Life of Mine | Mine Waste and Tailings Conference 2025

MINE WASTE AND TAILINGS VOLUME

29–30 July 2025
Brisbane, Australia

The Australasian Institute of Mining and Metallurgy
Publication Series No 3/2025

AusIMM

Published by:
The Australasian Institute of Mining and Metallurgy
Ground Floor, 204 Lygon Street, Carlton Victoria 3053, Australia

ISBN 978-1-922395-49-8

Advisory Committee

Life of Mine Committee

Claire Cote
Conference Advisory Committee Chair

Alex Thin
FAusIMM(CP)

Brad Radloff
MAusIMM

Ingrid Meek
MAusIMM(CP)

Jason Dunlop

Michael McLeary
MAusIMM

Sarah McConnell
MAusIMM

Sibasis Acharya
FAusIMM

Todd Bell
MAusIMM

Vanessa MacDonald

Maggie Ng

Mine Waste and Tailings Committee

David Williams
FAusIMM
Conference Advisory Committee Chair

Angélica Amanda Andrade

Lis Boczek

Edgar Salas

Hernan Cifuentes

Theo Gerritsen
MAusIMM(CP)

Symon Jackson
FAusIMM(CP)

Fernanda Maluly Kemeid
MAusIMM

Marcelo Llano

Allan McConnell
MAusIMM

Kathy Tehrani
MAusIMM

Chenming Zhang

Andy Fourie

Arun Muhunthan

Kristy Ellis

Kevin Spencer
MAusIMM

AusIMM

Julie Allen
Head of Events

Fiona Geoghegan
Senior Manager, Events

Raha Karimi
Program Coordinator, Events

Reviewers

We would like to thank the following people for their contribution towards enhancing the quality of the papers included in this volume:

Sibasis Acharya

Angelica Amanda Andrade

Todd Bell

Lis Boczek

Pascal Bolz

Kate Brand

Mark Chapman

Robynne Chrystal

Hernan Cifuentes

Amelia Corzo

Claire Cote

Manuel de Membrillera Ortuño

Jason Dunlop

Mansour Edraki

Andy Fourie

Philippe Garneau

Theo Gerritsen

Morteza Ghamgosar

Babak Hedayatifar

Stuart Henderson

Joe Hinton

Symon Jackson

Sina Kazemian

Fernanda Maluly Kemeid

Anjan Kundu

Scott Lines

Marcelo Llano

Vanessa MacDonald

Allan McConnell

Sarah McConnell

Karen McKenzie

Michael McLeary

Mauricio Medina Florez

Ingrid Meek

Liane Millington

Jack Moorhead

Camilo Moreno

Arun Muhunthan

Maggie Ng

Louisa Nicolson

Ina Prinsloo

James Purtill

Brad Radloff

Louisa Rochford

Edgar Salas

Mandana Shaygan

Kevin Spencer

Gideon Style

Kathy Tehrani

Alexander Thin

Ed Tuplin

Alex Walker

David Williams

Joseph Wu

Lauren Zappala

Chenming Zhang

Foreword

On behalf of AusIMM and the Sustainable Minerals Institute at The University of Queensland, welcome to the seventh Life of Mine (LOM) Conference.

Since its inception, LOM has grown to become a cornerstone of thought leadership in mine rehabilitation and closure, championing the importance of holistic thinking across the mine life cycle.

For the first time ever, LOM will run concurrently with the Mine Waste and Tailings Conference. This provides delegates a unique opportunity for cross-disciplinary dialogue, and a chance to explore strategic decisions shaping the future of mine waste management.

We are proud to present LOM's most extensive technical program to date, featuring 110 presentations across two streams. Attendees will benefit from a rich array of interactive panels and case studies, designed to encourage collaborative learning. A defining strength of this conference is the participation of AusIMM's traditional core professional disciplines – geology, mining engineering, and mineral processing – alongside those working in environmental, social, and sustainability-focused roles. The program has been carefully curated to foster cross-disciplinary engagement and learning.

A highlight of this year's conference is our opening keynote panel exploring the critical role of regional context in shaping effective closure strategies. Our keynote panellists will discuss how circular economy principles can turn environmental challenges into business opportunities, and the importance of engaging with supply chain actors to meet growing demands for transparency in environmental and social performance.

There is also a dedicated session on innovative rehabilitation practices, alongside a panel discussion and two must-attend technical sessions on building resilient landforms. Notably the LOM program includes a session dedicated to the Global Industry Standard for Tailings Management, reflecting the growing relevance of this topic for all professionals.

Importantly, the conference recognises that mine planners play a vital role in closure planning, and the perspectives of communities and Traditional Owners are also woven throughout the program, with a dedicated panel on strategies for building thriving, sustainable communities in mining regions.

I extend my heartfelt thanks to our presenting authors for their keen insights and our abstract reviewers for their time and diligence. We are also deeply grateful to our sponsors and exhibitors for their continued support.

Finally, I wish to acknowledge the Advisory Committee and AusIMM Management Team, whose dedication and expertise have been instrumental in bringing this event to life.

Welcome to the conference – we hope you find the program enriching and enjoy the opportunity to connect with your peers.

Yours faithfully,

Claire Côte

Life of Mine Conference Chair

The University of Queensland

Foreword

On behalf of the Advisory Committee, welcome to the 2025 Mine Waste and Tailings Conference, the fifth in this highly successful series first launched in 2015.

Co-hosted by AusIMM and The University of Queensland, this conference aims to continue as an industry-orientated benchmark for sharing knowledge and experience on all aspects of mine waste and tailings management, sustainable practice and closure. The co-location with Life-of-Mine will enrich the delegate experience and provide new learning opportunities.

As the demand for minerals and mine waste volumes continues to increase exponentially in the face of diminishing ore grades, the mining industry faces sustained threats to its financial and social licences to operate. Over the last decade, tailings dam failures have come to dominate discussion, with the longer term threat of acid and metalliferous drainage and erosion ever-present.

The mining industry accepts that the majority of the world's future minerals will come from ever more low-grade, high-tonnage, ultra-mechanised operations.

The industry is recognising that we are in the mine waste business, and this conference will present the latest ideas and tools to ensure we can better manage the responsibilities we have as professionals. With four pre-eminent keynote speakers, five industry expert panel discussions and numerous papers, I am confident that this conference will build on the noted success of its predecessors, as well as promoting networking to foster new connections.

I would like to thank the Advisory Committee, the AusIMM Management Team, authors, paper reviewers, attending delegates and all our sponsors and exhibitors, as well as our supporting partner, QTG.

We are delighted you are joining us at the conference, and we hope that you will find it an enjoyable and rewarding event.

Yours faithfully,

Emeritus Professor David John Williams FAusIMM

Mine Waste and Tailings Conference Chair

The University of Queensland

Sponsors

Major Conference Sponsor

BHP

Platinum Sponsors

GHD

Klohn Crippen Berger

WEIR
Mining technology for a sustainable future

Engineering Partner

red earth
engineering
A Geosyntec Company

Gold Sponsors

IGS
INSITU GEOTECH SERVICES
reducing geotechnical uncertainty

SGME
Geoforming
Solution Partners

srk consulting

Silver Sponsors

ATC WILLIAMS

BGC ENGINEERING

CONETEC

CSIRO

Deswik

SLR

Technical Session Sponsors

Ausenco

 **ERM**

xylem

Name Badge and Lanyard Sponsor

WEIR
Mining technology for a sustainable future

Coffee Cart Sponsor

McCOSKER
SAFETY FIRST. PRODUCTION ALWAYS.

Conference Proceedings Sponsor

MINE EARTH

Conference App Sponsor

 **kurloo**

Supporting Partner

QTG

Contents

Life of Mine — volume 1

Innovations in rehabilitation and closure

Integrated strategic mine planning

Life of mine waste and tailings management

Post-mining land use

Standards and reporting

Mine Waste and Tailings — volume 2

Case studies on operational aspects and closure

Design loadings and parameter selection, including BAT and BAP

Managing mine-affected water

MCA, risk assessment, governance and compliance

Mine closure and rehabilitation to accommodate site settings

Minimising and managing mine wastes, including dewatering tailings and comingling

Miscellaneous

Monitoring waste storages during operation and post-closure

Tailings and foundation characterisation

Tailings dam breach and runout analysis

Case studies on operational aspects and closure

Strength in waste – innovative tailings utilisation at the Kara Mine

C Cahill[1] and G Doherty[2]

1. Technical Director, GHD, Burnie Tas 7310. Email: clem.cahill@ghd.com
2. General Manager, Tasmania Mines, Burnie Tas 7310. Email: greg.doherty@tasmines.com.au

ABSTRACT

Tasmania Mines operates the Kara Mine, a magnetite and scheelite mining site located approximately 35 km south of Burnie in North-west Tasmania. The mine produces around 400 000 t of tailings annually, characterised as coarse material comprising 60–80 per cent sands, 7–19 per cent silts, and 12–21 per cent clays, and classified as Non-Acid Forming (NAF). The specific gravity of the tailings is approximately 3.6.

The most recent tailings storage facility (TSF) began construction in 2014, with the starter dam completed in 2019 across two distinct construction phases to meet evolving storage requirements. Tailings are deposited sub-aerially via spigots, with water managed using electric pumps. Originally, the facility design included 13 m of upstream raises. However, investigations revealed the tailings' susceptibility to liquefaction, prompting a shift to modified centreline raises.

Due to their NAF classification and high-strength characteristics, tailings were deemed suitable for use as embankment material. Comprehensive investigations and testing confirmed their viability, providing several advantages over conventional materials. Tailings could be sourced directly adjacent to the worksite, reducing transportation costs and time. Additionally, utilising tailings for embankment construction created extra storage capacity within the TSF, effectively extending the facility's lifespan.

To date, the embankment has been successfully raised by four metres, with an additional four metre raise currently underway. While construction has been largely successful, several unique challenges and lessons were identified during the use of tailings.

This paper presents a detailed case study of the Kara TSF, focusing on the innovative use of tailings as embankment material. It explores the TSF's design philosophy, the investigations and testing processes, the benefits of tailings utilisation, and the embankment's performance during deposition, offering valuable insights into the application of tailings in mining infrastructure.

INTRODUCTION

Tasmania Mines (Tasmines) owns and operates the Kara Mine, located near Hampshire in the North-west of Tasmania. Since operations began in 1978, the Kara Mine has produced a range of products from the Magnetite and Tungsten ores from the Kara Orebody. However, like many mining operations, the Kara Mine faces ongoing challenges in tailings management and storage. The tailings, which are susceptible to liquefaction, necessitated modifications to the Kara Tailings Storage Facility (TSF) No. 2 to meet design requirements.

Due to a limited mining lease footprint, there are limited opportunities to source materials for embankment construction from areas outside of the mine pit. The tailings are deposited into the TSF at a solids content of approximately 5 per cent, which is relatively low due to pumping constraints. Despite the dilute slurry, the tailings contain a significant sand fraction and exhibit rapid settling. Column testing indicates that approximately 92 per cent of total settlement occurs within the first hour, and 99 per cent within 24 hrs, relative to the total settlement measured over a 7-day period. The tailings are also non-acid-forming (NAF) which is seldom the case in Tasmania. These observations led to a concept study of utilising tailings for internal zones of the embankment. The use of tailings provided several potential positives for the Kara Mine:

- Tailings could be excavated and placed on the embankment with minimal hauling, increasing the efficiency of placement and reducing the unit cost of embankment material.

- An overall positive benefit to the site, by increasing the tailings storage volume per metre of raise (when compared to conventional earthfill), potentially reducing the total footprint and height of the facility when completed.

- Reduction in disturbance footprint as external borrow areas were not required.

The concept study found the use of tailings was viable, leading to laboratory testing to understand the strength parameters of the tailings. The detailed design considered a modified centreline cross-section, which could meet acceptable factors of safety under all stability load cases.

At the time of writing this paper, the embankment has been raised by a total of 8 m, in two stages, using tailings as one of the three primary materials. The construction was completed successfully.

A case study is presented in this paper on the novel use of tailings within the TSF embankment. The paper outlines the overall design of the TSF and embankment, describes the material testing undertaken, and provides an overview of the construction process, including the challenges encountered and how they were addressed.

SITE DESCRIPTION

The Kara Mine is located near Hampshire, approximately 30 km south of Burnie in North-west Tasmania. The location of the site is shown in Figure 1. The Kara Mine began operation in 1978 through the mining of the Kara No. 1 Pit. Currently, the life-of-mine is projected to be 2062, through mining of the Kara No. 1 Pit, Kara North Pit and the Rogetta Orebody. Ore is processed through the process plant, with the products transported by road to the Burnie port for shipment.

At the time of writing, tailings are produced at a rate of 400 000 t per annum, which is forecast to be consistent for the life of the mine. There are several tailings storage facilities at the Kara site, which are presented in Figure 2, currently, the tailings are stored in the Kara TSF No. 2, which is projected to provide tailings storage to approximately 2028 at the current ultimate design level.

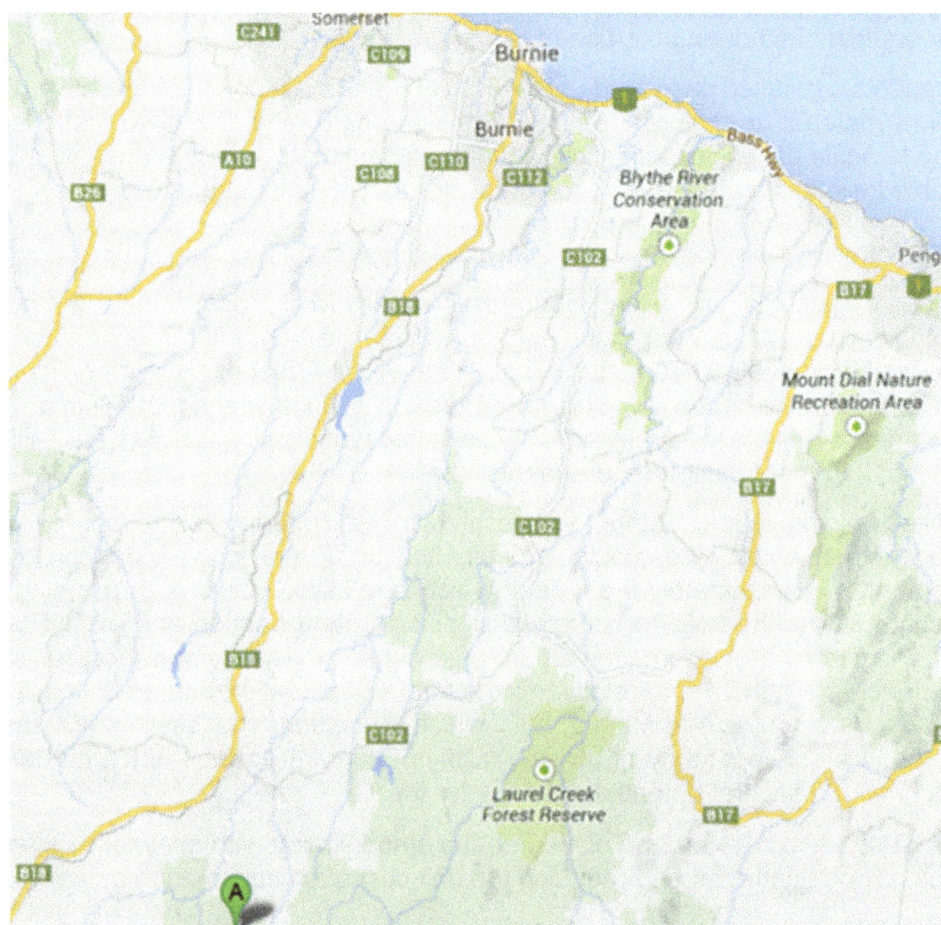

FIG 1 – Kara Mine location.

FIG 2 – Site layout.

The geology of the TSF No. 2 footprint shallow marine limestone, siltstone, sandstone, and Tertiary basalt. The TSF foundations comprise weathered sandstone and basalt in the west and metasomatised limestone (calc-silicate skarn) in the east. The TSF No. 2 embankment crosses contacts between basalt, sandstone, and calc-silicate skarn along the complete embankment alignment.

An example of the particle size distribution of the tailings is presented in Figure 3. The samples of Tailings 1 and Tailings 3 were sampled from the beach of the TSF close to the embankment, these samples are representative of the tailings adjacent to the embankment with 77–81 per cent sands, and 19–23 per cent fines. Tailings 2 was sampled at a distance further from the embankment, towards the decant pond, this sample displays typical tailings spatial distribution along the beach, with 60 per cent sands and 40 per cent fines.

The tailings have a specific gravity of approximately 3.6. Wet Cylinder Settled Density results indicate a Dry Density of 1.36 t/m^3 after 7 days. The settlement versus time results from this test are presented in Figure 4, with 92 per cent of settlement being achieved after 1 hr.

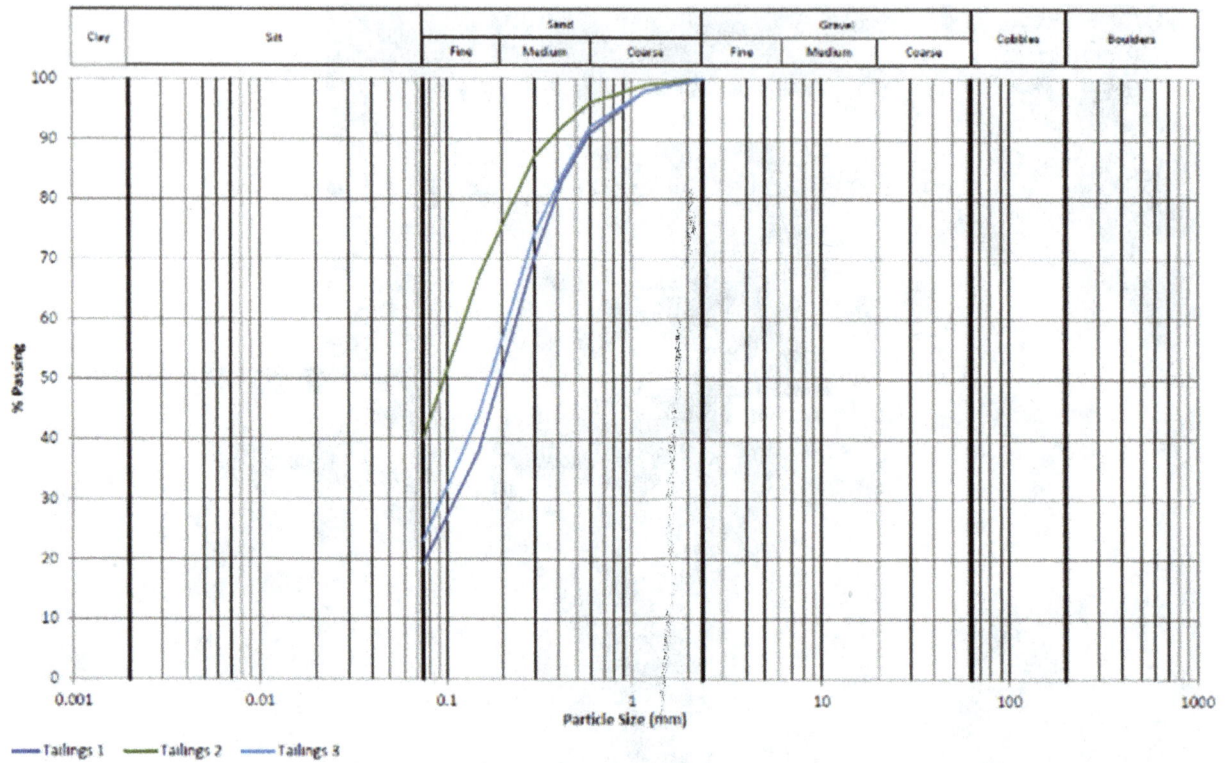

FIG 3 – Tailings particle size distribution.

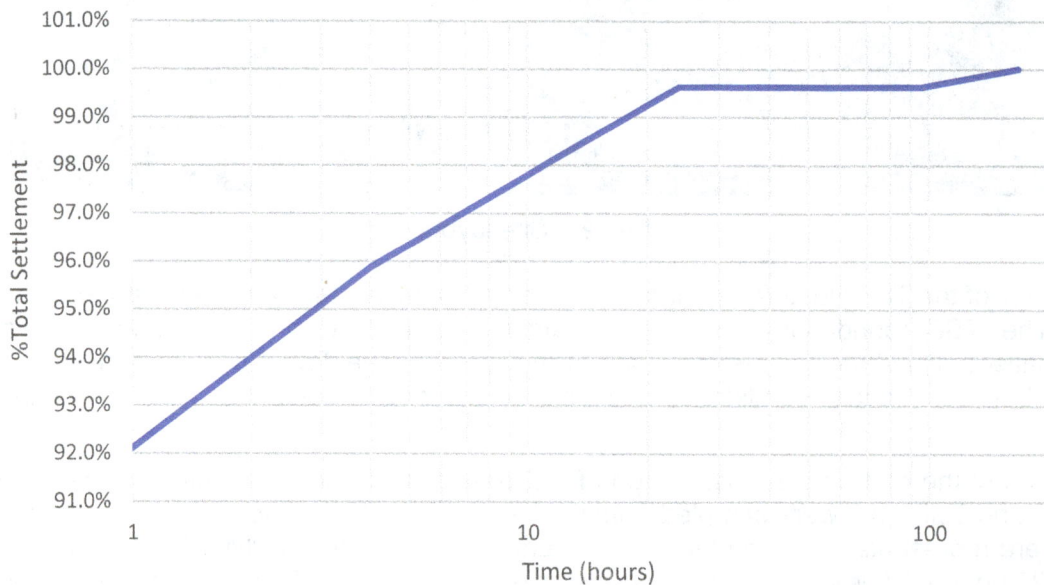

FIG 4 – Settling results.

TSF DESIGN AND INVESTIGATION

The Kara TSF No 2 design commenced in 2014 with the development of the concept design of the TSF to store approximately 30 years of tailings production. The TSF design has been developed in the following stages:

- Stage 1 – RL558 m

- Stage 1 Augmentation – RL558 m

- Stage 2 – RL562 m
- Stage 3 – RL566 m.

The overall design, investigation and construction stages through the life of the TSF are summarised in the following subsections.

Stage 1 and Stage 1 augmentation

Stage 1 of the TSF comprised an earthfill start embankment with a crest level of RL558 m. The embankment utilised general fill and Zone 1 clay materials for construction. A key feature of the Stage 1 embankment is the upstream toe drain constructed of coarse mine waste rock, which utilised the free-draining nature of the sandy tailings, to improve drainage in the tailings beach. The drain was sloped sufficiently to drain to the lowest point in the natural ground, where a flow-through zone of the embankment was detailed to allow drainage to a downstream sump. The Stage 1 embankment was a valley-type embankment along the northern extent of the TSF shown in Figure 2.

The typical section of the embankment is shown in Figure 5.

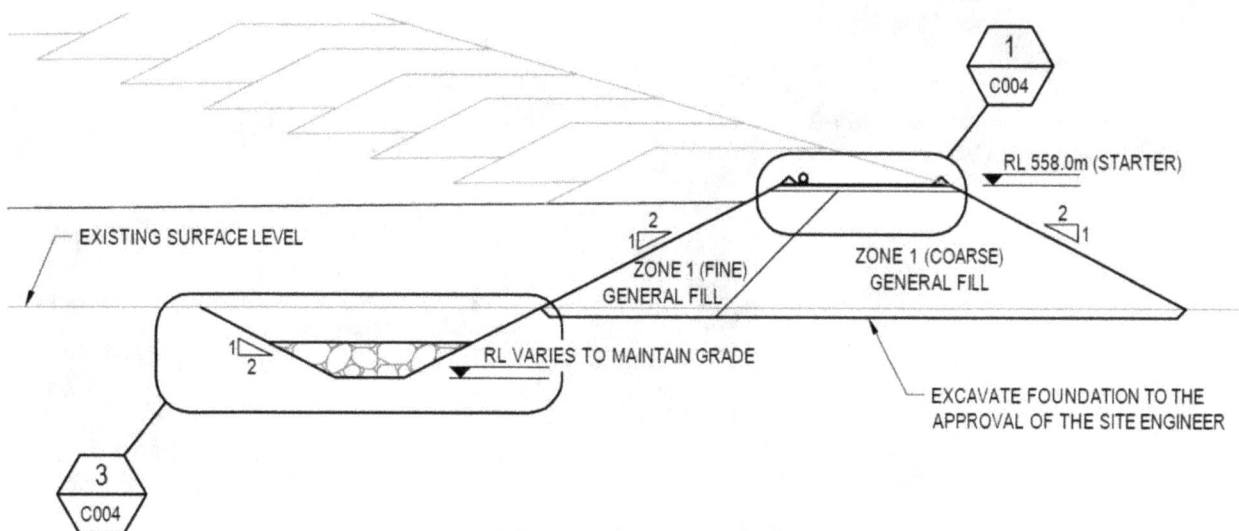

FIG 5 – Stage 1 typical section (GHD, 2014).

A concept design for the life of the facility was developed considering the use of upstream raises for all future raises to an ultimate level of RL580 m. A key assumption was made during the design phase that due to the coarse nature of the tailings and the drainage within the embankment, the tailings would not be susceptible to liquefaction. However, it was recommended that Cone Penetrometer Testing (CPT) be undertaken once the facility is constructed.

The Stage 1 starter embankment was constructed in 2015 and 2016.

The Stage 1 Augmentation essentially involved an extension of the embankment and upstream drain, using the Stage 1 design to complete the eastern and southern perimeters of the TSF at a crest level of RL558 m. The Stage 1 Augmentation was completed in 2018 and 2019.

2019 CPT investigation

A CPT investigation was undertaken in late 2019 to assess the engineering properties of the tailings deposited into the TSF since its commissioning. A total of eight CPTs were completed around the perimeter of the TSF as shown in Figure 6.

FIG 6 – CPT test locations.

Liquefaction triggering was assessed using a CPT-based approach in accordance with the procedures outlined by Idriss and Boulanger (2008), and Boulanger and Idriss (2014). CPTu data was analysed to calculate the Cyclic Stress Ratio (CSR) using the simplified procedure based on PGA and earthquake magnitude inputs derived from the 2018 National Seismic Hazard Assessment for Australia. The Cyclic Resistance Ratio (CRR) was determined from normalised CPT parameters with appropriate correction factors. A factor of safety (FoS) against liquefaction was determined at each depth as CRR/CSR, with values below 1.0 indicating potential liquefaction. The analysis considered both Operating Basis Earthquake (OBE, 1:475 AEP, PGA = 0.014 g) and Safety Evaluation Earthquake (SEE, 1:1000 AEP, conservatively using PGA = 0.044 g), with an assumed magnitude of 7.5 for both cases, in line with ANCOLD (2019) guidelines.

The key findings were summarised as follows (GHD, 2020):

- The Soil Behaviour Type (SBT) indicates that the sand-like tailings are susceptible to cyclic and flow liquefaction and the clay-like tailings are susceptible to cyclic softening and flow liquefaction.

- The majority of the tailings analysed appear to be contractive or in potentially contractive soil zones based on the assessment of the state parameter using the method proposed by Plewes, Robertson and Davies (1992).

- The majority of the tailings have been classified as sand-like.

- Liquefaction triggering assessments indicated that tailings are not likely to liquefy under a 1475 Annual Exceedance Probability (AEP) seismic event (OBE). However, it is considered that tailings are susceptible to liquefy under monotonic loading conditions (ie static liquefaction) for example, an increase in water pore pressure has the potential to reduce tailings' effective stress. It was recommended to investigate the implication of tailings strength reduction during the detail design of the next stage.

- Liquefaction triggering assessments indicated that tailings are likely to liquefy under a 1 in 2500 AEP event (SEE). A summary of the liquefiable zones is presented in Figure 7. A deeper liquefiable zone can be seen in CPT04 to CPT06 which may be due to the more recent deposition in this area at the time of the testing. Also CPT04 is situated near the end of the tailings pipelines and has been observed to have finer tailings typically deposited in this area.

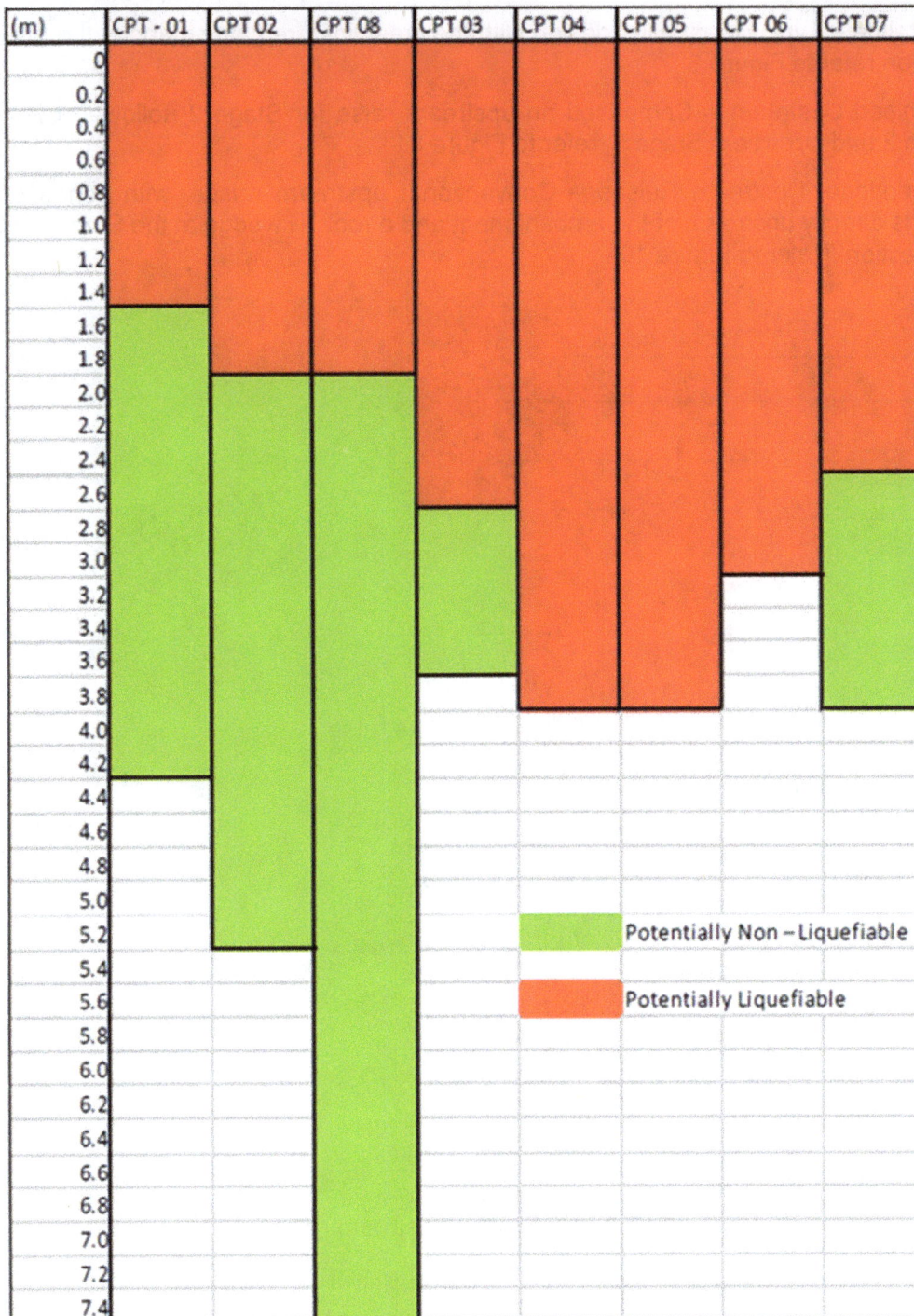

FIG 7 – Summary of liquefaction assessment for SEE.

The implications of the CPT investigation were clear, the assumption of the tailings being unlikely to liquefy was incorrect and post-liquefaction stability analysis for future raises would need to consider post-liquefied material strength parameters.

Stage 2 design and life of facility concept design

The outcomes of the CPT investigation necessitated revisiting the original life of facility design to confirm whether the use of the upstream raises was feasible with liquefiable tailings. A concept design was undertaken for the life of facility design involving raising the embankment by 11 m in three stages which was the maximum permitted height at the time of the design. The concept design considered three options for the embankment design:

- Centreline – Typical centreline design with the crest lines remaining above the previous crest position refer to Figure 8.

- Optimised Upstream – Comprised an upstream raise for Stage 2, followed by a centreline Stage 2 and Upstream Stage 3, refer to Figure 9.

- Conventional Upstream Raises – Conventional upstream raises, with zones of densified tailings directly upstream of the embankment and a rock fill wedge at the Stage 1 and Stage 2 connection. Refer to Figure 10.

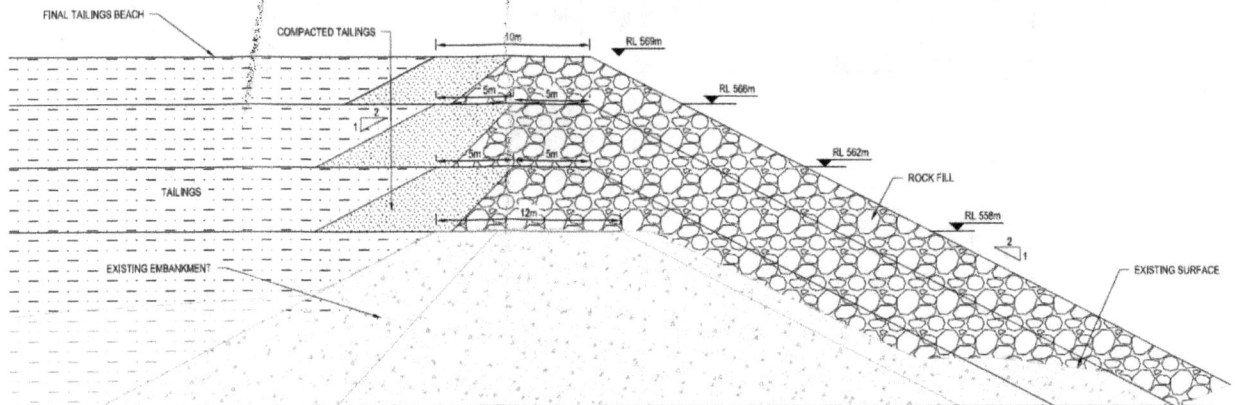

FIG 8 – Centreline raise.

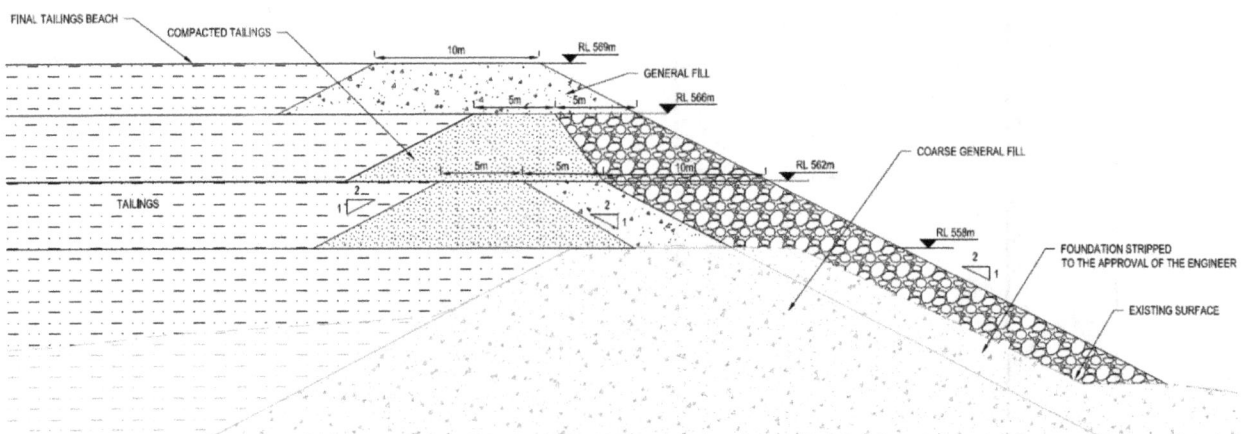

FIG 9 – Optimised upstream raise.

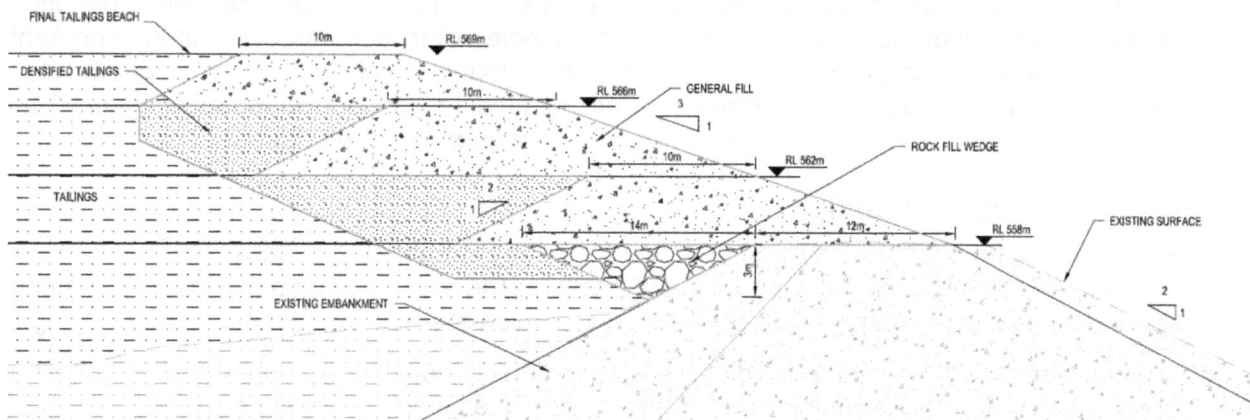

FIG 10 – Conventional upstream raise.

Both the centreline and the optimised upstream options adopt compacted tailings for embankment material. The tailings across the beach were observed to quickly develop strength post-deposition and once the beach has dried, it is trafficable by light vehicles, which provided confidence in the tailings performance as a construction material at this concept level. The CPT investigation found that the typical friction angle of the tailings in the static condition was approximately 30°. The use of tailings as an embankment material provided the following positive benefits:

- Tailings could be excavated and placed on the embankment with minimal hauling, increasing the efficiency of placement and reducing the unit cost of embankment material.

- An overall positive benefit to the site, by increasing the tailings storage volume per metre of raise (when compared to conventional earthfill), potentially reducing the total footprint and height of the facility when completed.

- Reduction in disturbance footprint as external borrow areas were not required.

The concept design found that all three options could meet current guidelines for stability, however, the upstream option had a factor of safety of 1.0 for the post-liquefaction case, representing a marginal factor of safety that would require increased rigour around material parameters and operations. The centreline options and the optimised upstream option achieved factors of safety greater than 1.5 for all loading cases, showing that these options are considerably more resilient in the post-seismic load cases.

The volumes showed that the optimised upstream alternative required a marginal increase in volume over the upstream options (considering the densified tailings zone). Therefore, the optimised upstream option was selected for the detailed design phase. Tailings samples were collected for triaxial testing to confirm the material properties once compacted and to determine if the tailings would be dilative when compacted.

Testing

Consolidated Undrained Triaxial Tests were performed on samples of the coarse tailings to determine the material strength properties of the tailings after compaction and whether the tailings would dilate during shearing.

A total of four triaxial tests were completed at densities as follows:

- Two tests completed at 90 per cent Standard Maximum Dry Density (SMDD).

- Two tests completed at 95 per cent SMDD.

The results of the triaxial testing indicated a friction angle of 36° for all tests completed and showed that the material was dilative at the effective stress of 150 kPa. Additionally, the permeability of tailings under the test conditions was found to be in the order of 1×10^{-8} m/s.

The results from the triaxial testing were favourable for utilising the tailings in the embankment and the detailed design commenced.

It is noted that the compaction process induced an apparent OCR below 150 kPa, hence resulting in a mechanical behaviour typical of dilative materials. Angles of friction can adequately represent the strength of dilative materials. Future work is required to understand how and at what overburden stress levels compacted tailings will transition from dilative to contractive material again. This transition from dilative to contractive is inevitable with subsequent upstream raises.

Stage 2 and 3 detailed design

The detailed design for both stages 2 and 3 involved adopting the updated material parameters from the CU testing for stability analysis. Updating and finalising the internal zoning of the embankment. The following items were updated in the detailed design which are relevant to the use of the tailings:

- Tailings were specified to be placed at a minimum density equal to or greater than 95 per cent SMDD, with the moisture content within 2 per cent of the Optimum Moisture Content (OMC).

- A general fill zone was detailed between the rock fill and tailings zone to provide a transition. Zone between the tailings and relatively coarse, free-draining rock fill.

- A rock fill protection zone was detailed on the upstream face of the embankment to protect the compacted tailings against erosion during active tailings deposition.

- A spillway was provided to safely pass the 0.1 per cent AEP critical duration flood event.

CONSTRUCTION

The embankment has been raised a total of 6 m using the optimised upstream design incorporating tailings as a construction material. Stage 2 comprised a 4 m raise and Stage 3 comprised two 2 m lifts, named Stage 3a and 3b, at the time of writing Stage 3a had been completed and Stage 3b was under construction. This section will focus on the general construction of the embankment with a focus on the tailings.

Construction is typically undertaken during the summer months on Tasmania's west coast to avoid the wet winter weather which slows construction considerably. Construction of the Kara TSF typically follows this cycle as well, with construction halted through the winter months. Construction was completed using an experienced earthworks and dam contracting company.

The specification for the tailings called for tailings to be placed in 300 mm layers and rolled with eight passes of a smooth drum roller to achieve a density equal to or greater than 95 per cent SMDD.

Tailings were borrowed by either one of the following methodologies. The borrowing methodology was varied typically due to the condition and trafficability of the tailings beach adjacent to the embankment.

- Direct borrowing of tailings adjacent to the embankment with an excavator and placed directly onto the embankment. This was used initially during construction but was typically not the most efficient means of placement due to the limited reach.

- Using dozers with low ground pressure tracks on the tailings beach to doze thin layers from the beach and stockpile adjacent to the embankment, where:

  o Tailings were placed on the embankment using an excavator and then spread with a dozer, or

  o Tailings were loaded into articulated dump trucks and hauled to the active construction area.

Density in excess of 100 per cent SMDD has been typically achieved. Dry density of placed tailings ranged from 1.95 t/m³ to 2.31 t/m³ with moisture contents ranging between 6.2 per cent and 15.9 per cent. However, due to the sandy nature of the tailings, achieving the moisture requirements of the tailings was found to be challenging. Test results are presented in Figure 11 for the Stage 3 construction. This is discussed further in the following section.

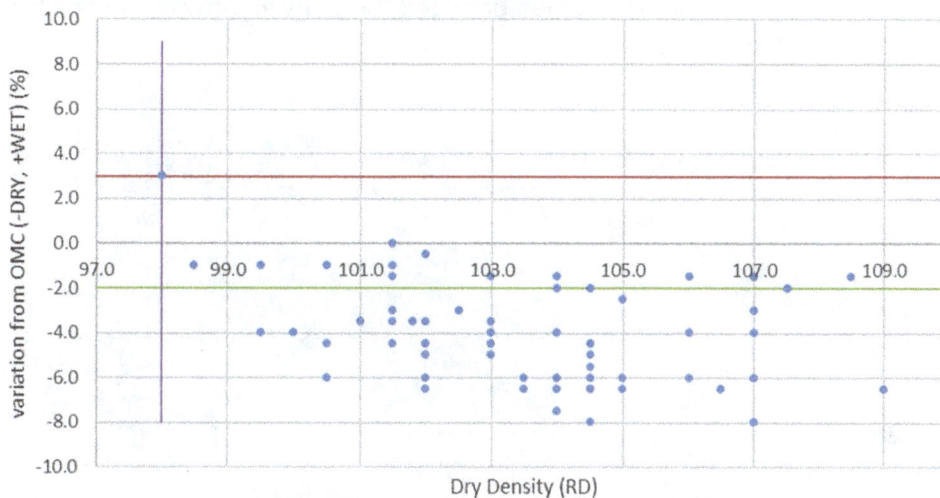

FIG 11 – Tailings compaction testing results.

The estimated total volume of tailings placed for each stage is as follows:

- Stage 2 = 65 400 m³
- Stage 3 = 47 000 m³.

CHALLENGES OF TAILINGS

Construction of the Kara TSF embankment utilising tailings presented several advantages as listed earlier, when compared to typical embankment materials but also challenges that are unique to the tailings (and similar materials).

Tailings were found to be easily compactable to the required minimum density, however, achieving moisture content specification was challenging. The tailings due to their sandy nature, dry quickly and do not retain moisture once excavated from the tailings beach. As shown in Figure 11, typically test results showed that the tailings were dry of optimum. A water cart was utilised for the addition of water to the tailings during placement and compaction. The addition of water was effective at raising the moisture content to approximately the OMC and the moisture was typically within specification when tested on the day of placement and compaction. If the material was tested on subsequent days, it was typically much drier than OMC. The out-of-specification results were accepted on the basis that the methodology of placement and compaction with the addition of water could consistently meet the required density. Further studies will be undertaken to assess the performance of the tailings when compacted at lower moisture levels to enable any adverse impacts on the design to be addressed in future design stages.

Borrowing the tailings from a TSF with active deposition provided challenges with scheduling between, deposition, drying and borrowing as well as the need to have different borrowing methodologies to deal with areas of the beach that weren't trafficable. Deposition occurred throughout the construction of the Stage 2 and Stage 3 embankment raises. The Kara TSF has two deposition pipelines with spigots, each of which covers roughly half of the TSF. The configuration allows for deposition on half of the TSF while tailings can be borrowed from the other half. During the summer, this methodology works particularly well because the tailings dry quickly and can be borrowed within a few weeks of discharge, which can be controlled by turning off the spigots in that area.

The trafficability of the tailings is generally adequate. Dozers with low ground pressure tracks can traffic most areas of the tailings beach. Three different methodologies were utilised to borrow and place tailings most efficiently. During construction, a few issues were encountered with weaker tailings, which were typically encountered when an area was nearly exhausted of usable tailings. These issues were highlighted by the increasing presence of sand boils, as can be seen in Figure 12.

The risk of equipment bogging was typically managed by using experienced operators. Due to the variable nature of the tailings beach, hard operating rules couldn't be implemented. The use of different borrow and placement methodologies provided the contractor with the flexibility to ensure construction occurred efficiently and safely.

FIG 12 – Sand boils during the borrowing of tailings.

The typical section shown in Figure 9 includes a rock fill protection layer on the upstream face of the embankment. The original intent of the protection layer was to prevent the erosion of tailings during active tailings discharge from the spigots. During the construction of Stage 2 it was decided not to construct the rock fill layer and to discharge tailings and monitor, repair and manage as required. Subsequent TSF operation revealed that exposed tailings batter were highly susceptible to erosion during intense rainfall events due to run-off. An example of significant erosion is shown in Figure 13. Initially, the eroded areas were repaired with rock fill. However, erosion would continue beside the repaired area in the following rainfall event. The rock fill layer was subsequently reinstated as originally designed. The adoption of the erosion protection layer has been effective at significantly limiting the erosion of the tailings. The observations of surface erosion show that the compacted tailings need to be effectively encapsulated after construction to limit erosion.

FIG 13 – Erosion of the tailings due to run-off.

Because the tailings are borrowed from the beach, the overall beach profile is altered, resulting in flatter sections and areas that are prone to ponding once deposition recommences. The impact of this on tailings strength has not been fully understood at the time of writing. However, it is anticipated that the strength may be negatively impacted due to the increased thickness of tailings required to dry, as well as the potential trapping of tailings slimes in lower areas. Future CPT testing will assess the changes in tailings strength, and future designs will include increased rigour around maximum tailings borrow depth to decrease the likelihood of ponded areas.

The challenges encountered were unique to construction with tailings due to the properties of the material and borrowing process on an active TSF.

CONCLUSION

The Kara TSF case study demonstrates that tailings can be successfully used as embankment material in a way that is both economically and operationally advantageous. The non-acid-forming nature of the tailings at Kara provided favourable conditions for their reuse as construction material. Laboratory testing and staged construction proved that the tailings met compaction and short-term strength requirements.

The optimisation approach led to several key benefits:

- Reduced reliance on external borrow material, minimising environmental disturbance.
- Cost and time efficiencies due to proximity of borrow source.
- Increased tailings storage volume per metre of raise, extending facility life.

The project also presented unique challenges, particularly in managing tailings moisture content, scheduling around active deposition, and addressing erosion risks. Each of these was overcome through adaptive construction methods, careful planning, and strong collaboration between mine operators and contractors.

Future stages of embankment construction should continue to assess tailings behaviour through CPT testing. For example, establishing the extent of the impact of the apparent OCR gained during tailings compaction. Further work is required to determine the necessary amount of overburden for the compacted tailings to behave contractively again. Areas of altered beach profile and potential ponding require careful work method statements that incorporate surface water management. More broadly, this case supports the viability of tailings reuse in embankment construction and encourages similar investigations at other mine sites where tailings characteristics permit.

ACKNOWLEDGEMENTS

The authors would like to acknowledge Tasmania Mines for their support throughout this innovative design and construction project, as well as their support for the preparation of this paper.

REFERENCES

Australian National Committee on Large Dams Incorporated (ANCOLD Inc), 2019. Guidelines for Design of Dams and Appurtenant Structures for Earthquake.

Boulanger, R W and Idriss, I M, 2014. CPT and SPT Based Liquefaction Triggering Procedures, Report No. UCD/CGM-14/01, Center for Geotechnical Modeling, Department of Civil and Environmental Engineering, University of California, Davis.

GHD, 2014. Kara Mine – Tailings Storage Facility – RL558 Raise Preconstruction Report.

GHD, 2020. Kara TSF – Stage 2 Raise – Pre Construction Report.

Idriss, I M and Boulanger, R W, 2008. SPT-Based Liquefaction Triggering Procedures, Report No. UCD/CGM-08/02, Center for Geotechnical Modeling, Department of Civil and Environmental Engineering, University of California, Davis.

Plewes, H D, Robertson, P K and Davies, M P, 1992. Assessment of the in-situ state of sands using the CPT, in *Proceedings of the 45th Canadian Geotechnical Conference*, pp 265–275.

The Life of Mine | Mine Waste and Tailings Conference 2025 | Brisbane, Australia | 29–30 July 2025

Glass reinforced plastic pipeline design for the life-of-mine of a tailings gravity decant system

C Han[1], L Day[2], W Ludlow[3] and O Dudley[4]

1. Principal Engineer, Red Earth Engineering – A Geosyntec Company, Brisbane Qld 4000.
    Email: chao.han@redearthengineering.com
2. Principal Engineer, Red Earth Engineering – A Geosyntec Company, Brisbane Qld 4000.
    Email: leis.day@redearthengineering.com
3. Senior Principal, Red Earth Engineering – A Geosyntec Company, Brisbane Qld 4000.
    Email: wade.ludlow@redearthengineering.com
4. Senior Engineer, Red Earth Engineering – A Geosyntec Company, Brisbane Qld 4000.
    Email: oliver.dudley@redearthengineering.com.au

ABSTRACT

The design and implementation of a Glass Reinforced Plastic (GRP) pipeline as a gravity decant system presents an attractive solution for managing tailings facilities in a mining operation over the life of the mine. This paper presents a design case study in which the engineering principles, material advantages, and sustainability aspects of utilising GRP pipelines embedded at up to 30 m depth below a tailings storage facility (TSF) are explored.

Tailings management is a critical aspect of mining operations, necessitating robust and long-lasting infrastructure to ensure environmental compliance and operational efficiency. The gravity decant system, which relies on the natural force of gravity to manage water decantation from the TSF, offers a low-energy, cost-effective solution. GRP, known for its high strength-to-weight ratio, corrosion resistance, and durability, emerges as an appealing material for constructing such systems.

In this study, the design criteria that underpin the effective use of GRP pipelines, including structural integrity, tolerable displacement and longevity, are discussed. Finite element analyses (FEA) simulations are undertaken to evaluate the soil-structure interactions and to further validate the standard based design assumptions and performance metrics. The paper also addresses the installation methodologies and maintenance protocols essential for optimising the life cycle cost and functionality of the GRP decant system.

Moreover, this paper presents a performance-based design approach in which the initial model and parameter assumptions will be further verified during construction via a quality control field and laboratory testing programme, as well as during operations via a sophisticated instrumentation system.

In conclusion, the adoption of a GRP pipeline as a gravity decant system in TSFs signifies a progressive stride towards enhancing the efficiency, reliability, and sustainability of tailings management. This paper aims to provide a technical guide and case-based insights for engineers and decision-makers in the mining industry, advocating for the integration of advanced materials and design innovations in critical infrastructure projects.

INTRODUCTION

The design of a gravity-fed decant dewatering system incorporating glass-reinforced plastic (GRP) pipe infrastructure presents a strategic approach to improving tailings water management within a Tailings Storage Facility (TSF). This system is intended to enhance decant capacity and operational reliability, particularly during wetter months, by maintaining pond levels and surface areas below the defined thresholds of the Trigger Action Response Plan (TARP) throughout the life of the facility. Effective control of pond levels enables the formation and preservation of a stable tailings beach slopes which supports subaerial deposition practices and facilitates enhanced drying and consolidation of the tailings mass. These outcomes align with the design philosophy recommended by the Independent Expert Engineering Investigation and Review Panel (IEEIRP, 2015), which advocates for the adoption of Best Available Technology (BAT) to ensure long-term safety and stability of TSFs. BAT, as defined by the IEEIRP from first principles of soil mechanics, is underpinned by three critical components:

1. Elimination of surface water from the impoundment.

2. Promotion of unsaturated conditions through drainage.

3. Achievement of dilative behaviour within the tailings through compaction.

GRP pipelines in the application of tailings decant systems mark an advancement in mining infrastructure. GRP specifically refers to composites using fibreglass as the reinforcement and various resins as the matrix. The GRP pipelines presented in this paper are manufactured by winding uni-directional fibres onto a mandrel and interspersing the layers with short random direction fibres and a filler such as sand to bulk up the wall thickness, which is implemented through the Flowtite continuous filament winding technology (Flowtite, 2025).

The GRP pipelines are known for their exceptional strength-to-weight ratio, corrosion resistance, and durability, and they are well-suited to the rigorous demands of tailings decant systems. The inherent properties of GRP allow for the construction of pipelines that can withstand significant depths and pressures, making them an ideal choice for embedding beneath tailings storage facilities (TSFs). However, limited literature regarding the successful applications of GRP pipelines in tailings facilities has been found so far.

This paper presents a design case study which covers the engineering rationale behind designing GRP pipelines for a gravity decant system. Finite element analyses (FEA) simulations were undertaken to evaluate the soil-structure interactions and to further examine the standard based design approach and to inform subsequent performance metrics. The simulation results were used to check the design against the design criteria that underpin the effective use of GRP pipelines., including structural integrity, tolerable displacement and longevity. The presented design has been integrated into a performance-based design approach in which the initial model and parameter assumptions are further verified during construction via a quality control field and laboratory testing programme, as well as during operations via a sophisticated instrumentation system. Additionally, the design has incorporated risk mitigation strategy where innovation Distributed Fibre Optic Sensing (DFOS) network would be installed at pipe levels along the alignment to monitor the operational design performance of the buried GRP pipe.

PROJECT BACKGROUND

This project includes the detailed design of a gravity decant system for an existing bauxite beneficiary TSF to better manage the decant water levels. The key design objective is to ensure the gravity decant system meets the design flow rate over the design life cycle without requiring any pumping systems, while the design components maintain structural integrity and design functionality over the design life cycle. To achieve the design objective, preceding hydraulic analyses were undertaken to determine the required combination of the alignments, gradients and associated diameters of the pipelines. Pipeline elevations were derived by fixing the weir inlet elevation and providing the minimum fall required to achieve self-cleaning velocities within the pipes at low flows. Pipeline alignments as shown in Figure 1 were selected with the aim of keeping the base of trench excavations within the controlled embankment fill to limit differential settlements. The pipeline was designed to be buried by up to 30 m of overburden fill and tailings during the life of the facility.

The main design components for the GRP pipeline alignments are summarised in Table 1. The GRP pipeline materials were selected following a two-step approach. A primary screening of commercially available pipe material options was undertaken in accordance with the Water Corporation's Pipeline Selection Guidelines (Water Corp, 2024). Subsequently, a secondary screening was undertaken based on project specific criteria. The considered material options include GRP, reinforced concrete (RC), Polymer Resin Concrete (RA) and Mild (Carbon) Steel (MSCL). The screening outcome indicates GRP pipes ranks higher than the other options due to durability and resistance to environmental factors during operation conditions over the design life as well as lower procurement and installation costs.

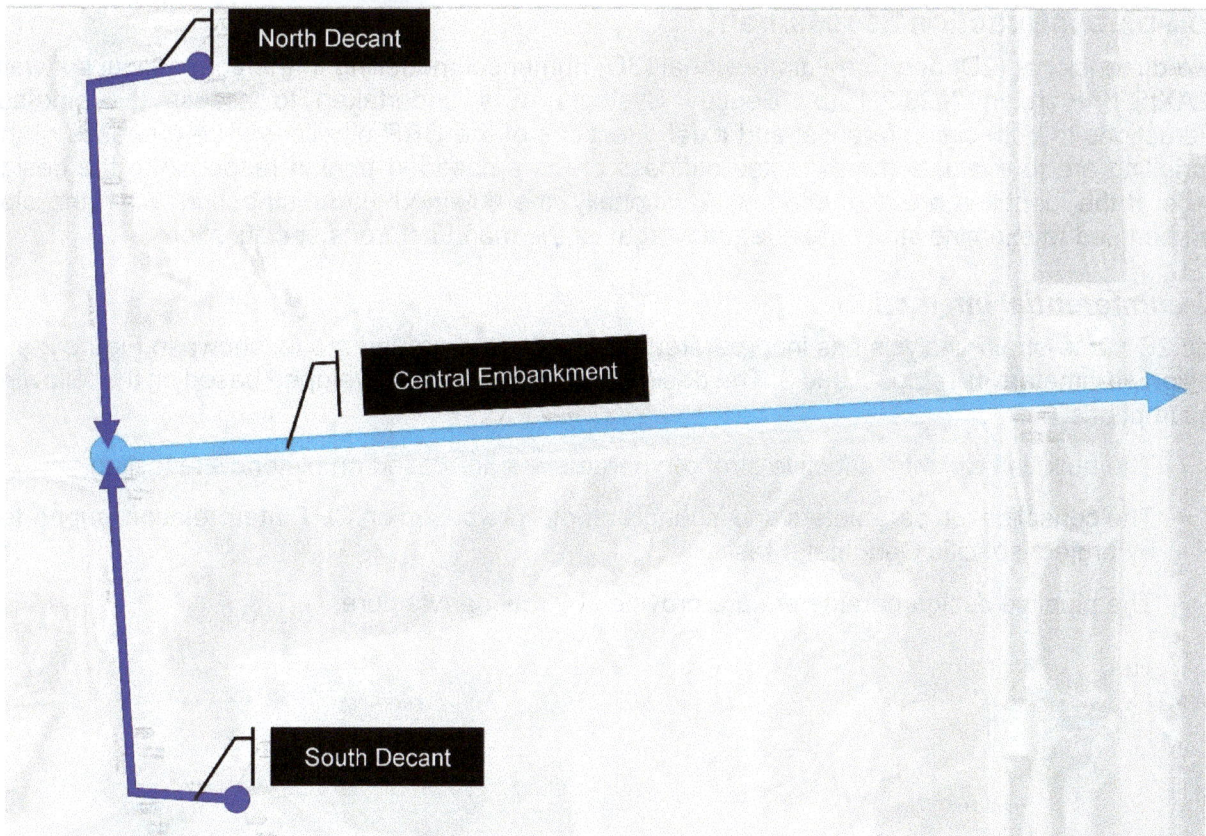

FIG 1 – GRP pipe alignments.

TABLE 1

Summary of gravity decant design components.

Alignment ID	Alignment length (m)	Main components
Central Embankment	1480	SN32000 PN6 DN1400 GRP pipeline
North Decant	640	SN32000 PN6 DN1100 GRP pipeline
South Decant	560	SN32000 PN6 DN1100 GRP pipeline

GRP PIPELINE DESIGN

The design has adopted a performance-based approach to improve the dewatering ability of the decant system to be better able to maintain the decant pond water levels below the trigger action response plan (TARP) levels throughout facility life cycle. The engineering design for the GRP pipelines includes the following key activities:

- Development of ground models to simulate the soil-pipeline interaction under the existing and future ground conditions informed by geotechnical investigation and laboratory testing data along the pipe alignments.

- The results from numerical modelling were used to inform the structural adequacy of design components and compared against relevant design standards and guidelines. The results also informed the expected movement ranges to ensure the design functionality.

- Forming part of the performance-based approach, the implemented design is subjected to regular review of construction, operational and surveillance data against key performance indicators.

Soil-pipe interaction assessment

Two-dimensional (2D) and three-dimensional (3D) numerical modelling in the commercial software PLAXIS™, version 2023.2.1 (by Bentley Systems) was undertaken to assess the pipe-soil interactions in both circumferential and axial directions of the GRP pipelines. The purposes of the modelling are to evaluate the expected stiffness of the embedded pipe in response to the design loads at the identified areas of interest. Additionally, the obtained load distribution within the pipe can be used to examine structural adequacy against the manufacturer's specifications.

Circumferential interaction

The 2D plane-strain analysis has incorporated the interpreted ground model shown in Figure 2 and design parameters in Tables 2 and 3. The design parameters were developed based on the following assumptions:

- The bulk unit weights, effective strength parameters are based on historic site testing.

- The consolidation parameters and Young's moduli are based on CPT interpretation, supported by project-specific Oedometer tests.

- The pipeline design parameters are provided by the manufacturer.

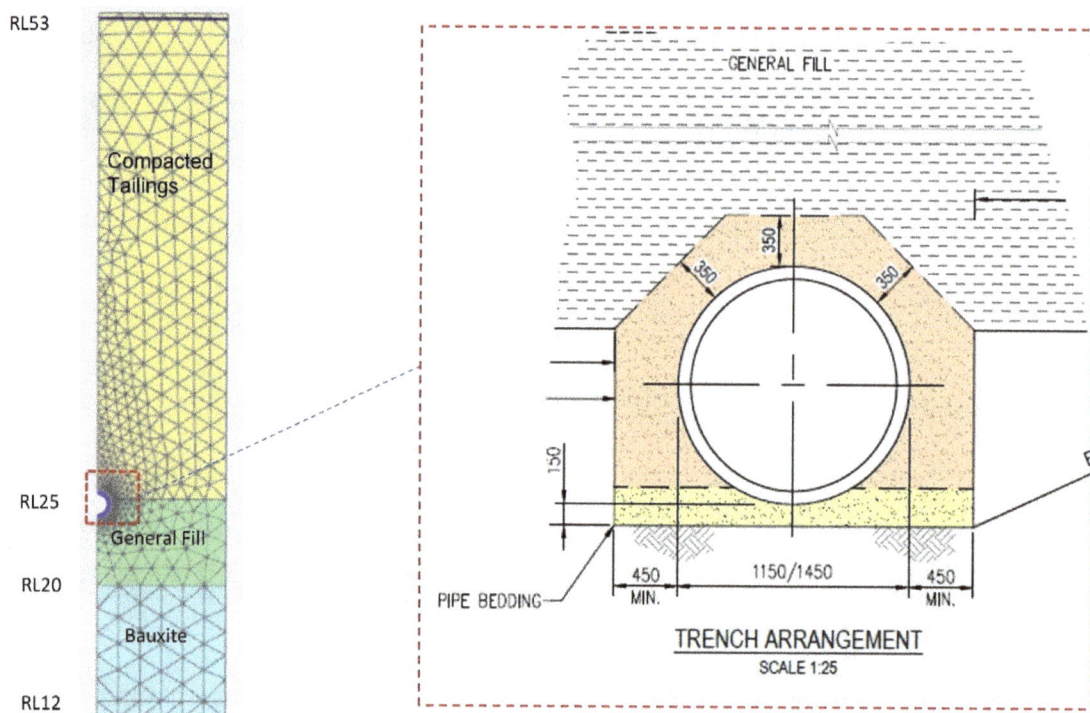

FIG 2 – Modelled cross-sectional details.

TABLE 2
Geotechnical parameters.

Unit	Bulk Unit Weight	Young's modulus	Poisson's ratio	Effective Cohesion	Effective Friction	Compression Index	Swelling Index	Initial Void Ratio	Over-consolidation Ratio
	γ (kN/m^3)	E' (MPa)	u	c' (kPa)	ϕ' (°)	Cc	Cs	e_0	OCR
General Fill	19	14–21	0.3	0	29	-	-	-	-
Compacted Fill	19	35	0.3	0	35	-	-	-	-
Bauxite	18	30–100	0.3	0	32	-	-	-	-
Mottled Zone Bulimba	18	15	0.3	0	29	0.15	0.015	0.65	2
Pallid Zone Bulimba	19	100	0.3	0	33	-	-	-	-
Trench Backfill *	20	50–200	0.3	5	36	-	-	-	-
Pipe Bedding	18	30	0.3	0	35	-	-	-	-
Compacted Tailings	17.5	25	0.3	0	35	-	-	-	-

Note: * flowable fill that has an estimated unconfined compressive strength (UCS) of at least 1 MPa.

TABLE 3
GRP pipe structural parameters.

Pipe type	External diameter	Wall thickness	Initial ring-bending modulus	Long-term ring-bending modulus	Allowable long-term hoop stress	Allowable long-term ring-bending strain	Poisson's ratio
	De (m)	t (mm)	E_b (GPa)	E_{bL} (GPa)	σ_{hall} (MPa)	ε_{hall}	v
DN1100	1.126	33	15.5	9.30	29.5	0.65%	0.21
DN1400	1.432	43	15.8	9.48	29.2	0.65%	0.21

Considering the uncertainties related to the key parameters of pipe surroundings, different design cases (Case A and B groups for DN1100 and DN1400 respectively) have been analysed to evaluate the pipe performance and to determine the critical case. The load distribution along the circumferential pipe wall is used to examine the proposed structural adequacy.

Axial interaction

A 3D PLAXIS™ model has been developed for the Central Embankment segment CH 2100 m to 2400 m, to assess soil-pipe interactions along the DN1400 axial direction where:

- The final design surface transition from the tailings impoundment and slopes towards the eastern perimeter embankment resulting in a reduction soil overburden loads above the pipeline.

- The presence of normally consolidated compressible Mottled Zone Bulimba materials may result in significant differential settlement at the overburden transition location.

The 3D section developed for the assessment is shown in Figure 3 along with the cross-sectional details. The parameters in Tables 2 and 3 have been assigned to the relevant materials and two phreatic surfaces were assessed based on the design operational scenarios, ie high and low representative of duration and after operations. The load and displacement distributions along the modelled pipe axis have been used to examine the adequacy of GRP pipeline strength, joint arrangement and design gradient arrangement.

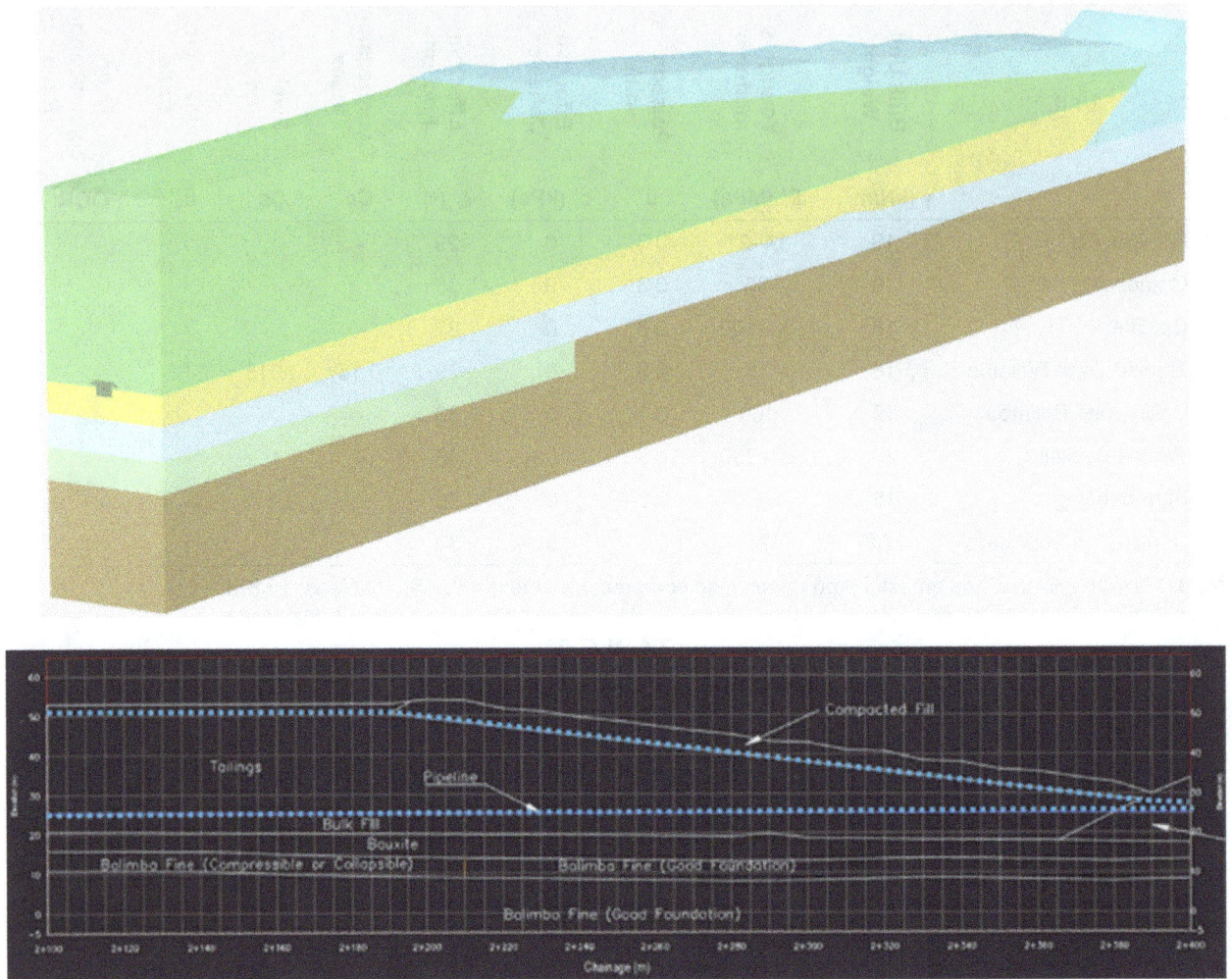

FIG 3 – 3D PLAXIS™ ground model and 2D cross-section between CH 2100 m and 2400 m.

Structural integrity

Design guidelines

The following guidelines have been referred to examine the structural capacity of the proposed GRP pipelines against the circumferential design actions.

- AS2566.1 (SAI Global, 1998) – R2018: note that for pipes buried with >10 m overburden, this method is conservative. Look and Cameron (2018) discusses the limitation of this method and recommends alternative values for the embedment soil modulus, E'_e, and native soil modulus, E'_n, as opposed to those provided in the standard. The recommended values are consistent with the design parameters in Table 2.

- FEMA (US Federal Emergency Management Agency, 2007) has been referenced to validate the design using AS2566.1. The considered failure modes align with those in AS2566.1. This manual acknowledges that deeply (> 50 ft) buried pipes are outside the scope of this document and the reader should consult other methods of analysis, such as by FEA.

Circumferential design

The results of structural design using AS2566.1-R2018 are summarised in Table 4. It is shown that the proposed GRP design satisfies the requirements for considered failure modes. As recommended in FEMA (2007) the PLAXIS™ modelling results are used to assess the mobilised hoop stress within the pipe wall. Figure 4 shows the combined axial compressions and bending moments for the analysed circumferential cases, compared to the allowable action envelope. The results indicate the estimated utilisations are less than 61 per cent of the allowable long-term hoop stress, which demonstrates adequate design safety as well as reasonable efficiency.

TABLE 4

Results of structural capacity check using AS2566.1-R2018.

Description	Case 1.1	Case 1.2	Case 1.3	Case 1.4	Unit
Pipe Size	1100	1100	1400	1400	mm
Cover	27.0	27.0	28.5	28.5	m
Native Soil Modulus	14.00	14.00	14.00	14.00	MPa
Backfill Modulus of Reaction	50.00	50.00	50.00	50.00	MPa
Width of trench at the spring line	2.00	2.00	2.40	2.40	m
Height of water surface above top of the pipe	26.0	-	27.5	-	m
Internal working pressure	0.25	0.00	0.27	0.00	MPa
Internal vacuum	-	-	-	-	MPa
Unit weight of trench fill	20.50	20.50	20.50	20.50	kN/m^3
Calculation results					
Deflection check	OK	OK	OK	OK	-
Strain check	OK	OK	OK	OK	-
Buckling check	OK	OK	OK	OK	-
Internal Pressure Check	OK	OK	OK	OK	-
Combined Loading Check	OK	OK	OK	OK	-

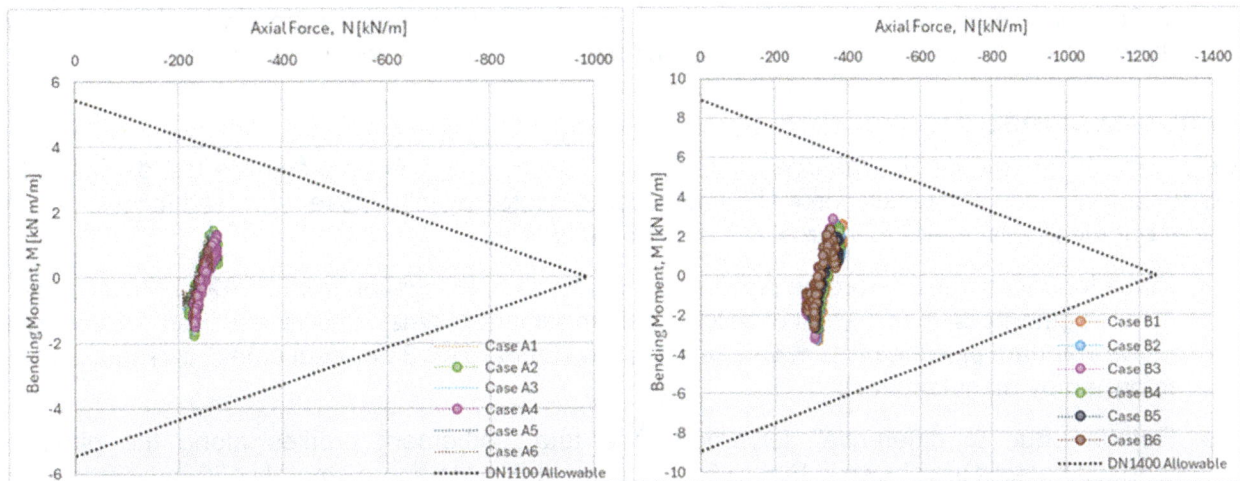

FIG 4 – Comparison of combined loads against allowable hoop stress surface (circumferential).

Axial design

The combined axial compressions and bending moments for the analysed cases are shown in Figure 5, compared to the allowable action envelopes. It is shown that the axial loads are less critical compared to the circumferential loads shown in Figure 4. The results also indicate the estimated utilisation of allowable compressive stresses is 52 per cent. Note that the modelling has conservatively ignored the contribution of joint centre stoppers to load redistribution, and hence the actual compressive loads are expected to be lower than presented herein.

The maximum shear stress from PLAXIS™ results is 160 kPa, which is approximately 15 per cent of the GRP pipeline lap shear strength of 1100 kPa, and 1.3 per cent of cross laminar shear strength of 12 300 kPa, as provided by the manufacturer. Since the utilisation is relatively low, the shear load would have marginal impact on bending capacity.

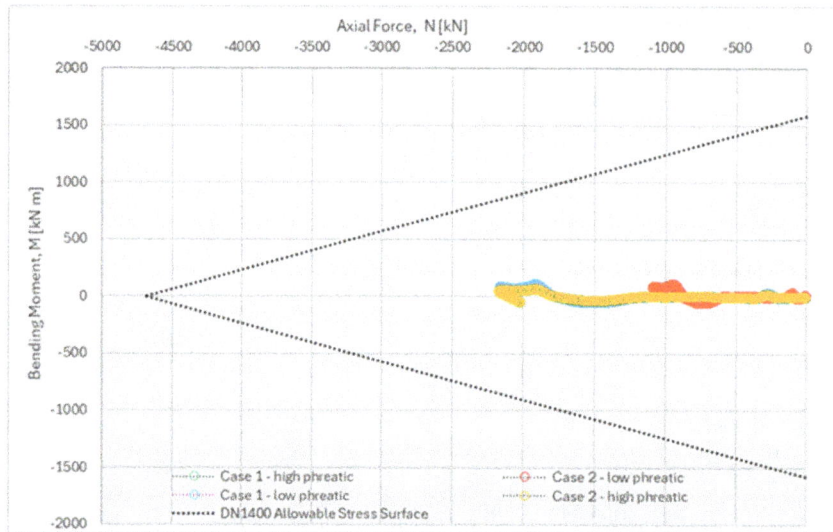

FIG 5 – Comparison of combined compression and bending against allowable hoop stress surface (axial).

A sensitivity study has also been undertaken to inform relevant risks associated with key uncertainties under extreme events over the life of operations which included:

- The absence of confining effect when water level is below the pipeline. This would increase the pipe ovality.

- The degradation of flowable fill would result in reduced backfill modulus.

- The combination of above.

The findings highlight that the long-term stiffness of the trench backfill is key consideration to pipe design. Selection of flowable fill mix design and composition will require the evaluation durability under long-term conditions and stiffness when compared to surrounding trench material.

Joint movement

A standard 12 m pipe length has been adopted for design. Since GRP pipes are not designed to take tensile stresses, the tensile force shall be accommodated by joint movement. The following pipe movement mechanisms are of interest for joint type selection.

- Pipe elongation: the interpreted pipeline maximum axial elongation from PLAXIS™ cases over 12 m segments is up to 10 mm of below final embankment crest. This is less than 0.1 per cent of the standard pipeline of 12 m and satisfies the standard <0.3 per cent pipe draw requirement informed by manufacturer.

- Rotation due to differential settlement: the total settlement profiles along the pipeline longitudinal axis are shown in Figure 6. It is shown that up to approximately 450 mm settlement is estimated below the tailings impoundment when the phreatic level is low. The differential settlement along the profiles would results in rotation of the pipeline, as interpreted in Figure 7. The maximum rotation is expected to coincide with the sudden changes of slope profile and the underlying materials, which is estimated to be up to 0.35 degrees.

- Combined joint movement: the inferred movement due to the combined elongation of 10 mm over 12 m segments and joint rotation of 0.35 degrees from 3D PLAXIS™ results is approximately 19 mm over 12 m segments, which is lower than the maximum elongation tolerance of 36 mm at joints. Therefore, no joint damage is expected over the design life.

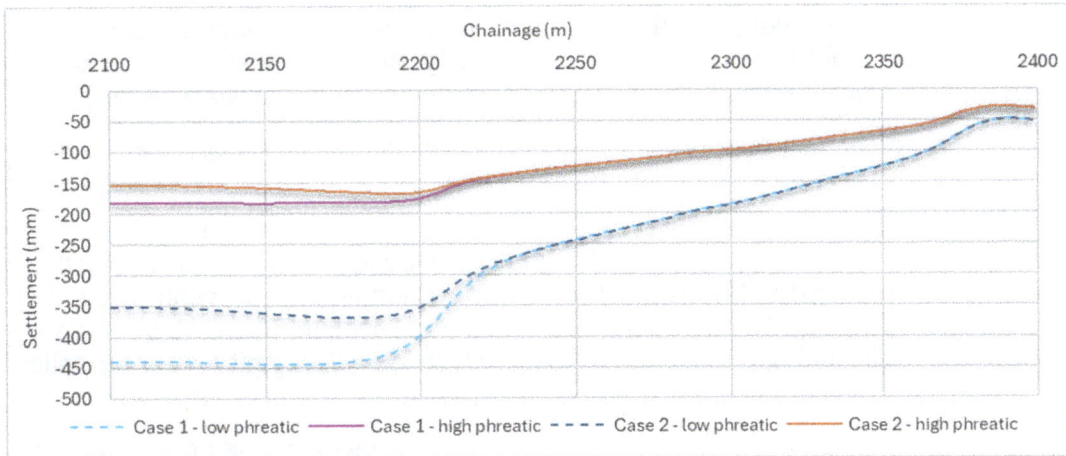

FIG 6 – Total settlement profile from PLAXIS™ 3D analyses.

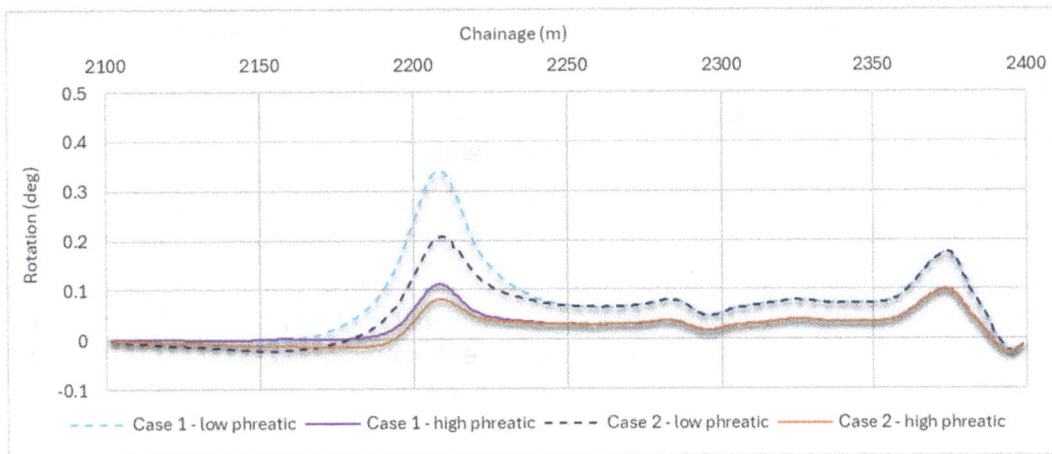

FIG 7 – Interpreted pipeline rotation profile from PLAXIS™ 3D analyses.

Durability

The GRP materials are expected to have equivalent or better performance than concrete. Therefore, durability risks of the pipeline associated with the loss of materials are expected to be low. Flowable fill's durability is primarily dependent on its designed compressive strength and is susceptible to degradation from repeated wetting and drying cycles, particularly in harsh environments with extreme temperature fluctuations. If it deteriorates *in situ*, it will continue to function as a granular fill. The flowable fill will utilise gravel as the filler to improve the stiffness (both intact and degraded) and the wet dry durability will be part of the mix design and trial considerations.

Construction

The quality control (QC) of the construction materials and the quality assurance (QA) of the construction product are of significant importance for the implementation of the design intents. Technical specifications included in the design have the following key elements:

- The procured GRP pipelines shall meet relevant industry quality standard with a valid certificate. The pipeline quality shall be maintained during transportation and storage period, subject to quality verification from qualified professionals prior to installation;

- The strength and modulus properties of the flowable fill for trench backfill shall be verified via laboratory testing by NATA accredited Laboratory testing facility against the design assumptions, regularly at the specified frequency;

- The strength and modulus properties of the *in situ* pipeline foundation shall be verified via *in situ* testing by geotechnical testing authority against the design assumptions, regularly at the specified frequency;

- The properties, placed densities and moisture conditions of compacted earthwork materials used for embankment backfill shall be verified by geotechnical testing authority against the design specifications, regularly at the specified frequency to ensure dam safety;

- Survey review of as-built components shall be conducted to meet the required tolerance limits.

Monitoring

To monitoring and validate the operational design performance of the buried GRP pipe, a DFOS network has been developed to be installed along the full length of the decant system to enable continuous monitoring along the entire length of optic fibre cables. The DFOS network includes measurements of strain, temperature and acoustic which provide monitoring information on the following:

- The temperature sensing can infer seepage and anomalies along the gravity decant system.

- The strain sensing can detect changes in internal deformation of the tailings embankment and pipework, complementing the continuous air measurement techniques of exterior deformation, such as using GPS survey monuments and satellite deformation estimates.

- The acoustic sensing has the potential to measure seismic wave velocity variations along the network to infer potential internal erosion, assess liquefaction risks due to seismic activities and map differential settlement of the embankment.

Review of Interrogators for assessment of the data from the DFOS will be developed and several extra DFOS cables have been installed to assess the data prior to completion of the full system.

In addition to the DFOS cable network, a series of conventional instrumentation such as piezometer, accel shape arrays, thermistors and geophones will be installed to validate the interpretation of data received via the DFOS. Existing surveillance instruments continue to be used to inform the dam performance.

CONCLUSIONS

This paper details a case study on designing GRP pipelines for a gravity decant system. FEA were used to evaluate soil-structure interactions, ensuring design criteria such as structural integrity, permissible displacement, and durability are met. The design parameters will be verified during construction via a quality control during field and laboratory testing. Operational performance will be monitored using an innovative DFOS network instrumentation system.

While a similar design approach could be taken to other forms of buried flexible pipe, only GRP was modelled as part of this design as it better met the project specific durability, abrasion and corrosion resistance and cost requirements.

In summary, the use of GRP pipelines in tailings storage facilities facilitates the efficiency, reliability, and sustainability of tailings management. Acknowledging the absence of documented successful applications of GRP pipelines within tailings management context, this paper serves as a case-based technical guide for engineers and decision-makers in the mining sector, advocating for advanced materials and innovative designs in essential infrastructure projects.

REFERENCES

Flowtite, 2025. GRP pipes with Flowtite continuous filament winding technology [online]. Available from: <https://www.flowtite.com/technology/> [Accessed: 11 April 2025].

Independent Expert Engineering Investigation and Review Panel (IEEIRP), 2015. Report on Mount Polley Tailings Storage Facility Breach, Independent Expert Engineering Investigation and Review Panel, Province of British Columbia, January 30, 2015.

Look, B G and Cameron, D A, 2018. Buried flexible pipes: Design considerations in applying AS2566 standard, *Australian Geomechanics*, 53(2):101–115.

Standards Australia & Standards New Zealand, 1998. AS\NZS 2566.1:1998 Buried flexible pipelines, SAI Global. Available from: <https://www.saiglobal.com/>

US Federal Emergency Management Agency (FEMA), 2007. Technical Manual: Plastic Pipe Used in Embankment Dams (FEMA P-675), November 2007. Available from: <https://damtoolbox.org/wiki/Technical_Manual:_Plastic_Pipe_Used_in_Embankment_Dams_(FEMA_P-675)>

Water Corporation, 2024. Pipeline selection guidelines - selection criteria for pipe, pipe fittings and interconnection with new and legacy corporation pipelines, ver 1, rev 2, February 2025. Available from: <https://pw-cdn.watercorporation.com.au/-/media/WaterCorp/Documents/About-us/Suppliers-and-contractors/Resources/Design-standards/Pipeline-Selection-Manual.pdf?la=en&rev=f4eba328fc3a49b4a644697b1eef601e&hash=7FA3E8A58216834F3B616D74483E495A>

Case study, development of integrated waste landforms, design and implementation

C Hogg[1] and P Atmajaya[2]

1. Senior Principal Tailings Engineer, CMW Geoscience, Perth WA 6155.
   Email: chrish@cmwgeo.com
2. Senior Geotechnical Engineer, CMW Geoscience, Perth WA 6155.
   Email: prasudia@cmwgeo.com

ABSTRACT

This paper explores the considerations and constraints in the implementation of an integrated waste landform type tailings storage (IWLTSF) required as part of a greenfields gold project in Western Australia. An existing tailings storage was designed and commissioned at the site. The existing TSF is an integrated waste landform type facility which comprises a tailings storage surrounded substantially by waste dumps. This type of facility was selected for the project due to a constrained project area and the ready availability of mine waste overburden. The design of TSF1 was based on limiting seepage to environmentally acceptable levels while reducing the cost of lining the facility.

TSF1 was designed to store 53.5 Mt of tailings with tailings densities ranging from an initial density of 1.25 t/m^3 (dry) to 1.4 t/m^3 (dry); due to the transition of storage from clay tailings to fresh rock tailings. TSF1 has a tailings area of approximately 100 ha and a maximum design height of 39 m. The existing tailings storage was performing well and the mine operator has decided to utilise the IWLTSF concept for future mine expansion.

Sites for the new IWLTSF were considered based on identifying areas away from sensitive creek lines with the other main constraint being siting the TSF at a reasonable distance from existing and proposed open pits. The site selected for TSF2 was a site immediately adjacent to TSF1 with a proposed nearby pit providing a source of construction materials.

TSF2 would have a similar area to TSF1 and has been designed to store approximately 27 Mt of tailings. Design challenges for TSF2 included a requirement to store water prior to tailings deposition, timing of construction and integration with the TSF1 landform.

This paper examines the challenges encountered in the design of the facilities, how these were overcome and presents some learnings from the design and implementation process of both TSFs.

INTRODUCTION

This paper explores the considerations and constraints in the implementation of an integrated waste landform type tailings storage (IWLTSF) required as part of a greenfields gold project in Western Australia. An existing tailings storage was designed and commissioned at the site. The existing TSF is an integrated waste landform type facility which comprises a tailings storage surrounded substantially by waste dumps. This type of facility was selected for the project due to a constrained project area and the ready availability of mine waste overburden. The case study examines a project in the Pilbara region of Western Australia. The project is located approximately 65 km south-east of Newman. A site plan of the project area is presented in Figure 1.

The planning and design of the IWLTSFs, considered life-of-mine storage, constructability and cost reduction, and operations. Whilst detailed closure studies were not part of the planning and design, the concept selected has several advantages for closure and this is discussed in the paper.

FIG 1 – Site plan.

BACKGROUND

Topography

The project area is located in the Bulloo Plains of the Hills Zone described as hardpan wash plains, stony plains, hills and ranges (with some sandplains) on sandstone and shale of parts of the Collier and Bresnahan Basins and granite of the Sylvania Inlier. Surface geology comprises typically red shallow loams (often with hardpans), red loamy earths, stony soils and red deep sands with some red shallow sands (refer Capricorn Metals Ltd (CMM), 2022). The project is located within the Lake Disappointment catchment, approximately 10–15 km south of the regional watershed boundary with the Upper Fortescue River catchment.

Climate

The project area has a semi-arid climate with hot and dry summers, and warm to cold and dry winters. Based on available records from the Bureau of Meteorology (<http://www.bom.gov.au>) for Newman Aerodrome) the annual rainfall is 232 mm and the average annual evaporation is 3733 mm. The site can experience intense rainfall, with a 1:100 year. Annual Exceedance Probability (AEP) event 72 hr duration, of 237 mm.

Hydrogeology

The project is located in the East Murchison Groundwater Area. The groundwater in the region typically occurs in fractured rock and weathered bedrock aquifers near the mine. Surficial thin calcrete aquifers occur to varying degrees along Savory Creek, south of the project area (refer GRM, 2017).

Groundwater quality in the region is typically fresh to brackish, with electrical conductivities (EC) generally <4000 µS/cm and neutral to slightly alkaline waters.

Geology and ground conditions

The project is located on a broad alluvial fan which obscures most of the local geology under a dominantly flat topography. A shallow near surface supergene horizon is hosted within a duricrust, which extends to approximately 10 m below ground levels (mbgl) on the east of the deposit. The depth of weathering at project deposit extends to approximately 60 mbgl.

The weathered rock (saprolite) at the project deposit was described as comprising of an upper saprolite and a lower saprolite (MHA, 2017). The upper saprolite consisted of sandy, mainly kaolinitic clay (60–70 per cent fines content by weight), which extended to the base of complete oxidation at between 25 mbgl and 35 mbgl. The underlying lower saprolite consisted of smectitic and kaolinitic clay (50 per cent fines), and sand-sized particles, which extended to the base of weathering at between 50 mbgl and 60 mbgl. At the base of the lower saprolite is a transition zone comprising highly friable saprolite with 20–30 per cent fines, becoming saprock.

The primary mineralisation has been found extending to approximately 270 mbgl, although it remained open down dip below this depth.

Sedimentary units of the Proterozoic Bangemall Group onlap the greenstone unit to the immediate south and west of the project deposit. These sediments comprise interbedded sandstone, siltstone, shale and dolomite of variable thicknesses, which are intruded by high level gabbro and dolerite sills. To the west and south-west of the deposit, an alluvial sequence of mixed gravel and silt up to approximately 30 m thick overlies the Bangemall Group sediments.

Tailings properties

Tailings test work was conducted in 2018 on samples of tailings from metallurgy testing, identified as 'Oxide Composite', 'Laterite Composite' and 'Fresh Composite'. In 2022, tailings samples termed 'Oxide Tailings' and 'Laterite Tailings' were obtained by CMM from the process plant. Table 1 presents a comparison of the results from the two sets of test work from 2018 and 2022:

TABLE 1

Tailings test work summary.

Engineering properties	Oxide tailings		Laterite tailings	
	2018[1]	2022	2018[1]	2022
% Fines Passing 75µ size	50% fines 9% passing 2 µ	61% 13% passing 2.8 µ	57% fines 8% passing 2 µ	57% 10% passing 2.5 µ
Plasticity	Medium plasticity	High plasticity PI 20%	Low plasticity	Low plasticity PI 9%
Air drying density		1.24 t/m^3 (dry) after approximately 15 days		1.44 t/m^3 (dry) after approximately 15 days
Undrained settling density		0.89 t/m^3 (dry) after 3 hrs		1.06 t/m^3 (dry) after 1 hr
Drained settling density		1.26 t/m^3 (dry) after 23 hrs		1.24 t/m^3 (dry) after 6 hrs
Hydraulic conductivity[2]	3.7x10^{-8} to 1.8x10^{-8} m/s	1.1 × 10^{-8} to 2.4 × 10^{-9} m/s	1.5x10^{-7} m/s	1 × 10^{-8} to 2.2 × 10^{-9} m/s

1) 2018 'Fresh Composite' sample was a non-plastic sandy silt with 53 per cent fines (passing 75-micron size) and 4 per cent passing the 2-micron size, permeability testing was not carried out on this sample.

2) The 2018 and 2022 tailings permeabilities were from oedometer tests and Rowe Cell tests, respectively.

The results of the test work indicated the tailings samples generally had adequate settling characteristics with relatively good consolidation characteristics. The 'Oxide Tailings' settled to lower densities than the 'Laterite Tailings'. The results of the air drying tests indicated maximum densities were achieved within 15 days under laboratory conditions. The 'Oxide Tailings' had an order of magnitude lower permeability (4.4 × 10^{-7} m/s) compared to the 'Laterite Tailings' (2.3 × 10^{-6} m/s) at densities around the settled density.

The inferred tailings densities from the results of the test work were 1.25 t/m^3 (dry) and 1.3–1.4 t/m^3 (dry) for the 'Oxide Tailings' and 'Laterite Tailings', respectively. 'Fresh Tailings', which would be produced after the 'Oxide Tailings' and 'Laterite Tailings', are expected to have higher densities ie conservatively estimated tailings density of 1.4 t/m^3 (dry).

Based on the densities and assuming a 'cycle' time of approximately 1.5 months, the estimated optimum tailings rate of rise was 2.5 m/annum. The rate of rise would allow a tailings throughput of approximately 3.3 Mt/a based on a tailings drying area of approximately 105 ha.

Based on the estimated tailings density of 1.4 t/m^3 (dry) and a 'cycle' time of around month, the estimated optimum tailings rate of rise for 'Fresh Tailings' was 3.0–3.5 m/annum. Approximately 4.4 Mt/a of 'Fresh Tailings' production would be feasible based on the expected rate of rise and a TSF storage (basin) area of 105 ha. This allowed planning for a second TSF based on tailings area and storage of future tailings production. An increased tailings area was ultimately required for future production ie 5.3 Mt/a.

TSF1 DESIGN AND IMPLEMENTATION

The project commenced in 2019. TSF1 was constructed as an IWLTSF in order to store 30 Mt of process tailings from the project CIL/CIP plant over ten years life. TSF1 was designed with a basal area of approximately 120 ha and a maximum embankment height of 25 m, which would be constructed as part of the mining operations with the facility being located within waste dumps. Subsequently the design of TSF1 was revised to accommodate an expected increase in storage demand to 53.5 Mt, with 14 m of embankment height added and a basal area that would extend over approximately 132 ha (CMW Geoscience, 2018).

TSF1 starter embankment was constructed to provide nominally one year's storage life, and afterwards was raised along with the surrounding waste dumps in order to store the LOM tailings. Downstream construction technique was used in the Stage 1 and Stage 2 embankment raises, and will be utilised for the future Stage 3 to Stage 7 embankment raises (CMW Geoscience, 2024a). Figure 2 shows the perimeter embankment staging and geometry. The use of downstream construction technique means a robust structure with the embankment foundational stability not relying on the liquefaction resistance of *in situ* tailings ie static liquefaction or earthquake induce liquefaction. The raising of the decant accessway and rock ring used the centreline construction technique.

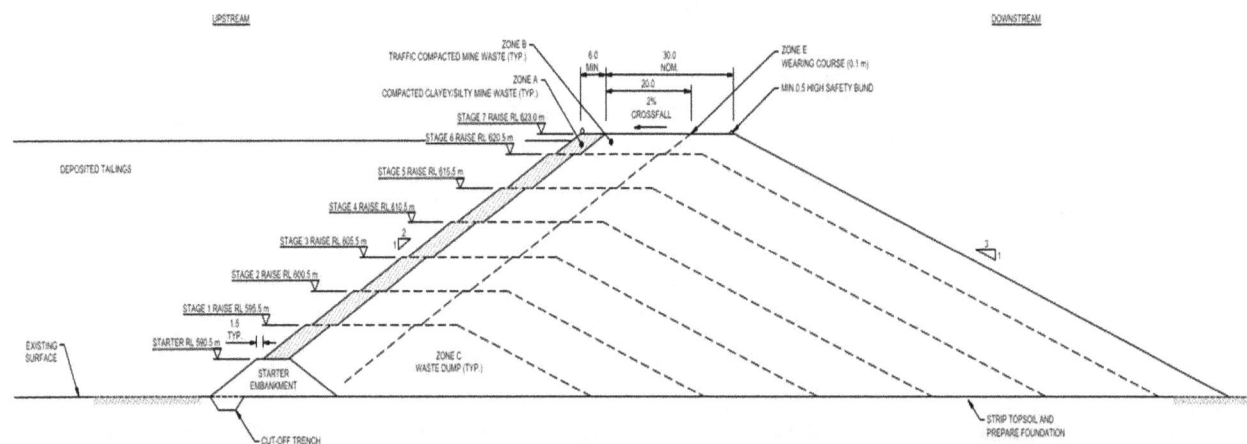

FIG 2 – Typical TSF1 embankment section (NTS).

The starter embankment of TSF1 was designed to be a zoned embankment comprising an upstream zone of roller compacted low permeability clayey/silty materials (Zone A) and a downstream zone of traffic compacted gravel material (Zone B). Zone A material was to be sourced from within the TSF site, and Zone B material which forms the bulk of the embankment was to be sourced from within the top of the nearby pit (CMW Geoscience, 2024a).

The embankment raises of TSF1 were designed to be zoned embankments similar to the starter embankment, with waste dumps surround (Zone C) which was to be built progressively over the life-of-mine. The raises were designed to be nominally 5 m in vertical height, each, and were to be constructed using materials sourced from the saprolite zones within the nearby pit for Zone A, and from adjacent waste dumps for Zone B (CMW Geoscience, 2024a).

The design of TSF1 was based on limiting seepage to environmentally acceptable levels while reducing the cost of lining the facility. The starter embankment incorporated a cut-off trench excavated to 'refusal' on cemented laterite gravels (Wiluna Hardpan) for reducing seepage losses. In conducting the seepage assessments, consolidation and storage of low permeability clayey tailings was allowed for in the design.

At implementation, the design of TSF1 evolved with some changes made prior to and during construction. One of the changes during the construction of the starter embankment involved the use of Zone A materials sourced externally from the TSF site as suitable clayey materials with higher fines contents and less oversizes were encountered from within the nearby pit; the specifications required fines content of >20 per cent and <66 per cent with PI of <30 per cent and a maximum particle size of 150 mm. Another change made during the starter embankment construction was dumping of mine waste outside the remit of Zone B, essentially integrating Zone C from inception rather than having a standalone starter embankment. A further change to the starter embankment was made to allow the mining fleet to construct the decant accessway and rock ring, with crest widths not bound by the 8 m and 6 m designed for the respective structures. The main design change made for the Stage 2 and Stage 3 embankment raises was the addition of 1.5 m bench on upstream slope to provide a pipe bench, and progressive raising of the decant accessway and rock ring to suit the rise in tailings level. Other design changes made during the embankment raises mainly related to QA/QC and operation of TSF1, with implementation of HILF tests to complement the standard compaction testing with determination of moisture content and dry density relationship using the standard and modified compactive efforts (AS1289; Standards Australia, 2017), and raising construction of the embankment in two halves to allow continuous operation of the tailings slurry pipeline.

Performance monitoring of TSF1 utilises a series of eight monitoring bores around the facility. Groundwater quality as indicated by monitoring bores needs to comply with strict limits for pH of 8 to 10, WAD Cyanide of <50 mg/L, and Copper, Chromium (Total) and Arsenic concentration limits of <1 mg/L in the supernatant pond during operation (DWER Licence No. L9324/2022/1; Department of Water and Environmental Regulation (DWER), 2022). Additionally, metals suite limits are set to comply with 95 per cent protection level for freshwater ecosystem (ref ANZECC, 2000).

The groundwater quality is reviewed quarterly and reported annually as part of the site annual environmental report and TSF annual audit and management review. Based on a recent annual audit (CMW Geoscience, 2024b), groundwater levels varied between approximately 7 mbgl and 27 mbgl. Non-compliances have been recorded and actioned for the bores and monitoring bores. Arsenic and copper levels are monitored closely, and dosing of the decant using Hydrogen Peroxide and Ferrous Sulfide are implemented to reduce the copper concentration in the supernatant water. The elevated levels of other metals (eg Boron, Cadmium, Zinc) are noted to be consistent with the background levels.

Overall, the existing TSF1 has performed adequately (CMW, 2024b) and the mine operator has decided to utilise the IWLTSF concept for future mine expansion ie a new TSF2.

TSF2 DESIGN

Sites for the new TSF2 were considered based on identifying areas away from sensitive creek lines with the other main constraint being siting the TSF at reasonable distances from existing and proposed open pits, and away from the existing and proposed process plants. The site selected for TSF2 was a site immediately adjacent to TSF1 with a proposed nearby pit providing a source of construction materials.

TSF2 would have a similar impoundment area to TSF1 and has been designed as a two-cell IWLTSF to store approximately 27 Mt of tailings over a target of up to 5.5 years of storage life. TSF2 will occupy a total footprint area of about 193 ha, providing a storage area of approximately 188 ha. The maximum embankment height will be approximately 17 m and the total crest length will be about 9.2 km. Design challenges for TSF2 included a requirement to store water prior to tailings deposition, timing of construction and integration with the TSF1 landform, and consideration to the impact of a dam break on the adjacent new pit.

TSF2 has been designed with zoned perimeter embankments comprising 6.0 m wide upstream zone of roller compacted clayey mine waste (Zone A) built on a 20 m wide transition traffic compacted zone (Zone B), which would be constructed upon waste dump (Zone C). As opposed to TSF1, the design of TSF2 implemented 'Lift' rather than 'Stage' in order to store more bulk waste from the adjacent new pit (refer to Figures 3 and 4).

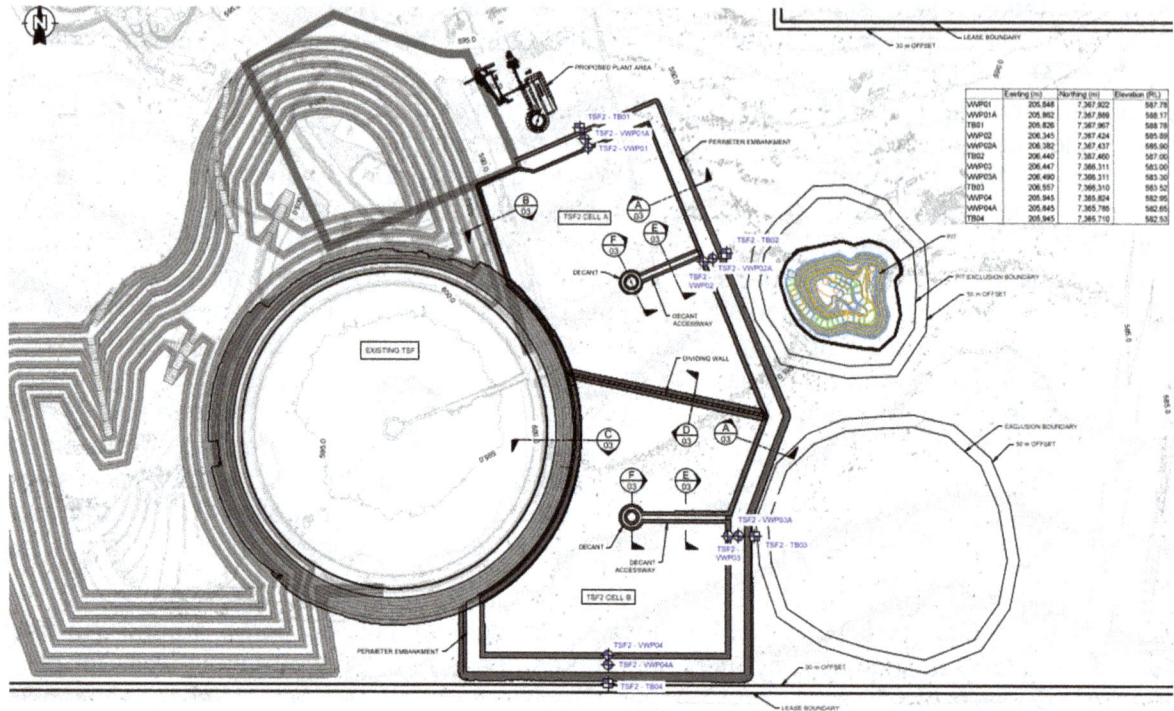

FIG 3 – TSF2 proposed layout.

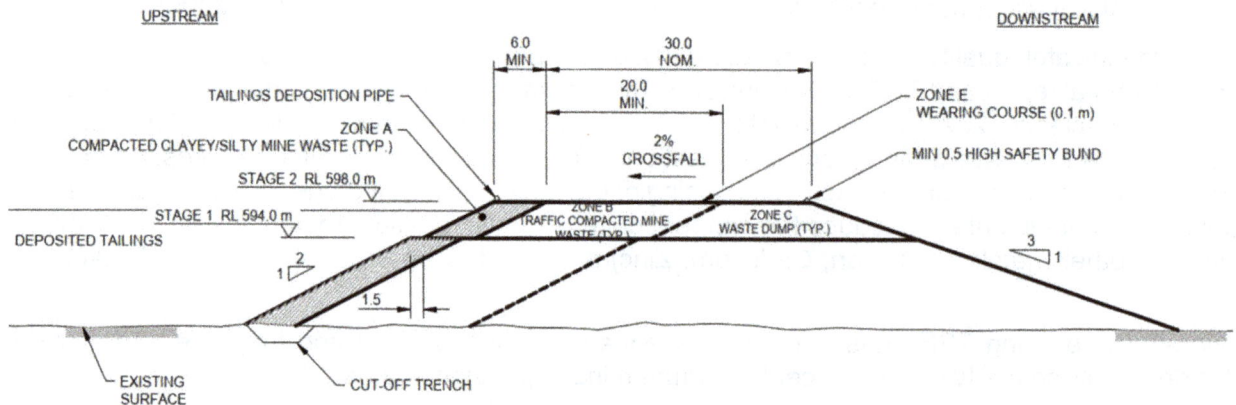

FIG 4 – Typical TSF2 embankment section.

Zone C will have a nominal width of 25.5 m upon the completion of TSF2 Stage 1 construction, which design has been adapted to the required amount of bulk waste. The Stage 1 embankment would have as a part of Zone A cut-off trench 1.5 to 2.5 m deep with 4 m wide base, founded on a Ferricrete layer in order to reduce seepage losses.

Embankment stability, seepage, and deformation analyses, plus dam break assessment and water balance estimations were performed to support the design of TSF2.

The seepage analyses indicated that approximately 100 m³/day and 150 m³/day of seepage can be expected during the operation of TSF2 Stage 1 and Stage 2 embankments, respectively. The stability analyses indicated adequate factors of safety for the drained, undrained, pseudo-static and post-seismic conditions when compared with the recommended minimum factors of safety in ANCOLD (2019). Both the seepage and stability analyses were performed assuming tailings deposition is managed so as to prevent prolonged ponding near the embankment, with pond of ≥100 m from the perimeter embankment at normal operating conditions.

The dam break assessment included a numerical modelling using a Eulerian numerical framework to assess the potential impacts of a TSF2 dam break on the adjacent new pit. The assessment comprised two breaching scenarios for the northern cell TSF2 Cell A, and one for the southern cell TSF2 Cell B. Estimated released volumes ranged from 2.7 Mm^3 to 4 Mm^3 for TSF2 Cell A and approximately 6 Mm^3 for TSF2 Cell B. Material releases were modelled over a four hour duration, capturing the upper bounds of peak flow rates (m^3/s), to evaluate conservative failure scenarios. The results demonstrated that in the most conservative scenario, the maximum flood depth around the adjacent new pit could reach 1.6 m, however, the maximum flood depths in the inundation area would range between 0.5 m and 15 m closer to the embankment.

The water balance estimations were carried out based on tailings area of 103 ha of the larger southern cell TSF2 Cell B, and using run-off coefficient of 0.4 for the tailings, a pool area equivalent to 40 per cent of the tailings area, running beaches of approximately 3 per cent of the tailings area, evaporation pan factor of 0.7, tailings residual moisture content of 33 per cent corresponding to the expected *in situ* density of 1.4 t/m^3 (dry), tailings slurry density of 46 per cent solids, tailings production rate of 4.5 Mt/a, and permeability through dam floor of 1 × 10^{-7} m/s. The estimations provided the base for the water recovery system, which decant, pumps and piping must be designed for a minimum recovery of ≥435 t/h constituting a potential for 25 per cent water return plus the additional capacity to recover water from design storm events of the 1:100 yr. AEP event 72 hr duration.

Additional water balance estimations were carried out to determine the expected dewatering from the adjacent new pit into TSF2, and it estimated that the project would have about 9 months of mining from the new pit before high water flows are predicted, and subsequently the need for temporary water storage and evaporation. Estimation was made for the amount of mine waste which would be required to construct an embankment to the level which will accommodate the P_{90} and P_{50} probabilities peak predicted water surplus of 1.5 Gl and 0.5 Gl, respectively. The total surplus volumes of 4.2 Gl and 6 Gl for the P_{50} and P_{90}, respectively, were modelled in the stability and seepage analyses to verify the impact of this temporary water storage in the northern cell TSF2 Cell A.

DISCUSSION

TSF1 and TSF2 would be classified, based on their respective designs up to date and in accordance with ANCOLD (2019) consequence rating, as 'Significant' and 'High C', respectively. The higher rating of TSF2, which has been designed with a shorter maximum embankment height as compared to TSF1, was largely due to its location near the adjacent new pit. In the event that a TSF is in imminent danger of failure and breach, the Emergency Response Action Plan in the existing TSF Operations Manual (CMM, 2023) would be enacted.

The probability of embankment failure during the life of TSF1 and TSF2 has been assessed as low provided construction and operation guidelines are adhered to by the mine operator, and ongoing monitoring and NATA-accredited QA/QC analyses are continued of the existing monitoring bores as outlined in the existing TSF Operations Manual and in the DWER licence conditions (DWER, 2022). Vibrating wire piezometers have in addition been installed around the perimeter of TSF1 to enable the phreatic surface within the embankment to be monitored thus allowing an early warning of seepage and providing data for future stability analyses.

A series of tailings test work provided valuable input into a TSF design. Tests were performed prior to commissioning and have been repeated during the operational life. The results of these tests have been used to inform the design engineer, specifically for water balance and tailings properties (*in situ* densities etc) for essential updates applicable to the existing design and for future storage capacity prediction.

Bulk mine waste samples were collected during a design study to determine both geotechnical and geochemistry properties. Grading, Atterberg limits, modified compaction, remoulded permeability, California Bearing Ratio (CBR) and remoulded shear box tests were undertaken to determine the geotechnical characteristics. Acid base analyses and determination of net acid generation results were undertaken to establish, among many of the geochemical properties, the potential for acid forming and major/minor-elements contents. Both sets of properties provided critical input to selection of materials for TSF construction and at closure.

Due to the need to accommodate TSF2 Cell A as a temporary water storage, a further review of the past and recent geotechnical investigations was made in order to refine the soil classification and permeability parameters used in the design. This work involved reviewing the stratigraphy of the ground encountered in the areas of TSF1, which surficial layer was in general logged as Silty Sandy GRAVEL, and tracing where the layer would intersperse with the Gravelly Clayey SAND that was generally encountered on the surface of TSF2 Cell A. In addition, unlike typical permeability tests the recent geotechnical investigation has accounted for differing depths with boreholes of 1 m, 5 m, 10 m and 15 m deep, allowing higher level of evaluations that provide a greater confidence in deriving suitable parameters for assessing the location of TSF2 Cell A.

Regular liaison and communications between design engineer and mine operator were conducted during the design process, by means of Teams meetings, email communications, reviews, etc. The design was developed such that it was 'fit for purpose'. Constructability, cost and closure were considered during the design by tapping into site mining and environmental experience along with expertise provided by the designer.

As with the past constructions of TSF1 starter, Stage 1 and Stage 2 embankments, future works to raise TSF1 and to build TSF2 will be monitored by the design engineer. QA/QC tests will form part of the construction, allowing both the identification of suitable materials for construction and the quality of the earthworks. Full-time supervision of the earthworks will be carried out by a civil engineer employed by the mine operator, who will communicate the technical and operational aspects of the construction, including the QA/QC tests, to the design engineer. Changes to the design of TSF1 and TSF2 could only be made upon discussion and approval from the design engineer.

The IWL concept lends itself to ease of TSF closure as the facility is in effect fully enclosed by waste dumps, which means there will be mine waste materials readily available for use in its final rehabilitation. Additionally, the concept also means a larger section of the perimeter embankment both for TSF1 and TSF2 and as such, allowed the designer to manipulate the stage-by-stage cross-section of the structure to adapt to the site requirements, not only for the current design but also when demands change during the life cycle of the mine including potentially also at closure.

CONCLUSIONS

The concept of IWLTSF offers notable environmental and economic advantages. Environmental benefits include rehabilitation of the embankments using nearby waste materials and reducing rehandling costs during final surface rehabilitation. Economically, it optimises mine waste for embankment construction, reducing capital and operational expenses, and closure costs. Tailings distribution lines can remain in place during construction, unlike conventional methods. Additionally, the design enhances embankment stability due to the waste mass surrounds.

An IWLTSF has proved to be an appropriate concept for TSF1, and is considered likewise for TSF2, given the constraints identified with sensitive areas, haulage distances, presence of adjacent infrastructure, and upfront availability of construction materials. The concept is therefore deemed appropriate for the project area considering the site locations, cost, engineering risks, environmental risks, social risks and closure requirements.

The technical reviews of the existing TSF1, carried out as part of an annual TSF audit, concluded that the adoption of the IWL concept for future storage TSF2 is appropriate provided that the facility is adequately managed with respect to tailings deposition and water management. Raising and expansion of TSF need careful consideration both from a designer and company perspective.

Regular liaison between design engineer and mine operator is essential for developing a design that fully meets the mining requirements. The design engineer maintains active communication with relevant lead mine personnel during TSF design activities by means of Teams meetings, email communications, reviews, etc. This enables the design to be 'fit for purpose' and improvements to be made to on-site systems and processes. Constructability, cost and closure were considered during the design by tapping into site mining and environmental experience along with expertise provided by the designer. The same principle is applied to the construction phase of TSF1, with this becoming crucial with variation of materials and operational challenges which became apparent during construction.

REFERENCES

Australian and New Zealand Guidelines for Fresh and Marine Water Quality (ANZECC), 2000. National Water Quality Management Strategy, paper (4):1, The Guidelines, Chapter 1–7, October 2000.

Australian National Committee on Large Dams (ANCOLD), 2019. Guidelines on Tailings Dam Planning, Design, Construction, Operation and Closure, ANCOLD.

Capricorn Metals Ltd (CMM), 2022. Karlawinda Gold Project – Environmental Licence Application M52/1070 Supporting Document, Greenmount Resources Pty Ltd, a 100% Owned Subsidiary of Capricorn Metals Ltd, ref, CAP-KARL-2022 Rev 1.

Capricorn Metals Ltd (CMM), 2023. Karlawinda Gold Project – Tailings Storage Facility Operation Management Plan, Capricorn Metals Ltd, document ID GRM-PLN-1000, Rev 1, 9 August 2023.

CMW Geoscience (CMW), 2018. Tailings Storage Facility, Karlawinda Gold Project, near Newman, WA, Design Report, ref PER2017–0251AB, Rev 4.

CMW Geoscience (CMW), 2024a. Tailings Storage Facility(TSF1), Additional Raises, Karlawinda Gold Project, WA, Design Report, ref PER2023–0242AE, Rev 1.

CMW Geoscience (CMW), 2024b. Annual Audit and Management Review, Tailings Storage Facility (TSF), Karlawinda Gold Project, ref PER2024–0007AB, Rev 1.

Department of Water and Environmental Regulation (DWER), 2022. Licence no, L9324/2022/1 in file DER2022/000042, for duration 29 August 2022 to 28 August 2042, Mining Lease – M52/1070, Capricorn WA 6642, Shire of Meekatharra, held by Greenmount Resources Pty Ltd, dated 29 August 2022.

Groundwater Resource Management (GRM), 2017. Karlawinda Gold Project Hydrogeological Feasibility Study, Groundwater Resource Management unpublished report prepared for CMM.

MHA Geotechnical (MHA), 2017. Geotechnical Site Investigation, Karlawinda Gold Project, Capricorn Metals, MHA Geotechnical ref, P03-16-RF.

Standards Australia, 2017. AS1289. Methods of testing soils for engineering purposes, Standards Australia, Sydney.

Considerations for design, construction and operation of protective levees and diversion drains for in-pit tailings storage facilities and mine infrastructure

W Ludlow[1], N Lumby[2], R Harrington[3] and A Goto[4]

1. Senior Principal, Red Earth Engineering, Brisbane Qld 4000.
   Email: wade.ludlow@redearthengineering.com.au
2. Tailings and Dams Specialist, Anglo American, Brisbane Qld 4000.
   Email: nicholas.lumby@angloamerican.com
3. Principal, Red Earth Engineering, Brisbane Qld 4000.
   Email: robert.harrington@redearthengineering.com.au
4. Senior Engineer, Red Earth Engineering, Brisbane Qld 4000.
   Email: anna.goto@redearthengineering.com.au

ABSTRACT

Levees and diversion drains are widely used to protect in-pit tailings storage facilities and mine infrastructure (including underground portals, population areas) to ensure suitable risk reduction and flood immunity from nearby channels and creeks and/or potential dam failure inundation extents. Despite the wide use of levees and diversion drains in the mining industry, a comprehensive industry guidance document has not been developed to cover the key aspects of design, construction and operation for Australian mining conditions. Levees and diversion drains designed or constructed without proper considerations inevitably turn into a maintenance burden for the owners over the mine life. This paper presents an assembly of key design criteria from various guidance documents and experience. The paper also presents key considerations for Australian conditions, including use and management of local soils, geomorphic approach to diversion drain design, mitigation of construction issues in dispersive soils and common operational issues such as caught catchments.

INTRODUCTION

Around the globe, levees and diversion drain structures are often used at mine sites to protect mine infrastructure, mine portals, mine pits and in-pit tailings storage facilities. Despite the wide use of levees and diversion drains in the mining industry, a complete industry guidance document has not been developed to cover all the key aspects of design, construction and operation for Australian conditions. There are several guidance documents that have been developed for levees and diversions around public infrastructure and communities, but these do not readily transfer to an Australian mining context (remote locations, varied local materials that don't meet standard specifications, different risk profiles etc). This paper provides details on design information and key considerations based on the authors' experience with all aspects of levee and diversion drain design, construction and operation in a mining context.

BACKGROUND

The International Commission on Large Dams (ICOLD) Bulletin 196 Preprint (2023) noted the following for levee practices in Australia (with a focus on public infrastructure and communities).

- Data for levees is not reported at a national level, and at a state/territory level, there are significant gaps in this information.

- Of the eight states and territories in Australia, three (Queensland, New South Wales and Victoria) are home to more than 90 per cent of Australia's at-risk properties vulnerable to flooding.

- In Victoria, levees are managed by local government agencies and Melbourne Water. Victoria has approximately 4000 km of levees.

- In New South Wales, levees are not regulated. A review conducted in 2013/2014 indicated there are approximately 350 km of urban levees.

- In Western Australia, levees are not regulated. There are approximately 240 km of levees in urban areas.

- In Queensland, a levee regulatory framework has been in place since 2014. However, there is an absence of information on levees and only two of the 77 local government organisations responded to a survey with information on levees. The two responding councils and a consultant design team reported a total of 15.4 km of levees and 1200 m of diversion drains.

- In Tasmania, the city of Launceston is protected by approximately 7 km of levees.

- No other information is currently available for South Australia, Northern Territory, and Australian Capital Territory.

- Of the information available, the height of levees above natural ground level in Australia varies considerably and can be as low as 0.2 m and as high as 8 m.

- Frequent and severe flood events have been reported across Australia, often with significant consequences including many fatalities. Most of this flooding was caused by floodwater overtopping the structures rather than levee failures. However, although not well documented, levee breaches are known to have occurred in the past.

- Some of the states have published guidelines. The *International Levee Handbook* (Ciria, 2013) and ICOLD bulletins are generally not directly applied within Australia.

The above shows there is not a complete levee design guidance document used in Australia for protective levees around public infrastructure and communities, and we suggest it is even less complete for the mining industry.

EXAMPLE MINING LEVEE FAILURE

On 19 January 2008, the Ensham Coal Mine in central Queensland had a flood event that exceeded the 1:100 annual exceedance probability (AEP). The levees protecting the mine were designed for the 1:100 AEP design event (Figure 1) and were overtopped and/or breached, resulting in large flooding of the pits and mine infrastructure. Large overtopping extents and breaches occurred at multiple levees, resulting in flooding of Pits B, C and D (Figures 2 to 4). This event was a critical reminder to the mining industry that levee and diversion designs need to be developed and managed to a high standard.

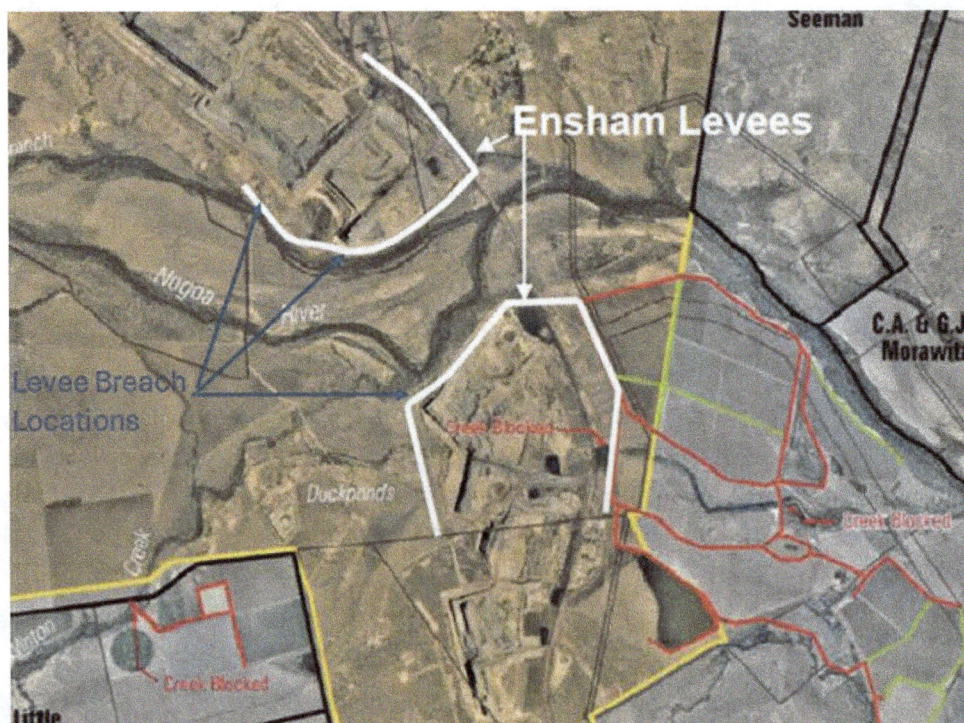

FIG 1 – Site plan Ensham Mine levees location and breach locations (Ensham Resources, 2008).

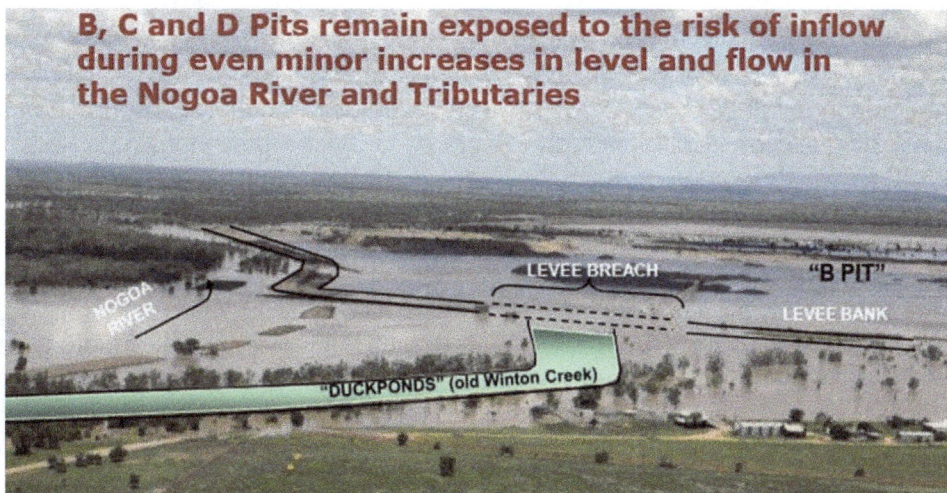

FIG 2 – Levee breach of B Pit (Ensham Resources, 2008).

FIG 3 – Floodwater commences levee overtopping (Ensham Resources, 2008).

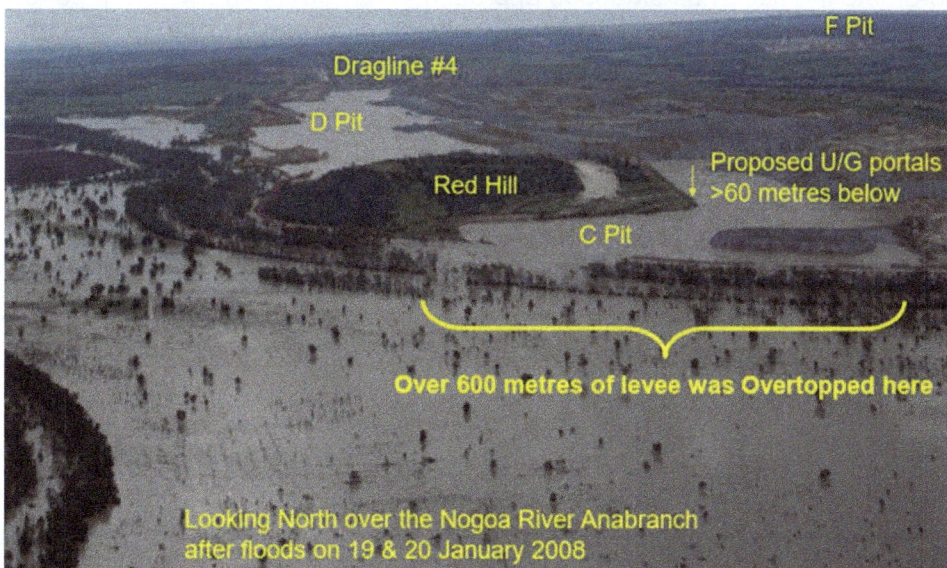

FIG 4 – Levee overtopping C Pit (Ensham Resources, 2008).

DESIGN RECOMMENDATIONS

For this paper, we compiled key design criteria with references to levee design guidance documents that have been primarily developed for public infrastructure and community protection. This compilation of criteria is provided below and we have provided additional commentary on how this can be applied in the mining industry.

Design flood event

In general, risk-based approaches are recommended for most civil structure design in mining – this extends to dam design, pit design, levee structures, etc. Below are various approaches for selecting the design event for the levee or diversion drain from several key references.

Reference 1: *The Manual*, Queensland government manual for assessing consequence categories and hydraulic performance of structures.

The Manual (Department of Environment, Science and Innovation, 2024), which was developed by the Queensland Regulator, provides guidelines for mine levees protecting mine pits from flood in-rush and for levees designed to divert contaminated waters or protect the structural integrity of a dam. The levees are subject to a consequence category assessment, and the minimum guidelines are provided in Table 1.

TABLE 1

Hydrological design criteria for levees (Department of Environment, Science and Innovation, 2024).

Consequence category for levee[*]	Design criteria: flood level for embankment crest levels[#]
Levee determined to be a regulated structures[^]	1:1000 AEP

[*] The design criteria identified in this table are relevant to section 2.3.1, Consequence assessment for levees. As such, they are not relevant to dam break failure modes where no overtopping occurs, such as failure caused by piping. Consideration by the suitably qualified and experienced person may need to be given to the appropriateness of the consequence category for the dam break scenario and the correct application of the design criteria in this table if there is a significant difference in consequence between the different failure modes.

[#] Crest Level should include a suitably designed freeboard.

[^] Refer definition of a levee. All regulated levees are required to provide a minimum of 1:1000 AEP flood protection.

Reference 2: Ciria (2013) International Levee Handbook

Ciria (2013) promote the use of a risk based approach for determining the return period of the flood event, based on the service life of the levee to calculate the probability of exceeding the flood event during the service life. An example of this approach is provided in Table 2, which we have found very useful in a mining context because it accounts for the service life of the infrastructure being protected by the levee.

TABLE 2

Likelihood of exceeding the design event based on the service life of the levee[*] (the likelihood that at least one storm event exceeds the design event during the design life of the structure).

		Design return period (1 in x years) for the levee								
		10	20	50	100	1000	10 000	20 000	100 000	1 000 000
	0.25	3%	1%	1%	0%	0.0%	0.00%	0.00%	0.000%	0.000%
	0.5	5%	3%	1%	1%	0.1%	0.01%	0.00%	0.001%	0.000%
	1	10%	5%	2%	1%	0.1%	0.01%	0.00%	0.001%	0.000%
Service life (years)	4	34%	19%	8%	4%	0.4%	0.04%	0.02%	0.004%	0.000%
	20	88%	64%	33%	18%	2%	0.20%	0.10%	0.020%	0.002%
	50	99%	92%	64%	39%	5%	0.50%	0.25%	0.050%	0.005%
	100	100%	99%	87%	63%	10%	1.00%	0.50%	0.100%	0.010%
	1000	100%	100%	100%	100%	63%	10%	5%	1%	0.100%

[*] The Probability P of an event having a given return period, T_r occurring at least once in N successive years:

$$P = 1 - \left(1 - \frac{1}{T_r}\right)^N$$

Reference 3: Australian Coal Industry Research Program (ACARP)

White *et al* (2014), shown in Table 3, provides the following detail for selecting a design flood event based on consequences of scouring of the diversion or levee for the mining industry and highlights a risk-based approach to most applications.

TABLE 3

Selection of design event based on consequence of failure (White *et al*, 2014).

Consequence of channel scour	Proposed design event	
	During mine life	Post mining
Scour that threatens mine infrastructure	To be determined by mine operator	NA
Scour that threatens public infrastructure	To be determined in consultation with relevant stakeholder (asset owner)	To be determined in consultation with relevant stakeholder (asset owner)
Scour that threatens capture of watercourse into open cut pit	1:1000 AEP	Probable Maximum Flood

DIVERSION DRAIN DESIGN

Diversion drain designs are often necessary to direct run-off away from receptors vulnerable to flood and can be supported by protective levee structures. There is a trend in the mining industry to develop increasingly natural diversion structures that mimic natural river systems and consider the geomorphology of the area. White *et al* (2014) states the following:

> In the past, watercourse diversions were viewed as engineered structures designed to transfer water from one part of the mine to another. The series of ACARP projects helped to change this perception by recognising that watercourse diversions are part of a wider river system. The clear message to the industry was that diversions need to function like natural river systems in order to achieve diversion licence relinquishment at mine closure.

The following attributes of natural watercourses have been identified for inclusion in a typical diversion design:

- Multistage channel with a low-flow, active and high-flow channel.
- Channel features such as benches and terraces.
- Diverse indigenous vegetation species and longitudinal coverage.
- Appropriate channel length and planform variability.

The ACARP series also identified that river processes such as erosion and sedimentation occur in natural watercourses and made the distinction between natural adjustment in the wider river system and accelerated erosion and depositional processes observed in some diversions. The series introduced the concept of dynamic equilibrium, enabling mine operators and industry regulators to understand that although a degree of channel adjustment is natural, particularly associated with flood events, major (or elevated rates of) adjustment should be considered a concern.

An example of a natural channel and a constructed diversion that provides more natural features such as a low-flow channel and flood benches is provided in Figures 5 and 6.

FIG 5 – Example of a natural channel (White *et al*, 2014).

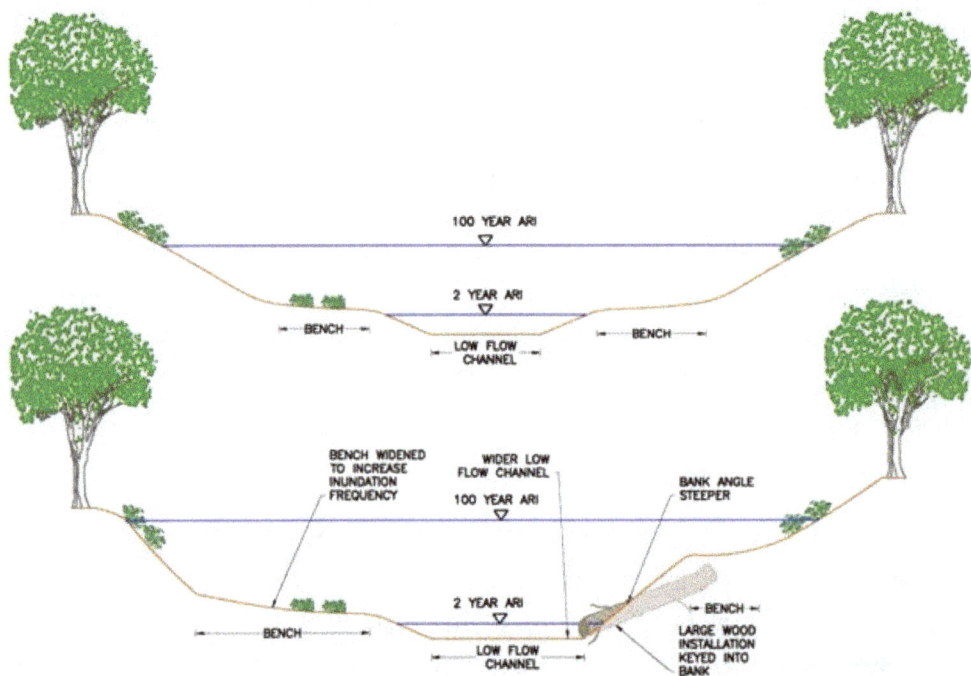

FIG 6 – Example of a constructed diversion with geomorphic features (White *et al*, 2014).

Hardie, White and Stewart (2001) presented the channel evolution graphic in Figure 7 of how a typical trapezoidal, engineered channel develops over time through erosion and sedimentation to form a more natural channel. The intent is to commence the engineering design at Stage 6 to 7 and mimic the natural system rather than design at Stage 2 with a typical engineered trapezoidal channel.

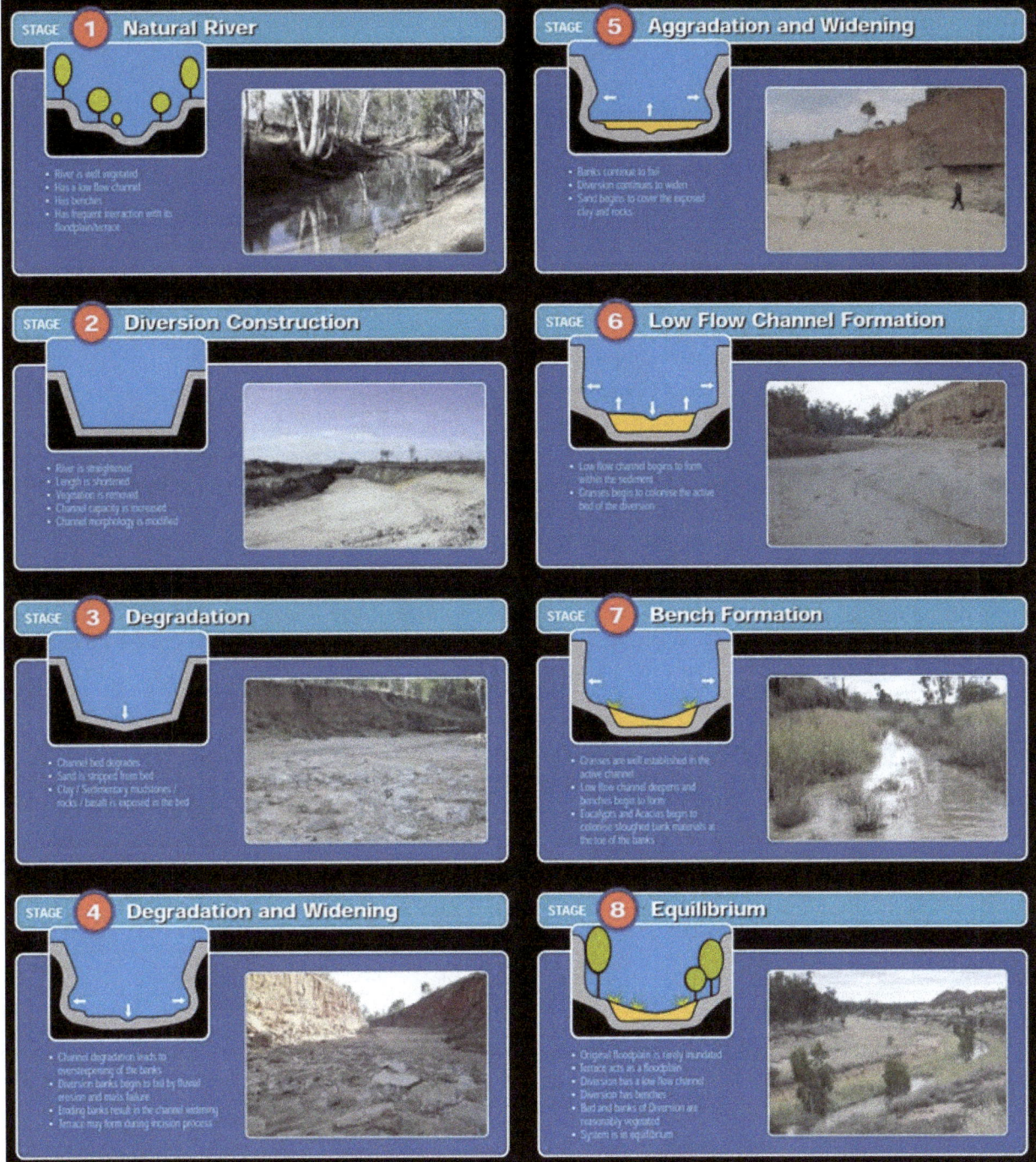

FIG 7 – Stages of diversion channel evolution (Hardie, White and Stewart, 2001).

White *et al* (2014) reviewed and assessed 60 diversion drains within the Bowen Basin in Queensland and were assessed using an Index of Diversion Condition (IDC) scoring system. The IDC was developed as a rapid method of assessing performance issues and potential management issues of diversion drains. The results of the White *et al* (2014) assessment are presented in Figure 8.

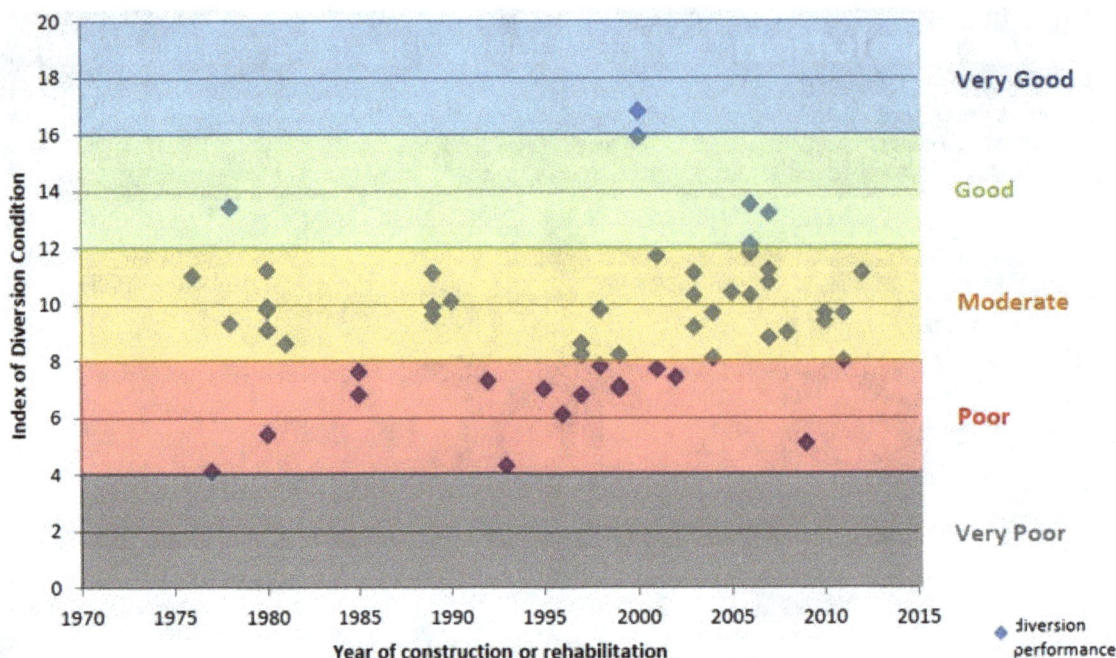

FIG 8 – Comparison of IDC results for diversions of different ages in 2012 (White *et al*, 2014).

The results above show that the majority of diversion drains constructed in the Bowen Basin are in the poor to moderate categories, and very few are in the very good category. There is more work to be done within the industry to achieve better performance for diversion drains.

Hardie and Lucas (2002) provide design criteria for assessing diversions using stream power, shear stress and velocity. This can be used to assess and compare the design parameters for natural streams and diversion drains to improve the performance of the diversion structures. Proposed stream power, shear stress and velocity values are detailed in Table 4.

TABLE 4

Typical values for stream reaches (adapted: Hardie and Lucas, 2002).

Stream type	Stream power		Velocity		Shear stress	
	1:2 AEP	1:50 AEP	1:2 AEP	1:50 AEP	1:2 AEP	1:50 AEP
Incised	20–60	50–150	1.0–1.5	1.5–2.5	<40	<100
Limited capacity	<60	<100	0.5–1.1	0.9–1.5	<40	<50
Bedrock controlled	50–100	100–350	1.3–1.8	2.0–3.0	<55	<120

White *et al* (2014) provides the following shear stress erosion thresholds for various materials to design the lining systems for diverison drains and levees.

TABLE 5

Erosion thresholds for different waterway materials (White *et al*, 2014).

Boundary category	Boundary type	Shear stress erosion threshold (N/m^2)
Soils	Fine colloidal sand	1.5
	Alluvial silt and silty loam (non-colloidal)	3
	Firm loam and fine gravels	4
	Stiff clay and alluvial silts (colloidal)	12
Gravel/Cobble	25 mm	16
	51 mm	32
	152 mm	96
	305 mm	192
Vegetation	Turf	45 to 177
	Long native grasses	80
	Short native and bunch grass	45
	Tussock and sedge	240
	Disturbed tussock and sedge, bunch grass 2–25 cm high	180
	Bunch grass 2–4 cm high	100
Biodegradable erosion control matting	Jute mat	22
	Coconut fibre with net	110
Non-degradable rolled erosion control products	Unvegetated	145
	Partially established	285
	Fully vegetated	380

Briaud (2013) proposed the following (Figure 9) for soil erosion subject to flow velocities which is also a commonly used reference for designing diversion drain lining systems.

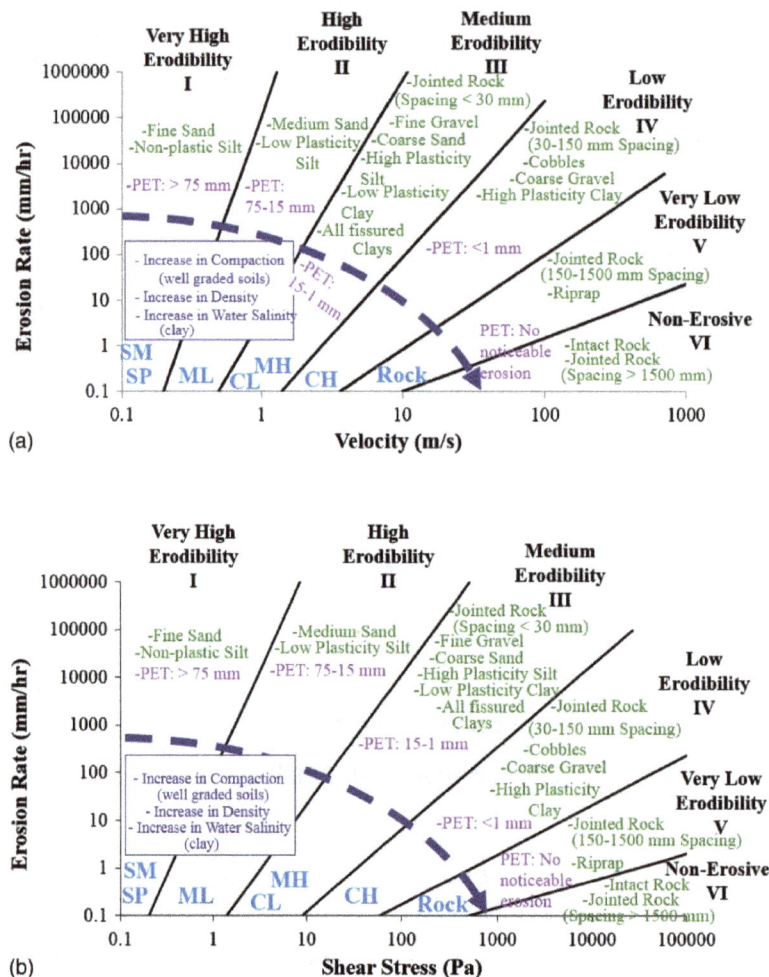

FIG 9 – Erosion thresholds for different materials (Briaud, 2013).

Levee geometry and construction materials

The Levee Management Guidelines (Department of Environment, Land, Water and Planning, 2015) provides a typical levee geometry for a zoned levee embankment with a central core and external shells (Figure 10). In our experience, mine levees are generally designed as homogenous embankments (not zoned) which enable more efficient construction and scheduling and reduce the potential need for filters. United States Army Corps of Engineers (USACE) also supports this approach and states the following:

> As a general rule levee embankments are constructed as homogeneous sections because zoning is usually neither necessary nor practicable. However, where materials of varying permeabilities are encountered in borrow areas, the more impervious materials should be placed toward the riverside of the embankment and the more pervious material toward the landside slope. Where required to improve under seepage conditions, landside berms should be constructed of the most pervious material available and riverside berms of the more impervious materials. Where impervious materials are scarce, and the major portion of the embankment must be built of pervious material, a central impervious core can be specified or, as is more often done, the riverside slope of the embankment can be covered with a thick layer of impervious material. The latter is generally more economical than a central impervious core and, in most cases, is entirely adequate. (USACE, 2000)

The cut-off is generally required but can be moved depending on the optimum location to cut-off potential seepage.

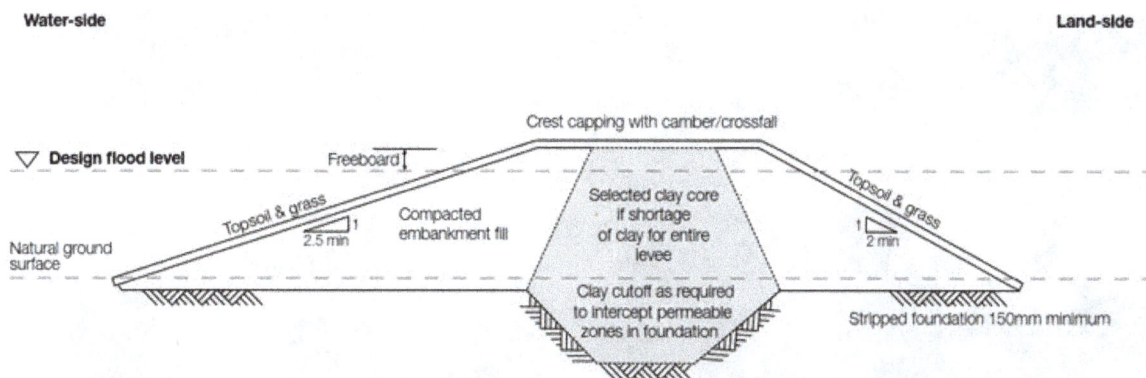

FIG 10 – Typical levee geometry from the Levee Management Guidelines (Department of Environment, Land, Water and Planning, 2015)

Furthermore, we recommend the following improvements in geometry:

- Camber the crest to the flood side at 3 per cent.

- Ensure that a good quality crest capping material is placed on the traffic to prevent rutting and provide all weather access for the inspections during rain events.

- Consider a minimum 3H:1V batters for safer constructability and promote vegetation growth on the slope. If there are footprint constraints, this may not be possible.

- Add safety bunds to the crest to prevent vehicles from driving off the crest.

Figure 11 shows the early rutting that is forming for a levee crest that was capped with topsoil rather than gravel.

FIG 11 – Topsoil and hydromulch levee crest suffering from rutting.

Mine levees often require crest safety bunds to meet safety standards and reduce the risk of vehicles driving off the crest. The bunds require regular breakthroughs to enable drainage off the crest and require batter drainage structures to reduce the erosion risk. Figure 12 shows erosion forming on batters where flow concentrations occur between bund breakthroughs with no batter erosion drain protection.

FIG 12 – Erosion forming in drainage down the levee batter.

For construction efficiency, a key objective is to source levee material locally and ideally cut material from local diversion drainage to use as fill in the levee with a short haul. The Levee Management Guidelines (Department of Environment, Land, Water and Planning, 2015) specify a plasticity index (PI) of more than 10 per cent and a particle size distribution with at least 25 per cent fines (passing the 0.075 mm sieve). The Dike Design and Construction Guide (Ministry of Water, Land and Air Protection, 2003) suggests the percentage of fines could be relaxed to 15 per cent to 30 per cent passing the 0.075 mm sieve. In the authors' experience, there does need to be some flexibility in detailing specifications to make use of local materials as much as possible.

Many Australian mines sites have consdierable quantities of dispersive soils which are highly erodible. The Levee Management Guidelines (Department of Environment, Land, Water and Planning, 2015) states that, ideally, dispersive soils having an Emerson Class of 3 or less should not be used. However, if non-dispersive soils are not available, dispersive soils can be used if appropriate measures are adopted in the construction. Such measures include tight control on compaction to a higher dry density, a higher compaction moisture and possibly the inclusion of filters to control piping erosion. Dispersive soils can also be treated with lime or gypsum (typically 2 per cent to 3 per cent by weight) to reduce the dispersiveness. In our experience, dispersive soils can be used in a homogenous embankment, but we recommend stabilising the soils with lime or gypsum. If lime is used, the lime stabilisation should be limited to the core of the embankment because the addition of lime increases the soil pH and reduces the likelihood of being able to successfully vegetate the batters. The outer shell materials could be treated with gypsum (which does not impact pH) enabling vegetation of the batters. It is good practice to check the percentage of lime/gypsum applied, via laboratory testing or a field trial.

Dispersive soil foundations also often have tunnel erosion features that require sealing to ensure an effective seepage barrier in the levee foundation. Some large features could be encountered which are best sealed with a suitable bentonite/cement grout mixture. Figure 13 provides photos of this application.

FIG 13 – Sealing of tunnel erosion features in highly dispersive soil foundation.

Levee freeboard

Levee freeboard is typically applied as a vertical distance above the design flood level. It accounts for potential waves from flood water, longer-term settlement of the levee crest, and uncertainty in the flood estimates. As such, we consider that the freeboard is a designer's decision based on their understanding of the design elements. Ciria (2013) notes that levee freeboard in the UK varies from 0.3 to 1.0 m, and we consider this to be a reasonable range, depending on design considerations. An alternate approach is to size the levee for design flood with freeboard set at the flood elevation of the next higher design flood with zero freeboard. Freeboard should include an allowance for climate change escalation over the design life of the structure.

Caught catchments

Construction of diversion drains and levees can result in caught catchments, that is, a catchment and/or creek lines on the upstream side of the levee that cannot drain away because the levee was constructed across the drainage line. This is very common in long length levee designs; in fact, these are somewhat inevitable due to the layout of levees and natural catchments (Figure 14). It is important to identify the caught catchments early in the design phase and develop methods to dewater the area to prevent excessive ponding against the levee or infrastructure and to ensure caught water does not impact mining. One method to dewater is to install a culvert through the levee to allow drainage during low-flow events. But this method requires also installing a one-way valve to ensure floods do not backwater through the culvert and into the protected area (Figure 15). Any pipe structure through a levee would require careful detailing because this will be the weak point of the structure. Standard dam engineering detailing is recommended. Another option consists of installing a pump system to pump the water over the levee (Figure 16).

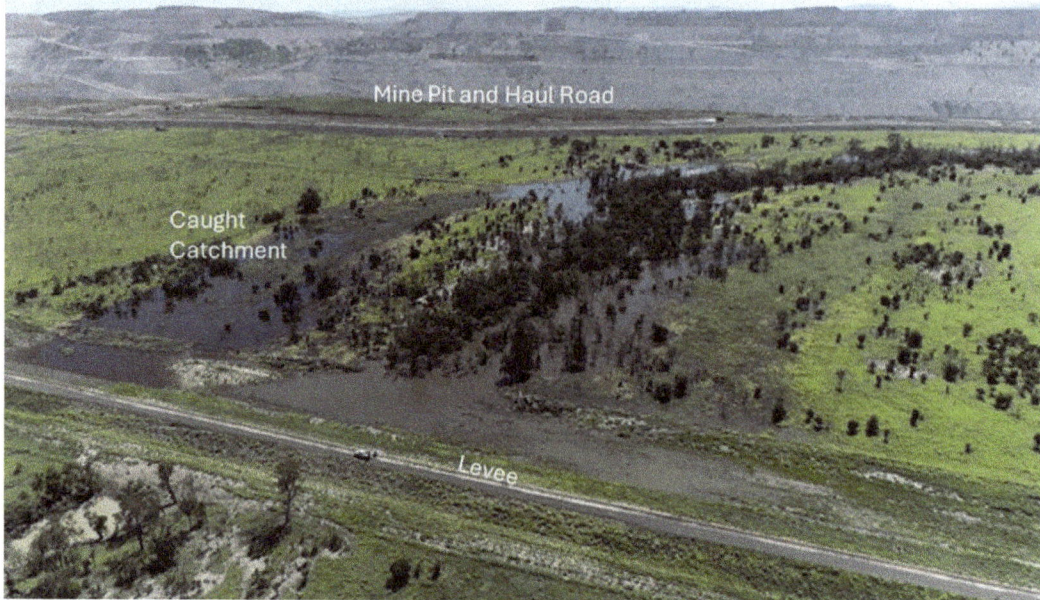

FIG 14 – Caught catchment upstream of a levee.

FIG 15 – One way valve on culvert through a levee.

FIG 16 – Pump and pipe system to dewater the caught catchment.

BLASTING NEAR LEVEES

Diversion levees around mine pits may be impacted by pit blasting during mining. Blasting results in high frequency, short duration blast waves that can result in deformation of earthen structures and there have been cases of soil liquefaction induced by blasting. Leeve's are often constructed over alluvial soils such as saturated, loose/soft sands and silts that are susceptible to liquefaction. Blast damage often results from the associated ground vibrations (which induce strains and increased stress in the soils), however it should be noted that embankment dams/levees generally have low natural frequencies and are not particularly susceptible to damage due to blast vibrations which have much shorter durations and much higher frequencies than an earthquake. For embankment levees, vibrations can induce several failure modes; including slope failures caused by large ground motions, slope failures due to liquefaction of foundation soils, and deformation of the levee crest (leading to loss of freeboard) and embankment cracking that could lead to a piping failure through the levee. When evaluating the impacts of blast vibration on a structure, it is typical to set a limit on the peak particle velocity (ppv). The ppv can be estimated at a particular structure for each blast and vibration monitors can be deployed to measure the value from the blast. A good discussion on blasting assessments for embankments and typical values of ppv for embankments is provided in Harrington (2016) and Figure 17 presents an assessment flow chart.

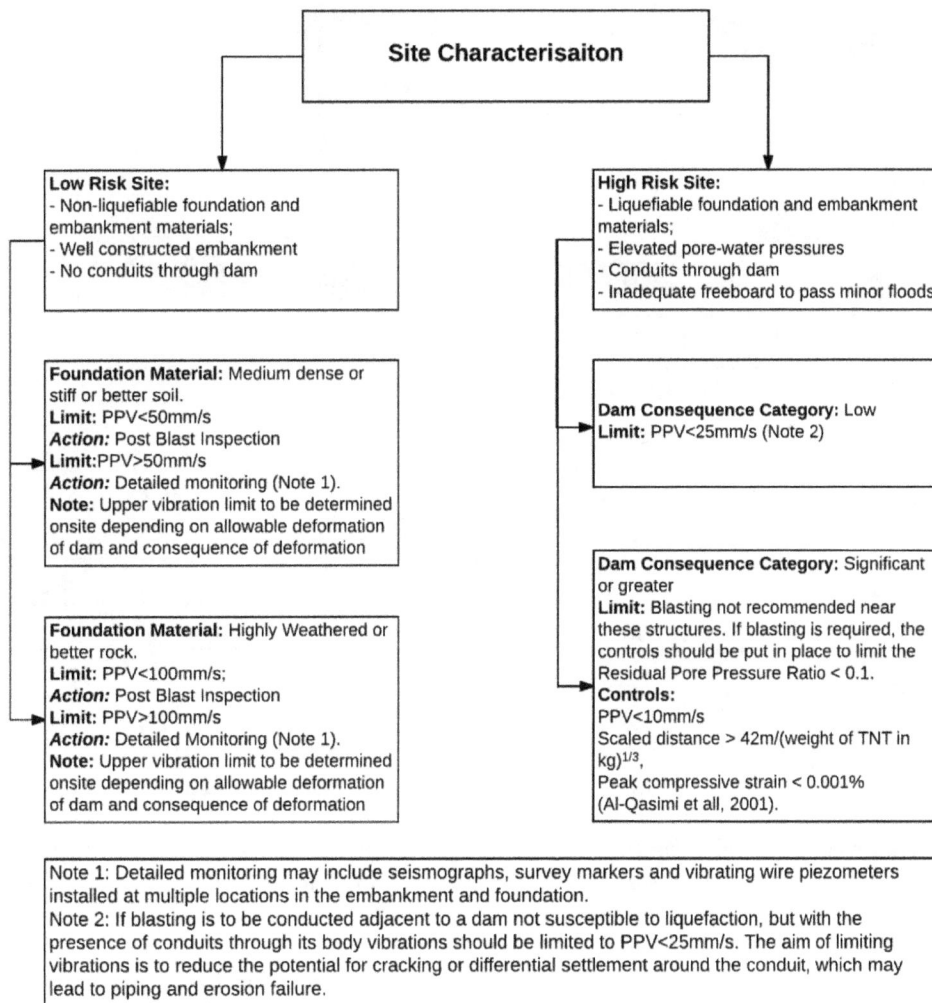

Site Characterisaiton

Low Risk Site:
- Non-liquefiable foundation and embankment materials;
- Well constructed embankment
- No conduits through dam

High Risk Site:
- Liquefiable foundation and embankment materials;
- Elevated pore-water pressures
- Conduits through dam
- Inadequate freeboard to pass minor floods

Foundation Material: Medium dense or stiff or better soil.
Limit: PPV<50mm/s
Action: Post Blast Inspection
Limit: PPV>50mm/s
Action: Detailed monitoring (Note 1).
Note: Upper vibration limit to be determined onsite depending on allowable deformation of dam and consequence of deformation

Dam Consequence Category: Low
Limit: PPV<25mm/s (Note 2)

Foundation Material: Highly Weathered or better rock.
Limit: PPV<100mm/s;
Action: Post Blast Inspection
Limit: PPV>100mm/s
Action: Detailed Monitoring (Note 1).
Note: Upper vibration limit to be determined onsite depending on allowable deformation of dam and consequence of deformation

Dam Consequence Category: Significant or greater
Limit: Blasting not recommended near these structures. If blasting is required, the controls should be put in place to limit the Residual Pore Pressure Ratio < 0.1.
Controls:
PPV<10mm/s
Scaled distance > $42m/(weight\ of\ TNT\ in\ kg)^{1/3}$,
Peak compressive strain < 0.001%
(Al-Qasimi et all, 2001).

Note 1: Detailed monitoring may include seismographs, survey markers and vibrating wire piezometers installed at multiple locations in the embankment and foundation.
Note 2: If blasting is to be conducted adjacent to a dam not susceptible to liquefaction, but with the presence of conduits through its body vibrations should be limited to PPV<25mm/s. The aim of limiting vibrations is to reduce the potential for cracking or differential settlement around the conduit, which may lead to piping and erosion failure.

FIG 17 – Proposed blast vibration limits (Harrington, 2016).

It is typical to set vibration limits (in terms of ppv) for a levee and confirm the actual ppv with multiple vibration monitoring locations during the blast. Supplemental monitoring should also be considered including:

- Piezometers to assess excess pore pressure generation in contractive soils.

- Survey monuments and GNSS survey monitoring to assess embankment movements.

- For critical locations, dynamic strain monitoring can also be employed and is best undertaken using specialists in this field.

SEEPAGE AND INTERNAL EROSION

Levees are often constructed over varying ground conditions that can have preferential flow paths subject to internal erosion and piping. This results in either sand boils along the toe or seepage discharge. Poor construction practices can also lead to internal erosion through the embankment during flood events. The design must address these potential issues. A good reference for this is Chapter D-6 (Best Practices) of *Internal Erosion Risks for Embankments and Foundations* (USACE, 2021). A typical foundation seepage issue is presented in Figure 18.

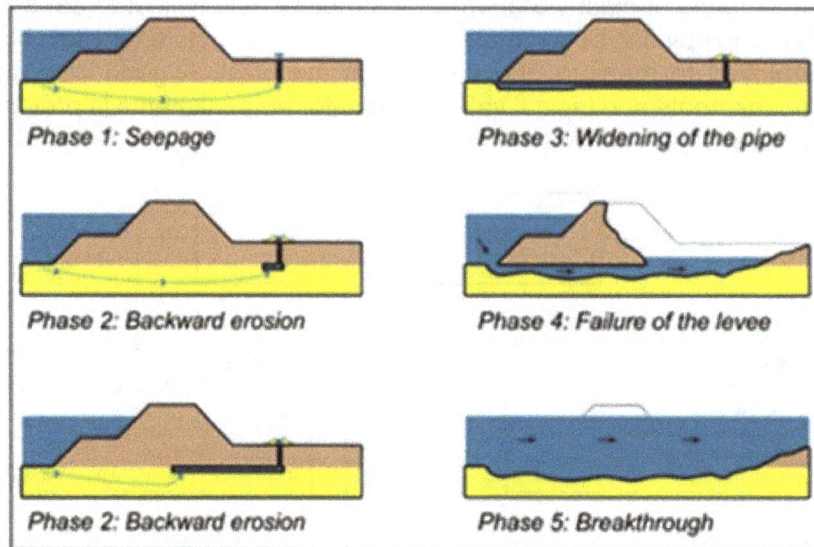

Figure D-6-9.—BEP (adapted from van Beek et al. 2011).

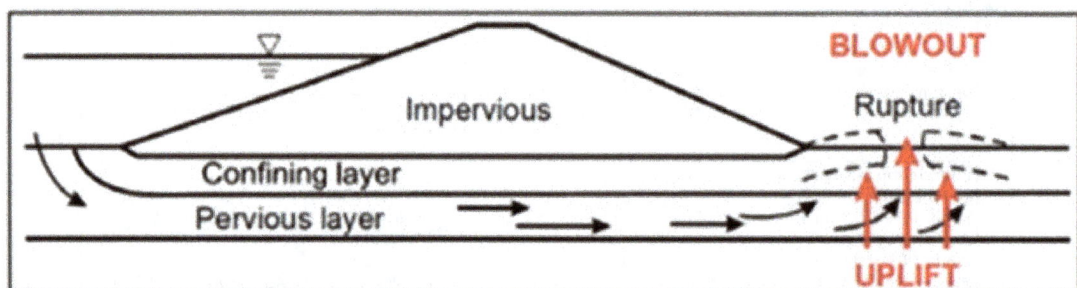

FIG 18 – Internal erosion scenarios for levees (USACE, 2021).

Piping risk assessments can also be undertaken with the aid of Fell *et al* (2024) to better understand the risk of a piping/internal erosion event. This guidance provides detailed inputs to establish event trees for piping/internal erosion failure events.

AFFLUX

It is vital to assess the afflux that occurs along the formation, including upstream and downstream of the levee/diversion drain system, to understand any impacts from the pre- and post-development. For example, many diversions have public road infrastructure upstream and downstream of diversions, and it is critical to understand any impacts to the existing infrastructure after development. Ideally, the design should try to limit afflux for the pre- and post-development conditions, but where this cannot be achieved, engagement with stakeholders and mitigation options need to be developed.

VEGETATION

It is important to consider which vegetation to use to stabilise levee or diversion structures, and the decision should be based on the purpose of the structure and the material that was used to construct it. The authors recommend that only grasses should be used for levee structures. Using only grasses avoids the potential for shrubs or trees to die back and cause potential weak areas in the levee. For diversion structures where mimicking the natural system is the goal, native trees and shrubs are usually used for stabilisation of the embankments and to create a riparian zone to blend into the natural system.

CONCLUSIONS

An overall guidance for design and construction of mine levee's does not exist. This paper provides key design guidance on levees and diversion drains for mine infrastructure from a number of references and presents some common issues with diversion levees in a mining context. This paper also provides key references that are useful for practitioners involved in levee design and management.

REFERENCES

Briaud, J-L, 2013. *Geotechnical Engineering: Unsaturated and Saturated Soils* (John Wiley and Sons: New York).

Ciria, 2013. *The International Levee Handbook* (Construction Industry Research and Information Association, French Ministry of Ecology and US Army Corps of Engineers). Available from: <http://www.ciria.org/ILH>

Department of Environment, Land, Water and Planning, 2015. Levee Management Guidelines, Dept of Environment, Land, Water and Planning, State Govt Victoria.

Department of Environment, Science and Innovation, 2024. *The Manual,* Manual for Assessing Consequence Categories and Hydraulic Performance of Structures, ESR/2016/1933, Version 5.03, Environmental Protection Act 1994, Qld Govt.

Ensham Resources, 2008. Presentation to Fitzroy Flood Forum, a Collaborative Approach to Recovery from an Unprecedented Natural Event.

Fell, R, Foster, M, Cyganiewicz, J, Davidson, R and Sills, G, 2024. Methods for Estimating the Probability of Failure of Embankment Dams by Internal Erosion and Piping, UNSW.

Hardie, R and Lucas, R, 2002. Bowen Basin River Diversions Design and Rehabilitation Criteria, Project C9068 Report for ACARP, by Fisher Stewart Ltd.

Hardie, R, White, K and Stewart, F, 2001. Monitoring Geomorphic Processes in Bowen Basin River Diversions, Project C9068, Australian Coal Industry's Research Program (ACARP).

Harrington, R, 2016. Assessing blasting impacts on embankment dams: Limitations with adopting an equivalent earthquake approach, ANCOLD conference.

International Commission on Large Dams (ICOLD), 2023. Levees Around the World – Characteristics, Risks and Governance, Bulletin 196, preprint.

Ministry of Water, Land and Air Protection, 2003. *Dike Design and Construction Guide*, prepared by: Golder Associates Ltd and Associated Engineering (BC) Ltd, British Columbia Flood Hazard Management Section, Environmental Protection Division, Province of British Columbia, Ministry of Water, Land and Air Protection.

United States Army Corps of Engineers (USACE), 2000. Design and Construction of Levees, EM 1110-2-1913, USACE.

United States Army Corps of Engineers (USACE), 2021. Internal Erosion Risks for Embankments and Foundations, Best Practices in Dam and Levee Safety Risk Analysis, Part D – Embankments and Foundations, Chapter D-6, USACE.

White, K, Moar, D, Hardie, R, Blackham, D and Lucas, R, 2014. Criteria for Functioning River Landscape Units In Mining and Post Mining Landscapes, Project C20017, Australian Coal Industry's Research Program (ACARP).

Design loadings and parameter selection, including BAT and BAP

Selection of critical earthquake loading and embankment sections for 2D seismic stress-deformation analyses

B Ghahreman Nejad[1,2] and M Sadeghipour[3]

1. Global Technical Advisor, ATC Williams, Mordialloc Vic 3195.
   Email: behroozg@atcwilliams.com.au
2. Principal Honorary Fellow, Department of Infrastructure Engineering, Melbourne University.
3. Senior Associate Engineer, ATC Williams, Mordialloc Vic 3195.
   Email: meris@atcwilliams.com.au

ABSTRACT

The principal seismic concern for design of an embankment dam is the development of displacement patterns that could lead to the embankment failure and uncontrolled release of water and/or tailings in the event of the design earthquake. The prediction of such movements is dependent on the accuracy and complexity of the analysis procedure used, and for plane strain (2D) analyses would also depend on the embankment section considered for the modelling.

This paper investigates the critical earthquake loading and variations in embankment natural frequencies and their effect on the predicted deformations for a 180 m high tailings retaining embankment dam. A simplified framework is proposed for selection of the most critical embankment sections for 2D seismic stress-deformation analyses.

INTRODUCTION

Seismic analysis is a critical component in the design of embankment dams, as it helps to evaluate potential ground and structural movements caused by earthquakes. According to ANCOLD (2019), designers should use dynamic deformation analyses to assess specific failure mechanisms that could compromise dam safety.

In a standard 2D limit equilibrium stability analysis, the most critical embankment section is usually adopted based on the maximum height (ie higher stresses) and/or foundation conditions. This approach may not be necessarily applicable to assessments of earthquake induced stresses and deformations. The critical embankment height for the seismic analysis that can experience the maximum deformation depends on the embankment natural frequency, which in turn is a function of its height and shear wave velocity of embankment and foundation materials. If the embankment natural frequencies approach the dominant frequencies of the design earthquake, the risk of resonation will increase and could potentially result in greater deformations and damages.

The natural frequencies of an embankment dam are not constant for several reasons including variations in its height due to elevation differences at the foundation level, dependency of the shear wave velocity of earth materials to confining stresses (ie depth). Furthermore, for a tailings dam its height and hence natural frequencies will vary over time given tailings dams are usually constructed in stages over the life-of-mine. In addition, cyclic softening and reduction in shear wave velocity of materials may also occur during earthquake loading which would in turn result in further uncertainties in calculation of the embankment dams natural frequencies.

Different approaches have been proposed to predict seismically induced deformations of earth dams, ranging from simplified Newmark (1965) based methods to complex stress–deformation numerical modelling (ie using finite element or finite different based programs such as FLAC etc).

The seismic response and deformation analysis method proposed by Makdisi and Seed (1978) can be used to identify the most critical embankment section for a rigorous and more time-consuming 2D deformation analysis. The responses of a cycloned sand embankment dam at different heights to four different earthquake loading cases have been analysed. The results are discussed and the estimated natural frequencies/periods and predicted earthquake induced deformations are compared.

METHODOLOGY

Newmark (1965) proposed a method for evaluation of permanent crest settlement of embankments subjected to earthquake loading. This method is based on sliding of soil mass along an inclined failure surface due to the inertia forces. Sliding would be initiated when the inertia forces exceed the resistance of the shearing sliding mass and would stop once the inertia forces are reversed. Makdisi and Seed (1978) extended the Newmark method by including the dynamic response of the embankment in the analysis process.

A series of seismic response and deformation analyses were conducted for a cyclone sand embankment dam based on the Makdisi and Seed (1978) approach using SHAKE2000 (Ordonez, 2005) and four design earthquake time histories.

Several failure surfaces were considered for the dam with the exit point of the failure surface gradually increasing in-depth, commencing in the upper quarter of the upstream embankment face and gradually increasing in-depth until the toe of the embankment slope was reached.

Makdisi and Seed method (1978) is based on iterations of shear strain to obtain strain compatible material properties, which is used to compute the maximum crest acceleration and the first three natural periods of the embankment. Then by using the embankment yield acceleration and natural periods, a range of permanent displacements are obtained.

TAILINGS DAM

The Tailings Storage Facility (TSF) is formed by a cross-valley cyclone sand embankment dam. The TSF has been in commission for over 24 years and is used to store tailings produced from a nearby mine. The embankment dam comprises the starter dam, start-up toe embankment and tailings sand stack, as schematically shown in Figure 1.

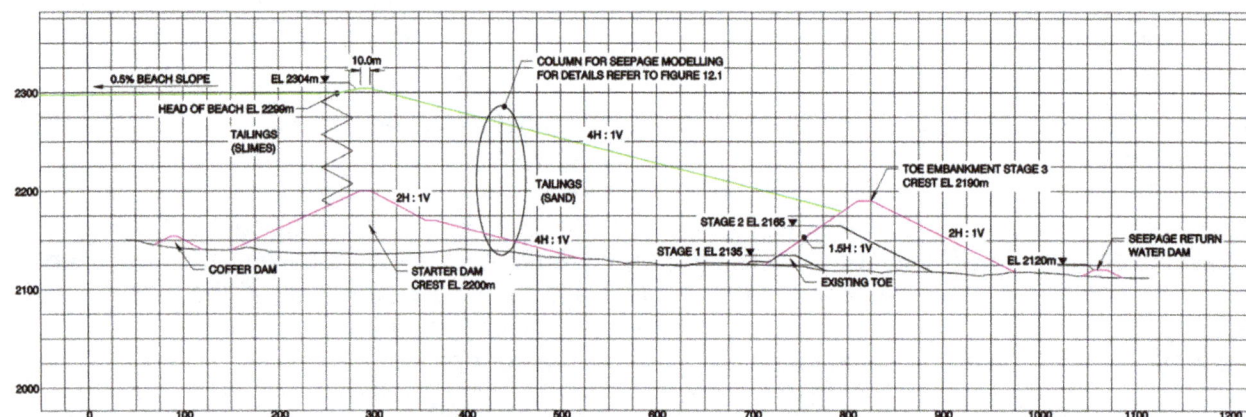

FIG 1 – Schematic embankment section.

The starter dam provides an embankment from which the tailings slimes (cyclone overflow) is deposited upstream and the sand (cyclone underflow) downstream via a number of cyclones mounted along its crest.

The start-up toe embankment's function is to form both a buttressing wall and a drainage element at the toe of the tailings sand fill which constitutes the overall embankment. The start-up toe embankment has been progressively raised, such that its crest is kept a suitable distance above the sand stack.

The TSF has been designed to store around 380 Mt of tailings (80 per cent slimes and 20 per cent sand tailings) over the 30 years mine life.

SELECTION OF DESIGN EARTHQUAKES

The dam is located in a region of high seismicity and has a consequence category of High C in accordance with ANCOLD Guidelines (2019). Hence, the Safety Evaluation or Maximum Design

Earthquake (SEE or MDE) was taken as an event having a peak ground acceleration (PGA) of 0.42 g corresponding to an annual exceedance probability (AEP) of 1:10 000.

Four acceleration-time histories (each with two horizontal accelograms) were used as input motions considered representative of the SEE event. They are reported in Table 1.

TABLE 1

Summary of design earthquakes.

Event name	Location	Date
Ardebil (Iran)	Kariq	28 February 1997
Hawaii	Hawaii	16 November 1983
Imperial Valley (USA)	Delta	15 October 1979
Imperial Valley (USA)	Compuertas	15 October 1979

Both horizontal components of the acceleration-time histories were considered. For each earthquake, a Fourier amplitude spectrum was derived to identify the dominant range of frequencies. The acceleration time history and Fourier amplitude spectrum derived for the earthquakes are presented in Figure 2.

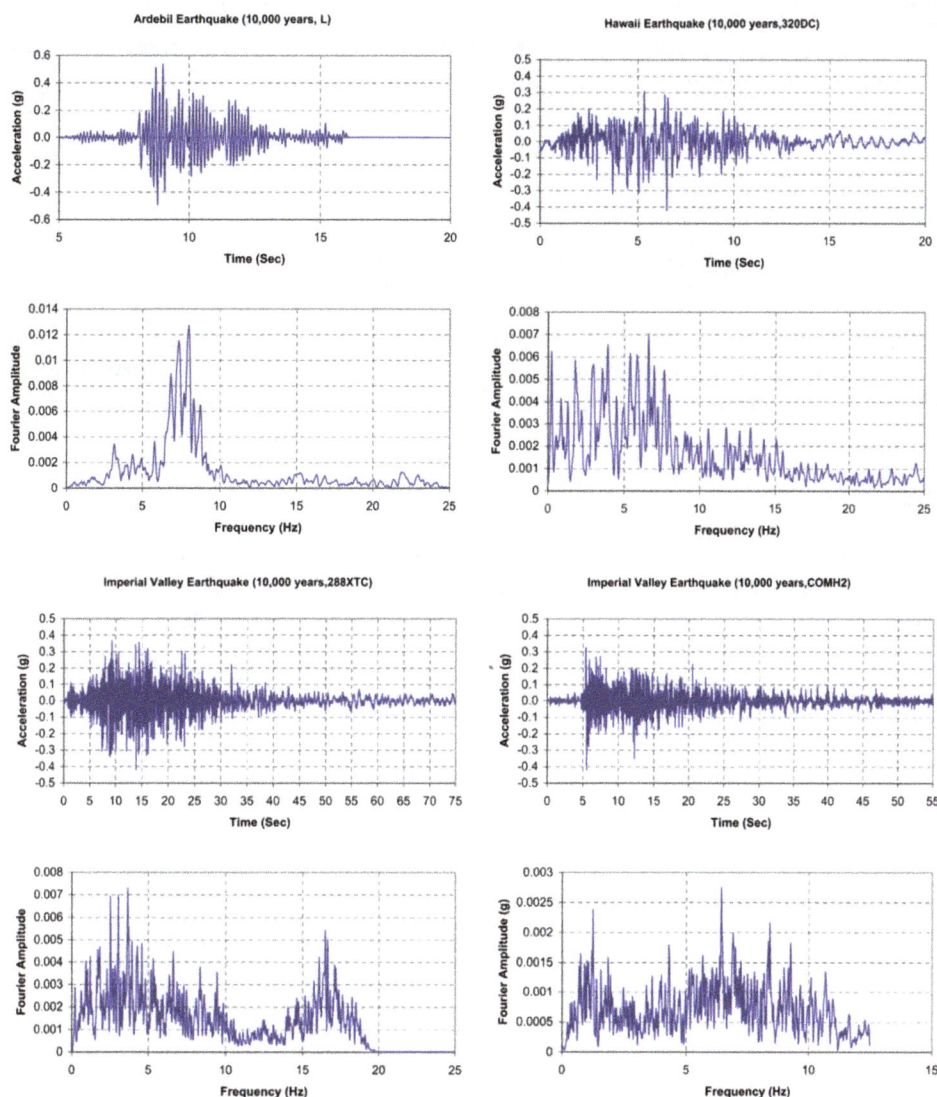

FIG 2 – Acceleration time histories and Fourier amplitude spectra of the design earthquakes.

MATERIAL PARAMETERS

Overview

The properties of materials used in different zones of the dam were obtained from site investigations, laboratory tests, published literature, established correlations and previous experience. A summary of the properties used in the analyses are presented in Table 2. The discussion following the table provides background information regarding the selection of some of these parameters.

TABLE 2
Material properties.

Material	Shear strength	K_0	G_{max} (MPA)	Modulus reduction and hysteretic damping	Density (KN/m^3)
Tailings slimes	$S_u/\sigma'_{vo} = 0.07$	0.4			20
Tailings sand	$\varphi' = 37$ $c' = 0$	0.40	Seed et al: $15.4\sigma'^{0.5}_m$	Deep cohesionless soils (EPRI, 1993)	19.4
Zone 3 rock fill	Leps Mean LB (capped at 46°)	0.3			22

c' is cohesion, ϕ' is friction angle, K_0 is coefficient of horizontal stress, G_{max} is small strain shear modulus, and σ'_m is mean effective stress.

Strength parameters

Tailings slimes

The particle size distribution for tailings slimes shows that 100 per cent of its particles are finer than 0.15 mm and it contains significant quantities of low plasticity fines.

The undrained shear strength under seismic loading was considered to be significantly lower than the strength under static condition as the tailings slimes were assumed to liquefy during earthquake. The strength parameter, shear strength as a function of the effective overburden stress, has been adopted.

Tailings sand

The tailings sand was the subject of significant laboratory investigation to establish the compaction level required to eliminate the risk of seismic liquefaction during the SEE event, if the sand becomes saturated. It was determined that a relative density (D_R) of 70 per cent needs to be achieved during construction (ie using compaction). This was considered to provide adequate protection against the occurrence of liquefaction (static and seismic). Significant differences between drained and undrained strength response are not expected for the sand material, unless it is susceptible to liquefaction. Consequently, drained strength parameters were used for stability analyses and determination of yield acceleration. A friction angle of 39° was derived from the results of triaxial tests but a lower bound value of 37° was adopted to include future variability in the material.

Rock fill (starter and toe embankment dam)

For simplicity, both Starter dam and Toe Embankment were considered to be uniform rock fill. Due to the size of the other zones (filters/drains and clay zones) compared to the size of final embankment, it was assumed that they have negligible effect on the overall stability of the embankment.

Leps (1970) has shown that shear strength of rock fill materials, as expressed by its friction angle, varies markedly as a function of the effective normal stress. The adopted strength function is between the Leps lower bound and average functions as the quality of rock is generally considered of medium quality. The friction angle has been capped at 46° in order to account for possible variations in the rock fill quality.

Stiffness and damping properties

The maximum Shear Modulus, G_{max}, is a measure of the stiffness of the material at very low strains, whilst the Modulus Reduction function describes how the stiffness decreases at larger strains.

Damping describes the hysteretic dissipation of energy within a material when is subjected to cyclic loading.

These parameters are required for the simplified Makdisi and Seed seismic response and deformation analyses. For simplicity and as these analyses are for the purpose of comparing the effect of embankment height on seismically induced deformations, it was considered that the embankment is homogeneous and represented by sand tailings properties, as it forms the main body of the embankment. G_{max}, for the sand tailings was taken as a function of mean effective stress, σ'_m, using the empirical relationship reported in Kramer (1996):

$$G_{max} = 220 K_{2\,max} \sigma_m'^{0.5} \text{ (kPa)} \tag{1}$$

K_{2max} depends on the quality and relative density of the material. For a gravel, K_{2max} ranges from 80 to 180. The sand tailings are considered to be of good quality and well compacted, consequently a K_{2max} value of 70 was adopted.

The modulus reduction and damping curves are from EPRI, recommended for deep cohesionless soils for the depth ranging from 35 to 75. The sand with an average depth of around 55 m (ranging from 0 to around 110 m lies within the provided EPRI range. The EPRI curves were obtained from the database included in the SHAKE2000 (Ordonez, 2005), as presented in Figure 3.

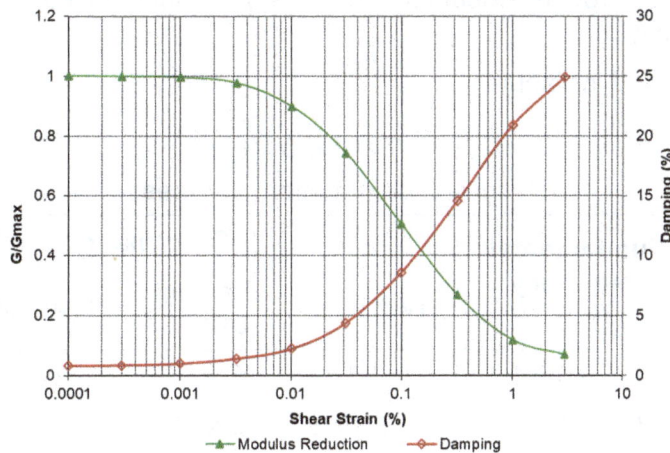

FIG 3 – Modulus reduction and damping curves (EPRI, 1993).

YIELD ACCELERATION

The strength parameters in Table 2 were used in SLOPE/W stability analysis software to determine the yield acceleration, a_y, for the downstream slope. The stability analysis was performed several times and each time the PGA was iterated until a factor of safety of around one was achieved. This acceleration was used as the yield acceleration in the Makdisi and Seed analyses. The critical slip surface exiting on the downstream shoulder at 0.165 times the dam height, h, above the downstream toe was analysed, as shown in Figure 4. This slip surface was found to generate the maximum deformations for the embankment at 180 m crest height during initial screening of the site response and deformation analyses.

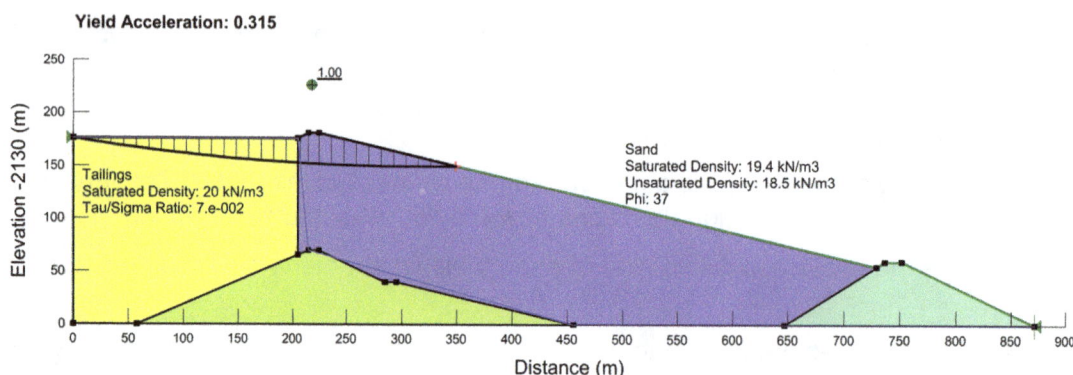

FIG 4 – Stability analysis section, yield acceleration (y/h=0.165).

Except for tailings (slimes), full soil strengths were used for different materials, and the phreatic surface was considered to be at the bottom of the downstream sand shell for the stability analyses.

DEFORMATION ANALYSIS

As the embankment is continuously raised, the site response and deformation analyses were carried out using the simplified Makdisi and Seed method in order to identify the critical embankment height that may lead to maximum deformation. It should be noted that the embankment natural periods/frequencies are a function of shear wave velocity and embankment height. If these frequencies approach the dominant frequencies of the design earthquake, the risk of resonation will increase, and consequently greater deformations might be resulted.

The final embankment height, 180 m, was reduced by 10 m increments to a minimum of 90 m. However, the depth of exit point for the slip surface on the downstream slope was kept constant at 30 m (ie y/h = 0.17 for a 180 m embankment). This was done to keep the yield acceleration unchanged at a_y = 0.315 g.

The results are summarised in Table 3.

TABLE 3

Results of deformation analyses.

Crest height, h (m)	y/h	Fundamental frequency (Hz)*	Deformation (cm)			
			IV-XTC	Hawaii-320DC	IV-COMH2	Ardebil-L
180	0.165	0.9,0.9,0.9,1.1	4 to 23.6	3.6 to 23.1	3.2 to 22.1	0.3 to 3.2
170	0.175	1.0,1.0,1.0,1.1	3.9 to 23.2	3.7 to 23.5		
160	0.186	1.0,1.0,1.0,1.2	5 to 28.9	4.6 to 27.5		
150	0.198	1.1,1.1,1.1,1.3	5.4 to 31.3	5.1 to 30.1	4.9 to 29.3	
140	0.212	1.2,1.2,1.2,1.4	5.4 to 31.5	5 to 29.4		
130	0.228	1.3,1.3,1.3,1.5	6 to 35.1	5.8 to 33.8	5.6 to 33.2	
120	0.248	1.4,1.4,1.4,1.6	6.5 to 37.6	6.2 to 35.9	6.2 to 36.4	3.7 to 22.4
110	0.27	1.5,1.5,1.5,1.7	7.6 to 44.3	9.6 to 53.5	6.7 to 38.4	6.2 to 36.1
105	0.283	1.6,1.6,1.6,1.8	7.3 to 42.8	9.2 to 51.6	6.5 to 37.6	9.4 to 54.9
100	0.297	1.6,1.6,1.7,1.9	6.8 to 39.5	8.7 to 48.9		12.9 to 75.2
90	0.33	1.8,1.8,1.9,2.1	6.6 to 38.2	7 to 39.8		15.1 to 88.3

NB: Maximum results are highlighted. *Fundamental frequencies associated with IV-XTC, Hawaii-320DC, IV-COMH2 and Ardebil-L earthquakes respectively.

As shown in the results, the maximum deformation generally occurs when the embankment height approaches 110 m, corresponding to a fundamental frequency of approximately 1.5 Hz, except in the case of Ardebil earthquake. For this event, the maximum deformation is observed at an embankment height of around 90 m, where the calculated fundamental frequency is 2.1 Hz.

CONCLUSIONS

For 2D stability and deformation analyses the maximum height sections (ie highest static shear stress) and those with unfavourable foundation conditions (weak zones, defects etc) should be always analysed.

The seismic response and deformation analyses method proposed by Makdisi and Seed (1979) may be successfully utilised as a precursor to identify the most critical embankment section for the more time-consuming numerical modelling. The critical height for seismic analyses is generally associated with the greatest amplification in ground accelerations, deformations or seismically induced shear stresses within the profile.

It is acknowledged that the Makdisi and Seed method is based on one-dimensional shear beam theory for seismic response analysis, assuming a horizontally layered soil profile. However, for assessing embankment stability and deformation, the method relies on 2D slope stability analysis. The application of a 1D seismic response to a 2D embankment geometry inevitably introduces some uncertainties in the predicted deformations. Nevertheless, the deformation trends produced by this approach are generally reliable for identifying the critical embankment height. Although the absolute deformations predicted using 2D and 1D seismic response analyses may differ significantly, both methods are expected to produce similar deformation trends with respect to changes in embankment height.

Despite its simplicity the Makdisi and Seed method provides a reasonable indication of embankment natural frequencies. The combination of natural frequencies relative to the earthquake dominant frequencies is the main factor affecting the deformations in the absence of liquefaction.

REFERENCES

Australian National Committee on Large Dams Incorporated (ANCOLD Inc), 2019. Guidelines On Tailings Dams - Planning, Design, Construction, Operation And Closure, revision 1, ANCOLD Inc. Available from: <https://ancold.org.au/product/guidelines-on-tailings-dams-planning-design-construction-operation-and-closure-revision-1-july-2019/>

Electric Power Research Institute (EPRI), 1993. Guidelines for Site Specific Ground Motions, Palo Alto, California, Electric Power Research Institute, TR-102293.

Kramer, S L, 1996. *Geotechnical earthquake engineering*, 1st edn (Prentice-Hall, Inc).

Leps, T M, 1970. Review of the shearing strength of rock fill, *Journal of the Soil Mechanics and Foundations Division*, 96(4):1159–1170.

Makdisi, F I and Seed, H B, 1978. Simplified procedures for estimating dam and embankment earthquake induced deformations, *Journal of the Geotechnical Engineering Division-ASCE*, 104:849–867.

Newmark, N M, 1965. Effects of earthquakes on dams and embankments, *Geotechnique*, 15:139–160.

Ordonez, G A, 2005. SHAKE2000 A Computer Program for the 1-D Analysis of Geotechnical Earthquake Engineering Problems, User's Manual, September 2005 Revision, GeoMotions LLC; Lacey, Washington, USA.

Effect of K_0 and bias on lateral extrusion propagation in upstream tailings dams – a conceptual numerical model

M Naeini[1] and B Ghahreman-Nejad[2,3]

1. Associate Engineer, ATC Williams, Mordialloc Vic 3195. Email: mahdin@atcwilliams.com.au
2. Global Technical Advisor, ATC Williams, Mordialloc Vic 3195.
   Email: behroozg@atcwilliams.com.au
3. Principal Honorary Fellow, Department of Infrastructure Engineering, The University of Melbourne.

ABSTRACT

Lateral extrusion in tailings dams refers to the horizontal displacement of soft materials, which can lead to a reduction in effective stresses and trigger liquefaction. This phenomenon may occur in layered tailings dams, where a tailings sand layer is underlain by a low-permeability finer zone, such as slimes. The possibility of this mechanism has been explored by researchers in previous failure cases, such as Fundão, Nerlerk Berm, and the Fort Peck Dam, under similar conditions in which strain softening of the underlying fine-grained soil led to increased shear stresses and reduced mean effective stresses in the overlying loose and coarser materials. This study aims to assess the effect of the coefficient of horizontal stress (K_0) and bias on the propagation of lateral extrusion in a hypothetical upstream tailings dam. This is done by performing a parametric study on K_0 and by adopting different downstream slopes to impose various bias conditions on the soft zones. The numerical modelling was conducted using the finite difference software FLAC (Itasca Group) with the NorSand constitutive model representing contractive materials. Lateral extrusion was simulated by applying horizontal deformation to the contractive zones. The propagation of lateral extrusion was assessed by adopting an instability triggering mechanism, in which the state parameter-dependent instability stress ratio was checked for each zone during the lateral extrusion. The modelling continued until a stabilised condition or failure was reached.

INTRODUCTION

Static liquefaction is a phenomenon in which a saturated, loosely placed material—such as tailings—undergoes a sudden and significant loss of strength. This strength loss can be triggered by several mechanisms, including undrained loading conditions (eg rapid construction), variations in pore water pressure, and reductions in lateral stress (Liu *et al*, 2024). As a result of static liquefaction, these materials may experience flow liquefaction, leading to large deformations and potential failure of the retaining structure (Jefferies and Been, 2015). Tailings dams, especially those constructed using the upstream raise method, are particularly vulnerable to this mechanism due to their construction history and the loose and saturated nature of deposited materials (Reid *et al*, 2021).

Static liquefaction has been identified as the primary cause of major tailings dam failures in recent years, including Fundão and Feijão tailings dams (Morgenstern *et al*, 2016; Jefferies *et al*, 2019; Robertson *et al*, 2019). Among these, the Fundão dam failure involved a notable case of lateral extrusion within a soft slimes layer (Morgenstern *et al*, 2016). Lateral extrusion refers to the deformation of a weak, saturated layer, typically composed of fine-grained tailings, which can lead to a reduction in effective stress in the overlying, denser materials. These denser layers may then exhibit brittle behaviour and experience a sudden loss of strength, resulting in a global failure.

Numerical modelling using constitutive models like NorSand is becoming a standard engineering practice for assessing static liquefaction in tailings dams. Some researchers have focused on the triggering potential of static liquefaction using NorSand (Ng *et al*, 2022; Liu *et al*, 2024) and on back-analyses of historical tailings dam failures (Morgenstern *et al*, 2016). However, some uncertainties remain in numerical modelling, perhaps the most significant being the influence of the coefficient of horizontal stress, K_0 (Reid *et al*, 2021).

For any given contractive soil under undrained loading conditions, failure is directly related to the initial K_0 state. A lower K_0 means the soil is closer to the instability line and can statically liquefy with a lower amount of shearing effort. The accurate field measurement of this parameter can be challenging and requires advanced *in situ* testing techniques, such as self-boring pressure metres, along with careful interpretation to account for stress history and anisotropy. The numerical modelling work could also overlook the importance of this parameter by adopting predefined values within the chosen constitutive model.

The other important factor that could affect static liquefaction is the bias, defined as the shear stress divided by the vertical effective stress. Bias tends to increase near slopes and decrease to zero in areas under flat surfaces and away from slopes. The geometry of a tailings dam, particularly the slope, can significantly alter the bias within the tailings zones. The combination of K_0 and bias can together influence how the tailings dam responds to static liquefaction triggering, under lateral extrusion, for example. This paper assesses the effect of these parameters on triggering of static liquefaction.

NUMERICAL MODELLING

Geometry and mesh

Numerical analyses were performed using the finite difference software FLAC2D V9 (Itasca Group). The nominal embankment model, shown in Figure 1, included a 16 m high starter embankment followed by five upstream raises, each 4 m in height. A slope of 1.5H:1V was applied to both the starter embankment and the upstream side of the embankment raises with the downstream slope of embankment raises varied. Three downstream slope arrangements including 2H:1V (Model_2H:1V), 3H:1V (Model_3H:1V), and 4H:1V (Model_4H:1V) were adopted for the parametric study.

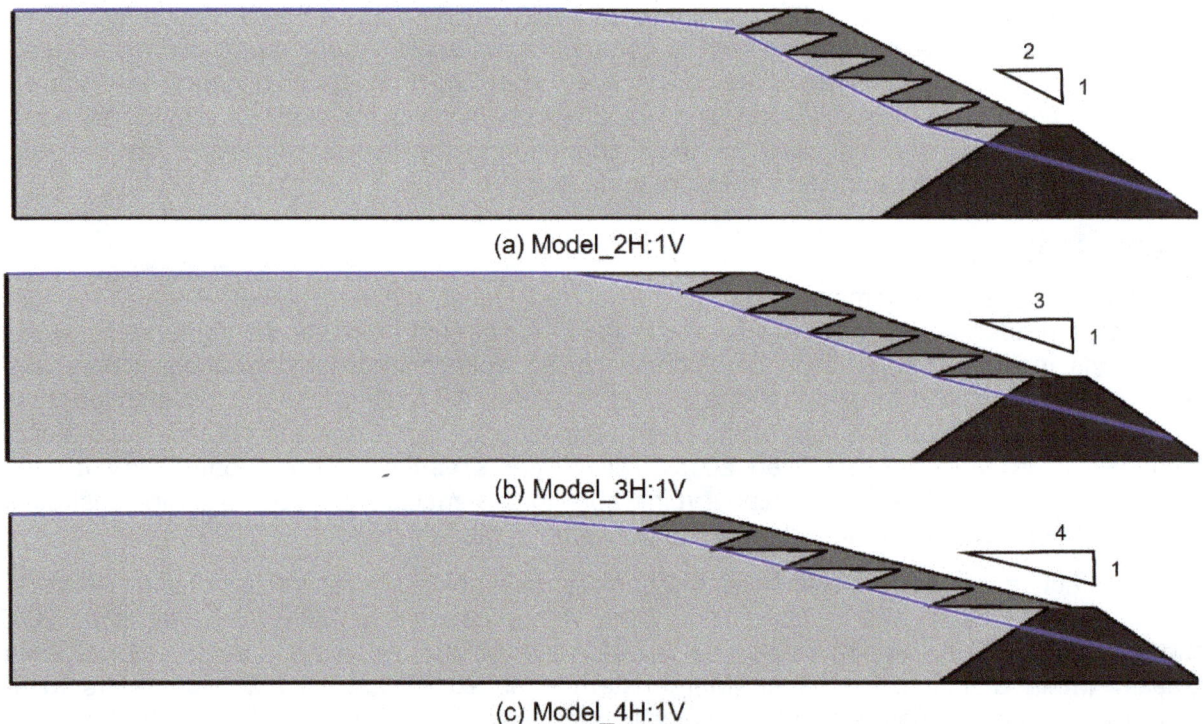

(a) Model_2H:1V

(b) Model_3H:1V

(c) Model_4H:1V

FIG 1 – Model geometries.

An unstructured mesh composed of quadrilateral zones with a global element size of 0.5 m was used. To minimise boundary effects, each model was extended 130 m from the upstream face of the last embankment raise.

Boundary conditions

The tailings storage facility, TSF, was assumed to be founded on a rigid bedrock layer (not modelled) to eliminate the influence of foundation deformations on the results. Accordingly, displacements at the base of the model were fixed in both horizontal and vertical directions. The left boundary was fixed in the horizontal direction, while vertical displacements were permitted within the tailings zones.

To ensure consistency across all models, an identical phreatic surface was defined as shown in Figure 1. The water table was assumed to pass through the upstream toe of each embankment raise and exit the model at a height of 4 m above the base of the starter embankment, for simplification and to ensure fully saturated tailings in all cases.

Modelling procedure and material parameters

The models were constructed in 1 m thick layers, with each lift allowed to reach equilibrium before the subsequent raise was added, resulting in a total of 72 model lifts. The phreatic surface was proportionally raised with the addition of each tailings layer. For each geometry model shown in Figure 1, four staged construction simulations were performed to initial K_0 of 0.5, 0.6, 0.7, and 0.8.

The adopted material parameters during the staged construction are summarised in Table 1. An elastic-perfectly plastic Mohr–Coulomb model was adopted for all materials during this phase. This approach allowed reaching the initial input K_0 values in the tailings zones (away from embankment slope), making it suitable for the parametric investigation presented in this study. A similar methodology has been adopted in previous studies (Reid *et al*, 2021). It should be noted that the term 'initial K_0' refers to the input *in situ* lateral stress condition assigned to each tailings layer at the start of staged construction.

TABLE 1

Material parameters adopted in staged construction.

Material	Young's modulus (MPa)	Poisson's ratio	Cohesion, C (kPa)	Friction angle, ϕ' (°)	Dry density (t/m^3)
Starter Embankment	27	0.35	50	40	2
Embankment Raises	18	0.35	5	40	1.8
Tailings	5	Varied from 0.335 to 0.445	5	40	1.6

Upon completion of the staged construction, the tailings materials were assigned the NorSand constitutive model, with parameters adopted from the literature by Morgenstern *et al* (2016) and discussed in Reid *et al* (2021). An overall state parameter, ψ, of -0.05 was used for the tailings. For the tailings layer located on the upstream side of the starter embankment crest, a ψ value of 0.05 was adopted. This 1 m thick layer served as the soft zone for lateral extrusion modelling. As shown in Figure 2, some zones beneath each embankment raise were modelled with a ψ of -0.1 to account for densification under embankment construction.

Upon completion of the stage construction and adaptation of NorSand constitutive model, a small deformation corresponding to approximately 2 per cent shear strain was applied to the central gridpoints of the soft layer (under undrained conditions), while fixing the displacement boundaries at the top and bottom of the layer. The lateral extrusion can be triggered by a combination of factors (eg concentrated loading during construction, ground movement etc), but has been assumed to have occurred in the soft tailings. The adopted approach allows to replicate undrained shearing conditions in the soft tailings layer that can lead to liquefaction of the upper sand tailings. The extent of shearing was varied across different models with 10 m, 30 m, 50 m, and 70 m of the soft layer laterally displaced. Upon completion of shearing, the imposed boundary conditions on the soft tailings layer were removed, and the model was allowed to run until reaching equilibrium.

(a) applied state parameter

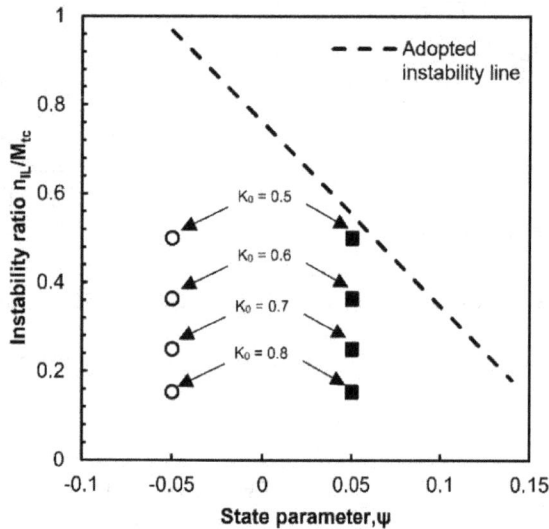

(b) adopted instability line (after Reid et al., 2021)

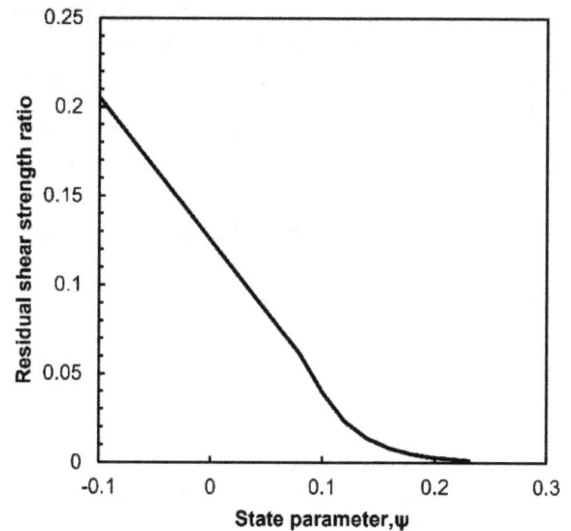

(c) adopted residual shear strength curve

FIG 2 – NorSand implementation.

After every 100 loading cycles, any tailings zone with a stress ratio (η) exceeding the instability line (η_{IL}), as defined in Figure 2b, was considered 'liquefied' and subsequently assigned the Tresca failure criterion. The cohesion of liquefied zones was determined using the best-practice line approach (Jefferies and Been, 2015), shown in Figure 2c. The shear modulus of liquefied zones was assumed to be 25 times their cohesion, while the bulk modulus was set to 100 times the shear modulus. This procedure was repeated until either equilibrium, or a global failure was achieved. The failure was defined as the occurrence of a maximum horizontal deformation of 1 m, as large shear strains were observed to develop at this level of deformation.

The implemented procedure aimed at simulating the strain softening behaviour in a soft tailings layer which in turn led to the increase in deviator stresses, reduction in mean effective stresses and increase in stress ratio in overlying sand tailings.

RESULTS AND DISCUSSIONS

Staged construction

The preliminary results of the staged construction analyses are presented in Figures 3 and 4. The stress ratio in the tailings significantly decreased with increasing initial K_0, as expected and demonstrated for Model_3H:1V in Figures 3(a–d). Flatter downstream slopes reduced the imposed stress ratio on the tailings, as observed by comparing Figures 3(b) and 3(e–f).

Bias, in the context of this study, is defined as the ratio of initial shear stress to initial vertical effective stress and serves as a proxy for assessing the impact of stress anisotropy in the static liquefaction process. Bias tends to increase near slope faces, where stress anisotropy is the highest, and approaches zero under relatively level grounds. Figure 4 illustrates how variations in geometry affect the spatial distribution of bias within the tailings mass. Moving away from the downstream, the variation in bias reduces, as shown in Figure 4(a). The soft tailings layer is shown with dashed lines.

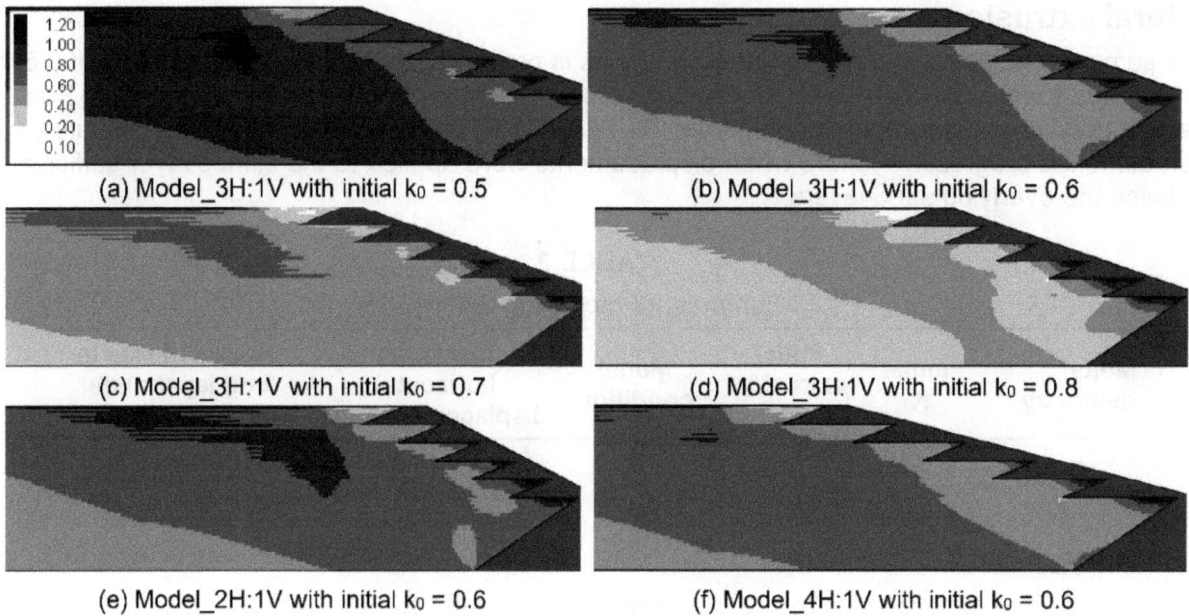

(a) Model_3H:1V with initial $k_0 = 0.5$

(b) Model_3H:1V with initial $k_0 = 0.6$

(c) Model_3H:1V with initial $k_0 = 0.7$

(d) Model_3H:1V with initial $k_0 = 0.8$

(e) Model_2H:1V with initial $k_0 = 0.6$

(f) Model_4H:1V with initial $k_0 = 0.6$

FIG 3 – Contours of stress ratio, η, at the end of staged construction.

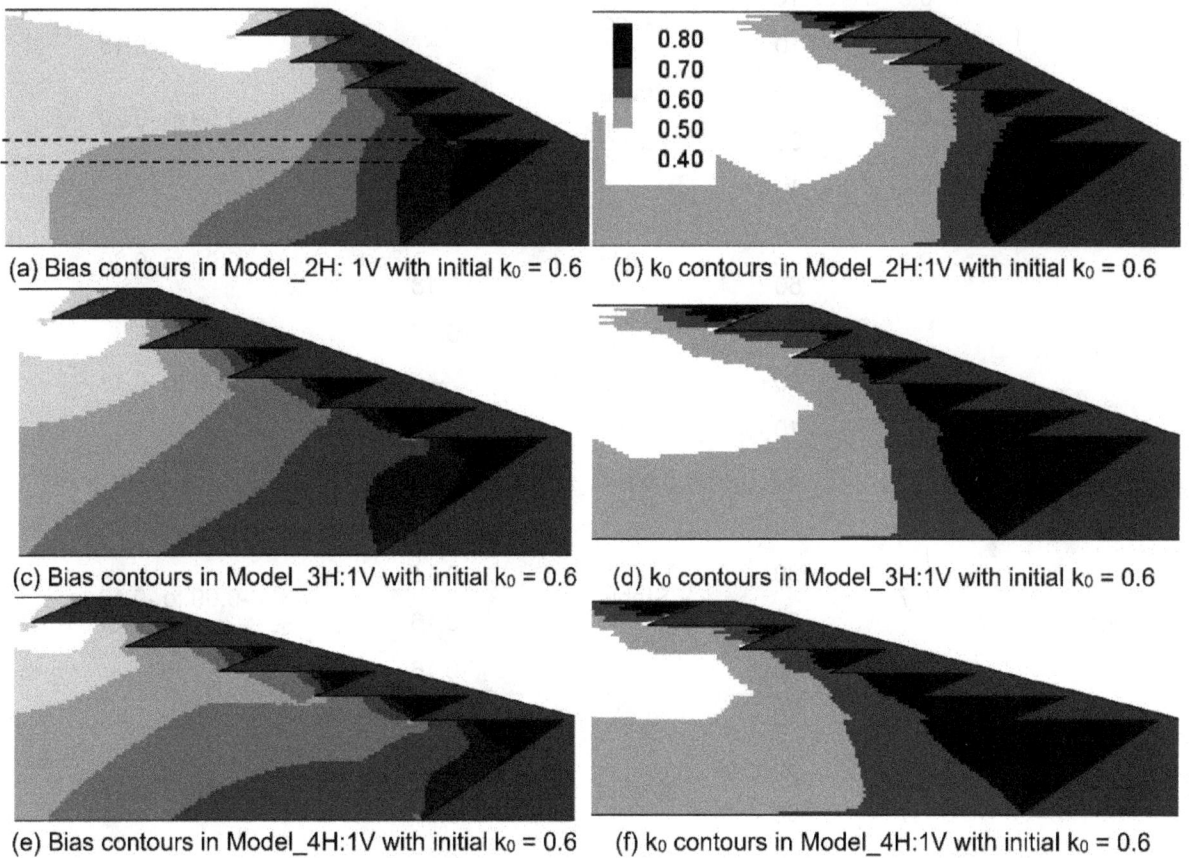

(a) Bias contours in Model_2H: 1V with initial $k_0 = 0.6$

(b) k_0 contours in Model_2H:1V with initial $k_0 = 0.6$

(c) Bias contours in Model_3H:1V with initial $k_0 = 0.6$

(d) k_0 contours in Model_3H:1V with initial $k_0 = 0.6$

(e) Bias contours in Model_4H:1V with initial $k_0 = 0.6$

(f) k_0 contours in Model_4H:1V with initial $k_0 = 0.6$

FIG 4 – Contours of bias and K_0 at the end of staged construction.

The K_0 value is calculated as the ratio of effective horizontal to vertical stresses for each zone (σ'_h/σ'_v). An initial K_0 was assigned to each tailings layer and allowed to change as the model was constructed in stages. As shown in Figure 4 (for an initial K_0 of 0.6), the K_0 distribution near the embankment slope significantly deviates from its initial value due to construction loading and changes in the stress level and orientation.

As illustrated in Figure 2(b), higher K_0 values are expected to reduce the potential for static liquefaction.

Lateral extrusion

The summary of lateral extrusion modelling results is presented in Table 2. At a target K_0 of 0.5, all three models were unstable under any applied lateral extrusion. Numerical modelling of the Fundão Dam, with downstream slopes ranging from 3H:1V to 5H:1V and adopting a K_0 of approximately 0.5, also exhibited progressive failure when displacements were applied to the slimes layer sufficient to mobilise the overlying sand layers.

TABLE 2

Summary of modelling results.

Model geometry	Initial K_0	Initial extrusion length (m)	Model condition	Maximum horizontal displacements (m)	Range of bias in soft layer under embankment raises
Model_2H:1V	0.5	10		—	-0.06 to -0.20
	0.6	10	Failed	—	-0.08 to -0.20
	0.7	10		—	-0.08 to -0.20
	0.8	10	Stable	0.14	-0.10 to -0.20
		30	Failed	—	
Model_3H:1V	0.5	10	Failed	—	-0.08 to -0.16
	0.6	10		—	-0.08 to -0.18
	0.7	10	Stable	0.14	-0.09 to -0.20
		30	Failed	—	
	0.8	10		0.14	-0.09 to -0.20
		30	Stable	0.16	
		50		0.18	
		70		0.17	
Model_4H:1V	0.5	10	Failed	—	-0.08 to -0.13
	0.6	10	Stable	0.14	-0.09 to -0.14
		30	Failed	—	
	0.7	10		0.13	-0.09 to -0.16
		30	Stable	0.14	
		50		0.16	
		70		0.16	
	0.8	10		0.14	-0.09 to -0.18
		30	Stable	0.16	
		50		0.17	
		70		0.17	

At an initial K_0 of 0.6, only the flattest model could tolerate 10 m of imposed extrusion without the liquefaction propagating. Model_4H:1V was resilient to liquefaction in the soft layer at K_0 values of 0.7 and 0.8. A similar response was observed in Model_3H:1V at initial K_0 of 0.8, whilst the applied extrusion of 30 m at K_0 of 0.7 resulted in failure. The only stable scenario for Model_2H:1V occurred at initial K_0 of 0.8 with a 10 m imposed extrusion.

Table 2 also presents the range of bias in the soft layer beneath the embankment. As the slope is flattened, the absolute maximum value of bias decreases at any given model with the same initial

K_0. This reduction, combined with the adopted and developed K_0 values near the embankment, influences the extent of failure propagation in the embankment.

To assess the effects of geometry, the liquefied zones and shear strain developments are presented in Figure 5 for the three geometries modelled, assuming an initial K_0 of 0.7, and 10 m of imposed shear strain in the soft tailings layer. A well-developed region of liquefied tailings and high shear strains is predicted for Model_2H:1V. The 10 m extrusion is sufficient to trigger liquefaction in the upper tailings zones, ultimately resulting in failure of the structure.

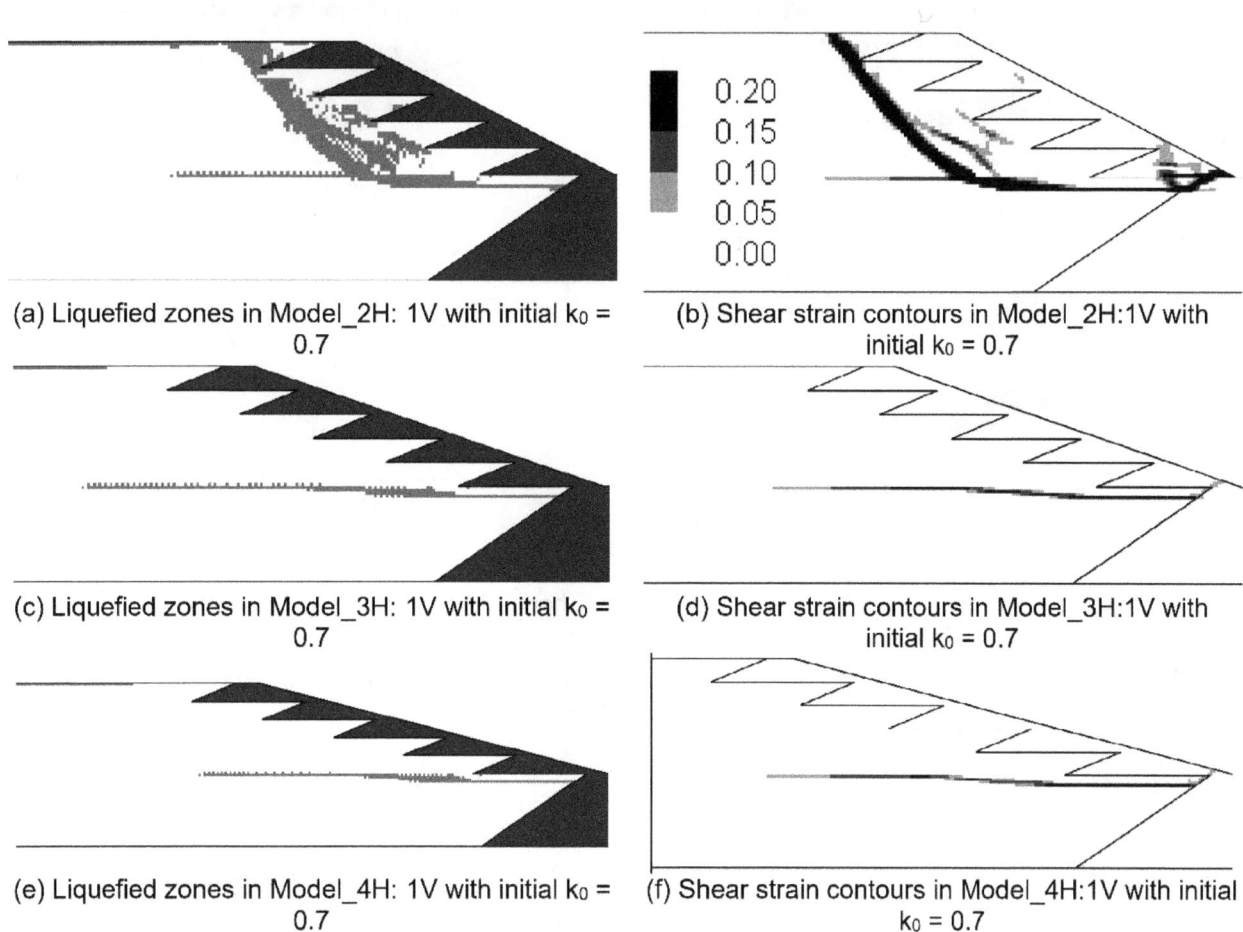

(a) Liquefied zones in Model_2H: 1V with initial k_0 = 0.7

(b) Shear strain contours in Model_2H:1V with initial k_0 = 0.7

Legend: 0.20, 0.15, 0.10, 0.05, 0.00

(c) Liquefied zones in Model_3H: 1V with initial k_0 = 0.7

(d) Shear strain contours in Model_3H:1V with initial k_0 = 0.7

(e) Liquefied zones in Model_4H: 1V with initial k_0 = 0.7

(f) Shear strain contours in Model_4H:1V with initial k_0 = 0.7

FIG 5 – Contours of liquefied zones (grey) and developed shear strain for 10 m extrusion.

The zones with maximum shear strain development in Figure 5(b) are directly related to the liquefied zones. Once a well-defined failure plane develops, shear strain progressively increases along the plane until complete structural failure occurs. In contrast, for the other two geometries, although liquefaction extends horizontally, it is not sufficiently developed to propagate through the upper tailings layers and cause a global failure.

The effects of K_0 on liquefaction development is illustrated in Figure 6 for Model_4H:1V with an initial 30 m of extrusion. In all three models, a relatively similar horizontal extent of soft tailings has liquefied, although the liquified zone is marginally thicker in the case of model with initial K_0 = 0.6. However, in both models with initial K_0 = 0.7 and K_0 = 0.8, the liquefaction does not extend into the upper tailings layer or form a failure zone similar to the K_0 = 0.6 case.

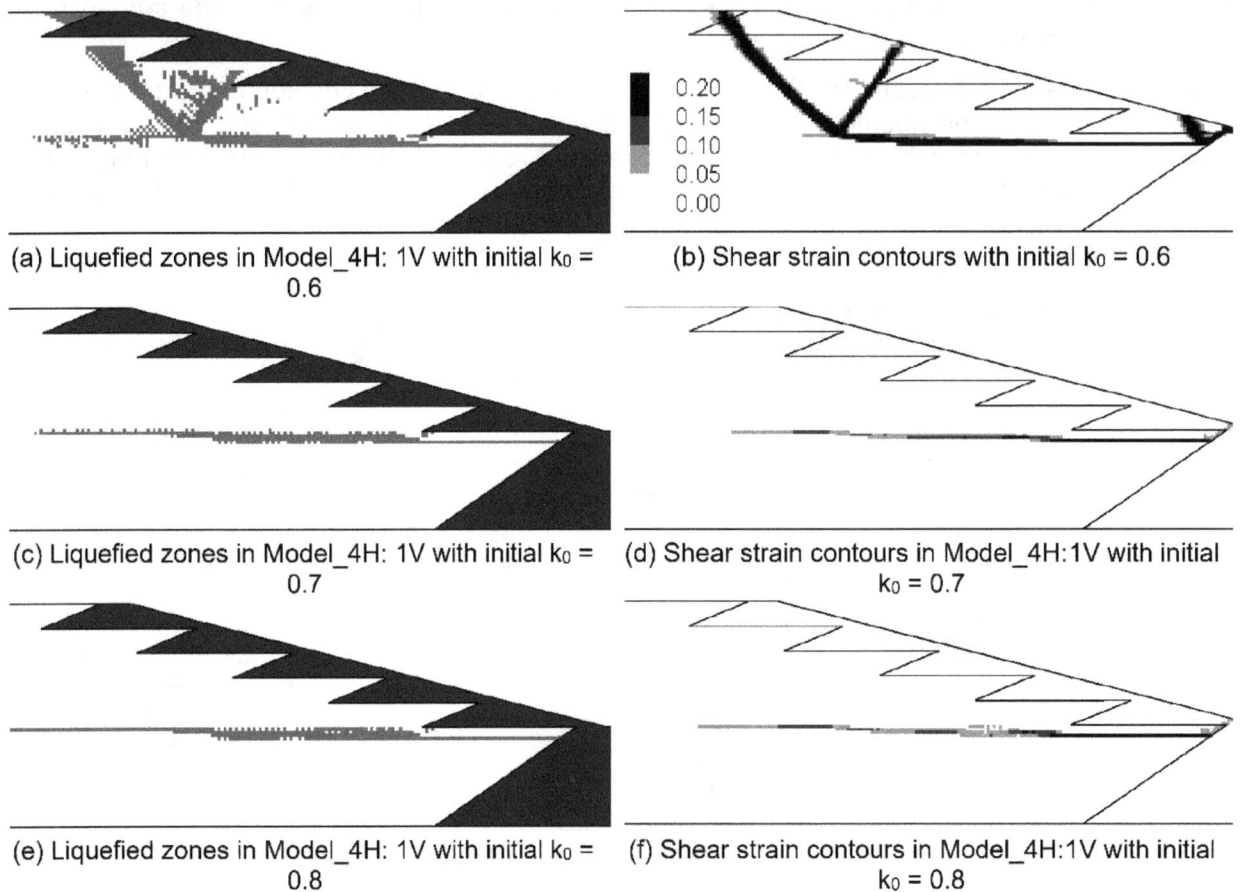

(a) Liquefied zones in Model_4H: 1V with initial k_0 = 0.6

(b) Shear strain contours with initial k_0 = 0.6

(c) Liquefied zones in Model_4H: 1V with initial k_0 = 0.7

(d) Shear strain contours in Model_4H:1V with initial k_0 = 0.7

(e) Liquefied zones in Model_4H: 1V with initial k_0 = 0.8

(f) Shear strain contours in Model_4H:1V with initial k_0 = 0.8

FIG 6 – Contours of liquefied zones (grey) and developed shear strain for 30 m extrusion in Model_4H:1V.

Modelling limitations

The modelling work conducted in the current study has some limitations as follows:

- The state parameter was assumed to be constant, although it should depend on the effective stress development during staged construction. This might be addressed by using the NorSand model during the model construction stages; however, proper control over the parameter K_0 cannot be adequately achieved in such cases.

- The initial K_0 was assumed to be constant across the soft tailings layer and other zones. Denser material or overlying sandy layers may potentially have different K_0 values compared to the soft slimes layer.

CONCLUSION

This paper investigated the effect of K_0 and bias on the flow liquefaction of a nominal upstream-raised tailings dam with different downstream slopes, after a lateral extrusion event is triggered. Three geometries with downstream slopes of 2H:1V, 3H:1V, and 4H:1V were assessed, using an initial K_0 value ranging from 0.5 to 0.8. The results of this study suggest:

- K_0 in combination with bias, significantly affects the development and propagation of liquefaction through the tailings mass.

- Flatter downstream slopes result in higher K_0 values across broader zones of tailings, and hence reduce the potential for static liquefaction by shifting the stress state further away from the instability line.

- At the lowest studied initial K_0 value of 0.5, all three geometries exhibited flow liquefaction and a global failure.

- The model with a downstream slope of 2H:1V, ie highest bias, could only tolerate the imposed shear strain (lateral extrusion) over 10 m of the soft tailings (initial K_0 of 0.8).

- Given the significant influence of K_0 it would be prudent to carry out rigorous investigation of this parameter for the assessment of static liquefaction. As demonstrated in this paper, for a given geometry, changes in K_0 can shift the model from being stable to experiencing flow liquefaction and a global failure.

REFERENCES

Jefferies, M and Been, K, 2015. *Soil liquefaction: A critical state approach,* 2nd edn (CRC Press).

Jefferies, M, Morgenstern, N R, Van Zyl, D V and Wates, J, 2019. Report on NTSF Embankment Failure, Cadia Valley Operations, for Ashurst Australia, Ashurst Australia, Sydney, Australia.

Liu, H, Nagula, S, Jostad, H P, Piciullo, L and Nadim, F, 2024. Considerations for using critical state soil mechanics based constitutive models for capturing static liquefaction failure of tailings dams, *Computers and Geotechnics*, 167:106089.

Morgenstern, N R, Vick, S G, Viotti, C B and Watts, B D, 2016. Fundão Tailings Dam Review Panel: Report on the Immediate Causes of the Failure of the Fundão Dam, Cleary Gottlieb Steen and Hamilton LLP, New York: USA.

Ng, C W, Crous, P A, Zhang, M and Shakeel, M, 2022. Static liquefaction mechanisms in loose sand fill slopes, *Computers and Geotechnics*, 141:104525.

Reid, D, Dickinson, S, Mital, U, Fanni, R and Fourie, A, 2021. On some uncertainties related to static liquefaction triggering assessments, in *Proceedings of the Institution of Civil Engineers – Geotechnical Engineering*, 175(2):181–199.

Robertson, P K, de Melo, L, Williams, D J and Wilson, G W, 2019. Report of the Expert Panel on the Technical Causes of the Failure of Feijão Dam, I, Skadden, Arps, Slate, Meagher and Flom LLP, New York: USA.

Yield stress characteristics of residual foundation soils – a tailings dam case history

M Walden[1], M Rust[2], N Rocco[3], M Shelbourn[4] and L Spencer[5]

1. Project Manager and Senior Geotechnical Engineer; NewFields Mining Design and Technical Services LLC, Lone Tree CO 80112, United States. Email: mwalden@newfields.com
2. Principal Geotechnical Engineer, PhD, NewFields Mining Design and Technical Services LLC, Lone Tree CO 80112, United States. Email: mrust@newfields.com
3. Principal Geotechnical Engineer, PhD, NewFields Mining Design and Technical Services LLC, Lone Tree CO 80112, United States. Email: nrocco@newfields.com
4. Director, Tailings and Dams, Newmont Corporation, Denver CO 80237, United States. Email: michael.shelbourn@newmont.com
5. Tailings and Water Treatment Superintendent, Newmont Corporation, Denver CO 80237, United States. Email: louise.spencer@newmont.com

ABSTRACT

Residual soils can pose unique challenges for geotechnical engineers due to their complex geological and stress history. Chemical and physical bonding mechanisms in residual soils contribute significantly to their strength and stiffness. However, even minor disturbances during sample collection, transportation, and testing can alter these properties, potentially leading to inaccurate geotechnical assessments.

The yield stress of residual soils differs significantly from the conventional frameworks developed for transported sedimentary soil types such as alluvial deposits.

The yield stress (or pre-consolidation stress) for a transported sedimentary soil formed by water, wind, or gravitational-driven movement is traditionally correlated to the maximum compressive stress experienced by the soil under self-weight or external loading, such as glaciation. In contrast, the yield stress of a residual soil is associated with the breaking down under compressive loading of weathering-induced bonding and fabric anisotropy, rendering traditional sedimentary soil models inadequate. Consequently, a unique approach is often needed to accurately characterise the behaviour of residual soils.

This paper presents a comprehensive case study on the yield stress of a residual soil derived from the *in situ* weathering of a low-grade metamorphosed sedimentary rock. The case study compares laboratory and *in situ* data to assess the consolidation characteristics of the residual soil. The degree of disturbance of intact laboratory samples is quantified and the impact of disturbance on the consolidation behaviour is investigated. Additionally, the laboratory strength testing results are directly compared to *in situ* soil strength estimates obtained through empirical correlations with data from seismic piezocone penetrometer testing with pore pressure measurements (SCPTu soundings). The study identifies discrepancies and correlations between the yield stress determined through laboratory oedometer testing and the *in situ* data correlations. The findings propose a refined approach for minimising sample disturbance during long-distance transport.

This research contributes to the development of more accurate and reliable methods for characterising residual soils, ultimately enhancing the safety and efficiency of geotechnical projects founded on such materials. The proposed approach has implications for the design and construction of tailings dams in regions with residual soils, and its adoption could lead to improved project outcomes and reduced costs.

INTRODUCTION

General

Residual soils, formed from the *in situ* weathering of parent rock, exhibit unique engineering properties that can significantly influence the performance of geotechnical structures. A substantial number of existing and planned mines and tailings storage facilities (TSFs) are situated in areas underlain by residual soils. Residual soils from varied regions often share similar traits due to common weathering processes. Characterising the engineering properties of these soils requires specialised approaches.

This study aims to investigate the yield stress of a low-plasticity residual soil deposit within the Guiana Shield, with a focus on an existing TSF subjected to embankment-induced foundation loading. The Guiana Shield, a vast geological province in South America, is characterised by its Precambrian basement rocks and deeply weathered near surface soil and rock profiles. The residual soils derived from these rocks often exhibit low plasticity and high sensitivity to disturbance. Understanding the yield stress of these soils is crucial for accurate geotechnical design and analysis, especially in working stress conditions that bound the estimated yield stress.

A comprehensive laboratory testing program was conducted, including index testing, scanning electron microscopy (SEM), one-dimensional oedometer testing, and consolidated undrained (CU) triaxial testing to evaluate the soil's behaviour and yield stress. These results were compared against *in situ* seismic cone penetration test (SCPTu) data to reconcile discrepancies between laboratory-derived consolidation parameters and field-measured responses.

Yield stress

Yield stress refers to the threshold stress level at which the soil begins to exhibit irreversible plastic deformation and significant changes in volume due to a reduction in stiffness. Unlike transported soils, residual soils have formed in place through weathering, and their behaviour under loading is commonly influenced by microstructure (eg cementation/bonding, and aging) in addition to past stress conditions. The yield stress in these soils is particularly important because it dictates the transition between predominantly elastic and predominantly plastic behaviour under loading. If a soil is stressed beyond its yield point, it may undergo significant settlement, strain, or structural reconfiguration, which affects its long-term stability and engineering properties.

Understanding yield stress is crucial in consolidation analysis, especially for predicting settlement behaviour under applied loads. Consolidation occurs when a saturated soil gradually expels pore water under sustained pressure, leading to a reduction in volume. If the applied stress remains below the yield stress, the soil deforms in a predominantly elastic manner, and consolidation occurs primarily due to gradual pore water drainage. However, exceeding the yield stress leads to predominantly plastic deformation and structural breakdown, significantly altering the compression behaviour of the soil. In residual soils, which often have variable structure and bonding due to weathering, the yield stress can vary widely, making it essential to determine through laboratory testing (eg oedometer tests) to ensure accurate settlement predictions in foundation design.

Yield stress also serves as a critical threshold for defining shear strength parameters in residual soils. The transition from predominantly elastic to predominantly plastic deformation degrades cementation bonds and fabric structure, resulting in shear strength reductions. Accurate yield stress determination is thus vital for selecting strength parameters that reflect both intact and post-yield states, preventing overestimation of soil resistance in geotechnical designs.

LABORATORY TESTING

Sampling

Methods

The majority of samples were collected in boreholes via traditional hydraulic push of thin-walled hollow steel tubes with a leading cutting edge (Shelby tube) pushed into the target unconsolidated formation to extract relatively undisturbed soil samples. For the project, some opportunities that were considered to increase specimen quality and mitigate disturbance included the following:

- Careful adherence to the guidance provided in ASTM D1587 (2015) for Shelby tube sampling.

- Larger diameter Shelby tubes (76 mm) were used to reduce disturbance. Disturbance may persist even after trimming, potentially compromising the reliability of laboratory test results.

- The samples were allowed time to partially bond to the tube after the Shelby tube push. Greater levels of bonding to the tube improve both the sample success rate and quality.

- Shelby tube specific sampling containers (ASTM D4220, 2014) were utilised for sample shipping. The samples were shipped from South America to the USA in these vibration resistant shipping containers.

- A band saw and piano wire method was trialled for sample extrusion but proved unsuitable due to the friable nature of the residual soil, which degraded during extraction. A piston extruder yielded better results.

- Block sampling via test pitting and ASTM D7015 (2018) was less commonly completed but was utilised in select cases to provide high quality laboratory samples and provided excellent sample quality.

Disturbance

Disturbance can significantly alter a soil's structure, leading to changes in its yield stress and compression behaviour. Any sample of soil being taken from the ground, transferred to the laboratory, and prepared for testing will be subject to disturbance (Clayton, Matthews and Simons, 1995). Collection of intact soil specimens can introduce disturbance through the drilling process (leading to extension), pushing the Shelby tube (compression), extraction of the tube (extension), transport and storage (vibration and temperature fluctuations), extrusion at the laboratory (compression), and during sample preparation for the specified laboratory test (Ladd and DeGroot, 2003).

Residual soils, in particular, are highly sensitive to disturbance due to their bonded microstructure and associated fabric, which can degrade under even minor handling. Accurate determination of yield stress from oedometer testing requires specimens that closely retain their *in situ* properties; otherwise, disturbed samples may underestimate yield stress, resulting in overly conservative designs or misrepresentation of settlement behaviour.

The effects of disturbance were quantified in this study by comparing the strain observed when the specimen is reloaded to its estimated *in situ* soil stress, per the designation scheme developed by Andresen and Kolstad (1979). The specimen quality designation was utilised to classify the sample quality in terms of disturbance and ranges from A (indicating minimal disturbance and negligible strain) to E (representing highly disturbed samples with significant strain). The specimen quality designation chart is presented in Table 1.

TABLE 1

Specimen quality designation chart.

Volumetric strain (%)	Specimen quality designation
<1	A
1–2	B
2–4	C
4–8	D
>8	E

Index properties

Index testing included particle size analyses (ASTM D6913, 2021; D7928, 2021) and Atterberg limits (ASTM D4318, 2017) using a Casagrande cup for the measurement of the liquid limit. The particle size distributions are presented on Figure 1. Some coarse-grained quartz vein materials were identified across the site that contained silty gravel and sand, and this data is also presented. Atterberg limit results indicate Plasticity Index (PI) ranges from non-plastic to 29. The average PI value is 9 and the standard deviation of the data set is 5. The Atterberg limits are plotted on Figure 1 with transparency, to assist in identification of areas of high data point concentration.

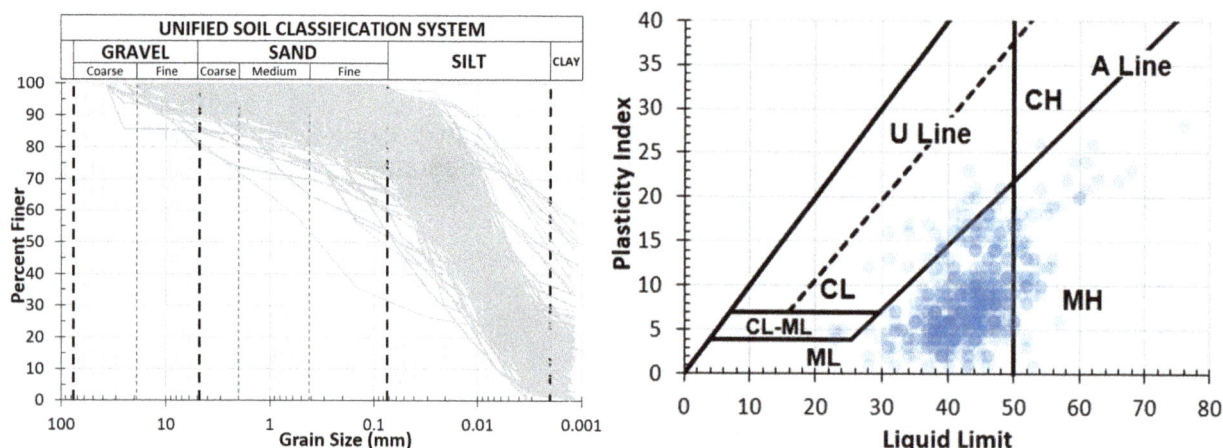

FIG 1 – Particle size distributions (left), Atterberg limits (right).

Results indicate that the materials encountered are fine-grained low to high-plasticity silt with varying amounts of clay, sand, and gravel particles.

Scanning electron microscopy

SEM was completed on intact samples of saprolite from the TSF foundation, and images were used to confirm the presence of bonding and microstructure within the soil. Figure 2 presents SEM imagery of an intact foundation sample at approximate magnifications of 60×, 1000×, and 5000× (left to right). The 5000× image depicts the bonding between particles and interlocking of platelets. Significant iron oxidation bonding is present in the sample, which is typical of saprolites. The bonding develops through a process known as laterisation, which occurs when iron-bearing minerals within the soil undergo oxidation and precipitation, leading to the formation of iron oxides that act as a natural cement, bonding soil particles together and increasing shear strength. The imagery presents the iron oxidation as the high contrast (brightly coloured) material that is cementing the platelets of quartz, mica, and kaolinite together.

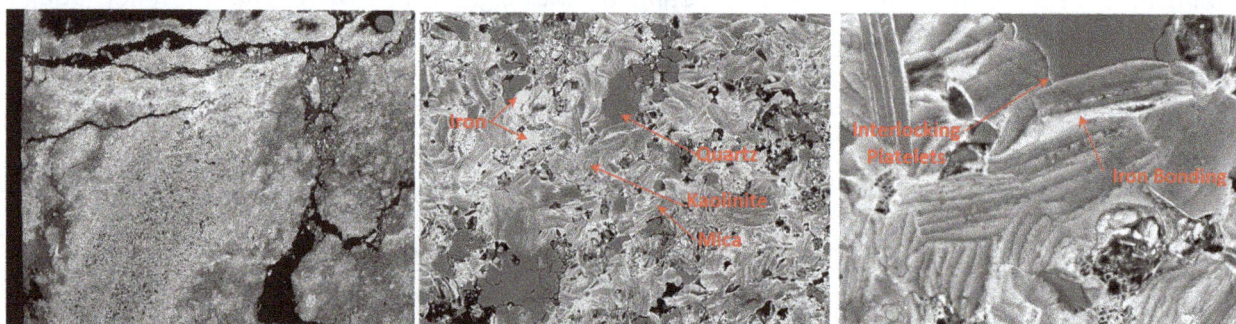

FIG 2 – Foundation SEM images at 60× (left), 1000× (middle), and 5000× (right).

Oedometer testing

Yield stress

One-dimensional oedometer tests were performed following ASTM D2435 (2004). The yield stress (σ_y) was determined using the cumulative work method as proposed by Becker et al (1987), which estimates yield by identifying the stress level at which the rate of cumulative plastic work begins to increase significantly, indicating the onset of yielding. Although Morin (1988) cautioned that the cumulative work method might overestimate yield stress in structured or cemented soils, it proved effective in providing definitive yield stress values where traditional approaches, such as the Casagrande construction, were unreliable or ambiguous.

Figure 3 presents the oedometer test data for a minimally disturbed specimen (quality designation of B) in terms of both void ratio (e) versus log σ_v' curve and cumulative work. The e versus log σ' curve is gradual, making identification of the point of maximum curvature difficult. In contrast, the cumulative work chart better identifies an inflection point and corresponding yield value.

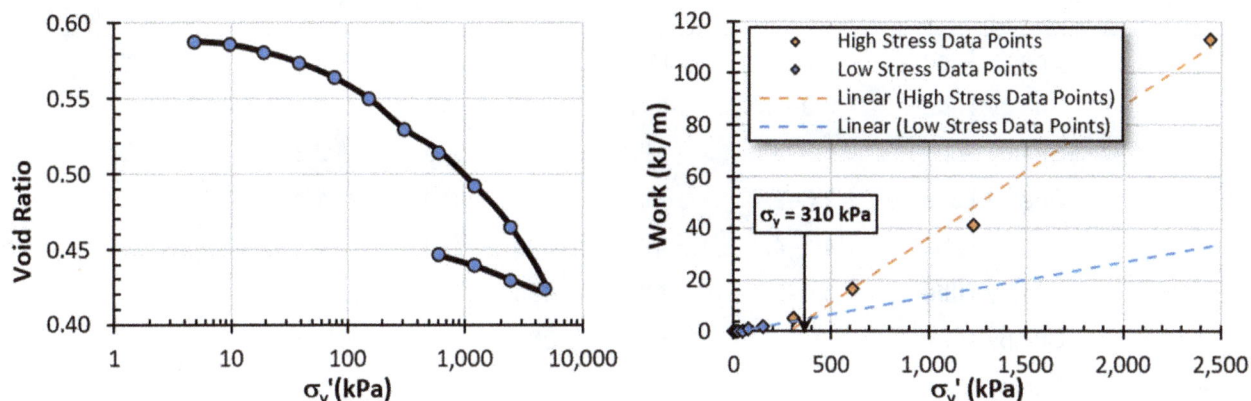

FIG 3 – Void ratio versus log σ_v' consolidation chart (left) and work per unit volume chart (right).

The work per unit volume method involves plotting the oedometer data on an arithmetic scale and fitting two distinct linear trends. The first line corresponds to a shallow-sloping trend in the low-stress region (blue data points and trend line defined in the right image on Figure 3), while the second line intersects data points in the high-stress regime (orange data points and trend line defined on the right image on Figure 3). The intersection of these two lines identifies the material's yield stress.

Results

The results, presented in Table 2, indicate that only two of the 27 specimens had quality designations lower than C, indicating that most laboratory tested soils were of reasonable quality.

The testing complements the SEM imagery and provides further indication that near-surface soil samples collected from the subject TSF site have bonding and cementation. The results were evaluated statistically, as presented in Table 3, which presents the average and 25th percentile yield stress values categorised by specimen quality designation. As presented in the table, both the average and lower-bound (25th percentile) yield stress values decrease as lower-quality specimens are included in the data set. This trend reinforces the expectation that sample disturbance reduces apparent soil strength.

To ensure reliability in the design, specimens with quality designations lower than 'C' were excluded from the statistical basis. This exclusion helps avoid incorporating data that may not accurately reflect *in situ* conditions due to excessive disturbance during sampling or handling. A conservative yield stress value was then selected for design purposes to mitigate the risk of underpredicting settlements or overestimating mobilised shear strength. This cautious approach reduces the potential for structural failure or excessive deformation, particularly in critical applications such as foundations, slopes, and retaining structures.

TABLE 2

Consolidation test results.

Sample #	Sample type	Depth (m)	USCS	e_0	σ_y (kPa)	Volumetric strain (%)	Specimen quality designation
1	Shelby Tube	3.5–4.2	MH	1.068	640	1.4	B
2	Shelby Tube	5.0–5.7	MH	1.105	515	1.7	B
3	Shelby Tube	6.0–6.7	ML	1.114	620	2.8	C
4	Shelby Tube	2.0–2.7	ML	0.938	310	1.4	B
5	Shelby Tube	4.1–4.8	ML	0.829	710	1.3	B
6	Shelby Tube	5.0–5.7	ML	0.903	300	3.3	C
7	Shelby Tube	3–3.7	ML	1.361	330	1.5	B
8	Shelby Tube	7–7.7	MH	1.224	340	8.2	E
9	Shelby Tube	6–6.7	ML	0.879	370	2.1	C
10	Shelby Tube	5.0–5.7	ML	0.799	390	0.5	A
11	Shelby Tube	3.0–3.7	ML	0.588	350	0.8	A
12	Shelby Tube	7.5–8.2	ML	1.228	350	0.8	A
13	Shelby Tube	10.5–11.2	ML	0.778	350	1.7	B
14	Shelby Tube	10.5–11.2	ML	0.959	350	2.3	C
15	Shelby Tube	12.0–12.7	ML	0.983	160	5.1	D
16	Shelby Tube	6.0–6.7	ML	1.051	340	2.3	C
17	Shelby Tube	6.50–7.20	ML	0.959	410	0.5	A
18	Shelby Tube	6.50–7.20	ML	1.254	270	3.9	C
19	Shelby Tube	7.00–7.25	ML	0.836	330	2.6	C
20	Shelby Tube	24.8–25.0	ML	0.758	380	2.3	C
21	Shelby Tube	7.4–7.8	ML	0.777	340	0.5	A
22	Shelby Tube	8.2–8.5	ML	0.723	220	1.6	B
23	Shelby Tube	7.2–7.4	ML	0.845	310	3.0	C
24	Shelby Tube	6.3–6.5	MH	0.673	370	2.1	C
25	Block Sample	2.0	ML	0.915	500	0.0	A
26	Block Sample	2.0	MH	1.271	460	0.0	A
27	Shelby Tube	3.7	ML	0.588	550	3.2	C

TABLE 3

Statistical oedometer test results.

Quality designation	Average σ_y (kPa)	25th percentile σ_y (kPa)
A	476	360
A+B	455	350
A+B+C	422	340
A+B+C+D	412	333
A+B+C+D+E	409	335

Consolidated undrained triaxial testing

The peak undrained shear strength was estimated for specimens that indicated both strain hardening and strain softening behaviour. More than half of all the CU triaxial tests (51/72 tests) present a slight strain-softening response, which is indicative of contractive shear behaviour. This behaviour may be linked to the relic microstructure apparent in the soil.

The peak undrained strength (Su_{peak}) for strain hardening specimens were selected at the point where the stress path reached critical state. The peak and residual undrained strength for strain softening samples were determined by selecting appropriate peak and residual values based on the stress-strain response during shearing.

Figure 4 presents the peak undrained strength ratio (Su_{peak}/σ') for each of the CU triaxial tests that have been performed. The results indicate that the undrained strength ratio decreases with increasing confining pressure to approximately 350 kPa. The results indicate that the 25th percentile yield stress (350 kPa) identified from oedometer testing is reasonable as the undrained strength ratio is relatively consistent beyond 350 kPa.

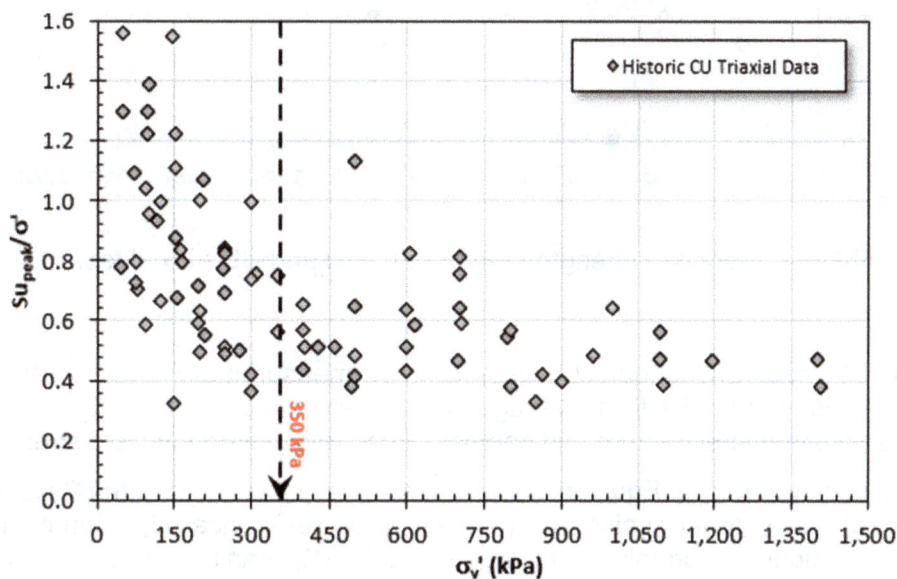

FIG 4 – Undrained strength ratio from CU triaxial testing.

IN SITU TESTING

Seismic piezocone testing

SCPTu testing was completed at the existing TSF (post embankment construction), and the data set encapsulates foundation material ranging no additional stress at surface to over 500 kPa of applied stresses from embankment loading.

Interpretation of CPTu data above the water table requires caution, as pore pressure measurements may be unreliable or absent in unsaturated conditions. However, elevated tip resistance and shear wave velocity in these zones can still provide indirect insights into bonding and cementation effects, especially in residual soils where structure plays a dominant role in strength. Where applicable, complementary methods such as SEM or microstructural analysis should be used to validate interpretations in the vadose zone.

Findings from the SCPTu investigations indicate that the upper foundation residual soils at the subject TSF may exhibit either contractive or dilatant behaviour depending on the effective stress conditions before shearing. This discovery is supported by the apparent exceedance of the yield stress and destruction of microstructure.

The SCPTu estimates of undrained shear strength were filtered to isolate data associated with expected contractive behaviour and plotted with transparency alongside triaxial test results in Figure 5. Both data sets demonstrate alignment, with the undrained strength ratio exhibiting a decrease as confining stress increases. However, due to refusal during penetration, the SCPTu data are constrained to confining pressures less than approximately 500 kPa, limiting its utility for direct yield stress estimation at higher stress levels. This truncation underscores the importance of integrating laboratory CU triaxial testing, which captures stress-strain behaviour across a broader pressure range to reliably extrapolate yield stress trends beyond *in situ* testing limitations.

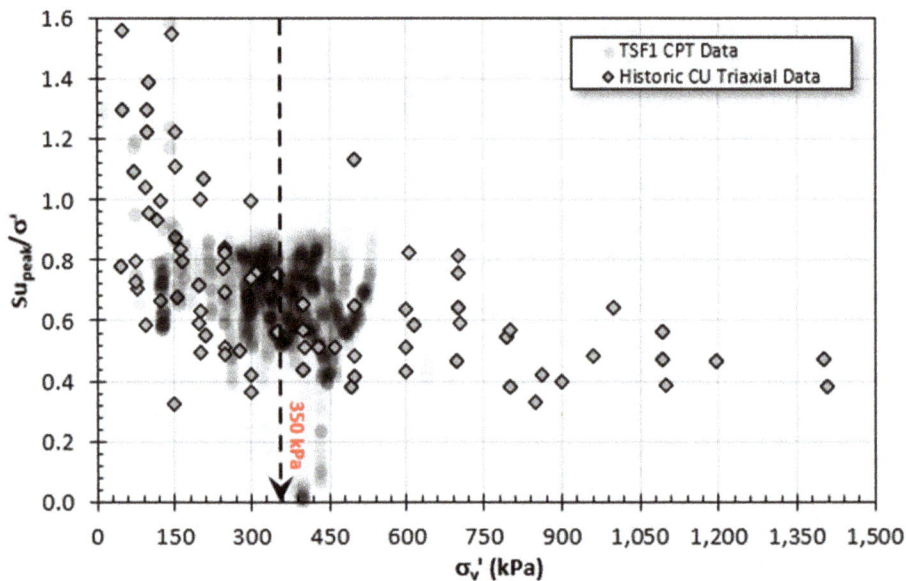

FIG 5 – Undrained strength ratio from CPTu and CU triaxial testing.

Seismic data

Compressional (P-wave) and shear (S-wave) velocity measurements were obtained at select locations during various TSF CPT field investigations. When paired with CPT, S-wave data can help identify the presence of microstructure and can be utilised to estimate the yield stress of the soil.

The small-strain rigidity index (I_G, Equation 1) was determined by comparing the S-wave velocity and inferred density of the material against the normalised tip resistance (Q_{tn}, Equation 2) to identify microstructure. The modified normalised small-strain rigidity (K_G) can be calculated using Equation 3.

$$I_G = \frac{G_0}{q_n} \tag{1}$$

Where:

q_n	is the net cone resistance (q_t-σ_v)
q_t	is the cone resistance corrected for water effects
G_0	is the small-strain shear modulus ($\rho(V_s)^2$)
V_s	is the shear wave velocity
ρ	is the soil mass density (γ/g)
γ	is the soil unit weight
g	is the acceleration due to gravity

$$Q_{tn} = \frac{q_t - \sigma_{v0}}{p_a} \times \left(\frac{p_a}{\sigma'_{v0}}\right)^n \tag{2}$$

Where:

$(q_t - \sigma_v)/p_a$	is the dimensionless net cone resistance
$(p_a / \sigma_{v0})^n$	is the stress normalisation factor
p_a	is the atmospheric reference pressure

n   is the stress exponent that varies with the normalised soil behaviour type (SBT$_n$)

$$K_G = I_G \times (Q_{tn})^{0.75} \qquad\qquad (3)$$

When K$_G$ is between 100 and 330 the corresponding soils are likely young and uncemented (ie have little or no microstructure) and can be classified as unstructured soils (Robertson, 2016).

The foundation soils appear to have microstructure, as expected for a residual soil. During analysis of the seismic data, a difference was observed between the behaviour of foundation soils under greater levels of applied stress from embankment loading, as presented on Figure 6. The data was screened at a variety of stress levels and the results indicate that the foundation soil microstructure remains intact provided that the applied stress at surface (eg from embankment loading) is less than 260 kPa. When the foundation soil is under stress conditions greater than 260 kPa the microstructure begins to yield. Plots of K$_G$ (comparing Q$_{tn}$ and I$_G$) for the foundation are presented on Figure 6.

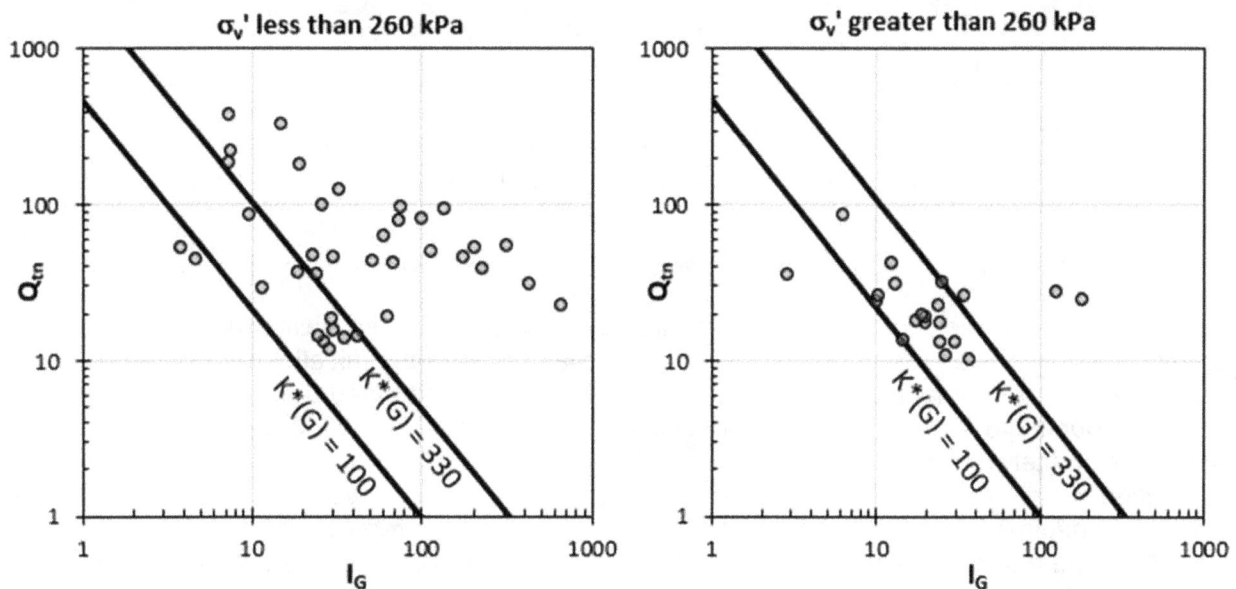

FIG 6 – Comparison of I$_G$ and Q$_{tn}$ in foundation soils.

SHEAR STRENGTH

Application in shear strength

While various methodologies exist for defining the shear strength, few effectively capture the non-linear behaviour of undrained shear strength associated with residual soils.

Shewbridge (2019) provides straightforward methodology for defining non-linear shear strength functions and offers a framework that can be utilised regardless of stress history, fabric anisotropy, and bonding effects. The approach can account for the progressive breakdown of cementation bonds during shearing, which is a critical process in residual soils where weathering-induced bonds dominate strength. The method can capture the transition from bonded to remoulded states by correlating peak and post-peak shear resistance with confining pressure, which is essential for characterising the nonlinear shear strength behaviour of residual materials. The model also incorporates parameters such as the friction angle at critical state and the rate of bond degradation, enhancing its applicability to field conditions. This methodology offers a significant advantage for residual soils over conventional models (eg Mohr–Coulomb, constant undrained strength, and constant undrained strength ratio) by addressing the dual influence of inherited structure and stress-dependent frictional resistance, thereby improving predictions of slope stability and foundation performance in geotechnical designs.

Figure 7 demonstrates the application of the shear strength methodology, integrating SCPTu data, triaxial test results, and the final non-linear strength envelope used for the facility design. The yield stress of 350 kPa marks the critical stress threshold where all bonding mechanisms are fully degraded, resulting in an expected transition from the critical state friction angle ($\phi' = 32°$) to a fully undrained peak strength ratio of 0.45.

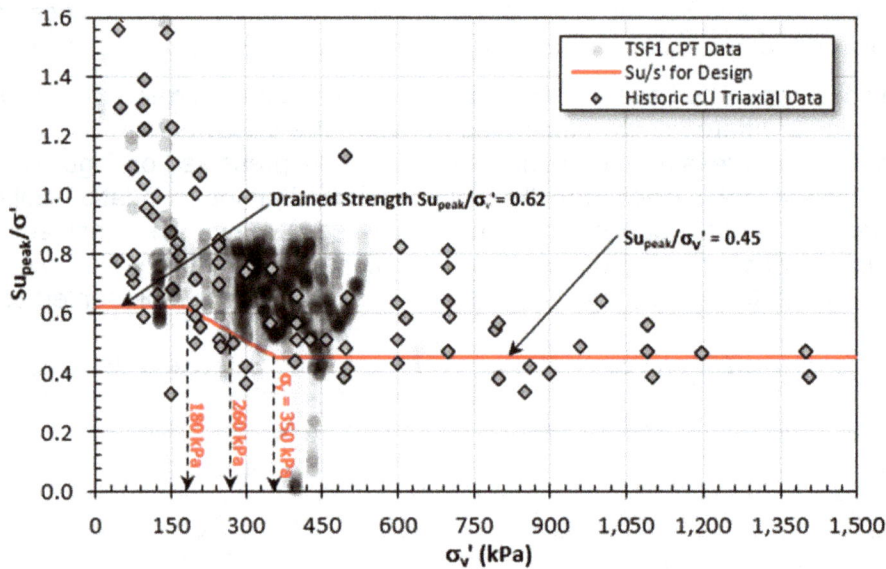

FIG 7 – Non-linear shear strength function.

Seismic shear wave velocity measurements indicate a breakdown in microstructure occurring at approximately 260 kPa. This threshold is interpreted as marking the initiation of progressive bond degradation, where partial loss of interparticle bonding allows undrained deformation mechanisms to begin to dominate, although some residual structure remains. The 260 kPa stress level appears to be approximately midway between the inferred yield stress of 350 kPa and the initial point at which a reduction in shear strength is observed. Accordingly, the onset of transitional soil behaviour is assumed to begin at around 180 kPa. This interpretation is supported by complementary evidence from CPTu and triaxial test data, both of which exhibit reductions in shear resistance at comparable stress levels.

Additional considerations

Residual soils typically exhibit a depth-dependent strength profile, with shear resistance increasing with depth due to reduced weathering intensity and enhanced preservation of the parent rock microstructure. This trend contrasts with traditional clay theory, such as the SHANSEP framework (Ladd and Foote, 1974) which correlates undrained shear strength with the pre-consolidation stress resulting in a general decrease in shear strength with depth. When defining yield stress in residual soils, consideration must be given to the thickness of the layer to which a single yield stress value is applied. Assigning a uniform yield stress across thick, heterogeneous profiles risks underestimating strength at depth, where bonding and fabric anisotropy can be more pronounced. Geotechnical analyses should account for this depth-strength relationship to avoid errors in slope stability assessments or foundation designs, particularly where residual soil horizons are common.

DISCUSSION

This study highlights the importance for characterisation of yield stress in residual soils through an integrated laboratory and *in situ* testing framework. Key findings reveal that residual soils derived from low-grade metamorphic rock in the Guiana Shield at the subject TSF site exhibit yield stress governed by weathering-induced bonding and fabric anisotropy, rather than historical stress conditions. The cementation mechanisms, particularly iron oxide bonding observed via SEM imagery, impart significant structural strength but indicate the soils are sensitive to strength loss when the bonding mechanisms are broken. The continued use of SEM to support a better understanding of soil behaviour for critical projects is highly encouraged.

The majority of the laboratory specimens were prepared from Shelby tube samples, which resulted in a range of soil disturbance. Specimens classified as moderately to highly disturbed (specimen quality designation ≥ C) demonstrated an average 24 per cent reduction in yield stress compared to minimally disturbed (specimen designation A) specimens, underscoring the necessity of stringent protocols for sample handling during long-distance transport. The maximum reduction in yield stress was approximately 66 per cent. Twenty-five (25) of the 27 specimens that underwent oedometer testing yielded quality designations ≥ C and 14 were ≥ B, indicating that the precautions taken to minimise sample disturbance were generally effective. It is important to highlight that laboratory test specimens obtained through block sampling methods (ASTM D7015, 2018) exhibited minimal disturbance, demonstrating the advantage of this method for preserving sample integrity. Similarly, alternative sampling techniques for stiff soils (eg Pitcher or Denison sampler) could be considered to improve sample quality in residual soils.

The application of the cumulative work method better defined the estimated yield stress values compared to traditional Casagrande construction methods, particularly in soils with gradual stress-strain curves. The laboratory-derived yield stresses (250–800 kPa) aligned with seismic shear wave velocity thresholds (260 kPa) and triaxial strength trends, revealing a critical transition from bonded to remoulded states. However, discrepancies persisted between oedometer results and SCPTu-based estimates, with the *in situ* data truncated at approximately 500 kPa due to penetrometer refusal. This highlights the importance of laboratory testing for extrapolating high-stress behaviour and validating field correlations.

The integration of Shewbridge's (2019) non-linear shear strength model showed strong potential for practical design applications. This model accounts for bond degradation and the transition to critical state friction, enhancing the accuracy of slope stability and foundation designs in residual soils. A notable finding was the identification of a microstructure breakdown threshold (~260 kPa), which lies between initial strength loss (180 kPa) and yield stress (~350 kPa), providing a valuable criterion for evaluating progressive failure in tailings dam foundations.

CONCLUSION

This study provides a robust framework for defining yield stress in residual soils, with direct implications for the safety and cost-effectiveness of critical infrastructure projects, particularly tailings dams. Based on the findings, the following recommendations are made for field investigation and design in residual soil environments:

- Use specimen quality designations routinely to assess sample integrity and preserve *in situ* microstructure.

- Adopt conservative yield stress values in design to avoid overestimating soil strength in residual formations.

- Leverage SCPTu S-wave data to define small-strain stiffness (K_G) and identify microstructural characteristics.

- Apply shear strength models that incorporate bonding degradation, especially in tropical or deeply weathered terrains.

These practices aim to enhance geotechnical reliability and contribute to the safe performance of structures built on residual soils.

ACKNOWLEDGEMENTS

The authors wish to express their sincere gratitude to Newmont Corporation for its support of this work. Specifically, we acknowledge Newmont's provision of essential data, without which this comprehensive analysis of residual soil behaviour would not have been possible. Newmont's commitment to advancing geotechnical understanding is gratefully recognised.

REFERENCES

Andresen, A and Kolstad, P, 1979. The NGI 54-mm Samplers for Undisturbed Sampling of Clays and Representative Sampling of Coarser Materials, Proceedings International Symposium of Soil Sampling.

ASTM D1587, 2015. Standard Practice for Thin-Walled Tube Sampling of Fine-Grained Soils for Geotechnical Purposes, ASTM International.

ASTM D2435, 2004. Standard Test Methods for One-Dimensional Consolidation Properties of Soils Using Incremental Loading, ASTM International.

ASTM D4220, 2014. Standard Practices for Preserving and Transporting Soil Samples, ASTM International.

ASTM D4318, 2017. Standard Test Methods for Liquid Limit, Plastic Limit and Plasticity Index of Soils, ASTM International.

ASTM D6913, 2021. Standard Test Methods for Particle-Size Distribution (Gradation) of Soils Using Sieve Analysis, ASTM International.

ASTM D7015, 2018. Standard Practices for Obtaining Intact Block (Cubical and Cylindrical) Samples of Soils, ASTM International.

ASTM D7928, 2021. Standard Test Method for Particle-Size Distribution (Gradation) of Fine-Grained Soils Using the Sedimentation (Hydrometer) Analysis, ASTM International.

Becker, D E, Crooks, J H A, Been, K and Jefferies, M G, 1987. Work as a criterion for determining in situ and yield stresses in clays, *Canadian Geotechnical Journal*, 24:549–564.

Clayton, C R I, Matthews, M C and Simons, N E, 1995. *Site Investigation*, second edition, Department of Civil Engineering, University of Surrey.

Ladd, C C and DeGroot, D J, 2003. Recommended Practice for Soft Ground Site Characterization: Arthur Casagrande Lecture, 12th Panamerican Conference on Soil Mechanics and Geotechnical Engineering, Massachusetts Institute of Technology Cambridge.

Ladd, C C and Foote, R, 1974. A new design procedure for stability of soft clays, *Journal of the Geotechnical Engineering Division*, 100(GT7):763–786.

Morin, P, 1988. Discussion of work as a criterion for determining in-situ and yield stresses in clays, *Canadian Geotechnical Journal*, 25:845–847.

Robertson, P K, 2016. Cone penetration test (CPTu)-based soil behavior type (SBT) classification system — an update, Canadian Science Publishing.

Shewbridge, S, 2019. Undrained Strengths and Long-Term Stability of Slopes, *ASCE Journal of Geotechnical and Geoenvironmental Engineering*.

Managing mine-affected water

Reactive transport and multiphase gas model of coal tailings to predict potential long-term seepage quality

J H du Toit[1], P J Fourie[2] and B H Usher[3]

1. Hydrogeochemist, KCB Australia, Brisbane Qld 4101. Email: jdutoit@klohn.com
2. Senior Hydrogeochemist, KCB Australia, Brisbane Qld 4101. Email: jfourie@klohn.com
3. Lead: Technical, KCB Australia, Brisbane Qld 4101. Email: busher@klohn.com

ABSTRACT

A geochemical reactive transport model was constructed to predict the long-term seepage quality from coal mining tailings backfilled into an opencast pit. The results of this model are intended to serve as a 'source term' for the tailings in a contaminant transport groundwater model.

Initially, a 1D gas model was developed to simulate atmospheric oxygen ingress, considering the moisture content and hydraulic head in the backfilled tailings. Critical aspects required for modelling gas migration into the tailings included the Oxygen Consumption Rate (OCR), Oxygen Diffusion Coefficient (ODC), Soil Water Retention Characteristics, and the hydraulic properties of the tailings material.

The mineral kinetic rate parameters were determined by simulating kinetic tests performed on tailings samples. These results, along with those from the gas transport model, were used as inputs for the reactive transport geochemical model, which was scaled to simulate site-specific conditions. The modelling domain was designed to allow for simulation of multiple output localities, representing seepage to the adjacent mining spoil backfill materials and the underlying aquifer.

The results of the reactive transport model indicated that sulfide mineral oxidation and leaching were restricted to the unsaturated zone exposed to atmospheric conditions. The modelling therefore also suggested that the neutralisation mineral content, especially in the unsaturated zone was crucial in providing buffering capacity against acidification, as assessed through sensitivity analysis. An additional modelling scenario was also undertaken where oxygen availability was limited by submerging the tailings underwater. The results of which indicated it as an effective acid-generating prevention strategy, regardless of the availability of carbonate minerals, to ensure no severe deterioration in the seepage quality from the tailings.

INTRODUCTION

In coal tailings, pyrite often occurs as one of the associated minerals. The weathering of pyrite is described by four commonly accepted reactions according to numerous studies, eg Stumm and Morgan (1981), which are summarised in Figure 1. The most important components required for this reaction generation are pyrite, oxygen and water, with additional factors such as pH, PO_2, specific mineral areas, morphology of the pyrite grains, absence of micro-organisms, carbonate minerals and hydrological system influencing the rate and outcomes of these reactions (Evangelou, 1995).

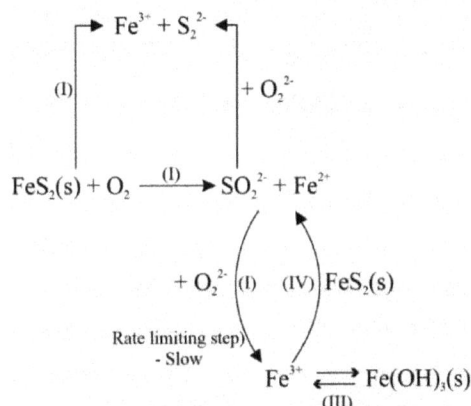

FIG 1 – Model for the oxidation of pyrite (modified from Stumm and Morgan, 1981).

To predict the rate at which materials generate acid Nicholson (1990) proposed the concept of 'limiting constituents'. In for example Figure 1, the oxygen supply limits the amount of pyrite that can be oxidised and therefore the amount of acid being generated. If the assumption is made that all the steps occur rapidly in comparison to the supply of oxygen, then all dissolved oxygen will be consumed and according to the stoichiometry of the reaction, an equivalent amount of sulfuric acid will be produced. The acid generated can be predicted from the amount of a limiting constituent (oxygen) present in a solution (in the absence of other oxidisers).

This principle was applied to predicting the drainage quality from backfilled tailings, where oxygen availability was used as the limiting factor.

METHODOLOGY

Tailings system under consideration includes a solid, water and gas phase. In general, when the solid phase (tailings) contains sulfide minerals, it is considered the reactive phase as it reacts spontaneously with oxygen and water to produce acidity. Figure 2 illustrates the setting that was modelled, where the tailings are to be partially submerged below water in the long-term. Therefore, the focus of the geochemical modelling is on the seepage quality that will be from the unsaturated zone where the most reactivity is expected.

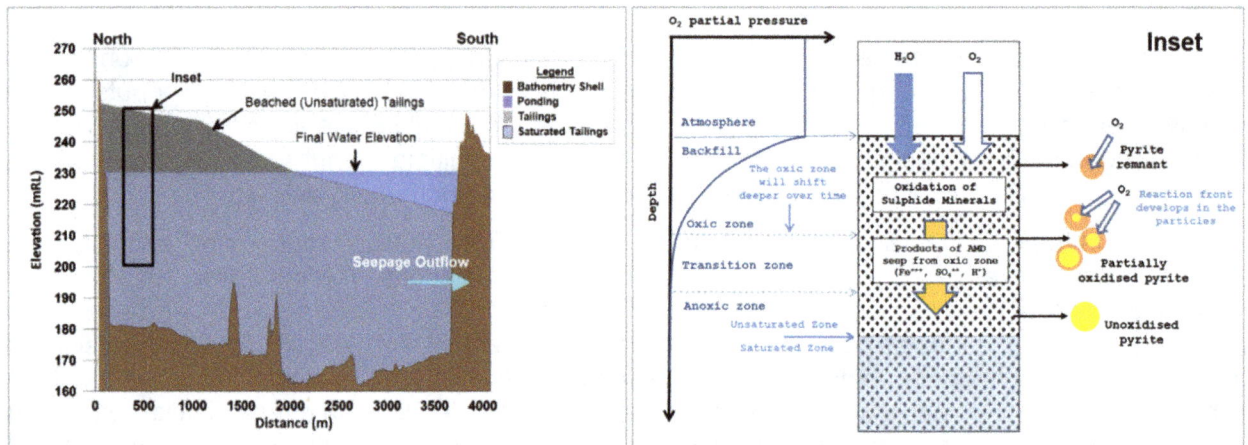

FIG 2 – Geochemical conceptual model of backfilled tailings.

In the model, the dissolution rates of primary minerals were controlled by the kinetic rate law as defined by the following equation:

$$r = \frac{dn}{dt} = (A_S k_+) \prod^{j} (a_j | m_j)^{P_j} \left(1 - \frac{Q}{K}\right)$$

Where:

A_S is the mineral surface area (cm^2)

k_+ is intrinsic rate constant (mol cm^{-2} s^{-1})

a_j, m_j is the function rate law's promoting and inhibiting species

P_j the species' power (+ is promoting, - is inhibiting)

Q the Activity product

K is the Equilibrium constant for dissolution reaction

The reaction surface areas were determined by reverse modelling on kinetic leaching testing that was undertaken on tailings using and the intrinsic rate constant were obtained from Oxygen Consumption Rates (OCR) that were also available for the tailings.

The interaction between the minerals, water, and gas phases was modelled using the Geochemist's Workbench Professional (GWB) and Chemplugin (Bethke, 2024). The Geochemist's Workbench is a set of interactive software tools for solving problems in aqueous geochemistry. This software allows the solving of hydrochemical and mineral reactions under equilibrium conditions, as well as the kinetic rate law for mineral dissolution. ChemPlugin is a GWB software object that can be used to create reactive transport simulations in the required configuration. A subroutine was compiled in Python 3.12 programming language and used ChemPlugin to convert a flow model into a multicomponent reactive transport simulator. In other words, each node in the modelling domain can be assigned a different flow rate and nodes can be assigned different flow destinations to better represent more intricate settings.

GWB and ChemPlugin have some implicit assumptions and limitations when used for Acid Mine Drainage (AMD) modelling. GWB relies on thermodynamic and kinetic data, assumes homogeneity, and often presumes chemical equilibrium, which may not fully capture the complexities of natural systems. ChemPlugin integrates with basic flow models and uses simplified representations of geochemical processes, which can affect accuracy. The accuracy of ChemPlugin's predictions is also sensitive to the specified boundary conditions, which may not always be well-defined and understood in natural systems.

MODEL INPUTS

A pseudo-two-dimensional geochemical model was constructed to model the expected water quality change of the interstitial water quality and seepage from the tailings into the adjacent spoils/aquifer (see Figure 2). The geochemical modelling domain was conservatively designed by assigning a source term for the tailings with a sufficient unsaturated zone to ensure adequate oxidation. However, in real-world conditions, a significant portion of the tailings will be submerged under water, preventing atmospheric oxidation of sulfidic materials.

The model had a total depth of 50 m, with the upper nodes simulated as unsaturated corresponding to the conceptual profile discussed above. Outflow from the domain was both in the horizontal (towards the adjacent spoils/aquifer) and downward direction (towards the underlying groundwater system).

Recharge to the system was based on larger-scale groundwater models, and for tailings source term models, the total horizontal flow was assumed to be 4 per cent of rainfall (which is 80 per cent of the inferred rainfall recharge infiltration (5 per cent recharge assumed)). Horizontal outflow was additionally also assumed to only occur in the saturated zone with the highest flow rate being at the top of the saturated zone, gradually declining with depth as indicated in Figure 3.

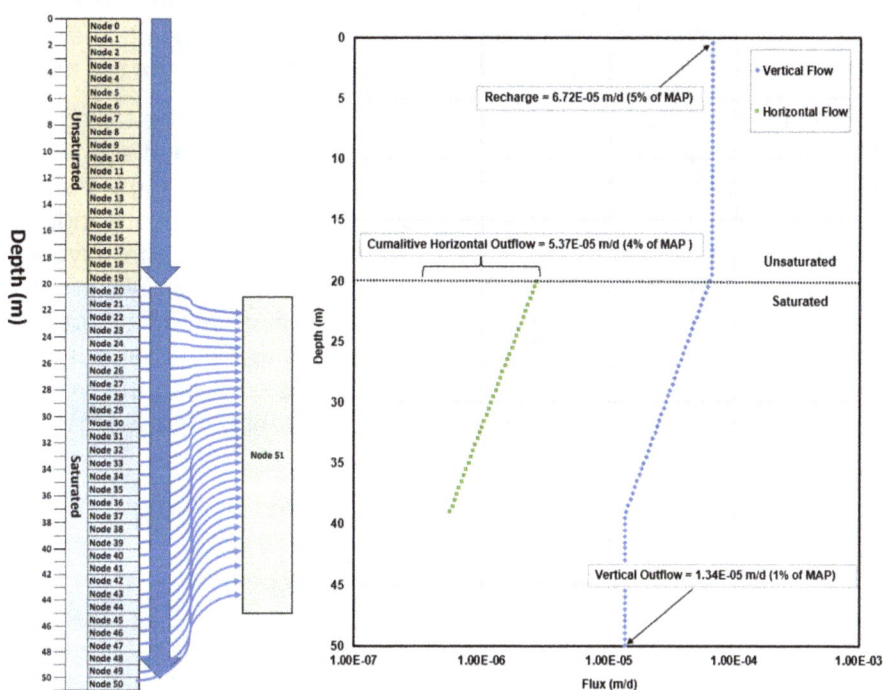

FIG 3 – Conceptual illustration of the model domain.

A base case model, using the mean Total S and acid neutralisation capacity (ANC) of the tailings was simulated with three additional sensitivity scenarios to assess the range of water quality outcomes based on the geochemical characterisation that the site had completed on the tailings. The additional scenarios had varying total sulfur content and/or acid neutralisation capacity, as shown in Table 1. The sensitivity scenarios considered the 75th centile of the Total S from a data set that consisted of 355 samples, as well as a conservative sensitivity scenario where only 50 per cent of the ANC may be effective (this aligned with a small subset of samples that showed this behaviour in acid-base characteristic curve (ABCC) testing.

TABLE 1

Geochemical model scenarios with major inputs.

Output name	Specifications	Description
Base Case	Total %S = 0.58, ANC (H_2SO_4/t) = 23.4	Mean Total %S, Mean ANC
Sensitivity 1	Total %S = 0.58, ANC (H_2SO_4/t) = 11.7	Mean Total %S, 50% Mean ANC
Sensitivity 2	Total %S = 0.71, ANC (H_2SO_4/t) = 23.4	75th Percentile Total %S, Mean ANC
Sensitivity 3	Total %S = 0.71, ANC (H_2SO_4/t) = 11.7	75th Percentile Total %S, 50% of Mean ANC

MODEL RESULTS

The oxygen availability was simulated using Geostudio 2024, with a fluid permeability of 3×10^{-6} m/s and a saturated volumetric water content of 0.45. Oxygen consumption testing was conducted to assess the rates of reaction, and the median rate of 4.5×10^{-11} kgO_2/kg/s from this testing was assigned as input to the multiphase model.

Under neutral pH conditions, without other oxidants, the presence of oxygen is critical for sustaining oxidation reactions of pyrite. A dissolved oxygen concentration of <0.5 mg/L is considered to be indicative of sub-oxic conditions in groundwater as described by McMahon and Chapelle (2008). At 25°C, oxygen in the air at 14 g/m^3 (1.2 per cent atm) will result in ~0.5 mg/L dissolved oxygen in the water. An oxygen concentration of 1 per cent atm was used to indicate the onset of sub-oxic conditions, and this criterion was used as background for the discussion of the model results.

Two (2) model scenarios were performed for a 20 year simulation for the tailing materials with the changing variable being the amount of rainfall recharge occurring in each scenario (5 per cent and 10 per cent of MAP). It was assumed that after the placement of the tailings, the oxygen in the gas phase within the material would be near atmospheric levels. The simulations indicated that oxygen in the model was quickly consumed in the upper unsaturated tailings with oxygen concentrations reaching a pseudo-equilibrium state (ie a stable depth of reaction) within 3–5 years and because of the low hydraulic conductivity of the tailings, the varying infiltration rate (5 per cent and 10 per cent of MAP) had a negligible effect on the oxygen concentrations in the upper portions of the tailings profile. Figure 4 provides the results of the 20 year simulations, showing the oxygen concentration with depth. These results were subsequently used as a fixed input into the geochemical reactive transport model.

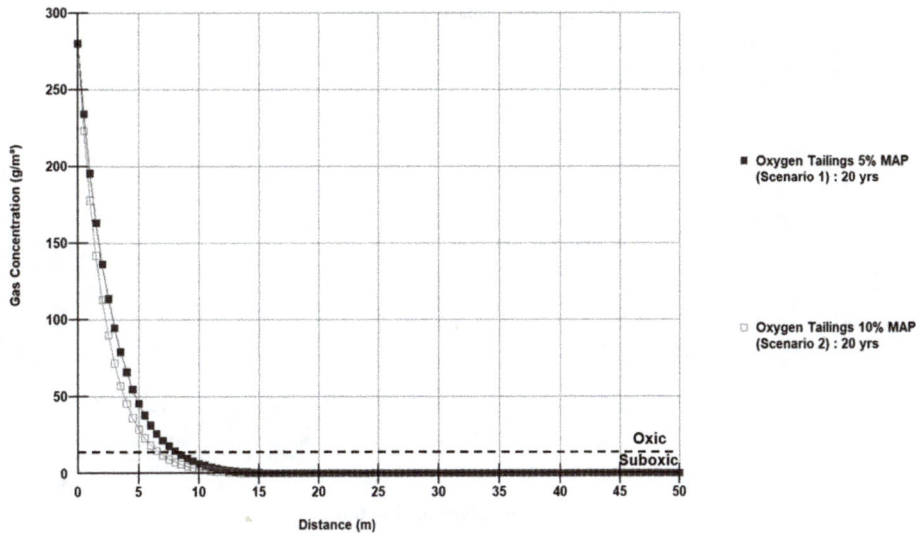

FIG 4 – Tailings – 1D oxygen profile after 20 years.

From the geochemical modelling, the impact of ongoing pyrite oxidation and depletion in the top ~9 m is shown in the modelling results in Figure 5. However, only the sensitivity scenarios with reduced neutralisation capacity (Sensitivity 1 and 3) indicate local acidification in the first couple of metres of the unsaturated zone (Figure 6). The base case using average tailings geochemical properties indicate that acidification is not expected within the 200 years of the simulation even with sufficient oxygen availability and results suggest near neutral pH throughout the tailings because of the sufficient neutralisation capacity.

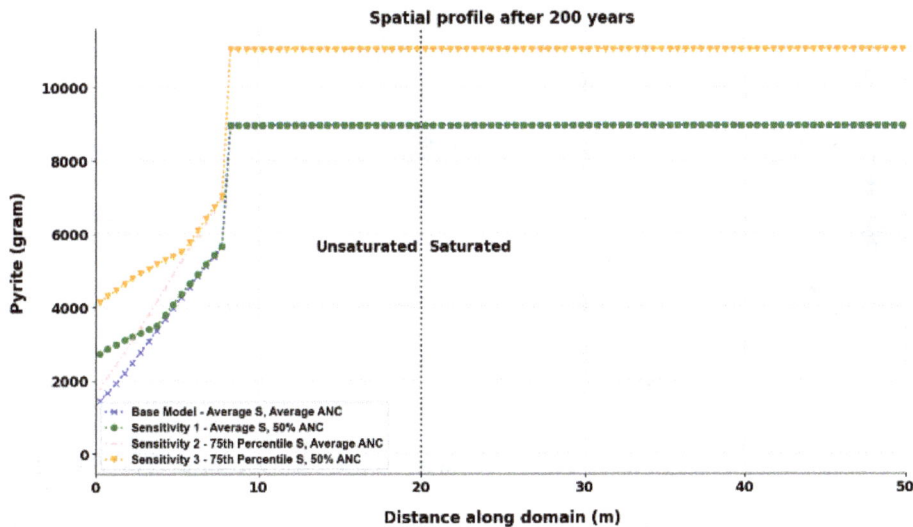

FIG 5 – Pyrite content in the tailings after 200 years.

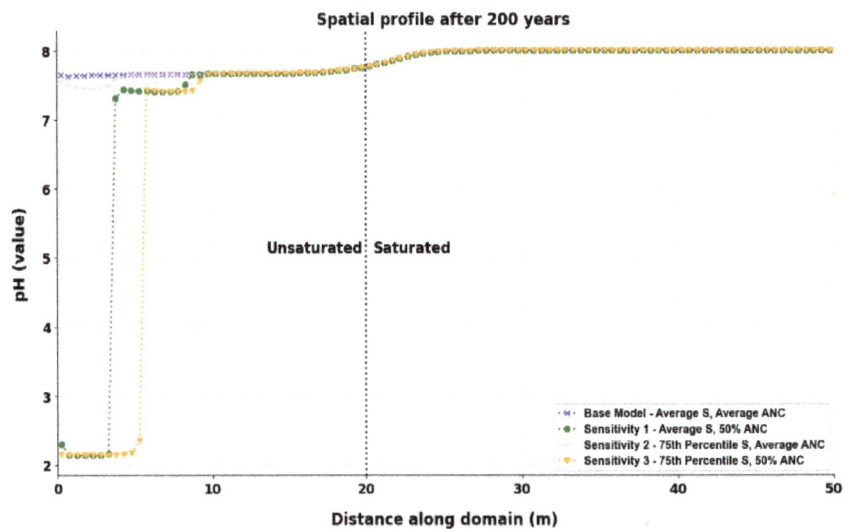

FIG 6 – pH values in the tailings after 200 years.

Sulfate was predicted to be the dominant anion in the tailings pore water. The sulfate in the Base Case model reaches about 2000 mg/L in the unsaturated zone but remains in the range of 500 mg/L in the saturated zone (Figure 7). The sensitivity scenarios with reduced ANC (Sensitivity 1 and 3), show some localised elevated concentration of sulfate in the acidified oxic zone.

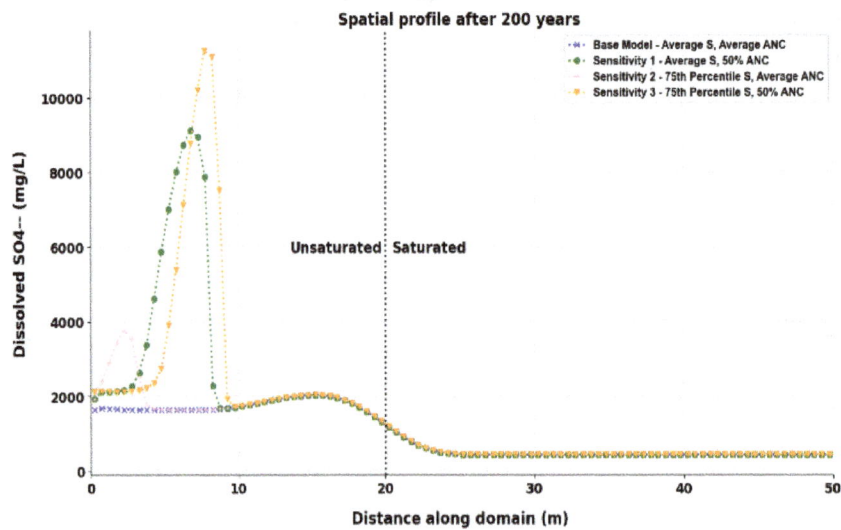

FIG 7 – Dissolved sulfate concentration in the tailings after 200 years.

As expected, neutral conditions in the base case result in low metal concentrations (Figure 8), while in the more conservative sensitivity scenarios with reduced ANC (Sensitivity 1 and 3), localised elevated metals are predicted as a result of the increased acidity in the oxic zone (because of pyrite oxidation).

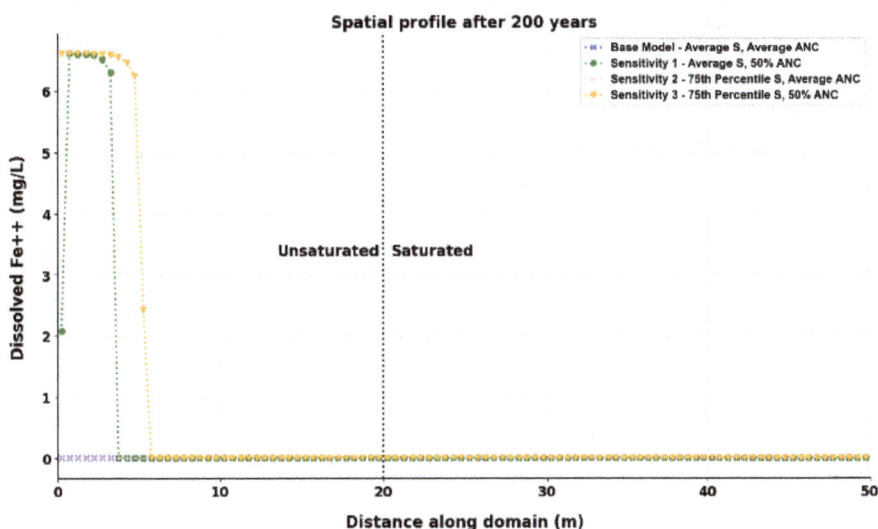

FIG 8 – Dissolved iron content in the tailings after 200 years.

Secondary minerals that were predicted to precipitate in the tailings will include mostly gypsum and goethite (Figure 9). Gypsum is expected to be saturated throughout the unsaturated zone and will act as the primary control of the sulfate concentrations. In the sensitivity scenarios with reduced ANC where acidification occurs in the oxic zone, the gypsum precipitation would cease once the calcium-containing carbonates are depleted.

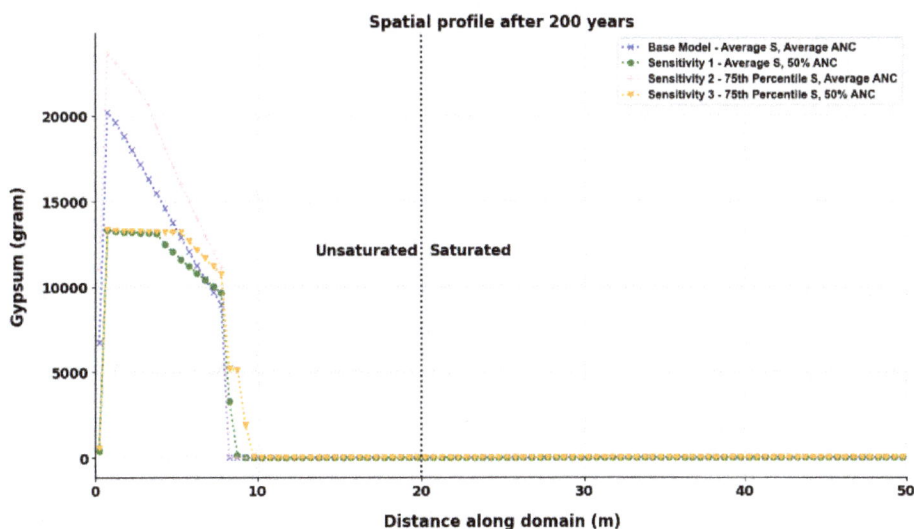

FIG 9 – Gypsum content in the tailings after 200 years.

From the previous model, it was shown that the sulfate concentration will be relatively higher in the unsaturated than saturated zone due to pyrite oxidation. The sulfate concentration in the horizontal flow of the tailings at the top of the saturated zone will however start to increase when the reacted mass from the unsaturated zone is periodically flushed into the saturated tails. Therefore, the thinner the unsaturated zone, the quicker this reacted mass may reach the saturated zone. However, if the unsaturated zone is too thin, it will also start to decrease the oxic zone where pyrite oxidation can occur.

Additional modelling was therefore performed using the Base Case and Sensitivity 1 inputs for various unsaturated depths, of 20 m (same as above), 5 m, 2.5 m and 0 m. The horizontal outflow was also changed to occur just below the saturation boundaries for each of these additional settings. The results of the sulfate concentrations and pH are provided as the cumulative horizontal flow Figures 10 to 13.

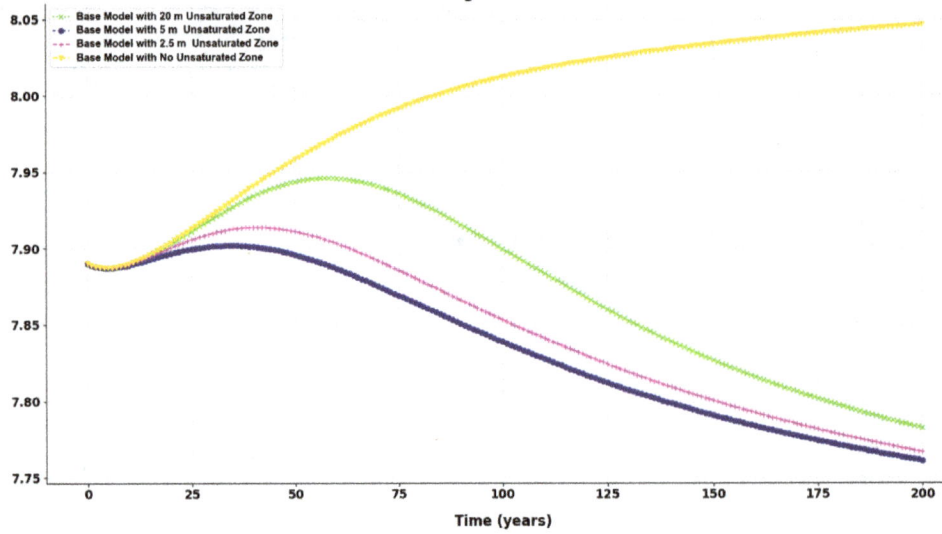

FIG 10 – Base model – pH in cumulative horizontal outflow (Node 51) from the tailings.

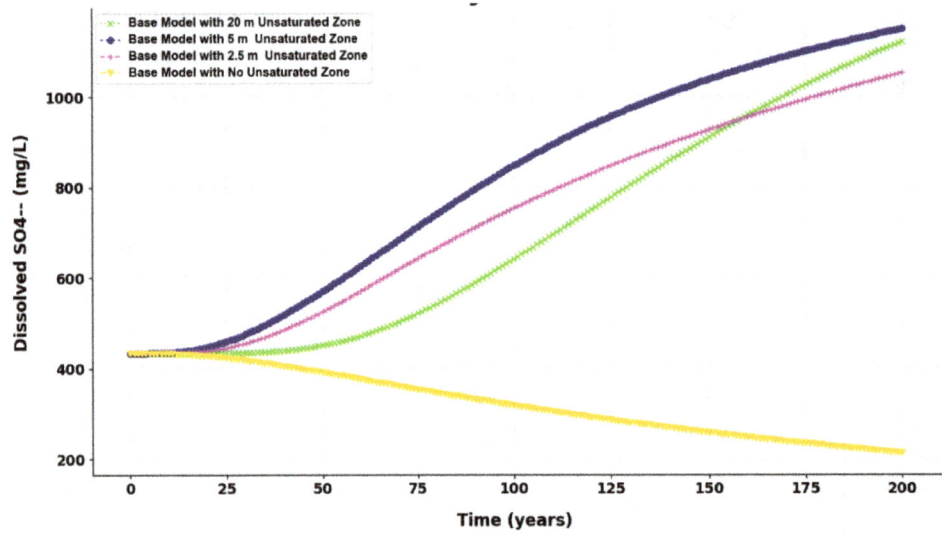

FIG 11 – Base model – sulfate in cumulative horizontal outflow (Node 51) from the tailings.

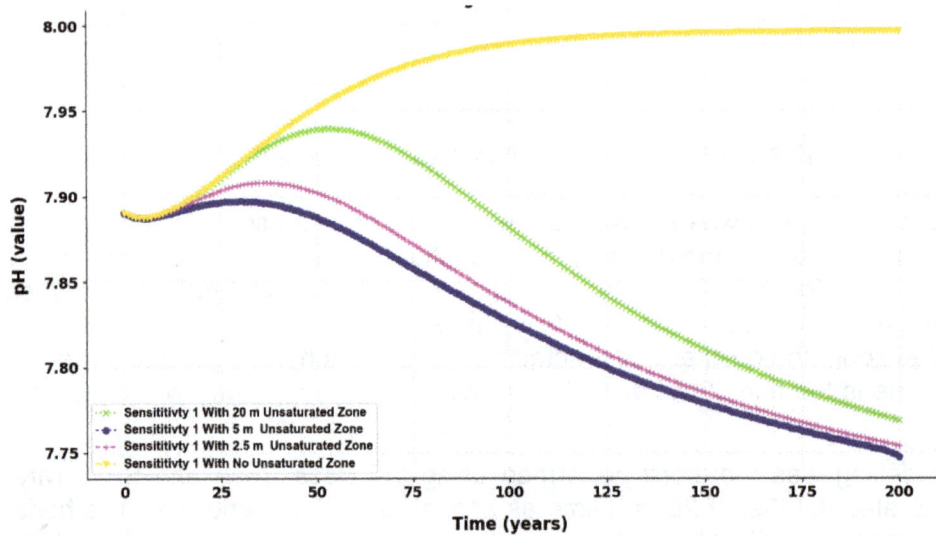

FIG 12 – Sensitivity 1 – pH in cumulative horizontal outflow (Node 51) from the tailings.

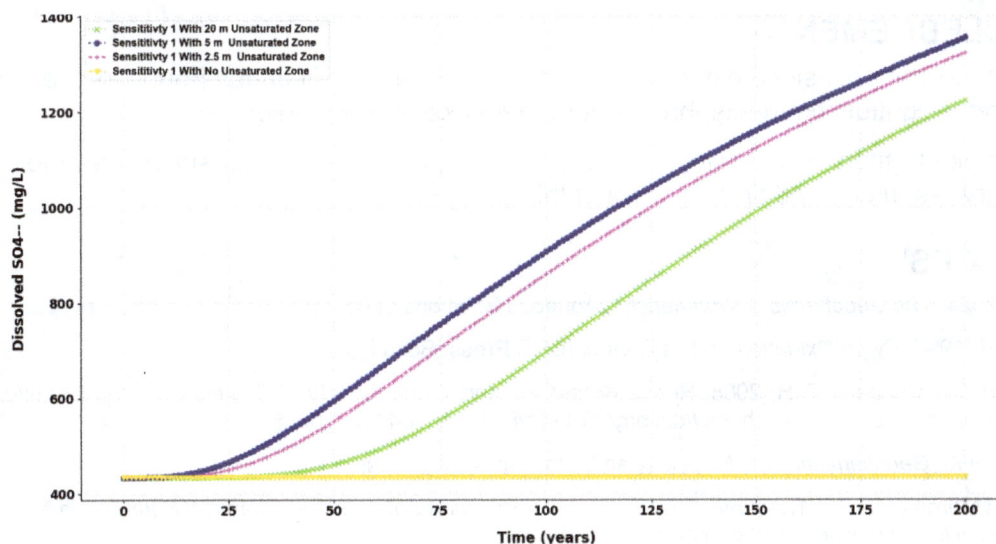

FIG 13 – Sensitivity 1 – sulfate in cumulative horizontal outflow (Node 51) from the tailings.

For both the Base Case and Scenario 1 (reduced ANC), the pH in the horizontal outflow remained neutral. This is because the saturated zone where the horizontal flow will be expected to occur is not predicted to acidify due to still available neutralisation minerals in this zone and the very limited depth of atmospheric oxygen penetration. Additionally, if no unsaturated zone is present, no acidification occurred, and the sulfate concentration will not significantly change over time.

Additionally, if no unsaturated zone is present, acidification does not occur, and the sulfate concentration will not significantly change over time. The horizontal outflow concentrations were not notably different between the Base Case and Scenario 1 (reduced ANC). This is expected given the location of the horizontal outflow node (Figure 3), which is dominated by flow from the saturated zone and the sulfate predicted in the saturated zone, as shown in Figure 7.

If the unsaturated zone is reduced to 5 m or 2.5 m, horizontal outflow would occur in the oxic zone, resulting in an increase in sulfate concentration in the horizontal outflow. However, this increase is limited due to the low hydraulic conductivity of the tailings. The elevated sulfates in the unsaturated zone migrate slowly into the saturated zone and seep horizontally out of the tailings into the adjacent aquifer.

CONCLUSIONS

A range of reactive geochemical simulations was completed for a coal tailings facility with moderate pyrite content and available ANC. Using the multiphase model results to provide oxygen profiles, the ChemPlugin code was employed to account for varying flow rates at different depths and to assign the diverse flow connections between nodal blocks in the modelling domain. This is especially useful in variable saturated conditions, such as for in-pit tailings disposal, where the tailings are pumped in as a slurry and allowed to settle and consolidate.

The geochemical model additionally also allowed consideration of more conservative scenarios (elevated sulfide content and lower availability of acid neutralisation).

From the results of the model, the simulated pH in the tailings remained neutral for the average tailings composition over 200 years. These results are consistent with kinetic testing results that show no signs of acidification after several years of testing. However, the modelling results indicate that if the ANC is not readily available, local acidification is likely to occur within the unsaturated zone at the top ~5 m of the tailings

For the model with mean tailings geochemical composition, the sulfate reached a concentration of about 2000 mg/L in the unsaturated zone (due to gypsum saturation), remaining in the order of ~500 mg/L in the saturated zone. No metals were elevated under neutral pH conditions. Salinity will initially be strongly dominated by the pore water trapped as the tailings consolidate. Where localised acidification is present in the oxic zone due to reduced ANC, elevation of TDS, sulfate and metals is expected.

ACKNOWLEDGEMENTS

I would like to express my sincere gratitude to Dr Brent Usher and Dr Johan Fourie for their invaluable guidance and insightful comments throughout the course of this study.

I would also like to thank my employer, KCB Australia (Pty) Ltd, for their support and for providing the necessary resources and time to conduct this study.

REFERENCES

Bethke, C M, 2024. *The Geochemist's Workbench*®, Aqueous Solutions LLC.

Evangelou, V P, 1995. *Pyrite Oxidation and Its Control* (CRC Press, Boca Raton).

McMahon, P B and Chapelle, F H, 2008. Redox Processes and Water Quality of Selected Principal Aquifer Systems, *Groundwater*, 46(2):259–271. https://doi.org/10.1111/j.1745–6584.2007.00385.x

Nicholson, K, 1990. *Geochemical Rate Models* (Cambridge University Press).

Stumm, W and Morgan, J J, 1981. *Aquatic Chemistry: An Introduction Emphasizing Chemical Equilibria in Natural Waters*, 2nd edn (Wiley-Interscience: New York).

Development and optimisation of a novel reactor for remediating acid mine drainage using natural substrates

D T Maiga[1], N A Tshikovhi[2], T T Phadi-Shivambu[3], L L Sibali[4] and T A M Msagati[5]

1. Scientist, Council for Mineral Technology (MINTEK), Johannesburg 2125, South Africa. Email: deogratius@mintek.co.za
2. Technician, Council for Mineral Technology (MINTEK), Johannesburg 2125, South Africa. Email: anastaciaT@mintek.co.za
3. Engineer, Council for Mineral Technology (MINTEK), Johannesburg 2125, South Africa. Email: terencep@mintek.co.za
4. Professor, University of South Africa, Johannesburg 1710, South Africa. Email: siballl@unisa.ac.za
5. Professor, University of South Africa, Johannesburg 1710, South Africa. Email: msagatam@unisa.ac.za

ABSTRACT

This study presents the development, optimisation, and evaluation of a novel passive treatment reactor designed to remediate acid mine drainage (AMD) using a combination of natural substrates. AMD, a by-product of mining operations, is characterised by its acidic pH and elevated levels of toxic metals and sulfates, posing significant threats to ecosystems and water quality.

The reactor was constructed using acid-resistant materials and incorporates locally available, low-cost materials, including low-quality limestone, organic waste from malt and hops, and activated sludge to enhance both chemical and biological treatment processes. The system operates under gravity-driven flow conditions, allowing AMD to percolate through layered substrates where contaminants are removed through a combination of pH neutralisation, adsorption, precipitation, and microbial activity. The limestone primarily acts to raise the pH, while the organic materials and sludge promote metal binding and biological sulfate reduction.

Preliminary results demonstrated the reactor's effectiveness in neutralising acidity and significantly reducing concentrations of metals such as iron (Fe) and manganese (Mn), as well as sulfate ions. This approach offers a sustainable and cost-effective alternative to conventional AMD treatment technologies, with the added benefits of low energy requirements and the potential for treated water reuse.

The reactor shows promise for application at mining sites as part of long-term environmental remediation strategies. The findings of this study suggest that leveraging natural substrates not only mitigates the impact of AMD but also contributes to more sustainable mining practices.

INTRODUCTION

Acid Mine Drainage (AMD) is a significant environmental concern resulting primarily from the oxidation of sulfide minerals, particularly pyrite (FeS_2). This process generates highly acidic water characterised by low pH levels and elevated concentrations of heavy metals and sulfates, posing serious risks to groundwater, surface water bodies, and surrounding ecosystems (Younger *et al*, 2002). In South Africa, where mining activities are prevalent, AMD represents one of the most challenging environmental issues faced by the industry. The detrimental effects of AMD can inflict extensive damage to aquatic life, soil quality, and overall ecosystem health, necessitating urgent and effective remediation strategies (Ighalo *et al*, 2022).

The economic implications of AMD are also substantial. Mining companies can incur significant financial liabilities related to environmental damage, regulatory compliance, and remediation efforts. Operational costs associated with AMD treatment technologies often include ongoing chemical treatments and maintenance, which can jeopardise the financial viability of mining operations (Wang *et al*, 2020). Consequently, there has been growing interest in developing more sustainable and cost-effective methods to treat AMD.

Passive treatment systems represent a promising alternative to conventional active treatment approaches. These systems utilise natural processes, relying on locally sourced low-cost materials to remediate AMD and offer several advantages, including lower operational and maintenance costs, reduced energy requirements, and the ability to leverage natural biogeochemical processes for pollutant removal (Shabalala, Ekolu and Diop, 2014). Among these processes, the use of natural or waste-derived substrates, such as organic materials, has gained popularity due to its ecological and economic benefits (Seervi *et al*, 2017).

Research has shown that conventional mineral-based treatments, such as limestone, are effective at neutralising acidity but often fall short in removing heavy metals (Zheng *et al*, 2018). The integration of organic material, such as malt and hops, can enhance both biological and chemical remediation processes by providing additional carbon sources that promote microbial growth (Skousen, Ziemkiewicz and McDonald, 2019). For instance, activated sludge has been demonstrated to improve the microbial community dynamics involved in sulfate reduction, which is critical for AMD treatment (Huang *et al*, 2020).

The primary objective of this study is to develop and optimise an innovative passive treatment reactor system that incorporates a combination of low-quality limestone, organic substrates (derived from malt and hops), and activated sludge. By utilising these materials, the study aims to create a sustainable solution for treating AMD, potentially enabling water reuse in mining applications and contributing to more sustainable mining practices. This research aligns with global trends in environmental remediation, emphasizing the importance of sustainable resource management and the utilisation of waste materials to mitigate environmental impacts.

EXPERIMENTAL PROCEDURES

Materials and equipment

Natural substrates included locally sourced low-grade limestone (Figure 1), malt and hops (industrial by-products), and activated sludge from wastewater treatment facilities. AMD samples were collected from affected sites in Mpumalanga (Figure 2). Analytical tools, including pH meters, spectrophotometers, ICP-OES, ICP-MS, FTIR, SEM-EDS, XRF, and XRD, were used for measurements. A laboratory-scale reactor was fabricated for controlled treatment experiments.

FIG 1 – Crushed limestone.

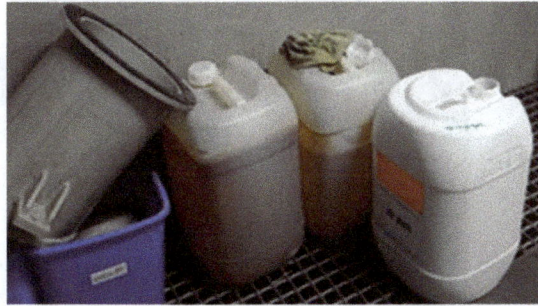

FIG 2 – AMD samples locally sourced.

Methodology

Substrate preparation

Limestone was crushed into granules of 1 mm, 2 mm, and 5 mm. Malt and hops were dried and milled to fine particles. Activated sludge was dewatered, calcined, and homogenised.

Scanning reactor set-up

Experimental reactors were loaded with various substrate compositions (1 kg, 2 kg, and 5 kg). AMD samples were introduced at a controlled percolation rate of 1 mL/min. Treated effluent was collected and analysed post-treatment (Figures 3–5).

FIG 3 – Crushed limestone (2 mm) loaded onto the reactor vessel.

FIG 4 – AMD being filled in the main tank of the reactor.

FIG 5 – Treated water after 30 mins of percolation.

Treatment process

The pH, electrical conductivity, biological oxygen demand (BOD), and concentrations of metals such as Fe, Mn, and Zn were monitored regularly throughout the experiments.

Data analysis

Ongoing data collection is being statistically evaluated to determine optimal substrate configurations and their treatment efficiencies.

RESULTS AND DISCUSSION

Substrate performance

Limestone effectively neutralised the acidic AMD but had a limited impact on metal removal. The addition of malt and hops promoted microbial activity and enhanced the removal of sulfate to sulfide, which precipitates as metal sulfides. Activated sludge played a significant role in the precipitation of metal ions, thereby enhancing overall treatment efficiency.

Treatment efficiency

- pH Neutralisation: The reactor achieved a consistent increase in pH levels, reaching near-neutral conditions within 30 mins of treatment.

- Metal Removal: Significant reductions in metal concentrations were observed:

  - Iron (Fe): Reduced by approximately 85 per cent.

  - Manganese (Mn): Reduced by approximately 70 per cent.

  - Zinc (Zn): Reduced by approximately 65 per cent.

- Sulfate Reduction: Sulfate concentrations decreased by approximately 60 per cent, indicating effective microbial sulfate reduction.

Observations

The combination of substrates demonstrated synergistic effects, with enhanced treatment efficiency compared to individual substrates.

CONCLUSION

This research demonstrates the potential of an integrated, low-cost, and sustainable approach for remediating acid mine drainage (AMD) using mixed natural substrates. Preliminary results are promising, showing substantial improvements in pH correction and metal removal efficiency. The use of locally sourced, low-cost materials aligns with sustainable and eco-friendly practices. The reactor system presents a viable alternative for the mining sector, offering opportunities for water reuse and a reduced environmental impact. Further optimisation and scaling-up studies are currently underway.

ACKNOWLEDGEMENTS

The authors gratefully acknowledge the Council for Mineral Technology (MINTEK) for funding and supporting this research. Special thanks to the LOM|MWT 2025 Conference organisers for providing a platform to share our findings.

REFERENCES

Huang, H, Biswal, B K, Chen, G-H and Wu, D, 2020. Sulfidogenic anaerobic digestion of sulfate-laden waste activated sludge: Evaluation on reactor performance and dynamics of microbial community, *Bioresource Technology*, 297:122396.

Ighalo, J O, Kurniawan, S B, Iwuozor, K O, Aniagor, C O, Ajala, O J, Oba, S N, Iwuchukwu, F U, Ahmadi, S and Igwegbe, C A, 2022. A review of treatment technologies for the mitigation of the toxic environmental effects of acid mine drainage (AMD), *Process Safety and Environmental Protection*, 157:37–58.

Seervi, V, Yadav, H L, Srivastav, S K and Jamal, A, 2017. Overview of active and passive systems for treating acid mine drainage, *Iarjset*, 4(5):131–137.

Shabalala, A N, Ekolu, S O and Diop, S, 2014. Permeable reactive barriers for acid mine drainage treatment: a review, *Construction Materials and Structures*, 1416–1426.

Skousen, J G, Ziemkiewicz, P F and McDonald, L M, 2019. Acid mine drainage formation, control and treatment: Approaches and strategies, *The Extractive Industries and Society*, 6(1):241–249.

Wang, Y, Cai, Z, Sheng, S, Pan, F, Chen, F and Fu, J, 2020. Comprehensive evaluation of substrate materials for contaminants removal in constructed wetlands, *Science of the Total Environment*, 701:134736.

Younger, P L, Banwart, S A, Hedin, R S, Younger, P L, Banwart, S A and Hedin, R S, 2002. Passive treatment of polluted mine waters, *Mine Water: Hydrology, Pollution, Remediation*, 311–396.

Zheng, W, Ma, X, Tang, Y, Ke, C and Wu, Z, 2018. Heavy metal control by natural and modified limestone during wood sawdust combustion in a CO2/O2 atmosphere, *Energy and Fuels*, 32(2):2630–2637.

Hourly rainfall from daily records – a method of fragments now and into the future

L Millard[1], M Batchelor[2] and H Hamer[3]

1. Principal Engineer, WRM Water and Environment Pty Ltd, Brisbane Qld 4000.
   Email: lindsay.millard@wrmwater.com.au
2. Senior Principal Engineer, WRM Water and Environment Pty Ltd, Brisbane Qld 4000.
3. Project Engineer, WRM Water and Environment Pty Ltd, Brisbane Qld 4000.

ABSTRACT

Continuous simulation run-off models are widely used to simulate and predict the future behaviour of operational mine site water management systems, final void lakes and landform evolution.

Most models are constrained to operate on a daily timestep by the limited availability of suitably representative long-term short-duration rainfall data sets. However, some hydrological and geomorphic processes are governed by long-term sub-daily rainfall variability.

Our paper presents an approach to generate hourly data sets from long daily records, based on sub-daily rainfall patterns extracted from surrounding 'donor' rain gauges to enhance the rainfall record and overcome data limitations using the regionalised method of fragments (Westra *et al*, 2012).

The approach uses 'IFD conditioning' to ensure consistency with the expected intensity-frequency-duration (IFD) characteristics for the site and incorporates the latest guidance from Australian Rainfall and Runoff (ARR) v4.2 and downscaled Coupled Model Intercomparison Project Phase 6 (CMIP6) climate model predictions to account for the expected impact of climate change on rainfall intensity and seasonality.

The paper provides an example of processing a daily rainfall data set and combining it with sub-daily records from surrounding areas to generate 136 years of hourly rainfall data that matches the Bureau of Meteorology IFD curves. The example includes adjustment to align with current and future CMIP6 climate horizons.

The resultant data sets can be more confidently used to make reliable predictions about future water management and mine closure challenges, to inform decisions about water storage design, pump capacity, and flood mitigation measures even where site data is scarce.

INTRODUCTION

Overview

Daily timestep continuous simulation (CSM) numerical models (Boughton and Droop, 2003) are widely used to simulate the behaviour of water management systems and landforms through the mine life cycle (Temme and Verburg, 2011; Hancock, Verdon-Kidd and Lowry, 2017).

However, some applications require sub-daily inputs to appropriately represent physical processes. For example, releases or overflows from water storage systems may be too short-lived to be adequately represented by daily run-off models. Similarly, Landform Evolution Models (LEMs) are increasingly used to assess the stability and evolution of landscapes over geological timescales (Hancock *et al*, 2025). Some models – such as CAESAR-Lisflood – can use sub-daily rainfalls to better represent the intensity and frequency of events, which are the key determinants of gully erosion extent, sediment transport and final landform. Studies by Hancock, Verdon-Kidd and Lowry (2017), Coulthard and Skinner (2016), and Skinner *et al* (2020) have demonstrated that, when using historical and stochastically generated rainfall sequences, sub-daily rainfall variability greatly affects landscape evolution.

While representative historical daily climate records of over 100 years duration can be drawn throughout Australia from nearby weather stations or gridded products (the Australian Bureau of Meteorology (BoM) or Queensland Government (SILO, Jeffrey *et al*, 2001) service), sub-daily rainfall records are rarely of sufficient quality or duration to capture the full climate variability, particularly in Northern Australia (McQuade, Arthur and Butterworth, 1996; Coulthard and Skinner, 2016).

The paper's objective is to present a methodology for deriving hourly rainfall data corresponding to daily rainfall data sets that are aligned with the expected Intensity Frequency Duration (IFD) characteristics of local rainfalls and can be adapted to incorporate projected rainfall changes under future climate scenarios.

DEFINING THE PROBLEM

Importance of representative rainfall data

Various studies have established that sub-daily rainfall data can lead to significantly different predictions in sediment yields and erosion patterns compared to daily, lumped, or time-averaged data.

Hancock, Verdon-Kidd and Lowry (2017) tested the sensitivity of modelled erosion rates to small changes in rainfall input at a study site located in the Northern Territory (NT) of Australia, At that site, only three complete (>85 per cent) long-term records of daily data were available from nearby recording stations, and only two sub-daily records were available within the entire NT with 40 years of 6 min pluviograph. In the absence of long-term data sets, A novel approach of employing stochastically generated daily rainfall data (derived using the DRIP model based on Darwin Airport records) was used to provides inputs to the CAESAR-Lisflood LEM to simulate 100 years of landform evolution on a proposed rehabilitated mine landform at Corridor Creek.

This research found that each stochastically generated rainfall scenario produced a unique pattern of erosion and sediment output, highlighting the sensitivity of landform evolution to small changes in rainfall sequences and the non-linear nature of these processes.

Limitations of available data sets

The method demonstrated its value as a risk-based approach, but underline the importance of using representative data sets reflecting local rainfall variability in predicting future sediment transport, erosion form and evolution.

This information is of particular importance for the design and testing of rehabilitated landscape systems such as post-mining landscapes.

Unfortunately, sufficiently well-conditioned rainfall pluviograph data sets are often either too short, not representative of the location, or are at a temporal resolution that is too coarse for use in these models.

Hancock, Verdon-Kidd and Lowry (2017) demonstrated that synthetic stochastic sub-daily rainfall can reproduce the very rare storm events that drive landform evolution and erosive processes. There is a need to ensure that data sets used in the modelling make the best use of the available information on rainfall, including capturing the IFD characteristics of the local area, both under current and future climate conditions, as discrepancies may have significant impacts on the modelled basin profile and shape from long-timescale simulations.

Impacts of climate change on rainfall

In March 2024, Engineers Australia (EA) in partnership with the Australian Government's Department of Climate Change, Energy, the Environment and Water (DCCEEW) released an update to the Australian Rainfall and Runoff (ARR) Guideline known as version 4.2 (Ball *et al*, 2019) which included a revision to the chapter relating to Considerations of Climate Change. The update focuses on the impact of climate change on design rainfalls of interest in flood modelling – and provides an approach to generating future rainfall IFDs based on projected global temperature increases. The update excludes guidance for incorporating climate change impacts into continuous simulation studies. The guidance recommends adjusting near-term rainfalls for temperature increases that have occurred in the period since the historical rainfall was collected.

In the method described in the present paper, the generated rainfall sequence can be conditioned to match a target's current or future climate IFD.

Further, the results of downscaled global climate modelling undertaken for the CMIP6, can be used to adjust the data set to be consistent with long-term seasonal rainfall projections for selected climate scenarios (Shared Socio-economic Pathways (SSPs) and time horizons).

METHODOLOGY

The following methodology outlines the authors' approach to generating hourly data sets from long daily records. The approach is based on sub-daily rainfall patterns extracted from surrounding 'donor' rain gauges to enhance the rainfall record and overcome data limitations. In this way, the authors' approach provides the necessary hydrologic input to enable the modelling of long-term processes that are governed by rainfall variability.

Conceptual overview

Figure 1 illustrates the methodology for generating climate-adjusted sub-daily rainfall data. This process is divided into two main steps:

1. *Disaggregation of daily rainfall totals using the regionalised method of fragments*: This step takes a daily rainfall record at the point of interest (POI) (as a long-term sub-daily record is often unavailable) along with sub-daily records at nearby gauges, and stochastically generates a sub-daily rainfall record matching the observed daily totals (Boughton, 1999; Westra *et al*, 2012; Dykman *et al*, 2024).

2. *Adjusting results to match predicted changes to seasonal rainfall totals and intensity in future climate scenarios:* In this step, the rainfall record is adjusted to match changes to seasonal totals predicted by downscaled global climate models and changes to intensity described in ARR 2019 (Ball *et al*, 2019) for a given climate scenario.

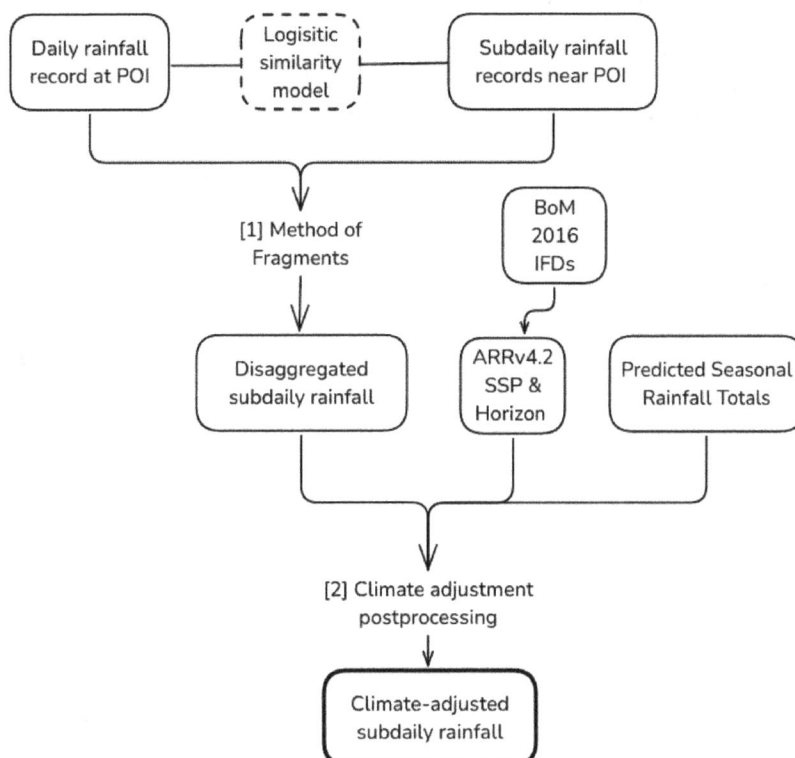

FIG 1 – Process to generate climate-adjusted, disaggregated sub-daily data from a daily rainfall record.

It is worth noting that the climate-adjusting techniques in step 2 are applicable to any sub-daily rainfall record – they do not necessarily require data produced by the method of fragments. The regionalised method of fragments was chosen because it does not require a long-term sub-daily rainfall record at the POI, but another method may be substituted. These steps and their respective inputs and outputs are discussed in further details.

Example application – generating future climate (2090) rainfall at Millmerran, Qld

This section provides a more detailed explanation of the proposed method by applying it to a real-world example: generating a climate-adjusted sub-daily rainfall series for the year 2090 (SSP3) at a point of interest in Millmerran, Queensland.

Step 1 – Disaggregating the daily rainfall record using the method of fragments

As is often the case, Millmerran does not have a long-term sub-daily rainfall record to work with directly. Therefore, the authors employed the regionalised method of fragments) to generate a sub-daily record matching known daily rainfall totals, using data from nearby sub-daily gauges. This process is often referred to as disaggregation.

Inputs

As input for this step, the authors used:

- 116 years of patched point daily rainfall totals (1889–2025) for the Millmerran Post Office BoM rainfall station (station number 41069) from the SILO rainfall database. The Millmerran Post Office station (27.874S, 151.271E) commenced operation in January 1900 and provides a 98 per cent complete daily record for 114 years. This station will be referred to as the 'Target' station. Where observed data is missing, it is infilled with data interpolated from surrounding daily data sets.

- Hourly rainfall records gathered from 101 stations throughout southern Qld. This data was obtained from the BoM and the Department of Regional Development, Manufacturing and Water (RDMW). These will be referred to as the 'donor' stations. Records ranged from 2 to 71 years in length, with a mean length of 28 years. Approximately 45 per cent of the data in these records was missing. These locations are shown in Figure 2.

- A latitude, longitude, elevation and distance to the nearest coast for all stations.

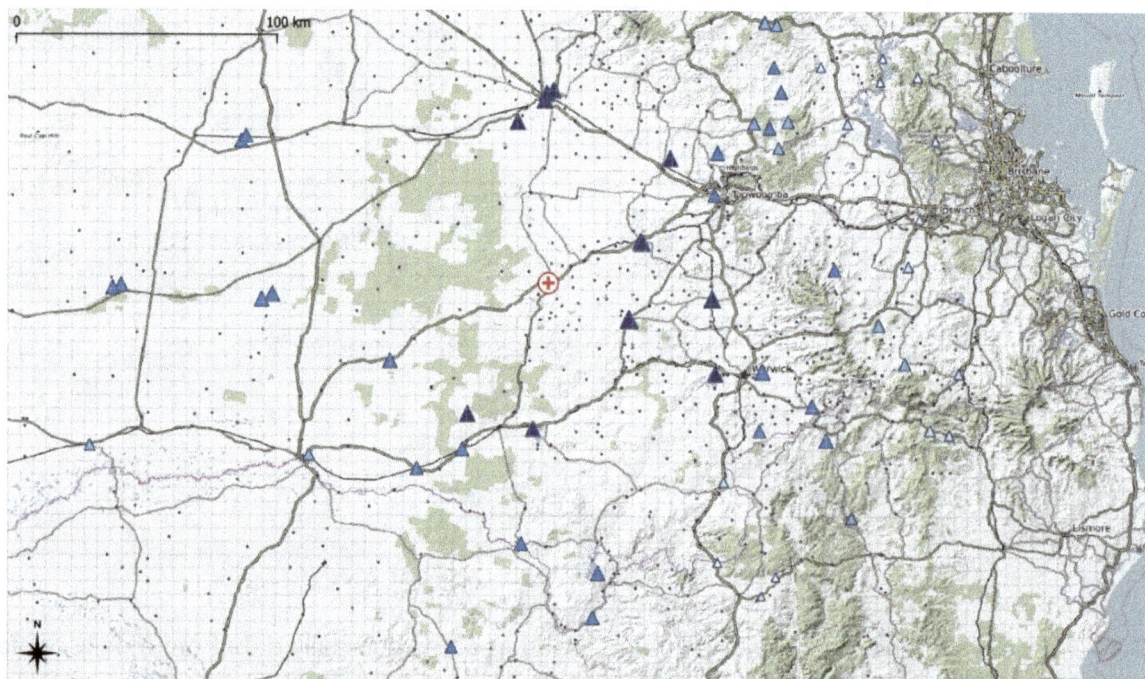

FIG 2 – Target and donor stations. NB: Target daily station at Millmerran PO is shown in red. Donor pluviographic stations shown as triangles. The similarity to target is shown by size and colour of donor triangle. Larger and darker triangles were calculated to be similar to target station. Other BoM daily stations are shown as points and BoM IFD grid is overlain for scale.

Process

Disaggregated data was obtained from the above inputs using a Python implementation of the Regionalised Method of Fragments (Westra *et al*, 2012). Below is a high-level overview of this process:

- Donor stations were ranked by their 'similarity' to the target station, based on a multivariate logistic regression model. The regression model considered latitude, longitude, the product of latitude and longitude, elevation and distance to the nearest coast. Book 2 ARR 2019 provides the logistic regression coefficient values in Table 2.7.7 (Ball *et al*, 2019).

- Data from donor stations was pooled to create a 50-year sampling data set of rainfall 'fragments' – hourly rainfall extracts from the donor station data, each 24 hrs in length. Fragments were preferentially drawn from donor stations by order of their similarity ranking, and fragments with any missing data were discarded. The sampling of fragments continues searching until sufficient donor locations are assembled to define the pool of fragments available in the database.

- For each daily total at the target station, a group of candidate fragments would be found that had a similar daily total, fell in the same part of the year and had similar rainfall levels on immediately preceding and succeeding days.

- These candidate fragments were sampled based on a harmonic distribution determined by the inverse of their similarity ranking. The sampled fragment is then scaled to match the daily total observed at the target station and becomes the sub-daily disaggregation for that day.

More details of this process are described in Westra *et al* (2012) or Book 2 of ARR 2019 (Ball *et al*, 2019).

Output

The method of fragments produced a 116-year simulated hourly rainfall record matching the daily totals at the Millmerran PO target station. The output is shown in Figure 3, zooming in on a randomly selected day in that series.

Because the sampling process introduces an element of randomness, the method is stochastic and can be used to generate any number of unique disaggregated records.

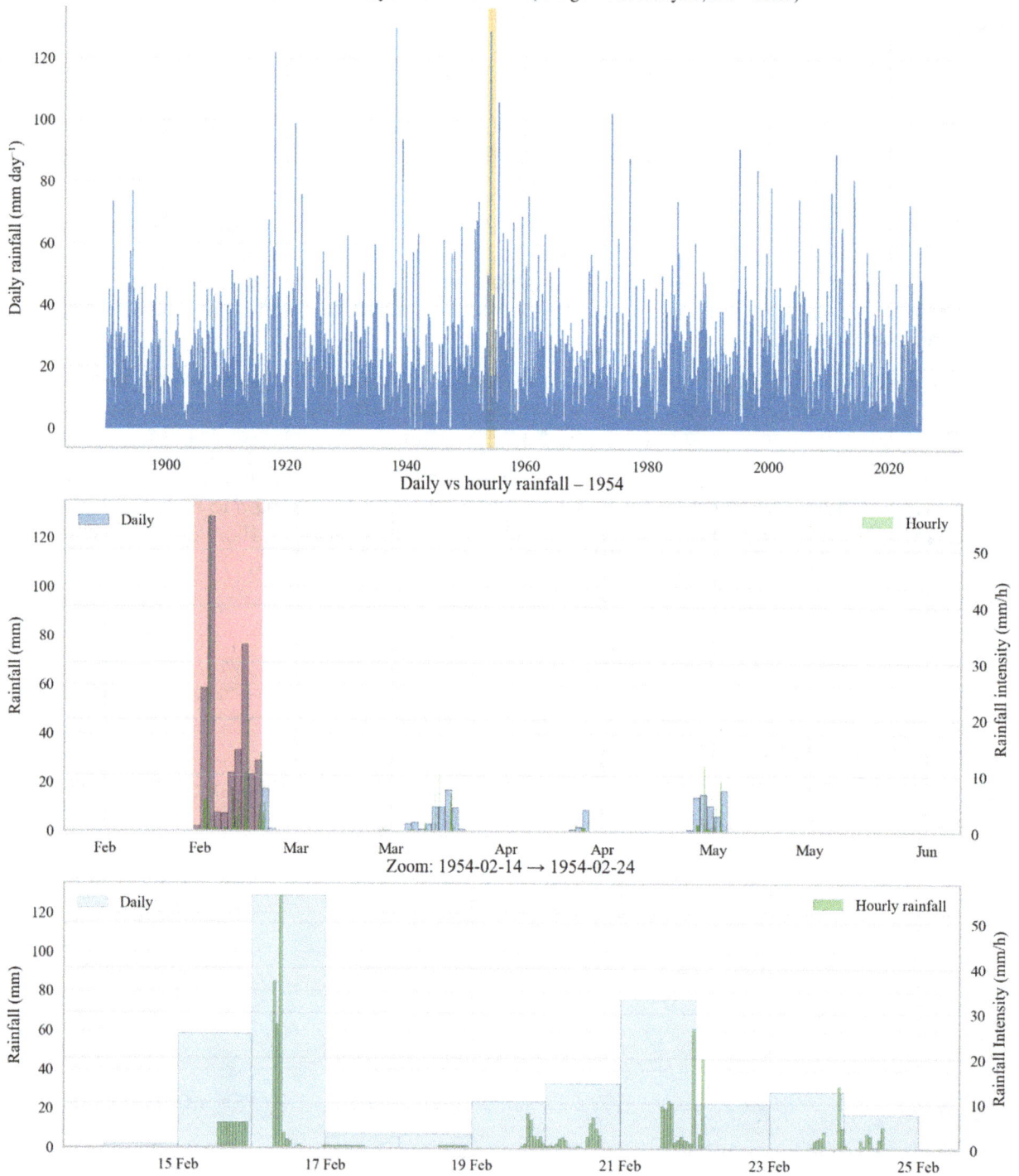

FIG 3 – Hourly rainfall data set disaggregated from daily rainfall data set using Method of Fragments showing progressive magnification of the series.

Step 2 – Adjusting rainfalls to match future climate predictions

This step adjusts rainfall based on predictions of how rainfall patterns will change in future climate scenarios (Visser *et al*, 2023).

Discrete events of each duration were recursively identified within the continuous series. Starting with short-duration events, each event was adjusted to meet the ARR 2019 uplift target for that duration. As longer-duration events encapsulate shorter-duration events, they require conditioning to ensure that the applied adjustment has not compounded overall and not exceeded the desired uplift in the IFD.

While global climate models predict increases in extreme rainfall intensity over Australia, seasonal totals may increase or decrease depending on location and time horizon. Rainfall events which were not identified for uplift were factored to match seasonal rainfall change predictions from downscaled CMIP6 climate modelling.

Two key sources were used to inform this adjustment:

1. ARR 2019 v4.2 Climate change factors are calculated using temperature increases from the IPCC AR6 Report (Ball *et al*, 2019; Wasko, 2024; IPCC, 2023; Stocker *et al*, 2013). These factors are used to scale Intensity-Frequency-Duration (IFD) characteristics based on predicted temperature change. In general, these factors predict that yearly rainfall maxima will increase exponentially with temperature according to Equation 1. This equation is applied to each interval/AEP to calculate future IFDs.

2. Seasonal precipitation predictions from the Queensland Future Climate Dashboard (Trancoso *et al*, 2024). These are gridded predictions derived from downscaled CMIP6 climate model results (Chapman *et al*, 2024) about how seasonal rainfall totals will change for a given time frame, climate scenario and location. In general, these predict that seasonal rainfall totals will remain the same or decrease in 2090 for SSP3 pathway across most of Queensland.

$$I_p = I \times \left(1 + \frac{\alpha}{100}\right)^{\Delta T} \tag{1}$$

where:

I	historical rainfall depth (taken from Australian Bureau of Meteorology (BoM, 2016) IFDs)
I_p	projected rainfall depth for the climate scenario
α	rate of change parameter (from tables provided by the ARR)
ΔT	change in global temperature projection for the design period (SSP3@2090)

Combined, these two sources tell us that (in general):

- Intense rainfall events (the yearly maxima represented by IFD statistics) will increase in severity as mean global surface temperatures increase in future climate scenarios.

- Total rainfall volumes in Queensland will mostly remain the same or decrease in future climate scenarios.

The method for adjusting sub-daily rainfalls presented here attempts to align with both predictions. At a high level, it scales intense events up to match predicted IFDs, while scaling overall rainfall volumes down to match predicted seasonal totals. The IFD statistics are calculated for the generated hourly record and compared with the BoM IFD statistics. An iterative post-processing step then conditions the synthetic hourly record to ensure that the frequency of different durations of storm intensity is accurately represented.

Inputs

The inputs for the authors' climate-adjusting method were as follows:

- The BoM (2016) IFD curve for the target location (with and without adjustment as per ARR 2019 V4.2 recommendations below).

- Climate change IFD uplift factors as per the ARR 2019 V4.2 (Ball *et al*, 2019) as shown in Book 2 Figure 2.7.13.

- Five simulated sub-daily records at Millmerran PO were obtained as outputs from Step 1 above. Only one daily record is necessary to perform this step, additional simulations were used to assess consistency in the results.

- Predicted percentage changes in seasonal rainfall totals for SSP3 in the year 2090, as per the Queensland future climate dashboard (Trancoso *et al*, 2024).

Process

Future climate adjustment was performed in two steps. First yearly maximum rainfalls were scaled to match the uplifted IFDs predicted by ARR V4.2. This will be referred to as the 'IFD constraining stage'. Non-maximal rainfalls (rain not part of a yearly maxima for any duration) were then scaled to align with predicted seasonal rainfall totals. This will be referred to as the 'seasonality adjustment stage'.

The IFD constraining stage used a modified version of the algorithm described by Woldemeskel *et al* (2016). The process at a high level is as follows:

- IFDs (rainfall depths for a set of durations and AEPs) are calculated for the generated sub-daily rainfall.

- These IFDs are compared with a set of target IFDs. In this case, these were SSP3 2090 statistics uplifted from the BoM (2016) using the ARR V4.2 factors. The metric used to compare two IFDs was the absolute relative percent difference (ARPD), described by Equation 2. See Table 1 for an example of this comparison.

  The duration with the highest average ARPD is selected as the 'priority duration'. This is the duration which least closely matches the target IFD statistics. In Table 1, two hrs is the priority duration with an average ARPD of 49 per cent.

- All rainfall that is part of a yearly maximum for the priority duration is rescaled to match the target IFDs for that duration. This rescaling is conducted according to the equation provided by Woldemeskel *et al* (2016).

- Steps 1–4 are repeated until the IFD statistics for the generated rainfall are acceptable, similar to the target IFDs. Figure 4 shows the convergence of the raw generated sub-daily data to the uplifted SSP3 2090 statistics over 15 iterations.

$$\text{ARPD}(a, b) = \frac{|a-b|}{\frac{a+b}{2}} \tag{2}$$

TABLE 1

Example comparison between IFD statistics using ARPD.

Duration	AEP (%)						
	63%	50%	20%	10%	5%	2%	1%
2 hrs	44%	49%	56%	56%	53%	46%	39%
6 hrs	37%	39%	42%	41%	40%	38%	35%
12 hrs	34%	34%	33%	33%	32%	33%	32%
1 day	29%	30%	29%	26%	23%	20%	17%
2 days	27%	28%	29%	29%	27%	26%	25%
4 days	27%	27%	27%	28%	29%	32%	34%
7 days	23%	23%	22%	21%	21%	20%	20%

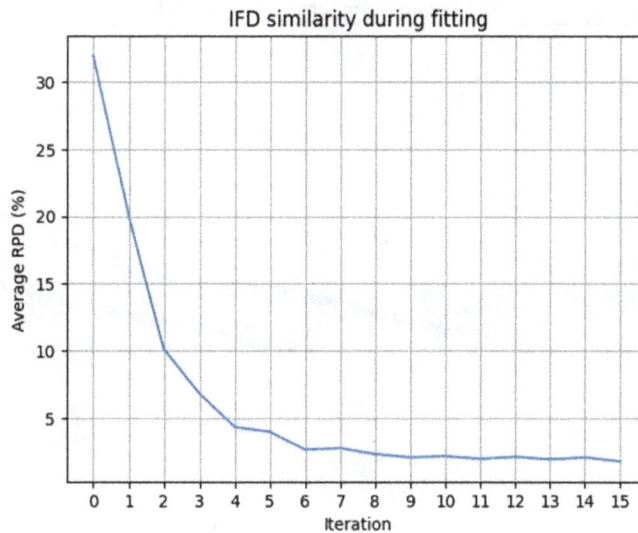

FIG 4 – ARPD convergence between generated and uplifted IFDs over 15 iterations.

The seasonality adjustment stage is simpler. Now that the yearly maxima rainfalls have been rescaled (usually increased), the seasonality adjustment stage rescales the non-maximal rainfall to match the predicted changes in seasonal totals – maximal rainfalls cannot be scaled at this stage as IFDs would be affected. This is achieved as follows:

- 'Expected' seasonal totals T_E are calculated by summing rainfall for each season (DJF, MAM, JJA, SON) across the original daily rainfall record and multiplying by seasonal adjustment factors.

- Maximal seasonal totals T_M are calculated by summing rainfall for each season across all rainfall that is part of a yearly maximum for some duration.

- Non-maximal seasonal totals T_{NM} are calculated by summing rainfall for each season across all rainfall that is not part of a yearly maximum.

- These totals are used to calculate a seasonal scaling factor S for each season, according to Equation 3. This factor represents the required scaling of non-maximal rainfalls to match predicted changes in seasonal totals.

- Non-maximal rainfalls are then scaled according to the rescaling factor for their season.

$$S = \frac{T_E - T_M}{T_{NM}} \tag{3}$$

Output

The output of this future climate adjustment is sub-daily rainfall that matches both the ARR V4.2 uplifted IFDs as well as the Queensland Future Climate Dashboard CMIP6 future climate seasonal rainfall changes. An example is shown in Figure 5 for a single duration. Table 2 presents the IFD results from the uplifted BoM IFD using ARRV4.2 compared to the value achieved within the continuous sequence. For ease of comparison, Table 3 provides the percentage discrepancy across the IFD.

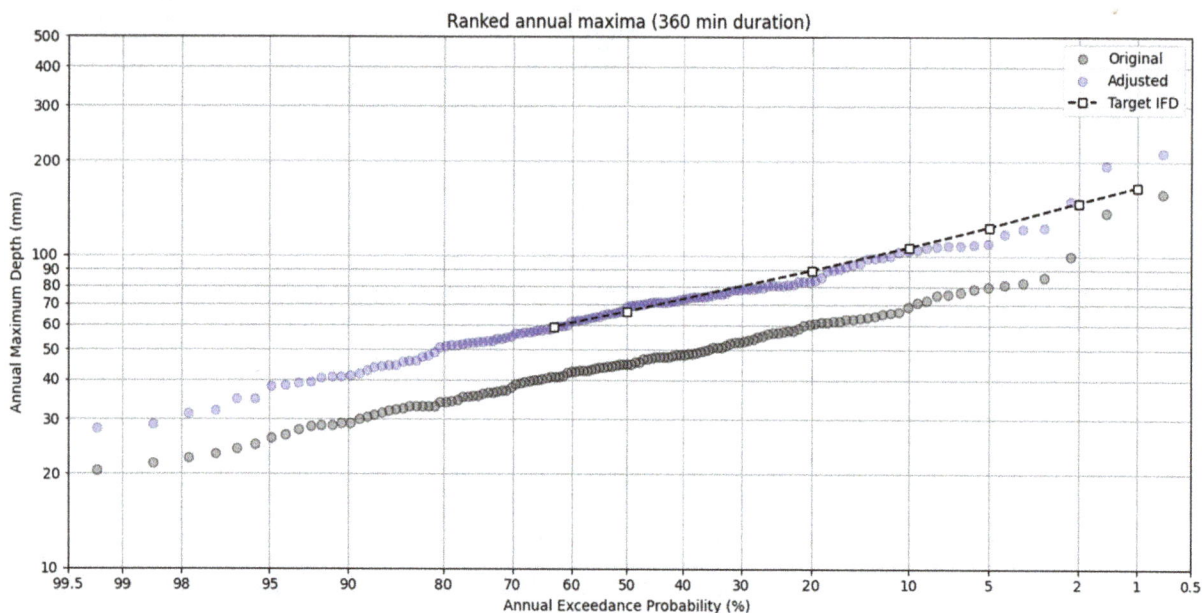

Ranked annual maxima (360 min duration)

FIG 5 – Adjusted IFD (single duration) showing adjustment to future climate IFD.

TABLE 2

Results of IFD conditioning MoF to ARR v4.2 SSP3 2090.

Duration (min)	AEP						
	63.2%	**50%**	**20%**	**10%**	**5%**	**2%**	**1%**
120	34.6 / 32.5	37.2 / 36.8	47.4 / 50.9	56.4 / 61.0	67.2 / 71.2	85.2 / 85.4	102.5 / 96.7
360	43.7 / 42.8	47.8 / 48.1	62.5 / 65.1	73.9 / 77.2	86.5 / 89.5	105.5 / 107.0	122.0 / 120.0
720	51.1 / 51.5	56.8 / 57.5	75.7 / 77.1	89.3 / 90.8	103.3 / 104.0	123.0 / 124.0	138.8 / 139.0
1440	61.6 / 62.1	68.4 / 69.6	92.2 / 93.1	110.6 / 109.0	130.5 / 125.0	160.1 / 148.0	185.5 / 166.0
2880	73.7 / 74.2	82.5 / 83.6	111.8 / 113.0	133.2 / 134.0	155.5 / 154.0	187.0 / 183.0	212.7 / 206.0
5760	84.1 / 85.5	96.4 / 96.9	134.5 / 134.0	159.8 / 159.0	184.0 / 185.0	215.4 / 222.0	238.9 / 251.0
10080	92.6 / 91.8	105.9 / 104.0	148.6 / 143.0	178.4 / 171.0	208.1 / 200.0	248.3 / 238.0	279.7 / 268.0

Note: MoF value/Future Climate Target value.

TABLE 3
Results of IFD conditioning MoF raw to BoM (BoM, 2016).

Duration	AEP						
	63.2%	50%	20%	10%	5%	2%	1%
3 hrs	6.30%	1.10%	-7.10%	-7.80%	-5.80%	-0.20%	5.80%
6 hrs	2.10%	-0.60%	-4.10%	-4.40%	-3.40%	-1.40%	1.70%
12 hrs	-0.80%	-1.20%	-1.80%	-1.70%	-0.70%	-0.80%	-0.10%
24 hrs	-0.80%	-1.70%	-1.00%	1.50%	4.30%	7.90%	11.10%
2 days	-0.70%	-1.30%	-1.10%	-0.60%	1.00%	2.20%	3.20%
4 days	-1.70%	-0.50%	0.40%	0.50%	-0.50%	-3.00%	-4.90%
7 days	0.90%	1.80%	3.80%	4.20%	4.00%	4.20%	4.30%

Discussion and application

Our approach describes how an hourly rainfall data set can be generated from surrounding sub-daily data sets, which is consistent with the IFD and a daily record for a point of interest.

Where a target site extends over a wide geographical area, the process can be repeated by adjusting the inputs to represent the daily rainfall and IFD based on gridded data or additional daily rainfall data.

The process can be repeated at a location to generate long IFD-conditioned millennial length rainfall sequences, or to generate any number of data sets at the same location to develop a risk-based understanding of the variance in model outcomes.

The MoF-generated rainfall approach addresses limitations identified in earlier studies on landform evolution modelling:

- Temporal Variability: The MoF-generated rainfall methods can assimilate nearby rainfall observations while anchoring the rainfall statistics to replicate the occurrence of short-duration, high-intensity events that drive nonlinear erosion responses, which daily rainfall would miss.

- Spatial Heterogeneity: MoF approaches may better capture the spatial rainfall variability that influences long-term channel network evolution by simulating convective rainfall cells and orographic effects.

- Avoiding the need to utilise LEM parameter calibration to 'correct' sediment yields from smoothed rainfall inputs.

- Preserving the geomorphic signature of rainfall variability, particularly for climate change projections where storm intensity/frequency may shift.

- Enabling more accurate simulation of sediment 'pulses' and their cascading impacts on valley floor deposition.

This highlights the potential for MoF rainfall generation to bridge the gap between LEMs' process representation and the spatiotemporal complexity of real precipitation patterns.

ACKNOWLEDGEMENTS

Code and data availability. The code described in this study was Python 3.11 and was adapted from the method described in Dykman *et al* (2024). The data presented in this paper can be made available on request from the corresponding author. The CAESAR-Lisflood model used is freely available under a GNU licence.

The data set of the Australian rainfalls can be obtained from the Australian Bureau of Meteorology. Daily and sub-daily rainfalls are available at <http://www.bom.gov.au/climate/data/>

Intensity Frequency Duration (IFD) design rainfalls were obtained from the Australian Bureau of Meteorology at <http://www.bom.gov.au/water/designRainfalls/revised-ifd/>

Projections of changes to mean annual rainfall from <https://www.climatechangeinaustralia.gov.au/en/projections-tools/climate-futures-tool/detailed-projections/>

REFERENCES

Australian Bureau of Meteorology (BoM), 2016. Design Rainfall Data System. Available from: <http://www.bom.gov.au/water/designRainfalls/revised-ifd/>

Ball, J, Babister, M, Nathan, R, Weeks, W, Weinmann, E, Retallick, M and Testoni, I (eds), 2019. Australian rainfall and runoff: A guide to flood estimation (Version 4.2), Commonwealth of Australia (Geoscience Australia).

Boughton, W and Droop, O, 2003. Continuous simulation for design flood estimation—a review, *Environ Model Software*, 18(4):309–318.

Boughton, W C, 1999. A daily rainfall generating model for water yield and flood studies Rep, Report 99/9, 21 p, CRC for Catchment Hydrology, Monash University, Melbourne.

Chapman, S, Syktus, J, Trancoso, R, Toombs, N and Eccles, R, 2024. Projected Changes in Mean Climate and Extremes from Downscaled High-Resolution CMIP6 Simulations in Australia. Available from: <https://papers.ssrn.com/sol3/papers.cfm?abstract_id=4836517>

Coulthard, T J and Skinner, C J, 2016. The sensitivity of landscape evolution models to spatial and temporal rainfall resolution, *Earth Surf Dynam*, 4(3):757–771. https://doi.org/10.5194/esurf-4-757-2016.2016

Dykman, C, Sharma, A, Wasko, C and Nathan, R, 2024. Pyraingen: A python package for constrained continuous rainfall generation, *Environmental Modelling and Software*, 175:105984. Available from: <https://pyraingen.readthedocs.io/en/latest/index.html>; https://doi.org/10.1016/j.envsoft.2024.105984

Hancock, G R, Nicolson, L, Purtill, J and Dunlop, J, 2025. Applying erosion and Landscape Evolution Models to assess post-mining landform stability: Technical Paper, Brisbane, Office of the Queensland Mine Rehabilitation Commissioner, Queensland Government.

Hancock, G R, Verdon-Kidd, D and Lowry, J B C, 2017. Sediment output from a post-mining catchment – Centennial impacts using stochastically generated rainfall, *J Hydrol*, 544:180–194. https://doi.org/10.1016/j.jhydrol.2016.11.027:2017

Intergovernmental Panel on Climate Change (IPCC), 2023. Sections, in Climate Change: Synthesis Report, Contribution of Working Groups, I, II and III to the Sixth Assessment Report of the Intergovernmental Panel on Climate Change (eds: H Lee and J Romero), IPCC, Geneva, Switzerland, pp 35–115. https://doi.org/10.59327/IPCC/AR6-9789291691647

Jeffrey, S J, Carter, J O, Moodie, K B and Beswick, A R, 2001. Using spatial interpolation to construct a comprehensive archive of Australian climate data, *Environmental Modelling and Software*, 16(4):309–330. https://doi.org/10.1016/S1364-8152(01)00008-1

McQuade, C, Arthur, J and Butterworth, I, 1996. Geobotany climate and hydrology landscape and vegetation ecology of the Kakadu Region, Northern Australia, vol 23.

Skinner, C J, Peleg, N, Quinn, N, Coulthard, T, Molnar, P and Freer, J, 2020. The impact of different rainfall products on landscape modelling simulations, *Earth Surf Process Landforms,* 45:2512–2523. https://doi.org/10.1002/esp.4894

Stocker, T F, Qin, D, Plattner, G-K and Tignor, M, 2013. The Physical Science Basis, Contribution of Working Group I to the Fifth Assessment Report of the Intergovernmental Panel on Climate Change (IPCC), Cambridge, United Kingdom and New York, USA.

Temme, A and Verburg, P H, 2011. Mapping and modelling of changes in agricultural intensity in Europe, *Agric Ecosyst Environ*, 140(1):46–56.

Trancoso, R, McGloin, R, Putland, D, Syktus, J, Ahrens, D, Chapman, S, Owens, D, Toombs, N, Eccles, R and Zhang, H, 2024. Queensland Future Climate Services – downscaled CMIP6 climate projections for Queensland's regions and locations. Available from: <https://www.longpaddock.qld.gov.au/qld-future-climate/>

Visser, J B, Wasko, C, Sharma, A and Nathan, R, 2023. Changing storm temporal patterns with increasing temperatures across Australia, *J Clim*, 1:26. https://doi.org/10.1175/jcli-d-22-0694.1

Wasko, C, Westra, S, Nathan, R, Pepler, A, Raupach, T H, Dowdy, A, Johnson, F, Ho, M, McInnes, L, Jakob, D, Evans, J, Villarini, G and Fowler, H J, 2024. A systematic review of climate change science relevant to Australian design flood estimation, *Hydrol Earth Syst Sci*, 28:1251–1285. https://doi.org/10.5194/hess-28-1251-2024

Westra, S, Mehrotra, R Sharma, A and Srikanthan, R, 2012. Continuous rainfall simulation, A regionalized sub-daily disaggregation approach, *Water Resources Research*, 48:W01535. https://doi.org/10.1029/2011WR010489.2012

Woldemeskel, F M, Sharma, A, Mehrotra, R and Westra, S, 2016. Constraining continuous rainfall simulations for derived design flood estimation, *J Hydrol*, 542:581–588. https://doi.org/10.1016/j.jhydrol.2016.09.028

MCA, risk assessment, governance and compliance

How should critical controls be applied to tailings risk management, if at all? And if not, how do we demonstrate adequate tailings risk governance?

J Coffey[1], G Trivett[2] and R Sisson[3]

1. Senior Principal, Red Earth Engineering – A Geosyntec Company, Perth WA 6000. Email: jarrad.coffey@redearthengineering.com.au
2. Principal, Trivett Risk Management, Brisbane Qld 4000. Email: greg@mytrivett.com
3. Senior Consultant, Geosyntec, California USA. Email: richard.sisson@geosyntec.com

ABSTRACT

Risk assessment of Tailings Storage Facilities (TSFs) has been a topic given much focus in recent years. In some cases, one of the outcomes of a risk assessment, whatever form is chosen, is to identify critical controls. This is discussed in several tailings management guides and, in the broader context for the mining industry, in International Council on Mining and Metals (ICMM; 2015). While neither the notion that some controls have greater impact on an assessed risk profile is not disputed, nor the benefits of simplifying risk management practices, the selection of what is critical can have significant impacts if the level of management is reduced for when controls inappropriately not selected as being critical. Couple this with an increasingly widely adopted need and/or goal to demonstrate that risk is managed As Low As Reasonably Practicable (ALARP), requiring all reasonably practicable controls to be implemented and maintained, and it appears that the identification of critical controls may in fact be adding unnecessary and distracting complexity, to the Responsible Design Engineer or Engineer of Record's whole of system governance requirements, particularly where an organisation has an entrenched Critical Risk Management (CRM) capability and CRM driven Key Performance Indicators (KPI). Further, the widely adopted example decision tree for the selection of critical controls within ICMM (2015) does not easily relate to geotechnical structures, which in the authors' experience has resulted in significant variation of outcomes. A final consideration is that the number of 'Controls' is relatively small compared to many other safety systems where many, multi-level controls are common. The overall control effectiveness also tends to be dominated by a relatively small number of controls for tailings facilities.

This paper objectively explores this topic and attempts to make practical suggestions for improved practice.

INTRODUCTION

The need to manage the risk associated with Tailings Storage Facility (TSF) failure has long been appreciated. However, until the significant industry response to failures in the past 10–15 years, there was limited published guidance on the topic. For example, the 2012 revision of the Australian National Committee on Large Dams Incorporated (ANCOLD Inc) Guideline on Tailings Dams (ANCOLD, 2012) does not provide explicit guidance on conducting a risk assessment but does reference the ANCOLD guideline on risk available at that time (ANCOLD, 2003), which in turn does not provide guidance on the assessment of tailings dams. In recent years, risk assessment and management has become a topic of keen interest, perhaps mostly due to the explicit inclusion of the need for risk assessment in the Global Industry Standard on Tailings Management (GISTM) (Global Tailings Review, 2020). Aside from the need to demonstrate that risk is regularly assessed and the necessary actions taken, conformance also requires that risk is managed ALARP and that critical controls are identified and used to manage risk in different contexts. While one might consider that tailings dam safety management should not significantly differ from that of water dams, for which there has been detailed guidance for some time (eg ANCOLD, 1994; International Commission of Large Dams (ICOLD), 2005), there appears to be consensus on the need for separate guidance documents across the tailings industry.

The mining industry has widely adopted Critical Control Management (CCM) for managing safety hazards over the past decade. Critical control management CCM has in effect been employed in other industries as well. One example is within the Safety Case regimes in process industries such as drilling and refining in Oil and Gas, an example being the National Offshore Petroleum Safety and Environmental Management Authority (NOPSEMA) Safety Case guidance (NOPSEMA, 2022), which requires an understanding of the factors that influence risk and the controls '...critical to managing risk'. Another example is the reliance of the Hazard Analysis Critical Control Point (HACCP) approach in the food safety industry on halting a failure pathway at a Critical Control Point that may lead to contamination. PPC Industries Inc. (2025), which is analogous to relying on preventative rather than mitigative controls. In the food industry this means implementing key controls to provide assurance that food is safe rather than relying upon testing prior to consumption.

In implementing CCM, many companies have adopted International Council on Mining and Metals (ICMM, 2015) guidance. As with GISTM, this document provides no guidance on how CCM fits within managing risk ALARP, leaving it to the asset owner and their consultant to determine the appropriate course of action.

Demonstration that risk is managed ALARP, or So Far As Is Reasonably Practicable (SFAIRP) which is considered an equivalent concept by Safe Work Australia (2013), relates to the actions one does or does not choose to take. All relevant matters are evaluated including the relative risk level compared to other threats and assets within the accountable party's control, benefit in terms of risk reduction, risks introduced to implement a proposed action and sacrifices including but not exclusively referring to cost. Guidance documents that delve into this topic further include Safe Work Australia (2013), Institution of Mechanical Engineers (iMechE; 2024), Engineers Australia (2014) and Health and Safety Executive (HSE; 2025). As the focus is comprehensive identification and implementation of actions, then it follows that control management is vital. However, some complications arise when applying CCM as outlined by ICMM (2015) to an asset where the risk is also managed ALARP, including:

- How to manage all controls effectively if focus is given to critical controls. Or in reverse, should one identify critical controls such that they are exhaustive for the safety of an asset and then solely focus on them?

- If a control is evaluated to be reasonably practicable, does it mean it must be critical?

- How does the example critical control selection decision tree in ICMM (2015) apply to geotechnical structures, such as a TSF?

In the authors' experience, it is appropriate that some controls are given greater focus, as it is without question that 'all controls are not created equal'. However, it has also been witnessed that the controls needing the greatest focus often change over time and too often f so many critical controls are identified that an overly complex system is created that is cumbersome to manage.

CRITICAL CONTROLS CONCEPT

Despite there being some recent conference papers that include discussion of critical controls (Martens and Kupper, 2024; Afriyie and Finke Morrison, 2023), they are not focused on how to select which controls should be defined as critical. As such, this section aims to provide a broad review of the concept to frame the discussion on the future use of CCM in the tailings industry.

ICMM guidance on CCM

The International Council on Mining and Metals (ICMM) have published three guides addressing critical control management focused on safety management across the mining industry that will be referred to heavily in this section to summarise their content relevant to the discussion in this paper:

1. Health and Safety Critical Control Management Good Practice Guide (ICMM, 2015).

2. Critical Control Management Implementation Guide (ICMM, 2016).

3. Leadership Matters: The Elimination of Fatalities (ICMM, 2009).

These documents describe CCM as means to manage risk that exceeds a threshold level, termed a Material Unwanted Event (MUE). Materiality is in turn defined by the owner of the asset as a threshold, that if breached would impact the ability of the business to meet its core objectives (ICMM, 2016) and thus warrants the highest level of attention (ICMM, 2015). As such, the threshold may vary between companies and for many reasons, including for example changes in management over time, community sentiment and market conditions.

The first steps in the CCM approach are to:

- Describe the MUE, including the relevant hazard, mechanism of release (often referred to as 'initiating events' in dam engineering) and the consequences.
- Assess opportunities to eliminate the MUE by improving design.

The next step is to identify both existing and new controls to prevent or mitigate the MUE. The controls must manage MUEs by focusing on what requires regular attention and communication of what action is required. The logic provided for identifying existing controls is not presented in this paper for brevity but is an important aspect of the process that should not be overlooked.

ICMM (2021) define critical controls as a sub-set of controls and clarify that they are risk management measures to mitigate credible failure modes that could lead to a catastrophic failure. One specific aspect of CCM that would probably be familiar to tailings professionals is identifying the trigger for an immediate stop, necessitating change of the operation, require remediation of one or more critical controls, or both. This is somewhat analogous to a Trigger Action Response Plan (TARP), as included in GISTM and other guidance documents. It likely is not a coincidence that the two ideas are blended in Requirement 7.4 of GISTM (Global Tailings Review, 2020).

It is also important to note the following also precede the identification of critical controls, which is akin to application of the hierarchy of control (ICMM, 2016):

- Assess opportunities to eliminate the MUE by improving design.
- Describe the MUE, including the relevant hazard, mechanism of release (often referred to as 'initiating events' in dam engineering) and the consequences.
- Identify existing controls and identify new controls, for which there is a logic provided that is not presented in this paper for brevity. However, it is an important aspect of the process that should not be overlooked.

ICMM (2015, 2016) provide substantial guidance on the definition of a critical control and how one might identify it. ICMM (2015) presents the following:

> The final set of critical controls for an MUE should represent the critical few that, when managed using CCM, can effectively manage the Material Unwanted Event risk.

> The following questions can help to determine if a control is critical:

> - Is the control crucial to preventing the event or minimizing the consequences of the event?
> - Is it the only control, or is it backed up by another control in the event the first fails?
> - Would its absence or failure significantly increase the risk despite the existence of the other controls?

To compliment and further develop this guidance an example decision tree is presented in ICMM (2015, 2016) and is reproduced in Figure 1. Applying the decision tree to TSFs is discussed throughout the remainder of the paper.

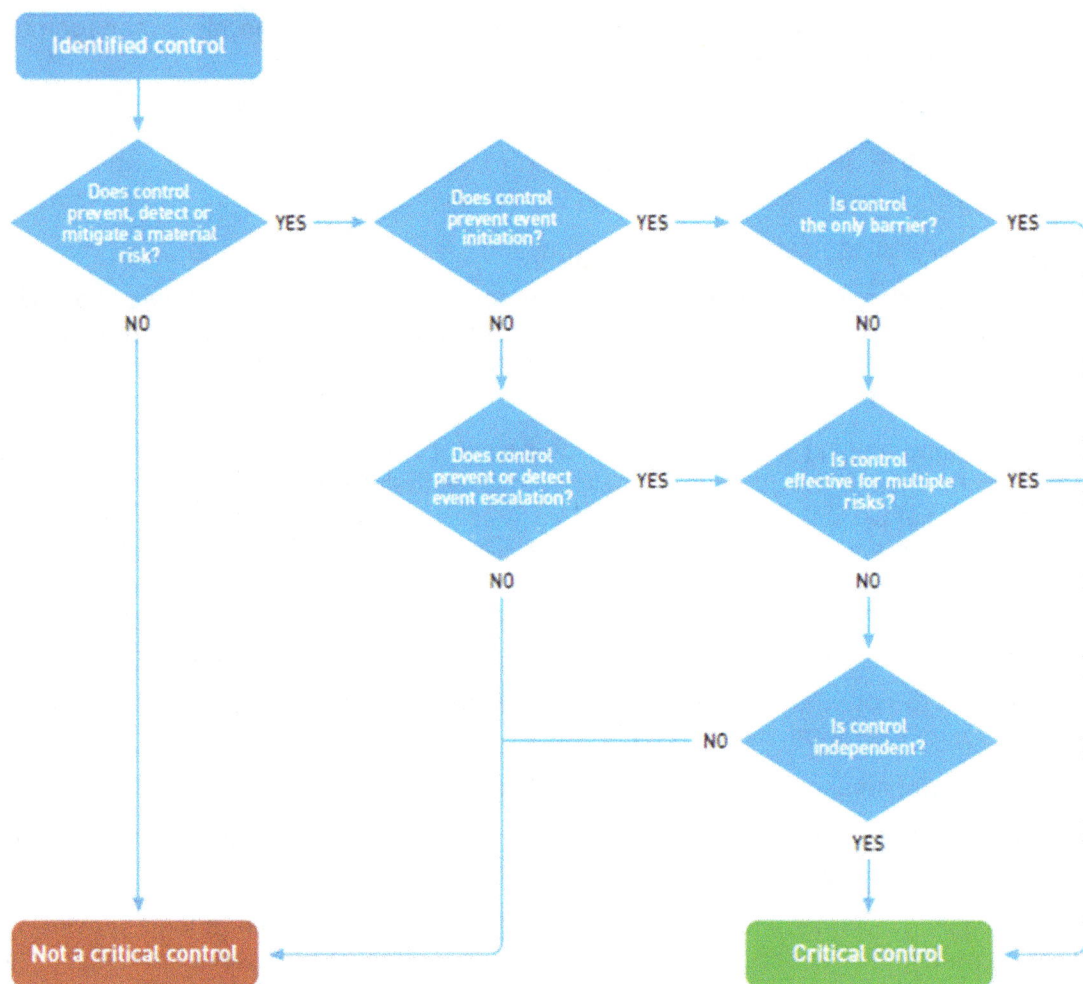

FIG 1 – ICMM (2015, 2016) example critical control decision tree.

ICMM (2015, 2016) include steps after critical control selection to define the objective and performance criteria for each critical control and then, after implementation, regularly review and report upon control effectiveness. This includes a review of critical controls summary reports on behalf of the MUE owner. Critical controls that are shown to be ineffective are recommended to be investigated and the outcomes actioned (ICMM, 2016), which may include improvements to the CCM process. In the authors' experience, it is also important to review the controls considering whether the risk level and events or failures that can initiate the MUE risk have changed over time. In the context of a TSF, an issue is often addressed by implementing a physical control such as additional drainage or slope flattening and these may materially change the risk profile and therefore the critical controls.

GISTM and supporting guides

The GISTM provides the following definition of a critical control, which is notably similar to that in ICMM (2015, 2016):

> *A control that is critical to preventing a potential undesirable event or mitigating the consequences of such an event. The absence or failure of a critical control would disproportionately increase the risk despite the existence of the other controls.*

Critical controls are in-turn noted as necessary elements of:

- Operations, Maintenance and Surveillance (OMS) manuals in providing for safe operations (see Requirement 6.4 and the Glossary definition of OMS).

- Trigger Action Response Plans (TARPs), which are noted in the Glossary as a tool to manage controls, including critical controls. Pre-defined trigger levels are also to be defined based on risk controls and critical controls. This is somewhat contradicted in Requirement 7.4, which states the performance outside expected ranges shall be addressed promptly through TARPs or critical controls. Regardless, the linkage between critical controls and safe operations is clear.

Given the scope of GISTM, this discussion appears consistent with ICMM.

Ongoing risk management, presumably including critical controls, is required in GISTM (Global Tailings Review, 2020) Requirements 10.1 and 10.2. These require risk assessment updates at minimum every three years (separate to requirements for update of ALARP evaluation, which is covered in Requirement 5.7) and review of the Tailings Management System and Environmental Social Management System, which presumably should provide the process and role assignments for CCM, every three years.

Mining Association of Canada

The Mining Association of Canada (MAC) provide perhaps the most detailed discussion of critical controls in their guide to the management of tailings facilities (MAC, 2021). The following definition and selection criteria are presented.

> *Critical controls are site-specific and governance-level risk controls that are crucial to preventing a high consequence event or mitigating the consequences of such an event. The absence or failure of a critical control would significantly increase the risk despite the existence of other controls. Critical controls may be technical, operational, or governance based.*

> *Risk controls are typically designated as critical controls if one or more of the following conditions are met:*

> - *implementation of the control would significantly reduce the likelihood or consequence of an unwanted event or condition that poses unacceptable risk;*

> - *conversely, removal or failure of the control would significantly increase the likelihood or consequences of an unwanted event or condition that poses an unacceptable risk, despite the presence of other controls;*

> - *the control would prevent more than one failure mode or would mitigate more than one consequence; or*

> - *other controls are dependent upon the control in question.*

MAC (2021) also state that critical controls should be included in the OMS and require verification, because otherwise they cannot be effectively maintained. Although perhaps implied by other guides, this is a key addition to the discussion. To manage risk to an accepted threshold critical controls must be designed so that they are never breached and to do this they need to be appropriately verified.

MAC (2021) includes discussion of management of critical controls including definition of performance criteria, role assignment, verification and reporting of deficiencies. This is quite similar to the recommendations of ICMM (2015, 2016), discussed above, but the process of periodic update of critical controls is recommended to be based upon updated risk assessment, risk management plans and past performance. This appears a practical addition and aligns well with the author's comments above on this topic.

ICOLD

ICOLD (2022) provide background to CCM in the mining industry similar to that reproduced in this paper from ICMM (2015). In discussing critical controls, links are again provided to the development of TARPs. The example decision tree (Figure 1) is reproduced from ICMM (2015, 2016). Also relatively well aligned with the other guides discussed above is the following (ICOLD, 2022):

> *Critical Controls are crucial to preventing an unwanted event or reducing the likelihood or consequence of the event. The Critical Controls inform the Trigger Action*

Response Plan (TARP). Critical Controls may include both the Preventative and Mitigative Controls and an assessment to determine if a given control is a Critical Control is related to the following:

- *Would the absence or failure of the control significantly increase the risk, despite the existence of other controls?*

- *Does the control address multiple causes or mitigate multiple consequences?*

Australian context – Work Health and Safety Act

The safety in design concept is explicitly included in the Model Work Health and Safety Legislation (Safe Work Australia, 2023). Of particular note, is the duty of a person (paraphrased for relevance to TSFs) designing plant, substance or structure to enable, so far as is reasonably practicable, that the plant, substance or structure is designed to be without risks to the health and safety of a person who at a workplace:

- Uses the plant, substance or structure for the purpose for which it was designed. Or

- Constructs the structure. Or

- Carrys out any reasonably foreseeable activity at a workplace in relation to the plant, substance or structure. Or

- Are at or in the vicinity of a workplace and who are exposed to the plant, substance or structure at the workplace or whose health and safety may be affected by a use or activity mentioned in the above.

The regulations that accompany the above text in the Act may vary between States and Territories. Although the term 'critical control' does not appear in the Western Australian WHS (Mines) Regulations (Government of Western Australia, 2022), they contain a relatively exhaustive list of matters must be considered in developing control measures to manage the risks of geotechnical structure instability. Consequently, extra diligence is required for Australian operations in developing, implementing and assuring controls.

Water dam industry

Whilst many references are discussed in this section; it is notable that none are from organisations that produce guidance for water dam safety management. The reason for a lack of adoption of the CCM concept in the water industry is not clear, although the authors speculate that it may be a result of:

- Greater consistency in types of structure and consequently, failure modes and controls that are consistent and well understood.

- Long-term adoption of a safety management system approach, where the tasks necessary to maintain a safe operating environment are embedded within the owner organisations, without the need for distinguishing individual controls.

- Water dams are designed for maximum operating pool levels and spillway flows, such that speculation with regard to maximum foreseeable loading is not required. This is to say that there is no space to design for anything but the worst-case loading with respect to pool level, whereas there is for a TSF, especially related to pond management that in turn relies on controls to ensure the pond does not become larger than intended or considered in the design, for example.

- For water supply companies, the dam is the central asset in the organisation and so there may be a culture that promotes all actions to manage its integrity, and in turn a distinction of controls is not desired or necessary.

GEOTECHNICAL STRUCTURES VERSUS OTHER ASSETS IN RISK ASSESSMENT

Geotechnical structures are heavily statically indeterminate and are inexorably linked with the natural and variable environments in which they exist. Consequently, there is significant dependence between the different elements that make up a geotechnical structure. For example, one cannot expect to make sense of long-term deformations looking only at one potential culprit, like changes in loading, in isolation of other candidates such as moisture content. This example can be significantly expanded by simply considering all natural and constructed variables related to those two potential culprits. This has relevance to this discussion as it helps illustrate how complex it is to determine how significant, or not so, that a given action is for a large geotechnical structure like a TSF. This is especially true when expanding beyond the idealised and often two-dimensional models that we often rely upon for analyses to consider the likely heterogeneity across the entire structure, including the geological setting in which it is founded. This is the level of complexity that one faces with when attempting to identify critical controls utilising the tools discussed above.

Failure modes are also often dependent within geotechnical structures. This means that the occurrence of one failure mode, or perhaps better termed development or onset, of one failure mode impacts the likelihood of another. The same can often be said of controls. For example, the erosion of the toe of an embankment due to flooding may trigger a slope failure. Again, such performance of a geotechnical structure is not easily forecast without considerable effort. A detailed discussion is provided in ANCOLD (2022).

TAILINGS STORAGE FACILITIES VERSUS OTHER GEOTECHNICAL STRUCTURES

Building on the preceding discussion, TSFs pose some further, somewhat unique, challenges that many other geotechnical structures do not face, further complicating the prediction of performance, defining controls, the impact that any given control may have:

- Significant distances and sometimes rapid run-out of material from a failure result in relatively higher consequences than many other earthen structures, especially since the materials may contain contaminants.

- In some cases, strain softening of materials significantly influence the stability of a TSF. In this instance, performance monitoring and forewarning of failure is difficult and often unreliable with current technology. This is addressed in GISTM with requirement for conservative design criteria (ie to eliminate or sufficiently manage the risk of triggering of a flow failure or unacceptable deformation).

- Although tailings are soil-like materials, they have been physically altered in a process plat and recently deposited in the context of geologic time. As a result, conceptual and constitutive soil models may not be able to be reliably directly applied, limiting the number of analysis tools available to inform risk analysis.

- Reliance on human actions during normal operation is often an unavoidable element of safe tailings management unless the most conservative of design criteria and assumptions are adopted, but generally this approach needs to be balanced with the pursuit of an economical solution.

THE ICMM EXAMPLE CRITICAL CONTROL SELECTION FLOW CHART

The example decision tree (Figure 1) has been observed to have been utilised increasingly frequently across the tailings industry. While the authors believe it has a sound basis and presents a well thought out and clear logic, some difficulties arise in its application to TSFs. These are presented as responses to the queries provided in the text supporting and within the example flow chart:

- *Would the absence or failure of the control significantly increase the risk, despite the existence of other controls?* As discussed above this may require a complex analysis for some controls to answer accurately, due to the dependence of elements that make up the TSF structure and also between failure modes and controls.

- *Does the control address multiple causes or mitigate multiple consequences?* Very often the answer is yes unless judgement is applied to provide some cut-off on the significance of impact to a cause or consequence

- *Is the control the only barrier?* This is almost never the case due to the issue of dependence, as discussed above.

- *Is the control effective for multiple risks?* Very often yes, although it does depend on how one identifies discrete risks. This is often confusing as the MUE is most often defined as uncontrolled release of catastrophic failure of the TSF having material impacts on people, the environment or the owner organisation. There are many failure modes and scenarios that may comprise this risk, but with a broad MUE selection they are all treated as the same risk.

- *Is the control independent?* As discussed above, almost never.

CRITICAL CONTROLS AND ALARP

As introduced above, managing risk to ALARP requires implementing all reasonably practicable controls, but in the literature known to the authors, no fixed process exists for doing so. A reasonable interpretation of the literature is that a reasonably practicable control is one where the sacrifice is not grossly disproportionate to the benefit it provides. This is consistent with legal origin in the case of Edwards versus National Coal Board from 1949 (Wikipedia, 2025) where the court ruling stated:

> *Reasonably practicable is a narrower term than 'physically possible' and implies that a computation must be made... in which the quantum of risk is placed in one scale and the sacrifice involved in the measures necessary for averting the risk (whether in time, trouble or money) is placed in the other and that, if it be shown that there is a great disproportion between them – the risk being insignificant in relation to the sacrifice – the person upon whom the obligation is imposed discharges the onus which is upon him.*

The application of critical controls at the same time as demonstrating risk is managed to ALARP introduces the following complications, which are expanded from the Introduction:

- CCM's objective is preventing or mitigating MUEs, which ultimately requires some definition of a threshold risk level, whereas managing risk ALARP is evaluated assessing each potential control on its merits to determine if it is reasonably practicable. Consequently, one could judge the controls required for ALARP to be less or more than as assigned from the CCM process.

- The success of implementing, or at least demonstrating, that all reasonably practicable controls are implemented, should not be compromised even when some controls are given a higher status. The management system employed must not focus solely on critical controls unless all reasonably practicable controls are designated as critical.

- Assessing the reasonable practicability of controls generally involved considering the relative risk reduction provided by each. With criticality also assigned to controls using a logic other than maximisation of resources to most efficiently reduce risk as expected in an ALARP evaluation, it may be that focus is wrongly given to a control that provides lesser impact on the total risk profile due to the selection logic discussed above. This could especially apply when selecting a critical control on the basis that it acts on multiple failure modes, when it may be that these various failure modes in aggregate pose a much lower risk than one risk-driving failure mode that is consequently given less attention.

- ALARP evaluations typically include identification of all potential risk reduction measures (ie controls), consider the hierarchy or control, alignment with good or current practices and precedent in similar situations. While many of these elements may also be applied in the CCM approach prior to assessing control criticality, they are not necessarily inherent in the process and may be missed, thus complicating an argument that management following CCM alone demonstrates that risk is ALARP.

- In practice, the level of risk assessment that accompanies selection of critical controls is often simpler compared to ALARP evaluations, although the authors are not suggesting this necessarily needs to be the case.

Whilst these are significant challenges, the author's do not consider them insurmountable and provide some suggested approaches for addressing them later in the paper.

IS THE DISTINCTION OF CRITICAL CONTROLS NECESSARY?

In introducing CCM, ICMM (2016) notes that traditionally a company would choose a reputable safety management system, undertake a gap assessment and then action the outcomes, however this produces a large amount of compliance tasks that we often experience hard to give focus to at site level and may not place emphasis on what materially reduces risk. Although there is no evidence that a framework such as the Dam Safety Management guideline (ANCOLD, 2003) is not functioning effectively in the water dam industry, it is interesting that the ICMM member companies felt such an approach was not working effectively enough in managing safety on mine sites. The interplay between risk and conformance-based approaches in tailings management has been explored previously, with the conclusion being that the two can and perhaps should function in a complimentary manner (Singh, Coffey and Herza, 2022).

At current, it appears CCM has been adopted by many mining companies and so is relatively embedded in the mining industry. Though it may evolve into something different, it will be present in the tailings industry as long as the term remains in GISTM.

Despite the nuances discussed in this paper, thoughtfully established controls typically have a unique impact on the hazard they seek to control and as such the management of any two controls is unlikely to be the same. It is also reasonable to give some controls more focus than others. Whether that necessitates a designation of a sub-set of controls to the extent that they are called out separately in GISTM is debatable and a matter that cannot be properly addressed until the broader integration of CCM and ALARP is resolved.

SUGGESTED UPDATED PRACTICE FOR CONSIDERATION BY INDUSTRY

It is undeniable that CCM is well-established and presumably well understood in some segments of the mining industry. As many well-travelled colloquial sayings suggest, changing something that is well established and has an adequate track record is not always wise. However, at current, it appears that disparate approaches are being taken due to differences in the details within, and maturity of, the broader risk/safety management systems employed by different owner organisations. Furthermore, this paper is intended to demonstrate that CCM deviates from management of risk to ALARP and potential perils arise due to the similarities in the process applied to identify controls in the two frameworks, especially when in the Australian legal context which explicitly refers to the latter. GISTM and ICMM (2015, 2016, 2021) do appear to provide a common point of reference. The paper does not provide a suggested resolution to the disparities and perils but is intended to foster discussion on this topic and lead to a clearer process. Suggestions are grouped within two categories. The first category of suggestions is those that relate to how critical controls could be implemented in parallel with managing risk ALARP. The second group of suggestions are focused on achieving a consistent methodology for selecting, designing, implementing and assuring critical controls The two concepts are not mutually exclusive, and the most practical solution may lay in between several of them.

Integrating ALARP and CCM

The following are potential pathways to integrating CCM into tailings management, whilst also managing risk ALARP:

- The risk threshold represented by the MUE is similar to societal and individual tolerability limits required to be demonstrated for a risk to be acceptable in ANCOLD (2022). As such, CCM could be implemented as a step in the process prior to evaluation of which controls are required to manage risk ALARP. Whilst this is simple in theory, some of the complications in control management highlighted earlier would remain, such as over-emphasis of critical controls in decisions making related to resource allocation.

- To overcome the complications alluded to above, all reasonably practicable controls could be treated as critical and then managed in accordance with ICMM (2015, 2016) provided this is confirmed to be adequate to maintain risk ALARP.

Establishing and implementing controls

As discussed above, direct application of the example decision tree (ICMM, 2015, 2016) faces some challenges for tailings storage facilities. The following suggestions are potential options to remedy this:

- In recognition of the wide application of the concept of critical controls and the processes for selection of them within the mining and tailings industry, some clarifications on the definitions of key terms within the example selection flow chart (ICMM, 2015) could be made per Table 1.

TABLE 1

Clarifications of key terms within the example critical control selection flow chart in ICMM (2015).

Query	Clarifying comments
Identified control	Background: a control selection flow chart is presented in ICMM (2015).
	Definition considerations: In order to align critical control selection with the application of ALARP, only controls that have already been shown to be reasonably practicable should be included in the process of selection of critical controls. This means that an exhaustive consideration of all potential controls would already been completed prior to selecting critical controls.
	Ideally, a critical control for a tailings system should entail a suite of requirements and actions. A summary of the author's collective experience from practice is presented in Table 2, as a suggested starting point or overall structure that could be considered.
Does the control prevent, detect or mitigate a material risk?	Background: Within health and safety management the control will almost invariably manage a specific single hazard eg high voltage power, Fall from height etc.
	Proposed definition: Material risk should be linked with the MUE, which in most cases would include catastrophic TSF failure, although the assessment team may choose to define it differently. Typically, a TSF failure is very likely to pose a material risk for any mining company, so this step may be superfluous in many cases.
Does the control prevent event initiation?	Background: Prevention of initiation is understood to be analogous to elimination of a hazard or the introduction of an engineering control to avoid interaction between hazard and human. These sit atop the hierarchy of controls.
	Proposed definition: Does the control act upon the physical initiating event(s) such that they would no longer trigger the initiation of the physical failure mechanism, or does it prevent the physical failure mechanism from manifesting given the occurrence of the physical initiating event?
	Design, construction, and operating controls are typically designed to prevent initiation. So, controls fitting under the definitions other than Emergency Response in Table 2 are likely to prevent initiation.
Does the control prevent or detect event escalation?	Background: In many systems early detection (lighting detection, pressure or temperature gauges etc) are intended to enable removal of the hazard and/or the persons.
	Proposed definition: Is the control an intervention after the physical initiating event has occurred and/or after the physical failure mechanism has been observed? If so, would it reliably perform this function in the range of conditions (eg during a significant storm, requiring high-risk access to a failing area etc) in which it may be called upon?
	An example may be detection of sudden saturation on an embankment triggering remediation works per the pre-developed plan as part of the implementation of the TARP an/or observation method.
Is control the only barrier?	Background: This would apply when a control with no redundancy, but high effect is employed, for example an E-stop on rail-mounted plant where administrative controls or personal protective equipment would be ineffective.
	Proposed definition: Would the applicable failure mode(s) be significantly more likely to manifest if the control was absent?

Query	Clarifying comments
Is control effective for multiple risks?	Background: this applies to actions that can be universally applied, such as pre-work Job Safety Analysis or a Safe Work Method Statements, that also have a proven methodology.
	Proposed definition: Does the control individually provide a significant reduction in the risk profile attributed to a risk-driving failure mode(s)? Or does the control solely provide, or in combination with a small number of other controls with which it is inter-dependent, a significant reduction in the risk profile attributed to the total risk?
	The group of example critical controls presented in Table 2 are examples that would likely pass this test.
Is control independent?	Background: an independent control is one that is not supported by any other control.
	Proposed treatment: this decision point is recommended to be omitted based on the preceding discussion regarding interdependence within geotechnical systems and the controls that apply to them, however there would be no harm if it were retained.

TABLE 2
Summary of outcomes from the Author's work in the identification of critical controls.

'Critical' controls	Primary intent of the control
Dam structural integrity	Make sure that components of the dam are designed, constructed and monitored in alignment with regulatory requirements, industry guidance, current engineering practice and precedent elsewhere in the operator's company or wider industry.
Tailings management	Manage tailings deposition such that it is controlled within the design intent of the dam and enables effective water management.
Water management	Manage supernatant water on the dam such that it is controlled within the design allowance, usually is maintained away from embankments and facilitates beach profiles and *in situ* density within the bounds assumed for the design if this is critical to the safety of the facility.
Emergency response capability	Enable the ability for the mine to respond in a timely and effective manner to any unplanned conditions whether they pose harm to persons or the environment or not.

- A different flow chart, or perhaps flow charts covering common failure modes with some specific detail pertaining to TSFs could be developed. However, it is possible this would result in an impractical amount of detail needing to be included which could inhibit application in practice.

- Instead of nominating a control as 'Critical', it may be more instructive to instead separate controls that relate directly to the functional integrity of the TSF System and nominate them as Integrity Controls. The other controls, necessary to facilitate effective 'integrity controls' could then be referred to as enablers or supporting controls. Regardless of nomenclature, the distinction of the two may facilitate improved safety outcomes.

- The hierarchy of control could be emphasised more in the process of selection of critical controls. The intent would be to limit the number of controls that require regular management in recognition that complex systems not only take more work to manage, but are typically also more common to fail if they involve a lot of human actions. For clarity on this approach; greater safety is not achieved by having more critical controls. In the author's experience, it is often the most resilient designs that require the fewest controls to be managed.

  o When emphasis is given to 'higher order' controls in the hierarchy, one could also consider if a critical control, or even any control requiring active management, is required if the design provides redundancy for a given or all failure modes that could result in catastrophic failure.

Management and governance of controls

Regardless of the process to assimilate CCM and managing risk ALARP, it is suggested risk management would be improved through better assimilation of the various elements in different guides, such as GISTM, and the risk assessment and management process. As suggested in most guides, this should be done through the implementation of a management system. In reference to previous papers related to tailings management (Herza, Coffey and Singh, 2022, 2023) and water dam management (Hunter, Foster and Arnold, 2022), the things required to maintain the chosen risk profile (ie elements of the TMS), especially when it pertains to managing risk ALARP, could be documented within a Safety Case similar to that required by NOPSEMA (2022) and discussed at length in iMechE (2024).

CONCLUSION

Critical controls are discussed in several guidance documents specific to tailings management and CCM is a mature process for many mining companies, as with other industries highlighted in this paper. The introduction of the GISTM increased focus on their adoption in tailings management, but there is limited guidance on what they are and how they interact with other elements of a management system, despite explicit requirements being included in the GISTM for integration with OMS and TARPs.

While it can be argued that critical controls need not be defined for tailings storage facilities, especially given the complications that it introduces when also applying the ALARP concept, for many companies CCM is so well entrenched within existing safety systems and culture that it could be a significant opportunity to facilitate safe and reliable operations. Some suggested approaches for integration of CCM and management of risk ALARP were presented in the hopes of catalysing further conversation and continual improvement of tailings management.

The decision tree presented by ICMM (2015, 2016) appears to be widely adopted in the mining industry, however there are several complications in applying it in its current form to tailings storage facilities. As such, some suggested updates and an example structure of critical controls were presented. These suggestions inevitably will not fit all facilities, but were presented in the hopes of assisting practitioners in simplifying and strengthening the application of the decision tree, and the broader CCM framework presented by ICMM (2015, 2016), to tailings storage facilities.

The recent focus and maturation of risk management theories into practice across the tailings industry is a positive development and the authors have developed this paper to assist further growth.

REFERENCES

Afriyie, G and Finke Morrison, K, 2023. Process safety approach for reviewing critical controls on tailings storage facilities, in Proceedings of Tailings and Mine Waste Conference 2023 (University of British Columbia).

Australian National Committee on Large Dams Incorporated (ANCOLD Inc), 1994. Guidelines on risk assessment.

Australian National Committee on Large Dams Incorporated (ANCOLD Inc), 2003. Guidelines on dam safety management.

Australian National Committee on Large Dams Incorporated (ANCOLD Inc), 2019. Guidelines On Tailings Dams - Planning, Design, Construction, Operation And Closure, revision 1, ANCOLD Inc. Available from: <https://ancold.org.au/product/guidelines-on-tailings-dams-planning-design-construction-operation-and-closure-revision-1-july-2019/>

Australian National Committee on Large Dams Incorporated (ANCOLD Inc), 2022. Guidelines on risk assessment.

Engineers Australia, 2014. Safety case guideline, Clarifying the safety case concept to engineer due diligence under the provisions of the model Work Health and Safety Act 2011, third edn.

Global Tailings Review, 2020. Global Industry Standard of Tailings Management (GISTM).

Government of Western Australia, 2022. Work Health and Safety (Mines) Regulations.

Health and Safety Executive (HSE), 2025. ALARP – As Low As Reasonably Practicable, Health and Safety Executive, Control of major accident hazards (COMAH). Available from: <https://www.hse.gov.uk/comah/alarp.htm>

Herza, J, Coffey, J and Singh, R, 2022. Key elements of tailings dam safety case, in Proceedings of Tailings and Mine Waste 2022 (Colorado State University).

Herza, J, Coffey, J and Singh, R, 2023. The tailings safety case – an example, in Proceedings of Tailings and Mine Waste Conference 2023 (University of British Columbia).

Hunter, G, Foster, M and Arnold, M, 2022. Risk and ALARP assessment for the Upper Yarra embankment upgrade selection, in Proceedings of ANCOLD/NZSOLD 2022.

Institution of Mechanical Engineers (iMechE), 2024. ALARP for engineers: a technical safety guide, Institution of Mechanical Engineers.

International Commission of Large Dams (ICOLD), 2005. Bulletin 130, Risk assessment in dam safety management.

International Commission of Large Dams (ICOLD), 2022. Bulletin 194 preprint, Tailings dam safety.

International Council on Mining and Metals (ICMM), 2009. Leadership matters. The elimination of fatalities.

International Council on Mining and Metals (ICMM), 2015. Health and safety critical control management good practice guide.

International Council on Mining and Metals (ICMM), 2016. Critical control management implementation guide.

International Council on Mining and Metals (ICMM), 2021. Tailings management good practice guide.

Martens, S and Kupper, A, 2024. Credible failure modes: considerations for assessment and application, in Proceedings of Tailings and Mine Waste 2024 (Colorado USA).

Mining Association of Canada (MAC), 2021. Tailings guide version 3.2, The Mining Association of Canada (MAC).

National Offshore Petroleum Safety and Environmental Management Authority (NOPSEMA), 2022. ALARP guidance note, National Offshore Petroleum Safety and Environmental Management Authority, Perth, Western Australia.

PPC Industries Inc., 2025. About HACCP. Available from: <https://ppcind.com/haccp/>

Safe Work Australia, 2013. How to determine what is reasonably practicable to meet.

Safe Work Australia, 2023. Model Work Health and Safety Legislation.

Singh, R, Coffey, J and Herza, J, 2022. Risk and conformance-based tailings management: two sides of the same coin, in Proceedings of ANCOLD/NZSOLD 2022, Sydney Australia.

Wikipedia, 2025. Edwards vs National Coal Board, last edited 2 May 2025. Available from: <https://en.wikipedia.org/wiki/Edwards_v_National_Coal_Board>

TailingsIQ – an AI-enhanced tool for improving tailings storage facility oversight and risk management

S Darmawan[1]

1. Principal Geotechnical Engineer, Geotesta Pty Ltd, Perth WA 6000.
   Email: sd@geotesta.com.au

ABSTRACT

Managing tailings storage facilities (TSFs) is becoming increasingly challenging as these structures grow in size and complexity, and as regulations such as the Global Industry Standard on Tailings Management (GISTM) demand higher levels of safety and oversight. One major difficulty is handling the enormous amount of data generated from many different sources – including sensor readings, inspection reports, and operational logs – which are often kept in separate 'silos' that do not communicate with each other. This lack of integrated data makes it hard for engineers to see the full picture, quickly assess risks, or easily verify compliance with standards.

TailingsIQ is a proposed software platform that uses Artificial Intelligence (AI) to solve these data integration problems. It offers a single, web-based system that brings together all types of TSF data in one place. Its key feature is an AI-powered Query Assistant that uses Natural Language Processing (NLP) – a branch of AI that enables computers to understand human language – to let users ask questions in plain English. The system will search through both the numerical sensor readings and the text of reports to provide a concise answer, drawing information from across the data sources. This capability is referred to as Cross-Data Synthesis (CDS) – the platform's ability to synthesize information from multiple data types, similar to how an experienced engineer would combine insights from instruments and reports. By breaking down data silos and making information easier to access, TailingsIQ is designed to improve situational awareness and support proactive risk management for TSF operations.

A proof-of-concept (PoC) implementation has been developed using simulated TSF data to validate the TailingsIQ approach. This paper presents the TailingsIQ concept, explains its architecture and features, describes the PoC results, and discusses how this AI-enhanced system can enhance safety, efficiency, and compliance in tailings management by helping engineers make better use of the data they already collect – all while keeping human experts in control of critical decisions.

INTRODUCTION

Tailings storage facilities (TSFs) – the large dam and pond systems that store mining waste – pose significant technical and safety challenges for the mining industry. As mining operations expand, TSFs are becoming larger and more complex, holding vast quantities of potentially hazardous waste under strict safety requirements. Ensuring the long-term stability of these structures is paramount for protecting downstream communities and the environment. Recent high-profile TSF failures, often caused by geotechnical instability or water management issues, have highlighted the catastrophic consequences of inadequate oversight and have led to heightened regulatory scrutiny worldwide. Notably, the Global Industry Standard on Tailings Management (GISTM) introduced in 2020 (Global Tailings Review, 2020) calls for sweeping improvements in TSF engineering design, monitoring, governance, and transparency in reporting performance and risks.

Meeting these elevated standards – particularly the GISTM requirements – an effective TSF management requires collecting and analysing huge volumes of diverse data from many different sources. Typically, a TSF team must deal with:

- Time-series monitoring data: frequent readings from geotechnical instruments (eg piezometers, inclinometers, settlement gauges), hydrological sensors (eg flow meters, water level sensors), and potentially environmental sensors, often delivered in proprietary formats from different logging systems (Qiu *et al*, 2024).

- Spatial data: surveys and remote sensing data such as satellite InSAR deformation measurements, drone photogrammetry, LiDAR scans, and conventional ground surveys.

- Operational records: daily logs of tailings deposition, pond water levels and balances, pumping and discharge operations, and construction or raise activities.

- Laboratory results: data from geotechnical lab tests on tailings and foundation samples, plus water quality analyses, usually stored in separate databases or spreadsheets.

- Design and as-built documentation: engineering design reports, drawings, and construction quality assurance records, often stored as PDF documents or CAD files.

- Inspection and review reports: routine inspection logs, Dam Safety Reviews (DSRs), Independent Tailings Review Board reports, regulatory audit findings, and compliance checklists, usually in narrative text (Word or PDF) format.

The core technical challenge is that all this information is heterogeneous and fragmented. Data resides in isolated silos – in different databases, file servers, spreadsheets, and legacy systems – with little to no connection between them. These silos often use incompatible formats and inconsistent data standards, making it hard to aggregate or compare information directly. As a result, engineers and managers spend an inordinate amount of time manually finding, extracting, cleaning, and compiling data from various sources. This manual process is time-consuming and error-prone. In short, fragmentation of data hinders the ability to identify risks in advance, demonstrate compliance (eg proving conformance with GISTM criteria), and make timely, well-informed decisions. This lack of integration severely hinders the ability to gain a holistic, near real-time understanding of the TSF's condition and performance trends (Mooder, Hawley and Williams, 2022). The sheer volume and speed of incoming data, especially from automated sensors, only adds to the difficulty, often overwhelming traditional analysis methods. The sheer volume and velocity of data, particularly from automated sensors, further exacerbate these challenges, overwhelming traditional analysis methods (Kruger, 2023).

Recognising these significant data management hurdles, the application of advanced digital technologies, particularly Artificial Intelligence (AI), presents compelling opportunities for transformative improvements (Li *et al,* 2021). AI encompasses computational systems capable of performing tasks typically requiring human intelligence, such as learning, problem-solving, and decision-making. Within mining, AI is already demonstrating value in exploration, predictive maintenance, process optimisation, and safety enhancement (Li *et al,* 2021). Specifically for TSFs and dam safety, AI offers substantial potential for revolutionising data management, interpretation, and risk assessment (Santos and Celestino, 2023), including methods for quantifying uncertainty. Machine Learning (ML), a subset of AI focused on algorithms that learn from data, can analyse vast data sets from monitoring systems to detect complex patterns, anomalies, or precursors to instability that might evade human detection (Australian National Committee on Large Dams Incorporated (ANCOLD Inc), 2022; Santos and Celestino, 2023). ML models show promise for predicting TSF behaviour, such as deformation or seepage patterns (ANCOLD, 2023; Lopes and Ramos, 2021), potentially leveraging advanced ML techniques. Furthermore, AI techniques can facilitate the integration of diverse data streams, including geospatial data from satellite or drone platforms, providing a unified operational picture.

This paper introduces TailingsIQ, a conceptual software platform created to address the critical data integration and accessibility challenges in TSF management through the smart use of AI. TailingsIQ envisions a unified, intelligent system that provides engineers, operators, and regulators with one-stop access to all relevant TSF data along with powerful analysis tools.

THE TAILINGS-IQ SOLUTION

TailingsIQ is conceived as a comprehensive web-based platform to improve how TSF data is managed and used. Its primary goal is to integrate heterogeneous TSF data streams and make them easily accessible through advanced AI tools, thereby helping engineers and managers make better decisions, anticipate problems earlier, and streamline regulatory compliance. In other words, it replaces the current patchwork of fragmented tools with a unified environment where all relevant data comes together and can be queried intelligently.

System architecture and accessibility

The TailingsIQ platform is designed as a cloud-supported web application that authorised users can access through a standard web browser. In practical terms, this means an engineer in the field or an off-site consultant can log into TailingsIQ via the internet and use its features immediately. The backend services (data storage and processing logic) are hosted in the cloud, which provides scalability and centralises the data management. This architecture was chosen to foster collaboration and information sharing, key aspects of modern integrated systems in mining. Instead of storing data on individual laptops or isolated local servers, all TSF information resides in one secure platform, always up-to-date and available to those who need it. This approach not only improves data consistency but also aligns with governance practices by controlling access and maintaining an audit trail of data usage.

Key components

TailingsIQ is structured into several core modules that reflect the main functional needs of TSF management:

- Integrated Dashboard: A centralised dashboard provides at-a-glance visibility into the TSF's status. It displays important metrics and Key Performance Indicators (KPIs) such as current piezometric levels, water pond elevation, recent rainfall, storage volume, and any active alerts. Graphs and charts show trends (eg piezometer readings over time) and status indicators highlight if something is outside normal or expected ranges. This gives engineers and managers an immediate sense of the facility's condition without having to manually compile data from different sources.

- Document Repository and Knowledge Base: All critical TSF documents are stored and managed in a unified repository. This includes design reports, inspection reports, monitoring reports, regulatory submissions, operational procedures, and so on. Documents are indexed so that users can quickly search for keywords or phrases. TailingsIQ uses AI-driven semantic search capabilities, meaning the search looks at the context and meaning of text, not just exact keywords. For example, a search for 'drainage issues' could find a section in an inspection report that discusses ponding water, even if the word 'drainage' isn't used.

- Monitoring Data Analysis Module: This module handles time-series sensor data (like readings from piezometers, inclinometers, flow metres etc). It provides tools to visualise and analyse these data streams. Users can pull up interactive charts of sensor readings over time, apply filters (for example, select a date range or a specific sensor group), and even overlay data (such as rainfall on top of piezometric levels to see correlation). By centralising sensor data and analysis, TailingsIQ makes it much easier to interpret the raw monitoring information and connect it with other context (like construction events or inspection notes).

- Compliance Management Module: This module is dedicated to tracking compliance with standards and requirements (for instance, GISTM guidelines or site-specific regulatory conditions). Users can define criteria or requirements (eg minimum freeboard, maximum rate of rise, required inspection frequency) and the system can help track whether those are being met. It can link each requirement to evidence in the platform – for example, linking a GISTM guideline about seepage monitoring to the relevant piezometer data and inspection reports that demonstrate compliance. The module can generate summaries or reports showing compliance status, which greatly simplifies preparing for audits or reviews. Instead of manually cross-referencing spreadsheets and documents, an engineer can use TailingsIQ to instantly retrieve proof that, say, freeboard was never below the required level in the past year, citing the exact data and reports as evidence.

- AI Query Assistant: This is the innovative heart of TailingsIQ. The Query Assistant allows users to interact with the system by asking questions in natural language. It is powered by advanced AI algorithms that understand the context of the question and retrieve information across all the integrated data sources to formulate an answer. For example, a user might ask, 'Have there been any unusual readings in the embankment piezometers since the last inspection?' The assistant would interpret this question, search the time-series database for anomalous piezometer readings, check the document repository for the last inspection report's observations, and then compile a response that might say: 'Yes, Piezometer PX-5 showed a sharp increase on March 3, which was noted in the March inspection report as a potential concern,' and provide links to the data plot and the report paragraph. This goes far beyond a simple keyword search – it attempts to synthesise data and text to answer the question directly. In TailingsIQ, this cross-source analytical process is referred to as Cross-Data Synthesis (CDS), indicating that the system can draw insights from both quantitative data and qualitative reports together. The aim is to mimic the way an experienced engineer would analyse a problem – by consulting various charts, notes, and reports – and to present that combined insight on demand.

Together, these components make TailingsIQ a powerful tool to address the specific technical challenges outlined earlier. The unified data hub inherently breaks down data fragmentation by providing one central access point for all information. The natural language Query Assistant makes it much easier to retrieve insights, essentially opening up the data to everyone on the team. By correlating data from sensors with observations from inspections, the system helps in early detection of potential issues, supporting proactive risk management rather than reactive response. And through the compliance module, it streamlines verification of standards like GISTM by directly linking requirements to supporting data and documents, making regulatory compliance checks more straightforward and transparent.

PROOF-OF-CONCEPT DEVELOPMENT AND VALIDATION

To ensure that the TailingsIQ concept would work in practice, a Proof-of-Concept (PoC) application was developed. This PoC is essentially a simplified version of TailingsIQ built to demonstrate the core ideas using test data. The primary objectives were to validate the concept and showcase the key features – especially the integrated data view and the AI query capability – in a working system. Building the PoC confirmed the technical feasibility of integrating very different types of data in one platform and provided a foundation for adding the AI-driven functions.

An agile, iterative approach was followed to build the PoC, focusing first on the most important features. Using a modern web application framework allowed rapid creation of a responsive interface that works in a browser. The use of a cloud-based backend was employed to simulate how the system would manage data centrally.

A crucial part of the PoC was assembling realistic test data to drive the system. Synthetic data sets that resembled what a TSF operation would produce were generated:

- Time-series sensor readings: Sample CSV files for a few piezometers and a pond level gauge, with readings over several months, were created. These were loaded into the system to emulate live monitoring data, allowing the dashboard graphs and anomaly detection logic to be tested.

- Document corpus: A collection of sample text documents imitating typical TSF reports was created – for example, a monthly inspection report noting various observations, an annual dam safety review summary, an emergency action plan excerpt, some design report snippets, and a regulatory audit checklist. These text files were used to populate the document repository, allowing the search and AI query functions to be tried out on them.

- Placeholder drawings and images: A few placeholder items (eg a dam cross-section drawing or a site map) were included (although the PoC did not focus on advanced geospatial integration) to ensure the structure was in place to add such content.

Having this synthetic data loaded into the PoC's database and file storage meant that using the PoC application simulated interaction with a real TSF's data – the dashboard would show a piezometer plot, the document list displayed titles of reports, and the AI assistant could attempt to answer questions based on the fake data provided.

The concept of an integrated TSF data platform with an AI query assistant is technically feasible and offers clear value. Notable outcomes and lessons from the PoC include:

- Validation of the Integrated Approach: Seeing all the disparate data (charts of sensor data, lists of documents, compliance checks etc) in a single interface strongly reinforced the benefit of integration. It was immediately clear how much easier it is to have everything in one place. This tangible demonstration aligns with industry observations that integrated data platforms greatly enhance efficiency and insight. For example, during testing, clicking on a point in a piezometer graph quickly retrieved the related inspection report entry, something that would be cumbersome without an integrated system.

- Positive User Experience Feedback: The initial user interface proved to be intuitive. Test users (colleagues familiar with TSF management) could navigate the dashboard and find information without extensive training, suggesting that the design is on the right track. This gives confidence that with further refinement and user feedback.

- Clarity on Full-Scale Requirements: By building the PoC, the requirements for a full production version were identified. For instance, the necessity for more sophisticated data processing pipelines – to automatically import and update data from real sensors and corporate databases in real-time – and robust APIs (Application Programming Interface) to integrate advanced AI models for tasks like complex query interpretation was recognised. The need for stronger security and user management for deployment in an actual company environment was also noted.

- Deployment Strategy Lessons: Running the PoC in a test environment revealed important deployment considerations. Some limits (for example, how much data could be handled in the initial set-up) were encountered, providing insight into how to configure a production deployment for reliability and performance. These lessons will guide how cloud services and resources (servers, databases etc) are set-up when moving to a larger scale, ensuring the live system can handle the load.

- Data Integration and Ingestion Challenges: The PoC underscored the importance of well-defined data ingestion processes to connect TailingsIQ with existing mine systems (like sensor databases or document management systems). It was confirmed that integrating with real operational systems is achievable, but it requires careful planning – for example, mapping out how frequently data updates, how to handle data quality issues, and ensuring that any data format differences are resolved on import. This is a standard challenge in any large system integration project, and the PoC provided a clearer picture of how to tackle it.

DISCUSSION

The development of the TailingsIQ proof-of-concept provides a basis to discuss how this AI-enhanced platform could transform tailings facility management and how it compares to current practices. The role of advanced AI techniques and the importance of human oversight in such a critical application are also considered.

Potential benefits for TSF management

A fully realised TailingsIQ platform promises several important benefits for those managing TSFs, including:

- Enhanced Situational Awareness: By unifying previously scattered data, the platform gives engineers a near real-time, comprehensive view of the TSF's status. Instead of flipping through multiple reports and files, they can observe all key indicators on one screen. This makes it easier to catch deviations or trends that might indicate emerging issues – for example, noticing that a certain sector of the dam is experiencing higher than usual piezometric pressures at the same time as increased rainfall.

- Proactive Risk Management: Faster access to integrated and synthesised information means potential problems can be identified and addressed sooner. For instance, the AI's ability to correlate data might reveal a subtle combination of factors (like a minor instrument anomaly coupled with an observation in an inspection log) that suggests a developing risk. In addition, as the system evolves, it could incorporate predictive analytics (machine learning models that forecast future behaviour) to anticipate issues before they occur (ANCOLD, 2023; Lopes and Ramos, 2021; Salih *et al*, 2025).

- Streamlined Compliance and Reporting: TailingsIQ's built-in compliance tracking greatly simplifies demonstrating adherence to standards like GISTM. Engineers can quickly produce evidence for each requirement (eg show the freeboard has been maintained above the threshold with supporting sensor data and inspection notes) because the system links requirements with actual data and documents.

- Improved Knowledge Management: By centralising all data and documents and even capturing the rationale behind decisions (through recorded queries and notes), the platform helps preserve institutional knowledge. With TailingsIQ, the information and insights live within the system, so new engineers or external experts can quickly get up to speed on the facility's history and current state.

- Broader Access to Information: The question-and-answer style interface means that even individuals who are not data experts or who are external to the core team (eg corporate oversight teams, independent reviewers) can retrieve information easily. This democratises access to TSF information, promoting transparency. For example, a corporate executive could query the system about 'the status of all critical instruments' and get a meaningful summary without having to call a meeting with the site engineers. It empowers a wider range of stakeholders to engage with the data directly.

Comparison to current practices

TailingsIQ represents a leap from the *status quo* of TSF data management. Currently, many sites rely on manual or semi-manual processes and disconnected tools:

- Reliance on Simplified Models: In geotechnical engineering, it's common to use empirical models and rules of thumb (derived from instruments like Cone Penetration Tests or Standard Penetration Tests) to predict how a dam might behave. These traditional models, while valuable, simplify reality and may not capture all factors at play. Major tailings dam failures often involve complex interactions (eg unusual seepage paths, changing material properties, operational factors) that may not be fully accounted for by static design models alone. TailingsIQ doesn't replace these engineering models but rather complements them by integrating real-world monitoring and observations. It can incorporate the results of empirical calculations (like a factor of safety from slope stability analysis) alongside live data from instruments and recent inspection findings, providing a more comprehensive understanding. In other words, it bridges the gap between theoretical predictions and actual performance.

- Manual Data Aggregation: Engineers often collect data from various sources (instrument databases, Excel sheets, lab reports) and manually combine them for analysis or reporting. This is labour-intensive and prone to errors or omissions. TailingsIQ replaces much of this manual effort with automated data integration.

- Fragmented Software Ecosystem: It's not uncommon to use one software for piezometer data, another for rainfall, another for geotechnical analysis, and basic tools like email and spreadsheets to tie things together. These disjointed systems create gaps where important correlations can be missed. TailingsIQ's unified platform overcomes these silos by having everything accessible in one place, which is a known best practice for improving safety and efficiency. As noted by Davies (2012), breaking down data silos can significantly improve dam management outcomes.

- Manual Document Review: Currently, finding information in old reports or correspondence can be like finding a needle in a haystack – one might have to read through dozens of PDF files to piece together a history of an issue. With TailingsIQ's AI-driven search and query, this process is accelerated by semantic search and direct answers. For example, if you want to know 'when was the last time there was erosion noted on the dam face,' the system can pinpoint the sections in past inspection reports that mention that, rather than you reading every report. This not only saves time but can reveal insights that might have been missed if someone didn't read the right document.

TailingsIQ's distinguishing feature is that it combines both quantitative data (numbers from instruments etc) and qualitative information (text from reports) into one interactive system. The Cross-Data Synthesis approach, powered by the AI Query Assistant, means that the platform can answer complex questions that involve both kinds of data. For example, existing anomaly detection software might alert users to an unusual piezometer reading, and a document system might let you search for 'piezometer' in reports; but TailingsIQ can connect the two, telling you which report discussed that specific piezometer's behaviour and what it said, all in one step. This level of synthesis is a novel advancement that current methods don't offer in an integrated way.

Ai techniques and engineering relevance

Under the hood, TailingsIQ utilises several AI techniques, but these can be described in straightforward terms as they apply to engineering use:

- Natural Language Processing (NLP): This is the technology that enables the Query Assistant to understand questions phrased in everyday language. Instead of forcing users to use specific keywords or rigid query formats, NLP allows the system to interpret the intent behind a question. For example, whether an engineer asks, 'Show me any abnormal readings in piezometers last week' or 'Were there any unusual piezometer trends recently?', the NLP component helps the system realise both queries are looking for anomalies in recent piezometer data. It does this by breaking down the sentence, recognising terms like 'abnormal' and 'piezometer,' and matching them to the data we have and the kind of answer needed.

- Machine Learning (ML) for Data Analysis: TailingsIQ can incorporate ML models to enhance its analysis of TSF data. In practice, this could mean using algorithms that learn from historical sensor data to forecast future values or detect anomalies. For instance, a machine learning model might learn the normal pattern of piezometric pressure changes over seasons and then flag when the current trend deviates significantly from that pattern. Similarly, ML could be used to classify the types of documents or even extract key parameters from a batch of reports (eg pulling out all recorded freeboard values from inspection documents).

- Contextual Retrieval (RAG) for Q&A: When the Query Assistant is answering a question, it uses a process known in AI as Retrieval-Augmented Generation (RAG). In simple terms, this means the system first retrieves relevant information from the database and document repository, and then generates an answer using that information as context. For example, if asked about 'piezometric trends versus design assumptions,' the system might retrieve the latest piezometer readings and the part of the design report that talks about expected pore pressures. It then composes an answer combining those pieces. The important point is that the answers are always backed by actual data or text from the TSF's records, not just a generic response. This makes the answers trustworthy and specific to the facility.

- Handling Uncertainty with Probabilistic Methods: In engineering, dealing with uncertainty is crucial – and AI is no different. TailingsIQ could integrate probabilistic approaches (like Bayesian methods) to quantify how confident the system is in a given prediction or alert. For example, if an ML model predicts that a certain instrument's behaviour is anomalous, the system could also calculate a probability (or confidence level) for that being a true issue versus a false alarm, taking into account prior knowledge such as the instrument's reliability or recent environmental conditions. This is analogous to how an engineer might weigh evidence, saying 'I'm 90 per cent sure this reading is problematic given the circumstances.' This aligns with established geotechnical practice of using factors of safety or reliability analyses – essentially providing a factor of safety for the AI's conclusions as well.

Human-in-the-loop oversight

Despite the advanced AI capabilities, TailingsIQ is explicitly designed to augment human expertise, not replace it. In the context of TSF management, human-in-the-loop (HITL) oversight is not just a preference but a necessity. The stakes are simply too high to allow fully automated decisions. Thus, the system is built with features that keep the engineer in charge:

- Transparency of Sources: Every answer or insight the AI provides comes with references to the source data or document. The system might show the actual sensor plot or quote the sentence from a report that it used to formulate an answer. This way, engineers can immediately verify the information. If something looks off, they can dive deeper or double-check the original source.

- User Feedback Mechanisms: TailingsIQ will allow users to give feedback on the AI's responses. If an answer was helpful or correct, they might confirm it; if it missed the mark, they can indicate that too. This feedback can be used to continuously improve the system's performance (for example, by adjusting the AI models or adding more training data for the NLP component). From an engineer's perspective, this is like coaching the AI to be a better assistant over time.

- Controlled Decision Points: The platform can be configured such that certain actions always require human confirmation. For instance, if the system were to generate a critical alert (say, 'possible impending slope failure'), it would be presented as a recommendation or warning to a human supervisor, who then validates it and decides on the action (like initiating an evacuation or inspection). It provides decision support, but humans provide the actual decisions.

- Audit Trails: Every query made, every AI response, and any subsequent human decision or override can be logged. This creates an audit trail for accountability and learning. In a post-incident analysis, one could review what the AI suggested and how the team responded. Or, during regular reviews, the team can look at patterns in what queries are being made and how effective the answers are, to identify areas for improvement. This logging is also useful for transparency with regulators, demonstrating that while AI is used, it is used under rigorous oversight.

By embedding human-in-the-loop oversight, TailingsIQ adheres to best practices for using AI in critical infrastructure management. It ensures that the technology serves as a powerful tool in the engineers' toolkit, enhancing their capabilities while respecting their judgment and experience. In essence, the system acts like a knowledgeable assistant and it defers to the human expert to interpret and act on that information.

CONCLUSION

Managing tailings storage facilities under today's conditions – with huge data volumes and strict standards like GISTM – is a formidable challenge. Traditional data management approaches in the mining industry are struggling to keep up, as vital information remains scattered and difficult to use effectively. TailingsIQ offers a new approach: an AI-enhanced platform that centralises diverse TSF data and provides intelligent tools to extract insights. By integrating everything from sensor readings to inspection reports into a unified system, it addresses the root problem of data fragmentation. Moreover, through its AI Query Assistant, which can understand context and retrieve relevant information (using methods like retrieval-augmented generation to combine data and documents), TailingsIQ enables a form of cross-data analysis that was not previously possible. This Cross-Data Synthesis capability means engineers can obtain answers that draw on the full breadth of the facility's knowledge base simply by asking questions in natural language.

The proof-of-concept implementation has shown that this concept is practical. Using modern web technology, the core architecture was built and tested with synthetic data, demonstrating that an integrated dashboard, document system, and AI assistant can work together as envisioned. The PoC validated the feasibility and highlighted the potential of TailingsIQ to improve how TSFs are overseen. Through examples in the PoC, improvements in situational awareness and efficiency were observed – confirming that such a platform can indeed help engineers spot issues earlier and perform their duties faster and more reliably.

It is important to emphasise that TailingsIQ is meant to support and enhance human decision-making, not replace it. The platform is designed with transparency and human control at its core, ensuring that engineers remain the ultimate decision-makers (consistent with the high responsibility and caution required in dam safety management). When AI is used, it operates under the guidance and review of qualified professionals, aligning with the industry's expectations for safety and accountability.

By transforming how critical TSF data is managed, accessed, and analysed, it helps bridge the gap between data and actionable knowledge. This directly supports better compliance with standards like GISTM and bolsters risk management practices, ultimately contributing to safer mine tailings management. The approach outlined in this paper – combining comprehensive data integration with AI-driven analysis in a user-centric platform – could redefine best practices in the field, enabling engineers and managers to make more informed decisions and protect people and the environment with greater confidence.

REFERENCES

Australian National Committee on Large Dams Incorporated (ANCOLD Inc), 2022. Guideline on the Use of Digital Technologies for Dam Safety Management, Australian National Committee on Large Dams.

Australian National Committee on Large Dams Incorporated (ANCOLD Inc), 2023. Guideline on Tailings Dam Monitoring, Australian National Committee on Large Dams.

Davies, M P, 2012. Tailings dam management for the 21st century, Proceedings of Tailings and Mine Waste 2012, Keystone, Colorado, USA.

Global Tailings Review, 2020. Global Industry Standard on Tailings Management (GISTM) [online], Global Tailings Review. Available from: <https://globaltailingsreview.org/global-industry-standard/>

Kruger, D, 2023. The digital transformation of tailings management, *CIM Magazine*.

Li, Y, Yang, C, Reis, R J and Zhang, L, 2021. Applications of artificial intelligence in mining and mineral processing: A review, *Minerals Engineering*, 173:107198.

Lopes, R and Ramos, L F, 2021. Machine learning applied to tailings dam monitoring: A review, *Geosciences*, 11(10):411.

Mooder, C, Hawley, M and Williams, D, 2022. Integrated data management systems for tailings storage facilities, Proceedings of Tailings and Mine Waste 2022, Denver, Colorado, USA.

Qiu, L, Zhang, K, Li, Z and Wu, J, 2024. Multi-source data fusion for tailings dam safety monitoring: A review, *Engineering Geology*, 331:107636.

Salih, B A, Jasim, A F, Al-Ansari, N and Khorshid, E A, 2025. Predicting Earth Dam Parameters Using Artificial Intelligence Techniques: A Review, *Sustainability*, 17(1):245.

Santos, T H and Celestino, T B, 2023. Artificial intelligence applications in dam safety: A state-of-the-art review, *Structures*, 54:104811.

The role of critical controls in safe tailings management

J Kilpatrick[1] and D O'Toole[2]

1. MAusIMM, Principal Engineer – Geotechnics and Mine Waste Engineering, SLR Consulting, Brisbane Qld 4000. Email: jkilpatrick@slrconsulting.com
2. MAusIMM, Technical Director – Geotechnics and Mine Waste Engineering, SLR Consulting, Townsville Qld 4810. Email: dotoole@slrconsulting.com

ABSTRACT

Risk management tools have become a common inclusion within the framework governing tailings management, translating technical risks into operational controls. These tools are used in the different stages of tailings management, including design, operation, governance, and closure. The identification of risks is accompanied by the identification of controls which may prevent or mitigate the risk from eventuating. In particular, 'critical controls' are defined by the Global Industry Standard for Tailings Management (GISTM; Global Tailings Review (GTR), 2020) as 'A control that is critical to preventing a potential undesirable event or mitigating the consequences of such an event'. A common challenge among industry is the balance of identifying the number of critical controls, developing performance standards for the controls that reflect technical and operational requirements, and resourcing to undertake the necessary risk management routines (review, verification, evaluation etc). This paper reviews the industry literature relating to critical control applications in the context of tailings management, connecting guidance material with industry applications. A reflection of the challenges associated with critical control development and management is also undertaken as part of this review. Further, it explores the processes relating to critical control identification and suggests their required implementation throughout each of the stages of a tailing's facility life cycle. Critical controls require ongoing review for the effectiveness of the control, and verification of its implementation on the ground. The paper concludes with practical recommendations for improving the integration of critical controls and tailings management, based on lessons learned from governance audits and reviews.

INTRODUCTION

The management of tailings storage facilities (TSFs) has become an integral component of a company's social license to operate after a number of global incidents which have occurred in recent years, resulting in catastrophic environmental impacts and fatalities. In response, increased standards and guidelines have been introduced to ensure that responsibility is undertaken by the asset owners to control the risks associated with TSF infrastructure and their operation. There are several key national and international guidelines that provide direction on risk management and critical controls specific to TSF management. These guidelines reflect evolving industry expectations and best practice principles for managing high-consequence risks associated with TSFs.

In accordance with the International Council on Mining and Minerals (ICMM) *Tailings Governance Framework* Position Statement (ICMM, 2016), the six key elements of management and governance necessary to maintain the integrity of TSFs and minimise the risk includes:

1. Accountability, Responsibility and Competency.

2. Planning and Resourcing.

3. Risk Management.

4. Change Management.

5. Emergency Preparedness and Response.

6. Review and Assurance.

As part of the governance framework, the Risk Management element is further defined to include requirements around risk identification, appropriate control regimes, and verification of control performance.

Control regimes can be identified as either preventing the failure from occurring or mitigating the impact if it does occur. As the mining industry continues to improve its approach to tailings and risk management, the concept of 'critical controls' has gained prominence as a key mechanism for managing the potential high-consequence events. Their role in tailings management is especially important given the complex and evolving nature of TSFs, which require robust systems to manage geotechnical, operational, and environmental risks across their life cycle.

Table 1 provides an overview of a subset of international and national guidelines that provide guidance on risk management and critical control impletion, including key principles and relevance to critical control management processes.

TABLE 1

Critical control literature.

Document/guideline	Key focus/relevant content	Application to critical control management (CCM)
GTR – Global Industry Standard on Tailings Management (GISTM) (Global Tailings Review (GTR), 2020)	Principle 10 requires implementation of quality and risk management systems across the TSF life cycle. Risk assessments must be updated at least every three years or when material changes occur.	Establishes requirements for levels of review, risk assessment frequency, and multidisciplinary involvement in TSF risk management. Defines 'critical controls'.
ICMM – Health and Safety Critical Control Management Good Practice Guide (International Council on Mining and Metals (ICMM), 2015a)	Outlines CCM process: control identification, performance requirements, verification, and governance. Originally developed for health and safety applications.	Framework widely applied to TSF risk management to structure critical controls and ensure verification and governance processes are in place.
ICMM – Critical Control Management Implementation Guide (ICMM, 2015b)	Provides detailed guidance for implementing CCM systems, expanding on risk assessment processes.	Practical guide for implementing critical control processes in TSF management, including monitoring and reporting.
MAC – Tailings Guide: A Guide to the Management of Tailings Facilities (Mining Association of Canada (MAC), 2021)	Framework for management of tailings facilities with emphasis on operational controls, monitoring, and response actions.	Strong alignment with CCM principles, focusing on-site-specific and governance-level controls, integrated into the tailings management system.
ANCOLD – Guidelines on Risk Assessment (Australian National Committee on Large Dams (ANCOLD), 2022)	Provides comprehensive guidance on the application of risk assessment methodologies for dam safety management.	While the guidelines do not explicitly use the term 'critical controls,' they emphasise the importance of identifying and implementing effective risk control measures to manage potential failure modes.
ICOLD – Bulletin 194: Tailings Dam Safety (International Commission on Large Dams (ICOLD), 2025)	Provides updated international guidance on tailings dam safety, incorporating lessons learned from recent failures. Emphasises a risk-based approach, governance structures, independent review, and performance monitoring. Highlights the importance of managing controls to address failure modes.	Strongly aligned with critical control management principles. Reinforces the need for clear control identification, implementation, verification, and ongoing governance processes across the TSF life cycle. Provides global benchmark practices for TSF risk and control management.

FAILURE MODE PROCESS AND RISK ANALYSIS

The identification and management of critical controls is a fundamental extension of the failure modes assessment and risk management process for TSFs.

A failure modes assessment is a structured risk identification process used to systematically identify the ways in which the facility could potentially fail, and to understand the consequences of those failures. There are a number of potential failure modes for TSFs, including slope instability, internal erosion and/or piping, surface erosion, seepage, structural failure, or overtopping (at a minimum). Understanding the potential failure modes that could result in adverse conditions of the facilities is an initial step required to influence the design, construction, and ongoing operation and management. This process typically involves identifying different components of the TSF, such as the embankment, decant structures, drainage systems, and monitoring equipment, and analysing how each component might fail under various conditions (eg overtopping, internal erosion, slope instability, equipment malfunction, or operational error).

The approach to the level of failure mode assessment and risk analysis varies with the stage or complexity of a project and/or facility and is typically managed through a Failure Modes and Effects Analysis (FMEA) or Failure Modes, Effects and Criticality Analysis (FMECA). The failure mode process can be extended to include risk analysis, where each identified failure mode is then assessed for its potential causes, likelihood of occurring, and the severity of its effects, or consequences, eg environmental damage, loss of life, operational disruption. This risk assessment process helps prioritise which failure modes pose the most significant threats and require the most attention.

Once the failure modes and associated risks are understood, controls can be identified to either prevent the failure from occurring (preventative controls) or to mitigate its impact if it does occur (mitigating controls). Among these, critical controls are highlighted as those that are essential to managing the highest-consequence risks. For example, a high-consequence failure mode like overtopping might lead to the identification of critical controls such as designing an adequate spillway or operating the facility in accordance with design (maintaining adequate freeboard). A summary of the processes for management of critical controls is outlined in Figure 1, adapted from the key elements listed within the Mining Association of Canada (MAC) *Tailings Guide: A Guide to the Management of Tailings Facilities* (MAC, 2021).

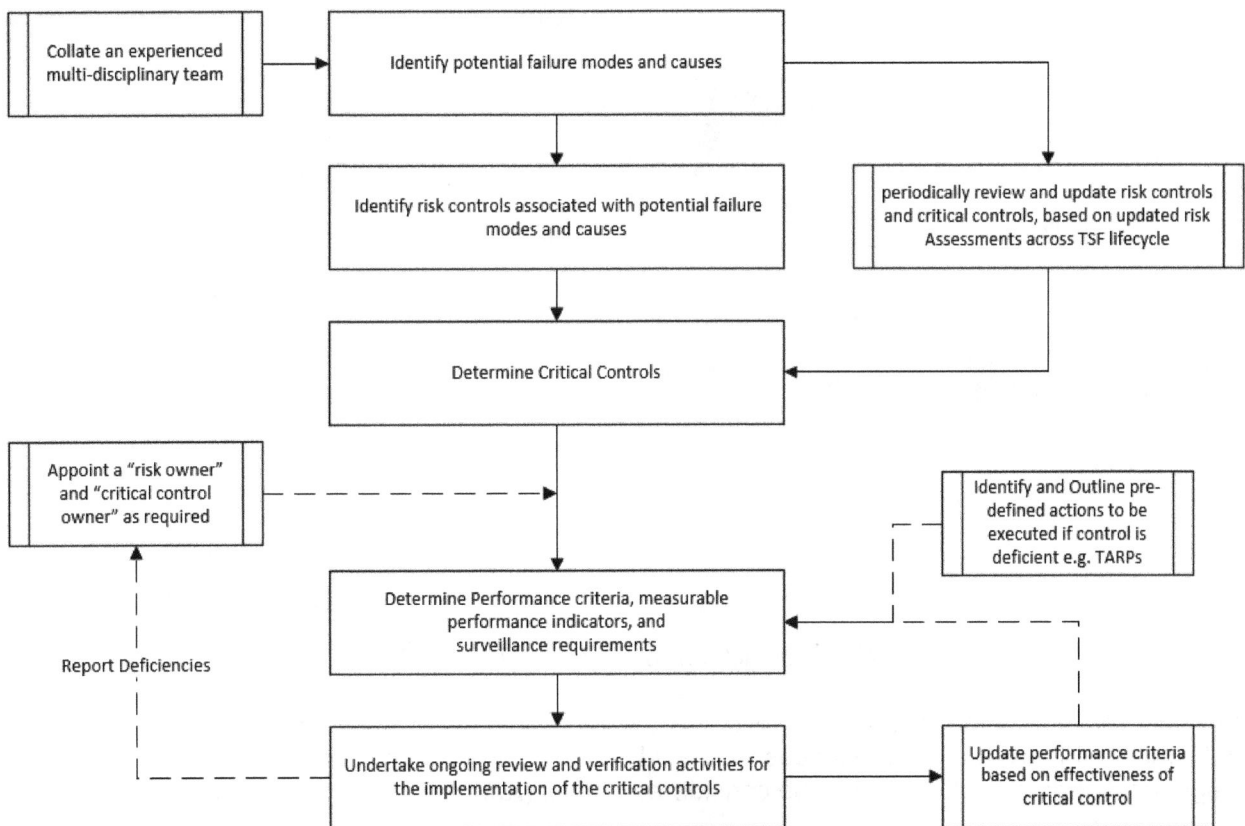

FIG 1 – Key elements relating to critical control management (adapted from MAC, 2021).

The framework emphasises ongoing review and verification activities to ensure that critical controls remain effective throughout the TSF life cycle. Performance criteria are updated based on the effectiveness of the controls, informed by monitoring, verification, and learnings from implementation. Periodic review and reassessment of risk controls and critical controls is also highlighted, particularly as the TSF evolves (eg through wall raises or changes in operational conditions), ensuring that risk assessments and controls remain current and fit-for-purpose.

IDENTIFYING AND MANAGING CRITICAL CONTROLS

According to the ICMM (2015a) *Health and Safety Critical Control Management Good Practice Guide* and the ICMM (2015b) *Critical Control Management Implementation Guide*, critical controls are those that must be implemented effectively to prevent a material unwanted event (MUE) ie dam failure, or to significantly reduce its consequences. The approach outlined by ICMM emphasises that not all controls are critical, and that identifying the few that matter most, allows for targeted implementation, monitoring, and assurance efforts. Clear criteria, multidisciplinary input, and context-specific analysis are vital to ensure that identified critical controls are fit-for-purpose and aligned with the facility's operational and risk profile. A flow chart is outlined within the guide to provide a structured and consistent approach to identifying which controls from the risk assessment should be labelled as 'critical' (Figure 2).

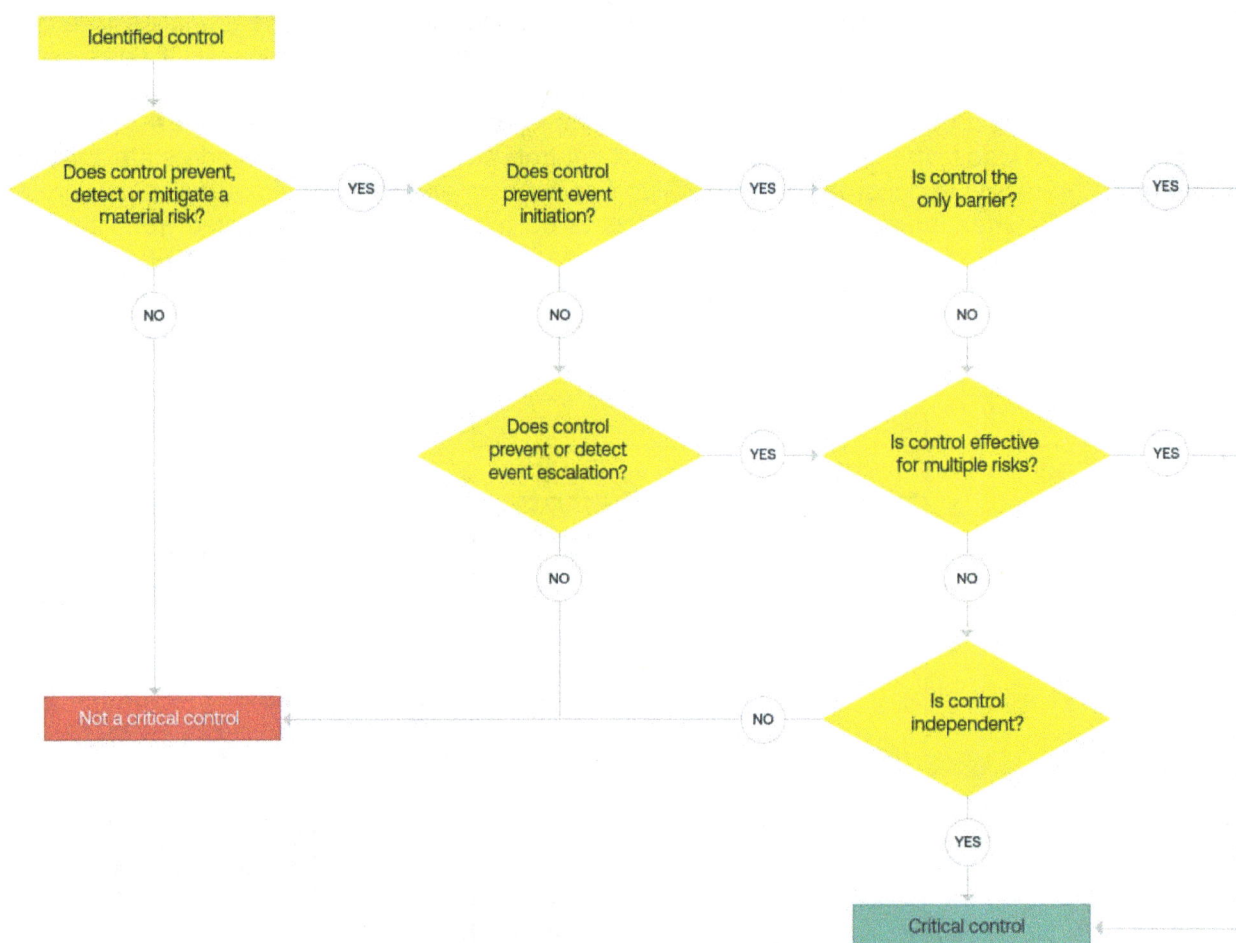

FIG 2 – ICMM Critical Control Decision Tree (ICMM, 2015b).

Although the flow chart is provided to guide the identification of the critical controls, the guide highlights that there are circumstances where additional questions may be asked to provide further insight on if certain controls should be considered critical (ICMM, 2015b), these include:

- Is the control crucial to preventing the event or minimising the consequences of the event?

- Is it the only control, or is it backed up by another control in the event that the first fails?

- Would the controls absence or failure significantly increase the risk despite the existence of the other controls?
- Does it address multiple causes or mitigate multiple consequences of the MUE?
- Can the critical control be implemented and actively monitored?

Monitoring activities associated with the implementation of critical controls include Critical Control Verification (CCV) and Critical Control Effectiveness testing (CETs). A CET evaluates whether the control remains effective at managing the risk event. It usually considers results from verification activities, potential control failures, and internal and external audits (which may highlight changes in industry standards). CCV's are the act of monitoring the control, which is required to verify that the critical control is being implemented as designed (ie outlined within the performance standard). These activities support the continuous improvement of the critical control, by identifying any deficiencies and planning remediation activities.

EXAMPLES OF CRITICAL CONTROL MANAGEMENT FOR TAILINGS FACILITIES

While industry guidelines provide structured processes for the identification and verification of critical controls, their practical application can vary significantly between sites, depending on facility-specific risks, operational contexts, and company governance practices. This section presents examples of critical controls observed during audits and reviews of TSFs across Australia associated with the risk of 'Dam Failure'. These examples provide insight into how operators are translating risk assessment outcomes into operational and governance controls, as well as the types of controls commonly identified as critical. The observations highlight both technical and operational controls, illustrating the diversity of approaches used to manage common failure modes and the importance of tailoring critical controls to site-specific conditions and evolving facility risks.

- Mitigating critical controls:
  - Effective emergency response (planning, procedures, and implementation).
  - Restricted Access and Exclusions Zones are in place to prevent unauthorised access to the facility or downstream areas.
- Preventative critical controls:
  - Site selection and review of tailings technology is undertaken to minimise impacts.
  - Facilities are designed and constructed in accordance with industry standards and guidelines.
  - Facilities are operated, maintained, and monitored in accordance with the Operations, Maintenance, and Surveillance (OMS) Manual.
  - Robust closure processes are in place to minimise impacts at the closure stage.
  - Routine audits and reporting processes are undertaken in relation to monitoring, instrumentation, and key facility performance indicators.
  - Competent, experienced, trained and appointed personnel are in place to manage the facility.
  - Change Management Processes.
  - A Tailings Governance Framework is in place to ensure a multi-layer approach to management is undertaken for the facility.
  - Undertake survey of the tailings beach.
  - Undertake tailings density monitoring, equipment, and reporting.
  - Undertake maintenance on tailings discharge pipelines.
  - Maintain the decant pond below determined level.

o Design of sufficient wall height and spillway capacity to accommodate extreme rainfall events.

Activities associated with reviewing the effectiveness of a control can include:

- Interviews with operational personnel or roles associated with critical control performance.

- Observations of the control implementation in field eg management and operation of dewatering/decant infrastructure.

- Review of incidents or hazards reported which may indicate potential deficiencies in the critical control.

- Review of external industry standards for potential changes or improvements in the critical control design.

- Review of internal standards, procedures, and audits which may indicate the need for potential updates to the critical control design.

Activities associated with verifying the implementation of the critical control are designed as part of performance standard. Typical examples are outlined in Table 2.

TABLE 2

Example critical controls and verification activities.

Example critical control	Example verification activities
Facilities are operated, maintained, and monitored in accordance with the Operations, Maintenance, and Surveillance (OMS) Manual.	• Verify that an OMS manual is available for the facility and routinely reviewed and updated as required. • Review subset of completed inspections undertaken by personnel in alignment with the OMS. • Review maintenance regimes and implementation in alignment with the OMS. • Review instrumentation and monitoring data management in alignment with the OMS.
Effective emergency response (planning, procedures, and implementation).	• Verify that an Emergency Action Plan (EAP) is available for the facility and routinely updated as required. • Review that desktop and field exercises are undertaken in accordance with the EAP, and the effectiveness of the exercises are discussed. • Review the in-field implementation of emergency evacuation (eg muster point) signage.
Restricted Access and Exclusions Zones are in place to prevent unauthorised access to the facility or downstream areas.	• Verify that site plans identify exclusion zones and restricted access areas. • Inspect that physical barriers (eg fencing, signage, locked gates) are installed, maintained, and fit-for-purpose. • Review records of site access control and incident reporting for breaches of restricted zones.
Facilities are designed and constructed in accordance with industry standards and guidelines.	• Review design reports and construction records to confirm compliance with relevant standards (eg ANCOLD, GISTM). • Verify that independent design reviews or third-party verifications have been completed. • Review quality assurance/quality control (QA/QC) documentation from construction activities.
Competent, experienced, trained and appointed personnel are in place to manage the facility.	• Verify that a training program and skills matrix is in place for all roles associated with the facility.

Example critical control	Example verification activities
	• Verify that key personnel (eg Responsible Tailings Facility Engineer (RTFE), Engineer of Record (EOR)) are appointed with clearly defined roles and responsibilities.
	• Review competency and training records to confirm technical qualifications and relevant experience are in place.
Change Management Processes.	• Review site-specific change management procedure is in place and implemented.
	• Verify that proposed changes (eg design changes, operational changes) are formally risk assessed and approved prior to implementation.
	• Review records of past changes, including lessons learned and close-out of actions from change processes.

IMPROVING THE IMPLEMENTATION AND MANAGEMENT OF CRITICAL CONTROLS FOR TAILINGS FACILITIES

While the identification of critical controls is a key outcome of the risk assessment and failure modes process, their effective implementation and ongoing management remains a common challenge in TSF governance. Improving the systems, processes, and behaviours that support critical control management is essential to ensure controls operate as intended and remain effective over the life of the facility. Common challenges include:

- Critical controls implemented lack connection with the underlying facility FMEA or are too generic in nature.

- Difficulty in the identification and prioritisation of critical controls during the risk assessment phase.

- Inadequate design of the critical control performance standards.

- Lack of planning relating to actions when the critical control is found to be deficient.

- Inadequate adequate ownership of critical controls.

- Verification activities are not appropriate.

- Lack of communications associated with control performance.

Clear connection to FMEA and re-evaluation in alignment with review of the FMEA

The identification of critical controls must be directly informed by the FMEA to ensure that controls are specifically targeted to manage the most significant failure modes and their causes. A key risk in critical control management is the tendency to identify controls that are generic, procedural, or disconnected from the technical failure mechanisms of the facility. Controls that are not clearly linked to the FMEA may fail to address the fundamental drivers of failure, reducing their effectiveness. The FMEA provides a structured framework to systematically identify potential failure modes, their initiating causes, and the consequences of failure. Critical controls should therefore map directly to these failure pathways. This traceability between the FMEA and the identified critical controls is essential for clarity, transparency, and defensibility of the risk management approach. It also enables effective performance monitoring and verification, as each control can be measured against its intended function within the failure mode analysis. An example may be where an FMEA has identified that critical controls for the risk of dam failure is an exclusion zone for blasting and blast monitoring. If this critical control is grouped generically with additional monitoring requirements and placed within the Operations, Maintenance and Surveillance (OMS) Manual, this can result in an overarching critical control of 'The facility is operated, maintained, and monitored in accordance with the OMS Manual'. It is important that the actual critical control of exclusion zone and blast monitoring is clearly documented in the OMS, and that the performance criteria for these critical control are appropriate, as identified initially within the FMEA.

While some controls will be specific to individual failure modes or site conditions, others may be more broadly applicable and embedded within corporate governance frameworks, such as 'A Tailings Governance Framework is in place to ensure a multi-layer approach to management is undertaken for the facility'. As outlined by the MAC (2021):

> The implementation of appropriate corporate governance, including the implementation of a tailings management system, is a form of critical control. However, most other critical controls are more specific to the risks associated with a given tailings facility. Thus, some critical controls can be implemented and monitored at a corporate level, while others are implemented and monitored at the site-specific level.

Collaborative approach to critical control identification

Identifying the critical controls is a function of the confidence of the risk assessment elements and outcomes. First, uncertainty regarding the likelihood or probability of failure for a specific failure mode can compromise the prioritisation of controls, especially where data is limited or failure mechanisms are not well understood. Second, uncertainty in the assigned consequences of failure may distort the perceived severity of risks, influencing which controls are deemed critical. Third, it is essential that all potential failure modes and hazards relevant to the facility are comprehensively identified and analysed to ensure no risks are overlooked. Fourth, the dynamic nature of TSFs such as wall raises, evolving deposition strategies, and exposure to changing environmental or climatic conditions can alter the risk profile over time, requiring that critical controls be periodically reassessed to remain effective and relevant. As such, identifying critical controls is not a one-off task but a dynamic process that must account for technical uncertainty, operational complexity, and the evolving context of the facility.

Due to the multidisciplinary nature of tailings management, the involvement of individuals from a range of roles and disciplines is essential to developing a risk management plan, particularly in the identification and implementation of risk controls and critical controls. Technical specialists, in fields such as geotechnical, civil, hydrology, hydrogeology, environmental science and more play a role in identifying potential failure modes, assessing associated consequences, and designing technical controls that address both structural and operational risks. In addition to the design and engineer of record professionals, operational personnel, including site managers, operators, and maintenance staff contribute practical insights into how controls are implemented and function on the ground. Their participation ensures that the identified controls are not only technically sound but also realistic, achievable, and aligned with day-to-day operations. Risk and governance professionals provide structure to the risk management process and oversee the documentation and tracking of controls through risk registers or critical control management systems. Management and leadership teams also have a responsibility to participate in and support the development of the risk management plan as their involvement ensures that adequate resources are allocated to implement and maintain critical controls. Having people of multiple experience levels, professional backgrounds, and roles involved in the process of critical control identification provides valuable context for those that will continue on in the process as risk control owners and undertaking verification activities.

Critical control performance designs

The objectives and performance requirements for each critical control require definition to allow for activities to occur which verify the critical control implementation. In the context of TSFs, where critical controls often manage complex geotechnical, hydrological, or operational risks, performance standards must be founded by engineering principles and site-specific design criteria. Performance standards define the specific objective for the critical control and describe the expected performance outcomes. For example, specifying a minimum freeboard level to prevent overtopping, a maximum pore pressure threshold to maintain slope stability, or evacuation exercises for emergency response plans. Without technically derived standards, there is a risk that critical controls become too subjective or administratively focused, lacking the precision needed for effective implementation. In addition, the industry needs to ensure these thresholds are measurable, technically justified, and reviewed periodically as conditions evolve, which is of particular importance for TSFs. In addition, GISTM (GTR, 2020) outlines that the 'OMS Manual should describe the performance indicators and *criteria for risk controls and critical controls, highlighting the link between OMS activities and critical controls management which should be developed to reflect site-specific conditions'* (tailored to the site).

Responding to critical control performance

Establishing critical controls is only one part of the risk management process; their effectiveness ultimately depends on how they are monitored, managed, and responded to in practice. A key aspect of critical control management is having clearly defined actions in place to respond when control performance deviates from expectations or thresholds are exceeded, to ensure a timely and effective response that mitigates potential consequences. The response must be well-structured, role-specific, and consistently understood across all parties involved in the operation of the tailings facility. To ensure effectiveness, these role-specific actions must be clearly described in operating procedures, training materials, and emergency response plans. Ambiguity in responsibilities or delayed responses can compromise the integrity of the facility and lead to escalating risks. The development and implementation of Trigger Action Response Plans (TARPs) are an important mechanism to formalise this process, providing a clear link between risk controls, performance monitoring, and operational response. TARPs provide pre-defined trigger levels for performance criteria that are based on the risk controls and critical controls of the tailings facility. The trigger levels are developed based on the performance objectives and risk management plan for the TSF.

Designated and trained critical control owners

An ongoing challenge in the effective implementation of critical control management for TSFs is the identification and appointment of suitably qualified personnel to act as control owners. Control owners are responsible for ensuring that critical controls are implemented, maintained, and verified over the life of the facility. However, given the complex and dynamic nature of TSFs, the number of critical controls identified for a single facility can be substantial in some cases. This can create resourcing restrictions or inappropriate appointments, particularly in an environment where the availability of experienced and competent personnel in tailings and dam engineering is acknowledged as limited. Without qualified and trained control owners, there is an increased risk that critical controls may not be effectively managed or verified, potentially undermining the integrity of the risk management framework. This highlights the need for the integration of risk training for roles which contain the potential and retain technical capability in tailings management, alongside clear governance structures to support control ownership where specialist resources are limited. Asset Owners are required to conduct regular training for staff, contractors, and consultants to ensure they understand critical control functions, triggers, and response requirements.

Verification activities

Critical control verification activities must be assigned, measurable, and integrated into routine operational practices to ensure that they are sustainable over the life of the facility. Good practice involves establishing clear performance criteria for each critical control, supported by objective evidence such as inspection records, monitoring data, or maintenance logs. For example, a well-designed verification activity for a critical control relating to seepage management may involve routine inspections of toe drains, combined with regular instrumentation data (eg groundwater bore level monitoring) to confirm that the risk of seepage is minimised and within acceptable limits. In contrast, poor verification practice may involve generic sign-offs such as 'checked as per procedure' without any supporting evidence or measurable parameters, making it difficult to demonstrate that the control is effective or to identify deterioration over time. Integrating verification activities operationally ensures that they become part of day-to-day site practices, embedded within the OMS Manual.

Communications

Identifying a control deficiency without clear and transparent communication mechanisms can result in delayed responses, increased exposure to risk, and missed opportunities for learning and improvement. To support this, many organisations are adopting dashboards or critical control monitoring technology that provide real-time visibility of critical control performance. This includes status of verification activities, and any deficiencies identified. These dashboards allow for the clear tracking of critical control health. Importantly, they provide a platform for escalation, enabling site personnel, management, and governance teams to monitor risk controls across multiple facilities or portfolios. The tracking of remediation actions including assigning responsibilities, due dates, and closure evidence is an important component of this process, ensuring that deficiencies are not only identified but also resolved. Without such systems in place, there is a risk that critical control failures are recorded but not effectively actioned, or that lessons are not shared across the organisation. Embedding these communication processes into governance frameworks supports a transparent safety culture and enhances critical control management system.

CONCLUSION

Critical controls are a key mechanism for translating technical risk assessments into operational risk management tools for TSFs. Their identification should be directly informed by failure modes and risk assessment processes to ensure they address the most significant failure scenarios and consequences. This paper has outlined that while identifying critical controls is important, the greater challenge often lies in their implementation, verification, and ongoing management. To be effective, critical controls must be measurable, integrated into operational systems, and supported by clear verification activities and response actions. The communication of control deficiencies, supported by dashboards and action tracking systems, is necessary to ensure timely response and accountability. Embedding critical controls throughout the TSF life cycle, from design to closure, requires clear ownership and defined processes, with an overall focus on continual improvement. The recommendations provided in this paper aim to support more consistent application of critical control management practices and improve confidence in their application for tailings facility risk management.

REFERENCES

Australian National Committee on Large Dams (ANCOLD), 2022. ANCOLD Guidelines on Risk Assessment, Australian National Committee on Large Dams, Australia.

Global Tailings Review (GTR), 2020. Global Industry Standard on Tailings Management (GISTM) [online], Global Tailings Review. Available from: <https://globaltailingsreview.org/global-industry-standard/>

International Commission on Large Dams (ICOLD), 2025. Bulletin No. 194: Tailings Dam Safety, International Commission on Large Dams, Paris.

International Council on Mining and Metals (ICMM), 2015a. Health and Safety Critical Control Management Good Practice Guide, International Council on Mining and Metals, London.

International Council on Mining and Metals (ICMM), 2015b. Critical Control Management Implementation Guide, International Council on Mining and Metals, London.

International Council on Mining and Metals (ICMM), 2016. Position Statement: Tailings Governance Framework, International Council on Mining and Metals, London.

Mining Association of Canada (MAC), 2021. A Guide to the Management of Tailings Facilities, version 3.2, Mining Association of Canada, Ottawa.

Towards the development of geotechnical engineers as TSF Engineers of Record

A H Kirsten[1], L H Kirsten[2], I Campello[3], V Gepilano[4] and D Johns[5]

1. Senior Geotechnical Engineer, KCB Australia Pty Ltd, Brisbane Qld 4000.
   Email: akirsten@klohn.com
2. Senior Geotechnical Engineer, KCB Australia Pty Ltd, Brisbane Qld 4000.
   Email: lkirsten@klohn.com
3. Senior Geotechnical Engineer, KCB Australia Pty Ltd, Brisbane Qld 4000.
   Email: icampello@klohn.com
4. Senior Geotechnical Engineer, KCB Australia Pty Ltd, Brisbane Qld 4000.
   Email: vgepilano@klohn.com
5. Senior Geotechnical Engineer, KCB Australia Pty Ltd, Brisbane Qld 4000.
   Email: djohns@klohn.com

ABSTRACT

The Engineer of Record (EOR) concept originated in North America and has been used in private and public works construction since the early 20th century. For mine process tailings dams and tailings storage facilities (TSFs), the EOR concept gained prominence over the past decade – a period marked by several catastrophic TSF failures. The role, function, and responsibilities of the EOR for TSFs were formalised in the Global Industry Standard on Tailings Management (GISTM; Global Tailings Review (GTR), 2020).

Following the introduction of the GISTM, there has been increased demand for engineering consulting firms to provide EOR services, including the assignment of suitably qualified and experienced engineers as EOR representatives. The current demand for EOR representatives greatly exceeds supply. As a result, EOR representatives are often accountable for multiple TSFs. The combination of high workloads and the long-term nature of EOR appointments compromises their ability to adequately meet duty-of-care obligations. A potential reluctance to accept the liability associated with EOR appointments further constrains the available pool of EOR representatives.

To address this supply constraint, in addition to recruiting talent into the mining and consulting industries, structured, purposeful, and continuous skills development of prospective EOR representatives is necessary. The strong demand and long life cycle of TSFs – often spanning several decades – together present a clear opportunity to develop aspiring TSF engineers into competent EOR representatives.

A framework to support the systematic development and training of geotechnical engineers to become suitably qualified and experienced TSF EOR representatives, is presented in this paper. Over time, industry-wide implementation of structured development programs such as the one proposed herein should help increase the supply and capability of EOR representatives and improve the quality and sustainability of EOR appointments.

The proposed framework also enables aspiring EORs to gauge their development towards EOR status and monitor their progress in fulfilling the necessary requirements.

The authors have developed and applied the framework to several TSFs since the publication of the GISTM. While the quantitative scores and weightings for the various criteria are not included in this paper, as they remain under development, the authors are prepared to share them on a personal basis to promote broader acceptance of the proposed framework and procedures.

INTRODUCTION

TSFs are unique, large-scale earth structures which construction, operation, and closure often span several decades. Their development and operation involve failure risk evaluations, ground investigations, environmental impact assessments, engineering design, water management, construction, operations, post-operational rehabilitation, closure, and post-closure maintenance, with performance monitoring and calibration throughout.

Due to their complexity and development needs, TSFs are typically designed by multidisciplinary engineering teams, comprising dedicated project, engineering, and construction management skillsets, under the direction of an appropriately qualified and experienced design or engineering manager.

Engineer of Record

The Engineer of Record (EOR) concept is widely recognised as having originated in North America, where it has been applied in private and public civil works construction (eg roads, bridges, buildings, water dams) since the early 20th century. In this domain, the design engineer and/or engineering manager and/or EOR were often considered equivalent roles (Tailings Engineer of Record Task Force (TEORTF), 2017).

In conventional mining practice, engineering consulting firms are engaged to provide EOR services and, in turn, to nominate suitably qualified and experienced engineers as EOR representatives.

A TSF EOR representative is typically, though not necessarily, a suitably qualified and experienced geotechnical engineer responsible for overseeing all aspects of a TSF's engineering requirement. This includes ensuring technical integration and quality assurance across all disciplines and stages of the facility's life cycle. Ideally, the EOR representative leads and directs a multidisciplinary team of engineers (eg geotechnical, hydraulic, civil, mechanical, structural, environmental, and mining) and scientists (eg geologists, seismologists, hydrogeologists, hydrologists, geochemists, and environmental scientists).

The EOR concept gained global prominence in the mining industry following the 2014 Mount Polley TSF failure. Notably, five different individuals reportedly occupied, in succession, the EOR representative role at this facility in the four years preceding the failure. This incident, along with the challenges of adequately staffing the EOR representative position, served as a catalyst for the mining industry and regulators to re-evaluate the EOR concept and strengthen governance frameworks for effective tailings management.

Since the Mount Polley failure, several other catastrophic TSF failures have occurred, further highlighting the need for improved and enduring TSF governance. The EOR's role has become pivotal in ensuring the safety and sustainability of TSFs, and increased scrutiny has prompted widespread reflection and efforts to clarify and strengthen this role.

Continued formalisation and standardisation of the EOR's role in TSF development and governance (TEORTF, 2017; Knight Piésold Consulting, 2021; Small and Witte, 2018) culminated in the publication of the GISTM.

Demand for engineer of record services

The events described above have resulted in a global shortage of TSF EOR representatives which may widen due to several widely encompassing demands, including:

- The introduction of the GISTM, which has intensified the need for adequate oversight at both operating and decommissioned TSFs.

- The ongoing energy transition, which is increasing demand for critical minerals and driving the development and commissioning of new mining operations.

- EOR representatives are often tasked with overseeing multiple TSFs, leading to workloads that may compromise their ability to adequately meet duty-of-care obligations. The enduring nature of EOR appointments adds further burden, straining both individual and collective capacity. A reluctance among some consulting firms and engineers to accept these demanding roles may exacerbate the shortage.

The demand for EOR services is amplified by mining industry-specific factors, such as:

- The persistent difficulty in attracting, suitably developing, and retaining enough candidates to meet growing demand within the mining industry.

- The significant time investment and professional commitment required to develop the experience, technical skills, and maturity needed to function as a TSF EOR representative. Industry consensus suggests that EOR representatives should possess at least a bachelor's degree (or equivalent) in civil or geotechnical engineering, along with approximately 15 years of relevant industry experience.

- A notable lack of universally applicable and transferable systematic training programs across jurisdictions, regions, and projects. In the absence of such programs, potential EOR candidates often rely on circumstantial opportunities and personal initiative to develop the required skills and experience. This not only results in professional development gaps but also limits the pipeline of suitably qualified and experienced professionals.

The first two industry specific factors underscore the need for strategic industry initiatives to attract talent and systemise EOR training and development. However, the systematic development of future EOR representatives presents an opportunity over which mining industry stakeholders hold the most effective immediate leverage.

Proposed engineer of record development pathway

This paper presents a development framework conceived to support the mining and consulting industries in progressively upskilling tailings dam engineers – specifically through the geotechnical engineering pathway to becoming TSF EOR representatives. While other pathways may exist, they are not considered in this paper.

In essence, the proposed framework comprises two key components: a development pathway and a skills and competency evaluation matrix.

The development pathway connects a TSF's failure risk category to the competency and experience level of EOR representatives. This ensures that less experienced representatives oversee facilities with lower failure consequences, and more experienced representatives oversee facilities with higher failure consequences. Regardless of an individual's relative progress along the pathway, each EOR representative is mentored or supervised by someone with more experience and expertise.

The accompanying skills and competency matrix outlines key technical and non-technical attributes required of EOR representatives. It provides a structured method for evaluating competencies based on subject exposure, complexity, responsibility, and duration. The goal is to guide the development of EOR representatives by addressing gaps over time and ensuring objective, transferable progression along the pathway.

By leveraging the demand for EOR services and the unique opportunities that EOR appointments provide, the adoption of a framework such as the one proposed herein should support the development of EOR representatives at all career stages – not only enabling continuous professional growth but also improving the supply rate and overall capability of prospective EOR representatives.

BASIS OF THE EOR DEVELOPMENT FRAMEWORK

This section outlines the principles and key elements underpinning the TSF EOR development framework. The framework aligns with a widely accepted understanding of the TSF EOR's purpose and function, incorporates the core principles and values associated with the role, and defines the skills and competencies required to fulfil it effectively.

EOR role and responsibility

Table 1 summarises the definition, responsibilities, and the purpose of the EOR representative role, as defined by industry standards including the GISTM and the ICMM Tailings Management Good Practice Guide (ICMM, 2021).

Building on the responsibilities outlined in the table, EOR representatives must demonstrate technical competence and experience commensurate with the operational complexity and risk profile of a TSF.

TABLE 1

Tailings storage facility EOR – definition, responsibility and purpose.

Definition	• TSF Owners are accountable for the TSFs as part their mineral extraction processes.
	• TSF Owners appoint TSF EORs in-house. Alternatively, suitably qualified engineering consultancies provide TSF EOR services (EOR).
	• If assigned to a consultancy, the EOR is a qualified engineering Company responsible for providing assurance that the tailings facility/dam is designed, constructed, and decommissioned with appropriate concern for the integrity of the facility, in line with the applicable regulations, guidelines, codes and standards. The Company is represented by an individual; however, the contractual and financial responsibilities for engineering activities is between Company and the TSF Owner. The EOR may delegate responsibility but not accountability (GTR, 2020).
	• The Company must nominate a senior engineer, approved by the TSF Owner, to represent the Company as EOR, and verify that the individual has the necessary experience, skills and time to fulfil the role and to assume accountability. The EOR must be competent and have experience appropriate to the Consequence Classification and complexity of the tailings facility (GTR, 2020).
EOR responsibilities	• EORs are responsible for the TSFs under their oversight and delegate the responsibility to suitably qualified and experienced engineers as their representatives (EOR representatives) (GTR, 2020).
	• EOR representatives must be competent and have experience appropriate to the Consequence Classification and complexity of the tailings facility (GTR, 2020).
	• The overarching responsibility for understanding the design concept and how it applies to successful construction and operation, resides with the individual appointed as EOR (ICMM, 2021).
EOR duty/role	• EORs ensure that the business and operational decisions made by TSF Owners about their TSFs are informed by engineers who understand the design principles and technical limitations of TSFs and the impact of changes on their safety and performance.
	• EOR representatives assure that TSFs are designed, constructed, operated, decommissioned and closed with due consideration for their integrity and in compliance with relevant regulations, guidelines and standards.
	• EOR representatives build and lead multidisciplinary TSF engineering teams.

EOR principles and values

For EOR representatives responsible for ensuring safety, regulatory compliance, and long-term integrity, professional conduct is fundamental to effective TSF management. Ethical principles serve as reference points to guide engineers through complex technical decision-making and stakeholder engagement. Engineering governing bodies, such as the Board of Professional Engineers Queensland, offer professional conduct guidance that emphasises values such as integrity, competence, public welfare, and environmental stewardship.

The proposed framework highlights that professional TSF management extends beyond technical competence to encompass values such as trustworthiness, accountability, and respect for societal and environmental impacts. By anchoring their actions in these principles, EOR representatives can build credibility, inspire confidence, and uphold the standards of engineering practice.

The guiding principles and associated values outlined in Table 2 provide a structured reference for both developing and maturing EOR representatives as they carry out their responsibilities with integrity and excellence. Each principle addresses a key aspect of professional conduct — from maintaining technical expertise to fostering collaboration and promoting sustainable practices. Together, these elements form a good foundation to support EOR representatives in ensuring the safety and reliability of TSFs throughout their life cycles.

TABLE 2

Tailings storage facility EoR – Principles and core values.

Principle 1: Execute work with integrity and adhere to professional norms

Core values:

- Trustworthiness and reliability
- Guided by an informed conscience (concern for public health and safety)

Principle 2: Apply knowledge and skills to protect public well-being and the environment

Core values:

- Respect and value affected communities
- Engage all stakeholders inclusively
- Balance the needs of current and future generations

Principle 3: Continuously improve skills and support the growth of colleagues

Core values:

- Commit to maintaining and developing knowledge
- Foster excellence and promote the advancement of the profession
- Represent areas of expertise with objectivity

Principle 4: Demonstrate professionalism in all aspects of work

Core values:

- Approach problems with objectivity
- Act as a problem solver
- Implement practical and effective solutions
- Take ownership and responsibility

Principle 5: Lead effectively and communicate with integrity

Core values:

- Uphold the honour and standing of the profession
- Communicate clearly, effectively, and honestly with all stakeholders

EOR knowledge base and technical tools

TSF EOR representatives must possess strong technical expertise across disciplines such as geology, geotechnical engineering, hydrology, and risk management. Their skills should align with the complexity and risk profile of the TSF to effectively guide all stages of its life cycle – including design, procurement, construction, operation, and decommissioning. Table 3 summarises the key technical competencies aspiring engineers must develop to become competent EOR representatives. These are grouped into five key areas: site characterisation, design, procurement and construction, operation and governance, and general skills.

TABLE 3

Tailings storage facility EoR – knowledge base.

Site characterisation	• Advanced geotechnical investigation techniques and state-of-practice testing methodologies and facilities and services
	• *In situ* or field testing technologies
	• Fundamental knowledge of general geology, structural geology, geophysics and seismology
	• Characteristic physical and mechanical properties of tailings particulate
	• Short and long-term weather and climate characteristics
	• Basic understanding of hydrogeological concepts
	• Familiarity with basic metallurgical concepts
	• Undermining potential
	• Natural and mining induced seismicity
Design	• State-of-practice approaches for TSF design and adherence to general design guidelines
	• Basic and advanced soil mechanics theory (eg critical state soil mechanics)
	• Engineering hydraulics and fluid dynamics, including pipe flow theory
	• Geotechnical modelling theories and techniques
	• Hydrology theory and related concepts
	• Statistical and probabilistic analysis
	• Options assessment techniques
	• Techniques for qualitative and quantitative risk assessments
	• Use of performance surveillance and monitoring instruments
	• General design philosophies
	• Seismic hazard assessment and its application to TSF design
Construction	• Best practices for construction monitoring, including quality control and quality assurance
	• General theories of engineering and construction management.
	• Understanding international and local contract law relevant to TSF projects
Operation	• State-of-practice approaches for TSF operations management
	• Compilation and application of operations manuals
	• Knowledge of environmental laws and regulations within the applicable jurisdiction
Governance	• Comprehensive understanding of the GISTM
	• Familiarity with practice guidelines, laws and regulations, and technical standards relevant to tailings dam ownership
	• Familiarity with TSF failure case studies
General	• Proficiency in report writing and public speaking
	• Competency in project management and computer technology
	• Skills in public liaison and stakeholder engagement

Site Characterisation requires a comprehensive understanding of both apparent and potential site conditions. EORs should be reasonably proficient in site investigation, field and laboratory testing, geology, geophysics, seismology, and metallurgical processes. This expertise enables them to evaluate site-specific factors that are critical to resilient TSF design. Unforeseen or poorly understood site conditions lead to compromised TSF integrity.

High-quality *Design and Construction* are critical for delivering safe and efficient TSFs. EORs must apply knowledge of soil mechanics, hydrology, short and long-term weather and climate, hydrogeology, dam hydraulics, statistical analysis, and seismic hazard assessment to develop designs that address operational needs and potential risks. During construction, they assure quality by monitoring compliance with design specifications. Their monitoring helps mitigate the risks associated with poor execution and supports the long-term safety and performance of the TSF.

Operation and Governance focus on maintaining TSF safety during ongoing operations and incremental construction phases. EORs monitor operations to ensure compliance with environmental regulations and global standards such as the GISTM. They also advise TSF owners on critical decisions that may affect safety or performance. Strong general skills — including report writing, communication, and engineering management — are essential for coordinating stakeholders and ensuring alignment among all parties involved in TSF management.

The EOR representative's technical toolbox comprises the practical and operational resources — tools, methodologies, equipment, and technologies — applied in fulfilling the EOR role. This toolbox naturally evolves with advances in technology and the availability of new resources, but it is effective only when paired with a robust knowledge base.

The knowledge base and technical toolbox form a complementary system that enables EOR representatives to conceptualise solutions, make informed decisions, and think critically. Together, they provide the foundation for evaluating an EOR representative's competency level as part of the development pathway.

Table 4 outlines some key elements of the technical toolbox, which are applied within the broader knowledge base.

TABLE 4

Tailings storage facility EoR – Technical toolbox.

General software	• GIS software
	• Geological data logging and modelling
	• CAD 2D and 3D
	• Operations planning and control
	• Project planning and resourcing
	• Database manipulation and management
	• Coding and/or spreadsheet proficiency
	• Geometric modelling
	• Quantitative risk assessment software
Equipment	• *In situ* materials testing techniques, apparatus and operator skill
	• Soils laboratory testing techniques, apparatus and operator skill
	• TSF performance monitoring apparatus, techniques and data transposition

Analytical methods/ software	• 2D and 3D seepage analyses software and user's skill
	• 2D and 3D limit equilibrium stability software and user's skill
	• Finite element/difference numerical software and user's skill
	• Dam break modelling software and user's skill
	• Flood routing software and user's skill
	• Water balance modelling software and user's skill
	• Pipe flow software and user's skill
	• Open channel flow software and user's skill
	• Mathematical software including risk assessment software and user's skill
	• Surface and piping erosion calculation software and user's skill
General	• Quantitative decision-making and risk management methods and user's skills
	• Application of guidelines, codes of practice, technical standards and regulations
	• Contract law fundamentals, application and user's skill
	• Project, contract and quality management fundamentals and application and user's skill

EOR DEVELOPMENT PATHWAY

Table 5 outlines a structured development pathway for engineers progressing towards the role of EOR for TSFs.

The development process ideally begins after a civil engineering degree is obtained, followed by the first three to four years focused on gaining broad experience in general engineering practice. Following this foundational period, the engineer transitions into tailings engineering, initially working as an EOR-in-training (EOR TR) under the technical guidance of an experienced EOR for a few years. The EOR TR then progresses to the role of Deputy EOR (DEOR) for a low-consequence TSF, serving in this capacity for some time.

During this phase, the DEOR supports the EOR in overseeing and leading all engineering and related aspects of either a new TSF development or an existing operation. The level of experience required for an EOR aligns with the TSF's failure consequence classification, as defined in the GISTM.

Upon completing a stint as DEOR, the individual transitions into the EOR role, while the outgoing EOR assumes a deputy role for a TSF in a higher failure consequence category. At the same time, a new DEOR steps in to support the EOR, creating a system of continuous mentoring. The incoming DEOR is paired with a more experienced EOR to receive ongoing guidance throughout the next two-year term. This structured progression enables engineers to steadily assume greater responsibility and refine both their technical and leadership skills.

Over approximately 16 years, this systematic approach cultivates the expertise, maturity, and confidence required for graduate engineers to eventually serve as EOR representatives for extreme-consequence TSFs. By fostering growth across progressive levels of competency and experience, the model provides a consistent development path while maintaining high standards and rigorous oversight at each stage.

For the pathway to be effective, it must be supported by a mechanism for evaluating the evolving competencies of EOR representatives. This need is addressed in the following section through a skills and competency assessment framework.

TABLE 5

Tailings storage facility EoR development pathway for geotechnical engineers.

Level #	Status/role	Technical responsibility	ANCOLD/CDA alignment	GISTM alignment	Practice level	Exposure/experience (yrs) Min[2]	Cum	EOR role[3]	Responsibility[4]	Technical mentor
3	Graduate civil engineer[1]	General civil engineering design and construction monitoring.	Building a foundational understanding of tailings management, geotechnical and hydrotechnical concepts, safety protocols and monitoring practices.	Familiarisation with the 15 principles of tailings management, focusing on principles related to understanding the TSF facility, emergency preparedness, and early-stage safety training.	Supervised	2	2	-	-	DEOR 1 [6]
4	TSF design engineer in training	Singular TSF design/assessment tasks. Construction monitoring. TSF performance monitoring.			Supervised	1–2	3–4	-	-	DEOR 2
5	TSF design engineer	Complex, singular TSF design/assessment tasks. Design criteria and, construction and performance monitoring specifications.	The responsible technical person must verify that all investigation, design and operation activities are appropriate and properly carried out (ANCOLD). The design engineer is responsible for the design of the dam up to and including construction (CDA).		Supervised complex tasks	1–2	5–6	EOR TR[6]	Significant/High	EOR 1
6[5]	Senior TSF engineer	Multiple TSF design/assessment tasks. Design criteria and, construction and performance monitoring specifications.			Unsupervised	1–2	7–8	DEOR 1	Significant/High	EOR 2
7	Senior TSF engineer	Complex, multiple TSF design/assessment tasks and internal review. Dam safety audits and reviews. *Holds ultimate responsibility for the design, construction and safety of TD.*		Facilitating/assuring the implementation of the 15 principles of tailings management.	Unsupervised	1–2	9–10	DEOR 2	High/Extreme	EOR 3
8	Lead TSF engineer	Complex, multiple TSF design/assessment tasks and internal review. Dam safety audits and reviews. *Holds ultimate responsibility for the design, construction and safety of TSF.*	The EOR must have appropriate knowledge and experience commensurate with the complexity and potential consequences of failure, in design, construction, performance, analysis and		Unsupervised	2–3	12–13	EOR 1	Significant	Principal

Level #	Status/ role	Technical responsibility	ANCOLD/CDA alignment	GISTM alignment	Practice level	Exposure/ experience (yrs)	EOR role [3]	Respons-ibility [4]	Technical mentor
7	Principal/ equivalent TSF engineer	Singular TSF project development and internal review. Dam safety audits, reviews and ITRB. Holds ultimate responsibility for the design, construction and safety of TSF.	operation of the subject dams and appurtenances – gained through directly related experience (CDA).	Facilitating/ assuring the implementation of the 15 principles of tailings management.	Unsupervised	2 14–15	EOR 2	High	Principal
9	Principal/ equivalent TSF engineer	Multiple TSF project development and internal review. Dam safety reviews and ITRBs. Holds ultimate responsibility for the design, construction and safety of TSF.			Unsupervised	2 16–17	EOR 3	Extreme	Principal
10	Manager –TSFs	Oversee and provide strategic management of multiple dam projects/portfolios.	Establish robust management systems for tailings dams (ANCOLD). Develop comprehensive tailings management programs (CDA).	Integrate management systems to ensure the safe operation of tailings facilities	-	16+	EOR 3	Extreme	-

Notes: 1. Minimum tertiary qualification: Bachelor of Engineering/BSc Engineer degree accredited by Engineers Australia; 2. Indicative minimum period; 3. Internally defined EOR roles; 4. TSF failure consequence classification defined by ANCOLD; 5. Minimum requirement to fulfill role of DEOR in Qld Aus is RPEQ; 6. DEOR – deputy EOR; EOR TR – EOR-in-training.

EOR SKILLS COMPETENCY MATRIX

Table 6 outlines a matrix for evaluating the skills and competencies of EOR representatives. The matrix incorporates both technical and non-technical criteria to assess the capability of EOR and Deputy EOR representatives at any stage of their development.

Technical criteria

The technical markers focus on:

- TSF engineering experience.
- TSF safety management (governance related) experience.

The engineering experience is further divided into four key subjects namely: characterisation; design; construction; and operation. Each subject is subdivided into key technical elements and sub-elements, which are appropriately weighted and assessed objectively within the matrix.

Non-technical criteria

The non-technical markers are grouped into four categories: *communication, independence, leadership and collaboration,* and *general attributes.*

These elements are similarly weighted and evaluated to provide a broad view of an EOR representative's non-technical competencies.

Evaluation and scoring

EOR representatives are evaluated based on their demonstrated levels of *exposure, responsibility* and *complexity* across each subject or element. Each of these is also weighted according to its importance within the evaluation matrix.

An EOR's company assigns the relevant scores and weighting factors in Table 6. The scoring system and weightings are adaptable and may be adjusted to meet specific company or project requirements, ensuring relevance and applicability across different contexts.

Tables 7–10 provide descriptive guidance for evaluating skills and competencies. They support consistent interpretation of the scoring system and its application.

The proposed matrix is intended as a guide, not as a definitive tool for assigning levels of competency such as EOR1 or DEOR1, 2, and so on. Such decisions remain subjective and should be made in consultation with the relevant parties.

The frequency of evaluation is determined at the discretion of the EOR's company. However, it is advisable to conduct a baseline evaluation and to repeat the assessment at least every two years (if not more frequently) thereafter. Notwithstanding formal assessments, EOR representatives are encouraged to use the matrix as a continual self-assessment and development tool, guiding their ongoing professional growth in relation to the matrix's key elements.

To ensure a robust evaluation, an EOR's company conducts the assessment of EOR representatives, who also complete self-assessments independently. When performed objectively on both sides, the results should converge, offering a balanced measure of skills and competencies. Figure 1 illustrates an example assessment conducted by a committee and six prospective EOR representatives. The further a point lies from the parity line, the greater the disparity between the evaluations (as seen with Representatives 1 and 6 in Figure 1). In such cases, a discussion between a committee member and the representative may be necessary to review the ratings and align perspectives and expectations.

The authors have developed a provisional scoring and weighting system for the matrix, which is currently trialled and calibrated for broader application.

TABLE 6

Skills and competency matrix.

Marker category	Experience category	Engineering aspect	Technical field	Technical element (TE)	Levels
Technical markers	Experience in TD/TSF engineering	Characterisation	Site geology and seismicity, foundations soils, embankment materials, tailings etc	Field testing (general)	Exposure level: Rating as per Table 7
				Field testing (tailings)	
				Laboratory testing	
			Site selection, optioneering	Data assessment and interpretation	
			Geotechnical elements	Data assessment and interpretation (parameterisation etc)	
				Design analysis	
				Feasibility/detailed design, technical specifications etc	
			Hydrological and hydraulic elements	Data assessment and interpretation	
				Design analysis	
				Feasibility/detailed design, technical specifications etc	
		Design	Hydrogeological elements	Overview of data assessment and interpretation	Responsibility level: Rating as per Table 8
				Overview of analysis	
				Overview of detailed design and technical specification	
			Environment	Data assessment and interpretation	
			Closure	Data assessment and interpretation	
				Design analysis	
				Feasibility/detailed design, technical specifications etc	Complexity level: Rating as per Table 9
		Construction	Commercial	Contract administration and management	
			Execution	Construction management, construction monitoring	
		Operation	Operation spec and performance limits	Compile, review OMS manual/code of practice/assess performance against TARPs	
			Performance monitoring	Data assessment, interpretation and reporting	
	Experience in TD/TSF management	Governance	Dam safety and governance related activities	Independent design review	
				Dam safety inspection	
				Dam safety review	
				Standards, guidelines and regulations	
				Independent technical review board collaboration	
				Industry activity	
Non-technical markers	Personal attributes and general skill			Communication (comprehension/writing/speaking)	Exposure level: Rating as per refer Table 10

TABLE 7
Exposure rating scale.

Numerical rating	Descriptive rating	Description
1	Very low	No to little experience with parts of this task or concept; limited in how to integrate information with other components of a project
2	Low	Performed tasks and applied related theory under supervision; demonstrates fair understanding of how to integrate information with other components of the project
3	Moderate	Repeated experience of task or applied related concepts to other projects; demonstrates understanding of how to generate and apply information
4	High	Consolidated experience with task and/or concept and has applied it to several projects; understands how to generate and use this information; started reviewing work by others
5	Very high	In-depth experience with this task and/or concepts; lead discussions on the topic

TABLE 8
Responsibility scale.

Numerical rating	Descriptive rating	Description
1	Very low	General exposure to rudimentary tasks; little to no responsibility
2	Low	Assisting; limited responsibility for work output
3	Moderate	Participating; full responsibility for supervised work
4	High	Contributing; full responsibility to supervisor for immediate work quality
5	Very high	Performing and leading; responsible with no support required but not accountable

TABLE 9
Complexity scale.

Numerical rating	Descriptive rating	Description
1	Very low	Routine tasks; well defined; minimal risk
2	Low	Known procedures; applying general technical knowledge; low risk
3	Moderate	Tasks with some uncertainty; required to solve problem; undertake analysing tools; moderate risk and impact
4	High	Significant levels of uncertainty; requiring advanced skills; high risk
5	Very high	High levels of uncertainty (ill-defined problems); requiring expert knowledge and innovation (derive solutions from 1st principles); multidisciplinary problems; lead with/develop new technology; very high risk

TABLE 10

Non-technical capability – capability scale.

Numerical rating	Descriptive rating	Description
1	Poor	
2	Average	
3	Good	
4	Very Good	
5	Advanced	

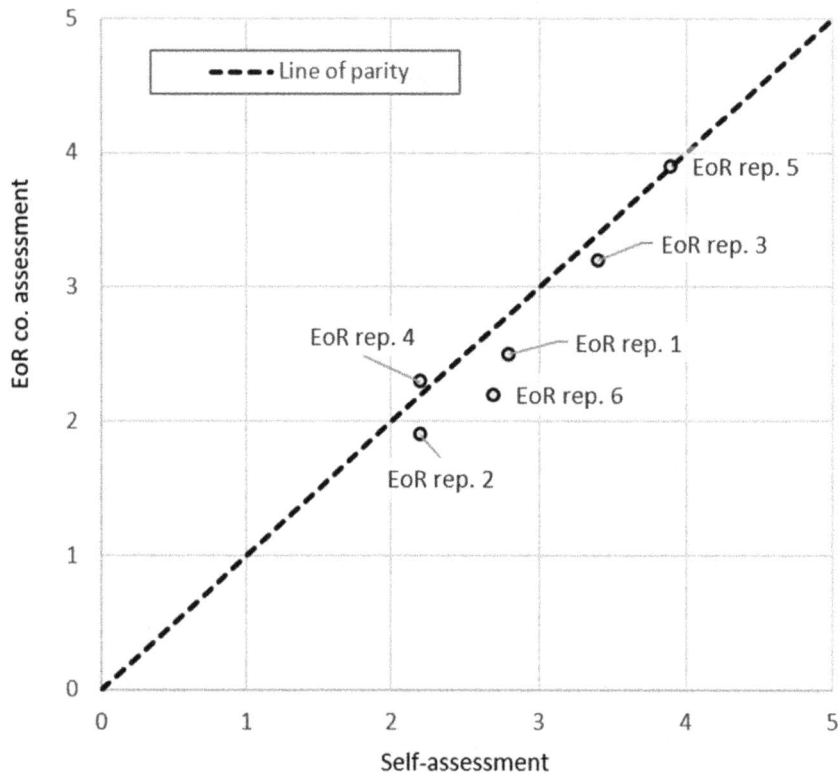

FIG 1 – EOR company and representative evaluation correlation.

TEAM BASED EOR EXECUTION

Complex or high-consequence TSFs are best managed through team-based EOR execution structures. No single engineer, regardless of experience, can realistically maintain comprehensive knowledge across all technical domains required for safe TSF oversight. A structured team approach ensures that critical areas – from site investigation and design to construction, operation, monitoring, and governance – are covered by specialists with the appropriate knowledge and skills.

A clearly defined EOR execution structure supports professional development and career progression within tailings dam engineering. Engineers may grow into the role of EOR-in-training (EOR TR), taking on defined technical responsibilities. As experience and competence increase, they can assume broader accountability. This structured pathway helps maintain continuity of knowledge and builds a pipeline of qualified professionals ready to take on EOR roles – addressing the industry-wide shortage of experienced tailings engineers.

A responsibility matrix within the EOR structure helps ensure that each functional area receives appropriate attention. By assigning primary and supporting roles across domains, industry organisations can reduce knowledge gaps and avoid critical oversights.

Implementing a team-based EOR model also strengthens the safety culture around tailings management. Teams that share EOR responsibilities foster collaboration, promote open technical dialogue, challenge assumptions, and support rigorous review. This reduces individual bias and oversight risk, while encouraging innovation and continuous improvement. Team structures also build resilience: if a team member departs, shared knowledge and clearly defined roles support smooth transitions without compromising oversight.

Table 11 presents a practical example of how EOR responsibilities can be distributed across career stages and functional domains. The matrix clarifies development pathways, supports comprehensive technical coverage, and highlights growth opportunities at each level.

TABLE 11

Example EOR technical execution/responsibility matrix for a TSF project.

Career stage	EOR role	Site investigation	Design Geotechnical	Water	Procurement and construction	Operation and monitoring	Governance
Principal geotechnical engineer	EOR	A	P	A	A	P	P
Senior geotechnical engineer	DEOR	P	A	A	P	A	A
Senior engineering geologist	EOR TR	P	A	-	-	-	-
Senior geotechnical engineer	EOR TR	A	P	-	-	P	A
Geotechnical engineer	EOR TR	-	A	-	-	-	P
Senior water engineer	DEOR	-	-	P	-	A	P
Senior resident engineer	DEOR	A	-	-	P	-	-

Key: EOR = Engineer of record; DEOR = Deputy EOR; EOR TR = EOR-in-training; P = primary responsibility; A – assisting.

ALTERNATIVE TSF EOR DEVELOPMENT PATHWAY

The tailings EOR development pathway presented herein follows the traditional geotechnical consulting development track. However, broadening the pool of potential TSF EORs may require expanding the concept to include alternative entry routes into the role, leveraging complementary experience as a foundation. Table 12 outlines the potential alternative entry routes and pathways, highlighting key areas of alignment and development needs.

TABLE 12
Example EOR technical execution/responsibility matrix for a TSF project.

Alternative starting point	Aligning elements	Development needs	Pathway support and integration
Water dam engineers	<ul><li>Familiarity with the EOR role and responsibilities for water dams</li><li>Water dam engineering (characterisation, design, construction, operations oversight)</li><li>Water dam operations/ performance management</li><li>Water dam risk management and governance</li><li>Stakeholder engagement</li></ul>	<ul><li>TSF specific knowledge</li><li>Integration into mining contexts (ie public to private transition)</li><li>TSF performance management</li><li>TSF life cycle technical management</li><li>TSF specific risk management and governance</li></ul>	Enabling factors:<ul><li>Existing roles and technical overlaps with TSF context</li></ul>Inhibiting factors:<ul><li>Limited exposure to tailings and mining contexts</li><li>TSF EOR mentoring and hierarchy</li><li>Limited exposure to TSF engineering design execution and/ management</li></ul>
TSF engineers – owner side	<ul><li>Embedded within mining operations</li><li>TSF life cycle planning</li><li>TSF technical (engineering) oversight</li><li>TSF operations management and oversight</li><li>TSF risk management and governance</li><li>Stakeholder engagement</li></ul>	<ul><li>TSF design</li><li>Professional independence (operational bias)</li><li>TSF life cycle technical management</li></ul>	Enabling factors:<ul><li>Embedded knowledge of TSF technical elements and mining context</li></ul>Inhibiting factors:<ul><li>Limited exposure to TSF engineering design and management</li><li>TSF EOR mentoring and structured opportunity</li></ul>
TSF engineers – TSF contractors and/or operators side	<ul><li>TSF construction</li><li>TSF operations execution</li></ul>	<ul><li>TSF design</li><li>TSF life cycle technical management</li><li>TSF risk management and governance</li></ul>	Enabling factors:<ul><li>Embedded knowledge of TSF sites and operations</li></ul>Inhibiting factors:<ul><li>Very limited exposure to TSF engineering design and management</li><li>TSF EOR mentoring and structured opportunity</li></ul>

CONCLUSION

This paper presents a framework to support the systematic development of tailings dam EOR representatives from a geotechnical engineering perspective, incorporating a structured development pathway and a complementary skills and competency matrix.

By systematising the development process, the framework provides a foundation for progressively building EOR capability and capacity, while aligning responsibility with experience and maintaining accountability across the TSF life cycle. When applied within a team-based EOR execution structure, it has the potential to strengthen continuity, promote knowledge sharing, and help address the industry-wide shortage of experienced tailings engineers.

Since the weighting and competency scoring of development markers are inherently subjective, this paper does not provide relative weights for development marker categories and competencies. Users should assign weightings according to their specific context and priorities when adopting the framework. The evaluation matrix is likewise not intended to result in an absolute measure of competency but should rather be used as a guide to be applied fairly and sensibly in assessment.

Finally, while the framework has been developed from a geotechnical perspective, it could be adapted or extended to reflect other entry routes to EOR development. Such alternatives were not explored to detail in the paper.

REFERENCES

Global Tailings Review (GTR), 2020. Global Industry Standard on Tailings Management (GISTM). Available from: <https://globaltailingsreview.org/wp-content/uploads/2020/08/global-industry-standard-on-tailings-management.pdf>

International Council on Mining and Metals (ICMM), 2021. Tailings Management – Good Practice Guide. Available from: <https://www.icmm.com/tailings-management>

Knight Piésold Consulting, 2021. Engineer of Record (EoR) Services for Tailings Facilities, August 2021.

Small, A and Witte, A, 2018. Guidance for Dam Safety Management, including the Engineer of Record, CDA 2018 Annual Conference.

Tailings Engineer of Record Task Force (TEORTF), 2017. Proposed Best Practices for the Engineer of Record (EoR) for Tailings Dams by The Tailings Engineer of Record Task Force (TEORTF), Geoprofessional Business Association, GBA.

Revisiting probability of failure versus factor of safety – a screening tool with limitations

J J Moreno[1] and S R Kendall[2]

1. Principal Consultant, Tailex Pty Ltd, Perth WA 6018. Email: pepe.moreno@tailex.com.au
2. Senior Consultant, Tailex Pty Ltd, Perth WA 6018. Email: sam.kendall@tailex.com.au

ABSTRACT

The use of simplified methods to establish a relationship between factors of safety and probability of failure (Silva, Lambe and Marr, 2008) has become more popular in tailings dam risk assessments as they offer a probabilistic perspective on stability that can be achieved with relative ease. While valuable as screening tools, their limitations warrant caution, especially when used to assess a TSF's ALARP position. This paper applies a probabilistic analysis to hypothetical cases, aiming to identify whether simplified methods could lead to misleading risk interpretations. A discussion on the limitations of Silva, Lambe and Marr (2008) is presented, and alternative pathways are explored.

INTRODUCTION

As part of evolving industry standards and heightened stakeholder expectations, mining companies are increasingly required to quantify risks related to tailings storage facilities (TSFs), not only to inform engineering decisions but also to support financial reporting through disclosure of potential failure liabilities. A key principle in this context is the ALARP (As Low As Reasonably Practicable) concept, which requires that risks be reduced to a level that achieves balance between safety, practicability and cost.

To demonstrate a TSF's ALARP position and enable clear communication of risk, a TSF's probability of failure (PoF) must be quantified across all relevant failure modes. Slope instability is a commonly observed failure mode (Stark, Moya and Lin, 2022) and often a major contributor to the PoF. To quantify TSF instability risk, practitioners rely on a range of methods, both subjective and objective, each offering strengths and limitations. These limitations must be diligently reviewed when comparing a TSF's ALARP position to tolerable risk thresholds: even risks considered well below societal limits carry potential for misrepresentation.

STATE OF PRACTICE

The current state of practice in assessing TSF stability typically relies on deterministic methods, using either limit equilibrium (LE) or stress-strain numerical modelling techniques, where a calculated safety factor (FoS) is measured against generic minimum criteria.

While industry guidance advocates the use of risk-based approaches, minimum FoS criteria are widely used, including by the authors, reflecting client/regulatory expectations and practical limitations in data availability to inform probabilistic analysis.

Slope stability is influenced by aleatoric uncertainty (arising from natural material variability) and epistemic uncertainty (arising from knowledge gaps or errors in understanding or measurement). Deterministic stability analyses typically use lower bound or characteristic strength values in addition to FoS criteria to manage aleatoric and epistemic uncertainty. This approach does not offer quantifiable indication of instability risk, as a compliant deterministic FoS could be underestimating risk and failing to meet tolerability thresholds—or overestimating risk, resulting in unnecessary conservatism and overdesign.

However, endorsed by GISTM, the industry is increasingly using risk—and PoF—as a means of communicating with stakeholders. Silva, Lambe and Marr (2008) (referred to herein as SLM) offers a practical bridge between deterministic analysis and PoF, allowing an annualised PoF to be inferred from empirical relationships.

Alternatives to SLM include more rigorous objective analyses; however, their complexity, cost and data requirements have impeded their widespread adoption in geotechnical engineering. While these approaches can offer additional precision in certain circumstances, they can also mask uncertainty and dilute accountability.

LIMITATIONS OF SLM

SLM presents a semi-empirical relationship that converts deterministic FoS into annualised PoF estimates. The underlying PoF estimates are informed by expert judgment using details from 75 projects. Notably, the method assumes the input FoS is derived from effective stress analyses and refines PoF estimates by categorising facilities (Class I–IV) according to the quality of investigation, design, construction and monitoring.

The SLM method is a well-founded approach that places value on expert judgment and geotechnical insight. It has useful applications, including screening-level assessments and portfolio risk comparisons. However, its application to quantifying TSF risk to inform TSFs' ALARP positions requires careful scrutiny This view is based on the following notable limitations.

Limited calibration for TSFs: The original development of SLM drew on a broad range of slope types but included limited calibration specifically for TSFs (two case histories), reducing its reliability in this context. SLM does not account for the wide range of TSF configurations—such as integrated waste landforms (IWLs) and filtered tailings stacks—that are often designed with operational objectives beyond stability, including material disposal efficiency or waste haulage/tailings transport logistics.

Loading conditions: The method assumes the input FoS is derived from effective stress analyses. Modern TSF design standards seek to address undrained loading conditions where stability is influenced by contractive materials (such as hydraulically placed tailings or weak foundations).

Foundation conditions: SLM treats foundation conditions indirectly through the input FoS and selection of engineering classes (I–IV) but is unable to explicitly assess or model complex foundation behaviours or heterogeneity—common in foundations over alluvial, colluvial, or highly weathered residual soils.

Conservatism in deterministic design: Standard deterministic design approaches often lead designers to employ lower bound or characteristic strength values to establish the deterministic FoS, which yield a higher PoF from the SLM curves (SLM notes that FoS using mean parameters should be applied). This can result in misguided concern and unnecessary expenditure.

QUANTIFYING POF SENSITIVITY TO SHEAR STRENGTH VARIABILITY

Uncertainty in shear strength is often the largest source of variability in slope stability analyses (Duncan, Wright and Brandon, 2014). This paper explores how PoF varies across a realistic range of material shear strength and the coefficient of variation (CoV) values, where the CoV captures aleatoric and epistemic strength uncertainties.

PoF has been estimated for various loading conditions and TSF configurations, as shown in Figure 1. Each TSF configuration featured a 30 m embankment height, 1V:3H downstream slopes and a 10 m crest width. The phreatic surface was modelled to simulate a steady-state seepage condition from a supernatant pond that is initiated 100 m upstream of the crest and exits at the downstream toe. Each scenario was assessed considering:

- competent (high strength) and weak (low undrained strength) foundation conditions

- drained strength for all tailings

- undrained strength for all tailings

- drained strength for tailings above, and undrained strength for tailings below the phreatic surface.

A probabilistic modelling approach using GeoStudio software was employed, applying the limit equilibrium method (Morgenstern-Price) with circular failure surfaces. Monte Carlo simulations with 10 000 samples were initially trialled to ensured convergence; however, it was observed that Latin Hypercube Sampling (LHS) with 1000 samples provided comparable accuracy with reduced computational time. Simulations were performed to compute a probability distribution of the FoS for a range of scenarios. The PoF was then estimated from fitted distributions where the FoS falls below unity or P[FoS < 1] (Harr, 1977; Christian, Ladd and Baecher, 1994; Baecher, 1984).

TSF Geometry

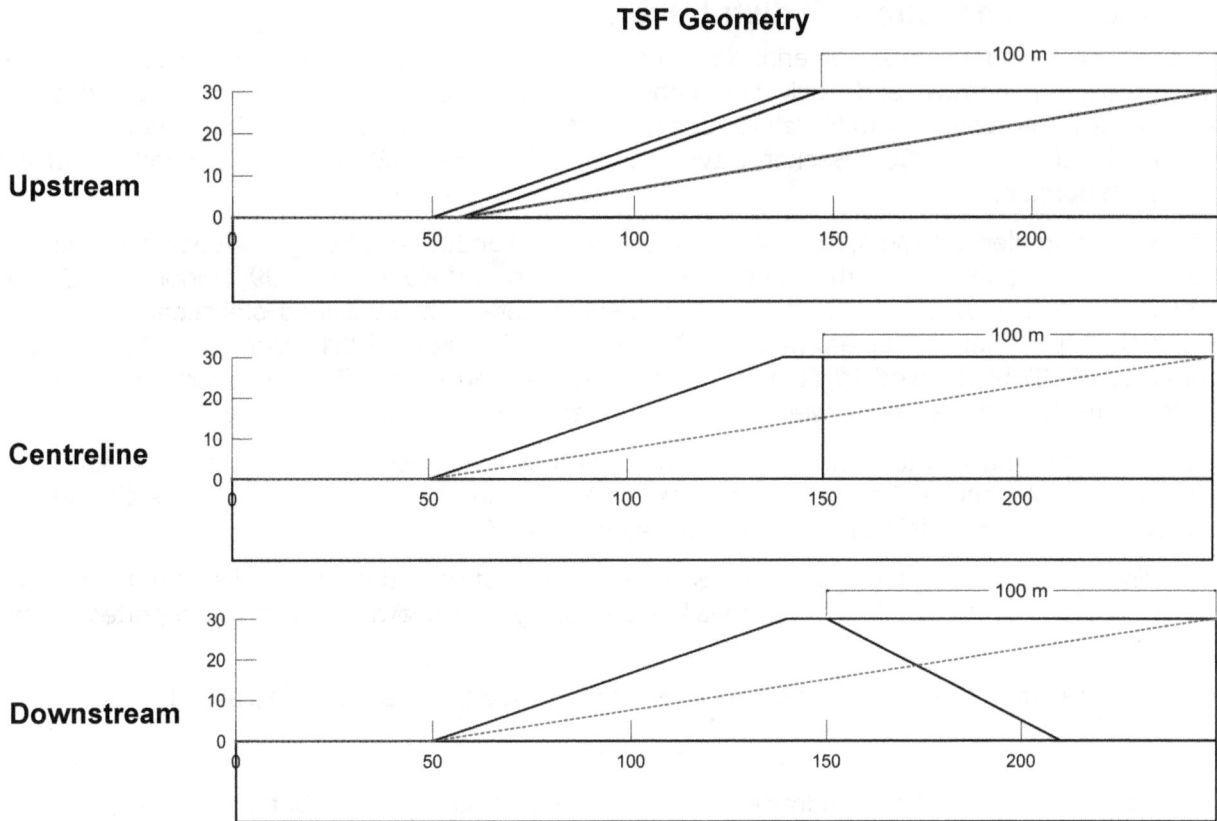

FIG 1 – Summary of analysed TSF stability geometries.

In steady-state conditions without a clear trigger or time-dependent degradation, the probability of failure from probabilistic slope stability analysis reflects uncertainty in the inputs—not the likelihood of failure over time. It is a measure of confidence in the design, not a prediction of failure rate. For this study, a 50-year period was assumed to represent the time during which the TSF remains in steady-state conditions prior to closure and reclamation. Under this assumption, the annual PoF can be approximated by dividing the total PoF by the selected time window. Key to this analysis is assessing the probability of drained and undrained behaviour and estimating their shear strength distributions using values derived from literature. A discussion of these aspects follows.

Drained versus undrained conditions

To capture a realistic range of potential failure mechanisms in TSF slopes, the analysis considers both drained and undrained conditions.

Under drained conditions, failure probability is largely governed by variability in effective strength parameters. Strength uncertainty in undrained scenarios is similarly present, though the mobilisation of undrained strength is often conditional on triggering events, including rapid loading or a consistent rise in phreatic surface.

To account for the conditional nature of mobilising undrained strengths, the LEM PoF was multiplied by the probability of a triggering event, using Barneich *et al*'s (1996) qualitative-to-quantitative mapping. This approach introduces the effect of time dependency, since the probability of triggering would be estimated on a yearly basis. The likelihood of triggering was considered to vary according to TSF status as follows:

- Certain: Actively raised TSFs (eg upstream, rapid rise) approach a trigger probability of 1.0, which is consistent with ANCOLD (2019).

- Unlikely: Intermittent deposition may reduce the probability to ~10^{-2}.
- Rare: Care and maintenance conditions may reduce probability further, to ~10^{-3}.

Probabilistic shear strength distributions

The analysis considers three material domains—tailings, foundation, and construction materials (low-permeability borrow and rock fill). Drained and undrained probabilistic shear strength distributions were derived from literature (Jefferies and Been, 2015; Christian and Baecher, 2011; Been and Li, 2009; Vick, 2002; Becker, Cavalcanti and Marques, 2023) to reflect strength variability within each domain.

To develop controlled data sets, representative drained and undrained strength values were selected as base cases. Literature-reported ranges of CoV (Phoon and Kulhawy, 1999; Phoon and Ching, 2015; Ching, 2021; Uzielli, 2008), as well as relevant data from published statistical analysis of geotechnical parameters (Linero-Molina, Contreras and Dixon, 2021; Vernengo Lezica and Contreras, 2023) were used to fit both normal and log-normal distributions. Three CoV levels (minimum, mean and maximum) were applied to assess sensitivity to uncertainty.

In general, lognormal distributions were preferred, as they better represent the skewness and non-negativity of geotechnical strength parameters. To avoid unrealistic sampling in Monte Carlo/Latin Hypercube simulations, distributions were truncated at ±1E-08.

Due to limited data, a key limitation of this study is the use of univariate distributions in the stability models. The choice of distribution type was based on typical skewness patterns reported in the literature.

The shear strength distribution parameters used in this study are presented in Table 1.

TABLE 1
Summary of drained and undrained probabilistic shear strength distribution parameters.

Material	Tailings						Clay foundation	Clay borrow		Rock fill
	Drained		Undrained					C	φ	φ
Strength	°	°	USR	USR	USR	USR	kPa	kPa	°	°
Mean	24	33	0.25	0.3	0.35	0.4	100	23.7	22.4	38
COV1 %	10	5	15	15	15	15	18	19	4	6
COV2 %	12.5	12.5	25	25	25	25	25	26	8	10
COV3 %	15	20	35	35	35	35	38	33	13	13
Skewness	>1	>1	>1	>1	>1	>1	>1	<1	<1	>1
Distribution	Log-Normal							Normal		Log-normal

RESULTS

Figure 2 illustrates the significant scatter observed in PoF across the range of FoS values. For FoS within the intervals [1.3, 1.4] and [1.4, 1.5], the PoF varied by more than 6 and 8 orders of magnitude, respectively, depending on the inherent variability of the foundation and embankment materials. For comparison, SLM curves (categories I–IV) span 4 to 5 orders of magnitude over the same FoS ranges. However, it is important to note that while SLM curves reflect increasing levels of engineering confidence, the probabilistic results presented here are independent of any predefined level of engineering, as CoV ranges used in the study are within expected natural inherent variability. While it may be possible to reduce epistemic uncertainty using better-quality data, a limit is reached where further investigation would not significantly reduce CoV.

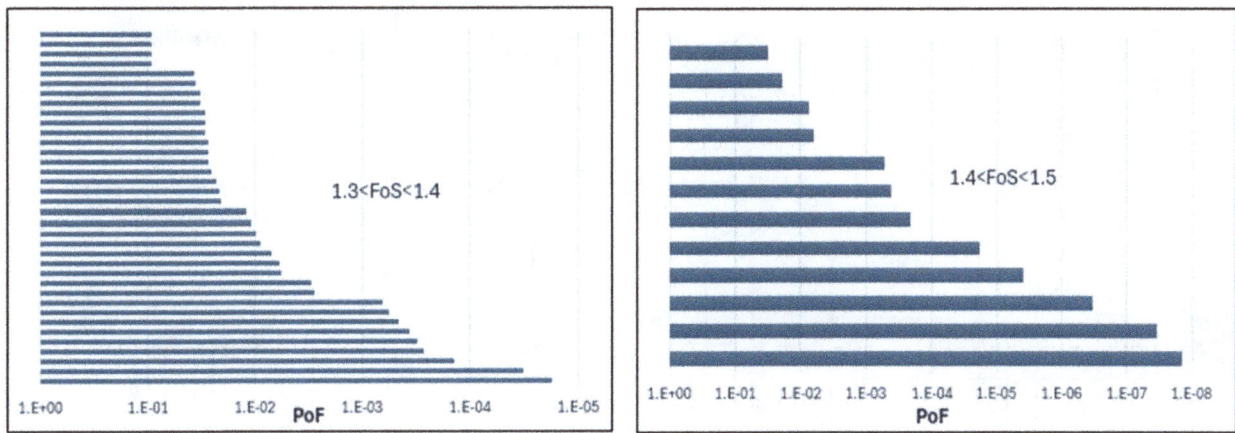

FIG 2 – PoF variability within FoS ranges.

A more detailed breakdown of PoF variability, focusing on the upstream raise case, is presented in Figure 3, showing PoF results for competent and weak foundations under varying mean values and CoV for undrained tailings strengths. In the competent foundation cases, variability in PoF increases with higher undrained strength, as the slip surface is governed primarily by the tailings. In contrast, results for the weak foundation do not display a clear trend, likely due to the interaction between the tailings and the weaker foundation materials influencing the slip surface.

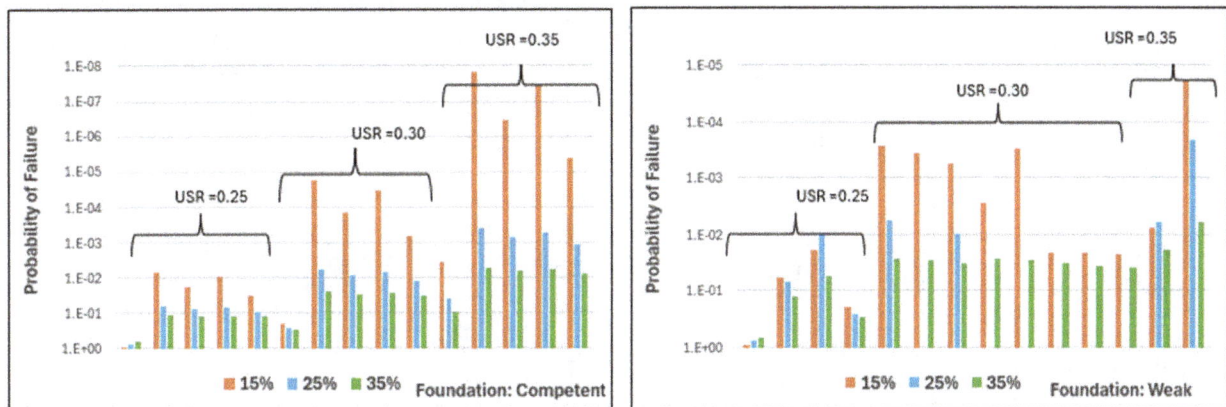

FIG 3 – PoF variability in upstream TSF cases considering competent and weak foundation conditions.

To enable direct comparison with the SLM framework, the probabilistic results were annualised as described earlier in the paper. Figure 4 presents these results across various operational cases, reflecting differing assumptions about the likelihood of triggering undrained loading conditions for each raise method as discussed earlier in the paper. A consistent trend is evident: most PoF values cluster around or above the SLM Category 1 curve—even in cases with relatively high CoV values and independent of raise method. This discrepancy suggests that case-specific probabilistic analyses, if interpreted without diligence, may yield PoF estimates that overstate the likelihood of failure, raising questions about how such results align with practical decision-making frameworks.

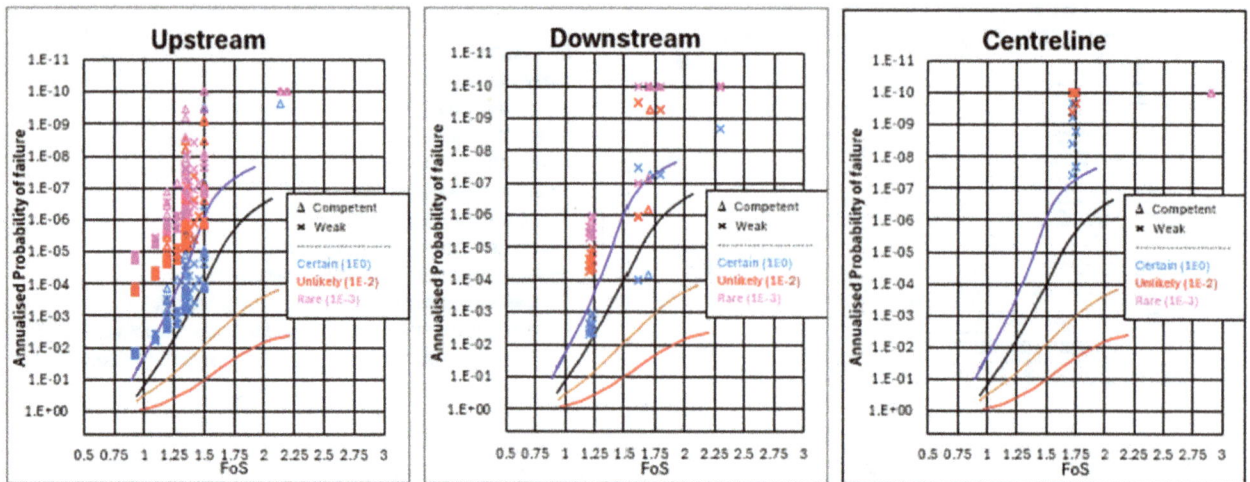

FIG 4 – Annualised PoF versus FoS varying likelihood of undrained triggers and foundation conditions.

For comparative purposes, a base case undrained strength ratio (USR) of 0.35 was selected for the tailings. The results, by foundation type and embankment raise construction method, considering that triggering an undrained response would have a likelihood of one in any given year, are shown in Figure 5. This plot offers a more granular view of PoF scatter and represents what might be encountered in a trade-off study. Interestingly, the results suggest that all raise construction types yield broadly similar PoF ranges under competent foundation conditions. Given that the critical slip surfaces analysed correspond to failure modes with severe consequences, the results imply that for this case, the downstream or centreline raise construction methods may not necessarily reduce PoF driven by undrained loading, provided foundation conditions are favourable.

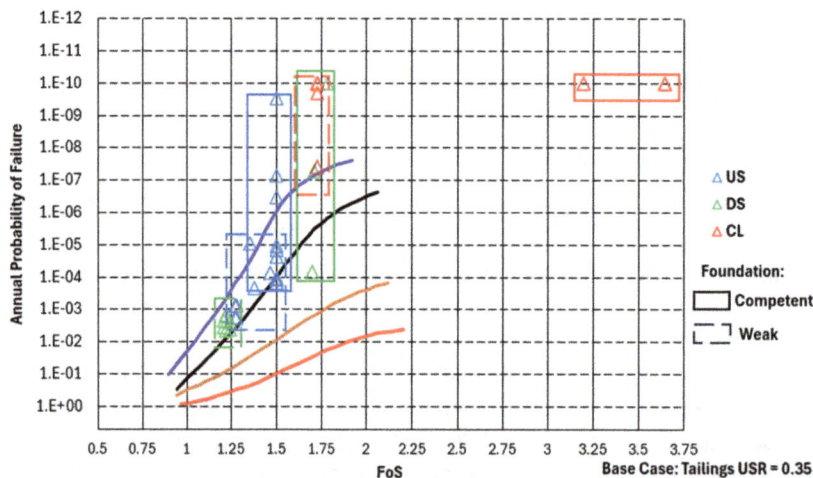

FIG 5 – Comparison of PoF between upstream, downstream and centreline methods of construction.

CONCLUSIONS

Selecting an approach to assess uncertainty in a geotechnical system is dependent on and constrained by regulatory, economic and technological factors.

This study highlights the need to exercise caution when using simplified methods such as SLM to assess TSF stability risk. Demonstrated misalignment between the FoS and PoF obtained from the SLM method can result in misrepresentation and overestimation of risk when used to establish a TSF's ALARP position. Where slope instability represents a major contributor to a TSF risk profile, costs to further investigate material variability would be justifiable. For many large-scale TSF operations, however, fully objective assessments remain challenging due to inherent limitations in data availability, precision, and accuracy.

Based on the findings of this study, the following conclusions are presented:

- The SLM method remains a useful screening tool in dam risk assessments; however, its applicability to TSFs is constrained by limited calibration, simplified treatment of foundation and loading conditions, and potential for misrepresentation of risk when deterministic FoS values are derived from conservative strength parameters.

- This study illustrates that PoF can vary by several orders of magnitude within narrow FoS bands, highlighting the significance of shear strength uncertainty particularly in undrained conditions. However, the application of probabilistic methods would require a critical view of the availability and potential variability of the geotechnical data.

- A practical way forward involves establishing bounded estimates of risk by identifying credible failure modes, defining realistic strength variability ranges through reasonable estimation of CoV, and applying a tiered investigation strategy. Literature-derived parameter distributions may offer meaningful input where site-specific data are limited. The aim is not to replace or downplay the importance of judgment; rather, to provide alternative framework for the rational application of geotechnical expertise.

- While conservatism is often applied as a safeguard, it can distort PoF estimates when probabilistic tools are used without adjusting input distributions accordingly. This may lead to either overstating risk or unjustified overdesign, both of which can undermine ALARP decision-making.

- Future work should aim to define practical guidelines or frameworks that identify minimum CoV targets that are achievable through standard testing and investigation methods, to help guide when a probabilistic analysis is meaningful and when it may be misleading.

- The authors are interested in collaboration using site-specific inputs to further refine a practical approach to quantifying TSF instability risks and informing ALARP assessments.

REFERENCES

Australian National Committee on Large Dams Incorporated (ANCOLD Inc), 2019. Guidelines On Tailings Dams - Planning, Design, Construction, Operation And Closure, revision 1, ANCOLD Inc. Available from: <https://ancold.org.au/product/guidelines-on-tailings-dams-planning-design-construction-operation-and-closure-revision-1-july-2019/>

Baecher, G, 1984. Geostatistics, Reliability and Risk Assessment in Geotechnical Engineering, *Geostatistics for Natural Resources Characterization*, Part II, pp 731–744.

Barneich, J, Majors, D, Moriwaki, Y, Kulkarni, R and Davidson, R, 1996. Application of reliability analysis in the environmental impact report (EIR) and design of a major dam project, Uncertainty '96, ASCE.

Becker, L D B, Cavalcanti, M d C R and Marques, A A M, 2023. Statistical Analysis of the Effective Friction Angle of Sand Tailings from Germano Dam, *Infrastructures*, 8:61. https://doi.org/10.3390/infrastructures8030061

Been, K and Li, A L, 2009. Soil Liquefaction and Paste Tailings, in *Paste 2009: Proceedings of the Twelfth International Seminar on Paste and Thickened Tailings* (eds: R Jewell, A B Fourie, S Barrera and J Wiertz), pp 281–290 (Australian Centre for Geomechanics: Perth). https://doi.org/10.36487/ACG_repo/963_32

Ching, J, 2021. State-of-the-art review of inherent variability and uncertainty in geotechnical properties and models, International Society of Soil Mechanics and Geotechnical Engineering (ISSMGE) – Technical Committee TC304 Engineering Practice of Risk Assessment and Management. https://doi.org/10.53243/R0001

Christian, J T and Baecher, G B, 2011. *Unresolved Problems in Geotechnical Risk and Reliability*, Geotechnical Special Publication.

Christian, J T, Ladd, C C and Baecher, G B, 1994. Reliability applied to slope stability analysis, *Journal of Geotechnical Engineering*, 120(12):2180–2207.

Duncan, J, Wright, S and Brandon, T, 2014. *Soil Strength and Slope Stability*, 2nd edn (John Wiley and Sons, Inc: Hoboken).

Harr, M E, 1977. *Mechanics of particulate media – a probabilistic approach* (McGraw-Hill Publishers: New York).

Jefferies, M and Been, K, 2015. *Soil Liquefaction: A Critical State Approach*, 2nd edn (CRC Press).

Linero-Molina, S, Contreras, L F and Dixon, J, 2021. Estimation of shear strength of very coarse mine waste, in *SSIM 2021: Proceedings of the Second International Slope Stability in Mining* (ed: P M Dight), pp 341–354 (Australian Centre for Geomechanics: Perth). https://doi.org/10.36487/ACG_repo/2135_21

Phoon, K-K and Ching, J, 2015. *Risk and reliability in geotechnical engineering* (CRC Press: Taylor and Francis Group).

Phoon, K-K and Kulhawy, F, 1999. Evaluation of geotechnical property variability, *Canadian Geotechnical Journal*, Nov.

Silva, F, Lambe, T W and Marr, W A, 2008. Probability and risk of slope failure, *Journal of Geotechnical and Geoenvironmental Engineering*, Dec:1691–1699.

Stark, T D, Moya, L J and Lin, J, 2022. Rates and Causes of Tailings Dam Failures, *Advances in Civil Engineering*, Hindawi.

Uzielli, M, 2008. Statistical analysis of geotechnical data, in *Reliability-based design in geotechnical engineering: Computations and applications* (ed: K K Phoon), (CRC Press). https://doi.org/10.1201/9780203883198.ch10

Vernengo Lezica, I and Contreras, L F, 2023. Application of Bayesian methods to the estimation of shear strength parameters for slope stability analysis of a TSF, in *Proceedings of the Mine Waste and Tailings Conference 2023*, pp 519–532 (The Australasian Institute of Mining and Metallurgy: Melbourne).

Vick, S G, 2002. *Planning, Design and Analysis of Tailings Dams*, 2nd edn (BiTech Publishers Ltd).

The challenge of GISTM adherence for smaller mining operations in remote parts of the world

G V Price[1]

1. Senior Staff Consultant, KCB Australia, Brisbane Qld 4000. Email: gprice@klohn.com

ABSTRACT

This paper, with special reference to Air Upas and Sandai Mines in west Kalimantan in Indonesia, describes the problems facing smaller mining operations in remote parts of the globe in acquiescing to the Global Industry Standard on Tailings Management (GISTM). The GISTM, as published by Global Tailings Review (GTR) (2020), references six topic areas namely Communities, Integrated Knowledge Base, Design Construction and Operation, Management and Governance, Emergency Response, and Disclosure. Each of these encompass 15 principles in turn incorporating 77 requirements and 219 criteria. Tailings storage facilities (TSF) are audited against these and though well-meaning and a very necessary requirement, are often difficult to adhere to in remote areas where flight access is arduous or impossible, road access is over long distances often along muddy tracks with deep rutting, satellite communications at times non-existent, specialised equipment necessary for geotechnical field testing unavailable or only accessed through difficult and costly establishment, senior geotechnical staff unwilling to locate to remote sites: these to mention but a few of the problems faced. The paper describes where adherence *is* possible as measured against the main GISTM topic requirements and principles, and perhaps more importantly – *where not*. These issues of compliance and non-compliance are integrated and presented in tabular format. The paper makes recommendations on how the most difficult adherence requirements can be overcome and concludes with suggestions of how GISTM audit requirements and processes can be softened in a 'GISTM-light' process where smaller remote operations can be audited against more appropriate and equitable targets.

INTRODUCTION

This paper, with special reference to mining operations at Air Upas and Sandai Mines in west Kalimantan, Indonesia, describes the problems facing smaller mining operations in remote parts of the globe in meeting GISTM requirements. Global Tailings Review (GTR, 2020), as co-convened by the International Council of Mining and Metals (ICMM), the United Nations Environment Programme (UNDP) and Principles for Responsible Investment (PRI) is structured around six topic areas, namely:

1. Communities.
2. Integrated Knowledge Base.
3. Design Construction and Operation.
4. Management and Governance.
5. Emergency Response.
6. Disclosure.

These six topic areas encompass 15 principles, which in turn include 77 requirements and more than 219 criteria. Separate from this, the ICMM has developed Conformance Protocols, which are mapped to the 77 requirements (ICMM, 2021a). The ICMM has also produced a Tailings Management Good Practice Guide (ICMM, 2021b). The Conformance Protocols aim to assist members in fulfilling their commitment to the GISTM implementation at TSF facilities, with a GISTM Classification of 'Very high' to 'Extreme' needing to conform to the standard by 5 August 2023. All other tailings facilities operated by members that are not yet in a state of safe closure, must conform by 5 August 2025. Time is running out.

Tailings storage facilities all over the world are audited against all of the GISTM requirements, and though well-meaning and necessary, one must question whether they are really applicable in difficult to access remote areas where operations are small, and environmental concerns often minor. Here a watered-down version of GISTM 'GISTM-light' may be more applicable as outlined in this paper.

REMOTE OPERATIONS IN INDONESIA AS CASE STUDY

GISTM audits were undertaken for Air Upas and Sandai Mines in west Kalimantan, Indonesia as published in December 2023 (KCB, 2023). These mines are very remote. It takes two days of travel from major centres such as Singapore and Australia, and even nearby Jakarta requires a full day's travel there, and another back. Exacerbating the problem are time-consuming visa requirements, flights required to bigger cities remote from these sites, and in many parts – Kalimantan included – many hours of 4×4 travel through jungle single-tracks, barely passable, at times flooded and sometimes inaccessible for long periods. Paved access is equally unappetising with many hours of driving across narrow undulating and busy road networks. Figure 1 provides the location of the two Kalimantan mines.

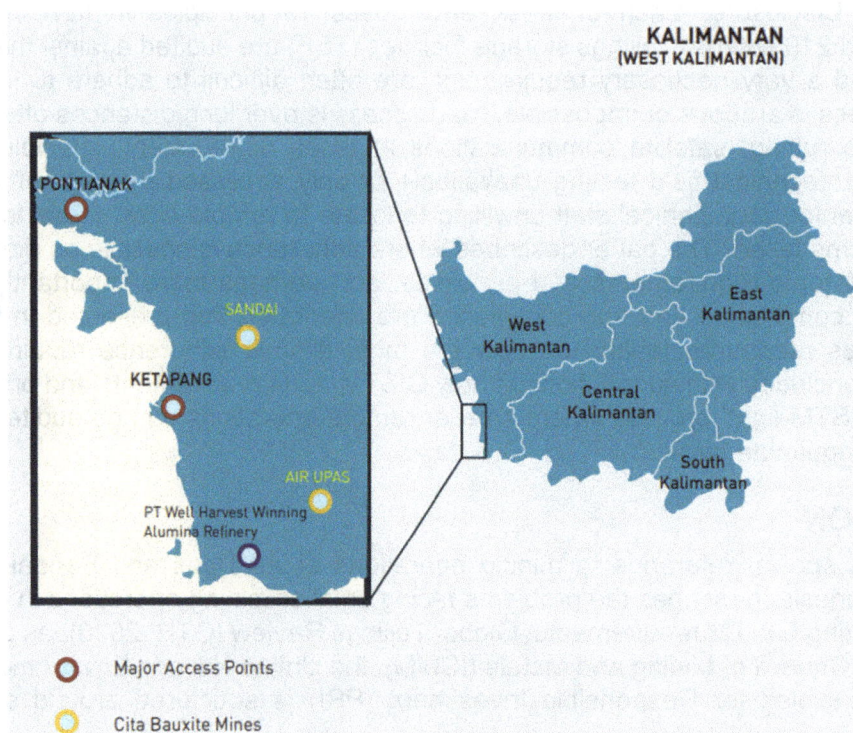

FIG 1 – Locality plan – Air Upas and Sandai Mines.

At both Sandai and Air Upas, bauxite-rich soil from near-surface mining is transported to nearby TSF washing stations where the bauxite-rich ore is separated from soil fines via rotating drum screens in a washing process, and the residue flushed out as tailings. Flushed residue draining from the washing process is treated with a flocculant before streaming into a primary trench and the thickened residue then excavated and further thickened in secondary, tertiary and quaternary sumps excavated for that purpose. The thickened tailings is eventually loaded onto trucks and transported to residue landfills. The remaining tailings is clarified through progressive overflow across a series of purpose-built residue tailings dams. Clear water from this process is then pumped back for reuse at the washing stations. Mining operations at these two facilities have been in existence for over three decades. The Metallurgical Grade Bauxite (MGB) washed and concentrated product is trucked from on-site stockpiles to a coastal refinery where it is used as raw material at the refinery plant to produce Smelter Grade Alumina (SGA).

The non-dumped tailings is captured in a series of relatively small in height (generally <15 m) TSF dams. The tailings is free from any added chemicals. These mine areas have few inhabitants given their low-lying swamp or jungle locations, and even should over-topping or breaching occur downstream areas are mostly devoid of people with no habitation as per mining requirements. Being flat lying also means that any over-topped or breached tailings would have only a limited flood footprint that would be mostly confined to the zone abutting the TSF facility. The rhetorical question begs then whether, under these circumstances, is it still necessary to conform to the full omnibus version of stringent GISTM requirements?

AIR UPAS AND SANDAI AUDIT RESULTS

A summary assessment of the conformance levels of the 77 audit GISTM requirements for these two sites is provided in Table 1. The table provides the number of requirements that align with each level of conformance.

TABLE 1

GISTM audit results.

Topic area	Principle	Compliant	Substantial Compliant	Partially Compliant	Not Compliant	Not Applicable
Communities	Principle 1 – Rights of project-affected people with meaningful engagement	2	2			
Integrated Knowledge Base	Principle 2 – Maintenance of a knowledge base to support a safe tailings management life cycle plus closure	1	2	1		
	Principle 3 – Use of all knowledge base elements – social, environmental, economic, technical – to support TSF life cycle	1		2		1
Design, Construction and Operation	Principle 4 – Develop plans and design to minimise all risk in all phases of the life cycle including closure		3	1	3	1
	Principle 5 – Provide robust design that integrates knowledge base and minimises risk through all life cycle phases	3	2	1		2
	Principle 6 – Plan, build and operate the TSF to manage risk through all life cycle phases		3	1	1	1
	Principle 7 – Design, implement and operate monitoring systems to manage TSF facilities throughout a life cycle	1	1	3		
Management and Governance	Principle 8 – Establish policies, systems and accountabilities to support TSF life cycle safety and integrity		1	2	2	2
	Principle 9 – Appointment and empowerment of an Engineer of Record			1	4	
	Principle 10 – Establishment and implementation of review levels to provide strong quality and manage risk for all phases		2	2	2	1
	Principle 11 – Develop an organisational structure that promotes learning, communication and early problem recognition	2	2	1		
	Principle 12 – Establish a process for reporting, addressing concerns and protecting whistleblowers	2				
Emergency Response	Principle 13 – Preparation of an emergency response to tailings failures		2			2
	Principle 14 – Preparations for long-term recovery in the event of catastrophic failure	1				4
Disclosure	Principle 15 – Publicly disclose and provide access to information in supporting public accountability		2	1		

Table 1 indicates the mine is achieving 45 per cent compliance in the top two compliance levels and only 15 per cent falling into the non-compliant category. This is considered a positive outcome given the remoteness of these sites and lack of sophisticated access to geotechnical testing mechanisms, monitoring, or very experienced geotechnical staff.

COMPLIANCE VERSUS NON-COMPLIANCE

This section provides a brief overview of compliance versus non-compliance items, with those that are compliant in terms of GISTM requirements at the Indonesia sites, and more importantly, those that are not since it is the latter that make compliance in remote areas for small operations so difficult.

Correct conformance compliance

- Communities surrounding the TSF areas, as well as all local public authorities, are fully engaged as part of the mining process and thereby compliant with the 'Communities' GISTM topic requirements.

- There is robust compliance with Integrated Knowledge Base and Design, Construction and Operation (DCO) topics including design, risk management and monitoring. All of which are understood and implemented. These have been mostly delivered by geotechnical consultants Tura Consulting Indonesia (2020) who have provided sound geotechnical principles for the TSF sites. There is also robust daily monitoring via a team of inspectors on motorbikes who, from remote stations at times, monitor the facility daily with immediate notification of any issues such as seepage from the dams, overtopping, bulging, erosion, tension cracking, or any other signs of slope instability. Their enthusiasm knows no bounds.

- Emergency Response plus Disclosure topics elicit a good response and whistleblowers are protected, but there are GISTM gaps in the Maintenance and Governance GISTM topic which require attention as presented in the section following.

These tailings facilities were previously monitored for general tailings deposition compliance by KCB (2019) in the period prior to GISTM. Even better, a second audit in 2023 provides ample evidence of efforts towards GISTM compliance, with a marked general improvement in management and monitoring of tailings at the mines. Mine management and staff are keen to ascribe to GISTM requirements, and it is this enthusiasm, as in other parts of the world, that must be encouraged and nurtured. It is also one of the prime reasons why smaller remote operations should be drawn into, and remain, within the GISTM system, and not allow them to be put off by creating targets that are difficult to achieve, if at all. Especially so given the significant costs relative to the level of conformance required.

Non-compliance

Two topic areas at Air Upas and Sandai mines provide all the non-compliant (KCB, 2023) items. These are as follows.

Design, construction and operation topic

Non-compliance for this topic include:

- No appointed Accountable Executive (AE).
- No standalone Design Basis Report (DBR).
- No detailed Tailings Management System (TMS) register or report.

Annexure 1 of GISTM (GTR, 2020) defines the requirements of an Accountable Executive to be 'accountable for the safety of tailings facilities' and further someone who 'may delegate responsibilities but not accountability'. This definition describes a very senior company position and is for someone who needs to be continuously in touch with day-to-day mine operations, and to all intents and purpose virtually residing on-site. Even remote AE with delegated authority on the ground will attract major costs. Can small distant and remote operations, as those described, afford such a senior person, and are there senior personnel willing to live/work at these relatively low-key operations? Probably not. There must therefore be a different way to provide such responsibility: perhaps one of a lower order but still acceptable position. This lack of a senior Accountable Executive and Engineer of Record (section following), as geotechnical council, mean that fulfillment of other GISTM requirements, such as DBR and TMS, are in turn difficult to ascribe to and at times virtually impossible.

Management and governance

Non-compliance for this topic includes:

- No appointed Accountable Executive (again).

- No appointed Engineer of Record (EoR).

- No designated Responsible Tailings Facility Engineer (RTFE).

- No documented evidence of any internal audits.

- No existing Closure Report or related documentation.

These, as before, all point to the appointments of senior executive personnel being one of the major hurdles to GISTM compliance. These persons are mostly unavailable or unwilling to fulfill senior posts in smaller remote operations. The EoR role, as for the earlier AE description, also requires the GISTM responsibility (GTR, 2020) of 'confirming that the tailings facility is designed, constructed, and decommissioned with appropriate concern for integrity of the facility, and that it aligns with and meets applicable regulations, statutes, guidelines, codes and standards'. There is also the repeated requirement for full 'accountability'. These stringent GISTM requirements, whilst well-meaning, appear not to fit the requirements of remote smaller mining operations. Here the appointment of a fulltime site RTFE to undertake this geotechnical work could be considered as a first step in addressing this impasse. This as expanded on in the sections following.

THE CASE FOR 'GISTM-LIGHT'

There looks to be sound argument for development of a 'GISTM-light' category that will still comply to international GISTM requirements, but one that will specifically cater for the TSF management and monitoring needs of smaller remote mining operations.

Reasons for a GISTM-light requirement

Reasons why the omnibus version of GISTM requirements does not meet small remote mining operations is summarised as follows:

- Mining operations are distant requiring difficult, time-consuming and costly logistics in terms of access.

- Sophisticated investigation techniques such as Seismic Cone Penetration (CPTu) testing, Vibrating Wire Piezometer (VWP) installations with telemetry, inclinometer installations with monitoring and requisite borehole drilling campaigns would be costly, and difficult, given the major logistics in establishment of machines on-site.

- The site investigation campaigns associated with the above techniques also require experienced and qualified geotechnical staff further inflating costs, which together raise costs by a significant order of magnitude.

- Electricity is often not available in remote areas and neither wi-fi connectivity.

- Senior executive staff, such as AE and EoR appointments, would probably be too costly for such small remote operations, whether on-site or remote with delegated authority.

- AE and EoR personnel are in any case probably unwilling to undertake such relatively low-key roles, especially since these could be regarded as unattractive opportunities for these core skills in a significantly constrained resource sector.

There is, notwithstanding the above, still every reason to encourage GISTM policies and adherence in even the remotest areas of the world. Even if partially so. It is recommended, for this reason alone and in order to provide GISTM continuity, that a GISTM-light category be considered as part of the broader set of GISTM requirements.

GISTM-light

GISTM-light could be structured as follows:

- Appointment of a Responsible Tailings Facility Engineer would be considered mandatory.

- This individual would need to be on-site full-time to cater to all geotechnical TSF requirements including technical geotechnical execution, management and monitoring.

- Sophisticated investigation and monitoring techniques could be substituted using more low-cost items such as Casagrande Standpipe Piezometers with daily monitoring instead of VWPs; daily land surveying of prism targets in areas suspected of being prone to lateral movements, satellite monitoring procedures, and continued daily inspections and monitoring by motorbike personnel of any potential dam safety issues.

- Consideration could be given to annual compilation of an internal geotechnical audit with concomitant reporting by the RTFE. This for presentation to a mine-appointed geotechnical engineering consultancy capable of checking for potential geotechnical gaps of on-site geotechnical systems.

- Local geotechnical on-site self-governance and on-site problem solving should be encouraged. The Air Upas and Sandai sites use, for example, a system of cantilever piles to check early detected lateral creep movement in dam slopes, and in so doing temporarily prevent major breach. Installing these provides time for installation of other geotechnical intervention such as slope flattening and/or additional subsurface drainage. For this piling large trees, which are plentiful, are stripped of their branches and banged into the ground at close spacing using an excavator. Simple but highly effective.

- TSF rehabilitation activities could be performed simultaneously with ongoing or new TSF operations. Large fallow dam areas can be planted up with timber, or fruit trees. This is possible at the Kalimantan sites audited given the chemical-free ore concentration washing process. Current rehabilitation operations in mined out areas already show excellent tree regrowth (CITA, 2022) even over short periods of a few years.

- A template for the GISTM-light audit would need to be compiled using geotechnical requirements as determined by the international GISTM group, with clear distinction between GISTM-light versus full GISTM requirements.

- GISTM-light could be included as an Annexure as part-fulfilment initially before upgrading, if considered necessary to full GISTM (GTR, 2020) requirements.

- An annual, or biannual, visit by the mine appointed geotechnical engineering consulting firm to check site geotechnical GISTM-light compliance and determine gaps within site operations and how these can be attended to, would be an enforced mandatory requirement.

- Accountability would reside with the RTFE.

- There will be a need to carefully formalise the GISTM-light to demonstrate clear delineation of responsibility and accountability, ensuring the system stands up to audit scrutiny.

CONCLUSIONS

The full suite of GISTM TSF audit requirements is difficult, time-consuming and costly to implement, which makes its use in remote small mining operations at times untenable. Geotechnical site investigation campaigns are also costly and difficult to undertake, and senior Accountable Executives and Engineers of Record difficult to come by, expensive and hard to entice to smaller remote operations. At the same time the enthusiasm and attempts at compliance by staff in these remote mining operations needs to be lauded and encouraged. It is recommended therefore, in efforts to ensure this, that a GISTM-light system be considered for use in remote small mining TSF operations. A GISTM-light system, as presented in this paper, could be included in the GISTM omnibus version as an Annexure, with protocols and direction as determined by the international GISTM body.

ACKNOWLEDGEMENTS

The author wishes to thank KCB for allowing publication of this paper and CITA Indonesia for giving KCB the opportunity to work with them on GISTM audits.

REFERENCES

Global Tailings Review (GTR), 2020. Global Industry Standard on Tailings Management (GISTM) [online], Global Tailings Review. Available from: <https://globaltailingsreview.org/global-industry-standard/>

International Council on Mining and Metals (ICMM), 2021a. Conformance Protocols: Global Industry Standard on Tailings Management, May 2021, ICMM Publication. Available from: <https://www.icmm.com/en-gb/our-principles/tailings/tailings-conformance-protocols>

International Council on Mining and Metals (ICMM), 2021b. Tailings Management: Good Practice Guide [online], International Council on Mining and Metals, pp 1–142. Available from: <https://www.icmm.com/tailings-management>

KCB, 2019. Geotechnical Review – KCB Preliminary Review Comments.

KCB, October 2023. CITA Indonesia, Global Industry Standard for Tailings Management Audit and Gap Assessment, pp 1–38.

Tura Consulting Indonesia, April 2020. Final Report – Geotechnical Investigation, Hydrology – Hydrogeology and Embankment Design Bauxite Residue Storage Facility, PT Cita Mineral Investindo, Tbk, Sadai Site and Upas Site, Ketapang Regency, West Kalimantan

Enhancing quality assurance in the construction of tailings storage facilities through automated workflows

M Rojas[1], M Teh[2], N Pereira[3] and M Llano-Serna[4]

1. Geotechnical Engineer, Red Earth Engineering a Geosyntec Company, Perth WA 6000.
   Email: martin.rojas@redearthengineering.com.au
2. Principal, Red Earth Engineering a Geosyntec Company, Perth WA 6000.
   Email: ming.teh@redearthengineering.com.au
3. Senior Engineer, Red Earth Engineering a Geosyntec Company, Perth WA 6000.
   Email: nicolas.pereira@redearthengineering.com.au
4. Principal, Red Earth Engineering a Geosyntec Company, Brisbane Qld 4000.
   Email: marcelo.llano@redearthengineering.com.au

ABSTRACT

Construction quality assurance (CQA), alongside quality control (QC) and construction versus design intent verification (CDIV), form a systematic process to ensure compliance with technical, contractual, and regulatory requirements, and ensure that the original design intent remains valid. In earth-structure projects such as tailings storage facilities (TSFs), CQA often requires manually handling large volumes of data—including field reports and laboratory results—organised in basic folder structures. This method is labour-intensive, requiring significant personnel hours, making data entry errors hard to detect. Transcription mistakes are common.

This paper presents an automated CQA workflow for tailings dam construction that integrates Artificial Intelligence (AI)-powered Optical Character Recognition (OCR), Python scripting, and email-based classification to extract, verify, and record test results, generate thickness reports, and organise other quality documents. The tool manages multiple test certificates in Portable Document Format (PDF), including particle size distributions (PSD), Atterberg limits, dynamic cone penetration, and density tests. recording results, verifying them against existing records, and flagging potential anomalies, like incorrect coordinates or repeated data. The script takes about 21 seconds to extract and compile raw data from each PDF into a spreadsheet, running in the background, freeing engineers for other tasks. This cuts transcription time to the spreadsheet by approximately 92 per cent, shifting manual effort to data verification.

The resulting improvements in data management and verification facilitate more efficient preparation of time-sensitive construction records reports. A case study is presented. The study involved a forensic geotechnical investigation. The project required processing around 500 compaction test certificates embedded in a report of more than 5000 pages. The data was processed and analysed in a few days. The efficiencies demonstrate the adaptability of the workflow. The outcome highlighting how the judicious integration of AI and scripts leads to more reliable data management in geotechnical engineering projects. This automated approach not only minimises transcription errors and reduces costs through higher efficiency but also allows engineers to focus on higher-value tasks, improving overall project quality and compliance.

INTRODUCTION

Poor construction of a Tailings Storage Facility (TSF) increases long-term risk and must be avoided (The Mining Association of Canada (MAC), 2021). To achieve this, appropriate quality management programs are needed, in accordance with global best practices for tailings management (Oberle, Brereton and Mihaylova, 2020) and aligned with the *Towards Zero Harm* goal of the Global Industry Standard on Tailings Management (GISTM, Global Tailings Review, 2020). A quality management program aims to ensure that the dam is built according to design principles and to mitigate risks associate to sub-standard construction (MAC, 2021). These programs typically involve two key different components: Quality Assurance and Quality Control (QA/QC).

Efficient management of the large volume of data generated through QC activities can be challenging for the Construction Quality Assurance (CQA) team. Data typically includes field and laboratory tests for compaction control, such as Standard and Hilf methods, as well as material characterisation tests, including particle size distribution and Atterberg limits. Topographic survey reports are important for verifying the as-built geometry against the design, including checks on lift thickness conformance. Manually processing, verifying, and validating this information is often time-consuming, repetitive, and prone to human error. This becomes particularly challenging during high-pressure scenarios, such as material investigations, design modifications, or non-conformance events, where large data sets must be quickly reviewed to support timely decisions and avoid delays in construction progress.

This paper presents a tool to facilitate QA that integrates automated workflows with AI-powered Optical Character Recognition (OCR). The system streamlines sorting, storage, and extraction of test data, and generates thickness reports for each constructed lift. It enhances documentation, traceability, and QA efficiency during TSF construction. Several application examples are described to illustrate the method's versatility and adaptability.

QUALITY MANAGEMENT IN TAILINGS DAM CONSTRUCTION

Effective quality management is essential to ensure a Tailings Facility is built according to the design intent, minimising the risk of construction-related failures. QA and QC are closely related yet distinct components of quality management. QA involves defining standards and procedures—such as material specifications and construction methodologies—established during the design phase. QC focuses on verifying compliance with these standards during construction. In simple terms, QA ensures tasks are performed correctly, while QC confirms the outcomes match expectations (MAC, 2021). The International Organization for Standardization defines QA as 'all planned and systematic actions necessary to provide confidence that a product or service will meet quality requirements' (ISO, 2015).

GISTM (Global Tailings Review, 2020) in the Requirement 6.2 mandates that operators manage the quality and adequacy of construction using QA, QC, and Construction versus Design Intent Verification (CDIV). The information related to QA/QC must be documented and included in the Construction Records Report (CRR). As outlined in Requirement 6.3, this report must be prepared whenever there is a material change to the TSF, its infrastructure, or monitoring systems.

The preparation of a CRR is often time-sensitive, particularly under constraints such as limited tailings storage capacity or tight commissioning deadlines. In some projects, CRRs have taken 12 to 18 months to be finalised after construction, according to industry feedback. The automated workflow presented in this paper has significantly improved the efficiency of managing the data required for the report. Approved documents and technical records are systematically stored as they are received, organised into structured directories, and automatically compiled into appendices for final reporting. This enhances traceability and supports a smoother closure of construction stages.

AUTOMATED WORKFLOW OVERVIEW

Automation is increasingly recognised in geotechnical engineering—as well as other disciplines—as tool to enhance efficiency, accuracy, and decision-making across data-intensive workflows. Several authors have explored the benefits of automating engineering processes to improve reliability and streamline repetitive tasks in data-driven environments (Stohr and Zhao, 2001; Farook, 2019; Reddy, 2017; Esiri, 2024).

In geotechnical design, digital workflows simplify subsurface data management, reduce manual input, and deliver faster, more reliable results (Reddy, 2017). In numerical modelling, scripting enables batch analyses, eliminates manual set-up, and accelerates iterations—freeing engineers to focus on higher-value decisions (Rocscience Inc, 2025). Across applications, automation consistently delivers technical and commercial benefits, including faster calculations, fewer errors, and increased productivity (Farook, 2019).

In the context of CQA, traditional workflows often involve repetitive and labour-intensive tasks—such as manual data entry, spreadsheet management, and filing systems—which increase the risk of error and inefficiency (Peng and Nadukuru, 2024). Automating these workflows using scripting languages such as Python can significantly improve consistency, traceability, and time management.

Python scripting

Python is widely adopted in automate engineering workflows due to its open-source nature, extensive ecosystem of scientific libraries, and flexibility in integrating with various tools and file formats. Its simplicity and adaptability make it well suited for automating repetitive tasks, streamlining complex workflows, and supporting efficient data processing across different stages of a project (Ogli, 2024). For more advanced analysis or large data set processing, platforms such as R or other specialised tools may offer additional advantages.

Optical character recognition

Several tools for information extraction (IE) from PDF documents have emerged since the format's inception, evolving from rule-based algorithms to statistical machine learning (ML) and, more recently, deep learning (DL) models—significantly improving extraction accuracy and flexibility (Meuschke *et al*, 2023). Optical Character Recognition (OCR) is one such IE method. It detects and converts printed or handwritten text in scanned documents and images into machine-readable digital formats and has followed a similar technological progression. Modern OCR systems now achieve high levels of accuracy and robustness across varied conditions (Llopart, Behjati and Aliguer, 2024; Meuschke *et al*, 2023; Meng and Wang, 2024).

A wide range of OCR tools is currently available, many of which are open source. Benchmark studies have assessed platforms such as Tesseract, Amazon Textract, Google Document AI, Adobe PDF Extract, and ABBYY Cloud OCR. Each tool offers different levels of accuracy, scalability, and compatibility with programmatic workflows, making them effective solutions for automating engineering processes (Farook, 2019; Hegghammer, 2022; Meuschke *et al*, 2023). For example, Adobe's PDF Extract API uses Adobe Sensei AI to analyse document structure and extract structured content such as text and tables with high precision (Adobe Inc, 2021; Meuschke *et al*, 2023). These benchmarking efforts help users select the most suitable OCR tool based on their data type, project needs, and desired outputs.

PROPOSED WORKFLOW SYSTEM

Figure 1 presents the proposed CQA workflow. The flow diagram outlines the logic followed by the automated scripts:

- Purple boxes represent tasks executed fully by scripts.
- Blue boxes represent script-driven tasks that require user confirmation for manual review.
- Green and red boxes indicate conformance and non-conformance outcomes, respectively.
- Rectangular boxes with sharp corners denote final workflow outcomes.

The streamlined workflow is based on the automated processing of incoming emails using Python scripting. The process can be triggered manually or scheduled using freeware tools such as Robocopy for Microsoft Windows. Once activated, the script filters emails according to user-defined relevance rules. Relevant messages are saved in the project inbox, and their attachments are automatically classified into the following categories:

- Documentation: Technical queries (TQs), inspection and test plans (ITPs), RFIs etc.
- Test reports: Typically, in PDF format.
- Survey reports: Usually DXF or CSV files from field surveys.

Each type follows a dedicated process stream, as shown in the Figure 1.

The following sections outline the general processes within the proposed workflow and suggest commonly used Python libraries. However, specific tools and libraries may vary depending on the user's environment or preferences. Likewise, the workflow structure can be adapted, refined, or expanded to meet project-specific requirements.

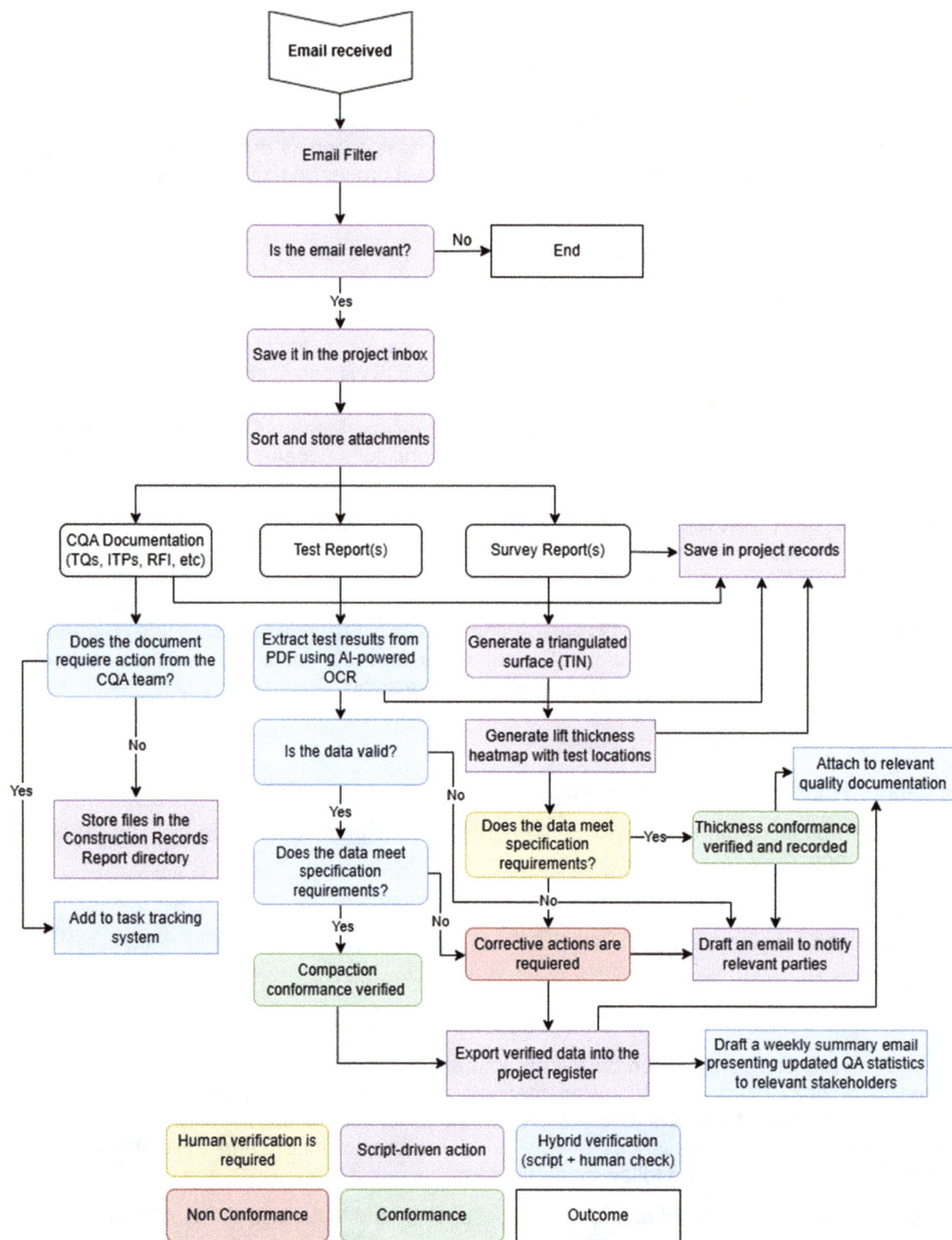

FIG 1 – Flow diagram showing the CQA automated workflows.

Document classification and handling

For quality documents, the script checks if action is required from the CQA team. Finalised and signed documents are stored in the relevant project directory and in the appendix folder of the Construction Records Report. Documents requiring review are added to the tracking system (*to-do list*) for CQA action.

Key libraries used include os, shutil, datetime, re, and win32com.client.

Processing test reports

Test certificates are saved in a temporary folder for initial manual verification. If they appear ready, user confirmation triggers a Python script that uses AI-powered OCR to extract raw data from the PDF files. The extracted data is stored in a structured format and recorded for the project database. A secondary script then checks for common errors, such as duplicated coordinates, values outside expected ranges, or inconsistencies in location assignments. A pop-up window alerts the user if the data meets project specifications or if potential issues are detected.

For example, the system can flag liquid limits exceeding specification thresholds, excessive fines in PSD tests, or dry density ratios below the required minimum. These checks occur almost in real time. The CQA engineer then validates the results and confirms conformance or identifies non-conformance. If the test is compliant, the associated reports are appended to the relevant quality documentation. If not, the script offers the option to draft a notification email to relevant parties and updates the project status.

Some key libraries used include pandas, and tkinter. The implementation can be adapted to suit different tools and preferences. For example, libraries such as pytesseract, PyMuPDF, or commercial OCR APIs can be used depending on the available set-up and user requirements.

Survey data and heatmap generation

Survey reports, such as point cloud files for a completed lift, are processed to create a triangulated surface (TIN). The system then generates a heatmap of thickness between layers and overlays the test locations based on the records to date, flagging any areas out of tolerance. This process can be completed in under a minute, allowing rapid identification of non-conformances and enabling timely corrective actions.

Key libraries used include ezdxf, numpy, matplotlib, and scipy.

Outcome

The outcomes of the automated workflow, as shown in the square-cornered boxes in Figure 1, include:

- QC results are assessed for conformance or flagged as non-conforming based on project technical specifications.
- Thickness heatmaps are generated to demonstrate conformance with lift design criteria, with test locations, failed results, and approved retests clearly marked.
- Project traceability is maintained through systematic file structuring and record-keeping.
- Verified test data is incorporated into the project records.
- Draft emails are automatically generated to inform relevant parties of non-conformances or actions required.
- All quality documentation and test results are saved in the project directory and appended to the Construction Records Report.

Practical examples

Particle size distribution test workflows

PSD test results are commonly used during tailings dam construction to confirm that fill material complies with the design specification. Each test result must be verified for location, sample ID, and plotted curve shape to ensure compliance with the project's Construction Technical Specification.

Manually processing a single PSD test can take around five mins, including file registration, data transcription and curve plotting. When multiple test reports are received—commonly ten to twenty at a time during material investigations—the total time investment becomes significant, particularly under tight reporting deadlines.

By contrast, when using the proposed automated workflow, the same set of tasks—receiving, extracting, verifying, and storing up to twenty test results from a single email—can be completed in approximately seven mins total.

The automation enables true multitasking. While the script runs in the background, the user is free to focus on other tasks. Verification of the processed data can be performed afterwards.

Survey data verification and lift approval

In another common scenario, the contractor submits survey data for a completed lift and requests release of the hold point to proceed with the next lift. With the automated system, a heatmap of thickness variation can be generated within seconds by comparing the current lift surface with the previous lift. This allows immediate identification of any out-of-tolerance zones or areas requiring corrective action.

The CQA engineer can quickly review the heatmap and draft an email to the construction team, confirming whether it is acceptable to proceed with the next lift or whether the submitted lift exceeds tolerance limits and must be addressed.

Construction records report

Once all documentation is signed by the responsible parties, the script automatically stores the quality documents in the appropriate project directories. Large volumes of files—typically PDFs such as Daily Reports—can be compiled using scripts, reducing strain on system resources since the code can run in the background while other tasks are performed.

Forensic investigation

A forensic investigation was initiated following the formation of sinkholes on a tailings dam. The assessment required rapid analysis of existing data, particularly Standard and Hilf compaction test results.

The client provided a report with over 500 test certificates embedded in the appendices as individual PDFs in an over 5000 pages CRR. Due to a tight deadline, an automated workflow was implemented to extract and process the data efficiently.

The main challenge was converting scanned PDFs—often digitised from physical sheets—into a spreadsheet format suitable for plotting and interpretation. The use of AI-powered OCR enabled rapid data extraction, significantly reducing the time required compared to manual methods.

The workflow followed a hybrid model: scripts extracted, cleaned, and organised the data using predefined rules, while the engineers applied their judgement to verify and interpret the results. This balance between automation and engineering oversight ensured reliable, high-quality outputs under tight time constraints—saving dozens of project hours.

Performance

The system significantly reduces processing time. On average, it takes approximately 21 seconds from script execution to extract test results from a single PDF file into an Excel spreadsheet—ready for verification.

This estimate is based on timing five individual extractions and two runs each for batches of five, ten, and twenty PDFs. For comparison, manually completing the same tasks—opening emails, saving attachments, creating the appropriate folders, and transcribing data into spreadsheets—takes in average 4.5 mins per test report. Based on these user-tracked measurements, the automated script reduces processing time by approximately 92 per cent per PDF file.

Additionally, generating a heatmap that displays lift thickness and plots all test locations to date takes approximately one min—from processing the survey report (DXF file) to producing the final output.

It is important to note that while automation significantly reduces data processing time, its implementation requires some initial set-up and coding proficiency. Although the gap between engineers and software developers can be substantial, recent advances in large language models

(LLMs), such as GitHub Copilot, now support the drafting and acceleration of scripts—further increasing process efficiency.

Implementation challenges

Due to variations in test formats and the specific QA/QC practices applied across projects, implementation of the proposed workflows must be case-specific. The tools and logic should be adapted, refined, and aligned with each project's documentation structure and operational needs to maximise their effectiveness.

Data security is a critical consideration. Client, company and project data must be safeguarded throughout any AI-based processing. Only secure, validated OCR or information extraction technologies—approved by the organisation—should be used to ensure compliance with internal governance policies and confidentiality requirements.

Importantly, AI tools cannot replace engineering judgement. While automation enhances efficiency, all outputs must be reviewed and validated by qualified professionals. Scripts must include error-checking routines to prevent inaccurate data from being used. Overreliance on automated processes without validation can undermine the intended benefits and introduce risk. AI should be seen as a support tool—not a substitute for critical thinking or professional responsibility.

CONCLUSIONS

This paper presents how automated workflow can improve digital data management in the Quality Assurance of tailings dam construction. From routine test verification to urgent forensic assessments, the integration of AI-powered OCR and scripted processes reduces manual handling time and enhances consistency and traceability.

The presented automated workflow cut the time required for repetitive daily tasks by up 90 per cent, enabling engineers to trace and record data and perform more efficient quality assurance. The system supports real-time decisions, early identification of non-conformances, and streamlined record-keeping.

Crucially, the hybrid approach ensures that automation complements—rather than replaces—engineering judgement. By combining scripted processing with human oversight, the methodology maintains both efficiency and accountability, aligning with project requirements.

As regulatory and operational expectations continue to evolve, practical, scalable automation of data handling processes offers a clear path to improve compliance, reliability, and overall performance in geotechnical engineering projects.

ACKNOWLEDGEMENTS

On behalf of all authors, the corresponding author states that there is no conflict of interest.

REFERENCES

Adobe Inc, 2021. Adobe PDF Extract API Technical Brief, version 1.0. Available from: <https://developer.adobe.com/document-services/docs/assets/268b4618cd5696a95ebf8cc01de5f310/Adobe_PDF_Extract_API_Technical_Brief.pdf> [Accessed: 8 April 2025].

Esiri, E, 2024. The Role of QA Automation in Eliminating Waste in Project Teams, MSc thesis, Harrisburg University.

Farook, M, 2019. Automation in geotechnical design – application and case studies, in *Proceedings of the 17th European Conference on Soil Mechanics and Geotechnical Engineering (ECSMGE 2019)*, 8 p. https://doi.org/10.32075/17ECSMGE-2019-0971

Global Tailings Review, 2020. Global Industry Standard on Tailings Management (GISTM), International Council on Mining and Metals (ICMM), the United Nations Environment Programme (UNEP) and the Principles for Responsible Investment (PRI). Available from: <https://globaltailingsreview.org>

Hegghammer, T, 2022. OCR with Tesseract, Amazon Textract and Google Document AI: a benchmarking experiment, *Journal of Computational Social Science*, 5(1):861–882.

International Organization for Standardization (ISO), 2015. ISO 9000:2015 Quality management systems – Fundamentals and vocabulary (definitions), Geneva: ISO.

Llopart, J, Behjati, P and Aliguer, I, 2024. AI-based digitization of legacy ground information, *Proceedings of the 7th International Conference on Geotechnical and Geophysical Site Characterization*. Available from: <https://www.scipedia.com/public/Llopart*_et_al_2024a>

Meng, F and Wang, C A, 2024. Artificial intelligence and machine learning approaches to text recognition: a research overview, *Journal of Mathematical Techniques and Computational Mathematics*, 3(3):1–5.

Meuschke, N, Jagdale, A, Spinde, T, Mitrović, J and Gipp, B, 2023. A benchmark of PDF information extraction tools using a multi-task and multi-domain evaluation framework for academic documents, *Information for a Better World: Normality, Virtuality, Physicality, Inclusivity*, Lecture Notes in Computer Science, 13972:383–405 (Cham: Springer Nature Switzerland). https://doi.org/10.1007/978-3-031-28032-0_31

Mining Association of Canada, The (MAC), 2021. *A guide to the management of tailings facilities* (Version 3.2). The Mining Association of Canada. Available from: <https://mining.ca/wp-content/uploads/dlm_uploads/2021/06/MAC-Tailings-Guide-Version-3-2-March-2021.pdf> [Accessed: 8 April 2025].

Oberle, B, Brereton, D and Mihaylova, A (eds), 2020. *Towards Zero Harm: A Compendium of Papers Prepared for the Global Tailings Review*, Global Tailings Review, London. Available from: <https://globaltailingsreview.org/wp-content/uploads/2020/08/towards-zero-harm.pdf> [Accessed: 23 June 2025].

Ogli, O K H, 2024. Python's Role In Revolutionizing Automation and Workflow Optimization, *Biologiya Va Kimyo Fanlari Ilmiy Jurnali*, 1(10):33–38.

Peng, X and Nadukuru S S, 2024. Transforming CQA Through Digital Innovation, *GeoStrata Magazine Archive*, 28(6):46–53.

Reddy, Y R, 2017. Automation of the digital information workflow in the geotechnical design process, *International Journal of Research and Analytical Reviews (IJRAR)*, 4(4):664–672.

Rocscience Inc, 2025. Why engineers use Python to automate geotechnical engineering in RS2. Available from: <https://www.rocscience.com/learning/why-engineers-use-python-to-automate-geotechnical-engineering-in-rs2> [Accessed: 8 April 2025].

Stohr, E A and Zhao, J L, 2001. Workflow automation: overview and research issues, *Information Systems Frontiers*, 3(3):281–296.

Standardising vulnerability assessments of tailings dams – advancing beyond trigger analyses

A O Sfriso[1] and M G Sottile[2]

1. Practice Leader and Group Chair, SRK Consulting, Buenos Aires, Argentina. Email: asfriso@srk.com.ar
2. Principal Consultant, SRK Consulting, Buenos Aires, Argentina. Email: msottile@srk.com.ar

ABSTRACT

The assessment of tailings dam vulnerability to flow liquefaction has evolved from deterministic factor-of-safety approaches to sophisticated numerical deformation analyses. Following the proposal introduced by the authors at PCSMGE 2024, this study refines the framework by integrating recent advancements in constitutive modelling, Eurocode 7 (Second Generation) reliability principles, and performance-based design concepts. The methodology aims at replacing traditional trigger analyses with a structured vulnerability assessment framework. It prescribes a uniform set of deteriorating actions—crest loading, toe contraction, and phreatic surface rise—regardless of site-specific conditions. These actions are applied systematically in pushover-style analyses to quantify the dam's susceptibility to flow liquefaction, moving towards a limit state-driven design philosophy akin to the Reliability Based Design approach of Eurocode 7. The methodology further introduces the construction of an Ultimate Limit State (ULS) surface, derived from multiple numerical simulations, and interpolated using Radial Basis Functions (RBF). A fuzzy transition zone is incorporated to account for modelling uncertainty, enabling the computation of a continuous probability of failure (PoF) across the action space using a Gaussian cumulative distribution. The framework includes the use of empirical or parametric probability density functions (PDFs) for each trigger, combined through Latin Hypercube Sampling in a Monte Carlo simulation to estimate annual PoF. By standardising vulnerability assessments, this methodology facilitates more objective risk management, improving dam resilience and regulatory compliance. The paper concludes with a case study illustrating the use of this procedure in a hypothetical tailings storage facility.

INTRODUCTION

Over the past several years, a comprehensive framework has been developed by the authors and collaborators to assess the vulnerability of upstream-raised tailings dams to failure induced by flow liquefaction. This body of work responds to regulatory shifts and technical imperatives arising after major tailings dam failures (eg Fundão, Brumadinho), and is grounded in the use of deformation modelling to simulate undrained strain-softening behaviour in saturated tailings.

The first numerical methodology for assessing the vulnerability of upstream-raised tailings dams to flow liquefaction was presented in Sottile, Cueto and Sfriso (2020). In that work, the authors proposed a simplified procedure that relies on deformation modelling using the Hardening Soil Model with Small-strain Stiffness (HSS model; Schanz, Vermeer and Bonnier, 1999; Benz, 2006), and a calibration method was introduced to adapt the HSS model for representing the strain-softening behaviour of tailings. The approach was validated against triaxial and direct simple shear tests and then applied to a real TSF to demonstrate how relatively small toe deformations can trigger liquefaction and progressive failure under realistic *in situ* stress conditions.

The foundations of the approach were formally laid out in Ledesma, Sfriso and Manzanal (2022), where a systematic procedure was proposed to assess dam vulnerability using three representative actions: a surface load, a toe displacement induced by a volumetric contraction of the starter dam, and a rise in phreatic surface. These actions represent different stress paths in the tailings body and are applied as numerical excursions to explore whether a stable configuration could evolve into instability under plausible adverse conditions (Figure 1). In Ledesma, Sfriso and Manzanal (2022), the procedure was tested on the Fundão Dam and demonstrated its ability to reproduce failure mechanisms consistent with those reported in forensic studies.

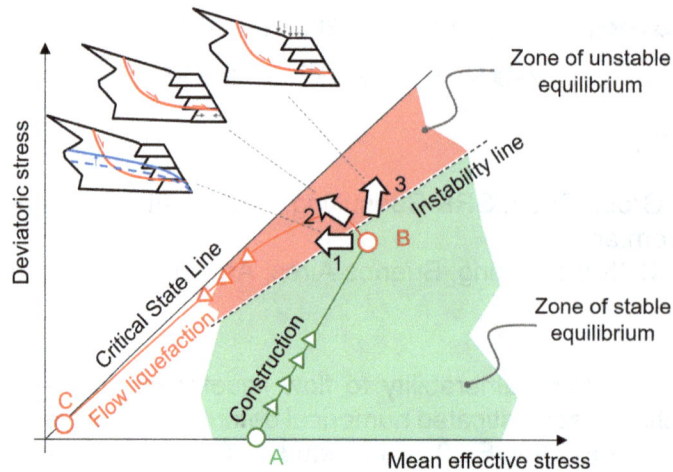

FIG 1 – Stress path of triggers proposed for assessing the vulnerability of tailings dams to flow liquefaction (adapted from Ledesma, Sfriso and Manzanal, 2022).

The numerical modelling framework is based on finite element deformation analysis and evolved into a systematic step-by-step procedure for the selection, calibration, and control of numerical and constitutive elements. Rivarola *et al* (2022) addressed numerical aspects required to guarantee enough computational robustness. Rivas *et al* (2023) benchmarked HSS against NorSand (Jefferies, 1993), a model widely accepted for modelling liquefaction. Sottile (2024) extended this benchmark exercise to include the CASM constitutive model (Yu, 1998) and calibrated the three models for a set of high quality triaxial tests done by Reid *et al* (2022). Rivarola *et al* (2023) presented a practical advancement in the implementation of these analyses by proposing a stepwise Soil Water Characteristic Curve (SWCC), allowing a simplified treatment of suction and partial saturation effects without compromising accuracy. The need for interpreting such analyses in a structured way, and some recommendations about the numerical aspects of the analysis were addressed in Sfriso *et al* (2023), who emphasised that vulnerability analysis is not about predicting a failure but about determining the conditions under which a system would be unable to resist a sufficiently adverse scenario. Rivarola and Tasso (2024) presented the convenience of employing fully coupled flow-deformation techniques to avoid mesh-dependency in strain-softening shear induced by the localised buildup of pore pressure. Sfriso, Ledesma and Sottile (2024a) presented a summary of these contributions.

The approach has been validated through comparisons with experimental data, different constitutive models, and sensitivity analyses of key numerical settings, and applied to several practical cases worldwide.

All the above contributions rely on deterministic deformation modelling to identify 'failure points'—values of the actions leading to system collapse. However, they do not quantify how likely these actions are to occur simultaneously in practice, nor do they provide a structured method for determining material parameters or aggregating uncertainty in the modelling.

Sfriso *et al* (2024b) introduced a standardisation framework to improve the reproducibility and scalability of vulnerability assessments. They proposed that the three trigger actions, each defined in terms of physical quantities with associated return periods, could serve as universal proxies for system deterioration. The standardisation also included guidance on interpreting model results to identify progressive failure, allowing comparison across different sites and dam geometries.

This paper further develops this framework. Drawing from the outcomes of deterministic simulations, an Ultimate Limit State (ULS) is determined in the three-dimensional space of triggering actions. This surface is generated using Radial Basis Function (RBF) interpolation, passing through six failure points obtained from simulations. Modelling uncertainty is then incorporated via a fuzzy margin, defined by a standard deviation field, also interpolated from input uncertainties at each point. A probability of failure (*PoF*) at each point in action space is computed using a cumulative Gaussian function, resulting in a continuous fragility surface. The full implementation is developed in a Python script which includes the fitting of empirical or parametric PDFs for each action based on return-period tables, and the use of Latin Hypercube Sampling to perform a Monte Carlo simulation. Each simulated triplet of actions is evaluated against the fuzzy surface to produce a *PoF* distribution.

This new step transforms the previously deterministic framework into a tool for probabilistic fragility assessment, while remaining fully compatible with the modelling practices and simulation workflows developed in Sottile, Cueto and Sfriso (2020), Ledesma, Sfriso and Manzanal (2022), and subsequent works. It provides a scalable and transparent foundation for integrating numerical deformation modelling into risk-informed tailings dam design and management. The paper concludes with a case study that illustrates its practical application.

PERFORMANCE-BASED, RISK-INFORMED DESIGN

International Committee on Large Dams (ICOLD, 2022) states: *'Risk-informed design using numerical models of particular failure modes can go beyond the traditional engineering and standards-based approach. In such cases, risk-informed design may be seen as an enhancement providing a defendable basis for decision making'*.

Morgenstern (2018) introduced the use of Performance-Based Risk-Informed Design to the design of TSFs. The approach combines two well established practices in geotechnical engineering: i) Risk-Informed Design, involving consideration of failure modes and Quantitative Risk Assessments (QRA); and ii) the Observational Method.

Canadian Dam Association (CDA, 2019) also places a strong emphasis on the application of the Observational Method. They however note that '... *the Observational Method is not applicable in cases where the failure mechanism is brittle and could evolve more rapidly than could be observed or responded to with contingency measures, or where other physical or economic constraints preclude the timely application of contingency measures'*.

Prescriptions by Eurocodes

EN 1990 (European Committee for Standardization, 2023) and EN 1997 (European Committee for Standardization, 2024) endorse the concepts of Performance-Based Risk-Informed Design (EN 1990 Section C.3.1) but states *'Risk-Informed and reliability-based approaches shall only be employed if uncertainties are represented consistently based on unbiased assumptions'*. EN 1990 Section C.3.2.3 states *'The quantification of uncertainties and their probabilistic representation should incorporate both relevant prior information and available new evidence, using Bayesian probability theory'*.

The procedure for determining material parameters in EN 1997 (European Committee for Standardization, 2024) presents a systematic way of addressing various sources of uncertainty, from spatial variability to model bias embedded in correlations, and is adopted here: measured values are subjected to QA/QC procedures to remove erroneous or non-representative results that might severely affect the outcome of statistical analyses. Derived values are obtained from empirical correlations, theoretical models (eg critical state soil mechanics), and back-analyses. From them representative values are computed, considering pre-existing knowledge, uncertainty due to the quantity and quality of data, spatial variability within the zone of influence and design criteria. Data from different sources are correlated to reduce uncertainty, and conversion factors are used to consider scale, variability of physical parameters such as humidity, aging and anisotropy.

Joint Research Commission (JRC, 2024) establishes three types of estimates for representative values: Type A refers to average values, and is applied when the limit state is insensitive to spatial variability of the ground; Type B represent lower or upper bounds (typically 5 per cent tail values) and is applied when the limit state is sensitive to spatial variability of the ground; finally, Type C is intermediate between Type A and Type B, and is computed taking into account the ratio of the scale of fluctuation to the extent of the failure surface is used in the determination of the characteristic values. In the procedure presented here, Type A is employed for unit weight, drained shear strength and stiffness, hydraulic conductivity and anisotropy, and stress field.

The undrained shear strength requires a special analysis. JRC (2024) establishes that, if a specific design situation/limit state is sensitive to weaker zones, identifiable weaker zones should be distinguished and treated as separate geotechnical units. Weak layers are very frequent in hydraulically deposited tailings and may control the overall performance of the TSF. The application of statistical measures of spatial variability to the determination of strength parameters of strain-softening tailings results in representative values that depend on the orientation of the potential sliding surface with respect to the horizontal layering of conventional tailings (would be different for filtered stacks), the out-of-plane extension of potential weak layers etc. The representative value of undrained shear strength at the back scarp of a potential sliding surface is normally higher than the representative value to be applied at the sub-horizontal base of the sliding surface and toe. The exact percentiles may vary from case to case and will depend on the sources of uncertainty of each section. If the position, thickness and size of the weak layers can be determined, they are identified as distinct geotechnical units and assigned Type A estimates (Figure 2). If the weak layers cannot be identified, their impact on the representative values is computed as sketched in Figure 3.

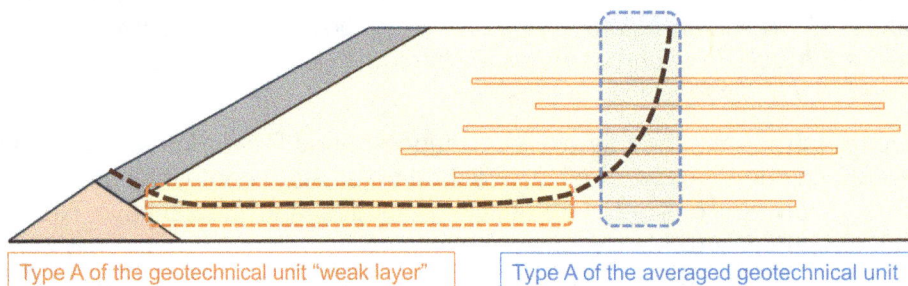

Type A of the geotechnical unit "weak layer"   Type A of the averaged geotechnical unit

FIG 2 – Weak layers as distinct geotechnical units.

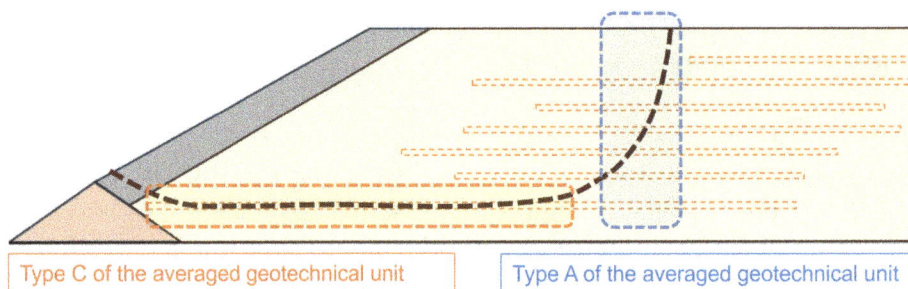

Type C of the averaged geotechnical unit   Type A of the averaged geotechnical unit

FIG 3 – Weak layers averaged in the tailings body.

Parameters computed with this procedure are random variables that can be expressed by a mean value and a standard deviation. For instance, the friction angle in the example below is expressed as $\phi' \sim Normal[\mu = 35.0°, \sigma = 1.2°]$. Input parameters that are random variables generate trigger values that are also random variables as described below.

Single-cause vulnerability assessments

The procedure proposed for the analysis of the vulnerability of tailings dams to flow liquefaction is presented elsewhere (Ledesma, Sfriso and Manzanal, 2022). Three triggers shown in Figure 1 are employed: a load at the crest q (resembling 'a rapid change in loading'); a toe contraction in the starter dam δ (resembling 'a deformation of the structure or foundation'); and an increase in the phreatic surface within the tailings body w (resembling 'a change in the state of drainage').

A triplet $\{q, \delta, w\}$ is defined for each set of values of these three actions. These actions start at zero and are increased separately as is standard in pushover analyses. Three triplets defining failure are produced: $\{q_1, 0, 0\}$, $\{0, \delta_1, 0\}$, $\{0, 0, w_1\}$, where q_1, δ_1 and w_1 are the values of the independent actions triggering failure, which are random variables because the input parameters are also random variables. Figure 4a shows the location of these three vertex triplets. For the sake of simplicity, in Figure 4 the triplets are shown as deterministic points instead of regions.

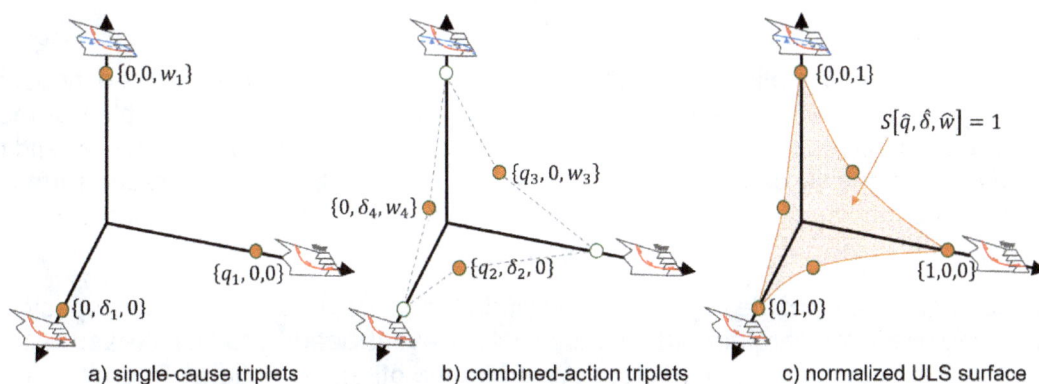

FIG 4 – The concept of a normalised ULS surface (Sfriso *et al*, 2024b).

Combined-cause assessments

Sfriso *et al* (2024b) introduced the idea of employing combined actions to define a ultimate limit state (ULS), or 'fragility' surface. Three additional triplets are computed by performing three additional pushover analyses, namely $\{q_2, \delta_2, 0\}$, $\{q_3, 0, w_3\}$, $\{0, \delta_4, w_4\}$, as shown in Figure 4b. For determining a combined-cause triplet (eg $\{q_2, \delta_2, 0\}$), the actions are increased consistently to their probability density functions (PDFs, described below). In the case of $\{q, \delta, 0\}$, load and the toe contraction are increased until failure is attained at $\{q_2, \delta_2, 0\}$, with q_2 and δ_2 having approximately the same return period.

Computing the ULS surface

To determine an analytic interpolation function for the ULS surface, triplets are normalised by the single-cause failure values to obtain the non-dimensional triplets $\{q_1/q_1 = 1,0,0\}$, $\{0, \delta_1/\delta_1 = 1,0\}$, $\{0,0, w_1/w_1 = 1\}$, $\{q_2/q_1, \delta_2/\delta_1, 0\}$, $\{q_3/q_1, 0, w_3/w_1\}$, $\{0, \delta_4/\delta_1, w_4/w_1\}$. A multiquadratic radial basis function (RBF) is then employed to produce a smooth interpolated ULS surface:

$$S[\hat{q}, \hat{\delta}, \hat{w}] = 1 \qquad (1)$$

where $\hat{q} = q/q_1$, $\hat{\delta} = \delta/\delta_1$ and $\hat{w} = w/w_1$ are the non-dimensional sampling variables normalised to the range 0 to 1 (Figure 4c).

RBF interpolation is a mathematical tool for estimating values at unknown locations based on a set of known values. Each of the known failure points creates a radial wave, which is strongest at the centre (the data point itself) and gradually fades out with increasing radial distance. To estimate the value at a new point in space — for example, a combination of load, contraction, and phreatic rise that hasn't been directly simulated —the influence of all the nearby waves is added up. Points that are closer have a stronger effect; points that are farther away contribute less. The result is a smooth, continuous surface that passes exactly through the known points, but also provides with reasonable values everywhere else. The RBF is chosen because it works well with a few scattered data points, produces a smooth surface, and it is straightforward to add new triplets, beyond the minimum of six, by performing additional runs of the numerical model.

Accounting for model uncertainty and computation of PoF

To account for the model uncertainty, the ranges of each triplet (eg $|q_1^{min}, q_1^{max}|$) is employed. The failure surface is defined as a fuzzy zone with Gaussian transition around $S[\hat{q}, \hat{\delta}, \hat{w}] = 1$ of the form:

$$PoF[\hat{q}, \hat{\delta}, \hat{w}] = \Phi\left[\frac{S[\hat{q}, \hat{\delta}, \hat{w}] - 1}{\sigma[\hat{q}, \hat{\delta}, \hat{w}]}\right] \qquad (2)$$

where $\sigma[\hat{q}, \hat{\delta}, \hat{w}]$ is the locally calibrated standard deviation of S, also computed by RBF interpolation from the ranges provided at the six data triplets, eg $\sigma_{q_1} = (q_1^{max} - q_1^{min})/4$. This formulation implies that a point lying exactly on the ULS surface has *PoF* = 0.50, points located close below the surface have 0.0 < *PoF* < 0.5 and points located close above the surface have 0.5 < *PoF* < 1.0 as dictated by their distance to $S = 1$.

PDFs of actions

The procedure requires estimating the probability density functions (PDF) of the three triggers. These can be obtained by risk assessment techniques and provided in the form of tables of magnitude versus return period for each action. The way they are defined, all triggers are positive, and physical upper bounds can be readily defined in most cases. For instance, the rise in phreatic surface cannot be higher than overtopping the dam. A truncated lognormal or Gumbel PDF is then calibrated from the data, so the triplet $\{\overline{Q}, \overline{\Delta}, \overline{W}\}$ of PDFs is obtained.

The estimation of the PDFs of the triggers must include the sources contributing to a specific hazard. For instance, the rise in the phreatic surface must include events leading to an increase in pore water pressure within the body of the tailings, not covered by the other two actions. Poor management of the pond location is an example of such an action. The toe contraction must include events leading to displacements at the toe of the dam. A toe erosion, starter dam excavation, creep of clay units at the foundation, are examples of such an action. The load at the beach must include events leading to loading at the crest of the dam. Rapid tailings deposition, fast flooding of the impoundment surface imposing transient downwards flow, are examples of such an action.

Monte Carlo simulation

The annual *PoF* of the structure is computed employing Montecarlo analysis. At a given iteration i, a sample $\{\hat{q}_i, \hat{\delta}_i, \hat{w}_i\}$ is chosen from the PDF set $\{\overline{Q}, \overline{\Delta}, \overline{W}\}$. The ULS $S[w_i, \delta_i, q_i]$ is evaluated and a contribution to the *PoF* is determined based on the fuzzy set. In most real applications, the probability of some of the actions will be small, and therefore efficiency can be highly improved by employing subset sampling. In this procedure Latin Hypercube Sampling (LHS; Helton and Davis, 2003) with $N \geq 100\ 000$ samples have been employed.

VALIDATION OF THE PROCEDURE

A typical upstream-raised tailings storage facility (TSF) is simulated using Plaxis 2D to investigate the combined impact of three flow liquefaction triggers and to numerically define the ULS of the system. The model geometry and the approach used to evaluate the TSF's susceptibility to flow liquefaction are consistent with those applied in earlier studies (Sottile, Cueto and Sfriso, 2020; Sottile *et al*, 2022; Rivarola *et al*, 2022; Rivas *et al*, 2023; Tasso *et al*, 2024; Sottile, 2024).

Mesh and geometry

The dam has a total height of 45 m, with external slopes of 1V:3H and a 30 m-wide berm at mid-height. The mesh, shown in Figure 5, comprises 5397 15-node triangular elements. Four geotechnical units are represented: tailings, embankment rises, foundation soil, and bedrock. Saturated tailings are modelled using the NorSand model, and assuming a constant state parameter $\psi = 0.07$, see Sottile (2024) for details on its calibration. Unsaturated tailings are modelled with the HSS model, assuming drained behaviour; this is done to overcome instabilities of the NorSand model on partially saturated zones, as discussed in Rivarola *et al* (2022). The embankment rises are simulated with HSS, the upper foundation with a Mohr–Coulomb model, and the bedrock as linear elastic; material properties are detailed in Sottile, Cueto and Sfriso (2020).

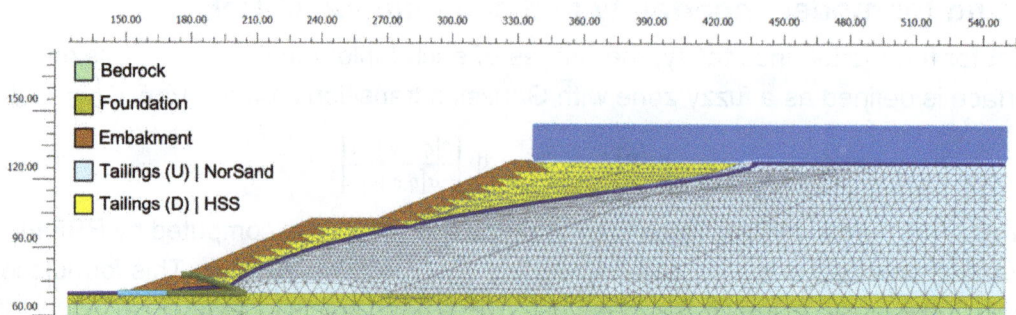

FIG 5 – Close view of the finite element mesh.

Modelling strategy

Conventional consolidation or undrained analyses were performed, ie inertia forces and transient seepage were neglected, and coupled flow-deformation modelling was not employed, as is usual practice in trigger analyses. The modelling sequence consists of two main stages: staged construction and triggering analyses and is explained in detail in Sottile (2024). After reaching the final configuration, dissipation of excess pore pressures is allowed and tailings below the phreatic surface are switched to NorSand model. The final modelling stage involves the combined application of triggers as explained above. A total of 650 scenarios were analysed by permutating five configurations of phreatic surface (PS-0 to 4, beach widths of 150, 100, 50, 20 and 5 m, see Figure 6), 13 values of horizontal contractions at toe (0 to 2.4 per cent, equivalent to 0 to 15 cm displacement at the crest of the starter dam) and ten values of surface load at the beach (0 to 180 kPa). In each case, failure or not failure scenarios were defined by considering whether static convergence was achieved or not.

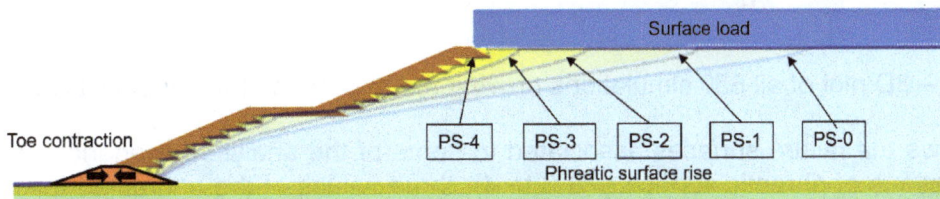

FIG 6 – Triggers applied to the dam.

The ULS surface

Figure 7 summarises the outcome of failure or non-failure for different combinations of triggers. As expected, the vulnerability of the dam is significantly influenced by the configuration of the phreatic surface: the scenario with the pond located furthest from the edge (150 m beach width, PS-0) can independently withstand toe contractions up to 2.4 per cent and beach loads of 180 kPa; these values progressively decrease as the pond approaches the edge, reaching minimums of 0.8 per cent and 20 kPa respectively in the case PS-4 (5 m beach width). Moreover, it is observed that the combination of load and contraction triggers have a significant effect on the dam vulnerability; for example, for PS-0 and considering a contraction of 1 per cent, the trigger load drops from 180 to 80 kPa.

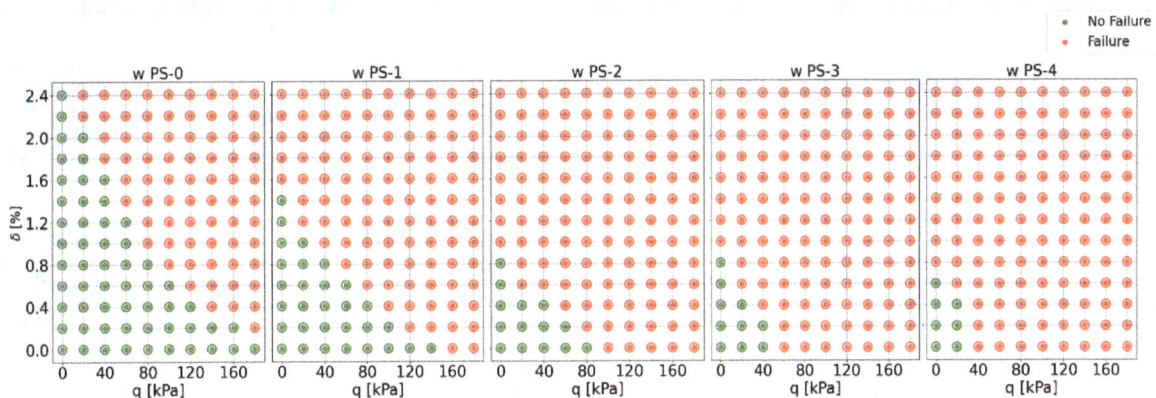

FIG 7 – Failure versus no failure scenarios for combined triggers.

The results shown above were used to produce a 3D plot that considers the combination of the three triggers (Figure 8). Results were normalised considering the maximum single-cause trigger values; this means: Load $q_1 = 180$ kPa, a contraction $\delta_1 = 2.4$ per cent (15 cm displacement at the crest of the starter dam) and width of beach $w_1 = 5$ m (pond located 5 m from the crest of the dam). The 'real' ULS surface defines the boundary in between the green and red dots; it is smooth enough to be interpolated by the RBF-based ULS interpolation presented above.

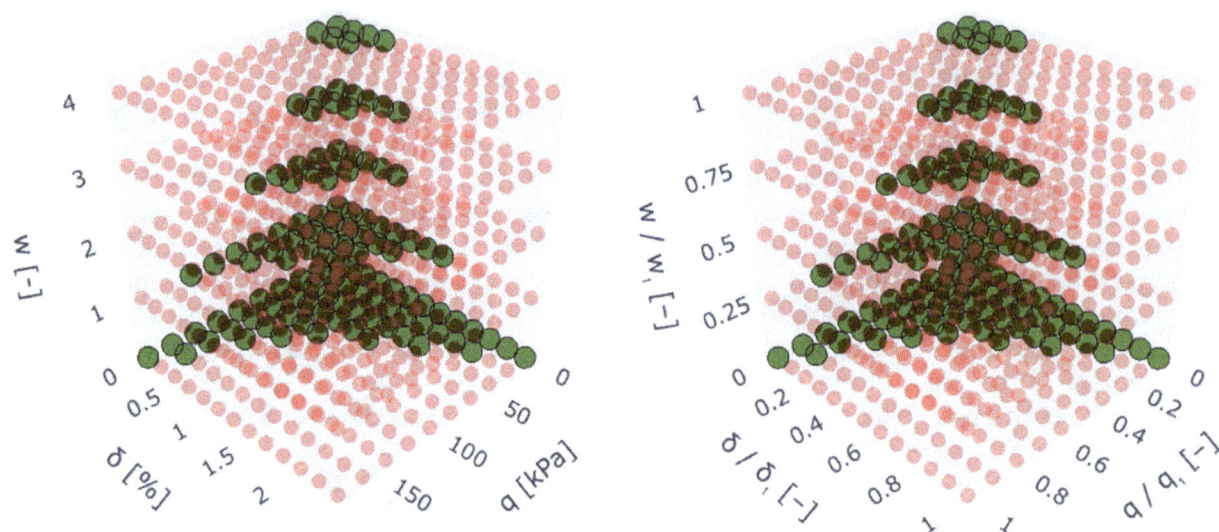

FIG 8 – 3D plot of all 650 simulations proving the existence of a smooth ULS surface.

Figure 9 shows the failure surfaces associated to some of the analysed scenarios; it includes the five configurations of phreatic surface (PS-0 to 4), three values of the undrained load (20, 60 and 100 kPa) and the corresponding contraction at toe that was needed to initiate the flow liquefaction triggering. It is observed that: i) for the further pond configurations (PS-0 and PS-1) all the failure surfaces are global and have a very similar shape, independent on the TA and TB combination; ii) when the pond gets closer to the dam slope (eg PS-3 and PS-4) and higher loads are applied, the failure shifts to a more local configuration on the upper portion of the dam.

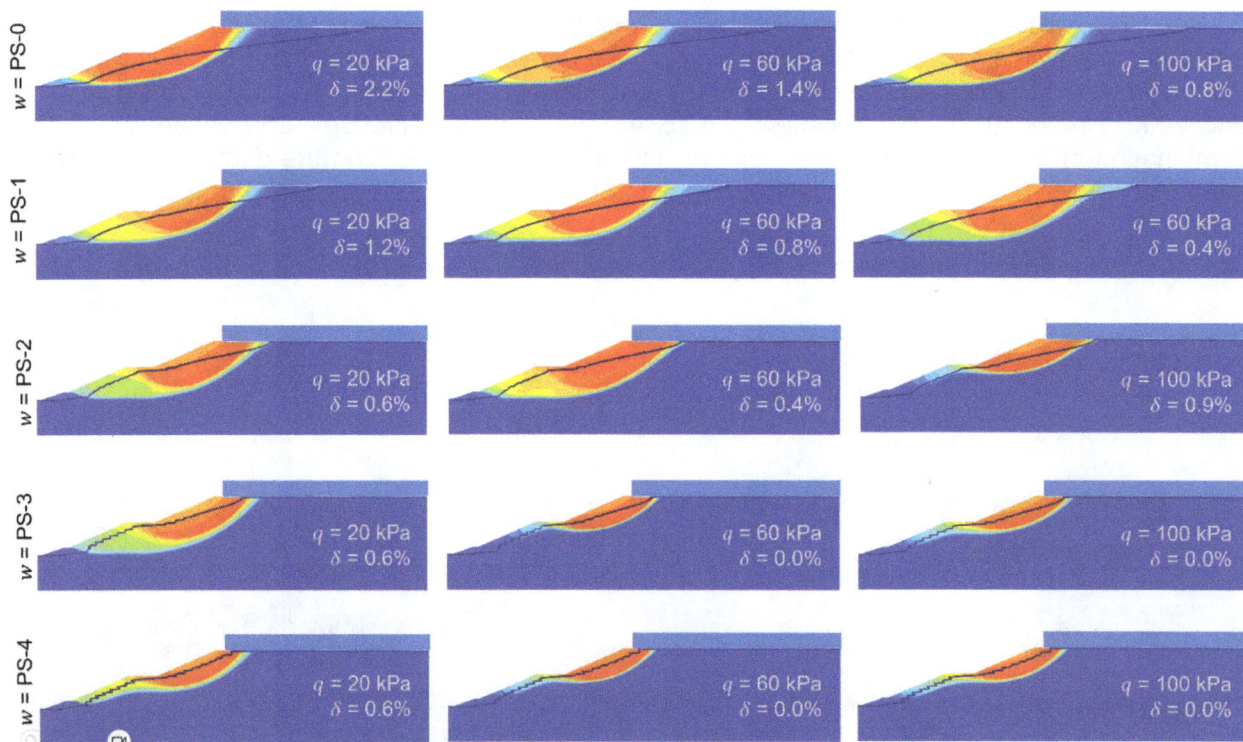

FIG 9 – Failure surfaces for selected scenarios.

Computing a Probability of Failure

To complete the exercise, the value of each action for a few return periods T_r is shown in Table 1. Two values are provided in each cell, representing the lowest and highest credible value for each trigger. These values are inserted here as an example; in a real case scenario, they would be estimated by quantitative risk assessments.

TABLE 1

Return periods for the three actions configuring the demand of the TSF.

Action	$T_r = 1\ yr$	$T_r = 10\ yr$	$T_r = 50\ yr$	$T_r = 100\ yr$
Load at crest (kPa)	10 | 15	20 | 50	40 | 80	60 | 100
Toe displacement (mm)	3 | 10	5 | 20	15 | 40	25 | 75
Width of beach (m)	125 | 50	100 | 25	50 | 20	25 | 10

With the demand defined by Table 1 and the capacity defined by the ULS surface, the following steps are conducted:

1. Lognormal distributions are computed for the triplet $\{\overline{Q}, \overline{\Delta}, \overline{W}\}$.

2. The ULS surface is computed as a fuzzy RBF interpolation $S[\hat{q}, \hat{\delta}, \hat{w}]$.

3. Latin Hypercube Sampling is employed to create 100 000 samples of $\{\overline{Q}, \overline{\Delta}, \overline{W}\}$.

4. For each sample, the *PoF* is computed employing Equation 2.

5. The total probability of failure is computed for the dam.

Adopting the most favourable values in Table 1, the computed annual probability of failure is *PoF* = 11.6 per cent if a deterministic ULS is employed, *PoF* = 18.0 per cent for a fuzzy ULS; employing the most pessimistic values in Table 1, *PoF* = 54 per cent. This is a reasonable outcome given the extreme vulnerability of this dam to toe displacement, as already highlighted by Sottile, Cueto and Sfriso (2020).

Informing TARPs

One contribution of this methodology is its ability to inform Trigger Action Response Plans (TARPs) with a probabilistic, multi-variable framework. Traditional TARP systems often rely on fixed, single-variable thresholds—such as a specific load or toe displacement—without fully accounting for the combined effect of multiple concurrent actions. The framework described here provides a means of evaluating stability considering different triggers acting together. This enhanced definition of the limit state enables TARPs to be informed by action combinations that, taken together, can pose a risk to the structural integrity of the dam, even if each individual action appears to remain within safe bounds. For example, a moderate load at the crest might be tolerable under static conditions, but when combined with a certain displacement at the toe, it could approach the failure envelope. This framework allows such combinations to be visualised and quantified, transforming the abstract concept of interaction into a tool for risk management.

Back to Figure 9, and assuming the phreatic surface is at its normal operational level (wide beach, first row of the figure), the 'red colour alert' would be triggered by any of the combinations $\{q = q_1 = 180\ kPa, \delta = 0\ cm\}$, $\{q = 100\ kPa, \delta = 5\ cm\}$, $\{q = 60\ kPa, \delta = 9\ cm\}$, $\{q = 20\ kPa, \delta = 14\ cm\}$, and $\{q = 0\ kPa, \delta = \delta_1 = 15\ cm\}$ and the interpolated values inbetween. In this example it's load at crest, toe displacement, and phreatic surface rise, but it can extended to other triggers that might be relevant to a given dam.

The vulnerability surface supports the automatic interpretation of sensor data. Continuous monitoring systems can map incoming measurements into the normalised action space and evaluate the position of the system relative to the vulnerability surface, triggering pre-defined responses, from increased surveillance to active intervention measures, based on the distance of the dam condition to the vulnerability surface. This application strengthens the observational method by incorporating the outcome of deformation modelling and probabilistic simulation into real-time operational decisions. A simple example of vulnerability analyses employed to inform the observational method applied to an upstream-raised tailings dam can be found in (Lino *et al*, 2024). Diagrams of the vulnerability surface, annotated with representative measurement paths, provide a powerful tool for building awareness among engineers, operators, and regulators, that can go far beyond opaque safety factors or abstract risk scores.

CONCLUSIONS

This paper builds upon previous contributions by the authors to propose a standardised, performance-based framework for assessing the vulnerability of tailings dams to flow liquefaction. The approach replaces conventional trigger analyses with a structured evaluation of system fragility, combining deformation modelling, probability density functions for key triggers, and statistical tools to quantify the probability of failure.

By defining a unique set of standardised actions—crest load, toe contraction, and phreatic surface rise—the method isolates the key stress paths that can induce instability in upstream-raised tailings storage facilities. A limit state surface is then constructed in the three-dimensional action space using Radial Basis Function interpolation of deterministic simulation results. Uncertainty in the location of the failure boundary is handled using a fuzzy margin defined by a Gaussian transition band, resulting in a continuous probability field rather than a binary outcome.

The methodology integrates risk-informed design principles from Eurocode 7 (Second Generation) and introduces a structured material parameter selection process aligned with European standards. It also incorporates performance-based evaluation criteria, extending the relevance of the method from forensic back-analysis to forward-looking design and dam safety management.

A Monte Carlo simulation using Latin Hypercube Sampling was used to assess the system's annual probability of failure, employing lognormal distributions calibrated from limited but realistic input data. The resulting fragility analysis showed that even with scarce input information, the method can produce meaningful risk metrics. In the illustrative case presented, the estimated annual probability of failure was approximately 50 per cent, driven largely by the system's sensitivity to toe contraction, as previously emphasised in Sottile, Cueto and Sfriso (2020).

This work demonstrates that deformation modelling, when standardised and embedded in a probabilistic framework, can provide actionable insight for dam safety evaluations and regulatory compliance. The procedure is scalable, robust, and fully compatible with existing numerical modelling workflows and Eurocode-aligned design protocols. The methodology extends the application of the observational method by incorporating a calibrated fragility surface into the design and implementation of TARPs. It allows real-time sensor data to be mapped against physically meaningful fragility thresholds while also fostering shared understanding among the professionals responsible for tailings dam safety. This dual utility makes the approach especially valuable for operational settings that require both automated alerts and human oversight.

ACKNOWLEDGEMENTS

This paper is a collaborative effort of SRK's international numerical modelling group, so all the names listed as co-authors in the reference participated, to some extent, in the creation of ideas that led to this paper. Appreciation also goes to clients and their reviewers for their support and advice.

REFERENCES

Benz, T, 2006. Small-Strain Stiffness of Soils and its Numerical Consequences, PhD, thesis, Universität Stuttgart.

Canadian Dam Association (CDA), 2019. Application of Dam Safety Guidelines to Mining Dams, Canadian Dam Association.

European Committee for Standardization, 2023. EN 1990:2023 – Eurocode 0 – Basis of structural and geotechnical design, European Committee for Standardization.

European Committee for Standardization, 2024. EN 1997-1/2/3:2024 – Eurocode 7 – Geotechnical design, European Committee for Standardization.

Helton, J C and Davis, F J, 2003. Latin hypercube sampling and the propagation of uncertainty in analyses of complex systems, *Reliability Engineering and System Safety*, 81(1).

International Committee on Large Dams (ICOLD), 2022. Tailings Dams Safety, Bulletin 194, Version 1.0.

Jefferies, M, 1993. NorSand: A simple critical state model for sand, *Géotechnique* 43(1):91–103.

Joint Research Commission (JRC), 2024. Determination of representative values from derived values for verification of limit states with EN1997, internal draft, Joint Research Commission.

Ledesma, O, Sfriso, A and Manzanal, D, 2022. Procedure for assessing the liquefaction vulnerability of tailings dams, *Computers and Geotechnics*, https://doi.org/10.1016/j.compgeo.2022.104632

Lino, E, Arquín, F, Ledesma, O and García, I, 2024. Mitigating risk through the observational method: A case study of an upstream tailings dam, 17th PCSMGE, La Serena, Chile.

Morgenstern, N, 2018. Geotechnical Risk, Regulation and Public Policy, The Sixth Victor de Mello Lecture, *Soils and Rocks*, 41:2.

Reid, D, Dickinson, S, Mital, U, Fanni, R and Fourie, A, 2022. On some uncertainties related to static liquefaction triggering assessments, in Proceedings of the Institution of Civil Engineers (ICE) – Geotechnical Engineering, 175(2):181–199 (ICE/Emerald Publishing Limited). https://doi.org/10.1680/jgeen.21.00054

Rivarola, F and Tasso, N, 2024. Analysis of flow liquefaction triggering in tailings dams considering coupled flow-deformation, in Proceedings of the 17th PCSMGE, La Serena, Chile.

Rivarola, F, Tasso, N, Bernardo, K and Sfriso, A, 2022. Numerical aspects in the evaluation of triggering of static liquefaction using the HSS model, 39 MECOM, Bahía Blanca.

Rivarola, F, Tasso, N, Bernardo, K, Sottile, M and Sfriso, A, 2023. Evaluation of triggering of static liquefaction of tailings dams considering the SWCC, in *Proceedings of the Mine Waste and Tailings Conference 2023*, pp 598–605 (The Australasian Institute of Mining and Metallurgy: Melbourne).

Rivas, N, Sottile, M, Rivarola, F and Sfriso, A, 2023. Comparing HS S and NorSand constitutive models for modelling flow liquefaction in tailings dams, 1 ICGTMW, Ouro Preto.

Schanz, T, Vermeer, P A and Bonnier, P G, 1999. The hardening soil model: Formulation and verification, *En Beyond 2000 in Computational Geotechnics*, pp 281–296 (Balkema: Rotterdam).

Sfriso, A, Ledesma, O and Sottile, M, 2024. Numerical modelling of tailings deposits and their interpretation, in First International Symposium on Tailings Deposits, Chihuahua.

Sfriso, A, Sottile, M, Rivarola, F and Ledesma, O, 2023. On the interpretation of trigger analyses of upstream-raised tailings dams, 1 ICGTMW, Ouro Preto.

Sfriso, A, Sottile, M, Terlisky, A, Rivarola, F, Lizcano, A and Ledesma, O, 2024. A proposal to standardize vulnerability analyses of flow liquefaction of tailings dams, 17th PCSMGE, La Serena, Chile.

Sottile, M, 2024. A comparison of advanced constitutive models to evaluate flow liquefaction of upstream raised tailings dams, PCSMGE, La Serena, Chile.

Sottile, M, Cueto, I and Sfriso, A, 2020. A simplified procedure to numerically evaluate triggering of static liquefaction in upstream-raised tailings storage facilities, XX COBRAMSEG, Brasil.

Sottile, M, Cueto, I, Sfriso, A, Ledesma, O and Lizcano, A, 2022. Flow liquefaction triggering analyses of a tailings storage facility by means of a simplified numerical procedure, XX ICSMGE, Sydney.

Tasso, N, Rivarola, F, de Santiago, O, Rivas, N and Sottile, M, 2024. Calibrating constitutive models for flow liquefaction: a word of caution, in First International Symposium of Tailings Deposits, Chihuahua.

Yu, H, 1998. CASM: a unified state parameter model for clay and sand, *International Journal for Numerical and Analytical Methods in Geomechanics*, 22:621–653.

NSW Resource Regulator tailings dam compliance priority assessment plan for 41 tailings storage facilities in NSW

J Stacpoole[1], X Hill[2], B Cullen[3], J Johnston[4], S Pegg[5] and A Margetts[6]

1. Inspector of Mines, NSW Resource Regulator, Orange NSW 2850.
   Email: john.stacpoole@regional.nsw.gov.au
2. Inspector of Mining Engineering, NSW Resource Regulator, Armidale NSW 2350.
   Email: xavier.hill@regional.nsw.gov.au
3. Senior Mine Safety Officer, NSW Resource Regulator, Newcastle NSW 2300.
   Email: brian.cullen@regional.nsw.gov.au
4. Senior Mine Safety Officer, NSW Resource Regulator, Newcastle NSW 2300.
   Email: john.johnston@regional.nsw.gov.au
5. Principal Inspector of Mining and Competency, NSW Resource Regulator, Newcastle NSW 2300. Email: shane.pegg@regional.nsw.gov.au
6. Chief Inspector, NSW Resource Regulator, Newcastle NSW 2300.
   Email: anthony.margetts@regional.nsw.gov.au

ABSTRACT

The NSW Resources Regulator is responsible for the regulation of both worker health and safety and planning conditions on mining operations throughout the state of NSW. This responsibility includes tailings storage facilities (TSFs) which often represent not just a significant, but often catastrophic consequence event to workers and the environment if they fail. The string of high-profile tailings dam failures across the world over the past 12 years – including the Cadia operation in NSW – highlights the real risk tailings dam failures pose and the need for their design, construction and management to be effectively regulated.

The NSW government recognises these risks and therefore has a specialist regulator to regulate mining across the state along with its own separate legislation. For Example, the NSW Workplace Health and Safety (Mines and Petroleum) Regulations (2022) require mine operators to manage 'Principal Hazards' which are hazards which may pose an unacceptable risk to large numbers of people if not appropriately managed such as hazards associated with tailings dam failure and the risks this poses to both workers and the environment.

To assess compliance with NSW legislation, the NSW Resources Regulator's specialist group in tailings dams conducted an assessment program across 41 mining operations and their active TSFs in NSW. The Tailings Dam Compliance Assessment assessed the safety management system for the TSF including dam break studies, failure modes, monitoring systems, design standard, risk assessments, monitoring links to trigger response plans, accountabilities, emergency response plans and the verification and effectiveness of the critical controls that control tailings dam embankment integrity. The assessment also reviewed how the tailings dam where constructed and operated against any recognised standards such as Australian National Committee on Large Dams Incorporated (ANCOLD, 2019) and Global Tailings Review (GTR), 2020.

The program found common deficiencies throughout NSW tailings dams, including Trigger Action Response Plans (TARPS) not being actioned, Walls not constructed to a recognised standard, Risk Assessments not considering necessary risks – particularly the risk to workers if the dam were to fail, no clear accountabilities for the management of the dams. There were other more operational failures commonly encountered, such as a lack of windrows at dams to provide edge protection to workers

This paper details the results and key findings of this program and provides learnings for not just mine operators in NSW but for dam operators across Australia.

INTRODUCTION

The NSW Resource Regulator undertake its regulatory activities in an industry sector with known extreme risks to both workers and the environment. These risks must be managed appropriately.

While the NSW mining industry is at the forefront of health, safety and environmental leading practice, the process of mining remains an inherently hazardous one. Instances of multiple worker fatalities arising from a single event and large-scale environmental degradation continue to occur globally. Notably, NSW has its own specific mine safety laws which specifically call out the existence of these risks as 'principal hazards' and requires that the high-risk activity such as tailings storage facility (TSF) construction and operation undergo assessment with a focus on the critical controls that verify tailings dam embankment integrity.

The existence of a specialist regulator for mining in NSW acknowledges the level of risk and highly technical aspects of NSW's modern mining sector, which allows for a special focus on the hazards and risks which are specific to the mining industry and tailings dam management, hazards which may pose an unacceptable risk to large numbers of people if not appropriately managed. Accordingly, the NSW Resources Regulator is different from more generalist regulators and prioritises compliance to risk profiles of the mining industry including the safe management of TSFs.

The Regulator's compliance priority assessment is a program, assessing a mine operator's management of controls to prevent Material Unwanted Events (MUE) and how they evaluate the effectiveness of controls in their safety management system. A safe tailings dam is dependent on effective and maintained controls.

There are a total of 360 TSFs in mine operations in NSW recorded in the NSW Mine rehabilitation portal. As part of the compliance priority assessment on NSW TSFs and emplacement areas, inspectors have reviewed the controls that maintain dam wall integrity for 26 mining operations that contain TSFs across NSW, with a further 15 planned throughout the program. This is to ensure the controls for monitoring and emergency response are sufficient to identify early signs of failure and to protect workers on and below the dam walls. Inspectors have attended tailings dams and emplacement areas focusing on principal hazard management plans, dam break studies, failure modes and effect analysis and monitoring system maintenance, verification and effectiveness.

CRITICAL RISK CONTROLS AND ASSESSMENT PROGRAM

The NSW Resources Regulator completed a broad-brush risk assessment and a program of bow-ties to review 'Principal Hazard' and 'Control Plan' topics. The bow-tie program identified material unwanted events (MUE) and critical controls to prevent serious injury or death of mine workers. The bow-tie for the 'Principal Hazard' topic of 'inundation or inrush' identified two MUE's associated with TSFs and developed critical controls for inclusion in the assessment program.

Following this assessment, the NSW Resources Regulator developed an assessment program focused on the identified critical controls used to prevent an uncontrolled inundation or inrush resulting from a failure of containment of a tailings dam, waste tip or water storage located on a mine site. These were identified as:

- The TSF (tailings emplacement, coarse reject or co-disposal areas) or water storage was designed to a recognised standard and managed to meet with best practice guidance information provided in a recognised industry standard, such as developed by ANCOLD.

- The mine operator had considered measures to reduce the magnitude of the source material (ie reduction of reject/waste and/or treatment of reject/waste). This is typically by tails harvesting such as drying the tails and emplacing them elsewhere, co-disposal with coarse reject or in waste rock dumps, use of pastefill for underground metalliferous mines or via splitting the tails with the course fraction being deposited closer to the TSF wall and the finer fraction being deposited closer to the TSF centre.

- The water management systems on-site effectively reduced surface water run-off from entering the TSF, that the TSF was managed in such a way that water was not ponding on the surface of the tailings and was directed away from the dam wall and that adequate drainage was in place at the toe of the TSF wall.

- Workers were effectively restricted from accessing the TSF.

- That the mine operator had an on-site emergency response plan that would be available and fit for purpose in the event of an unplanned failure associated with the TSF.

Each finding regarding a control used by the operation to manage the risks associated with the mine operation was assessed against the rating system shown in Table 1 with overall scores tabulated for each mining operation included in the assessment program.

TABLE 1

Criteria assessed ratings and points.

Assessed as	Rating	Points
Documented and implemented Compliant	4	4
Implemented but not documented Improvement needed	3	2
Documented but not implemented Significant improvement needed	2	1
Not documented and not implemented Non-compliant	1	0
Not applicable (N/A)		

Current status and preliminary results

The results described in this paper are an interim summary and are based on the results of 26 of the 41 active mining operations that contain TSFs across both coal and metalliferous mines in NSW.

The assessments were conducted between October 2024 and April 2025 with regulatory actions undertaken using provisions of the NSW Work Health and Safety (Mines and Petroleum Sites) Act (2013) at 17 of the 26 sites. These included:

- 130 individual findings, of which with enforcement actions recorded.
- 42 × S191 *Improvement Notices* issued.
- 11 × S195 *Prohibition Notices* issued.
- 18 × S23 *Matters of Concern Notices* issued.

Assessment findings results were calculated based on the total points allocated to the assessed ratings as a percentage of the maximum possible points for each critical control, and any findings rated as 'Not applicable' were excluded from the calculation.

Figures 1 to 3 summarise the overall assessment findings of the program.

The order of the mines listed in Figure 1 are the assessment findings providing a summary across all sites.

Figures 2–3 provide summaries of the notices and enforcement actions associated with the program, with Figure 3 identifying notices issued against the five critical controls, and Figure 2 summarising all notices issued during the program and the relevant principal hazard/control plan, MUE and category.

Critical control number	Critical control	Question	Chart
01	Tails facility (tailings emplacement, coarse reject or co-disposal) or water storage is designed to a recognised standard (eg ANCOLD)	How has the mine ensured that tails facilities are designed and constructed to the required standard?	18 / 5 / 3 = 26
02	Reduce magnitude of the source (ie reduction of reject or waste and or treatment of reject or waste)	Has the mine considered systems to reduce reject/waste material generation or treat reject/waste material?	21 / 1 / 1 / 1 / 2 = 26
03	Water management system	How does the mine ensure their water management system remains fit for purpose and operating correctly?	18 / 8 = 26
04	Workers and the public are restricted from accessing TSF		24 / 1 / 1 = 26
05	Onsite emergency response	Does the mine have adequate systems to response to an emergency related to tails and water storage facilities?	16 / 1 / 7 / 2 = 26

Legend:
- Not applicable
- Not documented and not implemented
- Documented but not implemented
- Implemented but not documented
- Documented & implemented

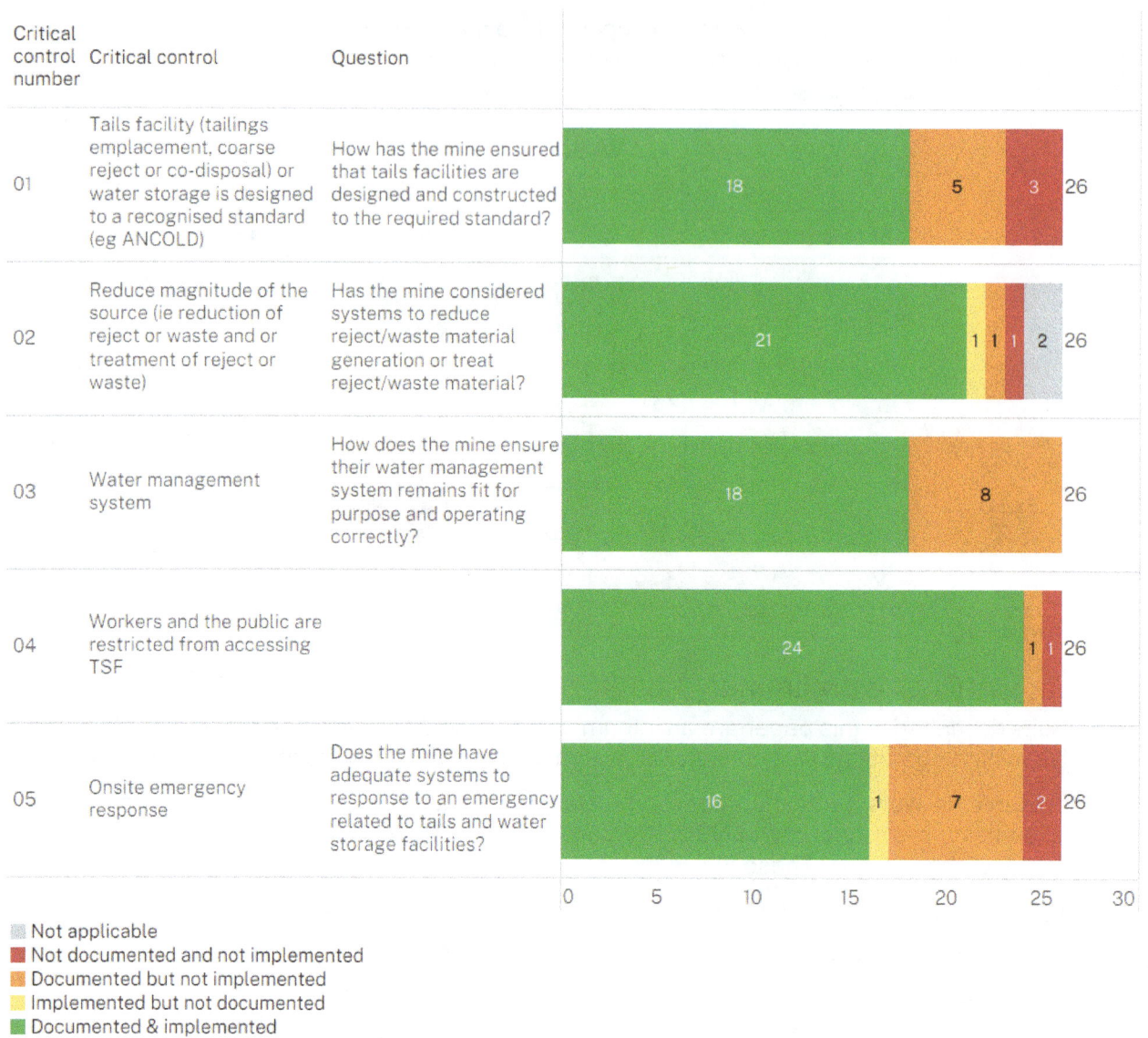

FIG 1 – Overall assessment findings ratings by critical control.

CDA or CIA	Critical control number	Critical control	Question	Concern (s23)	Contravention (s191)	Immediate risk (s195)	No Issue	Grand Total
						Enforcement action		
	01	Tails facility (tailings emplacement, coarse reject or co-disposal) or water storage is designed to a recognised standard (eg ANCOLD)	How has the mine ensured that tails facilities are designed and constructed to the required standard?	1	6	2	17	26
	02	Reduce magnitude of the source (ie reduction of reject or waste and or treatment of reject or waste)	Has the mine considered systems to reduce reject/waste material generation or treat reject/waste material?		1		25	26
CDA	03	Water management system	How does the mine ensure their water management system remains fit for purpose and operating correctly?	1	6	2	17	26
	04	Workers and the public are restricted from accessing TSF			2	1	23	26
	05	Onsite emergency response	Does the mine have adequate systems to response to an emergency related to tails and water storage facilities?		8	1	17	26
Grand Total				2	23	6	99	130

FIG 2 – Current notices issued by inspectors against the five critical controls.

Principal Hazard / Control Plan	Material Unwanted Event (MUE)	Category			Grand Total
		WHS(MPS)A s23 notice of concerns	WHSA s191 improvement notice	WHSA s195 prohibition notice	
Null	Null	18			18
01 Air quality or dust or other airborne contaminants	0101 Exposure to dust and airborne contaminants (including diesel particulate matter) in excess of exposure standards		5		5
04 Ground or strata failure	0402 Uncontrolled movement of ground or strata surface		1	1	2
05 Inundation or inrush of a substance	0501 Inundation or inrush of material or substance		6		6
	0502 Uncontrolled movement of tailings dams or water storages or waste tips		23	8	31
	0599 NA		1		1
07 Roads or other vehicle operating areas	0703 Collision of vehicle or mobile plant		2	1	3
	0799 NA		2	1	3
09 Subsidence	0999 NA		1		1
15 Mechanical engineering control plan	1503 Person in workbox or MEWP or IT or crane box		1		1
Grand Total		18	42	11	71

FIG 3 – All notices issued by principal hazard/control plan, MUE and category.

REGULATORY ACTIONS AND OBSERVATIONS

Regulatory notices and actions undertaken using provisions of the NSW Work Health and Safety (Mines and Petroleum Sites) Act (2013) during the program include Prohibition Notices (S195), Improvement Notices (S191) and Notices of Concern (S23).

Prohibition notices (S195)

Overall, 11 Prohibition Notices were issued during the program, relating to the following issues:

- 5 × due to insufficient edge protection.
- 1 × due to insufficient freeboard capacity in the TSF.
- 1 × due to tailings being placed in-pit prior to the finalisation of the bulkhead to protect underground workers.
- 4 × due to insufficient monitoring being undertaken that would allow the safe withdrawal of workers working on or directly below an embankment.

Examples of workers conducting an activity that involves or will involve a serious risk to the health and safety of a person emanating from an immediate or imminent exposure to a hazard requiring prohibition included:

- Workers working below a section of embankment without a risk assessment, an assessment of potential failure modes and insufficient monitoring in place to allow a trigger to be set in the TARP for withdrawal.

- Workers working below a section of embankment with a risk assessment and failure modes and effects analysis (FMEA) that required the use of monitoring equipment, but no monitoring equipment installed. No triggers for that monitoring equipment were linked to the TARP for withdrawal.

- In-pit TSF that required a barrier to underground workings, but without sufficient controls to protect underground workers from inrush.

Improvement notices (S191)

Details regarding the 42 Improvement Notices issued during the program as well as common issued are summarised here:

- 10 × mining operations where the mine operator had agreed to adopt the ANCOLD (2019) standard, however had failed to comply with section 2.7 of the ANCOLD (2019) Guidelines On Tailings Dams and the need to appoint persons to key roles such as the Engineer of Record (EoR) or the Responsible Site Person. There were also a number of examples where consultants were engaged by the mine operator to perform these roles, however had not provided any documentation saying that they had agreed to perform these functions.

- 6 × mining operations where Trigger Action Response Plans (TARP's) that were either not current or not relevant to the current operation of the tailings facility and therefore required review and update.

- 5 × mining operations with damaged and unserviceable instrumentation critical for monitoring deformation in the tailings dam wall, seismicity or pore water pressure change.

- 5 × sites with consequence category assessments with a significantly lower number of persons at risk than the actual number of people at risk on the ground. This was invariably due to not considering workers in the consequence analysis, with only residents downstream of the TSF being included.

- 4 × sites with emergency response plans requiring review and update.

- 4 × sites with insufficient controls to protect the embankment with excavations below or adjacent to the embankment.

- 4 × sites with embankments that have known factors of safety that were lower than the adopted design standard.

- 3 × sites with insufficient controls to protect workers from the risk of drowning or engulfment.

- 3 × sites with insufficient emergency training of workers including a failure to undertake mock emergency drills.

- 3 × sites with insufficient controls and insufficient monitoring to protect workers from respirable silica dust.

- 3 × sites with insufficient controls to protect workers from driving off an open edge.

- 2 × sites with the dam being operated outside the design limits.

- 2 × sites with insufficient positive communication between mobile plant.

- 2 × sites with significant cracks or erosion in TSF walls.

Specific examples of failing controls requiring improvement include:

- Geophones removed from a dam embankment to be used in another area of the mining operation.

- 8 × mining operations which the owners' engineer assumed the TARP to be in green trigger at the start of the assessment but on investigation of the instrumentation were found to be in amber or red.

- Dam break breach assessments undertaken on the opposite side to the site infrastructure and offices, containing a significant number of workers that were directly below an embankment.

- Areas of embankment with known low factors of safety in which insufficient (or no) monitoring was being undertaken.

Positive use of the Hierarchy of Controls

NSW mine operators must comply with Section 36 of the Workplace Health and Safety Regulations (2017) and to utilise the highest control to manage a hazard that is reasonably practicable. The highest order of control to manage the risks associated with tailings is to eliminate them. A number of mine operators were conducting innovative practices to not eliminate but significantly reduce the amount of tailings the operation produced.

There were four underground metalliferous operations in the assessment program who utilised pastefill, which in some cases reduced the amount of tails produced by 35 per cent. Pastefill also resulted in positive secondary effects, such as a reduced number of traffic movements and load on the ventilation circuit underground.

One mining operation harvests tailings by allowing them to dry to 8 per cent moisture, so that they are non-liquifiable prior to being co-disposed in an open pit. There are four coalmines in which a belt press or screw press was used to treat the tailings for subsequent placement in waste dumps and at least two operators that have undertaken upgrades or improvements to the washery circuit that has resulted in a greater recovery of the resource with a proportional reduction in tailings.

This represents good practice and provides additional benefits to the operations, such as reducing the TSF footprint, improving recovery and reducing traffic interactions.

CONCLUSIONS

The assessment program current being undertaken by the NSW Resource Regulator has provided insight into the current practices and management of risks associated with TSFs within the NSW Mining industry.

A number of initial observations relating to the management of tailings in NSW have been made, specifically as related to a mine operators responsibility to ensure there are sufficient controls in place to protect workers. These include ensuring the CCA, dam break and TARP are aligned and updated with the actual workers, process plant, offices and adjacent mine operations so the People at Risk (PAR) and Potential Lives Lost (PLL) is reflective of the actual risk profile. A consistent observation was that mine operations were not ensuring that there was adequate edge protection to prevent workers driving off or rolling over an open edge or into bodies of water.

Additionally, a clear understanding of the ownership and accountability for the critical controls relating to a tailings facility, as well as the roles of various positions within the management framework was poor at a number of sites. This confusion appeared to increase when the various tasks related to risk management of the TSF were being undertaken by multiple personnel or by different organisations, with instrumentation, monitoring, review and identification of triggers within a TARP requiring a large amount of coordination by the mine operator to remain effective.

For sites that have adopted the GISTM, the accountable executive, responsible owners engineer, designer and engineer of record roles are to be clearly delineated clearly identified in risk management documentation.

A large number of sites utilised embankment integrity monitoring as a key control in the triggering of associated TARP's and emergency response plans. For this to be effective, the mine operator must ensure that the monitoring of the critical control of embankment stability is able to be triggered sufficiently and that responsible personnel regularly review data trends to be able to provide adequate warning to evacuate workers and downstream effected communities on a shift by shift basis. A large number of sites had either inadequate or broken instrumentation that was critical in the safe operation of the dam, meaning that the instrumentation system was not fit for purpose and did not reflect the risk and consequence.

Mine operators should consider the MUE above as a minimum and ensure that the review of critical controls to prevent serious injury or death is included within the site principal hazard management plans and associated documentation. Guidance material is published on the Regulator's website about the best practice management of TSFs.

Further information can be found at: <https://www.resources.nsw.gov.au/resources-regulator/mine-rehabilitation/toolkit-and-guidance/tailings-storage-facilities>

REFERENCES

Australian National Committee on Large Dams Incorporated (ANCOLD), 2019. Guidelines On Tailings Dams - Planning, Design, Construction, Operation And Closure, revision 1, ANCOLD Inc. Available from: <https://ancold.org.au/product/guidelines-on-tailings-dams-planning-design-construction-operation-and-closure-revision-1-july-2019/>

Global Tailings Review (GTR), 2020. Global Industry Standard on Tailings Management (GISTM) [online], Global Tailings Review. Available from: <https://globaltailingsreview.org/global-industry-standard/>

NSW Government, 2022. NSW Work Health and Safety (Mines And Petroleum Sites) Regulation 2022. Available from: <https://legislation.nsw.gov.au/view/pdf/asmade/sl-2022-509>

NSW Government, 2023. NSW Work Health and Safety (Mines and Petroleum Sites) Act 2013 No 54, ver 15 December 2023. Available from: <https://legislation.nsw.gov.au/view/whole/html/inforce/current/act-2013-054>

NSW Government, 2025. NSW Work Health and Safety Regulations 2017, ver 1 March 2025. Available from: <https://legislation.nsw.gov.au/view/html/inforce/current/sl-2017-0404>

Case study – Engineer of Record transition process at a complex site

B Tiver[1], P Chapman[2], C Hatton[3] and J Schroeter[4]

1. Principal Tailings Engineer, WSP Australia, Adelaide SA 5000. Email: brad.tiver@wsp.com
2. Technical Director, WSP Australia, Perth WA 6000. Email: peter.chapman@wsp.com
3. Technical Director, WSP USA, Denver CO 80226, USA. Email: christopher.hatton@wsp.com
4. Senior Tailings Engineer, BHP, Adelaide SA 5000. Email: joachim.schroeter@bhp.com

ABSTRACT

This paper presents a case study of the transition of the Engineer of Record (EoR) role at a complex site. The site comprises six tailings storage facilities (TSFs), two receive tailings and four are inactive and transitioning to closure. The Owner had an EoR company at site since 2018, around the time when the role was popularly adopted at sites in Australia. In mid-2023, the EoR advised the Owner it was going to depart the role, giving approximately nine months of notice. The Owner requested bids from two different consulting firms to replace the EoR role, via a competitive tender process. There were two key influences in selection of the preferred EoR; the EoR had been involved at the site in various studies since 2014, and the Independent Tailings Review Board (ITRB) at the site had three out of four members from the same consultancy as the selected EoR, supporting transition of knowledge. The Owner undertook a risk assessment process in early 2024 for management of change, considering controls to be put in place for a successful transition. The transition process also included engaging a new ITRB to maintain the independence of that key governance function.

The new EoR undertook a Dam Safety Review (DSR) as a form of entry risk assessment and by April 2024 the transition had been completed. This paper highlights key items of the transition that went well as well presenting opportunities for improvement. A key challenge encountered was the recreation of models, such as slope stability models, which was necessary for the incoming EoR to take ownership of the assessments. Throughout the process it also became apparent that transition of some knowledge base items required more time and greater effort relative to others, in part due to different approaches and discussions with the Owner to gain agreement. Examples of aspects of the transition that worked well were the undertaking of a DSR and having a member of the ITRB being integrated into the EoR team and two others available for senior review.

INTRODUCTION

This paper presents a case study of the transition of the Engineer of Record (EoR) role at a complex site. The Owner had an EoR company at site since 2018, around the time when the role was popularly adopted at sites in Australia. In mid-2023, the EoR advised the Owner it was going to depart the role due to professional attrition, giving approximately nine months of notice. The planned transition date of the EoR role was April 2024. This paper presents some site background to give context on the complex nature of the site and then discusses the transition process with lessons learnt.

SITE BACKGROUND

The site comprises a series of upstream-raised paddock style tailings storage facilities (TSFs):

- Three of the TSFs were constructed in the late 1980s and early 1990s. These TSFs have not received tailings for over ten years.

- One TSF, which was constructed in the late 1990s, has recently reached its maximum capacity and is inactive.

- Two of the TSFs are active and are receiving tailings. The TSFs were commissioned within the last ten years and provide ongoing capacity in the medium-term.

The inactive TSFs are all in the order of 30 m high, and the active TSFs are anticipated to reach similar heights in time, at a rate of rise in the order of 1–2 m/annum. The TSFs are raised in the upstream method, contain ~50 per cent w/w slurry deposited through perimeter spigot discharge and have various states of buttresses for management of embankment stability.

The TSFs are founded on relatively flat terrain in an arid environment. The foundations generally contain sand dunes overlying stiff clays, common to desert areas in Australia.

TRANSITION PROCESS

Selection of new EoR

The EoR had been in the role at site for around six years when it advised the Owner it was going to depart the role. The Owner requested bids from two different consulting firms to replace the EoR role, which occurred through a competitive tender process. There was a high weighting on qualification-based selection and less so on cost criteria. The following were key reasons in selection of the preferred incoming EoR:

- The EoR consulting firm had been involved at the site in various tailings related studies since the mid-1990s.

- The EoR representing the firm in the role had been involved at the site in various tailings related studies since 2014. These studies included geotechnical investigations, design projects for raising TSFs and associated buttressing and design of appurtenant structures to the TSFs. Since 2018 when the EoR was appointed, these studies included interaction with the EoR.

- The EoR representing the firm is close geographical proximity to site such that a fast in-person response would be available if required.

- The Independent Tailings Review Board (ITRB) at the site had three out of four members from the same consultancy as the selected EoR, providing strong knowledge of the site and supporting knowledge transfer.

- The incoming EoR applied a team approach:

  o An overarching Deputy EoR role, which was a former member of the ITRB.

  o Three 'sub-deputy EoR' roles.

  o An internal technical peer review role, also a former member of the ITRB.

  o A project management team to allow the engineering team to focus on the EoR role.

The effective approach was designed to cover specialist skill sets required at the site, due to the wide variety of structures in various states of operation and closure.

Management of change processes

Overview

After selection of the incoming EoR, the Owner undertook a change risk assessment in accordance with its internal standards and the Global Industry Standard on Tailings Management (GISTM) for management of the change of EoR. The risk assessment identified a series of risks 'unwanted events' and ranked them using the Owner's internal risk assessment processes. As part of the risk assessment, a series of controls were identified and developed into an action plan. The controls and the way in which they were implemented are discussed in turn. The risk assessment was attended by 18 people, comprised of three staff from the outgoing EoR firm, two staff from the incoming EoR firm, and the Owner's team.

Completion of DSR

The incoming EoR undertook a DSR to familiarise itself with the site and facilities. The DSR had a requirement that where the incoming EoR identified risks that, if left unmitigated, would prevent them from taking on the role, agreement was attained from the Owner on mitigation of risks. The site inspection component of the DSR was completed in early 2024. After the site visit, the DSR analyses continued through the remainder of Q1 2024.

Notwithstanding, in January 2024 when the risk assessment was undertaken, a risk was identified that completion of a DSR in of itself does not immediately provide the incoming EoR with all required information to seamlessly take over completion of business-as-usual dam safety services at the site. These comprised of regular review of Critical Operating Parameters (COPs), review of Vibrating Wire Piezometer (VWP) data, and ad hoc slope stability checks.

The identified action was for the Owner to appoint the outgoing EoR for a crossover period of three months to provide parallel services to the incoming EoR for items such as the slope stability checks. As required during this time of crossover, obligations on the incoming and outgoing EoRs were clarified with contractual amending deeds, and clear communication of expectations.

The DSR was a useful process for the incoming EoR to develop its own slope stability models to be able to perform this task.

Slope stability models

The outgoing EoR had 20 slope stability models (cross-sections) across the six TSFs at site, and each cross-section had VWPs with trigger levels. While not a typical requirement of a DSR, as the incoming EoR, it was a requirement of the DSR scope to develop slope stability cross-sections to validate every model at site. The intent of this approach was to facilitate the delivery of business-as-usual tasks referred to earlier. The process of developing and validating comprehensive new models against existing models was an intensive and time-consuming task. Where differences in engineering approaches and application between the incoming and outgoing EoRs indicated changes in model outcomes, these aspects of the stability models were thoroughly reviewed and discussed to arrive at an agreed outcome.

Roles and responsibilities and transfer of information

It was identified that there was a risk that the EoR, both incoming and outgoing, may be unaware of their specific roles and responsibilities during the transition and upon completion of the transition. The identified actions were:

- A documented transition plan including familiarisation of the incoming EoR with COPs and Trigger Action Response Plans (TARPs). It was critical that the TARPs remained functional during the transition, to avoid losing valuable time or resulting in uncertainty, in the event adverse conditions were observed.

- The Owner provided the incoming EoR with a register (Excel) of regular dam safety activities and routines. This transfer of information with relation to regular routines provided the Owner with additional confidence that the likelihood of gaps in monitoring or reporting during the transition would be minimised.

- The Owner attained from outgoing EoR a list of documents in draft status that required review and publication to final status through the transition process.

- The incoming EoR provided a list of information required to complete the DSR. The intent was to compare this to the list of existing documents to identify any gaps that may exist between what exists, what was planned to be produced by the outgoing EoR and what may be missing.

- The incoming EoR was required to attend routine dam safety service meetings where relevant. These activities included:

  o fortnightly EoR meeting

  o one-monthly COPs review

  o three-weekly piezometer review

  o one-monthly construction inspections. Representatives from the outgoing EoR and incoming EoR teams undertook one parallel site inspection for the incoming EoR to be familiar with the required routine.

Gap in EoR services

The Owner identified a risk that there could a gap between the time when the outgoing EoR roles and responsibilities end and the incoming EoR roles and responsibilities commence. The agreed action was that before the outgoing EoR contract ends, and with the understanding that the incoming EoR makes no material adverse findings during completion of the DSR that would prevent them from taking on the responsibility of EoR, the incoming EoR is to deliver a 'letter of intent' to the Owner that outlines the recommendations that would need to be implemented. Upon acceptance from the Owner, the incoming EoR firm could then cover any time gap between the end of the outgoing EoR contract and the official finalisation of the DSR.

The incoming EoR issued a letter of intent in March 2024 to formally accept the role of EoR before the DSR was complete, which was finalised by July 2024.

Projects

The Owner identified that for projects in progress where the outgoing EoR and ITRB was engaged, there could be a risk that the project is not approved or endorsed by the outgoing EoR and ITRB by the transition date. Three key actions were identified:

1. The Owner provided the incoming EoR with a register including project details and gate status of current projects, as well as an approximate estimate of the foreseeable level of EoR input.

2. The Owner developed a communication plan to notify the project teams of the EoR change.

3. Project teams communicated the relevant required EoR input.

Stakeholders

There was a risk identified that the incoming EoR would not be familiar with stakeholders, for example the Accountable Executive (AE), that they are required to know as part of the role. The action to control the risk was that the Owner shared the most recent AE engagement discussion outcomes.

It was also identified that site stakeholders might not be familiar with the new EoR and their contact details in order to escalate issues. The Owner updated the Operations Maintenance and Surveillance Manual with new EoR and Deputy EoR details and developed a communication plan to notify site and operational stakeholders of the new EoR.

COMMENTARY

Overall, the process was a success. It was a nine-month process from the outgoing EoR notification to the incoming EoR officially taking the role. We make the following comments about the successes of the transition:

- The process was an integrated and collaborative approach between the Owner, the outgoing EoR and incoming EoR. Following the selection of the incoming EoR and the management of change risk assessment, each action was followed through on and each party was respectful of the process. This should be commended and be a common practice for transitions of EoRs.

- Undertaking a DSR as the incoming EoR was essential. It allowed the EoR to gain knowledge quickly, even though they had years of prior experience at the site. It also allowed time to generate the required slope stability models for the role. It should be considered industry best practice to undertake a form of risk assessment as the incoming EoR, as has been done during this transition. As this was a complex site, DSRs had been occurring every two years since 2018, and it was convenient timing that the DSR occurred for the incoming EoR.

- Having a former member of the ITRB integrated into the EoR team was useful for knowledge transfer and guidance on some of the historical issues identified at site. The particular former ITRB member had been on the ITRB for approximately three years before the transition. This person has also become the Deputy EoR, which provides confidence to stakeholders such as the RTFE and AE.

Alongside the successes, the authors identified the following considerations. These are not necessarily items that caused detriment to the transition, but should be considered by others for future EoR transitions:

- The procurement process took a long time, with six of the nine months of the transition taken up by procurement. This resulted in an intense transition during the DSR, and business as usual processes were interrupted. A smoother procurement process would have allowed more time to spend on the transition itself.

- The process for transfer of documents could be improved, noting the challenge of transferring such a large amount of information. While the incoming EoR was provided with documents at the commencement of the DSR, the incoming EoR found themselves often requesting more information, and have continued to do so one year on from the transition.

- Not all correspondence between the previous EoR and the Owner was formally documented in memos/reports. As much as possible, it should be EoR practice to record key decisions in memoranda, reports or meeting minutes. Site knowledge is paramount to a successful EoR role, and sharing knowledge between the Owners team, outgoing ITRB and EoR, and incoming EoR was a key success in the transition.

- The incoming EoR had gained some knowledge of the outgoing EoRs slope stability models in the DSR. Following the appointment, the incoming EoR team undertook a geotechnical investigation at site in mid-2024, which occurs on a two-yearly frequency. The interpretation of the newly gathered data amongst the historical information and interpretation has been a key challenge. Nuances in different approaches between the outgoing and incoming EoRs resulting in different outcomes has taken time to explain and gain agreement.

- Changes in EoR/ITRB bring with it changes of professional opinion and this risk needs to be considered upfront, so as not to cause unforeseen impacts later. The ITRB governance function is one of high-level reviews, to judge whether the Owners team and EoR are undertaking good practice. The former ITRB reviewed approaches by the former EoR, which may have been viewed as agreement. Whilst there was a great advantage in having former ITRB members on the incoming EoR team, when the EoR wanted to change an approach, sometimes it was met with caution by the Owners team. An example of this was changing software codes for deformation analysis. Future EoR transitions could make proactive steps in the transition around discussion of approaches to modelling and selection of software codes.

CONCLUSIONS

This paper presents the steps of a successful transition of the Engineer of Record role at a complex site. The key conclusions drawn from the process that should be considered in future transitions at complex sites are:

- Allow time. The procurement process takes a long time, and at complex sites, there is a lot of knowledge to gather.

- Undertake a selection process that places a high degree of importance on qualifications and previous site experience.

- The incoming EoR should undertake a DSR as an essential process for gathering knowledge and being informed of the risks at site, and completion of a risk assessment is strongly recommended

- Foster a progressive, integrated and collaborative environment between the Owners team, the outgoing EoR and incoming EoR.

- Acknowledge change and continuous improvement, making time for the incoming EoR to justify differences in approaches.

ACKNOWLEDGEMENTS

The authors would like to acknowledge the support of the RTFE and his team at site, and the approval of the Owner to publish this paper.

A competency based development framework for current and future Engineers of Record

M B Willan[1], P Chapman[2] and D Bleiker[3]

1. Principal Geotechnical Engineer, WSP Canada, Vancouver, Canada.
   Email: martyn.willan@wsp.com
2. Technical Director, WSP Australia, Perth WA 6000. Email: peter.chapman@wsp.com
3. Fellow, WSP Canada, Mississauga, Canada. Email: david.bleiker@wsp.com

ABSTRACT

The development of professionals, such that they are ready to take on the responsibilities of Engineer of Record (EoR), continues to be an industry challenge due to a perception that the role carries high professional risk. This has been exacerbated by the shortage of potential competent professionals within the mining industry.

To date industry initiatives have focused on defining competency requirements by establishing a minimum number of years of experience and/or the technical knowledge of the professional. However, the authors and WSP believe that the demonstration of escalating levels of competency define important milestones on a professional's career journey. As such, a comprehensive EoR development program can attract more professionals and develop more effective future EoRs. This has the added benefits of building individual confidence in the role over time and allowing individuals to develop the required knowledge and, just as importantly, behaviours at their own pace while maintaining required standards.

In this paper the authors will present an overview of the EoR development program, which is being implemented globally within WSP, with the aim of achieving the following key elements:

- Provide a framework for development by establishing a set of core competencies/behaviours for individuals to demonstrate. These are defined at various competency levels to allow for progression throughout an individual's career.

- Provide an assessment process by which an individual's competency, relative to the requirements, can be assessed, providing for consistency across the organisation.

- Provide formal training so that EoRs understand and appreciate their responsibilities, the risks involved and key outcomes from past case studies.

- Provide ongoing support to EoRs to assist with decision-making and in navigating their responsibilities. This helps to ensure that no individual is or feels like that are making decisions in isolation.

INTRODUCTION

Alongside the Responsible Tailings Facility Engineer (RTFE) and Accountable Executive (AE), the Engineer of Record (EoR) is one of the three named roles in the Global Industry Standard for Tailings Management (GISTM; Global Tailings Review (GTR), 2020) and is a critical position of responsibility in the safe management of tailings storage facilities (TSFs). The International Council on Mining and Metals (ICMM), Tailings Management Good Practice Guide (ICMM, 2025) states:

> The purpose of the EoR role should be understood as a means to ensure that business and operational decisions made by the operator [owner] are informed by an engineer who understands the design principles and technical limitations of the tailings facility and the impact of changes on its safety performance.

Unlike the RTFE and AE, which are roles typically assigned within an operator's company, the EoR role can, and is often, undertaken by external consultants. This puts the EoR in a unique position where they function as a partner with the client, but do not have direct decision-making authority. This relationship is also significantly different to the standard consulting one and often includes the EoR having influence on a client's overall strategy and stewardship as well access to senior members of the client's corporate team, through a trusted advisor role. It may also require that the EoR be willing to risk long-term client relationships to protect facility safety and in some jurisdictions, under certain circumstances, may require an EoR to report a client to regulatory bodies.

It therefore follows that the development of EoRs is also unique and cannot be solely based on a minimum number of years of experience or amount of technical knowledge, which has generally been the focus of initiatives adopted to date. The authors believe there are clear benefits to the establishment of a more formal development plan for EoRs, including:

- Supporting EoRs in executing their accountabilities and responsibilities.

- Providing a framework for competency, which provides additional resources to practitioners when competency is challenged, such as by licensing bodies following a failure or incident (eg Engineers and Geoscientists of British Columbia (EGBC) between 2014 and 2022 following the Mount Polley Failure (EGBC, 2022).

- Providing a road map for practitioners at all levels that can be integrated with career development planning activities, including progression from supporting an EoR team, to a Deputy EoR, before taking on a full EoR role.

- Identifying attributes and core competencies for undertaking EoR and associated roles.

- Achievement of a consistent level of competency across a company and potentially the industry.

- Assisting EoRs with the management of risks related to increased public/regulatory visibility.

- Establishing and supporting an approval process for EoR identification (internal and external to the EoRs organisation).

- Providing confidence to EoRs, EoR teams, clients and the EoR's organisation, that those undertaking the role have demonstrated competency.

- Attracting more professionals to the role and developing the future generation of EoRs.

As a global leader in the provision of EoR (and design) services to the mining industry, WSP has established a global comprehensive EoR development program, the outline of which is provided in this paper. The authors believe that this framework is of benefit to the wider industry in promoting good governance and consistency in EoR competency.

EOR CAREER DEVELOPMENT FRAMEWORK

Being an EoR is a challenging leadership position. The EoR leads a multi-disciplinary team of professionals from within and sometimes outside of their organisation including within the client's organisation. It is therefore imperative that potential EoRs develop, practice and strengthen key attributes and competencies. The required minimum experience for an EoR is substantial and the range of experiences required are highly varied. Career progression and increasing levels of competency are important performance indicators and the focus should be in the quality of experience rather than the speed in which it is acquired. However, the intent is to allow career planning and ensure that future EoRs can effectively gain the competencies and experiences required.

EoR competencies

The EoR competencies identified in the WSP EoR career development framework are shown in Figure 1.

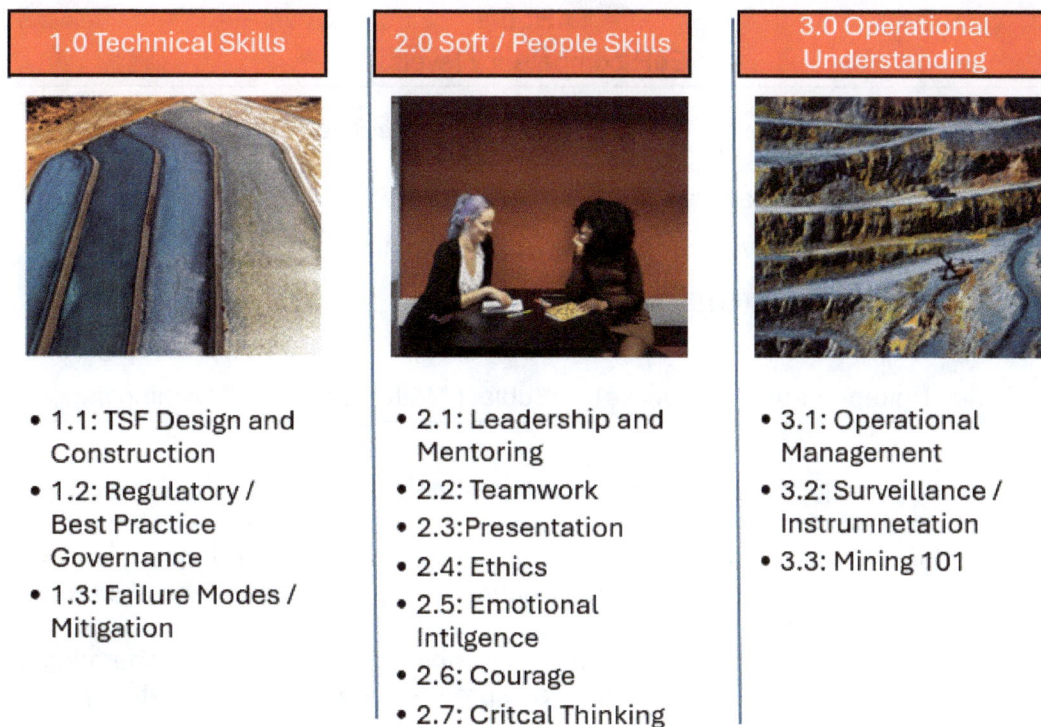

| 1.0 Technical Skills | 2.0 Soft / People Skills | 3.0 Operational Understanding |

- 1.1: TSF Design and Construction
- 1.2: Regulatory / Best Practice Governance
- 1.3: Failure Modes / Mitigation

- 2.1: Leadership and Mentoring
- 2.2: Teamwork
- 2.3: Presentation
- 2.4: Ethics
- 2.5: Emotional Intilgence
- 2.6: Courage
- 2.7: Critcal Thinking

- 3.1: Operational Management
- 3.2: Surveillance / Instrumnetation
- 3.3: Mining 101

FIG 1 – EoR core competencies.

These are intended to provide a general knowledge/behaviour base for the 'typical' EoR role but are not intended to cover all the knowledge areas that may be required for individual facilities, clients, geographical/geological locations and/or regulatory environments.

Each of the competencies identified in the WSP program is supported by specific typical experiences at each competency level. These provide examples to participants of the types of activities that would meet the specific competency levels. However, it is important to note that these are not intended as a 'tick-box' exercise and it is expected that each participant's journey in the programme will be unique, in terms of achieving the competencies and the timing thereof.

EoR competency levels and attributes

Progressive competency levels are defined, that is participants starting in the program at the commencement of their professional careers will likely advance through each of the levels in time. However, the program was developed to allow participants to join at any time, in which case the attributes can be used for gap analysis. They are aligned with, but not reliant, on an individual's years of experience (although guidance on the minimum years of experience to assume the role of EoR at Level 4 is established, taking cognisance that many site owners also have guidance on this). As such, an individual will generally become increasingly responsible/accountable for their actions and inactions as they advance through the program. An overview of the five competency levels, as well as the typical knowledge level, project role and stage of an individual's career are provided in Table 1.

TABLE 1

EoR competency levels.

Competency level	Knowledge level	Typical project role	Typical stage of career
1	Interested (Reactive)	Staff Engineer	Exploring
2	Aware (Reactive)	Project Engineer	Establishing
3	Transitional (Engaged)	Deputy EoR	Transitioning
4	Collaborative (Proactive)	EoR	Performing
5	Entrepreneurial (Proactive)	Subject Matter Expert	Teaching/supporting

A key component of developing a consistent level of competency for EoRs is the identification and focus on general attributes for those undertaking the EoR and Deputy EoR role. These allow for local variation/knowledge requirements that are inherent in a complex global industry while aiming to foster the same behaviours amongst professionals undertaking EoR roles. General attributes at each of the five competency levels are provided below and are intended to provide overall guidance to participants and their mentors in assessing the level at which they are currently operating at, as well as the attributes they should seek to demonstrate to progress to the next competency level.

- Level 1 – Exploring EoR as a Potential Career Path:
  - Willingness to learn the potential impacts and consequences of their work/actions.
  - Developing an understanding of the requirements of the core competencies.
  - Willingness to learn about tailings facility design/operations/closure.

- Level 2 – Establishing Understanding of EoR Role:
  - Awareness of facility design concepts used to determine construction and operational requirements.
  - Awareness of applicable regulations, standards and guidelines.
  - Development of initial understanding of how tailings facilities are designed, constructed, operated, and closed.
  - Are aware that their work/actions can have adverse effects but may have limited understanding of consequences.

- Level 3 – Transitioning into EoR Role:
  - Demonstrate knowledge of the need for proactive tailings stewardship.
  - Demonstrate working knowledge of applicable regulations, standards and guidelines.
  - Understand they can affect change (positive and negative).
  - Understand the tools they have or do not have but need.
  - Own their role.
  - Understand that the role carries significant liability and risk (to self, organisation, and client).
  - Understand their actions can have long-term impacts.
  - Understanding of the concepts and engineering principals behind tailings facility design.
  - Starting to understand how tailings systems work in their entirety.
  - Be named as a Deputy EoR.

- Level 4 – Performing EoR Role:
  - Demonstrate proactive tailings stewardship.

- o Strong understanding of the concepts and engineering principles behind tailings facility design, operations and closure.
- o Make thoughtful contributions and manage facility safety with intent.
- o Functions as a team member/leader.
- o Demonstrates a culture of pride and ownership.
- o Understand impacts of changes on the facility through experience or access to others with required knowledge.
- o Seeks collaboration and innovation (internal and external).
- o Uses predictive tools to anticipate behaviour.
- o Demonstrate advanced knowledge of applicable regulations, standards and guidelines.
- o Communicates effectively and transparently with all stakeholders.
- o Proactively identifies opportunities to improve facility safety.
- o Maintain awareness of external resources and understand current state of practice.
- Level 5 – Defining and Supporting the EoR Role:
  - o The individuals opinions and assistance are sought across various geographies and/or industry groups.
  - o The individual contributes meaningfully to working groups; typically, industry associations or regulators.
  - o Clients seek longer-term strategic advice for planned expansions or other sites.
  - o Mentor/coach and support EoRs and is sought after as a mentor/coach.

EOR QUALIFICATIONS

Specific qualifications for EoRs have been developed with the intent of providing clarity on the requirements and potential career path. It should be noted that these guidelines describe the general knowledge/behaviour base for the 'typical' EoR role. However, each Participant's journey through the program will be unique based on individual facilities, clients, geographical locations and/or regulatory environment.

Formal education

In most cases it is expected that an EoR will have an engineering or equivalent degree in a discipline related to mine waste engineering. In addition, if the participant does not have a Masters or PhD degree, additional experience or formal mine waste training may be required.

If required by the jurisdiction within which the individual works, the participant will hold a professional license to practice engineering (eg Professional Engineer).

Minimum years of experience

It is expected that the typical EoR will have a minimum of 10 years of relevant project experience. This minimum experience may exclude non-tailings work or closely related work. For example, someone who has spent 50 per cent of their time undertaking non-dam or mine waste work, may require 20 years to accumulate 10 years of relevant project experience. Similarly, more relevant experience may be needed for complex, extreme consequence facilities.

For Deputy EoRs, we would target a minimum of seven years of experience, but less may be acceptable for lower consequence facilities.

Peer validation

ICMM and industry stakeholders recognise the need for an EoR to have leadership skills and behaviours. An EoR can face pressure from owners, contractors and others within their organisation and it is important that the EoR is able to manage that pressure to minimise public, personal, corporate and client liabilities/risks.

An EoR also requires professional maturity and emotional intelligence. These qualities cannot be measured by educational level or years of experience and as such must be demonstrated through the performance of the participant to manage complex tasks, interact with clients, other consultants, contractors, regulatory agencies, etc.

Consequently, individuals are to be vetted by their peers and references from a minimum of three people with whom they have worked. It is expected that at least two of these references will have reviewed and taken professional responsibility for the applicant's work. These references will typical be EoRs of tailings facilities (ie at a 4 or 5 level of competency).

Multi-disciplinary leadership experience

EoRs rely on other experts within their own organisations, the client's organisation and third parties. As such, the EoR must be capable of understanding these roles and evaluating if the expertise and experience of others is suitable. Since the EoR is responsible for the entire team, they must demonstrate their ability to:

- Identify the required disciplines to evaluate and steward the facility.

- Understand and approve/accept other disciplines' scopes of work as complete and adequate to meet the project needs.

- Evaluate the qualifications of others on their team.

- Incorporate comments, recommendations, and conclusions from other disciplines into the facility design and operation.

- Review the team structure and performance regularly, with a view to making sure the right people are engaged on the project.

TRAINING

As participants gain experience they will benefit from 'on-the-job' training. Nonetheless there is also a need for formal training programs. In general, these training programs are grouped into technical and non-technical.

Technical training can be formal or informal and may be related to types of analyses, analytical software, technologies, case studies, etc. These may take the form of internal resources, courses at academic institutions (eg college or university), etc.

Non-technical training relates to role requirements, procedures, and obligations of an EoR. While there are some academic institutions offering such courses, organisations will likely need to develop their own training. The authors recommend that training in some/all of the following areas is required:

- Current guidelines, standards, and decisions related to the role of EoR.

- Roles, accountabilities, responsibilities, and levels of authority within the teams stewarding a tailings facility.

- Expectations for assuming the EoR role for an existing site.

- Succession planning.

- Termination of EoR services and dispute resolution.

- Understanding EoR contracts, specialist design services contracts and general engineering contracts.

- Understanding, using, and developing leading business practices.

ASSESSMENT PROCESS

An assessment process is required and should align with the aim of allowing for continuous progressive development of competencies/attributes. Formal assessments are also incorporated at key progression milestones. The WSP assessment:

- Progressive Assessment (Competency Levels 1 to 4):
  - o Progressive documenting of experience and self-assessment.
  - o Regular review of experience and self-assessment with mentors.
- Formal Assessment:
  - o Competency Level 3 (Deputy EoR Role).
  - o Competency Level 4 (EoR Role).

Progressive Assessment (Levels 1 to 4)

As described above, development of an EoR is a progressive process and as such individuals are responsible for periodic documentation of their experience, and their progress towards the key attributes, competencies and experience. To assist with this process the WSP development framework incorporates a self-assessment tool, which is aligned with both the core competencies, and attributes presented.

Following self-assessment, participants are required to provide a copy of their completed self-assessment along with their documented experience to their mentor, who complete a parallel scoring assessment including seeking feedback from others, as required. Results are then discussed and agreed and gaps/development areas identified.

Progression to the next level is based on achieving a minimum average score in each area/competency, as confirmed/agreed with the mentor.

Competency Level 3 – Formal Assessment (Deputy EoR Role)

To progress beyond competency Level 3, it is anticipated that participants will need to undertake a Deputy EoR role in which they shadow/are mentored by an existing EoR over an extended period of time. As such, in addition to meeting the minimum scoring requirements for their progressive assessment, a suitable role must have been identified and a Deputy EoR candidate will be interviewed by the EoR under whom they will be working and an independent EoR peer.

These interviews are intended to provide an opportunity to discuss experience to date, areas for development, the nature of the work, client relationship and structure and the expectations of the Deputy EoR role. Overall, the objective is to ensure a successful working/mentoring relationship can be achieved. Acceptance to the Deputy EoR role requires acceptance by the EoR.

These interviews are also intended to provide the Participant with experience of the formal interviews required for progression to Competency Level 4 (EoR role).

Competency Level 4 – Formal Assessment (EoR Role)

Competency Level 4 is a significant step and the assessment effort from both participants and their organisation is likely to be significant. In the WSP system, participants are formally interviewed by a minimum of two existing EoRs from within the organisation. The interviewing EoRs will have been briefed by participant's mentors prior to the interview as well as had the opportunity to review participants project experience summaries. It is also expected that interviewers will reach out to references and other team members that participants have worked with.

The objective of the interview is to provide an independent verification that participants have developed the key attributes required to successfully undertake an EoR role. As such, interviews focus on both technical and non-technical skills as well as a participant's knowledge of the overall mine waste industry and governance practices. These interviews also provide experience of the formal interviews required as part of many client's EoR approval processes as well as ITRB/ governance review, or similar.

During this interview EoR candidates are required to discuss their qualifications pertaining to the EoR role, as outlined earlier in this paper. These are formally assessed and confirmed as part of the Competency Level 4 assessment process.

It should be noted that meeting the minimum EoR qualifications does not qualify a professional to be an EoR at any specific facility. The EoR, their team and the structure all need to be carefully matched and approved, including by the client. As such, it may take additional training/assessments to match an EoR with their first facility, as well as each new facility going forward.

ONGOING SUPPORT TO EORS

It is recognised that the EoR role is a long-term and challenging role. Guidelines and regulations for the role are also evolving as is engineering science and available technologies for tailings management. To assist EoRs in confidently executing the EoR role, WSP provides ongoing support to EoRs including but not limited to:

- EoR Advisory Group – EoRs may find themselves in difficult situations with clients, reviewers, their own management or others related to the execution of the role. The authors recommend that a group of senior practitioners be established to provide advice, intervene, and/or advocate on behalf of a project, EoR, or the companies EoRs in general. It is important to note that the decision-making authority for a facility still reminds with the appointed EoR.

- Informal EoR Meetings – These meetings provide an opportunity for EoRs to discuss current trends or concerns amongst peers.

- Ongoing training – As detailed earlier in this paper.

INITIAL VALIDATION OF THE PROCESS

As part of a validation process, current Deputy EoRs, EoRs and former EoRs (now Senior Reviewers and occasional Independent Tailings Review Board members) went through the self-assessment process. The purpose of the validation process was multi-faceted:

- To review the process by using it.

- To identify any gaps in the experience base for the EoRs currently assigned to facilities.

- To identify the path forward for engineers transitioning from other technical disciplines to focus on tailings (eg foundation geotechnics, or water dam engineering).

- To validate the 'bar' that should be met for each level.

A total of 93 people completed the survey, of which 40 were Deputy EoRs and 53 were EoRs or Senior Reviewers. The survey included a qualitative scoring scale for each attribute, ranging from 'A clear strength', scoring 100 per cent to 'No experience', scoring 0 per cent. The results of the validation process are presented as average values in Figures 2 and 3.

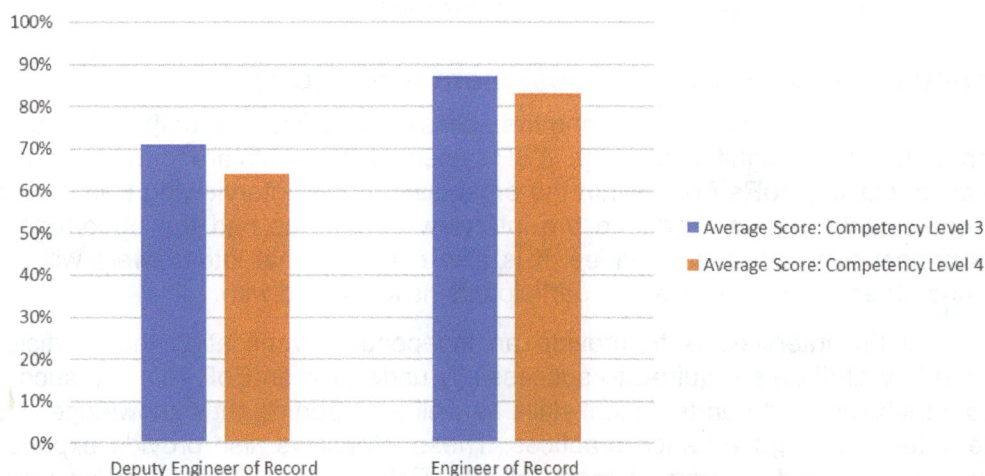

FIG 2 – Overall summary of validation process.

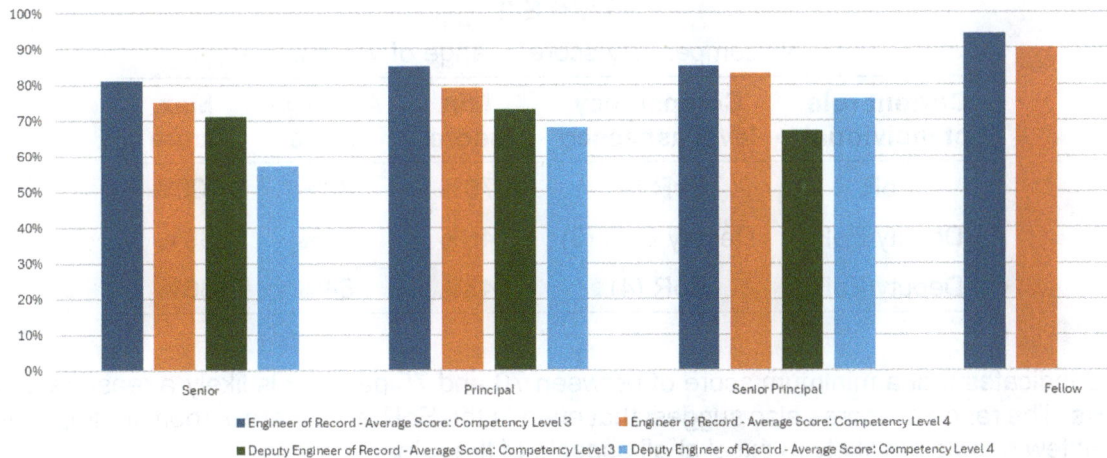

FIG 3 – Detailed summary of validation process.

It is evident from the summary data presented in Figure 2 that:

- The current EoRs (Competency Level 4) scored extremely well for both the Level 3 and Level 4 competency levels, with Level 3 slightly higher, as anticipated.

- The Deputy EoRs (Competency Level 3) scored well for Level 3, as anticipated, but there was a notable drop in their scores at competency Level 4. This supports the assessment that there are gaps which allows for a focused career conversation on what is needed to progress to Competency Level 4.

It is evident from the more detailed data presented in Figure 3 that:

- For practitioners in Deputy EoR roles, there is strong alignment with the minimum score at Competency Level 3 but, as noted above, there is a notable drop in score for self-assessments at Competency Level 4. This supports the assessment that there are gaps, as discussed above. At the Senior Principal level the values indicate there are individuals who are likely to be ready to take on an EoR role, if they are not currently doing so.

- Practitioners currently in EoR roles scored much higher than those in Deputy EoR roles, for both Level 3 and Level 4 competencies. It can also be seen that this was true even when the individuals were at the same level of career seniority. This validates the assumption that those completing EoR roles have varied skillsets compared to their non-EoR peers and that this is reflected in the EoR competency framework. It is also noted that some of these practitioners have transitioned from a related technical discipline area (especially those Senior Principals in Deputy EoR positions), or may be progressing at a different rate than others.

- For our most experienced practitioners (Fellows), who often sit on review boards as well as provide senior review, it is clear that the competencies are well covered.

The authors believe that the information presented above validates the overall process, however it is also noted that only the average scores are shown. For completeness, and to support meaningful conversations with our practitioners, as part of developing their career paths, the range of scores was also considered and is presented in Table 2.

TABLE 2

EoR competency score – range of values.

Current role of individual	Competency level assessed	Min score	Average score	Max score
EoR	EoR (4)	70%	82%	100%
Deputy EoR	Deputy EoR (3)	46%	71%	89%
Deputy EoR	EoR (4)	43%	64%	86%

Table 2 indicates that a minimum score of between 70 and 75 per cent is likely a reasonable value for EoRs. The range of scores also suggest that even at the EoR level there is room for improvement and that few people are likely to excel at all aspects of the role.

For Deputy EoRs, a minimum score of between 60 and 70 per cent is likely a reasonable value. The range of scores also suggest that some Deputy EoRs may be operating beyond their competency, and as such additional review is required. However, it is also noted that depending on the site and when combined with a strong EoR, this may be acceptable (and even deliberate).

The max score for Deputy EoRs assessing at EoR level was above minimum score for EoR. However, it is noted that this was based on a single Senior Principal doing a Deputy EoR role on an extreme consequence facility, which is likely appropriate given the consequence classification for the facility and the time required to effectively manage such a facility. It also indicates that experienced EoR for lower consequence facilities may have to 'step back' from EoR roles and take on deputy EoR roles for higher consequence facilities in order to advance their long-term development.

NEXT STEPS

As part of advancing the EoR development program, WSP is currently in the process of:

- Developing/identifying training modules to support the development of professionals.
  - o Internal development is expected to cover the minimum training identified within this paper, which are generally focused on internal guidelines, best practices and approaches to the provision of EoR services.
  - o The authors note that significant recourses already exist to support professionals soft/people skills and that individual learning styles differ. As such, it is not WSP's intent to replicate resources that already exist but to develop a list of resources that may be beneficial.
- Initiating the EoR Advisory Group – including the development of a mandate for the group to define clear responsibilities.
  - o This advisory group will also serve to help train the most senior EoRs for roles on ITRBs.

CONCLUSIONS

The authors and WSP believe that the comprehensive framework presented in this paper provides a basis for the assessment of the competency of EoRs as well as provides a road map for developing professionals. This supports with attracting more professionals and develop future EoRs. As such, it is recognised that this framework could be useful to the wider industry in helping to manage the risks associated with undertaking EoR roles, especially when competency is challenged by others, within and external to the industry.

However, it is recognised that professional protections would be enhanced by the establishment of international guidelines for EoR competency which the authors feel should be developed by a cross-industry team of EoR practitioners, mine owners and other key stakeholders. In publishing the outline of the WSP framework, it is the authors intent to provide a 'starting-point' to facilitate such an international effort. We welcome interested parties and organisations to join us in efforts to provide increased good practice governorship for the undertaking of the EoR role with the aim to meet the Global Industry Standard on Tailings Management (GTR, 2020), namely: 'achieve the ultimate goal of zero harm to people and the environment with zero tolerance for human fatality'.

ACKNOWLEDGEMENTS

The authors wish to thank the WSP Mine Waste Practice Area Network members for their participation in the global process including guidance and feedback provided by them during the development of the process. In particular, the authors would like the thank the following people who were key authors and contributors to the program: Ben Wickland, Laura Fidel, Chris Hatton, Andy Haynes and Paul Bedell.

REFERENCES

Engineers and Geoscientists of British Columbia (EGBC), 2022. Mount Polley Investigation Concludes; Three Engineers Disciplined [online]. Available from: <https://www.egbc.ca/News/News-Releases/Mount-Polley-Investigation-Concludes-Three-Enginee> [Accessed: 29 March 2025].

Global Tailings Review (GTR), 2020. Global Industry Standard on Tailings Management (GISTM) [online], Global Tailings Review. Available from: <https://globaltailingsreview.org/global-industry-standard/>

International Council on Mining and Metals (ICMM), 2025. Tailings Management: Good Practice Guide [online], International Council on Mining and Metals. Available from: <https://www.icmm.com/tailings-management>

The Life of Mine | Mine Waste and Tailings Conference 2025 | Brisbane, Australia | 29–30 July 2025

What role should controls play when declaring a failure mode non-credible?

J Willis[1], I Gillani[2] and K Spencer[3]

1. Practice Lead Tailings (Acting), BHP, Adelaide SA 5000. Email: justin.willis1@bhp.com
2. Principal Advisor, Tailings and Dams, Rio Tinto, Montreal QC, Canada. Email: imran.gillani@riotinto.com
3. Principal Tailings, BHP, Brisbane Qld 4000. Email: kevin.spencer@bhp.com

ABSTRACT

Assessing when a failure mode is credible versus non-credible is an aspect of risk management that has come to the fore following the release of the GISTM, and that standard's focus on credible failure modes. There is no clear definition of credible versus non-credible failure modes available either in the standard or in common practice in the industry, leading many mining companies to develop their own approaches. Additionally, beyond the strictly technical credibility and non-credibility divide are approaches for assessing the credibility of failure modes by their levels of likelihood and/or consequence.

Often the assessment of credibility is made with explicit or implicit assumptions about external conditions, tailings storage facility (TSF) behaviour and control effectiveness. This paper explores the role that controls, and the underlying assumptions that support them, play when assessing the credibility of a failure mode. This is done by considering approaches in common use in the industry for identifying failure modes, assigning controls, and assessing technical credibility and credibility based on risk. Key elements are identified for consideration when assessing credibility of failure modes to avoid mis-stating the risk and a false sense of assurance at the TSF. This paper highlights the need to understand well the underlying limitations of the methods, the assumptions that underpin the failure modes, and the explicit and implicit reliance on controls when declaring failure modes as non-credible.

INTRODUCTION

The Global Industry Standard on Tailings Management (GISTM) released in 2020 by the Global Tailings Review (GTR) increased the focus placed on the credibility or non-credibility of failure modes, by linking a number of requirements to credible failure modes. This has created ongoing debate in the industry on how to define non-credibility, the best approaches to define this, and several papers on the topic (Martens and Kupper, 2024; Thomson and Williams, 2024).

Determining the credibility of a failure mode relies upon a sound understanding of the tailings storage facility (TSF) design, characteristics, environment and operation. This reliance can incorporate the work of others (such as hydrological or geotechnical studies), assumptions that conditions don't change (such as tailings characteristics or upstream conditions), and the operation of the TSF or associated controls (such as responding to pond size, or maintenance of a spillway). These considerations apply to all TSFs, and are typically managed through controls associated with design, operation, performance monitoring and review.

This reliance then needs to be factored into the credibility of the failure mode. Teams that are involved in the assessment of failure modes will need to consider how many controls does a non-credible failure mode rely upon, and at what point would that failure mode transition from non-credible to well-controlled?

LITERATURE REVIEW

Global Industry Standard on Tailings Management (GISTM)

The GISTM (2020) uses credible failure modes or scenarios to qualify the following requirements:

- Human rights risks of credible failure scenarios (Req 1.1).

- Breach analysis for credible failure modes (Req 2.3).

- Human exposure and vulnerability to credible failure scenarios (Req 2.4).

- Consequence classification based on credible failure modes (Req 4.1).

- Design criteria that minimise risk for credible failure modes (Req 4.4).

- Resettlement where reduction of credible failure modes has been exhausted (Req 5.8).

- Emergency Preparedness and Response Plan/s (EPRP) based on credible failure scenarios (Req 13.1).

- Shared state of readiness for credible failure scenarios (Req 13.3).

- Engaging with public sector agencies and other organisations on credible failure scenarios (Req 14.1).

- Summary of impact assessments, exposure and vulnerability to credible failure scenarios (Req 15.1.B.4).

- Summary version of the EPRP for credible failure modes (Req 15.1.B.8).

For a failure mode to be credible, GISTM requires the failure mode to be technically feasible given the materials present in the structure and its foundation, the properties of these materials, the configuration of the structure, drainage conditions and surface water control at the facility, throughout its life cycle. The GISTM also refers to the credibility of a failure mode not being associated with the probability of an event occurring nor with facility safety. The ICMM (2021) Conformance Protocols do not provide any further detail and refer to the GISTM definition of credible failure modes.

The credible failure modes qualification related to the Requirements identified above makes identifying credible failures modes an important task in applying the GISTM. It also provides an incentive to reduce/eliminate credible failure modes, providing a reduction in workload to implement and maintain GISTM conformance. If a TSF can be declared as having no credible failure modes, it is unclear if GISTM applies to that TSF, with some companies adopting that approach. This has driven the debate within the industry as to the best approach to determine credible versus non-credible.

The GISTM refers to Critical Controls only in relation to the Operations, Maintenance and Surveillance manual (Requirement 6.4) and addressing performance outside of expected ranges (Requirement 7.4). The Glossary identifies a control as critical where the absence or failure would disproportionately increase the risk despite the existence of other controls.

ICMM

The ICMM (2025) Tailings Management Good Practice Guide relies on the GISTM definition of credible, and provides additional detail on determining credibility, based on breaking a credible failure mode into three components:

- credible mechanism

- credible initiating event

- credible failure process.

A credible failure mode needs all three elements to be credible. The Good Practice Guide proceeds to define a credible failure scenario as a credible failure mode plus credible consequence(s).

According to the Good Practice Guide, a failure mode with a credible mechanism and failure process may be deemed non-credible if it does not have a credible initiating event (also referred to as a trigger). Note it is unclear how this interacts with Requirement 4.6 of the GISTM that describes identifying and addressing failure modes independent of trigger mechanism.

The ICMM (2015b) Health and Safety Critical Control Management Good Practice Guide defines a control as an act, engineered object or system (combination of object and act) intended to prevent or mitigate an unwanted event. The Tailings Management and Health and Safety Critical Control Management guides both provide definitions for a critical control the same or similar to the GISTM.

ANCOLD, CDA and ICOLD

Australian National Committee on Large Dams (ANCOLD) tailings guidelines (2019) do not define credibility or use it to describe failure modes in the guidelines. Potential failure modes are referred to in relation to determining the consequence category of the TSF. It is similar for the 2022 ANCOLD Risk Guidelines, that do not define credibility or describe failure modes as credible. The failure modes assessment process in the Risk Guidelines describes considering all conceivable failure modes for a dam and removing only those that can be clearly stated to have a negligible level of risk, with negligible not defined. An example is given where a dam that has a factor of safety (FoS) >1.5 using reasonable but conservative assumptions and monitoring indicates no unusual movement the likelihood of slope failure can be taken as negligible. The authors have difficulty agreeing with this example.

Canadian Dam Association (CDA, 2013) is silent on credible failure modes, making no reference to either credible or failure modes generally. The Tailings Dam Breach Bulletin (CDA, 2021) refers to 'critical and credible failure mechanisms' for selecting failure scenarios to model but does not provide a definition of credible, while the Dam Safety Analysis and Assessment Bulletin (CDA, 2007) refers to 'hazards relevant to a particular dam', but also does not directly address credible versus non-credible failure modes or how to determine these.

International Committee on Large Dams (ICOLD, 2022) describes failure modes as needing to be physically possible, and eliminating potential failure modes as not physically possible must have a rigorous basis and rationale developed and documented. ICOLD further, similar to the Good Practice Guide (ICMM, 2025), requires an initiating event, failure mechanism and process. No reference to the credibility of these is made in the guidelines.

ICOLD (2022) identifies two types of control: critical controls that prevent, detect or mitigate a material risk; and supporting controls.

Other literature

Thomson and Williams (2024) conclude that there are three defining traits of credible failure modes, following their review of the available literature:

- technically feasible, conceivable and/or physically possible
- not associated with probability of occurrence
- not a reflection of facility safety.

Based on this, they recommend an approach that screens credible failure modes further into three categories, considering the technical feasibility and probability of occurrence, which are each then managed in a different manner. No further detail is given beyond what is provided in other industry guidance to determine non-credibility, rather they believe that the focus on defining failure modes as credible or non-credible distracts from the intent of the exercise. They provide an example of a non-credible failure mode as containment failure of an in-pit TSF.

Martens and Kupper (2024) look in detail at the relationship between credibility, likelihood of occurrence and consequence. They note that the definition of credible cannot be based on likelihood alone and must consider consequence. They also note that there is a poor track record of estimating likelihoods, typically arriving at values lower than reality, even in technologically sophisticated organisations. As a result, they state that the bar to demonstrate a credible failure mode presents negligible risk must be set very high, as not recognising the existence and credibility of potential failure modes can lead to inadequate design, construction, and operational controls. Martens and Kupper also note that any potential failure mode that requires monitoring is a credible failure mode.

Martens and Kupper (2024) reinforce the point that having a credible failure mode is not an indication of the likelihood of that failure mode occurring; and that structures can be very safe and have credible failure modes, such as aircraft that have multiple failure modes yet operate with a very high level of safety.

In 2021 and 2023, two workshops were convened to discuss interpretations and applications of credible failure modes (Small *et al*, 2023). These workshops identified two approaches to determining credibility, based on physically possible and determining that the probability of failure was non-negligible. Both approaches were noted as satisfying the intent of GISTM. Angela Kupper was a member of the panel formed to develop the GISTM and presented at both workshops. She made the following point in the workshops, that if a potential failure mode needs action, monitoring, human judgement, procedures to minimise human error, or maintenance of field conditions to manage the likelihood of failure, it is likely a credible failure mode (Small *et al*, 2023).

FAILURE MODE ASSESSMENT

The primary tool for identifying failure modes is the Failure Modes Assessment or Analysis (FMA) or variations such as the FMEA (Failure Mode Effects Analysis), PFMA (Potential Failure Modes Analysis) or FMECA (Failure Modes Effects Criticality Analysis). ANCOLD Risk Guidelines (2022) describes in detail the variations and use of these analysis tools.

Uses and limitations

The FMA is a systematic approach for identifying failure modes, beginning with the splitting of the TSF into components and then analysing in turn the effect if that component were to fail. The output of the FMA is a list of potential failure modes, that can be used to inform further risk assessments (such as quantitative risk assessment), design, preventative and mitigating controls and potential failure consequences.

ANCOLD Risk Guidelines (2022) lists the following limitations with an FMA:

- Difficulty in dealing with redundancy and the incorporation of repair actions to a failure progression.

- Focus on single component failures, with the interdependency of components and functions not easily able to be analysed.

- Difficulty incorporating human error into the failure progression (but can identify elements sensitive to human error).

- Requires an in-depth knowledge of the conditions to adequately account for environmental factors in the failure mode.

Using the ANCOLD (2022) guidelines as the most recent example of risk assessment guidance for TSFs, the recommendation is that all conceivable failure modes are included in the risk assessment until they can be excluded. This is an approach supported by Thomson and Williams (2024).

PFMA

Morgenstern (2018) highlights the value of PFMA for risk assessment of tailings dams. This approach adds a third dimension to the FMEA approach, which is typically comprised of a likelihood rating and a consequence/severity rating. The third dimension is a detection rating that is an assessment of the <u>effectiveness of controls</u> employed to avoid failure. Among other uses, PFMA is a good tool to assess the credibility of a failure mode based on hazard, likelihood, and effectiveness of controls in a cumulative manner.

The three ratings/criteria used in PFMA are defined as:

1. Likelihood rating – Likelihood of occurrence of the failure mode.

2. Severity rating – Estimate of the severity of failure (Consequence rating).

3. Detection rating – Likelihood for the failure mode to be detected in time to avoid failure (Control effectiveness).

Likelihood, Severity and Detection criteria are scored with the scores multiplied to calculate the 'Risk Priority Number' (RPN). Each criterion is scored independently of the other eg when scoring likelihood, the user must not consider the detection rating score and *vice versa*. RPN is a widely adopted method for risk assessment and is used often in the automotive industry.

The RPN is used to make decisions regarding the failure mode, such as its treatment or the implementation of mitigative measures. A high RPN means high risk; therefore, additional treatment such as mitigation should be considered to minimise the failure mode exposure. A failure mode with a low RPN may be argued to be a low contributor to the risk.

In the authors' opinion, a scale of 1 to 100 for RPN is convenient and logical. For this scale to work, authors suggest using the following rating scales for the likelihood, severity, and detection ratings:

Likelihood rating rated on a scale of 1 to 5, with 5 indicating very high likelihood of occurrence. Likelihood/probability descriptors, published by leading organisations, could be adopted to develop a rating table for this purpose. In the authors' opinion, the annual probability descriptors, with five categories, published by the U.S. Bureau of Reclamation and US Army Corps of Engineers is a good reference for this purpose.

Severity rating rated on a scale of 1 to 5, with 5 indicating extreme consequences of failure. The consequence classification table provided in GISTM, with five categories, could be adopted for the severity rating.

Detection rating rated on a scale of 1 to 4, with 1 indicating good control effectiveness and 4 indicating weak control effectiveness. Published descriptors for control effectiveness could be adopted to develop a table for this purpose.

Using the above rating scales, a RPN between 1 and 100 could be calculated using the following formula.

$$RPN = Likelihood\ rating \times Severity\ rating \times Detection\ rating$$

The RPN value is then used to make decisions regarding implementation of mitigative measures. In the authors' opinion, the following table may be used for this purpose.

RPN	Description
1 to 8	Minimal or No Action Required
9 to 40	Action Recommended
41 to 100	Action Required

RPN could be used for all or any one of the following purposes:

- Assess contribution to risk of a failure mode: a failure mode with a very low RPN may be argued to be a low contributor to the risk.

- Portfolio risk ranking: the higher the RPN value, the higher the risk.

- Prioritisation of resources: Mining companies do not have unlimited resources and hence need to prioritise. RPN can be a good basis for prioritising attention and resources: More attention and resources allocated to facilities with relatively high RPN values.

- Site management may use RPN to justify capital expenditure to senior management/corporate. A mining company may institute a rule that no RPNs can be above a certain level.

- Show progress in risk reduction: RPN is calculated before and after implementation of mitigative actions and the reduction in RPN is used as a measure of risk reduction.

DEFINING TECHNICAL NON-CREDIBILITY

Without a clear definition of what is meant by technical non-credibility, or not physically possible, the determination of technical non-credibility relies on engineering judgement, the experience of the risk assessment team, and the level of residual risk the TSF owner is willing to tolerate when declaring a failure mode non-credible.

Clear examples

It is possible to provide clear examples of technical non-credibility that can be used to provide some perspective. For example, it is straightforward to determine that a landlocked TSF is unable to fail by impact from a ship. Similarly, if there is no railway adjacent, impact by train can be equally discounted.

Thomson and Williams (2024) provide the example of containment failure of an in-pit TSF, while other examples that have been known to be used include piping failure where there is no source of water on the TSF, overtopping on a landform that has been shaped to shed incident water, or liquefaction in a free draining material.

There is some question of the value provided in these examples, as they are unlikely to organically arise in a FMA where that component does not exist. Including failure of a drain in the TSF as a non-technically credible failure mode when there is no drain installed may be viewed as being suitably comprehensive, however it risks making the exercise cumbersome and providing little value in the minds of stakeholders. Often, we discount failure modes intuitively, and listing every single non-credible failure mode provides little value.

That notwithstanding, considering and documenting the reasons for discounting the most common failure modes would make an FMA more defensible. ICOLD (2000) grouped TSF failures into the following categories (from the greatest number of incidents to least):

- slope stability
- earthquake
- overtopping
- foundation
- seepage
- structural
- erosion
- mine subsidence.

Less clear examples

Less clear examples typically rely to a greater extent on engineering judgement and risk tolerance. Considering the example above of impact by ship or train, if there is an airport nearby to the TSF, and noting that many remote mine sites do have purpose built airfields, an impact by an aircraft then may be a technically credible failure mode. That notwithstanding it is rare to see aircraft impact on an FMA.

One example of varying technical credibility is to consider a 40 m high TSF embankment that is supported by waste rock. If the waste rock layer is 3 m thick, then most practitioners would be likely to agree that slope failure is credible. However, if that waste rock layer is 300 m thick, then most practitioners would say slope failure is non-credible. Within those two extremes is a point where the TSF failure transitions from credible to non-credible, but defining this point is difficult and ultimately likely to be based on judgement.

This can also be related to rainfall and overtopping failure modes. If our TSF in the paragraph above does not have a spillway and can store a 1 in 100 year event, then most practitioners would be likely to agree that overtopping is credible. However, if the storage volume is many multiples of the PMF, then most practitioners would say overtopping is non-credible, with credibility transitioning at some point between.

If we consider our hypothetical TSF again and look at conditions for liquefaction, we could state with confidence that an unsaturated TSF won't liquefy, but how about if the saturation level is at 82 per cent? The degree of saturation that separates materials that are vulnerable to static liquefaction is commonly taken as 85 per cent but that doesn't mean that liquefaction immediately can't occur below that point.

Ability to control a failure mode

Another consideration is what can be done to control the failure mode. The often-used example in this case is failure by impact from a meteorite. While this is technically credible or physically possible, there is no reasonable active control that can be implemented to reduce the risks associated with this event. Incorporating it into an FMA may serve to complete the record, but there is little else that is done with that failure mode. Other examples include volcanic activity such as the Paricutin volcano that formed in a corn field in Mexico (Wikipedia, 2025).

Industry examples

Oroville Dam

The Oroville Dam, the tallest in the United States of America, had a service spillway and an emergency spillway, yet in 2017 a rainfall event below the design event and less than the largest event in the dam's history triggered the evacuation of the downstream population when the service spillway failed and unexpected erosion occurred below the emergency spillway (France et al, 2018). The failure of the service spillway and the failure of the emergency spillway had both been identified within an FMA and assessed as non-credible. No failure mode involving both spillways failing at the same time was identified.

Edenville

The Edenville dam failure in Michigan in May 2020, shares some characteristics to the Oroville Dam incident. A smaller than design rain event resulted in a water level higher than had previously been experienced at the dam that caused static liquefaction in a section of the embankment (France et al, 2022). A failure mode similar to the actual cause was identified in a 2005 FMA, and determined to be non-credible. This was on the basis of a 1991 stability assessment indicating a FoS of 1.8 using drained parameters. Two subsequent FMAs in 2010 and 2015 did not challenge this assessment. The failure report notes that elements of three other identified failure modes – each identified as credible – were involved in the failure.

NON-CREDIBILITY ON THE BASIS OF NEGLIGIBLE CONTRIBUTION TO RISK

Besides identifying technical non-credibility of a failure mode, there is the negligible contribution to risk approach. This approach is noted in ANCOLD's Risk Guidelines (2022) and appears in the Federal Energy Regulatory Commission (FERC) Engineering Guidelines (2021). This approach considers the likelihood and/or consequence of the failure mode, allowing those failure modes with very low likelihoods, and/or consequence to be deemed non-credible. Adopting this approach requires setting a threshold of likelihood and/or consequence below which the risk contribution can be considered negligible. FERC sets the likelihood threshold at 1 in 1 000 000, and this value has been adopted by some operators within the tailings industry (Small et al, 2023). Note the authors are not recommending the use of this value, any threshold for negligible contribution should be based on a well thought through and defensible position considering the specific aspects of the TSF or site.

Clear examples

A combination of extreme events will often result in a likelihood so remote as to be negligible, such as combining the maximum credible earthquake with the probable maximum precipitation (PMP). This combination of events while physically possible, will typically produce a very low likelihood, and is generally excluded. Another example is the meteorite strike, described previously. This failure mode while physically possible could be ruled non-credible on the basis of negligible contribution to risk.

Within slope stability analysis, shallow failure surfaces – typically those less than 1 m in depth – will often be excluded. These can be an artefact of the modelling, and if such a failure were to occur the consequences would be low. These types of failure are rarely carried through into FMAs, being intuitively screened out.

Similar to technically non-credible, the value provided in these examples is questionable, as they are unlikely to organically arise in an FMA. Listing every single remote likelihood or low consequence failure mode provides little value.

Less clear examples

Consider the example provided by ANCOLD of a dam with a FoS >1.5 using reasonable but conservative assumptions and monitoring indicating no unusual movement the likelihood. In this scenario the following is still being relied upon:

- The variability in the foundation materials is known and has been accounted for.

- The foundation is sufficiently well understood and there are no unidentified layers or weak zones.

- The material properties are geochemically stable and will not change over time.

- The monitoring instruments are installed in locations that will detect movement that would be expected ahead of a failure.

- The monitoring of the instruments is of a sufficient frequency and accuracy to identify precursor movements.

In this example, the buffer that accommodates the unknowns is the FoS >1.5 and the conservative assumptions. There is insufficient detail in the example to identify if these are sufficient to warrant identifying the failure mode as non-credible based on negligible contribution to risk.

INTERACTION OF ASSUMPTIONS, CONTROLS, AND TSF STATUS WITH CREDIBILITY

Non-credibility may still be associated with reliance on underlying assumptions or reliance on explicit or implicit controls. As described by GISTM, breach analysis, consequence category, design criteria and emergency response are all informed by credible failure modes. Mischaracterising credible failure modes can result in inappropriate design criteria, governance, controls and emergency responses.

Underlying assumptions and controls

Acknowledging and understanding the assumptions and controls that underpin an assessment of non-credibility is critical, because if the assumptions cease to be valid or the controls fail, then the failure mode may become credible. Coming back to the question raised in the introduction section of this paper, at what point does a failure mode move from being non-credible to well controlled?

There may be further controls to address some of the assumptions, such as a trigger action response plan (TARP) that acts if the pond size grows too large, or inspections that would identify erosion on the crest. Even these though, have underlying assumptions that may not be valid when put under stress. Particularly important is the consideration of a response in the event of a large scale event that affects a wide area and assumes resources can be mobilised for the TSF. A large flood may restrict access and personnel, and create multiple demands on resources.

If the role that the control plays in the assessment of credibility is not well understood and the failure mode is assessed as non-credible, the criticality of this control may be missed.

Reducing likelihood and consequence of failure considering controls

Failures that are declared non-credible based on negligibility may have explicit or implicit reliance on controls affecting either the likelihood or consequence.

Quantitative Risk Assessments (QRAs) can include explicit or implicit probabilities about intervention that reduce the overall likelihood of failure. These may not hold true where a potential failure triggering event can affect the response, such as access to the TSF in the event of a large flood described in the section above. These probabilities also may not hold true in circumstances without an initiating event, such as the beach, freeboard, and pond management all deviating at the same time, as all require consistent proactive operational management. Also, once one or more of those start to deviate, the system may be far more vulnerable to the others deviating. Lastly the incremental nature of the deviation could reduce the urgency of the intervention, such as the normalisation of the erosion at the Oroville Spillway (France *et al*, 2018).

The consequence of failure may be reduced through reliance on explicit mitigating controls such as Emergency Preparedness and Response Plans (EPRPs) that rely on the failure being identified in a timely manner, or implicitly through selection of fatality rates that incorporate an element of warning, or an assumed rate of progress of the failure.

Absent controls for non-credible failure modes

ICOLD (2022) directly links controls to failure modes identified from the risk assessment. The logical extension of this is that if there are no failure modes, there are no controls required at the TSF.

Martens and Kupper (2024) note that controls can be seen to be unnecessary if there are believed to be no credible failure modes. They use the example of the Titanic when describing the risk associated with declaring no credible failure modes. Declaring the ship unsinkable was the direct cause in the loss of life through actions such as not installing sufficient lifeboats.

Non-credibility for active TSFs

Assumptions often relate to conditions remaining the same, while for active TSFs at active mine sites, conditions are often constantly changing. Declaring a failure mode as non-credible requires confidence in the future condition, environment, and performance of the TSF, at least until the next FMA review. Active TSFs typically are receiving additional loading, may be receiving additional liquor, may be interacting with nearby infrastructure such as open pits and underground mines, may be impacted by changes in the upstream or downstream environment, and are reliant on the ongoing operation of the TSF in accordance with the design.

Examples

Non-credible overtopping failure mode

Consider an above ground paddock style TSF with no emergency spillway that can store a 72 hr duration PMP rainfall. It could be argued that the TSF has no credible-overtopping failure mode, based on the PMP being the largest conceivable single rain event. However, there are a number of underlying assumptions that accompany the ability to store the PMP:

- The beach slope remains the same.
- The pond location and size remain the same.
- The TSF geometry remains the same and there isn't a step in or set back.
- The walls are raised at the same rate and one wall is not advanced and operated ahead of the others.
- The PMP is the scenario that results in the greatest volume of water on the TSF (there are examples where smaller events over a number of months in succession provide more volume and are more likely than the PMP).
- Wind doesn't result in wave set-up and/or wave action that overtops the TSF.
- The PMP isn't exceeded, and the estimated return period is correct (see ANCOLD (2022) for comments on PMP reliability).
- The facility can also overtop if the embankment drops in elevation, through erosion, settlement, embankment failure or human intervention. While these mechanisms may be captured as separate failure modes, the limitations of an FMA for considering combination failure modes means that these may not be considered in conjunction with a wet weather event or upset pond condition.

If a series of smaller weather events occur that cumulatively lead to an excess of water in the system, then the following weakness may be encountered when trying to implement a pond size TARP:

- Water return to plant may reduce as plant manages the surplus of water in the system.

- There may be management pressure to store excess water on the TSF because of this excess water in the system.

- The TARP likely assumes there is somewhere to pump water to, however all other storages may be full and discharge to the environment may not be permitted.

- The decant and pumps may become submerged, rendering them inoperable.

- Obtaining additional pumps may not be possible, with pumps on-site or in the region also running at full capacity, possibly combined with shortages in spare parts.

- Access to the TSF or even the site may be restricted in wet events, through washing out of ramps, flooding of roads, or flooding of camps and airports.

On this basis a series of wet events that may not even deviate significantly from the normal patterns could put a TSF into a position where it is at risk of overtopping despite having declared this failure mode to be non-credible.

Non-credible liquefaction failure mode

Let's consider the saturation level of this TSF in relation to liquefaction. We may have a tailings saturation level of 50 per cent and confidence that liquefaction is not possible at that level, hence rule liquefaction failure modes as non-credible. However, there are again assumptions that could accompany this determination:

- The saturation state remains at 50 per cent, and the TSF doesn't re-saturate over time, with further reliance that:
    - Pond size remains in target.
    - Tailings solid concentration remains the same.
    - Drains under the TSF continue to function.
    - Seepage from adjacent TSFs doesn't recharge the TSF.
    - Artesian pressure doesn't recharge the TSF.

- The TSF doesn't have zones or layers of higher saturation, either laterally or vertically within the TSF that could liquefy.

In this scenario, the following controls may be implicitly relied upon to support the assumptions above:

- Monitoring of overall saturation levels in the TSF.

- Pond control (with the associated issues described above).

- Monitoring of drainage system efficacy/recovery.

- Tailings properties that avoid creating zones of higher saturation.

- Deposition control to avoid creating zones of higher saturation.

The team operating the TSF is unlikely to have the appreciation that changes in these controls may have an impact on a failure mode deemed non-credible, and would be relying on Engineer of Record (EoR) reviews, Independent Tailings Review Board (ITRB) reviews, or Dam Safety Reviews (DSR), or an update to the FMA to capture this.

Industry examples

Baia Mare

In January 2020 the Baia Mare TSF in Romania overtopped releasing 100 000 m³ of material. The TSF stored material from the reprocessing of an existing TSF and was constructed using coarser particles separated from the tailings by hydrocyclones. The coarse content of the tailings varied from the design and resulted in insufficient production for the embankments. This was exacerbated by the hydrocyclones being unable to operate in very low temperatures. These two factors meant that the freeboard was insufficient leading to the overtopping of the TSF (Garvey *et al*, 2000).

Mount Polley

The Mount Polley TSF failed in Augst 2014 due to the failure to recognise how the consolidation state of a foundation layer changed over time with embankment loading, moving from over consolidated to normally consolidated (Morgenstern, Vick and Van Zyl, 2015). The assumption that the material was over consolidated led to incorrect stability assessments that informed the design. Contributing factors were the geotechnical investigation spacing that meant that the location of this foundation layer was not well understood, the TSF slope had been steepened due to a shortage of fill, and the TSF stored increasing volumes of water due to discharge constraints.

RELIANCE ON CONTROLS

General effectiveness of controls

Comon ways controls may cease to be effective include weaknesses in design, mechanical breakdown, and human factors.

Control design should flow from the risk assessment used, such as being directly linked to a failure mode (ICOLD, 2022) if using an FMA, or breaking the linkage between cause and event if using a Bow tie (ICMM, 2015a). A poor understanding of the failure mode or linkage between cause and event could lead to incorrect control design.

Design elements that perform a control function, engineering controls, and instruments that verify control performance are susceptible to mechanical failure. A drain installed at the base of a TSF can crush or silt up, a reverse filter installed on an active seepage may not have been constructed properly, and a VWP can corrode over time.

There is a broad field of study related to human factors and human reliability assessment (HRA). Kirwan (1994) notes that human error is extremely commonplace, and human error in complex systems can have a large impact. Impact from human error can arise from a requirement for reliable human performance in response to an event, where the error is then failed to be corrected or compensated for by the system (Kirwan, 1994).

Industry examples

Merriespruit TSF

Tailings deposition at the northern cell of the Merriespruit TSF was suspended in March 1993 due to ongoing issues with excessive seepage and embankment sloughing (Fourie, Blight and Papageorgiou, 2001). Despite this, for reasons that were attributed to poor communication, sporadic deposition continued at the cell. In addition, spillage from the adjacent cells occurred, and water was being actively pumped into the cell the night of the failure. Lastly there was a failure to maintain the required freeboard in the TSF. The reliance on operational controls, which failed, resulted in the overtopping of the TSF and the loss of 17 lives.

Oroville Dam

Considering the Oroville Dam spillway incident again, this TSFs had three mechanisms for reducing water level in the dam – the powerhouse, the service spillway and the emergency spillway – and the risk of overtopping, as noted above, was considered non-credible. All three controls however, were at risk following a single rain event with the powerhouse at risk of flooding from tailwaters, and both the service and emergency spillways heavily eroded.

DISCUSSION

Technically non-credible failure modes

There are a few things that are apparent from the consideration of technically non-credible failure modes:

- it is more difficult than is generally acknowledged to declare failure modes as technically non-credible.

- there are few clear cut cases where this can be done in practice for active TSFs, with more opportunities for TSFs that are inactive or closed.

- unless the failure modes are common in the industry, there is little value in identifying and listing failure modes that don't apply to a TSF, and typically these are instinctively ruled out.

Ability to define non-credible failure modes

The limitations of the FMA approach need to be kept in mind when attempting to determine credible and non-credible failure modes. An FMA output of non-credible failure modes for a TSF only indicates that the FMA process and team have not been able to identify any credible failure modes for the TSF, not that the TSF cannot fail. As Martens and Kupper (2024) point out, this is akin to calling the Titanic unsinkable.

It could be argued that most TSF failures have occurred due to a failure mode that was either not identified, identified but not considered credible, or was mistakenly considered well controlled. Given the limitations of the FMA process, and the ability for a non-credible failure mode to rely on assumptions and implicit controls, an explicit control may still be required for non-credible failure modes, at the very least such as a process for regularly checking the FMA against any changes at the TSF.

Where the failure mode is not a clear example and relies on judgement, extensive technical analysis, or the application of controls, the more appropriate approach is to be conservative and ensure the relevant design and operational controls are carried forward. Effort focused on determining a failure mode as non-credible may be better spent ensuring the controls are robust and effective. Given the tendency to reduce focus on controls that manage non-credible failure modes, declaring the failure mode credible ensures these controls receive higher levels of attention. It would also ensure the failure mode is carried forward for review in FMA updates or when conditions change. The PFMA technique described earlier in this paper can highlight the role that controls play when assessing failure modes.

Where a failure mode is deemed technically non-credible and yet this determination is reliant on a single control, according to ICOLD (2022) this control should be designated as a critical control. This creates a seeming contradiction in terms defining a critical control for a failure mode that is technically non-credible.

Non-credible on the basis of negligible risk

This approach is attractive, as it presents a route that is easier to achieve than technically non-credible. It uses established tools such as QRA and dam break assessment and provides a quantified output to support the determination of non-credibility. There is limited reproducibility for QRAs, and as noted by Martens and Kupper (2024) a poor track record for accurately identifying the risk even for sophisticated organisations. It is possible that a likelihood is several orders of magnitude out, and such uncertainties should be reflected in the threshold defined at which level the risks are considered negligible. Noting Martens and Kupper's concerns, this threshold should include both likelihood and consequence considerations.

Similar to technically non-credible failure modes, where the determination of negligible risk is based on explicit or implicit reliance of a control, then this control should be classified as a critical control, with the associated contradiction.

Combinations of failure modes

The primary tool for identifying credible failure modes, the FMA, has difficulty assessing a combination of events or a complex system of interactions (ANCOLD, 2022; France *et al*, 2018, 2022). The Edenville failure report expressed this well noting '... the challenges involved in postulating a set of PFMs which is sufficient to reasonably capture all the ways in which a dam might fail' (France *et al*, 2022).

If we revisit the example earlier of a TSF able to store the PMP event, but this time assume it is a cross valley TSF, the risk of overtopping could be from a large rain event, or a landslide into the TSF creating a sufficient wave to overtop the crest. Individually these failure modes may be low likelihood events and considered non-credible. However a single event could increase the likelihood of both events, such as a large rainfall that increased the pond level and destabilises a section of the adjacent hillside, such as in the Vajont failure in Italy in 1963 (Norbert, 2009). This would also apply for TSFs adjacent a large waste dump.

At Oroville Dam, an underlying assumption about rock strength of the natural soils was the reason behind the erosion/failure of the spillways. This can similarly apply where the non-credibility of the failure mode is reliant on controls that could be disrupted by an event that also increases the likelihood of failure. An extended power outage or fuel shortage could mean that an active dewatering system is inoperable at the same time as the monitoring system.

In these instances, the likelihood of failure converges to the likelihood of the single initiating event.

Disadvantages of assessing all conceivable failure modes as credible

The most conservative approach when assessing failure modes is to assess all conceivable failure modes as credible and carry them through for management. This approach may present downsides such as:

- Increased workload for managing the failure modes, associated controls and verification.
- Increased workload defining design criteria, consequences and classification, emergency response plans (noting that the these may be reduced if the failure modes or failure scenarios have similar consequences).
- Reduced focus on the higher likelihood and consequence failure modes.
- Carrying population at risk and the corresponding consequence classification in cases where the probability of failure is vanishingly low, and therefore the related ongoing effort of maintaining emergency preparedness and response for that population at risk.

Disadvantages of not assessing all conceivable failure modes as credible

If an approach is adopted that discounts controlled or low consequence/likelihood failure modes as non-credible, then the following downsides may present:

- Reduced focus on the failure modes, and associated explicit or implicit controls.
- Monitoring and surveillance that may not be sufficient to identify changes in conditions or assumptions or when failure modes become credible.
- False sense of security or increased risk taking (such as the Captain of the Titanic not slowing down when entering the iceberg field as described by Martens and Kupper (2024)).
- An inappropriate assessment of the consequences of the TSF failing, leading to inappropriate classification of the risk and level of governance.
- Subsequent FMAs don't critically assess the failure modes and instead adopt the same credibility assessment as previous FMAs.

- Emergency response plans for the TSF that don't alert all the population at risk.

CONCLUSION

When assessing failure modes of a TSF with a view to determine if they are non-credible or not, these are some key aspects of which to be mindful.

- Accept that FMAs may miss failure modes and not cover all combinations of events that could lead to failure. Controls may still be required for non-credible failure modes.

- It is important to understand the controls and assumptions that are implicitly or explicitly relied upon within the FMA, and ensure these get the appropriate level of attention. These should be captured, and a PFMA can assist with identifying and recording the underlying controls and assumptions.

- The authors advise caution when declaring a failure mode non-credible when there is a reliance on controls, and for active TSFs where conditions and assumptions can change. Non-credibility in these circumstances may be challenging to justify.

- There are a number of ways controls can fail, and non-credible determination that is heavily reliant on controls may be better left as credible, with a focus instead on ensuring the controls are effective. The detection rating within the RPN can assist identifying the extent to which a failure mode is reliant on controls.

- Where the failure mode is reliant on a single control or a control that meets the definition of a critical control, this should be captured as such, regardless of the credibility of the failure mode.

- The FMA process should be revisited frequently to check the controls are still effective and the assumptions unchanged. The FMA process should revisit failure modes declared as non-credible each time.

REFERENCES

Australian National Committee on Large Dams (ANCOLD), 2019. Guidelines on Tailings dams, Planning Design, Operation and Closure, Revision 1, July 2019.

Australian National Committee on Large Dams (ANCOLD), 2022. Guidelines on Risk Assessment, July 2022.

Canadian Dam Association (CDA) 2013. Dam Safety Guidelines 2007 (2013 Edition).

Canadian Dam Association (CDA), 2007. Technical Bulletin: Dam Safety Analysis and Assessment.

Canadian Dam Association (CDA), 2021. Technical Bulletin: Tailings Dam Breach Analysis.

Federal Energy Regulatory Commission (FERC), 2021. Chapter 17 – Potential Failure Mode Analysis, Engineering Guidelines for the Evaluation of Hydropower Projects, Washington.

Fourie, A B, Blight, G E and Papageorgiou, G, 2001. Static liquefaction as a possible explanation for the Merriespruit tailings dam failure, *Canadian Geotechnical Journal*, 38:707–719.

France, J W, Alvi, I A, Dickson, P A, Falvey, H T, Rigbey, S J and Trojanowski, J, 2018. Independent Forensic Team Report: Oroville Dam Spillway Incident, 5 January 2018.

France, J W, Alvi, I A, Miller, A C, Williams, J L and Higinbotham, S, 2022. Independent Forensic Team Final Report Investigation of Failures of Edenville and Sanford Dams, May 2022.

Garvey, T, Barlund, K, Mara, L, Marinov, E, Morvay, K, Verstrynge, J F and Weller, P, 2020. Report of the International Task Force for Assessing the Baia Mare Accident, December 2020.

Global Tailings Review (GTR), 2020. Global Industry Standard on Tailings Management (GISTM) [online], Global Tailings Review. Available from: <https://globaltailingsreview.org/global-industry-standard/>

International Committee on Large Dams (ICOLD), 2000. Tailings Dams Risk of Dangerous Occurrences, Lessons learnt from practical experiences, Bulletin 121.

International Committee on Large Dams (ICOLD), 2022. Tailings Dam Safety Bulletin No. 194, Preprint submitted for publishing, November 16 2022.

International Council on Mining and Metals (ICMM), 2015a. Critical Control Management, Implementation Guide, April 2015.

International Council on Mining and Metals (ICMM), 2015b. Health and Safety Critical Control Management, Good Practice Guide, April 2015.

International Council on Mining and Metals (ICMM), 2021. Conformance Protocols, Global Industry Standard on Tailings Management, May 2021.

International Council on Mining and Metals (ICMM), 2025. Tailings Management, Good Practice Guide, February 2021.

Kirwan, B, 1994. *A Guide to Practical Human Reliability Assessment*, 1994 (Taylor and Francis Publishers).

Martens, S and Kupper, A, 2024. Credible failure Modes: Considerations for Assessment and Application, Proceedings of Tailings and Mine Waste 2024, Colorado USA.

Morgenstern, N R, 2018. Geotechnical Risk, Regulation and Public Policy, *Soils and Rocks, São Paulo*, 41(2):107–129, May–August 2018.

Morgenstern, N R, Vick, S G and Van Zyl, D, 2015. Independent Expert Engineering Investigation and Review Panel Report on Mount Polley Tailings Storage Facility Breach, 30 January 2015.

Norbert, J, 2009. *Beyond Failure, Forensic Case Studies for Civil Engineers* (ASCE Press).

Small, A, Kupper, A, Johndrow, T and Al-Mamun, M, 2023. Credible Failure Modes – Summary of 2021 and 2023 Workshops, Proceedings of Tailings and Mine Waste 2023.

Thomson, H and Williams, H, 2024. Defining 'Credible' in the Context of Tailings Storage Facility Failure Modes, in Proceedings of ANCOLD Conference 2024, Adelaide, Australia, 11–14 November 2024.

Wikipedia, 2025. Parícutin, Mexico. Available from: <www.en.wikipedia.or/wiki/Paricutiin> [Accessed: 4 July 2025].

Mine closure and rehabilitation to accommodate site settings

Validation of landform evolution modelling using remote sensed erosion data

H Crisp[1], N Nazarov[2], E Smedley[3] and S Gregory[4]

1. Environmental Scientist, Mine Earth Pty Ltd, O'Connor WA 6163.
   Email: harry@mineearth.com.au
2. Principal Engineer, Mine Earth Pty Ltd, O'Connor WA 6163. Email: nickolai@mineearth.com.au
3. Chief Executive Officer, Mine Earth Pty Ltd, O'Connor WA 6163. Email: elis@mineearth.com.au
4. Chief Strategy Officer, Mine Earth Pty Ltd, O'Connor WA 6163.
   Email: stacey@mineearth.com.au

ABSTRACT

A key requirement for landform closure design is to characterise the long-term erosion risk. Landform evolution modelling (LEM) is often conducted prior to landform construction to predict erosion impacts over extended time periods. Increasingly, regulatory authorities require LEM to predict post-closure erosion performance of landforms during the project approvals phase. The purpose of this presentation is to investigate the predictive accuracy of LEM on a mining landform in comparison to observed erosion at a Goldfields mine, in Western Australia (WA).

Due to the development of modern survey hardware, including unmanned aerial vehicles (UAV) and high-resolution mounted surveying sensors, high-resolution surveys of rehabilitated landforms can be readily undertaken. This case study takes advantage of UAV captured LiDAR data that was collected to conduct landform-scale erosion monitoring and classified to filter out vegetation. This data was used to measure the geometry of erosion features including length, depth, slope and volume at the landform scale.

LEM was used to predict erosion on a reconstructed as-built model of the landform. Initially, the LEM was parameterised using only information available for the greenfield project such as geological information, analogue conditions, climate, local soils, landscapes and vegetation data. This represents a scenario that might be common for the approval stage of a Project. The LEM was run to match the duration of the current rehabilitation age to enable comparison of the predicted and actual erosion. The LEM was then re-calibrated using measured erosion geometries and rerun to assess predicted and measured erosion over a 300-year time frame.

Conclusions from this study will demonstrate methods to use LiDAR data to assess the scale of erosion features and discuss the effectiveness of LEM at predicting erosion in multiple project stages and assess long-term risk.

EXTENDED ABSTRACT

Erosion poses a significant risk to the rehabilitation of mining landforms after closure. Various factors influence erosion, including mine waste and growth media properties, climate conditions, local drainage patterns, landform construction and rehabilitation design, surface treatments, vegetation establishment, post-closure land use, and exposure to extreme weather events.

A crucial aspect of landform closure design is characterising long-term erosion risk. Landform evolution modelling (LEM) is often performed before landform construction to predict erosion impacts over extended periods. Regulatory authorities increasingly require LEM to predict post-closure erosion performance during the project approval phase. While LEMs are valuable for informing rehabilitation designs, they depend on well-defined parameters. Without available as-mined waste rock and rehabilitation trials, practitioners may lack sufficient information to develop suitable parameters for LEMs. Using overly conservative parameters can impose unnecessary constraints on landform design, leading to additional disturbances, higher mining costs, and unfavourable outcomes. Conversely, using overly optimistic parameters can result in designs that promote excessive erosion, potentially harming the environment and failing to meet stakeholder expectations, requiring costly rework.

The growing use of whole-of-dump scale approaches like airborne LiDAR or photogrammetry surveys in post-closure monitoring enables high-density and high-quality data collection to assess mining landform performance. However, current practices often emphasise data collection without adequately informing management responses based on the available data. Drainage reviews can identify high-risk erosion areas due to ineffective water management, but deciding whether to intervene for erosion features on sloped surfaces can be challenging. A well-calibrated LEM helps understand the long-term risks posed by erosion features and determines where remedial actions may be necessary. It is important to remember that erosion risk of materials can change over time, potentially decreasing (eg via self-armouring or vegetation growth) or increasing (eg via material weathering).

This presentation investigates the predictive accuracy of LEM on a mining landform compared to observed erosion at a Goldfields mine in Western Australia (WA) and how uncertainty about physical material properties can be considered through sensitivity analysis. The presentation then demonstrates how a well-calibrated LEM can inform risk understanding and guide remedial management activities post-closure. It also discusses the importance of episodic post-closure monitoring to understand changes in erosion risk over time.

This case study uses UAV-captured LiDAR data collected for landform-scale erosion monitoring and classified to exclude vegetation. This data measured the geometry of erosion features, including length, depth, slope, and volume. Key erosion feature dimensions were verified with field measurements.

The SIBERIA LEM predicted erosion on a reconstructed as-built model of a landform rehabilitated over 20 years ago. An as-constructed survey of the original rehabilitation surface was unavailable, so to create an assumed starting surface, the survey digital elevation model was modified to remove erosion features and incorporate ripping surface features, similar to those observed in other high-stability areas of the landform. Initially, the LEM was parameterised using only greenfield project information, such as geological data, analogue conditions, climate, local soils, landscapes, and vegetation data—common for the project approval stage. Sensitivity testing accounted for uncertainty using lower and upper bounded parameters of likely material behaviour.

The LEM matched the duration of the current rehabilitation age to compare predicted and actual erosion. The LEM was then calibrated by modifying parameters to recreate erosion features of similar size to those observed on the current surface. An averaged erosion rate indicated unrealistically high future predicted erosion rates, so historical aerial imagery was reviewed to understand the evolution of key erosion features over time. This review showed that most erosion occurred shortly after rehabilitation work completion, possibly due to settlement of the initially loose material and self-armouring of erosion features. As a result, two sets of erosion parameters were developed: one for the initial high-erosion period and another for stabilised material.

Once calibrated LEM parameters were established, SIBERIA was used to model the predicted behaviour of the current surface over a 200-year period using the stabilised material parameters to inform the long-term trajectory of the landform and to compare the predicted erosion rates with threshold values.

Conclusions from this study will demonstrate methods to use LiDAR data to assess erosion feature scale and discuss LEM effectiveness in predicting erosion across multiple project stages and long-term risk assessment. The presentation will emphasise the importance of regular monitoring during different phases of post-closure establishment for mining landforms to understand long-term risks and guide early interventions when needed. It shall discuss some opportunities and limitations associated with LEMs in assessing the long-term trajectory of mining landforms.

The remediation of an abandoned tailings storage facility – Collingwood Tin Mine, Far North Queensland

A Friend[1], T Hall[2], A Grabski[3], M Thompson[4] and D Lewis[5]

1. Project Manager, Department of Natural Resources and Mines, Manufacturing, and Regional and Rural Development, Brisbane Qld 4000. Email: andrew.friend@resources.qld.gov.au
2. Director Resources Remediation, Department of Natural Resources and Mines, Manufacturing, and Regional and Rural Development, Brisbane Qld 4000. Email: tania.hall@resources.qld.gov.au
3. Executive Director Technical Services, Department of Natural Resources and Mines, Manufacturing, and Regional and Rural Development, Brisbane Qld 4000. Email: andrew.grabski@resources.qld.gov.au
4. Senior Project Officer, Department of Natural Resources and Mines, Manufacturing, and Regional and Rural Development, Brisbane Qld 4000. Email: mitchell.thompson@resources.qld.gov.au
5. Manager Reporting and Engagement, Department of Natural Resources and Mines, Manufacturing, and Regional and Rural Development, Brisbane Qld 4000. Email: damian.lewis@resources.qld.gov.au

ABSTRACT

The Collingwood Tin Mine (COLT), located within the Annan River catchment 35 km south of Cooktown, presented unique environmental and public health challenges due to its proximity to drinking water sources, the Wet Tropics World Heritage Area and Great Barrier Reef catchment. The site was operational from 2005 to 2008 followed by abandonment in 2015; and has been managed since by the Queensland Government's Abandoned Mine Lands Program (AMLP), which aims to mitigate risks to public health, safety, and the environment. The main risk sources present at the mine were the uncapped tailings storage facility containing accumulated mine-affected water, an associated decant water dam, a processing plant and associated infrastructure, and an open portal and vent shaft to underground mine workings.

This paper focuses on the remediation of the tailings storage facility (TSF) and decant water dam (DWD) – identified as the most technically complex site domains. Extensive assessments were conducted, alongside engagement with Traditional Owners and other stakeholders, to define remediation objectives and approaches. Earthworks planning prioritised use of existing site materials to avoid new land disturbance or the need for bulk-fill material imports or exports. The remediation plan also emphasised local economic opportunities, with significant involvement from local businesses, small-to-medium enterprises, and Indigenous-owned businesses. The remediation works were completed over two years (FY 2021/22 and 2022/23) at a cost of approximately $6.8 M, delivering substantial environmental and community benefits.

This paper provides an in-depth overview of the remediation activities, sequencing, challenges, outcomes, and lessons learned, offering valuable insights for future projects in abandoned mine management and decommissioning of TSFs.

INTRODUCTION

Mining has played a significant role in the history of Queensland for more than 150 years. However, over time some mines have been abandoned, leaving risks to public health, safety and the environment. In Queensland, a mine is considered 'abandoned' when there is no longer a current mining tenement nor environmental authority (Department of Resources, 2024).

The Abandoned Mine Lands Program (AMLP) is a state government initiative to assist with mitigating risks posed by abandoned mines in Queensland. The overall objectives of the program are outlined within the Abandoned Mine Management Policy (Department of Resources, 2019):

- preventing potential exposure of the surrounding community to hazards on an abandoned mine site by removing or mitigating hazards (Safe).

- implementing control measures to limit the level of adverse impacts to the surrounding and downstream environments (Secure).

- minimising the ongoing maintenance and monitoring requirements for a site—this includes geotechnical and geochemical stability (Durable).

- investigating opportunities to re-commercialise abandoned mines and/or repurpose the land for future use, where appropriate, considering the economic, community, cultural, and conservation values (Productive).

The AMLP is delivered by the Technical Services team within the Department of Natural Resources and Mines, Manufacturing, and Regional and Rural Development (DNRMMRRD) (the department), with resources allocated based on the Risk and Prioritisation Framework for Abandoned Mine Management and Remediation (Department of Resources, 2021a).

Site background

The Collingwood Tin mine (the site) operated from 2005 to 2008 before entering an extended period of care and maintenance. The mine operator subsequently entered receivership, followed by liquidation in 2015, which resulted in the disclaiming of the mining leases and environmental authorities (Department of Resources, 2021b). No remediation or rehabilitation had been undertaken prior to disclaiming. Since then, the site has been classified as an 'abandoned mine' under the *Mineral Resources Act 1989* (Department of Resources, 2024) and incorporated into the AMLP.

The site is located 35 km south of Cooktown, within the Annan River catchment and adjacent to the Wet Tropics World Heritage Area and Great Barrier Reef catchment, as shown in Figure 1. The Collingwood Tin mine area has been subject to historic mining dating back to the 1890s, followed by the sinking of a 1.7 km exploratory adit in the 1980s. The most recent mining operation, involving underground mining and surface processing, was commissioned in late 2005 and the first commercial shipment of concentrates was produced in early 2006.

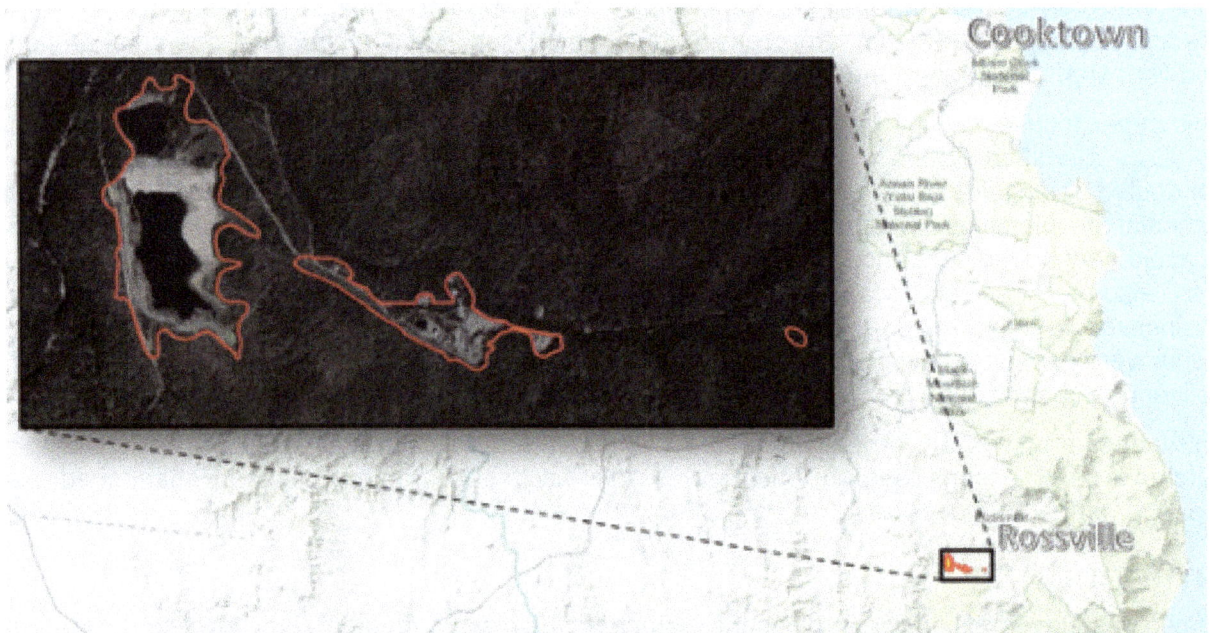

FIG 1 – Location of the Collingwood Tin Mine, in Far North Queensland (Source: Department of Resources, 2021b).

Approximately half of the mine footprint is located on Aboriginal Freehold Land, held by the Jabalbina Yalanji Aboriginal Corporation (JYAC) as the Registered Native Title Body Corporate (RNTBC) for the Eastern Kuku Yalanji. The remainder of the site is located on a State Timber Reserve with an overlying pastural Occupational Licence for grazing (Department of Resources, 2021b).

Site remediation overview

The department facilitated extensive consultation and collaboration with Traditional Owners to understand future land use aspirations and establish appropriate remediation objectives that reflected the site's constraints, risks and opportunities. This stakeholder engagement process is outlined further within McGuire *et al* (2022).

The site was split into eight domains, and a fact sheet was developed to share with all stakeholders, summarising agreed remediation objectives for each domain. A condensed overview is presented in Table 1 including the approximate timing that remediation was carried out.

TABLE 1

Overview of Collingwood Tin remediation works per domain (Department of Resources, 2018).

Domain	Primary remediation activities	Calendar year
Workshop Area	Remove drill core, remove scrap and waste, remove contaminated soil and retain concrete hardstand slabs	2021
Processing Area	Demolish and remove processing infrastructure, decommission process water pond, retain and reprofile hardstand pads for potential future commercial use	2018
Underground Portal and Vent	Cap ventilation shaft, seal public access to the portal while retaining bat access	2019–2020
Buildings	Retain buildings (acquired by JYAC)	-
Power line	Refurbish and transfer power line infrastructure to electrical utility	2020
Revegetated Area	Minimise disturbance	-
Decant Water Dam (DWD)	Remove sediments from impoundment, lower embankment crest, install long-term closure spillway	2021–2023
Tailings Storage Facility (TSF)	Reprofile tailings impoundment and encapsulate as free-draining landform, construct long-term closure spillway	2021–2023

The full site remediation was carried out over multiple years and phases due to factors such as risk prioritisation, logistics and resourcing. Initial management efforts focused on essential care and maintenance activities, such as managing water inventory within the TSF and other water storages. This was followed by immediate clean-up and make-safe works across the site, including process plant demolition. Closure planning for the more complex TSF and DWD domains progressed in parallel, with remediation then completed by late 2022.

TSF background

Configuration

The tailings dam was designed as a cross-valley tailings impoundment with a maximum design storage capacity of 1325 ML; however, only 830 ML of tailings had been deposited prior to abandonment. All embankments were constructed using zoned earth and rock fill, using locally sourced materials, with the configuration shown in Figure 2.

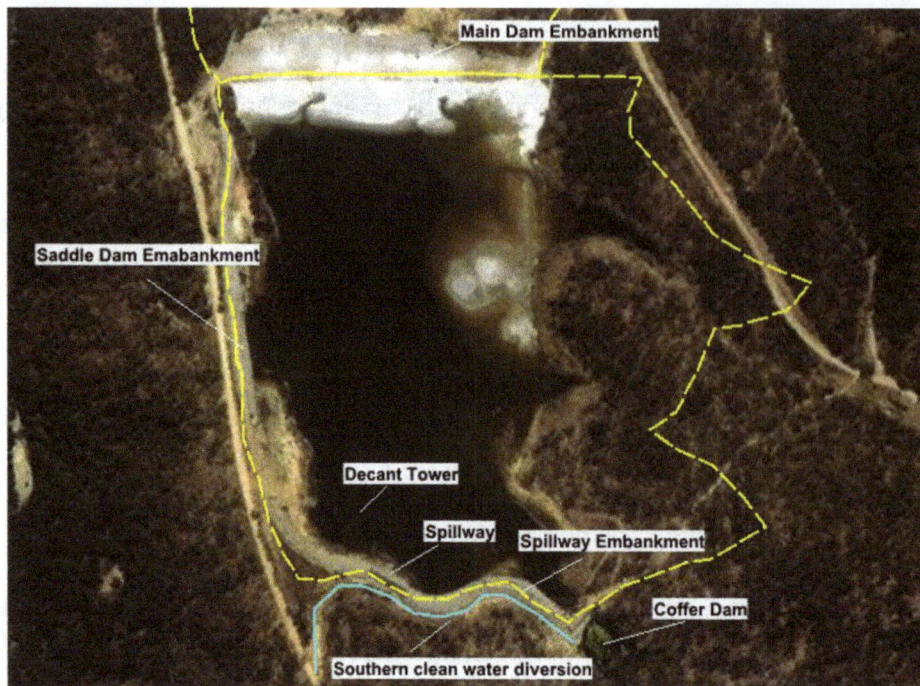

FIG 2 – The TSF configuration in 2017 prior to remediation (Engeny Water Management, 2017).

The TSF main embankment, located on the northern side, measured 300 m in length and consisted of a starter embankment with maximum height of 15.5 m followed by an additional 3.5 m crest raise using upstream construction methods. The raised embankment had an 8 m wide crest as well as downstream and upstream batter slopes of 2.75H:1V and 2.25H:1V respectively.

Tailings were primarily deposited along the main embankment, resulting in a tailings profile sloping southwards. However, the significant coarse-grained fraction within the tailings caused significant quantities of material to settle-out immediately after deposition, with only the much smaller quantity of fine-graining tailings flowing into the central and southern sections of the tailings impoundment. As a result of the localised tailings beach surface reaching the embankment crest level, additional end-of-pipe tailings deposition points were added along the eastern edge of the impoundment area.

The Stage 2 raise also included construction of a small saddle embankment along the western and southern perimeter of the impoundment area, approximately 800 m in length and averaging 2 m in height. The southern embankment also provided two additional functions, firstly to divert external clean catchment areas to the south-east of the TSF away from the impoundment area, towards the south-west. Secondly, the southern embankment housed the 15 m wide concrete crested spillway, discharging into the southern clean water diversion.

The TSF included two staged decant towers located at the centre and southern edge of the facility, comprising slotted 1.5 m diameter concrete pipes stacked vertically and surrounded by coarse rock fill. These towers connected to a single 315 mm HDPE pipeline, installed into the foundation of the TSF with continuously falling gradient towards the DWD, including passage through the main embankment foundations, with downstream valving used to regulate flow. The design and construction of this outlet appeared effective, with no ingress of tailings material into the pipeline, nor seepage around the pipe penetration through the TSF main embankment.

Water management

The site is located within the wet tropics, with closely matched average annual rainfall and evaporation of 1946 mm and 1984 mm respectively. Approximately 80 per cent of average rainfall occurs during the December to April monsoon season. The tropical region is also at high risk of Tropical Cyclones that bring extreme wind and rainfall over a compressed period (Klohn Crippen Berger (KCB), 2007).

The site discharges into the Annan River as shown in Figure 3. Seven km downstream of site is the Lions Den, a popular tourist destination with a range of nature-based activities, including recreational use of the Annan River. It is also likely local residents use the Annan River for a range of informal and unregulated activities, such as water extraction, fishing, hunting or swimming. Downstream of the mine 15 km, is the intake used to supply Cooktown's drinking water. The Annan River ultimately flows out to the Coral Sea.

FIG 3 – The site catchment areas and proximity to the Annan River (Engeny Water Management, 2017).

Water stored within the TSF and DWD generally met downstream environmental standards, with occasional exceedances against aquatic ecosystems (95 per cent protection) and drinking water limits for Copper, Arsenic, Antimony and Fluoride. These exceedances could be eliminated by managing outflow rates to achieve a dilution ratio in the order of 50:1, which was readily available due to significant year-round baseflow within the Annan River. A release protocol was adopted that enabled low-risk and low-volume discharges, inclusive of a water monitoring program to monitor upstream, on-site, and downstream water quality (Engeny Water Management, 2018). Through these small-scale water releases, pond inventory and free-board levels within the TSF and DWD could be effectively managed, reducing the risk of wet season overtopping as well as reducing the risk of other embankment failure mechanisms.

TSF and DWD risks prior to remediation

The primary long-term risk was uncontrolled discharge of mine-affected water through the spillway, due to insufficient storage capacity following water accumulation from progressive seasonal rainfall or a specific extreme event, such as a Tropical Cyclone. This was particularly acute as the site was generally unmanned and located in a sensitive ecosystem with minimal catchment disturbance, as well as being located upstream of Cooktown's drinking water supply intake.

Other key risks included spillway failure due to erosion (particularly the DWD), and embankment failure from seepage or geotechnical instability. All failure modes had increasing probability and consequences as water storage levels within the impoundment area increase.

Wind-blown tailings also posed a risk. Coarse-grained tailings material deposited along the main embankment was highly susceptible to movement by wind, and this was most evident by the visible volume of material that had migrated from the TSF impoundment area onto the downstream batter of the main embankment, as shown in Figure 4. While tailings movement by wind reduced significantly upon reaching the surrounding tree lines, any material outside of the TSF impoundment area remained susceptible to further migration through rainfall erosion. As such, tailings material was observed within the DWD pond.

FIG 4 – Tailings material on the upstream (left) and downstream (right) batters of the main embankment (Engeny Water Management, 2017).

Finally, the site's remoteness and absence of a regular site workforce increased the risk of unauthorised access and subsequent exposure to contaminants, particularly for high-risk activities such as camping or interacting with mine-impacted waters for personal use, consumption or recreational purposes.

TSF CLOSURE PLANNING

Objectives and constraints

The overall remediation works were focused on achieving risk reduction, specifically making the site safe, secure, durable and productive as per the Abandoned Mines Management Policy. However, in relation to the TSF, further definition of objectives was required.

This included:

- Ensuring secure long-term containment of all tailings contaminants, including physical encapsulation, and preventing migration of contaminants through surface or groundwater movement.

- Achieve a completely free-draining landform, with negligible annual operational requirements or active water management measures necessary.

- Ensuring embankments were geotechnically stable (minimum static factor of safety 1.5 and maintain integrity under seismic OBE earthquake loading).

- TSF spillway structures to maintain containment integrity from Probable Maximum Precipitation (PMP) flood events, and minimal long-term maintenance requirements arising from more common storm events.

- Downsize the DWD to achieve a tolerable risk profile, while still retaining some sediment control and flood attenuation functionality.

- Achieve a landform generally compatible with a commercial industrial future land use, despite no foreseeable intention for future development across the capped TSF surface.

- Meet these objectives with minimal disturbance of new areas (land clearing), as well as minimising the import or export of earth and rock fill materials as far as practically possible.

Environmental studies (environmental values, animals etc)

As the remediation works were constrained to the existing disturbed areas, this significantly reduced the risk of causing impacts to identified protected flora and fauna.

During the stakeholder engagement phase, it was identified that some of the local stakeholders expressed a general desire to retain the existing TSF configuration indefinitely, to retain the artificial wetland area which some fauna had adapted to and observed to be using as habitat. These views were considered, although a risk assessment framework ultimately demonstrated that the risks associated with retaining the current arrangement, namely potential dam failure or future release of contaminants to the downstream environment, were intolerable.

An Environmental Values and Impact Assessment study concluded that the TSF and DWD were classified as artificial lacustrine wetlands but were not ecologically or legally significant. They were not mapped as Referable Wetlands and had been used for mineral waste storage, limiting their ecological value. No local, state, or federal listings identified them as important wetlands (AARC, 2019a).

A threatened fauna breeding places assessment study was undertaken, which identified that the Northern Quoll (*Dasyurus hallucatus*) occurs within the local area inclusive of the TSF footprint, namely within rock voids on the embankment downstream batters seen in Figure 5 (AARC, 2019b). However, impacts were able to be subsequently managed by minimising disturbance to the existing rock layers as far as practical, undertaking the essential construction activities outside of breeding season, and through engagement of qualified fauna spotter-catchers when undertaking disturbance works in those areas.

FIG 5 – Northern Quoll (left) and large rock fill occasionally used as breeding habitat (right) (AARC, 2019b).

Geochemical sampling

The TSF design by KCB (2007) concluded that as there was no credible geochemical risk associated with the tailings, and that it is typically representative of a sand-like product with elevated concentrations of Arsenic, Copper and Antimony above ANZECC guideline limits for sediments.

During the detailed design phase, a further 17 tailings samples and 22 waste rock samples were taken from across site, to verify that *in situ* materials were geochemically benign as expected. Analysis of mineralogy from XRD analysis indicated a slightly alkaline, chemically inert, granitic material at various stages of weathering (GHD, 2020).

Static geochemical testing was also undertaken for all samples, indicating average total sulfur concentrations of 0.005 per cent in waste rock and 0.1 per cent in tailings samples. The results were also assessed against classification criteria from both AMIRA International Ltd (2002) and International Network for Acid Prevention (INAP, 2009), as shown in Figure 6, conclusively identifying the material as Non-Acid Forming (NAF) at low risk of acid generation.

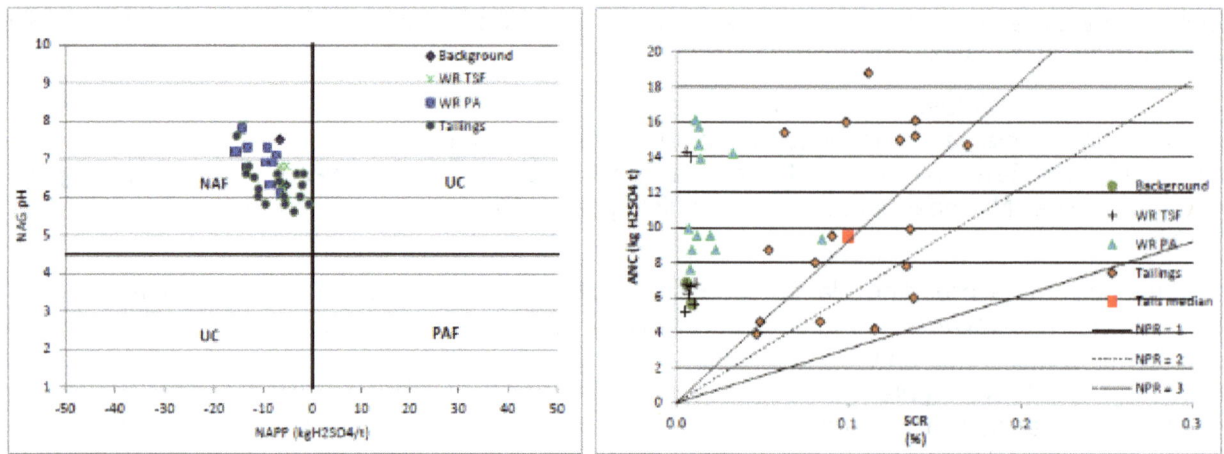

FIG 6 – Geochemical classification of materials against INAP and AMIRA criteria (GHD, 2020).

Remediation materials sampling

During the detailed design process, an additional site investigation was undertaken to address identified data gaps, as shown in Figure 7. This included borehole drilling via a track-mounted DB525 drill rig, test pitting with a 10 t excavator, Cone Penetration Test (CPT) testing via an excavator mounted rig, and vane shear testing on the tailings carried out using hand-held equipment.

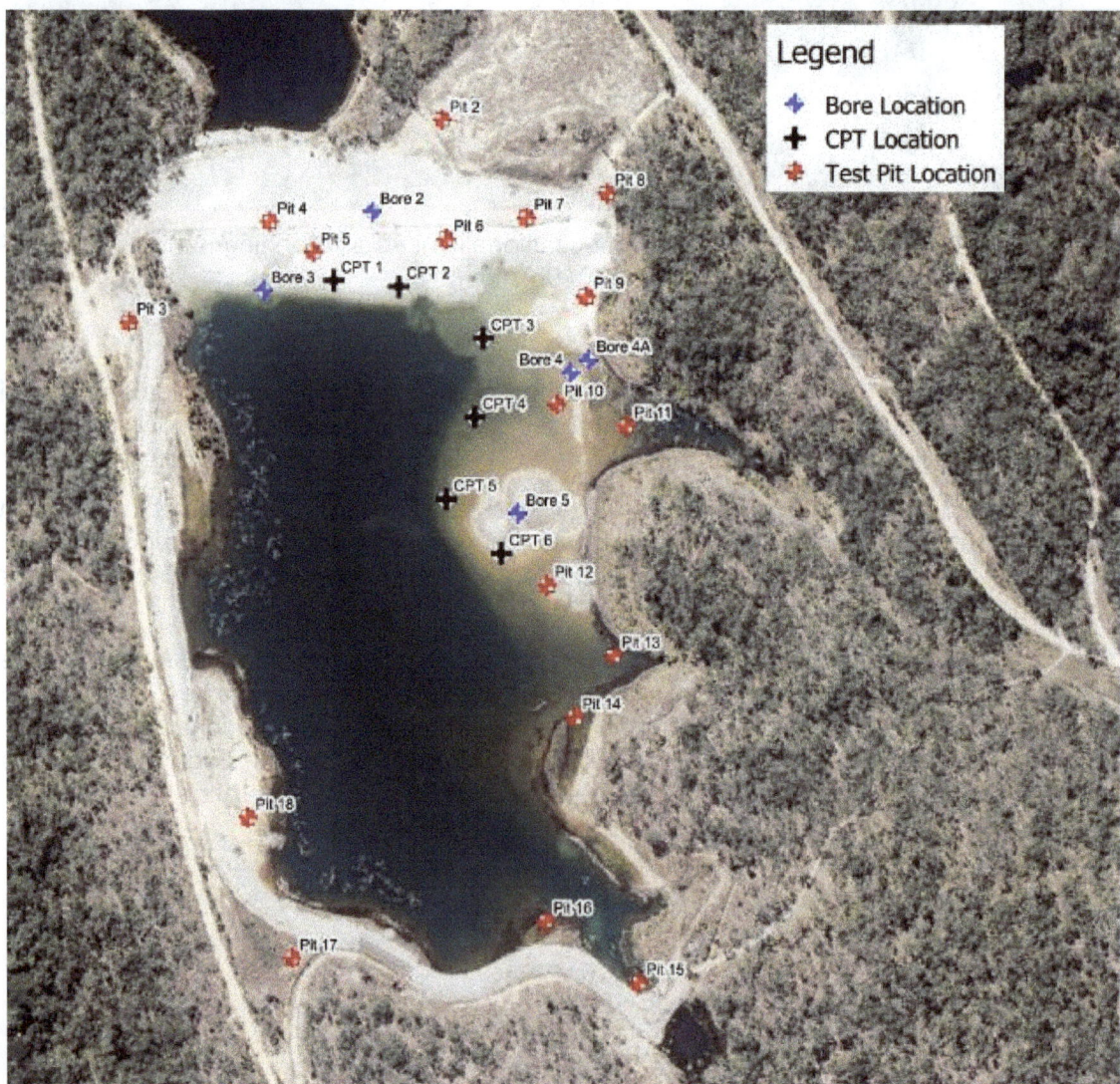

FIG 7 – Sampling undertaken in the detailed design phase (excludes earlier sampling campaigns) (Douglas Partners, 2020).

A series of ten shear vane tests were undertaken across the upper tailings beach area, with results summarised in Table 2, based on information from Douglas Partners (2020).

TABLE 2

Overview of average near surface shear vane results on the upper tailings beaches (Douglas Partners, 2020).

Depth (mm)	Peak (kPa)			Residual (kPa)		
	Min	Average	Max	Min	Average	Max
100	6	18	28	3	7	12
500	22	30	36	9	13	17

Other laboratory classification tests included particle size distribution and hydrometer analysis, Atterberg limits, oedometer, pinhole dispersion, triaxial test, Emerson class, chemical dispersion and moisture content.

Results from the CPT testing on the upper tailings beaches demonstrated consistently loose tailings throughout the upper zones that were likely to excavated during tailings re-profiling, as illustrated in Figure 8.

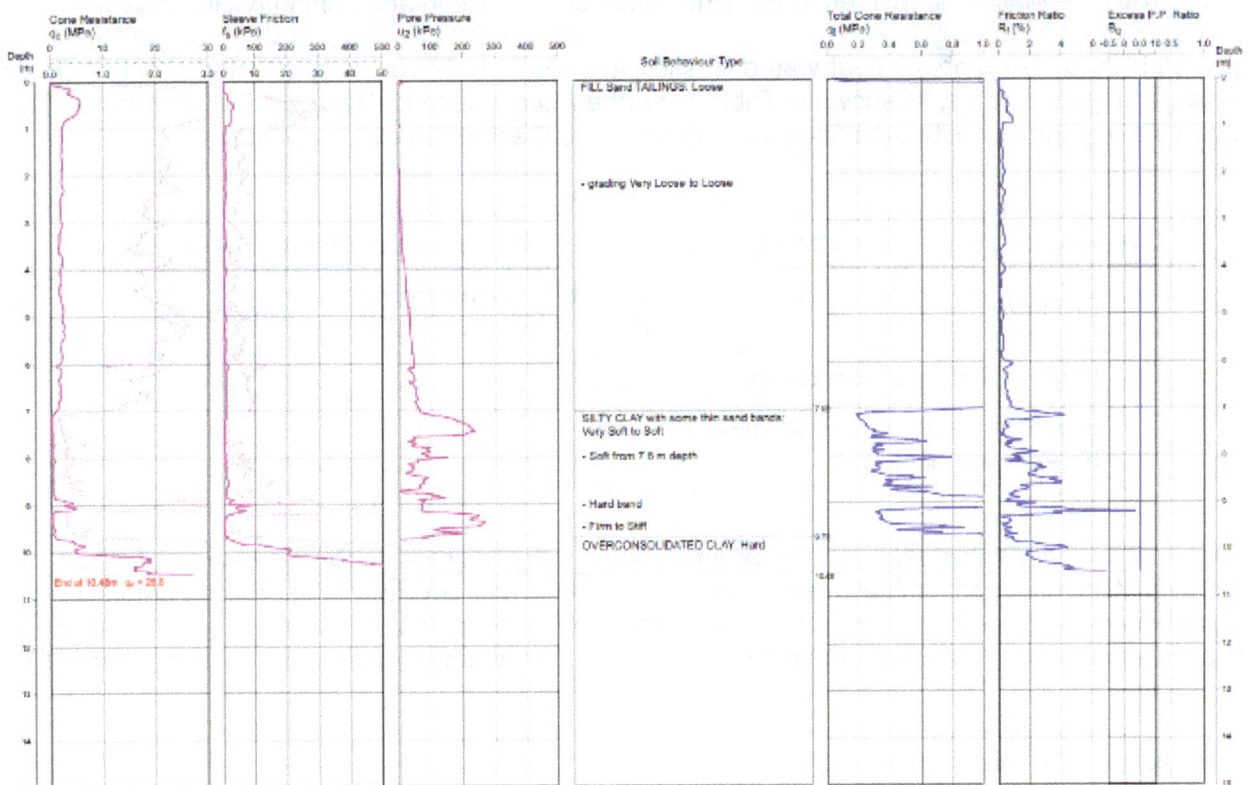

FIG 8 – CPT sampling results obtained on the upper tailings beach (Douglas Partners, 2020).

Embankment stability

The proposed TSF remediation included lowering the main embankment crest to completely remove the upstream raise, while retaining the original starter embankment. The embankment cross-section, as well as zoning and material parameters, is shown in Figure 9.

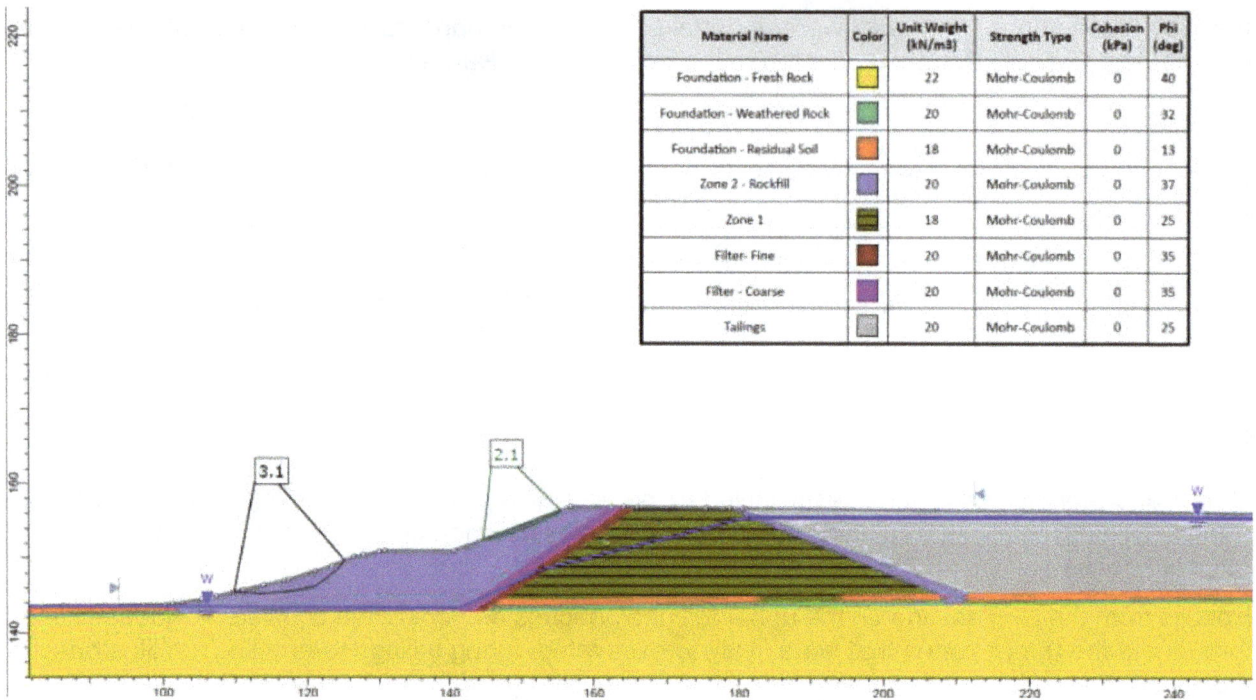

FIG 9 – Main embankment post remediation cross-section and zoning (GHD, 2020).

The stability of this embankment was assessed and found to have a static factor of safety (FoS) exceeding requirements, as shown in Table 3. Furthermore, the critical failure surfaces were shallow failures along the downstream batter, with no impact to crest or the clay core integrity.

TABLE 3

Overview of geotechnical stability results for the TSF main embankment (GHD, 2020).

Scenario	Circular min FoS	Non-circular min FoS	Satisfactory
Long-term drained conditions	2.1	2.1	Yes
Short-term flood condition (high phreatic surface)	2.1	2.1	Yes

Further detailed assessment of tailings liquefaction was deemed to be unnecessary given the removal of the upstream embankment raise and the robust self-stability of the retained embankment. In addition, deformation risks were considered negligible given the design and construction specifications of the retained embankment, and the free-draining nature of the remediated TSF landform.

Reduction to DWD capacity

An assessment was undertaken for downsizing the DWD in order to lower the long-term risk profile of the dam, while also retaining some sediment control and storm attenuation functions. Another consideration was the surrounding topography and providing a configuration that facilitated construction of a significant closure spillway that can pass annual operational flows as well as extreme storm events – without posing risk to the DWD embankment integrity.

The outcome of this assessment was lowering the DWD embankment crest by 3 m, to a revised maximum embankment height of 6.5 m, as shown in Figure 10. The spillway invert was set at 145 mRL, enabling permanent ponded volume of 40 ML below the spillway invert, and a 2 m vertical freeboard-to-embankment crest to prevent overtopping due to spillway flows and wave run-up. Importantly, the DWD embankment reduction also provided a source of high-quality clay material that could be utilised for tailings capping.

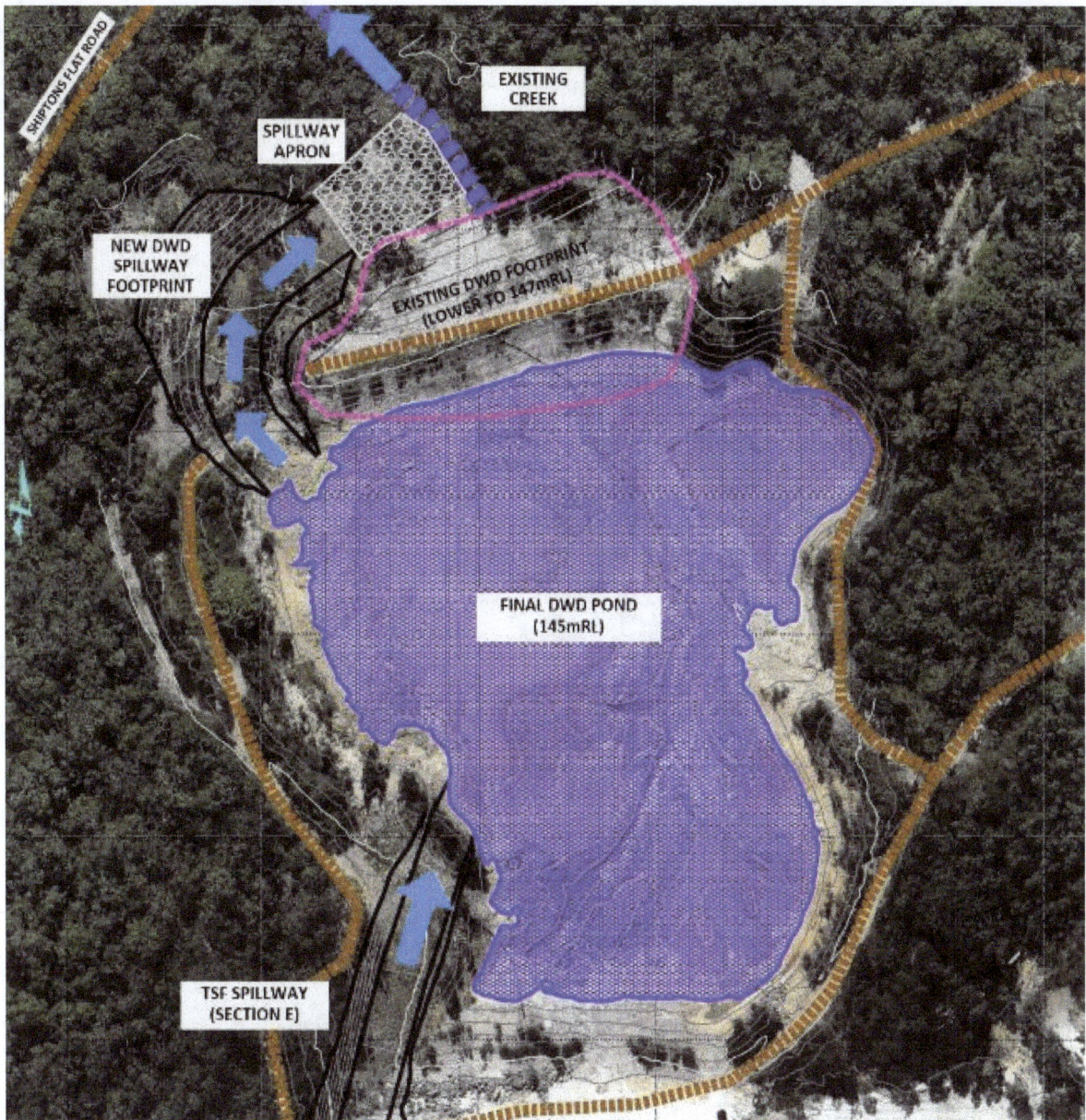

FIG 10 – Overview of DWD embankment modifications and spillways (Source: Department of Resources, 2022).

The DWD spillway was designed immediately to the west of the reduced DWD embankment, however founded entirely within excavated *in situ* ground with no connection to embankment fill. The spillway consisted of a 15 m base width and used a 3.3 per cent gradient over 90 m before discharging onto a flat rock apron area downstream of the DWD embankment. Rock fill with d50 of 400 mm was used ranging in thickness from 600 mm to 800 m. Where available, large boulders were also placed selectively in the rock apron.

Landform design and optimisation

Due to the coarse-grained composition of the tailings, significant beaching of tailings occurred immediately adjacent to the main embankment on the northern side, with only a small proportion of finer-grained tailings flowing to the central and southern ends of the impoundment area. As such, significant earthworks were identified as necessary to achieve a regular landform with gentle gradients sufficient to ensure it remains free-draining, but also so that any run-off velocity will remain low enough to limit long-term erosion to the capping layers.

Initial closure designs for the TSF included an eastern diversion drain that would convey water to the southern bypass, however this option was subsequently discarded due to topographic limitations that required raising significant sections of the drainage channel above natural ground level in order to achieve sufficient flow gradients. To avoid constructing raised drainage channels over tailings, the decision was taken to combine the relatively modest eastern catchment flows with run-off from the tailings capping area, and discharge via the TSF spillway into the DWD.

However, the limited deposition of tailings within the southern end of the facility also created an opportunity to consolidate the tailings footprint, and in doing so, create a perimeter drain that diverts water from the eastern side to the spillway located to the north-west of the impoundment, without traversing or interacting with the final tailings.

Similarly, a small volume of tailings in the eastern gullies were removed and replaced with general fill, including construction of two clay cut-off walls, to ensure incoming surface water could not interact with or saturate the tailings, instead being forced to flow into the TSF diversion drain.

The consolidated footprint of the tailings impoundment is illustrated in Figure 11, in addition to the perimeter diversion drain that was able to be excavated into predominantly natural ground.

FIG 11 – Overview of the TSF tailings impoundment area before and after consolidation (Source: Department of Resources, 2022).

The TSF remediation works also involved lowering the main embankment crest by three metres, to tie in with the final tailings level following reprofiling of the tailings. This reduction completely removed the Stage 2 upstream lift and subsequently eliminated any associated longer-term geotechnical risks.

The entire earthworks landform design was optimised to achieve a cut-to-fill balance, eliminating the need for any additional material to be imported – aligning with Traditional Owner aspirations. There was sufficiently abundant access to clay material and rock armouring materials, however careful sequencing was required to ensure excavations and capping activities were aligned. Removal of the western saddle embankment was excluded from the landform design as retaining it posed no risk, however that allowed it to act as a 'backup' material source in case a shortfall of material arose mid-construction, such as due to unfavourable shrink/swell changes, loss of material and/or construction settlement, or other issues.

Drainage design

The TSF had a catchment area of 23 hectares, comprising 17 ha of tailings impoundment footprint and three external sub-catchments to the east comprising 6 ha. The TSF drainage configuration includes a spillway chute on the western abutment, discharging into the DWD pond, as shown in Figure 12.

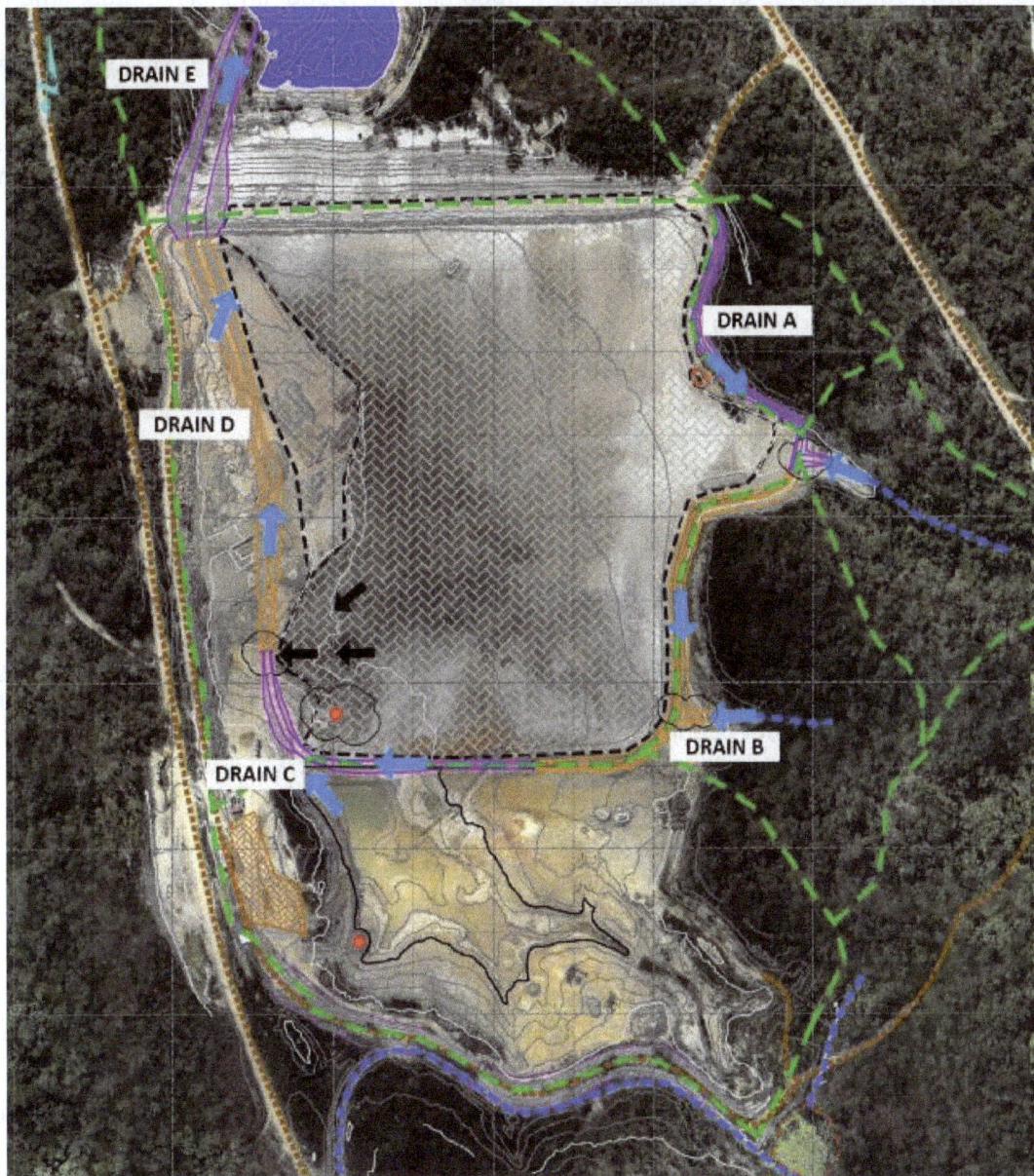

FIG 12 – Overview of the TSF surface drainage design arrangement (Source: Department of Resources, 2022).

The DWD pond has its own catchment area of seven ha, in addition to the TSF spillway inflows. The DWD has a spillway on its western abutment, discharging into a dry creek-line that flows to the Annan River.

Despite the significant depth of many sections of the TSF diversion drain and spillways, which was necessary to achieve free-draining gradients, channel widths were also designed with the intention of limiting design flood flow heights with a nominal target of not exceeding one-metre depth. The overall intent was to limit discharge flow velocities and minimise the occurrence of high-energy turbulent flow conditions.

TSF CLOSURE IMPLEMENTATION

Governance and objectives

Following funding approval, the TSF Remediation Project was put into execution phase, using the department's Project Management Framework. This included a Project Control Group (PCG) comprising the Project Sponsor (representing business risk considerations), Senior Supplier Representative (representing Project Delivery considerations), and Senior User Representative (representing external stakeholder considerations). A monthly PCG meeting was held, where the project team provided updates on completed and upcoming activities, financial status, time frames, issues encountered, project risks, and any other queries raised by the PCG members or invited guests. The PCG also provided review and approvals between project phases (tollgates) as well as for any scope change requests.

The project objectives outlined the remediation outcomes necessary to achieve a safe and secure TSF, however were also extended to include additional social considerations (Department of Resources, 2021c), specifically:

- active engagement with Traditional Owners (via Jabalbina Yalanji Aboriginal Corporation) throughout the remediation project, with opportunities for involvement and employment explored and pursued where practical and appropriate.

- to explore and pursue employment opportunities for local businesses, small/medium businesses, and indigenous owned businesses, throughout the site remediation project, where practical and appropriate.

Sequencing

The project received funding approval in June 2021, with funds split across two financial years; $3.5 M in FY2021/22 and $3.9 M in FY2022/23.

A significant consideration when developing the project schedule was the wet season period between December and May each year where a significant number of rain days occur, and ground conditions become saturated and at times unpassable. While there are narrow periods within the wet season where rain does not occur and some work can be undertaken, it is highly unpredictable, and as such, undesirable to undertake construction works during that wet season period. As a result, the two-year project only contained three suitable construction periods:

- July to December 2021 (6 months)
- May to December 2022 (8 months)
- May to June 2023 (2 months).

Vegetation would need to be established in the final wet season (early 2023) as well as performance monitoring, corrective actions if any issues were identified, and final close-out activities. Therefore, working backwards, bulk earthworks would need to be completed by December 2022.

While it was generally understood that a significant portion of the bulk earthworks would need to be undertaken in the middle dry season period, there was also a risk if any delays or unforeseen issues arose during that period it could significantly compromise the project – especially if the remediation works were not fully completed by the onset of the 2022/23 wet season.

Therefore, with a view to de-risking the project, it was decided to focus on opportunistically progressing as many of the higher-risk activities in the limited 2021 construction period as practically possible. This included:

- Establishment of on-site construction facilities. This included construction of parking laydown areas, establishment of demountable site offices, crib facilities and toilets, and installation of a cel-fi antenna to provide partial mobile phone coverage.

- Dewatering of the tailings dam, which was holding approximately 40 ML in mid-2021.

- Removing all tailings from the southern valley area and consolidating it into the main tailings area to the north, as shown in Figures 13 and 14. This including post-works verification sampling to confirm the stripped southern valley was free of contaminated material.

- Removing a small volume of tailings from the eastern gullies, and construction of a cut-off wall to avoid incoming flows from interacting with the tailings mass.

- Reprofiling of the tailings material within the TSF, including significant cut-to-fill activities, to achieve the preferred final tailings surface and with sufficient bearing capacity to permit trafficability by construction machinery, including fully-loaded articulated dump trucks.

- Completing the limited areas of vegetation clearing that would be necessary in the next phase of works, primarily focused around the new TSF and DWD spillways.

FIG 13 – Photograph from 2021 showing tailings material being removed from the TSF southern valley (source: DNRMMRRD).

FIG 14 – Photograph from 2021 showing completion of tailings removal from the TSF southern valley (source: DNRMMRRD).

Delivery approach for enabling works

In order to practically implement the initial dry season program of works, the project team adopted a collaborative approach with a number of local contractors. All identified local earthworks contractors were approached to obtain equipment lists and hourly wet hire unit rates, which were then assessed and used to develop a value for money construction workforce that could work directly with the department's site supervisor.

The tailings reprofiling work was undertaken with a fleet comprising, at its peak, three D6 equivalent bulldozers, three 20 t equivalent excavators, two 25 t articulated dump trucks, a grader and a 5T roller.

Equipment was ultimately sourced from several local suppliers, as no single supplier could fully meet demand. This approach worked effectively, with all machinery operators working constructively as an integrated team under the direction of the site supervisor.

This approach not only enabled a relatively quick commencement of activities within the limited dry season construction period but was also well-suited to the inherent uncertainty of ground conditions and therefore methodology and productivity rates for carrying out the tailings reprofiling. Those factors lead to an inability for contractors to reliably estimate their costs and subsequently develop a bill of quantities (BOQ) unit price to use in a tender process.

This method ultimately worked effectively for all parties. The project team was able to complete this activity, that had a high level of associated engineering and construction uncertainty, at an early stage and at a fair and reasonable cost that was reflective of the machine hours expended. The local suppliers were able to access a significant package of works, that would otherwise have been beyond their capability if bundled into a complete TSF remediation package and could also do so without carrying significant financial risk (due to the schedule of rates invoicing basis rather than BOQ).

Carrying out tailings re-profiling

Reprofiling tailings is an inherently difficult activity, with a high level of uncertainty upon project kick-off regarding what methodology would be successful, in addition to the level of resourcing that would be required. The remediation designs included some commentary about constructability aspects of tailings reprofiling, however understandably it was mostly general information, for two reasons. Firstly, no factual information was available about *in situ* tailings conditions in the middle and lower-lying areas of the tailings impoundment, because the extent of ponding during the site investigation phase meant only sampling of the upper tailings beaches could occur. Secondly, even on the upper areas where factual geotechnical data could be obtained, it had limited ability to be translated into specific and reliable constructability advice.

The project team identified that developing a suitable methodology for tailings reprofiling would need to be carried out through at-scale field trials using construction equipment, followed by monitoring and optimising throughout implementation.

The TSF surface was able to be dewatered by mid-2021 for the first time since mine abandonment. While the upper tailings beaches consisting of coarse sands were trafficable, it was quickly identified that the substantial lower-lying areas of the TSF, comprising saturated loose and fine-grained sands, were not trafficable. Furthermore, following surface dewatering, those areas did not appear to benefit from air-drying (such as clay tailings desiccation observed at other sites) or desaturation and consolidation under self-weight. Therefore, a number of active measures were trialled, with the aim of improving bearing capacity.

Firstly, pushing out causeways 'fingers' of drier tailings sands out onto the soft low-lying tailings areas using bulldozers was attempted. To ensure safety of dozers, causeway fill heights of 1.0– 1.5 m were used, as shown in Figure 15, consuming a lot of material and dozer effort. Despite the weight surcharge imposed by these causeways, no heave 'bow-wave' occurred and negligible consolidation or expulsion of water from the tailings was evident. Several variants of this method were also trialled, including using greater causeway heights, as well as pushing slotted PVC pipes into the surrounding tailings with a view to assisting pore pressure dissipation, however negligible improvements were achieved.

FIG 15 – Photograph from 2021 of tailings causeways pushed out onto the soft low-lying areas (source: DNRMMRRD).

The greatest improvement to the low-lying tailings areas was encountered by dynamic excavator activities along the advancing 'fill' edge, rather than static surcharge efforts. In particular, using a methodology of imposing repetitive dynamic loading to the tailings by hitting ('pumping') with an excavator bucket, under controlled conditions. This caused consolidation and densification of the tailings material followed by a rapid and substantial release of previously entrained water from the tailings mass. Water released to the surface could then readily run-off across the tailings surface downslope towards the TSF decant, as shown in Figure 16.

FIG 16 – Photograph from 2021 of an excavator undertaking localised tailings densification activities (source: DNRMMRRD).

Once the excess pore pressures associated with this liquefaction-induced consolidation of the tailings had sufficiently dissipated and surface outflow had ceased, ranging between 6 to 24 hrs later, a noticeable improvement in bearing capacity was observed. This then allowed drier sand to be pushed-out with sufficient trafficability for excavators to progressively advance.

The process could be repeated as many times as necessary, with diminishing returns each time. The process could also be aided by displacing bucket loads of dry tailings sand directly into the underlying tailings, forcing an increase to the tailings density, and in the process releasing further entrained water from the tailings mass.

A system was developed where excavators would rotate and cycle along a long advancing edge, allowing slow and steady creation of new tailings footprint areas that could now be safely trafficked by tracked construction machinery. Bulldozers were used to provide a steady supply of dry, coarse-grained tailings sand to the excavators from the upper tailings beach areas several hundred metres away. This arrangement is shown in Figure 17.

FIG 17 – Excavators in 2021 undertaking localised tailings densification activities while dozers supply bulk material movement from upper tailings beach areas towards the lower lying areas (source: DNRMMRRD).

Ultimately, through improving the geotechnical properties of the *in situ* tailings, and reprofiling drier coarse-sand tailings across the full tailings footprint, we were able to achieve a final landform that was completely free-draining as well as trafficable by construction equipment on-site, including fully loaded 25 t ADT trucks.

The enabling works program successfully resolved several of the key project uncertainties, allowing the remaining major remediation works to be tendered with a controlled scope.

Delivery approach for bulk earthworks

A major tender process was conducted throughout early 2022, with most of the scope of work activities being clearly defined and with minimal uncertainty. The tender received a positive response rate, with evaluations and contract award completed in May 2022, followed by mobilisation in June 2022. The successful contractor was a joint venture between two local businesses, an earthmoving company and an engineering firm. The contractor carried out the works under a Principal Contractor arrangement, with departmental staff providing contract management.

Bulk earthworks phase sequencing

The Contractor was able to achieve a sequence for the works that balanced the timing of excavation material types to when that material type was required within the tailings cap, as well as traffic flow considerations.

The initial activity was removal of all sediments from within the DWD, as shown in Figure 18. Given the potential for contaminants, as well as the known presence of wind-blown tailings, all material removed from the DWD was placed and incorporated into the tailings impoundment area, prior to the commencement of capping. As the tailings were already at a suitable and regular gradient, the tailings were spread as a thin layer across the surface, rather than being concentrated in any one area. Because this material was saturated, it was initially paddock dumped and then later loose spread to allow air drying, as shown in Figure 19, followed by final grading and compaction.

FIG 18 – Photograph from 2022 showing completion of sediment removal within the DWD basin (source: DNRMMRRD).

FIG 19 – Photograph from 2022 of paddock dumped DWD sediments spread across the reprofiled tailings surface to dry (source: DNRMMRRD).

The TSF main embankment was then lowered, providing valuable clay capping material (Figure 20).

FIG 20 – Photograph from 2022 of the main embankment crest being lowered, with clay material being hauled for use in the tailings capping (source: DNRMMRRD).

The lowering of the DWD embankment crest and excavation of the DWD closure spillway were prioritised next as they also provided significant quantities of clay material necessary to complete the tailings clay capping.

The DWD excavation works were completed over a three-week period, immediately followed by installation of geofabric and erosion protection rock fill in the DWD spillway so that it was ready and available if earlier than expected rainfall was encountered. The final DWD spillway can be seen in Figures 21 and 22.

FIG 21 – Photograph from 2022 of the completed DWD spillway through the western abutment of the lowered DWD embankment (source: DNRMMRRD).

FIG 22 – Photograph from 2022 of the constructed DWD spillway apron comprising 800 mm thick rock fill (source: DNRMMRRD).

Sufficient clay fill was sourced from lowering the TSF crest (11 000 m^3) and the DWD crest lowering and spillway excavation (10 000 m^3), that only 4000 m^3 of additional clay was needed to complete the tailings capping layer. That shortfall was made up by partial removal of the western saddle embankment.

A further 2000 m^3 of material excavated from the TSF and DWD excavations contained non-clay components and was deemed unsuitable for use in the clay capping layer, so was instead utilised in the upper protective capping layer, as shown in Figure 23.

FIG 23 – Photograph from 2022 showing the progressive placement of the clay capping layer, followed by the overlying protective layer of rock-dominated material (source: DNRMMRRD).

The remaining 22 000 m^3 of material required for the upper protective capping layer was sourced from the TSF diversion drain and main spillway chute excavations. This material consisted of a competent rock-dominated matrix, however still contained a sufficient soil fraction to be free of visible voids when placed and compacted. While less favourable for vegetation establishment, its erosion resistant characteristics ultimately reduced the necessity to establish a vegetative cover.

Extremely hard rock was encountered during excavation of the TSF diversion drain and spillway chute, requiring the use of excavator-mounted rock hammers and rippers to dislodge or break the *in situ* rock prior to removal. This had been included as a provisional activity within the Contract, avoiding Contract disruption and delays, however it did cause costs to increase as the pre-determined unit rate was much higher than the general free-dig excavation rate.

Ultimately the presence of hard rock within the TSF drains and spillway chute was regarded as beneficial, as it enhanced long-term channel erosion resistance along those sections and provided additional quantities of coarse durable rock fill for use in other drainage sections. The final TSF spillway chute can be seen in Figure 24.

FIG 24 – Photograph from 2022 of the completed TSF spillway through the western abutment of the TSF embankment (source: DNRMMRRD).

The final earthworks activity involved backfilling surface depressions within the southern valley, where tailings had been previously removed, to prevent water ponding in line with Traditional Owner preferences for a free-draining landform.

Revegetation

The site lacked substantive topsoil stockpiles, and the limited material that was present was assessed as low quality and weed-affected. However, as the capped tailings surface comprised rock-dominated material, a vegetative cover was not required for erosion control.

Nonetheless, hydro-mulching of the capped tailings surface, as well as drain channel batter slopes, was undertaken to stabilise any exposed fine-grained material, reduce run-off sediment loads, improve visual amenity, and discourage unauthorised land use. The revegetated tailings cap, a year after site remediation and hydro-mulching, is shown in Figure 25.

FIG 25 – Photograph from 2024 of light vegetation cover across the TSF capping area one year after remediation (source: DNRMMRRD).

POST-REMEDIATION REVIEW

Costs and expenditure outcomes

The project delivery phase was successfully completed under budget, despite several unforeseen costs arising. This was attributable to significant prior 'front-end' works during the planning phase, realistic budget allowances, and various cost efficiencies that were realised throughout the works. A summary of major cost items is presented in Table 4.

TABLE 4

Overview of costs for the delivery phase of the TSF remediation project (data source: DNRMMRRD).

Activity	Expenditure (ex GST)	Key items
Owners costs	$1.03 m	Project staff salaries, travel costs, vehicles, site facilities and comms, stakeholder engagement.
Enabling works	$0.70 m	Dewatering, water monitoring, road maintenance, survey, consultants, cultural heritage.
TSF works 2021	$0.91 m	Tailings reprofiling, southern valley tailings removal
TSF works 2022	$2.95 m	Tailings capping, diversion drain and spillway construction, backfill southern valley
DWD works	$0.44 m	DWD reduction and spillway construction
Revegetation	$0.53 m	Revegetation across TSF and DWD
Other works	$0.21 m	Miscellaneous activities, site laydown pads and roads
Total	**$6.78 m**	**Delivery phase (TSF)**

Monthly expenditure was relatively consistent across the two-year construction period, reflecting the project's approach of bringing forward activities whenever practical, and utilising a small, sustained workforce whenever conditions permitted. This strategy helped maximise workforce continuity and reduce mobilisation overheads. The project S-curve is presented in Figure 26.

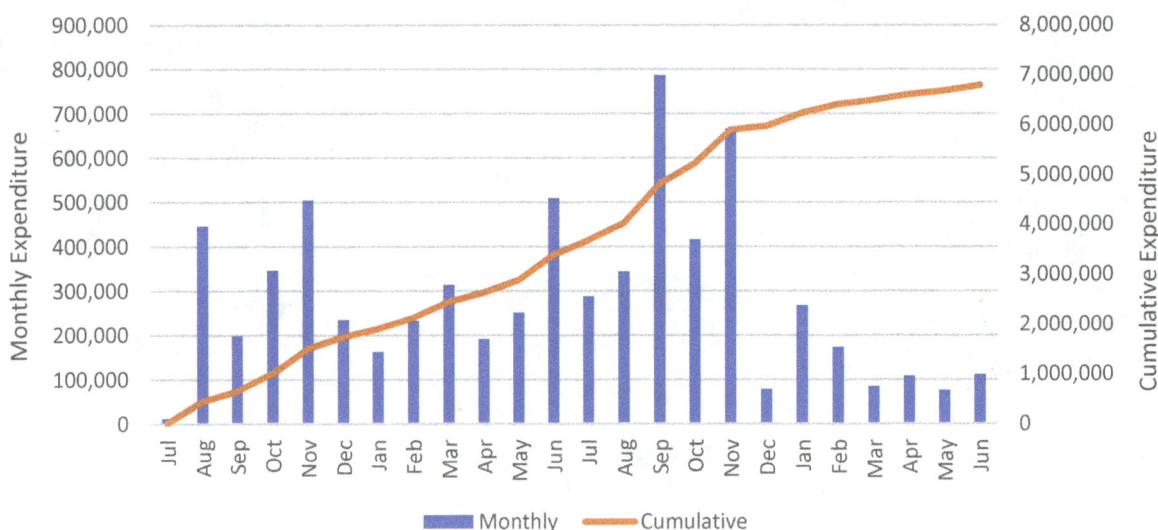

FIG 26 – Monthly and cumulative expenditure across the two-year delivery phase (data source: DNRMMRRD).

In terms of social procurement outcomes, a total of $5.2 million – equating to 80 per cent of total external expenditure – was with local suppliers located within a 125 km radius of the site. A further 11 per cent was spent with suppliers based elsewhere across the Far North Queensland regional, and the remaining 9 per cent with other Queensland-based suppliers. All local suppliers were classified as small-to-medium enterprises, with predominantly local workforces.

Additionally, an indigenous-owned earthworks company was engaged during early enabling works and was later a partner in the joint-venture tender that was won and delivered the major remediation works package.

These outcomes were achieved through deliberate procurement strategies made by the project team, in particular a focus on unbundling the remediation works into smaller packages that were within the capability of local contractors, taking a collaborative risk-sharing approach to uncertainties during the enabling works, and simplifying administrative and commercial arrangements where practical. These actions removed participation barriers for local businesses without compromising delivery on safety, quality, time or budget.

Project review and lessons learnt

The project close-out review considered the project outcomes, and factors that contributed a positive or adverse impact on the project. The leveraging of internal resources and expertise, and willingness to pursue opportunities such as design optimisations and unbundling of work packages, led to improved construction deliverables as well as social benefits, while remaining within original budget limitations. It was noted that this is conditional on sufficient capability and experience within the project team, as well as ongoing external governance and oversight.

In terms of lessons learnt (continual improvement), it was also identified that there may be lower-risk strategies for dealing with hard rock occurrence beyond just including a provisional one-size-fits-all cubic metre rate, given the inherent variability of *in situ* rock properties and therefore variability and uncertainty for excavation requirements and productivity rates. As such it becomes inherently impossible for contractors to accurately determine a fixed unit rate, leading to the necessary inclusion of substantive risk loadings, and a high rate. It was proposed that, for similar future projects, a time-charge cost basis could be used for the pre-loosening/pre-breaking of hard rock to reflect the actual machine/equipment used and hours expended. The subsequent handling of that material would then be treated under the general free-dig excavation rate.

It was also recommended that in future projects, additional drilling be undertaken along the final drainage alignment to assist with assessing rock strength and fracturing; particularly in segments with significant excavation depths or proximity to *in situ* rock foundations. This would assist with identifying the presence of hard rock, quantifying the range of properties and characteristics, and confirming it is feasible to be mechanically removed without blasting.

Performance during ex-TC jasper floods

In late 2023, Tropical Cyclone Jasper formed in the Coral Sea, making landfall near Wujal Wujal – approximately 30 km south of the site – as a Category 2 tropical cyclone on December 13. Intensive rainfall persisted across Far North Queensland from 11 to 18 December, resulting in widespread flooding and substantial damage to towns, roads and other public infrastructure in the area (Vujkovic and Eaton, 2023).

The site received 1360 mm of rainfall over a seven-day period, recorded by the on-site weather station. Analysis of the hourly rainfall data enabled maximum rainfall over various time periods to be plotted against Intensity Duration Frequency (IDF) graphs for the site obtained from the Bureau of Meteorology, as shown in Figure 27. This indicated the peak rainfall over shorter time periods up to six hours corresponded to a 1-in-50 year recurrence event, while the overall four to seven day rainfall totals equated to a 1-in-450 year event.

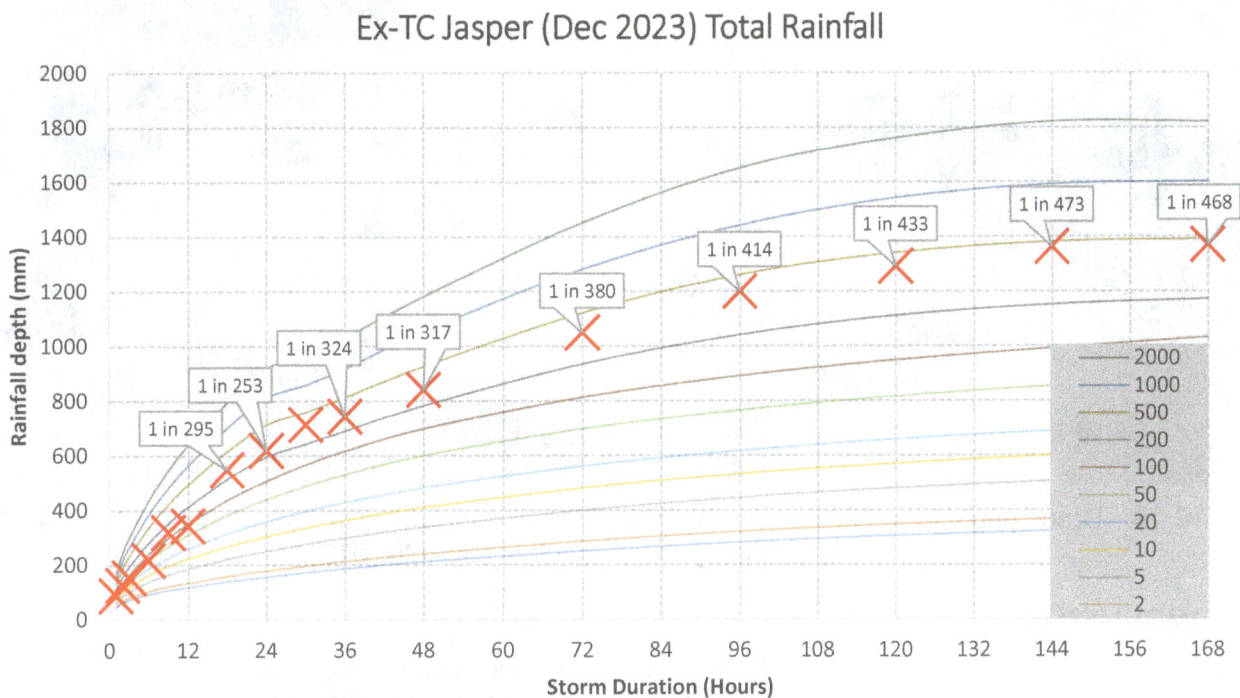

FIG 27 – Observed peak rainfall intensities throughout the ex-TC Jasper floods, plotted against the site's Intensity-Frequency-Duration graph from Bureau of Meteorology (Department of Resources, 2023).

Access to the site was cut-off for approximately two weeks due to inundated river crossings and multiple road washouts in the Rossville area. Upon gaining access to site, inspections were able to confirm that that all TSF and DWD drains and spillways had operated as designed, with no erosion or displacement of rock fill observed. The spillways' robust construction – particularly the wide base and substantial rock fill depth – ensure flow remained within the rock fill layer, even during peak rainfall.

Additionally, no scouring or damage was observed on the tailings capping surface or embankments.

Downstream water quality

Across the 2023/24 wet season, water quality within the DWD pond as well as downstream in the Annan River was positive, with metal concentrations below guideline values. However occasional sporadic exceedances were detected in some samples taken in the drainage channel between the DWD spillway outlet and the Annan River – suggesting historical contaminants and sediments were still present within that watercourse and could be re-suspended and flushed downstream during rain events, albeit in low concentrations.

In contrast to the reductions in metal concentrations, turbidity and suspended sediment concentrations increased over the wet season following remediation, due to the significant rainfall encountered as well as clay and other exposed soil across the remediation areas, where vegetation had not yet been able to fully establish. The DWD pond following ex-TC Jasper is shown in Figure 28. Flushing of historical sediments and/or erosion of the drainage channel downstream of the DWD spillway outlet (beyond the project limits) was also identified as another potential contributing factor (Howley Environmental Consulting, 2024).

FIG 28 – Drone photo from 2024 of the DWD pond taken January 2024 (following ex-TC Jasper floods the previous month) (source: DNRMMRRD).

CONCLUSION

The remediation of the Collingwood Tin TSF and DWD was a complex project; but set-up for success in the early stages by a thorough stakeholder engagement process that identified agreed outcomes, followed by a broad range of data gathering and technical assessment work. Management supported requests from the project team to optimise the remediation plan throughout the project, as new information became available, leading to improved outcomes such as footprint consolidation and improvements to long-term drainage structures. The project team was empowered to pursue construction and procurement strategies that encouraged participation of small-to-medium sized local contractors, local workforce, and indigenous-owned corporations – and significant positive expenditure outcomes were realised, while ensuring the project also met safety, quality, time and cost objectives. Confidence that the remediated site is now safe, secure and durable has increased following a significant seven-day storm event that caused no damage to drainage structures.

ACKNOWLEDGEMENTS

The authors would like to acknowledge the valuable contributions of the many individuals and organisations that contributed to this project. From the external stakeholders, whose engagement and inputs were critical to establishing the pathway to remediation, and the technical specialists whose insights and expertise guided the development of a robust and practical remediation strategy.

We are especially grateful to the dedicated workforce who carried out the on-ground works at this remote site, often with uncomfortable weather conditions and a range of challenging site conditions — your persistent efforts and commitment were instrumental to the project's success.

Finally, we wish to thank the department and management team for their ongoing support of the project, for empowering the project team to innovate and use a flexible delivery approach, and for enabling the sharing of our experience and learnings through this publication.

REFERENCES

AARC Environmental Solutions (AARC), 2019a. Collingwood Tin Mine Environmental Values and Impact Assessment, AARC, Brisbane.

AARC Environmental Solutions (AARC), 2019b. Collingwood Tin Mine Threatened Fauna Breeding Places Assessment, AARC, Brisbane.

AMIRA International Ltd, 2002. ARD Test Handbook: Project P387A Prediction and Kinetic Control of Acid Mine Drainage, AMIRA International Ltd, Melbourne.

Department of Resources, 2018. Collingwood Tin Mine Remediation Fact Sheet, Queensland Government, Brisbane.

Department of Resources, 2019. Abandoned Mines Management Policy, Queensland Government, Brisbane.

Department of Resources, 2021a. Risk and Prioritisation Framework for Abandoned Mine Management and Remediation, Queensland Government, Brisbane.

Department of Resources, 2021b. Collingwood Tin Abandoned Mine Overview, Queensland Government, Brisbane [online], The State of Queensland. Available from: <https://www.qld.gov.au/environment/land/management/abandoned-mines/remediation-projects/collingwood-tin> [Accessed: 2 March 2024].

Department of Resources, 2021c. Collingwood Tin Mine Remediation Project Brief, internal document, Queensland Government, Brisbane (unpublished).

Department of Resources, 2022. Collingwood Tin Mine Site Remediation Construction Drawings, internal document, Queensland Government, Brisbane (unpublished).

Department of Resources, 2023. Review of December 2023 Storm Event, internal document, Queensland Government, Brisbane (unpublished).

Department of Resources, 2024. *Mineral Resources Act 1989*, September 2024.

Douglas Partners, 2020. Collingwood Tin Factual Geotechnical Investigation Report, Douglas Partners, Brisbane.

Engeny Water Management, 2017. Collingwood Tin Mine Regulated Structure Inspection 2017, Engeny, Brisbane.

Engeny Water Management, 2018. Collingwood Tin Mine Concept Design Report, Engeny, Brisbane.

GHD, 2020. Collingwood Tin Disclaimed Mine TSF Closure Detailed Design Report – Rev 1, GHD, Brisbane.

Howley Environmental Consulting, 2024. Water Quality Monitoring Report for Collingwood Mine Remediation Project, Howley Environmental Consulting, Cooktown.

International Network for Acid Prevention (INAP), 2009. Global Acid Rock Drainage (GARD) Guide, INAP, Vancouver. Available from: <http://www.gardguide.com> [Accessed: 01 April 2025].

Klohn Crippen Berger (KCB), 2007. Surface Water Management TSF Embankment Raise, KCB, Brisbane.

McGuire, G, Warren, T, Smalley, T, Grabski, A and Hall, T, 2022. Traditional Owner Engagement for Collingwood Tin Mine Closure Planning, in Proceedings of the 4th International Congress on Planning for Mine Closure (Gecamin: Santiago).

Vujkovic, M and Eaton, M, 2023, 13 Dec. Tropical Cyclone Jasper crosses the coast near Wujal Wujal, north of Cairns [online], ABC News. Available from: <https://www.abc.net.au/news/2023-12-13/qld-tropical-cyclone-jasper-weather-pattern-warning-bom/103220130> [Accessed: 01 April 2025].

Capping and landform construction over an in-pit TSF for mine closure

K Koosmen[1], S Christian[2], A Inderbitzen[3], J Huelin[4] and C Anderson[5]

1. Principal Geotechnical Engineer, PSM, Brisbane Qld 4000. Email: kai.koosmen@psm.com.au
2. Geotechnical Engineer, PSM, Brisbane Qld 4000. Email: samuel.christian@psm.com.au
3. Senior Geotechnical Engineer, PSM, Brisbane Qld 4000.
   Email: andrew.inderbitzin@psm.com.au
4. Director, Practical Mining Consultants (PMC), Wamuran Qld 4512.
   Email: joe@practicalminingconsultants.com
5. Mining Engineer, Practical Mining Consultants (PMC), Wamuran Qld 4512.
   Email: craig@practicalminingconsultants.com

ABSTRACT

The subject site is a large open pit mining complex in the Bowen Basin, Queensland, Australia. The last coal was mined on-site in March 2023 and the site is now being rehabilitated, including several in-pit tailings storage facilities (TSFs) with a combined surface area of more than 100 hectares. As part of the closure plan it is a requirement to place a soil cap and construct landforms over all TSFs to meet the mine closure objectives and satisfy the post mining land use.

Pit X is an in-pit TSF where the final landform has now been constructed. Geotechnical stability during and after construction has been a major focus for closure planning. Consideration of landform settlements has also been a major focus for Pit X as the landform is required to remain free draining after mine closure, yet, the placement of fill over the TSF was predicted to cause large settlements which could allow water to pool.

This paper presents a case study of the Pit X in-pit TSF focusing predominantly on geotechnical aspects relating to: landform design criteria, landform concept selection, geotechnical data collection, geotechnical stability, settlement assessments, construction implementation, and post-construction monitoring.

INTRODUCTION

The subject site is in the Bowen Basin in Queensland, Australia. Mining commenced at the site in the early 1980s with the operations spanning across three mining areas with multiple pits in each area. Open pit strip mining was the primary extraction method, although, underground longwall mining was also employed to a lesser extent. Coal was processed on-site and tailings produced from processing were deposited into several old pit voids to form in-pit TSFs.

Coal mining ceased in March 2023 and all activities are now focused on rehabilitating the site for mine closure. As part of the rehabilitation plan it is a requirement to cap and construct landforms over the four in-pit TSFs which have a combined surface area of more than 100 hectares.

The purpose of this paper is to describe the geotechnical studies that were undertaken to facilitate the design and construction of a final landform over the Pit X TSF. The main aspects which are discussed in the paper include:

- project considerations and landform design criteria
- geotechnical investigations and data analysis
- geotechnical stability assessments for construction and final landform
- settlement assessments
- construction implementation
- post-construction monitoring.

HISTORY OF PIT X

Historic details of Pit X as follows (Figure 1):

- Prior to 1995: mining was undertaken using a dragline and truck/shovel fleet to extract coal.

- 1997 to 2002: tailings were discharged into the facility from the southern end.

- 2003 to 2007: rehabilitation began by dozing fill from adjacent dragline spoil piles over the TSF surface using the displacement method. This created a large bow wave as works advanced northward, causing tailings to bulge up by several metres at the northern end of the facility (Figure 2). Perimeter embankments (~5 m high) were constructed to contain the bulged-up tailings.

- August 2007: around 4 ha of tailings remained uncapped at the northern end when a large section of the platform broke away and sank into the tailings (Figure 2). This triggered further upwelling and minor release of tailings to the east and north-west (see Figure 1).

- After August 2007: rehabilitation activities ceased, and the facility remained dormant for many years. A low point and large pond developed over the capped surface at the southern end of the facility (see 2008 and 2012 images in Figure 1).

- 2018: mining commenced in Pit Y immediately north of Pit X. As part of these works the dragline dumped additional spoil along the northern boundary of Pit X, partly covering the overflow tailings (see locations in 2012 image in Figure 1).

- 2021 onwards: various works have been undertaken including ground investigations and geotechnical assessments to inform closure studies.

- September 2024 to February 2025: the final landform was constructed over the TSF.

FIG 1 – Historic aerial images of Pit X.

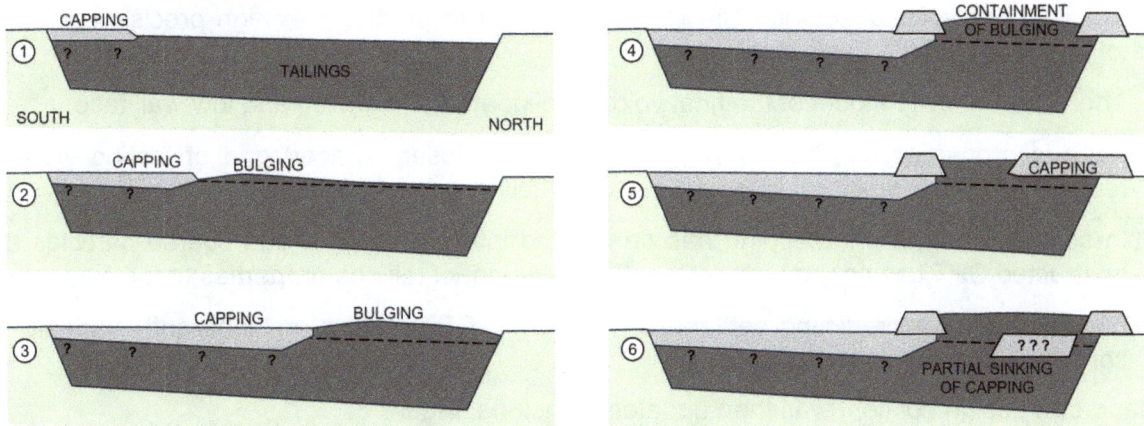

FIG 2 – South to North schematic long section showing historic capping sequence (not to scale).

Some of the complications associated with the historical aspects were:

- there was no available survey model for the final pit void prior to tailings filling

- there were no records of capping fill thickness in areas that had been capped

- there were no details on the depths and extents of the waste rock platform that sank into the tailings at the northern end of the facility.

LANDFORM DESIGN CONSIDERATIONS

General criteria

The final landform for Pit X was designed considering several key criteria that were developed to satisfy requirements of the mine closure plan. These included:

- The post-mining land use will be grazing, and the landform must support vegetation that is consistent with the surrounding environment.

- The landform must remain free draining after mine lease relinquishment (ie no ponding).

- Limited ponding is tolerable during the post-construction monitoring phase provided any low points can be filled and remediated prior to mine lease relinquishment.

- Slope gradients across the final landform must be designed to be 10 per cent or less.

- A minimum cover of 2 m is required over all tailings materials.

- The landform must be constructed using locally sourced fill and placed using the existing mining fleet.

- Fill placement volumes must be minimised as much as reasonably practicable.

- Potential for future rework must be minimised as much as reasonably practicable.

- Geotechnical stability must satisfy corporate and legislative standards, both during construction and after the final landform is complete.

Landform concept optioneering

An initial landform concept was developed for Pit X in 2020 and involved:

- minimal fill volumes placed across the entire facility

- a landform slope gradient of 1 per cent to drain the landform from south to north

- this landform specified less than 2 m of cover over the exposed tailings in some areas.

Given the profile was almost flat it was highlighted that differential settlements may occur in the tailings over time, which may generate low points where ponding could develop. It was also highlighted that differential settlements would be difficult to predict with high precision across all areas of the facility on the basis that:

- There is no survey model of the final void to delineate the highwall and low wall face.

- Tailings properties can be highly variable, and a closely spaced grid of testing would be required to give a reasonable indication of variability across the facility.

- In some areas, historic capping has penetrated into the tailings, and coarse particles have obstructed CPT testing, preventing the measurement of tailings properties to full pit depth.

- The methods for predicting settlements based on CPT testing are inherently variable with considerable scatter.

The landform design concept was then updated to include, Figure 3:

- Shedding run-off across the landform surface from west to east, achieved by placing more fill along the western side of the facility so that water drains towards the east.

- Collecting the run-off water in a surface water drain that runs south to north with 1 per cent gradient. It was planned to excavate the drain into the *in situ* ground (ie not over tailings) to minimise potential for differential settlements to develop along the drain alignment.

- The outflow point for the drain is in the north-east corner of the TSF.

- The landform would be constructed from mine waste rock with no requirements for internal structural layers, filters or drains.

With this updated design it was highlighted that if the landform is constructed with relatively flat cross fall, then it was still plausible for differential settlements to form low points and facilitate ponding of water on the final landform. With this in mind, two options for the landform design were then considered:

1. A conservative design which placed larger fill volumes as part of the initial construction works, to ensure that long-term gradients would be maintained and provide a 'walk-away' solution. This option would implement larger cross gradients in the west to east direction so that the landform would remain free draining even if large settlements were to occur. Or,

2. A less conservative design based on predicted and observed settlement behaviour, accepting some margin for error and the possibility that minor low points could develop but could be identified by monitoring then remediated prior to mine lease relinquishment.

After evaluating the trade-offs between larger initial capping volumes and potential future remediation, the second option was selected as the preferred approach.

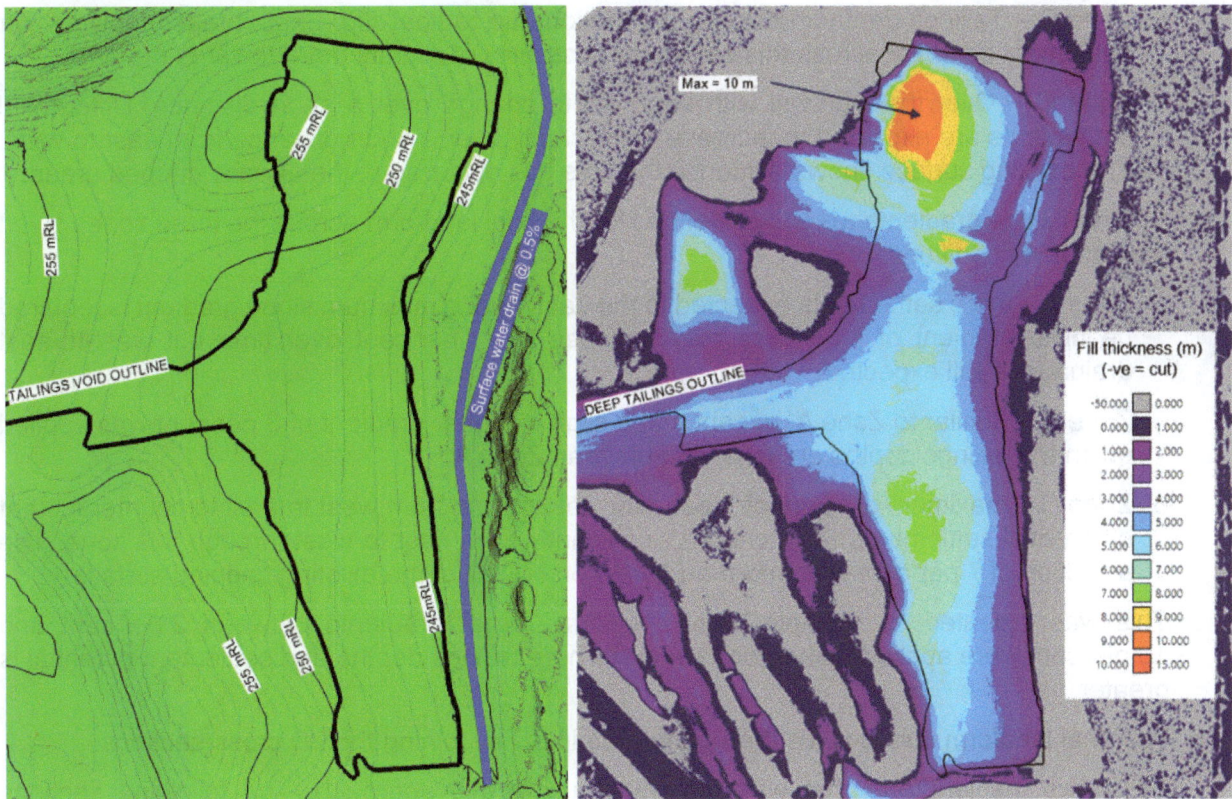

FIG 3 – Final landform design: (left) 1 m contours of final surface; (right) fill thickness above existing surface.

Landform design approach

Complications for designing the landform to account for differential settlements were that: (1) differential settlements depend on the tailings depths; and (2) there was no available void model to provide details of the tailings depths across all areas. Consequently, it was decided that an approximate void model would be defined for the study using the following approach:

- Use the coal seam floor triangulation to define the base of the pit.

- Use historic aerial photos to define the perimeter of the impounded tailings.

- Along the eastern side, project a line from the tailings limit down to the coal seam floor at 70° from horizontal. This is a typical highwall angle applied elsewhere across the site.

- Along the western side, project a line from the tailings limit down to the coal seam floor at 40° from horizontal. This is a typical dragline low wall angle applied elsewhere across the site.

With the above method used to define the highwall and low wall face positions, the following method was then adopted to design the landform profile, Figure 4:

1. Define a series of east–west cross-sections through the TSF including the following information on each of the sections:

    o survey model of the existing surface

    o coal seam floor based on the site geological model (to define the base of tailings)

    o eastern and western limits of deep tailings (based on the review of aerial photos).

2. Draw the highwall and low wall face positions on the section using the method noted above.

3. Define three settlement zones: A, B and C:

    o Zone A – Over the low wall with variable tailings depths and high potential for differential settlements. Differential settlements are expected to increase from west to east with increasing tailings depth.

- Zone B – Tailings depth shows slight variation due to floor dip. Expect reasonably similar settlement potential across this zone with possibility of slightly more towards the east.

- Zone C – Over the highwall with large depth change over a short distance. Differential settlements are expected to increase significantly over a short distance from east to west. The existing surface shows evidence that this has occurred in the already capped areas.

4. Different slope gradients were then defined for the final landform across the three zones given the variable conditions:

- Zone A – more settlements expected to the east, hence minimum slope gradient is required (ie initial gradient of 1 per cent may increase to >1 per cent over time but west-to-east drainage is still expected).

- Zone B – Similar to Zone A, except that a slightly higher gradient of 3 per cent was applied given the tailings depths are less variable than Zone A.

- Zone C – maximum gradient of 7 per cent was applied from west to east to counteract that differential settlements will increase significantly from east to west through this zone. The selection of 7 per cent was informed based on review of the existing capping surface.

5. This was repeated for multiple cross-sections. Larger gradients between 2 per cent and 10 per cent were applied at the northern end where tailings depths and settlement potential is greater.

6. The final landform surface was then generated in 3D by joining the 2D cross-sections.

Figure 4 shows the landform that was generated using this approach. The following sections describe the studies that were undertaken to provide a geotechnical evaluation of this landform concept.

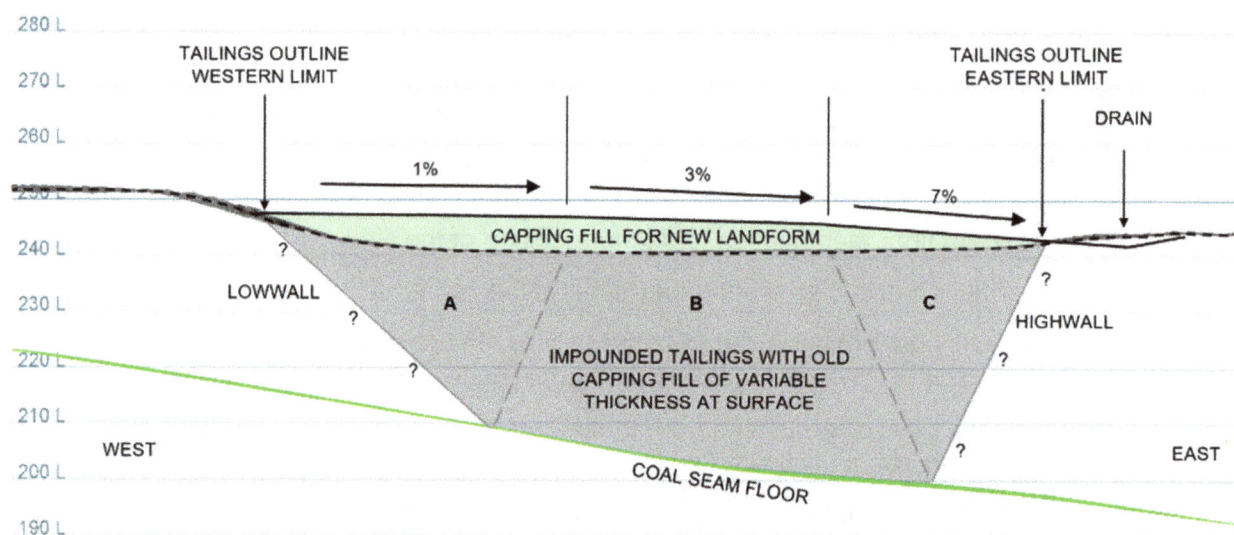

FIG 4 – Example of cross-section approach that was used to define the final landform profile.

GEOTECHNICAL DATA COLLECTION

2021 CPT testing

In situ tests were conducted in 2021 to inform the feasibility level study and included (Figure 5):

- Seven CPT tests at various locations across the facility (red dots in Figure 5)

- 17 pore pressure dissipation tests conducted across five of the sites

- 12 vane shear tests conducted at two of the sites

- Push tube samples were also collected from two of the sites for laboratory testing.

Site ID	Test Depth (m)	Predrill Depth (m)	Lab Tests	Vane Shear Tests	Dissipation Tests
CPT19	45.3	0			
CPT20	14.2	0			
CPT20A	10.9	0			
CPT21	23.2	0		8	2
CPT22	51.4	0		12	4
CPT23	5.5	0			
CPT23A	7	0			
CPT23B	2.4	0			
CPT23C	14.1	0			
CPT23D	6.8	0			
CPT23E	39.7	0			
CPT24	1.1	0			
CPT24A	7	3			
CPT24B	10.1	3			
CPT25	0.4	0			
CPT25A	26.8	3.5			
CPT26	0.6	0			
CPT26A	6.4	4.5			
CPT26B	5.1	3.5			

Site ID	Test Depth (m)	Predrill Depth (m)	Lab Tests	Vane Shear Tests	Dissipation Tests
PX-01	36.5	2.7			4
PX-02	24.3	0			
PX-03	52.7	0	Yes	7	4
PX-04	3.9	2.9			
PX-05	32.8	2.7	Yes	5	4
PX-06	19.9	3			
PX-07	35.6	1.5			4

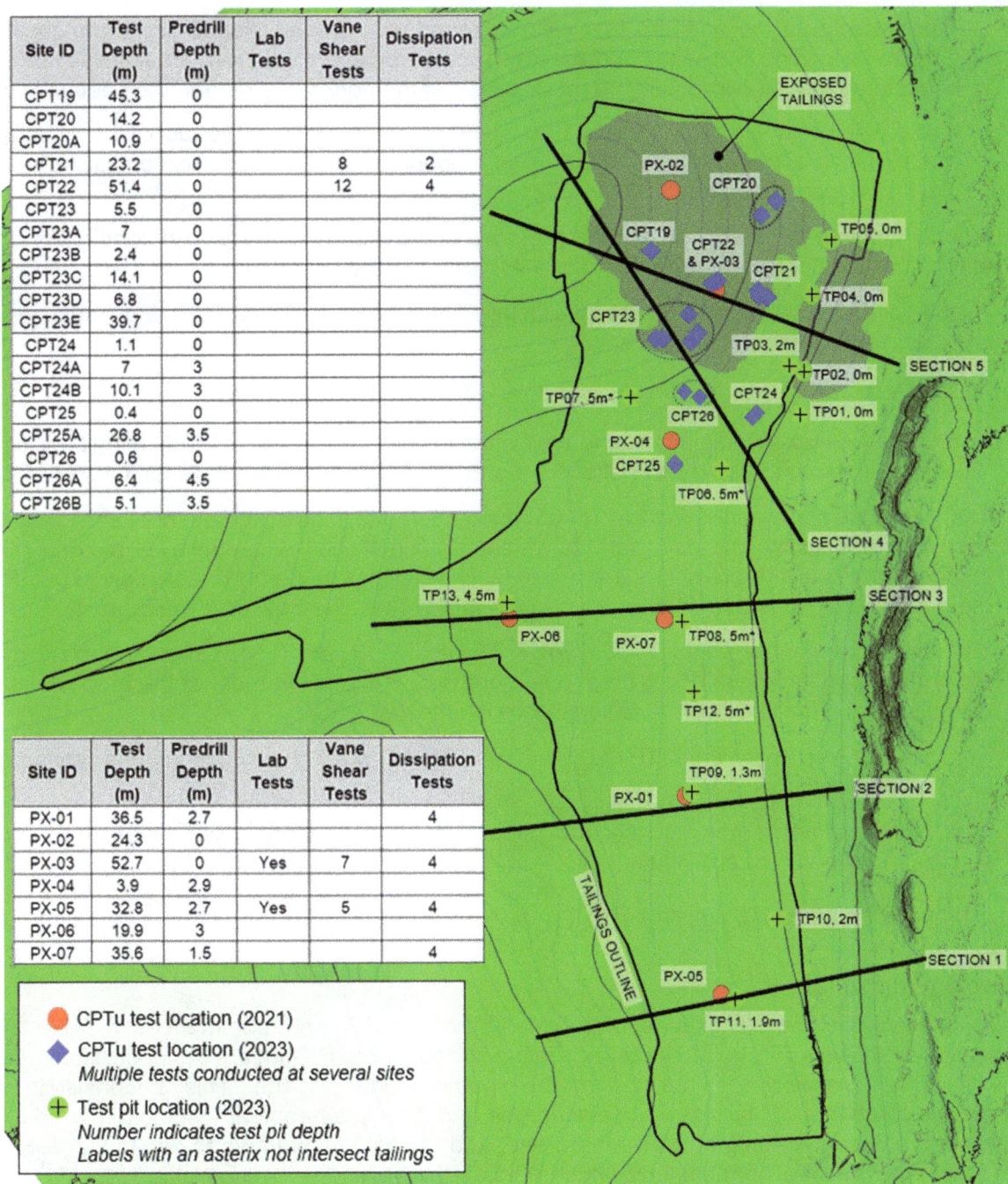

- ● CPTu test location (2021)
- ◆ CPTu test location (2023)
  Multiple tests conducted at several sites
- + Test pit location (2023)
  Number indicates test pit depth
  Labels with an asterix not intersect tailings

FIG 5 – Overview of geotechnical data collection points.

The 2021 campaign also included testing across the other TSFs at the mine site which provided a reasonably large data set.

2023 CPT testing

Further testing was conducted in 2023 for the detailed capping design and landform assessment study. The 2023 testing included:

- CPT testing was conducted across eight sites at the northern end of the TSF. Many of these sites required multiple testing attempts due to premature termination of tests on hard objects (presumably submerged mine spoil, or the edges of the pit void).

- The reasons for concentrating the tests at the northern end were:

  o The 2021 campaign indicated poorer tailings properties with higher clay contents towards the northern end of the TSF, and there were only two existing data points in the northern end following the 2021 campaign (PX-02 and PX-03).

- o The landform in the northern end was designed with greater slopes and therefore more differential loading across the surface when compared to the southern area. It was therefore considered that stability would be more critical at the northern end than the southern end.

- o It was also considered that the 2021 tests at the southern end of the facility would still give a reasonable approximation of the tailings properties in those areas as conditions had not significantly changed from 2021 to 2023.

- Twenty vane shear tests and six pore pressure dissipation tests were also conducted at two of the sites to supplement previous data that was collected in 2021.

- A multipoint VWP with four sensors was installed into the tailings at CPT22. Monitoring results are discussed at the end of the paper.

2023 test pits

Thirteen test pits were excavated across the area with the primary objective of confirming the existing cap thickness and depth to tailings. Locations are shown in Figure 5 with details as follows:

- Test pits in the north-east (TP-01 to TP-05) encountered tailings at surface or at shallow depth. In several test pits the tailings at surface were underlain at shallow depths by fill materials. This confirms previous interpretations that a thin layer of tailings in the north-east flowed out of the main TSF during previous attempts to cap the facility (see overflow area in Figure 1).

- Test pits in the central area (TP-06, TP-07, TP-08, TP-12 and TP-13) show the existing cap has a minimum thickness of 4.5 m, and is more than 5 m thick in most of the area where the test pits were terminated before tailings were intersected.

- Test pits in the southern area (TP-09, TP-10 and TP-11) show the existing cap is between 1.3 and 2.0 m thick in this area.

- Test pits were not required to confirm capping thickness where tailings were exposed at surface in the northern end of the facility.

GEOTECHNICAL CHARACTERISATION

Tailings properties

Physical properties of the tailings measured from laboratory testing are summarised in Table 1 with grading curves shown in Figure 6. The tailings in the northern end (PX-03) are generally finer grained and with higher plasticity than tailings at the southern end (PX-05). This is consistent with expectations given that: (1) tailings were discharged and formed a beach at the southern end of the facility with the decant forming at the northern end; and (2) historic attempts to cap the facility were more problematic at the northern end which implies locally poorer tailings characteristics. High variability in the tailings characteristics is also observed from the CPT data shown in Figure 7.

TABLE 1

Physical properties of the tailings measured from laboratory testing.

Test ID	Depth	LL	PL	PI	GMC	Silt + clay	Wet density (t/m^3)
PX-03	8.35	49%	28%	21%	43%	90%	1.44
PX-03	18.40	45%	25%	20%	42%	85%	1.49
PX-03	38.39	34%	26%	8%	32%	30%	1.47
PX-05	10.41	Not plastic			38%	20%	1.45
PX-05	23.20	Not plastic			40%	15%	1.47

LL = Liquid Limit, PL = Plastic Limit, PI = Plasticity Index, GMC = Gravimetric Moisture Content.

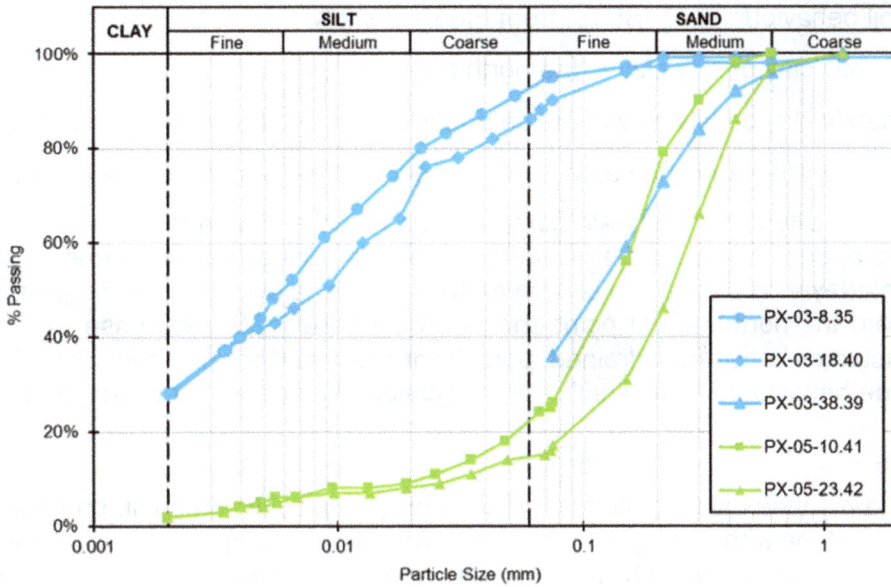

FIG 6 – Grading curves measured by sieve and hydrometer.

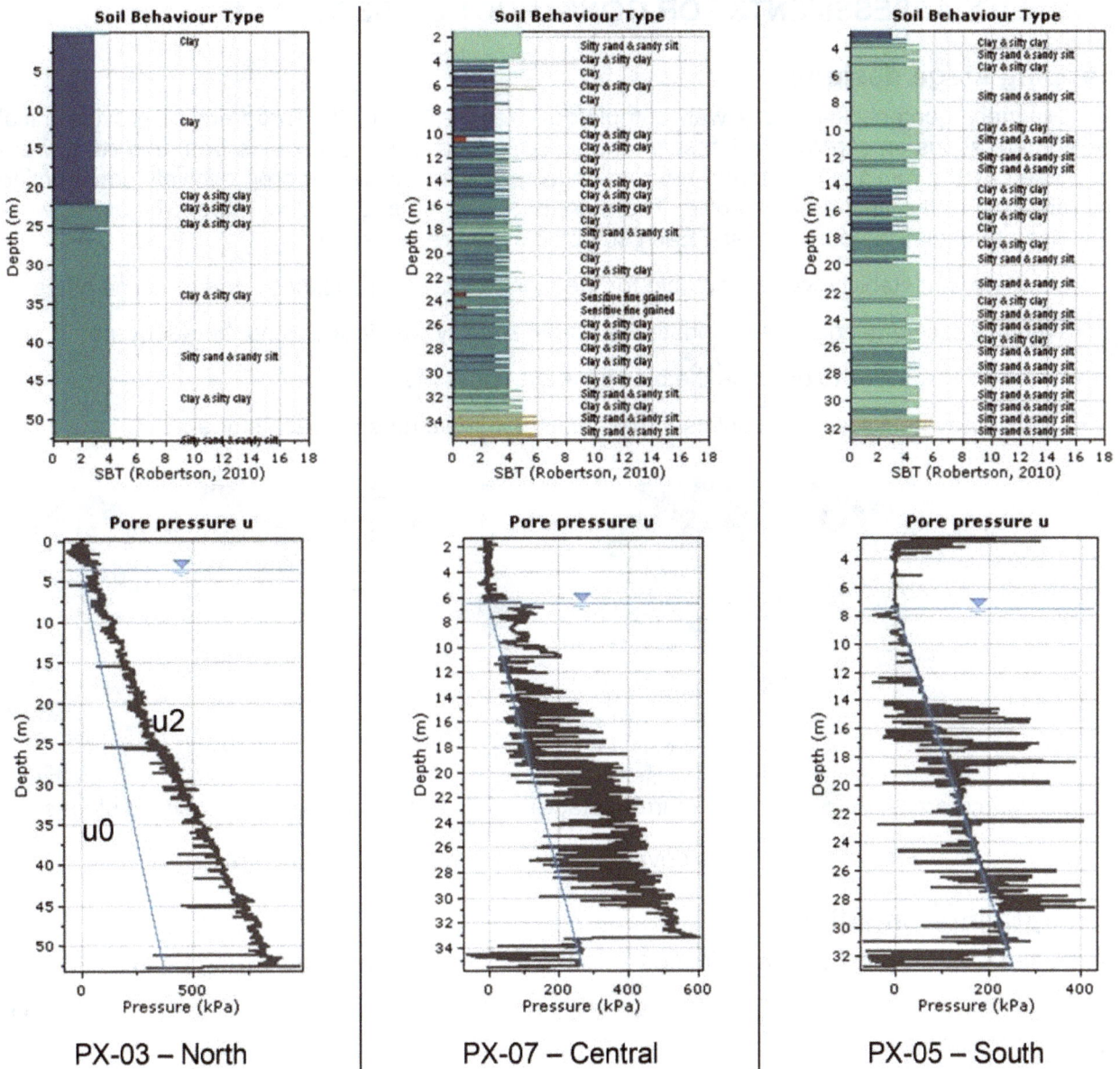

FIG 7 – Tailings characteristics based on CPT data collected in different areas of the TSF.

The different soil behaviour types (SBTs) from Figure 7 show:

- Predominantly clay or silty clay in the north (PX-03).

- Predominantly clay or silty clay in the central area (PX-07) but with some sandy lenses.

- Predominantly silty sand and sandy silt in the south (PX-05) but with some clay lenses.

Although clays and silty clays dominate compositions in the central and northern areas, tailings in the central zone appear to have better drainage characteristics and may be due to slightly coarser gradations or interlayering of fine and coarse materials. This is evident from the pore pressure data in Figure 7 where the northern test point shows an undrained CPT response ($u_2 > u_0$), while the central point fluctuates between undrained, partially drained, and drained conditions ($u_2 \approx u_0$). These variable drainage behaviours have significant implications for settlement times, as discussed later.

Fill materials

The existing and planned capping fill is comprised of typical coalmine spoil, primarily derived from sandstone and siltstone with varying size fractions. This fill was sourced from spoil piles surrounding the TSF which were placed during historic mining activities. With pit depths around 50 m and *in situ* weathering to about 25 m, the fill comprises a mix of fresh and weathered material.

STABILITY ASSESSMENTS FOR CONSTRUCTION SEQUENCE

Capping stage designs

A preliminary geotechnical study was completed after the 2021 CPT program to provide initial recommendations on capping thicknesses and suitable machinery. Outcomes from this study were then used by the mining engineering team to develop stage designs for progressively constructing the landform to the final design profile. The extents and fill source for each stage (Figure 8) were optimised to minimise haulage costs. Each stage was then reviewed to:

- confirm that stability was acceptable for the planned capping thickness and machine types

- provide recommendations to update the plan if stability was found to not be adequate

- seek opportunities to optimise the plan wherever possible

- define any necessary hold points or risk mitigation measures for each stage.

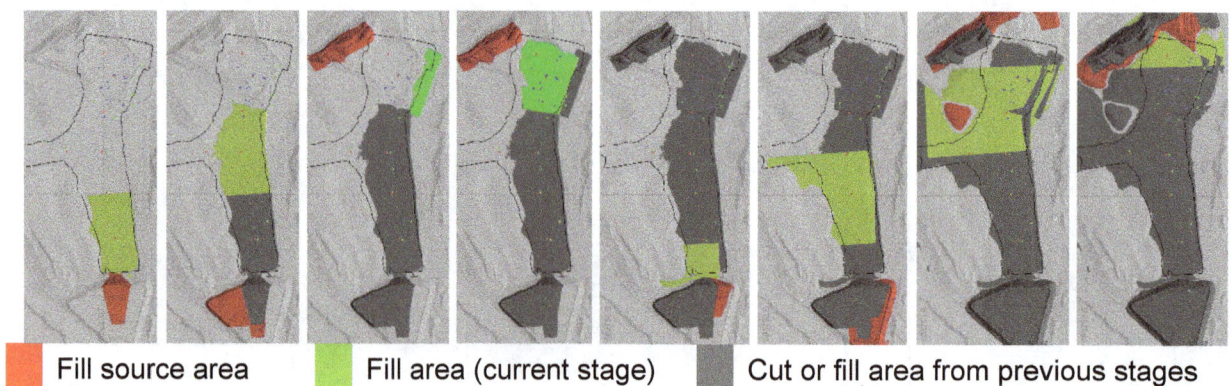

■ Fill source area   ■ Fill area (current stage)   ■ Cut or fill area from previous stages

FIG 8 – Stage plans developed by the mining engineering team.

Strength interpretations

Undrained shear strengths were estimated from the CPT data at each point using Equation 1:

$$S_u = (q_t - \sigma_v) / N_{kt} \qquad (1)$$

Where:

S_u is the undrained shear strength

q_t	is the net cone resistance corrected for excess pore pressure
σ_v	is the vertical total stress
N_{kt}	is the empirical cone factor

The tailings density was assumed to be 14.5 kN/m^3 for the purpose of calculating the vertical total stress.

N_{kt} values were estimated using correlations with the pore pressure parameter (B_q), developed from *in situ* vane shear testing of coal tailings at the site and other coal mining operations (Figure 9). The adopted N_{kt}-B_q relationship plots near the upper bound of the data set meaning it yields higher-than-average N_{kt} values. Since N_{kt} is in the denominator of Equation 1, this results in more conservative (ie lower-than-average) estimates of undrained shear strength.

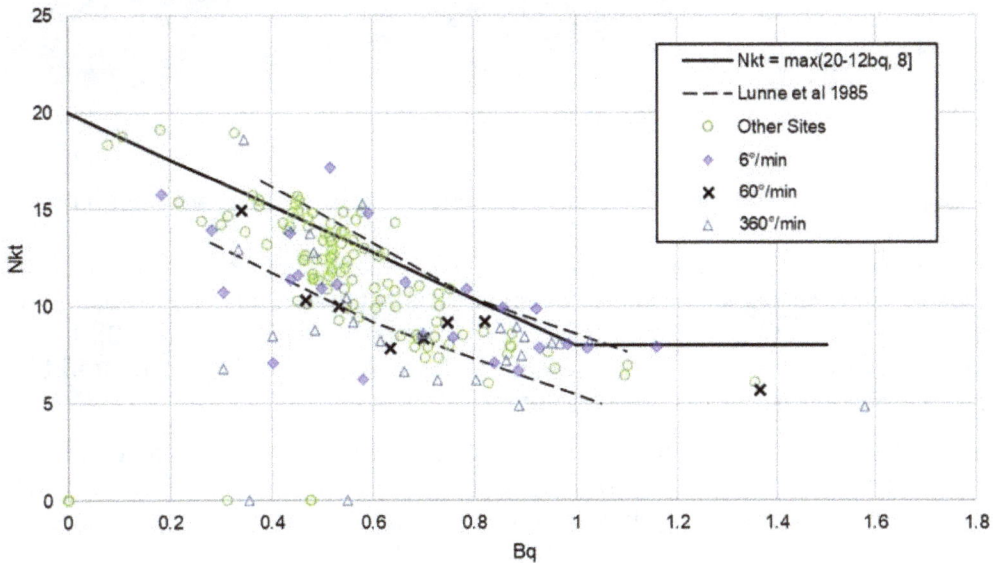

FIG 9 – Relationship between N_{kt} and B_q.

Undrained shear strengths were calculated using Equation 1 across the full depth of each CPT profile. Each strength profile was then used to define tailings layers in the stability models, with each layer assigned a representative strength as either a constant S_u or increasing from a minimum value at the top (Figure 10). This approach reflects an unconsolidated undrained (UU) shear strength mode for short-term construction assuming no strength gains from consolidation following fill placement. Whilst this approach varies from conventional practice where undrained shear strengths apply a vertical stress ratio (VSR), it is conceptually similar to using a VSR with a B-bar of one. While this may be conservative for coarser tailings that could consolidate between lifts, this approach avoids the need for a VWP network to monitor pore pressures, which would be required to verify conditions if the design strengths had assumed that some degree of consolidation had to be achieved. Not having to manage construction around a VWP network has significant practical benefits. Furthermore, the schedule was also de-risked from a possible scenario where consolidation is not being achieved in the field to the extents or over the time frames assumed in the design.

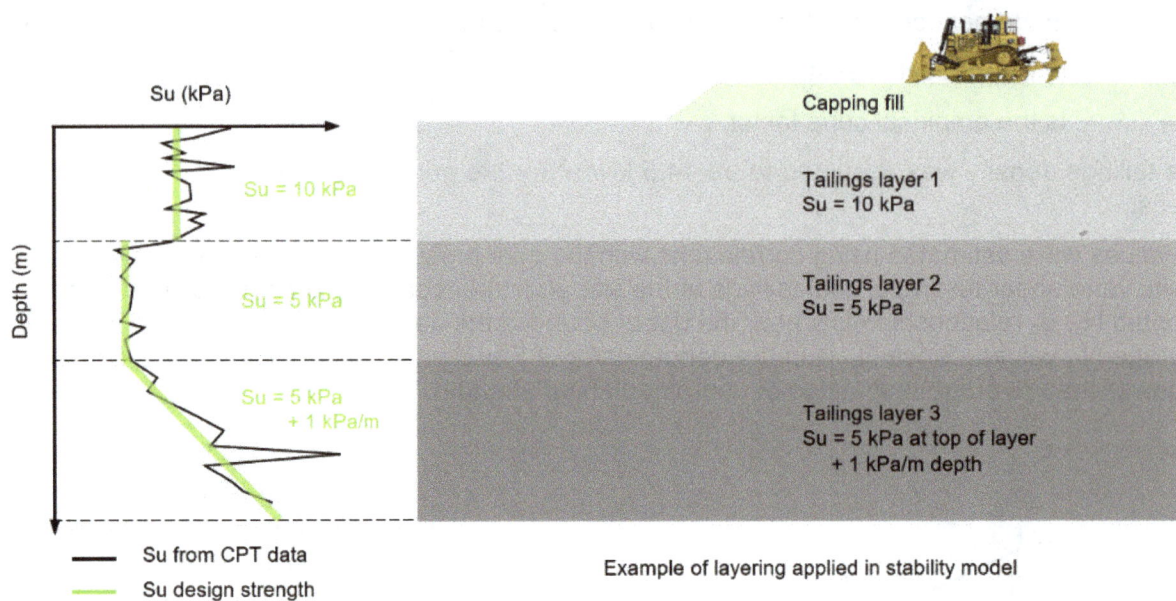

FIG 10 – Concept sketch showing strength assignment for tailings layers based on CPT data.

Stability simulations were also performed assuming drained shear strengths for some of the tailings layers at some of the CPT locations. Friction angles were estimated from CPT data using published methods by Robertson and Campanella (1983) and Kulhawy and Mayne (1990). In all cases, the models applying undrained strength parameters predicted lower stability compared to the models which applied drained strength parameters to some of the layers.

The fill materials were assumed to have a drained shear strength with cohesion of 3 kPa and friction angle of 30°. These values were selected considering: (1) the presence of variably weathered fill materials; and (2) it was planned to place the fill in layers that were several metres thick with no engineering compaction. The fill density was assumed to be 17.5 kN/m³.

Machine loadings

Equipment surcharge loads were included in the stability analyses using machine operation weights for a CAT D11 dozer, CAT D10 dozer and CAT 785 haul truck, based on manufacturer specifications. Some factoring of the loads was applied to account for load dissipation through the capping platform which is not accurately captured by plane-strain 2D limit equilibrium modelling. Scenarios with the dozer at the tiphead included both the machine load and an additional load for the spoil heap in front of the blade, estimated using the dozer's maximum blade capacity.

Assessment methodology

A representative stability model was developed for each capping stage considering the lowest CPT strength profile and variable thicknesses of the existing and planned capping within each stage. In some cases, multiple models were developed for each stage considering different strength profiles or different thickness for the capping materials.

The stability was assessed for each model using 2D limit equilibrium techniques adopting the GLE/Morgenstern-Price Method. The minimum acceptable factor of safety (FOS) was defined as being 1.3 or greater, in accordance with corporate standards. The stage plans were modified for any cases where the FOS was calculated to be less than 1.3.

Examples showing typical results from stability assessments are provided in Figures 11 and 12.

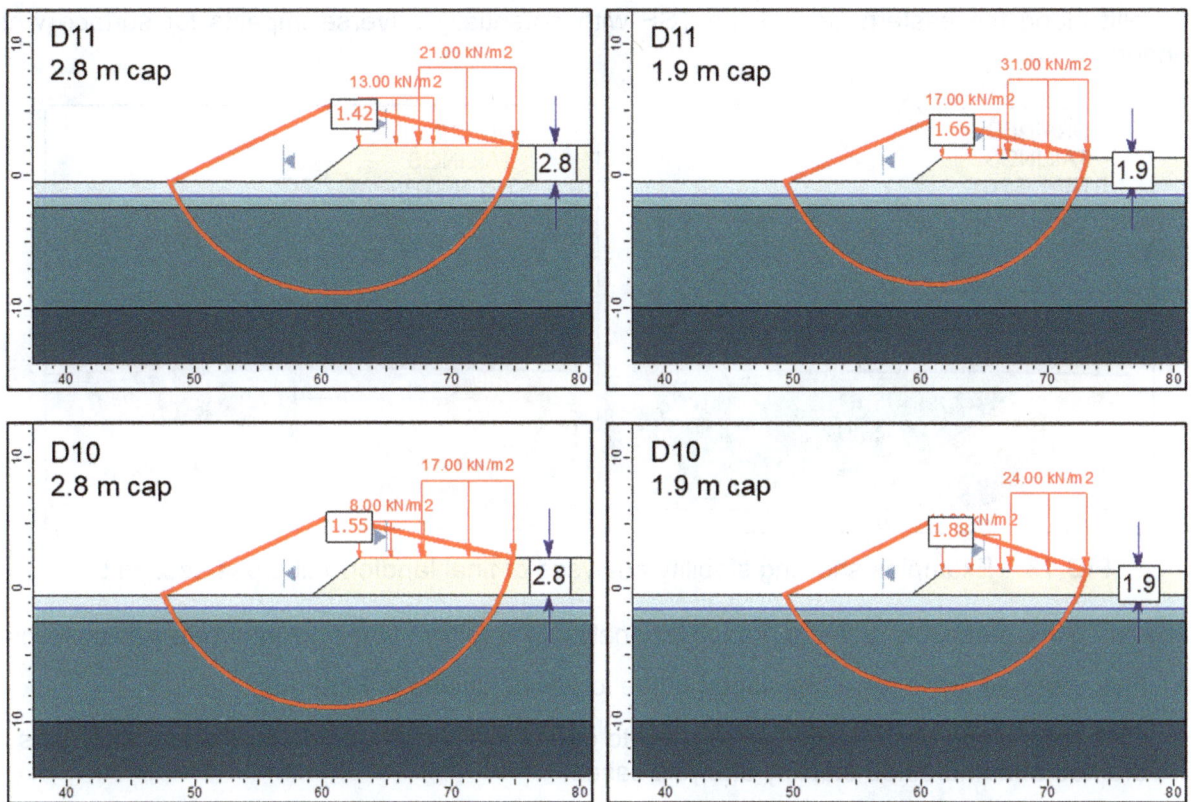

FIG 11 – Stability results for Stage 3 at northern end considering variable lift thickness and dozer types.

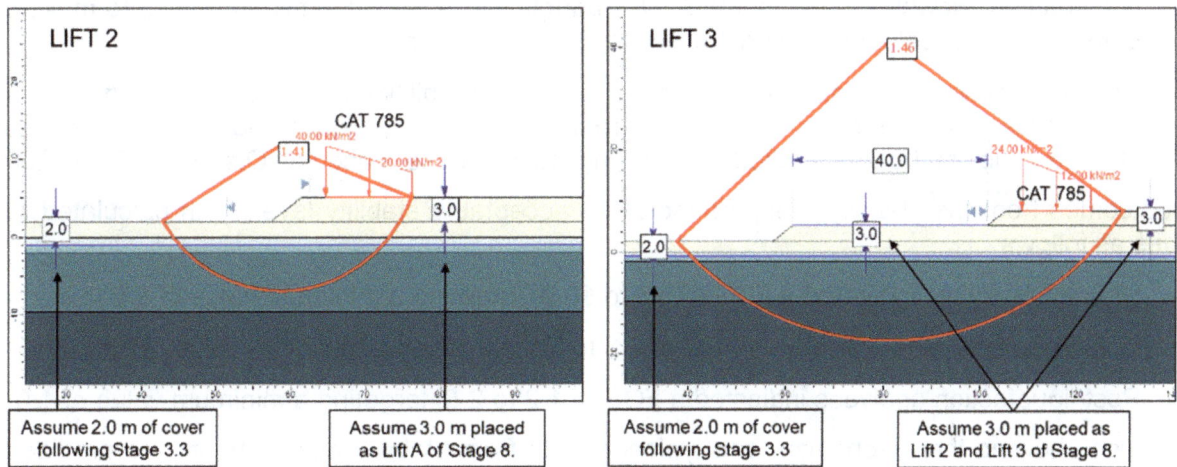

FIG 12 – Stability results for Stage 8 at northern end considering second and third lifts.

STABILITY ASSESSMENTS FOR FINAL LANDFORM

The overall stability risk for the final landform is considered relatively low given that:

- the tailings are not retained by an embankment structure
- most of the tailings are retained below ground level, except at the northern end where the tailings bulged up to 3 m higher but are fully contained on all sides by mine fill
- all tailings within the facility will be covered by several metres of fill.

Due to the greater fill depths along the western side of the facility, the key failure mechanism involves a deep-seated rotational surface from west to east extending down through the tailings (Figure 13). Although this mechanism is unlikely to cause loss of containment after the landform is constructed, development of this mechanism could result in subsidence of the landform along the western side and uplift along the eastern side of the TSF with potentially adverse impacts for surface water drainage.

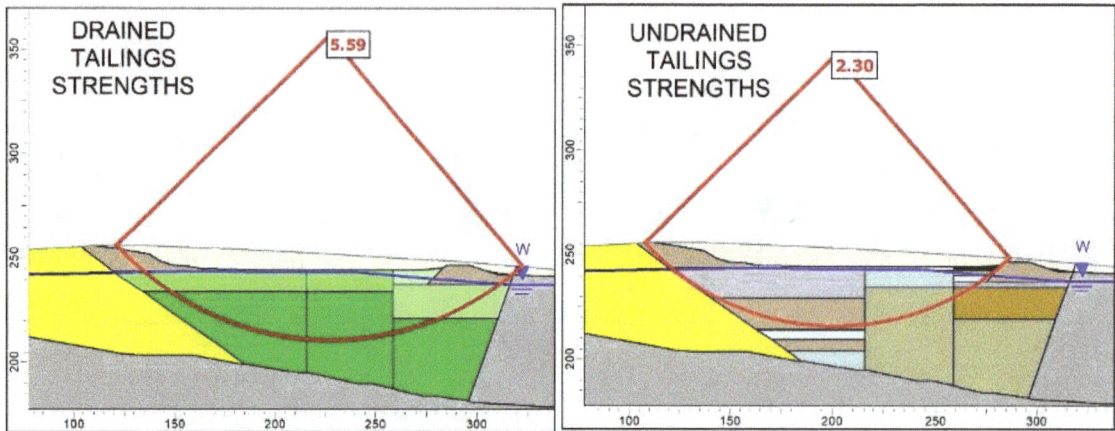

FIG 13 – Examples showing stability analyses of final landform at cross-section 5.

Considering this mechanism, the approach for analysing stability of the final landform was as follows:

- Five cross-sections were developed at the locations shown in Figure 5.

- CPT tests along each section were used to define drained and undrained shear strengths for drained and undrained stability analysis cases.

- Drained and undrained strengths were derived from CPT data and applied in the stability models using the methods outlined previously. Applying UU strengths for the undrained case is likely conservative, as it neglects potential strength gains from consolidation after landform construction. However, this approach was adopted as a conservative first-pass to make use of the models developed for construction-phase stability analyses.

- Post-liquefaction analyses were carried out to assess stability under seismic loading. As a first-pass it was assumed that an earthquake could trigger tailings liquefaction and a lower-bound strength ratio of 0.05 was applied to the tailings for this case (based on Olson and Stark, 2002).

All outcomes from the stability analyses indicated acceptable stability levels with calculated FOS values as follows:

- Drained analyses indicated a FOS of 5.6 to 10.3, targeting a minimum value of 1.5.

- Undrained analyses indicated a FOS of 2.3 to 5.5, targeting a minimum value of 1.5.

- Post-liquefaction analyses indicated a FOS of 1.2 to 2.5, targeting a minimum value of 1.0.

Examples showing the mechanism and typical results from stability assessments are provided in Figure 13.

The high FOS values and conservative strength assumptions suggest that the landform could support steeper cross falls whilst still maintaining acceptable stability levels. This could be achieved by placing more fill on the western side or less fill on the eastern side. However, neither option is desirable given the crossfall is optimised to balance long-term settlement risk with earthworks costs.

SETTLEMENT ASSESSMENTS

Introduction

The landform is required to remain free draining and not allow water to pool following mine closure. It is therefore important that gradients are maintained from west to east, and the landform does not settle to an elevation lower than the surface water drain along the eastern side of the facility. Whilst the landform has been designed primarily to address the settlement risks, checks were still required to confirm the expected settlement magnitudes.

Reliable 3D predictions of the settled landform shape were not possible given the spacing of testing, the inability to test to full pit depth in several areas, and without a detailed survey of the final void before tailings filling. In view of this, 1D settlement predictions have been made at CPT test locations to confirm if the landform surface at those locations remains higher than the drain further to the east.

Methodology

Settlement predictions were made factoring in primary and secondary consolidation settlements in the tailings, and post-construction creep settlements in the capping fill materials. The method and assumptions that were used to calculate the settlements were as follows:

- Primary and secondary consolidation settlements in the tailings were estimated using the method developed by Koosmen and Gerridzen (2023) which involved:

  o Collecting push tube samples at CPT sites across the different TSFs at the mine.

  o Performing oedometer tests on push tube samples to determine constrained modulus (M) and secondary consolidation coefficients (α_M).

  o Deriving equations based on a comparison of the laboratory and CPT data, so that M and α_M can be predicted at all depths throughout a CPT profile based on the soil behaviour type index (Ic).

  o Settlements were then calculated for each CPT profile in response to a change in effective stress, factoring in the change in surcharge load due to capping placement and the change in soil properties (M and α_M) as the tailings consolidate.

- Long-term settlements of the capping fill were estimated to be 10 per cent of the fill layer thickness, accounting for:

  o Creep settlements likely to be in the order to ~1 per cent per log cycle of time.

  o Possibility for larger collapse settlements if the base of the fill becomes saturated or experiences a significant increase in the moisture content.

- A duration of 500 years was assumed for predicting creep settlements.

- Upper-probable-bound (UPB) settlements were calculated by summing the estimated fill settlements with 150 per cent of the predicted tailings settlements (primary + creep).

- Settlements were generally only calculated for CPT profiles reaching the base of the tailings. Where profiles were incomplete, some 'full-depth' profiles were manufactured by combining the upper portion of an incomplete profile with the lower portion of a nearby full-depth profile.

Settlement predictions

Settlement predictions calculated at different CPT locations are shown in Table 2.

TABLE 2

Long-term settlement predictions at different CPT locations.

CPT test location	Final landform thickness (m)	Tailings settlement (m)	150% tailings settlement (m)	Fill settlement 10% of cap (m)	UPB total settlement (m)**
CPT19	9.4	2.6	3.9	0.9	4.8
CPT22A	6.9	2.3	3.6	0.7	4.3
CPT23E*	6.8	2.0	3.0	0.7	3.7
CPT25A*	4.7	1.1	1.7	0.5	2.1
PX-06	5.5	0.3	0.5	0.6	1.1
PX-07	5.9	0.7	1.1	0.6	1.7
PX-01	6.8	1.0	1.4	0.7	2.1
PX-05	4.9	0.6	0.9	0.5	1.4

* Tests not completed to full pit depth. Tailings settlement through the interval from base of testing down to pit floor are added from nearby test sites to give total settlements.

** Upper Probable Bound (UPB) for Total Settlement = (Tailings Settlement + 50 per cent) + (Fill Settlement = 10 per cent of fill thickness).

For each CPT location, the UPB settlements were then subtracted from the 'as-designed' landform elevation to determine the 'settled' landform elevation (Figure 14). Using this approach, it was found that the settled elevations at each CPT point remained above the elevation of eastern drain with average gradients from west to east of 1.4 per cent or greater (Figure 15). Although these results suggest the final landform will remain free-draining across the broader area, small and localised low points may still be possible due to differential settlements in isolated areas (Figure 14). Any such areas are expected to become apparent over the next three to five years through ongoing monitoring and will be locally regraded prior to mine lease relinquishment if this is required.

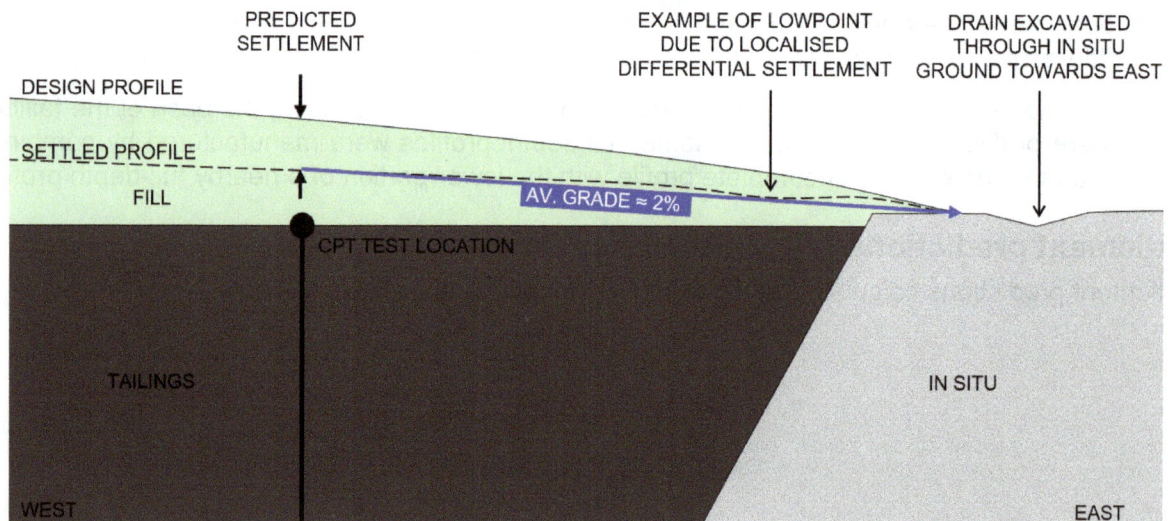

FIG 14 – Schematic to illustrate concept of 'settled profile' and calculation of average drainage gradient.

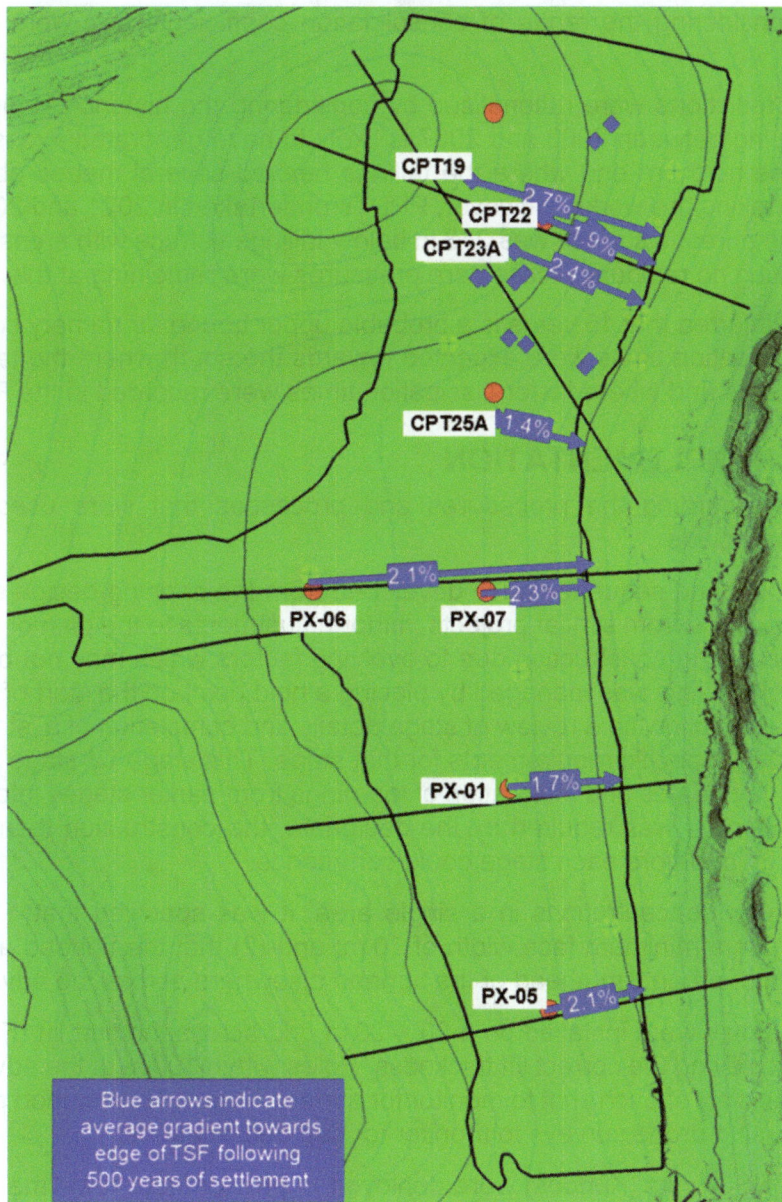

FIG 15 – Average drainage gradients from west-to-east after accounting for UPB settlements.

It is also worth mentioning that this method implicitly assumes that settlements only begin after the landform is constructed up to final elevation. In reality, some settlements may occur during placement of the capping, and where this does occur, the magnitude of the post-construction settlements will be reduced after the landform is constructed up to the design profile. This introduces some degree of conservatism in areas where a substantial portion of consolidation takes place during construction.

Primary consolidation times

Most settlements within the tailings are expected to occur from primary consolidation. Understanding the approximate timing for primary consolidation was important to understand if a significant portion would occur in the next five years so that any potential low points could be identified in time to be corrected before mine lease relinquishment.

The maximum primary consolidation times were expected at the northern end of the facility, where the lowest consolidation coefficients were inferred from pore pressure dissipation testing (PPDTs) around PX-03 and CPT-22. Terzaghi's 1D consolidation theory was used as a first pass to calculate consolidation times assuming one way drainage with a drainage path length of 50 m (approximate pit depth). From this the time taken to achieve 95 per cent consolidation ranged widely from 30 to

200 years when considering the range of consolidation coefficients that were inferred from the PPDTs.

These theoretical predictions were rationalised by considering the historic capping works and the PPDT results. Capping between 2003 and 2007 is likely to have generated excess pore pressures and especially at the northern end where the tailings heaved several metres above their original levels prior to undertaking the works. However, PPDTs undertaken in 2021 and 2023 (at PX-03 and CPT-22) showed the phreatic surface was 2–3 m below tailings surface with pressures ~85 per cent hydrostatic – indicating no residual excess pore pressures were remaining at this time.

On this basis it was inferred that 15 years is a probable upper bound for primary consolidation in the north. Shorter consolidation times were expected towards the south where the tailings have better drainage characteristics and where faster dissipation times were recorded in the PPDTs.

CONSTRUCTION IMPLEMENTATION

Some brief details regarding the procedures and processes that were used to manage the construction process include:

- The capping plan and sequence were developed from the capping design studies. However, based on experience from similar projects, minor adjustments to the plan or stage sequencing were expected during construction due to evolving factors which may not be apparent at the start of the project. This was managed by placing a hold point on the start of each stage which was only released following a review of stage details and completion of a 'stage sign-off sheet' which outlined the specific requirements for that stage. This stage-by-stage approach allowed for interim stage updates and integration of learnings from earlier stages into the work plan for later stages. Sign-off was required by the designers, the construction team, and the project management team before each stage could commence.

- To minimise load concentrations in a single area, it was specified that: (1) the tiphead be advanced across a minimum face width of 70 m; and (2) that the tiphead at any location not be advanced more than 3 m ahead of the tiphead elsewhere across the advancing face.

- White marker poles were installed on a 20 × 20 m grid across the first lift (Figure 16) to guide operators in achieving the correct lift thickness. Poles within 100 m of the advancing face were surveyed at the start of each shift to monitor for surface heave. This method proved more time-efficient than using drone surveys to monitor for cap deformations.

- Five mobile GNSS units (Kurloos) were deployed to provide near real-time monitoring ahead of the advancing face in select areas to supplement the white marker poles. The units were moved regularly as the works progressed.

- Trigger Action Response Plans (TARPs) were developed to identify risk indicators and guide appropriate responses if adverse conditions were detected. Risk indicators included abnormal deformations (ie slumping, heaving, cracking), high rainfall, ponding of water, and design non-conformance.

- Regular design conformance checks were undertaken, and corrective actions were included in the TARP for any significant non-conformance.

- There were no requirements to compact the fill and the allowable material types for capping fill were kept broad (this was accounted for in the design studies by using relatively low strength assumptions). This was to ensure that the operation could remain flexible, given that fill material of varying quality was expected from the adjacent spoil piles.

- A workforce was allocated specifically for the tailings capping project so that workers on the facility would be familiar with the conditions and experienced in the procedural requirements. The mine site operators have an internal short course for tailings risk management and all workers within the tailings team were required to undertake the training.

Photographs which show the operation during the capping are provided in Figures 16 and 17.

FIG 16 – Progress photo showing tiphead advanced over exposed tailings by dozer push across stage 3B. Photo is taken from east looking towards the west. Note the dozer in the top left.

FIG 17 – Progress photo showing tiphead advanced south to north as part of stage 5 being constructed over stage 2. Photo is taken from north looking south. Note the truck in the middle dumping material at the active face. The excavator in the back right is at the fill source location, not over tailings.

POST CONSTRUCTION MONITORING

Pore pressures

A multipoint VWP was installed at CPT-22 at the northern end of the TSF whilst the tailings were still exposed at surface and before capping activities were undertaken. Tips were installed 5, 10, 20 and 40 below surface. Results from the VWP are shown in Figure 18, where:

- Conditions are initially less than hydrostatic, with deeper units having lower total heads.

- Three clear steps on 7 Nov, 5 Dec and 10 Jan correspond with capping lift placement.

- The maximum head changes range from 16 m (CPT22_20) down to 10 m (CPT22_05).

- A small amount of pore pressure dissipation is observed between the placement of lifts.

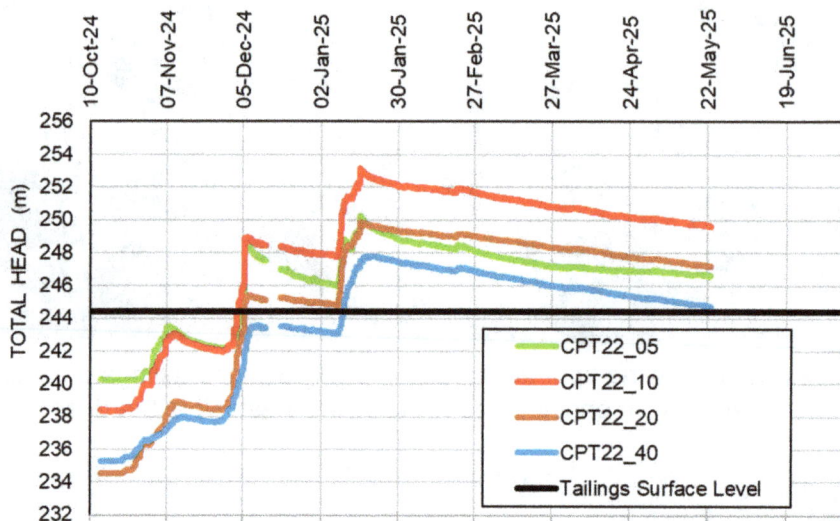

FIG 18 – Results from VWP monitoring at CPT-22 at the northern end of the TSF.

Pore pressures are expected to continue dissipating for several years as the tailings consolidate. It is also plausible that pore pressures may equilibrate to some level higher than the pre-capping levels with rising of the water table due to: (1) compression of the deeper tailings forcing water to migrate upwards; and/or (2) the development of a water table at the base of the fill material, where previously, ponded water on the exposed tailings would have been removed by evaporation.

Settlement plates

Settlement plates were installed at CPT-22 (in the north) and PX-01 (in the south) to monitor tailings settlements (see locations in Figure 5). At CPT-22, the plate was placed directly on the tailings surface, while at PX-01 the plate was seated onto tailings at the base of a 1.5 m deep test pit excavated through the historic cap. Screw-in rods were used to extend the units upward as fill was placed. Surface monitoring pegs were also added to track total landform settlements (fill + tailings) after the landform was construction up to final profile. Current results are compared with predictions in Figure 19, showing:

For the southern plate (PX-01):

- An initial lift was placed in September 2024, causing 0.2 m of settlement in the tailings. Survey data was not collected at regular intervals during this time, although negligible settlement from October onwards indicates primary consolidation was complete in a few weeks.

- A larger second lift was placed at the start of November 2024. The data indicates that primary consolidation was achieved within about two weeks following 0.8 m of settlement. This compares very well with the predicted primary consolidation settlements of 0.78 m.

- Thereafter, it is interpreted that the tailings entered the secondary consolidation or creep phase where very little settlement has occurred.

- Total settlements of 0.3 m have been measured at surface since the start of February 2025. These are predominantly localised in the fill given that little settlement has been measured in the tailings over this period.

For the northern plate (CPT-22):

- Primary consolidation is interpreted as still ongoing and ~25 per cent complete with ~0.5 m of settlement measured in the tailings since December 2024.

- Assuming that primary consolidation is 25 per cent complete after 5 months then a first-pass estimate for the time to achieve 99 per cent primary consolidation can be made using Terzaghi's 1D consolidation theory. Assuming one-way drainage over a 50-m-long drainage path with time factors of 0.05 for 25 per cent consolidation and 1.8 for 99 per cent consolidation, the primary consolidation period is estimated to be around 15 years at CPT-22. Calibration of numerical models against field data is planned in the future to improve confidence in this estimate.

- Around 0.4 m of total settlement has been recorded at surface since February 2025. During this same period the tailings settlements were around 0.2 m which implies that 0.2 m of additional settlement has occurred in the fill.

Regarding a comparison of the settlement data from both locations:

- Primary consolidation has occurred much faster at the southern plate. This because: (1) the tailings in the south are coarser grained with higher consolidation coefficients measured from CPT testing; and (2) tailings in the south are more interlayered (Figure 7) so that coarser grained layers can serve as drainage paths.

- Significantly larger consolidation settlements are predicted in the north, despite both locations having a similar cap thickness (Table 2). This is due to the northern areas having greater tailings depths and higher clay contents with higher compressibility.

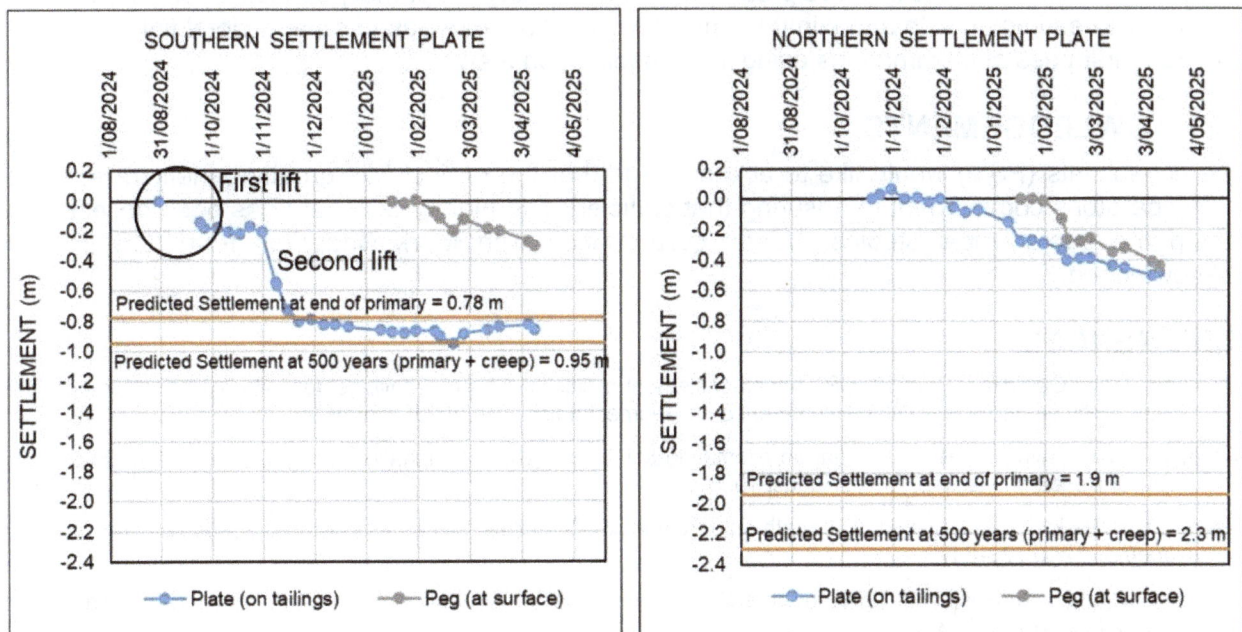

FIG 19 – Results from settlement monitoring at PX-01 in the south, and CPT-22 in the north.

Drone LiDAR surveys

Drone Lidar surveys are specified to be undertaken every 6 months. These will be used in the future to provide broad area monitoring and identify any areas where a more targeted monitoring approach may be warranted through installation of additional instruments.

CONCLUSIONS

Geotechnical assessments have been carried out to assess the performance of the final landform at the Pit X TSF, both during construction and following mine closure. The works have been underpinned by detailed site investigations, spanning several campaigns, which have enabled a detailed understanding of the material properties throughout the facility.

The capping and construction work for Pit X, in many respects, was similar to TSF capping projects that have been implemented elsewhere. Notwithstanding, some of the unique aspects for Pit X included the requirements to use historic aerial photographs to define the TSF extents, requirements to consider and investigate the existing cap that was placed during historic capping activities, and the use of site-specific B_q-N_{kt} relationships to predict the undrained shear strength of the tailings from the CPT data.

A major aspect of the Pit X project (and less common for most other capping projects) was the requirement to design the landform considering long-term settlement risks to ensure the landform would remain free draining after mine closure. Due to limitations with historic information, reasonable assumptions were applied to identify zones with varying potential for adverse impacts from differential settlements. The landform was then designed to minimise fill volumes for cost efficiency, whilst also intending to maintain west-to-east drainage even after differential settlements occurred. The method for predicting tailings settlements from Koosmen and Gerridzen (2023) was applied to estimate long-term settlements at different CPT locations throughout Pit X due to landform construction. The results from this analysis indicated that the landform would remain free draining following mine closure in accordance with the landform design intent.

The landform has now been constructed up to final level and post construction monitoring is underway using VWPs, settlement plates and LiDAR. Current data indicates that the landform is performing as expected. Settlement data has also helped to verify the reliability of the method that was used to predict tailings settlements. Monitoring will continue, and if needed, remediation may be carried out before relinquishment to address any emerging low points. Numerical modelling is also planned in the future to improve predictions for estimating the primary consolidation times, which are currently estimated with a maximum time of around 15 years based on anecdotal field evidence and from first pass approximations using 1D consolidation theory.

ACKNOWLEDGEMENTS

The consultants (PSM) would like to acknowledge that the project has greatly benefited from the mine operator's commitment to ensuring that sufficient and high-quality data has been collected to inform the geotechnical studies. Useful collaboration with Mark Chapman from IGS is also acknowledged.

REFERENCES

Koosmen, K and Gerridzen, P, 2023. CPT-based settlement predictions for coal tailings deposits, in Proceedings of the 14th Australia and New Zealand Conference on Geomechanics (ANZ2023), Cairns 2023.

Kulhawy, F and Mayne, P, 1990. Manual on estimating soil properties for foundation design, Report EL-6800 Electric Power Research Institute, EPRI, August 1990.

Olson, S and Stark, T, 2002. Liquefied strength ratio from liquefaction flow failure case histories, *Canadian Geotechnical Journal*, 39(3):629–647.

Robertson, P and Campanella, R, 1983. Interpretation of cone penetration tests – Part I: Sand, *Canadian Geotechnical Journal*, 20(4):718–733.

Risk could escalate during transition from 'care and maintain' to 'active closure' – owners beware

A Kundu[1]

1. Senior Technical Director and Business Group Leader, GHD, Brisbane Qld 4069.
   Email: anjan.kundu@ghd.com

ABSTRACT

When a Tailings Storage Facility (TSF) closes operation, there is likely a time gap between the stoppage of operations and commencement of active closure. This gap could be between a few months to multiple years. This gap is typically mentioned as a state of 'care and maintain' when the asset may lose adequate surveillance from the operations team. All engineering input and asset maintenance are also likely to be reduced significantly during this stage.

When active closure commences on the asset, the author has noted three possible changes:

1. The asset gets transferred to a different asset owner team (eg closure or rehabilitation team).

2. The asset closure design is undertaken by a different designer (different from the engineer responsible for the asset development design).

3. As preparation of closure construction and profiling, material movements (eg storage of closure material on top of TSF/around the TSF) occurs.

While changes 1 and 2 enhance risk due to knowledge discontinuity, however, a key area the author would like to focus on this paper is item 3 – linkage between storage of closure material inside TSF footprint and integrity of the embankment. This is critical as construction material, when stored within the zone of influence of slip circle of the embankment, could create additional destabilising force and may lead to slope instability and failure of the embankment. This could be more critical to the upstream raised TSFs. The author would also like to present thoughts on how such risk can be alleviated and where Global Industry Standard on Tailings Management (GISTM; Global Tailings Review (GTR), 2020) can assist in attaining safe closure construction without compromising on asset integrity.

INTRODUCTION

There are many tailings dams currently either going through the process of closure or developing their closure strategies. Closure/rehabilitation can be undertaken in many ways by profiling the closure surface. A dry closure and capping (with a convex profile) has frequently been seen as a closure strategy for tailings storage facilities undertaken over several years to address the consolidation of tailings. Other than that, author has also observed plans for beneficial reuse of the tailings surface, such as the installation of solar panels to generate energy in the area close to embankment crest.

Tailings dams are critical structures, and any failure of any component can be disastrous to humans, sensitive vegetation and riparian corridors in the run-out zone, where tailings are flowable. Any tailings retaining embankment is designed to a certain design basis and material parameters derived through geotechnical investigation, material testing and material strength assessment through various interpretations. Tailings retaining structures are typically designed to a minimum required Factor of Safety (FoS) in different loading conditions following specific industry standard (eg Australian Standard – ANCOLD Guidelines (Australian National Committee on Large Dams Incorporated (ANCOLD Inc), 2019); Canadian Standard – Dam Safety Guidelines (Canadian Dam Association (CDA), 2013); International Standard – ICOLD Guidelines: Tailings Dam Safety (International Committee on Large Dams (ICOLD), 2024). During the operational stage of the TSFs, the facilities are also designed based on certain assumptions (eg freeboard, distance of water surface from the embankment) and governance (eg regular inspections, third-party reviews).

However, when the closure/rehabilitation process progresses, the facility is expected to see multiple changes while the facility moves to closure profile.

CHANGES DURING TRANSITION

When an asset moves from operations to closure stage, multiple things can happen, for example:

- Part of design basis may change (eg seismic criteria to MCE – maximum credible earthquake).

- Closure material may be stored/stacked close to the embankment crest (where capping is to be undertaken) through vehicle movement on crest.

- Equipment movement and associated vibration increases near the zone of influence (the zone covered by the slip circle in the upstream area of the embankment footprint with tailings cover).

All the above-mentioned activities are generating loads that may influence the stability of the embankment, which may be easily overlooked as part of closure preparation or preliminary closure activities. It is possible that significant volume of material be stored prior to undertaking any capping/closure design without an engineering assessment.

IMPACT ON STABILITY

The author has tried to understand the impact of such capping construction on the stability FoS values under static loading. A typical example of such impact on an upstream raised TSF is presented in Figure 1. The example assumes all design parameters remain unchanged. Due to confidentiality, the author is unable to present any design details or photographs.

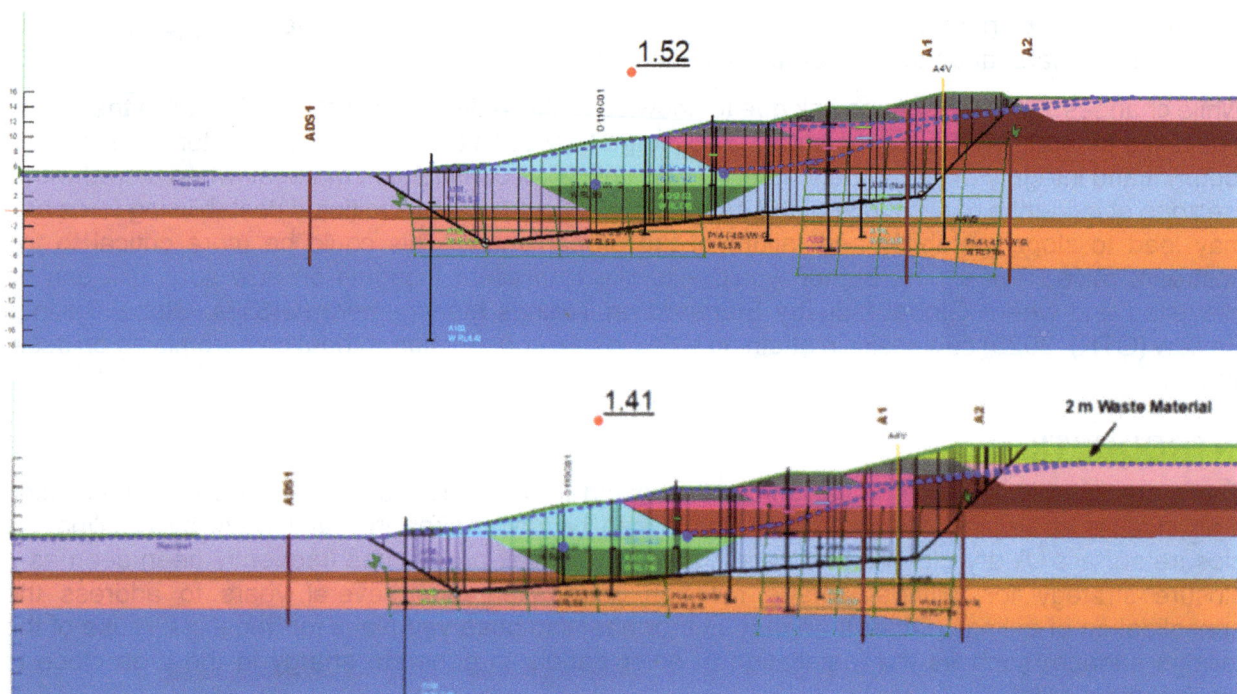

FIG 1 – Impact on Factor of Safety (FoS) values with 2 m of capping material within the zone of influence for an upstream raised TSF (with all parameters remaining unchanged).

The analyses noted a significant difference in FoS values that will likely enhance risk profile and require mitigation measures eg buttresses ahead of closure.

KEY UNDERSTANDING

Considering the sensitivity associated with the TSFs around the globe, any additional loading outside of the design loading (eg in the form of material storage, additional installation for a purpose unrelated to the tailings storage) on the TSF can pose additional risk to TSF stability. The risk, if not appropriately understood, evaluated and addressed, could lead to unwanted incidents.

WHAT CAN OWNERS DO?

An owner organisation owns all the risks associated with the TSF. With the principle of 'any work within a reasonable distance from the TSF footprint will require a careful and risk-assessed approach', the owner organisation is able to adopt a robust process to mitigate such risks. Some suggested measures are:

- Undertake an assessment of the benefit of the construction of a downstream buttress/toe berm prior to undertaking any material storage or vehicle movement upstream of the crest, with a principle that 'the facility is required to be stable at any point in time' to align with 'fundamentally stable landform'.

- Engage a qualified engineer (preferably one with previous understanding of the asset) to undertake an engineering assessment and carefully plan the various construction stages.

- Undertake a surface water assessment to understand if the storage of material is going to impact any storage capacity or have any impact on surface water hydrology. Not only this surface water management needs to be considered with the presence of any stockpiled material, but also as the cover is progressively constructed. Keep all records of change through a 'management of change (MoC)' process with all required approvals.

- Undertake a trial to understand constructability, change in design conditions due to required compaction effort, tailings consolidation characteristics, etc. The trial should also assess the performance of the cover against its design objectives, which may require instrumentation to monitor, eg moisture level, phreatic surface, settlement etc).

- Undertake a risk assessment to capture and mitigate any risk associated with any change outside of the operational design.

- Establish an independent review process to validate the design and execution process.

GISTM RELATED REFERENCES IN RELATION TO ENGINEER OF RECORD

GISTM (GTR, 2020) requires engagement of a competent Engineer of Record (EOR) organisation, development of a design basis and capture any change of the tailings facility through various Requirements as follows. For the owners, who have not adopted the GISTM, the following can be considered as good practice:

- **Requirement 4.8:** The EOR shall prepare a Design Basis Report (DBR) that details the design assumptions and criteria, including operational constraints, and that provides the basis for design of *all phases of the tailings facility life cycle.*

- **Requirement 6.3:** *Prepare a detailed Construction Record Report ('as-built' report) whenever there is a material change to the tailings facility, its infrastructure* on its monitoring system. The EOR and the Responsible Tailings Facility Engineer (RTFE) shall sign this report.

- **Requirement 6.5:** *Implement a formal change management system that triggers the evaluation, review, approval and documentation of changes to design, construction, operation or monitoring during the tailings facility life cycle. The change management system shall also include the requirement for the EOR to prepare a periodic Deviance Accountability Report (DAR), that provides an assessment of the cumulative impact of the changes on the risk level of the as-constructed facility.*

- **Requirement 9.1:** *Engage an engineering firm* with expertise and experience in the design and construction of tailings facilities of comparable complexity *to provide EOR services for operating the tailings facility and for closed facilities* with 'High'. 'Very High' and 'Extreme' Consequence Classification, that are in active closure space.

CONCLUSIONS

Any change in tailings facility footprint (irrespective of its' temporary or permanent status, either in construction or during operational stage) is to be supported by an engineering assessment by competent engineers. This becomes more important during the transition phase when the asset is awaiting closure as the level of governance may reduce. Any intended change is also recommended to be recorded through an appropriate change management process and reviewed thoroughly prior to implementation. Asset owners need to take the leadership in making the process effective.

REFERENCES

Australian National Committee on Large Dams Incorporated (ANCOLD Inc), 2019. Guidelines On Tailings Dams - Planning, Design, Construction, Operation And Closure, revision 1, ANCOLD Inc. Available from: <https://ancold.org.au/product/guidelines-on-tailings-dams-planning-design-construction-operation-and-closure-revision-1-july-2019/>

Canadian Dam Association (CDA), 2013. Dam Safety Guidelines, 2007 edition. Available from: <https://cda.ca/sites/default/uploads/files/CDA_Dam_Safety_Guidelines_TOC-Preface.pdfhttps://cda.ca/publications/cda-guidance-documents/dam-safety-publications>

Global Tailings Review (GTR), 2020. Global Industry Standard on Tailings Management (GISTM) [online], Global Tailings Review. Available from: <https://globaltailingsreview.org/global-industry-standard/>

International Committee on Large Dams (ICOLD), 2024. Tailings Dam Safety, Bulletin No. 194, 1st edition. Available from: <https://routledgetextbooks.com/textbooks/icoldportal/9781032871554.php>

Characterisation and potential application – an investigation on *Haliotis midae* shells for tailings rehabilitation

N A Tshikovhi[1], D T Maiga[2], T T Phadi-Shivambu[3] and T A M Msagati[4]

1. Technician, Council for Mineral Technology (MINTEK), Johannesburg 2125, South Africa. Email: anastaciat@mintek.co.za
2. Scientist, Council for Mineral Technology (MINTEK), Johannesburg 2125, South Africa. Email: deogratiusm@mintek.co.za
3. Engineer, Council for Mineral Technology (MINTEK), Johannesburg 2125, South Africa. Email: terencep@mintek.co.za
4. Professor, University of South Africa, Johannesburg 1710, South Africa. Email: msagatam@unisa.ac.za

ABSTRACT

This study explores the potential use of South African Abalone (*Haliotis midae*) shells for tailings rehabilitation, addressing a critical environmental challenge posed by mining operations. Tailings, composed of crushed rock and processing fluids, often contain hazardous heavy metals that pose significant risks to ecosystems. Traditional rehabilitation techniques rely heavily on costly chemical reagents, highlighting the need for innovative and sustainable alternatives.

A comprehensive characterisation of the *Haliotis midae* shells was conducted through advanced techniques, including Scanning Electron Microscopy (SEM-EDS), X-ray diffraction (XRD), Brunauer-Emmett-Teller (BET) analysis, and Fourier Transform Infrared Spectroscopy (FTIR). The SEM-EDS analysis revealed rough, porous structures conducive to heavy metal adsorption, while XRD confirmed the presence of aragonite and calcite, both of which enhance the shells' stability and buffering capacity in acidic environments. The BET analysis indicated mesoporous characteristics, with an average pore size of 13.68 nm and a surface area of 2.50 m²/g, positioning these shells as competitive with conventional synthetic adsorbents. Furthermore, FTIR spectra confirmed the high calcium carbonate content, underscoring the shells' utility in neutralising acidic leachates and enhancing soil quality. Additionally, the angular, porous structure of *Haliotis midae* shells enhances shear strength, compaction, and moisture retention, thereby improving geotechnical stability. Their chemical composition also supports vegetation growth and minimises surface run-off, thereby aiding in erosion control.

This research contributes to the ongoing dialogue on waste management and sustainable practices in mining, advocating for the repurposing of organic waste materials like *Haliotis midae* shells. The findings suggest that these natural materials offer a viable and sustainable approach for mitigating the environmental impacts of tailings, aligning with principles of the circular economy. By leveraging local resources, this project aims to improve ecological restoration efforts and facilitate the development of effective waste management strategies in South Africa, ultimately supporting broader environmental sustainability initiatives.

INTRODUCTION

Tailings can be defined as crushed rock and processing fluids which remain after the extraction of economic metals and minerals from the mining mills, washers or concentrators (Kossoff *et al*, 2014). Tailing facilities from mining operations pose significant environmental risks due to their potential to release toxic contaminants (Karaca, Cameselle and Reddy, 2018). Traditional rehabilitation methods often involve high costs and extensive use of chemical reagents (Edraki *et al*, 2014; Sun *et al*, 2018). The mining industry is invested in tailings management initiatives because once the contaminants are removed, the land has potential for land development.

Marine shells, primarily composed of calcium carbonate, offer a natural and sustainable alternative for contaminant reduction through ion exchange processes (Cheng *et al*, 2023); Topić Popović *et al*, 2023). Previous studies have highlighted the effectiveness of calcium carbonate in neutralising acidic environments and removing heavy metals (Li, Zhang and Yang, 2020; Zeng *et al*, 2020). This project builds on these findings, aiming to utilise marine shells for potential land reclamation. Cheng *et al*

(2023) shows that with compositions of 95 per cent inorganic and 5 per cent organic phases, seashells are valuable for exploiting both organic and inorganic materials. They should not become waste but should be used as new raw materials for their best potential as a major by-product; their inherent characteristics lay a good foundation for application in different fields.

The alkaline nature of $CaCO_3$ enhances the pH of acidic tailings, fostering conditions conducive to heavy metal adsorption and stabilisation, which is vital for effective environmental restoration (Phillips *et al*, 2016). Research by Atikah *et al* (2015) and Dhami, Reddy and Mukherjee (2013) shows that calcium carbonate can efficiently bind to heavy metals through mechanisms such as adsorption and co-precipitation, making it an effective solution for contaminated water and soil. However, while the use of *Haliotis midae* shells offers significant ecological benefits, challenges remain. The variability in the composition of seashell wastes, particle size, and surface area can affect their effectiveness in adsorption compared to conventional chemical methods, such as lime treatment, which is widely used to neutralise acidic mine drainage. A study by Phillips *et al* (2016) shows that, although lime is typically more immediate in its effects, its use can lead to increased salinity and alter the ecological balance. Moreover, conventional methods often involve significant capital and operational costs that can be prohibitive, especially for small-scale operations. In contrast, the use of natural seashells presents a more sustainable and cost-effective alternative, aligning with the principles of a circular economy by repurposing waste materials that would otherwise contribute to environmental degradation Zhu, Dittrich and Dittrich (2016).

By characterising *Haliotis midae* shells through advanced techniques such as Scanning Electron Microscopy with Energy-Dispersive X-ray Analysis (SEM-EDS), X-Ray Diffraction (XRD), Brunauer-Emmett-Teller (BET) analysis, and Fourier Transform Infrared Spectroscopy (FTIR), this study aims to provide critical insights into the material properties and potential applications of these shells in sustainable tailings rehabilitation.

Additionally, the knowledge generated from this study will enhance waste management practices in South Africa, in line with the Waste Act 59 of 2008 (NEMA, 2008), which aims to reform waste management laws to protect public health and the environment by implementing effective pollution prevention measures and reducing ecological degradation (Waste Act 59; NEMA, 2008). Moreover, the study aligns with the objectives of the United Nations World Summit on Sustainable Development, held on August 26, 2002, in Johannesburg, which underscored the need for governmental initiatives to implement integrated coastal area management plans and develop infrastructure to reduce ocean pollution.

EXPERIMENTAL PROCEDURES

Apparatus and equipment

The apparatus and equipment utilised in this study included polythene plastics for the collection of *Haliotis midae* shells, a 2 L beaker for washing the shells, an oven for drying the washed shells, a crusher for breaking the dried shells into a homogeneous texture, and a sieve for ensuring consistent particle size among the crushed material. For characterisation purposes, the following instruments were employed: X-Ray Diffraction (XRD) for crystallographic analysis, Fourier Transform Infrared Spectroscopy (FTIR) for identifying functional groups and molecular structures, the Brunauer–Emmett–Teller (BET) Surface Area Analyzer for measuring surface area and porosity and Scanning Electron Microscopy with Energy Dispersive X-ray Spectroscopy (SEM-EDS) for morpho structural characterisation and elemental composition assessments. Safety was prioritised using personal protective equipment (PPE), which included lab coats, closed shoes, gloves, and goggles.

Study site and sampling

The purpose of this study, a total of 20 kg of South African abalone *(Haliotis midae)* shells were collected from Jacobsbaai Sea Products (JSP), located in Gonnemanskraal, Western Cape, South Africa (see Figure 1). This site was selected due to its established reputation in the aquaculture industry and its operational capacity to provide large amounts of abalone shell samples. Following the collection, the samples were securely transported to MINTEK, where they underwent thorough sample preparation and subsequent analytical procedures. The transportation was conducted under conditions that minimised contamination risks and ensured the integrity of the samples. Figure 2 displays the *Haliotis midae* shells as sourced from Jacobsbaai Sea Products, providing a visual reference for the type and condition of the shells utilised in this study.

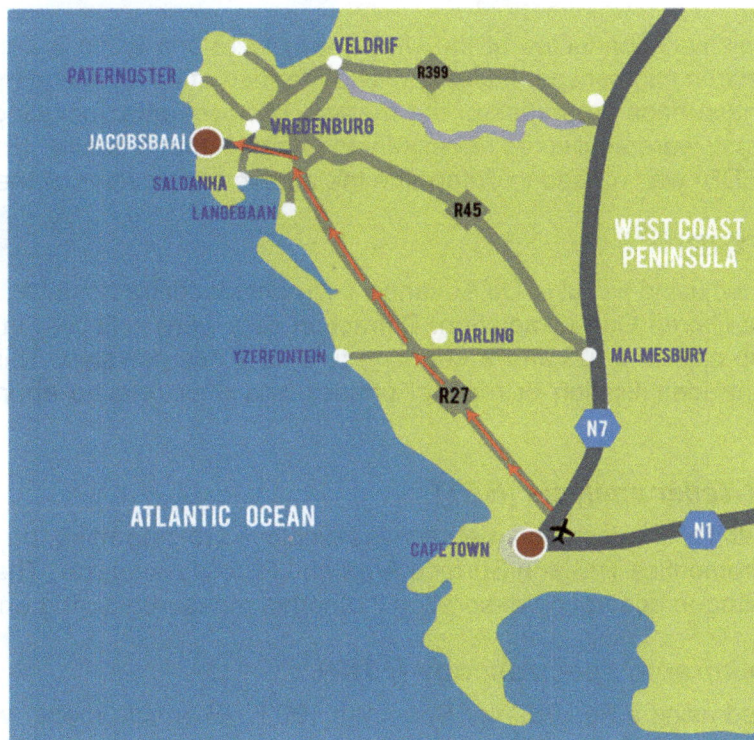

FIG 1 – Google map of a sampling location of the South African Abalone <https://jacobsbaai.co.za/maps/directions-map/>.

FIG 2 – *Haliatos midae* shells in Jacobsbaai Sea Products.

Sample preparation

The sample preparation process for *Haliotis midae* shells involved several critical steps to ensure the integrity and suitability of the specimens for subsequent characterisation. Initially, 1 kg of *Haliotis midae* shells was carefully weighed. The shells were then thoroughly washed three times in a 2 L beaker with deionised water to eliminate contaminants, including salt and organic material adhering to their surfaces. Once cleaned, the shells were dried in an oven at 100°C for 24 hrs. Following the drying process, the shells underwent mechanical crushing using a crusher to achieve a homogeneous texture, which is vital for standardisation in characterisation. Finally, the crushed shell fragments were sieved to ensure uniform particle size, thereby promoting consistency across samples for effective analysis.

Characterisation of *Haliotis midae* shells

Scanning electron microscopy and energy dispersive X-ray spectroscopy (SEM-EDS)

Morphological analysis was conducted at the University of Pretoria using a Zeiss Ultra Plus 196–000106 scanning electron microscope. Imaging was performed at an accelerating voltage of 2 kV to assess the shape and surface properties of the material. Adjustments, including enhancements in brightness and contrast, were applied as necessary to improve image quality. Energy Dispersive X-Ray Spectroscopy (EDS) was utilised to determine the atomic composition of the shells.

X-ray diffraction (XRD)

Analysis was performed using a Bruker D8 Advance X-ray powder diffractometer, where the samples were subjected to Fe-filtered Co-Kα radiation. Diffraction data were collected in the 2θ range from 4° to 80°, with a step size of 0.02° and a counting time of 1 sec per step. This semi-quantitative analysis facilitated the identification of mineral phases and their relative abundances within the samples.

Brunauer-Emmett-Teller analysis (BET)

The Brunauer-Emmett-Teller adsorption method was employed to determine the surface area of the shells, utilising a Micromeritics TriStar II Surface Area and Porosity Analyzer. The texture properties were derived from nitrogen adsorption/desorption isotherms measured during analysis.

Fourier transform infrared spectroscopy (FTIR)

Spectra were obtained using a PerkinElmer Spectrum 100 FTIR spectrometer, with measurements taken within the range of 650 to 4000 cm^{-1}. This method was instrumental in investigating the functional groups present in the *Haliotis midae* shells.

RESULTS AND DISCUSSION

Scanning electron microscopy combined with energy-dispersive X-ray analysis (SEM-EDS)

SEM-EDS analysis was used to investigate shells' surface characteristics and chemical composition. High-resolution SEM images revealed rough, angular structures with layered and fractured surfaces, as observed in Figure 3a. These morphological features suggest the presence of micropores and irregular edges, indicative of a high surface area that may facilitate the adsorption of heavy metals, thereby reducing their mobility in tailings and enhancing chemical interactions during remediation, as revealed by Letshwenyo and Mokokwe (2021) and Gong *et al* (2011). Additionally, the fragmented structure, as observed in Figure 3b suggests ease of integration into tailings, potentially improving soil structure and aeration, which is supported by the findings of Daubert and Brennan (2007) and Liu *et al* (2013).

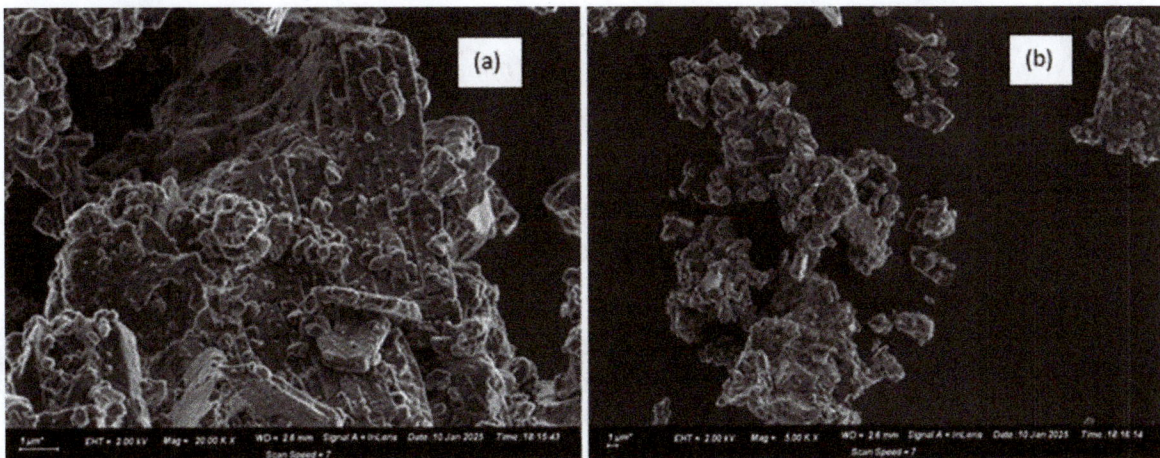

FIG 3 – SEM images of *Haliotis midae.*

The EDS spectrum in Figure 4 confirmed the chemical composition of the shells, revealing significant elemental contents of calcium (Ca) and oxygen (O). Specifically, the shell composition was approximately 50 per cent Ca and 50 per cent O, which supports the predominant presence of calcium carbonate ($CaCO_3$) in the shell structure. This high calcium content suggests that *Haliotis midae* shells could play a beneficial role in neutralising acidic environments and facilitating the removal of heavy metals, as demonstrated by Kızılkaya and Yıldız (2024).

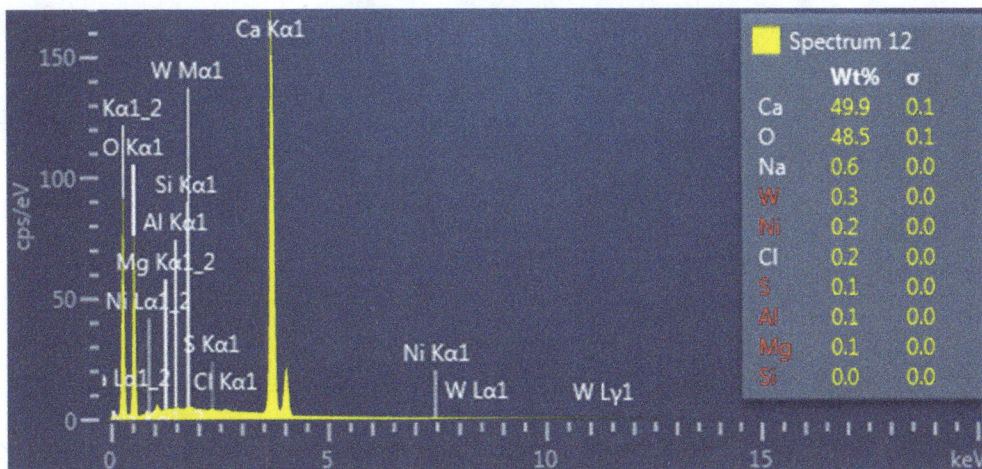

FIG 4 – EDS spectrum of *Haliotis midae.*

Compared to other materials used in environmental remediation, previous studies indicate that the high surface area and porous structure of these shells make them competitive with conventional adsorbents. For instance, Gong *et al* (2011) reported the efficacy of nanoparticles for heavy metal ion adsorption, while Liu *et al* (2013) highlighted the favourable characteristics of metal oxides based on their nanoscale structures and high BET surface areas for effective ion exchange and adsorption. Similarly, Masukume, Onyango and Maree (2014) discussed the potential of composite materials in heavy metal adsorption processes, underscoring the comparable potential of biogenic materials, such as *Haliotis midae* shells, in environmental applications.

X-ray diffraction (XRD)

The X-ray diffraction (XRD) method was employed to examine the crystal structures of *Haliotis midae* shells, providing insights into their mineral composition. The analysis revealed two primary mineral phases present in the shells, aragonite, and calcite. The relative abundance of these phases was observed, with aragonite being the predominant phase and calcite classified as major (Table 1).

TABLE 1

Compound composition of *Haliotis midae* sample by XRD.

Compound name	Formula	Relative abundance
Calcite	$CaCO_3$	Major
Aragonite	$CaCO_3$	Predominant

Predominant (>50 mass%), major (30–50%), intermediate (15–30%), minor (5–15%) and trace (<5%).

The XRD spectrum of the *Haliotis midae* shell sample demonstrates distinct peaks corresponding to the aragonite and calcite phases (Figure 5). The most intense peaks are associated with aragonite, indicating its predominance in the shell's mineralogical composition. In contrast, calcite appears as a major phase within the sample. The differing peak positions and intensities between aragonite and calcite can be attributed to their unique crystal structures; aragonite typically exhibits a more intense diffraction pattern owing to its higher crystallinity, reflecting a greater degree of structural order (Kızılkaya and Yıldız, 2024).

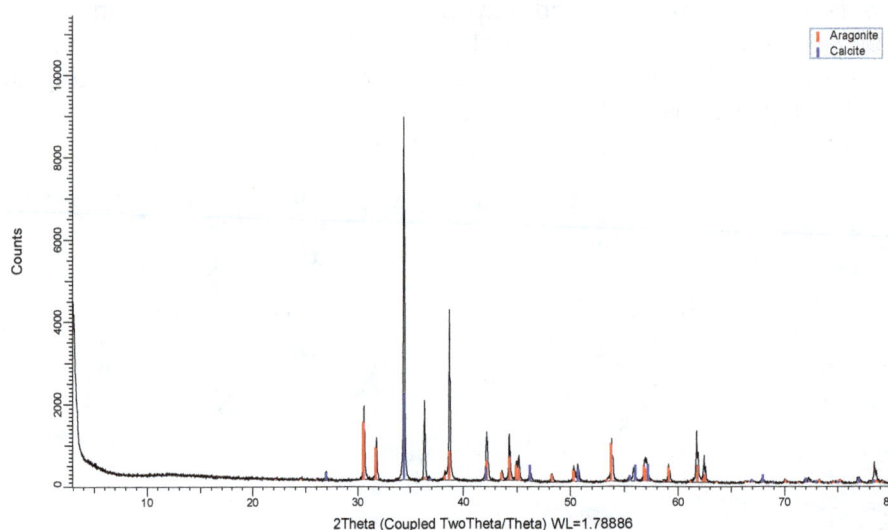

FIG 5 – XRD spectrum of calcite, aragonite in *Haliotis midae*.

The dual-phase composition of these biominerals is significant for environmental applications as both aragonite and calcite contribute to the overall stability of the material. This stability is essential, as it may enhance the shells' ability to buffer acidic environments and facilitate the adsorption of heavy metals, an important consideration for tailings rehabilitation (Gong *et al*, 2011; Liu *et al*, 2013). Previous studies have highlighted the utility of natural materials, such as biogenic carbonates, for remediation purposes due to their structural properties and chemical compositions (Letshwenyo and Mokokwe, 2021). The biomineral composition offers unique advantages when comparing *Haliotis midae* shells to other conventional adsorbents used for environmental remediation, such as activated carbon or synthetic zeolites. For instance, conventional adsorbents like activated carbon, although effective for capturing a range of pollutants, often arise from non-renewable sources and can involve significant processing energy (Gisi *et al*, 2016; Guo *et al*, 2024; Liu *et al*, 2013). In contrast, natural adsorbents, such as those derived from molluscan shells, present a sustainable alternative.

Notably, the use of *Haliotis midae* shells could alleviate some typical drawbacks associated with synthetic materials, such as environmental dependency and cost, while also leveraging the high surface area and porous nature of the shells to enhance adsorption efficiency and stability in diverse environmental conditions (Shikuku *et al*, 2015). Therefore, the XRD analysis underscores the complex biomineralisation process within *Haliotis midae* shells and highlights their potential role in environmental remediation efforts. The unique properties of these shells, combined with their inherent abundance and sustainability as a natural resource, position them as promising candidates for applications in tailings rehabilitation and heavy metal remediation.

Brunauer-Emmett-Teller (BET) analysis

The Brunauer-Emmett-Teller (BET) method was utilised to determine the surface area and porosity properties of *Haliotis midae* shells. The results indicate a Type IV isotherm with a hysteresis loop, as depicted in Figure 6, which is consistent with the IUPAC classification of adsorption isotherms (Grana, 1955). This hysteresis loop suggests that *Haliotis midae* shells exhibit mesoporous characteristics, a finding that aligns with the observations of Leng *et al* (2021). The interconnected pores implied by the hysteresis loop play a critical role in adsorbing heavy metals and other pollutants from mining water, as previously reported by Babel and Kurniawan (2003) and Park *et al* (2011).

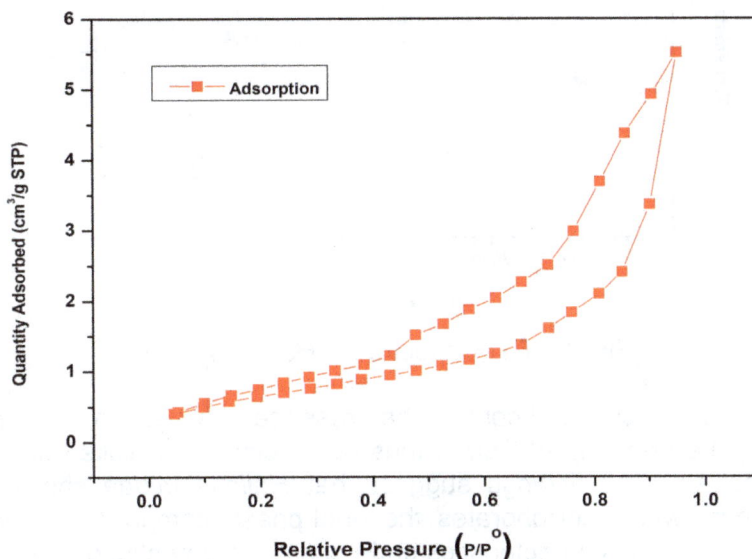

FIG 6 – Adsorption isotherm of *Haliotis midae*.

The BET analysis reveals that the average pore size of *Haliotis midae* shells is approximately 13.68 nm, which is consistent with findings by Barrera-Zapata (2017), who reported that the average pore diameter of similar samples ranged from 10 to 50 nm, classifying them as mesoporous. Additionally, Figure 6 illustrates that the quantity of gas adsorbed steadily increases at lower pressures, with a significant rise observed near a relative pressure of approximately 0.8–1.0. According to Chuyingsakuntip and Tangsathitkulchai (2013), this behaviour reflects a moderate to high surface area, which could potentially enhance the adsorption of contaminants in tailings.

The BET surface area of *Haliotis midae* shells was determined to be 2.50 m^2/g. Although this value is lower compared to many synthetic adsorbents, such as activated carbon, the unique mesoporous structure and chemical composition of these seashells may offer competitive advantages in tailings rehabilitation applications. The natural abundance, low cost, and sustainability of *Haliotis midae* shells position them as a promising alternative to conventional adsorbents (Gong *et al*, 2011; Liu *et al*, 2013). Moreover, the porosity and surface properties of *Haliotis midae* shells allow for increased binding sites for contaminants. Similar results have been demonstrated with other natural materials; for instance, Barus *et al* (2021) observed that marine biomaterials could effectively adsorb heavy metals compared to conventional materials. The porous structure of *Haliotis midae* shells potentially enhances their adsorptive capacity, as indicated by the effective adsorption of pollutants reported by Fan *et al* (2021).

Fourier transform infrared spectroscopy

The Fourier Transform Infrared Spectroscopy (FTIR) method was employed to elucidate the bond structures of molecules within *Haliotis midae* shells, providing key insights into their chemical composition and potential applications. Figure 7 presents the FTIR spectra for *Haliotis midae*, revealing a prominent peak at 1457 cm^{-1}, which corresponds to the asymmetric stretching vibration of the carbonate ion (CO_3^{2-}). Additionally, the peak observed at 861 cm^{-1} is associated with the in-plane bending vibrations of the carbonate group, while a peak at 706 cm^{-1} corresponds to out-of-plane bending vibrations of the carbonate group. These findings align with those of Kızılkaya and

Yıldız (2024), who also identified these peaks as indicative of calcium carbonate (CaCO$_3$), the primary component of seashells.

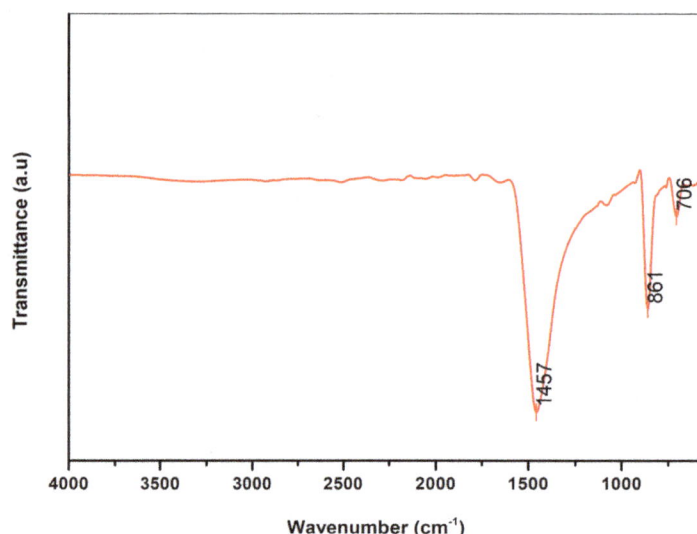

FIG 7 – FTIR spectra for *Haliotis midae*.

The strong peaks at 706 and 861 cm^{-1} confirm the presence of aragonite, while the calcite phase is primarily indicated by the peak at 1457 cm^{-1}, thus reinforcing the results obtained from the X-ray diffraction (XRD) analysis. The findings suggest that *Haliotis midae* shells are predominantly composed of aragonite, which corroborates the dual-phase composition identified in the XRD analyses (see Figure 5). These characteristics underscore the potential of *Haliotis midae* seashells as a sustainable material for tailings rehabilitation and environmental remediation. The rich calcium carbonate content in the shells enhances their capacity to buffer acidic environments, offering an effective means of mitigating the adverse effects of metal leachates in mining operations (Amenabar *et al*, 2013; Masukume, Onyango and Maree, 2014).

Furthermore, a recent study highlighted the adsorption capacity of shell-based adsorbents for heavy metals in mining effluents. Fernández Pérez, Ayala Espina and Fernández González (2022) demonstrated significant adsorption efficiencies for various metal ions, indicating that seashells could serve effectively for treating contaminated waters. This reinforces the argument for utilising *Haliotis midae* shells, given their inherent structural properties and chemical advantages over traditional materials.

In addition to their chemical buffering and heavy metal adsorption capabilities, *Haliotis midae* shells exhibit geotechnical and erosional stability properties that enhance their suitability for mine tailings rehabilitation. Their angular, porous structure can improve shear strength, compaction, and moisture retention, contributing to the mechanical stability of amended tailings. Compared to traditional stabilisers, these shells may also offer reinforcement benefits while remaining eco-friendly (Dejong *et al*, 2014). Furthermore, the high calcium carbonate content promotes soil cohesion and supports vegetation establishment, which aids in erosion control. While the specific anti-erosive effect of calcium carbonate in soil amendments requires further examination, the presence of calcium carbonate is pertinent in studies related to erosion control and soil stability (Scandiffio *et al*, 2018). Their porous texture also enhances water infiltration, reducing surface run-off and particle detachment, making them effective in mitigating both wind and water erosion (Wang *et al*, 2021).

CONCLUSION

In conclusion, this study presents significant findings regarding the characterisation and potential application of South African *Haliotis midae* shells for tailings rehabilitation. The results from advanced analytical techniques such as Scanning Electron Microscopy with Energy-Dispersive X-ray Spectroscopy (SEM-EDS), X-Ray Diffraction (XRD), Brunauer-Emmett-Teller (BET) analysis, and Fourier Transform Infrared Spectroscopy (FTIR) provide critical insights into the physical and chemical properties of these shells. The predominance of aragonite and calcite mineral phases, along with the characteristic mesoporous structure, highlight their potential for heavy metal adsorption and acid neutralisation, addressing the urgent need for sustainable solutions in tailings management.

Our findings suggest that *Haliotis midae* shells possess the necessary alkaline properties to enhance pH levels in acidic tailings, thus facilitating the adsorption of heavy metals and other contaminants. This aligns well with the growing body of literature that emphasises the effectiveness of calcium carbonate in environmental remediation, as demonstrated in various studies focusing on biogenic materials and their applications in pollutant absorption and neutralisation processes (Tshilate, Ishengoma and Rhode, 2024; Vosloo and Vosloo, 2017). The inherent sustainability and cost-effectiveness of *Haliotis midae* shells, derived from aquaculture waste, present a compelling alternative to synthetic adsorbents or traditional chemical methodologies, which can be expensive and ecologically disruptive.

In addition to their chemical effectiveness, the geotechnical and erosional stability properties of *Haliotis midae* shells further enhance their applicability in tailings rehabilitation. Their angular, porous structure improves shear strength, compaction, and water retention, while the high calcium carbonate content promotes soil cohesion and supports vegetation growth. These characteristics help reduce surface run-off and particle detachment, making the shells effective in mitigating both wind and water erosion.

Moreover, the research underscores the importance of utilising agricultural and marine waste materials, as mandated by legislative frameworks such as the Waste Act 59 of 2008 (NEMA, 2008) in South Africa, which advocates for improved waste management practices and the promotion of circular economy principles. Ultimately, the characterisations performed in this study not only benefit local environmental remediation efforts but also contribute to a broader understanding of how renewable resources can be harnessed to reclaim and restore contaminated landscapes.

The insights gained from this investigation pave the way for future studies aimed at optimising the application of *Haliotis midae* shells as an eco-friendly approach to mitigate the environmental impacts of mining activities. Continued research in this area will further elucidate the mechanisms underlying the adsorption processes and enhance the efficiency of utilising these shells in large-scale remediation projects, ultimately supporting the principles of sustainable development and responsible resource management.

ACKNOWLEDGEMENTS

The authors express their sincere gratitude to the Council for Mineral Technology (MINTEK) for their generous funding, which made this research possible. Special thanks are extended to Jacobsbaai Sea Products (Pty) Ltd for supplying the South African Abalone (*Haliotis midae*) shells used in our study. We would also like to acknowledge the organisers of the LOM|MWT 2025 Conference for their diligent efforts in hosting this event, providing a valuable platform for dialogue and knowledge exchange among researchers and industry professionals.

REFERENCES

Amenabar, I, Poly, S, Nuansing, W, Hubrich, E H, Govyadinov, A A, Huth, F, Krutokhvostov, R, Zhang, L, Knez, M, Heberle, J, Bittner, A M and Hillenbrand, R, 2013. Protein complexes by infrared nanospectroscopy. https://doi.org/10.1038/ncomms3890

Atikah, N, Latiffi, A, Maya, R, Radin, S, Apandi, N M and Mohd, H, 2015. Application of Phycoremediation using Microalgae *Scenedesmus sp.* as Wastewater Treatment in Removal of Heavy Metals from Food Stall Wastewater, pp 1168–1172. https://doi.org/10.4028/www.scientific.net/amm.773-774.1168

Babel, S and Kurniawan, T A, 2003. Low-cost adsorbents for heavy metals uptake from contaminated water: A review, *Journal of Hazardous Materials*, 97(1–3):219–243. https://doi.org/10.1016/S0304-3894(02)00263-7

Barrera-Zapata, R, 2017. Morphological and physicochemical characterization of biochar produced by gasification of selected forestry species (Caracterización morfológica y fisico-química de biocarbones producidos), *Revista Facultad de Ingeniería*, 26(46):123–130. https://doi.org/10.19053/01211129.v26.n46.2017.7324

Barus, B S, Chen, K, Cai, M, Li, R, Chen, H, Li, C, Wang, J and Cheng, S Y, 2021. Heavy Metal Adsorption and Release on Polystyrene Particles at Various Salinities, *Frontiers in Marine Science*, 8:671802. https://doi.org/10.3389/fmars.2021.671802

Cheng, M, Liu, M, Chang, L, Liu, Q, Wang, C, Hu, L, Zhang, Z, Ding, W, Chen, L, Guo, S, Qi, Z, Pan, P and Chen, J, 2023. Overview of structure, function and integrated utilization of marine shell, *Science of the Total Environment*, 870:161950. https://doi.org/10.1016/j.scitotenv.2023.161950

Chuyingsakuntip, S and Tangsathitkulchai, C, 2013. Adsorption of Natural Aluminium Dye Complex from Silk-Dyeing Effluent Using Eucalyptus Wood Activated Carbon, *American Journal of Analytical Chemistry*, 04(08):379–386. https://doi.org/10.4236/ajac.2013.48048

Daubert, L N and Brennan, R A, 2007. Passive remediation of acid mine drainage using crab shell chitin, *Environmental Engineering Science*, 24(10):1475–1480. https://doi.org/10.1089/ees.2006.0199

Dejong, J T, Soga, K, Kavazanjian, E, Burns, S, Van Paassen, L A, Al Qabany, A, Aydilek, A, Bang, S S, Burbank, M, Caslake, L F and Chen, C Y, 2014. Biogeochemical processes and geotechnical applications: progress, opportunities and challenges, in *Bio- and Chemo-Mechanical Processes in Geotechnical Engineering: Géotechnique Symposium*, pp 143–157 (Ice Publishing).

Dhami, N K, Reddy, M S and Mukherjee, A, 2013. Biomineralization of calcium carbonates and their engineered applications : a review, *Frontiers in Microbiology*, 4(Oct):1–13. https://doi.org/10.3389/fmicb.2013.00314

Edraki, M, Baumgartl, T, Manlapig, E, Bradshaw, D, Franks, D M and Moran, C J, 2014. Designing mine tailings for better environmental, social and economic outcomes: a review of alternative approaches, *Journal of Cleaner Production*, 84:411–420.

Fan, T, Zhao, J, Chen, Y, Wang, M, Wang, X, Wang, S, Chen, X, Lu, A and Zha, S, 2021. Coexistence and Adsorption Properties of Heavy Metals by Polypropylene Microplastics, *Adsorption Science and Technology*, 4938749. https://doi.org/10.1155/2021/4938749

Fernández Pérez, B, Ayala Espina, J and Fernández González, M d L Á, 2022. Adsorption of Heavy Metals Ions from Mining Metallurgical Tailings Leachate Using a Shell-Based Adsorbent: Characterization, Kinetics and Isotherm Studies, *Materials*, 15(15):5315. https://doi.org/10.3390/ma15155315

Gisi, S, De, Lofrano, G, Grassi, M and Notarnicola, M, 2016. Characteristics and adsorption capacities of low-cost sorbents for wastewater treatment : A review, *SUSMAT*, 9:10–40. https://doi.org/10.1016/j.susmat.2016.06.002

Gong, J, Liu, T, Wang, X, Hu, X and Zhang, L, 2011. Efficient Removal of Heavy Metal Ions from Aqueous Systems with the Assembly of Anisotropic Layered Double Hydroxide Nanocrystals @ Carbon Nanosphere, *Environmental Science and Technology*, 45(14):6181–6187.

Grana, A, 1955. Some aspects of the serology of congenital syphilis [in Italian: Alcuni aspetti della sierologia della sifilide congenita], in Proceedings of the Italian Society of Dermatology and Syphilography and of the Interprovincial Sections [Atti Della Società Italiana Di Dermatologia e Sifilografia e Delle Sezioni Interprovinciali], Italian Society of Dermatology and Syphilography [Società Italiana Di Dermatologia e Sifilografia], 6(4):710–713.

Guo, C, Jiang, X, Guo, X and Ou, L, 2024. An Evolutionary Review of Hemoperfusion Adsorbents: Materials, Preparation, Functionalization and Outlook, *ACS Biomaterials Science and Engineering*, 10(6):3599–3611. https://doi.org/10.1021/acsbiomaterials.4c00259

Karaca, O, Cameselle, C and Reddy, K R, 2018. Mine tailing disposal sites: contamination problems, remedial options and phytocaps for sustainable remediation, *Reviews in Environmental Science and Bio/Technology*, 17:205–228.

Kızılkaya, B, Yıldız, H and Vural, P, 2024. Shell composition analysis of European flat oyster (Ostrea edulis, Linnaeus 1758) from Marmara Sea, Türkiye : Insights into Chemical Properties, *Marine Science and Technology Bulletin*, 13(2):142–150. https://doi.org/10.33714/masteb.1493896

Kossoff, D, Dubbin, W E, Alfredsson, M, Edwards, S J, Macklin, M G and Hudson-Edwards, K A, 2014. Mine tailings dams: Characteristics, failure, environmental impacts and remediation, *Applied Geochemistry*, 51:229–245. https://doi.org/10.1016/j.apgeochem.2014.09.010

Leng, L, Xiong, Q, Yang, L, Li, H, Zhou, Y, Zhang, W, Jiang, S, Li, H and Huang, H, 2021. An overview on engineering the surface area and porosity of biochar, *Science of the Total Environment*, 763:144204. https://doi.org/10.1016/j.scitotenv.2020.144204

Letshwenyo, M W and Mokokwe, G, 2021. Phosphorus and sulphates removal from wastewater using copper smelter slag washed with acid, *SN Applied Sciences*, 3(12). https://doi.org/10.1007/s42452-021-04843-7

Li, X, Zhang, Q and Yang, B, 2020. Co-precipitation with CaCO3 to remove heavy metals and significantly reduce the moisture content of filter residue, *Chemosphere*, 239:124660.

Liu, L, Yang, L, Liang, H, Cong, H, Jiang, J and Yu, S, 2013. Bio-Inspired Fabrication of Hierarchical FeOOH Nanostructure Array Films at the Air–Water Interface, Their Hydrophobicity and Application for Water Treatment, *ACS Nano*, 7(2):1368–1378.

Masukume, M, Onyango, M S and Maree, J P, 2014. Sea shell derived adsorbent and its potential for treating acid mine drainage, *International Journal of Mineral Processing*, 133:52–59. https://doi.org/10.1016/j.minpro.2014.09.005

National Environmental Management (NEMA), 2008. Waste Act 59 of 2008, Government of South Africa.

Park, J H, Lamb, D, Paneerselvam, P, Choppala, G, Bolan, N and Chung, J W, 2011. Role of organic amendments on enhanced bioremediation of heavy metal(loid) contaminated soils, *Journal of Hazardous Materials*, 185(2–3):549–574. https://doi.org/10.1016/j.jhazmat.2010.09.082

Phillips, A A J, Cunningham, A B, Gerlach, R, Hiebert, R, Hwang, C, Lomans, B P, Westrich, J, Mantilla, C, Kirksey, J, Esposito, R and Spangler, L, 2016. Fracture Sealing with Microbially-Induced Calcium Carbonate Precipitation: A Field Study, *Environmental Science and Technology*, 50(7):4111–4117.

Scandiffio, P, Mantilla, T, Amaral, F, França, F, Basting, R and Turssi, C, 2018. Anti-erosive effect of calcium carbonate suspensions, *Journal of Clinical and Experimental Dentistry*, 10(8):e776.

Shikuku, V O, Kowenje, C O, Donato, F F, Zanella, R and Prestes, O D, 2015. A Comparison of Adsorption Equilibrium, Kinetics and Thermodynamics of Aqueous Phase Clomazone between Faujasite X and a Natural Zeolite from Kenya, *South African Journal of Chemistry*, 68(1):245–252.

Sun, W, Ji, B, Khoso, S A, Tang, H, Liu, R, Wang, L and Hu, Y, 2018. An extensive review on restoration technologies for mining tailings, *Environmental Science and Pollution Research*, 25:33911–33925.

Topić Popović, N, Lorencin, V, Strunjak-Perović, I and Čož-Rakovac, R, 2023. Shell waste management and utilization: Mitigating organic pollution and enhancing sustainability, *Applied Sciences*, 13(1):623.

Tshilate, T S, Ishengoma, E and Rhode, C, 2024. Construction of a high-density linkage map and QTL detection for growth traits in South African abalone (*Haliotis midae*), *Animal Genetics*, 55(5):744–760. https://doi.org/10.1111/age.13462

Vosloo, D and Vosloo, A, 2017. Short Postspawning Recovery Time Affects DNA Integrity and Fertilization Success of South African Abalone (*Haliotis midae*) Oocytes, *Journal of Shellfish Research*, 36(1):169–174. https://doi.org/10.2983/035.036.0117

Wang, Z, Zhang, N, Jin, Y, Li, Q and Xu, J, 2021. Application of microbially induced calcium carbonate precipitation (MICP) in sand embankments for scouring/erosion control, *Marine Georesources and Geotechnology*, 39(12):1459–1471.

Zeng, C, Hu, H, Feng, X, Wang, K and Zhang, Q, 2020. Activating CaCO3 to enhance lead removal from lead-zinc solution to serve as green technology for the purification of mine tailings, *Chemosphere*, 249:126227.

Zhu, T, Dittrich, M and Dittrich, M, 2016. Carbonate Precipitation through Microbial Activities in Natural environment and Their Potential in Biotechnology: A Review, 4(January):1–21. https://doi.org/10.3389/fbioe.2016.00004

Monitoring and modelling of hydrological and geochemical processes in large *in situ* waste rock leaching columns

C Zhang[1], Z Zhao[2], C Ptolemy[3], S Quintero[4], L Tan[5] and D Williams[6]

1. Senior Research Fellow, Geotechnical Engineering Centre, University of Queensland, St Lucia Qld 4072. Email: chenming.zhang@uq.edu.au
2. Hydrological Engineer, ATC Williams, Newmarket Qld 4051. Email: zichengz@atcwilliams.com.au
3. Senior Environmental Officer, Grange Resources, Burnie Tas 7320. Email: carl.ptolemy@grangeresources.com.au
4. Senior Laboratory Manager, University of Queensland, St Lucia Qld 4072. Email: s.quintero@uq.edu.au
5. Senior Research Assistant, Geotechnical Engineering Centre, School of Civil Engineering, University of Queensland, St Lucia Qld 4072. Email: l.tan@uq.edu.au
6. Professor, Geotechnical Engineering Centre, School of Civil Engineering, University of Queensland, St Lucia Qld 4072. Email: d.williams@uq.edu.au

ABSTRACT

A comprehensive *in situ* monitoring and geochemical modelling was conducted to investigate the hydrological and geochemical behaviour of waste rock materials, with a focus on addressing environmental challenges posed by potentially acid-forming (PAF) waste. The research evaluated alternative waste co-disposal strategies and the effectiveness of various cover materials. Six large, vertically aligned columns (with a diameter of 1 m and height of 4 m), simulating different configurations of neutral and acid-forming waste types, were instrumented to monitor parameters such as moisture content, suction, oxygen levels, and leachate chemistry over five years. Both Monitoring and geochemical modelling results demonstrated that compacted waste significantly improved moisture retention and reduced oxygen ingress, thereby limiting pyrite oxidation and maintaining near-neutral pH levels in most configurations. Geochemical analysis revealed that calcite dissolution and muscovite weathering were critical in buffering acidity and sustaining alkalinity. Long-term monitoring indicated equilibrium in moisture dynamics and stable leachate hydrochemistry, with pH levels between 6 and 9. The study highlights the importance of material selection, compaction, and cover design in mitigating acid mine drainage, offering valuable implications for sustainable waste management in mining operations.

INTRODUCTION

Managing acid mine drainage (AMD) is an ongoing challenge within the mining industry, particularly due to its potential long-term environmental impacts on soil, water resources, and ecosystems. The generation of acid mine drainage occurs primarily when sulfide-bearing minerals, such as pyrite, are exposed to oxygen and water, resulting in oxidation processes that produce sulfuric acid. If unmanaged, AMD can significantly degrade water quality, affect biodiversity, and lead to costly remediation efforts. Consequently, understanding and controlling the geochemical and hydrological processes in waste rock storage facilities is essential for sustainable mining practices.

This study focuses on the Savage River Mine in Tasmania, Australia, operated by Grange Resources. The mine previously proposed a Pit Rim Crushing and Conveying (PRCC) system designed to handle future waste rock, aiming to integrate potentially acid-forming (D-Type waste) and non-acid-forming (NAF) waste types (A-Type and B-Type) within a structured co-disposal scheme. Although the initial proposal for PRCC was later reconsidered due to economic constraints, the underlying principles and interest in innovative waste management approaches continued to inform ongoing research at the mine.

To evaluate the long-term behaviour and effectiveness of alternative waste management and co-disposal strategies, Grange Resources, in collaboration with the University of Queensland, initiated a comprehensive monitoring and modelling study using large-scale column leaching tests. These tests aimed to replicate realistic field conditions, offering insights into the complex interactions between hydrological flow, oxygen ingress, and geochemical reactions over an extended period.

METHODOLOGY

Experimental set-up

As shown in Figure 1, six large concrete columns, each with a diameter and height of 1 m per segment and stacked to reach a total height of 4 m, were constructed between December 2017 and February 2018. These columns were designed to represent different waste rock scenarios typically encountered at the Savage River Mine:

- Column 1: Loosely placed, coarse-grained A-Type waste (NAF).

- Column 2: Loosely placed, coarse-grained B-Type waste (NAF with high muscovite content).

- Column 3: Loosely placed, coarse-grained D-Type waste (PAF).

- Column 4: 3 m of loosely placed D-Type waste capped with 1 m of compacted A-Type waste.

- Column 5: 3 m of loosely placed D-Type waste capped with 1 m of compacted B-Type waste.

- Column 6: Loosely placed mixture of A-Type, B-Type, and D-Type wastes.

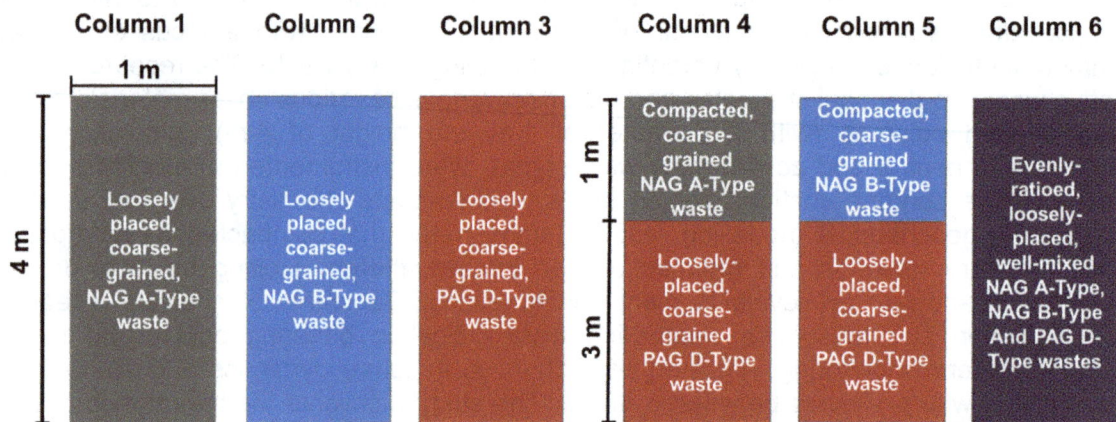

FIG 1 – Schematic representation of column design illustrating waste configurations and photograph showing the actual set-up of the large-scale leaching columns (column 1 to 6 from left to right).

Each column was constructed on a levelled gravel base and sealed internally to minimise air infiltration. A U-bend system at the base ensured no atmospheric venting, mimicking conditions of restricted oxygen ingress in field scenarios. Leachate was collected at the base and sampled quarterly for water quality analyses.

Instrumentation and monitoring

Columns were instrumented with sensors positioned at various depths to continuously monitor moisture content, suction, oxygen concentration, and temperature profiles, as illustrated in Figure 2. Moisture sensors provided real-time data on volumetric water content, while suction sensors measured soil-water potential. Oxygen sensors deployed were based initially on electrochemical methods and later replaced by optical sensors for improved long-term reliability. Temperature sensors recorded the thermal dynamics within each column.

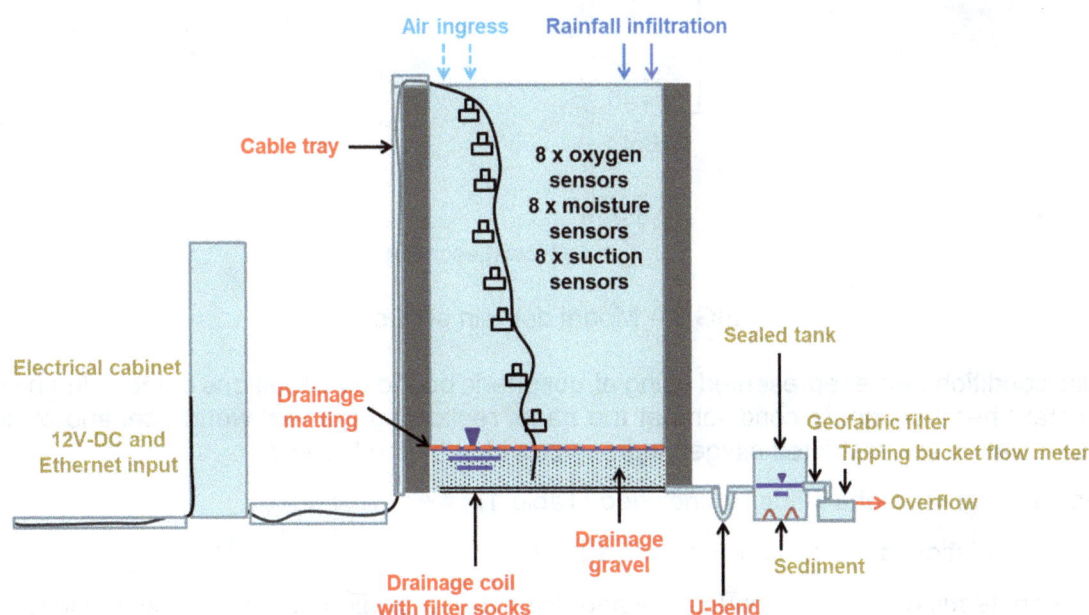

FIG 2 – Schematic design and instrumentation layout of each column.

Leachate flow was continuously recorded using tipping buckets, allowing accurate quantification of drainage rates. Water samples were collected and analysed monthly for chemical constituents, including sulfate, calcium, magnesium, and pH. Climatic conditions, including rainfall, solar intensity, temperature, humidity, and wind data, were monitored by two weather stations installed adjacent to the column set-up to correlate external environmental conditions with internal column dynamics.

Waste material characterisation

Detailed characterisation of waste materials was conducted before and after the leaching experiment using X-ray diffraction (XRD) to determine mineralogical composition and assess mineral transformations. Particle size distribution and soil-water characteristic curves (SWCC) were also determined to evaluate the hydraulic properties of the waste materials. This comprehensive characterisation facilitated interpretation of hydrological and geochemical behaviours observed during the leaching tests.

Hydrogeochemical modelling

Numerical modelling was conducted using TOUGHREACT, a reactive transport simulator designed to capture coupled hydrological and geochemical processes in porous media (Xu *et al*, 2006). The model domain replicated the physical dimensions and boundary conditions of the experimental columns, using 40 discretised vertical layers to capture detailed variations, as shown in Figure 3.

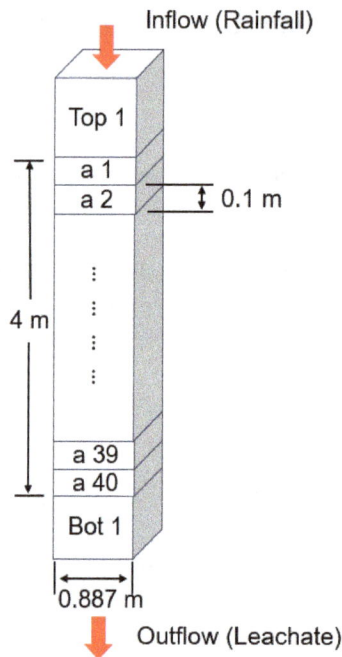

FIG 3 – Model domain set-up.

Hydraulic conditions were represented using atmospheric boundary conditions at the column surface and constant head boundary conditions at the base, replicating the real-world scenario of gravity-driven percolation and restricted oxygen ingress provided by the U-bend.

Key geochemical reactions modelled included (Table 1):

- pyrite oxidation, generating acidity and sulfate
- carbonate mineral dissolution (calcite and dolomite), providing acid-neutralising capacity
- muscovite and albite dissolution, contributing to alkalinity and silica release
- gypsum precipitation, capturing calcium and sulfate ions.

TABLE 1

List of minerals considered in the geochemical model and associated equations.

Mineral name	Reaction equation
Pyrite	$FeS_2 + 3.5O_2 + H_2O \rightarrow Fe^{2+} + 2SO_4^{2-} + 2H^+$
Calcite	$CaCO_3 + 2H^+ \rightarrow Ca^{2+} + HCO_3^-$
Dolomite	$CaMg(CO_3)_2 + 2H^+ \rightarrow Ca^{2+} + Mg^{2+} + 2HCO_3^-$
Muscovite	$KAl_2(AlSi_3O_{10})(OH)_2 \rightarrow K^+ + 3SiO_2 + 3AlO_2^- + 2H^+$
Albite	$NaAlSi_3O_8 \rightarrow Na^+ + 3SiO_2 + AlO_2^-$
Gypsum	$CaSO_4 \rightarrow Ca^{2+} + SO_4^{2-}$
Quartz	Assumed non-reactive

Mineral volume fractions were derived from pre-leaching XRD analyses, and reaction kinetics parameters were sourced from established literature, with subsequent calibration against observational data. The calibration involved adjusting kinetic parameters such as dissolution and precipitation rates and surface area calculations to align simulated and observed leachate chemistry trends, particularly sulfate, calcium, magnesium concentrations, and pH.

Through iterative model calibration and validation against five years of observational data, the hydrogeochemical model effectively simulated the coupled hydrological and chemical processes within the column tests. This robust modelling approach validated the conceptual understanding of mineral reaction dynamics and solute transport, thereby enhancing confidence in predicting long-term AMD behaviour in mine waste management.

RESULTS

Hydrological behaviour and oxygen dynamics

This paper focuses on the hydrogeochemical modelling and mineralogical transformations associated with leaching processes. This section provides only a brief summary of the hydraulic behaviour and oxygen dynamic results. For further details, please contact the corresponding author.

The monitoring data collected over five years provided detailed insights into the hydrological behaviour within the experimental columns. Initial moisture conditions of all columns were relatively dry upon construction; however, significant moisture infiltration was observed during the first rainy season, marking a rapid shift towards equilibrium moisture conditions within several months. Moisture content profiles indicated consistent equilibrium states, suggesting that rainfall infiltration was effectively balanced by evaporation and bottom drainage in each column.

Columns containing compacted materials showed enhanced water retention capabilities compared to loosely placed counterparts. Specifically, Column 5, capped with compacted B-Type waste, maintained higher moisture content within its uppermost layers compared to Column 2, which contained loosely placed B-Type waste. This increased water retention was attributed to reduced pore sizes and improved moisture retention characteristics associated with compaction. Furthermore, Column 6, which contained a mixture of A-Type, B-Type, and D-Type wastes, demonstrated the greatest moisture retention across its entire profile. This phenomenon was primarily due to the bonding and cementation processes facilitated by mixing diverse waste types, resulting in reduced hydraulic conductivity and increased water-holding capacity.

Distinct zones of moisture hold-up and hydrostatic conditions were identified in each column, with water hold-up zones typically occurring between depths of 1 to 3 m, characterised by relatively stable, elevated moisture contents and low suction values. In contrast, hydrostatic conditions were clearly observed near the base of Columns 3, 4, 5, and 6, indicative of limited vertical drainage. However, Columns 1 and 2 exhibited minimal hydrostatic conditions, attributed to larger voids and enhanced drainage pathways within loosely placed, coarse-grained materials.

Oxygen dynamics were monitored extensively throughout the experiment, revealing critical interactions between oxygen ingress, waste material properties, and moisture conditions. Initial oxygen concentrations in all columns approximated atmospheric levels (~21 per cent), but quickly decreased due to oxidative reactions and restricted oxygen ingress from the surface. Over the first two years, oxygen levels progressively declined, particularly within PAF materials and beneath compacted cover layers.

Compacted layers significantly reduced oxygen ingress due to reduced permeability, effectively creating an oxygen barrier that protected underlying reactive materials from oxidation. For example, the compacted B-Type cover in Column 5 showed substantially lower oxygen levels compared to its loosely placed counterpart in Column 2, confirming the efficacy of compaction in limiting oxidative processes. Furthermore, Column 6, containing mixed waste types, consistently demonstrated the lowest overall oxygen concentrations, reinforcing the effectiveness of mixed-waste bonding and cementation in minimising oxygen availability.

Seasonal fluctuations were observed in oxygen concentrations, with modest increases during dry periods and substantial decreases during wet seasons due to enhanced water saturation limiting oxygen diffusion. Optical oxygen sensors used in later columns provided reliable, long-term data, confirming lower oxygen levels stabilised after approximately two years, further supporting the effectiveness of structural waste management strategies in controlling acid generation.

Mineralogical change before and after leaching

Table 2 shows the XRD results before and after column leaching test. Post-leaching results represent the average values obtained from the four segments of the column. The key acid-generating minerals and acid-neutralising minerals are highlighted in red and green, respectively. Dolomite is excluded from the list of acid-neutralising minerals as its content remained unchanged (for A-Type) and increased (for B-Type) before and after leaching.

TABLE 2
XRD Results Before and After Leaching.

Mineral	Units	A-Type		B-Type		D-Type
		Before	After	Before	After	Before
Quartz	%	17	16.5	31	28.5	9
Mica/Muscovite	%	2	0.6	42	21	4
Chlorite	%	31	21	12	15.3	17
Calcite	%	9	4	1	0.5	<1
Dolomite	%	4	4	5	7.5	2
Albite	%	18	25	6	9.4	15
Pyrite	%	2	0.5	-	0.9	4
Rutile	%	1	-	1	-	1
Amphibole	%	7	-	1	-	12
Magnetite	%	2	2.3	-	0.3	3
Hematite	%	-	-	2	-	-
Talc	%	1	-	-	-	17
Serpentine	%	4	-	-	-	4
Augite	%	2	8.2	-	1.4	2
Vermiculite	%	-	-	-	-	10
Hornblende	%	-	12.5	-	1.2	-
Biotite	%	-	0.5	-	9.4	-
Microcline	%	-	3.6	-	3	-
Kaolinite	%	-	1.2	-	0.5	-
Siderite	%	-	-	-	0.4	-
Total	%	100	99.9	101	99.3	100

While the XRD-derived mineralogical compositions approached 100 per cent in total mass balance, the analytical method does not quantify non-crystalline or amorphous phases reliably. Therefore, although no amorphous phases were explicitly identified, their potential presence at trace levels is acknowledged. Such components may still participate in geochemical reactions during leaching, albeit their influence was inferred indirectly via leachate chemistry.

The key findings from the changes in mineralogical compositions observed in A-Type and B-Type waste before and after leaching are summarised as follows.

For A-Type waste:

- Reduction in Pyrite: The pyrite content decreased from 2 per cent to 0.5 per cent, indicating substantial oxidative dissolution during the leaching process. This suggests that pyrite oxidation was the key process contributing to acid generation.

- Reduction in Mica/Muscovite, and Chlorite: Although present in low quantities in A-Type waste, these minerals showed reductions, suggesting some degree of dissolution or alteration during the leaching process.

- Decrease in Calcite: Calcite experienced significant reductions, consistent with their high solubility under acidic conditions. Calcite decreased from 9 per cent to ~4 per cent. These carbonates played a major role in neutralising the acid generated by pyrite oxidation.

For B-Type waste:

- Appearance of Pyrite: This observed increase in pyrite content from 0 per cent to 0.9 per cent after leaching could be attributed to sampling or analytical artefacts rather than actual geochemical processes (ie no formation of new pyrite during leaching). Post-leaching sampling may have captured a portion of the waste material with a higher pyrite concentration that was not included in the pre-leaching sample. This can occur due to natural heterogeneity in the distribution of pyrite within the waste material.

- Weathering of Muscovite: The muscovite content was reduced by 50 per cent over five years (from 42 per cent to 21 per cent), reflecting its gradual weathering and dissolution. This process contributed to the release of potassium, silica, and potentially aluminium into the leachate. Over time, muscovite became more important in controlling leachate composition as faster-reacting minerals were depleted.

- Decrease in Calcite: Calcite content decreased from 1 per cent to 0.5 per cent, suggesting partial dissolution of calcite during the leaching process.

The comparison of waste types revealed that A-Type waste exhibited higher reactivity due to its elevated initial proportions of reactive minerals such as chlorite, calcite, and pyrite, which were significantly depleted during leaching. Additionally, A-Type waste exhibited a stronger reliance on carbonate buffering, while B-Type waste demonstrated sustained alkalinity through aluminosilicate weathering.

Leachate chemistry

In the context of mine waste leachate, drainage waters are generally classified as:

- acid drainage (pH < 5.5), typically associated with rapid oxidation of sulfide minerals without adequate buffering

- neutral drainage (pH 5.5–7), where partial buffering mitigates acidity

- alkaline drainage (pH > 7), often observed in systems dominated by carbonate or aluminosilicate weathering.

In this study, pH values consistently ranged from 6 to 9 across all column configurations, indicating that drainage conditions were dominantly neutral to mildly alkaline. This outcome reflects effective acid buffering from both carbonate dissolution and aluminosilicate weathering under limited oxygen ingress.

The chemical analyses of leachate samples collected from each column, shown by Figure 4 provided critical insights into the geochemical behaviour of waste materials, particularly regarding acid generation and buffering dynamics.

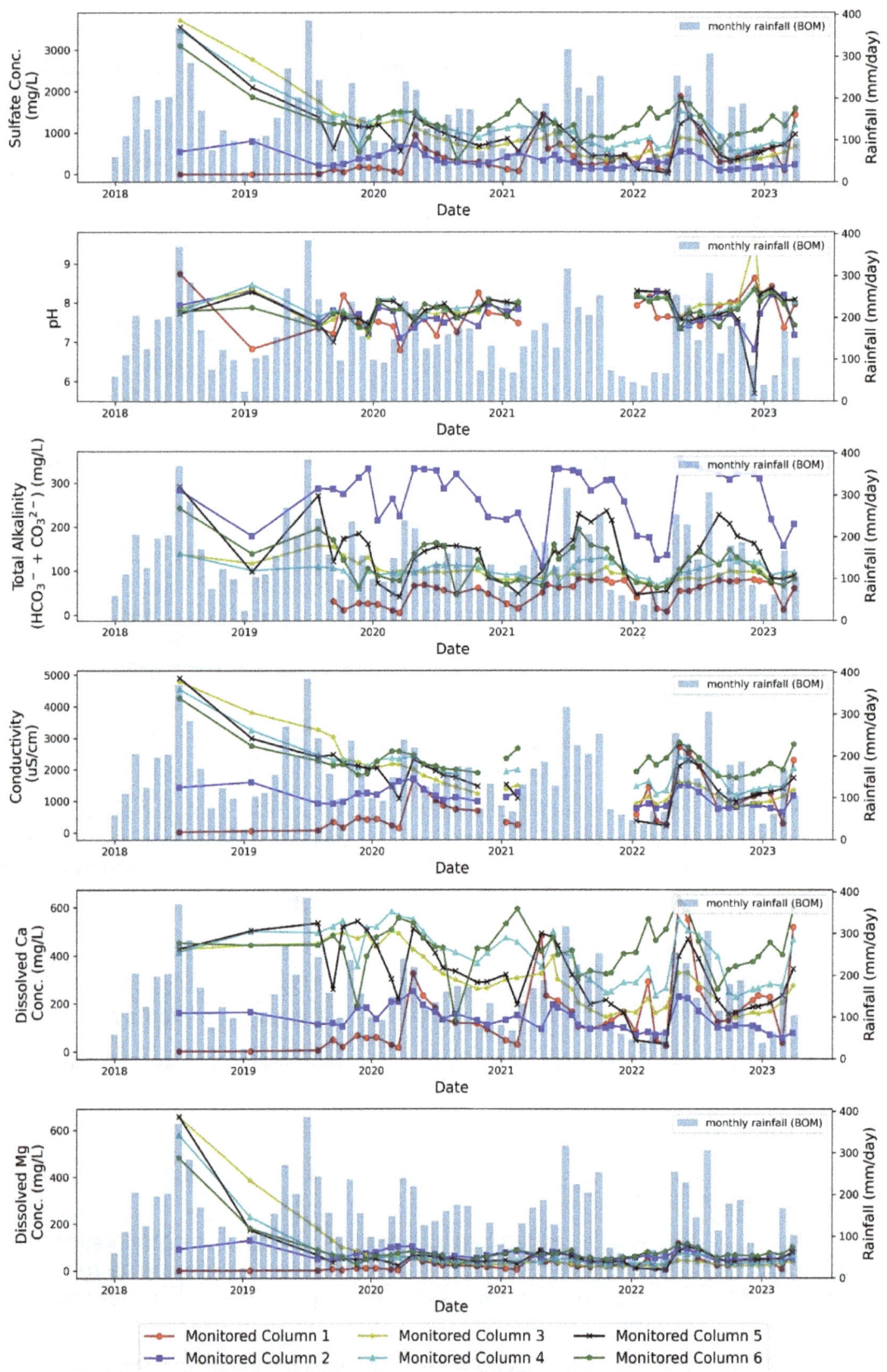

FIG 4 – Temporal variations in leachate water quality parameters.

Columns containing D-Type waste initially showed elevated sulfate concentrations indicative of rapid pyrite oxidation. Column 3, composed entirely of loosely placed D-Type waste, exhibited the highest initial sulfate concentrations, demonstrating its highly reactive nature. However, these sulfate concentrations sharply decreased over the first two years, indicating the progressive depletion of easily oxidisable pyrite. Similar trends were observed in Columns 4 and 5, which included layers of

D-Type waste beneath compacted NAF covers. The compacted covers effectively slowed oxygen ingress, reducing oxidation rates and subsequently lowering long-term sulfate release compared to Column 3.

Leachate from Column 2 (B-Type waste) consistently showed low sulfate levels but maintained high alkalinity throughout the monitoring period, despite its limited carbonate mineral content. This sustained alkalinity was primarily due to the weathering of muscovite minerals, which released basic cations such as potassium (K^+) and sodium (Na^+), contributing to the neutralisation capacity of the system (White and Brantley, 2003). Consequently, Column 2 demonstrated a stable, near-neutral pH between 7 and 9 over the entire experimental period.

Columns containing mixed waste types (Column 6) demonstrated a dynamic balance between sulfate production and buffering capacity. While initial sulfate release was elevated due to partial pyrite oxidation, the leachate chemistry stabilised after two years, indicating the establishment of a steady-state balance between acid production and neutralisation reactions driven by both carbonate dissolution and aluminosilicate weathering.

Calcium and magnesium concentrations mirrored sulfate trends, reflecting the interplay between sulfate mobilisation, carbonate mineral dissolution, and precipitation processes such as gypsum formation. The calcium-to-sulfate ratio analysis supported the occurrence of gypsum precipitation within the columns, particularly where sulfate concentrations approached stoichiometric limits for gypsum formation. This precipitation effectively controlled sulfate and calcium mobility, contributing to the stabilisation of leachate chemistry in the longer term (Nordstrom, 2011; Blowes et al, 2003).

Hydrogeochemical modelling

Model calibration

The numerical model developed using TOUGHREACT (Xu et al, 2006) was calibrated rigorously using five years of observed leachate chemistry and hydrological data. Calibration focused primarily on sulfate concentrations, pH, calcium, and magnesium levels, as these were critical indicators of pyrite oxidation and buffering reactions.

Initial mineral volume fractions, derived from XRD analyses, were adjusted minimally during calibration, highlighting the accuracy of initial mineralogical characterisation. However, minor adjustments to reaction kinetics, such as dissolution and precipitation rates and reactive surface areas, were necessary to closely align model predictions with observed data.

Model calibration effectively replicated observed sulfate dynamics across columns (Figure 5), accurately predicting the initial spike and subsequent decline associated with pyrite oxidation. Similarly, simulated pH levels closely matched measured values, confirming the accuracy of buffering reactions involving carbonate and aluminosilicate minerals (Figure 6). The model also reliably captured the trends in calcium and magnesium concentrations (Figures 7 and 8), validating the representation of mineral precipitation processes, notably gypsum formation.

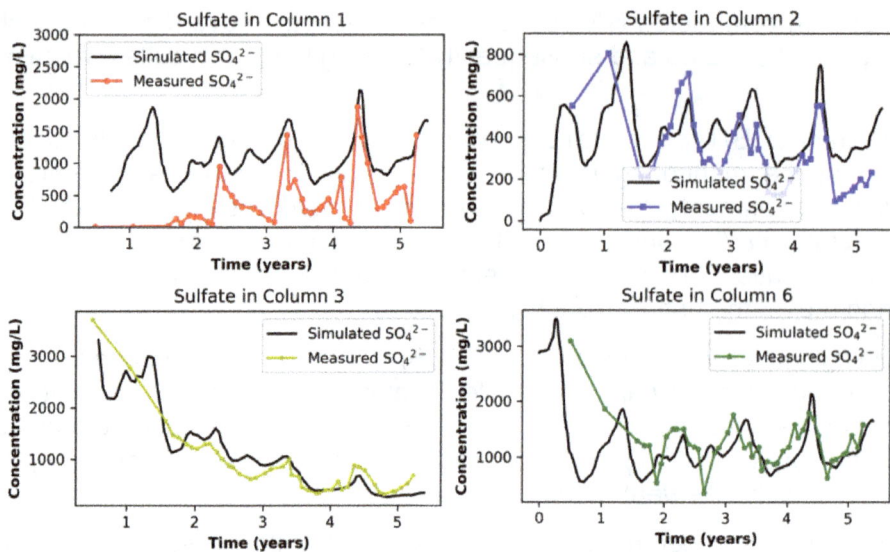

FIG 5 – Model calibration results – sulfate.

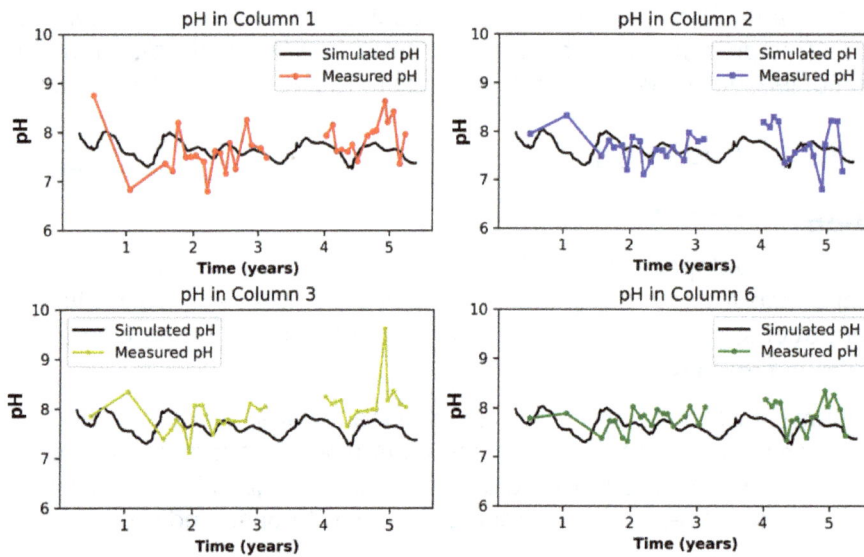

FIG 6 – Model calibration results – pH.

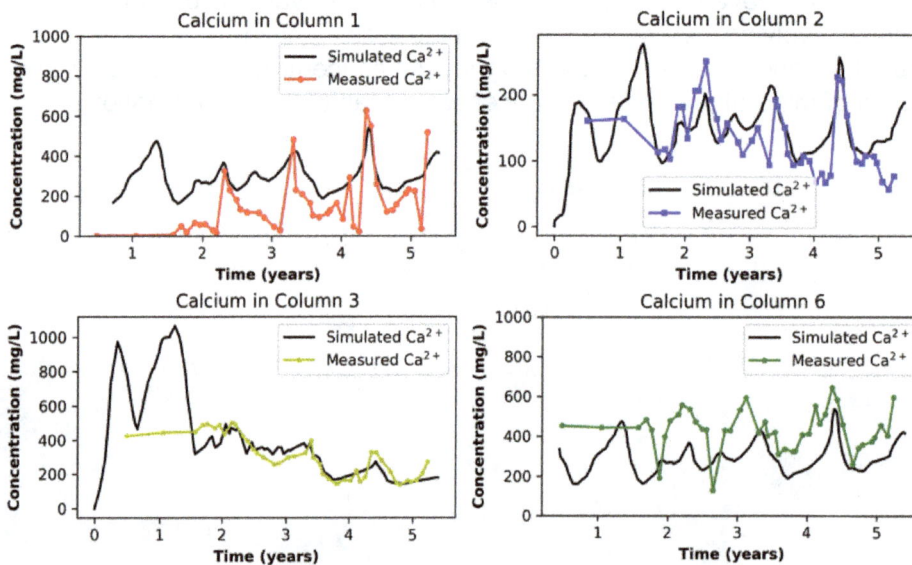

FIG 7 – Model calibration results – calcium.

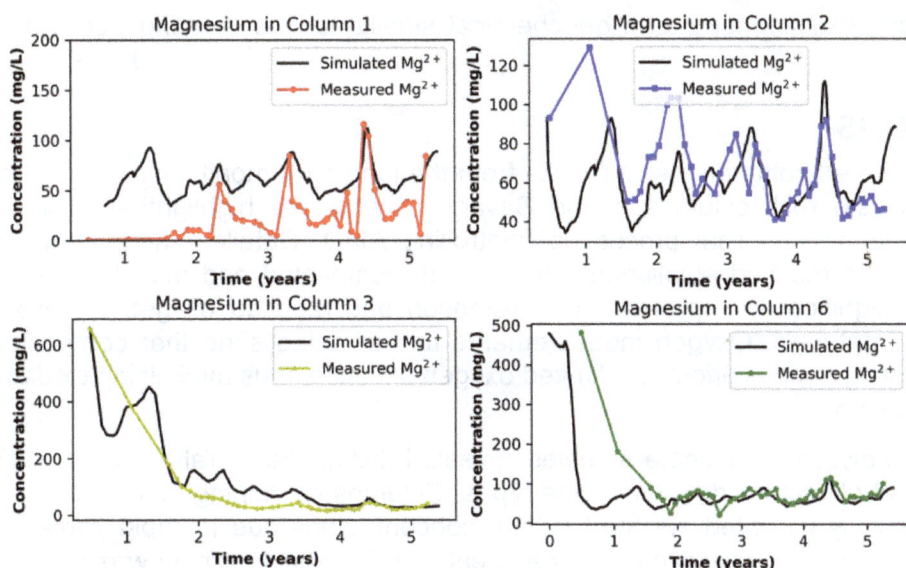

FIG 8 – Model calibration results – magnesium.

Note the TOUGHREACT model was constructed using initial mineral volume fractions derived directly from pre-leaching XRD characterisation. Although amorphous phases were not detected in XRD results, sensitivity adjustments were implemented during calibration. These included modifying reactive surface areas and kinetic parameters of silicate minerals to account for potential reactivity associated with undetected, fast-dissolving components. This approach was employed to bracket uncertainty arising from uncharacterised phases and helped ensure robust alignment with observed leachate chemistry.

Model results summary

The results of the hydrogeochemical modelling using TOUGHREACT provided detailed insights into the key geochemical processes driving acid generation and neutralisation in the columns. The calibrated model successfully simulated the temporal evolution of sulfate concentrations, pH, calcium, and magnesium levels, effectively capturing the complex interplay of hydrological and geochemical dynamics observed during the five-year monitoring period.

Simulated sulfate concentrations closely matched observed values, accurately capturing the initial sharp increase due to pyrite oxidation and the subsequent decline as reactive sulfides were progressively depleted. The model indicated that sulfate release was largely governed by oxidation kinetics, influenced significantly by oxygen availability and moisture content. The simulations showed reduced sulfate production in columns with compacted cover materials, highlighting their effectiveness in limiting oxygen ingress and pyrite oxidation.

The pH simulations confirmed effective acid buffering provided primarily by carbonate dissolution, especially calcite and dolomite, which were accurately represented in the model. Columns with limited carbonate content, such as the B-Type waste column, showed consistent maintenance of pH through aluminosilicate weathering, particularly muscovite dissolution. The modelling further confirmed that this aluminosilicate weathering released essential cations (K^+, Na^+) into the leachate, sustaining alkalinity and buffering pH effectively over time.

The simulation of calcium dynamics reflected a close coupling with sulfate concentrations, indicative of gypsum precipitation processes within the columns. Model outputs suggested gypsum precipitation was particularly active in columns containing higher sulfate and calcium concentrations, effectively immobilising both ions and reducing their mobility in leachate. Magnesium dynamics were similarly well-represented, with the model accurately simulating gradual declines in magnesium concentration due to precipitation processes and interactions with carbonate minerals.

Overall, the modelling results provided robust validation of the conceptual model developed from field observations, confirming the critical role of mineralogical composition and waste management strategies in controlling AMD processes. The calibrated TOUGHREACT model thus served as a reliable tool for predicting long-term geochemical stability and evaluating potential management interventions.

CONCLUSIONS

This study presented comprehensive results from the long-term monitoring and hydrogeochemical modelling of waste rock columns at the Savage River Mine, highlighting critical insights into hydrological and geochemical processes controlling AMD. Detailed monitoring confirmed that moisture dynamics reached equilibrium rapidly, with compacted and mixed-waste configurations demonstrating significantly improved water retention and reduced oxygen ingress compared to loosely placed materials. Oxygen measurements further established that compacted layers and mixed waste configurations effectively limited oxygen diffusion, thus minimising oxidative processes and acid generation.

Geochemical analyses of leachate samples revealed distinct temporal trends in sulfate, calcium, magnesium, and pH across different waste types. Columns containing potentially acid-forming (D-Type) waste initially exhibited elevated sulfate concentrations due to rapid pyrite oxidation, with subsequent decreases as reactive pyrite was depleted. Non-acid-forming waste types, particularly B-Type waste, maintained consistently high alkalinity, driven predominantly by aluminosilicate weathering processes.

The calibrated hydrogeochemical model developed using TOUGHREACT effectively replicated observed geochemical trends, validating the conceptual understanding of mineral reactions and buffering mechanisms. Modelling results emphasised the critical role of mineralogical composition, waste layering, compaction, and mixing strategies in controlling AMD processes. The simulation outcomes clearly demonstrated effective sulfate immobilisation through gypsum precipitation, robust carbonate buffering, and sustained alkalinity from aluminosilicate weathering. While XRD analysis did not detect amorphous phases, their potential presence below detection thresholds was acknowledged and indirectly addressed during model calibration via sensitivity adjustments to aluminosilicate reaction kinetics.

Ultimately, this study provides valuable practical insights and robust predictive tools that support the implementation of innovative waste management strategies at mining operations. The validated hydrogeochemical modelling approach offers reliable forecasting capabilities to inform decision-making processes, enhancing environmental stewardship and operational sustainability in the mining industry.

REFERENCES

Blowes, D W, Ptacek, C J, Jambor, J L and Weisener, C G, 2003. The geochemistry of acid mine drainage, in *Environmental Aspects of Mine Wastes* (eds: J L Jambor, D W Blowes and A I M Ritchie), 31:149–204 (Mineralogical Association of Canada).

Nordstrom, D K, 2011. Mine waters: Acidic to circumneutral, *Elements*, 7(6):393–398. https://doi.org/10.2113/gselements.7.6.393

White, A F and Brantley, S L, 2003. The effect of time on the weathering of silicate minerals: Why do weathering rates differ in the laboratory and field?, *Chemical Geology*, 202(3–4):479–506. https://doi.org/10.1016/j.chemgeo.2003.03.001

Xu, T, Sonnenthal, E, Spycher, N and Pruess, K, 2006. TOUGHREACT – A simulation program for non-isothermal multiphase reactive geochemical transport in variably saturated geologic media: User's Guide, Lawrence Berkeley National Laboratory Report LBNL-55460.

Two-year *in situ* monitoring of water balance and vegetation growth on soil covers for tailings on a slope

C Zhang[1], L Tan[2], A Taylor[3], S Quintero[4], D Williams[5] and C Gimber[6]

1. Senior Research Fellow, Geotechnical Engineering Centre, School of Civil Engineering, University of Queensland, St Lucia Qld 4072. Email: chenming.zhang@uq.edu.au
2. Senior Research Assistant, Geotechnical Engineering Centre, School of Civil Engineering, University of Queensland, St Lucia Qld 4072. Email: lieven.tan@uq.edu.au
3. Environmental Specialist, Incitec Pivot Ltd, Phosphate Hill Qld 4825. Email: amy.taylor@incitecpivot.com.au
4. Senior Laboratory Manager, University of Queensland, St Lucia Qld 4072. Email: s.quintero@uq.edu.au
5. Professor, Geotechnical Engineering Centre, School of Civil Engineering, University of Queensland, St Lucia Qld 4072. Email: d.williams@uq.edu.au
6. Partner, ERM, Brisbane Qld 4000. Email: chris.gimber@erm.com

ABSTRACT

To meet mine closure objectives, the design of cover systems for tailings in a semi-arid climate must satisfy three major criteria:

1. Safe – the cover must not collapse, ensuring the protection of people and animals.

2. Stable – the cover must resist erosion over time.

3. Non-polluting – the cover should support vegetation, store-and-release precipitation, shed excess water, and effectively segregate the tailings.

To evaluate the efficacy of different potential cover designs, four trials were established on dried gypsum tailings with a 1V:3H slope. The trials were each 120 m long, 20 m wide and 1.6 m deep. Each trial was configured using locally available materials, with overlying topsoil as a seedbank, followed by a layers of varying material properties and thickness above tailings.

Soil moisture and suction sensors were installed at various depths to monitor moisture migration and net percolation conditions. Cameras were installed on the cover surface to capture images, enabling the observation of vegetation growth and gully formation (erosion).

This paper presents the findings from a two-year monitoring period since installation, highlighting the range of infiltration depths, the volume of water percolating to the tailings, and the presence and persistence of erosion on the slopes. This study provides valuable insights into the rehabilitation of sloped tailings storage facilities in semi-arid environments, highlighting the importance of material selection in ensuring long-term stability and ecological sustainability.

INTRODUCTION

Mine rehabilitation is a crucial aspect of sustainable mining, aimed at restoring disturbed landscapes and minimising long-term environmental impacts. As part of planning for the closure of a tailings storage facility (TSF), mine cover trials are conducted to assess strategies improving slope stability, moisture movement management, and mitigating risks associated with run-off, gully formation, and erosion. These trials focus on the performance of store-and-release cover systems, which are designed to control the infiltration of water into tailings, a critical factor in ensuring the long-term stability and minimising erosion risks (Dunlop, Nicolson and Purtill, 2025).

The TSF is located at Phosphate Hill in western Queensland. The site is in a semi-arid climate zone experiencing hot summers and mild winters, with low annual rainfall averaging between 400 mm and 500 mm. The trials provide valuable knowledge for future cover design by evaluating how these systems perform under local hydrological and climatic conditions. Excessive moisture percolation into tailings can compromise the integrity of the cover, leading to erosion, instability, and potential contamination risks.

Store-and-release covers aim to mitigate these issues by capturing and storing water during wet periods and releasing it gradually during dry periods, thus reducing moisture infiltration and preventing destabilisation (Dunlop, Nicolson and Purtill, 2025). Vegetation growth, supported by the moisture retention properties of the cover, plays a vital role in slope stability by reducing erosion and stabilising surface soils. The success of vegetation establishment depends on the cover system's ability to regulate moisture and protect the surface from degradation. This paper does not attempt to analyse vegetation spread and the density of distinct species.

The trial site selected was along the slope of an existing TSF with a slope gradient of 1:3 due to wet stacking. Four connected cells were constructed for the cover trial, which will be discussed in this paper and will focus primarily on the moisture profile. The insights gathered from these trials will directly inform future cover designs at Phosphate Hill, guiding the optimisation of water infiltration management and slope stability as part of the rehabilitation process.

COVER AND INSTRUMENTATION SET-UP

Four connected cover trial cells were constructed along the side of an existing TSF in April 2023 (Figure 1). Each trial cell has a footprint of 20 m (width) × 120 m (slope length) × 1.6 m (depth).

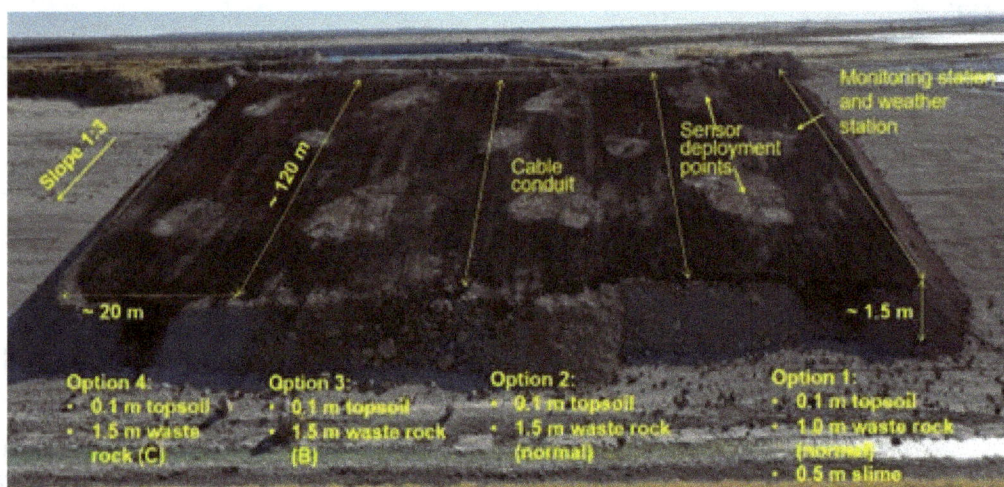

FIG 1 – Aerial image of trial cell post construction.

Each cell was instrumented post cover construction, with an upper and lower slope instrumentation group. Each group, summarised in Table 1 consists of METER Group Teros-12 moisture sensors, heat dissipation suction sensors, and gaseous oxygen sensors. A weather station was installed to monitor and record rainfall, wind, and solar intensity. Vegetation and surface condition monitoring were performed with remote cameras located on each cell, capturing daily images. All telemetry and camera images were transmitted over 4G cellular network to provide remote monitoring. The degree of saturation was obtained by linearising the sensor reading between an upper and lower bound limit.

TABLE 1

Summary of sensor locations and cover trial cell stratification layer.

Approx thickness (m)	Cell 1	Cell 2	Cell 3	Cell 4	Moisture sensor depth (m)
0.1	Topsoil	Topsoil	Topsoil	Topsoil	0.05
1	Normal waste rock	Normal waste rock	Waste rock B	Waste rock C	0.6
					1.0
0.5	Slime				n/a
N/A	Tailings	Tailings	Tailings	Tailings	2.2

Cover design and profile

The following materials were used for the construction of the trial cells with a summary provided in Table 1.

- Topsoil: Growth medium to promote vegetation growth.

- Waste rock (Normal): Primary cover armouring, store-and-release, consisting of fines to medium sized rocks, prone to weathering.

- Waste rock (B): Alternate cover armouring, store-and-release, material consisting of fines to large boulders and is less prone to weathering.

- Waste rock (C): Alternative cover armouring, store-and-release, limited supply from normal mining, material consisting of fines to small rocks, least prone to weathering.

- Waste (Slime): Hydraulic barrier, fines from clay silt and silt sand obtained from ore waste.

- Tailings: Waste byproduct post extraction of minerals of interest from ore.

OBSERVATIONS

Rainfall

Major rainfall events marking the rainy season were observed starting from the period of November 2023 to March 2024; this cycle was further repeated in 2024–2025 albeit with lower rainfall as observed in Figure 2. Notably, there was a large single rainfall event which recorded a daily accumulation of 76.6 mm in July 2023, one month post installation. Over the observational period, a maximum of 83.4 mm of daily rainfall was recorded in January 2024, with a monthly total of 182.2 mm. Over the observational period, the average daily rainfall during the rainy season is greater than 2.5 mm, in contrast, the daily average is less than 0.5 mm during the dry season.

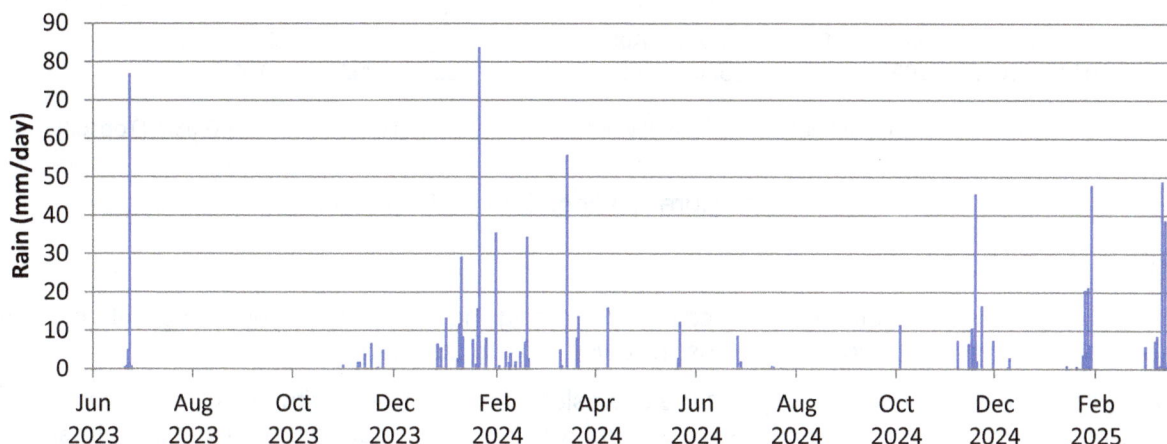

FIG 2 – Daily rainfall over the two-year monitoring period.

Cell 1

Cell 1 construction as summarised in Table 1 consisted of a slime layer (0.5 m thick) capping the tailings layer, followed by normal waste rock (1 m thick) and topped with topsoil.

The moisture data for the downslope and upslope instrumentation groups are included in Figures 3 and 4 respectively. At both the upslope and downslope sections, there is significant activity within the topsoil layer, which strongly correlates with rainfall events. At depths of 0.6 m below ground level (BGL) and deeper, excluding the upslope measurements, which maintained a relatively stable moisture level with minimal variation; the moisture profile, including the tailings, was observed to be gradually decreasing over the observational period.

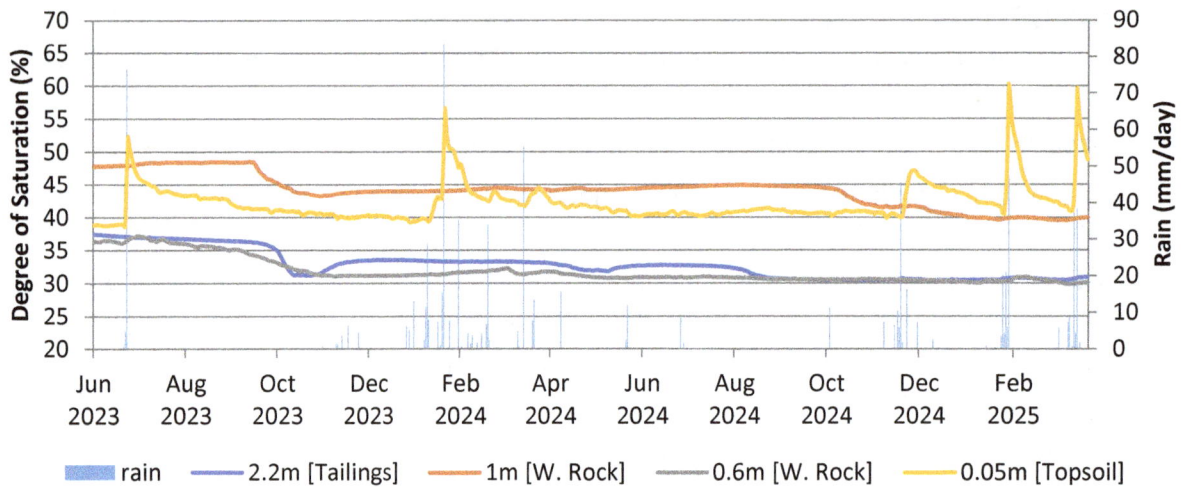

FIG 3 – Moisture readings for cell 1 downslope.

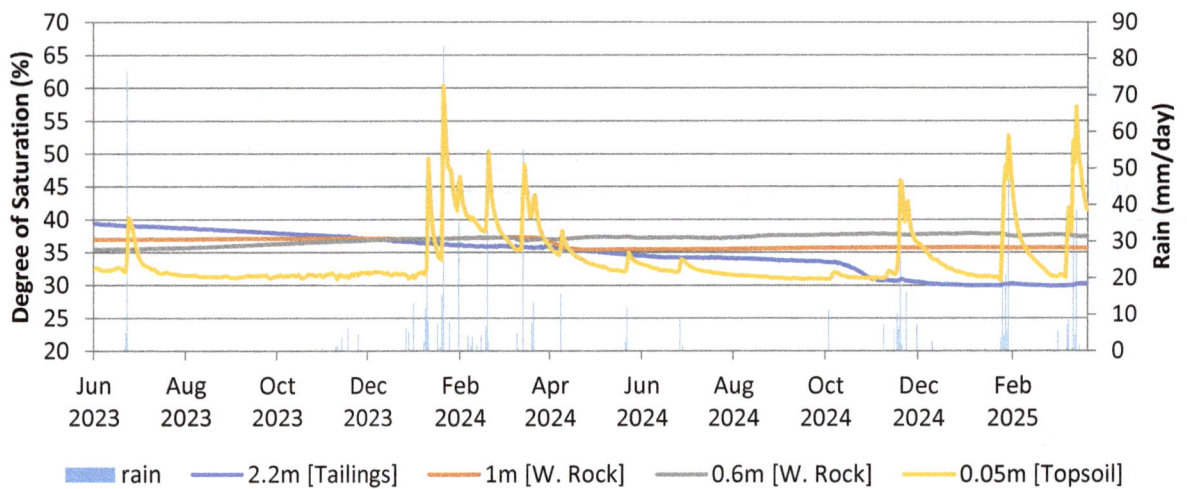

FIG 4 – Moisture readings for cell 1 upslope.

Cell 2

The construction of cell 2 differed from cell 1 by omitting the slime layer, consisting only of normal waste rock (1.5 m thick) to cap the tailings, followed by a topsoil layer.

Figure 5 shows the moisture readings for the downslope sensor group, with the sensor located within the topsoil at 0.05 m BGL recording a subdued delayed response during the July 2023 rainfall event and moderate activity during the November 2023 to March 2024 rainfall period. In contrast, a significant increase in response during the 2024–2025 wet season was observed. Such contrast is likely caused by the micro topography of the cover surface and moisture in the cover gradually reaching a steady state that facilitates stormwater infiltration.

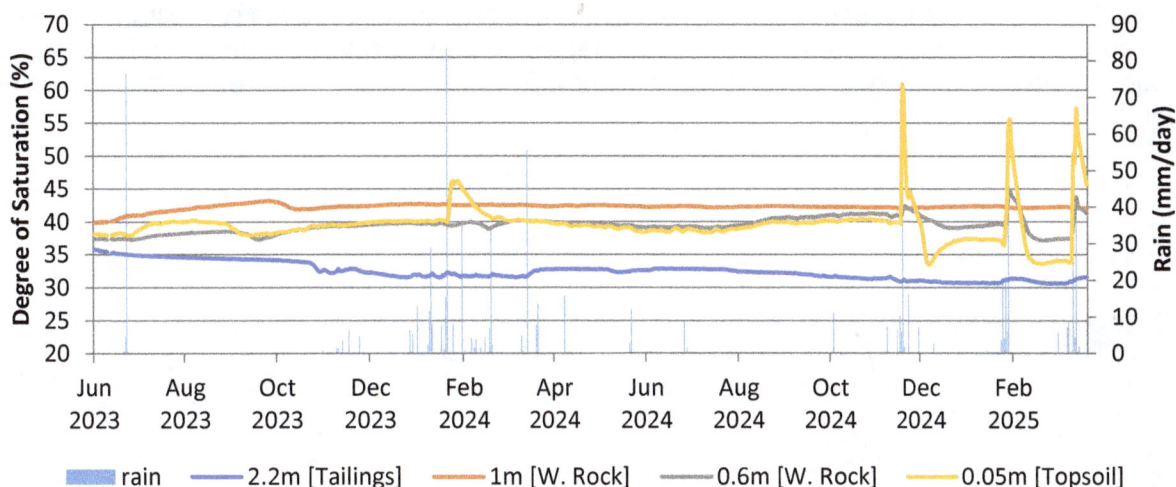

FIG 5 – Moisture readings for cell 2 downslope.

The upslope moisture sensor readings, as depicted in Figure 6 exhibited pronounced activity corresponding to rainfall events from November 2023 to March 2024 for the topsoil sensor and did not register any significant response to the rainfall event in July 2023, similarly to the downslope sensor.

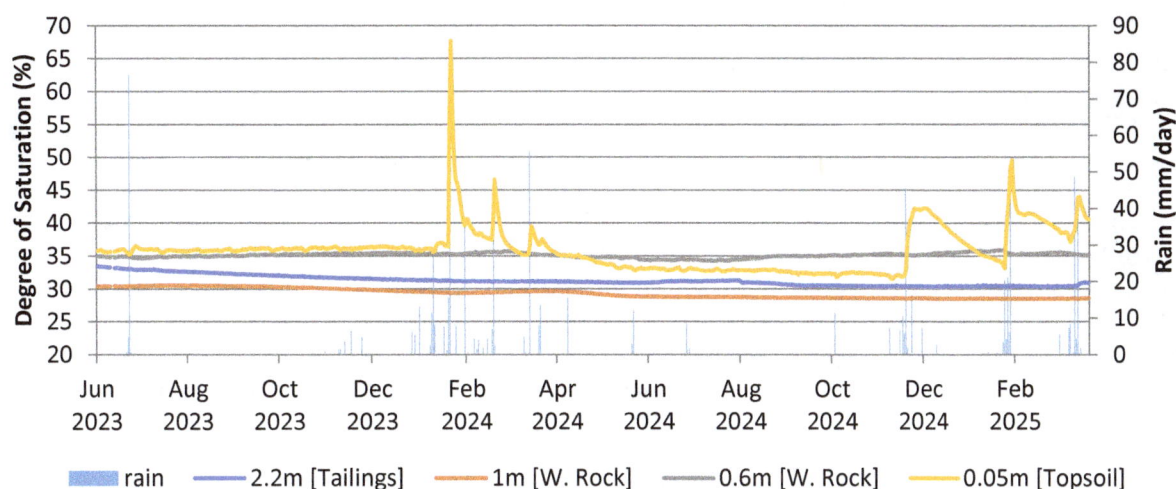

FIG 6 – Moisture readings for cell 2 upslope.

At 0.6 m BGL, the downslope moisture sensor observed a minor increase in the baseline measurement with moderate activity observed during the 2024–2025 wet season. In contrast, the upslope sensor did not observe significant variability throughout the observational period, only showing a very weak wet-dry cycle variation.

At 1 m BGL and deeper, upslope moisture sensors reported a slow and stable decrease in moisture readings. In contrast, the downslope moisture sensors recorded a minor increase in moisture at 1 m BGL during the initial four months post installation and remained stable for the rest of the recording period. The downslope moisture sensor located within the tailings observed a gradual decrease over the observational period, however appeared to have a delayed and minimal response to rainfall.

Cell 3

Analogous to the design of cell 2, the construction of cell 3 omits the slime layer above the tailings. The armouring layer was constructed using waste rock B, which is predominantly characterised by large sized boulders.

Figures 7 and 8 shows the downslope and upslope moisture measurements respectively for cell 3. Once again, the topsoil sensor at 0.05 m BGL depicts strong correlation with rainfall events, with variation of the upslope sensor appearing to not show a reaction to the March and November rainfall events of 2024.

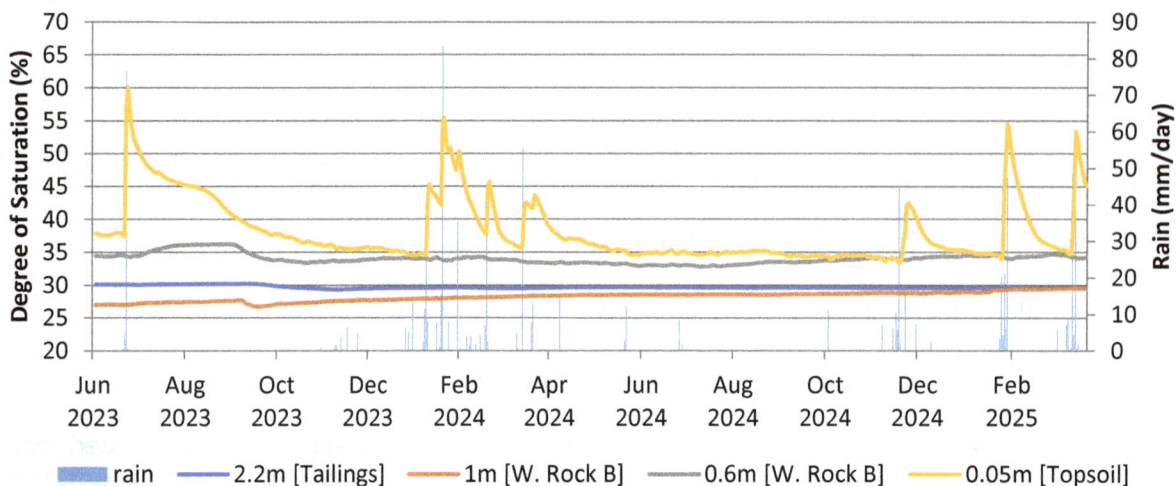

FIG 7 – Moisture readings for cell 3 downslope.

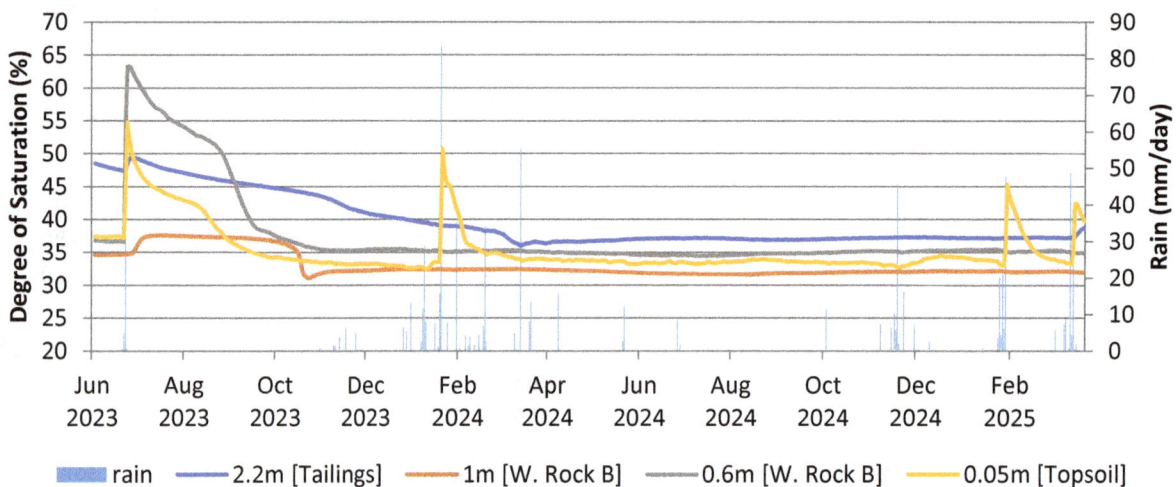

FIG 8 – Moisture readings for cell 3 upslope.

At the upslope instrumentation group, moisture sensors reported moisture activity after the first initial rainfall in July 2023, with a large sharp spike observed at 0.6 m BGL. A subdued and slower response was recorded by the sensors located at 1 m BGL and within the tailings layer. Subsequent measurements at the upslope location at 0.6 m BGL and deeper observed the moisture levels to decrease over time, appearing to reach a low state by April 2024.

In contrast, only the moisture sensor located at 0.6 m BGL for the downslope section recorded any activity for the July 2023 rainfall. All deeper moisture sensors recorded stable measurements over the monitoring period with insignificant variation.

Cell 4

Analogous to cell 2 and cell 3, the armouring layer consists of the limited supply of waste rock C. Figures 9 and 10 depicts the downslope and upslope moisture readings respectively. Following the July 2023 rainfall, moisture sensors within the topsoil observed an increase in moisture level and only observed a reduction one month later. Topsoil sensors located downslope further showed strong reactivity to rainfall events during the rainy season in contrast to the upslope sensors which showed a very subdued response.

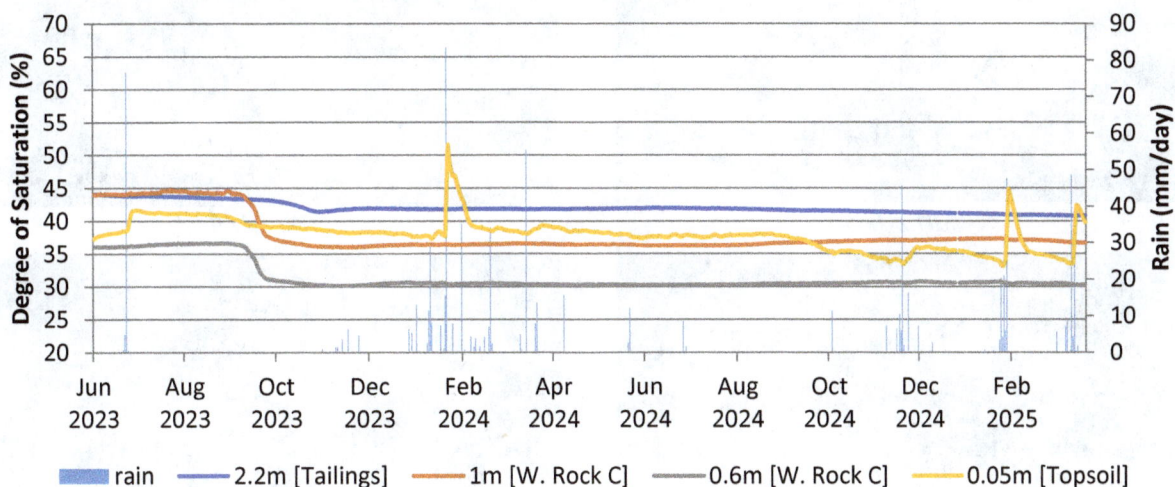

FIG 9 – Moisture readings for cell 4 downslope.

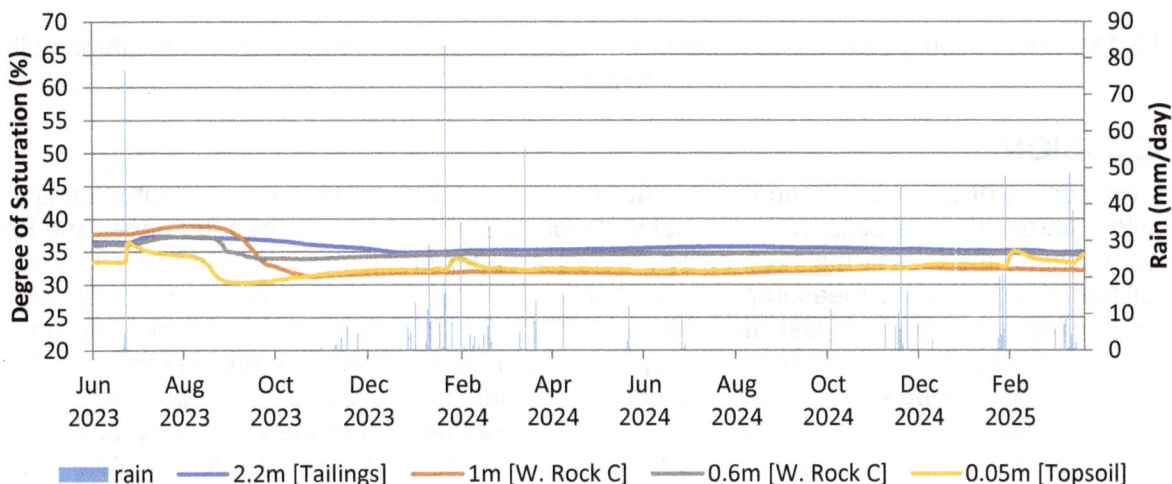

FIG 10 – Moisture readings for cell 4 upslope.

Further observations of both upslope and downslope sensors located at depths greater than 0.6 m BGL inclusive of tailings, notes decreased moisture levels by December 2023. With further gradual decrease in moisture levels recorded by moisture sensors located within the tailings over the observational period.

Vegetation growth

Figure 11 illustrates the vegetation growth at various stages throughout the monitoring period. There was no vegetation at the start of monitoring in June 2023, and by April 2024, vegetation cover had increased significantly in response to the 2023–2024 wet season. In comparison to the April 2024 observations, vegetation cover in January 2025 was reduced, and by March 2025 had appeared to increase slightly following rainfall. A small gully can be seen to have formed on the surface.

FIG 11 – Vegetation growth as assessed from surface timelapse images with highlighted gully formation.

DISCUSSION

Over the course of the observation period, the moisture profile generally demonstrated consistent behaviour across the trial cells, with minimal variation of significance between the upper and lower slopes. Of note, the topsoil moisture activity of cell 4 upper slope appears to be less reactive to rainfall than the downslope measurements, and cell 3 recorded significant moisture activity down to the tailings layer following the first rainfall in July 2023. Cell 3 consists of large rocky and boulder sized material, thus the upslope sensors may have been localised within a preferential path during the initial rainfall contributed by the disturbance of material during the installation process, and with consolidation assisted by the movement material due to rainfall and or the weathering of the waste rock, this further alters the local microenvironment thus creating a new equilibrium state. Not all moisture sensors reported a similar response to the July 2023 rainfall event; the observations occurred one month post installation and does not represent the steady state of the cover and latest observations from January 2025 further supports that a steady state has yet to be reached. The variance observed between the upper and lower slope of cell 4 by the topsoil is most likely attributed to differing microenvironments, with the possibility of a more preferential path taken by surface flow on the upper slope of cell 4.

Over the observation period, moisture infiltration was restricted to the surface layer, with limited activity observed beyond 0.6 m BGL. No evidence was found of upward movement of pore water from the tailings, nor of significant rainfall infiltrating into the tailings as the observational period continued. Subsequent measurements indicated a progressive decline in moisture content, suggesting that the material may have continued to self-settle. This process, along with the breakdown of material and subsequent clogging of pore spaces by fines, likely reduced the hydraulic conductivity, thereby limiting further infiltration.

Time-lapse imagery analysis indicates that vegetation growth closely followed rainfall events, with vegetation cover persisting throughout the first wet season of 2023–2024 and remaining detectable in the subsequent months. During drier periods, a decline in vegetation cover was observed, suggesting water stress or possible vegetation physiology cycle, as seen in the January 2025 image (Figure 11) Subsequent rainfall events led to renewed vegetation growth, as observed in March 2025.

Gullies were observed to have formed from surface run-off. No formal gully assessment has been performed. It is too early to comment on the efficacy of erosion resistance across the trial cells, and a slope surface study is warranted to identify the extent of erosion that has occurred.

Observations indicate that the trial cover effectively limited water infiltration into the tailings. The cover appears to retain moisture within the surface layer, supporting sustained vegetation growth throughout the current observation period. While long-term vegetation survival remains to be confirmed over time, these initial findings suggest positive performance. Ongoing long-term monitoring is essential to comprehensively evaluate the performance of the different trial cover systems.

REFERENCES

Dunlop, J, Nicolson, L and Purtill, J, 2025. Mine waste cover system trials – a comparative review of case studies: Technical Paper 2, Office of the Queensland Mine Rehabilitation Commissioner, Brisbane, Queensland Government.

Minimising and managing mine wastes, including dewatering tailings and comingling

Reprocessing and upgrading cassiterite from tailings – an assessment of the fourth generation Reflux™ Classifier

M E Amosah[1], J Zhou[2] and K P Galvin[3]

1. ARC Centre of Excellence for Enabling Eco-Efficient Beneficiation of Minerals, University of Newcastle, Callaghan NSW 2308. Email: margaret.amosah@uon.edu.au
2. ARC Centre of Excellence for Enabling Eco-Efficient Beneficiation of Minerals, University of Newcastle, Callaghan NSW 2308. Email: james.zhou@newcastle.edu.au
3. ARC Centre of Excellence for Enabling Eco-Efficient Beneficiation of Minerals, University of Newcastle, Callaghan NSW 2308. Email: kevin.galvin@newcastle.edu.au

INTRODUCTION

Tailings reflect the limitations of older methods of beneficiation, often associated with poor recovery of relatively fine particles. These limitations include the effects of slimes, the need for further liberation, and the inability to recover down to ~10 μm (Das and Sarkar, 2018). Most tailings dams, left idle for years, exhibit varying characteristics that affects reprocessing techniques. This variability can pose a significant challenge, impacting separation efficiency, recovery, and overall beneficiation performance.

Therefore, innovative methods of extraction are needed. The Reflux™ Classifier, shown in Figure 1, features a system of inclined channels above a conventional fluidised bed. This design significantly enhances throughput compared to conventional gravity separators by greatly increasing the effective sedimentation area. A channel spacing of 1.8 mm was used in this study, replacing the previously tested wider spacings of 3 mm and 6 mm (Galvin, 2021).

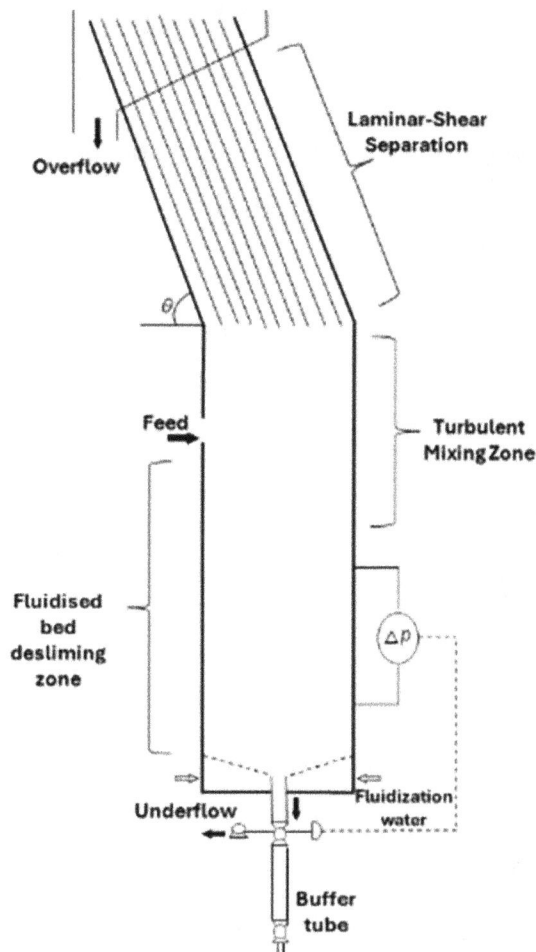

FIG 1 – Schematic representation of the Reflux™ Classifier.

This study highlights the application of the Reflux™ Classifier (RC) for reprocessing fine cassiterite (SnO_2) from a NSW tailings dam, tackling the long-standing, 40-year-old challenge of tin recovery from tailings. Specifically in this study the cassiterite tailings contains only ~0.3 wt per cent tin, with a significant portion finer than 20 µm and a wide size range up to 300 µm.

Earlier research on this feed showed that the RC effectively concentrates ultrafine tin, enhancing recovery and provides desliming and preconcentration of feed (Amosah et al, 2024). More recently, (Amosah, Zhou and Galvin, 2025) introduced a fourth-generation Reflux Classifier with a novel fluidisation arrangement and a slow-moving rake, enhancing gangue rejection across the full-size range and fine tin retention due to the promotion of a dense medium.

This paper extends the previous work by examining the separation performance for feed sourced from the same tailings dam, but with a much stronger concentration of the tin in the finer sizes. Of interest is the potential to maintain the recovery and the upgrade achieved previously.

FEED CHARACTERISATION

The methodology used here is described by Amosah, Zhou and Galvin (2025). They detailed the errors associated with the approach, while also confirming the strong mass balance reconciliation across the full-size range. The experiments were conducted using the fourth generation RC on these feeds at a constant pulp density of 10 wt per cent solids. The solids feed flux was ~6 t/m²/h, though it is noted this level can readily be increased by increasing the feed pulp density.

Table 1 shows the mass distribution and tin grades of two tailings feed streams from different sections of the same cassiterite tailings site. Notably, the original feed (Run A) has about 46 per cent of the tin below the liberation size of 38 µm while the new feed (Run B) has 71 per cent of its tin below the liberation size of 38 µm.

TABLE 1

Summary of balanced feed data.

Sieve range (µm)	Run A – feed		Run B – feed	
	Mass (%)	Grade (% Sn)	Mass (%)	Grade (% Sn)
300–150	14.71	0.20	4.69	0.15
150–90	30.02	0.21	19.24	0.14
90–45	15.36	0.40	26.56	0.18
45–38	3.37	0.62	1.52	0.37
38–20	8.80	0.49	12.93	0.55
-20	27.73	0.39	35.08	0.41
Feed Head Grade		0.33		0.30

RESULTS AND DISCUSSION

Table 2 shows the product grade and upgrade versus particle size for the two runs. The overall recovery achieved by the two runs was very similar, 48.2 per cent for feed A and 45.7 per cent for feed B, with uncertainties in the recovery of ~±5 per cent. The overall upgrade achieved for feed A of 19.7, however, is nearly double that achieved for feed B of 10.7, with the solids yield for feed A (2.4 per cent) nearly half that of feed B (4.3 per cent). This result suggests that with the tin concentrated much more in the relatively fine sizes for feed B compared to feed A, maintenance of a similar recovery requires the extraction of a higher solids yield for feed B, leading to the lower upgrade. If the solids yield for feed B was reduced further, the question remains whether the recovery could be maintained. In general, the recovery would be expected to decline, however, Amosah et al (2024) showed that the new rake delivers significant increases in grade, with relatively little loss in recovery, a consequence of the dense medium effect.

Very high upgrades were achieved at the finer sizes for both feeds, attributed to the new fluidised rake system in the RC. The dense medium effect enhances the concentration and retention of fine tin, down to below 20 µm, a size typically lost to tailings with conventional technologies. To address future variability, we therefore propose targeting a 5 per cent solids yield, similar to that achieved for feed B. The material coarser than 53 µm would then be milled to achieve liberation. This pre-concentrated feed (5 per cent of the original feed) would then be sent for final upgrading in a much smaller Grade Pro to produce a concentrate with over 50 wt per cent tin, maximising recovery, grade, and minimising fine tailings.

The Reflux™ Classifier offers significant advantages over conventional gravity-based separators, especially under conditions of high feed variability. Most conventional separators such as spirals, though widely used, face several critical limitations. These include the need for multi-staging, inefficient recovery of fine particles (particularly below ~50 µm), low throughput per unit area, requiring large operational footprints, and a high level of gangue entrainment which leads to reduced product grades.

The combination of a fluidised bed, a slow-moving rake system, and inclined plates in the RC delivers plug-flow efficiency under laminar flow conditions, counter current washing of the concentrate, and hence highly efficient separations with 1:1 scalability. This design is specifically engineered for fine particles down to ~20 microns and finer, making it ideally suited for tailings reprocessing. The use of inclined plates significantly expands the effective settling area, enabling much higher throughput.

In a recent study by Galvin, Zhou and Rodrigues (2025), the Reflux™ Classifier was evaluated for its ability to upgrade ultrafine iron ore, with results directly compared to spiral separation at the optimised splitter position and profile geometry. The spirals achieved a separation efficiency (hematite recovery minus quartz recovery) of 77 per cent recovery, while the RC achieved an efficiency of 91 per cent across a broader particle size range. This demonstrates the ability of the Reflux™ Classifier to deliver enhanced recovery and sharper separation performance in just a single processing stage.

TABLE 2
Summary of balanced product data.

Sieve range (µm)	Results					
	Run A – Product			Run B – Product		
	Grade (% Sn)	Recovery (%)	Upgrade	Grade (% Sn)	Recovery (%)	Upgrade
300–150	1.28	63.72	6.4	0.30	78.58	2.0
150–90	6.06	67.45	28.9	1.30	73.69	9.3
90–45	17.87	86.23	44.7	4.73	78.03	26.7
45–38	42.80	87.26	69.0	10.70	86.89	28.9
38–20	50.59	57.38	103.2	32.60	73.06	59.3
-20	25.53	2.60	65.5	61.77	14.72	150.7
*Product Head Grade	6.30			3.22		
*Overall Recovery	48.2			45.7		
*Solids Yield	2.4			4.3		
*Upgrade	19.7			10.7		

* Results on product head grade, overall recovery, solids yield and upgrade are calculated based on balanced head samples.

CONCLUSIONS

Reprocessing tailings provides a more sustainable mining approach but presents challenges such as feed variability and low mineral concentrations, making recovery difficult. This paper examined the performance of a fourth-generation Reflux™ Classifier for tin recovery from two distinct cassiterite tailings compositions. This work shows the potential for a two-stage system consisting of a pre-concentration stage with a yield of ~5 per cent which would be followed by classification, grinding of the coarse, and then a final upgrade to ~50 per cent tin. Overall, the Reflux™ Classifier delivers strong operational performance in terms of energy efficiency, throughput capacity, and scalability. Its energy efficiency stems from effective pre-concentration ahead of grinding, reducing the load on downstream circuits. It also supports exceptionally high throughput compared to conventional units while maintaining a 1:1 scalability, making it well suited for both pilot and full scale operations.

ACKNOWLEDGEMENTS

The authors acknowledge the funding support from the Australian Research Council for the ARC Centre of Excellence for Enabling Eco-Efficient Beneficiation of Minerals, grant number CE200100009, contributions from the University of Newcastle, and the contributions of industrial partner organisation FLSmidth and Australian Tin Resources (Ardlethan Mine).

The authors also thank Josh Sutherland, Josh Starrett, Desire Awuye and Jessie Cummings for their kind assistance during the experiments.

REFERENCES

Amosah, M E, Yvon, M, Zhou, J and Galvin, K, 2024. The role of enhanced desliming and gravity separation as a precursor to flotation in the upgrading of cassiterite from tailings, *Minerals Engineering*, 208:108581.

Amosah, M E, Zhou, J and Galvin, K P, 2025. Fourth generation gravity separation using the Reflux Classifier, *Minerals Engineering*, 224:109216. https://doi.org/10.1016/j.mineng.2025.109216.

Das, A and Sarkar, B, 2018. Advanced gravity concentration of fine particles: a review, *Mineral Processing and Extractive Metallurgy Review*, 39(6):359–394.

Galvin, K P, 2021. Process intensification in the separation of fine minerals, *Chemical Engineering Science*, 231:116293.

Galvin, K P, Zhou, J and Rodrigues, A F, 2025. Single Stage Production of Ultra-High-Grade Iron Ore using a Novel Fluidisation Arrangement in a Reflux Classifier, *Minerals Engineering*, manuscript under review.

Use of waste rock fill and tailings in the construction of a tailings dam lift

E Baxter[1] and M Medina[2]

1. Dams Engineer, GHD, Melbourne Vic 3000. Email: ella.baxter@ghd.com
2. Senior Tailings Engineer, GHD, Brisbane Qld 4000. Email: mauricio.medina@ghd.com

ABSTRACT

The Henty Gold Mine's Newton Pond Leach Residue Storage Facility (NLRSF) raise project demonstrates innovative approaches to tailings management in challenging conditions. The construction of Lift 9 focused on sustainable practices by utilising on-site waste materials, including waste rock fill and tailings, reducing the operation's environmental footprint.

Throughout the project, several significant challenges arose. The waste rock fill and tailings exhibited considerable variability, requiring geotechnical characterisation to confirm their suitability as construction materials. Tasmania's wet climate presented ongoing challenges, with frequent rainfall affecting material handling and compaction activities. These conditions necessitated flexible construction scheduling and adapted methodologies. Additionally, site conditions prompted design refinements, particularly regarding slope stability parameters and seepage management systems.

The success of this project hinged on strong collaboration between the engineering team and constructor. Regular site meetings enabled swift resolution of technical challenges, ensuring construction proceeded safely while maintaining project momentum.

The completion of Lift 9 delivered numerous benefits for the operation. By utilising on-site waste materials, we significantly reduced the project's environmental impact compared to traditional construction methods requiring imported materials or use of borrow areas. This approach generated cost savings through reduced development of new borrow areas and reducing transport requirements. Moreover, the project has contributed valuable practical insights into waste material utilisation in tailings dam construction, advancing industry knowledge in sustainable mining practices.

This experience offers valuable lessons for tailings engineers, particularly regarding construction efficiency and environmental performance optimisation within the context of mine waste management. The project demonstrates how reflective engineering, and strong teamwork can transform operational challenges into opportunities for sustainable practice.

INTRODUCTION

Historically, tailings dam raises in Australia have predominantly relied on fine, low-permeability soils sourced from dedicated borrow areas. This conventional approach leads to the development of extensive borrow pits, creating significant land disturbance that requires rehabilitation efforts. The availability of suitable materials is often uncertain, with borrow areas sometimes found at considerable distances from construction sites. Consequently, material haulage not only increases project costs but also generates additional environmental impacts through fuel consumption and emissions.

These industry-wide materials challenges were particularly evident at Henty Gold Mine which is located approximately 22 km north of Queenstown on the west coast of Tasmania Australia. The mine deposits tailings in an existing tailings storage facility (TSF) located approximately 2 km south of the mine, named the Newton Pond Leach Residue Storage Facility (NLRSF).

Henty Gold Mine had consistently used clays to construct their lifts however, during construction of Lift 8 it was identified that the existing clay borrow was going to be exhausted for the construction of Lift 9. Therefore, during the design phase (GHD, 2023a) an evaluation of importing material versus utilising materials available on-site was done, highlighting the potential of using existing waste materials, including tailings and waste rock.

In addition to the shortage of materials available another major challenge during construction of Lift 9 was the significant rainfall that this region experiences. The west coast of Tasmania experiences some of Australia's highest precipitation levels, with data from the Bureau of Meteorology at the Rosebery climate station (14 km from the NLRSF) showing annual rainfall averaging approximately 2000 mm (Figure 1). These significant rainfalls were accounted for in the design and construction planning process, however also resulted in a number of complications during the construction process.

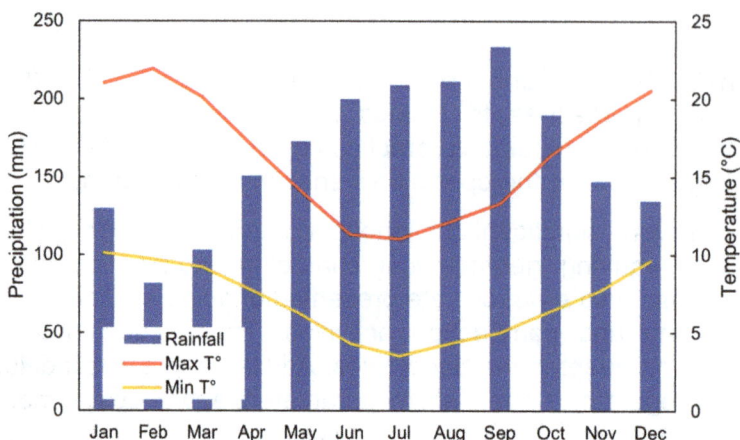

FIG 1 – Annual rainfall in Rosebery, Tasmania (BOM, 2024).

This paper examines the innovative approach of using mine waste materials in the construction of the NLRSF Lift 9, which raised the embankment 1.5 m from RL 2523 m to RL 2524.5 m using both downstream and centreline methodologies. It explores the technical and practical challenges encountered, the solutions developed, and the valuable lessons learned that may benefit similar projects facing material constraints in challenging environments.

CONSTRUCTION HISTORY

The embankment design for NLRSF has evolved significantly from its initial conception. Previous designs predominantly utilised an upstream method, where new embankments were constructed directly on tailings. While this approach was beneficial regarding material volumes use and cost, it presented challenges regarding stability and seepage control, particularly in light of the site's limited availability of low-permeability materials.

The initial NLRSF embankment was constructed from rock fill and clay in early 2001 to a crest height of RL 2505 m. As shown in Figure 2, the embankment has been raised eight times since then. The NLRSF is now at Lift 9 with a crest level of RL 2524.5 m.

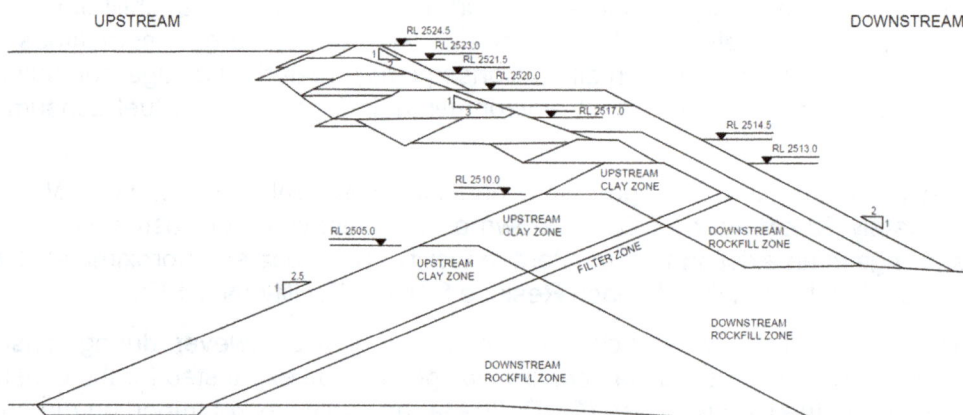

FIG 2 – NLRSF main embankment historical lifts.

MATERIAL SOURCING STRATEGY

Due to previous raises of the NLRSF materials available on-site had depleted during construction of Lift 8, prior to design and construction of Lift 9. A solution to this problem was to use existing waste materials on-site including tailings and waste rock.

The existing on-site borrow area, which had previously supplied low-permeability clay-like material for earlier lifts, was depleted during the construction of Lift 8 of the NLRSF. A proposed alternative for sourcing low permeability materials was to repurpose tailings from the beaches of the NLRSF. CPT testing was undertaken around the upstream perimeter of the dam to assess the condition and consistency of the tailings material. Material which had been deposited within 50 m of the upstream edge of the embankment (top 1 m) was deemed appropriate due to the particle distribution and moisture content being close to their optimum moisture content.

During the period between the completion of Lift 8 of the NLRSF and Lift 9, Henty Gold Mine stockpiled their excess waste rock from underground operations in the on-site quarry to the south-east of NLRSF as well as another borrow to the north of the mine. The waste rock was tested by HGM and was found to be non-acid forming (NAF), and the general particle distribution of the waste rock was undertaken during site investigations to confirm the material was appropriate as a construction material.

CONSTRUCTION OF THE LIFT

Design

The design of the lift (Figure 3) involved a rock fill downstream portion of the raise which overlaid the existing embankment (with a geotextile liner) with a 3 m wide upstream tailings zone, with a geosynthetic clay liner (GCL) on the upstream face to provide an additional low permeability barrier. The GCL confining layer was also constructed from tailings (GHD, 2023a).

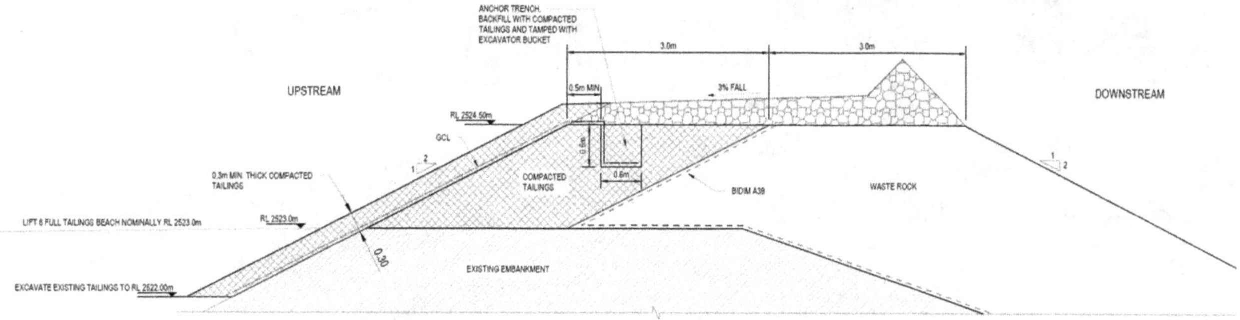

FIG 3 – General design cross-section (GHD, 2023a).

Construction methodology

Waste rock

Construction began with the placement of the downstream rock fill buttress. Waste rock material was sourced from multiple stockpiles around the Henty Gold Mine. The primary stockpile was in the existing quarry approximately 0.5 km south-east of the NLRSF, which provided approximately 65 per cent of the required rock fill volume. When this source was depleted, additional material was sourced from a secondary stockpile situated 7 km north of the site near the mine's operational area. This greater distance significantly increased haulage requirements, with trucks traveling an additional 13 km round-trip compared to the primary source. Despite these logistical challenges, utilising both stockpiles enabled the project to meet material requirements without the need for additional drill and blast operations, which would have extended the construction timeline and increased environmental disturbance.

Waste rock was placed in uniform 500 mm layers using a tracked excavator and compacted with eight passes of a 20-tonne smooth drum roller. This compaction methodology was determined by an initial trial conducted at the start of construction, which assessed the compaction level under different number of passes (GHD, 2023b).

Quality assessment of the waste rock was mainly done by visual inspection rather than laboratory testing once construction commenced. While Henty Gold Mine undertook assessment of the waste rock to confirm NAF characteristics, ongoing quality control during placement depended on the site engineer's visual assessment of each load. This approach was adopted due to the impracticality of frequent laboratory testing given the project's timeline constraints and the remote location. As per technical specifications material containing more than 10 per cent by weight passing the 4.75 mm sieve would be rejected as too fine, as would any material containing particles exceeding 400 mm in diameter. Non-compliant loads were immediately identified, rejected by the site engineer, and removed by the excavator operator before incorporation into the embankment.

Tailings

Tailings were initially won and stockpiled around the upstream edge of the dam, where the tailing has the lowest moisture content. A long reach excavator formed shallow (<1 m) borrow pits along the whole perimeter of the dam to prevent pooling of water (Figure 4). The ability to stockpile material around the whole perimeter of the dam reduced cartage costs. Once the tailings were deemed to be at an appropriate moisture content, they were placed in 300 mm layers and compacted using a pad foot roller to assist in breaking up any clumps in the material.

FIG 4 – Stockpiling of tailings around dam using long reach (GHD, 2023b).

Approximately 14 000 m^3 of tailings were placed and compacted in the embankment for Lift 9 (Figure 5). All lots of tailings placed underwent quality assurance testing during construction including compaction and moisture testing and characterisation testing. The tailings were required to be compacted at 95 per cent Maximum Dry Density (MDD) and within -2 per cent to +4 per cent wet of the Optimum Moisture Content (OMC). The moisture requirement was kept as wide as possible, whilst also ensuring the quality of the final product, due to the unpredictable weather on the West Coast of Tasmania and its potential impact on the construction programme.

FIG 5 – Placement and compaction of tailings portion of the batter against the waste rock and geotextile (GHD, 2023b).

The majority of the results were compliant with the tailings batter having an average compaction of 102 per cent MDD and an average moisture variation of 1 per cent wet of OMC, Figure 6 provides the results of all tests. Where the tailings were not found to meet the specified limits, the lot would be remediated and retested as per technical specifications. There were only six lots which failed, one for not meeting the required compaction and five for not meeting moisture requirements. The contractors strategic staging of the construction, based on rainfall trends, allowed the tailings batter to be constructed in the drier months, ultimately reducing the amount of rework required, which assisted in meeting budget and time restraints.

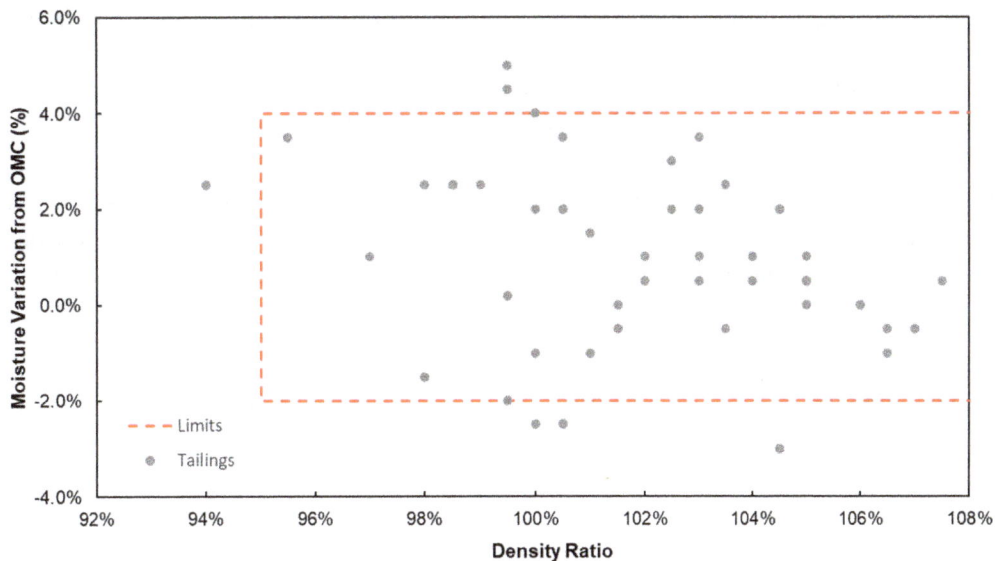

FIG 6 – Construction Quality Assurance testing results for tailings (GHD, 2023b).

GCL was used as an additional low permeability layer on the upstream face of the tailings, and tailings were used as a confining layer. Initially, a thin 300 mm layer of tailings was placed over the GCL however the layer thickness was increased, and a flatter slope to prevent damage during rainfall events or blow outs of tailings pipeline.

ENVIRONMENTAL AND SUSTAINABILITY BENEFITS

The use of waste materials in construction not only addressed the lack of available construction materials on-site but also had a number of environmental and sustainability benefits.

The use of the waste rock from existing stockpiles increased the future storage capacity of these areas for Henty Gold Mine without extending the footprint of the stockpiles. Therefore minimising impact to the surrounding environment.

Sourcing and stockpiling tailings around the perimeter of the dam eliminated the need for carting of the material to construction areas. This reduced not only cartage costs but also emissions from use of the dump trucks. It also provided additional tailings storage space in the dam, extending the time it would take to fill the dam.

CHALLENGES AND SOLUTIONS

During the construction, multiple challenges arose due to weather conditions and variability in the waste materials available for use.

A frequent challenge encountered with the waste rock, as it was sourced from underground and came from changing geological units, was its variability. Some material which had been stockpiled for longer was also more conducive to breaking down during compaction. The breakdown of the waste rock material resulted in an increase in the proportion of fines, resulting in non-conforming material. To address this issue the site engineers undertook regular checks of the material both pre and post compaction. Layers of rock fill would be inspected as it was being spread and then following compaction test pits dug to allow for visual inspection. The site engineer would assess the overall particle distribution and if required would instruct operators to remove portions of a layer to be removed if the material had broken down and was outside of tolerance.

The frequent wet weather on Tasmania's west coast also presented significant challenges for stockpiling and placing for the tailings. Heavy rainfall events caused rutting and erosion of the fine portion of the tailings, therefore, to minimise these issues, it was determined that tailings placement and compaction should be done during dry weather windows. Additionally, it was decided that rather than trimming the slopes to their final surface, the tailings were over placed ('fat') in the slopes along the entire perimeter. These over-placed slopes were only trimmed to their final design profile prior to GCL installation, minimising exposure time. This approach proved effective, as any subsequent rainfall-induced scouring affected only the sacrificial outer layer of tailings. Compaction testing confirmed that despite the challenging weather conditions, the core tailings material maintained its required engineering properties throughout construction.

As the mining operations continued during the construction period and the NLRSF is the only active tailings storage facility at the site, HGM were continuously depositing tailings in the dam. This was a challenge when trying to win tailings from the beach. If operations deposited tailings near stockpiles or borrow areas it would cause the moisture content of the tailings to increase and the ability to use them in construction decrease. As a result, in some areas the tailings could not be won and had to be carted from elsewhere. To address this issue the embankment was constructed in stages to allow the critical/lowest storage capacity areas to be raised first to allow for deposition to continue without overfilling. This was successfully in allowing construction to continue however highlighted the importance of the relationship between the mine and contractors to allow this to be a seamless process.

LESSONS LEARNED AND RECOMMENDATIONS

Use of waste materials is highly dependent on the site and materials available at the time of construction. It is highly recommended that testing be undertaken on the tailings to understand their permeability and the compaction requirements and whether an additional synthetic layer is required. It is also recommended that compaction trials be undertaken to assess the condition on the rock fill after compaction and whether any issues may arise during this stage.

Construction timing is also an important aspect as tailings batter construction should not be undertaken during inclement weather. Moreover, if using tailings a confining layer, time to filling should be considered as any additional inclement weather could cause the tailings to wash away and expose the liner.

CONCLUSIONS

The NLRSF Lift 9 construction project at the HGM represents a forward-thinking perspective in tailings storage facility design and construction. The project team successfully adapted to material and environmental challenges and demonstrated flexibility and innovation. The shift from traditional upstream raises to a combined approach of centreline and downstream raises facilitated the efficient use of limited materials and ensured the facility's structural integrity and environmental compliance. Introducing a GCL and strategically using compacted tailings and rock fill exemplify the project's commitment to sustainable practices and efficient resource management.

This project underscores the critical importance of adaptability in changing conditions and constraints, setting a precedent for future tailings storage facility constructions in such conditions. By balancing the demands of material availability, environmental considerations, and structural stability, the NLRSF Lift 9 project extends the facility's lifespan and enhances its capacity, ensuring the HGM's continued operation and environmental safety.

Through proactive planning, creative design, and dedicated execution, the project team contributed significantly to the field of tailings management, offering valuable insights and strategies for similar future endeavours.

ACKNOWLEDGEMENTS

We sincerely thank Henty Gold Mine for their invaluable collaboration and support in the development and execution of the NLRSF project and for their openness to sharing data, insights, and operational challenges. This paper stands as a testament to the progressive vision and commitment of Henty Gold Mine towards sustainable mining practices and environmental stewardship.

The collaboration between Henty Gold Mine and GHD highlights the importance of partnership in achieving engineering excellence and environmental sustainability in the mining sector.

REFERENCES

Bureau of Meteorology (BOM), 2024. Climate statistics for Australian locations – Rosebery (HEC Substation). Available from: <http://www.bom.gov.au/climate/averages/tables/cw_097073.shtml> [Accessed: 2024].

GHD, 2023a. Newton Pond Leach Residue Storage Facility, pre-construction report for Lift 9 (RL 2524.5m), internal document.

GHD, 2023b. Work as Executed Report, Newton Residue Storage Facility Raise to RL 2524.5 m, internal document.

Clearing the muddy waters – the do's and don'ts of tailings filtration technology

H Cifuentes[1]

1. Principal Tailings Consultant, Tailings HC, Brisbane Qld 4068. Email: hernan@tailingshc.com

ABSTRACT

The pressing need for sustainable mining practices has propelled the development and implementation of innovative solutions, among which tailings filtration technology stands out as one of the main options. Tailings, the byproduct of mining and mineral processing, present considerable safety and environmental risks if not managed appropriately. Tailings filtration technology plays a crucial role in mitigating these risks, underscoring its importance in water conservation, enhancing physical stability, reducing the environmental footprint, etc.

This paper will provide an overview of tailings filtration technology, detailing its operational mechanism and the environmental benefits. It delves into the operational mechanisms and environmental advantages of tailings filtration and filtered tailings stacking compared with conventional approaches. The paper will highlight some of the 'do's' to understand tailings characteristics, strategic integration of filtration systems within broader tailings management practices, and a proactive approach to regulatory compliance and stakeholder engagement. Furthermore, the paper will draw attention to some common pitfalls in implementing tailings filtration technology— the 'don'ts' to avoid. These include underestimating the technological complexity and economic investment required, neglecting filtration systems' scalability and maintenance needs, and overlooking the importance of adapting to evolving environmental standards and community expectations. Discussing these pitfalls is meant to serve as a cautionary guide and acknowledge the importance of meticulous planning, continuous evaluation, and adaptive management in successfully adopting tailings filtration technology.

INTRODUCTION

The mining industry is undergoing a structural shift in how it manages tailings—a shift accelerated by escalating ESG expectations, water scarcity, climate change, and an intensifying focus on risk-based design standards such as the Global Industry Standard on Tailings Management (GISTM; Global Tailings Review (GTR), 2020). Traditional slurry tailings storage facilities (TSFs) are increasingly seen as vulnerable to failure, regulatory challenge, and social opposition—especially in regions with high seismicity, limited land availability, or scarce water resources.

In this context, tailings filtration and dry stacking have emerged as transformative technologies that promise to decouple tailings storage from water retention. By mechanically dewatering tailings into a stackable, trafficable 'cake,' filtration systems enable geotechnically stable landforms, reduce closure liabilities, and enhance water recovery, often exceeding 80 per cent (Amoah, 2024; Coghill et al, 2024; Fitton, 2024). These benefits are particularly significant in arid environments, high-altitude operations, or jurisdictions moving toward dry closure mandates.

However, the implementation of filtration is not without complexity. As outlined in the BHP Rio Tinto *Filtered Stacked Tailings: A Guide for Study Managers* (Coghill et al, 2024), success depends not only on selecting the right filter but also on integrating filtration into the mine's water balance, haulage strategy, closure plan, and stakeholder expectations. Industry data and case studies suggest that underperformance often results from treating filtration as a stand-alone process rather than part of a broader tailings management system (Davies, 2011; Grohs et al, 2024).

This paper seeks to 'clear the muddy waters' by demystifying the core principles of tailings filtration, distilling key 'Do's' and 'Don'ts' based on recent operational learnings, and highlighting emerging best practices that align with long-term sustainability goals. It integrates technical reviews, field case studies, and strategic guidance to provide a roadmap for implementing filtration solutions that are robust, adaptive, and ESG-aligned.

UNDERSTANDING TAILINGS FILTRATION TECHNOLOGY

Technology overview – filtration methods

Tailings filtration can be achieved through a range of mechanical dewatering technologies, selected based on throughput needs, particle size distribution, and target moisture content. The most used systems include:

- Vacuum Disc Filters: Suitable for coarse or moderately fine tailings with good drainage properties. These systems offer continuous operation and relatively low energy demand but are limited in achieving very low moisture levels.

- Drum Filters and Belt Filters: Applicable in some intermediate cases but less common in modern high-capacity mining operations due to maintenance intensity and footprint constraints.

- Pressure Filters (Plate-and-Frame, Membrane Presses): The dominant choice for fine-grained or high-clay-content tailings. These systems deliver lower final moisture contents (often below 18 per cent), making them suitable for trafficable dry stacking in a range of climates. However, they require higher CAPEX and OPEX and are sensitive to feed variability.

The selection of a filtration method must consider throughput, filterability, operational flexibility, and downstream stacking logistics. Pilot-scale testing and tailings characterisation are essential to align technology choices with site-specific requirements (Meneses *et al*, 2024; Sommacal *et al*, 2024).

Filtered tailings characteristics

Filtered tailings typically exhibit solids contents between 80–85 per cent, resulting in a dense, 'cake-like' material. This improves both trafficability and stack stability over conventional slurry tailings. Filtered tailings can be conveyed or trucked and are suited to dry stacking arrangements, which significantly reduce risks associated with water-retaining embankments.

The mechanical and hydraulic behaviour of filtered tailings is influenced by the mineralogy, grain size distribution, and degree of saturation. Amoah (2024) reports that filtered tailings deposited at low moisture content tend to maintain an unsaturated state, reducing the likelihood of internal water table development and pore pressure buildup. However, if deposited at higher moisture, tailings may remain saturated, increasing geotechnical risks if not well drained or compacted.

Grohs *et al* (2024) detail how filtered tailings at the Martabe gold mine achieved 12–14 per cent moisture using plate-and-frame Diemme filters, enabling co-disposal with mine waste and forming a compactable, trafficable material. Similarly, Masengo *et al* (2024) demonstrate successful stacking of filtered tailings over a pre-existing slurry TSF at LaRonde by using a 2 m waste rock bridging layer, supported by instrumentation monitoring to validate stability assumptions.

To assess the filterability of tailings, Meneses *et al* (2024) propose a screening method based on geotechnical properties aligned with the ICOLD (2022) Bulletin 181 classification. This allows practitioners to quickly evaluate whether a given tailings material can feasibly be filtered and dry stacked, based on parameters such as plasticity index, fines content, and specific gravity.

Comparison with conventional tailings

Unlike traditional tailings deposited as high-water-content slurries into tailings storage facilities (TSFs), filtered tailings offer:

- Improved geotechnical performance due to lower pore pressures and better shear strength.

- Reduced environmental risk, particularly in seismic or high-rainfall regions.

- Enhanced water recovery, with some operations recovering up to 85 per cent of process water.

- Lower long-term liability, allowing for progressive reclamation and potential dry closure strategies.

Filtered tailings eliminate the need for large impoundments and decant ponds, mitigating key failure mechanisms such as overtopping, internal erosion, or liquefaction. Fitton (2024) contrasts thickened discharge and filtered stack systems, noting that dry stacks have demonstrated strong performance in terms of both structural integrity and closure.

Amoah (2024) points out that over 95 per cent of current TSFs operate with unthickened or thickened slurries, and that filtered tailings, while not universally applicable, are especially valuable in high-risk or low-rainfall settings. For example, in desert regions where water sourcing is critical and land is limited, filtered tailings allow for smaller footprints and water recycling. Nevertheless, these benefits come with significant trade-offs:

- High capital and operating costs for filter plants, conveyors, and haulage.

- Increased energy use, especially with pressure filters or large tonnage throughput.

- Sensitivity to tailings variability and operational downtime (Hahn, 2024; Coghill *et al*, 2024).

A holistic approach—considering site conditions, water balance, closure objectives, and life cycle costs—is essential to determine whether filtered tailings offer net value. For example, this can be achieved by reverse-engineering the design, starting from the final landform and working backwards through the filtration and transport systems to optimise trade-offs across the tailings continuum (Crystal and Jansen, 2024).

THE 'DO'S' OF TAILINGS FILTRATION

Do understand tailings characteristics

Tailings composition is critical in determining filterability, cake structure, and achievable moisture content. Key factors such as particle size distribution, clay content, plasticity index, and mineralogy should be assessed early in the design process. Meneses *et al* (2024) presents a geotechnical screening method—based on the ICOLD classification system—that enables preliminary evaluation of filtration potential using standard laboratory data. While this approach provides a valuable first-pass tool for early decision-making, it is important to note that the study is based primarily on data sourced from iron ore tailings, which typically exhibit coarser particle size distributions and different filtration behaviour than more complex tailings streams, such as copper or gold, which often contain higher clay fractions and exhibit lower permeability. In general, coarse, free-draining tailings may be effectively dewatered using vacuum filters, whereas fine-grained or high-clay-content tailings typically require high-pressure filtration to achieve target moisture levels for trafficability and stack stability.

Table 1 presents a summary of the key tailings characteristics that influence filterability and filtration system performance, based on practical guidance from recent studies and adapted from the BHP Rio Tinto *Filtered Stacked Tailings: A Guide for Study Managers* (Coghill *et al*, 2024). These parameters should be systematically evaluated during the early stages of filtration design to ensure appropriate technology selection and reliable operational outcomes.

TABLE 1

Key tailings filterability parameters. Adapted from Coghill *et al* (2024).

Tailings characteristic	Relevance to filtration
Particle size distribution (PSD)	Directly affects permeability, drainage, and filter cycle time.
Clay content	High clay content reduces permeability and increases cake moisture content.
Plasticity index (PI)	Indicates plastic behaviour that impacts cake consolidation and trafficability.
Mineralogy	Certain clay minerals (eg smectite) are more problematic for dewatering.
Slurry consistency (solids %)	Higher solids concentrations may improve throughput but affect pumpability.
Chemical composition/solute load	Influences corrosion, cloth blinding, and cake integrity; relevant for long-term performance.
Slurry rheology (viscosity, yield stress)	Affects pumpability and resistance to flow; essential for equipment selection.
Solution chemistry (pH, salinity, redox)	Impacts flocculant performance, cloth scaling, and cake detachment.
Geochemical risks (eg AMD Potential)	High sulfide or reactive minerals may lead to post-deposition oxidation and instability.
Temperature/thermal behaviour	Alters cake consolidation and drying behaviour; affects filter design and maintenance.
Water source and compatibility	Variable water chemistry can impact flocculation, scaling, and filtration efficiency.

Do integrate with the whole TSF life cycle

Filtration should not be treated as an isolated unit operation. Its effectiveness depends on its integration within the broader tailings storage facility (TSF) life cycle—from placement and compaction to water management, closure, and post-closure performance. A life cycle-based approach ensures that filtered tailings contribute meaningfully to long-term geotechnical stability, reduce water retention risks, and support progressive rehabilitation efforts (Morrison, 2022).

Designing a filtered tailings system must therefore begin with a clear understanding of the final landform and closure objectives, rather than focusing solely on filtration efficiency. Crystal and Jansen (2024) advocate for a reverse-engineered approach, starting with the TSF's end-state requirements to guide the selection of dewatering technologies, stacking geometry, and haulage strategies. This integrated methodology enables more cost-effective design decisions, improves environmental and social performance, and ensures alignment with ESG expectations throughout the mine life.

Figure 1 shows how filtration should be integrated across the full TSF life cycle to support long-term stability, efficient closure, and alignment with ESG goals (International Council on Mining and Metals (ICMM), 2021).

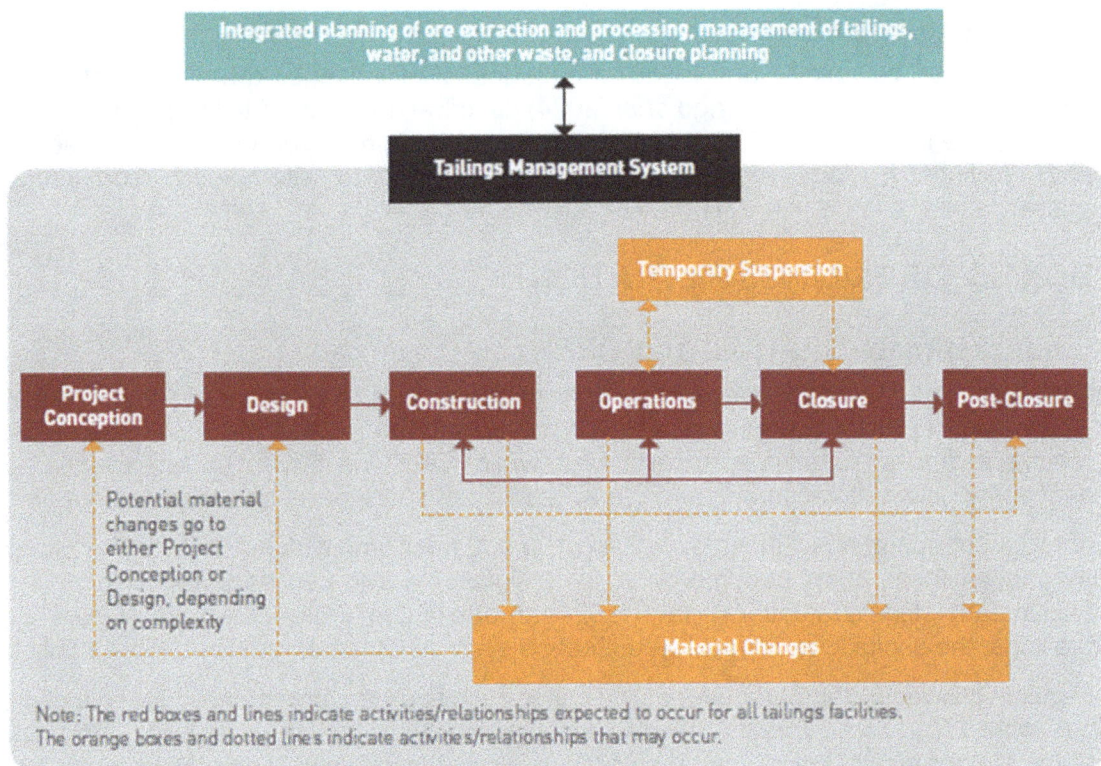

FIG 1 – Filtration must be integrated with broader life cycle planning. Source: ICMM (2021).

Do design for water reuse and recovery

In water-scarce or high-cost regions, filtration systems are increasingly valued for their potential to maximise water recovery and reduce reliance on external sources. Filtered tailings can enable water recovery rates exceeding 80 per cent, making them a key component in achieving sustainable water management objectives. Notable operations such as Karara and Martabe have reported substantial gains in process water recirculation using high-capacity plate-and-frame filter presses (Amoah, 2024; Grohs *et al*, 2024).

To fully realise these benefits, filtration must be integrated into the broader water management strategy. This includes coordinated design between the dewatering circuit, thickener underflow systems, return water infrastructure, and plant process demands. A system-wide approach ensures efficient recovery, reduces water losses, and supports operational continuity and environmental performance targets.

Do engage stakeholders early

The successful implementation of filtered tailings systems depends on early and meaningful engagement with a broad range of stakeholders, including regulators, local communities, and investors. As ESG considerations become more central to project evaluations, factors such as visual impact, water stewardship, and long-term landform stability are increasingly linked to a mine's social license to operate. Proactive stakeholder involvement not only facilitates smoother permitting processes but also builds trust and long-term accountability (Martinez, 2019).

Likewise, as noted by Fitton (2024), social license and risk mitigation are increasingly strong drivers of filtration adoption, even where costs are high.

Internally, tailings filtration projects intersect with multiple functional areas across the operation, from mine planning and metallurgy to closure design, logistics, and procurement. Early coordination with operations teams, closure specialists, and finance departments helps align performance expectations, uncover interdependencies, and manage cross-functional risks.

Do consider site-specificity

Topography, climate, seismicity, and legacy infrastructure all influence the viability of the filtered tailings stacking. For example, Masengo *et al* (2024) describe how filtered tailings were successfully stacked on top of a legacy slurry TSF at LaRonde, using a bridging layer and real-time geotechnical monitoring. Tailings management plans must be tailored to site-specific constraints and opportunities.

THE 'DON'TS' OF TAILINGS FILTRATION

Don't underestimate CAPITAL and OPEX

Implementing tailings filtration represents a significant financial undertaking, involving substantial capital and operating expenditures. Upfront CAPEX is required for filter presses, conveyors, haul trucks, stackers, and related infrastructure. Meanwhile, OPEX is driven by energy use, labour, maintenance, filter cloth replacement, and the availability of spare parts and technical support.

In high-throughput operations—those exceeding 100 000 t per day, which are still in the planning or feasibility phase—these costs become particularly significant and can escalate rapidly if not fully considered in early-stage evaluations. Hahn (2024) reports that in large-scale projects involving fine-grained tailings, the capital cost for pressure filtration systems alone can exceed EUR 1 billion.

While filtration provides clear advantages in water recovery, environmental risk reduction, and alignment with ESG commitments, these benefits must be weighed against their financial implications. A rigorous life cycle cost analysis is essential to assess trade-offs, ensure economic viability, and support informed decision-making at both operational and strategic levels.

Don't let short-term NPV thinking undermine long-term value

A frequent barrier to the adoption of tailings filtration systems lies not in technical feasibility, but in how projects are financially evaluated. Conventional Net Present Value (NPV) models, often used during early-stage decision-making, heavily discount future costs such as closure, long-term monitoring, and post-closure liabilities. This tends to favour low-CAPEX, slurry-based tailings systems that appear more cost-effective in the short-term, even though they defer substantial long-term risks and expenditures (ICMM, 2021; Williams, 2021).

Filtration systems, while more capital-intensive upfront, offer long-term benefits such as enhanced stability, reduced water demand, and smaller closure footprints. These advantages are often underrepresented in traditional financial assessments that prioritise near-term cash flow over life cycle cost optimisation. Figure 2 illustrates this dynamic, showing how cumulative costs associated with slurry tailings escalate significantly in the closure phase, while filtered systems maintain a flatter, more predictable cost profile over the life of the project.

To address this bias, some operations are beginning to explore alternative valuation frameworks—such as whole-of-life cost modelling or real options analysis—that better reflect long-term value creation and ESG alignment. Unless such methods become standard practice, filtration technologies risk being undervalued or dismissed during feasibility, despite their strategic and environmental advantages.

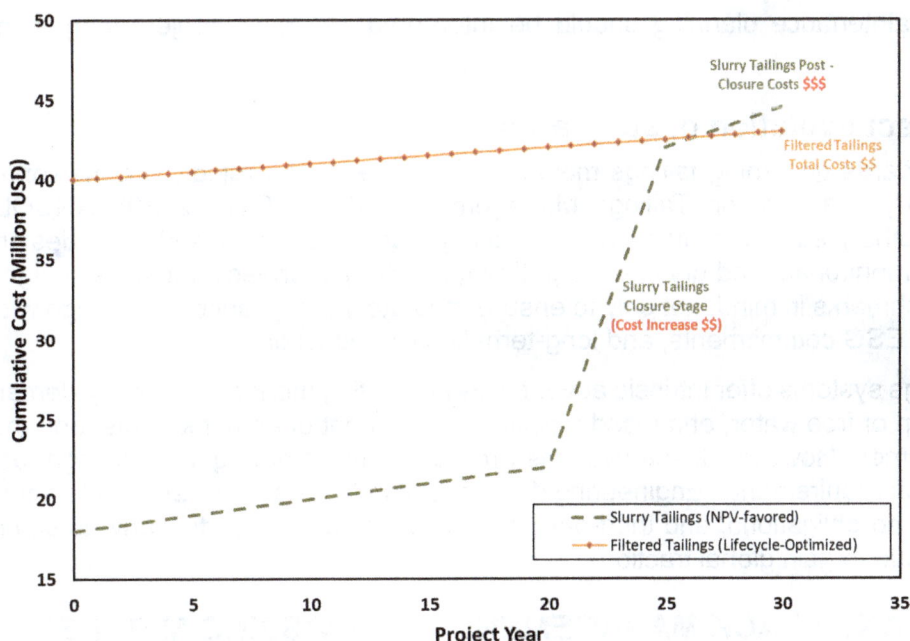

FIG 2 – NPV versus life cycle cost: slurry versus filtered tailings.

Don't assume 'one size fits all'

Each tailings stream and mine site present distinct technical, environmental, and logistical challenges. Relying solely on vendor specifications or laboratory-scale data, without site-specific trials, can lead to underperformance, operational inefficiencies, or costly retrofits. A common pitfall is skipping pilot testing, often due to time or budget constraints. However, pilot-scale trials are critical for validating assumptions related to filter sizing, cycle times, achievable cake moisture, throughput, and equipment behaviour under real process and climatic conditions (Baron van Wassenaer, 2021).

What works at one operation may fail at another due to differences in tailings mineralogy, clay content, throughput rate, water balance, and haulage constraints. Amoah (2024) highlight the importance of testing and modelling actual tailings behaviour at scale, factoring in downtime, variability in tailings feed, and the effects of blending multiple streams. Tailings filtration must be approached as a site-specific engineering solution, not a plug-and-play technology.

Don't forget ongoing maintenance and filter downtime

Filtration systems are mechanical workhorses that require diligent maintenance and operational discipline. Components such as filter cloths, seals, hydraulic actuators, and pumps are subject to wear and must be routinely inspected, serviced, and replaced. Unexpected failures can halt production and lead to significant financial losses without adequate redundancy, such as standby units or a well-stocked inventory of critical spare parts. Downtime risk must be incorporated into both capital planning and operational readiness strategies.

Common challenges include:

- cloth degradation
- hydraulic malfunction
- filter blinding, especially in tailings with variable or fine-grained characteristics.

Grohs *et al* (2024) emphasise that extended pilot testing at Martabe was critical to understanding these maintenance cycles and fine-tuning operational parameters such as blow pressure and squeeze time. Overlooking these issues during the feasibility and design stages can lead to:

- reduced throughput
- suboptimal cake quality
- escalating maintenance costs.

Therefore, maintenance planning should be integrated into early project stages, not left as an afterthought.

Don't neglect evolution of standards

Industry standards governing tailings management are evolving rapidly. Frameworks such as the Global Industry Standard on Tailings Management (GISTM; GTR, 2020), ANCOLD guidelines (2019), and other regional regulations increasingly promote risk-based TSF designs, enhanced performance monitoring, and dry closure pathways. Filtration projects must be designed with these evolving benchmarks in mind, not only to ensure regulatory compliance but to align with stakeholder expectations, ESG commitments, and long-term liability reduction.

Filtered tailings systems offer intrinsic advantages in meeting modern regulatory demands, including the elimination of free water, enhanced stability independent of embankments, and more controlled closure outcomes. However, these systems are not exempt from rigorous design, documentation, and verification requirements. Engineering decisions should proactively anticipate future compliance norms, reporting obligations, and third-party review expectations, particularly as standards like the GISTM continue to gain global traction.

HAULAGE AND STACK MANAGEMENT – THE MISSING MIDDLE

Despite being central to the success of filtered tailings operations, haulage logistics and stack management are often underestimated in terms of complexity, cost, and long-term impact. While much attention is typically given to filtration technology and water recovery, the *'missing middle'*— comprising how filter cake is transported, placed, and managed on-site—can become a dominant cost driver and operational bottleneck if not fully addressed.

Transporting filter cake involves balancing terrain, distance, climate, and material consistency, influencing trafficability, equipment wear, emissions, and scheduling. These challenges intensify at sites with steep topography, remote stack locations, or high climatic variability. As shown in recent studies, effective stack design requires close coordination between dewatering performance, transport system selection, and surface stability considerations (Lara and León, 2011; Grohs *et al*, 2024; Sommacal *et al*, 2024). Yet, in many projects, stack management is treated as a downstream afterthought, leading to costly retrofits, safety risks, and inefficiencies.

Addressing these gaps requires integrating haulage and stack management into early design stages, right alongside filtration system selection. This includes evaluating cake moisture control, terrain constraints, transport method feasibility, and real-time geotechnical monitoring strategies. As operations scale up and ESG expectations tighten, overlooking this 'middle link' in the filtered tailings value chain may erode the very benefits that filtration technologies are meant to deliver.

Filter cake transport options

The transport of filtered tailings, typically in the form of a 'damp cake', is a key design consideration influenced by distance, terrain, moisture content, and daily tonnage. In challenging geographies such as the Andes in South America, topography significantly impacts equipment selection, haul route design, and operational efficiency. Trucking remains the most flexible and widely used option, allowing for adaptive routing and phased development, but it carries high OPEX due to fuel, maintenance, and labour costs. Conveyor systems, by contrast, offer higher efficiency and automation over shorter, more consistent routes, but require higher upfront CAPEX and, in many cases, substantial terrain modifications (Cacciuttolo and Pérez, 2022).

The choice between haul trucks, conveyor belts, and mobile spreaders depends on balancing filter cake properties (eg moisture, cohesion), topographic constraints, and stacking geometry. At Martabe, for example, haul trucks were selected due to the flexible geometry of the filtered tailings stack and the short distance to the plant (Grohs *et al*, 2024). Emerging alternatives such as in-pit stacking and tramline systems are also under evaluation in some large-scale projects, particularly where long-term cost optimisation is a priority.

Stacking equipment and trafficability

Effective stacking operations depend heavily on the trafficability of filtered tailings, which is governed by moisture content, particle gradation, and deposition method. Equipment selection—typically involving haul trucks, dozers, and spreaders—must be tailored to the site's terrain, stacking geometry, and tailings consistency. Moisture content must be sufficiently low to prevent rutting, slippage, and equipment bogging, especially when operating on newly placed lifts. Uniform deposition and adequate compaction are also essential for ensuring slope stability and safe working conditions.

At LaRonde, Masengo *et al* (2024) report successfully using a 1.5–2 m thick rock platform to improve surface trafficability and support equipment movement during placement over a legacy slurry TSF, see Figure 3. This bridging layer was key in mitigating geotechnical risks associated with soft, saturated foundations. Such site-specific adaptations are often required to maintain operational continuity and reduce safety hazards during stack construction.

FIG 3 – Bridgelift construction over TSF (left); filtered tailings stack during construction in 2023 (right). Source: (Masengo *et al,* 2024).

Moisture control versus trafficability

Maintaining the appropriate moisture content in filtered tailings is critical to balancing trafficability, stability, and operational efficiency. Excess moisture can result in soft, untrafficable surfaces, increasing the risk of grooving, slumping, and equipment immobilisation. Conversely, overly dry filter cakes may lead to dust generation, material segregation, and compaction difficulties during placement. The ideal moisture target must therefore account for both geotechnical performance and site-specific operating conditions.

In high-altitude Andean environments, natural desiccation during the dry season can assist in further reducing moisture content after deposition. However, this effect is highly seasonal and should not be relied upon year-round; filtration systems must be designed to meet performance targets even during periods of limited evaporation (Lara, Pornillos and Muñoz, 2013).

Sommacal *et al* (2024) highlight that modern filter presses are increasingly equipped with automated control systems that optimise pressing and blow cycles to maintain consistent moisture levels. These systems help operators manage trade-offs in real time, ensuring the filter cake meets stability and handling requirements under variable processing conditions.

Climate considerations

Climatic conditions are critical in filtered tailings performance, influencing moisture control, stack stability, and operational scheduling. In cold climates, filter cakes may freeze, making them brittle and difficult to spread or compact. In contrast, high rainfall can re-saturate stacks, reducing shear strength, increasing erosion potential, and posing stability risks. Mitigation measures—such as engineered surface drainage, stack covers, and seasonal scheduling—are essential to maintaining safe and efficient operations (Oldecop and Rodari, 2021).

In arid or high-altitude environments, dry and windy conditions can enhance natural desiccation of tailings, aiding moisture reduction. However, they also elevate dust generation risks, which can have environmental and health implications. In wet or tropical climates, filtered stacks must be designed to actively shed water and avoid ponding or infiltration. Protective strategies such as progressive capping, filter cake layering, slope management, and cover systems help maintain stack integrity.

Adaptive design that accounts for seasonal variability in rainfall, freeze-thaw cycles, and wind exposure is crucial. As Amoah (2024) and Crystal and Jansen (2024) emphasise, filtration systems must be robust enough to maintain performance across a wide range of climatic extremes, ensuring resilience, regulatory compliance, and long-term stability.

THE FUTURE OF FILTRATION

Circular economy models

Filtered tailings are increasingly being redefined from inert waste to potential secondary resources. As demand for critical minerals such as rare earth elements, cobalt, and lithium accelerates, particularly in support of the global energy transition, interest in tailings reprocessing is growing. Emerging technologies now enable the extraction of residual metals from historical and current tailings, aligning with circular economy principles and offering the dual benefit of environmental footprint reduction and extended value generation (Das, van Hullebusch and Akçil, 2024).

One landmark initiative is the BHP Tailings Challenge, launched in 2020 to identify technologies capable of reprocessing and repurposing copper tailings (BHP, 2023). The competition attracted over 150 proposals globally, with two solutions advancing to pilot-scale testing. This initiative aims to reduce the volume of tailings requiring long-term storage by converting waste into usable construction and backfill materials.

Filtered tailings stacks offer unique advantages for reuse due to their improved geotechnical stability, uniformity, and accessibility. Reuse pathways currently under evaluation include:

- aggregate recovery for construction and road base

- engineered fill blending for mine or civil infrastructure

- paste backfill integration in underground operations

- capping and cover material for closure landforms.

In Chile and Australia, several projects are actively exploring the crushing and blending of filter cakes to produce products that meet internal or market-based specifications. Unlike traditional slurry impoundments, which are challenging to access post-closure, filtered stacks remain modular and accessible, facilitating future reprocessing or recovery if economic or regulatory drivers change.

Efforts by major miners such as BHP and Rio Tinto further illustrate the industry's commitment to innovation and waste valorisation. Their 2023 joint call for collaboration on novel tailings dewatering and reuse technologies emphasises the critical role of filtration in enabling sustainable tailings strategies (BHP and Rio Tinto, 2023).

Integration with closure planning

Filtered tailings are increasingly recognised as strategic assets in mine closure planning. Their physical characteristics—low moisture, compactability, and structural stability—make them well suited for forming engineered, geo-stable landforms and supporting dry closure strategies. Unlike conventional slurried tailings, filtered materials can be shaped, compacted, and integrated into long-term rehabilitation plans with greater control and predictability. This facilitates progressive reclamation, reduces long-term water management needs, and aligns with evolving regulatory and ESG requirements (Jose et al, 2024).

Dry stacked tailings enable visual and functional integration into the surrounding landscape, supporting landform design principles such as contouring, erosion resistance, and vegetation establishment. Masengo et al (2024) describe how LaRonde's filtered tailings stack incorporates staged capping and revegetation based on real-time monitoring of in situ behaviour—an approach that reduces post-closure risks and long-term monitoring obligations. When properly planned, filtered tailings contribute directly to landform stability, water management, and socio-environmental legacy objectives.

Data-driven optimisation

As tailings filtration systems scale up and operational complexity increases, data-driven tools play a growing role in optimising performance. Advances in real-time monitoring, sensor integration, and process automation enable more consistent filter operation, moisture control, and predictive maintenance, particularly in high-throughput environments where small inefficiencies can have significant cost implications.

Modern filter presses are increasingly equipped with smart control systems that monitor key parameters such as cake moisture, cloth wear, blow cycles, and energy consumption. These systems support dynamic adjustments to pressing conditions, reducing downtime and improving throughput. As highlighted by Hahn (2024), tracking moisture variability between shifts or across filter units can help operators fine-tune cycle settings and proactively schedule maintenance.

Integration with plant-wide data systems, including digital twins, allows for more responsive planning, simulation of performance under varying conditions, and better alignment with water balance and stacking logistics. While the implementation of such technologies still varies across sites, their role in reducing variability and improving life cycle efficiency is becoming increasingly clear.

CONCLUSION

Tailings filtration has evolved from a niche engineering solution into a central pillar of responsible mine waste management. Its ability to reduce water use, enhance physical stability, and support progressive closure makes it increasingly relevant in a mining landscape shaped by ESG pressures, climate resilience goals, and regulatory shifts. Yet, filtration is not a universal remedy—it requires tailored design, life cycle thinking, and cross-disciplinary coordination. This paper has highlighted the importance of understanding tailings characteristics, aligning systems with closure outcomes, and engaging stakeholders from the outset. The risks of underperformance—whether due to short-term economic bias, inadequate maintenance planning, or overlooked haulage constraints—underscore the need for integrated planning and adaptive management. When applied strategically, filtration offers not just operational benefits but a long-term path toward safer, more sustainable tailings practices that align with societal expectations and corporate responsibility.

REFERENCES

Amoah, N, 2024. The place for filtered tailings and stacking in the search for safe and sustainable tailings management, in *Paste 2024: Proceedings of the 26th International Conference on Paste, Thickened and Filtered Tailings* (eds: A B Fourie and D Reid), pp 3–24 (Australian Centre for Geomechanics: Perth). https://doi.org/10.36487/ACG_repo/2455_0.01

Australian National Committee on Large Dams Incorporated (ANCOLD Inc), 2019. Guidelines On Tailings Dams - Planning, Design, Construction, Operation And Closure, revision 1, ANCOLD Inc. Available from: <https://ancold.org.au/product/guidelines-on-tailings-dams-planning-design-construction-operation-and-closure-revision-1-july-2019/>

Baron van Wassenaer, D, 2021. The reprocessing of historic mine tailings, Aalto University. Available from: <https://aaltodoc.aalto.fi/items/4f7c1b42-51c2-4a4b-aefd-b09fb7322043>

BHP and Rio Tinto, 2023. Joint Innovation Call on Tailings Technologies, BHP. Available from: <https://www.bhp.com/news/media-centre/releases/2023/05/bhp-and-rio-tinto-invite-collaboration>

BHP, 2023. BHP Tailings Challenge: Innovating Tailings Reuse, BHP. Available from: <https://www.bhp.com/news/case-studies/2023/08/bhp-tailings-challenge>

Cacciuttolo, C and Pérez, G, 2022. Practical experience of filtered tailings technology in Chile and Peru: An environmentally friendly solution, *Minerals*, 12(7):889. https://doi.org/10.3390/min12070889

Coghill, M, Dressel, W, Liggins, G, Rogers, J, Staines, R and Wisdom, T, 2024. *Filtered Stacked Tailings: A Guide for Study Managers* (ed: R Jansen), (BHP Rio Tinto Tailings Management Consortium).

Crystal, C and Jansen, R, 2024. A holistic approach to large-scale alternative dewatered tailings management: lessons from case studies, in *Paste 2024: Proceedings of the 26th International Conference on Paste, Thickened and Filtered Tailings* (eds: A B Fourie and D Reid), pp 305–320 (Australian Centre for Geomechanics: Perth). https://doi.org/10.36487/ACG_repo/2455_25

Das, A, van Hullebusch, E and Akçil, A, 2024. *Sustainable Management of Mining Waste and Tailings: A Circular Economy Approach* (Springer).

Davies, M, 2011. Filtered Dry Stacked Tailings – The Fundamentals, in *Proceedings of Tailings and Mine Waste 2011 Conference*, 9 p (University of British Columbia).

Fitton, T G, 2024. Avoiding dam failures: is filtration the best solution?, in *Paste 2024: Proceedings of the 26th International Conference on Paste, Thickened and Filtered Tailings* (eds: A B Fourie and D Reid), pp 51–64 (Australian Centre for Geomechanics: Perth). https://doi.org/10.36487/ACG_repo/2455_03

Global Tailings Review (GTR), 2020. Global Industry Standard on Tailings Management (GISTM) [online], Global Tailings Review. Available from: <https://globaltailingsreview.org/global-industry-standard/>

Grohs, K, Liu, M, Satriawan, A, Kunadi, S and Simanjuntak, B, 2024. Application of wet tailings pressure filtration for filtered tailings stack and co-disposal with mine waste at various sites including upstream and downstream of the tailings storage facility, in *Paste 2024: Proceedings of the 26th International Conference on Paste, Thickened and Filtered Tailings* (eds: A B Fourie and D Reid), pp 77–94 (Australian Centre for Geomechanics: Perth). https://doi.org/10.36487/ACG_repo/2455_05

Hahn, J, 2024. How tailing characteristics affect capex and opex in filtration: two case studies, in *Paste 2024: Proceedings of the 26th International Conference on Paste, Thickened and Filtered Tailings* (eds: A B Fourie and D Reid), pp 105–114 (Australian Centre for Geomechanics: Perth). https://doi.org/10.36487/ACG_repo/2455_07

International Commission on Large Dams (ICOLD), 2022. Bulletin 181: Tailings Dam Design – Technology Update, ICOLD.

International Council on Mining and Metals (ICMM), 2021. *Tailings Management: Good Practice Guide*, first edition, ICMM. Available from: <https://www.icmm.com/en-gb/guidance/environmental-stewardship/tailings-good-practice-guide>

Jose, S A, Calhoun, J, Renteria, O B, Mercado, P, Nakajima, S, Hope, C N, Sotelo, M and Menezes, P L, 2024. Promoting a Circular Economy in Mining Practices, *Sustainability*, 16(24):11016. https://doi.org/10.3390/su162411016

Lara, J L and León, E, 2011. Design and operational experience of the Cerro Lindo filtered tailings deposit, in *Paste 2011: Proceedings of the 14th International Seminar on Paste and Thickened Tailings* (eds: R Jewell and A B Fourie), pp 25–37 (Australian Centre for Geomechanics: Perth). https://doi.org/10.36487/ACG_rep/1104_03_Lara

Lara, J L, Pornillos, E U and Muñoz, H E, 2013. Geotechnical-geochemical and operational considerations for the application of dry stacking tailings deposits – state-of-the-art, in *Paste 2013: Proceedings of the 16th International Seminar on Paste and Thickened Tailings* (eds: R Jewell, A B Fourie, J Caldwell and J Pimenta), pp 249–260 (Australian Centre for Geomechanics: Perth). https://doi.org/10.36487/ACG_rep/1363_19_Munoz

Martinez, J, 2019. Reutilization, recycling and reprocessing of mine tailings, considering economic, technical, environmental and social features, a review, Montanuniversität Leoben. Available from: <https://pure.unileoben.ac.at/en/publications/reutilization-recycling-and-reprocessing-of-mine-tailings-conside>

Masengo, E, Ingabire, E P, Huza, J and Julien, M R, 2024. Impact of the construction of a filtered tailings stack on top of an existing slurry tailings storage facility at LaRonde gold mine, in *Paste 2024: Proceedings of the 26th International Conference on Paste, Thickened and Filtered Tailings* (eds: A B Fourie and D Reid), pp 321–334 (Australian Centre for Geomechanics: Perth). https://doi.org/10.36487/ACG_repo/2455_26

Meneses, B, Llano-Serna, M, Dressel, W, Coffey, J P and Gerritsen, T, 2024. A geotechnically derived screening method to assess the filterability of tailings, in *Paste 2024: Proceedings of the 26th International Conference on Paste, Thickened and Filtered Tailings* (eds: A B Fourie and D Reid), pp 115–128 (Australian Centre for Geomechanics: Perth). https://doi.org/10.36487/ACG_repo/2455_08

Morrison, K, 2022. *Tailings management handbook: A life cycle approach* (Society for Mining, Metallurgy and Exploration).

Oldecop, L A and Rodari, G J, 2021. Unsaturated mine tailings disposal, *Soils and Rocks*, 44(3). https://doi.org/10.28927/SR.2021.067421

Sommacal, E, Brum, J C, Doveri, F and Boriello, A, 2024. Tailings filtration toward smaller filters with higher efficiency, in *Paste 2024: Proceedings of the 26th International Conference on Paste, Thickened and Filtered Tailings* (eds: A B Fourie and D Reid), pp 95–104 (Australian Centre for Geomechanics: Perth). https://doi.org/10.36487/ACG_repo/2455_06

Williams, D J, 2021. The Role of Technology and Innovation in Improving Tailings Management, in Towards Zero Harm – A Compendium of Papers Prepared for the Global Tailings Review, pp 64–67 (Global Tailings Review / ICMM).

Bioleaching of sulfur enriched coal discards applied to circular economy

M Gcayiya[1], J R Amaral Filho[2] and S T L Harrison[3,4]

1. PhD candidate, Centre for Bioprocess Engineering Research (CeBER), Dept of Chemical Engineering, University of Cape Town, Rondebosch 7701, South Africa. Email: gcymsi001@myuct.ac.za
2. Chief Scientific Officer, Centre for Bioprocess Engineering Research (CeBER), Dept of Chemical Engineering, University of Cape Town, Rondebosch 7701, South Africa. Email: j.amaralfilho@uct.ac.za
3. Executive Dean, Faculty of Engineering, Architecture and IT (EAIT), University of Queensland, St Lucia Qld 4072. Email: sue.harrison@uq.edu.au
4. Professor emerita, Centre for Bioprocess Engineering Research (CeBER), Dept of Chemical Engineering, University of Cape Town, Rondebosch 7701, South Africa. Email: sue.harrison@uct.ac.za

ABSTRACT

Coal waste discards are currently treated as materials of no value despite containing minerals of interest such as pyrite, aluminium silicates, carbonates and rare earth elements (REEs). Under natural weathering conditions and where sulfidic minerals are embedded, these materials generate acid rock drainage (ARD), leading to environmental issues. Isolating pyrite from coal waste can reduce the residual waste's acid generation potential while creating a stream for valorisation. This study explores bioleaching as a method to accelerate sulfur removal from these discards under controlled conditions, mimicking the ARD process in a controlled setting. The bioleaching process enhances sulfur removal efficiency, with sulfur removal exceeding 90 per cent, while generating an iron-rich solution from which we have produced ferrous sulfate crystals with potential for other valuable by-products. Additionally, this approach creates a safer disposal pathway for the residual material, reducing its long-term environmental impact through risk removal rather than delaying risk. The results highlight the potential of integrating bioleaching into circular economy strategies, transforming coal waste into a useable resource while mitigating ARD-related risks. The implementation of bioleaching can contribute to more sustainable mining practices, reducing environmental footprints while extracting value from previously discarded materials. This research underscores the potential for bioleaching to play a key role in the transition toward a circular economy in the mining sector.

INTRODUCTION

In 2018 in South Africa, approximately 70 per cent of primary energy needs are still derived from coal (Bahrami et al, 2018). While renewable energy is increasing in South Africa, coal is expected to continue to play the major role in electricity generation for many years and continues as an export product, hence its handling to minimise environmental burden is key. The run-of-mine (ROM) coal contains a mix of combustible coal, ash forming minerals, impurities of iron and silicates and different sized mining fragments. In order to meet the quality standards required by the market, ROM coal is typically beneficiated, separating usable coal from the waste materials. This waste or discard material is typically disposed in mine waste storage facilities such as dump deposits and tailings dams with no economic value recovered. This contributes to major environmental challenge, particularly for sulfidic coal deposits, as the exposure of these sulfur-containing residues, in particular those carrying pyrite, to water and oxygen, results in the pyrite being oxidised to produce a highly acidic and saline effluent carrying dissolved toxic elements mobilised and high sulfate concentrations, commonly referred to as acid rock drainage (ARD).

A variety of methods have been established to either manage or prevent the formation of ARD. Active treatment methods such as lime neutralisation are extensively used and aim to neutralise acidity using chemical reagents such as sodium hydroxide or lime for inducing metal precipitation (Skousen, Ziemkiewicz and McDonald, 2019). Although effective in reducing acidity, these methods consume large amounts of costly chemicals while generating significant volumes of sludge which require further processing for final disposal, and some of which may re-dissolve. These methods are only effective in reducing ARD risks in the short-term. In contrast, preventive methods to control ARD generation are widely studied and can be used to limit the presence of oxygen and water in waste discard storage facilities. Examples are the use of dry covers, and the co-disposal of coarse waste rock with non-acid forming fine tailings, arranged in either layered or blended (Kotsiopoulos and Harrison, 2017; Machado and Schneider, 2008; Mjonono, Harrison and Kotsiopoulos, 2019; Soares et al, 2009). Most effective risk mitigation is expected from methods that focus on the prevention of ARD generation by the removal of the sulfidic fraction from the coal waste and discards prior to disposal. These approaches have demonstrated potential for effective minimising or elimination of ARD formation while opening possibilities for the use of the sulfidic portion as raw material for other industries (Harrison et al, 2013; Kazadi Mbamba et al, 2012; Tambwe, Kotsiopoulos and Harrison, 2020).

The removal of sulfide from mine wastes can be accomplished through chemical, biological or physical methods, and studies have demonstrated several benefits in terms of ARD management and control. By processing a coal waste discard stream rich in pyrite using dense medium separation, at least 50 per cent of the sulfide contained in the feed bulk discards could be recovered with a mass recovery varying from 2 to 10 per cent depending on the characteristics of the stream (Amaral Filho et al, 2017, 2022; Weiler, Amaral Filho and Schneider, 2016). After pyrite isolation, the ultimate discards have a reduced acid generation potential, associated with enhanced environmental considerations, improved potential for re-purposing and consequently reduced long-term liability, together presenting socio-economic and environmental benefits (Harrison et al, 2013; Weiler, Amaral Filho and Schneider, 2014; Weiler and Schneider, 2019). In addition to reducing ARD risk, the integration of a sulfide removal step into the beneficiation circuit offers the opportunity for value recovery and enhanced resource efficiency (Amaral Filho et al, 2013; Harrison et al, 2013). This approach would reduce the environmental damage caused by coal waste storage facilities, the volume of coal waste disposed and the reagent requirement for ARD treatment (Amaral Filho et al, 2017; Weiler, Amaral Filho and Schneider, 2016).

Several studies have explored the potential of producing sulfur and iron-based materials such as, but not limited to, ferrous sulfate heptahydrate, sulfuric acid and iron oxides (goethite, hematite and magnetite) directly from pyrite contained in mine wastes (Colling, Menezes and Schneider, 2011; Harrison et al, 2013; Lopes, 2017; Runkel and Sturm, 2009; Stander, Harrison and Broadhurst, 2022; Vigânico, 2014; Villetti, 2017; Harrison et al, 2020). While the production of sulfuric acid via pyrite roasting is a well-established industrial process, its economic feasibility is highly dependent on the grade of the pyrite. According to Lauriente, Deboo and Sakota (1999), sulfur content must lie between 30 per cent and 52 per cent to ensure cost-effective sulfuric acid production. When the ash content in pyrite is high, the material is unable to sustain the oxidation reaction during roasting. In contrast, the production of iron-based products such as ferrous sulfate heptahydrate through bioleaching of pyrite present in coal discards does not require high-grade material (Tambwe, Kotsiopoulos and Harrison, 2020). Bioleaching uses micro-organisms to regenerate the lixiviants necessary to extract metals from low-grade ores. This method is relatively simple and environmentally friendly. It can be used to process mine waste and concentrates for extraction and recover of metals, but may be associated with long residence times.

Ferrous sulfate heptahydrate ($FeSO_4 \cdot 7H_2O$) is widely used in the field of agriculture as an iron supplement for the treatment of chlorosis in plants (Katyal and Sharma, 1980), as coagulant in treatment of textile dye wastewaters through oxidative processes eg Fenton's reaction (Lopes, 2017), as an iron supplement to treat anaemia in humans (Zaim et al, 2012). Crystals of ferrous sulfate, if further processed, have potential to be used to produce ferric-based coagulants (Mazza et al, 2020), pigments and other material of value such as nano particles of magnetite (Lopes, 2017).

The aim of this study focuses on two aspects: (i) risk removal to prevent acid rock drainage through bioleaching of pyrite present in high sulfur-containing coal discards; and (ii) the recovery of value from the generated iron and sulfate rich leachate by producing ferrous sulfate heptahydrate. Proof of concept experiments were conducted over a period of 1300 days using laboratory-scale glass leaching reactor columns. Both inoculated bioleach and non-inoculated control were evaluated to assess the efficiency and feasibility of bioleaching as an environmentally sustainable method for pretreatment of sulfide-rich coal waste prior to disposal.

MATERIALS AND METHODS

Experimental approach

The experimental approach for this study is illustrated in Figure 1. A leach column was packed with pyrite-enriched coal discards and irrigated with nutrient solution inoculated with iron and sulfur-oxidising micro-organisms to enhance the solubilisation of iron and sulfur from pyrite (Vigânico *et al*, 2011; Tambwe, Kotsiopoulos and Harrison, 2020). The ferric iron rich solution generated from microbial oxidation of ferrous iron was recirculated back into the column using peristaltic pump. The process continued until a leachate rich in ferric iron and sulfate is achieved, along with a desulfurised solid residue that was less prone to acid rock drainage (ARD) generation. Subsequently, the desulfurised solid residue was characterised for total sulfur and carbon content using CHNS analyser, and its acid generation potential is assessed through ARD static tests. The reduction of Fe^{3+} to Fe^{2+} was performed by contacting ferric iron solution with pyrite-enriched material (reducing agent) in a packed bed reactor under anaerobic conditions. Subsequently, the ferrous iron-rich solution produced was mixed with ethanol to precipitate the dissolved ferrous iron as ferrous sulfate heptahydrate. The resulting ferrous sulfate heptahydrate crystals were then purified using ethanol, followed by drying, weighing.

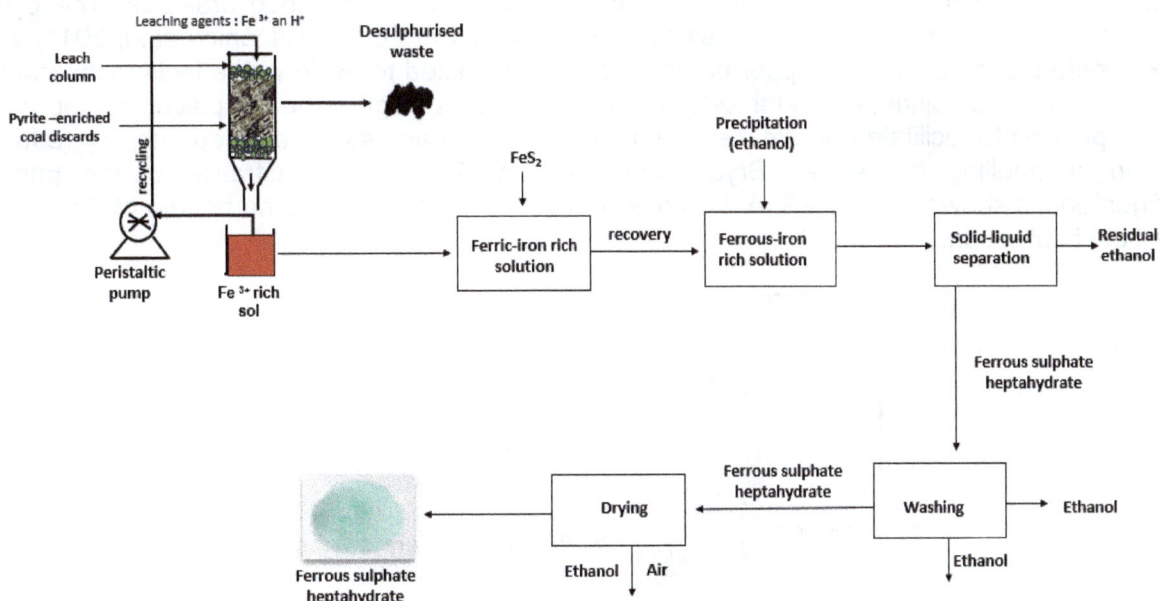

FIG 1 – Process flow diagram for the production of ferrous sulfate heptahydrate and desulfurised solid residue through leaching of pyrite present in coal waste.

Sample preparation and characterisation

The pyrite-rich coal discards with diameter in the range 2 and 50 mm were collected from a dump deposit in the Emalahleni area, Mpumalanga, South Africa. The sample was crushed to liberate pyrite from the surrounding coal matrix, as shown in Figure 2.

(a)   (b)

FIG 2 – Images showing: (a) unliberated pyrite-rich coal discards; and (b) crushed pyrite-rich coal discards to liberate pyrite.

The crushed material was screened to achieve a +2–6 mm size fraction and split representatively into aliquots of 500 g for sample characterisation and column leaching tests. For sample characterisation, a representative sample was pulverised to a particle size less than 37 μm and homogenised. The total sulfur content was determined using PerkinElmer 2400 CHNS/S analyser. The mineralogy analysis was carried out to quantity sulfide minerals and associated gangue materials present in the waste sample using X-rays Diffraction (XRD) analysis.

Acid base accounting tests were conducted to characterise the ARD potential of the waste sample before and after bioleaching according to the methods presented elsewhere (Miller, Robertson and Donahue, 1997; Skousen, Ziemkiewicz and McDonald, 2019). The MPA was calculated from the total sulfur content of the sample based on assumption that all the sulfur present in the sample was sulfidic. The net acid producing potential (NAPP) was determined by estimating the difference between the maximum potential acidity (MPA) and the acid neutralising capacity (ANC): NAPP = MPA – ANC (Skousen, Ziemkiewicz and McDonald, 2019).

Column leaching experiments

The laboratory scale glass reactor column, of 7 cm diameter and 30 cm working height was used in this study. A layer of glass beads measuring 4 mm in diameter was placed at the bottom of the reactor to prevent the loss of smaller discard particles and ensure good drainage. The pyrite-enriched coal discards with particle sizes ranging from 2 mm to 6 mm (Vigânico *et al*, 2011) were packed onto these beads. The upper particle size was selected to avoid wall effects in the packed bed, based on the column diameter selected. Further layers of glass beads placed on top of the discard packing to facilitate uniform distribution of the irrigant across the ore bed, thereby reducing solution channelling (Govender, Bryan and Harrison, 2013). An illustration of the packing configuration is shown in Figure 3. A 1 L collection vessel was placed at the bottom of the column for effluent collection.

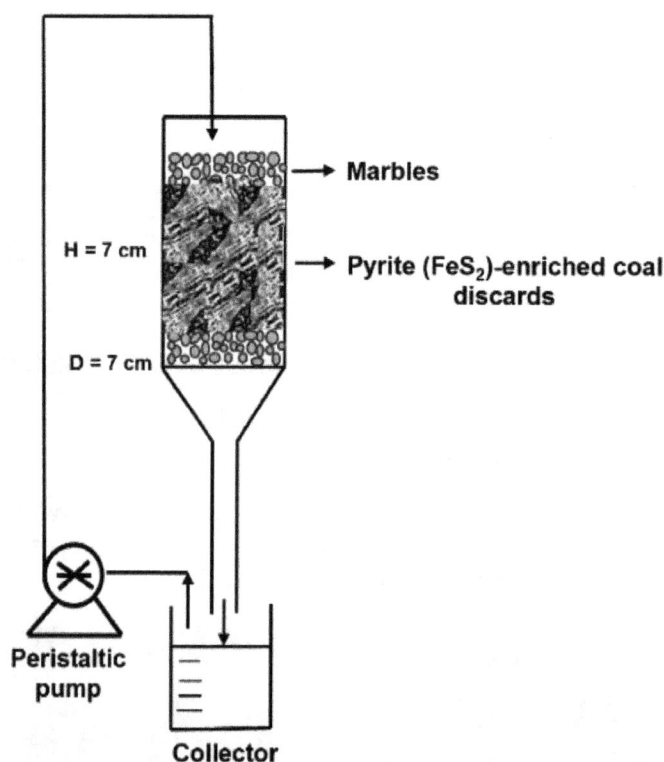

FIG 3 – Experimental reactor column used for column leaching.

The two reactor columns used in this study were initially irrigated with a 1 L of deionised water and the collected leachate was continuously recirculated to the column using a peristaltic pump. The leachate was periodically replaced with fresh feed solution to remove acid neutralising materials. This was repeated until the leachate reached a pH of 4.43 for column A1, and 5.52 for column A2, on day 351. The irrigation with water was selected here to allow demonstration and characterisation of the natural acidification process of these coal discards. It is recognised that, where bioleaching of discards is intentional, this stage of acidification can be accelerated by irrigation with an acidic solution as is typical in the start-up of bioleaching processes.

On day 352, 50 per cent of the leachate was removed in column A1 and replaced with 50 per cent nutrient solution comprised of 183.3 mg/L $(NH_4)_2SO_4$, 60.5 mg/L $NH_4H_2PO_4$, and 111.2 mg/L K_2SO_4 in deionised water with the solution pH was adjusted to 1.6 using sulfuric acid (H_2SO_4), inoculated with 10^9 cells/g of mixed iron- and sulfur-oxidising community of micro-organisms described in *Microbial culture* section, supplemented with 0.5 g/L of Fe^{2+} in the form of $FeSO_4.7H_2O$ to accelerate pyrite dissolution. For column A2, 50 per cent of the leachate was removed and replaced 50 per cent nutrient solution (pH 1.6) of same composition to that in column A1; however, no micro-organisms were added, and this was treated as an un-inoculated control. The system was corrected for evaporation with deionised water. Daily effluent solutions were collected and analysed for pH, redox potential, ferric and total iron concentration, sulfate and cell concentration.

Microbial culture

The microbial consortium used for inoculating column A1 was made up of a 1:1 mixture of two stock cultures, each fully characterised by quantitative polymerase chain reaction (qPCR). The first mixed mesophilic culture comprises an approximately 1:1 ratio of the iron oxidising species *Leptospirillum ferriphilum* and the sulfur oxidising species *Acidithiobacillus caldus*. The culture was maintained in 1 L continuously stirred tank reactor (CSTR) at approximately 35°C with weekly feeding by removing a fifth of the reactor volume and replacing it with fresh nutrient solution and 1 per cent (w/v) pyrite concentrate. This second culture was made up of *Sulfobacillus spp, Sb. thermosulfidooxidans* and *Sb. benefaciens* as shown in Figure 4.

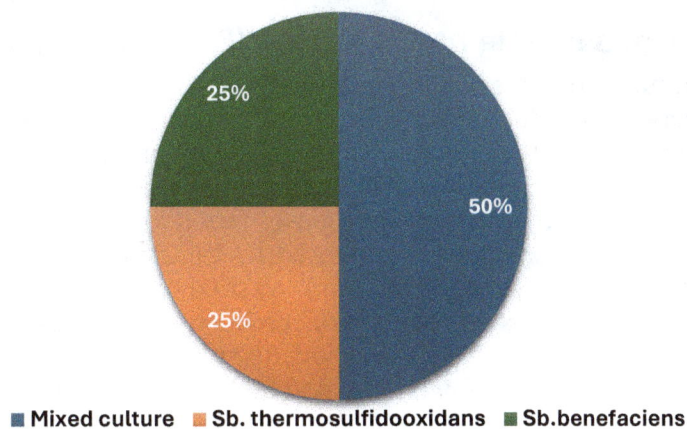

FIG 4 – Composition of mixed microbial consortium used in this study.

Chemical reduction of ferric to ferrous iron in packed bed reactor

The conversion of Fe^{3+} to Fe^{2+} was carried out under the hypothesis that, in the absence of iron-oxidising micro-organisms under anoxic conditions, the ferric iron-rich solution in contact with pyrite-enriched material is reduced to ferrous iron according to Equation 1.

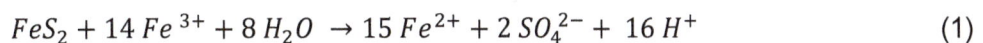

$$FeS_2 + 14\,Fe^{3+} + 8\,H_2O \rightarrow 15\,Fe^{2+} + 2\,SO_4^{2-} + 16\,H^+ \tag{1}$$

The reduction was performed using 50 mL of ferric iron solution generated according to the *Column leaching experiments* section, which was introduced into a packed bed reactor containing 92 g of pyrite-enriched material, as shown in Figure 5. The system was maintained under anaerobic conditions to promote the reduction of ferric iron to ferrous iron. After seven days of contact between ferric iron and pyrite-enriched material, the solution was analysed for pH, redox potential, ferric iron, and total iron concentrations. The ferrous iron concentration was determined by calculating the difference between total iron and ferric iron.

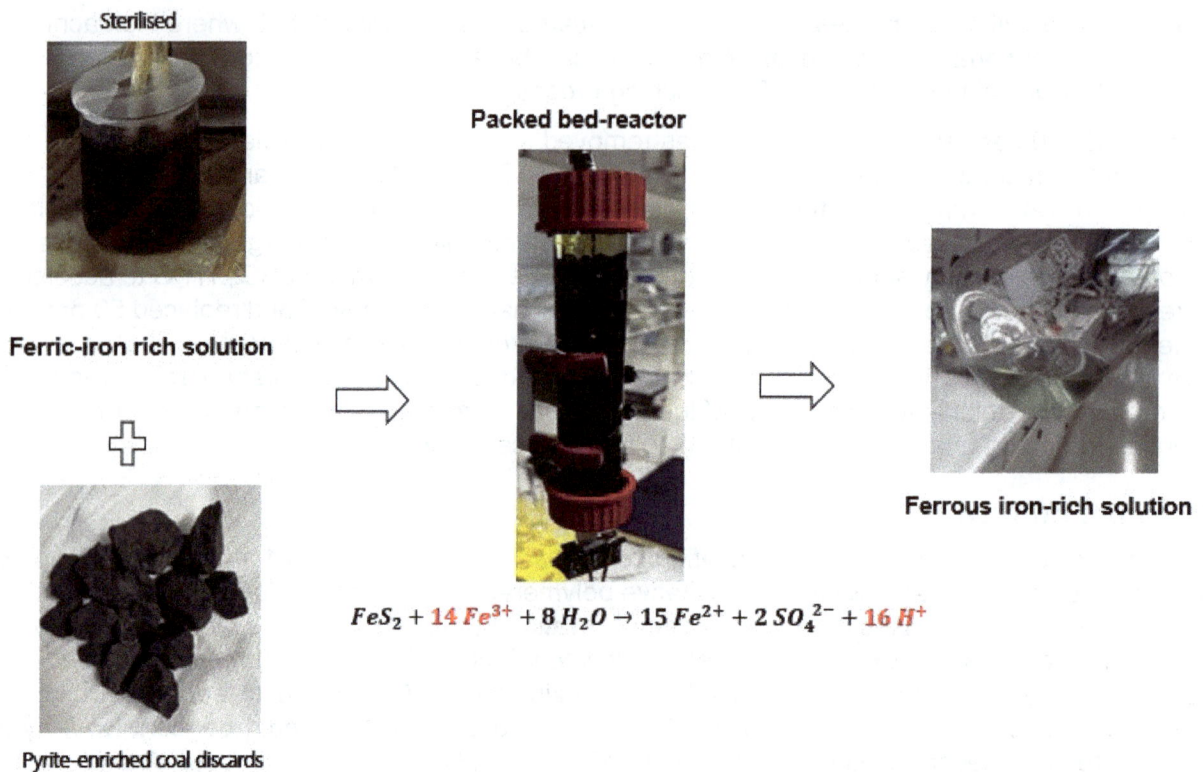

Sterilised

Ferric-iron rich solution

✛

Pyrite-enriched coal discards

Packed bed-reactor

Ferrous iron-rich solution

$$FeS_2 + 14\,Fe^{3+} + 8\,H_2O \rightarrow 15\,Fe^{2+} + 2\,SO_4^{2-} + 16\,H^+$$

FIG 5 – Reduction of ferric to ferrous iron in a packed bed reactor with pyrite-enriched material.

Precipitation of ferrous sulfate heptahydrate (FeSO₄.7H₂O) crystals

A 34 mL aliquot of ferrous iron-rich solution described in the previous section was mixed with an equal volume of absolute ethanol to achieve a solution comprising 1-part ferrous sulfate and 1-part absolute ethanol (99.5 per cent v/v). Precipitation experiment was conducted in a 250 mL Erlenmeyer flask for 20 mins at room temperature, with agitation using a magnetic stirrer bar to enhance contacting and accelerate the precipitation of ferrous iron in solution as ferrous sulfate heptahydrate crystals. The precipitate was recovered by filtration using standard Whatman filter paper with a pore size of 11 µm (diameter 150 mm), washed twice with ethanol, air-dried at room temperature, and subsequently weighed. The overall process flow sheet for precipitation of ferrous sulfate heptahydrate from ferrous iron rich solution using ethanol is provided in Figure 6. The composition of the resultant precipitates was analysed by X-ray diffraction (XRD).

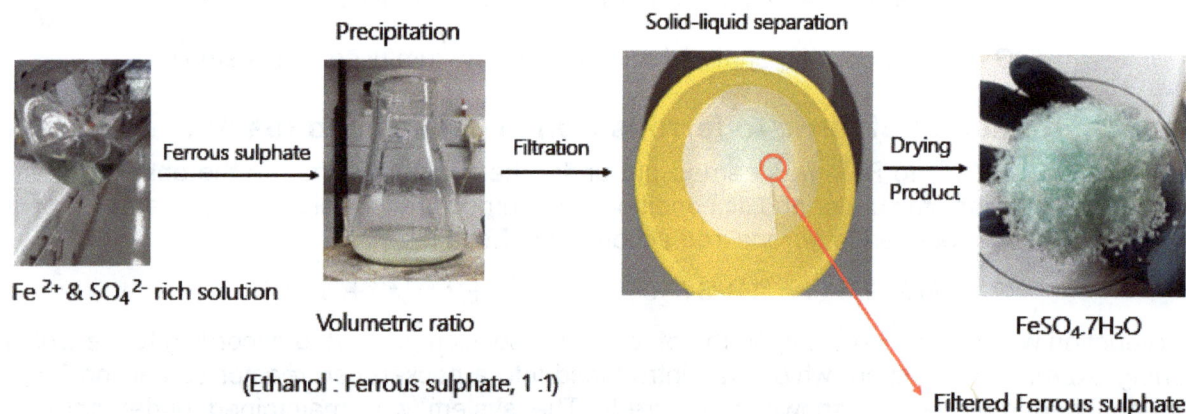

Precipitation

Solid-liquid separation

Ferrous sulphate → Filtration → Drying / Product

Fe^{2+} & SO_4^{2-} rich solution

Volumetric ratio

$FeSO_4.7H_2O$

(Ethanol : Ferrous sulphate, 1 :1)

Filtered Ferrous sulphate

FIG 6 – Steps followed for production of ferrous sulfate crystals from a solution ferrous iron rich solution.

RESULTS AND DISCUSSION

Sample characterisation

Figure 7 shows the XRD diffractogram of the mineralogical composition of the pyrite-enriched coal discards used in this study. The material was analysed using X-ray diffraction according to the method described in the *Sample preparation and characterisation* section. The semi-quantitative results showed that pyrite was the major sulfide mineral present in the waste, accounting for about 51.1 wt per cent (Table 1). The carbonate minerals calcite (8.4 wt per cent), and dolomite (7.4 wt per cent) were the most abundant acid neutralising minerals. The sulfate minerals included gypsum, bassanite and jarosite, accounted for 2.4, 6.1 and 3.8 wt per cent, respectively. The gangue minerals quartz (14.3 wt per cent), kaolinite (3.3 wt per cent), feldspar (1.2 wt per cent), and muscovite (2 wt per cent) were also found. The total sulfur content of the pyrite-enriched coal discard obtained according to the ASTM (2012) method using a LECO sulfur analyser presented approximately 22.3 wt per cent S. Assuming that the majority of sulfur present in the sample is derived from pyrite, accounting for > 96 per cent of the sulfur present (Amaral Filho *et al*, 2022), the calculated pyrite content of the waste material is 42 per cent, indicating high pyrite content. The reason for difference between the calculated pyrite content and the reported content from X-ray diffraction (XRD) analysis can be attributed to the semi-quantitative nature of XRD analysis. XRD only detects crystalline constituents and may underestimate the presence of coal and amorphous components within the waste material.

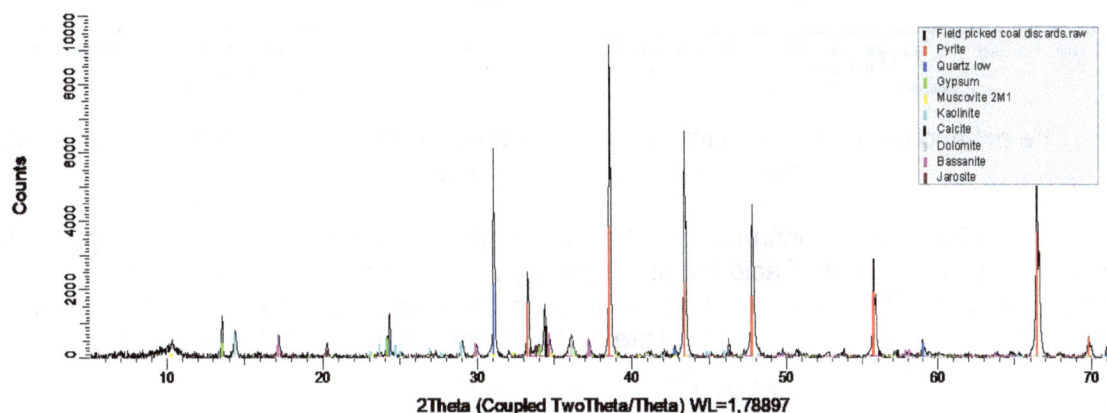

FIG 7 – X-ray diffractogram of the liberated pyrite-enriched coal discards (analysed using XRD).

TABLE 1

Semi-quantitative results of the mineral phases present in the pyrite-enriched coal discards.

Mineral	Chemical composition	Concentration (wt %)
Quartz	SiO_2	14.3
Muscovite	$KAl_2(AlSi_3O_{10})(F, OH)_2$	2.0
Kaolinite	$Al_2(Si_2O_5)(OH)_4$	3.3
Feldspar	$KAlSi_3O_8 - NaAlSi_3O_8 - CaAl_2Si_2O_8$	1.2
Calcite	$CaCO_3$	8.4
Dolomite	$CaMg(CO_3)_2$	7.4
Pyrite	FeS_2	51.1
Gypsum	$CaSO_4 \cdot 2H_2O$	2.4
Bassanite	$Ca(SO_4) \cdot 0.5H_2O$	6.1
Jarosite	$KFe^{3+}_3(OH)_6(SO_4)_2$	3.8

To characterise the ARD generation potential of the sample, static acid-base accounting (ABA) tests were conducted according to methods presented elsewhere (Miller, Robertson and Donahue, 1997; Skousen, Ziemkiewicz and McDonald, 2019; Stewart, Miller and Smart, 2006). With an MPA of 682.4 kg H_2SO_4/t and an ANC of 127.7 kg H_2SO_4/t, resulting in a positive net acid producing potential (NAPP) of 554.7 kg H_2SO_4/t, the pyrite-enriched material used in this study was classified as acid forming (AF).

Column leaching results

A leaching performance was evaluated by analysing the leachate in terms of pH, redox potential, iron and sulfate concentration throughout the experimental run. Figure 8 shows the change in pH and redox potential over time.

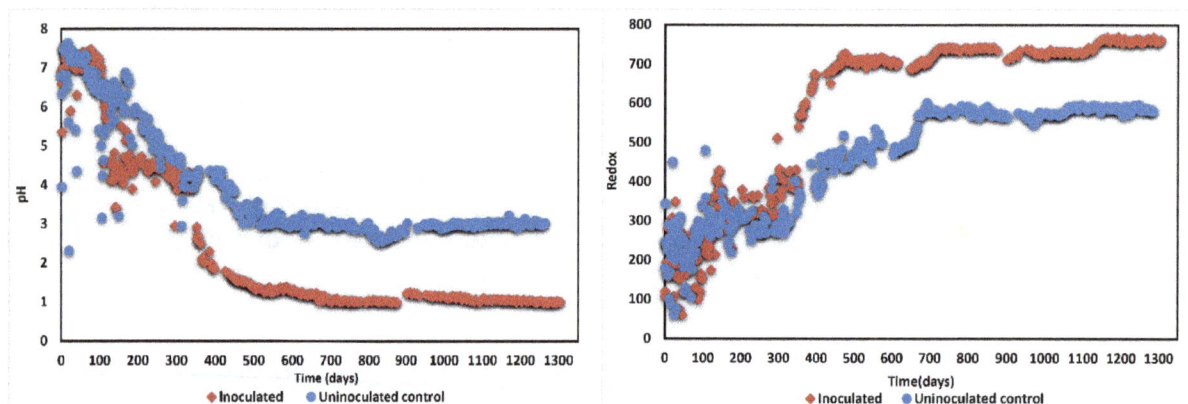

FIG 8 – The pH and redox (Ag/AgCl reference electrode) profiles for inoculated and uninoculated columns over 1300 days of operation.

The characterisation assays confirmed that the coal sample used in this study was rich pyrite. It also contained significant amount of acid consuming gangue minerals, as shown in Table 1. The latter resulted in an initial increase in solution pH in both columns, owing to the ready dissolution of carbonate containing minerals such as calcite and dolomite, as shown in Equations 2 and 3.

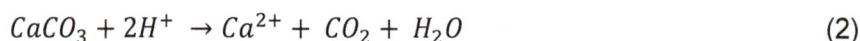

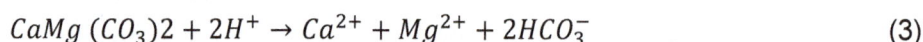

$$CaCO_3 + 2H^+ \rightarrow Ca^{2+} + CO_2 + H_2O \tag{2}$$

$$CaMg\ (CO_3)2 + 2H^+ \rightarrow Ca^{2+} + Mg^{2+} + 2HCO_3^- \tag{3}$$

The high level of acid consuming minerals significantly reduced the efficiency of acidic or ferric leaching by buffering any acid produced from the dissolution of the sulfide mineral and thereby restricting solubility of ferric iron. From day 45, the leachate was replaced periodically with fresh deionised water in both columns, to remove the solubilised acid neutralising materials, simulating an open leach environment. The water replacement in the system and associated removal of alkalinity can be observed from day 50 (inoculated and uninoculated control), reaching pH levels under 5 after 260 days of leaching. The observed reduction in solution pH is due to dissolution of accessible water-soluble acid producing materials. At day 180, solution pHs of 4.46 in the inoculated column and 5.2 in the non-inoculated column were recorded. For the inoculated column (A1) pH remained relatively constant over a long period (to > 300 days). For the non-inoculated column (A2), the pH decreased further to approximately pH 4.7 at day 303 where it remained stable till around day 400. These trends indicated a depletion of water-soluble acid producing materials.

On day 352, 50 per cent of the leachate was removed from column A1, replaced with 50 per cent nutrient solution (pH 1.6), and inoculated with 10^{12} cells/kg of the mixed community of acidiphilic iron-and sulfur-oxidising micro-organisms supplemented with 0.5 g/L of Fe^{2+} in the form of $FeSO_4.7H_2O$ to accelerate pyrite dissolution. For column A2, 50 per cent of the leachate was removed and replaced 50 per cent nutrient solution (pH 1.6) but not inoculated; this served as un-inoculated control. After inoculation of column A1, a notable decrease in the solution pH was observed and redox potential increased to around 700 mV, indicating ferric iron as the dominant form of iron in solution, and remained relatively constant thereafter (Figure 8). Under this environment, microbial growth and activity was enhanced, contributing to effective regeneration of ferric iron and associated accelerated leaching reactions. The increase in redox potential was linked to a decrease in Fe^{2+} concentration, from an initial supplemented concentration of 500 mg/L Fe^{2+} to less than 4 g/mL This was coupled with an increase in ferric iron concentration, owing to its active microbial regeneration as shown in Figure 9, and dominated throughout the remainder of the experimental run. The iron concentration on the non-inoculated column A2 remained below 40 mg/L, due to its high pH (Figure 8). At pH > 2, ferric iron precipitated out of solution. This is evident in the iron profile of non-inoculated column shown in Figure 9.

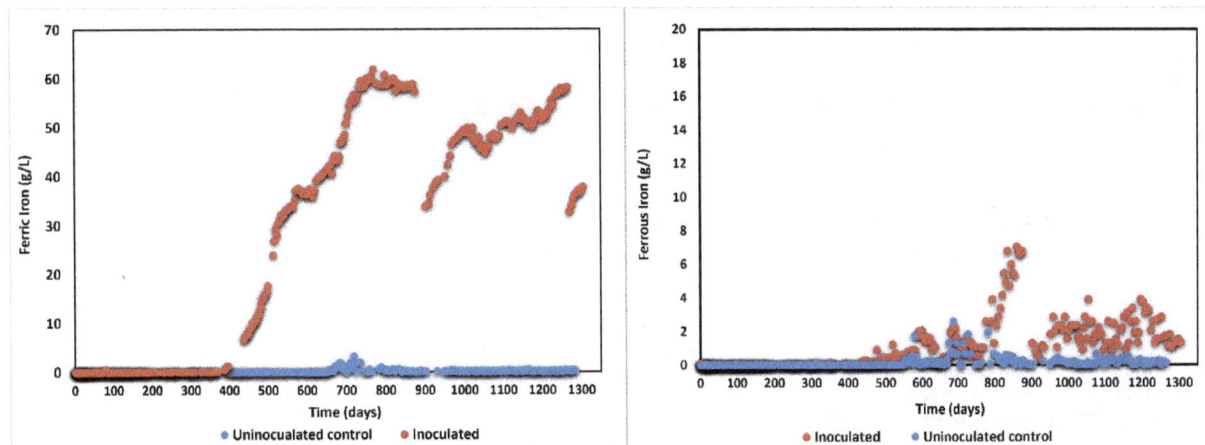

FIG 9 – Ferric and ferrous iron concentration profiles generated through bioleaching of (between inoculated and uninoculated control) of pyrite-enriched coal discards over 1300 days of operation.

Table 2 presents the elemental composition of leachate collected on day 876. Iron was the dominant element in the solution from the inoculated column, primarily resulting from the oxidation of pyrite. Calcium and magnesium from the dissolution of dolomite and calcite (as shown in Equations 2 and 3), were also detected. In the uninoculated column, iron concentrations were below 200 mg/L, likely due to higher pH values (>2), which promote ferric iron precipitation.

TABLE 2

Major elemental composition of inoculated and uninoculated collected on day 876, as determined by XRF analysis.

Major elements	Concentration (mg/L)					
	Fe	Ca	K	Mg	Na	P
Inoculated	62340.0	1678.4	<LOQ	910.6	140.0	185.4
Uninoculated	132.2	649.2	11.3	0.0	388.5	34.6

On Day 700, a ferric iron concentration of approximately 60 g/L was reached in Column A1 whereafter a decrease in microbial activity occurred, and the rate of ferric iron reduction exceeded than microbial ferrous iron oxidation, with ferrous iron showing a sharp increase in solution to around 7 g/L (Figure 9). This suggests inhibition of the microbial community by the high concentration of ferric iron and possibly a depletion of nutrients. After removing the 50 per cent of leachate in inoculated column and replacing it with 50 per cent feed fresh on Day 902, a notable increase in ferric iron and decrease in ferrous iron was again observed, indicating restored microbial activity.

The sulfate concentration in inoculated column increased rapidly after inoculation, suggesting good activity of sulfur oxidising micro-organisms assisting in solubilisation of sulfur in pyrite (Figure 10). This increase in sulfate concentration coincided with an increase in total iron concentration, further supporting pyrite oxidation.

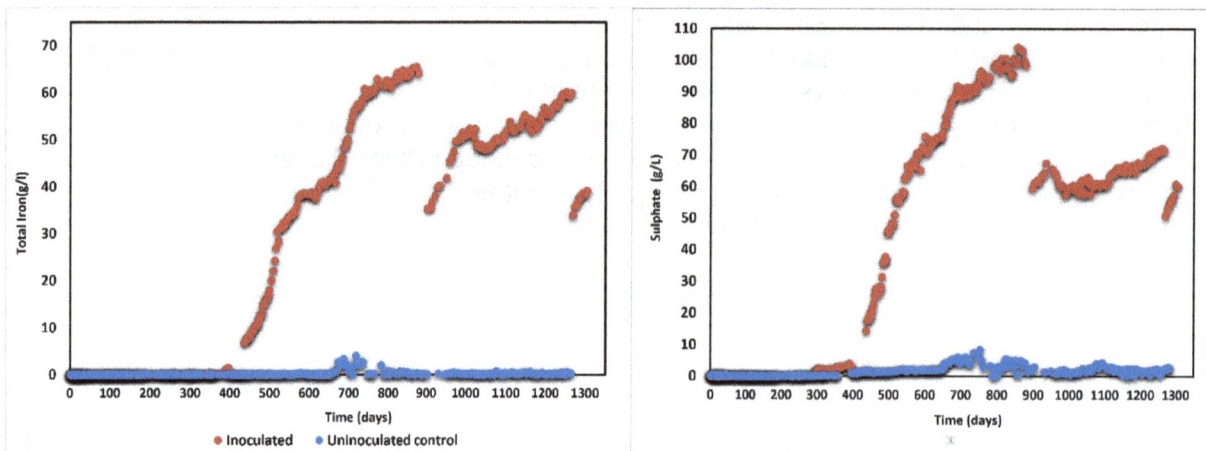

FIG 10 – Total Iron and Sulfate concentration as a function of time generated over 1300 days through abiotic acid leaching (Inoculated and uninoculated control) of South African pyrite-enriched coal discards.

Figure 11 shows the extent of iron and sulfur removal from Columns A1 and A2 over the duration of experiment. Results indicated more than 90 per cent removal of Iron and sulfur from coal discards containing 22.3 per cent total sulfur.

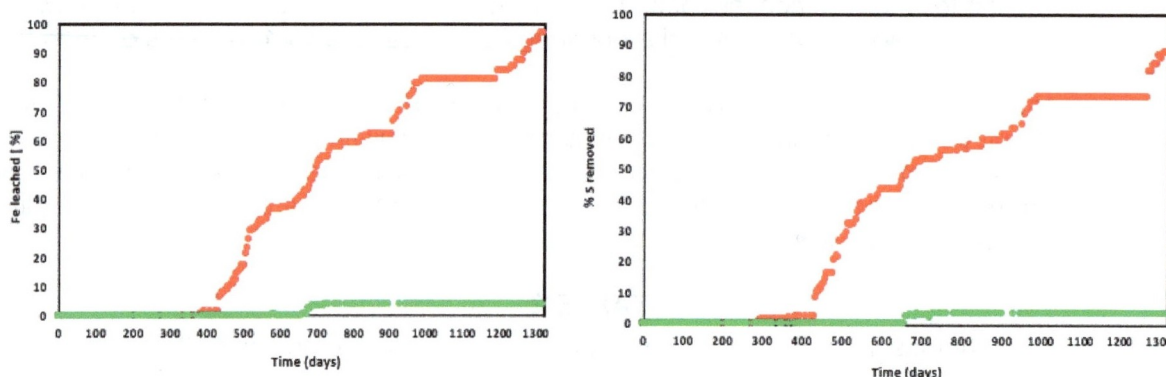

FIG 11 – Extent of Iron and sulfur removal from column leaching test.

This confirms the potential of bioleaching as a viable strategy for desulfurisation of high sulfur-containing coal discards crushed to a particle size of <6 cm during life-of-mine. The bioleached coal discards have a net acid producing potential of -155.3 kg.H_2SO_4/t after 1300 days, hence are classified as non-acid forming, mitigating ongoing liability. On the other hand, the non-inoculated column has a NAPP of 455 H_2SO_4/t and the sample is classified as acid forming.

Reduction of Fe^{3+} to Fe^{2+} and crystals characterisation results

Figure 5 showed the set-up for ferric iron reduction in a packed bed reactor with pyrite-enriched material. The abiotic reduction of Fe^{3+} to Fe^{2+} in the reactor was accompanied by a decrease in redox potential from 717 mV to 311 mV (Table 3), in line with the Nernst equation, which shows a decrease in redox potential as the Fe^{3+}/Fe^{2+} ratio decreases (Equation 4):

TABLE 3

Reduction of ferric to ferrous iron in a packed bed using pyrite-enriched material as the reductant.

Time (days)	pH	Eh	Fe^{3+}	Fe^{2+}	SO_4^{2-}
1	1.23	717	40.9	2.98	63.5
7	2.18	311	0	49 g/L	97 g/L

$$E = E° - \frac{0.059}{n} \log \left(\frac{Fe\,2+}{Fe\,3+}\right) \qquad (4)$$

The ferrous iron increased from 2.98 g/L to 49 g/L on day 7, coinciding with increased sulfate concentration from 63.5 g/L to approximately 97 g/L and this shows a chemical reduction of Ferric to Ferrous accompanied by oxidation of sulfur in pyrite to produce sulfate according to the reaction described in Equation 1.

The precipitation of dissolved ferrous ion in solution as ferrous sulfate heptahydrate using a 34 mL of ferrous iron solution and 34 ml of 95 per cent ethanol to get 1:1 ration, resulted in the recovery of 7.8 g of ferrous sulfate heptahydrate crystals ($FeSO_4 \cdot 7H_2O$), with a precipitation efficiency of 91.3 per cent. Figure 12 shows the recovered ferrous sulfate heptahydrate crystals by filtration and air-dried at room temperature.

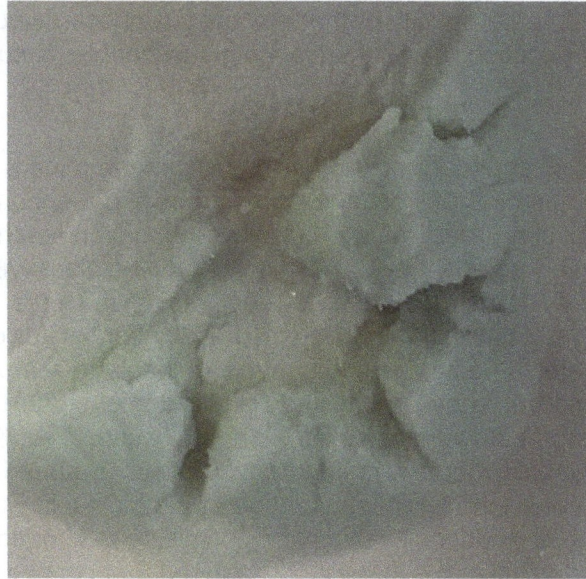

FIG 12 – Precipitated ferrous sulfate heptahydrate crystals from ferrous iron rich solution using ethanol.

Figure 13 shows the XRD diffractogram of the precipitated crystals, showing that the final crystals consist of predominantly ferrous sulfate heptahydrate, comprising approximately 90 per cent of the material.

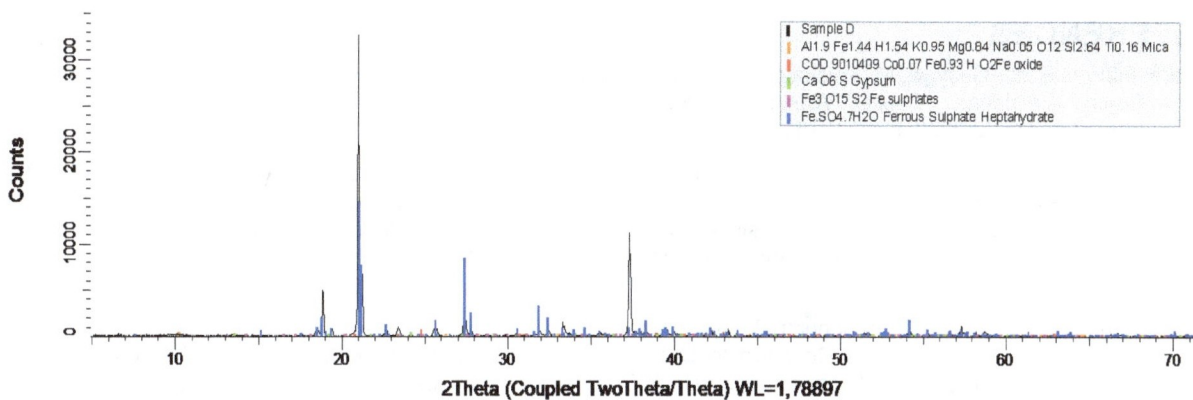

FIG 13 – XRD diffractogram of the precipitated ferrous sulfate heptahydrate crystals.

Characterisation of the solid residue after leaching

The particle size distribution (PSD) of the solid residues was evaluated by sieving and compared with pre-leaching PSD test. Prior to bioleaching, the pyrite enriched discards used in this study had a particle size of +2 mm and -6 mm and post leaching characterisation showed that approximately 10 per cent of the residual material was -2 mm, suggesting that pyrite oxidation weakens the structural integrity of the particles, thus reducing the size. Approximately 90 per cent of the residual material after leaching was within the 2 mm and 6 mm size range.

Post-leaching solid residues were evaluated by static acid–base accounting to determine their ARD risk and reuse potential. The starting material, a raw pyrite-enriched discard, contained 22.3 wt per cent sulfur and exhibited a net acid producing potential (NAPP) of approximately 582.67 kg H_2SO_4/t, classifying the head sample as acid-forming with high ARD liability. In the uninoculated control column, the sulfur present remained essentially unchanged (21.5 wt per cent), and the NAPP stayed elevated (485.6 kg H_2SO_4/t), so the material remained acid-forming. By contrast, microbial oxidation in the inoculated column removed over 90 per cent of the sulfur (estimated total S of 2.8 wt per cent), decreasing the NAPP to -155.3 kg H_2SO_4/t and reclassifying the residue as non-acid-forming. The results indicate an excess of neutralising minerals over residual sulfides, confirming that bioleaching potentially eliminates ARD risks. The treated residue, not only minimises long-term environmental liability of the residual coal discards but also opens avenues for its beneficial reuse in soil fabrication, thereby advancing sustainable, circular-economy outcomes.

CONCLUSION

Despite the buffering effect of abundant acid-neutralising gangue minerals, inoculated bioleaching achieved over 90 per cent removal of iron and sulfur from pyrite-enriched coal discards initially containing 22.3 wt per cent sulfur. After about 1300 days of column leaching, the residual material exhibited a net acid-producing potential of -155.3 kg H_2SO_4/t—reclassifying it as non-acid forming and effectively eliminating ARD risk. Ferric-to-ferrous conversion in a packed-bed reactor was complete within seven days, enabling the efficient precipitation of ferrous sulfate heptahydrate crystals, yielding approximately 245 kg per cubic metre of leachate – a product with an estimated market value of ZAR 1.2 million. These promising results demonstrate the potential viability of a bioleaching-based valorisation pathway for sulfide-rich coal wastes, yet the current reaction times are still long for industrial applications. Clear potential exists to speed up both the preliminary acidification stage and to shorten the ferric leach stage of the bioleach. Future work must therefore focus on accelerating microbial oxidation kinetics and optimising reactor configuration and operation to shorten leaching time and achieve the throughput required for economic scale-up.

ACKNOWLEDGEMENTS

The authors acknowledge, with thanks, the South African Water Research Commission (WRC) through WRC 2761, the Carnegie Fellowship (DEAL/HUMA) administered through the University of Cape Town and National Research Foundation (NRF) of S.A for funding through the SARChI in Bioprocess Engineering held by STLH.

REFERENCES

Amaral Filho, J R do, Gcayiya, M, Kotsiopoulos, A, Broadhurst, J L, Power, D and Harrison, S T L, 2022. Valorization of South African Coal Wastes through Dense Medium Separation, *Minerals*, 12(12). https://doi.org/10.3390/min12121519

Amaral Filho, J R do, Schneider, I A H, de Brum, I A S, Sampaio, C H, Miltzarek, G and Schneider, C, 2013. Characterization of a coal tailing deposit for integrated mine waste management in the Brazilian coal field of Santa Catarina [Caracterização de um depósito de rejeitos para o gerenciamento integrado dos resíduos de mineração na região carbonífera de San], *Revista Escola de Minas*, 66(3). https://doi.org/10.1590/S0370-44672013000300012

Amaral Filho, J R do, Weiler, J, Broadhurst, J L and Schneider, I A H, 2017. The Use of Static and Humidity Cell Tests to Assess the Effectiveness of Coal Waste Desulfurization on Acid Rock Drainage Risk, *Mine Water and the Environment*, 36(3):429–435. https://doi.org/10.1007/s10230-017-0435-7

ASTM, 2012. ASTM D4239-11 Standard Test Method for Sulfur in the Analysis Sample of Coal and Coke Using High-Temperature Tube Furnace Combustion, *Annual Book of ASTM Standards*, 552:2–5. https://doi.org/10.1520/D4239-11.2

Bahrami, A, Ghorbani, Y, Mirmohammadi, M, Sheykhi, B and Kazemi, F, 2018. The beneficiation of tailing of coal preparation plant by heavy-medium cyclone, *International Journal of Coal Science and Technology*, 5(3):374–384. https://doi.org/10.1007/s40789-018-0221-6

Colling, A V, Menezes, J C S D S and Schneider, I A H, 2011. Bioprocessing of pyrite concentrate from coal tailings for the production of the coagulant ferric sulphate, *Minerals Engineering*, 24(11):1185–1187. https://doi.org/10.1016/j.mineng.2011.04.003

Govender, E, Bryan, C G and Harrison, S T L, 2013. Quantification of growth and colonisation of low grade sulphidic ores by acidophilic chemoautotrophs using a novel experimental system, *Minerals Engineering*, 48:108–115. https://doi.org/10.1016/j.mineng.2012.09.010

Harrison, S T L, Broadhurst, J L, Opitz, A, Fundikwa, B, Stander, H, Mostert, L, Amaral Filho, J and Kotsiopoulos, A, 2020. An Industrial Ecology Approach to Sulphide containing Mineral Wastes to Minimise ARD Formation; Report to the Water Research Commission, Pretoria. Available from: <https://scholar.google.com/scholar?oi=bibs&cluster= 16351987838301511644&btnl=1&hl=en>

Harrison, S T L, Franzidis, J-P, Kazadi Mbamba, C, Stander, H, van Hille, R P, Mokone, T, Broadhurst, J L, Mbamba, C, Opitz, A, Chiume, R, Vries, E and Jera, M K, 2013. Evaluating approaches to and benefits of minimising the formation of acid rock drainage through management of the disposal of sulphidic waste rock and tailings: Report to the Water Research Commission, Pretoria. Available from: <https://wrcwebsite.azurewebsites.net/wp-content/uploads/mdocs/2015-1-13.pdf>

Katyal, J C and Sharma, B D, 1980. A new technique of plant analysis to resolve iron chlorosis, *Plant Soil*, 55:105–119. https://doi.org/10.1007/BF02149714

Kazadi Mbamba, C, Harrison, S T L, Franzidis, J P and Broadhurst, J L, 2012. Mitigating acid rock drainage risks while recovering low-sulfur coal from ultrafine colliery wastes using froth flotation, *Minerals Engineering*, 29:13–21. https://doi.org/10.1016/j.mineng.2012.02.001

Kotsiopoulos, A and Harrison, S T L, 2017. Application of fine desulfurised coal tailings as neutralising barriers in the prevention of acid rock drainage, *Hydrometallurgy*, 168:159–166. https://doi.org/10.1016/j.hydromet.2016.10.004

Lauriente, D H, Deboo, A and Sakota, K, 1999. Sulfuric Acid. Available from: <https://www.researchgate.net/publication/ 265277876_Sulfuric_Acid>

Lopes, F A, 2017. Hydrometallurgical Production of Magnetic Oxides from Pyrite Concentrate from Carbon Mining [in Portuguese: Producao Hidrometalurgica de Oxidos Magneticos a partir de Concentrado de Pirita Proveniente de Rejeitos da mineracao de carvao], DSc thesis, The Federal University of Rio Grande do Sul [UFRGS].

Machado, L A and Schneider, I A H, 2008. Static and kinetic tests for the prevention of acid mine drainage generation from coal mining mines with steel slag [in Portuguese: Ensaios estaticos e cineticos para a prevencao da geracao de drenagem acida de minas da mineracao de carvao com escoria de aciaria], *School of Mines Magazine [Revista Escola de Minas]*, 61(3):329–335. https://doi.org/10.1590/S0370-44672008000300011

Mazza, V B, Teixeira, L A C, Martins, A R F D A and Santos, B F dos, 2020. Process optimization for the production of ferric sulfate coagulant by the oxidation of ferrous sulfate with hydrogen peroxide, *Chemical Product and Process Modeling*, 15(3). https://doi.org/10.1515/cppm-2019-0091

Miller, S D, Robertson, A and Donahue, T, 1997. Advances in acid drainage prediction using the Net Acid Generation (NAG) test, in *Proceedings of the 4th International Conference on Acid Rock Drainage*, pp 533–549.

Mjonono, D, Harrison, S T L and Kotsiopoulos, A, 2019. Supplementing structural integrity of waste rock piles through improved packing protocols to aid acid rock drainage prevention strategies, *Minerals Engineering*, 135:13–20. https://doi.org/10.1016/j.mineng.2019.02.029

Runkel, M and Sturm, P, 2009. Pyrite roasting, an alternative to sulphur burning, *Journal of the Southern African Institute of Mining and Metallurgy*, 109(8):491–496.

Skousen, J G, Ziemkiewicz, P F and McDonald, L M, 2019. Acid mine drainage formation, control and treatment: Approaches and strategies, *Extractive Industries and Society*, 6(1):241–249. https://doi.org/10.1016/j.exis.2018.09.008

Soares, A B, Ubaldo, M de O, de Souza, V P, Soares, P S M, Barbosa, M C and Mendonça, R M G, 2009. Design of a dry cover pilot test for acid mine drainage abatement in southern Brazil, Materials characterization and numerical modeling, *Mine Water and the Environment*, 28(3):219–231. https://doi.org/10.1007/s10230-009-0077-5

Stander, H M, Harrison, S T L and Broadhurst, J L, 2022. Using South African sulfide-enriched coal processing waste for amelioration of calcareous soil: A pre-feasibility study, *Minerals Engineering*, 180. https://doi.org/10.1016/j.mineng.2022.107457

Stewart, W A, Miller, S D and Smart, R, 2006. Advances in Acid Rock Drainage (Ard) Characterisation of Mine Wastes, *J Am Soc Min Reclam*, pp 2098–2119. https://doi.org/10.21000/JASMR06022098

Tambwe, O, Kotsiopoulos, A and Harrison, S T L, 2020. Desulphurising high sulphur coal discards using an accelerated heap leach approach, *Hydrometallurgy*, 197:105472. https://doi.org/10.1016/j.hydromet.2020.105472

Vigânico, E M, 2014. Pilot-scale prototype for the production of ferrous sulphate from pyrite concentrate from coal mining [in Portuguese: Protótipo em escala piloto para produção de sulfato ferroso a partir de concentrado de pirita da mineração de carvão], DSc thesis, The Federal University of Rio Grande do Sul [UFRGS].

Vigânico, E M, Colling, A V, Silva, R D A and Schneider, I A H, 2011. Biohydrometallurgical/UV production of ferrous sulphate heptahydrate crystals from pyrite present in coal tailings, *Minerals Engineering*, 24(11):1146–1148. https://doi.org/10.1016/j.mineng.2011.03.013

Villetti, P I C, 2017. Ferric sulphate coagulant production from the leaching of pyrite concentrate from coal mining via ferrous sulfate crystallization/solubilization: a comparative study between tailings from two deposits [in Portuguese: Produção de coagulante férrico a partir da lixiviação de concen- trado de pirita da mineração de carvão via cristalização/solubilização de sulfato ferroso: estudo comparativo entre rejeitos de duas jazidas], MSc thesis, The Federal University of Rio Grande do Sul [UFRGS]. Available from: <http://hdl.handle.net/10183/174405>

Weiler, J and Schneider, I A H, 2019. Pyrite utilization in the carboniferous region of Santa Catarina, Brazil - Potentials, challenges and environmental advantages, *REM: Int Eng J*, 72(3):515–522. https://doi.org/10.1590/0370-44672018720139

Weiler, J, Amaral Filho, J R and Schneider, I A H, 2016. Coal waste processing to reduce costs related to acid mine drainage treatment - Case study in the Carboniferous District of Santa Catarina State, *Engenharia Sanitaria e Ambiental*, 21(2). https://doi.org/10.1590/S1413-41522016116411

Weiler, J, Amaral Filho, J R and Schneider, I A, 2014. Coal Tailings Processing and Environmental Impact Reduction [in Portuguese: Processamento de rejeitos de carvão e redução do impacto Ambiental], *Augm Domus*, 6(51):80–94. Available from: <http://revistas.unlp.edu.ar/domus/issue/view/99>

Zaim, M, Piselli, L, Fioravanti, P and Kanony-Truc, C, 2012. Efficacy and tolerability of a prolonged release ferrous sulphate formulation in iron deficiency anaemia: a non-inferiority controlled trial, *Eur J Nutr Mar*, 51(2):221–229. https://doi.org/10.1007/s00394-011-0210-7

Applying hydrocyclone classification for tailings management in Australia

J Penman[1] and A Mayot[2]

1. Senior Geotechnical Engineer/Associate, KCB Australia Pty Ltd, Perth WA 6005.
   Email: jpenman@klohn.com
2. Senior Geotechnical Engineer/Associate, KCB Australia Pty Ltd, Brisbane Qld 4000.
   Email: amayot@klohn.com

ABSTRACT

The Global Industry Standard on Tailings Management (GISTM; Global Tailings Review, 2020) has a stated goal for new tailings storage facilities (TSFs) to 'minimise the volume of tailings and water placed in external tailings facilities'. One method that can be used to achieve this goal is to use the sand fraction of the whole tailings to construct part (or all) of the perimeter containment dams which reduces the volume of 'wet' tailings to be contained in the TSF.

Hydrocyclone technology can be used to 'classify' the whole tailings into coarse (or sands) and fines (or slimes) fractions with the coarse fraction subsequently used as fill to build earthfill dams to contain the fines fraction. This technique has been used since the 1950s for tailings management and continues to be in widespread use in several of Australia's peer mining jurisdictions including in North America, South America, and Southern Africa. The resulting 'cyclone sand dams' have been successfully built to modern design standards with consideration for a range of extreme conditions including high-seismicity sites in Chile to high-rainfall sites in Panama. Despite being a well-established technique, the application of tailings classification to operations in Australia has been very limited with the only significant example of its use being at Alcoa's Western Australian bauxite processing operations to contain residue mud.

The objective of this paper is to identify and critically examine potential reasons for the lack of use in Australia through a comparison to conditions at comparable operations where tailings classification is used. The basic considerations for use of hydro-cycloning such as tailings characteristics and the management of the coarse and fine fractions are also discussed to provide context. This assessment is based on the author's experience in the application of this technique to operations in the Americas and more recently in the detailed study of its application to existing and proposed TSFs at Australian projects. This paper is intended to support tailings professionals to consider alternatives to more typical tailings management techniques practices in Australia.

INTRODUCTION

The Global Industry Standard on Tailings Management (GISTM; Global Tailings Review, 2020) states that the goal for a tailings management solution should be one that: *(1) minimises risks to people and the environment throughout the tailings facility life cycle, and (2) minimises the volume of tailings and water placed in external tailings facilities*. Achieving this goal requires the consideration and application of a range of tailings management technologies. One such technology with clear potential to achieve the stated goal of GISTM is 'classification' of tailings classification using hydrocyclones which separates the tailings stream into coarse and fine fractions. The coarse 'sand' fraction can be used to construct robust, free-draining embankments (forming so-called cyclone sand dams) that safely contain the remaining fine tailings (slimes). This method inherently reduces the volume of wet fine tailings stored and can improve water recovery for reuse.

Despite the use of hydrocyclones for tailings management being a mature, well-established technology, its operational application in Australia is limited to a handful of operations. The Authors' believe that conditions are broadly favourable for application of this technology in Australia; however, a paradigm shift in attitudes will be required to realise the potential benefits. This paper aims to present an assessment of potential reasons for the lack of adoption in Australia and address these reasons.

The Life of Mine | Mine Waste and Tailings Conference 2025 | Brisbane, Australia | 29–30 July 2025   763

BACKGROUND ON HYDROCYCLONE CLASSIFICATION

A hydrocyclone – as illustrated in Figure 1 – is a mechanical device used to separate or 'classify' particles in a liquid suspension based on their size and density. The hydrocyclone works by feeding a slurry into the device at high pressure which leads to vortex formation and centrifugal force generation. The centrifugal force within the vortex leads to the lighter materials – including smaller particles and water – moving towards the centre. The conical shape of the device results in the slurry accelerating as it moves down the device further encouraging separation of lighter and heavier materials.

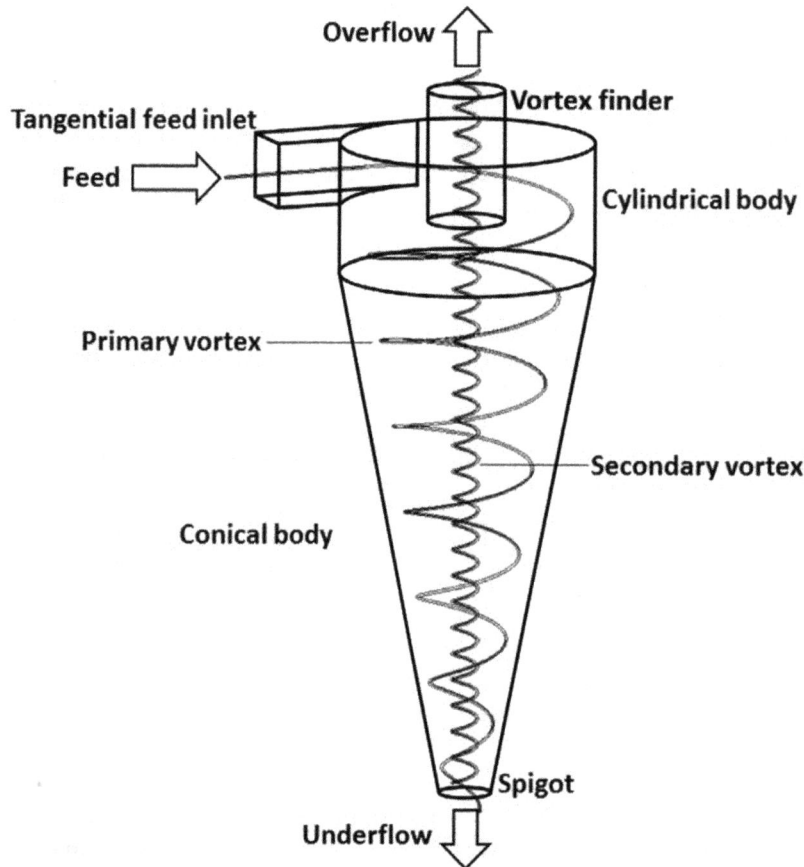

FIG 1 – Schematic of hydrocyclone operation (from Vega-Garcia, Brito-Parada and Cilliers, 2018).

Once the heavier material reaches the apex of the cone it is discharged as the 'underflow'. The lighter material is pushed to the top of the cone and discharged as the 'overflow'. The underflow typically contains the majority of the larger (and therefore heavier) particles along with some carrier fluid. The overflow typically contains the majority of the smaller (and therefore lighter) particles along with most of the carrier fluid. The classification process can be repeated several times on the underflow fraction to further separate the heavier (coarser) particles from the lighter (finer) particles until the desired specification is achieved. Notably, the hydrocyclone has no internal moving parts and does not require energy inputs other than the feed pressure allowing the technology to be 'bolted-on' to an existing slurry distribution system with relatively minor effort relative to other technologies.

With respect to tailings, the underflow and overflow are significantly different in terms of geotechnical and mechanical properties. Where the underflow is typically characteristic of a coarse-grained (sand) material and the overflow is typically characteristic of a fine-grained (silt or clay, colloquially known as the 'slimes' fraction) material. This difference in properties presents opportunities for tailings management techniques outside of the conventional slurry deposition such as management in separate facilities (eg co-mingling with waste rock) or, more typically, use of the underflow (sand) fraction as fill for construction of the containing embankments (commonly known as 'cyclone sand dams').

SPECIFICATION OF UNDERFLOW FOR EMBANKMENT CONSTRUCTION

Typical cycloned sand embankments (using the underflow fraction as a structural fill) rely on drained, unsaturated conditions being present in the sand fill and this requires several orders of magnitude difference in hydraulic conductivity (K_{sat}) between the cycloned sand and the stored overflow/tailings as well as adequate underdrainage of the embankment. Vick (1990) concluded that the greater the contrast in permeability between cycloned sands and the impounded tailings, the lower the phreatic level that will be maintained in the facility. A well-draining material is also required to support hydraulic placement of the underflow as part of the hydrocyclone process.

Whitehead and Witte (2019) reviewed the characteristics of selected cyclone sand embankment fills and concluded that hydraulic conductivity (K_{sat}) of sand at successful operations typically ranged from 10^{-5} m/s to 10^{-4} m/s. The hydraulic conductivity is primarily a function of the 'fines' content (or fraction particles passing 75 µm in the material). The angularity of tailings and associated impact on porosity further supports the ability of the underflow sand to drain.

For cyclone sand dam fills, the typical 'rule of thumb' is to specify ≤15 per cent of the particles (by weight) to be passing 75 µm in the underflow. Laboratory testing of underflow at a higher fines content may yield hydraulic conductivities within the target range; however, it is prudent to target a lower fines content in the underflow to account for variability following placement in the embankment.

McLeod and Bjelkevik (2017) collated hydraulic conductivity data for a range of tailings and grouped the tailings to show typical characteristics. As shown in Figure 2, only some 'coarse' tailings would inherently achieve this performance. The data shown supports the benefits from 'upgrading' a portion of the whole tailings with a high fines content to achieve the performance of 'coarse' tailings.

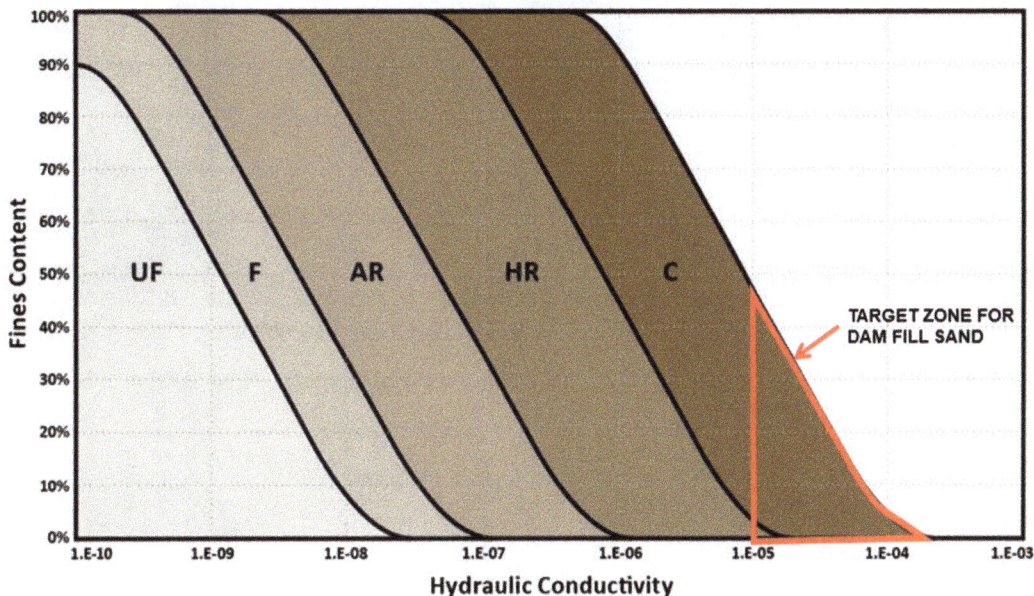

FIG 2 – Generalised tailings hydraulic conductivity versus fines content.

Where a single-stage of classification is unable to achieve the target fines content (eg due to high clay fraction) a second-stage can be added in series to further classify the underflow from the first stage. Advances in cyclone design, such as the development of integrated double-stage cyclones (eg Weir's Cavex® DE), are improving underflow yields with comparable infrastructure requirements to a single-stage cyclone arrangement.

A well-designed internal drainage is equally important. Drains or filter zones (eg chimney drains, toe drains, blanket/finger drains) are installed within or beneath the sand embankment to rapidly discharge water that seeps through the sand. Underdrainage systems coupled with wide dry beaches of fine tailings can maintain a phreatic surface away from the embankment. This design approach – large cycloned sand shell with internal drains, backed by a fine tailings beach – promotes stable, drained conditions in the dam.

ASSESSING HYDROCYCLONE APPLICABILITY

Assessing applicability of hydrocyclone technology to tailings can be done with relative ease at an early stage for most projects using key data of particle size distribution (PSD), particle specific gravity, and pulp density (or solids concentration) of the whole tailings slurry.

When provided with a typical PSD, specific gravity, and slurry solids concentration, Original Equipment Manufacturers (OEMs) such as Weir Minerals or Krebs can quickly run desktop simulations of potential hydrocyclone performance which – for tailings applications – should focus on the fines content and mass balance. These simulations include assessment of mass balance, solids content, particle size distributions etc, for the underflow and overflow as illustrated in Figure 3. The simulated hydrocyclone process can be adjusted, eg through trialling different cyclone sizes, feed concentrations, and/or the addition of further classification stages, to achieve the desired specification in the underflow or overflow.

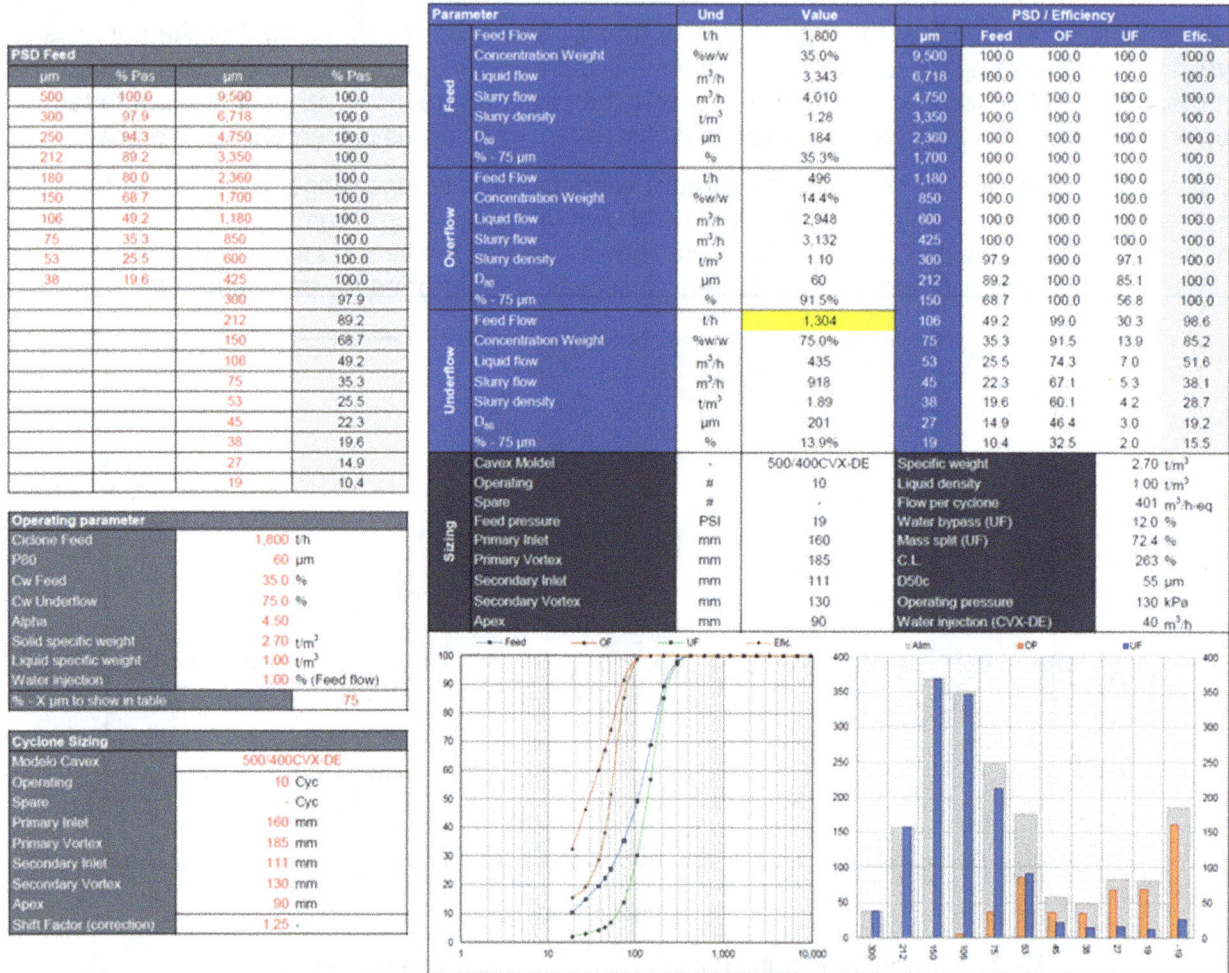

PSD Feed

µm	% Pas	µm	% Pas
500	100.0	9,500	100.0
300	97.9	6,718	100.0
250	94.3	4,750	100.0
212	89.2	3,350	100.0
180	80.0	2,360	100.0
150	68.7	1,700	100.0
106	49.2	1,180	100.0
75	35.3	850	100.0
53	25.5	600	100.0
38	19.6	425	100.0
		300	97.9
		212	89.2
		150	68.7
		106	49.2
		75	35.3
		53	25.5
		45	22.3
		38	19.6
		27	14.9
		19	10.4

Operating parameter

Ciclone Feed	1,800 t/h
P80	60 µm
Cw Feed	35.0 %
Cw Underflow	75.0 %
Alpha	4.50
Solid specific weight	2.70 t/m³
Liquid specific weight	1.00 t/m³
Water injection	1.00 % (Feed flow)
% - X µm to show in table	75

Cyclone Sizing

Modelo Cavex	500/400CVX-DE
Operating	10 Cyc
Spare	- Cyc
Primary Inlet	160 mm
Primary Vortex	185 mm
Secondary Inlet	111 mm
Secondary Vortex	130 mm
Apex	90 mm
Shift Factor (correction)	1.25 -

Simulation parameters

	Parameter	Und	Value
Feed	Feed Flow	t/h	1,800
	Concentration Weight	%w/w	35.0%
	Liquid flow	m³/h	3,343
	Slurry flow	m³/h	4,010
	Slurry density	t/m³	1.28
	D80	µm	184
	% - 75 µm	%	35.3%
Overflow	Feed Flow	t/h	496
	Concentration Weight	%w/w	14.4%
	Liquid flow	m³/h	2,948
	Slurry flow	m³/h	3,132
	Slurry density	t/m³	1.10
	D80	µm	60
	% - 75 µm	%	91.5%
Underflow	Feed Flow	t/h	1,304
	Concentration Weight	%w/w	75.0%
	Liquid flow	m³/h	435
	Slurry flow	m³/h	918
	Slurry density	t/m³	1.89
	D80	µm	201
	% - 75 µm	%	13.9%
Sizing	Cavex Moldel	-	500/400CVX-DE
	Operating	#	10
	Spare	#	-
	Feed pressure	PSI	19
	Primary Inlet	mm	160
	Primary Vortex	mm	185
	Secondary Inlet	mm	111
	Secondary Vortex	mm	130
	Apex	mm	90

PSD / Efficiency

µm	Feed	OF	UF	Efic.
9,500	100.0	100.0	100.0	100.0
6,718	100.0	100.0	100.0	100.0
4,750	100.0	100.0	100.0	100.0
3,350	100.0	100.0	100.0	100.0
2,360	100.0	100.0	100.0	100.0
1,700	100.0	100.0	100.0	100.0
1,180	100.0	100.0	100.0	100.0
850	100.0	100.0	100.0	100.0
600	100.0	100.0	100.0	100.0
425	100.0	100.0	100.0	100.0
300	97.9	100.0	97.1	100.0
212	89.2	100.0	85.1	100.0
150	68.7	100.0	56.8	100.0
106	49.2	99.0	30.3	98.6
75	35.3	91.5	13.9	85.2
53	25.5	74.3	7.0	51.6
45	22.3	67.1	5.3	38.1
38	19.6	60.1	4.2	28.7
27	14.9	46.4	3.0	19.2
19	10.4	32.5	2.0	15.5

Specific weight	2.70 t/m³
Liquid density	1.00 t/m³
Flow per cyclone	401 m³/h-eq
Water bypass (UF)	12.0 %
Mass split (UF)	72.4 %
C.L	263 %
D50c	55 µm
Operating pressure	130 kPa
Water injection (CVX-DE)	40 m³/h

FIG 3 – Example output from simulation of hydrocyclone performance.

With these outputs, the facility designer can assess if classification should be explored further and, if so, the best way to manage the separate streams with consideration for the mass balance of solids and water, eg adequate quantity of sand to construct the perimeter embankment.

The underflow characteristics are most sensitive to the PSD of the whole tailings and – as this is often difficult to modify – this may be a significant impediment to implementation. Data from operational tailings classification projects generally show the D80 and D50 of the whole tailings to be >100 µm and >55 µm, respectively. The whole tailings PSD is a function of the grind size required to efficiently liberate the metals from the ore during processing and this may be problematic to change with materially affecting project economics. Emerging processing methods such as coarse-particle flotation presents potential for a coarser tailings to be produced without comprising the metal recovery. The application of CPF and similar methods, eg ore sorting, should be assessed wholistically alongside potential downstream benefits for tailings management (lower CAPEX, OPEX).

HYDROCYCLONE TAILINGS MANAGEMENT IN PRACTICE

Hydrocyclone classification of tailings is a well-established technique being first implemented at large scale in the 1960s and 1970s for copper mines in British Columbia, Canada. From there, the technology has been implemented in a wide range of physio-graphic conditions, jurisdictions, and commodity types as shown in Table 1. The list includes both historical and modern operations, with tailings production rates ranging from modest (~20 000 tpd) to very large (~220 000 tpd), annual rainfalls from arid (~35 mm) to tropical (>2000 mm), and dam heights up to 250 m.

TABLE 1

Selected projects utilising hydrocyclone classification of tailings.

Project	Commodity	Start-up year	Location	Annual average rainfall (mm)	Total tailings production rate (tpd)	Peak U/F sand production rate (yield) (tpd)		Maximum dam height (design) (m)
Gibraltar	Copper	1972	Canada	500	55 000	27 500	50%	100+
Kennecott	Copper	1999	USA	500	100 000	30 000	30%	60+
Highland Valley	Copper	1962	Canada	350	130 000	22 000	17%	150+
Copper Mountain	Copper	1972	Canada	452	40 000	16 000	40%	170+
Cerro Verde	Copper	1994	Peru	35	120 000	37 000	31%	250+
Cobre Panama	Copper	2019	Panama	2500	220 000	66 000	30%	150+
Wagerup	Bauxite	1984	Australia	1200	30 000	9000	30%	50+
Pinjarra	Bauxite	1972	Australia	1100	25 000	7500	30%	45+
Kwinana	Bauxite	1963	Australia	900	20 000	6000	30%	40+

These examples demonstrate that hydrocyclone technology has been successfully applied for projects with characteristics that should cover most (if not all) climatic and production rate scenarios encountered in Australia. The required cyclone sand embankments are among some of the highest such structures in the world which demonstrates the favourable stability characteristics of this technique.

One key area of difference that potentially impacts the viability of the technique in Australia is the topography of the project site and the required quantity (yield) of underflow sand to construct the confining embankment(s). In the examples shown above, the underflow yield does vary (from 17–50 per cent) but is most commonly around 30 per cent of the total tailings produced. The non-Australian projects listed are all 'valley-fill' style facilities which generally have a favourable geometry allowing for centreline- or downstream-raised embankment construction with the available underflow sand. In contrast, the three Australia projects listed are 'paddock' style facilities and these have been built using the upstream-raised method as the quantity of sand produced is likely insufficient to construct centreline- or downstream-raised sand embankments.

Notwithstanding the vintage of this technique, classification of tailings is also being used as an integral part of novel approaches to wholistic tailings management such as Anglo-American's 'Hydraulic Dewatered Stacking' (HDS) technique being trialled at the El Solodado copper mine in Chile. This system combines coarse particle recovery (to generate sand-size material) with hydrocyclone separation and alternate stacking of the sand and fines to facilitate drainage of the deposit. Anglo reports that HDS can recover over 80 per cent of the water from the tailings (approaching the performance of filtered tailings), while producing a dense, compacted tailings stack that remains stable and accelerates rehabilitation (International Mining, 2023).

APPLICATION IN AUSTRALIAN CONTEXT

Clearly tailings management using hydrocyclone classification is a well-established and successful practice in places with more challenging geohazards, similar commodity types and production rates, similar cost bases (in the case of North America), and comparable regulatory environments as Australia. Many mining companies that operate some of the largest and most complex hydrocyclone operations have extensive portfolios within Australia.

Despite this context, the large-scale adoption of this technology in Australia has been limited to Alcoa's Kwinana, Pinjarra, and Wagerup bauxite refineries. Alcoa's successful long-term operation of cyclone sand dams for bauxite residue containment is a strong local proof-of-concept that this technique can succeed in Australia.

Furthermore, it is important to note that, fundamentally, the use of tailings to construct containment embankments is a well-used method in Australia, particularly 'harvesting' of the coarser fraction from a tailings beach for use as fill. In principle, there appear to be no fundamental technical barriers justifying the lack of widespread adoption in Australia.

The authors' opinion on some of the core reasons for lack of adoption in Australia include:

- prevalence (and preference) of upstream-raised construction
- fear of the 'leaky dam'
- strict battery limits – 'siloed thinking'.

The following sections challenges the validity of these reasons in an Australian context.

Prevalence of upstream-raise construction

Expanding tailings storage capacity through upstream-raised embankments represents the *status quo* for most operations in Australia. The cost benefits of upstream-raising are well known with this method representing the lowest capital cost for surface facility expansion. Additionally, physiographic conditions in Australia are broadly favourable for upstream-raised facilities with typically low-rainfall and low-seismicity, and large surface areas to limit rate-of-rise.

Notwithstanding the favourable environment, the increased focus on mitigating risk of residual-strength failure modes has led to a widespread program of buttressing which has essentially negated much of the capital cost benefits (excluding early-life capital deferment) of upstream raise construction.

Applying hydrocyclone technology can effectively mitigate the risks and uncertainties of upstream-raised construction through using the underflow fraction to construct centreline- or downstream-raised embankments. The risk mitigation can be achieved with the additional benefit of an overall reduction in the volume of the stored tailings. This theoretical benefit is conceptually illustrated in Figure 4 which shows an equivalent volume of tailings stored in a simplified 'paddock-style' TSF with the sand used to build an inherently more stable centreline-raised configuration dam within the same footprint as an upstream-raised dam.

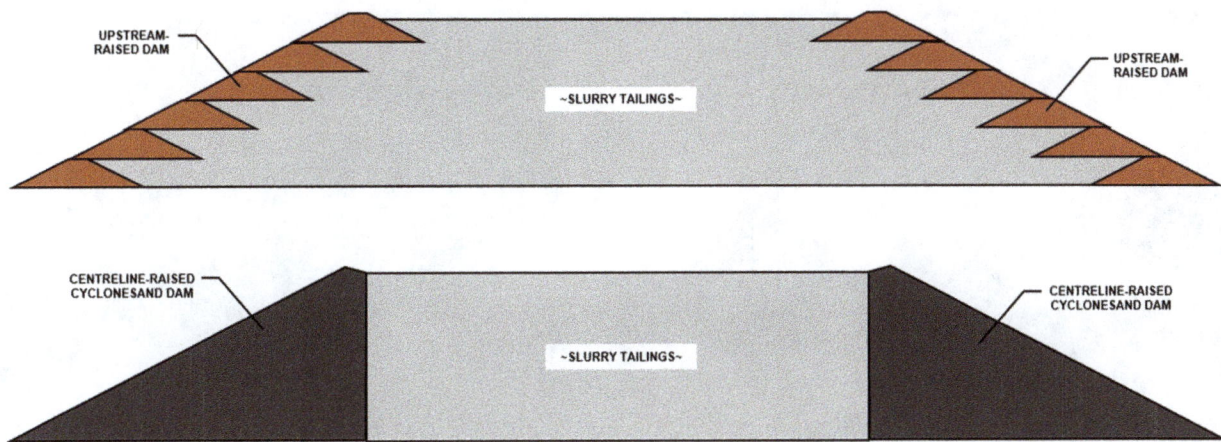

FIG 4 – Upstream-raised TSF versus centreline-raised cyclone sand dam TSF.

As previously discussed, the limiting factor in realising the potential structural benefits is the quantity of underflow sand required to build the desired dam configuration matching what can reliably be produced. This is more likely to occur for a paddock style facility as observed for Alcoa's residue storage areas. A buttressed (or hybrid) upstream raise may be required where the quantity of sand is likely to be insufficient for a centreline- or downstream-raised embankment.

A further limitation in applying this technique is the intensity of operational support required. Constructing cyclone sand embankments requires a permanent, fulltime crew and associated earthwork fleet to match sand (fill) production and tailings deposition. This contrasts with the more common approach in Australia of engaging a civil works contractor to construct raises of the embankments as a discrete works package. For this technique to work and the potential benefits realised, a paradigm shift will be required to integrate the construction works with facility operations as the fill material is directly supplied from processing operations.

This 'permanent crew' requirement is often viewed as a cost downside. However, it must be weighed against the costs that are avoided: for example, one large copper mine (Copper Mountain in Canada) feeds raw tailings directly to cyclones and was able to omit a tailings thickener, saving a substantial capital expense and operating costs. Likewise, if adopting cycloning prevents the need for future buttresses or remedial works, that is a significant cost and liability avoidance that should be considered.

Fear of the 'leaky dam'

Australian regulators and mine operators have – with good intentions – traditionally exhibited a strong aversion to seepage from tailings facilities. Many project approvals and designs focus on containing tailings pore water as much as possible, eg through lining facilities, installing cut-off walls, or designing impoundments to minimise seepage gradients. This mindset, while well-intentioned, creates a challenge for cyclone sand dams because they deliberately encourage seepage through the embankment to keep the dam in a drained, unsaturated state.

The fine tailings (slimes) forming against the sand act as a natural core or barrier, substantially reducing the flow rate and acting as a filter. As a result, the seepage quantities can be managed and captured with proper engineering controls. For instance, underdrain systems and toe drains can intercept seepage and direct it to collection sumps where it can be pumped back to the pond or direct to the plant. Many cyclone facilities around the world operate under environmental regulations that require seepage collection – for example, in Canada and the US, it's common to have seepage return systems at the downstream toe to protect aquifers as illustrated in Figure 5 (from Klohn Crippen Berger Ltd (KCB), 2025). Australian operations often adopt similar features as a reactive response to observed seepage rather than a proactive inclusion in the design. The perception that a cycloned sand dam might uncontrollably contaminate groundwater is not borne out if appropriate controls are in place.

FIG 5 – Operational cyclone sand dam with toe seepage collection pond (KCB, 2025).

Caution is required where the tailings may present acid metalliferous drainage risks which may necessitate additional controls including water treatment. This is particularly important for the sand dam fill which will, by design, be unsaturated and exposed to oxidation and regular 'flushing' from seepage. Geochemical characterisation is a critical early piece of data to collect and analyse for risk.

Battery limits and 'silos' (integrated plant and tailings design)

A less tangible but critical barrier to adopting hydrocyclones lies in organisational and design-process silos. In many mining projects, the processing plant design and the tailings storage design are handled separately, often by different teams or consultants, with minimal integration.

Processing engineers tend to focus on maximising ore throughput and recovery which is their key performance indicator. Their scope typically ends at the plant 'battery limits' – usually the tailings thickener underflow pump or the pipeline to the TSF. Anything beyond that is left to geotechnical/tailings specialists. In this traditional set-up, there is little incentive for the plant designers to consider alternative tailings handling technologies that might introduce perceived risk, complexity to the plant, reduce ore recovery, or increase plant CAPEX and OPEX. For example, suggesting a coarser grind to increase cyclone sand yield could be met with resistance, as it might reduce metallurgical recovery. The safer route for plant designers is to stick to conventional assumptions: a standard high-rate thickener generating a tailings slurry to be managed 'by others'.

This compartmentalised approach often means opportunities for better tailings solutions are missed. Hydro-cycloning, in particular, must be conceived early in design because it can influence upstream decisions, eg the grind size in the mill, the need (or not) for a thickener, the layout of pump and pipeline systems etc, If tailings engineers are only brought in after the process flow sheet and plant design are fixed, it may be too late or too costly to retrofit further tailings treatment circuit. Alternative dewatering or tailings classification technologies tend to be considered only when there is a strong driver such as water scarcity or space constraints forcing high-density thickening or filtration to be considered. Known examples of potential opportunities from introducing tailings classification include:

- Omitting a tailings thickener altogether (a significant capital cost saving) as is done at Copper Mountain and feeding the hydrocyclones directly with 'raw' tailings.

- Pursuit of a coarser grind size (or coarse partial flotation) which should support a higher yield of underflow sand leading to tailings management cost benefits that could offset a reduction in recovery (if any). This approach is being used with success as part of the HDS technique at Anglo American's El Solodado copper mine.

- Tailoring the tailings classification to produce an overflow suitable for use as feedstock for an underground paste backfill plant.

To overcome this, a cultural shift in project development is needed, ie tailings planning must be integrated with plant design from the earliest stages and not be seen as 'the tail wagging the dog'. A holistic evaluation might show, for instance, that accepting a slightly lower recovery could reduce tailings management costs so much (by producing needing smaller TSF, omitting thickeners etc) that the net project value is neutral (or even higher) compared to the *status quo*. These kinds of trade-offs rarely surface when teams work in silos.

CONCLUSIONS

Classification of tailings via hydrocyclone is a well-established (70+ years) and proven technique to reduce the volume of tailings to be stored in surface TSFs. Furthermore, there are no fundamental technical reasons for the application of this technology in Australia. In the Authors' opinion, the lack of adoption – outside of isolated cases like Alcoa's WA operations – is indicative of an overly conservative and compartmentalised approach to tailings management in Australia. Due to the long history of the technology being implement, the risks associated with tailings classification are well established and can be considered in the engineering of a TSF.

To achieve the stated goals of the GISTM, we must be more willing to pursue alternative technologies and techniques. Given the relative ease with which the applicability of tailings classification can be assessed at even the earliest stage of a project (even if it ends up being unviable), failure to consider this technology is incompatible with the goals of the GISTM and the drive to zero harm from tailings management.

REFERENCES

Global Tailings Review, 2020. Global Industry Standard on Tailings Management.

International Mining, 26 April 2023. Anglo American proves hydraulic dewatered stacking value at El Soldado. Available from: <https://im-mining.com/2023/04/26/anglo-american-proves-hydraulic-dewatered-stacking-value-at-el-soldado/> [Accessed: 1 June 2025].

Klohn Crippen Berger Ltd, 2025. Highland Valley Copper Mine. Available from: <https://klohn.com/projects/highland-valley-copper-mine/> [Accessed: 1 June 2025].

McLeod, H and Bjelkevik, A, 2017. Tailings Dam Design: Technology Update (ICOLD Bulletin), in Proceedings of the 85th Annual Meeting of International Commission on Large Dams.

Vega-Garcia, D, Brito-Parada, P R and Cilliers, J J, 2018. Optimising small hydrocyclone design using 3D printing and CFD simulations, *Chemical Engineering Journal*, 350:653–659. https://doi.org/10.1016/j.cej.2018.06.016

Vick, S G, 1990. *Planning, Design and Analysis of Tailings Dams* (Canada: BiTech Publishers Ltd).

Whitehead, J and Witte, A, 2019. Cycloned Copper Tailings Sands: Practical Methods for Benchmarking Design Parameters, in Proceedings of Tailings and Mine Waste 2019.

Dewatering behaviour of filtered red mud

A Ranabhat[1], C Zhang[2], S Quintero[3] and D Williams[4]

1. PhD candidate, School of Civil Engineering, University of Queensland, St Lucia Qld 4072. Email: a.ranabhat@uqconnect.edu.au
2. Research Fellow, Geotechnical Engineering Centre, School of Civil Engineering, University of Queensland, St Lucia Qld 4072. Email: chenming.zhang@uq.edu.au
3. Senior Research Technologist, School of Civil Engineering, University of Queensland, St Lucia Qld 4072. Email: s.quintero@uq.edu.au
4. Professor of Geotechnical Engineering, Director, Geotechnical Engineering Centre, School of Civil Engineering, University of Queensland, St Lucia Qld 4072. Email: d.williams@uq.edu.au

ABSTRACT

The high-water content of red mud in bauxite residue tailing storage facilities (TSF), combined with its composition, presents challenges such as reduced structural stability, environmental risks from toxic seepage, and the necessity of larger disposable land. Filtration is a widely employed dewatering method that can increase the solid content of red mud to 50–75 per cent, depending on the technique used, thereby improving TSF storage efficiency and structural stability, while reducing seepage risks. However, once deposited, red mud continues to undergo settling, consolidation, and desiccation, often achieving a dry density beyond what filtration alone can accomplish. Moreover, climatic factors such as frequent rainfall and limited solar exposure can lead to rewetting and reduced evaporation, potentially diminishing the long-term effectiveness of filtration. Therefore, understanding the impact of natural weather on the dewatering and densification of filtered red mud is essential.

To investigate this, a comprehensive study was conducted, integrating characterisation tests on both filtered and unfiltered red mud, followed by basin tests to examine their desiccation behaviour. An outdoor instrumented column experiment was performed, where two columns, each equipped with moisture and suction sensors, were filled with filtered and unfiltered red mud and subjected to natural drying. Finally, consolidation tests were conducted on both samples to assess their behaviour after disposal in the storage facility. This study presents a long-term comparative analysis of dewatering and densification behaviour of filtered and unfiltered red mud under natural environmental conditions to further understand the effectiveness of filtration on red mud

INTRODUCTION

The growing demand for aluminium-based products has driven a substantial increase in aluminium production globally, resulting in a corresponding rise in the production of red mud as a by-product. Annually, approximately 180 million tons of red mud is produced, adding to the existing global stockpile of over 4 billion tons (Jovičević-Klug et al, 2024).

Red mud, also known as bauxite residue, is predominantly generated as a by-product during the extraction of alumina from bauxite using caustic soda. It is commonly transported to TSF by hydraulic deposition at high water content with a solid content of 25 per cent to 30 per cent by weight (Avery et al, 2022). The residue generated consists of a coarse fraction, primarily quartz (red sand), and a fine fraction (red mud), rich in iron oxides. The high water content in the red mud stored in the TSF combined with its composition poses significant challenges including decreased structural strength, environmental impacts due to seepage of toxic chemicals in the tailings, and the necessity of larger disposable land (East and Fernandez, 2021; Fourie, 2009).

Due to the presence of residual caustic soda, red mud is highly alkaline, with a pH exceeding 11 and contains other toxic elements such as water-soluble fractions of Al and Cr as $[Al(OH)4]-$ and Cr(VI) (Snars and Gilkes, 2009; Milačič, Zuliani and Ščančar, 2012). These toxic chemicals in red mud can dissolve with water and transport contaminants to surrounding soil, groundwater, and surface water (Cacciuttolo et al, 2023).

The high water content of red mud disposed of in tailings storage facilities (TSFs) can also contribute to hydraulic erosion and liquefaction of the tailings (Cacciuttolo and Atencio, 2022). Tailings disposed as slurry potentially remain vulnerable to seismic activities and liquefaction unless the tailings are compacted or consolidated and desiccated (Williams, 2021).

There are numerous approaches to managing water both before and after TSF deposition. The method or methods selected will rely on several variables, including the type of tailings and its capacity to retain water, environmental concerns, the amount of investment required, and the local water supply. The methods selected may also be influenced by local regulations, which may limit the amount of water available for mining.

There has been growing interest in adopting filtered and dry stacked tailings over conventional disposal methods using hydraulic deposition. This shift is driven by the water efficiency and enhanced geotechnical stability of filtered tailings, as well as advancements in large-scale pressure filtration technology, reducing the cost per unit area, making the method more economically viable (de Kretser, 2018).

Filtration utilises either pressure to force water through the filter medium, effectively separating red mud solids, or a vacuum to draw water through the medium, facilitating dewatering (Davies, 2011). Flocculants are generally used before filtration to enhance dewatering by reducing consolidation and increasing cake porosity, leading to higher residual moisture content. However, it is crucial to use an optimal amount of flocculant, as low dosage may result in ineffective flocculation while excessive dosing can increase filtration resistance (Alam *et al*, 2011).

The effectiveness of tailings filtration is largely influenced by their particle size distribution and mineralogical composition. A high concentration of fine clay-sized particles (<74 µm) or clay-rich mineralogy can significantly hinder filtration efficiency (Davies, 2011).

Filtered tailings are generally transported to storage location using conveyor belts or trucks and can be stacked and compacted to meet the geotechnical requirements. There are several advantages of storing filtered tailings over conventional methods of impounding, such as (Kaswalder *et al*, 2018):

- Improved stability and reduction in seepage potential due to lower moisture content.

- Progressive reclamation and site closure is easier and can be planned early.

- Land use can be optimised, reducing the footprint while increasing storage capacity.

- Enhanced water and soda recovery, often making it the driver to choose filtration over conventional methods.

Depending on the target moisture content, filtration effectively lowers the moisture content of the tailings. In contrast, in the conventional method of red mud deposition, the amount of water is significantly higher, with a solids content of approximately 25 per cent (Xue *et al*, 2016). The reduced moisture content in filtered tailings enhances the mechanical strength of the deposits, rendering them more resilient to seismic events (Gallardo Sepúlveda, Sáez Robert and Camacho-Tauta, 2022).

A key factor influencing the stability of tailings deposits is the initial void ratio at the time of deposition. Filtered tailings generally exhibit lower void ratios compared to those deposited using conventional methods. According to Li *et al* (2012), the initial void ratio is crucial in determining the consolidation behaviour and long-term stability of TSF. Figure 1presents consolidation curves for tailings prepared at varying initial void ratios. The results indicate that samples with lower initial void ratios consistently achieve lower final void ratios. The parallel nature of the consolidation lines highlights the influence of initial void ratio on the overall consolidation behaviour of the tailings.

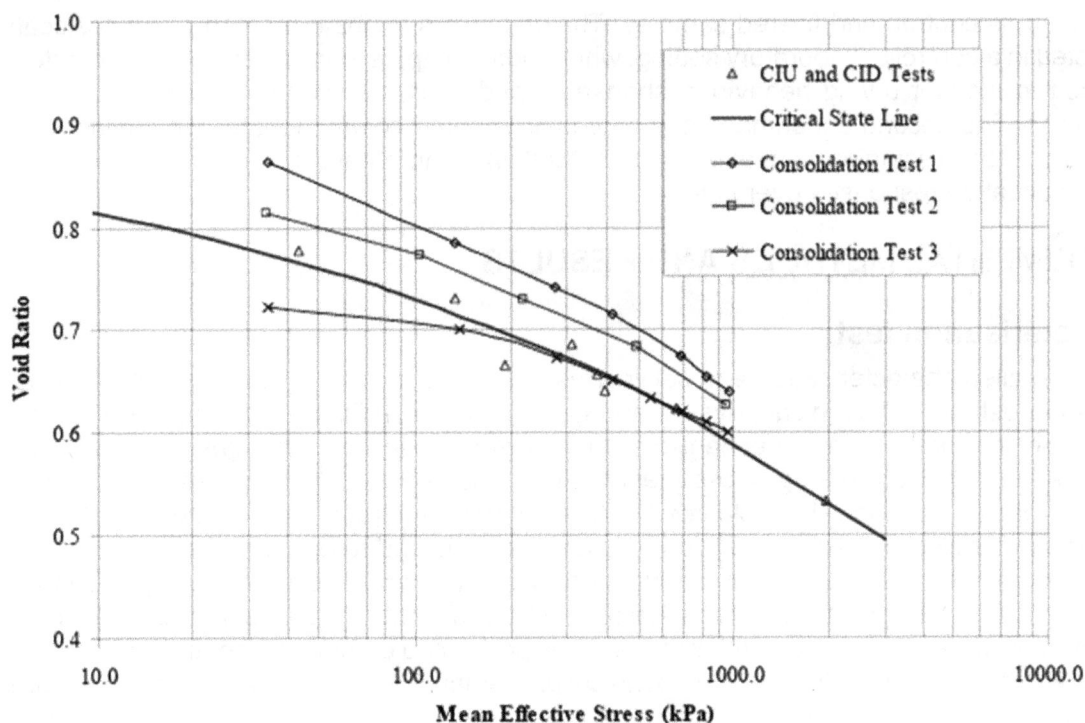

FIG 1 – Triaxial compression and consolidation tests (Li *et al*, 2012).

In addition to consolidation, desiccation significantly contributes to the densification of subaerially deposited tailings. This process is driven by surface evaporation, which generates capillary pressures that expel pore water from the tailings matrix, thereby enhancing both densification and mechanical strength (Fujiyasu, 1997; Zhang *et al*, 2024). Rodríguez *et al* (2007) demonstrated that the hydraulic properties of tailings, including water retention, saturated hydraulic conductivity, and permeability, are strongly influenced by porosity and vary with changes in the degree of saturation. Building on this Daliri, Simms and Sivathayalan (2016) conducted a laboratory simulation on gold tailings and categorised into two phases: phase I, characterised by high water content, rapid drainage, and settlement; and phase II, marked by lower water content and slower evaporation. In phase I, desiccation was accelerated due to dominant downward water flow and stronger capillary forces within the underlying tailings. In contrast, Phase II experienced slower desiccation because limited upward water movement reduced moisture recharge at the surface. Notably, the allowance of drainage during phase I further enhanced the dewatering process. However, it is important to note that drainage is allowed for phase I, enhancing dewatering. In a study using a large-scale consolidometer, Shokouhi, Zhang and Williams (2018) found that coal tailings consolidated with a consolidometer desiccated more rapidly compared to tailings allowed to consolidate with self-weight weight alone due to increased capillary forces and higher permeability. The results also showed that approximately 88 per cent of the total settlement occurred during the initial settling phase of the slurry.

The disposal of red mud typically involves three interconnected processes: sedimentation, consolidation, and desiccation. While these mechanisms occur simultaneously in a TSF, laboratory testing often isolates each process under controlled conditions using separate samples. Such an approach may not be able to represent field conditions, where the interactions between processes and environmental factors are more complex. A comparative analysis of filtered and non-filtered red mud tailings is essential to fully understand how filtration influences settling, consolidation, and desiccation behaviour of red mud and its interaction with additional layers deposited over the previously desiccated layer in the natural environment. To achieve this, a well-designed experimental program is required to evaluate the long-term effects of filtration on red mud behaviour in realistic field-like settings.

Based on the objective of developing a detailed understanding of the long-term effects of filtration on red mud behaviour, including self-weight consolidation, desiccation, settlement, drying, shrinkage, and crack formation, two major experiments were conducted using both unfiltered (referred to as control) and filtered samples. The first test is an instrumented basin desiccation test conducted in a controlled laboratory setting, where both samples were placed in instrumented basins equipped to monitor drying behaviour, shrinkage, and crack development and dry density during desiccation. The second experiment is a large instrumented column test for the control and filtered sample, where the samples were exposed to natural environmental conditions to evaluate settlement and consolidation responses over time.

EXPERIMENTAL METHODS AND RESULTS

Characterisation test

Initially, detailed characterisation and settling tests were conducted on both the control and filtered samples to establish their initial physical and geochemical conditions. This baseline information is critical for interpreting the mechanisms driving the observed changes in desiccation and consolidation behaviour during subsequent testing. The red mud tailings used in this study were sourced from an alumina refinery located in Queensland, Australia. The control sample underwent sea water neutralisation and flocculation but was not subjected to any dewatering processes, while the filtered sample was mechanically dewatered by filtration and subsequently repulped to facilitate transport to the tailings storage facility along with sea water neutralisation and flocculation. A comprehensive series of characterisation tests was performed on both the control and filtered tailings to evaluate their initial physical and geotechnical properties. The results of these tests are presented in Table 1.

TABLE 1

Characterisation results of the control and filtered sample.

Sample ID		Control	Filtered
Simplified descriptor		Neutralised flocculated unfiltered	Neutralised flocculated filtered and repulped
Initial state	GMC (%)	258.7	143.6
	% Solids	27.9	41.0
	pH	8	11
PSD	% Clay size	52	79
	% Silt size	45	18
	% Sand size	3	3
Atterberg limits	Liquid limit (%, Casagrande)	80.7	64.3
	Liquid limit (%, Falling cone)	85.6	69.6
	Plastic limit (%)	55.3	45.1
	Plasticity index (%)	25.5	19.3
	USC	MH	MH
Settling from 25% solids	Settling time (days)	16	13
	Settled % solids	29.2	31.1
	Settled dry density (t/m^3)	0.361	0.375
Settling from 25% solids	Settling time (days)	16	12
	Settled % solids	29.2	39.5
	Settled dry density (t/m^3)	0.361	0.534

The filtered sample shows significantly lower moisture content and higher solids content compared to the control sample, primarily due to water removal during filtration. Both samples are composed predominantly of fine particles, with high proportions of clay and silt, and only around 3 per cent sand. In the control sample, the proportions of clay and silt are relatively balanced. However, the filtered sample contains a notably higher percentage of clay-sized particles. This difference is likely due to the breakdown of flocculated aggregates during filtration. Despite the higher clay content in the filtered sample, both its liquid limit and plastic limit are lower than those of the control. Nevertheless, both samples are classified as high-plasticity silt (MH) according to the Unified Soil Classification System (USC).

Settling tests were carried out on both samples at initial solids contents of 25 per cent and 35 per cent. The tests were concluded once two consecutive daily measurements showed no change in settled height. The filtered sample reached final settlement more rapidly, settling three days earlier than the control at 25 per cent solids and four days earlier at 35 per cent solids. Furthermore, the filtered sample exhibited a higher final dry density after settling, especially at 35 per cent solids, where its dry density was approximately 47 per cent greater than that of the control sample.

Basin desiccation test

The basin desiccation test was conducted to simulate and monitor the drying behaviour of red mud tailings under controlled laboratory conditions that approximate natural field desiccation. This test is essential for evaluating the evolution of cracking patterns, shrinkage, matric suction, and degree of saturation in both filtered and control samples as they progressively lose moisture.

Basin test design and procedure

The drying behaviour of the control and filtered red mud when it is allowed to dewater by evaporation is observed using instrumented basins. As illustrated in Figure 2, separate basins are allocated for each sample. Each basin was equipped with three calibrated dielectric moisture sensors and three suction sensors, with measurements taken every 30 mins.

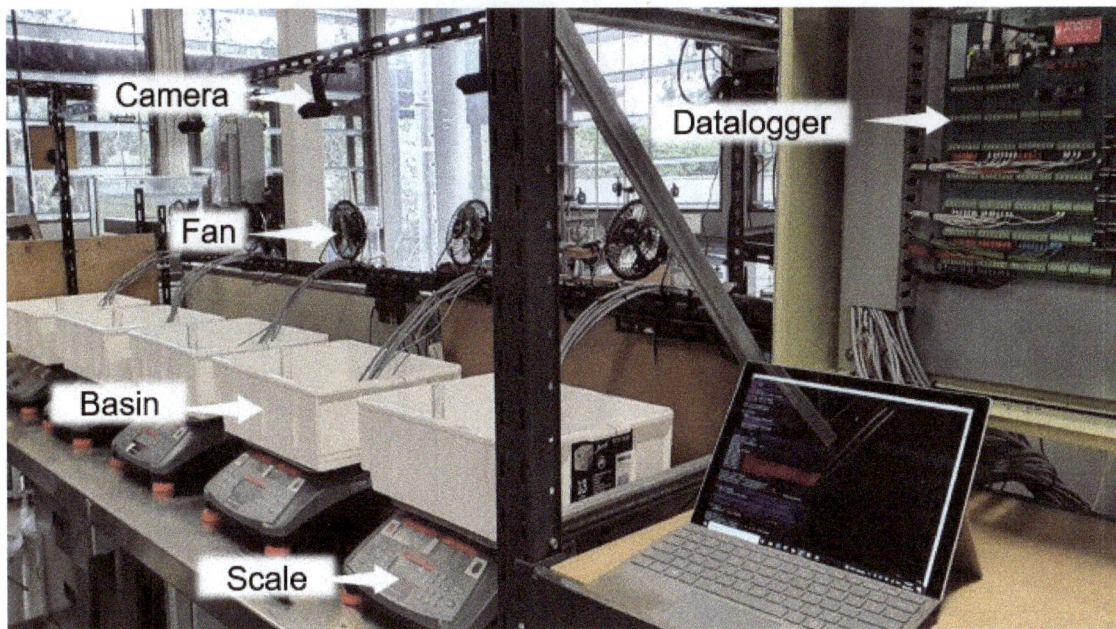

FIG 2 – Basin test configuration.

The instrumented basins were placed on digital scales. As the samples were tested at their initial moisture content, the change in weight over time was continuously monitored. This set-up allowed for the observation of moisture loss throughout the desiccation process, providing a quantitative assessment of drying behaviour.

To simulate a natural drying environment and promote desiccation, an electric fan was installed above each basin. These fans were positioned carefully to ensure that airflow passed above the surface of the red mud without directly impinging on it. This arrangement aimed to replicate the effect of a gentle breeze, encouraging evaporation while minimising disturbance to the surface features and natural drying mechanisms.

Additionally, a high-resolution camera was mounted above each basin to capture key surface phenomena, including crack development, salt precipitation, and crust formation. This visual record offered valuable insights into the physical responses of the tailings to moisture loss. A datalogger was used to coordinate and collect data from all installed sensors, ensuring synchronised and continuous monitoring of the desiccation process.

Following the completion of the basin test set-up, as illustrated in Figure 2 both the control and filtered red mud samples were thoroughly mixed to ensure uniform consistency throughout each sample. Since the test required both materials to be evaluated at their respective initial moisture contents (258 per cent for the control sample and 143 per cent for the filtered sample), small subsamples were extracted and tested to confirm that the control sample and the filtered sample met target requirements. Once verified, each sample was poured into its designated basin until a uniform height of 50 mm from the base was achieved. The sensors were positioned so that they were fully embedded within the red mud while remaining suspended above the basin floor, ensuring accurate monitoring of internal conditions throughout the desiccation process.

Result and discussion

Figure 3 illustrates the images of the control and the filtered sample at various gravimetric moisture contents throughout desiccation. The control sample and the filtered sample had an initial moisture content of 258 per cent and 143 per cent, respectively. As the water content decreases, notable differences in surface behaviour and crack formation can be observed between the two samples.

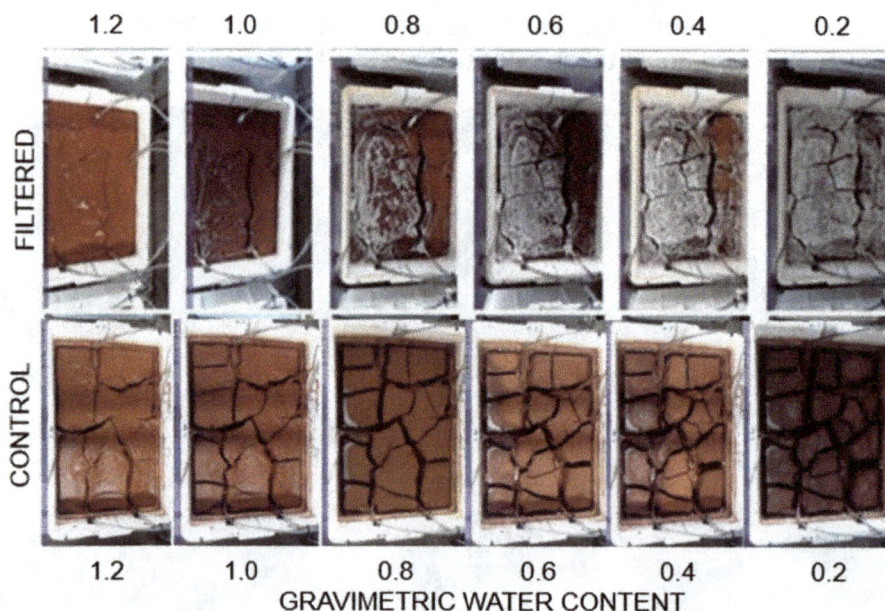

FIG 3 – Images of the control and filtered sample at various moisture content.

During drying, both the control and filtered samples undergo shrinkage due to the loss of moisture. Vertical subsidence results from particle rearrangement, while tensile stresses generated by shrinkage lead to horizontal crack formation (Vo *et al*, 2017). Cracks appeared earlier in the control sample at a gravimetric moisture content (GMC) of approximately 130 per cent, compared to 105 per cent in the filtered sample. As shown in Figure 4, matric suction in the control sample begins to rise around a moisture content of 150 per cent, close to its air entry value. The intrusion of air into the pore spaces increases suction and effective stress, initiating shrinkage and cracking earlier in the control sample (Kindle, 1917; Kodikara and Costa, 2013). In contrast, the filtered sample developed finer, narrower, and more closely spaced cracks (Figure 3), which suggests lower volumetric shrinkage (Figure 4). This is likely due to its lower plasticity index (PI = 19.3) and higher solids content.

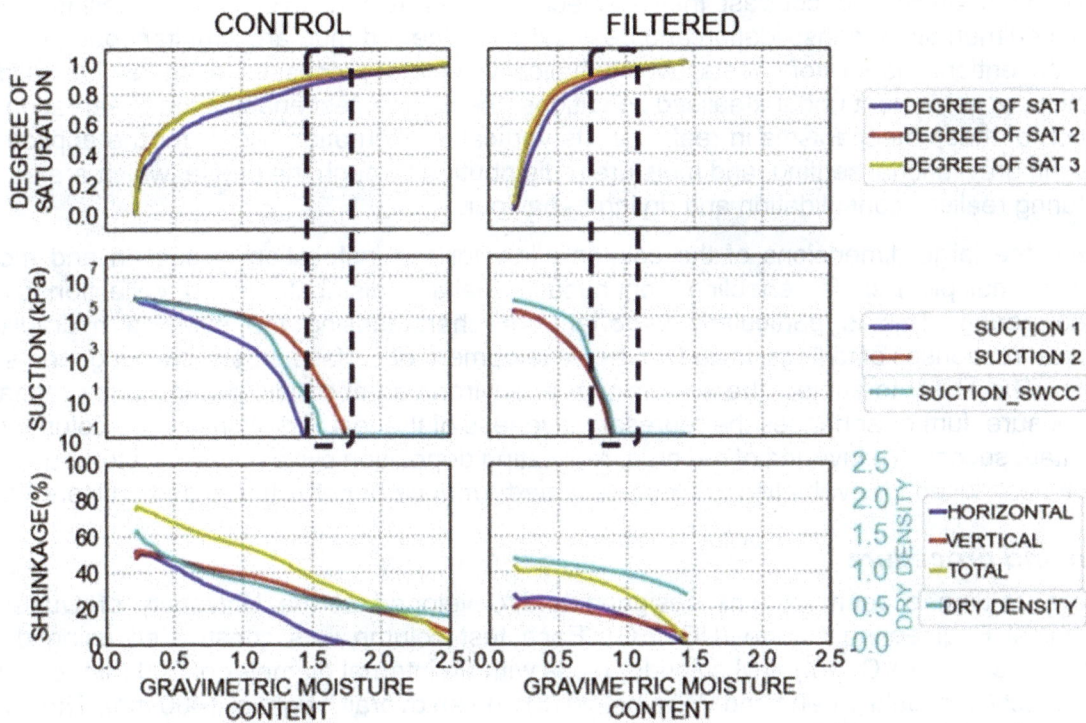

FIG 4 – Comparison of the degree of saturation, matric suction, shrinkage, and dry density during desiccation of control and filtered red mud samples.

The filtered sample exhibits a highly alkaline pH of 11, compared to pH 8 in the control sample, due to the presence of elevated sodium salt concentrations. This high alkalinity is visually evident in the filtered sample through the formation of surface efflorescence during drying.

The experiment commenced with fully saturated control and filtered samples. Desaturation initiated earlier in the control sample, at approximately 200 per cent moisture content, while the filtered sample began desaturation at around 100 per cent. The control sample, with lower solids content and density, exhibits higher void ratio and more interconnected pore spaces, making it easier to lose water and desaturate with early air entry. In contrast, the filtered sample has a denser particle arrangement and lower void ratio, allowing full saturation even at lower water content (Kodikara and Costa, 2013).

The evolution of dry density during desiccation reveals distinct behaviours between the control and filtered samples. The filtered sample maintained a consistently higher dry density than the control sample across higher moisture contents, particularly from the start of the experiment to 50 per cent moisture content. This behaviour is attributed to its initial condition following filtration, which resulted in higher solids content, lower void ratio, and a denser particle structure. These characteristics reduce the potential for volume change, leading to lower shrinkage. Hence, the improvement in dry density is gradual and steady throughout drying.

In contrast, the control sample had higher initial water content and lower dry density. During initial desiccation, the shrinkage in the control sample was vertical only with evaporation, thus the dry density was improving slowly. However, once horizontal shrinkage commenced, the sample experienced significant volume reduction, which led to a steeper rise in dry density. The total shrinkage observed in the control sample was substantial, approximately 78 per cent, compared to about 40 per cent in the filtered sample. This large volumetric reduction allowed the control sample to ultimately achieve a higher final dry density of approximately 1.6 t/m^3, exceeding that of the filtered sample, which reached around 1.2 t/m^3.

Large column test

This test is a continuation of the basin desiccation test with the same two samples, filtered and unfiltered, to quantify and contrast their geotechnical behaviours over time. In tailings storage facilities, sedimentation, consolidation and desiccation of the red mud are simultaneous processes. Unlike conventional laboratory tests, which typically isolate processes such as consolidation, desiccation, or settlement under idealised conditions, the column test enables the observation of the interaction of these mechanisms in real time. Its vertical configuration allows for the application of self-weight, gravitational settling, and moisture redistribution through the profile, which are essential for capturing realistic consolidation and drying behaviour.

Moreover, the large dimensions of the column allow for the installation of suction and moisture sensors at multiple depths, enabling continuous, spatially distributed data collection over an extended period. This is particularly important for characterising the evolving hydraulic and mechanical responses of tailings, including the development of surface crusts, cracking, and suction gradients. The ability to subject the system to real environmental conditions, including rainfall and solar exposure, further enhances the representativeness of the test. Additionally, the column set-up can facilitate successive layering of red mud, replicating deposition cycles of tailing storage facilities, making it best suited for evaluating the long-term performance of both filtered and unfiltered tailings

Design and procedure

The large column experiment was designed and developed at the University of Queensland Geotechnical Engineering Centre (UQGEC). Each test column was constructed using durable, chemically resistant PVC (polyvinyl chloride) pipes with an internal diameter of 200 mm to minimise the wall effect (Shokouhi, Zhang and Williams, 2018) and an overall height of 1400 mm. The columns were formed by joining a 780 mm upper section and a 620 mm lower section using a metal coupling ring to allow easy access for instrumentation. The bottom was sealed to restrict dewatering to the exposed surface at the top (Figure 5).

FIG 5 – Set-up of large column test.

Ten dielectric moisture sensors and eight thermal sensors developed at UQGEC were employed for each column. The dielectric moisture sensors are calibrated to measure the volumetric water content of red mud due to its variation in dielectric permittivity, from which the degree of saturation can be calculated. This type of moisture sensor is capable of providing continuous and reliable measurements even at high salinity levels. The thermal conductivity sensors use a thermocouple and porous cylinder to measure how quickly heat dissipates from a small heater embedded in a porous ceramic tip that equilibrates with the surrounding material. The rate of heat dissipation reflects the moisture conditions, allowing estimation of matric suction based on thermal conductivity. The sensors were installed on the walls of each column along the depth with sensor density higher near the surface of the column (ie 15, 50, 100, 150, 200, 285, 385, 485, 700, 850 mm from surface) where more variations were expected to occur.

A high-resolution camera is mounted above the columns to observe settlement, cracking, crust formation, and salt precipitation throughout the experiment. Two dataloggers collect the sensor data, and the system operates independently using a solar panel and battery set-up housed in a weather-resistant enclosure beside the columns.

After the experimental set-up was completed, an additional 100 mm section was attached to the top of each column to collect supernatant water following tailings placement. The columns were then filled with water and sealed at the top to verify waterproofing. Each of the control and filtered red mud samples was thoroughly mixed using a handheld mixer to ensure uniform consistency. The control sample was prepared at its initial moisture content of 258 per cent, and the filtered sample at 143 per cent. After mixing, small portions of each sample were taken to confirm the moisture content through oven-drying. The slurries were then carefully poured into their respective columns using a funnel to avoid disturbing the sensors. Both columns were filled to the top (1.5 m including the added 100 mm section to collect and remove supernatant water), and the total weight of material placed in each was recorded.

Once the supernatant water had accumulated, the top section and water were removed to expose the columns to natural environmental conditions, including sunlight, wind, and rainfall. Moisture and suction data were recorded at 30 min intervals, while weather data were logged every 8 mins. Surface images of the tailings were captured twice daily, at 12:00 pm and 5:00 pm, to monitor settlement, cracking, and crust development.

Result and discussion

The large column experiment was conducted for 15 months, capturing both wet and dry seasonal cycles to simulate field-like weather exposure. Both the control and filtered columns were subjected to direct exposure to environmental conditions, including rainfall, solar radiation, and wind. Figure 6 illustrates the progressive evolution of the two samples throughout the monitoring period.

FIG 6 – Progressing images of the large column test

Following the initial disposal, both samples underwent settlement, self-weight consolidation, and surface desiccation. On 19 May, a second layer of red mud was deposited in each column after the initial material had settled by approximately 200 mm. Distinct differences in surface response emerged over time. The filtered sample exhibited pronounced efflorescence upon drying, consistent with its higher alkalinity (pH 11) and results observed in the basin desiccation test. This efflorescence is indicative of soluble salt migration and crystallisation during evaporation.

Although surface cracking on both samples began at approximately the same time, with the control sample showing cracks around 20 December and the filtered sample around 27 December, the extent and severity of cracking were notably different. The control sample, which had a higher initial moisture content (258 per cent) and plasticity index (PI = 25.5), developed deeper and wider cracks. This reflects greater shrinkage and tensile strain during desiccation. In comparison, the filtered sample, with a lower initial moisture content (143 per cent) and plasticity index (PI = 19.3), exhibited finer and more closely spaced cracks. These narrower crack patterns suggest reduced volumetric change and lower shrinkage potential. The visual distinctions between the two samples are consistent with the initial characterisation results and highlight the role of filtration in modifying the surface behaviour of red mud under natural drying conditions.

Both the samples undergo settlement consolidation and desiccation simultaneously in the natural environment. Figure 7 presents the temporal evolution of dry density, settlement, degree of saturation, and suction in the control and filtered red mud columns over a 15 month monitoring period. The experiment captures the response of each sample to natural weather cycles, including wet and dry seasons.

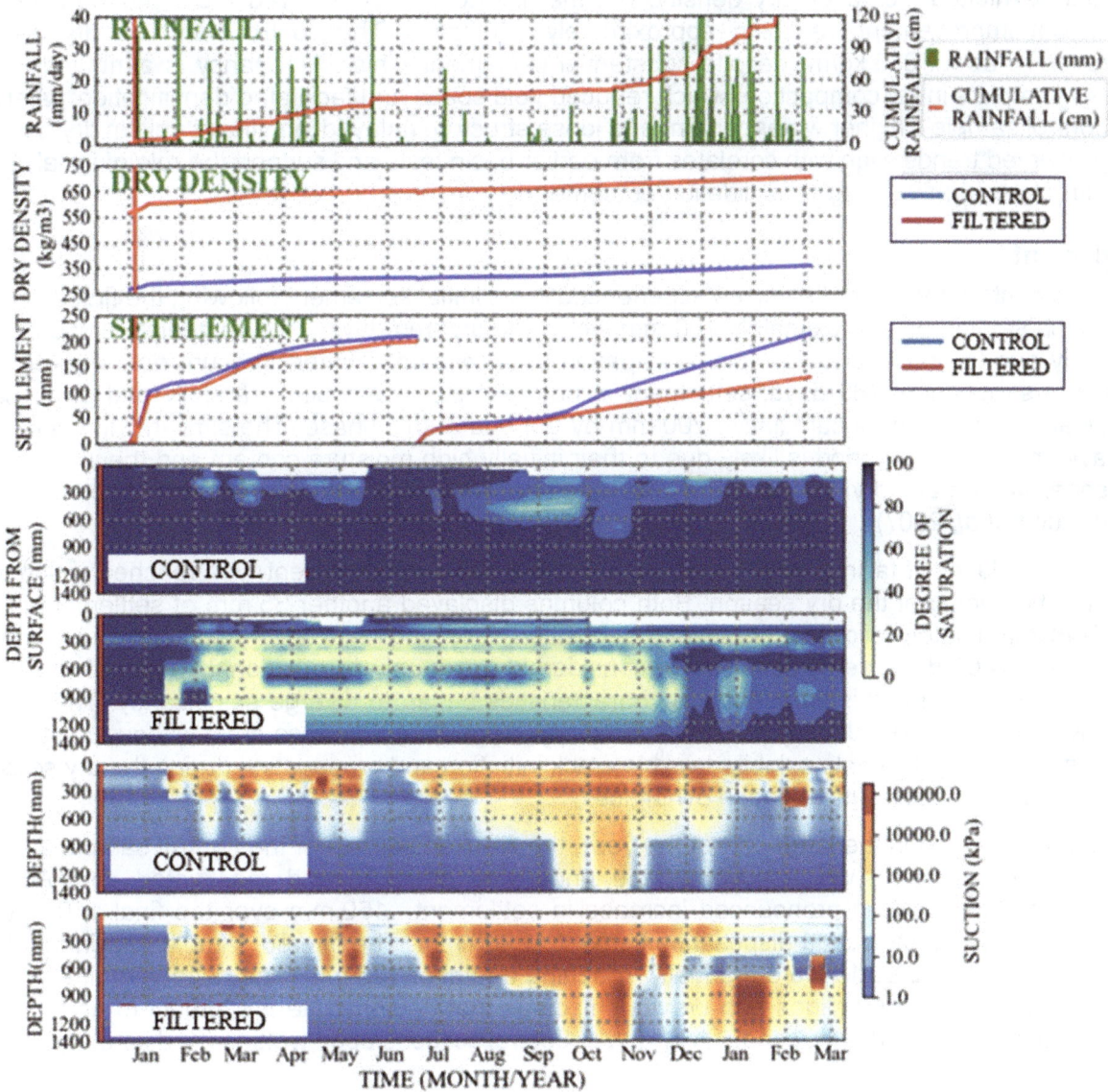

FIG 7 – Temporal variation of rainfall and changes in geotechnical parameters (dry density, suction, degree of saturation, and settlement) in control and filtered red mud columns during the experiment period.

In Figure 7, first plot shows the rainfall per day and cumulative rainfall. Rainfall occurred predominantly between January and April, with cumulative rainfall exceeding 80 cm during this period. After mid-April, a dry season followed with sporadic rainfall events below 25 mm per day, lasting until November. Rainfall resumed in late November, initiating another wet phase. These cycles of wetting and drying played a key role in influencing the degree of saturation and suction response in both samples, particularly affecting the surface and upper layers where desaturation and rewetting occurred more rapidly.

Detailed analysis of each geotechnical parameters and the comparative analysis of the behavioural response of the control and filtered samples are explained in this chapter.

Dry density

The control sample, with an initial solids content of 27.9 per cent and a gravimetric moisture content (GMC) of 258.7 per cent, began with a low dry density of approximately 250 kg/m³. In contrast, the filtered sample, which had undergone mechanical dewatering, started with 41 per cent solids, 143.6 per cent GMC, and a much higher dry density of 550 kg/m³. Throughout the experiment, both samples exhibited increased dry density, but the filtered sample showed more pronounced and consistent improvement, reaching approximately 700 kg/m³. The control sample's dry density improved to about 350 kg/m³ despite substantial settlement. This discrepancy is attributed to the filtered sample's initial compaction, which reduced void space and facilitated densification, whereas the control sample's higher water content and loose structure delayed significant gain in dry density. The observed trends align with corelates from earlier basin tests and support the role of initial solids and structure in influencing consolidation outcome.

Settlement

Both the control and filtered samples experienced rapid initial settlement following the first disposal, with each settling by approximately 100 mm within the first ten days. This immediate settlement is primarily due to gravity and the quick dissipation of excess pore water pressure near the surface. Over the subsequent 169 days, settlement in both samples continued gradually and at a nearly identical rate, reaching a cumulative 200 mm by the end of this phase. The similarity in settlement behaviour during this period is likely due to their initially high moisture content and the influence of seasonal wetting and drying cycles that alternated between accelerating and slowing consolidation (Rodríguez et al, 2007).

The second layer of tailings was deposited once the 200 mm settlement was reached which was also the beginning of the dry season. Both columns displayed another 35 mm of settlement in the ten days immediately following this second deposition, reflecting rapid adjustment under self-weight. Over the next 60 days, settlement slowed considerably, with only an additional 15 mm of change in both samples. During this time, surface characteristics began to diverge significantly. The filtered sample, with higher dry density and lower plasticity, developed a surface crust without cracking. In contrast, the control sample exhibited visible crack initiation and propagation during the dry season, which drastically altered its drainage dynamics.

As rainfall resumed in the latter part of the year, the cracked surface of the control sample allowed rapid water infiltration, which softened the underlying structure and reactivated consolidation processes. This led to a pronounced increase in settlement—150 mm over the final 110 days—bringing its total post-second-disposal settlement to 200 mm. The filtered sample, lacking surface cracks, continued to settle more gradually, reaching a total of 125 mm. The observed difference in settlement post-cracking aligns with previous findings that demonstrate the impact of desiccation-induced fissures on hydraulic conductivity and structural collapse upon wetting (Kodikara and Costa, 2013).

Although the control sample exhibited higher settlement, the filtration allowed 35 per cent more solids to be stored within the same column volume, demonstrating the efficiency of filtration in enhancing storage capacity compared to tailings without dewatering.

Degree of saturation (Dos)

The evolution of degree of saturation in both the control and filtered samples closely reflects natural climatic conditions, particularly the alternating wet and dry seasons. In the control sample, frequent rainfall during the first 60 days maintained a fully saturated profile throughout the column. By mid-May (around Day 130), during the transition into the dry season, the upper 400 mm began to desaturate, with saturation levels reducing to approximately 70 per cent. However, the deposition of a second layer resaturated the profile, primarily due to the moisture content of the newly deposited red mud and intermittent rainfall.

Following the second deposition, dry weather induced further desaturation, with the Dos decreasing to 50 per cent up to a depth of 800 mm. The observed drying trend in the control column is consistent with greater pore connectivity and lack of surface sealing, all of which enhance moisture loss. As the wet season resumed, the upper 300 mm of the control column exhibited repeated drying and rewetting cycles, while the profile below remained saturated. The presence of wide desiccation cracks in the control sample Figure 6 allowed seepage to rewet the deeper layers.

In contrast, the filtered sample remained fully saturated for the first 45 days due to rainfall. Afterwards, the upper 200 mm began to desaturate, reaching 65 per cent, but subsequent rain events rewet the tailings. The second disposal also rewets the tailings; however, it is not sufficient to saturate. During the dry season, dewatering of filtered tailings is prominent, with occasional rainfall only rewetting up to 450 mm. Unlike the control sample, the filtered red mud formed a surface crust during the dry season (Figure 6). This crust limited vertical infiltration, resulting in unsaturated deeper levels with Dos approximately 40 per cent. the accumulation of water in the surface due to limited infiltration can be seen in Figure 6.

From December to March, saturation levels in the filtered sample were distinctly stratified. Figure 7 shows that between December and March, saturation reached approximately 80 per cent only within the upper 600 mm, while the layer below (600–1200 mm) maintained a lower and stable degree of saturation around 60 per cent. This behaviour reflects the lower initial void ratio and dense structure of the filtered sample, which restricts water movement and as well as the barrier effect of the crust that impedes rewetting in underlying layers.

These patterns demonstrate how initial moisture content, structural properties, and surface conditions govern the hydrological response of tailings to natural weather cycles. The filtered sample's restricted desaturation depth and crust-induced water retention contrast with the more permeable and reactive behaviour of the control sample, underscoring the long-term advantages of filtration for controlling moisture dynamics in tailings deposits.

Suction

Suction (negative pore water pressure) behaviour in both samples reflected the seasonal climatic conditions and tailings properties. In the control sample, suction remained low during the first 60 days due to frequent rainfall and resulting saturation. As drier conditions set in, suction gradually increased, reaching approximately 100 kPa up to a depth of 800 mm by mid-May. A subsequent heavy rainfall event reduced the suction to negligible before the second deposition. From July through November, the dry season caused significant moisture loss, and suction increased sharply, peaking near 10 000 kPa close to the surface and approximately 2000 kPa in the layers below. This sharp increase was consistent with the development of cracks that enhanced evaporative loss and air entry. However, due to these large cracks, some suction sensors became misaligned or fell into cracks, requiring careful interpretation of data.

In the filtered sample, suction remained low during the initial phase due to full saturation and closely followed the wetting and drying cycles. During the dry season, from mid-June to November, suction increased significantly near the surface. Values reached approximately 10 000 kPa within the upper 300 mm, likely due to limited rainfall and surface drying. Between 300 and 600 mm, suction peaked around 50 000 kPa, reflecting progressive desaturation in the mid-depth zone. Below 600 mm, the tailings remained relatively moist, with suction ranging between 1000 and 5000 kPa. In the subsequent wet season, from December to March, rainfall led to rewetting of the 300–600 mm layer, reducing suction to around 10 kPa, while the deeper layers remained unsaturated, with suction increasing up to approximately 4000 kPa.

CONCLUSION

This study presents a detailed comparative evaluation of the geotechnical behaviour of filtered and unfiltered red mud through laboratory desiccation tests and a long-term large column experiment under natural weather conditions. This research quantitatively examines the long-term effect of filtration on the performance of red mud in terms of densification, structural stability, and moisture control and storage capacity.

In the basin desiccation test, filtered red mud, prepared with flocculation, sea water neutralisation, and filtration, exhibited lower initial moisture content (143.6 per cent) and higher dry density (0.55 t/m^3) compared to the unfiltered control sample (258.7 per cent moisture and 0.25 t/m^3 dry density). During drying, filtered samples showed reduced shrinkage (40 per cent compared to 78 per cent in control), delayed cracking, and finer, denser crack patterns, largely attributed to lower plasticity index (PI = 19.3 versus 32.6) and improved particle packing. Matric suction developed earlier in the control sample, starting near 150 per cent GMC, while the desaturation in the filtered sample begins at a lower moisture content, close to 100 per cent GMC, due to its denser structure and lower void ratio.

In the large column test, similar trends were confirmed under field-like conditions. The filtered sample maintained higher dry density throughout (reaching 0.7 t/m³ versus 0.35 t/m³ in the control) despite settling less (125 mm versus 200 mm). The storage capacity for the same volume could accommodate 35 per cent more solids upon filtration, which is quite significant for the tailings industry.

Control sample also remained highly saturated for longer, with a degree of saturation consistently above 80 per cent during much of the test, whereas the filtered sample desaturated more effectively, reaching around 60 per cent at depth, indicating better dewatering and densification performance which increases the stability of tailings storage as well as reduces the seepage potential.

Collectively, these results indicate that filtration not only improves initial material properties but also leads to more stable, predictable long-term behaviour. Filtered tailings resist volumetric shrinkage, reduce cracking, sustain higher dry density, and respond more gradually to environmental changes. These advantages are critical for increasing storage efficiency, improving TSF stability, and reducing environmental risks.

This study adopted a single 1.5 m lift to accelerate consolidation and drying, whereas field deposition generally proceeds in 0.1–0.3 m layers that undergo multiple wet–dry cycles; future studies may incorporate progressive thin-layer placement to capture these cyclic effects. The filtered residue obtained for this study was repulped to 41 per cent solids, a condition that simulates a rain-wetted filter cake. Parallel tests on the filtered cake without repulping could be conducted for future studies. In addition, the filtered sample displayed higher soluble-salt content and pH, variables known to alter surface tension, air-entry value, and hence evaporation kinetics (Nachshon et al, 2011); controlling or systematically varying salinity and alkalinity would clarify their separate influence.

ACKNOWLEDGEMENTS

I would like to express my deepest gratitude to Professor David Williams, Dr Chenming Zhang, Sebastian Quintero, and Lieven Tan for their invaluable contributions to this research. Their continuous support, guidance, and thoughtful feedback were instrumental in shaping the direction of this study and ensuring its successful completion. Their assistance in experimental design, data collection, analysis, and interpretation were immensely valuable, and their dedication to academic excellence has been truly inspiring. It has been a privilege to work with such knowledgeable and supportive mentors and colleagues, and I sincerely thank each of them for their unwavering commitment throughout this project.

REFERENCES

Alam, N, Ozdemir, O, Hampton, M A and Nguyen, A V, 2011. Dewatering of coal plant tailings: Flocculation followed by filtration, *Fuel*, 90:26–35.

Avery, Q D, Bach, M, Beaulieu, S, Cooling, D, Evans, K and Schoenbrunn, P F, 2022. Bauxite Residue/Red Mud, *Smelter Grade Alumina from Bauxite: History, Best Practices and Future Challenges* (Springer).

Cacciuttolo, C and Atencio, E, 2022. An Alternative Technology to Obtain Dewatered Mine Tailings: Safe and Control Environmental Management of Filtered and Thickened Copper Mine Tailings in Chile, *Minerals*, 12:1334.

Cacciuttolo, C, Pastor, A, Valderrama, P and Atencio, E, 2023. Process water management and seepage control in tailings storage facilities: Engineered environmental solutions applied in Chile and Peru, *Water*, 15:196.

Daliri, F, Simms, P and Sivathayalan, S, 2016. Shear and dewatering behaviour of densified gold tailings in a laboratory simulation of multi-layer deposition, *Canadian Geotechnical Journal*, 53:1246–1257.

Davies, M, 2011. Filtered dry stacked tailings: the fundamentals, in *Proceedings of Tailings and Mine Waste Conference 2011*, pp 1–9 (University of British Columbia: Vancouver).

de Kretser, R, 2018. Tailings filtration: risk reduction through understanding and designing for variability, in *Proceedings of the 21st International Seminar on Paste and Thickened Tailings, Paste 2018*, pp 63–74 (Australian Centre for Geomechanics: Perth).

East, D and Fernandez, R, 2021. Managing water to minimize risk in tailings storage facility design, construction and operation, *Mine Water and the Environment*, 40:36–41.

Fourie, A B, 2009. Preventing catastrophic failures and mitigating environmental impacts of tailings storage facilities, *Proceedings of the International Conference on Mining Science and Technology (ICMST 2009)*, 1:1067–1071.

Fujiyasu, Y, 1997. Evaporation behaviour of tailings, Doctor of Philosophy, The University of Western Australia. https://doi.org/10.26182/5ca3072c34624

Gallardo Sepúlveda, R, Sáez Robert, E and Camacho-Tauta, J, 2022. Assessment of the self-compaction effect in filtered tailings disposal under unsaturated condition, *Minerals,* 12:422.

Jovičević-Klug, M, Souza Filho, I R, Springer, H, Adam, C and Raabe, D, 2024. Green steel from red mud through climate-neutral hydrogen plasma reduction, *Nature,* 625:703–709.

Kaswalder, F, Cavalli, D, Hawkey, A and Paglianti, A, 2018. Tailings dewatering by pressure filtration: process optimisation and design criteria, in *Proceedings of the 21st International Seminar on Paste and Thickened Tailings, Paste 2018*, pp 427–438 (Australian Centre for Geomechanics: Perth).

Kindle, E, 1917. Some factors affecting the development of mud-cracks, *The Journal of Geology,* 25:135–144.

Kodikara, J and Costa, S, 2013. Desiccation cracking in clayey soils: mechanisms and modelling, *Multiphysical testing of soils and shales* (Springer).

Li, A, Been, K, Wislesky, I, Eldridge, T and Williams, D, 2012. Tailings initial consolidation and evaporative drying after deposition, in *Proceedings of the Proceedings of the 15th International Seminar on Paste and Thickened Tailings, Paste 2012*, pp 25–42 (Australian Centre for Geomechanics: Perth).

Milačič, R, Zuliani, T and Ščančar, J, 2012. Environmental impact of toxic elements in red mud studied by fractionation and speciation procedures, *Science of the Total Environment,* 426:359–365.

Nachshon, U, Weisbrod, N, Dragila, M I and Grader, A, 2011. Combined evaporation and salt precipitation in homogeneous and heterogeneous porous media, *Water Resources Research,* 47.

Rodríguez, R, Sánchez, M, Ledesma, A and Lloret, A, 2007. Experimental and numerical analysis of desiccation of a mining waste, *Canadian Geotechnical Journal,* 44:644–658.

Shokouhi, A, Zhang, C and Williams, D J, 2018. Settling, consolidation and desiccation behaviour of coal tailings slurry, *Mining Technology,* 127:1–11.

Snars, K and Gilkes, R, 2009. Evaluation of bauxite residues (red muds) of different origins for environmental applications, *Applied Clay Science,* 46:13–20.

Vo, T D, Pouya, A, Hemmati, S and Tang, A M, 2017. Numerical modelling of desiccation cracking of clayey soil using a cohesive fracture method, *Computers and Geotechnics,* 85:15–27.

Williams, D J, 2021. Lessons from tailings dam failures—where to go from here?, *Minerals,* 11:853.

Xue, S, Zhu, F, Kong, X, Wu, C, Huang, L, Huang, N and Hartley, W, 2016. A review of the characterization and revegetation of bauxite residues (Red mud), *Environmental Science and Pollution Research,* 23:1120–1132.

Zhang, W, Zhang, C, Lei, X, Quintero Olaya, S, Zhu, Y, Zhao, Z, Jensen, S and Williams, D J, 2024. Instrumented column testing on long-term consolidation and desiccation behaviour of coal tailings under natural weather conditions, *Acta Geotechnica,* 19:1891–1909.

Advanced seiche hazard modelling for in-pit tailings storage facilities using FLOW-3D™

A P Resende[1], N Moon[2] and H Duarte[3]

1. Water Resources Engineer, WSP, Perth WA 6000. Email: adolfo.resende@wsp.com
2. Principal Water Resources Engineer, WSP, Brisbane Qld 4006. Email: nigel.moon@wsp.com
3. Senior Water Resources Engineer, WSP, Perth WA 6000. Email: helvecio.duarte@wsp.com

ABSTRACT

The Seiche Hazard Assessment can serve as a tool for evaluating the risks posed by mass movements, such as landslides, into tailings storage facilities (TSFs). This external factor should be considered in TSF risk assessments, as it has the potential to cause uncontrolled releases or even lead to failure through overtopping. This paper presents a case study of the potential hazards posed by seiche waves in an in-pit tailings storage facility (IPTSF), with focus on three-dimensional (3D) numerical simulations.

This study assessed eighteen failure scenarios identified based on pit wall stability and tailings deposition, considering tailings as both solid and liquid. Numerical simulations provided detailed insights into wave characteristics, such as wave propagation and run-up effects, with the most critical scenarios involving liquid tailings, revealing significant wave heights and potential overtopping. Key findings indicate that overtopping risks could be linked to maximum tailings deposition and extreme rainfall events. Comparing numerical and empirical approaches, both produced similar results, but numerical simulations offered higher accuracy by considering complex phenomena. Empirical methods served as quick validation tools for intermediate conditions.

The study highlights the importance of advanced modelling techniques like 3D simulations for predicting seiche hazards and informing risk mitigation strategies. The results broadened the understanding of risks surround the IPTSF operation and informed the development of emergency response plan. Continuous monitoring and model updates are essential for improving predictive accuracy, contributing to safer and more efficient tailings storage facility management.

INTRODUCTION

Seiche modelling is crucial for understanding and managing the oscillatory behaviour of water bodies, which can have significant implications across various contexts. Seiches, standing waves in enclosed or semi-enclosed bodies of water, are driven by factors such as wind, atmospheric pressure changes, and seismic activity.

In the context of large lakes and reservoirs, seiche modelling helps predict water level fluctuations that can impact shoreline infrastructure and aquatic ecosystems. For instance, Hamblin *et al* (1999) highlighted the importance of understanding seiche dynamics in the Great Lakes to manage coastal ecosystems effectively. Similarly, Ichinose *et al* (2000) demonstrated the potential hazards of seiches in Lake Tahoe, emphasising the need for accurate modelling to mitigate risks to lakeside communities.

Seiches triggered by earthquakes can lead to significant water movement, affecting dam stability and water quality. Hewitt, Scolan and Balmforth (2011) explored the instability of seiches in reservoirs with movable dams, demonstrating the complex oscillatory behaviour resulting from fluid-structure interactions, which is vital for designing resilient hydraulic structures.

In environmental impact assessments, seiche modelling provides insights into the propagation of energy and its effects on aquatic environments. Wang *et al* (2014) discussed the application of seiche modelling in offshore wind farm assessments, illustrating how these models help understand underwater acoustic propagation and its impact on marine life.

Overall, seiche modelling is a multidisciplinary tool that enhances our ability to predict and manage the dynamic behaviour of water bodies under various natural and anthropogenic influences. By integrating knowledge from different contexts, we can develop more comprehensive and effective strategies for water resource management and environmental protection.

IPTSFs are generally considered safe structures with low risks of uncontrolled releases, which often leads to the neglect of seiches as credible scenarios for such events. However, this hazard was identified in the case study due to high levels of ponded water during the early life of the IPTSF. Several pit wall failure scenarios into the IPTSF were considered, varying landslide volume, location and orientation. Numerical modelling was performed using the hydraulic modelling package FLOW-3D™, which were validated against empirical equations for landslide-generated impulse waves in reservoirs (Evers *et al*, 2019).

METHODOLOGY

The adopted methodology and assumptions for each step of the seiche modelling were divided into three main steps, as follows:

- Assessed failure scenarios: to establish and identify the failure scenarios.

- Numeral simulations: to conduct numerical simulations using FLOW-3D™.

- Empirical correlations: to perform empirical analysis to validate the numerical results.

Assessed failure scenarios

Based on stability assessment results, the failure pit wall sections were defined, as shown in Figure 1. Different scenarios were assessed considering variable pit conditions resulting from the tailings deposition strategy. The following criteria were used to select the failure pit wall sections to be assessed as triggers to the seiche hazard modelling, which are also cross-referenced in Table 1:

- Pit wall failure volume based on results of two-dimensional (2D) stability analysis, which may dislocate a greater volume and yield greater wave heights.

- Pit wall areas with low Factor of Safety (FoS), which may be more susceptible to failures.

- Locations near the IPTSF spill point, which may be more susceptible to overtopping risks and consequently greater impacts on the surrounding work areas and critical infrastructure.

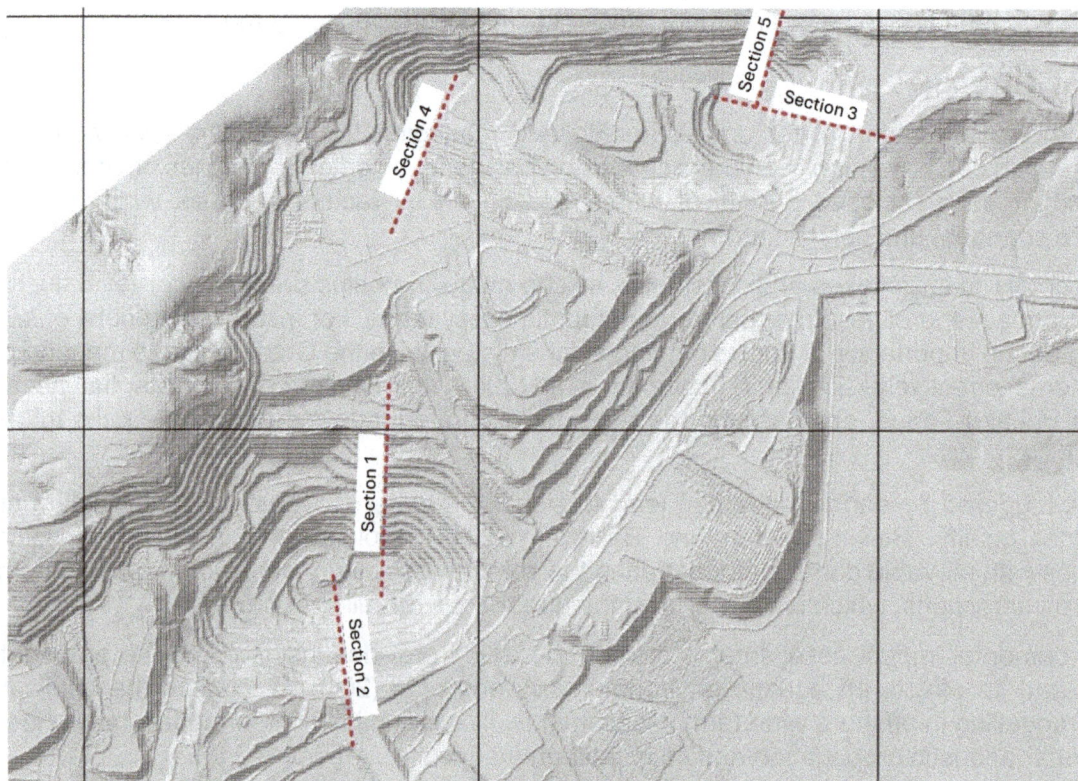

FIG 1 – Pit wall stability analysis sections.

TABLE 1

Selected scenarios.

Scenario	Stability section	Failure volume (m³)	Bottom level (RL m)	Tailings level (RL m)	Water level (RL m)	Decant pond (m)
1b		68 000		-	518.0	58.6
2a	1	14 662	459.4	561.0	564.0	3.0
2b				-	564.0	104.6
3b		41 080		-	518.0	58.6
4a	2	8 649	459.4	561.0	564.0	3.0
4b				-	564.0	104.6
5a		29 000		527.0	530.0	3.0
5b	3		479.7	-	530.0	50.3
6a		33 100		555.5	558.5	3.0
6b				-	558.5	78.8
7a		1 271 725		527.0	530.0	3.0
7b	4		515.0	-	530.0	15.0
8a		648 580		555.5	558.8	3.0
8b				-	558.5	43.5
9a		38 264		527.0	530.0	3.0
9b	5		479.7	-	530.0	50.3
10a		73 379		555.5	558.5	3.0
10b				-	558.5	78.8

The deposited tailings behaviour within the IPTSF were considered as either solid not able to mix with water (identified by letter 'a' in scenario name) or as water (identified by letter 'b' in scenario name). This approach was adopted due to the lack of rheological properties of deposited tailings and supported the coverage of the range of conditions, from tailings with high solids concentration to those with low solids concentration.

After screening all potential scenarios according to the criteria mentioned above, a total of eighteen scenarios were selected (as specified in Table 1).

Numerical simulations

The following steps and assumptions were adopting in the numerical simulations.

- 3D pit wall failure surfaces:
  - The Slide 2D stability model sections were georeferenced within the 3D geological model (Leapfrog). The 3D failure surfaces were estimated based on the approximated failure widths provided in the stability model and the orientations of geological units in the geological model. The pit wall failure volumes were calculated using the failure surfaces.
  - There were small discrepancies between the Slide 2D volumes, and the volumes calculated. Slide 2D employs a straightforward process extrapolating the 2D area across the expected failure width. This approach does not consider the 3D geometry, while Leapfrog does. The failure volume difference is limited to 20 per cent. To construct the 3D seiche model, it was required to produce a 3D failure surface using a Leapfrog model.

- There were minor discrepancies between the volumes calculated by Slide 2D and those determined using Leapfrog. Slide 2D employs a method of extrapolating the 2D area across the anticipated failure width, which does not account for the 3D geometry. In contrast, Leapfrog incorporates the 3D geometry, resulting in a failure volume difference limited to 20 per cent. The 3D failure surfaces calculated by the Leapfrog model were used as inputs in the 3D seiche model.

- 3D seiche model:

  - FLOW-3D™ HYDRO was used as the 3D computational fluid dynamics software package, with an inbuilt water-focused user interface.

  - Volume of Fluid (VOF) and Variable Density models were activated.

  - The model assumed the landslide behaves as a fluid, with a density equivalent to that of the failed rock mass.

  - The failed rock mass descends under the influence of gravity.

  - As the rock mass enters the decant pond, it displaces water, resulting in wave propagation.

  - Wave propagation within the pond extends to the perimeter, resulting in the wave run-up effect.

- 3D and shallow water model:

  - The shallow water model was enabled in the 3D model for critical scenarios where overtopping is observed, particularly when the initial water level is at the spill point elevation. A combined mesh was created under these circumstances, to simulate the seiche wave along surrounding areas of the IPTSF.

- Model inputs:

  - Rock mass density of 2.6 t/m³, consistent laboratory testing of pit wall samples.

  - Surface roughness height: A surface roughness height of 0.21 was adopted (Brisbane City Council, 2023). The roughness height for bedrock formation can vary significantly depending on factors such as the type of rock, weathering, and surface conditions. The assumption adopted for this study considers a roughness of bedrock with a minor degree of irregularity (upper limit).

  - Tailings surface: The tailings surface was modelled as a solid surface and assumed to be levelled (ie flat).

  - Decant pond depth: Assumed as 33 m deep on top of the levelled tailings surface, according to the water balance model.

  - Additional 'water-only' scenarios (named as 'b'): There is some uncertainty regarding the behaviour of the saturated tailings underlying the decant pond, including the moisture content, particularly in the event of a pit wall failure. Therefore, the authors completed additional model runs assuming the tailings behave as water (ie no tailings surface was input into the model).

  - Topography: The same topographic survey (1 m grid resolution) used for the stability analysis and the Leapfrog model was adopted.

  - Grid size: The optimisation of the mesh is a crucial phase in the modelling process. The goal is to enhance model accuracy, computational efficiency, and reliability of the results. This phase involved refining and adapting the mesh to balance between resolution and computational demand, considering that a CFD model requires significant computational resources.

Scenarios 2a, 2b, 4a, and 4b will use the 3D and shallow water modelling approach, with a 2 m grid size for 2a and 4a, a 3 m grid size for 2b and 4b, and all other scenarios will be modelled using the 3D approach with a 3 m grid size.

Empirical correlations

A landslide-generated impulse wave (and its effect on the pits) was estimated using the methodology presented by Evers *et al* (2019). This method bases its empirical equations on granular slide material, This method bases its empirical equations on granular slide material, providing an understanding of wave behaviour across the three wave phases Figure 2):

- Slide impact and wave generation.

- Wave propagation.

- Wave run-up and overtopping.

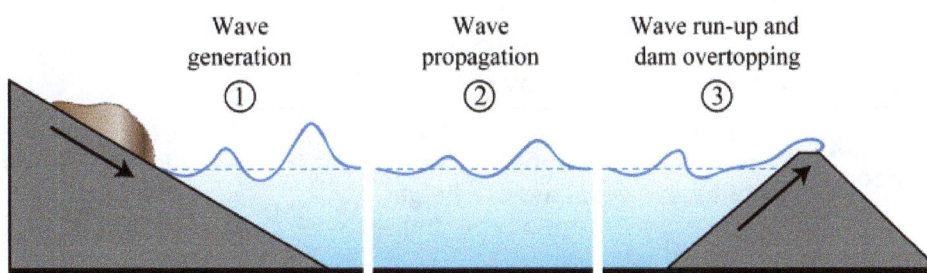

FIG 2 – Governing parameters schematisation (source: Evers *et al*, 2019).

The relevant governing parameters required to perform the calculations are provided in Table 2. This approach contrasts with the wave generation methods used in Phases 1 and 2 of the progression Figure 2). Notably, Phase 3 is excluded from this study, primarily because it relies on two-dimensional investigations to derive equations for wave run-up and overtopping. This contrasts with the methodology used in Phases 1 and 2. Furthermore, Phase 3 is omitted due to the presence of complex wave phenomena—such as dispersion, diffraction, and refraction—which limit the accuracy of wave run-up predictions. Lastly, it is noted that significant wave run-up is absent on the main axis because the pit wall's elevation surpasses the amplitude of the waves.

TABLE 2

Relevant governing parameters required to perform step 1 (source: Evers *et al*, 2019).

Governing parameter	Symbol	Unit	Description
Slide impact velocity	V_s	m/s	Slide velocity at the point of impact (function of the drop height and bed friction angle)
Bulk slide volume	$\forall_s$	m³	Volume of the displaced material
Slide thickness	s	m	Maximum thickness of the slide measured perpendicularly to the slide slope
Slide width	b	m	Average width during impact
Bulk slide density	ρ_s	kg/m³	Density of the displaced material
Bulk slide porosity	n	%	Porosity of the displaced material
Slide impact angle	α	°	Hill slope angle from the horizontal at the impact location
Still water depth at the impact zone	h	m	Average depth in the slide impact zone
Bed friction angle	δ	o	Friction angle of the hill slope material
Distance	r	m	Propagation distance from the impact location to the point of interest
Wave propagation angle	γ	°	Propagation angle measured from the slide axis to the point of interest

RESULTS

This section summarises the results of seiche modelling, illustrating the numerical simulations, empirical correlations, and comparisons between both approaches.

Numerical simulations

The seiche simulation produced waves that demonstrate several key processes, including non-linearity, frequency dispersion, diffraction, refraction, shoaling, and wave breaking. Table 3 presents the wave characteristics obtained from the FLOW-3D™ simulations.

Figure 3 presents the total hydraulic head for each location for the critical scenarios.

TABLE 3

Summary of the results.

Scenario	Decant pond (m)	Wave crest amplitude (m)	Wave trough amplitude (m)	Max wave height (m)	Spill elevation (m)	Max. elevation reached (m)	Max overtopping depth (m)
1b	58.6	~ 6.0	~ 2.4	~ 8.4	~ 564.6	~ 524.0	-
2a	3.0	~ 1.2	~ 0.3	~ 1.5	~ 564.6	~ 565.0	~ 0.4
2b	104.6	~ 0.7	~ 0.7	~ 1.4	~ 564.6	~ 564.7	~ 0.1
3b	58.6	~ 6.7	~ 3.8	~ 10.5	~ 564.6	~ 524.7	-
4a	3.0	~ 4.2	~ 1.3	~ 5.5	~ 564.6	~ 565.0	~ 0.4
4b	104.6	~ 5.2	~ 3.9	~ 9.1	~ 564.6	~ 564.8	~ 0.2
5a	3.0	~ 3.9	~ 0.9	~ 4.8	560.2	533.9	-
5b	50.3	~ 2.9	~ 3.1	~ 6.0	560.2	532.9	-
6a	3.0	~ 4.5	~ 0.5	~ 5.0	560.2	560.1	-
6b	78.8	~ 8.8	~ 1.9	~ 10.7	560.2	560.0	-
7a	3.0	~ 11.2	~ 1.4	~ 12.6	560.2	541.2	-
7b	15.0	~ 10.6	~ 3.3	~ 13.9	560.2	540.6	-
8a	3.0	~ 6.4	~ 0.5	~ 6.9	560.2	564.2	~ 4.0
8b	43.5	~ 10.6	~ 3.4	~ 14.0	560.2	566.5	~ 6.3
9a	3.0	~ 7.5	~ 2.2	~ 9.7	560.2	537.5	-
9b	50.3	~ 6.2	~ 12.3	~ 18.5	560.2	536.2	-
10a	3.0	~ 1.7	~ 0.8	~ 2.5	560.2	559.0	-
10b	78.8	~ 2.2	~ 2.2	~ 5.0	560.2	559.0	-

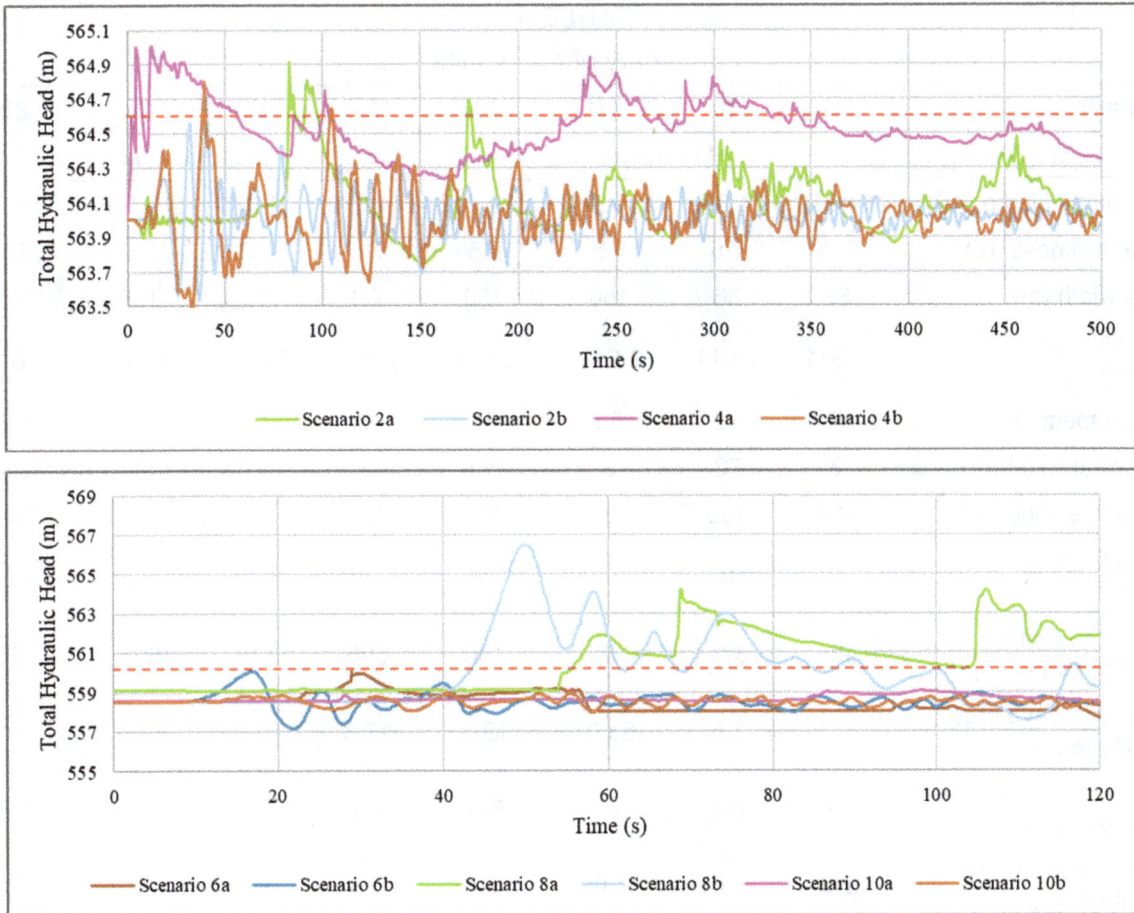

FIG 3 – Comparison of overtopping in Scenarios 2, 4, 6, 8 and 10.

Empirical correlations

An empirical validation of the two critical overtopping scenarios was conducted, with the main results presented in Table 4. The results indicate that the best-validated outcomes were found for intermediate waves rather than shallow waves.

TABLE 4

Summary of the results.

Scenario	2a	2b	8a	8b	2a	2b	8a	8b
Item	Main axis				Radial axis			
Bulk slide volume (m³)	14 379	14 379	539 280	539 280	14 379	14 379	539 280	539 280
Slide thickness (m)	7.6	7.6	8.5	8.5	7.6	7.6	8.5	8.5
Slide width (m)	66	66	150	150	66	66	150	150
Bulk slide density (kg/m³)	2 611	2 611	2 611	2 611	2 611	2 611	2 611	2 611
Slide impact angle (°)	48	48	55	55	48	48	55	55
Still water depth (m)	3	72	3	30.9	3	52.1	3	21.7
Radial distance [m]	463	463	207	207	426	426	940	940
Wave propagation angle (°)	0	0	0	0	-39	-39	82	82
Surrogate radial distance (m)	436	375	166	106	396	348	865	835
Initial first wave crest amplitude (m)	17.8	3.6	53.0	16.1	17.8	4.3	53.0	19.6
First wave crest amplitude (m)	1.1	0.4	15.1	6.6	1.1	0.3	2.7	0.9
Second wave crest amplitude (m)	0.3	3.0	1.0	4.1	0.3	1.8	0.3	1.2
Shallow (S), intermediate (I), or deep water (DW)	S	I	S	I	S	I	S	S

The outputs are divided between the initial first wave crest amplitude and the first and second wave crest amplitudes (see Table 4). The main difference in these results is the location where the maximum wave is estimated to occur. The initial first wave crest amplitude is expected somewhere within the surrogate radial distance, while the first and second wave crest amplitudes are expected within the total radial distance. It is important to note that the wave at the spill point will be the greatest value from either the first or the second wave.

Additionally, the methodology differentiates the waves between shallow, intermediate, and deep, as this factor affects the ratio of wavelength to still water depth (L/h). The calculations for celerity in shallow water are different due to the large ratio between L/h, making the method used in this analysis more appropriate for intermediate and deep waves. As mentioned by Evers *et al* (2019), an impulse wave generated by a landslide usually creates a large wave at the impact location, which typically separates into smaller waves. This process, called frequency dispersion, is one of the reasons why it is so challenging to accurately predict waves using empirical techniques.

Ma *et al* (2024) present a numerical model to simulate wave generation, propagation, and run-up on dams caused by landslide motion. The results were validated against published physical model experiments. The authors demonstrated that the numerical model is capable of accurately simulating waves generated by landslides.

Ke *et al* (2025) present a 3D analysis of landslide-generated impulse waves in a narrow reservoir. Wave characteristics were investigated through 180 trials across 36 cases and compared with predictive run-up equations. The study also includes a real-world case of a landslide that occurred on 4 December 2007, in Canada, demonstrating that the application of empirical equations closely matched the observed event.

Comparison between numerical simulations and empirical correlations

The results from the FLOW-3D™ numerical simulations were cross-referenced and compared with the empirical correlations for validation purposes. The empirical correlations are more suitable for intermediate and deep water models. Hence, this comparison is focused on scenarios labelled as 'b', where the tailings were assumed to have a significantly low solids concentration, resembling water behaviour, and leading to the creation of an intermediate/deep water model. The outcome of this comparison, for the worst cases scenarios, is summarised in Table 5.

TABLE 5
Comparison between FLOW-3D™ and empirical methodologies.

Scenario	2b	2b	8b	8b
Location	Spill	Main	Spill	Main
Failure section	39G	39G	11G	11G
Numerical simulation – generation	3.2	3.2	0	21.5
Numerical simulation – propagation	1.5	1.9	17.5	26.9
Numerical simulation – at radial distance	0	0.1	8.4	25.6
Empirical correlation – generation	4.3	3.6	19.6	16.1
Empirical correlation – propagation				
Empirical correlation – at radial distance	1.8	2.9	1.2	17.0

The results were compared at two main locations and include the maximum wave at generation, over both the propagation distance and at the radial distance. As observed in Table 5, the results converged to values with the same orders of magnitude, regardless of several physical phenomena that occur during wave propagation, such as diffraction, refraction, shoaling, and wave breaking.

CONCLUSIONS

This seiche hazard assessment was primarily completed using numerical simulations with FLOW-3D™ software. While this method requires a high level of computational effort, and therefore higher cost and time, it ultimately provides greater accuracy in its results. This is because it considers the topographic complexity of both pits, which is essential for accurately estimating seiche impacts and the complexity of wave propagation.

Importantly, the numerical simulation estimated run-out extents for the two critical scenarios where a wave might overtop the IPTSFs, inundating an area adjacent to the facility. As described above, this analysis has identified that at elevated water levels, there is a risk of flooding outside of the pit because of a seiche event. This risk is linked to specific operational conditions for the facilities when they are at full capacity and following an extreme rainfall event. The risk is not present in everyday operational conditions. This risk will need to be considered in the emergency response plan and trigger action response plans for the IPTSFs.

One limitation of the study is that empirical correlations were applied to certain scenarios classified as shallow water. These scenarios may exhibit different celerity values than those calculated using this methodology, potentially affecting the overall results.

Another limitation is that the assessed scenarios considered two extreme cases: (1) tailings deposited as a solid with a decant pond on top (scenarios labelled 'a'); and (2) tailings treated as entirely water (scenarios labelled 'b'). Additionally, the tailings were not modelled as a non-Newtonian fluid, and mixing effects were not incorporated into the simulation.

Future work could address these limitations to enhance the accuracy and robustness of the results.

ACKNOWLEDGEMENTS

The authors would like to thank everyone who contributed to this research. Special thanks to WSP for their support and incentives for innovative approaches. Finally, we acknowledge the AusIMM for their feedback on the research.

REFERENCES

Brisbane City Council, 2023. Natural Channel Design [online]. Available from: <https://s3.ap-southeast-2.amazonaws.com/docs.brisbane.qld.gov.au/Technical+Documents/Natural+Channel+Design/Appendix+C+-+Manning%27s+roughness+(pages+142-153)+(PDF+-+443kb).pdf> [Accessed: 20 November 2023].

Evers, F M, Heller, V, Fuchs, H, Hager, W H and Boes, R M, 2019. Landslide-generated impulse waves in reservoirs: Basics and computation, VAW-Mitteilungen, 254, Research Institute for Hydraulic Engineering, Hydrology and Glaciology (VAW), ETH Zürich. https://doi.org/10.3929/ethz-b-000413216

Hamblin, T J, Davis, Z, Gardiner, A, Oscier, D G and Stevenson, F K, 1999. Unmutated Ig V (H) genes are associated with a more aggressive form of chronic lymphocytic leukemia, *Blood*, 94:1848–1854.

Hewitt, I J, Scolan, H and Balmforth, N J, 2011. Flow-destabilized seiches in a reservoir with a movable dam, *Journal of Fluid Mechanics*, 673:1–26.

Ichinose, G A, Anderson, J G, Satake, K, Schweickert, R A and Lahren, M M, 2000. The potential hazard from tsunami and seiche waves generated by large earthquakes within Lake Tahoe, California-Nevada, *Geophysical Research Letters*, 27(8):1203–1206. https://doi.org/10.1029/1999GL011119

Ke, C, Miao, F, Wang, Y, Yang, B, Wu, Y, Liu, J, Zhang, Y and Li, X, 2025. Three-dimensional experimental analysis of landslide-generated impulse waves in narrow reservoirs with variable local water depths, *Landslides*. https://doi.org/10.1007/s10346-025-02523-w

Ma, H, Wang, H, Xu, W, Zhan, Z, Wu, S and Xie, W C, 2024. Numerical modeling of landslide-generated impulse waves in mountain reservoirs using a coupled DEM-SPH method, *Landslides*, 21:2007–2019. https://doi.org/10.1007/s10346-024-02243-7

Wang, X, Holden, C, Power, W, Liu, Y and Mountjoy, J, 2014. Seiche effects in Lake Tekapo, New Zealand, in an Mw8.2 Alpine Fault earthquake, *Pure and Applied Geophysics*, 177:5927–5942.

Challenges and opportunities in implementing combined paste and aggregate fill systems – lessons learned and potentials

M Ryan[1]

1. Senior Backfill and Tailings Engineer, Perth WA 6028. Email: masseyryan999@gmail.com.au

ABSTRACT

The Combined Paste and Aggregate Fill (PAF) system represents an innovative approach to sustainable waste management in underground mining. It is designed to integrate the simultaneous placement of cemented paste and aggregate for improved ground stability and reduced surface waste. This paper evaluates the implementation of a PAF system in an underground mining operation, where the technical feasibility of concurrent disposal was validated. Still, operational success was hindered by design flaws and unanticipated challenges.

Key issues impacted the system's overall performance, including inefficiencies in process integration, equipment reliability concerns, and inadequate adaptation to site-specific conditions. Despite these obstacles, the project highlighted the potential of the PAF system to optimise underground backfill operations, reduce surface waste disposal, and support sustainable mining practices.

This paper provides a detailed analysis of the challenges encountered during implementation, the opportunities identified for improving system performance, and the lessons learned from the project. It emphasises the importance of proper design, robust maintenance strategies, and process adaptability to achieve the full potential of combined paste and aggregate systems.

This study aims to encourage continued research and development in this field by offering actionable recommendations and demonstrating the value of the PAF concept. The findings underscore the importance of advancing sustainable waste management practices in mining operations, paving the way for future innovations in backfill technology.

INTRODUCTION

One of the significant challenges faced by deep underground mining operations is the management of waste rock. Waste rock is typically generated during the excavation of main declines and level access drives in areas that lack economically viable mineral deposits. Given the spatial constraints inherent in underground environments, the waste rock must either be hauled to the surface for disposal or deposited in underground voids, such as abandoned stopes or drives. As reported by Yilmaz and Guresci (2017) Surface haulage of waste rock imposes substantial costs and adversely impacts the mine's overall ore haulage capacity, creating inefficiencies in resource extraction and transportation logistics.

Stacking waste rock in ore drives can lead to several adverse effects that compromise both operational efficiency and long-term mine viability. One primary concern is restricting access to remaining ore reserves, as backfilled drives may obstruct future mining activities or exploration opportunities. Geotechnical instability can arise if the waste is not compacted correctly, leading to uneven settling and potential ground collapse. Additionally, as reported by McPherson (1993) such practices can disrupt ventilation systems, causing reduced airflow and accumulation of harmful gases, which pose risks to worker safety. From an environmental perspective, waste rock can leach contaminants into underground water systems, complicating water management and potentially violating environmental regulations. Furthermore, emergency response and escape routes may be compromised, as backfilled ore drives limit access and manoeuvrability during critical situations.

A widely practised method for waste disposal in underground mining involves tipping waste rock from the stope top access into the void until it is entirely filled with backfill material (Behera, Mishra and Das, 2023). Two primary approaches are employed for this process. The first approach is to completely fill the stope with waste rock, which is a viable option if the filled stope will not have any future exposure, either horizontally or vertically. While this method reduces surface waste disposal and enhances geotechnical stability, it also has adverse effects. As reported by Xue et al (2021) one significant concern is that it can obstruct the natural flow of groundwater through the stope, potentially leading to water pooling and the washing out of finer materials, which creates mud inrush hazards. This scenario can compromise mine safety and geotechnical stability, leading to operational challenges and increased maintenance costs. Careful planning and hydrological assessments are essential to mitigate these risks effectively.

Backfilling methods have been developed to address the challenges of mud inrush hazards and to enhance geotechnical stability, incorporating waste rock with cementitious materials such as grout, cemented hydraulic fill, and cemented paste backfill (Mitchell and Hassani, 2018). One widely utilised method is Cemented Rock Fill (CRF), which involves mixing waste rock with grout or shotcrete before tipping it into the stope using a loader. This approach improves stability by creating a more cohesive backfill material that resists water ingress and structural collapse. However, the method is notably slow, as only small volumes of 50 m^3 to 100 m^3 can typically be achieved in a single shift using a loader and delivery agitator. Due to its low production rate and high labour and equipment costs, this method proves inefficient and uneconomical for mines with large stopes, especially those with significant dips.

To enhance the efficiency of stope filling and reduce associated costs, a co-disposal method has been developed that integrates waste rock with cemented hydraulic fill (CHF) or cemented paste fill (Lee and Gu, 2017). This approach involves tipping waste rock into the stope from a top access point or backfill filling rise, with the tipping point carefully selected to prevent waste material from reaching future exposures, such as horizontal or vertical surfaces, thereby safeguarding the integrity of planned mining operations (Figure 1). Following the disposal of waste rock, CHF or paste fill is poured over the waste rock cone, forming a covering layer. The cementitious medium percolates through the void spaces within the waste rock and, upon curing, consolidates the bulk of the disposed material into a stable, cohesive mass. Once a roughly flat backfill layer is achieved in the stope, the process is repeated by disposing of the next lift of waste rock and covering it with CHF or paste fill. This cycle continues until the height of the waste rock reaches the tipping point. The remaining void in the stope is then tight-filled with CHF or paste fill, ensuring complete stope stabilisation.

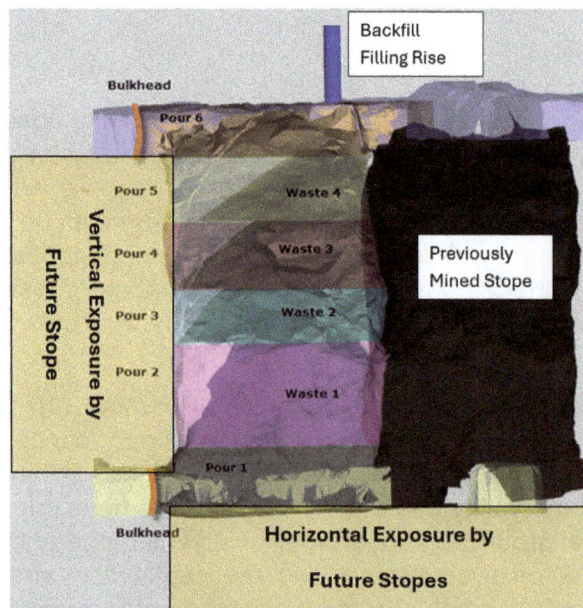

FIG 1 – Illustration of the co-disposal method for waste rock and cemented paste or CHF in stopes.

The co-disposal method offers significant advantages in underground mining by accelerating the filling rate and improving geotechnical stability, primarily by reducing risks associated with groundwater infiltration and mud inrush. This approach is particularly well-suited for large-scale stopes, combining efficiency with structural reliability to provide a cost-effective and sustainable waste disposal solution. However, several limitations need to be addressed. A critical challenge is ensuring the availability of a tipping point on top of the stope or a backfill tipping rise, as developing such infrastructure can be costly and is often constrained by ground stability issues. Another drawback is the limited percolation of cementitious material into the waste rock cone. With only 4–6 per cent of voids filled when using paste and 11–15 per cent when using CHF, a significant portion of the waste rock remains unbonded despite being surrounded by cementitious material. This lack of bonding presents risks for future exposures, particularly in the event of a massive overbreak. If a future stope intersects the waste rock bulk, a large portion of the unbonded material may spill into adjacent stopes, leading to ore dilution or, in worst-case scenarios, rendering the next stope uneconomical to mine.

To address the challenges associated with traditional co-disposal methods, this paper explores the implementation of Combined Paste and Aggregate Fill (PAF) Systems, a more advanced and efficient backfilling technique. This approach produces a homogeneous mixture of crushed aggregate and paste in a controlled environment, ensuring consistent quality and bonding. This mixture is then delivered to the stope using a network of filling pipes, enabling precise placement and minimising operational inefficiencies. Combining paste and aggregate at the source eliminates the issues of limited cementitious percolation and unbonded waste rock, ensuring a fully consolidated backfill mass. The Combined PAF System offers enhanced geotechnical stability and reduces the risks of ore dilution or stope failure caused by unbonded material entering adjacent stopes. Additionally, it optimises the filling process by increasing efficiency and achieving a uniform backfill density, making it a viable solution for tackling the limitations of conventional methods in large-scale mining operations.

COMBINED PASTE AND AGGREGATE FILL SYSTEM DESIGN

The Combined Paste and Aggregate Fill (PAF) System is designed to produce a homogeneous mixture of crushed aggregate and cemented paste delivered to stopes through the reticulation system.

The process begins with clean waste rock being tipped into a Waste Rock Reservoir Stope via the trucking fleet from a designated tipping rise. A loader feeds the waste rock into a crushing system on the Crushing Level. The crushing circuit consists of a grizzly and apron feeder, a jaw crusher, and a cone crusher. The grizzly and apron feeder manages waste rock flow from the reservoir and pre-screening material while ensuring a consistent feed rate for the crushing process. The jaw crusher reduces the primary size of large waste rock, breaking it into smaller, manageable pieces. Following this, the cone crusher performs secondary crushing, producing uniform aggregate particles of the required size (2–12 mm) suitable for mixing with cemented paste. The crushed aggregate is transported to a Crushed Aggregate Reservoir Rise, which is stored temporarily before being fed into the mixing system. Figure 2 provides a schematic flow diagram of the Combined Paste and Aggregate Fill (PAF) System, illustrating the aggregate preparation, mixing, and delivery to the underground stopes.

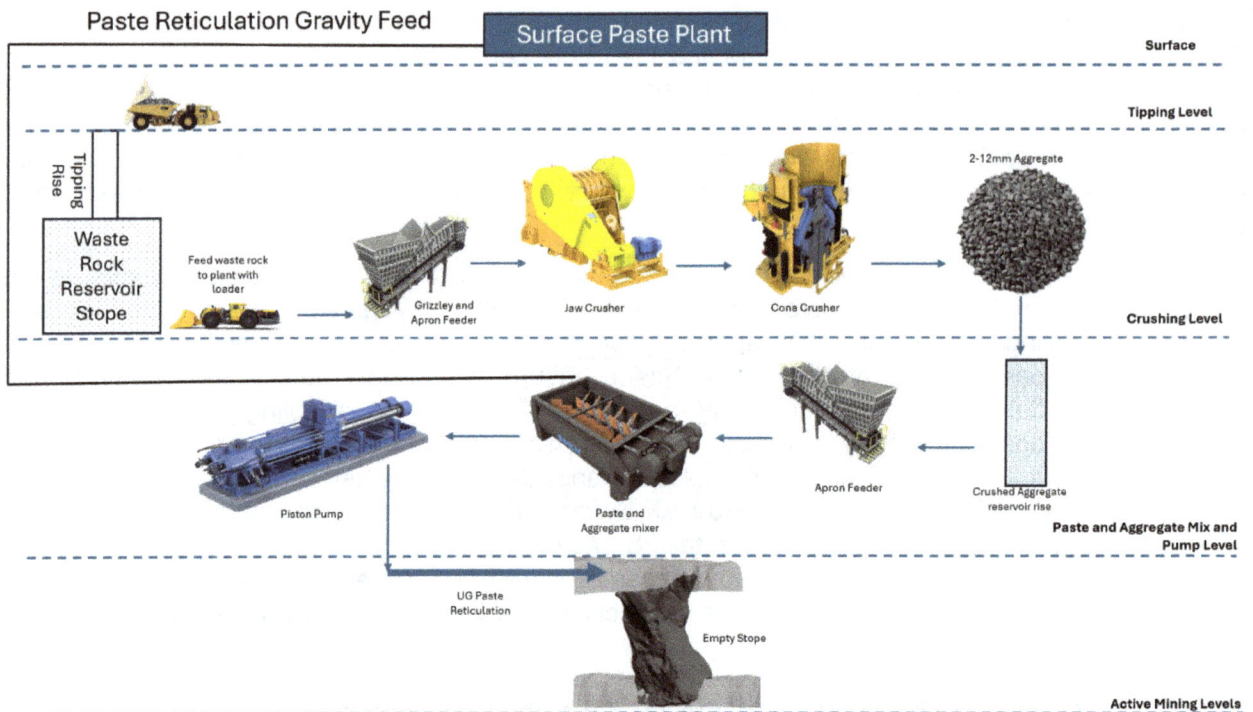

FIG 2 – Schematic flow diagram of combined paste and aggregate fill (PAF) system with gravity feed.

At the Paste and Aggregate Mix and Pump Level, the crushed aggregate is fed through an apron feeder into a continuous mixing plant, combined with cemented paste. The paste is delivered to the mixing plant from the surface paste plant via a gravity-fed reticulation system. Within the mixer, the paste and aggregate are thoroughly blended to create a homogeneous material that meets the geotechnical and operational requirements of the stope. Once the mixing process is complete, the paste-aggregate mixture is transferred to a piston pump, which propels the material through an underground reticulation system to the designated stoping area.

Performance and characteristics of PAF backfill material

Figure 3 illustrates the Unconfined Compressive Strength (UCS) Performance of PAF and Paste Materials. The comparison of the UCS between Cemented Paste Fill (CPF) and Combined Paste and Aggregate Fill (PAF) over a 60-day curing period highlights the better performance of the PAF material. The PAF system demonstrates faster strength development, achieving approximately 600 kPa within 10 days compared to 510 kPa for CPF. By 30 days, the PAF material reaches 1580 kPa, surpassing the 1250 kPa of CPF, and continues to outperform, achieving 1800 kPa by day 60, while CPF plateaus at 1200 kPa. This enhanced performance is due to the inclusion of crushed aggregate in the PAF, which improves structural integrity, reduces void spaces, and enhances bonding within the backfill mass.

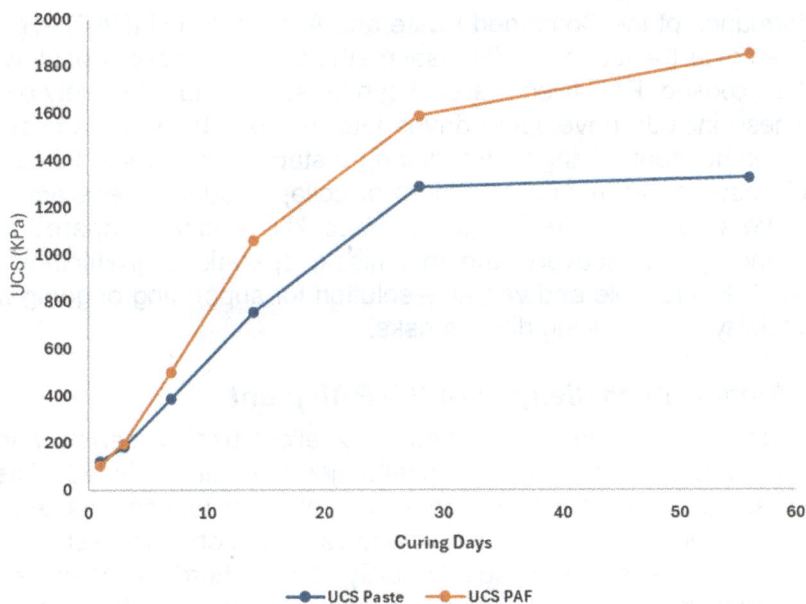

FIG 3 – Unconfined compressive strength (UCS) performance of PAF and paste materials, 5.5 per cent LH binder, 76 per cent solid content.

Figure 4 provides a visual comparison between the paste-only and PAF samples after 58 days of curing highlighting the unique structural characteristics of the two materials. The paste-only sample shows a uniform texture with a fine-grained, dense structure typical of cemented paste fill (CPF). While this structure provides adequate compressive strength, it lacks the reinforcement offered by larger, more rigid aggregate particles. In contrast, the PAF sample exhibits a heterogeneous composition, with clearly visible crushed aggregate particles embedded within the paste matrix. This composite structure results in better durability and strength, as the aggregate particles contribute to the overall stability of the material. The paste binds the aggregates together and fills the void spaces, creating a cohesive and robust backfill material. Including aggregate in the PAF sample enhances its geotechnical performance, making it less prone to shrinkage or deformation than the paste-only sample. These visual differences directly correlate with the mechanical properties, such as higher unconfined compressive strength (UCS), observed in the PAF material, making it a more effective solution for stope backfilling in underground mining operations.

FIG 4 – Comparison of paste-only and PAF samples after 58 days of curing.

The exposure performance of the Combined Paste and Aggregate Fill (PAF) system demonstrates significant advantages over traditional co-disposal methods. Unlike co-disposal, where the fill mass is not intended to be exposed, PAF-filled material can be safely and effectively exposed for various mining activities. These include developing drives into the PAF backfill, vertical firing against the backfilled material, and horizontal firings, all requiring a stable and durable fill mass. The structural integrity of the PAF ensures minimal risk of failure or collapse during these activities. Additionally, PAF-filled stopes have shown a marked reduction in backfill dilution compared to paste-only filled stopes, further enhancing ore recovery and minimising operational inefficiencies. This superior performance makes PAF a reliable and versatile solution for supporting ongoing mining operations while maintaining stability and reducing dilution risks.

Operational and long-term challenges of the PAF plant

Despite achieving good results and demonstrating excellent backfill performance, the long-term operation of the PAF plant encountered major challenges that ultimately led to its discontinuation. While the backfill material consistently met geotechnical expectations and effectively supported mining activities, persistent operational inefficiencies, maintenance issues, and unforeseen complications undermined the system's sustainability. These hurdles, combined with escalating costs and logistical difficulties, overshadowed the initial success of the PAF system, forcing a revaluation of its viability for daily and long-term use. In the following paragraphs, we will explore these daily and long-term hurdles in detail, highlighting the factors contributing to discontinuing the plant after a few years of operation.

- The first hurdle observed in the daily operation of the PAF plant was the mandatory shutdown during underground firing times, which occurred twice per day. This practice led to a total of 10 hrs of operational downtime daily, significantly reducing the plant's efficiency and overall output. Such a challenge, while impactful, could have been addressed by positioning the PAF plant within an independent firing zone. This arrangement would have allowed plant workers to continue operations safely during underground firing activities, maintaining productivity and minimising interruptions in the backfilling process.

- The second hurdle that substantially hampered the plant's output was the direct coupling of the crushing level with the mixing level, compounded by the limited capacity of the crushed aggregate reservoir, which held only 500 t. With the aggregate disposal rate at 85 t/h, the reservoir could be depleted in less than a single shift if the crushing level did not continuously supply material. This dependency required a loader, its operators, and additional personnel to remain consistently on the crushing level to maintain a steady aggregate supply, leading to increased labour demands and logistical inefficiencies. This challenge could have been mitigated by implementing multiple crushed aggregate reservoir rises, all feeding the mixing level. By increasing the crushed aggregate capacity to 3000–4000 t, the crushing level could have operated independently of the mixing level, reducing operational pressure and improving the overall efficiency of the PAF plant.

- The third hurdle encountered over the long-term was the accelerated rate of depreciation in the pump and mixing levels caused by extensive saltwater ingress. This issue led to severe corrosion of the steel structures, ultimately rendering the maintenance of the plant impractical. Over time, the corrosion significantly weakened critical components, causing frequent breakdowns and escalating repair costs. The extent of the damage made it increasingly difficult to sustain the plant's operations, contributing to its eventual shutdown. As reported by Richards (2018) this challenge could have been avoided through a comprehensive hydrogeological study during the planning phase, which would have allowed for the selection of a plant location less susceptible to saltwater ingress, ensuring the long-term integrity of the infrastructure.

- Another main hurdle was the complexity of the plant's instrumentation, which was over-engineered and, in practice, became a significant obstacle to efficient operation and maintenance. For example, the oversized pump design, intended to handle high-capacity throughput, often caused operational inefficiencies due to its impractical scale and maintenance demands. Additionally, limited access to certain critical parts of the plant made routine inspections and repairs unnecessarily complex, leading to prolonged downtime and increased labour requirements. Simplifying the instrumentation and designing the plant with accessibility in mind could have reduced these challenges, ensuring smoother daily operations and easier maintenance in the long run.

Future improvements and development strategies for PAF systems

Although the PAF plant faced operational and structural limitations, the backfill material it produced consistently demonstrated high performance. To build on this success, future developments must address both the practical and engineering aspects of PAF system design, with a focus on simplifying infrastructure and enhancing material behaviour.

- Rheological behaviour and mixing efficiency:

    A critical area for future research is the rheological behaviour of the PAF mixture, particularly under pipeline flow conditions. As the paste-aggregate mix travels through the reticulation system, turbulence significantly influences the distribution and homogeneity of the material. Studies should explore how shear thinning and thixotropic properties of the paste interact with the aggregate particles during turbulent flow. A better understanding of this interaction can lead to refinements in pipe design and flow velocity control, improving uniformity and reducing the risk of segregation or blockage.

    Additionally, optimising the in-line mixing dynamics—especially where turbulence is highest—can help maintain consistency in aggregate distribution throughout the pipe length. Computational fluid dynamics (CFD) modelling could be used to design mixing sections that enhance dispersion and reduce wear.

- Optimised mixing and pumping designs:

    The original plant's pumping and mixing set-up was overengineered and challenging to maintain. Future systems should focus on more compact, modular, and energy-efficient designs that reduce mechanical complexity while maintaining throughput. Research should investigate alternative pump types better suited to handle coarse, abrasive mixtures without sacrificing flow stability or pressure head.

    Designs should also consider introducing real-time rheological monitoring, enabling operators to dynamically adjust mix ratios and flow parameters based on feedback from sensors measuring viscosity, density, and pressure loss within the system.

- Simplified aggregate injection and dosing systems:

    The aggregate injection system was one of the most labour-intensive components of the original plant. Future improvements should aim to simplify the dosing mechanism by integrating controlled feed systems that are responsive to paste flow rate and mix demand. Pneumatic or gravity-fed surge hoppers, equipped with variable speed feeders and real-time load cells, can provide a more manageable and predictable aggregate dosing process.

CONCLUSIONS

The Combined Paste and Aggregate Fill (PAF) system showcased a high-performing backfill solution that delivered exceptional geotechnical stability, durability, and reliability, significantly supporting underground mining operations. Despite these successes, the PAF plant faced numerous challenges, including operational inefficiencies, maintenance demands, and long-term structural issues, ultimately leading to its discontinuation after four years of use. These hurdles, such as the coupling of crushing and mixing levels, saltwater-induced corrosion, complex instrumentation, and dependency on firing schedules, highlighted critical areas for improvement in the plant's design and operation.

However, the backfill material produced by the PAF system consistently met and exceeded performance expectations, underscoring the value of this technology for future mining applications. Lessons learned from this project emphasise the importance of addressing logistical, structural, and operational shortcomings to ensure the long-term sustainability of PAF systems.

Future developments should focus on optimising plant location, increasing storage capacity, using corrosion-resistant materials, simplifying instrumentation, and implementing advanced automation systems. By applying these improvements, the next generation of PAF plants can overcome past challenges and unlock the full potential of this innovative backfill technology, paving the way for more efficient, sustainable, and cost-effective mining operations.

REFERENCES

Behera, B K, Mishra, B and Das, S K, 2023. Progress and prospects of mining with backfill in metal mines: A review, *International Journal of Minerals, Metallurgy and Materials*, 30(3):266–280. Available from: <https://link.springer.com/content/pdf/10.1007/s12613-023-2663-0.pdf>

Lee, C and Gu, F, 2017. Co-disposal of waste rock with backfill, in *Proceedings of the First International Conference on Underground Mining Technology (UMT 2017)* (eds: M Hudyma and Y Potvin), pp 353–362 (Australian Centre for Geomechanics: Perth). https://doi.org/10.36487/ACG_rep/1710_27_Lee

McPherson, M J, 1993. Subsurface Ventilation and Environmental Engineering, *Subsurface Ventilation and Environmental Engineering*.

Mitchell, D and Hassani, F P, 2018. Enhancing the mechanical properties of backfill using mixed aggregate and paste: A case study, *Journal of Mining and Geotechnical Engineering*, 12(4)245–260.

Richards, L, 2018. Hydrogeological impacts on underground plant corrosion, *Environmental Geoscience Reports*, 22(3):117–130.

Xue, Y, Kong, F, Li, S, Qiu, D, Su, M, Li, Z and Zhou, B, 2021. Water and mud inrush hazard in underground engineering: Genesis, evolution and prevention, *Tunnelling and Underground Space Technology*, 114:103987.

Yilmaz, E and Guresci, M, 2017. Design and Characterization of Underground Paste Backfill, *Paste Tailings Management*, pp 115–136. Available from: <https://link.springer.com/chapter/10.1007/978-3-319-39682-8_5>

Enhancing tailings management through optimised *in situ* dewatering and consolidation

O Santiago[1], R Menezes[2] and W McAdam[3]

1. Global Tailings Principal, Phibion Pty Ltd, Lytton Qld 4178. Email: oscar.santiago@phibion.com
2. Specialist Engineer, Phibion Pty Ltd, Lytton Qld 4178.
3. Specialist Engineer, Phibion Pty Ltd, Las Condes, Santiago, Chile.

ABSTRACT

The effective management of tailings remains a critical challenge in the mining industry, requiring innovative solutions to improve TSF stability, minimise storage footprints, and optimise embankment construction. This paper presents a combined methodology utilising *in situ* dewatering and optimised consolidation techniques to enhance the mechanical properties of deposited tailings, enabling cost-effective and safe management. The study focuses on complementary approaches that accelerate dewatering of tailings through *in situ* mechanical consolidation to improve *in situ* strength for embankment raising or alternative reuse of the stored material to reduce reliance on extended settlement periods while improving tailings storage rates.

Controlled mechanical consolidation techniques, including precise and tailored dewatering cycles, increased shear strength from <5 kPa to >30 kPa in <60 days, providing construction benefits for subsequent raises, improving capital cost savings with improved storage rate, stable denser material compared to conventional methods. The process overall is defined by optimal tailings deposition depths, machine pass frequencies, and coverage strategies, demonstrating that mechanical consolidation and compaction could reduce overall tailings settlement, consolidation and compaction time independent of initial solids content. Field data from multiple operations confirm that adopting a structured, data-driven framework for mechanical consolidation application prevents crust formation, enhances drainage, and sustains predictable dewatering rates. Additionally, the implementation of modelling tools for real-time tailings monitoring supports operational adjustments to maximise efficiency. These results underscore the importance of integrating geotechnical investigation, real-time monitoring, and targeted dewatering techniques to achieve significant economic and environmental benefits.

The findings presented provide a scalable, adaptable framework applicable across mining tailings storage sites globally. By minimising storage rate of tailings, reducing required disposal footprints, and improving tailings strength, this approach offers a sustainable and cost-efficient alternative to traditional tailings management practices. The conclusions highlight the potential for industry-wide adoption of optimised dewatering and consolidation techniques to improve tailings stability while enhancing operational efficiency.

INTRODUCTION

The International Council on Mining and Metals (ICMM) signatories are committed to advancing the safe and sustainable management of tailings storage facilities (TSFs) through innovative dewatering and stabilisation techniques (ICMM, 2022, 2021). Tailings, the finely ground by-products of mineral extraction, are typically deposited as high-moisture slurries in TSFs, posing risks to structural integrity and the environment. To mitigate these risks and challenges, the processes of settlement, consolidation, and compaction are key to enhance tailings stability, reduce water content, and ensure long-term safety and stability. These methods are pivotal in transforming tailings from a vulnerable waste slurry into a dense, manageable material. Traditional tailings management practices are being challenged by emerging dewatering technologies aimed at enhancing safety and preparing TSFs for closure (McPhail, Ugaz and Garcia, 2019; Gerritsen *et al*, 2024). While scaling and adoption of solutions such as high-pressure filtration and/or 'dry-stacking' presents obstacles—including high capital costs, operational complexity, and limited applicability to legacy impoundments (CGI, 2025), there is an industry wide need for transformative tools, offering precise control over dewatering, consolidation and compaction to revolutionise tailings management and ensure the industry's environmental and operational resilience.

One of the emerging technologies which has been tested, implemented and scaled, and significantly improves tailings management issues in new and existing dams, is the *in situ* mechanical dewatering, consolidation and compaction (known as Accelerated Mechanical Consolidation or AMC™) methodology using an advanced and customised amphibious tractor equipped with Archimedean screws (Smirk, Santiago and Pardon, 2022; Smirk and Jackson, 2010; Munro and Smirk, 2018, 2012) that effectively dewater, enhance consolidation and compaction of tailings in existing or new TSFs. The methodology uses the principles of consolidation and compaction while delivering superior results than other industry practices for the management of tailings (Woolston and DiDonna, 2020). The methodology originates with settlement which initiates this transformation as solid particles within the tailings' slurry settle under gravitational forces following deposition. This natural process it is enhanced by the amphibious tractor (amphirol) which displaces water upward, gradually expelling it from the tailings matrix and increasing the material's density. By reducing pore water pressure, settlement enhances TSF stability, though its performance is influenced by settling rates of fine-grained and complex clay-minerals, often necessitating additional intensity with the amphirol to accelerate water removal. This process densifies the tailings, elevating the shear strength reducing vertical infiltration and reducing susceptibility to geotechnical failures, particularly where tailings are utilised for upstream embankment construction or safer storage. The final process is compaction of the tailings, which in contrast, involves active mechanical compression using amphibious and civil equipment for mechanical compaction, applied to dewatered tailings to eliminate residual voids and gain maximum dry density. When applied appropriately to the tailings it enhances resistance to erosion, subsidence, fugitive dust, among other benefits, producing a robust, cohesive mass suitable for long-term storage or alternative use.

Collectively, these processes, enhanced with mechanical aid improve water management by recovering excess moisture that is reused, minimising seepage-induced groundwater contamination and alleviate hydraulic stress on containment structures. Environmentally, consolidated and compacted tailings exhibit reduced dust generation and leaching of potential damaging substances (eg heavy metals), mitigating the ecological impact of mining operations. Furthermore, volume reduction optimises storage capacity—a critical consideration given the voluminous tailings output, finite land resources in mining regions, and maximisation of capital invested. For site closure, these stabilised tailings provide a solid foundation for capping with soil or vegetation, ensuring compliance with regulatory standards and facilitating post-mining rehabilitation.

BACKGROUND

Typically, tailings are stored as slurries in tailings storage facilities (TSFs), these materials pose risks to safety, operational efficiency, and the environment if not managed properly. Effective consolidation and compaction are critical to reduce water content, increase density, and ensure TSF stability, yet conventional methods often fall short, exacerbating these issues (Figure 1). As global demand for minerals rises, so does tailings production, amplifying the need for robust management strategies to mitigate financial and ecological consequences.

Consolidation, the process of expelling water from tailings to enhance density, relies on applying a load to release interstitial water trapped within the material matrix. Self-weight consolidation, driven by the tailings' own mass, is a common approach in theory. However, its low load often proves ineffective, particularly with fine-grained and clay-rich tailings, leading to slow water expulsion. This can cause surface drying and crust formation, trapping substantial water between deposition layers, which resists extraction by environmental mechanisms such as evaporation. Compaction, intended to mechanically densify tailings and eliminate voids, also struggles when initial dewatering is inadequate, limiting its efficacy. Low consolidation rates force operators to prematurely raise TSF heights, construct new dams or other management infrastructure, incurring significant costs and expanding operational footprints.

FIG 1 – Examples of fresh and natural consolidated tailings of: (a) zinc, (b) alumina, (c) bauxite and (d) copper.

To address these challenges, mechanical consolidation and compaction emerges as a solution. Unlike passive self-weight methods, AMC™ applies targeted mechanical disturbance to tailings *in situ*, accelerating water egress and reducing pore spaces on fresh tailings to boost density rapidly.

Mechanical consolidation and compaction in practice

Mechanical consolidation and compaction, also known as AMC™ (Figure 2), present an efficient and cost-effective solution to mechanically dewater, consolidate and compact *in situ* tailings with comparable and in some cases superior results to filtered and stacked tailings facilities (Woolston and DiDonna, 2020; Santiago *et al*, 2024). The theory of mechanical enhancement implies that *in situ* mechanical perturbation of a deposited material accelerates its consolidation process. In the case of suspensions, this technique overcomes the effects of hindered settlement and facilitates the exit of water from the tailings profile. When applied to highly flocculated or porous materials, it collapses the pores and rapidly increases density to levels equivalent to those achieved by filtration or centrifugation (Munro and Smirk, 2018). Reduction of interstitial pore spaces and water is by far the most important consideration for the design and operation of TSFs. Significant efforts are made to maximise the water recovered from tailings to allow continued tailings transportation and process water.

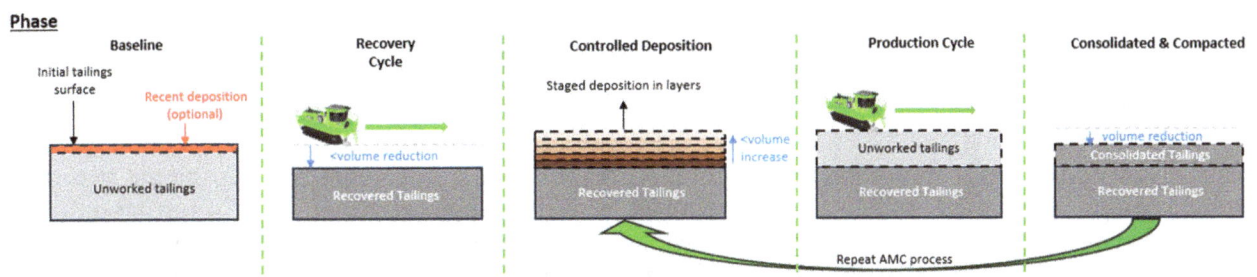

FIG 2 – AMC™ Operational phases of implementation.

The AMC™ process involves operational cycles, starting with an initial period the 'recovery cycle' where interstitial tailings water is compressed from the tailings to a point where the amphibious machine has consolidated the tailings into a dense material upon which the machines weight is bearing. This stage is critical for operational effectiveness as machines then gain adequate traction in the consolidated tailings to traverse at greater speeds as well as creating a dense base upon which fresh tailings can be deposited and compressed against. Following the recovery phase, the operation moves into the production phase (operational cycles) where dewatering, consolidation and compaction are optimised. A specialised operational plan must be developed to optimally drain excess water effectively and work in conjunction with tailings placement and construction activities since control the deposition depth, location and timing of fresh tailings discharge are critical to the process.

To investigate and demonstrate the effectiveness of mechanical consolidation and compaction a test plan is developed with a combination of *in situ*, sampling and laboratory analysis for both altered and unaltered tailings as shown in Table 1.

TABLE 1

Suggested geotechnical testing for altered and unaltered tailings.

Test type	Parameter
Granulometry	Granulometric distribution – percentage fines
Atterberg limits	Liquid limit (LL), plastic limit (PL) and plasticity index (PI)
USCS classification	-
Sand cone test	*in situ* dry density
Shear vane test	*in situ* undrained shear strength
Modified proctor	Dry maximum compacted density (DMCS) and optimum moisture
Humidity	Percent humidity
Particle density	Particle density solids – specific gravity (GS)
Porchet test	Infiltration rate
DCP PANDA test	Tip resistance
Survey – LiDAR	Fill and consolidation rate

AMC™ in action

To enhance and accelerate dewatering and consolidation, mechanical enhancement has been undertaken in copper concentrate and zinc refinery tailings. This method involved the use of an amphirol that traverse the surface of deposited tailings. These repeated loads created a rapid dewatering effect and expediting the removal of interstitial water from the material matrix. In parallel, the mechanical loading also compacted the tailings, enabling effective particle rearrangement and densification, which in turn improved the material's strength and resistance properties (South32 2023; McPhail, DiDonna and Ugaz, 2021). To quantify the changes in volume and strength, a monitoring program was established, consisting of fortnightly aerial surveys to detect surface level changes and calculate volumetric changes. Additionally, *in situ* undrained shear strength testing was carried out at various depths across the work area to evaluate strength gains over time.

Mechanical treatment of tailings was conducted at a large copper mining operation located in Chile, where both copper and molybdenum tailings are stored at a high intensity of up to 180 000 t per day and spanning a total area of approximately 600 Ha. Though, a concentrated operational trial was completed in 90 Ha. The tailings are discharged from the concentrator plant at a targeted solids concentration of 54 per cent by weight, following hydro-cycloning that separates coarser fractions used in dam construction. Copper tailings are classified as non-plastic ranging from low plasticity silt and silty sand.

At the second mine operation zinc tailings discharged into the facility at approximately 1.15 t/m^3 (bulk density) or 20 per cent solids (w/w) in an area of 27 Ha. Particle size distribution (PSD) indicates the material is made up of largely silt (80 per cent); equal clay and sand (<10 per cent each) like particles with a d60 = 14.5 µm. The noticeable difference between both materials is the initial discharge density and the larger clay size and mineral present in zinc tailings.

Recovering TSF storage capacity through AMC™

The topographic survey monitoring produced biweekly volumetric data at each location as shown in Figure 3. The estimated volume reduced is the net volume reduced which it is an indication of the overall performance of dewatering and consolidation, this includes interference and influx of fresh tailings in the monitoring area as TSFs were active during the trial. This volume can assess the metrics to estimate the total volume of water liberated by mechanical consolidation although it is proxy for total water liberated by the *in situ* mechanical consolidation.

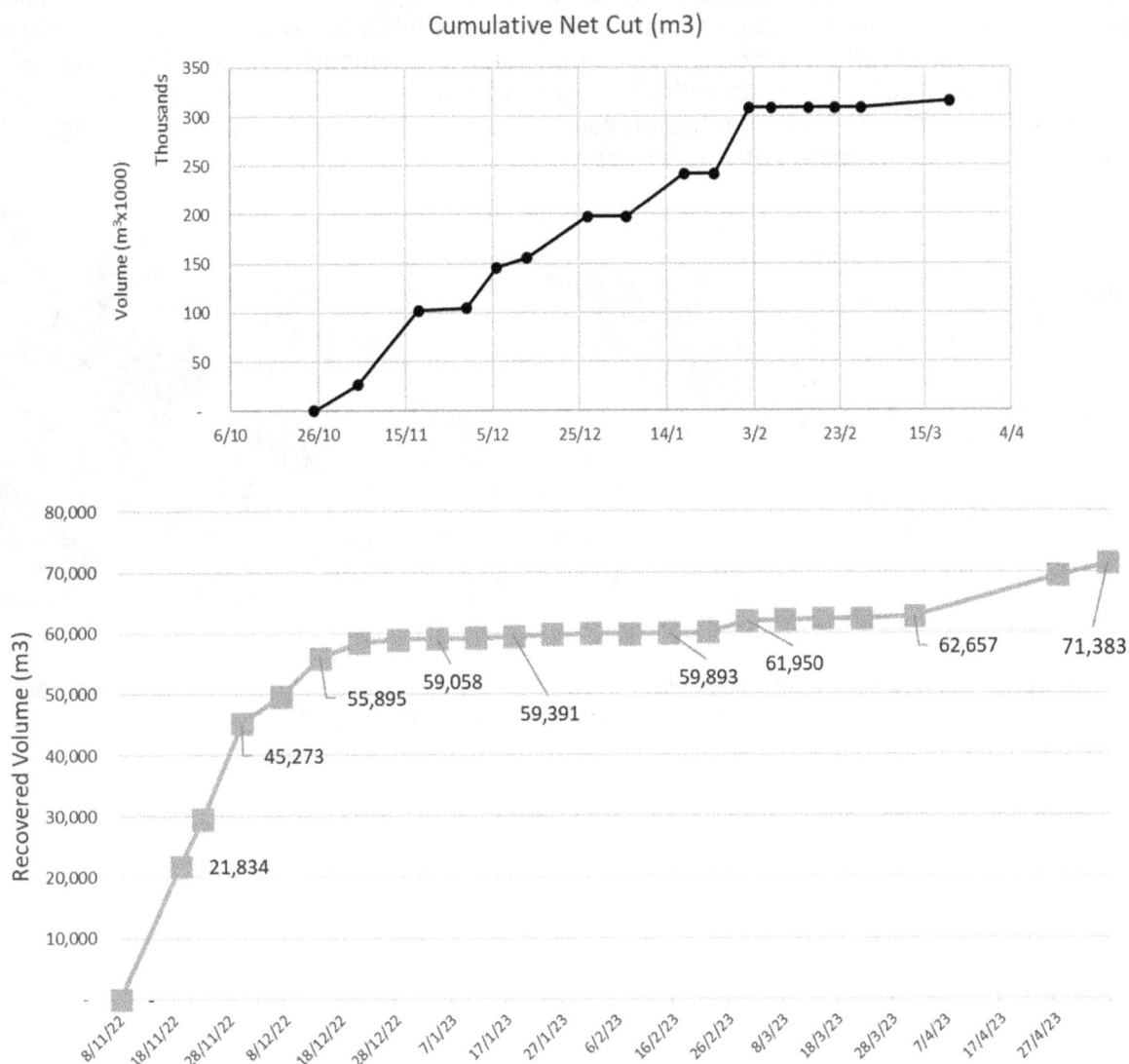

FIG 3 – Recovered storage capacity measured as cumulative volumetric change for: (top) Copper and (bottom) Zinc tailings.

The *in situ* mechanical dewatering was applied to both materials through a consolidation and compaction cycle, highlighting the fact that both scenarios showed an improvement in the storage capacity. The mechanical consolidation of copper tailings achieved a recovered tailings volume of 308 499 m³ over 4 months of continuous work producing a total of 6692 m³/Ha.

On zinc tailings, *in situ* mechanical dewatering was applied over a period of 4 months in order to achieve maximum surface strength. Nonetheless, the similarly applied dewatering methodology created additional tailings storage of 71 383 m³ which is equivalent to 2643 m³/Ha. This additional volume generated in the facility with an improved *in situ* undrained strength was achieved despite the last full deposition cycle at the TSF occurring seven years prior to commencing works. Further highlighting the prolonged settlement periods required to naturally consolidate and dewater these materials.

Improving tailings surface strength with AMC™

Tailings surface strength is a critical path towards generation of stable stored tailings in dams during continuous operation and for facility closure. *In situ* mechanical consolidation and compaction demonstrated superior results when compared with years of natural consolidation and self-weight. For mechanical consolidation dewatering cycle under normal continuous operation is considered to be complete when shear strength is >35 kPa which could sustain low-ground pressure equipment for construction or further compaction activities. Closure works require that mechanical dewatering, consolidation and compaction delivers higher shear strength (>50 kPa) to enable traffic of conventional civil machinery. *In situ* strength testing results often follow a direct relationship with application and intensity of *in situ* mechanical consolidation and compaction increasing incrementally from an initial strength of 0–10 kPa to >50 kPa for both examples, copper and zinc tailings. Figures 4 and 5 show the results obtained from testing campaigns and observations in copper tailings, while Figures 6 and 7 show the results and outcomes in zinc tailings.

(a)   (b)

FIG 4 – Copper tailings condition of the work area prior mechanical treatment (a), and approaching maximum *in situ* undrained shear strength (b).

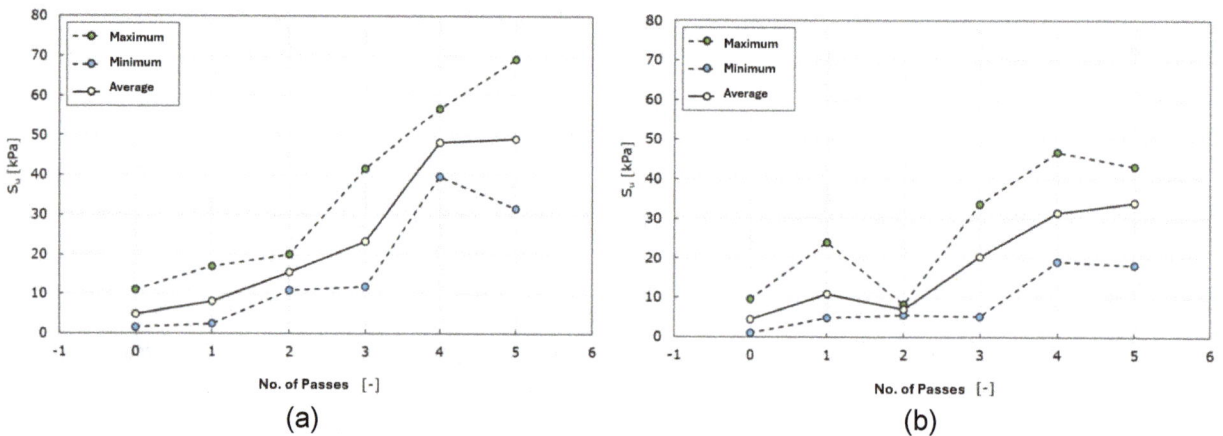

(a)   (b)

FIG 5 – Undrained shear strength versus number of passes for copper tailings: (a) shows testing at 0.5 m deep, and (b) shows at 1.0 m deep, data represents results across all test areas.

FIG 6 – Zinc tailings condition of the work area prior mechanical treatment (a), and approaching maximum *in situ* undrained shear strength (b).

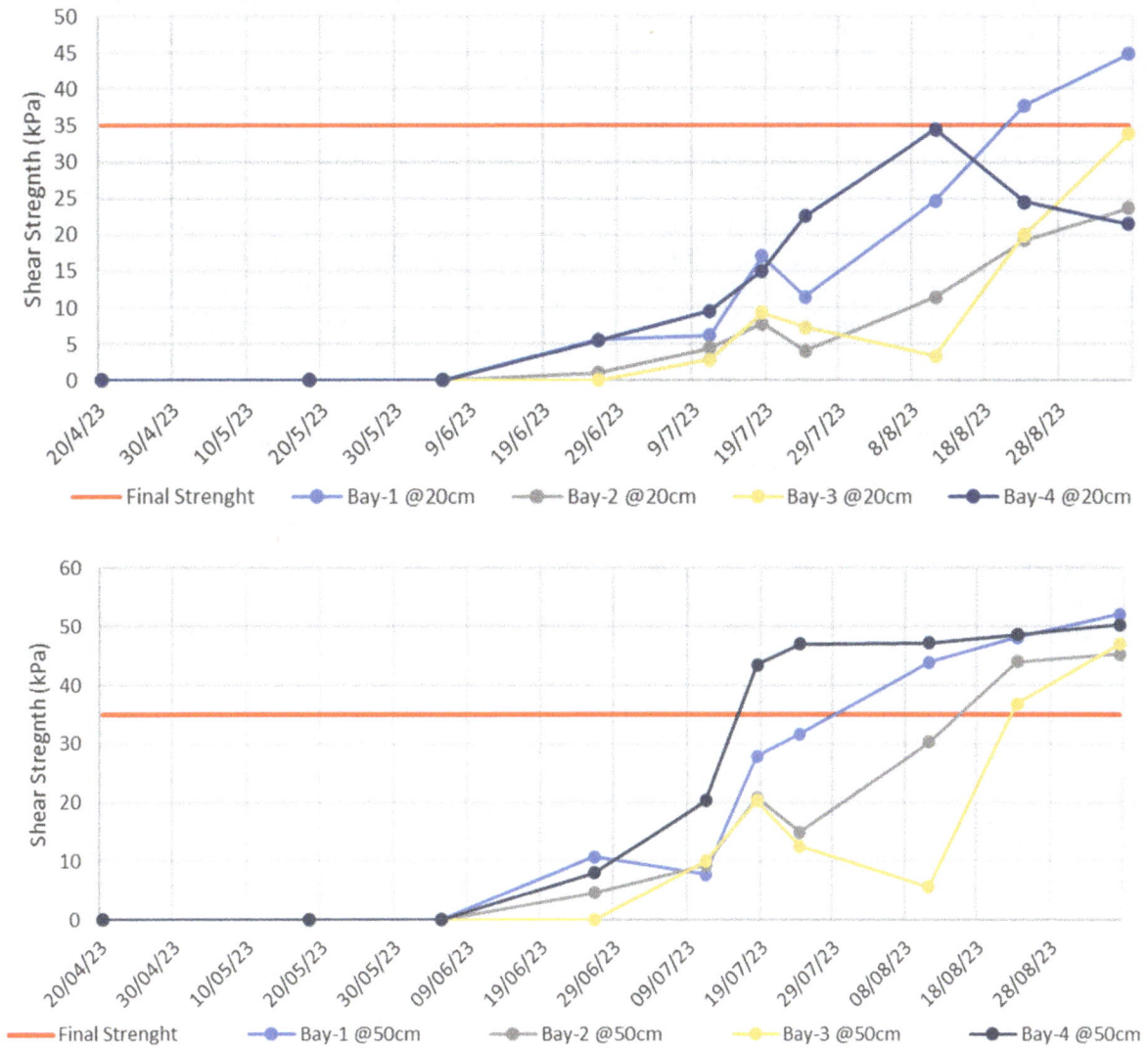

FIG 7 – Average *in situ* undrained shear strength (kPa) of Zinc TSF. (top) Results obtained at 0.2 m deep, (bottom) average results at 0.5 m. Note 35 kPa is the required strength.

As expected, material strengthening during the first few amphirol passes is not largely improved since material is rapidly draining excess water. After excess water is removed, strengthening starts to take effect, further application of *in situ* mechanical consolidation and compaction homogenises the strength across the surface until it reaches the desired strength (Figures 5 and 7).

DISCUSSION

Tailings management has advanced with new tools and enhanced industry practices, elevating the performance and standards of storage facilities. However, adaptive management remains vital throughout a facility's life cycle, despite technological innovations and mining improvements. The ongoing challenge is to dewater, store, and minimise tailings sustainably and cost-effectively. Tailings management methods often involve significant investment, high operational costs, and scaling limitations, making them impractical for some operations. While these technologies address certain tailings storage issues, they fail to independently enhance waste management or resolve challenges posed by legacy facilities.

A compelling alternative to other tailings management solutions is AMC™, implemented by Phibion. This approach offers mine operators a practical, flexible method to safely store tailings, reducing facility footprints and extending asset lifespans. *In situ* mechanical consolidation and compaction integrates swiftly into new or existing facilities with minimal infrastructure changes, requiring only operational process alignment. This eliminates the need for expensive infrastructure or hardware, maximising the use of current facilities, cutting costs, and potentially extending tailings storage facility (TSF) longevity. Benefits include safer storage, stabilised materials, reduced risks, and support for facility closure. As shown by the results, consolidation and compaction are essential for effective tailings management, as they significantly enhance the stability, safety, and long-term performance of storage facilities. These processes reduce water content and increase material strength, lowering the risk of dam failures and improving the structural integrity of the deposit. By densifying tailings, storage capacity is optimised, operational risks are reduced, and water recovery is improved for reuse in processing. Additionally, consolidated and compacted tailings provide a stable surface for equipment access and support progressive rehabilitation and closure. Overall, they are critical for achieving safe, cost-effective, and environmentally responsible tailings management.

CONCLUSIONS

This study demonstrates that *in situ* mechanical dewatering, consolidation, and compaction using the AMC™ methodology offers a viable, scalable, and cost-effective alternative for improving tailings management across several types of tailings, including copper and zinc. Field testing conducted over several months in active TSFs showed substantial increases in undrained shear strength—often exceeding 50 kPa—and notable volumetric reductions, with up to 6692 m³/Ha recovered for copper and 2643 m³/Ha for zinc. These improvements enhance operational efficiency by reducing the storage footprint, increasing TSF life, and supporting safer, more stable conditions for both ongoing operations and closure activities.

Mechanical consolidation significantly shortens consolidation timelines compared to natural settlement, accelerating surface preparation for embankment raises or final capping. This approach minimises reliance on costly infrastructure, extensive land use, and time-intensive traditional methods such as filtration or dry stacking. Additionally, it supports water recovery and reduces infiltration risks, thereby strengthening environmental performance and compliance with global standards such as the ICMM's Global Industry Standard on Tailings Management.

Broader adoption of AMC™ could drive a paradigm shift in tailings management, enabling the mining industry to meet growing expectations for safer, more sustainable, and more resilient waste practices.

REFERENCES

Gerritsen, T, Wood, R, Llano-Serna, M, Meneses, B and Dressel, W, 2024. An alternative approach to developing compaction specifications for tailings materials, in *Paste 2024: Proceedings of the 26th International Conference on Paste, Thickened and Filtered Tailings* (eds: A B Fourie and D Reid), pp 293–304 (Australian Centre for Geomechanics: Perth). https://doi.org/10.36487/ACG_repo/2455_24

International Council on Mining and Metals (ICMM), 2021. Tailings Management – Brief, August 2021.

International Council on Mining and Metals (ICMM), 2022. Mining with Principles – Tailings Reduction Roadmap, September 2022.

International Geotechnical Centre (CGI), 2025. Filtered and Dry Stacking Tailings: Universal Solution or Costly Mirage?. Available from: <https://www.linkedin.com/pulse/filtered-dry-stack-tailings-universal-dezee/?trackingId=SSuoCAnERCyfRo4gqbRhGA%3D%3D>

McPhail, G I, DiDonna, P and Ugaz, R, 2021. Dam break analysis for BRDA 5 at Worsley Alumina Refinery, in *Paste 2021: Proceedings of the 24th International Conference on Paste, Thickened and Filtered Tailings* (eds: A B Fourie and D Reid), pp 177–200 (Australian Centre for Geomechanics: Perth). https://doi.org/10.36487/ACG_repo/2115_16

McPhail, G, Ugaz, R and Garcia, F, 2019. Practical tailings slurry dewatering and tailings management strategies for small and medium mines, in *Paste 2019: Proceedings of the 22nd International Conference on Paste, Thickened and Filtered Tailings* (eds: A J C Paterson, A B Fourie and D Reid), pp 235–243 (Australian Centre for Geomechanics: Perth). https://doi.org/10.36487/ACG_rep/1910_15_McPhail

Munro, L D and Smirk, D D, 2012. Optimizing Bauxite Residue Deliquoring and Consolidation, in Proceedings of the 9th International Alumina Quality Workshop, AQW Inc. Available from: <https://aqw.com.au/papers/item/optimising-bauxite-residue-deliquoring-and-consolidation>

Munro, L D and Smirk, D D, 2018. How thick is thick enough?, in *Paste 2018: Proceedings of the 21st International Conference on Paste and Thickened Tailings* (eds: R J Jewell and A B Fourie), pp 22–34 (Australian Centre for Geomechanics: Perth). https://doi.org/10.36487/ACG_rep/1805_01_Munro

Santiago, O, Menezes, R, McAdam, W and D, Smirk, 2024. Why accelerated mechanical consolidation delivers equal or greater benefits to other tailings management solutions, in *Paste 2024: Proceedings of the 26th International Conference on Paste, Thickened and Filtered Tailings* (eds: A B Fourie and D Reid), pp 349–360 (Australian Centre for Geomechanics: Perth). https://doi.org/10.36487/ACG_repo/2455_28

Smirk, D D and Jackson, S, 2010. In situ foundation improvement for upstream raising of embankments using dried tailings, in *Mine Waste 2010: Proceedings of the First International Seminar on the Reduction of Risk in the Management of Tailings and Mine Waste* (eds: R J Jewell and A B Fourie), pp 251–260 (Australian Centre for Geomechanics: Perth). https://doi.org/10.36487/ACG_rep/1008_22_Smirk

Smirk, D, Santiago, O and Pardon, H, 2022. Operation, Engineering and safety in tailings management, Lima, Peru, in 7th Relaves Peru Congress 2022.

South32, 2023. GISTM Requirement 15.1, Public Disclosure, Worsley Alumina.

Woolston, J S and DiDonna, P, 2020. A case study comparison of best available technology (BAT) for the stacking of bauxite residue: Accelerated Mechanical Consolidation (AMC) and filtration and stacking (F&S), in Paste 2020 – Proceedings of the 23rd International Conference on Paste, Thickened and Filtered Tailings (ed: H Quelopana), (Gecamin Publications: Santiago).

Optimising filtration in coal processing – the Glencore Collinsville case study: proof of concept

E Sommacal[1], F Doveri[2] and M Riboni[3]

1. Director, Matec Pacific, Australia. Email: sommacal@matecpacific.com
2. Chemical Engineer, Matec Industries, Italy. Email: francesco.doveri@matecindustries.com
3. State Manager, Matec Pacific, Australia. Email: m.riboni@matecpacific.com

ABSTRACT

The filtration of tailings, especially in recent times given the general increase in prices, the decrease in available areas, and the ever-increasing drive towards recovery and reclamation, is becoming a race towards greater efficiency and productivity to reduce the dimensions of treatment plants and their overall costs. So, what are the improvements that can reduce machinery sizes and the project's overall costs?

First, as every site has its own characteristics, the plant must be studied in its scenario so the machines are optimised to treat the site-specific material. Test studies on cycle times and other pressing parameters are the key to designing the perfect plant for each application, reducing unnecessary losses in time and performance.

Filtration pressure is one of the most impactful variables in the process parameters. Moving to higher pressures, directly and indirectly, impacts the filtration times and therefore the overall dimensions of the machinery because it permits obtaining cake that is also better formed and less moist in less time. This change alone can significantly reduce the volume needed to treat the same amount of sludge, thus resulting in a reduction of the space required for the installation.

Another process parameter that significantly impacts the dimensions of the machinery is the mechanical time, ie the time that the machine needs to stay idle to perform all the required steps for its opening/closing, discharging, washing etc. Reducing these dead times is essential to improve equipment efficiency.

Other than this, the sheer amount of daily (and hourly) dry solids, which easily reach multiple thousands of tonnes per hour, is another factor that impacts this type of application. Capex and opex, as well as land consumption, are to be kept within tight limits, while at the same time, productivity, efficiency and quality of the final cake have to be as high as possible.

This is where technology plays a crucial role. Treating lots of material without losing performance is what leads to the future of a well-designed plant.

INTRODUCTION

The demand for metals continues to grow as new technologies and infrastructures develop to meet market requests. Even though material recycling is expanding, this alone is insufficient to cover the demand. If there is one thing we can be sure about, mining businesses will keep operating for many years ahead of us (Furnell *et al*, 2022).

More stringent environmental constraints, driven by water scarcity and community issues, make water recycling as crucial as possible.

Because of this, systems that dewater sludge to the minimum water content, like high-efficiency filter presses, are unequivocally supplanting other types of systems that do not allow for such levels of dehydration. Compared to other technologies (eg vacuum filters and belt filters), filter press filtration can obtain very high dewatering levels, producing cakes with residual values as low as 20–15 per cent w/w and even lower. Cakes obtained with this process are compact, perfectly transportable, stackable, and disposable at lower costs.

However, in current large-scale mining (updated to 2022) in dry climate areas, most typical tailings disposal schemes still consist of conventional or slightly thickened at modest levels, with tailings solids weight of up to concentration of 52–48 per cent w/w (Cacciuttolo Vargas and Pérez Campomanes, 2022).

This approach leads to problems such as large area requirements to store the thickened sludge, hydrogeological risks, high disposal costs, low water recycling and high environmental impact.

Sludge solidity stages

The operations of solid–liquid separation are usually related to water recovery for reuse in the process, adequacy of solids percentage on pulp required by the subsequent unit operations, reduction of a product's moisture for transport and sale, preparation of waste for transportation and disposal.

Figure 1 shows the correlation between using different dewatering technologies versus increasing yield stress and solid content.

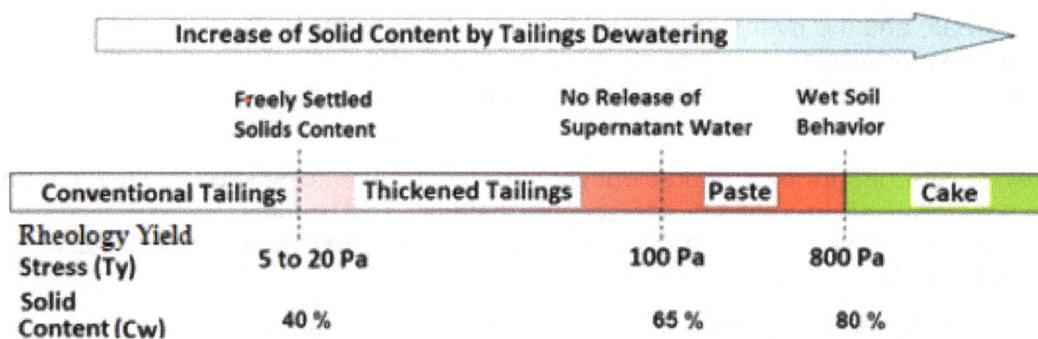

FIG 1 – Levels of dehydration of sludges.

Conventional tailings processing delivers slurry waste to settling ponds, which require large accessible areas and volumes. Very low yield stress is obtained.

Moving on to thickened tailings, high-rate rake thickeners are utilised for dewatering. This technology delivers a mud-like consistency slurry sent to ponds, but the volume utilised is exceptionally reduced from the previous case. Adding flocculating agents can increase the solidity (meant as solid content (% w/w)), but never above 55 per cent of solids. Yield stress remains under 100 Pa.

Deep cone paste thickeners deliver a paste-like consistency (maximum 75 per cent solids) so mine waste can be stored in tailings storage facilities. However, even in this case, the yield stress is still low to prevent risks of catastrophic failures.

In Figure 2, some of the most tragic calamities regarding tailing dams are displayed, both in past and present days (Williams, 2021). Furthermore, in Figure 3, we can see the real effect of this calamities on the environment with the collapse of the Feijão dam's collapse 2019.

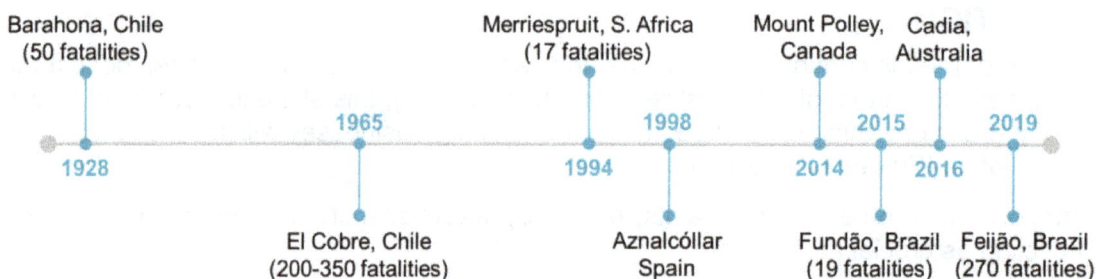

FIG 2 – Major calamities caused by tailings dam failures.

FIG 3 – Feijão dam's collapse 2019.

The 2019 Mariana disaster in Brazil brought a new focus to mining tailings disposal. The collapse of the Feijão dam released around 50 million cubic metres of sludge, killed 19 people, destroyed villages and travelled more than 600 km to the Atlantic Ocean, where it left a reddish-brown plume visible from space.

This incident raised questions about the safety of dams, leading to more significant difficulties in licensing areas for the construction of new dams, and was associated with growing public pressure and increasing proximity to residential areas in Brazil and other countries.

New challenges for the sector in Australia

In countries like Australia, with a high risk of drought and such a high concentration of natural resources and mining activities, water recovery is a decisive factor in the feasibility of new projects. Energy consumption, process efficiency, and reduction of environmental impact are generally considered the main drivers for mining companies in selecting optimised dewatering equipment for their operations.

With the increasing water scarcity, access to water has become a concern. New projects have encountered difficulties in acquiring grants for water use without a tailings management plant.

As depicted in Figure 1, filter press is the only technology to deliver a yield stress of over 800 Pa and as such, cakes are so firm and compact that they can be dry stacked and stored in a much more secure way than paste.

These cakes, which can sometimes reach a solidity of over 90 per cent, can be transported at a meagre cost as their volume and weight are primarily solids. At the same time, all the recovered water is reintegrated into the process, reducing further operational costs. The reduction of operational costs fully recovers the price of the machinery required.

PROCESS IMPROVEMENTS

So, what improvements can be made in the industry to improve filter press technologies, reduce machinery sizes, and be more attractive for mining operations?

Studies on cycle times and other parameters are the key to projecting the perfect plant for each application, reducing unnecessary losses in time and performance.

Filtration pressure is one of the most impactful variables in the process. The higher the pressure, the lower the cycle time and therefore, the overall dimensions of the machinery are lower because it permits the cake to be obtained in less time, which is also better formed and less moist.

First of the most critical and exclusive technical developments introduced by Matec equipment is the high pressure technology®. Matec filter presses work at high pressures (16–21 bar to 1.6–2.1 MPa) ensuring maximum water recovery, minimum moisture on cakes, and lower cycle time. The following figures show the effect of pressure on dehydration (Figure 4) and cycle time (Figure 5) for a given material.

Dehydration

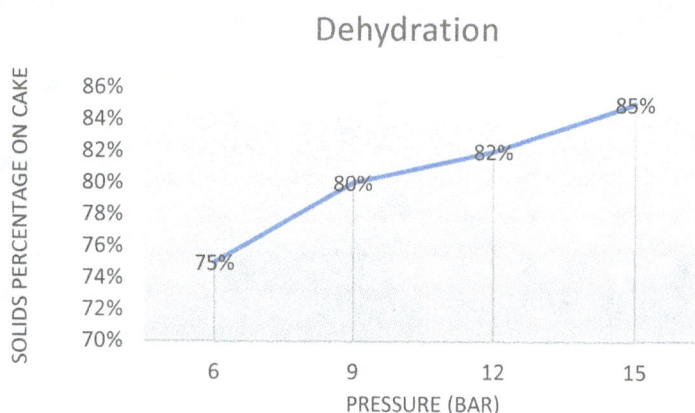

FIG 4 – Pressure effect on cake dehydration.

Cycle time

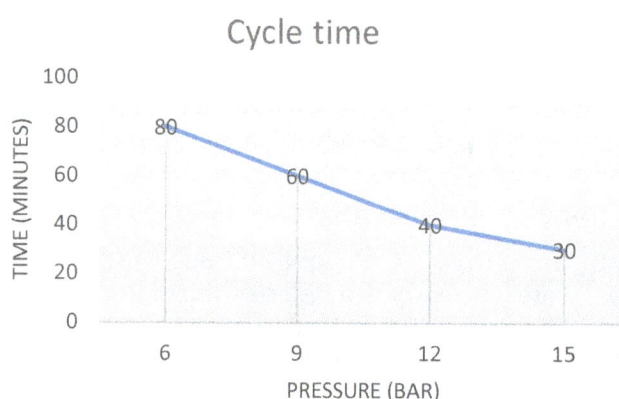

FIG 5 – Pressure effect on cycle time.

It is clear that for a given throughput and target TM%, a filter press operating at higher pressure will require a shorter cycle time.

This change alone can significantly reduce the volume needed to treat the same amount of sludge, thus reducing the space required for the installation. The overall benefits are as follows:

- Smaller equipment size and a smaller number of units are needed to achieve an equivalent given production throughput.

- Processing complex materials with a high proportion of ultrafine and clay without relying on the additional complexity and cost of membrane plate systems or adding chemical aids.

- Higher performance and recovery rate with a resulting TM% reduction in the cake.

High-pressure technology is specifically suitable for applications with a high proportion of ultra-fines and clay commonly associated with tailings and increased by the crescent exploitation of deposits with lower metal content.

Ultrafine particles in the range of 0–30 µm and clay are challenging in the filtration process, as they tend to blind filtering media, creating an impermeable layer that decreases or even limits the dewatering process. Thanks to the higher pressure over 16 and 21 bar, it is possible to break through this material layer, forming into the filter chamber at the very start of the filtration phase and thus completing the dewatering process and producing dry stackable cakes (Figure 6).

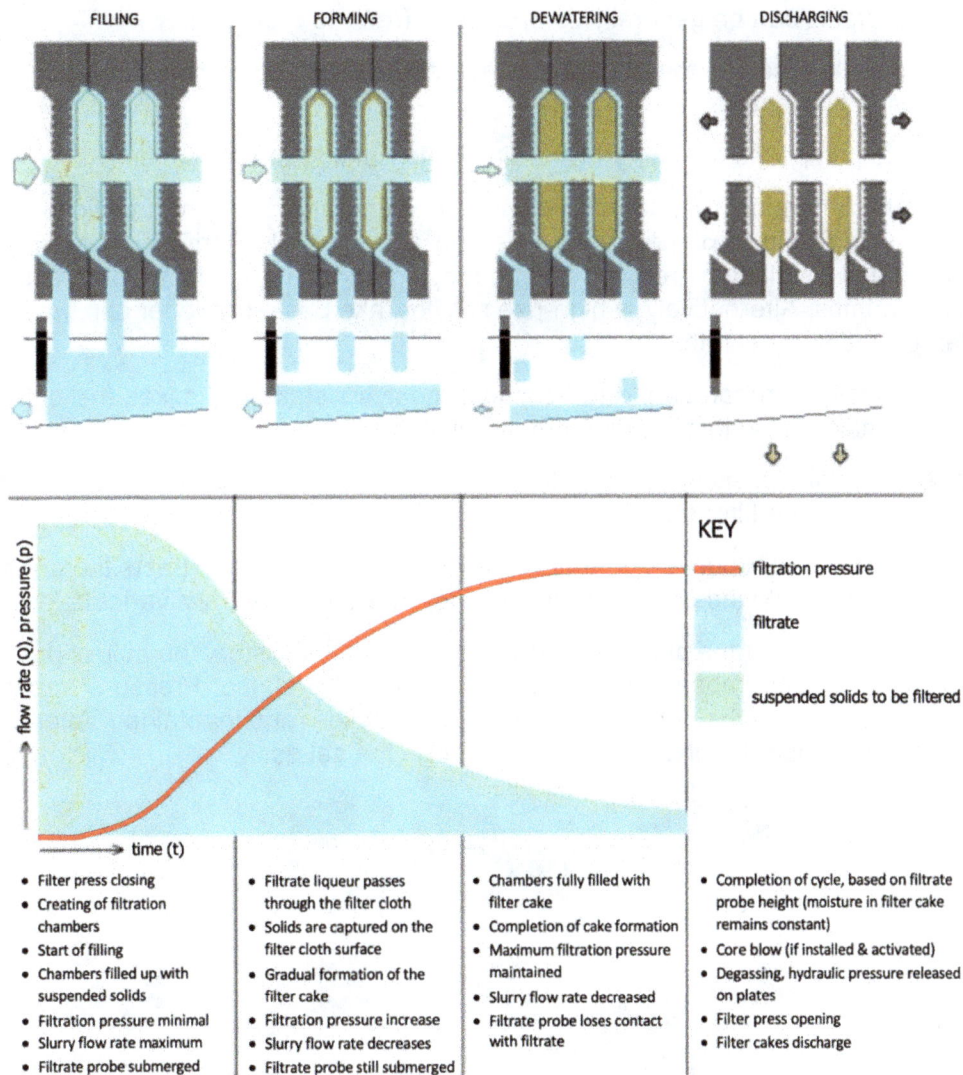

FILLING	FORMING	DEWATERING	DISCHARGING

KEY

— filtration pressure

▨ filtrate

▨ suspended solids to be filtered

flow rate (Q), pressure (p) → / time (t) →

• Filter press closing	• Filtrate liqueur passes through the filter cloth	• Chambers fully filled with filter cake	• Completion of cycle, based on filtrate probe height (moisture in filter cake remains constant)
• Creating of filtration chambers	• Solids are captured on the filter cloth surface	• Completion of cake formation	
• Start of filling		• Maximum filtration pressure maintained	• Core blow (if installed & activated)
• Chambers filled up with suspended solids	• Gradual formation of the filter cake	• Slurry flow rate decreased	• Degassing, hydraulic pressure released on plates
• Filtration pressure minimal	• Filtration pressure increase	• Filtrate probe loses contact with filtrate	• Filter press opening
• Slurry flow rate maximum	• Slurry flow rate decreases		• Filter cakes discharge
• Filtrate probe submerged	• Filtrate probe still submerged		

FIG 6 – Matec plate filter dewatering process phases (2018).

Another process parameter that significantly impacts the dimensions of the machinery is the mechanical time, ie the time that the machine needs to stay idle to perform all the required steps for its opening/closing, discharging, washing etc. Reducing these dead times is essential by studying new ways to operate the machine and discovering new technologies that speed up the process without losing performance.

The second of the most critical and exclusive technical developments introduced by Matec equipment is the TT2 Fast Opening®. The filter plates' fast opening system ensures lower dead time between cycles, reducing the opening time of the machine from the typical 20–25 mins to approximately seven mins even for the larger 220-plate machines and consequently achieving more productivity. The systems are automated, controlled by PLC and easily integrated into the existing plant.

A faster opening of the plates, for example, will lead to a shorter cycle time and a smaller filter press for the same application. By opening 10 plates at a time instead of one by one, the discharging time is reduced by one-third and if, while opening 10, the 10 before them are being closed, this time will reduce even further. This technology employs lateral hydraulic cylinders that pull on the first plate of every pack, effectively opening the whole pack and closing the one before it in just 20 seconds (shaking the plates to ensure a proper detachment of the cakes included).

Matec filters recover the maximum amount of water for reuse and reduce moisture to minimum levels in the concentrates and tailings, ensuring the lowest possible operating cost and lower power consumption.

Matec filtration systems can be generally composed of (refer to Figure 7):

- Raw water tank: receives the tailings and slimes from the beneficiation plant and conditions and pumps to thickeners.

- Submersible pump: lifts the slurry in the dirty water tank to the thickener.

- Matec deep cone thickener: combining the thickening and clarifying functions, it delivers an underflow with 50 up to 65 per cent solids and a reasonably clarified overflow. It requires a relatively small installation area and has no moving parts inside. No motors, rake, bearings or extraction pumps. Alternatively, a high-rise rake or rake paste thickener can be deployed for a very large process flow rate.

- Automatic dosing, preparation, and flocculant analysis station. It mixes the polymer powder with water, makes tube tests and automatically adjusts the dosing of flocculant into thickener.

- Homogeniser tank: receives the underflow slurry from the thickener and keeps it homogeneous until pumped into the filter press.

- High-pressure single, double, and triple cavity centrifugal pump: extracts the dense pulp from the homogeniser tank and presses it into the filter. It is controlled by a variable-frequency drive.

- Matec filter press: after the plates are closed by a hydraulic piston, the pulp is pumped into the filter, filling all the chambers formed by the union of the plates. Pressure forces the filtrate through the cloths and the drainage ducts while the solids remain retained. After the cycle, the feed pump stops and the plates open to discharge the cakes.

FIG 7 – Matec filtration plant in Brazil 2019.

TEST WORK AND CUSTOMISATION

Testing the material is crucial in filtration as every mine has its characteristics and needs. Therefore, a customised plant accounts for various factors determined before plant sizing and scaling.

Results obtained with pilot equipment are fully scalable to every type of plant, regardless of its production and dimensions, because it can recreate all the conditions of the plant and has all the options that a full-scale machine could have.

Figure 8 shows a typical test work regime and procedure including placing thickened slurry into a homogeniser tank that keeps the solid particles in suspension.

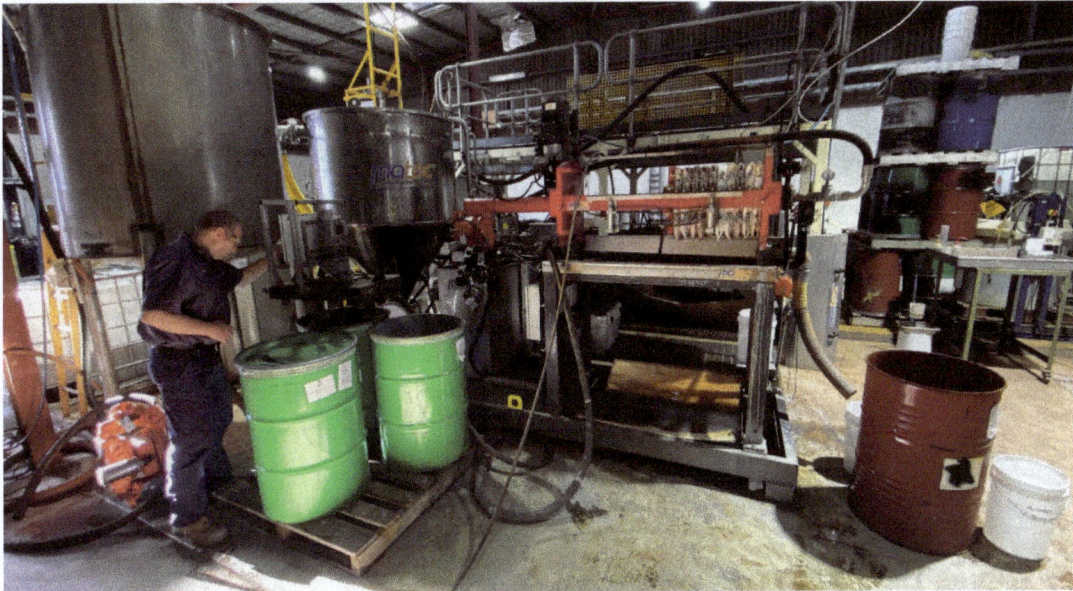

FIG 8 – Matec FP400/10 test rig in Perth (WA).

A review of process conditions, particle size distribution, and solid gravity is completed to decide the type and set-up for the filter press system to be utilised. The main parameters measured during test work are:

- cake thickness
- filtration time
- mass of filtrate over time
- cake moisture.

Based on test work results, the plant's design can finally commence.

CASE STUDIES

Besides test work campaigns, experience is at the base of projecting a plant that works perfectly. In this section, we will report three case studies related to existing plants. Three different but equally important aspects of tailing filtration will be highlighted.

The Companhia Siderurgica Nacional (CSN) plant in Brazil is a plant that underwent three expansions and is perfect for understanding the relationship between machinery and production, as well as comprehending how a project develops.

Consorcio Peña Colorada in Mexico is presented to describe how implementing larger machines with latest generation systems reduces mechanical times, increases productivity and reduces the space needed for the installation.

Lastly, a plant in Brazil will be described from a cost-related point of view to analyse the benefits of installing filter presses and producing cake-like consistency waste versus using only high-rate rake thickeners and managing huge volumes of mud.

Case study – Glencore Collinsville Coal Mine

Objective: Demonstrate the effectiveness of Matec's filter press system in coal dewatering operations.

Phase 1 – Lab testing:

- Samples tested across a range of solid concentrations (4–48 per cent).
- Achieved target cake moistures below 25 per cent.
- Cycle times ranged from 15 to 182 mins, depending on feed composition.

- The best result achieved a final cake moisture of 23.2 per cent at 14 bar pressure with a 28 per cent solid feed concentration.

Phase 2 – Full-scale installation:

- Equipment: FP1520/100 HPT21 CH35 CUBE modular system (Figure 9).

- Installed in 2023 at the CHPP reject area.

- Set-up included homogeniser tanks, underbelly conveyor, and robowash systems for efficient handling.

FIG 9 – Matec FP1520/100 HPT21 CH35 CUBE.

Phase 3 – Data collection and analysis

- Over 1950 cycles completed between Nov 2023 and Feb 2024.

- Water recovery: ~91 per cent per cycle.

- Production: ~16–17 tons/cycle.

- Average cycle time: 26–32 mins.

Expanded results:

- Consistently low cake moisture (~23 per cent) was achieved across multiple feed types and operating conditions.

- Filtration rates reached up to 202.17 $kg/m^2/h$, significantly outperforming traditional technologies.

- The system demonstrated strong adaptability, with reliable performance across feed densities ranging from 4–48 per cent solids.

- Higher solid feed concentrations resulted in shorter cycle times and denser cakes, enhancing throughput and drying efficiency.

- The filter system operated without the need for filter aids, which significantly reduces consumable costs and environmental impact.

- Filtrate clarity exceeded expectations, with values under 10 ppm after 1 min of cycle time, ensuring compliance with discharge standards.

- No downtime reported due to equipment failure; remote diagnostics via Mara (diagnostic remote services) contributed to prompt troubleshooting and optimal uptime.

The CSN plant in Brazil

Starting with the CSN case study, this project was divided into three stages to increase production and keep up with the mining operations. First, the plant needed to treat 550 t/h waste tailings from an iron ore mine. Matec installed two vertical high deep cone thickeners and four filter presses FP2000/190 plates for this first phase. The plant was then expanded by adding five more filter presses of the same size and three more vertical thickeners to cover a further 650 t/h production, bringing the overall plant total throughput to 1200 t/h. For the final target of 35 000 t per day of tailings to be dewatered, an additional 40 m diameter high-rate thickener and two more filter presses were also implemented in the plant, bringing the total to 11 filter presses with an hourly output of almost 1500 t/h, making it one of the largest mine waste treatment plants in the world (Figure 10).

FIG 10 – Companhia Siderurgica Nacional plant completed with the final expansion.

This project has allowed several square kilometres to be cleared of mining sludge ponds, making it safer for those who work there and limiting the environmental impact of the occupation of land for these activities for the benefit of personnel and mine operations.

Consorcio Peña Colorada

A tailings management plant can sometimes occupy space. As such, it is of great interest to explore solutions that can also reduce the machinery footprint, effectively decreasing the land consumption for dewatering operations.

the rendering of the plant design for the Consorcio Peña Colorada tailing plant is shown in Figure 11. This project aimed to dewater sludges from an iron ore mine, roughly 450 t/h. The solution proposed for this plant was the installation of six filter presses FP2526/190 plate size with five machines on duty and one on standby to cover the maintenance schedules.

In this case, using much larger machines with high-pressure technology and fast opening cycle time has improved the efficiency of the overall project, thus reducing the real estate required from the initial scope to nearly 30 per cent, minimising cost and avoiding expansion of the site.

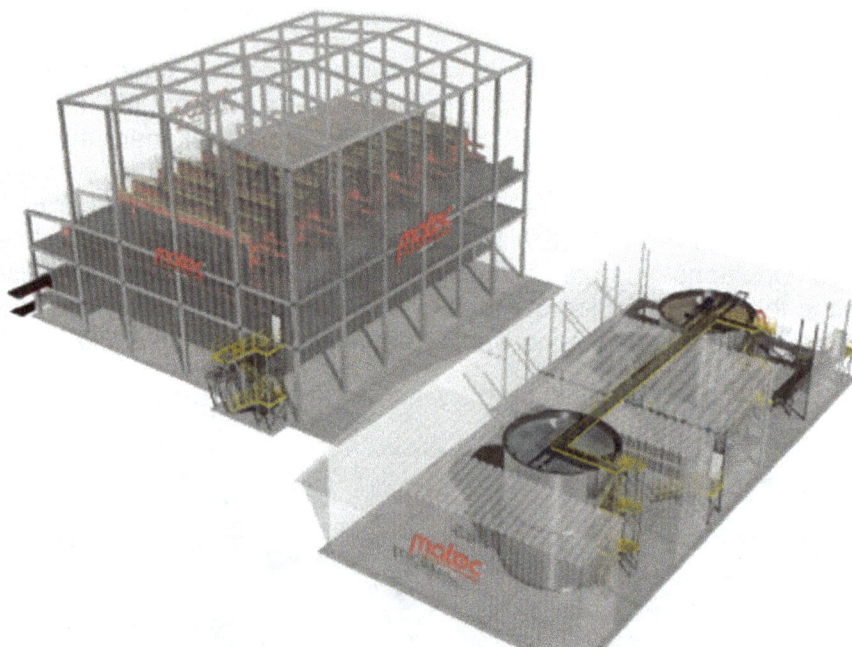

FIG 11 – Rendering of the plant under construction with six high beam filter presses 2500 × 2600 and 190 plates.

In addition, the solution has been perfectly optimised from the point of view of mechanical time and cake result, implementing systems such as drying with compressed air, which reduces residual moisture in the cakes (and therefore the volume required for the same production), automatic washing of the cloths, which minimises the time needed for washing the entire machine, which is effectively dead time, provided it is necessary for the correct operation of the equipment, and very high flow loading pumps, which influence the filling time before filtration itself.

Given that the filtration and air-drying time of this type of material is extremely short, usually no more than six to seven mins in total, the reduction of machine time in all its phases (opening/closing, unloading, washing etc) is significant, especially for machines of this size, that usually take more than 10 mins for all the above-listed operations. It's easily understandable that reducing idle time to the minimum possible, which is longer than the production phase, namely filtration, is crucial to increase efficiency and reduce cost.

MML Mineracao

The last case study presented is a plant again located in Brazil, which is symbolic of deeply understanding the different operational costs of running a plant that produces waste in the form of thickened sludge versus dry stacking.

This original plant design consisted of several paste thickeners, which could not treat all the material from the mining operation. Additionally, paste thickeners could only reach a paste consistency of the resulting mud, still containing a high moisture value of 30–35 per cent and limited competency.

Two Matec HP filter FP2000/120 were installed to replace the thickeners, which are capable of treating up to 125 t/h of dry solids. Considering that the present thickeners were capable of only 6 t/h each, 21 units would have been required to achieve the same productivity.

Comparing the operating costs of the units alone, then those related to the electrical supply, it is possible to calculate the mere annual savings that the installation of the two filter presses provided compared to the thickeners. This calculation also doesn't take into account the fact that the cake residual moisture coming out of the filter press is way less than that of the mud coming out of the thickeners, so there was also an improvement in sludge transportability/stack ability, reducing, therefore, costs related to the operation.

Data on the power consumption and its relative costs and yearly savings are reported in Table 1.

TABLE 1

Power consumption costs of the two solutions.

Equipment	FP option	Paste thickener option
No.	2	21
Total power (kW)	526	3366
Power consumption (kW/h)	287	2111
Total annual cost (R$)	723.182	5.319.864
Yearly savings (R$)	4.596.682	

CONCLUSION

This paper presents technological advancements and improvements in tailing filtration that are being implemented to reduce costs and land requirements while preserving the environment from the increasing production of mineral wastes.

Reducing cycle time has the dual benefit of reducing the size of the equipment required to achieve a target throughput and, as such, reducing both the capital cost for the equipment as well as the footprint and overall operational cost per ton of material, compared to lower feed pressure technologies and other technologies generally in use in the mining dewatering operations.

Matec's high-pressure filter press technology, showcased through the Glencore project, sets a new benchmark for safe, sustainable, and efficient tailings dewatering. The ability to reduce water usage, improve cake dryness, and enable land reuse makes Matec a leader in dry stacking innovation.

REFERENCES

Furnell, E, Bilaniuk, K, Goldbaum, M I, Shoaib, M, Wani, O, Tian, X, Chen, Z, Boucher, D and Bobicki, E R, 2022. Dewatered and stacked mine tailings: a review, *ACS EST Engineering*, 2:728–745.

Cacciuttolo Vargas, C and Pérez Campomanes, G, 2022. Practical experience of filtered tailings technology in Chile and Peru: an environmentally friendly solution, *Minerals*, 12(7).

Williams, D J, 2021. Lessons from tailings dam failures – where to go from here?, *Minerals*, 11(8).

Utilising iron ore-sand for sustainable mine haul road construction

G Vizcarra[1], K Tehrani[2], M Ghamgosar[3], J Segura-Salazar[4] and C Andrade[5]

1. Specialist Geotechnical Engineer, Vale S.A., Belo Horizonte, Brazil.
   Email: ginocalderon@hotmail.com
2. Senior Tailings Engineer, ATC Williams, Australia. Email: kathyt@atcwilliams.com.au
3. Senior Associate, ATC Williams, Australia. Email: mortezag@atcwilliams.com.au
4. Research Fellow, The University of Queensland, Global Centre for Mineral Security,
   Sustainable Minerals Institute, Brisbane Qld 4000, Australia. Email: j.segurasalazar@uq.edu.au
5. Geotechnical Projects Manager, Vale S.A., Belo Horizonte, Brazil.
   Email: claudio.rodrigues.andrade@vale.com

ABSTRACT

The mining industry produces large amounts of tailings, which are typically deposited in tailings dams. However, this method overlooks the potential value of silicate minerals, posing economic, social, and environmental risks. Utilising iron ore-sand, a byproduct of iron ore processing, in the construction of mine haul roads presents a promising solution to the challenges of sustainable construction and resource management in mining operations. Additionally, adopting iron ore-sand aligns with sustainable mining practices and promotes the efficient use of resources by preventing and reducing mine waste generation and generating an alternative source of sand at scale. This study investigates the feasibility and potential benefits of utilising ore-sand sourced from iron ore mine in the Iron Quadrangle region in Minas Gerais, Brazil. as an embankment material for mine haul roads. A series of laboratory tests were performed to understand the physical properties, compaction behaviour, mechanical characteristics of the ore-sand such as shear strength under long- or short-term stability scenarios. This material exhibits high shear strength and favourable compaction properties. Moreover, the stability of iron ore-sand embankments is further enhanced by its stiffness. These properties make it particularly suitable for use in areas under the heavy loads and dynamic conditions typical of mine haul roads. The aim is to promote the use of ore-sand as a viable and sustainable alternative for constructing mine haul roads. Furthermore, the material is readily available at mining sites, reducing transportation costs and the need for external sourcing, which can lead to significant cost savings for mining companies.

INTRODUCTION

The disposal of mine tailings, particularly iron ore tailings, remains a critical environmental challenge for the mining industry (Bastos *et al*, 2016). Additionally, the vast quantities of tailings generated necessitate large-scale storage solutions, such as tailings dams, which carry substantial environmental and social risks, including structural failures and long-term land degradation (Innis *et al*, 2022). Repurposing mine waste, including tailings, as construction materials presents a dual opportunity: reducing the volume of waste requiring disposal while simultaneously decreasing reliance on natural *in situ* resources. The successful application of sand from waste in civil engineering demonstrates the potential for such alternatives to preserve natural sand reserves, which are increasingly depleted due to overexploitation (Salim and Prasad, 2020).

This study investigates the feasibility of using iron ore-sand sourced from the Iron Quadrangle region in Minas Gerais, Brazil, as an embankment material for mine haul roads. Through a series of laboratory tests, we evaluate the material's physical properties, compaction behaviour, shear strength, and stability characteristics. Our findings indicate that iron ore-sand exhibits strong resistance to shearing forces, and excellent compaction properties making it particularly suitable for heavy-load applications under dynamic applied loads, such as those encountered in mine haul roads. By promoting the use of iron ore-sand as a sustainable alternative, this research aims to contribute to cost-effective and environmentally responsible mine waste management. The material's on-site availability reduces transportation costs and external sourcing needs, offering significant economic benefits to mining operations.

Production of iron ore sand

Iron ore processing involves crushing, milling, and classification (via sieves or cyclones), followed by concentration through gravity separation, flotation, or magnetic separation, and finally dewatering. The ore is separated into four size fractions: coarse (+8 mm), medium (+1–8 mm), fine (+0.15–1 mm), and very fine (-0.15 mm), yielding products like lumps, sinter feed, and pellet feed fines. Fines require agglomeration (sintering or pelletising) to enhance their properties and increase iron content, making them more valuable. Sand can be co-recovered during processing by further classifying and treating silica-rich rejects with hydrocyclones before dewatering and storage, rather than discarding them as tailings.

Environmental characterisation

Golev *et al* (2022) characterise iron ore sand as environmentally safe for construction, with neutral pH (6.9–7.3), low salinity (sulfates/chlorides <10 mg/kg), and trace metals (eg Cr 10 ppm, Pb 2–3 ppm) below thresholds; its composition is predominantly silica (90 per cent) and iron oxides (9 per cent), primarily as angular quartz (90–91 per cent) and hematite/goethite, while TCLP leaching tests confirm non-toxicity (only trace zinc detected).

METHODOLOGY

This study evaluates the suitability of iron ore-sand as an embankment material for mine haul roads by assessing its geotechnical properties and comparing them with a locally available natural soil. a comprehensive laboratory testing program was conducted to characterise the material's behaviour under various conditions, ensuring its compliance with engineering requirements for haul road construction.

Sample collection

- Iron ore-sand was sourced from the process plant at an iron ore mine in the Iron Quadrangle region, Minas Gerais, Brazil.

- Natural soil (for baseline comparison) was collected from a site adjacent to the planned haul road construction area.

Laboratory testing program

The following geotechnical tests were performed to evaluate the physical and mechanical properties of the materials:

- Grain size distribution – determined via sieve analysis and hydrometer testing to classify the material and assess its suitability for compaction.

- Specific gravity – measured using a pycnometer to understand particle density.

- Atterberg limits – evaluated the plasticity characteristics (liquid limit, plastic limit, plasticity index) to assess the material's behaviour under varying moisture conditions.

- Compaction tests (Standard Proctor) – established the optimum moisture content (OMC) and maximum dry density (MDD) for effective embankment construction.

- Permeability tests – determined hydraulic conductivity to evaluate drainage characteristics.

- Shear strength tests (triaxial) – quantified the friction angle to ensure stability under heavy haul vehicle loads.

Data analysis and suitability assessment

The test results were analysed to:

- Compare the iron ore-sand properties with those of the natural soil.

- Verify compliance with industry standards for mine haul road embankments regarding stability.

- Identify potential advantages (eg higher shear strength, better compaction) or limitations.

RESULTS AND DISCUSSIONS

Laboratory testing results

As shown in Table 1, the iron ore-sand has a specific gravity of 2.89 and a fines content of 28 per cent, with no liquid or plastic limit. In contrast, the Gneiss residual soil has a specific gravity of 2.66, a fines content of 60 per cent, a liquid limit of 55 per cent, and a plasticity index of 27 per cent. Figure 1 presents the granulometric analysis of the iron ore-sand sample used in this study and compares it to the Gneiss residual soil.

TABLE 1

Properties of testing materials.

Properties	Iron ore sand	Gneiss residual soil
Specific gravity, Gs	2.89	2.66
Fines content (%)	28	60
Liquid limit (%)	-	55
Plastic limit (%)	-	28
Plasticity index (%)	-	27
Unified Soil Classification System (USCS)	SM	CH
Permeability (cm/s)	6.7×10^{-4}	6.6×10^{-5}
Grain size distribution (%)		
Sand (0.075 mm < d < 4.75 mm)	72	37
Silt (0.005 mm < d < 0.075 mm)	27	8
Clay (d < 0.005 mm)	1	52

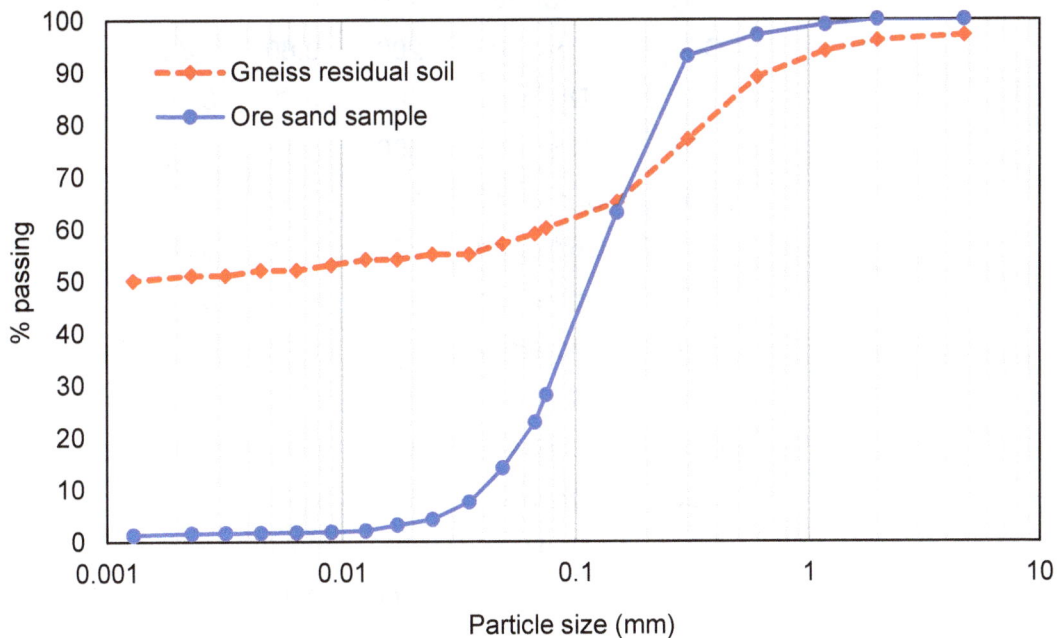

FIG 1 – Particle size distribution of the materials.

Figure 2 illustrates the compaction curves for both materials. Each material has two curves – one depicting dry unit weight and the other showing void ratio. The data reveals that the residual soil has a higher optimum water content (approximately 26.4 per cent versus 18 per cent), while the iron ore sand exhibits a higher maximum dry unit weight (1.6 g/cm^3 versus 1.4 g/cm^3) and a lower void ratio.

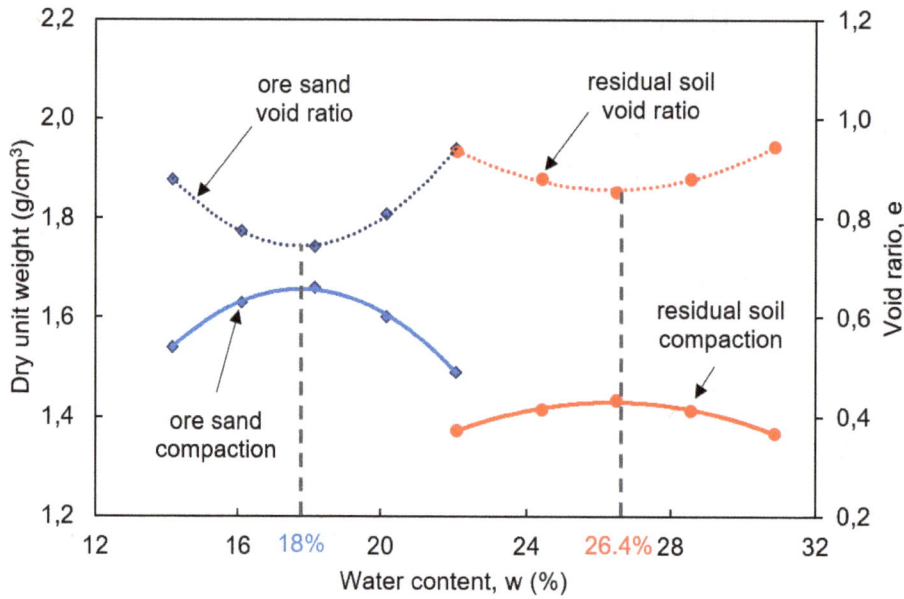

FIG 2 – Compaction results for ore sand and residual soil.

The shear behaviour of both materials was evaluated through a comprehensive series of 18 triaxial tests, comprising eight tests on iron ore sand and ten on gneiss residual soil under both drained and undrained conditions. Table 2 presents the complete tested specimen characteristics, including the void ratios before consolidation (e_e) and after shearing (e_f).

TABLE 2

Characteristics of specimens of triaxial testing.

Material	Testing type	Drainage condition	σ'_0 (MPa)	e_c	e_f
Iron ore sand	CD1	Drained	100	0.81	0.85
	CD2	Drained	200	0.80	0.82
	CD3	Drained	400	0.78	0.79
	CD4	Drained	800	0.76	0.76
	CU1	Undrained	100	0.81	0.81
	CU2	Undrained	200	0.80	0.80
	CU3	Undrained	400	0.78	0.78
	CU4	Undrained	800	0.77	0.77
Gneiss residual soil	CD1	Drained	200	0.78	0.73
	CD2	Drained	400	0.75	0.68
	CD3	Drained	800	0.70	0.62
	CD4	Drained	1600	0.61	0.50
	CD5	Drained	3200	0.48	0.36
	CU1	Undrained	200	0.87	0.87
	CU2	Undrained	400	0.75	0.75
	CU3	Undrained	800	0.65	0.65
	CU4	Undrained	1600	0.62	0.62
	CU5	Undrained	3200	0.55	0.55

Triaxial tests on iron ore sand (Figures 3–6) show a high friction angle (ϕ'=32°) with pronounced dilatancy, reaching a critical state stress ratio q/p'=1.27. This exceptional strength, maintained even at high stresses, reflects the material's favourable particle characteristics for embankment construction. The friction angle (ϕ') was determined from the critical state stress ratio (M = q/p') using the following relationship: $\phi' = \sin^{-1}(3M/(6 + M))$

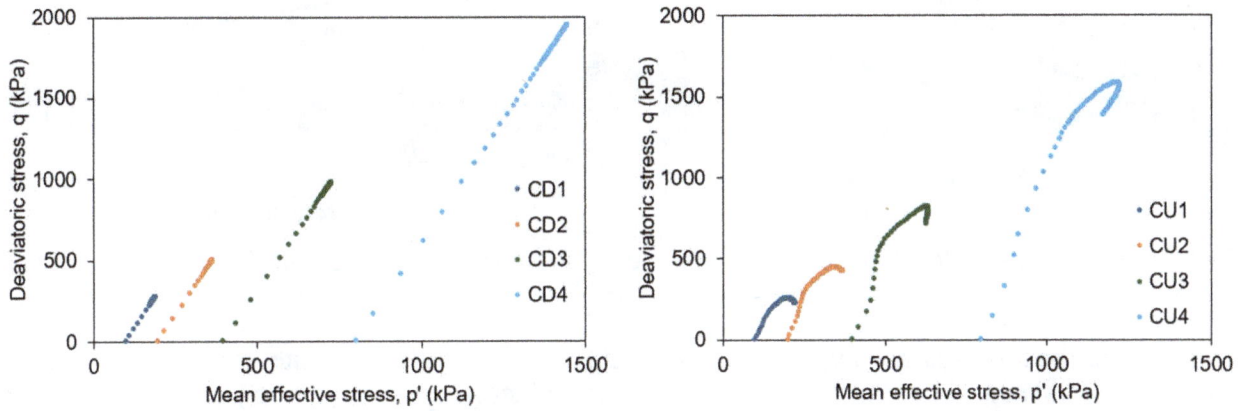

FIG 3 – Iron ore sand – stress paths: (a) drained tests; (b) undrained tests.

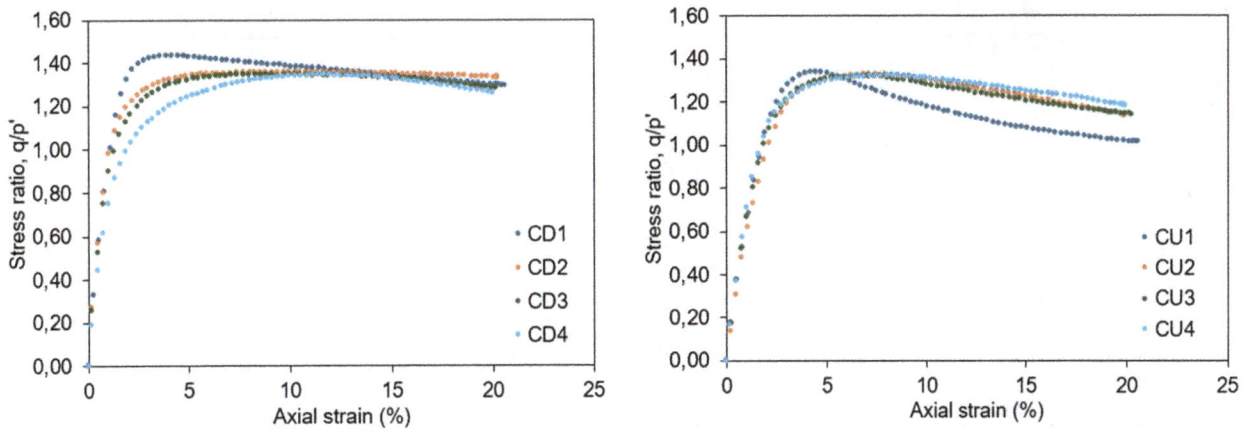

FIG 4 – Iron ore sand – stress ratio versus axial strain: (a) drained tests; (b) undrained tests.

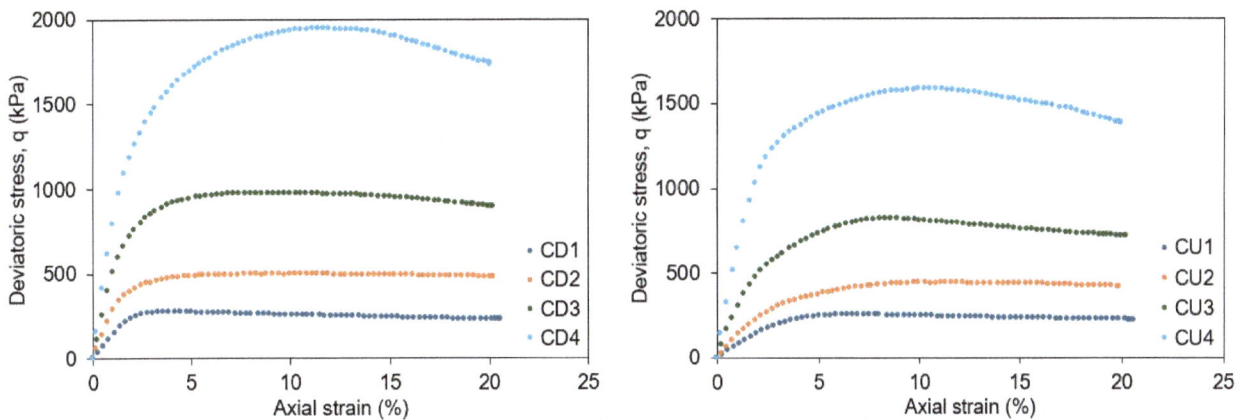

FIG 5 – Iron ore sand – deviatoric stress versus axial strain: (a) drained tests; (b) undrained tests.

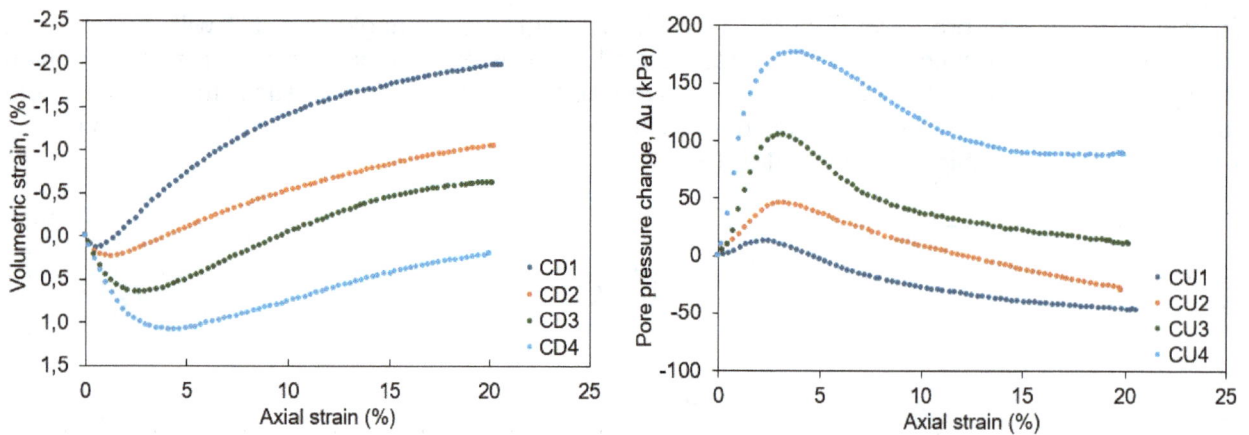

FIG 6 – Iron ore sand: (a) drained tests – volumetric strain versus axial strain; (b) undrained tests – pore pressure changes versus axial strain.

The Gneiss residual soil exhibits characteristic critical state behaviour with $\phi'=23°$, as shown in Figures 7–10. While all specimens converge to a critical state stress ratio (q/p') of 0.91, undrained tests (Figure 7b) reveal an important stress-dependent transition – contractive response dominates above 800 kPa mean effective stress, while dilative behaviour prevails at lower stresses.

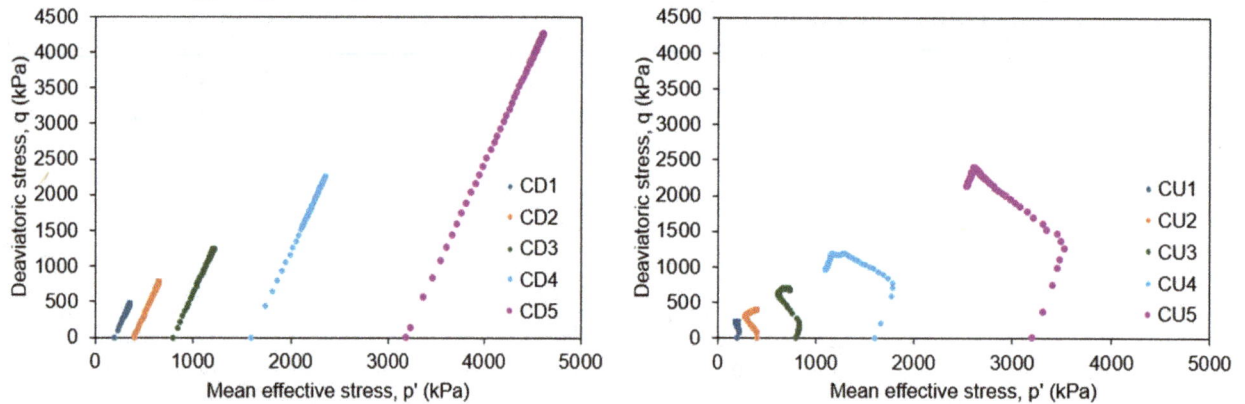

FIG 7 – Gneiss residual soil – stress paths: (a) drained tests; (b) undrained tests.

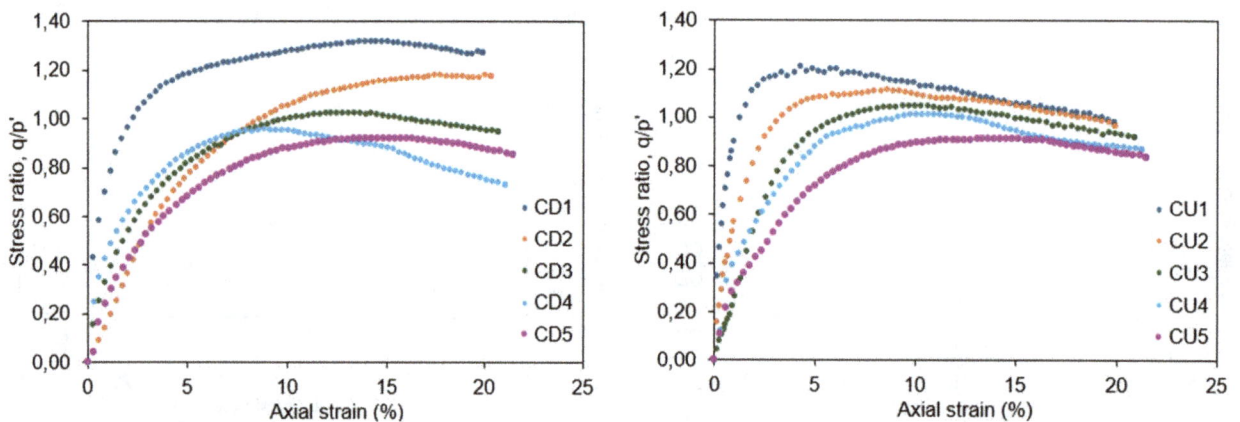

FIG 8 – Gneiss residual soil – stress ratio versus axial strain: (a) drained tests; (b) undrained tests.

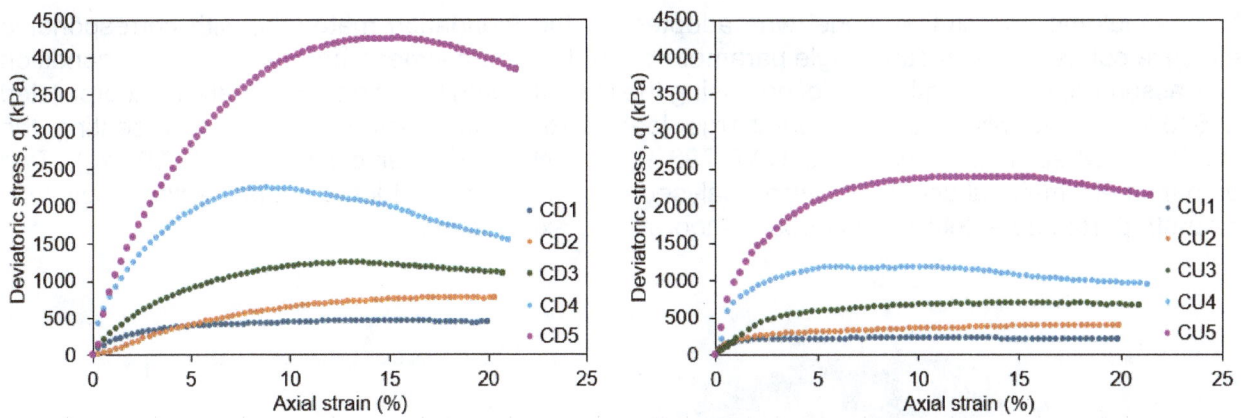

FIG 9 – Gneiss residual soil – deviatoric stress versus axial strain: (a) drained tests; (b) undrained tests.

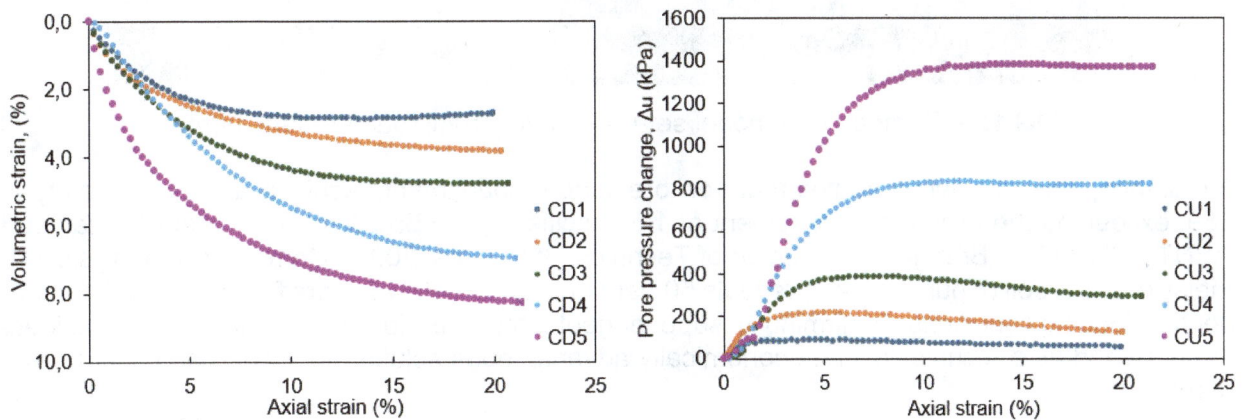

FIG 10 – Gneiss residual soil: (a) drained tests – volumetric strain versus axial strain; (b) undrained tests – pore pressure changes versus axial strain.

Figure 11 presents the mean effective stress versus void ratio behaviour of the materials. The iron ore sand exhibits a flatter critical state line compared to the gneiss residual soil, indicating lower sensitivity to high stress conditions.

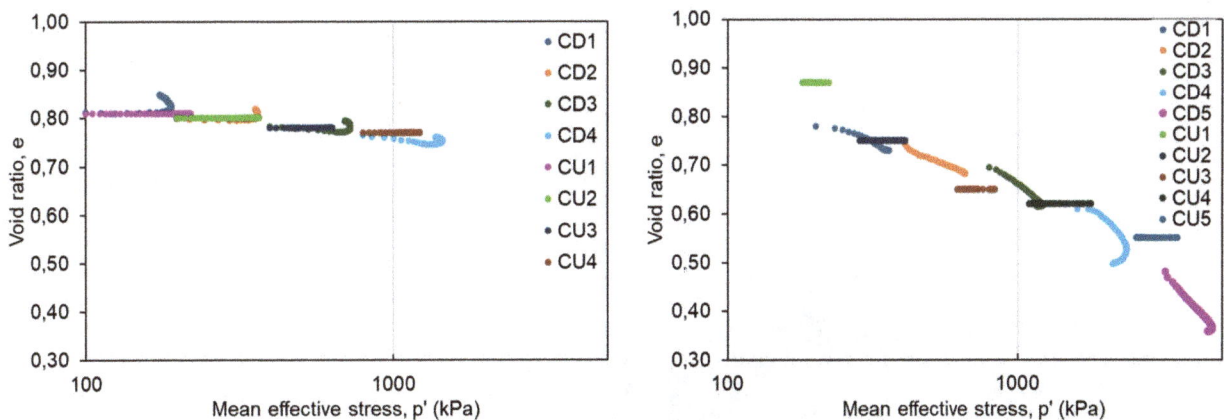

FIG 11 – Mean effective stress versus void ratio: (a) iron ore sand; (b) gneiss residual soil.

Embankment slope stability

The embankment slope stability was evaluated using Spencer's method of limit equilibrium analysis, as implemented in Rocscience Slide 2D. The analysis incorporated the geometry and material properties illustrated in Figure 12. Two primary zones were considered in the model: the foundation layer and the embankment. The critical cross-section was selected based on the maximum elevation of the embankment crest to capture the most critical condition for the stability assessment. The

Mohr–Coulomb constitutive model was adopted for the foundation materials, with corresponding effective cohesion and friction angle parameters. For the embankment material, a drained condition was assumed, and strength was defined using the C/Φ Strength Function. The analysis incorporated a 540 kPa uniformly distributed surcharge load across the running surface, representing the maximum wheel line loading of a CAT 793 haul truck (gross vehicle mass: 623 690 kg). Two embankment material scenarios were analysed: iron ore sand and local gneiss residual soil, with strength parameters determined through laboratory triaxial testing.

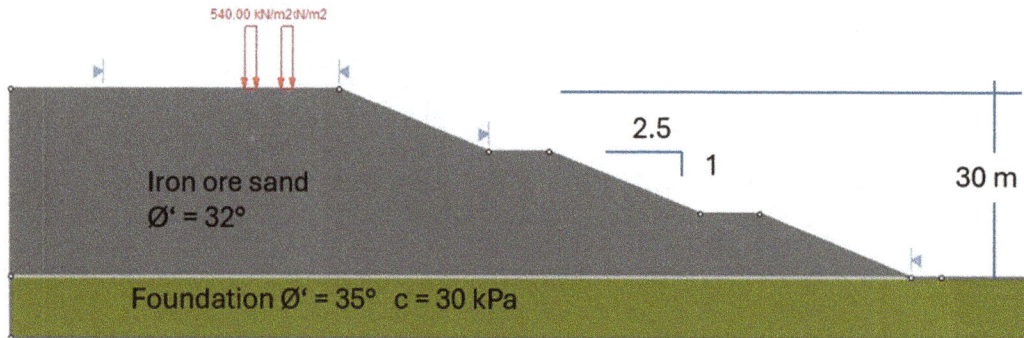

FIG 12 – Geotechnical model set-up including haul road embankment.

The stability analysis demonstrates that iron ore-sand embankments achieve a factor of safety of 1.59, exceeding the minimum requirement of 1.5 specified in the Brazilian Slope Stability Standard (ABNT NBR 11682; Brazilian Association of Technical Standards, 2009). Comparative analysis with gneiss residual soil (Figures 13–14) reveals 50 per cent higher safety factors for iron ore-sand under identical loading conditions, confirming its superior geotechnical performance. These results validate iron ore-sand as a technically and economically advantageous solution for haul road construction, offering:

- enhanced stability: 50 per cent higher safety margins compared to conventional materials

- operational reliability: consistent performance under dynamic truckloading (540 kPa surcharge)

- economic benefits: elimination of imported material costs through on-site utilisation

- sustainability: 100 per cent tailings utilisation, reducing waste storage requirements.

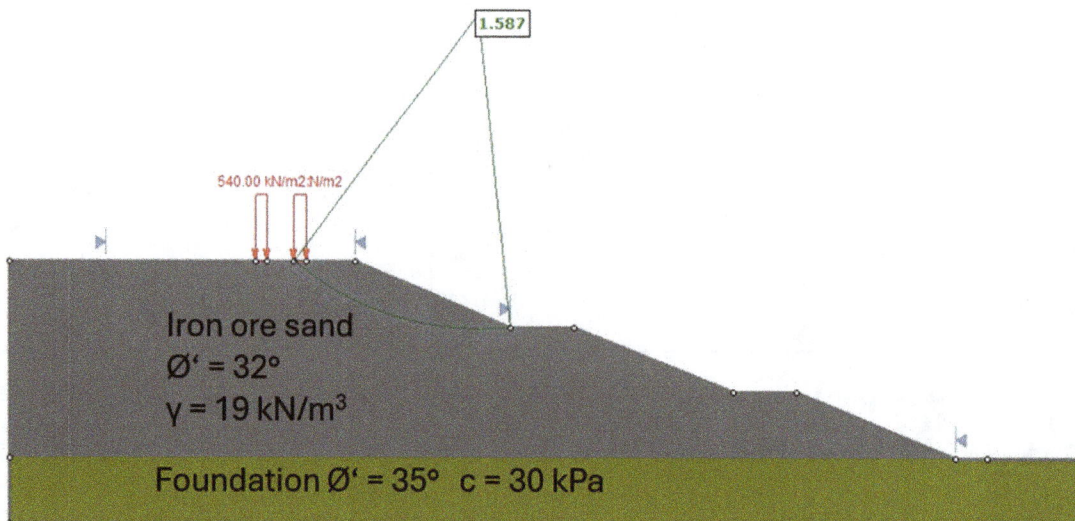

FIG 13 – Slope stability results for Iron ore sand embankment.

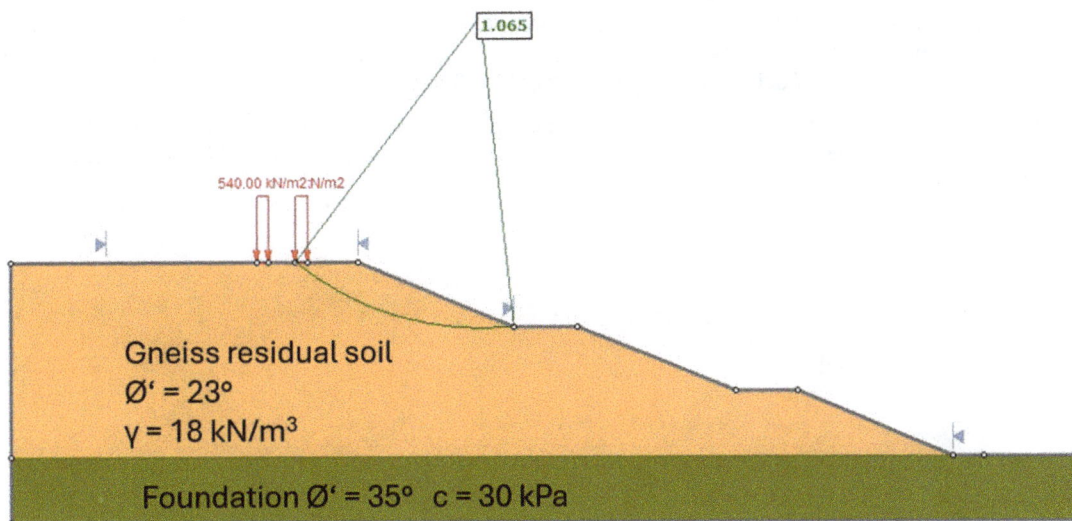

FIG 14 – Slope stability results for Gneiss residual soil embankment.

These findings align with recent research by the University of Queensland (Golev *et al*, 2022), which confirmed that ore-sand from tailings exhibits geotechnical properties suitable for construction, offering a sustainable alternative to natural sand. The use of iron ore-sand not only enhances haul road performance but also contributes to the mining industry's circular economy goals by repurposing waste, reducing tailings storage pressures, and preserving natural resources (Golev *et al*, 2022). While this study demonstrates the viability of iron ore-sand, further research—such as long-term abrasion resistance and binder optimisation—could build on global efforts, like those by the University of Queensland, to standardise ore-sand applications (Golev *et al*, 2022).

CONCLUSIONS

This study demonstrates that iron ore-sand, a byproduct of iron ore processing in Brazil's Iron Quadrangle region, exhibits adequate geotechnical properties for use in mine haul road embankments. Laboratory tests confirmed that the material possesses compatible shear strength, optimal compaction characteristics, and favourable permeability, ensuring superior stability under the loads imposed by heavy mining trucks. Comparative analysis with locally sourced gneiss residual soil revealed that iron ore-sand consistently exceeds minimum stability requirements, achieving a higher factor of safety and validating its technical viability as a sustainable construction material.

From an environmental standpoint, repurposing iron ore-sand significantly reduces tailings storage volumes, mitigating risks associated with tailings dam failures while conserving natural soil resources. Economically, its on-site availability eliminates the need for external material sourcing, reducing transportation costs and providing mining companies with a cost-effective, circular economy solution that aligns with global sustainability goals.

Future research should systematically evaluate the stress-strain behaviour of iron ore sand through advanced laboratory testing (eg cyclic triaxial shear) and field monitoring, while optimising construction techniques (compaction methods, layer thickness, stabilisation) to enhance its performance in critical mining applications under diverse environmental conditions.

REFERENCES

Bastos, L A, de C, Silva, G C, Mendes, J C and Peixoto, R A F, 2016. Using Iron Ore Tailings from Tailing Dams as Road Material, *Journal of Materials in Civil Engineering*, 28(10). https://doi.org/10.1061/(asce)mt.1943-5533.0001613

Brazilian Association of Technical Standards, 2009. ABNT NBR, 11682 2009. Slope stability – Procedure [Original title in Portuguese: Estabilidade de encostas – Procedimento], Brazilian Association of Technical Standards, Rio de Janeiro.

Golev, A, Gallagher, L, Vander Velpen, A, Lynggaard, J R, Friot, D, Stringer, M, Chuah, S, Arbelaez-Ruiz, D, Mazzinghy, D, Moura, L, Peduzzi, P and Franks, D M, 2022. Ore-sand: Apotential new solution to the mine tailings and global sand sustainability crises, final report, version 1.4, March 2022, The University of Queensland and University of Geneva.

Innis, S, Ghahramani, N, Rana, N M, McDougall, S, Evans, S G, Take, W A and Kunz, N C, 2022. The Development and Demonstration of a Semi-Automated Regional Hazard Mapping Tool for Tailings Storage Facility Failures, *Resources*, 11(10):82. https://doi.org/10.3390/resources11100082

Salim, P M and Prasad, B S R K, 2020. A Review on the Usage of Recycled Sand in the Construction Industry [Review of A Review on the Usage of Recycled Sand in the Construction Industry], *IntechOpen eBooks*, IntechOpen. https://doi.org/10.5772/intechopen.92790

A practical approach to tailings repurposing

C Vuillier[1]

1. Technical Director Tailings and Repurposing, GHD, Melbourne Vic 3000.
   Email: charles.vuillier@ghd.com

ABSTRACT

This paper presents a fresh and practical approach to re-mining, recycling and repurposing tailings, addressing the significant challenge of mine waste management. The mining industry generates the largest volume of mineral waste globally, yet this vast resource remains largely undervalued. With the right approach, tailings can present a substantial opportunity for other industries to access a wide range of valuable materials

The objective of this paper is to encourage the mining industry to rethink the way they perceive and manage tailings. By adopting a practical, systematic approach, companies can explore viable pathways to minimise the volume of tailings stored on-site while unlocking new value streams for other industries. Many sectors are increasingly interested in the minerals contained within tailings, offering mutual benefits through collaboration

The paper reviews different types of tailings, focusing on common categories such as coarse and fine tailings, including those containing excess contaminants. We propose a structured methodology based on a detailed assessment of fraction-based chemical and mineralogical properties of tailings. Simple sorting techniques aligned with the intrinsic value of tailings will be discussed, followed by an approach to progressively create multiple material streams that can be utilised by external industries

Furthermore, guidance will be provided on how to effectively present sorted tailings to potential industry partners and initiate collaborative ventures that could lead to external funding opportunities. The proposed strategy demonstrates how tailings can be transformed into valuable resources at any stage of the mining life cycle, whether during ore processing, operational phases, or post-closure, when tailings can effectively become a new ore source for other industries.

INTRODUCTION – A HIDDEN OPPORTUNITY IN PLAIN SIGHT

Using just a simple high-level mass balance, combining waste rock and tailings, the global mining industry produces about 100 billion tonnes (Bt) of mine waste every year. On the other side of the equation, the global need for raw minerals to produce construction material and to replace fertile soil lost through climate-driven washout exceeds 100 Bt annually (Vuillier, 2024).

It looks like a good match. However, despite this obvious correlation, there are still very few synergies between the mining industry and the building and construction or agricultural sectors. This paper presents a practical, implementation-focused strategy to support this necessary collaboration.

HOW DO OTHER INDUSTRIES SEE MINE WASTE?

This is a fundamental question, because these industries are the potential users.

Mine waste, which is mostly composed of mineral matter, could easily be viewed as a valuable resource across various sectors. In fact, almost all of it is a potential resource. But the focus, especially in the public domain, is overwhelmingly on the perceived risk posed by mine tailings and tailings dams and the potential for contamination. News coverage is often dominated by failures, leaks, and spills. As a result, the perception of mine waste is frequently negative, more about liability and risk than potential or opportunity. This negativity, and fear of association should something go wrong, poses a barrier to other industries entering the market of tailings repurposing.

HOW CAN WE HELP OTHER INDUSTRIES SEE MINE WASTE AS A RESOURCE?

There have been numerous mining-led trials that have attempted to repurpose tailings, pilot projects producing bricks, road base, or using tailings as aggregates for concrete manufacturing. These initiatives are well-intentioned, but they often miss a key point: they're built from a mining perspective, not from the perspective of the end user, which is essential for a large-scale application.

It is a bit like a copper mine deciding to manufacture copper wire in-house, without engaging with wire manufacturers or understanding their specifications. Should it not be the end user role to test and define what they need?

The same principle applies here. Construction companies and manufacturers should be the ones defining what they require, and what materials they're willing to accept, not the mining companies trying to reverse-engineer a use.

To truly make progress, we need to prioritise cross-industry collaboration. Some concrete manufacturers are already aware of these opportunities, but progress has been slow, partly because communication between sectors is fragmented. The approach outlined here offers a structured way to address this.

PRACTICAL APPROACH

The author has already been involved in multiple projects involving the re-mining, reprocessing, and repurposing of tailings. Before diving into the technical details, it is worth reinforcing how similar this approach is to the development of a typical mining or quarrying project.

The first key step for a mining project is exploration, finding and defining an orebody, and analysing its economic potential. In a quarry, it is the same process, but with a focus on minerals instead of metals.

Repurposing tailings or waste rock requires the same mindset: instead of identifying a new orebody, explore the mine waste to see what value (metal and minerals) it still contains. A critical question is: who sees this material as a resource? The potential user becomes the target.

Process

In a mining project focused on a specific metal, the company usually understands the market and tailors its extraction and recovery to suit that need. The same logic applies for repurposing, except the market could be for construction materials, geopolymers, ceramic tiles, soil products, or cement additives.

From the perspective of a brick or tile manufacturer, tailings are a manufactured raw material. If a mining company wants to repurpose its mine waste, it should involve the relevant downstream professionals as early as possible, whether consultants, product designers, or material manufacturers.

Ask: what should this material look like for the market to see it as a valuable resource? That's where the process starts, not from the tailings dam, but from the end-user's product specification.

This is almost a reversed engineering process (see Figure 1):

- start from the end user
- identify how the product must perform
- work backwards through processing, classification, and storage
- finally, design the tailings management to enable this outcome.

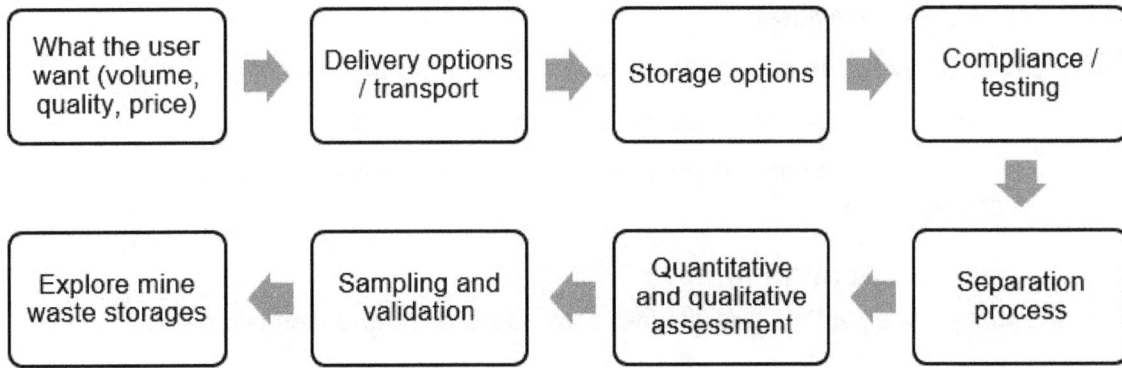

FIG 1 – Reverse engineering process example.

Understand your tailings

The ore/waste rock/tailings should be seen as an aggregate of valuable minerals and metals. Typically, tailings contain large volumes of silicates and alumino-silicates, and a smaller portion of unrecovered metal-bearing minerals, usually sulfides and oxides (and sometimes chlorides and fluorides). There may be other components, carbonates, phosphates, or in the case of many gold operations, arsenopyrite. If the goal is to create a cost-efficient and low risk mine, reducing the volume of waste is a logical strategy. It reduces bond requirements, closure liability, dam risks, and long-term rehabilitation and management costs.

The targeted metal values in the ore are usually well-known. However, the silicate and alumino-silicate content, often more than 80 per cent of the total mass, offers opportunities for repurposing, especially if the mine is near an urban area (raw material for construction) or in a remote area (mine site-based production of precast elements, geopolymers).

Each option has a development pathway and standards. To make it practical, it's essential to include a materials specialist alongside the metallurgists, process engineers, and tailings experts. The material specialist will help make sure the process addresses not just metals, but also mineral quality and compliance with end-user standards.

ESSENTIAL INGREDIENTS FOR A SUCCESSFUL OUTPUT?

Let's begin with the principle: All ore extracted from underground can be reused. There should be no waste (see Figure 2).

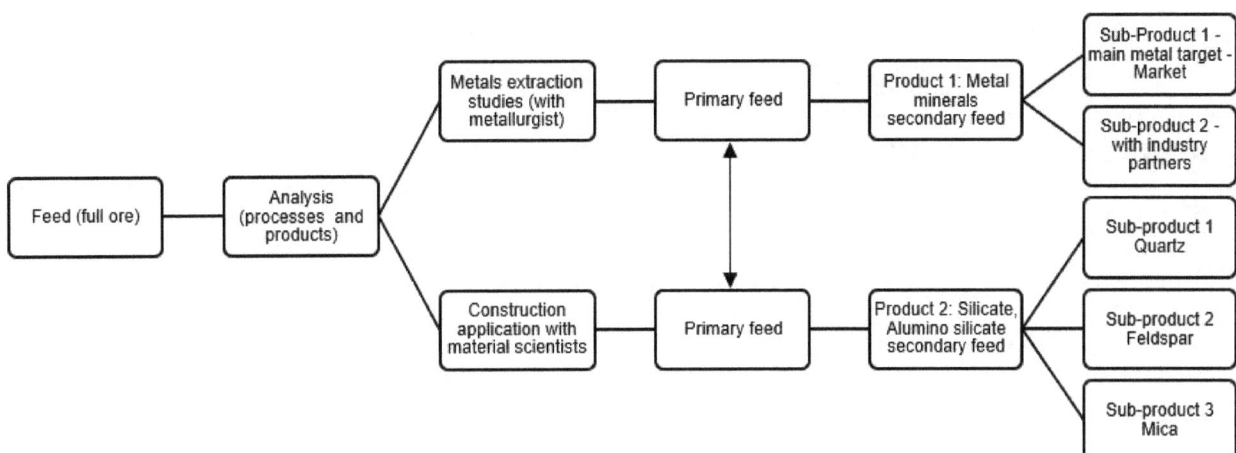

FIG 2 – Upstream captured full mined ore value.

Treat the ore as a value aggregate

Most tailings originate from ore that has been crushed, ground, and processed, yet over 90 per cent of the original mass is often discarded. That discarded portion still holds value, not just environmentally, but economically, if re-evaluated.

What's needed is a shift in mindset:

- view tailings as a mixture of potential products, not just as inert material
- integrate repurposing pathways into the initial design of the tailing's strategy
- ensure the chemical and mineralogical analysis of tailings considers industrial applications, not just environmental risk.

Technical and market alignment

Working with materials scientists and downstream users early is essential (see Figure 3). These parties can:

- define performance criteria (compressive strength, reactivity, durability)
- link tailings properties to construction product standards
- highlight necessary testing regimes and product certification pathways.

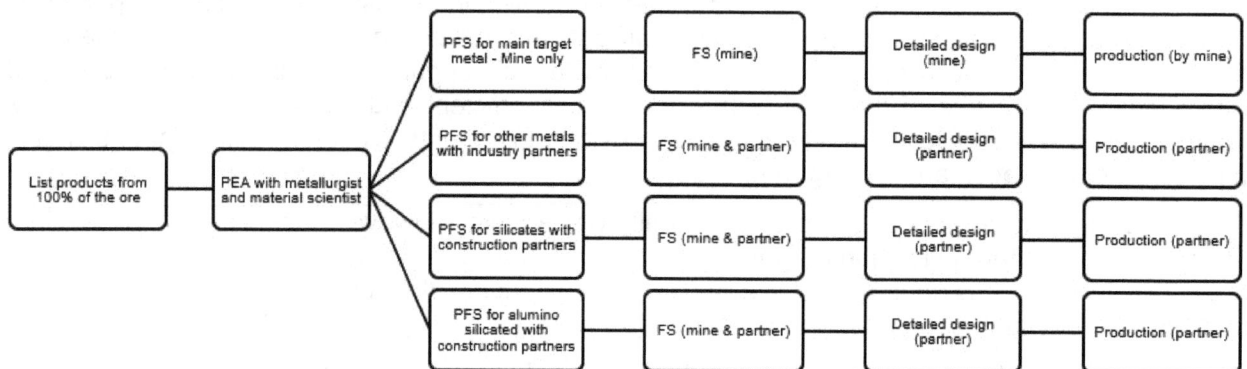

FIG 3 – Study process for identified product and collaboration.

For example, compliance with AS 1141.11.1 (Standards Australia, 2020b; grading of fine aggregates) or AS 2758.1 (Standards Australia, 2020a; aggregates for concrete) can dramatically improve product credibility.

It is critical that repurposed products:

- meet or exceed industry expectations
- are presented professionally (with certificates, lab reports, technical datasheets)
- are shown to be functionally and economically viable.

Engage end users and diversify product streams

Engagement with downstream users, whether in construction, ceramics, soil mixing, or geopolymer production, needs to happen during feasibility, not after.

Tailings volumes are too large for a single market. That's why parallel product pathways should be developed:

- Bricks and precast blocks (silica + alumina, with low sulfides).
- Geopolymer cement (reactive fines, sodium-activated).
- Tiles or ceramics (feldspar/mica-rich fractions).
- Industrial filler (clean quartz in fine sizes).
- Polymetallic concentrates (residual sulfides, recovered via flotation or bioleaching).

Quantify a cost comparison

Repurposing tailings should be benchmarked against the long-term cost of storage for the following items:

- wet storage and filtered tailings stacking
- closure and post-closure monitoring
- environmental and regulatory liability
- tailings repurposing (screening, testing, product delivery).

If 30–40 per cent of tailings can be removed and sold, the avoided cost alone can fund the required processing, with the added benefit of income from sales.

IMPORTANT CONSIDERATIONS

Location and market access

The feasibility of tailings reuse is highly location dependent. Considerations include:

- distance to urban centres
- transport infrastructure (rail, haul road, port)
- energy availability (especially for geopolymer activation or drying)
- climate conditions (which affect product curing, storage, and dust control).

Products like precast panels or geopolymer bricks can justify longer transport, while raw sand substitutes typically require local markets to remain viable.

Contamination and compatibility

Construction materials must meet strict standards for:

- durability
- chemical stability
- absence of deleterious substances.

Sulfides, arsenic, and salts (ie sodium, potassium, chloride) must be addressed:

- Sulfides generate sulfuric acid, which reacts with calcium hydroxide $Ca(OH)_2$ in concrete to form gypsum ($CaSO_4$), causing long-term expansion and cracking.
- Mica content weakens bonding in cement paste.
- Arsenic may leach under alkaline conditions and must be stabilised or removed.

Handing over material responsibility

Rather than the mining company attempting to create a full downstream operation, it may be more efficient to transfer ownership of the clean material to a third party (eg quarry operator, construction group, precast company).

This creates a clean interface:

- mining company focuses on material preparation and quality assurance
- partner manages processing, marketing, compliance, and sales.

Volume opportunity

To illustrate scale:

- 1 Mt of fine silicates = ~285 million bricks

- 20 000 bricks/home = material for 14 000 homes

- 2000 t of precast panels = enough for 500 ten-storey buildings.

Meanwhile, the global demand for aggregates exceeds 40 Bt per annum. Tailings are already ground, which means they may have a lower embodied energy (carbon footprint) than quarried products.

WHERE TO FROM HERE

Silicates and alumino-silicates, which constitute up to 80 per cent of the ore, are in fact valuable products. They just need to be prepared and presented to potential end users to meet their standards. A mine would not present a low-grade concentrate to a smelter or investor as a final product. The same applies here. Mines should not present raw tailings and expect the construction industry to be impressed.

Both mining and quarrying are extracting rock from the ground. The difference is that a quarry turns 90 per cent of its extracted rock into saleable products. Mines discard the majority as waste. Quarries are therefore more material-efficient than mines, yet their approval processes are often just as rigorous, sometimes more so.

If a mining operation can combine with a quarrying strategy, everyone wins. Tailings become raw material, not a waste management problem. The cost of studying, processing, and marketing this material ranges from $15/t to $20/t, comparable to many quarry operations (see Figure 4). For a concrete manufacturer, accessing already-ground silicate material is a significant benefit.

FIG 4 – Similar life cycle for mines and quarries.

SOME CONSIDERATIONS REGARDING MANUFACTURED SAND (M-SAND)

Tailings that meet the particle size requirements and contain suitable minerals can be used as manufactured sand for concrete. But there are technical standards that must be met (Cement Concrete and Aggregate Australia, 2008).

Key parameters for M-sand include:

- Grading/particle size distribution. This can affect workability and strength and must conform to AS 1141.11.1 (Standards Australia, 2020b; fine aggregate grading) and AS2758 (Standards Australia, 2020a).

- Fineness modulus. Indicative of average grain size and important indicator or future concrete performance.

- Shape and texture. Angularity can increase strength, but reduce workability.

- Fines content. Excess fines increase water demand and shrinkage of concrete.

- Water absorption. Affects mix water and concrete strength.

- Durability/soundness. Resistance to weathering (AS 1141.24).

- Clay lumps / mica / deleterious material. Weakens cement bond (AS 2758.1; Standards Australia, 2020a).

- Alkali-Silica Reactivity (ASR). Long-term concrete durability (AS 1141.60.1).

These standards are not optional. They are essential for acceptance in the construction industry.

OTHER PRODUCTS

Manufactured products

- Bricks and Tiles: fine M-sand and cement can be used to make high-quality bricks.

- Geopolymers: activated with sodium hydroxide, can also be used for bricks and tiles. They offer good strength and better resistance to chemicals and heat than traditional concrete. Many of these can be formed at low temperature or room temperature.

As always, purity matters. For example:

- Sulfides, mica, high iron content, sodium, potassium salts all reduce long-term concrete quality and performance.

- Arsenic is a serious issue; it leaches and poses regulatory risks due to its toxicity.

Metals

Removing these contaminants also creates the opportunity to produce a polymetallic concentrate, adding revenue, while decontaminating the silicates for construction use. These metal fractions often contain critical elements, including highly valued rare earth, gallium and germanium.

WHAT IS THE BEST PRACTICAL APPROACH FOR MINING COMPANIES?

Combine a mining approach with a quarrying strategy. The mining company doesn't need to become a quarry operator, this is not their core business. But they can partner with quarry operators or construction product manufacturers.

A dual strategy

The partners know:

- what technical specifications the construction industry needs

- what to test for

- how to screen, wash, and blend materials

- how to ensure compliance with AS standards.

This is how tailings stop being a liability and start being a product.

The simplified approach:

- characterise the tailings

- clean (separate metals, remove harmful components)

- sort by size and mineral content

- match to downstream user specifications

- blend or refine as needed

- present product samples that comply with standards

- work with end users to validate

- avoid wasting time and cost on standalone R&D—collaborate instead (see Figure 5).

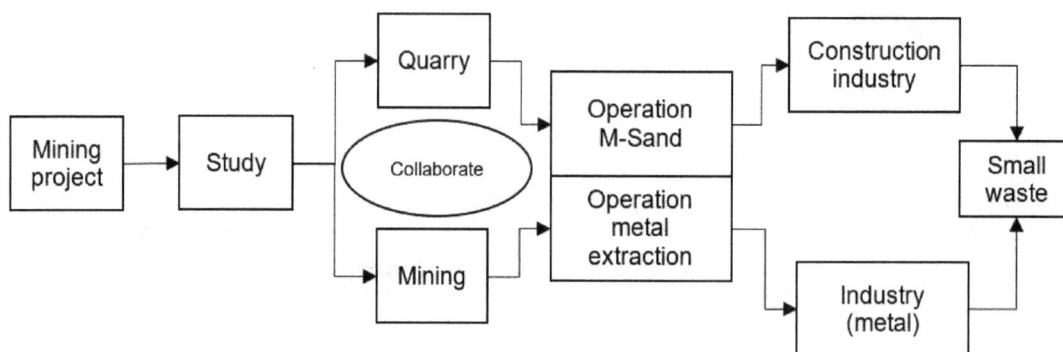

FIG 5 – Visual summary of the development strategy.

CHALLENGES AND LIMITATIONS OF TAILINGS REPURPOSING

While the opportunities are substantial, tailings repurposing is not without challenges. Understanding and addressing these limitations is essential for success and credibility.

Technical challenges

Heterogeneity of tailings

Tailings from the same facility can vary significantly in composition across the storage facility, over time, depending on ore variability and processing changes. This variability complicates product consistency and may require blending or sorting.

Particle size

Tailings are often ultra-fine. While this benefits geopolymer or ceramic applications, it can reduce suitability for structural aggregates unless coarser fractions are separated. Very fine particles (<75 μm) may need densification, drying, or flocculation.

Contamination and deleterious components

As discussed, sulfides, chlorides, sodium, potassium, arsenic and other heavy metals need to be separated from the non-metallic fraction to produce a valuable metal-rich product.

Economic and market barriers

Transport costs

Unless located close to users, tailings may not compete with traditional aggregates on cost. The economic case improves with:

- local demand (eg mine-adjacent infrastructure or housing)
- high-value niche products (tiles, geopolymer panels, ceramics)
- availability of shared logistics.

Lack of familiarity or trust

Construction users are often conservative and risk averse. If tailings-derived products are unfamiliar or unproven, adoption will be slow unless:

- technical testing aligns with national standards
- products are introduced incrementally
- full traceability and compliance documentation are provided.

Capital investment

Upfront investment is required to:

- pilot test
- install separation or washing circuits
- set-up partnerships or co-located plants.

Joint ventures or public-private funding can help bridge the gap (see NEMO Project).

Regulatory and liability risks

Waste classification and liability

In some jurisdictions, even cleaned tailings may still be legally classified as waste, making off-site transport or repurposing and sale more difficult. Clear regulatory pathways and engagement with environmental agencies are essential at a very early stage.

Leachate and durability risks

Even if tailings meet structural specifications, long-term leaching under wet conditions must be ruled out or mitigated. This is particularly important for products used in housing, road base, or public infrastructure. As discussed before, thorough preparation and testing are essential.

Social and cultural perception

Mine waste is often seen as dangerous, even if technically inert. Public and stakeholder acceptance is vital, especially for projects involving housing or public works. Success depends on:

- transparency of testing and certification
- engagement with local communities
- demonstrating visible, safe, high-quality end products.

In summary: these challenges do not negate the opportunity; they shape the roadmap. Each challenge becomes more manageable when projects are phased, partnered, and grounded in compliance and testing.

INTERNATIONAL CASE STUDY – THE NEMO PROJECT

The NEMO Project (New Mining and Remediation Technologies), funded by the European Union's Horizon 2020 programme, is a landmark case in tailings repurposing. It tackled sulfidic mining waste, some of the hardest to reuse, and turned it into a model of circular economy innovation.

NEMO focused on three main sites:

1. Sotkamo Mine, Finland: Bioleaching of Ni-Co-Zn-Cu tailings with residues used in cement and aggregate.
2. Luikonlahti Processing Plant, Finland: Tank bioleaching and reuse of residues in plant flow sheets or construction.
3. Tara Mine, Ireland: Tailings slimes (fines) were converted into composite cements, granulated aggregates, ready-mix and precast concrete.

Their approach included four integrated pilot systems:

1. Bioleaching (ponds, heaps, tanks).
2. Metal and critical mineral recovery.
3. Sulfur removal and element stabilisation.
4. Residue upcycling into certified building materials.

Key takeaways:

- The project featured a strong collaboration between mining, companies, engineering partners, and construction material producers.

- Bioleaching enabled recovery while preparing solids for reuse.

- European and national standards (EN) were met for building materials.

- Transparent processes enhanced community trust and license to operate.

Implication for mines: Even sulfidic tailings can be reused, with the right technology, planning, and stakeholder engagement. Australian projects can draw from this blueprint, especially if aligned to AS testing standards and proximity to construction markets.

CONCLUSION

- All tailings are re-purposable. Large quantities of tailings are silicates and alumino-silicates, Re-using these is the easiest way to reduce tailings volumes significantly. The construction industry is already consuming equivalent materials, including M-sand and crushed fines.

- Quarries face high costs to open new pits. Mines are already operating, and the tailings are already crushed and ground.

- A crucial step is to ensure these materials meet technical standards, especially for concrete, bricks, tiles, or geopolymer products. This means removing sulfides and harmful minerals, and understanding grading, absorption, alkali-silica reaction (ASR) potential, and deleterious material limits.

If implemented correctly, the ore becomes a dual-feed source:

- one stream for metals

- one for clean silicates and alumino-silicates.

Each product has its own market and each reduces environmental impact.

The result is less waste, lower cost, stronger investor interest, and a cleaner operation.

The mining industry doesn't and shouldn't have to do this alone. There must be collaboration between partners who can co-finance, build, and operate the downstream processes.

It is time to move from managing tailings in isolation to designing tailings as an additional product stream. The future is circular, and it's practical.

REFERENCES

Cement Concrete and Aggregate Australia, 2008. Guide to the Specification and Use of Manufactured Sand in Concrete.

European Union Horizon, 2020, 2019–2023 NEMO Project (New Mining and Remediation Technologies). Available from: <https://nemo2020.eu>

Standards Australia, 2018. AS 1141.24 – Aggregate soundness — Evaluation by exposure to sodium sulphate solution.

Standards Australia, 2019. AS 1141.60.1 – Methods for sampling and testing aggregates – Potential alkali-silica reactivity.

Standards Australia, 2020a. AS 2758.1 – Aggregates and rock for engineering purposes – Concrete aggregates.

Standards Australia, 2020b. AS 1141.11.1 – Methods for sampling and testing aggregates.

Vuillier, C, 2024. Turning Mine Waste into Value – Sustainable Strategies for Modern Mines, *The AusIMM Bulletin*.

Improved predictions of acid and metalliferous drainage for greenfield projects

W Zhang[1], A Corzo Remigio[2], R Green[3] and M Edraki[4]

1. Research Fellow, The University of Queensland, St Lucia Qld 4072.
   Email: wenqiang.zhang@uq.edu.au
2. Research Fellow, The University of Queensland, St Lucia Qld 4072.
   Email: amelia.corzoremigio@uq.edu.au
3. Principal Environmental Geochemist, Rio Tinto, Brisbane Qld 4000.
   Email: rosalind.green@riotinto.com
4. Professor, The University of Queensland, St Lucia Qld 4072. Email: m.edraki@uq.edu.au

ABSTRACT

Extrapolating predictions of acid and metalliferous drainage (AMD) from laboratory to the field face significant challenges due to the heterogeneity of waste rock dumps (WRDs), encompassing variations in particle size distribution, geochemical and hydrological properties of rock fragments, control of experimental conditions and duration, and lack of input parameters reference for numerical models. For greenfield mine sites where the design of WRDs is not finalised, these challenges are further compounded beyond the shortcomings of industry-standard procedures for AMD predictions. Therefore, it is necessary to develop appropriate methods for systematic AMD predictions for greenfield sites. This study investigates how closely the hydrological conditions of waste rock dumps, such as intrinsic permeability, liquid-to-solid ratio, and residence time, can be simulated in the laboratory to derive WRD design parameters. The case study focuses on a greenfield copper-gold deposit in a semi-arid region. Comprehensive mineralogical and geochemical characterisations have been conducted using drilled core samples. The main lithologies consist of metasediments (sandstone, siltstone) and mafic rocks with varying mineralogy. Silicates, such as muscovite, the biotite group, hornblende, and chlorite, provide some neutralising capacity at low reaction rates, while carbonates may or may not offer neutralising capacity. To address the limitations of conventional kinetic testing methods, such as humidity cell and AMIRA funnel leaching, this study introduces newly designed kinetic leaching columns that integrate direct oxygen consumption measurements with leachate chemistry analysis and can evaluate AMD risk of the samples at different compaction rates. Comparison of the results will help reduce the duration of kinetic tests, improve understanding of particle size and mineral liberation effects on reaction rates, and enable distinguishing the role of carbonates versus silicates in the neutralisation of acidity through monitoring of CO_2 generation. The findings will assist in identifying key parameters affecting the geochemical and hydrogeological performance of WRDs, aiming to provide optimum dump designs before mining activities.

INTRODUCTION

Extrapolating predictions of acid and metalliferous drainage (AMD) from laboratory to the field face significant challenges due to the heterogeneity of waste rock dumps (WRDs), encompassing variations in particle size distribution, geochemical and hydrological properties of rock fragments, control of experimental conditions and duration, and lack of input parameters reference for numerical models. For greenfield mine sites where the design of WRDs is not finalised, these challenges are further compounded beyond the shortcomings of industry-standard procedures for AMD predictions. Therefore, it is necessary to develop appropriate methods for systematic AMD predictions for greenfield sites. This study investigates how closely the hydrological conditions of waste rock dumps, such as intrinsic permeability, liquid-to-solid ratio, and residence time, can be simulated in the laboratory to derive WRD design parameters.

METHODS

The case study focuses on a greenfield copper-gold deposit in a semi-arid region. Comprehensive mineralogical and geochemical characterisations have been conducted using drill core samples. The main lithologies consist of metasediments (sandstone, siltstone) and mafic rocks with varying mineralogy. Silicates, such as muscovite, the biotite group, hornblende, and chlorite, provide some neutralising capacity at low reaction rates, while carbonates may or may not offer neutralising capacity.

Conventional kinetic testing methods, such as humidity cell and AMIRA funnel leaching, are often too simplified with respect to experimental conditions. They typically lack the capability to monitor for gas (eg oxygen) variations, provide limited temperature control and particle sizes, and do not account for density changes of materials. To address those limitations, this study introduces newly designed kinetic leaching columns that integrate direct oxygen consumption measurements with leachate chemistry analysis and can evaluate AMD risk of the samples at different compaction rates. The column leaching apparatus was improved to withstand compaction energy, allowing the target dry density to be directly achieved within the column rather than transferring the compacted samples from other devices. The dimensions of the column are 150 mm in diameter and 200 mm in height. Oxygen consumption is always a key parameter to assess the oxidation of pyrite, which is a dominant reaction in AMD generation. While previous studies also indicated the importance of silicates and carbonates in providing acid-buffering/neutralising capacity within waste materials (Schoen *et al*, 2023), which can be evaluated by the release of CO_2 in the waste products. So, the improved column leaching apparatus was equipped with a sensor case that contained an O2-A3 oxygen sensor and an IRC-A1-CO_2-A-NDIR sensor from Alphasense® to monitor variations in both O_2 and CO_2 concentrations in the weal-sealed column. The two types of sensors were calibrated using sealed bottles containing known concentrations of O_2 and CO_2, and temperature compensation was applied to the readings when the experimental temperature changed following the sensor manual. The sensor case was connected to a three-way ball valve which was integrated with a timer. Gas concentrations in the column were measured at four-time intervals over a 24-hour period: 3:00 am, 9:00 am, 3:00 pm, and 9:00 pm, with each measurement lasting 45 mins. During the measurements, the channel between the sensor case and the column was connected, allowing gas from the column to diffuse into the sensor case. After each measurement, the ball valve rotated to release the moist gas in the sensor case to the atmosphere. This process ensures that the sensors are effectively protected from humidity in the column. According to the standard proctor compaction curve, three levels of compaction (ie max compaction, medium compaction and less compaction) were achieved by compacting the crushed samples with different initial moisture contents of 3 per cent, 4 per cent and 5.1 per cent (AS 1289.5.2.1, Australian Standards, 2017). For each column, the pre-prepared sample was compacted in five equal layers using a hammer, with 25 blows applied for every 20 mm height. Once the compaction of the final layer was completed, the instrumented lid was screwed onto the top of the column to ensure it is water- and air-tight. Once all the samples were packed into the columns, silicone heat pads were applied around the column wall to control the experimental temperature, which was set to be constant at 35°C to accelerate the chemical reaction inside the column and mitigate the impact of temperature fluctuations on sensor readings. Figure 1 shows the schematic diagram of the improved kinetic leaching column. At the beginning of each leaching cycle, half a pore volume of deionised water was pumped into the column through the inside nozzle. Additional deionised water was added after a 14-day reaction, and leachates were collected one day after to provide enough time for solute dissolution. Then the leachates were analysed for electrical conductivity, pH, alkalinity/acidity, major cations and anions and metals concentrations.

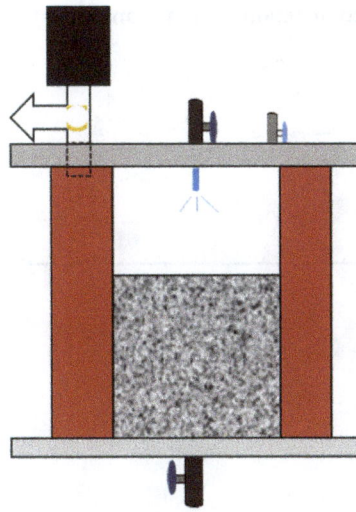

FIG 1 – Schematic diagram of the improved kinetic column leaching apparatus for different compaction rates with oxygen and carbon dioxide concentrations monitoring.

RESULTS

The experiment has been running for over ten leaching cycles. The gas monitoring results after ten leaching cycles indicated that the sample with maximum compaction had insignificant changes in oxygen consumption and minimal CO_2 generation over leaching cycles, whilst the sample with less compaction achieved the highest cumulative O_2 consumption of 4.8 per cent and CO_2 generation of 1.8 per cent (Figure 2). From the leachate chemistry in Figure 3, we found that higher compaction resulted in higher electrical conductivity in leachates, but pH fluctuated in a similar manner for the three samples. It is noted that the sample with maximum compaction had the highest sulfate release in leachates, and it's Fe^{2+} concentration was 5~6 times higher than the other two samples after three leaching cycles.

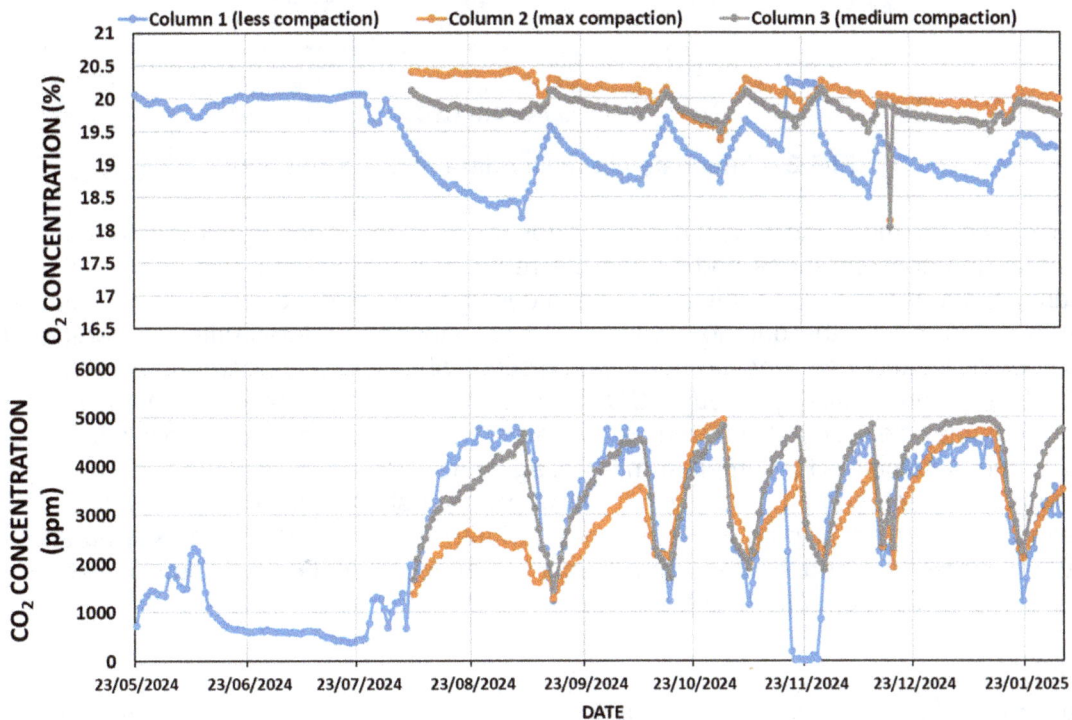

FIG 2 – Temporal changes in oxygen and carbon dioxide concentration in three columns with different compact rates.

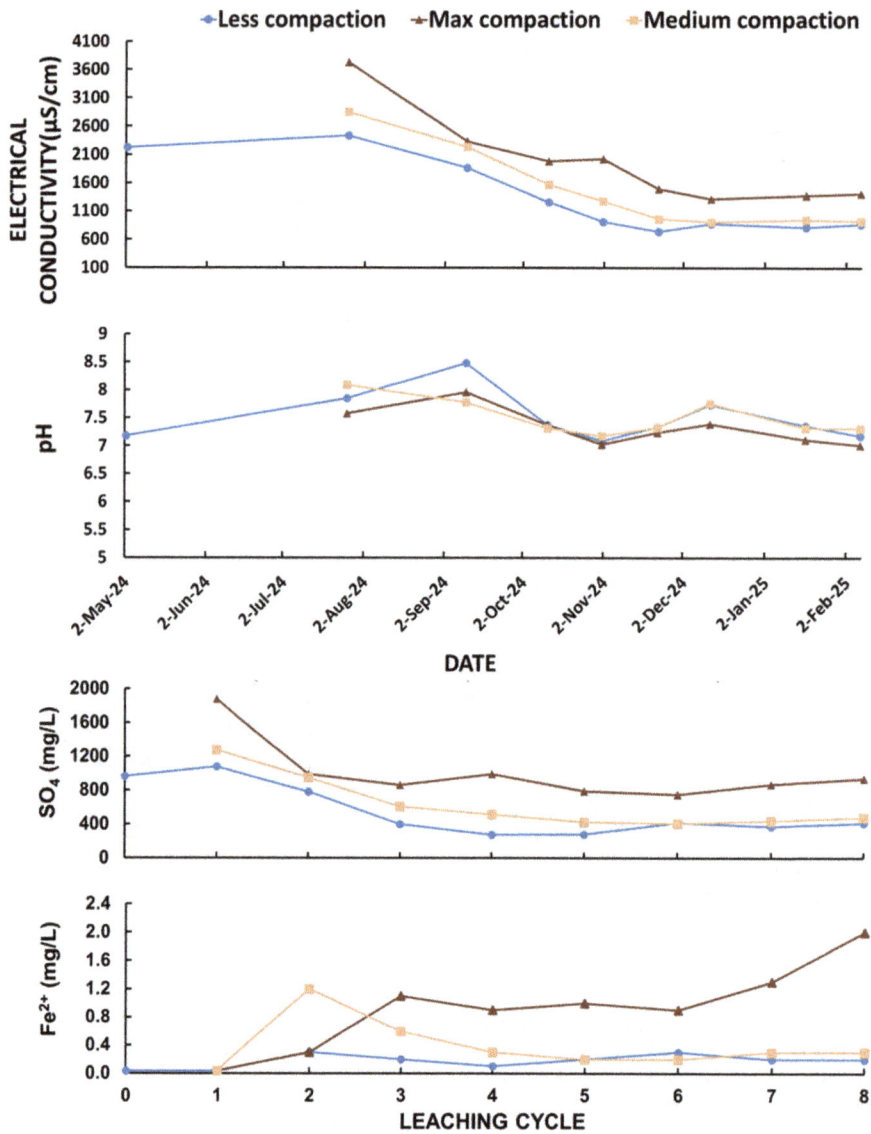

FIG 3 – Leachate chemistry data of three columns.

The new kinetic column leaching test demonstrates the capacity of the apparatus for both capturing variations in gas concentrations and tracing the release of metals and acid. It also highlights the significant impact of compaction rates on oxygen consumption and CO_2 generation. Comparison of the results helps reduce the duration of kinetic tests, enhance understanding of particle size and mineral liberation effects on reaction rates, and distinguish the roles of carbonates versus silicates in the neutralisation of acidity through CO_2 monitoring. Although the experiment does not fully represent complex field conditions, the results are still able to provide reliable reference values regarding input parameters for numerical models, eg hydraulic properties of the material and reaction rates under different compaction levels, unlike using estimations solely based on empirical parameters or mineralogical data. The findings contribute to not only identifying key parameters that affect the geochemical and hydrogeological performance of WRDs, but also aiming to provide optimum dump designs prior to mining activities.

REFERENCES

Australian Standards, 2017. AS 1289.5.2.1 – Methods of testing soils for engineering purposes Soil compaction and density tests – Determination of the dry density/moisture content relation of a soil using modified compactive effort, Methods of Testing Soils for Engineering Purposes, Australian Standards.

Schoen, D, Savage, R, Pearce, S, Shiimi, R, Gersten, B, Roberts, M and Barnes, A, 2023. A novel empirical approach to measuring pore gas compositional change in mine waste storage facilities: A case study from northern Europe, in Mine Closure 2023: Proceedings of 16th International Conference on Mine Closure (eds: B Abbasi, J Parshley, A Fourie and M Tibbett), (Australian Centre for Geomechanics: Perth). https://doi.org/10.36487/ACG_repo/2315_070

Miscellaneous

A risk-based approach to bauxite residue disposal at Worsley Alumina

A J Schteinman[1] and J Wu[2]

1. Tailings and Dams Engineer, South32, Worsley WA 6225.
   Email: alexis.schteinman@south32.net
2. Superintendent Governance and Standards, South32, Worsley WA 6225.
   Email: joseph.wu@south32.net

ABSTRACT

Worsley Alumina's Bauxite Residue Disposal Area (BRDA) operations have grown increasingly complex since commissioning in 1984 driven by higher production rates, diminishing land for additional storage areas, and intensified construction activities.

In response to these demands, Worsley developed a Residue Deposition Strategy (RDS) to manage consolidation, strength gain and dewatering incorporating improved deposition practices, mudfarming and residue testing regimes. However, the limited construction season—constrained by regional climate—introduces risks of misalignment between deposition and construction schedules.

To address these challenges, Worsley developed an integrated Construction and Deposition Plan (CDP) using in-house modelling tools. The CDP aligns operational execution with the RDS philosophy, enabling early identification and mitigation of medium- and long-term risks, and supporting life-of-mine planning. The CDP is used across a 10-year planning horizon, allowing Worsley to re-sequence construction, optimise deposition and defer non-critical works.

Integration of the CDP into Worsley's operation has reduced planning inaccuracies which placed RDS compliance at risk, and ensured fallow period compliance over the 10-year plan. Previous generic assumptions on construction sequencing have been replaced with a data-driven model and accurate plan aligned with residue deposition.

INTRODUCTION

Worsley Alumina Pty Ltd (WAPL) operates several Bauxite Residue Storage Areas (BRDAs) divided into the areas shown in Figure 1. A 10.5 Mt/a of residue mud is delivered via four pipelines to the BRDAs, which are constructed using the upstream method in 5 m raises. Two of the residue delivery lines transport slurry to the northern BRDAs (Areas 1–3) while the other two deliver to the southern areas (Areas 5–9). Areas 4 and 10 are not currently operational. In the BRDAs, residue undergoes accelerated mechanical consolidation (AMC) with amphi-scrolling machines which ploughs the surface maximising evaporation and increase the overall deposited density of the residue.

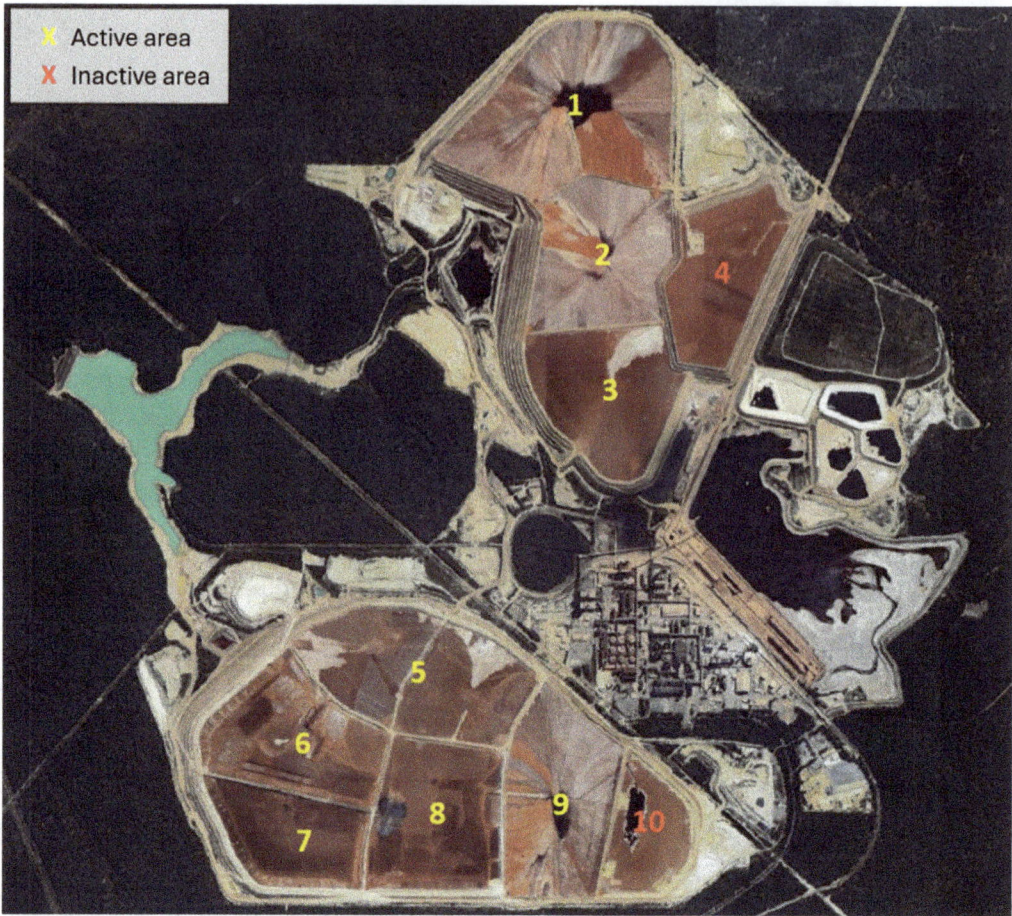

FIG 1 – Worsley Alumina residue area layout.

The use of upstream construction requires the underlying residue to be sufficiently consolidated. As such, WAPL developed a Residue Deposition Strategy (RDS) which guides the planning of BRDA operation and construction. Adherence to the RDS secures the operational life of its BRDAs, as historical operation demonstrated that its implementation results in rapid residue consolidation and strength-gain to design specifications, on which an upstream raise can be safely constructed. The rotational nature of deposition at WAPL requires construction to be scheduled during long-term fallow periods in the BRDAs. Figure 2 shows a graphical concept of the operational philosophy of the BRDAs.

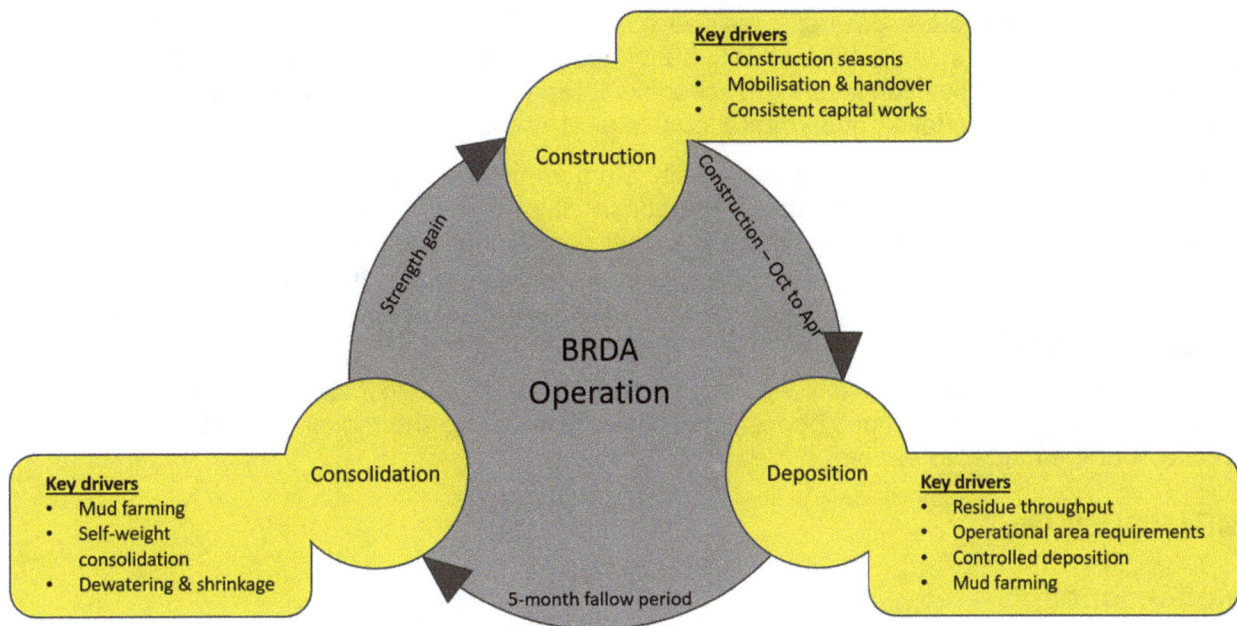

FIG 2 – BRDA operation life cycle.

While the RDS has been implemented and continuously improved at WAPL since its conception in 2012, substantial risks to medium- and long-term deposition and construction still exist. Conflicting construction and deposition plans were identified as the main risk, driven by increased production, limited construction seasons due to regional climate, and a historically used generic sequence for BRDA construction—not based on residue storage demand. The separate development of construction and deposition plans could only address short-term risks to construction, causing friction between contractual obligations and conformity to the RDS. The lack of data-driven planning incapacitated WAPL's ability to assess long-term risks to construction—the time frame during which changes to the plan can easily be made and risks to deposition be managed.

Reducing these risks required the construction and deposition plans to be harmonised, prompting WAPL to develop an optimised construction and deposition model integrating BRDA storage volumes, production rates and the RDS key performance criteria. The resulting model was termed the combined Construction and Deposition Plan (CDP). A conceptualised case study of the CDP is presented in this paper to demonstrate its effect on risk management in the BRDAs. Naming convention of the BRDAs is for illustration purposes only.

RESIDUE DEPOSITION STRATEGY

WAPL's current RDS consists of the following key performance criteria:

- Residue with consistent solids content, limited to a design range, balancing adequate residue consolidation and operational viability.

- Controlled deposition layers, limited to 1.1 m depth to ensure amphi-scroller traction with previous lift (1.2 m scrolls). These layers undergo significant consolidation to 600 mm depth.

- Allowance for fallow periods between depositional cycles in a cell to facilitate consolidation. Consolidation is achieved via initial shrinkage and mud-farming with amphi-scrollers. Enabling these fallow periods requires a minimum of 240 ha to be available for deposition.

- Allowance for an extended pre-construction fallow period of 5 months to ensure the design strength is achieved and residue surface is well prepared to serve as foundation for an upstream raise.

- Vane shear testing following inter-cycle and pre-construction fallow periods to verify the required residue strengths have been achieved.

Historical adherence to the desirable residue solids has been well managed, and while sporadic spikes in residue solids would theoretically benefit BRDA capacity, it can cause BRDA Operations to lose control over deposition layering, exceeding 1.1 m (overpouring). Close monitoring of residue solids is therefore undertaken at an operational level to avoid overpours.

The duration of fallow periods varies seasonally, from approximately 30 days in summer to 60 days in winter, with an average of 42 days. Pre-construction fallow period is extended to 5 months between May and October for the residue to further consolidate and achieve higher strength as per design specifications, minimising consolidation settlement during construction.

SCOPE OF CDP

The CDP serves as an early indicator of medium- and long-term risks to BRDA operation by integrating the constructional and depositional plans, forecasting shortcomings in RDS performance criteria over the modelled period.

METHODOLOGY

Model concept

A conceptual representation of the CDP model is shown in Figure 3. The CDP model requires two base inputs – BRDA storage capacity and a preliminary construction schedule. Storage capacity can be estimated using any CAD software, but in this case was modelled using Muk3D. Capacity volumes were converted to available days of residue deposition (ADRD) based on forecast refinery production from the survey date.

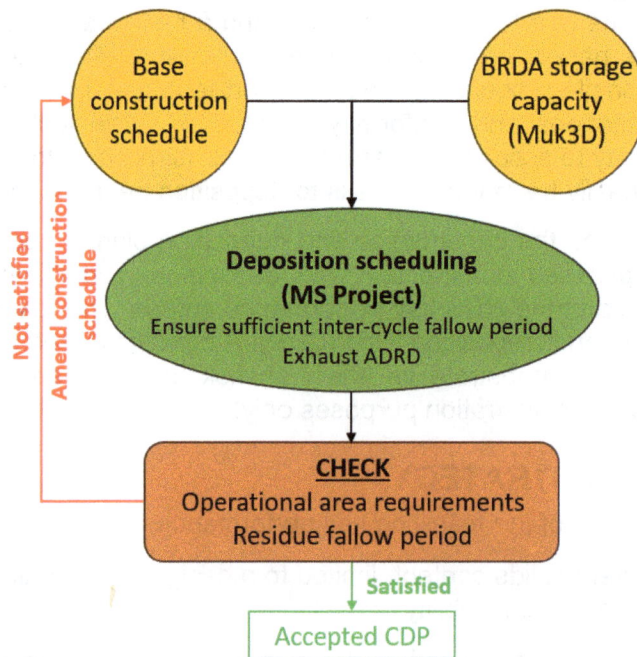

FIG 3 – Conceptualised CDP model.

The base construction schedule and ADRD were then used to develop a preliminary deposition plan in Microsoft Project, the suitability of which was evaluated against key RDS monitoring criteria – sufficient residue operational area (240 ha), inter-cycle fallow period (42 days) and pre-construction fallow period (5 months).

If the RDS criteria are not satisfied, the construction and/or deposition sequences are revised depending on-site priorities. This process is repeated until the key RDS criteria are satisfied and facilities are de-risked. The resulting construction and deposition plan is then risk-assessed against the relevant critical risks and accepted once these are sufficiently controlled or accepted by relevant stakeholders.

Key inputs

The model requires the key data listed in Table 1. These are critical in quantifying the ADRD, which are calculated using the following equation for every BRDA residue area raise:

$$ADRD = V_{raise}/P$$

Where:

V_{raise} is the storage capacity for a given raise in a BRDA residue area

P is the volumetric residue production rate

TABLE 1

Model key inputs.

Input	Value	Use
Residue production rate	Nominal refinery, dt/day	Used to calculate ADRD alongside BRDA storage volume obtained from Muk3D.
Residue *in situ* density	Measured / calculated, t/m^3	
Consolidated cycle depth	600 mm	Used to calculate number of cycles required to fill BRDA residue area.
Min. fallow period [1]	42 days	Min. number of days between deposition cycles into the same area to allow for consolidation.
Min. active deposition area [1]	180	Calculated operational area requirements for ample fallow period between cycles as per the RDS.
Min. emergency deposition area [1]	60	
Min. total deposition area [1]	240	
Min. pre-construction fallow period	5 months	Model requires area deposition to be completed by end of April to begin construction in October.

[1] RDS standard calculated based on amphi-scrolling productivity and coverage.

Maximum cycle duration must also be calculated for every residue area using the following equation:

$$D_{cycle} = (ADRD*A*CD_{cycle}) / V_{raise}$$

Where:

A is the surface area of disposal

CD_{cycle} is the consolidated cycle depth = 600 mm in the case of WAPL

D_{cycle} is used to guide the deposition scheduling as described further below

Initial construction sequence

If not already available, a construction sequence for the residue areas must be developed as a primary input to the CDP model. The construction sequence is used to determine disposal area availability, which is then used as a guide to deposition scheduling. Since construction must be completed by the end of April of a given FY, it is assumed that a BRDA is available on the subsequent 1st of July (beginning of the following FY).

The initial construction sequence is listed in Table 2.

TABLE 2

Initial construction sequence.

Financial year (FY) constructed	Residue area and stage
FY2025	Area 1 Stg 1
	Area 9 Stg 1
FY2026	Area 2 Stg 1
FY2027	Area 3 Stg 1
FY2028	Area 5 Stg 1
FY2029	Area 6 Stg 1
	Area 7 Stg 1
FY2030	Area 8 Stg 1
FY2031	Area 9 Stg 2
FY2032	Area 1 Stg 2
	Area 2 Stg 2
FY2033	Area 3 Stg 2
FY2034	Area 5 Stg 2
	Area 6 Stg 2
	Area 7 Stg 2
FY2035	Area 8 Stg 2

Deposition scheduling

The deposition schedule is developed in Microsoft Project, but any scheduling program can be used. The schedule integrates actual deposition since the aerial survey date and projected deposition based on operational priorities. Care is needed to ensure that time between cycles in the same residue area is a minimum of 42 days as per the RDS, and that total deposition duration achieves the ADRD.

Once completed, the CDP model combines both the deposition and construction schedules and graphically presents the plan for the modelled duration, used to identify key risks and gaps.

Construction sequence optimisation

The following figures show the resulting output of the CDP model.

Figure 5 shows the summary of the next 10 years of BRDA operation, with the vertical axis representing storage capacity in ADRD, and the shaded regions highlighting wet seasons (no construction).

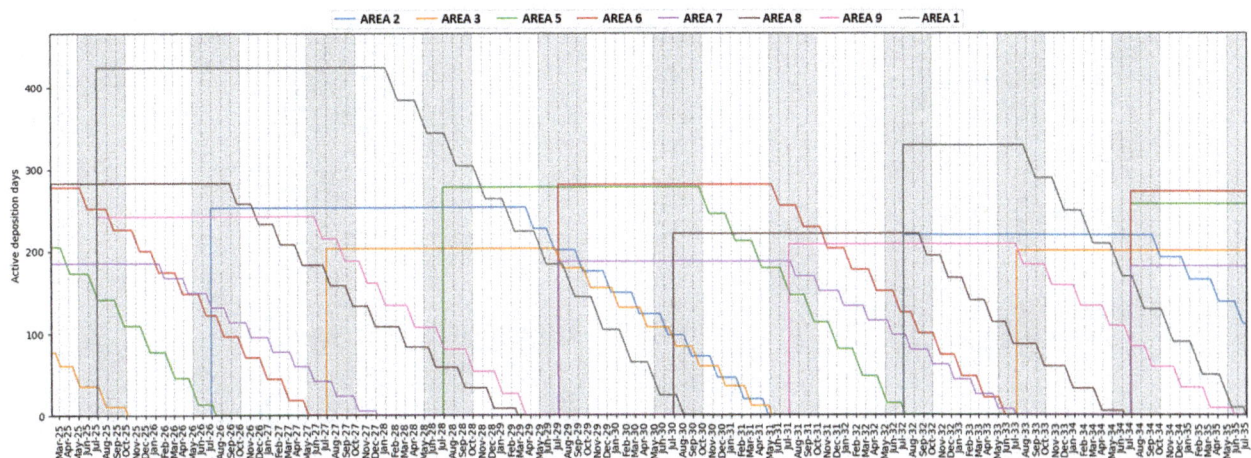

FIG 5 – Initial 10-year BRDA operational summary.

Figure 6 presents the pre-construction fallow period availability in each BRDA stage, with the minimum required 5 months outlined. The model shows significant fallow period availability in the short-term, but substantial risk to the RDS due to pre-construction fallow period shortfall from FY2034 onward. To rectify this, either deposition must be re-prioritised or the relevant BRDA areas need to be constructed later. As no deposition is projected in the range of 10-years for these stages in the model, their construction can be delayed, and re-evaluated once deposition in these areas becomes part of the 10-year plan. Deposition can be optimised by prioritising deposition in residue areas with shortfall at the expense of areas with significant fallow period available at the same time, such as Area 3 Stage 3 and Area 5 Stage 3.

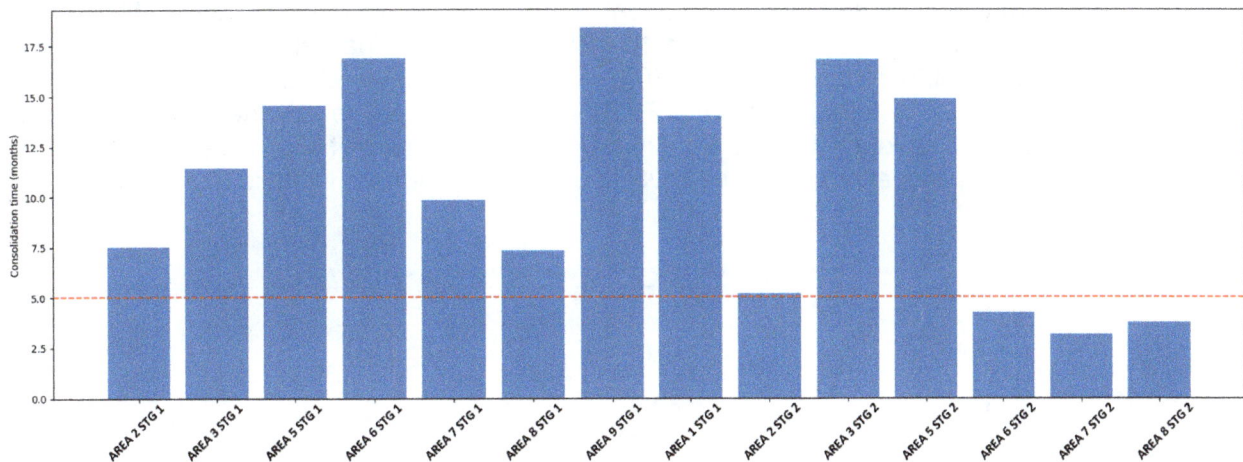

FIG 6 – Initial 10-year BRDA pre-construction fallow period.

Figure 7 displays the available operational area over the next 10 years alongside the minimum area required by the RDS for active, alternate and total area. While some periods exhibit active areas below 180 ha, sufficient alternate storage area is available to mitigate operational risks. Most importantly, a minimum of 240 ha is available throughout the 10-year plan.

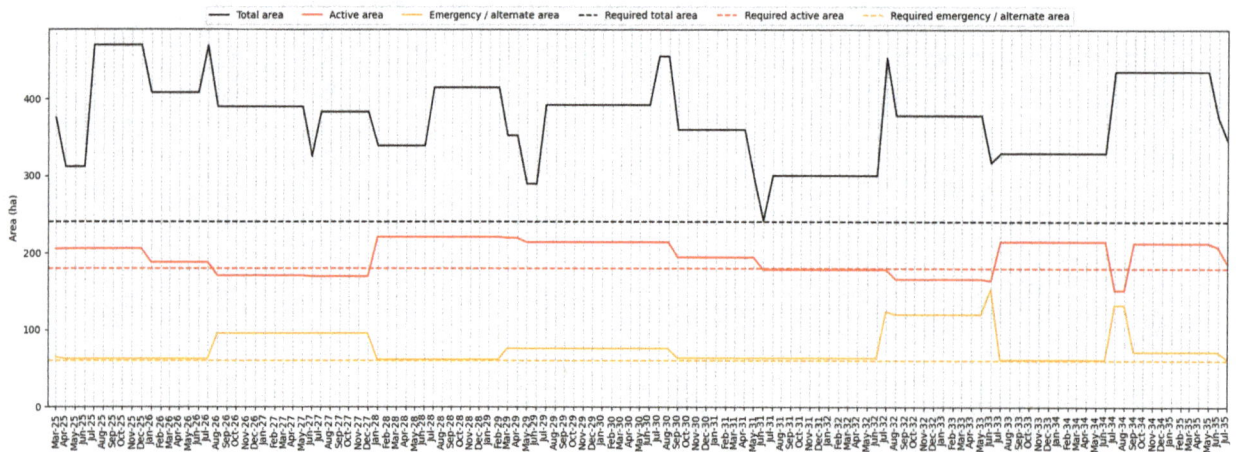

FIG 7 – Initial 10-year BRDA operational area availability.

Feedback loop

Where challenges to achieve the required fallow periods were observed during the modelling, modifications to construction and deposition schedules were undertaken until all criteria were satisfied. For example, advanced deposition to Areas 6, 7 and 8 required a delay in deposition to Area 5, which could only be accommodated by delaying Area 5 Stage 3 construction to FY2035.

CDP RESULTS

Optimised 10-year plan

Figures 8, 9 and 10 present the refined CDP model. Figure 9 shows all residue areas achieving the target pre-construction fallow period. Figure 10 however, shows periods of significant active operational area shortfall, primarily around FY2031 and FY2034. Evidently, the FY2031 shortfall is due to paused deposition in Area 3, while the FY2034 deficit is explained by the operation of only Area 8 and Area 9 for several months. Substantial alternative depositional area is available to mitigate risks posed by this shortage to the RDS and the total area available is maintained above 240 ha.

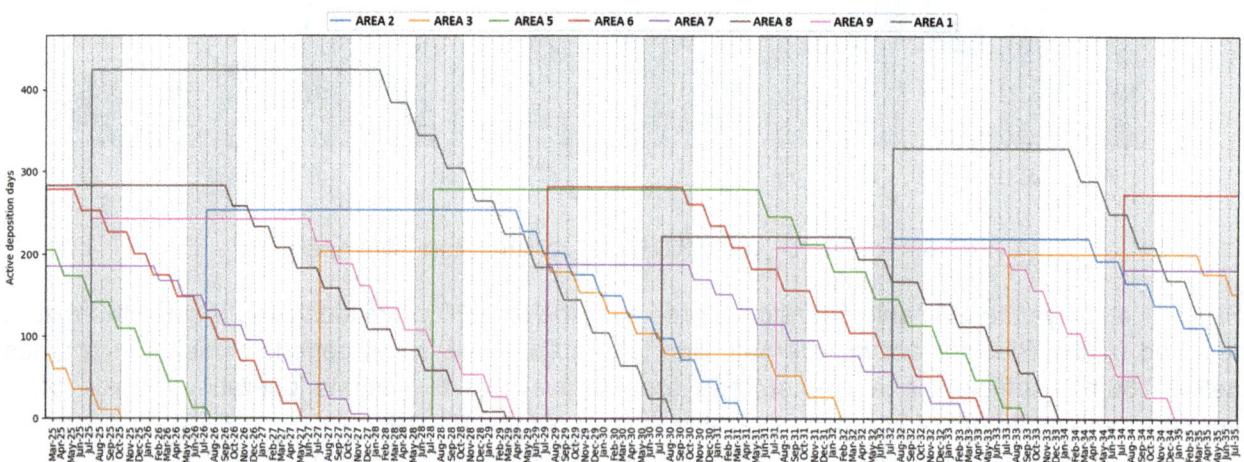

FIG 8 – Optimised 10-year BRDA operational summary.

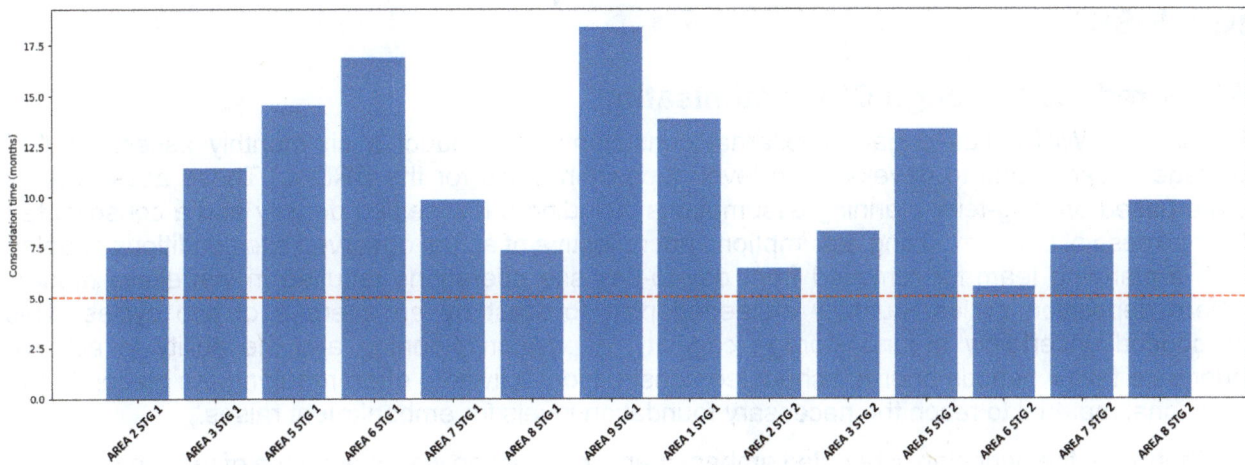

FIG 9 – Optimised 10-year BRDA pre-construction fallow period.

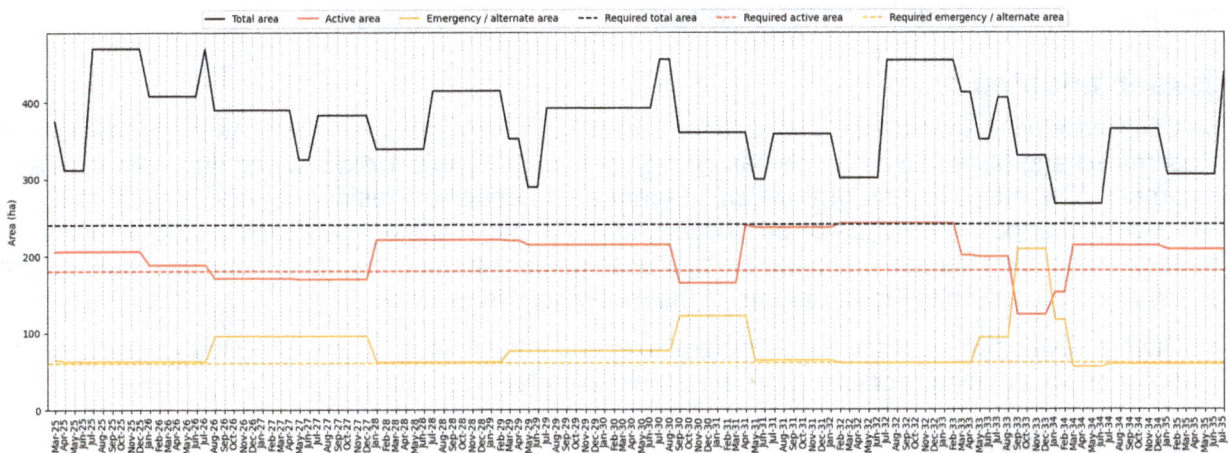

FIG 10 – Optimised 10-year BRDA operational area availability.

Figure 11 compares the adopted optimised plan to the initial plan, demonstrating the optimised plan clearly meets the RDS performance criteria where the initial plan falls short.

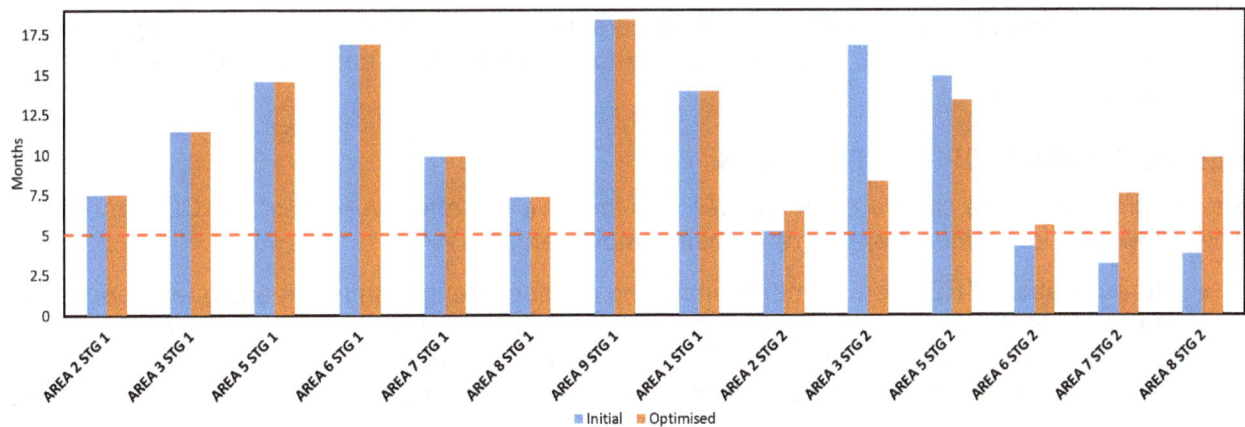

FIG 11 – De-risked pre-construction fallow period.

The resulting 10-year plan significantly improved WAPL's capacity to predict and manage risks to construction and production, permitting construction deferral of several residue areas and easing pressure on residue deposition conformance by BRDA Operations. When compared to the initial plan, significant effort and compromise to RDS compliance would be needed to achieve the required fallow periods for construction across the 10-year horizon. Adopting a data-driven plan secures the future of BRDA construction and refinery operation through flexibility in long-term planning.

KEY RISKS

Risks reduced through CDP optimisation

Historically, WAPL has engaged external consultants to conduct a six-monthly assessment of storage capacity and to develop high level deposition plans for the BRDAs. These assessments were based on long-term planning assumptions including a low settled density and a consolidated lift thickness of 650 mm. Using assumptions not reflective of actual observed site conditions coupled with a planning team far removed from day-to-day site operations resulted in ineffective models where deposition cycles routinely exceeded plan forecast by an average of two cycles. This introduced uncertainty around storage capacity, deposition planning and the ability to achieve adequate fallow periods prior to scheduled construction activities, often requiring the placement of additional material to reach the necessary foundation levels for embankment raises.

Furthermore, the prior plan scheduled embankment construction well in advance of required storage, which increased maintenance costs to the operation and introduced medium-term contractual risks due to unavailability to construct. This presented opportunities to simultaneously reduce contractual and operational risks by deferring the construction of several residue areas.

Risks in adopted plan

The CDP introduces several operational risks during its implementation phase. As the model is built on estimated and observed data, the parameters outlined in Table 1 require ongoing validation and refinement. The effectiveness of the CDP depends on consistent residue characteristics and any unexpected changes in ore type or production rates will necessitate reforecasting and adjustment. To ensure the plan remains accurate, regular updates informed by field data and survey are essential, particularly in the initial stages of the model as it is calibrated.

The CDP is developed in-house and is integral to execution planning. Maintaining technical expertise and succession planning is critical to support mid-term planning.

CONTROL VERIFICATION

The CDP undergoes control verification by complete re-calibration and re-optimisation under the following conditions:

- **Periodical:** an annual control verification is conducted at minimum to ensure all construction, changes to deposition, and verified key RDS criteria are captured.

- **Ad hoc:** as-required control verifications are undertaken when actual RDS input values deviate from the forecast, or when changes to design or major changes to deposition due to incidents occur.

IMPROVEMENT OPPORTUNITIES

Several improvements can be made at WAPL to improve the effectiveness of the CDP and RDS. The following improvements are currently being implemented:

- Re-evaluation of RDS requirements, including residue strength gain rate and operational area. Recent introduction of autonomous amphi-scrolling to improve AMC is expected to have reduced the required fallow period and operational area requirements. Density data will be collected over the initial implementation phase to refine WAPL's understanding of the residue consolidation rate.

- Revision of residue input data and deposition factors such as layer thickness, solids content, delivery rates and durations.

- Implementation of a structured approach to residue operational process reviews, to ensure stakeholder engagement from operational personnel to engineering.

- Addition of mudline flow measurement capabilities to capture exact volume disposal into the different BRDA areas and understand the split between the mudlines.

- Global residue consolidation rates tracking to ensure accurate volumes are accounted for in the model, and validating assumptions regarding consolidation uniformity in the residue areas.

CONCLUSIONS

- The CDP has been successfully integrated into BRDA operational planning and execution.

- It has significantly reduced operational risk by optimising the deposition and construction plans.

- The CDP is sensitive to key input data, and therefore requires consistent revalidation of site data and alignment with recent deposition tracking.

- Further improvements could be made to improve tracking of key inputs on-site to improve the accuracy of the CDP.

ACKNOWLEDGEMENTS

The authors would like to acknowledge South32 for permitting the publication of this paper and WAPL Superintendent – Execution Cam Golding for providing valuable data and access to modelling software.

Monitoring waste storages during operation and post-closure

Evaluating unexpected piezometer trends in a tailings dam – insights from dissipation testing

R Dellamea[1], J Rola[2], M Sottile[3] and J Eldridge[4]

1. Senior Consultant, SRK Consulting (Australasia) Pty Ltd, West Perth WA 6005.
   Email: rdellamea@srk.com.au
2. Principal Consultant, SRK Consulting (Australasia) Pty Ltd, West Perth WA 6005.
   Email: jrola@srk.com.au
3. Principal Consultant, SRK Consulting (Argentina) S.A., Buenos Aires, Argentina.
   Email: msottile@srk.com.ar
4. Consultant, SRK Consulting (Australasia) Pty Ltd, West Perth WA 6005.
   Email: jeldridge@srk.com.au

ABSTRACT

The monitoring of deformation and pore water pressure (PWP) within a tailings dam is crucial for ensuring structural integrity and operational safety. Unexpected trends are sometimes observed within monitoring instrumentation, and careful consideration of these trends is important as actions may range from 'do nothing' to 'evacuate'. This paper presents and discusses unexpected trends in vibrating wire piezometer (VWP) readings within a heavily instrumented tailings dam during construction of a stabilisation solution. Observation of these unexpected trends resulted in a stoppage of works and prompted a detailed investigation to evaluate the validity of the VWP readings. The investigation included cone penetration testing (CPT), dissipation testing and installation of additional VWPs adjacent to those showing unexpected trends. The testing methodologies involved saturating the CPT pore pressure disc with both silicone oil and then glycerine after observing an unusual dissipation curve. The results at various locations showed differences based on the saturation fluid used. Findings using glycerine generally aligned with the expected trends and corroborated the data from the additional VWPs. This paper highlights and discusses the potential impact of saturation fluids in dissipation tests and the importance of assessing PWP trends and ensuring instrumentation accuracy for geotechnical monitoring.

INTRODUCTION

The monitoring of PWP and deformation within tailings dams is fundamental to ensuring their stability and operational safety. VWPs play a crucial role in this process by providing real-time pore water pressure data within the dam structure and its foundation. However, anomalies in VWP readings can raise concerns regarding their accuracy and the potential implications for dam stability. Misinterpretation of unexpected trends can lead to unnecessary actions and/or construction delays as well as potentially failing to recognise critical conditions, all of which pose significant risks.

This paper presents a case study from a heavily instrumented tailings dam where unexpected trends in VWP readings were observed during the construction of a stabilisation solution. These anomalous trends led to a temporary halt in construction activities and prompted a detailed investigation to assess the reliability of the readings. The investigation included CPT, dissipation testing, and the installation of additional VWPs to evaluate data consistency.

During the investigation, it was observed that the fluid used to saturate the CPT pore pressure disc appeared to influence dissipation behaviour. While this was not the primary focus of the study, the findings suggest that the type of saturation fluid may affect how PWP stabilises during dissipation tests. This observation merits further investigation to better understand its implications for geotechnical monitoring.

By presenting the methodology and findings of this investigation, this paper highlights the importance of critically assessing unexpected VWP readings and refining dissipation test procedures. The paper concludes with insights into improving instrumentation reliability and proposes further research into the effects of saturation fluids in dissipation testing.

BACKGROUND

The tailings dam discussed in this study was constructed over liquefiable materials – see Rola *et al* (2024) for further details. Due to the assessed low factor of safety (FoS), the dam was undergoing stabilisation construction works and was heavily instrumented to monitor its performance. The instrumentation included inclinometers, settlement points, Casagrande piezometers, and VWPs, all of which were strategically placed across the structure to monitor deformation and PWP throughout the works allowing real-time warnings of unexpected conditions.

Two different sets of alerts were defined for the instrumentation: i) Construction alerts, that served as early warnings if readings exceeded anticipated values due to construction activities; and ii) Dam Safety Emergency Plan (DSEP) alarms, that were directly associated with the FoS/stability of the dam.

Historically, DSEP alarms for this dam were based on data from Casagrande piezometers. As part of the instrumentation upgrade process, these piezometers were replaced with VWPs connected to a telemetry system to enhance the frequency of monitoring. The VWPs were installed adjacent to the Casagrande piezometers, within the historical tailings geotechnical unit, to allow direct comparison of readings. Many of these newly installed VWPs exhibited unexpected trends, recording PWP values significantly above expected levels. Investigation by the instrumentation subcontractor indicated that the instruments appeared to be functioning correctly with no identified instrument or electronics errors. Therefore these instruments were maintained and their trends regularly tracked.

ANOMALIES IN VWP READINGS

VWPs installed for the construction works were monitored hourly to identify stabilised trends and compared to the expected PWP based on existing instrumentation, historical data, and site investigation results, including CPTu with dissipation tests and pumping test data from monitoring wells.

Several VWPs showed unexpected trends in the PWP readings. Many of the VWPs installed adjacent to historical Casagrande piezometers (CP-VWPs) recorded unexpectedly high PWP values that were inconsistent with the dip measurements from the Casagrande piezometers and other VWPs installed nearby within the same unit. As an example, Figure 1 shows the installed VWPs in section CH825, along with the projected position of nearby CP06-VWP and CP07-VWP.

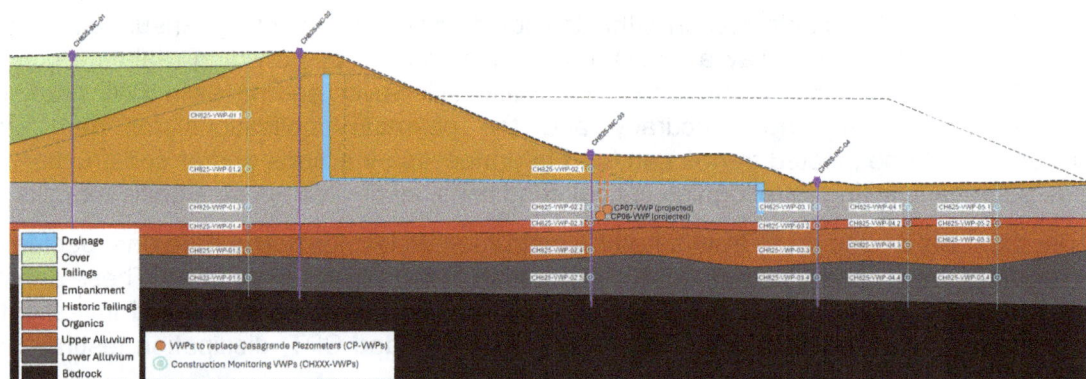

FIG 1 – Installed VWPs in section CH825 and projected nearby CP06-VWP and CP07-VWP.

A noticeable difference was observed between CP06-VWP and CP07-VWP in comparison with CH825-VWP02.2, despite all of the piezometers being very close to each other and located in the same geotechnical unit. Figure 2 presents a comparison of six VWPs installed in the toe berm at that location: the two CP-VWPs displayed unexpected trends, whereas four other VWPs (CH825-VWP-02.2 to CH825-VWP-02.5) followed the expected trends. This pattern was observed in several sections across the dam. Initially, these anomalous readings were disregarded, but daily monitoring continued to track their evolution.

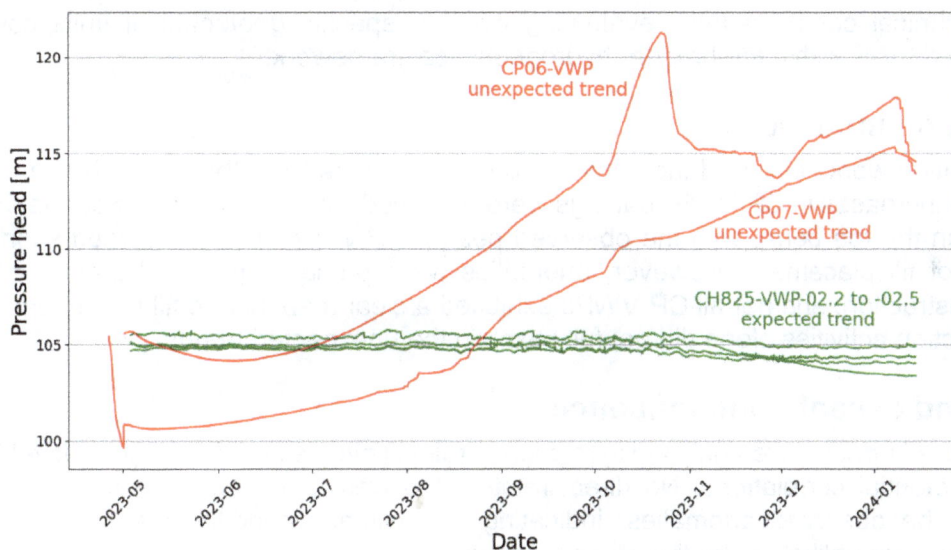

FIG 2 – Comparison of some VWPs with unexpected trends.

A sudden rise in PWP in CH980-VWP-1.3 (Figure 3) was observed during construction work and triggered a construction alarm. This instrument was installed within the historical tailings below the dam crest, where no significant PWP increase was expected. The increase in PWP at this location raised doubts in the reliability of VWPs installed within the historical tailings, a critical unit for dam stability. As many VWPs were recording unexpected trends in this geotechnical unit, the alarm led to a temporary halt in construction activities in the area to allow investigation into these anomalies.

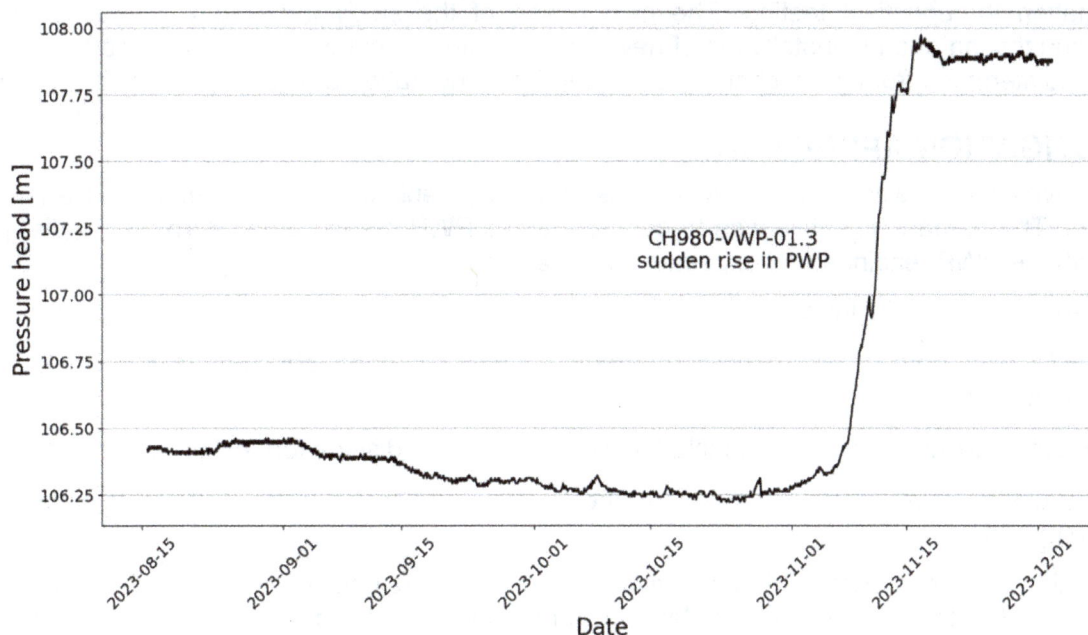

FIG 3 – Sudden rise in PWP in CH980-VWP-1.3.

DESKTOP REVIEW

A comprehensive desktop review was conducted to evaluate potential factors contributing to the unexpected trends and high PWP readings within the historical tailings unit. The review included:

- construction activity timeline analysis: comparing the elevated PWP readings with construction activities to identify potential triggers

- hydrological influences: assessing the impact of rainfall events and fluctuations in the dam's decant pond level on the VWP readings through cross-comparison of data

- geotechnical considerations: evaluating whether specific geotechnical units could support confined PWP within the broader hydrogeological framework.

Construction influences

The stabilisation works included the construction of a buttress fill at the toe of the dam and its toe berm. Some increases in CP-VWP readings were observed within a time frame corresponding to fill placement on the toe berm, with the observed peak PWP values being recorded shortly after the completion of fill placement. However, anomalies were present prior to the commencement of buttress construction, and not all CP-VWPs exhibited a clear response to fill placement, suggesting that construction activities alone did not fully explain the unexpected trends.

Rainfall and decant pond influence

The unexpected trends were analysed in relation to rainfall events and decant pond level fluctuations to identify potential correlations. No direct relationship was found between these environmental factors and the observed anomalies, indicating that climatic conditions and operational water management were unlikely to be the primary causes.

Conceptual considerations

The measured PWP values suggested that water levels were significantly above the dam's drainage blanket, raising concerns about the reliability of the readings. For these unexpectedly high PWPs to persist, a continuous aquitard would need to exist across the dam footprint to isolate pressure zones from the drainage system. However, this condition was not anticipated based on site stratigraphy and historical investigations.

Despite these desktop assessments no definitive explanation emerged, necessitating further investigation through field testing. The next phase of the study involved CPT soundings with dissipation testing and the installation of new VWPs to directly compare readings and assess if the anomalies stemmed from geotechnical behaviour, instrumentation errors, or procedural factors.

INVESTIGATION APPROACH

A field investigation was designed to assess PWP adjacent to the VWPs exhibiting anomalous readings. The primary objective was to compare static PWP measurements from CPT dissipation tests with the VWP readings and expected PWP values.

The investigation plan included:

- CPTs with pore pressure measurements at five locations adjacent to the VWPs under investigation

- a total of 57 dissipation tests, including seven overnight dissipation tests

- installation of additional VWPs using the CPT rig and the 'push-in' method to obtain direct comparative readings.

The location of the conducted CPTs and investigated VWPs are shown in Figure 4. Table 1 presents the tests conducted in chronological order, the corresponding investigated VWPs, and the saturation fluid used for the dissipation tests.

FIG 4 – Test locations and investigated VWPs.

TABLE 1

Summary of CPT conducted.

CPT ID	Target VPW	Saturation fluid
CPT23-CP6	CP06-VWP	Silicone
CPT23-CP10	CP10-VWP | CP10.2-VWP	Silicone | glycerine
CPT23-CP6B	CP06-VWP	Silicone | glycerine
CPT23-CP7	CP07-VWP	Glycerine
CPT23-CP10B	CP10-VWP | CP10.2-VWP	Glycerine
CPT23-CH980–01	CH980-VWP-01.3	Glycerine
CPT23-CH725–01	CH725-VWP-01.3	Glycerine
CPT23-CH980–01B	CH980-VWP-01.3	Glycerine

The CPTs were conducted from a 20 t track-mounted rig equipped with a high-resolution compression cone with a 15 cm^2 tip area, a u_2 pore pressure filter location, and a net end area ratio of 0.8. Testing followed standard procedures and maintained a penetration rate of 20 mm/s.

Each CPT location was first advanced using a high-capacity cone (150 MPa tip resistance and 1500 kPa sleeve friction) to reduce the risk of pore pressure disc desaturation when passing through unsaturated embankment fill. A lower-capacity cone (37.5 MPa tip resistance and 1000 kPa sleeve friction) was then used for dissipation testing at target depths.

Saturation of the CPT pore pressure sensor was performed using either silicone oil or glycerine, with on-site vacuum chamber saturation applied before each test. Overnight dissipation tests were conducted at selected depths to observe long-term trends.

The results from the CPT dissipation tests and the push-in VWPs were then compared to existing VWP data to assess the validity of the observed anomalies and to assess if the unexpected trends were instrumentation-related or reflected actual geotechnical conditions.

SATURATION FLUID INFLUENCE

Initially, silicone oil was used as the saturation fluid for the PWP discs. The first location investigated was CPT23-CP6. At this location, the dissipation tests were completed once the expected static PWP values were reached, and a reasonably stable dissipation curve was presented.

At the second location (CPT23-CP10), the initial dissipation tests in the coarse historical tailings above the groundwater level proceeded as expected. However, during the dissipation test at elevation RL 104.30 m in the fine historical tailings, the PWP exhibited an atypical trend. Instead of stabilising, a minimum PWP was reached, followed by a gradual increase in PWP (Figure 5a). The anticipated pressure value at this depth was 5 kPa. Initially, the pressure remained stable for the first 20 mins at 12.6 kPa, but then began to rise steadily. The test was terminated after 113 mins, with the final recorded pressure reaching 21.8 kPa.

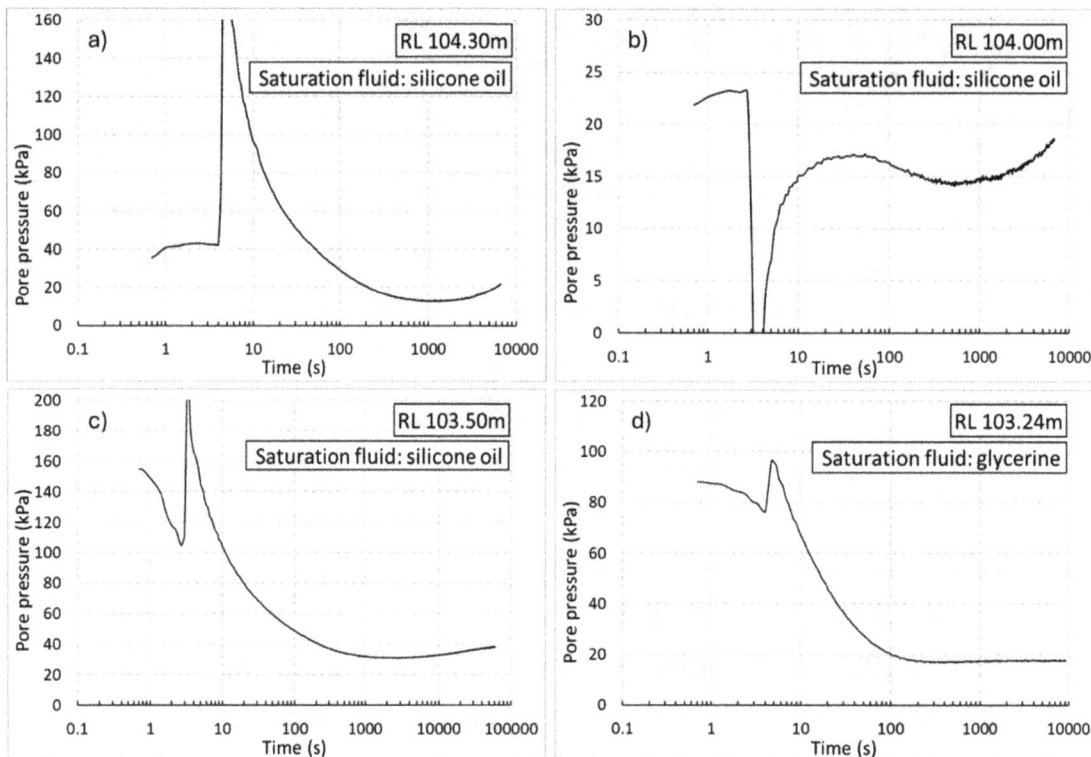

FIG 5 – Dissipation tests in CPT23-CP10 with different saturation fluids.

Two additional dissipation tests were performed at greater depths within the same fine historical tailings unit, using silicone oil as the saturation fluid (Figure 5b and 5c). Both tests displayed a consistent atypical trend of a continuous gradual increase in PWP. The first test, conducted at an elevation of RL 104.0 m, was terminated at 115 mins, with a recorded PWP of 18.6 kPa. The second test, conducted at elevation RL 103.50 m, was terminated after 17 hrs (overnight dissipation), with a final recorded PWP of 38.3 kPa. In both cases, no static or stable PWP was achieved when the test was terminated.

A final dissipation test was conducted at CPT23-CP10 using a glycerine-saturated disc. The pressure stabilised within the first 8 mins at approximately 18 kPa, and the test was run for a total of 154 mins without any upward trend (Figure 5 d).

Three additional dissipation tests using glycerine as the saturation fluid were performed at CPT23-CP10B, located approximately 500 mm south of CPT23-CP10. These tests were conducted at the same depths where the atypical PWP response had been observed with silicone oil saturation fluid at CPT23-CP10 (Figure 6). The first two tests were run for 90 mins, and the third test was run for 14 hrs (overnight dissipation). The PWP stabilised at the expected values in all three tests.

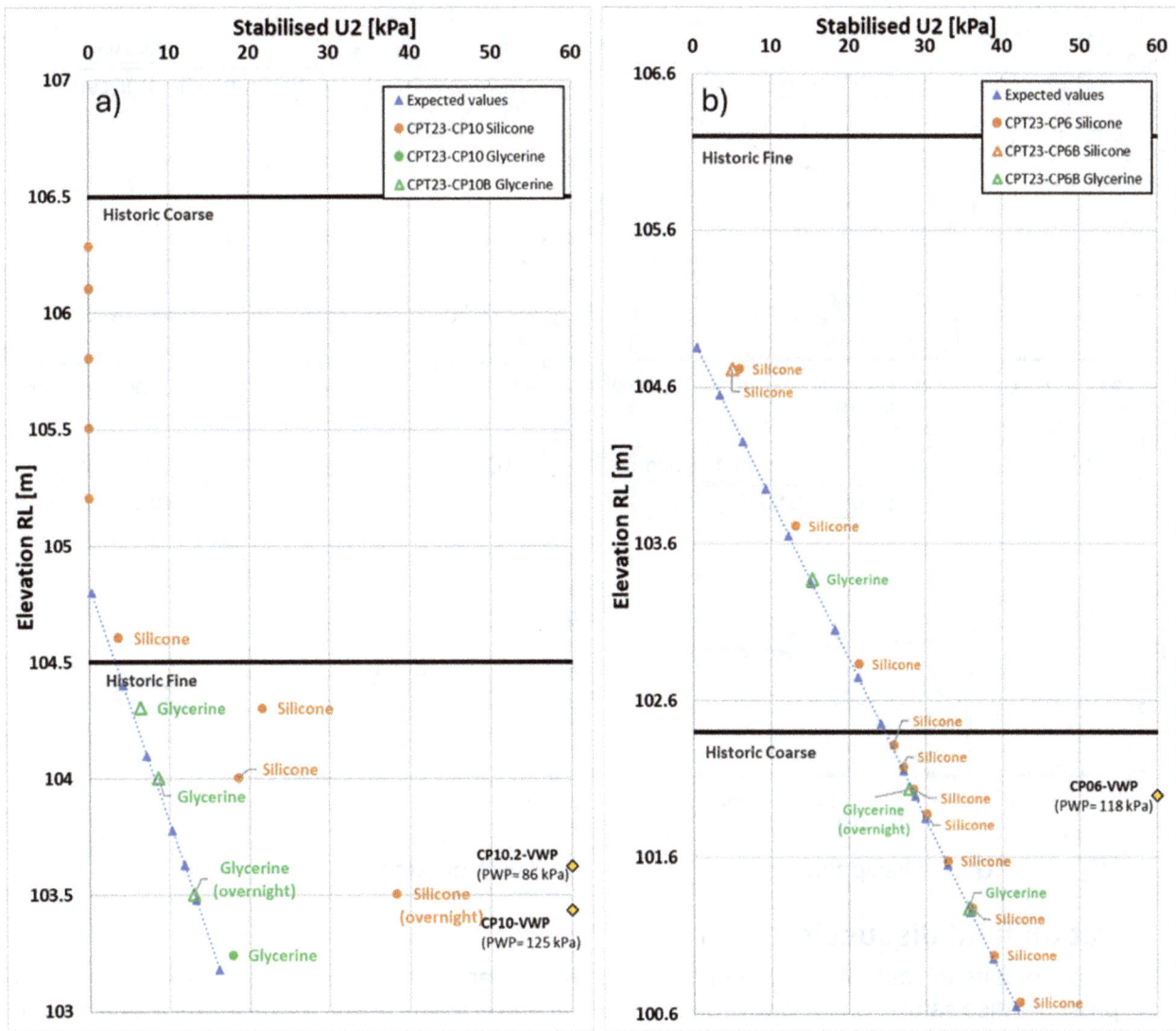

FIG 6 – Dissipation tests: a) CPT23-CP10 and CPT23-CP10B; b) CPT23-CP6 and CPT23-CP6B.

Following this outcome in CPT23-CP10, the tests at the initial location (CPT23-CP6) were re-run to evaluate if a PWP rise would be evident in the historical tailings for tests run over longer periods. A similar set of tests was conducted at CPT23-CP6B. The first dissipation test, using silicone oil as the saturation fluid at elevation RL 104.71 m in the coarse historical tailings, was run for 95 mins and exhibited an atypical upward trend. Three additional dissipation tests using glycerine as the saturation fluid were conducted at elevations RL 103.37 m, RL 102.03 m, and RL 101.27 m. The test at RL 102.03 m ran overnight, and in all cases, the PWP stabilised at the expected values within the first 5 mins (Figure 7).

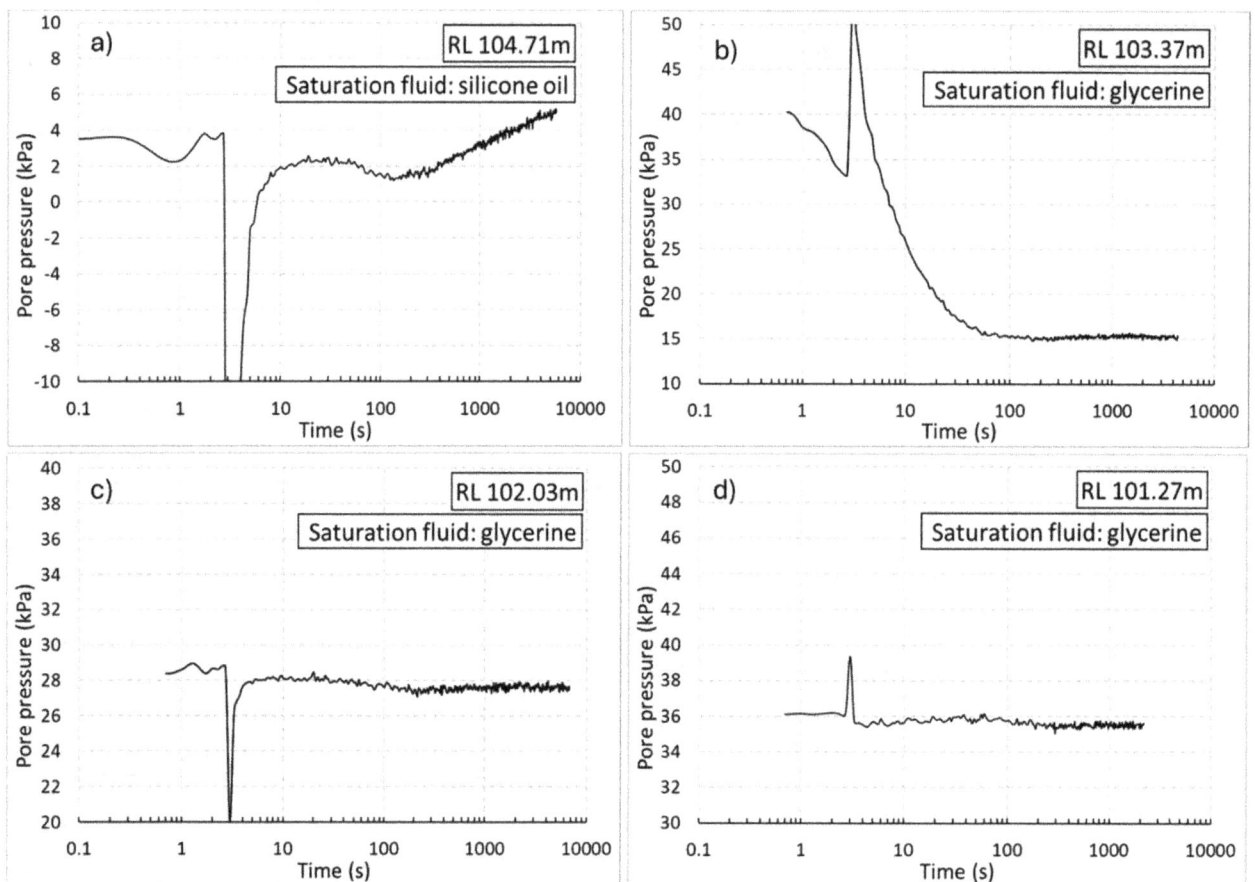

FIG 7 – Dissipation tests in CPT23-CP6B with different saturation fluids.

Saturation fluid discussion

The above results indicate that varied PWP responses were evident at the site depending on the pore pressure disc saturation fluid, with glycerine providing results consistent with the expected value based on measurements from other instruments. Vermeulen and Archer (2023) have also reported varied PWP response within tailings during dissipation tests using silicone oil and glycerine saturation fluids. Similar to this case study, they found unexpected dissipation responses when using silicone oil and that glycerine provided similar static PWP readings comparable to measurements from adjacent VWPs. Vermeulen and Archer (2023) note that while the poor performance of silicone oil is not fully understood, it is believed that its higher viscosity, stronger surface tension, and water insolubility contribute to the formation of a persistent meniscus at the porous disc interface, which hinders the complete dissipation of excess pore pressure.

In the authors' experience, dissipations tests in tailings typically follow the following types (Figure 8) as described by Sully *et al* (1999):

- Type 1: Dynamic PWP greater than equilibrium PWP, with a monotonic decrease of initial dynamic PWP.

- Type 3: Dynamic PWP greater than equilibrium PWP, with initial PWP increasing after commencement of the dissipation test followed by a monotonic decrease of PWP.

- Type 5: Dynamic PWP less than equilibrium PWP, with a monotonic increase of initial dynamic PWP.

However, similar to this case study, the authors have experienced atypical dissipation responses across many tailings storage facilities and tailings types when aiming to estimate static PWP.

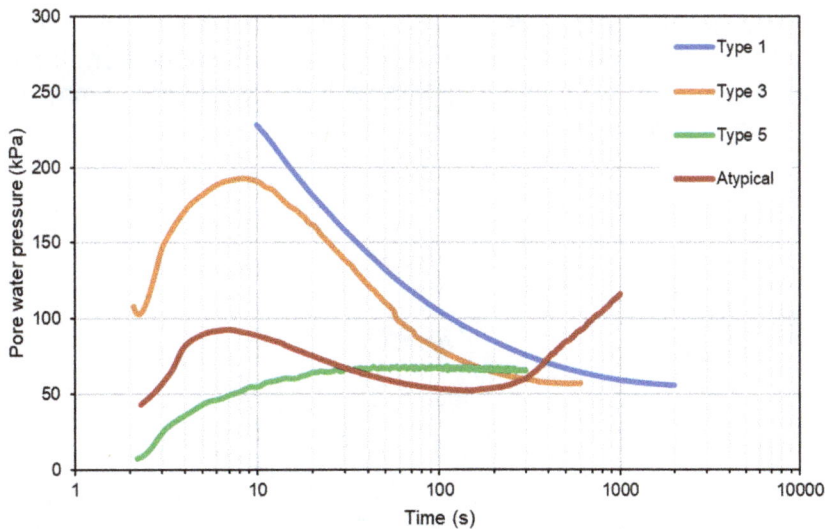

FIG 8 – Dissipation test types.

Substantial discussions and testing with the CPT contractors regarding varied saturation fluids and pore pressure discs (sintered metal filter) have not provided a clear answer on what is driving the atypical dissipation response. Although the response in this case study appears to be related to the silicone oil, the authors have experienced atypical responses when using glycerine as a saturation fluid. Consistent with Vermeulen and Archer (2023), research is encouraged to investigate this phenomenon as the authors consider it is being experienced by many practitioners with only limited discussion in the literature.

ADDITIONAL VWPS

A VWP was installed at the location of CPT23-CP10 using the push-in method as part of the PWP investigation. This VWP (CP10.3-VWP-A) was positioned between the elevations of CP10-VWP and CP10.2-VWP, corresponding to the depth range where the overnight dissipation tests were conducted at CPT23-CP10 and CPT23-CP10B.

Figure 9 compares the pressure head values from this newly installed push-in VWP with historical data from the standpipe CP10 and VWPs CH980-VWP-02.3 to CH980-VWP-02.5, which were installed nearby. The data show that the measurements from the push-in VWP align closely with the expected values and the results from the dissipation tests conducted with glycerine at CPT23-CP10 and CPT23-CP10B.

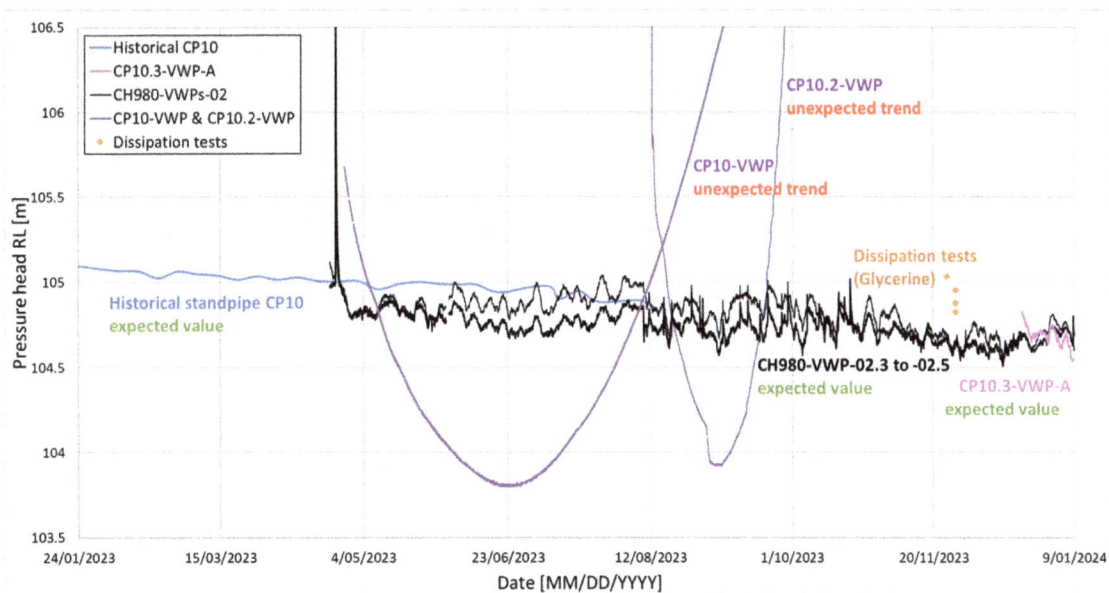

FIG 9 – CP10.3-VWP-A pressure head comparison.

In total, 16 additional push-in VWPs were installed. Figure 10 displays data from these VWPs in relation to other existing VWPs. The results show that the PWP values from the push-in VWPs fall within the expected range, further supporting the decision to disregard any VWPs that exhibit unexpected trends or anomalous values.

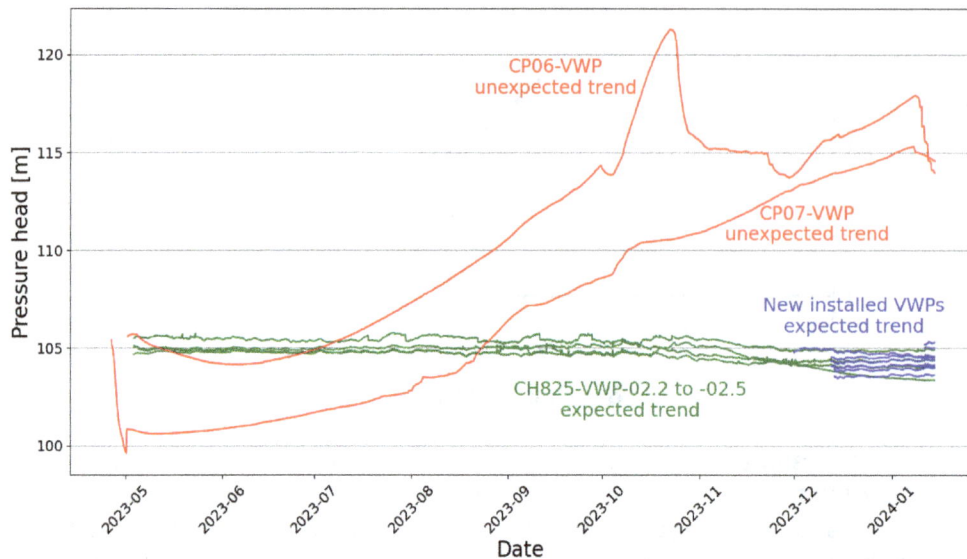

FIG 10 – Newly installed push-in VWPs compared with some previously installed VWPs.

CONCLUSIONS

The instrumentation of tailings dams is crucial for monitoring their stability and ensuring safe construction and operational practices. A comprehensive approach to data interpretation is essential. Relying solely on individual instruments without considering the broader data set may lead to misinterpretations and incorrect assessments of dam performance, which can result in unnecessary actions and/or construction delays as well as potentially failing to recognise critical conditions.

In this study, unexpected trends in VWP readings prompted an investigation using CPT dissipation tests to verify the reliability of piezometer measurements. The investigation was complemented by the installation of additional VWPs using the push-in method. Both the dissipation tests and the additional VWPs confirmed the initial interpretation of PWP behaviour within the dam that was based on existing instrumentation, historical data, and previous site investigations.

During the investigation, the results from the dissipation tests suggested that the type of fluid used to saturate the CPT pore pressure disc may influence PWP dissipation behaviour. These findings aim to highlight potential responses so that practitioners can be prepared to investigate if they experience similar conditions at their projects. Further research is necessary to fully understand the implications of saturation fluid selection on CPT dissipation test results.

ACKNOWLEDGEMENTS

The authors would like to acknowledge Menard Oceania for its support and collaboration on this project. The authors also would like to extend their gratitude to the dam owners for enabling the investigative work and for their willingness to share the learnings documented in this paper.

REFERENCES

Rola, J, Sottile, M G, Rivas, N A, Roldan, L and Sfriso, A, 2024. Development of a 3D ground model to design the stabilisation of a dam founded on weak liquefiable ground, in *Proceedings of the 7th International Conference on Geotechnical and Geophysical Site Characterization* (International Center for Numerical Methods in Engineering: Barcelona).

Sully, J P, Robertson, P K, Campanella, R G and Woeller, D J, 1999. An approach to evaluation of field CPTU dissipation data in overconsolidated fine-grained soils, *Canadian Geotechnical Journal*, 36:369–381.

Vermeulen, N J and Archer, A, 2023. Brittleness of iron ore tailings – fact or artifact, a case study, in *Proceedings of Tailings and Mine Waste 2023*, pp 1457–1467 (The University of British Columbia and C3 Alliance Corp: Vancouver).

Tailings dam monitoring instrumentation system design and maintenance for improved functionality

R F Gleeson[1] and J White[2]

1. FAusIMM(CP), Principal, Spectrum Mining Consultants, Adelaide SA 5000.
   Email: rohan.gleeson@spectrummining.com.au
2. MAusIMM, Principal, Spectrum Mining Consultants, Adelaide SA 5000.
   Email: joshua.white@algona.com.au

ABSTRACT

Tailings storage facilities (TSFs) represent one of the highest-risk assets in the mining industry, where failures can result in catastrophic environmental, social, and economic consequences. Although dam safety is increasingly recognised as a critical component of responsible mining, monitoring systems often lack the design rigour and investment afforded to other safety-critical functions.

This paper applies a systems engineering approach to the design, management, and life cycle maintenance of dam monitoring instrumentation systems. It outlines a management framework that incorporates critical element identification, redundancy planning, documentation, cyber security integration, preventative maintenance, and disaster recovery planning.

A case study from a large-scale operational site is presented, demonstrating measurable improvements achieved through structured implementation of the framework. Within 12 months, the site achieved a 67 per cent reduction in operational downtime and a 30 per cent reduction in alarm occurrences, significantly improving operational efficiency and risk governance.

The results highlight that even complex, organically grown monitoring systems can be transformed into resilient, reliable, and transparent safety-critical systems through deliberate, structured management practices.

INTRODUCTION

Tailings storage facilities (TSF) around Australia, both operational and decommissioned represent a portfolio of structures that require careful and considered management. TSF's incorporate multiple and varied risks that have the potential to affect environmental, social, and economic outcomes of a region, often with catastrophic effect (Australian National Committee on Large Dams Incorporated (ANCOLD Inc), 2019). The significant collapses at Cadia, Brumadinho, and Samarco forced the global mining industry to confront the serious inadequacies in its TSF management frameworks and performance standards. This recognition led to the release of the Global Industry Standard for Tailings Management (GISTM) (Global Tailings Review (GTR), 2020). The GISTM calls for a greater level of rigour and transparency in the management framework of tailings storage facilities as defined by the six key areas shown in Figure 1.

Among these six areas, monitoring and surveillance is fundamental to ensuring the early identification of abnormal dam behaviour. However, in practice, monitoring systems often lack the design rigour, life cycle planning, and integration afforded to other critical safety functions. Monitoring instrumentation systems frequently evolve organically, leading to reduced reliability, operational inefficiencies, and governance challenges.

This paper addresses the monitoring and surveillance of tailings facilities and suggests management practices that can assist operators in meeting best practice. Furthermore, the monitoring framework presented is designed to meet the requirements of monitoring across the life cycle of the facility and ensuring the monitoring system is robust and reliable, thereby engendering confidence in decision-makers. A systems engineering approach has been applied to the dam monitoring instrumentation system to ensure high levels of reliability, redundancy, and documentation. Learnings from high-reliability industries such as aviation have been incorporated into the design philosophy. This paper presents key principles addressed in the system design philosophy and presents an anonymous case study to highlight the benefits and improvements that can be realised by implementing this approach.

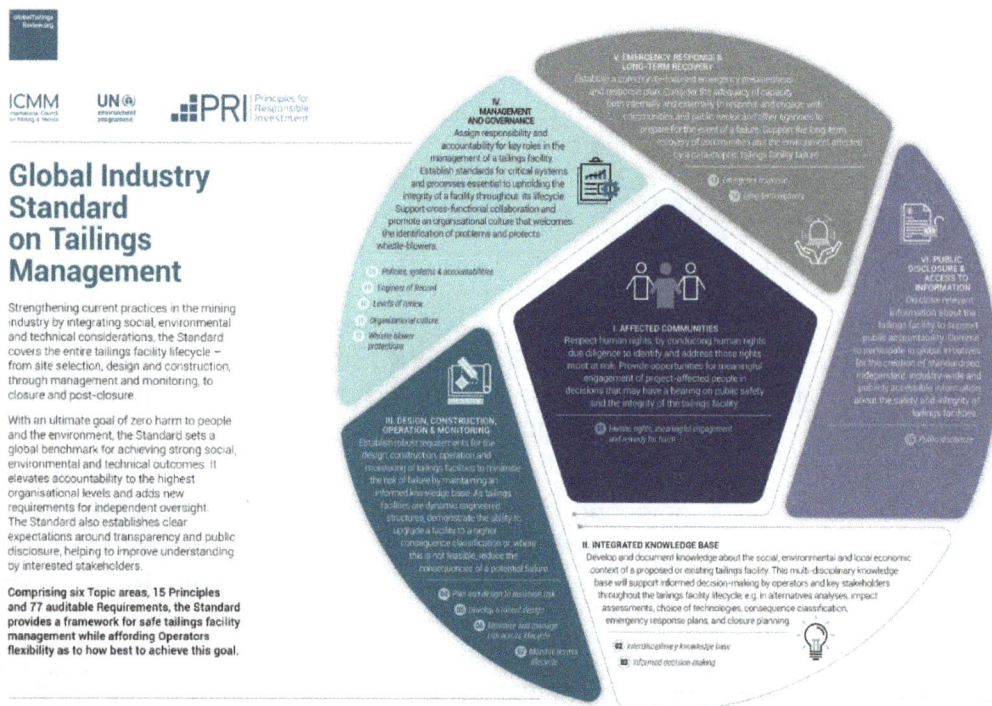

FIG 1 – Global Industry Standard on Tailings Management (GISTM) – Management Framework. Source: GTR (2020).

DESIGN PHILOSOPHY AND CRITERIA

The dam monitoring instrumentation (DMI) system is a safety-critical system that must be deliberately designed, maintained, and governed throughout the life of a facility (International Commission on Large Dams (ICOLD), 2001). To ensure its effectiveness, a systems engineering approach has been applied, drawing on lessons from high-reliability industries. The design philosophy is underpinned by four core principles:

1. Reliability

A reliable DMI system provides consistent, accurate data to decision-makers and enables early detection of abnormal conditions. It must remain operational during critical phases such as embankment construction, post-seismic events, or following heavy rainfall.

To achieve this, the system should:

- Target ≥98 per cent availability, allowing for reasonable unplanned maintenance periods.
- Include redundancy (minimum N+1) for all safety-critical elements (SCEs).
- Be simplified to reduce points of failure and interdependencies.
- Be hardened against environmental damage (eg fauna, moisture, UV exposure).

The design must also consider the expected operational lifespan of each instrument. Where sensors are inaccessible (eg buried VWPs or SAAVs), a minimum OEM-rated lifespan of 20 years should be targeted. Documentation of all existing instrumentation should be reverse-engineered to ensure whole-of-system coherence and minimise legacy performance issues.

2. Safety

The DMI system plays a fundamental role in risk mitigation by verifying the performance of dam design and construction. It must provide early warnings with minimal delay.

To maximise safety:

- Alert times must be minimised, particularly during construction or periods of elevated downstream population at risk.

- Instrumentation should be physically protected against damage from ongoing or future site activities.

- Sensor cabling must be designed to prevent mechanical, moisture, and fauna-related degradation.

- Instrument placement and trigger thresholds should align with identified failure modes from FMEA studies.

In addition, critical elements must be identified and documented in the Instrument Register, with clear linkage to the facility's Trigger Action Response Plan (TARP).

3. Cost-efficiency

A well-designed DMI system reduces total financial cost of ownership while maintaining critical performance. Cost-efficiency is achieved by:

- Integrating instrumentation planning into future embankment designs to reduce relocation and retrofit costs.

- Standardising equipment types (eg dataloggers, sensors, communications systems) to lower spare part inventory, simplify technician training, and reduce complexity.

- Carefully setting trigger thresholds to prevent unnecessary operational interruptions due to false positives.

Proactive investment in system design avoids reactive costs and operational disruptions later in the facility's life.

4. Documentation and transparency

Robust documentation and transparency are essential to ensure system governance (Pells, 2017), facilitate knowledge transfer, and support regulatory compliance. The DMI system must include:

- Comprehensive as-built documentation of all existing and new instrumentation.

- Clear records of system design rationales, maintenance schedules, and calibration requirements.

- Active management of instrument registers, network addresses, maintenance logs, and performance tracking.

Thorough documentation ensures that instrumentation data is trustworthy, and that system performance can be independently verified, particularly as facility management teams evolve over time.

INTEGRATION WITH FACILITY DESIGN REQUIREMENTS

The DMI system design philosophy complements the surveillance and monitoring requirements identified by the facility designer through risk classification, failure modes effects analysis (FMEA), and regulatory obligations. It extends beyond specifying data types and trigger thresholds by ensuring that instrumentation selection, installation, maintenance, and redundancy are deliberately engineered to deliver safe, reliable, and cost-effective monitoring across the facility's entire operational and closure life cycle.

DMI MANAGEMENT FRAMEWORK

The DMI Management Framework, referenced in Figure 2, includes the development of a hierarchy of documentation that defines all aspects of the design, procurement, installation, management, and decommissioning of the DMI system. This framework guides the development of the management system and identities specific documentation required. The intention of this framework is to integrate with and support the legislated Operation, Maintenance and Surveillance Manual.

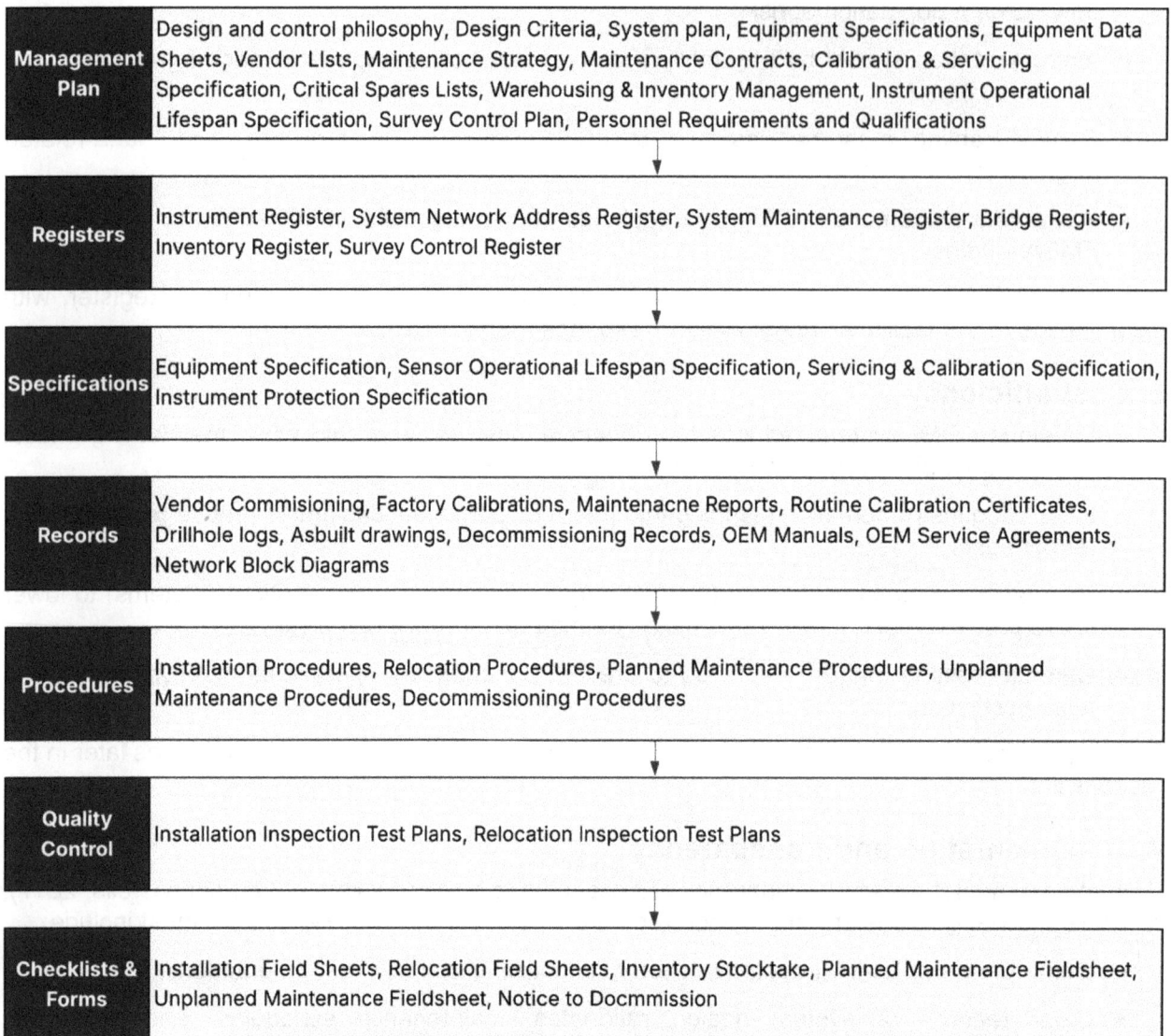

Management Plan	Design and control philosophy, Design Criteria, System plan, Equipment Specifications, Equipment Data Sheets, Vendor LIsts, Maintenance Strategy, Maintenance Contracts, Calibration & Servicing Specification, Critical Spares Lists, Warehousing & Inventory Management, Instrument Operational Lifespan Specification, Survey Control Plan, Personnel Requirements and Qualifications
Registers	Instrument Register, System Network Address Register, System Maintenance Register, Bridge Register, Inventory Register, Survey Control Register
Specifications	Equipment Specification, Sensor Operational Lifespan Specification, Servicing & Calibration Specification, Instrument Protection Specification
Records	Vendor Commisioning, Factory Calibrations, Maintenacne Reports, Routine Calibration Certificates, Drillhole logs, Asbuilt drawings, Decommissioning Records, OEM Manuals, OEM Service Agreements, Network Block Diagrams
Procedures	Installation Procedures, Relocation Procedures, Planned Maintenance Procedures, Unplanned Maintenance Procedures, Decommissioning Procedures
Quality Control	Installation Inspection Test Plans, Relocation Inspection Test Plans
Checklists & Forms	Installation Field Sheets, Relocation Field Sheets, Inventory Stocktake, Planned Maintenance Fieldsheet, Unplanned Maintenance Fieldsheet, Notice to Docmmission

FIG 2 – Dam Monitoring Instrumentation Management System.

SAFETY CRITICAL ELEMENTS AND INSTRUMENT CRITICALITY

Process safety literature defines a safety critical element (SCE) as a control measure, the failure of which could lead to a material unwanted event. In the context of a tailings storage facility a simple bow-tie analysis identifies the material unwanted event as embankment failure, the controls being the design and construction verification. The DMI is thus the control verification as it monitors the ongoing performance and health of the facility. For the purpose of this paper, Critical dam monitoring instruments and/or portions of the DMI system can be considered SCE's. SCE's can include sensors as well as power supplies and communications links that are vital to the performance of the system and are determined by the system architecture and instrument criticality. Where possible, SCEs should be reduced to only include critical sensors by ensuring adequate redundancy in the power and communications systems. This can also be achieved by decentralising power and communications wherever possible. A simplified system with reduced layers of complexity thus achieves are higher level of reliability and availability.

Furthermore, not all sensors are necessarily considered critical. As a facility ages, additional sensors are often installed following additional construction phases or deformation events. This organic growth of the system should always occur following the DMI management framework to ensure integration of new sensors into the existing system. This long-term evolution of the system can often lead to a situation where not all sensors are considered critical to the monitoring of the facility. All sensors provide value and can be used to provide a broad understanding of the facility's health; however critical instruments are defined as those directly monitoring failure modes identified during the FMEA that have been classified as possible or likely. These sensors should be identified in the Instrument Register and should inform the TARP. Thus, the determination of critical instruments should play a key role in the subsequent identification of SCE's. An item of hardware that has been identified as an SCE is subject to greater requirements for lifespan, maintenance efforts and redundancy.

INSTRUMENT LIFESPAN

Instrument lifespan is determined by instrument criticality and any SCE designation. Where sensor replacement is restricted, such as those that are buried, a minimum OEM lifespan of 20 years should be targeted. End-of-life replacement of critical instruments should be scheduled and budgeted at the time of system design. Sensor replacement should occur on schedule or on failure, whichever occurs first.

Sensor lifespan can be impacted by installation methodology and the quality of the installation. All care should be taken during installation to ensure the maximum lifespan can be achieved. Detailed record keeping of installations assists with identifying installation practices that may be leading to reduced instrument lifespan.

REDUNDANCY AND STANDARDISATION

At a minimum, all SCEs should be designed and installed with redundancy level of N+1. Where redundant hardware is installed, the determination of standby configuration is determined by an analysis of a hypothetical failure event.

Where an outage of the primary hardware will result in delays or expense that is not acceptable to the business, it is recommended to operate redundant hardware in the Hot Standby configuration. In this situation both the primary and redundant systems are powered on and operating simultaneously, with the backup system continuously mirroring or tracking the primary system's performance. If the primary unit fails, the redundant system immediately takes over with minimal or no interruption. This approach shortens the operational lifespan of the redundant hardware but avoids any data loss or delay downtime.

The preference for standby configuration is cold standby. As seen in Figure 3, a cold standby configuration, the redundant system remains powered off or in a passive state until the primary system fails. Upon failure, the backup system is manually or automatically activated. This maximises the lifespan of the redundant hardware, minimises cost, reduces complexity and has a lower risk of impacting data integrity. However, slightly longer recovery time compared to hot standby, as activation processes are required.

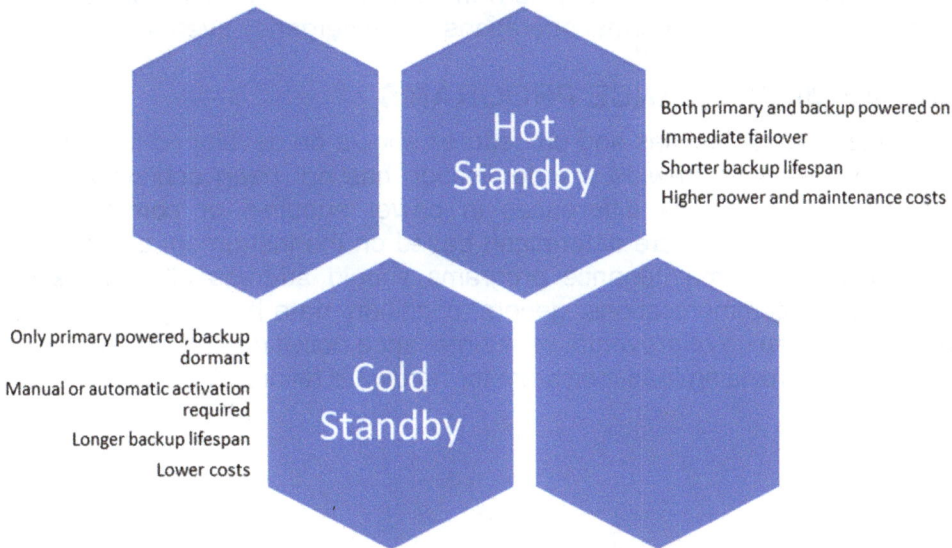

Both primary and backup powered on
Immediate failover
Shorter backup lifespan
Higher power and maintenance costs

Only primary powered, backup dormant
Manual or automatic activation required
Longer backup lifespan
Lower costs

FIG 3 – Hot standby and cold standby assessment.

Standardisation of the DMI system is a highly effective method to reducing complexity and minimising cost. For example, by standardising dataloggers, the number of models required to be stored as crucial spares is reduced and the total stock inventory is reduced. Further benefits include reduced training requirements and improved response times from maintenance technicians as the hardware will always be recognised and familiar. Standardisation can be difficult, or impossible to realise for older facilities. Strategies to better achieve standardisation in the DMI of older facilities can include:

- New-for-old replacement strategy – as hardware malfunctions and requires replacement, it should be replaced with a standard model.

- Targeted upgrade strategy – as budget becomes available conduct targeted upgrade programs to achieve standardisation where possible.

- Consistency – avoid the temptation to buy the newest models. If an installed system is functioning maintain it with consistent sensor models, rather than the newest model.

SURVEY CONTROL

A well-designed and maintained survey control network is fundamental to the performance of dam monitoring instrumentation systems, supporting both geotechnical surveillance and construction activities. The quality of the survey control network directly determines the accuracy, precision, and reliability of instruments such as robotic total stations and monitoring prisms.

Survey control networks should be established by experienced surveyors with a strong understanding of geotechnical monitoring requirements. The design of the network must consider the scale of the facility, the precision required for monitoring deformation, and the operational conditions across the site.

Best practice involves designing a survey network with:

- Cross-braced triangles between control points to strengthen geometric integrity and reduce cumulative errors.

- Redundant backsights to ensure reliable positioning even if a control point is lost or damaged.

- Extended coverage to accommodate future facility expansions and monitoring needs.

All survey control points should be tied into the broader mine survey network and, where practical, linked to a known State Map Grid to ensure spatial consistency and to assist in emergency response scenarios.

Redundancy in the control network is critical. Loss or damage to a single control point should not impair the accuracy of the monitoring system. By implementing cross-bracing, redundant backsights, and multiple fixed control points, the survey network remains resilient to localised failures, ensuring ongoing data integrity during both normal operations and emergency events.

PREVENTATIVE MAINTENANCE PROGRAMS

Preventative maintenance schedules and procedures should be developed for each type or model of sensor in use. Where system-wide standardisation has not been achieved, this may require additional procedures to address differences in power supplies or communications set-ups. Preventive maintenance regimes are determined based on the instrument criticality, power supply and exposure. Preventative maintenance programs should address all aspects of the sensor including power supply, communications, general condition, data handling and software. Detailed and consistent record keeping of preventative maintenance activities can be used to identify trends in sensor performance, providing lead indicators for hardware failure.

AUTOMATION, REMOTE ACCESS, CYBER SECURITY AND TRIGGER LEVELS

Remote access and automation

To maximise system availability and operational efficiency, all sensors, dataloggers, and communication systems should allow for secure remote access. Remote connectivity minimises downtime during unplanned maintenance, facilitates rapid troubleshooting, and reduces personnel exposure to hazardous areas.

Where feasible, it is recommended to automate sensor data collection, reporting, and alarm generation. Automation improves response consistency, reduces decision fatigue, and eliminates manual errors in interpreting data streams. A well-designed automated system supports timely decision-making while freeing operational teams to focus on higher-level analysis.

Cyber security considerations

The increased use of remote access and automation introduces heightened cyber security risks. Given the safety-critical nature of dam monitoring instrumentation systems, cyber security must be treated as integral to system design and maintenance. Key protective measures include:

- Implementing secure access controls and multifactor authentication.
- Encrypting all data transmissions.
- Regularly updating and patching system firmware and software.
- Isolating operational monitoring systems from corporate IT networks where practical.
- Developing a specific cyber incident response plan for the monitoring system.

Maintaining strong cyber security protects the integrity of monitoring data, preserves system availability, and ensures that alarms and trigger actions are not compromised by malicious interference.

Trigger level design

Trigger thresholds must be carefully configured to reflect the facility's risk profile, expected behaviour, and historical instrument performance. Thresholds should:

- Be set above the known precision limits of the monitoring equipment.
- Account for natural noise, environmental variability, and seasonal fluctuations.

For example, prism monitoring thresholds must consider instrument precision, baseline distances, and atmospheric influences on measurement. If the required detection sensitivity cannot be achieved with existing instrumentation, alternative or supplementary monitoring methods should be explored to maintain early warning capability.

RECORD KEEPING, REGISTERS AND PERFORMANCE TRACKING

Consistent and thorough record keeping is highly valuable and can assist in developing the integrated knowledge base. Further, thorough record keeping at all levels allows the operation to analyse the performance of the DMI system at multiple levels. A data driven approach to identifying poor performance in different instrument types, or different segment of the system can assist with the diagnosis and rectification of underlying problems.

The Instrument Register records key details regarding the instrument installation and trigger thresholds. This document is a live document and is used to maintain an active record of the commissioned instruments in use. This is the primary document that informs the TARP.

The Network Address Register is used to record the network addresses associated with each instrument, datalogger and communications node. This document is primarily used to maintain a record of network addresses and is useful in diagnosing issues in the system.

The Maintenance Register records all instances of planned and unplanned maintenance. It is useful for identifying nodes in the system that are displaying ongoing poor performance as well as ensuring that maintenance requirements are complied with.

The Bridge Register records events where a Safety Critical element has been bypassed, either intentionally during maintenance or unintentionally during an outage. This Register can also be used to record TARP trigger events. These events are not classified as bridge events; however, it is useful to record these events. Analysis of unintentional bridge events can develop a lead indicator of instruments or nodes in the system that may be nearing end of life and require replacement. Analysis of past TARP trigger events is important for the facilities governance team to understand the health and ongoing performance of the facility.

The Survey Control Register records the details including location, accuracy, type, and installation details of each survey control point.

The above registers are most powerful when maintained regularly as live documents. Ensure to maintain a complete metadata record of all changes to the registers for transparency purposes. Furthermore, these registers are ideally maintained in a relational database or geomatic systems to assist with ease of data lookup and to facilitate simple and efficient reporting. This approach also allows the operator to establish automated alerts or alarms based on frequency of events for individual or groups of instruments. Thus, it can become a powerful tool for identifying trends of change in the facility performance that may otherwise be missed.

WAREHOUSING AND INVENTORY MANAGEMENT

Maintaining an inventory of critical spares is effective at minimising downtime and improving the efficiency of the technicians maintaining the system. It is crucial to consider items such as software or programming scripts as critical spares as well. Duplicates of each datalogger script should be maintained in a centralised storage location. This can avoid significant downtime and data loss during the replacement of dataloggers.

Maintenance of a stock of non-critical consumables is important, especially during times of construction and instrument relocations.

Regular stocktakes of critical spares and non-critical consumables ensures the DMI system can be adequately maintained and managed at all times during its life cycle.

PERSONNEL COMPETENCIES

Personnel involved in the design, installation, management, and maintenance of the DMI system should have a sound understanding of embankment dam monitoring practices and understand the principles regarding how and what each sensor is monitoring. Furthermore, these personnel should have a strong understanding of earthen embankment visual inspection and know the tell-tale signs of an embankment in distress. The DMI personnel should work closely with the governance team to ensure the best outcome for the facility is achieved.

Further competencies that are highly useful for the DMI team include:

- A sound understanding of drilling and investigative techniques used in and around embankment dams.

- A sound understanding of extra low voltage electrical systems.

- An understanding of wireless communications architecture and a basic understanding of IT network architecture.

- A moderate to strong understanding of programming languages such a python or C++.

- Competency in the use of UAVs to conduct inspections is desirable.

- Engineers involved in the design, planning, and management of DMI systems should have a strong understanding of earthen embankment dam design principles and considerations as well as a strong understanding of tailings management.

Training and upskilling of the DMI team should be addressed as a priority to ensure the required skillsets are captured within the team. The utilisation of project based contractors and consultants should be minimised to ensure a complete understanding of the system is maintained within the DMI team.

DISASTER PREPAREDNESS AND RECOVERY

A Disaster Preparedness and Recovery Plan (DPRP) for the DMI system should be established as a matter of priority. A large portion of the value that the DMI system provides is the data records that are maintained. This data provides valuable insights into the performance of the facility during phases of its life cycle and provides a record that transcends the movement of personnel in and out of the business. A well-established database and DMI system leave a clear and traceable record of facility performance thereby providing assurance to the business that the risks associated with the facility are managed now and into the future.

In the first instance, the DPRP should identify and document how the DMI system is prepared for and hardened against a system wide disaster. This process, similar to the SCE identification process, identifies key infrastructure in the system that poses a point of failure and where data integrity can be lost.

The DPRP should address a number of key scenarios, identified by a panel of subject matter experts and deemed as credible, that could lead to total or partial loss of the DMI system or data. A thorough understanding of the DMI system, and complete documentation of the system, is key to developing an effective Disaster Recovery Plan.

The DPRP should encompass all aspects of the DMI system, including all hardware and software on-site as well as all third-party suppliers that provide a service or host data. Data hosting suppliers should be requested to provide the principal their own Disaster Recovery Plan.

Further, the DPRP should identify and define a series of steps to be undertaken immediately and through to full recovery following a DMI disaster event. These steps should address items such as:

- How the facility will continue to be monitored during any extended period of system loss?

- How the DMI system will be managed during a partial or complete outage?

- How the DMI system will be recovered during a partial or complete outage?

- How the assurance of the data will be undertaken following the recovery of the DMI system?

CASE STUDY

This case study examines the application of the DMI Management Framework at a large-scale operational site in Australia. A snapshot of the monitoring system is provided to illustrate the scale, complexity, and unique challenges faced during implementation. Baseline system performance metrics prior to the framework rollout are also presented to highlight the opportunities for improvement and set the context for the results achieved.

System description and performance

The DMI system monitored five declared facilities spanning over 17 km^2 and 11 km of earthen embankments. It comprised more than 773 sensors, including robotic total stations, monitoring radars, doppler radars, shape array accelerometers, monitoring prisms, moisture sensors, vibrating wire piezometers, settlement systems, CCTV and thermal cameras, seismic arrays, earth pressure cells, and smart markers.

Installed organically over the preceding eight years, the system lacked a cohesive design philosophy or management strategy. This led to excessive complexity, with numerous interdependencies between nodes, poor documentation, and a reliance on a small number of personnel for system knowledge. Repairs were often delayed, and overall system reliability was poor.

By 2023, the DMI system was experiencing significant performance issues, averaging 93.3 hrs of downtime per month and totalling 1119 hrs across the year. During this period, 127 TARP events were recorded — with 75 per cent classified as grey alerts linked to power or communications failures, highlighting the critical impact of system instability on operational efficiency.

Management framework implementation

Challenges

The implementation of the dam monitoring instrumentation (DMI) management framework at the case study site encountered several notable challenges, reflective of the site's complex operational history and organically developed monitoring system.

Legacy system complexity

The existing DMI system had evolved over an eight-year period without a unified design philosophy, resulting in a highly complex and interdependent network of sensors, dataloggers, and communication nodes. Integration efforts required extensive reverse engineering and documentation to uncover undocumented installations and dependencies. The lack of a coherent as-built record significantly delayed the early phases of the framework roll-out.

Knowledge and documentation gaps

Knowledge of the system's configuration and functionality was concentrated among a small group of individuals, many of whom were involved in original installations. Due to incomplete documentation and inconsistent record-keeping, accurately mapping the system and assessing criticality of individual components proved challenging. Bridging this knowledge gap required significant stakeholder engagement and verification of historical installations.

Resource and budget constraints

Resource availability and budget prioritisation presented early hurdles. While the value of proactive maintenance, standardisation, and critical spares management was clear to the project team, securing funding and resourcing required extensive advocacy. Initial hesitation was evident to allocate resources toward long-term reliability goals when immediate operational pressures demanded attention.

Resistance to change

Cultural resistance also emerged as a major challenge. Maintenance teams were accustomed to reactive troubleshooting rather than structured preventative maintenance routines. Shifting practices to a proactive approach with greater emphasis on documentation, standardised procedures, and governance integration required not only training, but ongoing leadership support and reinforcement.

Cyber security risks

The increased focus on remote access and automation introduced heightened cyber security concerns. The existing system lacked standardised access controls, encrypted communications, and formal network segmentation. Retrofitting appropriate cyber security measures onto the legacy infrastructure demanded careful planning to avoid operational disruptions while ensuring data integrity and system resilience.

Data governance and integration

Consolidating multiple diverse data streams into a unified relational database was a technically complex undertaking. Differences in sensor types, communications protocols, and data handling practices had to be rationalised into a consistent structure. Establishing rigorous metadata standards and integrating historical data into the new system was critical to ensuring long-term data integrity and performance analysis capability.

Although implementation challenges were considerable, the application of the DMI Management Framework ultimately delivered significant improvements in operational performance, system reliability, and governance transparency.

This case study illustrates that while transitioning from an organically developed system to a structured, life cycle-oriented management approach requires perseverance and careful change management, the long-term benefits to dam safety and business performance are substantial. The realised improvements are outlined below.

Realised improvement

The structured implementation of the DMI management framework delivered significant and measurable operational improvements at the case study site within just 12 months.

Key outcomes included:

- A 67 per cent reduction in average monthly operational downtime associated with dam monitoring alarms and system faults.

- A 30 per cent reduction in monthly TARP (Trigger Action Response Plan) occurrences, significantly decreasing operational disruptions and the frequency of unnecessary alarm escalations.

- Reduction in total annual operational downtime from 1119 hrs to 369 hrs (evident in Figure 4), translating to a recovery of over 750 productive hrs across the calendar year.

- Reduction in total annual TARP occurrences from 127 to 89, reducing the number of events by 38 over the calendar year.

FIG 4 – Case study TARP performance metrics.

In addition to these quantifiable improvements, the site also realised several qualitative benefits:

- Increased confidence among operational and governance teams in the reliability of instrumentation data and alarm systems.

- Enhanced responsiveness to maintenance issues through improved remote access, preventative maintenance scheduling, and system documentation.

- Improved resource efficiency, with reduced reliance on specialist troubleshooting and a broader base of personnel trained in system maintenance and interpretation.

- Strengthened alignment with the Global Industry Standard on Tailings Management (GISTM) guidelines (GTR, 2020), positioning the site ahead of regulatory compliance timelines.

Critically, the reduction in false positive alarms and avoidable TARP escalations helped to refocus site attention on genuine dam performance risks rather than on system malfunctions. This strengthened the overall risk management framework, improved stakeholder assurance, and contributed to a demonstrable uplift in the site's dam safety culture.

These results highlight that even highly complex, organically grown monitoring systems can be transformed into reliable, resilient, and transparent safety-critical systems with the application of a disciplined management framework.

CONCLUSION

The management of tailings storage facilities demands not only technical excellence in dam design and construction but also a disciplined, proactive approach to ongoing monitoring and surveillance.

This case study demonstrates that the implementation of a structured Dam Monitoring Instrumentation (DMI) Management Framework can deliver significant improvements in system reliability, operational efficiency, and overall governance performance — even within complex, organically developed monitoring environments.

By applying systems engineering principles, identifying and protecting Safety Critical Elements, and embedding strong maintenance, cyber security, and data management practices, the site was able to recover over 800 productive hours annually and achieve a substantial reduction in false alarms and unnecessary operational disruptions.

Beyond measurable operational gains, the framework strengthened stakeholder confidence, improved regulatory alignment with standards such as the Global Industry Standard on Tailings Management (GISTM), and contributed meaningfully to the site's broader dam safety culture. Organisations that invest early in the design, documentation, and life cycle management of their dam monitoring systems are better positioned to manage risk, maintain compliance, and meet the evolving expectations of regulators, communities, and investors.

The mining industry must continue to shift from reactive to structured, proactive monitoring approaches if it is to sustainably manage tailings storage facilities and protect the communities, environments, and businesses that depend on their safe operation.

ACKNOWLEDGEMENTS

The authors would like to acknowledge the contributions of the site Operations and Maintenance teams for their support and collaboration throughout the implementation of the DMI Management Framework. Their operational insights and commitment to improvement were critical to the success of the project.

Appreciation is also extended to the Dam Governance Team and site Leadership for their support in resourcing and prioritising the framework rollout.

Thanks are given to the survey and instrumentation personnel whose efforts in system mapping, reverse engineering, and data verification enabled the foundation for successful integration.

Finally, the authors acknowledge Spectrum Mining Consultants for supporting the development of the framework and providing technical resources during implementation.

REFERENCES

Australian National Committee on Large Dams Incorporated (ANCOLD Inc), 2019. Guidelines On Tailings Dams - Planning, Design, Construction, Operation And Closure, revision 1, ANCOLD Inc. Available from: <https://ancold.org.au/product/guidelines-on-tailings-dams-planning-design-construction-operation-and-closure-revision-1-july-2019/>

Global Tailings Review (GTR), 2020. Global Industry Standard on Tailings Management (GISTM) [online], Global Tailings Review. Available from: <https://globaltailingsreview.org/global-industry-standard/>

International Commission on Large Dams (ICOLD), 2001. Tailings dams: risk of dangerous occurrences – Lessons learnt from practical experiences, *Bulletin 121,* International Commission on Large Dams.

Pells, D L, 2017. The missing link: Benefits realisation management – and welcome to this edition [editorial], *PM World Journal,* 6(8). Available from: <https://pmworldlibrary.net/wp-content/uploads/2017/08/pmwj61-Aug2017-Pells-missing-link-benefits-realization-editorial-welcome2.pdf>

Legacy tailings dams – asset or liability? A case study on sub-aqueous tailings reclamation

M Jones[1]

1. Senior Associate, Enable Advisory, Brisbane Qld 7007.
   Email: marie.jones@enableadvisory.com

ABSTRACT

Legacy tailings storage facilities exist within operating mine sites across Australia. These tailings storage facilities are often viewed as rehabilitation liabilities but with considered assessment may prove to be a resource. Extraction of such a resource might add value to operations by providing cash flow during the development of new mines; providing supplementary mill feed for existing mines undergoing step changes in operations; and supporting mine closure activities.

Hellyer Gold Mines is successfully extracting and reprocessing sub-aqueous material from legacy tailings storage facilities at their Hellyer operation in north-west Tasmania. The Hellyer tailings storage facilities contain poly-metallic material from the Hellyer and Fossey underground deposits dating back to the 1980s as well as material from tailings reprocessing activities by other operators of the Hellyer site. Hellyer Gold Mines is reclaiming sub-aqueous material using cutter-suction dredgers and reprocessing in the existing plant on-site.

Hellyer is the only operation of its type in Australia and few such operations exist worldwide. Key learnings around the estimation of a resource for legacy tailings storage facilities have been acquired at Hellyer that can be applied to the assessment of other tailings storage facilities.

The application of dredging as a mining method at Hellyer is delivered through collaboration with experts from outside the mining industry under the direction of a mining engineer. Key elements required for the planning and delivery of safe and efficient dredging inside tailings storage facilities have been identified at Hellyer and are presented to provide a road map to the development of other dredging operations.

INTRODUCTION

There are over 750 legacy tailings storage facilities on active and inactive mine sites across Australia. Inside these tailings storage facilities are valuable metals which, with advances in minerals processing techniques, it may now be possible to extract. Considered assessment of these tailings storage facilities may unlock new resources on existing mine sites.

Hellyer Gold Mines (HGM) is successfully extracting and reprocessing sub-aqueous material from legacy tailings storage facilities at their Hellyer operation in north-west Tasmania to produce lead, zinc and pyrite concentrates. The lessons learned from HGM's operations are documented here with a view to informing the development of future mineral resource and ore reserves estimates and the safe extraction of sub-aqueous tailings from other legacy tailings storage facilities.

HISTORY OF HELLYER TAILINGS STORAGE FACILITIES

The Hellyer mine site in the north-west of Tasmania has been active since 1983 as both underground and tailings reprocessing operations (Table 1). The Hellyer site comprises two separate underground voids; Hellyer and Fossey, and six sub-aqueous tailings storage facilities (TSFs) (Figure 1).

TABLE 1

Hellyer operations history.

Years	Entity	Operation type
1983–2000	Aberfoyle Resources Limited Western Metals Resources Limited	Underground (Hellyer Mine)
2006–2008	Polymetals Pty Ltd	Tailings reprocessing
2010–2012	Bass Metals Ltd	Underground (Fossey Mine)
2017–2025	Hellyer Gold Mines Pty Ltd	Tailings reprocessing

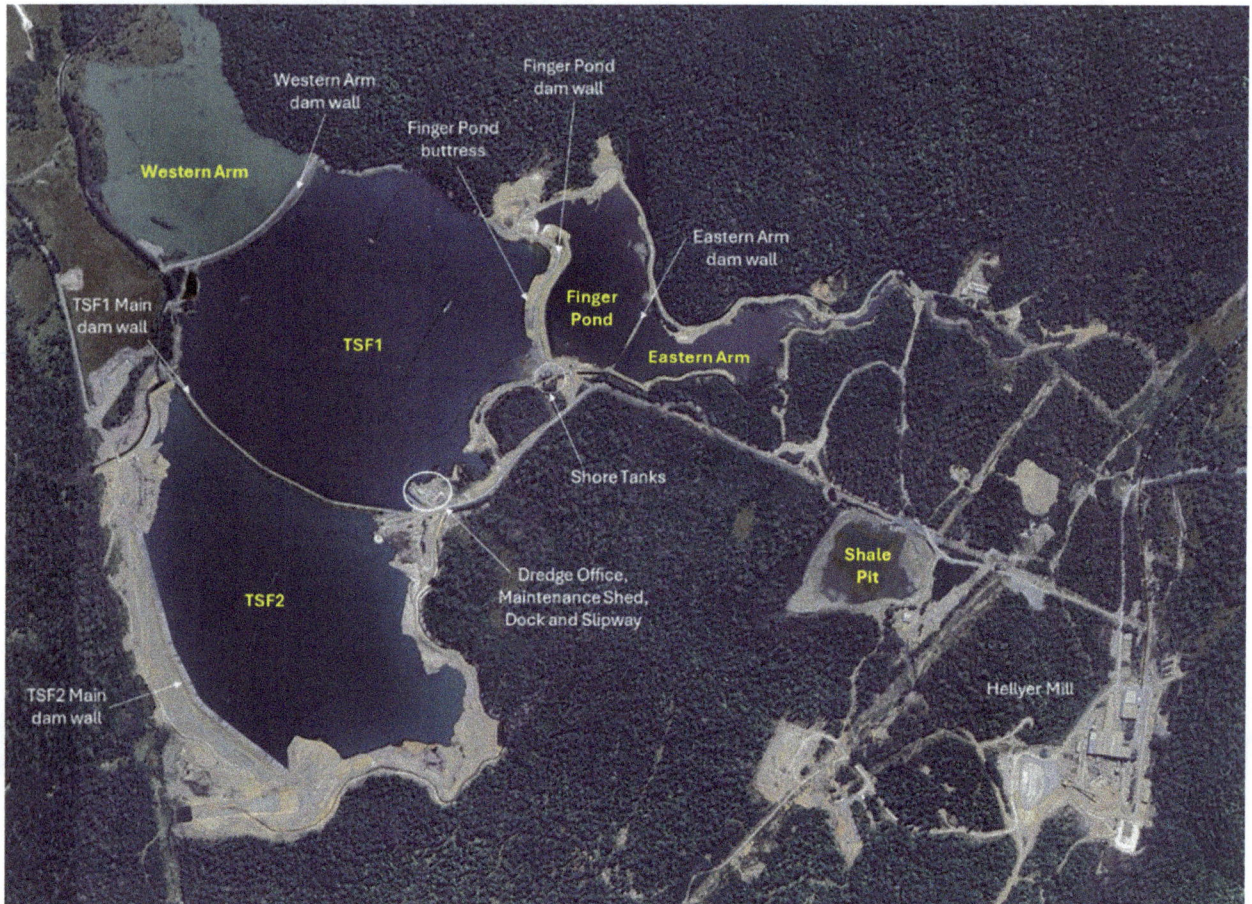

FIG 1 – Hellyer tailings storage facilities, November 2023.

Until the year 2000 all tailings were deposited sub-aqueously into TSF1 which at that time was a single body comprising the current TSF1, the Western Arm, the Fingerpond and the Eastern Arm (Figure 2).

The intermediary embankments: Western Arm Wall, Eastern Arm Wall, Fingerpond Wall and Buttress (Figure 1) were constructed between 2006 to 2008 over unknown quantities of *in situ* tailings. This was done to increase storage capacity without lifting the main TSF1 embankment (TSF1 wall). Consequently, the elevation of the Western Arm, Fingerpond and Eastern Arm TSFs are up to five metres above the waterline of TSF1.

During this period a small open pit (Shale Pit) was repurposed as a TSF to avoid returning tailings to TSF1. TSF2 was constructed during Hellyer Gold Mines' tailings reprocessing operations after capacity in the Western Arm, Fingerpond, Eastern Arm and Shale Pit TSFs was reached.

FIG 2 – Hellyer tailings storage facilities, March 1999.

Lessons learned – developing a mineral resource estimate for tailings

Resource boundaries

An accurate understanding of the resource boundaries is important not only for the mineral resource estimate but also for the safety of any future dredging operations. The data required to determine the resource boundaries is:

- an accurate dam floor survey
- as constructed models of dam walls
- bathymetry surveys (to capture the tailings surface underwater)
- drone surveys (to capture the tailings surface of any exposed tailings).

For legacy TSFs an accurate dam floor survey and as constructed models of the dam walls can be difficult to obtain. A detailed survey of the area may never have been undertaken; boundaries and depths of any borrow pits within the dam may not have been recorded or surveyed; and the dam floor may not have been cleared prior to commencing tailings deposition. Potential sources of data for legacy tailings dams include physical plans from the operations archives, local surveying firms and the mining regulator.

Bathymetry surveys establish the sub-aqueous surface of the tailings and are undertaken from small craft that can access areas with water depths as shallow as 0.33 m. The accuracy of bathymetry surveys is influenced by water turbidity and any tailings deposition activities should cease for a minimum of 12 hrs prior to the survey. Drone surveys can accurately capture the surface of any beached or exposed tailings.

Modelling the in situ tailings

A significant amount of data about the tailings within a legacy TSF is available without any drilling. In addition to the resource boundaries, the data which can assist the development of a mineral resource estimate is the operations historical:

- daily, weekly and monthly processing records, and specifically:
  - o the feed mineralogy
  - o the tail grade and tonnes
  - o tailings deposition points.
- process plant design
- mining production records (ie source orebodies)
- aerial photographs and surveys.

The above data together with the resource boundaries can be used to build spatial models of deposition periods with discreet tailings mineralogy and/or grade to guide the design of drilling programs and modelling constraints.

Broadly, there are two drilling methodologies available for tailings: vibracore and direct push. The selection of drilling methodology influences the quality of the data derived. Both methodologies have been employed in the Hellyer tailings: vibracore in 1998, 2000 and 2017; and direct push in 2021. The advantages and disadvantages of each in relation to sample recovery in the tailings at Hellyer are described in Table 2.

TABLE 2

Tailings drilling methodologies for sample recovery in the Hellyer tailings.

Methodology	Sample recovery	Data derived	Comment
Vibracore	Disturbed	Assay	<ul><li>Cost-effective</li><li>Quick</li><li>Maximum depth of sample recovery: 188 m</li><li>Water addition required to extract samples from tubes making them unsuitable for *in situ* bulk density determination</li><li>Larger samples of representing up to four vertical metres were recovered</li></ul>
Direct push	Disturbed and Undisturbed	Assay, CPT, *In situ* Bulk Density, PSD	<ul><li>Expensive</li><li>Time consuming (especially at depth)</li><li>Maximum depth of sample recovery: 222 m</li><li>Undisturbed samples can be recovered suitable for *in situ* bulk density determination</li></ul>

Based on the data derived from the Hellyer drilling programs, a combination of both drilling methodologies is recommended to reduce costs whilst maximising the available data points in legacy TSFs. Important considerations for drilling program design in tailings are the selection of:

- collar point location
- sample size (in vertical metres)
- sample interval
- sample recovery type.

The developed spatial models of deposition periods can be used to guide the selection of collar point location and sample size and interval to ensure that sufficient samples are recovered for each deposition period. The sample recovery type can be guided by the data required: disturbed samples are suitable for assay whereas only undisturbed samples are suitable for the determination of *in situ* bulk density.

The determination of the *in situ* bulk density (IBD) of tailings is challenging. A global calculation from the resource boundaries and historical records of deposited tailings can provide an average IBD. However, testing and reconciliation of the extracted Hellyer tailings indicate a correlation between depth and IBD and metal grade and IBD. It is recommended that the design of drilling programs in tailings provide for recovery of sufficient undisturbed samples in each deposition period to understand any correlation between depth, metal grade and IBD.

Drilling program design in tailings should also provide for Vane Shear, Cone Penetration (CPT), particle size distribution (PSD) and geochemistry testing for each deposition period. These tailings material properties are required for the development of an ore reserve estimate: Vane Shear and CPT for the modelling of tailings stability, CPT for dredge anchor selection, PSD for tailings slurry transport calculations and geochemistry to understand the impact of the tailings acidity and water quality on the wear of dredging equipment and infrastructure.

HELLYER DREDGING OPERATIONS

The Hellyer tailings are reclaimed using cutter-suction dredgers (CSDs) and the extracted tailings slurry is pumped via floating pipelines to holding tanks at the waterline of TSF1. The elements that make up the Hellyer dredging operation and their purpose are laid out in Table 3 and Figure 3.

TABLE 3

Hellyer dredging operation elements.

Element	Function
Dredger	Extraction of tailings – See Table 4, Figure 4
Dredge anchors	Controls the position of the dredger on the TSF and influences the extraction arc (sweep) Dredge anchors are: • located at the TSF waterline, including on TSF embankments (shore anchor) • placed on or in the tailings within the TSF (in-dam anchor)
Anchor lines	Steel cables that connect the dredger to the dredge anchors
Product lines	Floating pipelines to transport the extracted tailings slurry from the dredgers to the holding tanks
Trailing cable	Floated electrical cable supplying power to the dredge pump and dredger
Workboat	Support craft with the capacity to: • supply diesel to the dredge pump engine • lift and place in-dam anchors • transport crew and materials to the dredgers • re-locate the dredgers on the TSF
Holding tanks	Storage of extracted tailings slurry
Support craft	Transport of crew and materials to the dredgers; adjustment and repair of anchor, mooring and product lines; floating infrastructure inspections
Mooring lines	Controls the position of product lines and the trailing cable on the TSF
Gates	Allows passage of support craft over dredge anchor lines

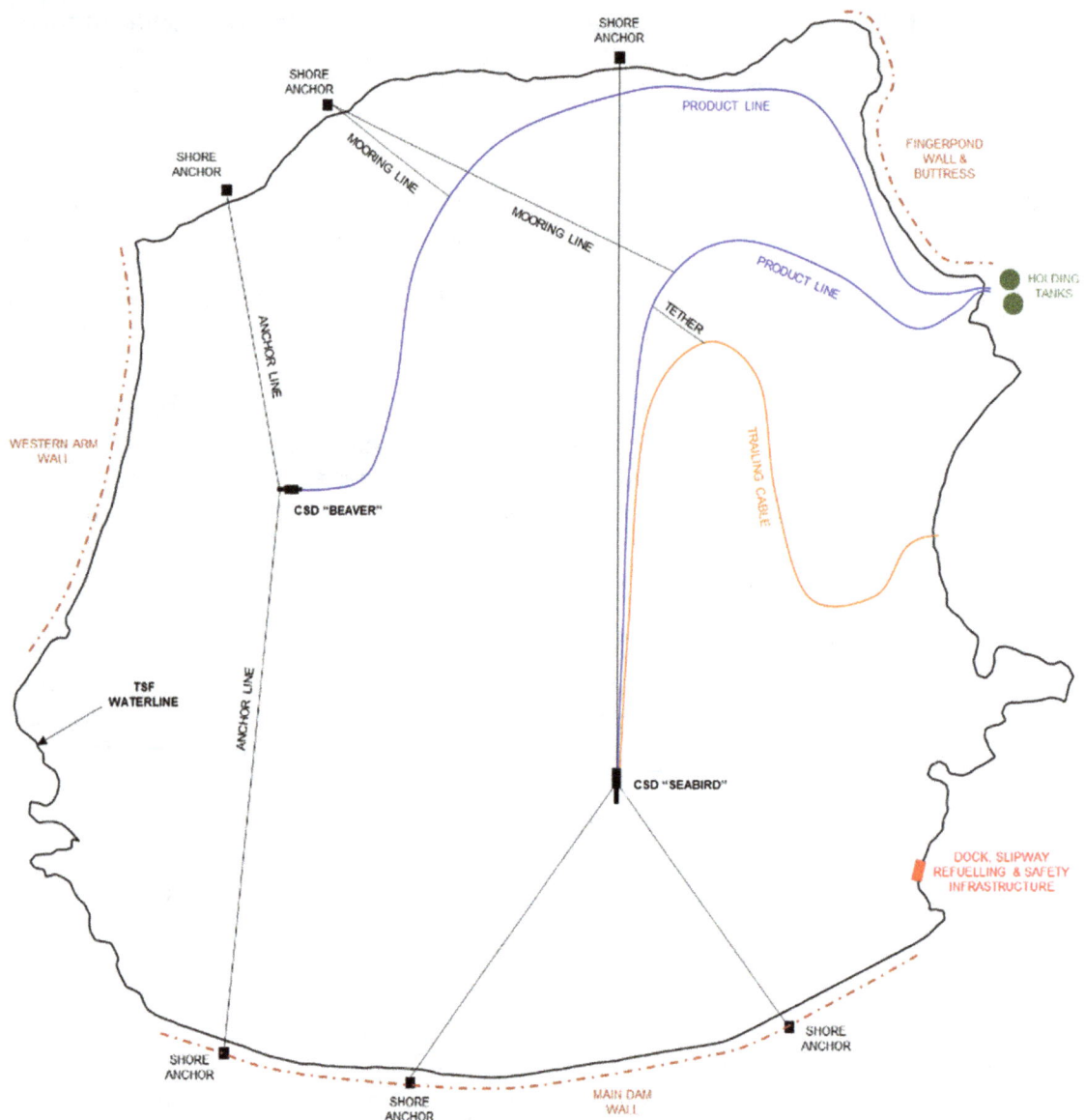

FIG 3 – Indicative Hellyer TSF1 dredging operations layout.

The Hellyer operation has two dredgers with different operating characteristics in service (Table 4, Figure 4). Both are barges without propulsion and anchors and anchor lines are used to control the position of the dredgers on the TSF. Shore anchors are 5 t and 9 t weights located at the waterline around the TSF, including on the TSF embankments. In-dam anchors are 360 kg mushroom type anchors, and the spuds at the stern of the Beaver dredger placed on and in the tailings to be extracted. Relocation of the dredgers and in-dam anchors on the TSF is achieved using the workboat and support craft.

The dredgers breakup the tailings using a rotating cutterhead positioned at the end of a pivoting boom (the ladder) with the dredge suction pump inlet located directly behind the cutterhead. The dredge pump transports the tailings slurry from the cutterhead to the Holding tanks via a floating pipeline.

The dredge operator extracts the tailings by gradually lowering and raising the ladder whilst sweeping the cutterhead from port to starboard. During this process and as the dredge operator moves the dredger forward a face, walls and benches are created. Dredge operators progress the dredger forward until the limit of the anchor lines has been reached. The anchor lines are then adjusted or disconnected and reconnected to allow the workboat and support craft to reposition the dredge at the next location.

TABLE 4

Hellyer dredger operating characteristics.

Operating characteristic	CSD 'Beaver'	CSD 'Seabird'
Ladder type	Fixed length	Variable with the addition or removal of ladder sections and hull pontoons
Maximum recovery depth	100 m	233 m
Stern anchor	Spud	Wire line to shore
Port and starboard anchors	Wire line to anchor in tailings or Wire line to TSF shore	Wire line to TSF shore
Dredge operation	Ladder sweep pivots from stern spud and is controlled by winches on port and starboard anchor lines Forward movement controlled by the raising and lowering of stern spuds Reverse movement not possible unassisted	Ladder sweep controlled by port and starboard winches on slew ropes passing through blocks located in the port and starboard anchor lines Forward and reverse movement controlled by winches on stern, port and starboard anchor lines
Dredge sweep schematic	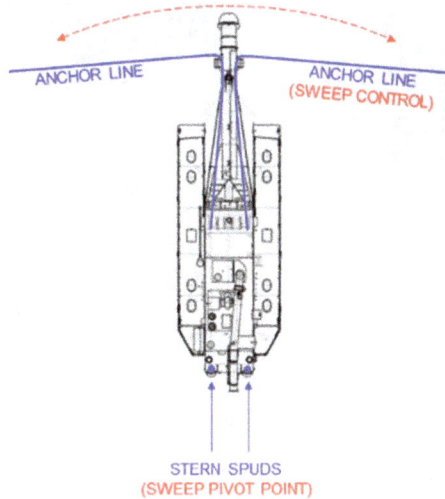	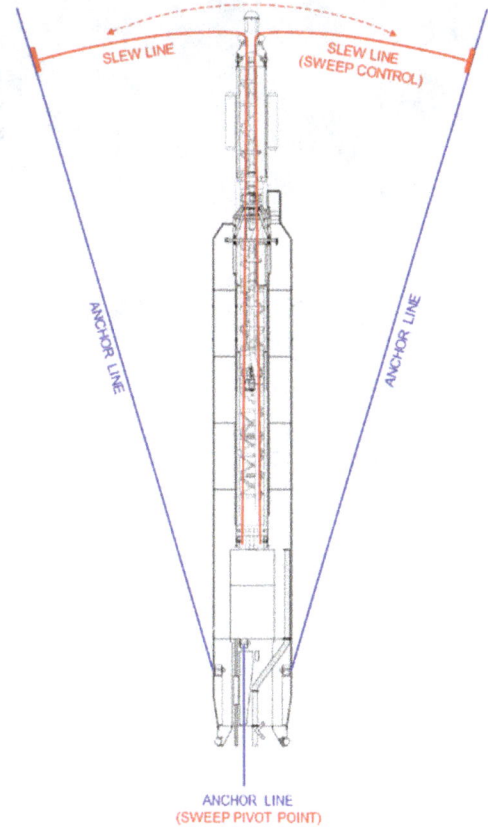

FIG 4 – CSD 'Seabird' and CSD 'Beaver'.

The principal hazards of dredging operations are water, weather conditions, floating infrastructure, and tailings stability.

Dredging operations require personnel to work on water and some tasks require dredge operators to work over the side of support craft. Weather conditions such as high winds, low temperatures, and low visibility due to fog, rain and darkness impact the ability to safely carry out work on water. Anchor lines, mooring lines, product lines and the trailing cable impact the ability to safely transit support craft over the TSF, particularly in low visibility conditions. Once disturbed the tailings within the TSF are highly mobile and movement of large volumes of tailings can trigger waves that impact the position and stability of the dredgers on the TSF.

Lessons learned – developing an ore reserve estimate for tailings

Addressing tailings stability

The critical parameters for addressing the principal hazard of tailings stability in the development of an ore reserve estimate are:

- maximum tailings bench height
- maximum tailings face angles
- minimum tailings berm width such that tailings benches behave as discreet entities
- extraction exclusion zones for TSF embankment stability.

Numerical modelling using the tailings material properties drawn from Vane Shear and CPT testing during resource development determines these parameters. Bench height, berm width, face angles and exclusion zone boundaries guide the development of a safe extraction schedule.

Determination of achievable production rates

The maximum throughput rate of the processing plant (at average resource grade) sets the lower limit for the dredger production rate. This allows for shift change, planned and unplanned dredger maintenance activities. The following parameters guide dredge pump selection:

- the relative RL differential between the TSF maximum depth and the holding tank inlet
- product line length required to access the entire surface area of the TSF
- product line diameter
- tailings material properties.

The specifications of the selected dredge pump are a required input into the dredge asset selection process. The pump curve of the selected dredge pump dictates the maximum achievable production rates from different tailings benches (differentiated by distance below the TSF waterline). These production rates guide the development of an achievable extraction schedule.

Selection of appropriate anchoring methodology

Dredge anchors are used to control the position of the dredgers on the TSF and influence the extraction sweep. The characteristics of the TSF that influence the selection of anchoring methodology are:

- the accessibility of the waterline around the perimeter of the TSF
- tailings material properties
- the depth of water cover over the TSF
- the size of TSF floor area.

Shore anchors are positioned at the water line and access is required for anchor installation and relocation, the connection of anchor lines and anchor inspections. Ideally this requires a perimeter access road that can provide for two-way traffic. Placement of shore anchors on embankment wall crests that provide for one-way traffic only is possible where anchors can be installed below the crest in the downstream embankment wall.

A water cover of less than 1.55 m over the tailings prohibits the use of in-dam anchors. Workboat access to the port and starboard in-dam anchor locations is required and workboats capable of lifting in-dam anchors generally have a minimum draft of 1.55 m. Dredgers with a stern spud as an in-dam anchor also have a minimum draft of 1.55 m.

The material properties of the *in situ* tailings determine the expected penetration of spud and other in-dam anchors. In-dam anchors are unsuitable where the expected spud and anchor penetration exceeds the capacity of the dredger spud and workboat anchor raising systems.

Observations from Hellyer operations show recovery of tailings from the TSF floor to be greater when a shore stern anchor is utilised. In this configuration the dredge operator controls the reverse movement of the dredger via winches. This facilitates manoeuvrability around hazards such as rocks, tree stumps and fallen trees on the dam floor resulting in greater tailings recovery. Reverse movement of dredgers with a stern spud (such as the Beaver dredger) is not possible without assistance from the Workboat and support craft.

The selected anchor methodology is required as an input into the dredge asset selection process.

Selection of appropriate dredging asset

The selection of the appropriate dredging asset is influenced by the following parameters:

- dredge pump specifications
- anchor methodology
- power availability

- maximum depth of the TSF (distance between the waterline and the TSF floor).

Electric dredgers have comparatively lower operating costs but require substantial electrical infrastructure and a floated trailing cable of sufficient length to facilitate access across the TSF. Losses across the trailing cable limits the maximum operational area of an electric dredger. Diesel dredgers have comparatively higher operating costs and require shore based refuelling infrastructure and watercraft (generally a Workboat). Over a TSF with a large surface area a diesel dredger provides greater flexibility.

The maximum depth of the TSF determines the required ladder length. Where this is greater than the ladder lengths of available dredgers two options are possible:

- A custom modular dredger where the ladder length can be increased and decreased with the addition and removal of ladder sections and hull pontoons (such as Hellyer operations' Seabird dredger).

- Staged extraction where after the removal of an upper layer of tailings across the TSF, the water level of the TSF is lowered to facilitate access to tailings beyond the reach of the ladder of available dredgers.

The selected dredge asset(s) guide the cost estimation process for the development of an ore reserve estimate.

Ancillary selection

Ancillary equipment and infrastructure required to support safe dredging operations are:

- support craft
- workboat
- product line
- holding tanks
- slipway and dock
- electrical infrastructure
- diesel storage and refuelling infrastructure
- safety equipment and infrastructure.

Considerations for the selection of support craft include propulsion and steering type, fuel type, hull material and size. Diesel outboard propulsion with central steering control is recommended. Hull material is guided by the TSF water quality. Support craft size is guided by manoeuvrability on water, number of crew to be carried and the ease of trailering/launching for maintenance.

Workboat selection is guided by anchor methodology, product line diameter and dredger diesel consumption, the aim being to refuel a diesel dredger over water in daylight hours only.

Consideration of tailings material properties is required to ensure that tailings remain in suspension throughout the product line from the dredge pump to the holding tank inlet. Buildup of settled tailings in the product line causes product line blockages that are difficult to safely repair on water. Larger diameter product lines reduce friction losses but increase the potential for tailings settlement and are subject to greater wind action on the TSF.

Holding tanks at the TSF shoreline are required to manage dredge pump pressures and provide surge capacity to facilitate planned maintenance activities. Holding tank size is guided by the production rate with a minimum of 12 hrs capacity recommended.

A slipway in which to conduct dry-dock maintenance of dredgers, workboat and support craft is required. Slipway size and length is defined by the selected dredge asset. A dock is required to safely load and unload personnel and materials from the support craft and workboat.

The slipway, dock and refuelling and safety infrastructure all require lighting and power. Safety infrastructure for dredging operations includes communications, a heated drying room and first aid room. The selected ancillary equipment and infrastructure guide the cost estimation process for the development of an ore reserve estimate.

Lessons learned – managing dredging operations

Hellyer dredging operations has safely managed simultaneous dredging by two dredgers and associated ancillary infrastructure on tailings storage facility of which two of the three embankment walls are constructed on top of an unknown quantity of tailings. Key learnings regarding managing regulatory compliance as well as the principal hazards of water, weather conditions, floating infrastructure and tailings stability have been drawn from this operating experience.

Regulatory compliance

Commercial operations conducted from vessels falls under the purview of the Australian Maritime Safety Authority (AMSA). The legislation under which AMSA operates is highly prescriptive and covers certification and compliance of vessels, safety management systems for marine operations, and vessel operator certification and training requirements. To assist with understanding compliance obligations, it is recommended that a prospective dredging operation establish relationships with:

- an AMSA accredited marine surveyor (vessel certification and compliance)
- a marine operations specialist (safety management system, safe operating procedures and emergency response training)
- a maritime industry training provider (vessel operator certification and training).

Water and weather conditions

Robust risk assessments are required to determine what tasks can safely be performed on water and the weather conditions in which those tasks can occur. Engagement with a marine operations specialist will ensure that all risks associated with conducting work over water have been identified and addressed in the safety management system.

The development of trigger action response plans (TARPS) for weather conditions such as high winds, low temperatures, low visibility (due to rain and fog) and electrical storms is recommended to guide the daily execution of the production plan.

Floating infrastructure

Infrastructure located on and in the TSF presents a hazard to both the anchored dredgers and the traverse of support craft over the TSF. Wind action on floated product lines and trailing cables can move their position on the TSF such that it interferes with the dredger anchor lines and extraction arc (sweep). Modelling by marine operations specialists is required to determine the placement and type of mooring lines to prevent movement of product lines. Tethers are utilised to protect a trailing cable from damage.

Anchor lines, mooring lines and tethers are rope or steel cables that lie just under the water surface of the TSF. Support craft traversing over the TSF can become snagged on these lines causing injury to personnel and damage to the support craft. Visibility can be improved through the installation of buoys and reflective markers on the lines and tethers. Sections of the anchor and mooring lines can be weighted to increase the distance of the line below the water to form a gate over which support craft can safely pass.

Tailings stability

Numerical modelling to define the maximum bench height, minimum berm width, face angle and exclusion zones from TSF embankments guides the development of a safe extraction schedule.

During dredging operations, regular bathymetry surveys are required to monitor tailings stability and manage compliance to the extraction schedule. Monthly surveys are recommended to:

- identify non-compliant zones to address as soon as practicable in production planning
- facilitate production reporting and reconciliation.

The accuracy of bathymetry surveys is influenced by water turbidity and any dredging activities should cease for a minimum of 12 hrs prior to the survey. Bathymetry surveys are a specialist service not normally within the scope of mining and cadastral surveyors.

Water turbidity resulting from dredging activities prohibits any real-time monitoring of face angles however real-time visualisation of benches and face angles based on the movement of the cutterhead is provided by dredge management software such as DredgePack to the dredge operator.

CONCLUSION

Legacy tailings storage facilities contain valuable metals which, with advances in minerals processing techniques, it may now be possible to extract. Understanding the potential value within these tailings storage facilities requires the development of mineral resource and ore reserve estimates. This paper provides a roadmap for the development of a mineral resource and ore reserve estimates for legacy tailings dams. The principal hazards associated with dredging operations are identified and steps to address those principal hazards during the development of an ore reserve estimate and in the management of dredging operations are defined.

ACKNOWLEDGEMENTS

I gratefully acknowledge *Hellyer Gold Mines* for the continuing opportunity to support the Dredge Crew and the dredging operation and for permission to publish the lessons learned through the provision of this support in this paper.

I thank *Sub41* for their invaluable marine operations experience and willingness to apply this in support of the Dredge Crew and the Hellyer operation.

Use of a conductive multi-linear drainage geocomposite to enhance the performance of a double-lined pond enabling a high sensitive leak location survey on the primary liner

M Leroux[1], P Saunier[2] and C Charpentier[3]

1. General Manager, CSP Fidelio, Nouméa 98800, New-Caledonia. Email: mleroux@csp.nc
2. Business Development Manager North-America and Pacific, AFITEX-Texel, Québec G6E 1G8, Canada. Email: psaunier@afitextexel.com
3. ELL Director and Tech Development, Alphard, Montréal Québec H2T 1X9, Canada. Email: ccharpentier@alphard.com

ABSTRACT

Leachates are one of the most critical effluents to be managed in mining operations. Thus, the regulation is generally very strict about how leachates and pregnant solution are contained, requiring double-lined ponds to maximise the performance of the leak-proofing. Reducing the risk of an environmental contamination through a potential leak in the system can only be achieved when the control of the primary liner integrity is as accurate as the one executed on the secondary liner. However, using a non-conductive element between the two liners might lead to poor leak detection sensitivity, and therefore fail to detect small and undetected defects on the floor. Using a conductive multi-linear geocomposite will solve this issue in a very efficient way in comparison with other techniques such as using a conductive geomembrane or a conductive geotextile installed over a geonet single-sided composite, or even worse, flooding between the two liners. This paper presents a recent case study of a double-lined pond project located along the second biggest coral reef in the world.

INTRODUCTION

Process water or leachate ponds are the most critical components of containment facilities, as they are intended to completely and reliably contain materials that are highly loaded with pollutants. Literature indicates that typically geomembranes leak at a rate of seven defects per hectare under standard installation conditions, even when installed according to best practices (Forget, Rollin and Jacquelin, 2005). One way to reduce this defect density is by implementing a third-party quality assurance system and using standardised electronic leak location methods.

Geoelectrical surveys are well established in the industry. ASTM standards D7002 (2022; Water Puddle) and D7953 (2020; Arc Testing) govern survey techniques on exposed geomembranes. ASTM standard D7007 (2016; Dipole Method) governs survey techniques on covered geomembranes. In all cases, it is necessary to have a conductive medium beneath the geomembrane being tested.

In the case of double-lined ponds, it is therefore necessary to introduce a current carrying system beneath the primary geomembrane (ie between the primary and secondary geomembranes) in order to apply these techniques. Several approaches are possible, one of which is to install a multi-linear conductive drainage geocomposite, which serves three functions: protecting the geomembranes, draining seepage through the leak detection system, and providing a conductive path for electrical current beneath the primary geomembrane.

LINERS DO LEAK

The simple act of using geosynthetics on a project does not guarantee imperviousness of the barrier layer. Granted proper fabrication practices, a geomembrane by itself is a fully impermeable material. However, every operation required to transform a manufactured geomembrane roll into an installed liner exposes the geomembrane to potential damage (mostly mechanical damage, failed welds or chemical degradation resulting from improper storage, for instance).

Therefore, wherever a high level of uncompromised impermeability is demanded from the barrier layer, the use of third-party quality assurance and leak location services is imperative.

The relationship between third-party quality assurance (QA) and electrical leak location (ELL) is that of mutual dependency. QA is in part dependent on ELL for detecting breaches in geomembranes which have been overlooked, are invisible to the naked eye, or have occurred following the installation of subsequent system layers. Whereas the ELL party relies on the QA party for ensuring the adequacy and traceability of all installed materials in addition to minimising the number of geomembrane defects through invigilation of the storage, handling and installation phases of the work. Thus, ELL ensures that the geomembrane is uncompromised at the time of the inspection, while QA ensures that the geomembrane will continue to fulfil its function for the entire life expectancy of the project.

CQA third-party control

Internal Quality control is typically a flawed process due to the inherent conflict of interest existing between the Installation team and the Quality team: both parties operate under an authority whose underlying interest is generally the fastest-possible installation of geosynthetics on a given project. Thus, the employment of an independent Quality assurance party introduces an unbiased stakeholder whose underlying interest is the best-possible installation of geosynthetics (more specifically: the mandate of a third-party QA inspector is to ensure the work is carried out as per project plans and specifications).

An expert third-party Quality assurance personnel will oversee all aspects related to geocomposite materials on-site: transport, handling and storage of geocomposite rolls; ensuring all material has undergone appropriate factory QC testing, sampling to verify shipped materials are compliant with project specifications; subgrade approval, visual inspection of installed panels and seams, validation of welding machine calibration, non-destructive testing and intermittent destructive testing, including coordination with a testing laboratory.

In addition to collection and verification of all factory issued documentation (factory QC testing, mill test certificates, datasheets etc), QA personnel also keep daily logs of panel installation sequence and the respective *in situ* testing results, thus ensuring full traceability of geosynthetic material from the factory floor to their final resting place on the project site.

Additionally, much like the influence of a leak location operation, the sole existence and on-site presence of independent QA personnel has the effect of raising the diligence and operational discipline of the internal QC personnel and installation team.

Leak detection

Most commonly, electrical leak location is carried out using two standardised methods: the water-puddle method and the dipole method.

The water-puddle is the more direct, more effective method of testing for breaches in the barrier layer, and is carried out directly on top of the exposed geomembrane. The concept behind this methodology is that a continuous, fully impermeable assembly of geomembrane panels will not allow surface water to come into contact with the underlying substrate layer. When such contact is made, the water acts as an electrical bridge between overlay and underlay media. A ground electrode is connected to the underlay layer and voltage is applied to the water coming from the waterlance, and when that connexion is made, a signal indicates to the field technicians that a defect has been reached.

In order to successfully carry out dipole leak detection, the geomembrane must be covered by a single layer of homogeneous, electrically-conducting material (eg wet granular material). The concept is identical to that of the water-puddle method: a continuous, impermeable assembly of geomembrane panels will not allow current propagation from above the geomembrane to the underlying substrate. The presence of a breach in the barrier layer is indicated by a typical leak signal in electrical current detected by the dipole apparatus.

When possible or if required, both leak location methods are used on the same barrier layer. In addition to a redundancy check, this allows for separation of liability: holes detected via the water-puddle methods are necessary the responsibility of the geomembrane installer (which should cover repair costs), whereas holes detected via the electrical dipole method are the responsibility of the civil-works contractor (who should also cover those repair costs).

Average leakage per hectare

Typically, on a project in North-America involving third-party QA oversight of geomembrane installation, a leak location operation will locate an average of seven leaks/ha via the water-puddle method, and between one and four leaks/ha via the dipole method (contingent upon whether a water puddle survey had been conducted first). Unsupervised projects typically exhibit a much larger presence and range of breaches in the barrier layer. In some scenarios, the presence of breaches is so prevalent that it effectively nullifies the purpose of employing geomembrane panels in the first place (Forget, Rollin and Jacquelin, 2005).

Under favourable conditions, a thorough leak location survey will reduce the average presence of leaks down to two or fewer holes per hectare. It is important to keep in mind that the relationship between a hole and a leakage rate is dependent upon the size and the location of the hole within the containment basin (Charpentier *et al*, 2017).

In addition, in its history, the geomembrane system transformed from single to double lined (Peggs, 2009) because damage is unavoidable for a geomembrane during construction. The purpose of a double lined system in which leakage through the primary geomembrane (with a constant hydraulic head on it) is collected by the secondary geomembrane and removed so there is no head on the secondary. Therefore, the double lining system doesn't leak – just as double hulled ships do not sink.

Today, the 'most often applied primary action leakage rate (ALR) for water impoundments is 500 gallons per acre per day (gpad)' as specified in the 'Recommended Standards for Wastewater Facilities' (Florida Department of Environmental Protection, 2004) by ten northern states and on Canadian provinces for liners under 6 ft of water in wastewater treatment plants. 'Right sizing' the ALR to match the technical capabilities of the current leak-location methods is necessary (Darilek and Laine, 2011). Specifying a leakage rate that is too low can be a disaster if the source of the leakage cannot be located by current technology. If the source of the leakage cannot be located, then the only alternative is to reline the facility and hope that the new geomembrane does not also exceed the specified ALR. According to the Quality Based Action Leakage (QBAL) method based on Giroud's equation for calculating flow-through defects, with a good quality installation, a geomembrane could be expected to have between one and four defects per acre and a poor-quality installation that could have 10–20 defects per acre. This could equate to an ALR of 720 gpad to 3600 gpad. Furthermore, according to Peggs (2009), a regulatory agency with a zero-leakage policy is not being practical which can lead to arguments, wasted time and efforts, and unnecessary expenses that benefit no one.

GLOBAL QUALITY

In this document, the overall quality of a project is defined as the multiplication of the quality level of the three principal components of a double lined pond project (Figure 1), which are:

1. The design, or conception stage.
2. The construction stage.
3. The operational stage.

FIG 1 – The three components of the Global Quality of a project.

The Global Quality of the entire project can be calculated as follows:

$$GQ = QDesign \times QConstruction \times QOperations$$

Design

Two of the most important design criteria when determining the liner materials are, first, the long-term operational requirements of the system ie fresh water versus produced water, circulation versus non-circulation, types of pumping and piping being used within the system and, second, the client's operational knowledge of liners; ie does the operator's understanding of the fragility of the liners. In this example, CSP understood the liner fragility and their operational requirements during design. The leachate pond was originally designed with a composite cross-section with a GCL and a 2 mm HDPE primary geomembrane. In order to increase the overall quality of the project and also due to geometrical constraints, it has been decided that the design would change from a composite lining system to a double lining system, using the existing primary geomembrane as the new secondary liner. Ultimately, this decision led to a design that included a secondary and primary liner system comprising 2 mm HDPE liners as shown in the detail below. The original HDPE (new secondary) was cleaned after the pond was drained and leak tested before installing the new components.

The client's design guidelines were as follows: fit an operational treatment area within a small rig anchor pattern, meet State design regulations, maximise the amount of produced water/leachate storage within the permitted boundary (Figure 2).

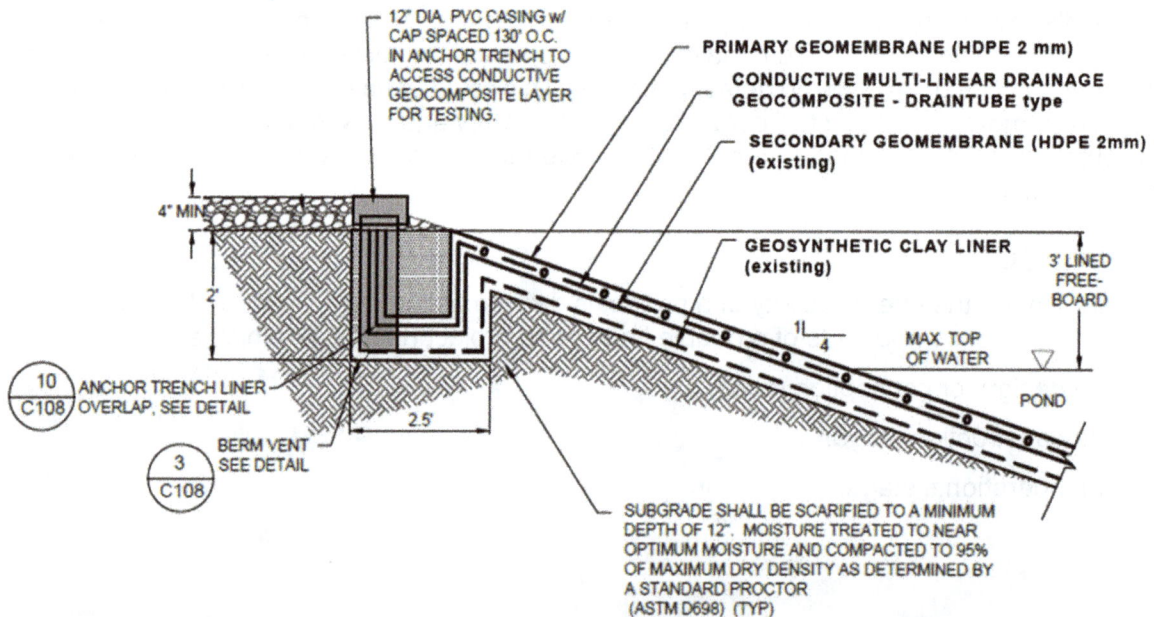

FIG 2 – Typical cross-section on the slopes and anchor trench.

The Conductive multi-linear drainage geocomposite comprises 20 mm corrugated polypropylene perforated pipes spaced on 1 m centres between two non-woven polypropylene geotextile layers (Figure 3). A conductive grid made of stainless steel mesh is inserted between the two geotextiles to offer long-term electrical conductivity below the primary liner.

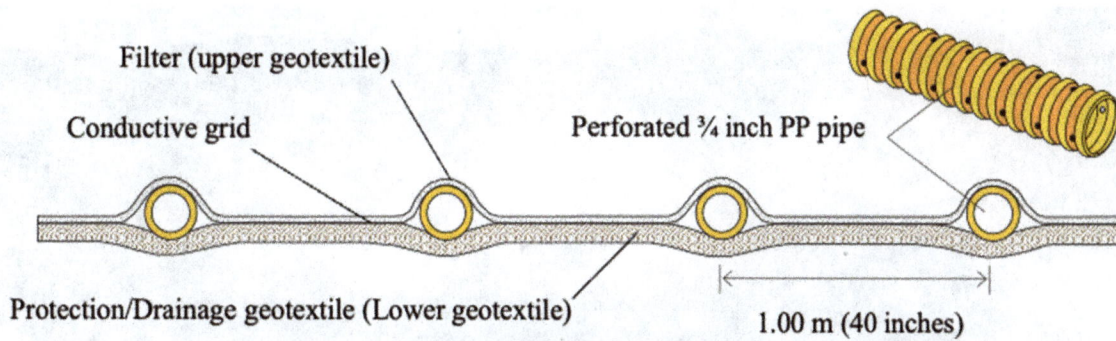

FIG 3 – Conductive multi-linear drainage geocomposite description.

Multi-linear drainage geocomposites have been used in landfill and mining (ponds) applications in Europe and Africa for 25 years. An important characteristic of those drainage geocomposites is that they maintain their transmissivity under significant normal stresses (Saunier, Ragen and Blond, 2010) because they don't experience geotextile intrusion into the primary high-flow component (Figure 4). Therefore, for most of the applications, the applied combined reduction factors for multi-linear drainage geocomposite are almost half of those applied to standard geonet geocomposites (Diamiano and Steinhauser, 2020; Maier *et al*, 2013). These type of products have also a good behaviour in extreme conditions (Fourmont and Saunier, 2015).

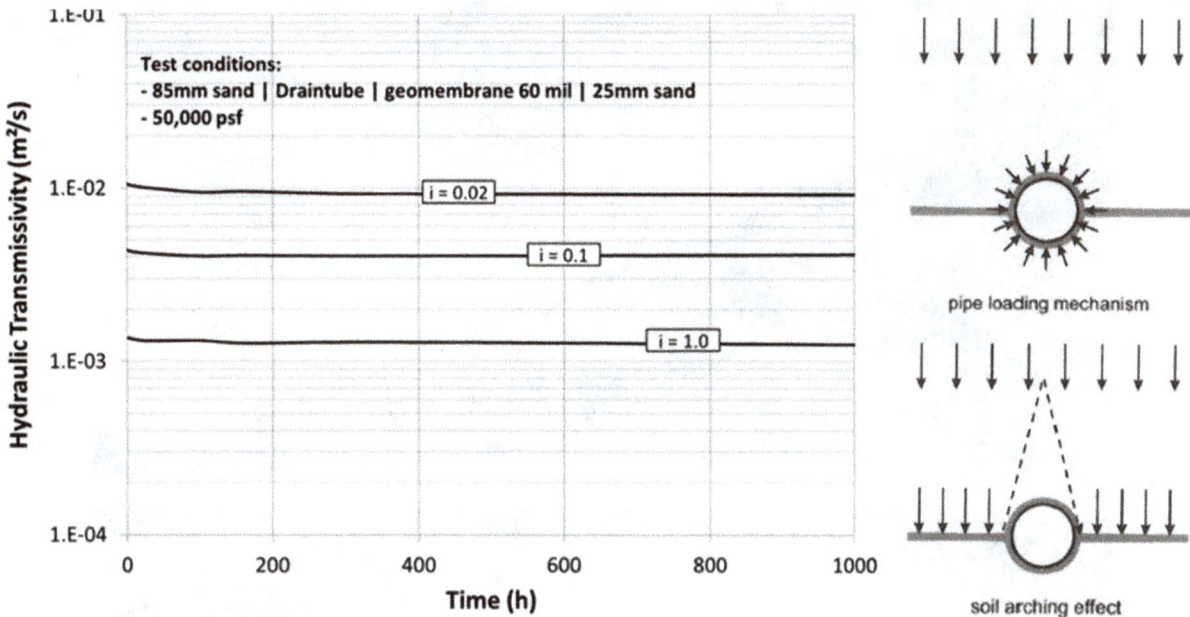

FIG 4 – Measurement of transmissivity over time under high load.

Construction

One of the major selling points of the design was using the actual primary liner as a new secondary liner and multi-linear drainage geocomposite for extremely high installation speed. The first step was to empty the pond from its leachate, clean the surface of the exposed liner from mud and dust. Then the new exposed secondary liner has been prospected using the water puddle method before the installation of the upper components. In that situation, because the liner is lying on top of GCL and native soil, the electrical current is able to be properly established below the liner to be prospected without the need of having an additional conductive layer.

Figure 5 shows the existing liner cleaned after having been prospected with the water puddle leak location survey. Leaks, most likely generated during the cleaning process, have been found and repaired before installing the drainage geocomposite.

FIG 5 – Picture of the pond with the secondary liner cleaned and checked.

Figures 6 shows the lightness of the multi-linear drainage geocomposite. The rolls can be handled like geotextiles, unrolled from the top of the slope and their structure is soft enough against the liner to avoid any risk of perforation due to heavy load or harsh angles of plastic. Also, the installation is accelerated (in the order to 20 to 30 per cent more productivity) and is safer for the crew than the installation of a geonet type drainage geocomposite and a conductive geotextile.

FIG 6 – Deployment of the Conductive Multi-Linear DRAINTUBE from the top of the slope.

The electrical contact between the panels is critical in order to ensure the continuity of the electrical path within the entire area of the pond. A proper overlap of the conductive grid of the geocomposite is observed (Figure 7) and a continuity test is operated every 20 to 30 m of longitudinal joint (Figure 8).

FIG 7 – Overlap of the conductive grid in the machine direction.

FIG 8 – Electrical continuity test in the length of the connection.

The installation of the primary liner starts (Figure 9) as soon as the Conductive multi-linear drainage geocomposite is mechanically connected to the main header positioned along the low points in the bottom of the pond (Figure 10).

FIG 9 – Deployment of the primary liner on top of the Conductive multi-linear drainage geocomposite.

FIG 10 – Mechanical connection to the main header and protection with a geotextile.

Leak Location surveys

Leak location surveys have been conducted on the entire surface of the secondary and the primary liner using the Water Puddle method (as per ASTM D7002) (Figure 11).

During the leak location of the secondary liner, the electrical current passed below the liner into the soil and through the saturated GCL. Seven leaks were detected, mostly in the extrusion welds or knife cuts (Figure 12).

During the leak location of the primary liner, the electrical current passed below the liner through the Conductive multi-linear drainage geocomposite. Three leaks were detected in the extrusion weldings (Figure 13). Considering the area of the pond, the estimated leakage rate was between five and eight leaks per hectare.

FIG 11 – Water Puddle in the slopes and the bottom.

FIG 12 – Location of the leaks found and picture of a knife cut.

FIG 13 – Location of the leaks found and picture of a typical leak in the extrusion welding.

Cost comparison

Defects on a double lined pond can lead to tremendous consequences on costs and delays. An estimate from a project in New-Mexico in 2015 has been evaluated and is presented in Table 1. Originally, the pond, after its completion, had not been leak tested before operation. This catastrophic scenario, based on this real story, was as follows:

- Day 1: First filling of the pond.

- Day 1 + 1 month: Shutdown the pond due to observed leaks.

- Day 1 + 2 months: Investigation, complete discharge of the pond, cleaning of the liner and run an ELL program.

- Day 1 + 3 months: Relining of the primary liner, line a third layer and complete a second ELL program.

- Day 1 + 4 months: Pond re-filled and now operates without unacceptable leakage.

TABLE 1

Comparison of costs between with and without an ELL campaign and third-party QA/QC.

Good practices (ELL + 3rd Party QA/QC)		Bad practices (no external QA/QC nore ELL)	
Leak location survey program	$25k	$720k	Pond shutdown for 3 months
3rd party QA/QC	$25k	$50k	Leak location survey program × 2
		$50k	3rd party QA/QC × 2
		$250k	Relining the primary
Total expected costs	**$50k**	**$1.07M**	**Total estimated costs**

In this cost comparison, each day the pond is unavailable equates to US$80 000 of lost revenue for the operation.

Table 1 shows that poor quality management may lead to significant cost increases overall, compared to including a proper quality assurance program, including ELL and third-party QA if considered at the beginning of the project.

CONCLUSIONS

This case study shows that, despite a good design and a normal construction, the high overall quality of a project is linked to details. There is absolutely no way to meet stringent performance requirements in a geosynthetic installation without having, in addition to a high-performing team of sub-contractors and owner representative control, a strong QA program including a third-party QA and a complete Electrical Leak Location survey campaign. Conductive Multilinear drainage geocomposites have demonstrated their efficiency in this project by helping the operators to find a reasonable amount of leaks, even very tiny ones (1 mm diameter) that kept the overall project at a maximum level of Quality and performance.

REFERENCES

ASTM International, 2016. ASTM D7007 – Standard Practices for Electrical Methods for Locating Leaks in Geomembranes Covered with Water or Earthen Materials, ASTM International, West Conshohocken. https://doi.org/10.1520/D7007-16

ASTM International, 2020. ASTM D7953 – Standard Practice for Electrical Leak Location on Exposed Geomembranes Using the Arc Testing Method, ASTM International, West Conshohocken. https://doi.org/10.1520/D7953-20

ASTM International, 2022. ASTM D7002-22 – Standard Practice for Electrical Leak Location on Exposed Geomembranes Using the Water Puddle Method, ASTM International, West Conshohocken. https://doi.org/10.1520/D7002-22

Charpentier, C, Bremmer, H, Jacquelin, T and Rollin, A, 2017 Is electrical leak location helping us getting better with geomembranes?, in Proceedings GeoAfrica '17, Marocco, 8–11 October.

Darilek, G T and Laine, D L, 2011. Specifying allowable leakage rates, *Geosynthetics Magazine,* August 2011.

Diamiano, L and Steinhauser, E, 2020. A guide for specifying drainage geocomposites, *Geosynthetics Magazine*, 8–10 April.

Florida Department of Environmental Protection, 2023. Recommended Standards for Wastewater Facilities (2004). Available from: <https://floridadep.gov/water/domestic-wastewater/documents/>

Forget, B, Rollin, A and Jacquelin, T, 2005. Lessons Learned from 10 years of Leak Detection Surveys on Geomembranes, 9 p. Available from: <https://evolui.eco.br/wp-content/uploads/2020/10/Lessons-Learned-from-10-years-of-leak-detection-surveys-on-geomembranes.-Quebec-Canada-2005.pdf>

Fourmont, S and Saunier, P, 2015. Behavior of Drain Tubes planar drainage geocomposite under extreme cold temperatures, in Proceedings GeoQuebec 2015 Conference, 20–23 September, Quebec, Canada.

Peggs, I, 2009. Geomembrane Liner Action Leakage Rates: What is Practical and What is Not?, Geosynthetica.com, July/August P47.

Saunier, P, Ragen, W and Blond, E, 2010. Assessment of the resistance of drain tube planar drainage geocomposites to high compressive loads, in Proceedings 9th International Conference on Geosynthetics, vol 3, Guarujá, Brazil.

Machine learning (ML)-based analysis of significant surface elevation changes in operational tailings storage facilities

W Lu[1], R Shirani Faradonbeh[2], H Xie[3] and P Stothard[4]

1. Postgraduate Student, Curtin University, Kalgoorlie WA 6430.
   Email: wang.lu@postgrad.curtin.edu.au
2. Lecturer, Curtin University, Kalgoorlie WA 6430.
   Email: roohollah.shiranifaradonbeh@curtin.edu.au
3. Lecturer, Curtin University, Bentley WA 6102. Email: hui.xie@curtin.edu.au
4. Adjunct Associate Professor, Kalgoorlie WA 6430. Email: phillip.stothard@curtin.edu.au

ABSTRACT

Tailings Storage Facilities (TSFs) are complex and dynamic infrastructures that require continuous monitoring to prevent geotechnical failures. Traditional manual monitoring approaches lack the spatial and temporal resolution needed for effective risk assessment. This paper proposes a Machine Learning (ML) framework to predict significant elevation changes in operational TSFs using high-resolution Digital Surface Models (DSMs) derived from Unmanned Aerial Vehicle (UAV) photogrammetry. A multi-stage methodology was implemented, including DSM detrending, DoD (DSMs of Differencing) generation, feature engineering, and the development of ML models focused on zones of significant deformation. Among several ML algorithms evaluated, XGBoost achieved the highest predictive performance, with R^2 values of 0.708 and 0.643 respectively, for the negative and positive elevation change for the test data. SHapley Additive exPlanations (SHAP) analysis revealed the importance of spatial coordinates and time-related interaction features. These results highlight the potential of data-driven models to enhance proactive TSF risk management by enabling predictive insights into deformation behaviour.

INTRODUCTION

Tailings storage facilities (TSFs) are critical mining infrastructures, that pose inherent risks due to the properties of structures (Mwanza, Mashumba and Telukdarie, 2024; Sun, Zhang and Li, 2012). Traditional monitoring methods heavily rely on manual, point-based measurements (Adamo *et al*, 2020; Chovan *et al*, 2021; Witt *et al*, 2004). These methods lack the necessary spatial and temporal understanding to fully capture the dynamic surface evolution of these large, complex structures, which potentially miss early indicators of instability. Recent advancements in Unmanned Aerial Vehicle (UAV) photogrammetry provide a flexible, high-resolution approach for monitoring TSF stability (Rauhala *et al*, 2017; Zubícek *et al*, 2024). On the other hand, the generation of Digital Surface Models (DSMs) based on aerial images and subsequent DSM differencing (DoD) enables detailed mapping of surface and contour detections over time (Wang *et al*, 2018). While these approaches compensate for the limitations of field investigations, such as high labour costs and lack of retrospective data in local areas, it lacks predictive capabilities for the stabilities based on the active deposition.

A key challenge lies in transitioning from observing past changes to proactively predicting future significant surface deformation. This is complicated when working with DoD data, as minor changes may be influenced by background variability or measurement uncertainties, making it challenging to distinguish true physical changes from noise (Cavalli *et al*, 2017; Wheaton *et al*, 2010). This study addresses this gap by proposing a Machine Learning (ML) framework specifically designed to predict significant surface elevation changes in operational TSFs (Luo *et al*, 2022; Wang *et al*, 2018). These frameworks utilise time-series UAV-derived DSMs and focus the ML model on areas exhibiting statistically significant past changes (specifically, where elevation change falls beyond the positive and negative third quartiles of the DoD distribution) to mitigate noise and concentrate on potentially higher-risk zones.

Using extracted topographic features, including slope, curvature, time interval etc, alongside recent significant DoD values as input, multiple ML algorithms are trained to predict the magnitude of future elevation change. This paper presents the full methodology, evaluates model performance using R-squared (R^2), Mean absolute error (MSE), and Root Mean Squared Error (RMSE) metrics, and discusses how this ML-based approach enhances proactive TSF monitoring and complements existing risk management strategies.

MATERIALS

This research focuses on the Mungari mine site (30.76°S, 121.24°E), operated by Evolution Mining Co. Mining operations, primarily focused on gold and silver extraction, commenced at the site in 1999 and are projected to continue until 2041. The TSF comprises four cells (refer to Figure 1 for site layout and location). Following the approval of the initial two cells in 2015, the facility was expanded in 2020 with the construction of additional cells. This study specifically investigates Cell 3, an area designated for deposition volume estimation, which encompasses approximately 0.7 km² and has a designed storage capacity of over 12 million m³. The operational deposition cycle for Cell 3 is typically nine months.

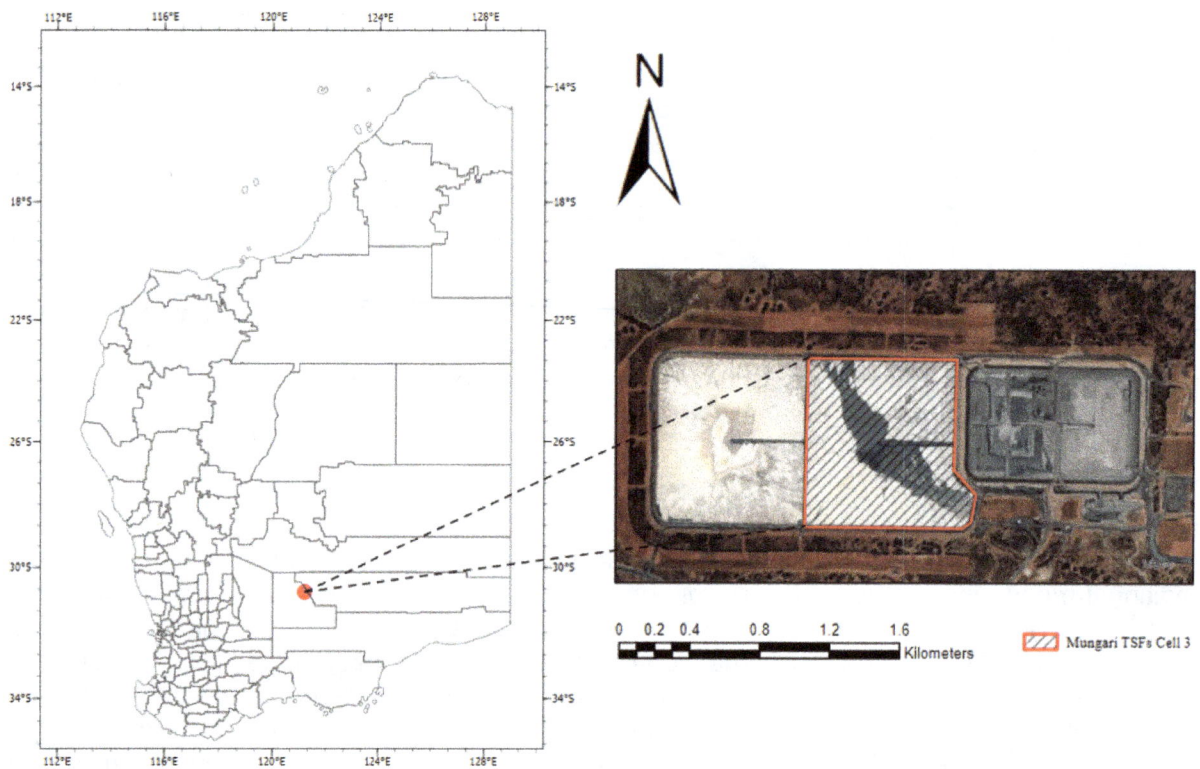

FIG 1 – Studied area: Cell 3 in Mungari mine site in Western Australia.

The primary data set for this research consists of 24 time-series DSMs provided by Evolution Mining. These DSMs were acquired via routine UAV-based photogrammetry surveys conducted over a two-year monitoring period, spanning from 29 January 2023 to 26 January 2025. All data sets are georeferenced to the EPSG:28351 coordinate system (MGA94, Zone 51S). The spatial resolution of these DSMs varies from 37.12 mm/pixel to 40.16 mm/pixel, resulting in raster dimensions ranging between 20 397 × 19 596 and 22 067 × 21 201 pixels. Elevation values within the DSMs range from 339.29 m to 354.14 m. Intra-model standard deviations of elevation, varying between 0.65 m and 0.84 m, reflect moderate spatial variability across the tailings surface. This comprehensive DSM time series forms the basis for assessing temporal patterns of tailings deposition and surface elevation changes within the TSF. Six sampled dates of selected DSMs are summarised in Table 1.

TABLE 1

Properties of key date DSM files.

Date	Dimension (pixel × pixel)	Resolution (mm/pixel)	Min elevation (m)	Max elevation (m)	Mean elevation (m)	SD elevation (m)
29/01/2023	22067 × 21201	37.12 × 37.12	344.34	351.32	346.04	0.82
30/07/2023	20397 × 19596	40.16 × 40.16	344.08	350.52	346.06	0.83
28/01/2024	20732 × 19918	39.51 × 39.51	342.07	353.37	347.26	0.76
28/07/2024	20412 × 19611	40.13 × 40.13	341.46	353.51	347.82	0.65
26/01/2025	21844 × 20986	37.50 × 37.50	339.29	354.14	348.23	0.84

METHODOLOGY

This study employed a structured workflow (see Figure 2) to accurately monitor significant elevation changes and understand the deformation of relative parameters. Generally, as seen in Figure 2, the methodology can be divided into three main steps of Data Preprocessing, Feature Extraction and Data set Construction, and ML Prediction and Metric Evaluation, which are discussed in more detail in the next section.

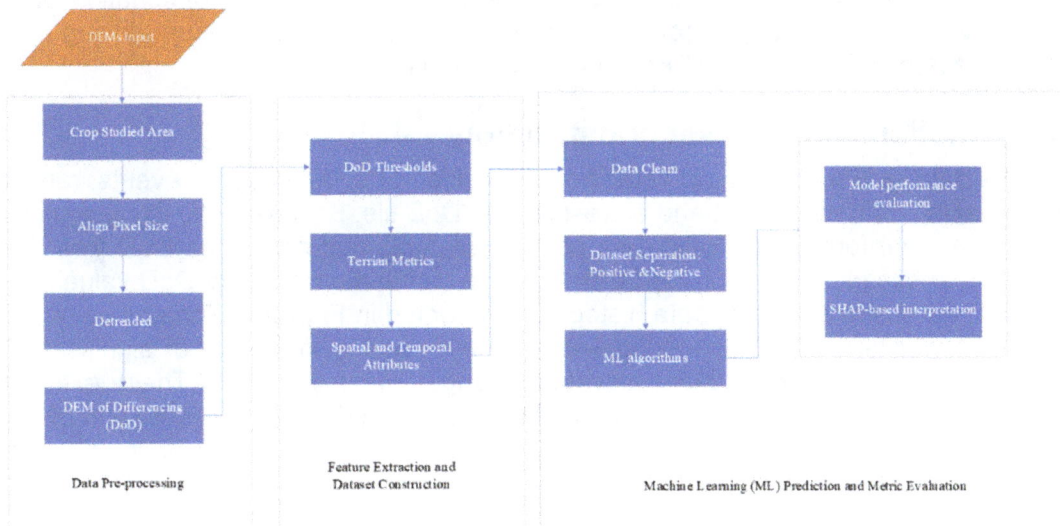

FIG 2 – Workflow of elevation change prediction using DSMs and machine learning.

Data preprocessing

The preprocessing workflow encompasses several sequential steps designed to assure the consistency and reliability of elevation change analysis. Initially, the DSMs were spatially subsetted using predefined shapefiles delineating the boundary of Cell 3. This procedure excluded the decanting tower areas, thereby preserving only the active deposition zone for analysis. Following spatial processing, a Gaussian filter was applied to each DSM to detrend the surface. This filter effectively mitigates large-scale topographic trends and low-frequency noise associated with the underlying terrain morphology. The detrending process was implemented via Gaussian filtering, which intended to remove large topographic trends and enable robust parameterisation of local surface deformation related to active tailings deposition (Erdoğan, 2010; Groom, Bertin and Friedrich, 2019).

Figure 3 illustrates a comparative example: the left panel shows a DoD derived from the original DSMs, while the right panel presents the corresponding DoD generated from detrended DSMs. The detrended DoD reveals spatial patterns of elevation change, with less interference from background terrain morphology. The effectiveness of filtering in isolating operational changes from broader topographic structures, which reduce the uncertainty from the DSM generations, and improve the reliability of the deformation signals captured.

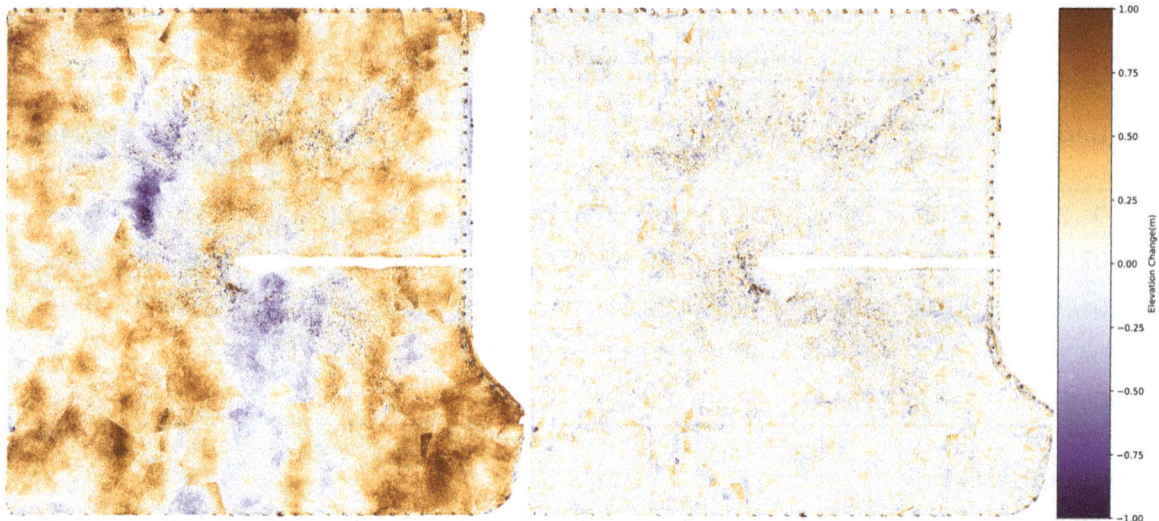

FIG 3 – Heatmap comparison of DSM of differencing (DoD) with and without detrending: original DoD (left) and detrended DoD (right).

Additionally, terrain derivatives such as slope and aspect were extracted from the detrended DSMs to serve as explanatory variables in the machine learning models. The combination of raw and processed elevation data enabled comprehensive understanding of the underlying processes influencing surface dynamics within TSFs active depositions.

Feature extraction and data set construction

To construct a data set representing outstanding deposition or sedimentation events, random pixel data were extracted from each DSM and corresponding DoD file. Following the exclusion of missing or invalid entries, over four million valid data points were obtained across 23 DoD files. Statistical thresholds were subsequently established based on the distribution of these DoD values to identify extreme cases, as illustrated by the data histogram presented in Figure 4. Thresholds representing the upper and lower tails of the distribution—specifically, values indicative of significant positive (deposition) or negative (sedimentation) elevation change—were identified. These extreme value instances, constituting approximately 1 per cent of the total valid data points, were compiled to form the ML data set.

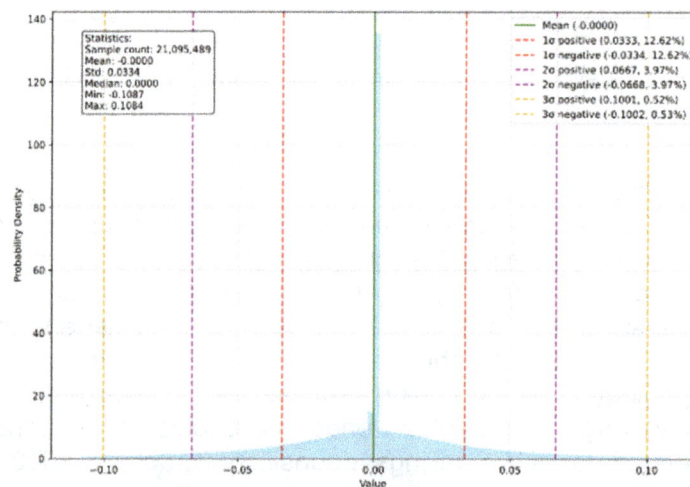

FIG 4 – DoD statistical distribution.

Furthermore, due to the limitations of computational memory with the preservation of spatial and temporal variability, a maximum of one million randomly sampled valid data points were retained from each DoD file. These resulting curated data sets were subsequently integrated into ML workflows for the purposes of model training and testing.

In parallel, relevant topographic derivatives were extracted from the DSMs to facilitate investigation into the terrain factors influencing observed elevation changes. A suite of relative terrain attributes—encompassing coordinates, slope, curvature, aspect, roughness, and Topographic Position Index (TPI), time interval—were computed utilising standard Geographic Information System (GIS) algorithms. These derived parameters serve to characterise local surface terrain and specific geomorphological features relating to the dynamics of tailings deposition.

Machine learning (ML) prediction and metric evaluation

For the machine learning (ML) training and prediction tasks, a comprehensive data set was constructed, comprising over five million pixel-level instances. Each instance was characterised by 13 input features derived from UAV-based DSMs and their associated terrain derivatives. These features included spatial coordinates (X, Y), time intervals between DSM acquisitions, and surface morphological indicators such as slope, aspect, curvature, roughness, and topographic position index (TPI). To capture the spatiotemporal dynamics of elevation change, several augmented features were introduced through interaction terms—for example, the multiplication of spatial coordinates with time-related variables. These cross features were designed to enhance the model's ability to learn localised temporal trends in deposition and erosion processes.

The data set was stratified into two subsets representing significant positive and negative elevation changes, corresponding to zones of deposition and subsidence, respectively. This stratification was based on statistical thresholds derived from the distribution of DSM differencing (DoD) values, focusing on extreme cases beyond the upper and lower third quartiles. The resulting subsets were randomly divided into training and testing sets using an 80:20 ratio to facilitate robust out-of-sample performance evaluation. Five regression algorithms were selected for model development: Linear Ridge Regression, Random Forest (RF), XGBoost, Gradient Boosting, and Histogram-based Gradient Boosting. These models were chosen to represent a range of complexity—from linear to non-linear ensemble learners—capable of capturing both simple and complex relationships between features and target variables. Given the structured nature of the data and the importance of model interpretability, tree-based models were prioritised.

Model parameters were optimised using hyperparameter tuning strategies aimed at minimising a predefined loss function while maximising prediction accuracy. Techniques such as randomised grid search and cross-validation were employed to identify optimal values for key hyperparameters, including maximum tree depth, learning rate, number of estimators, and regularisation terms. This process ensured a balance between model bias and variance, reducing the risk of overfitting and improving generalisation to unseen data.

Following training, the predictive performance of the resulting models from each algorithm was quantitatively assessed on the unseen testing subset using standard evaluation metrics, namely the R-squared (R^2), Mean Squared Error (MSE), and Root Mean Squared Error (RMSE).

RESULTS

Table 2 and Figure 5 present a comparative evaluation of the predictive performance across the five machine learning algorithms considered in this study. Among them, XGBoost exhibited the highest performance, achieving an R^2 value exceeding 0.84 on the training data sets for both positive (deposition) and negative (erosion) elevation change predictions. However, a performance drop was observed on the testing data sets, where R^2 values decreased to 0.643 and 0.708 for positive and negative changes, respectively. This discrepancy suggests a degree of overfitting, where noise or unmodelled mechanisms in the training data compromise the model's generalisation performance. While overfitting reflects the limitation of generalisation, it can be somewhat constrained when training on large and complex data sets using tree-based models such as XGBoost. This is especially the case when the input feature space is high-dimensional and incorporates complex engineered spatiotemporal interaction features, as in this study. To mitigate overfitting, future work could explore regularisation techniques, cross-validation strategies, or feature pruning to enhance model robustness and interpretability.

TABLE 2

Training and testing performance metrics for various machine learning algorithms applied to negative and positive elevation change models.

	ML algorithms	Training			Testing		
		R2	MSE	RMSE	R2	MSE	RMSE
Negative elevation change model	Linear Ridge	0.0161	0.1977	0.4447	0.0150	0.1994	0.4465
	Random Forest (RF)	0.5708	0.0863	0.2937	0.5280	0.0955	0.3091
	XGBoost	0.8422	0.0317	0.1781	0.7080	0.0591	0.2431
	Gradient Boosting	0.5485	0.0907	0.3012	0.5121	0.0988	0.3143
	Hist-gradient Boosting	0.3409	0.1325	0.3639	0.3349	0.1346	0.3669
Positive elevation change model	Linear Ridge	0.0145	0.2429	0.4929	0.0148	0.2442	0.4941
	Random Forest (RF)	0.5170	0.1191	0.3450	0.4617	0.1334	0.3652
	XGBoost	0.8425	0.0388	0.1971	0.6427	0.0886	0.2976
	Gradient Boosting	0.6375	0.0893	0.2989	0.5309	0.1163	0.3410
	Hist-gradient Boosting	0.2474	0.1855	0.4307	0.2481	0.1863	0.4317

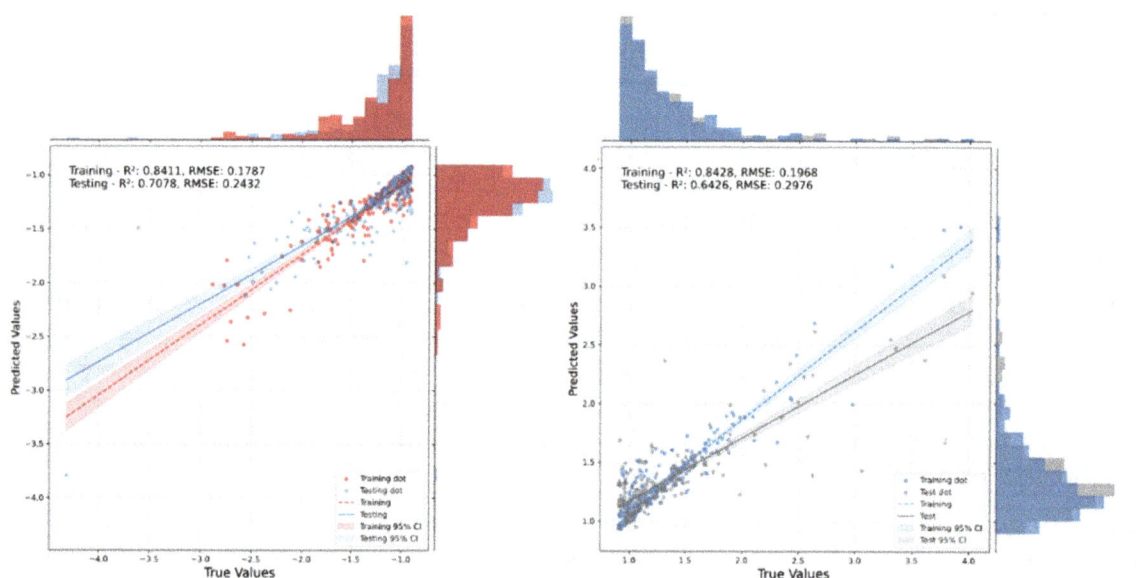

FIG 5 – Model performance evaluation showing predicted versus true values for two prediction tasks: negative DoD model (Left) and positive DoD model (Right). Results differentiate training and testing data by different colours, including 95 per cent confidence intervals.

In comparison, other algorithms such as Random Forest and Gradient Boosting performed moderately well but consistently underperformed relative to XGBoost in both training and testing phases. For instance, Random Forest achieved R^2 values of 0.528 (negative) and 0.462 (positive) on testing data, significantly lower than XGBoost. Linear Ridge Regression yielded minimal predictive power, with R^2 values close to zero, reflecting its inability to model the non-linear and spatial-temporal complexities inherent in TSF deformation processes. These comparisons present the advantage of utilising XGBoost algorithm, which effectively capture complex patterns while maintaining relatively strong generalisation ability.

To interpret the predictions of the best-performing XGBoost models, SHAP analysis, a game-theoretic approach, was employed to quantify the contribution of each input feature to the predicted output variables. Figures 6 and 7 provide further insight into the internal decision-making processes of the XGBoost models through SHAP analysis. Figure 6 presents a SHAP summary plot, illustrating the distribution of feature contributions across the data set for both positive (left) and negative (right) elevation change predictions. Each point represents a SHAP value corresponding to a single observation, with colour denoting the original feature value. Notably, features such as spatial coordinates, time interval, and their interaction terms consistently exhibit high SHAP values, indicating that spatial positioning and its interplay with deposition timing are critical determinants of significant elevation changes. This pattern implies that the temporal dynamics of deposition are spatially heterogeneous across the TSFs, likely reflecting operational practices such as localised discharge or directional slurry flow.

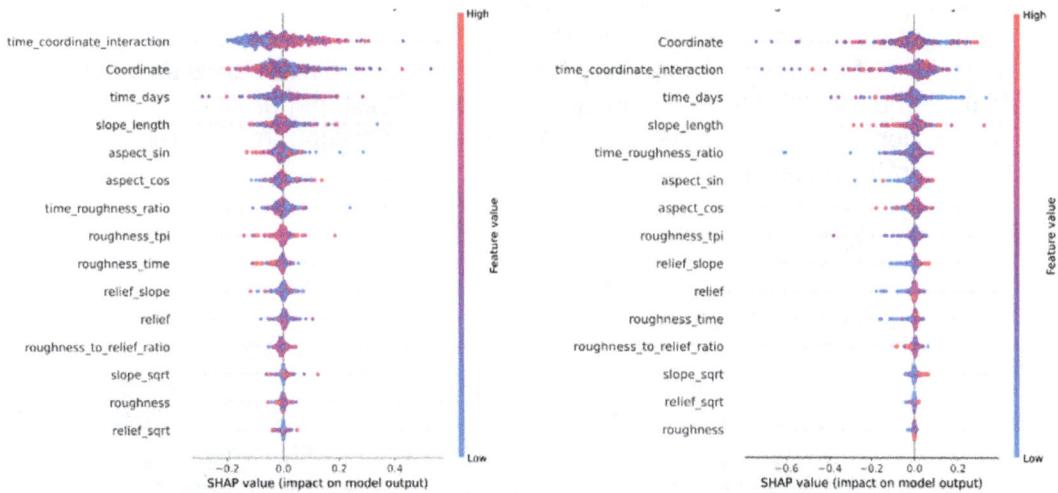

FIG 6 – SHAP summary plot, positive SHAP value (left) and negative SHAP value (right).

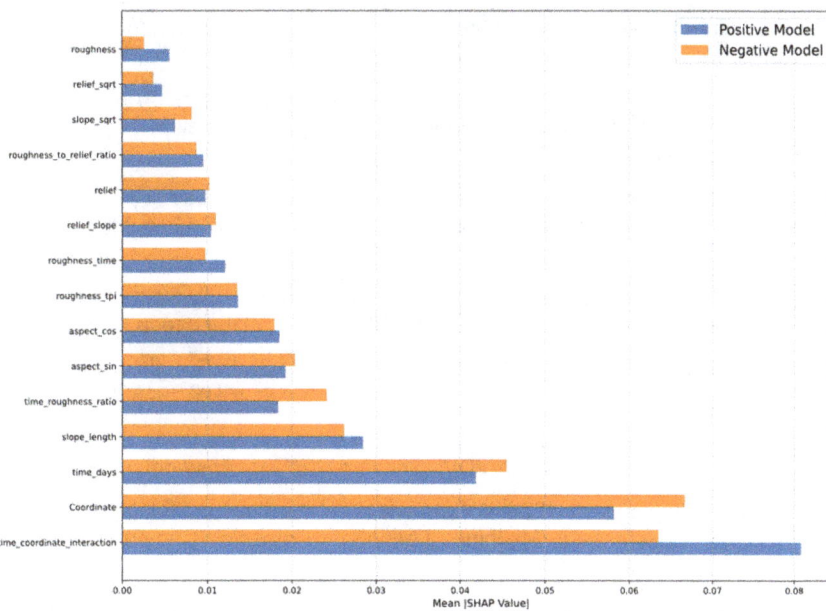

FIG 7 – Features importance comparison SHAP values.

Figure 7, which depicts a bar plot of mean SHAP values, enables a comparative analysis of feature importance between the positive and negative elevation change models. The results confirm that coordinate and time-coordinate interaction feature consistently dominate the importance rankings, collectively accounting for approximately 30 per cent of the model's predictive influence. This finding underscores the pivotal role of spatiotemporal coupling in driving TSF surface evolution, exceeding the predictive power of traditional topographic descriptors such as slope or curvature. These insights show the presence of structured operational cycles and directional deposition mechanisms in active cells.

CONCLUSION

This study presents a promising ML-based framework to predict significant elevation changes in TSFs, utilising UAV-derived DSMs for enhanced spatial and temporal resolution. The predictive performance of the XGBoost algorithm outperformed other four ML models across both deposition (positive change) and sedimentation (negative change) cases. Although training R^2 values exceeded 0.84 for both models, the lower prediction performance on test data, (ie R^2 values of 0.64 and 0.70 for positive and negative elevation changes, respectively), indicates moderate overfitting. Future work can explore regularisation techniques, hybrid models, or additional data augmentation strategies to mitigate this. Lastly, the framework's reliance on UAV data makes it scalable and adaptable to other TSFs globally. However, further validation across diverse sites with varying geomorphological and climatic conditions is essential to establish generalisability. Future directions include the integration of near-real-time DSM updates, incorporation into digital twin systems, and deployment for operational decision-making support.

ACKNOWLEDGEMENTS

We gratefully acknowledge Evolution Mining for providing the high-resolution DSM data set from the Mungari site, which served as the foundation of this study. We extend our special thanks to the Geotechnical Engineering team, particularly Dr. Mohammad Ali (Moe) Moridi and Mr. Anthony Kerr, for their technical support and valuable insights into the practical engineering aspects of this research. We also wish to express our sincere appreciation to the Western Australian School of Mines (WASM) at Curtin University for their institutional support and for providing the computational infrastructure essential to the development and analysis of the proposed models.

REFERENCES

Adamo, N, Al-Ansari, N, Sissakian, V, Laue, J and Knutsson, S, 2020. Dam Safety: Monitoring of Tailings Dams and Safety Reviews, *Journal of Earth Sciences and Geotechnical Engineering*, 249–289. https://doi.org/10.47260/jesge/1117

Cavalli, M, Goldin, B, Comiti, F, Brardinoni, F and Marchi, L, 2017. Assessment of erosion and deposition in steep mountain basins by differencing sequential digital terrain models, *Geomorphology*, 291:4–16. https://doi.org/10.1016/j.geomorph.2016.04.009

Chovan, K M, Julien, M R, Ingabire, E P, James, M, Masengo, É, Lépine, T and Lavoie, P, 2021. A risk assessment tool for tailings storage facilities, *Canadian Geotechnical Journal*, 58(12):1898–1914. https://doi.org/10.1139/cgj-2020-0329

Erdoğan, S, 2010. Modelling the spatial distribution of DEM error with geographically weighted regression: An experimental study, *Computers and Geosciences*, 36(1):34–43. https://doi.org/10.1016/j.cageo.2009.06.005

Groom, J, Bertin, S and Friedrich, H, 2019. Moving-Window Detrending for Grain-Roughness Parameterization, *Journal of Hydraulic Engineering*, 145(6):06019009. https://doi.org/10.1061/(ASCE)HY.1943-7900.0001612

Luo, W, Gan, S, Yuan, X, Gao, S, Bi, R and Hu, L, 2022. Test and Analysis of Vegetation Coverage in Open-Pit Phosphate Mining Area around Dianchi Lake Using UAV–VDVI, *Sensors*, 22(17). https://doi.org/10.3390/s22176388

Mwanza, J, Mashumba, P and Telukdarie, A, 2024. A Framework for Monitoring Stability of Tailings Dams in Realtime Using Digital Twin Simulation and Machine Learning, *Procedia Computer Science*, 232:2279–2288. https://doi.org/10.1016/j.procs.2024.02.047

Rauhala, A, Tuomela, A, Davids, C and Rossi, P M, 2017. UAV Remote Sensing Surveillance of a Mine Tailings Impoundment in Sub-Arctic Conditions, *Remote Sensing*, 9(12). https://doi.org/10.3390/rs9121318

Sun, E, Zhang, X and Li, Z, 2012. The internet of things (IOT) and cloud computing (CC) based tailings dam monitoring and pre-alarm system in mines, *Safety Science*, 50(4):811–815. https://doi.org/10.1016/j.ssci.2011.08.028

Wang, K, Yang, P, Hudson-Edwards, K A, Lyu, W S, Yang, C and Jing, X F, 2018. Integration of DSM and SPH to Model Tailings Dam Failure Run-Out Slurry Routing Across 3D Real Terrain, *Water*, 10(8):1087. https://doi.org/10.3390/w10081087

Wheaton, J M, Brasington, J, Darby, S E and Sear, D A, 2010. Accounting for uncertainty in DEMs from repeat topographic surveys: improved sediment budgets, *Earth Surface Processes and Landforms*, 35(2):136–156. https://doi.org/10.1002/esp.1886

Witt, K, Schönhardt, M, Saarela, J, Frilander, C E R, Csicsak, J, Csővari, M, Várhegyi, A, Georgescu, D, Radulescu, C and Zlagnean, M, 2004. Tailings management facilities–risks and reliability, Report of the European RTD project, 76.

Zubícek, V, Hudecek, V, Orlíková, L, Dandos, R, Jadviscok, P, Adjiski, V and Dlouhá, D, 2024. Thermal signatures of relict coal spoil, waste and tailing dumps, *International Journal of Mining Reclamation and Environment*, 38(3):267–279. https://doi.org/10.1080/17480930.2023.2263264

Integrating complementary technologies for tailings facility topographic surveying

V Nell[1]

1. VP Products and Solutions, PhotoSat, Vancouver V4E 1T2, Canada.
   Email: veronique.nell@photosat.ca

ABSTRACT

Ensuring the safety, compliance, and efficient management of tailings storage facilities (TSFs) demands robust surveying techniques. In his context, the integration of complementary technologies and data sources offers a holistic approach to TSF surveying. This paper explores the synergies between imagery and surveys from drones, high-resolution optical satellites, rapid revisit optical satellites, bathymetric sensors, and LiDAR for comprehensive TSF surveying.

While LiDAR excels at capturing elevation data, it faces constraints related to accessibility, expense, and revisit intervals. Drones, in comparison, can offer excellent flexibility and accessibility for capturing detailed video, imagery and topography of TSFs. However, their limited collection area can lead to challenges in the merging of data from multiple flights, as well as the time to acquire and process data over an entire TSF.

High-resolution and rapid revisit optical satellites provide complementary data to drones and LiDAR by capturing a whole TSF in a single acquisition at revisit rates ranging from days to weeks. They offer a supplementary, cost-effective solution with rapid delivery times. In addition to ongoing surveying, archived satellite imagery can retroactively provide time stamped data to determine the deposition history of a TSF when as-built records are unavailable.

Finally, bathymetric surveys provide essential information about underwater topography. By combining bathymetric data with surface topography, operators can gain a comprehensive understanding of TSF geometry, construction, and deposition plan advancement.

Integrating data from these complementary sources allows the TSF operator to overcome the limitations of each technology and enhances the overall accuracy and completeness of TSF surveying.

This paper highlights the collaborative potential of diverse technologies, emphasizing their collective impact on TSF survey quality and accuracy. The integration of these complementary technologies provides operators the data for safer and more efficient management of tailings facilities, aligning with industry standards and sustainability goals.

INTRODUCTION

Ensuring the safety, compliance, and efficient management of tailings storage facilities (TSFs) demands robust topographic surveying techniques. In this context, the integration of complementary technologies and data sources offers a holistic approach to TSF surveying.

No two sites are identical. Each location presents unique challenges, influenced by factors such as size, ease of access, operating requirements, reporting requirements. TSF operators often face recurring challenges that affect the overall design and implementation of the survey program.

Frequency of data acquisition – Different data is needed at different frequencies and for different tasks. Month end reporting, weekly operational updates, and integration with multiple sensors such as bathymetry collections, all require data collections at different frequencies, resolutions, accuracies and over different areas. Another consideration in this category is how quickly data can be collected and how fast the topography can be delivered post collection so that a site has the data they need when they need it.

Offset errors and data integration – Once the data from multiple sources and sensors is collected, it is important to ensure that all the data sets are coherent for all teams to be working from the same baseline. Correcting any offset errors is a key step in achieving effective data integration.

Lack of data continuity – With many TSFs having been in operation for years, if not decades, changing regulations, staff turnover, and changes of ownership can lead to incomplete records and lack of topographic data continuity. With the implementation of new global standards for TSF management, these data gaps can hinder current operators managing their TSFs within the accepted standards.

In this paper, we will look at the strengths and limitations of four surveying methods, or tools: ground surveyors, manned airborne LiDAR, drones, satellites (high resolution optical and rapid revisit optical). We will compare their performance in six main categories:

1. Area coverage: how large of an area can this tool survey in a single pickup?

2. Accuracy: what is the absolute vertical accuracy of the survey?

3. Repeatability: how repeatable are the results from one survey to the next?

4. Flexibility: how quickly can data be captured from request?

5. Cost: what is the cost of the survey?

6. Processing time: how quickly after the raw data is capture can the survey be delivered?

The following sections will detail how these different tools can be used to address different surveying needs and how, by leveraging their complementary characteristics, their individual limitations can be offset.

DIFFERENT TOOLS FOR DIFFERENT NEEDS

Ground surveying

Ground surveying is the original surveying method and Ground Surveyors provide unmatched accuracy on discrete areas of change – sub-centimetre horizontal and vertical accuracy in some cases. The cost of these surveyed points can also be low if work is being performed by existing site survey teams.

The presence of personnel on-site can present challenges, as the surveying work needs to be conducted in a way that does not interfere with operation, while ensuring safety of all personnel. Also, with a limited number of on-site surveyors, the demands of multiple groups can lead to scheduling issues. When scheduling challenges can be avoided, ground surveyors provide a very flexible way to collect data points over small areas.

Survey points from ground surveyors provide valuable and accurate survey points. Their value can be increased exponentially when paired and merged with the other survey methods described below, where they can be used as ground control or to get as small number of highly accurate survey points over a critical area.

Figure 1 offers a visual representation of the strengths and limitations of ground surveying across the six categories described.

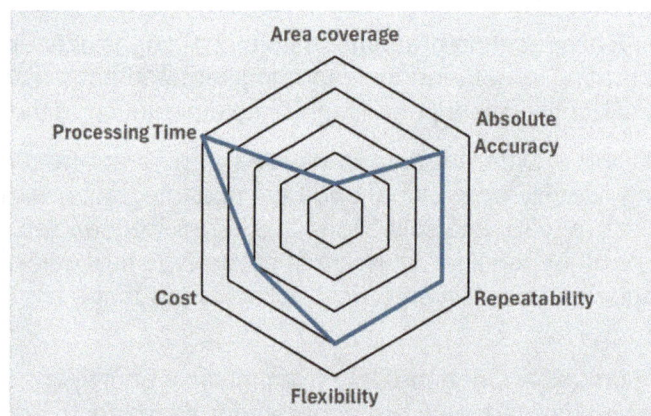

FIG 1 – Strengths and limitations of ground surveying.

Fixed wing manned airborne LiDAR

Manned Airborne LiDAR is the traditional benchmark for high accuracy surveys over mine site scale areas. With vertical accuracies as low as 5 cm (with ground control) and the ability to capture an entire TSF in a single flight, LiDAR surveys can be a great option for a baseline survey that all subsequent surveys are matched to. LiDAR's ability to penetrate vegetation means that the resulting surface is a bare ground surface – also referred to as a DTM, Digital Terrain Model – another key feature of a baseline survey.

LiDAR surveys are large data sets with very high data density. These can be long to process and hard to manipulate. The need to dispatch a manned aircraft means this is a relatively high-cost option, with some regional variation, and can also present challenges for more remote sites or in certain parts of the world where this type of service may not be readily available.

Figure 2 offers a visual representation of the strengths and limitations of LiDAR across the six categories described.

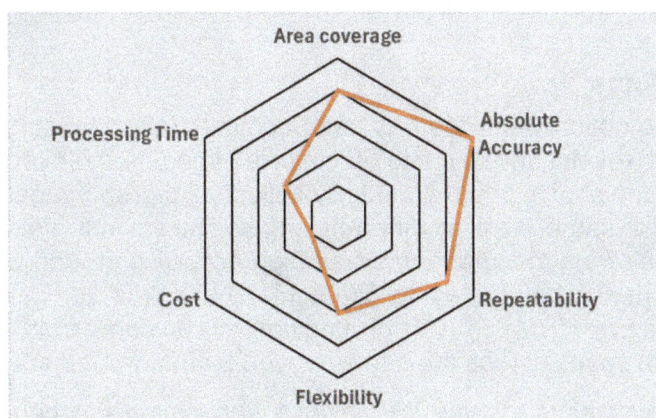

FIG 2 – Strengths and limitations of fixed wing manned airborne LiDAR.

Drones

Drones have come a long way in the last few years; significantly increasing their payload capacity and autonomy. In the context of topography, the key sensors are optical (including multispectral) and LiDAR. Despite the improvements, the payload capacity is still limited so high-end LiDAR and multispectral sensors may require larger, more expensive drones.

Since they fly at lower elevation than traditional manned aircraft, drones can avoid some atmospheric conditions and capture clear data even when clouds are present. The lower flight altitude can also lead to higher resolution imagery and topography with vertical accuracy down to the centimetre level. They are nevertheless sensitive to weather conditions such as high winds, rain, fog, or extreme temperatures that can affect flight stability and data quality. Adverse weather may limit the operational window for surveys.

In-house drone programs provide incredible flexibility for regular data acquisition if scheduling requests from multiple teams on a site can be managed.

Limited flight times mean that the whole TSF can often not be captured in a single flight, which can lead to offsets between different pickups. Also, surfaces derived from drone data are typically Digital Surface Models (DSM), which still include features such as vegetation and buildings. These surfaces need to be post processed to remove these features and produce a bare ground survey.

Figure 3 offers a visual representation of the strengths and limitations of drones across the six categories described.

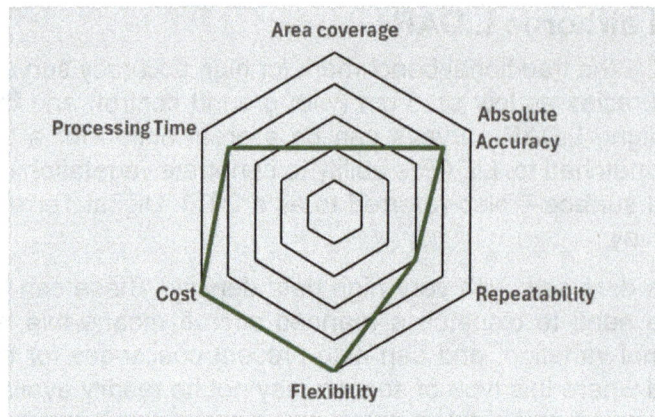

FIG 3 – Strengths and limitations of drones.

Satellites

High-resolution satellites

Data from high resolution optical satellites has been commercially available since the early 2000s. Their ability to collect high resolution (30 cm to 50 cm pixel) imagery over very large areas (hundreds of square kilometres) in one shot is unparalleled. By collecting stereo imagery of an area of interest, that is two images of the same area at different angles, taken only seconds to minutes apart, topography can be derived from the imagery. When high accuracy ground control over areas of no-change is available this method of surveying can achieve absolute vertical accuracies better than 15 cm Root Mean Square Error (RMSE). This ground control can come from existing LiDAR surfaces or discrete ground control points across the site surveyed by the ground surveying teams

When applied to TSFs, this means a whole TSF, even a whole mine site, can be collected in a single, instantaneous snapshot. Since the topography is derived from a single acquisition, the consistency of the data over the whole area of interest is excellent. The challenges of correction for errors and offsets that can come when multiple collections need to be stitched together are avoided. The data is self-consistent which leads to high precision topography. As subsequent collections come from the same sensor and go through the same processing systems, there is excellent repeatability between surveys over a same area and the results from surveys over different areas are easily comparable.

Figure 4 offers a visual representation of the strengths and limitations of high-resolution satellites across the six categories described.

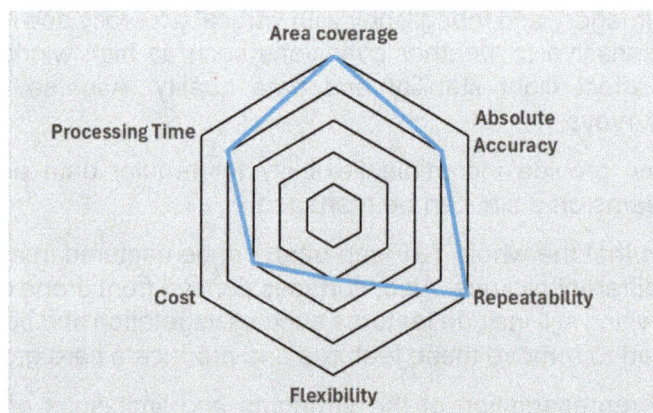

FIG 4 – Strengths and limitations of high-resolution satellites.

Rapid-revisit satellites

Rapid Revisit Satellite constellation, and specifically ones from which topography can be derived, are a new development in surveying. These growing constellations of ten or more satellites can provide collection opportunities weekly or even daily anywhere in the world, with collection occurring as little as 3 hrs after tasking.

Their collection area offers larger coverage than a single drone pickup but smaller than high-resolution satellites and manned airborne LiDAR. The imagery resolution and survey accuracy from these satellites is currently lower than what is achieved with the other tools described above but the sensor technology and processing is rapidly improving. These characteristics make them an attractive option in areas where drones are unavailable or where on-site drone availably is limited.

Figure 5 offers a visual representation of the strengths and limitations of rapid revisit satellites across the six categories described.

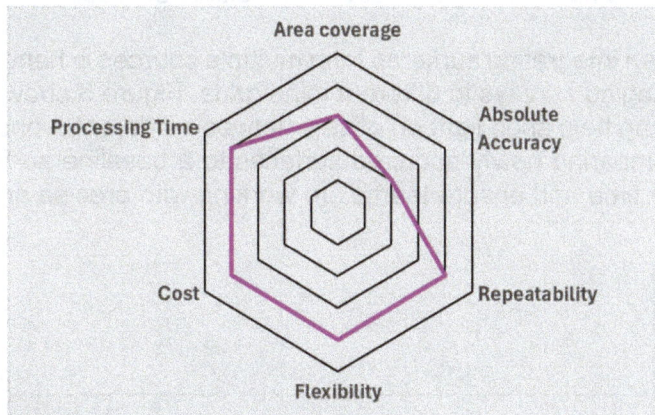

FIG 5 – Strengths and limitations of rapid revisit satellites.

Archived satellite imagery

There is a continuous archive of commercially available optical satellite imagery dating back to the 1960s. Figure 6 shows a non-exhaustive list of satellites for which archive data is commercially available. Topography can be generated from most of the sensors on this list.

Accessing archived imagery means that current operators can fill potential data gaps left by change of ownership, personnel, and regulations, to better understand how a site was operated in the past in order to plan for safer and more efficient operations moving forward.

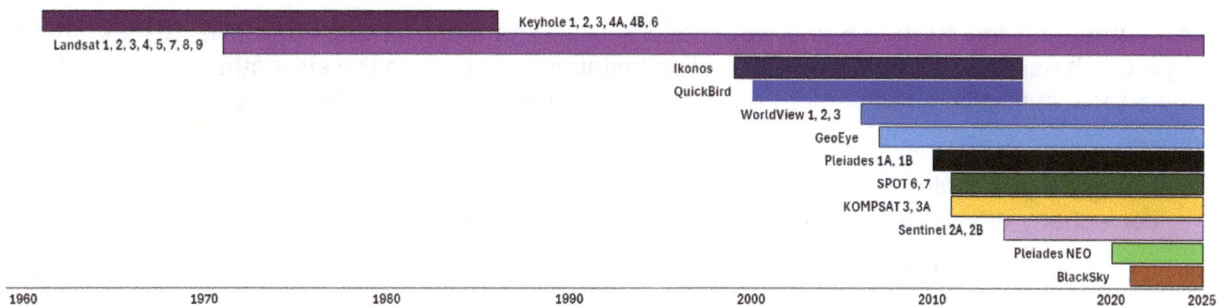

FIG 6 – A timeline of commercially available satellite imagery.

LEVERAGING SYNERGIES

Integrating data from multiple sources

The surveys derived from all the acquisition tools described above are valuable on their own, but to fully leverage their synergies, the various surfaces can be integrated with one another to provide a single, up to date surface for all the site teams to work with. This is also true for other sensors such as bathymetric sensors whose data can be integrated with the above water surface.

Figure 7 illustrates how topographic data from multiple sources can be merged into a single surface. This means teams can work with a single, cohesive and coherent surface that covers the whole area while also integrating the most up to date data.

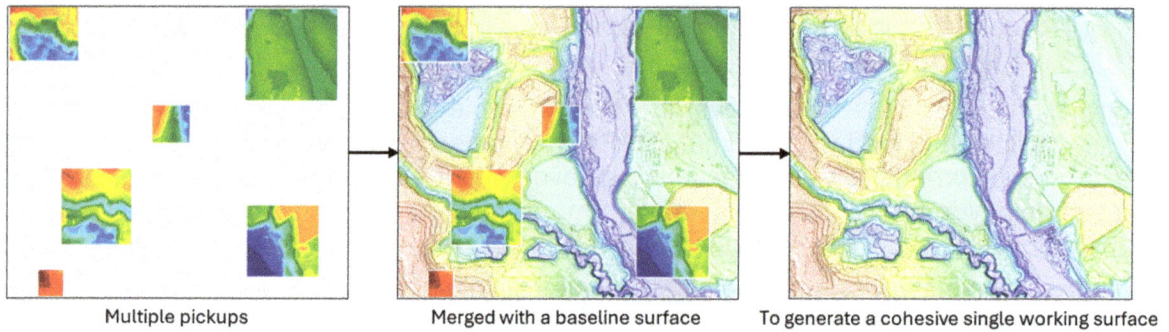

| Multiple pickups | Merged with a baseline surface | To generate a cohesive single working surface |

FIG 7 – Merging data from multiple sources.

A common challenge when integrating surfaces from multiple sources is handling offsets in data and, and in some cases, managing surveys in different mine grids. Figure 8 shows how using a baseline surface of a whole TSF can help shed light on offsets between different drone pickups and therefore help to correct them. Comparing newly acquired surfaces to a baseline surface will help avoid any propagation of error over time and ensure teams are working with precise and reliable data.

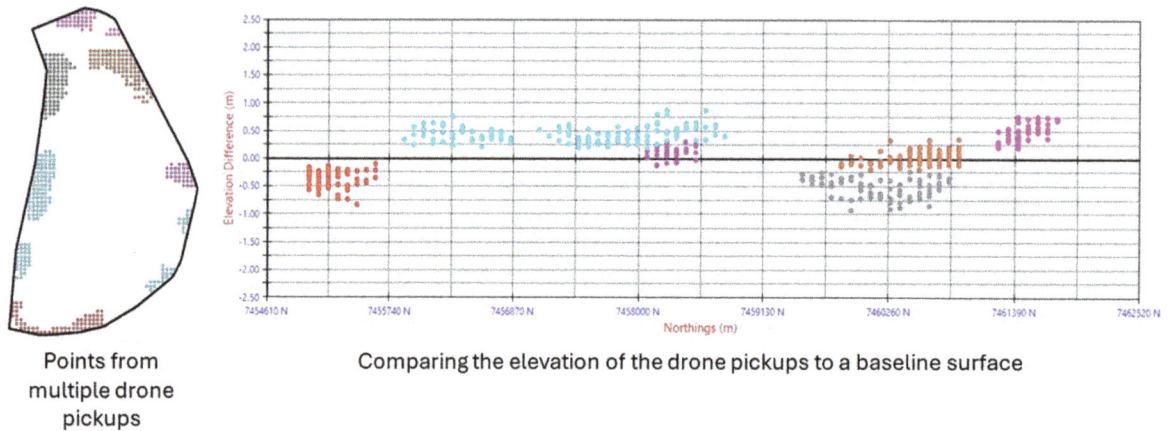

Points from multiple drone pickups

Comparing the elevation of the drone pickups to a baseline surface

FIG 8 – Correcting offsets using a baseline survey.

Leveraging strengths to offset limitations

When comparing the strengths and weaknesses of all the tools discussed above, we see that by using a combination of them, we can offset the limitations of one with the strengths of another. Using multiple tools supporting each other can increase the value of each of those tools on their own. For example:

- Using a high-resolution satellite or LiDAR survey as a baseline to cross-check drone data and correct any offsets makes the survey from drone data more precise and reliable.

- Using highly accurate points surveyed by a ground surveyor to tie any other kind of survey to, will greatly improve the absolute accuracy of that survey.

- By integrating an above water surface with a bathymetric survey, operators can generate a single TSF survey to work from, improving overall understanding of the TSF surface and supporting efficient operations.

Figure 9 illustrates how the limitations of each tool can be offset using the strength of a different tool.

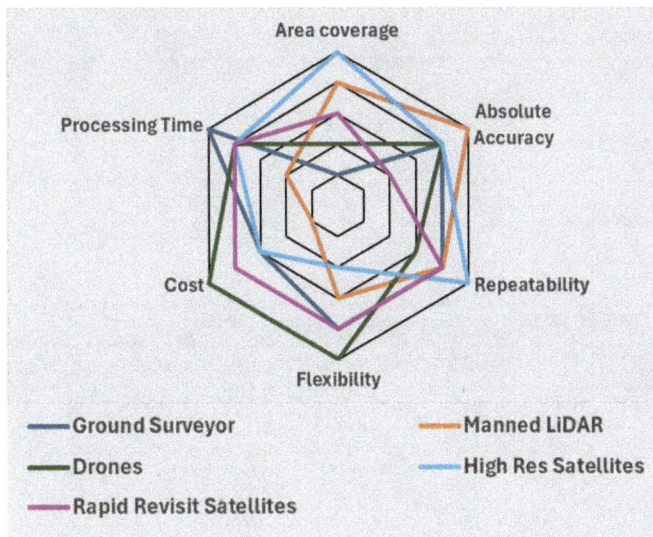

FIG 9 – Leveraging strengths to offset limitations.

The right data at the right frequency

Each tool described is better suited to different frequency of acquisition. By leaning into the strengths of each tool, TSF operators can ensure they have the right, reliable data when they need it. The cycle shown in Figure 10 is an example of a possible acquisition frequency for each tool.

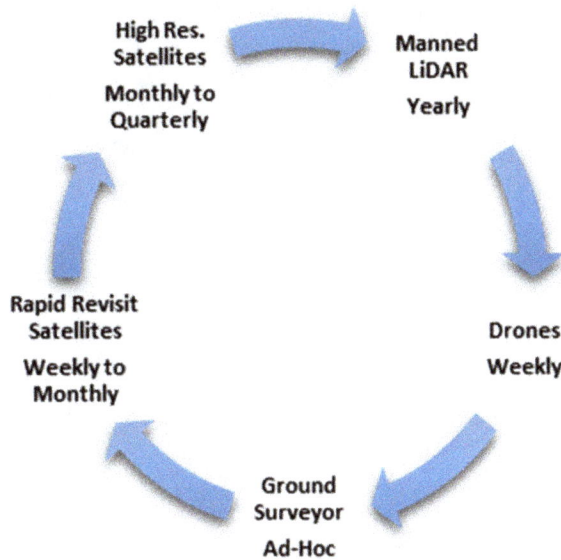

FIG 10 – The right data at the right frequency.

CONCLUSION

Designing and implementing a complete survey program is less about getting more data and more about getting the right data at the right time and efficiently integrating survey data from multiple sources.

Trade-offs will always need to be made around safety of personnel on-site, collection logistics, cost over time and accuracy. By optimising collection frequency to capitalise on the strength of the tools at their disposal, TSF managers will be able to acquire the survey data necessary to make informed decision for efficient operations.

Monitoring over the TSF life cycle – a structured approach

W Weinig[1], R Rekowski[2], T Gerritsen[3], J Rogers[4] and B Wang[5]

1. Discipline Leader Hydrogeology and Mine Closure, Stantec, Brisbane Qld 4006.
   Email: walter.weinig@stantec.com
2. Senior Principal Tailings Engineer, Stantec, Brisbane Qld 4006.
   Email: robert.rekowski@stantec.com
3. Principal Tailings Engineer, Rio Tinto Copper, Brisbane Qld 4000.
   Email: theo.gerritsen@riotinto.com
4. Principal Tailings Engineer, Rio Tinto Copper, Salt Lake City UT, USA.
   Email: josh.rogers@riotinto.com
5. Senior Principal Tailings Engineer, Stantec, Santiago Chile. Email: bing.wang@stantec.com

ABSTRACT

The mining industry is experiencing a rapid change in how TSFs are instrumented and monitored. A methodical, well-documented approach to TSF instrumentation and monitoring can help satisfy requirements of GISTM, International Committee on Large Dams (ICOLD, 1996, 2023), Australian National Committee on Large Dams Incorporated (ANCOLD Inc, 2019) and Canadian Dam Association (CDA, 2013) as well as corporate governance programs. Instrumentation and data management are the core tools available to evaluate the success of tailings governance policies as they are applied to individual facilities.

Instrumentation and monitoring programs at TSFs often develop haphazardly over time. Instruments may be installed based on a perceived need without evaluating the long-term usefulness of the data in achieving the monitoring program objectives. Closed TSFs may have monitoring systems based on operational requirements that are no longer applicable. Meanwhile each instrument carries ongoing costs for maintenance, monitoring visits, data management, and eventual replacement or decommissioning.

The authors have developed a playbook for instrumentation and monitoring system design that can be applied to a TSF at any point in its life cycle. The playbook takes advantage of the accumulated knowledge and experience of tailings practitioners across the globe and the guidance offered by industry consortiums. During design, the playbook can be used to optimise the planned monitoring and reporting systems given the critical factors identified from risk assessments. Existing systems at operating facilities can be assessed against current and future practices and areas for improvement identified. The changing monitoring needs for facilities in closure and post-closure stages can be assessed and appropriate long-term monitoring practices identified.

Applying the playbook offers the opportunity to standardise TSF monitoring systems. Once implemented, operational staff and corporate-level managers can benefit from an improved understanding of TSF conditions and risks across a portfolio of TSFs.

INTRODUCTION

The importance of effective and comprehensive monitoring programs at tailings storage facilities (TSFs) has become evident in recent years. Industry consortia such as the International Council on Mining and Metals (ICMM) and the Society for Mining, Metallurgy and Exploration (SME) provide recommendations on appropriate monitoring approaches for TSFs throughout their life cycles from construction through closure and post-closure care (eg ICMM, 2019, 2021; Morrison, 2022; amongst others). Key industry guidelines such as the Global Industry Standard on Tailings Management (GISTM; Global Tailings Review (GTR), 2020) and those produced by the Australian National Committee on Large Dams (ANCOLD Inc, 2019), the Canadian Dam Association (CDA, 2013) and the International Committee on Large Dams (ICOLD, 1996, 2023) describe the need to monitor TSFs. Major mining companies provide their own guidelines as part of increasingly robust tailings governance frameworks.

Data from TSF monitoring programs play key roles in safety evaluations and early detection of potential issues. The number of monitoring points and the sophistication of instrumentation continue to grow. The result is a trend toward increased data needs, both in number and variety, to satisfy the wide range of requirements for TSF monitoring and reporting.

While industry guidance is many and varied, some aspects of a TSF monitoring program are not well covered. Elements that are not typically addressed include:

- Guidance on design – usually provided by manufacturers.
- Information on installation, such as construction quality assurance (CQA).
- Pitfalls and common flaws in the overall system.
- Data processing from logger to actual data and dashboard.
- Dashboard implementation.
- Threshold levels for Trigger Action Response Plans (TARPs).
- Communications protocols.

These gaps can contribute to TSF monitoring programs growing haphazardly over time as the TSF evolves from construction into operation and eventually into closure.

OBJECTIVES

Why do we monitor TSFs throughout their life cycles? Ultimately, we need data to take appropriate actions. Figure 1 illustrates the path from raw data collection to action using data from vibrating wire piezometers (VWPs) as an example. The sequence includes all aspects of the monitoring system, from the physical connections through calculations and ultimately comparison to relevant TARP limits. A similar set of connections and calculations exists for each type of measurement – VWPs, inclinometers, accelerometers, open standpipe water levels, weather, remote sensing, etc.

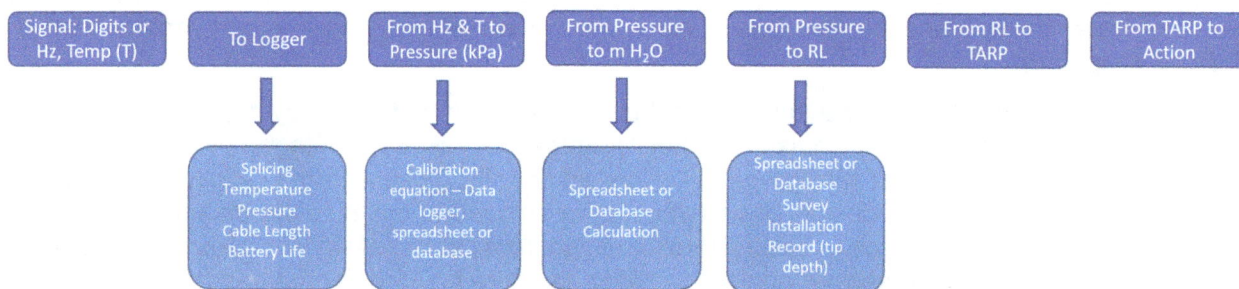

FIG 1 – The path from VWP data to action.

The overarching goal of the work presented in this paper was to develop a consistent framework for the design, installation, operation and data management aspects of TSF instrumentation and monitoring considering the full sequence from installation and raw data collection through decision-making based on the data. Specific objectives included:

- Evaluate current practices and installations at TSFs in different phases of their life cycle.
- Identify state-of-the-practice systems that could be applied across a full range of TSFs.
- Develop a workflow to ensure TSF instrumentation, monitoring and data-management practices meet industry standards.
- Identify emerging technologies that show potential for incorporation into existing or future TSF monitoring programs.

PLAYBOOK DEVELOPMENT

The TSF Monitoring Playbook takes advantage of the accumulated knowledge and experience of a range of tailings practitioners located in Australia and in North America. The development process included:

- Site visits to three mining operations with TSFs at different stages of development:

- Ongoing operations, recently designed and commissioned.
- Ongoing operations, legacy design.
- New design, not yet in construction.
- Closed and reclaimed.

- Review of tailings and foundation conditions to ensure that Observational Methods are valid.
- Discussions with operations personnel to understand their priorities and challenges.
- Literature and guidance review to identify current practice.
- Interviews with practitioners and vendors to identify promising near-term and long-term emerging technologies.

Consideration of data quality is key to developing robust, defensible TSF monitoring programs. Ljunggren, Logan and Fløystad (2023) presents the idea of the Data Quality Cycle (DQC) illustrated in Figure 2 to describe the importance of decision-making in each step of the monitoring program.

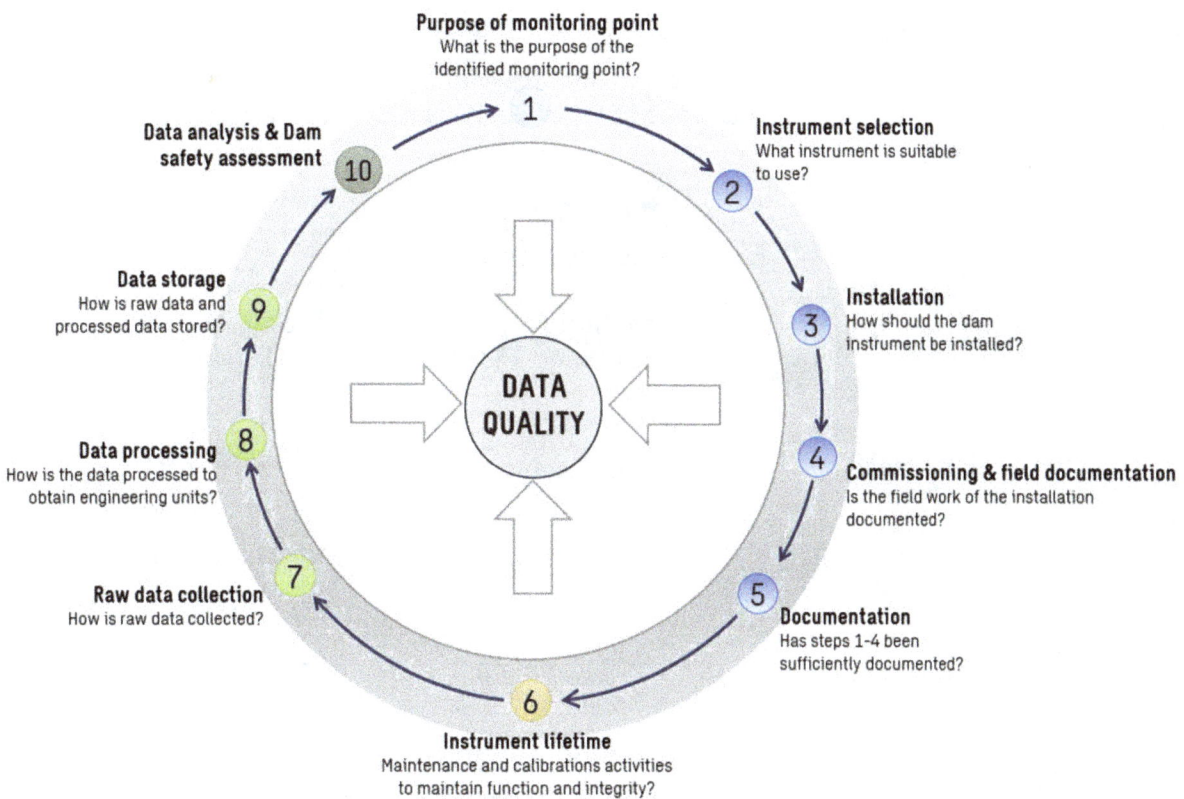

FIG 2 – The data quality cycle (DQC, after Ljunggren, Logan and Fløystad, 2023).

Top-level index categories

The Playbook is implemented in Microsoft Excel® as a living quick-reference guide. The top-level index categories summarised in Table 1 allow the user to work through a structured process similar to the ten steps of the DQC. The user can enter the process at any stage and link from the index page to topical pages that describe current practices, considerations, and potential issues that may be encountered across the range of topics. The Playbook includes a bibliography that is categorised against the topics shown on the index page for further exploration on individual issues.

TABLE 1

Playbook top-level index categories.

Section 1: Type of Tailings Storage Facility	Section 5: Climate and Setting Considerations	Section 9: Construction Instruments
Section 2: Risk Assessments	Section 6: Design	Section 10: Closure and Decommissioning
Section 3: Dam Consequence Rating	Section 7: Instrumentation Types	Section 11: Troubleshooting
Section 4: Failure Modes	Section 8: Data Mgmt and Dashboards	Section 12: Bibliography

Examples of topical page structure

The top-level index categories are broken down into individual topics that are implemented as tabs or pages in the Excel workbook. Topical-page titles for Sections 6, 7, and 8 are shown in more detail in Table 2 as examples.

TABLE 2

Example topical categories.

Section 6: Design	Section 7: Instrumentation Types	Section 8: Data Mgmt and Dashboards
8 Essential Steps	Pore Water Pressure Measurement	Processing Data
Life of Facility Considerations	Lateral Movement Sensors	Critical Elements to Display
Purpose of Instrument	Vertical Movement Sensors	Timeliness of Data Display and/or Processing
Common Instrumentation Combinations	Remote Sensing	Data Management Tools
Installation and Calibration	Pond Level and Precipitation	Routine Performance Reporting
Redundancy of Critical Instruments	New and Emerging Technologies/Trials	Trigger Action Response Plans (TARPs)
Common Flaws in Overall System	Key Notes About Suppliers	Data Resolution
	Key Notes About Manufacturers	

Drilling down further into the Playbook, the topical pages from Section 7, Instrumentation Types, includes the types of content summarised in Table 3. As seen in Table 3 there is some overlap in content that allows for users to reach the information through different paths in the Playbook.

TABLE 3

Examples of content for topical categories in Section 7, Instrumentation Types.

Section 7: Instrumentation Topical Category	Content

Pore Water Pressure Measurement	• Introduction: Physical quantity and rationale
	• Descriptions:
	  o Standpipe piezometers
	  o Vibrating wire piezometers
	  o Pneumatic piezometers
	• Pore Water Pressure Instrument Summary: Characteristics of different types of instruments

Lateral Movement Sensors	• Introduction: Physical quantity and rationale
	• Descriptions:
	  o Slope Inclinometer
	  o Terrestrial Radar
	  o Total Stations (Survey)
	  o Interferometric Synthetic Aperture Radar (InSAR)
	  o Borehole Extensometers
	  o Accelerometers

Vertical Movement Sensors	• Introduction: Physical quantity and rationale
	• Considerations: General characteristics of major measurement categories
	• Instrumentation and Monitoring Considerations:
	  o Extensometers
	  o Settlement gauges
	  o Global Positioning System (GPS)
	  o Light Detection and Ranging (LiDAR)
	  o Photogrammetry
	• Examples

Remote Sensing	• Introduction: Physical quantity and rationale
	• Descriptions:
	  o InSAR
	  o Hyperspectral monitoring
	  o Photogrammetry
	  o Unmanned Aerial Vehicle (UAV) surveys

Pond Level and Precipitation	• Introduction: Physical quantity and rationale • Considerations: General characteristics and key concerns of which to be aware • Pond Level and Water Balance Schematic • Instrumentation and Monitoring Considerations: Categories of measurement with advantages and disadvantages
New and Emerging Technologies/Trials	• Introduction: Changing landscape of technologies • Considerations: General characteristics and key concerns of which to be aware • Fibre-optic Technology • Electrical Resistivity Tomography (ERT) • Muon Vision
Key Notes about Suppliers	• List of suppliers and areas of expertise
Key Notes about Manufacturers	• List of manufacturers and major product categories

Example content – remote sensing page

The following sections reproduce the content from the Remote Sensing page within Section 7, the Instrumentation Types topical category. While not an exhaustive review, the content allows the Playbook user to quickly assess key concerns related to the topic and understand the types of instruments that may be applicable to the monitoring system under consideration.

Introduction

Remote sensing can enhance tailings dam monitoring, particularly in improving safety and minimising personnel exposure to risk. Although the initial set-up for some remote sensing technologies might be costly, over time, these methods can be more cost-effective than traditional monitoring. They reduce the need for frequent on-site visits and physical inspections, thereby saving resources and reducing risk exposure. Remote sensing data can be integrated with data from ground-based monitoring systems, such as piezometers and tiltmeters, to create a comprehensive monitoring strategy. This integration enhances the overall understanding of the dam's condition, leading to better-informed decision-making regarding safety measures Remote sensing enables the monitoring of the entire dam and its surrounding areas, not just specific points. This broad coverage ensures that no part of the dam is overlooked, providing a comprehensive understanding of its condition and potential risks.

InSAR

InSAR can be a powerful tool for tailings dam monitoring by providing high-resolution, precise measurements of ground deformation over time. InSAR's non-intrusive nature, ability to cover large areas, and high precision make it a valuable tool in the ongoing effort to ensure the safety and integrity of tailings dams.

- Surface Deformation Monitoring: InSAR can detect and measure very small surface movements over large areas, including the subtle deformation of tailings dams, which may indicate potential stability issues or failure risks.

- Early Warning System: Continuous or periodic InSAR monitoring can serve as an early warning system for detecting changes in the dam's structure or integrity. This allows for proactive measures to be taken before a minor issue becomes a critical failure.

- Operational Safety: By identifying areas of potential risk, InSAR data can inform operational decisions, such as where to place or relocate critical infrastructure, how to adjust tailings deposition practices, and when to perform targeted inspections or maintenance.

- Historical Deformation Analysis: InSAR can provide a historical record of deformation over the lifespan of a tailings dam. Analysing this data can help understand the dam's behaviour under various load conditions, contributing to improved design and management practices for existing and future facilities.

- Integration with Other Monitoring Systems: InSAR data can be integrated with information from other monitoring systems, such as piezometers, GPS, and visual inspections, to provide a comprehensive view of the dam's health and stability.

- Water Management: InSAR can help monitor the phreatic surface within the tailings, informing water management practices that can influence the dam's stability.

- Regulatory Compliance and Reporting: The precise, quantifiable data provided by InSAR can support compliance with regulatory requirements for dam safety monitoring and facilitate transparent reporting to stakeholders.

Site specific studies should be performed where applicable to calibrate InSAR data and interpretations. Tailings facilities offer a unique advantage because settlement of the tailings facility is usually measured by other sensors and can be used to calibrate the InSAR data with the vendor.

Hyperspectral monitoring

Hyperspectral photo analysis can enhance tailings dam monitoring through its advanced capabilities for detailed and precise material characterisation over large areas.

- Identification of Minerals and Chemicals: Hyperspectral imaging can detect specific minerals and chemicals in the tailings dam area, including potentially hazardous substances. This allows for the monitoring of chemical changes or leakages that could indicate the presence of contaminants seeping into the environment.

- Moisture Content Analysis: By analysing the spectral signatures, hyperspectral imaging can assess moisture content in the dam. Variations in moisture content can indicate potential weak spots or areas where structural integrity might be compromised due to saturation.

- Erosion and Sedimentation Monitoring: Hyperspectral data can help identify areas of erosion or sedimentation build-up. These phenomena can alter the dam's stability and effectiveness, signalling when maintenance or corrective actions are needed.

- Vegetation Health Monitoring: The health and distribution of vegetation around tailing dams can be indicators of water quality and soil stability. Hyperspectral imaging can detect stress in vegetation, which might be due to contaminated groundwater or soil, providing indirect clues about the dam's impact on the surrounding ecosystem.

- Thermal Anomalies Detection: While traditionally associated with multispectral rather than hyperspectral imaging, combining these approaches can help detect thermal anomalies indicative of leaks or internal processes affecting the dam's integrity.

- Structural Integrity Analysis: Changes in the spectral signature of materials can also indicate alterations in the physical composition or structure, helping to identify areas that may require further geotechnical investigation.

- Leakage and Seepage Detection: Hyperspectral imaging can identify variations on the dam's surface or nearby areas that suggest the presence of unexpected water flow paths, seepage, or leakage, which are critical for maintaining the dam's safety and environmental protection.

By integrating hyperspectral photo analyses into routine tailings dam monitoring programs, operators can gain a more comprehensive understanding of the dam's condition and its environmental impact. The key to its effective application lies in the ability to process and interpret the vast amount of data generated, requiring specialised software and expertise in hyperspectral data analysis.

Photogrammetry

Photogrammetry is a technique used to measure distances, dimensions, and volumes, and to create maps or 3D models of real-world objects and environments from photographs. It involves capturing images from different angles and then using software to analyse these images to extract spatial data. This process relies on the principles of stereoscopy, where two or more photographs taken from different perspectives are used to reconstruct the 3D structure of the subject.

- Topographical Mapping: Photogrammetry can create detailed 3D models of the tailings dam and surrounding area, providing accurate topographical data.

- Volume Calculations: By generating precise 3D models, photogrammetry allows for accurate volume calculations of the tailings material. This is useful for managing the capacity of the dam.

- Change Detection: Regular photogrammetric surveys can detect changes in the dam's geometry over time, such as settling or erosion. Early detection of these changes can prompt timely interventions to prevent dam failure.

UAV surveys

UAV (Unmanned Aerial Vehicle) Surveys: Drones equipped with cameras or other sensors can conduct detailed inspections of areas that are difficult to access, providing high-resolution data for monitoring dam conditions.

CONCLUSIONS

The mining industry is experiencing a rapid change in how TSFs are instrumented and monitored. The project team developed a structured Playbook to guide design and implementation of TSF instrumentation and monitoring programs. Taking a stepwise, thoughtful approach to monitoring is an important step in translating the large volume of data collected at a typical TSF into actionable information. The documented process can also help satisfy requirements of GISTM (GTR, 2020), ICOLD (1996, 2023), ANCOLD Inc (2019) and CDA (2013) as well as corporate governance programs.

The Playbook is designed as a living quick-reference guide that can be updated as new requirements and new methodologies emerge. It is not intended to be a complete standalone document or replace good governance, good design and good operational function and good closure at each facility. It is instead intended to help the end user gain an insight into common configurations and warn of potential pitfalls at key decision points.

ACKNOWLEDGEMENTS

The authors acknowledge the contributions of project team members from both organisations who participated in the site visits and monitoring-program reviews. We appreciate the input of tailings, instrumentation, and data-management experts who helped make the Playbook a useful and living tool for improving TSF instrumentation and monitoring programs.

REFERENCES

Australian National Committee on Large Dams Incorporated (ANCOLD Inc), 2019. Guidelines On Tailings Dams - Planning, Design, Construction, Operation And Closure, revision 1, ANCOLD Inc. Available from: <https://ancold.org.au/product/guidelines-on-tailings-dams-planning-design-construction-operation-and-closure-revision-1-july-2019/>

Canadian Dam Association (CDA), 2013. Dam Safety Guidelines 2007, 2013 edn, CDA.

Global Tailings Review (GTR), 2020. Global Industry Standard on Tailings Management (GISTM) [online], Global Tailings Review. Available from: <https://globaltailingsreview.org/global-industry-standard/>

International Committee on Large Dams (ICOLD), 1996. *Bulletin 104 – Monitoring of Tailings Dams – Review and Recommendations* (International Commission on Large Dams).

International Committee on Large Dams (ICOLD), 2023. *Bulletin 194 – Tailings Dam Safety* (International Commission on Large Dams, Committee L – Tailings Dams and Waste Lagoons).

International Council on Mining and Metals (ICMM), 2019. Integrated Mine Closure Good Practice Guide, 2nd edn, ICMM.

International Council on Mining and Metals (ICMM), 2021. Tailings Management: Good Practice Guide [online], International Council on Mining and Metals. Available from: <https://www.icmm.com/tailings-management>

Ljunggren, M, Logan, T and Fløystad, A, 2023. The instrument data quality cycle, in Symposium 'Management for Safe Dams', 91st Annual ICOLD Meeting, Gothenburg, Sweden, 13–14 June 2023.

Morrison, K F (ed), 2022. *Tailings Management Handbook: A Life cycle Approach*, 1024 p (Society for Mining, Metallurgy and Exploration).

Non-invasive and real-time monitoring of tailings slurry density in transmission pipeline using the electrical resistance tomography method

Y Xiong[1], C Zhang[2], S Quintero[3], N Racha[4], T Bore[5] and D Williams[6]

1. PhD candidate, The University of Queensland, St Lucia Qld 4072. Email: yue.xiong@uq.edu.au
2. Senior Research Fellow, The University of Queensland, St Lucia Qld 4072.
   Email: chenming.zhang@uq.edu.au
3. Senior Research Technologist, The University of Queensland, St Lucia Qld 4072.
   Email: s.quintero@uq.edu.au
4. Coal Chain Superintendent, Stanwell Corporation Limited, Brisbane Qld 4000.
   Email: naresh.racha@stanwell.com
5. Senior Research Fellow, The University of Queensland, St Lucia Qld 4072.
   Email: t.bore@uq.edu.au
6. Emeritus Professor, The University of Queensland, St Lucia Qld 4072.
   Email: d.williams@uq.edu.au

ABSTRACT

Monitoring and maintaining the density of tailings slurry during pipeline transport is essential for optimising transportation efficiency and preventing blockages. This study proposes using electrical resistance tomography (ERT) to measure the cross-sectional density distribution of slurry tailings in a transmission pipeline as an alternative to the conventionally used nuclear densitometer. A pipeline system was constructed at the Long Pocket Campus, the University of Queensland, featuring a 200 mm-diameter pipeline, the same size as that used *in situ*. The system can be configured for a dam break test, in which 1 m³ of tailings slurry is discharged through a monitored pipeline at time-varying velocities ranging from 0 to 2 m/s, as well as a pipe loop test, where tailings slurry is circulated inside the loop at a constant velocity of up to 0.7 m/s. Sensitivity analyses were conducted by varying the solids density of the slurry and the electrical conductivity of the process water. In addition to laboratory tests, the ERT system was trialled *in situ* alongside a gamma-ray-based nuclear densitometer. Comparisons were made between density measurements obtained from ERT and those from the conventional method. This paper presents the results obtained to date.

INTRODUCTION

Australia is rich in mineral resources such as gold, bauxite, iron ore, lead, and coal, making it one of the world's leading mining nations. While the mining industry has significantly contributed to the country's economic development, its environmental impact cannot be overlooked. Mine tailings—residual materials left after extracting valuable metals or minerals—can cause serious environmental issues including water contamination, soil degradation, air pollution, and damage to vegetation. Therefore, the transport and management of tailings require careful handling.

Currently, the most effective and safest method of tailings management involves transporting tailings slurry via pipelines to a tailings storage facility (TSF) for centralised containment or to a subsequent mechanical dewatering process to produce dry cake for co-disposal with coarse rejects. Compared to traditional truck transport, pipelines can operate continuously 24 hrs a day, emit no exhaust fumes, avoid damage to road infrastructure, and reduce labour costs associated with truck operations. However, during transport over several kilometres—or even tens of kilometres—solid particles in the slurry may settle and accumulate, potentially leading to pipeline blockages (Kotzé *et al*, 2019). High solids concentration and low flow velocity are the main causes of sedimentation and blockage. Although increasing flow velocity or reducing solids concentration can prevent this issue, it also increases water and energy consumption and operating costs. To strike a balance between reducing the risk of pipeline blockage and minimising operational costs, it is essential to monitor the solids concentration within the pipeline and adjust operational parameters accordingly.

In recent decades, pipeline monitoring equipment has evolved from traditional intrusive probes to non-intrusive sensors. This is due to the difficulty of maintaining intrusive probes in highly abrasive slurries. Non-intrusive sensors are designed as ring-shaped units that become part of the pipeline itself. These sensors are not only more durable, but also do not disturb the flow, enabling more accurate and stable measurements. Currently, gamma-ray computed tomography (CT) is the most widely used technology for monitoring solids concentration in the mining industry. It provides high-resolution cross-sectional density images of the slurry in the pipe (Bieberle *et al*, 2013; Krupička and Matoušek, 2014). However, it involves radiation exposure, posing health risks to on-site personnel.

To eliminate radiation-related risks, alternative technologies that do not rely on radioactive sources must be explored. Electrical Resistance Tomography (ERT), a well-established and non-intrusive technique for real-time monitoring of fluid properties in pipelines, presents a promising solution. This study aims to evaluate the effectiveness of ERT in monitoring solids concentration in tailings slurry. The evaluation will commence with laboratory-scale experiments using a dam break set-up and a pipeline loop system and will subsequently be validated through field trials at a mine site. Results from ERT measurements will be compared with those obtained from existing gamma-ray density systems to assess performance. The study aims to demonstrate ERT's capability in monitoring solids concentration, reduce reliance on nuclear-based methods, minimise radiation risks and operational costs, and support the broader adoption of safer, more sustainable pipeline monitoring technologies. In addition, the feasibility of using a single-plane ERT system to monitor flow velocity will also be explored as part of the study's broader objectives.

FACTORS AFFECTING ERT MEASUREMENTS OF TAILINGS SLURRY

The ERT system primarily consists of three components: the sensor, the data acquisition system, and the image reconstruction system, as illustrated in Figure 1. In pipeline applications, the ERT sensor is typically designed in a ring shape, with 16 electrodes evenly distributed along the inner wall of the sensor. This creates a smooth internal surface that minimises any disturbance to the flow. Since different fluids have different electrical conductivities, ERT captures the conductivity distribution within the sensor's cross-section by systematically injecting current through selected electrodes and measuring the resulting voltage differences across others.

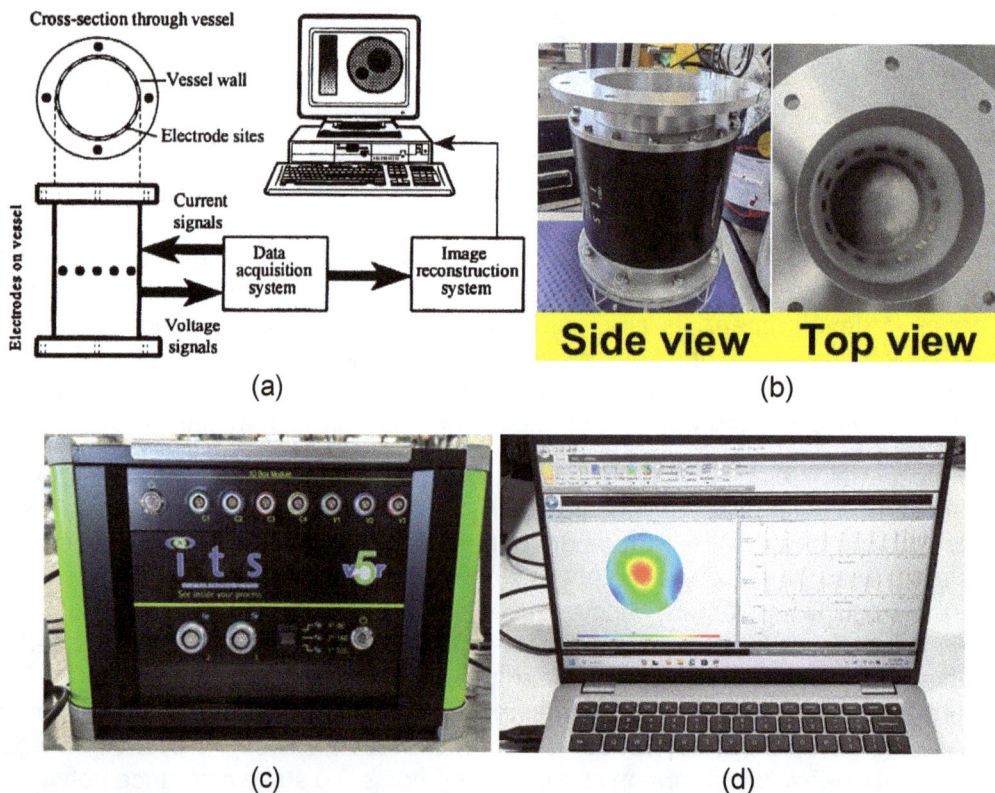

FIG 1 – Schematic and photos of the ERT System Components: (a) schematic diagram of an ERT system (Dyakowski, Jeanmeure and Jaworski, 2000); (b) ERT sensor; (c) data acquisition system; (d) image reconstruction system.

When tailings slurry is considered as the fluid under investigation, the primary focus is to examine the relationship between solid concentration and bulk electrical conductivity—this is the main objective of this study. In addition, the effects of factors such as the electrical conductivity of process water, clay content, and temperature on the overall conductivity also need to be analysed.

Impact of process water electrical conductivity

In 1942, an empirical formula known as Archie's law was proposed based on experimental observations. It has since become a standard method for describing the electrical conductivity of mixtures such as sand–water and rock–water systems. Archie's law relates the electrical conductivity of a saturated soil to the porosity and conductivity of the pore water. The formula is expressed as follows (Glover, 2010):

$$\sigma_{mix} = \varphi^m \cdot \sigma_w = \frac{1}{F} \cdot \sigma_w \tag{1}$$

where σ_{mix} is the electrical conductivity of the mixture, σ_w is the electrical conductivity of the water, φ is the porosity of the mixture, and m is the cementation factor. F is called formation factor, describing the effect of rock matrix. This implies that tailings slurry can be regarded as a mixture of liquid and solid phases, where the bulk electrical conductivity is primarily determined by the electrical conductivity of the liquid phase (ie process water) and the volumetric water content. The cementation factor m is related to the arrangement of solid particles and the complexity of the pore structure. Its value varies across materials such as sand and rock and is typically determined by fitting after obtaining σ_w and φ through laboratory measurements. It is important to note that the original equation does not account for the influence of temperature on electrical conductivity, and assumes that solid particles (eg rock) are non-conductive. However, in practice, the effect of temperature cannot be ignored, and a certain degree of conductivity may exist between solid particles.

Impact of temperature

The influence of temperature on electrical conductivity has been widely reported. An increase in temperature enhances the thermal motion of ions in water, leading to a rise in solution conductivity. This change typically follows an approximately linear relationship when the temperature is below 100°C. For aqueous mixtures such as tailings slurry, temperature fluctuations can cause significant variations in conductivity measurements. To minimise temperature-related effects, some studies have maintained a constant testing temperature (Fu et al, 2021), while others have applied temperature compensation formulas for correction (Giguère et al, 2008), such as Equation 2. Since the temperature in tailings slurry pipelines is subject to continuous variation, using a temperature compensation formula is considered a more suitable approach.

$$\sigma_w(T_{ref}) = \frac{\sigma_w(T)}{1 + \alpha_T(T - T_{ref})} \tag{2}$$

Impact of clay content

The electrical conductivity of soil particles primarily arises from surface conduction, whereby the electrical double layer at the particle surface provides an additional current pathway beyond that of the pore water. This phenomenon is particularly significant in soils containing a high proportion of clay minerals. Previous studies have indicated that tailings typically contain substantial amounts of clay minerals, including kaolinite, illite, and montmorillonite (De Kretser, Scales and Boger, 1997). These components are likely to influence the conductivity measurements obtained through ERT, and it is therefore essential to investigate the effects of clay minerals on the measurement results. Electrical conduction through pore water and along mineral particle surfaces is considered to occur in parallel. Accordingly, the electrical conductivity of tailings slurry containing clay minerals can be expressed by Equation 3.

$$\sigma_{mix} = \varphi^m \cdot \sigma_w + \sigma_{surf} = \frac{1}{F} \cdot \sigma_w + \sigma_{surf} \tag{3}$$

INVESTIGATION OF THE INFLUENCE OF CLAY CONTENT AND PROCESS WATER CONDUCTIVITY ON ELECTRICAL CONDUCTIVITY

The subject of this study is coal tailings slurry sourced from the Meandu mine site. First, the tailings slurry collected from the mine site was oven dried. The resulting solid blocks were then ground and sieved through a 1 mm mesh to obtain fine particle samples, as shown in Figure 2a. XRD analysis indicated that the clay mineral content in the sample was approximately 31.1 per cent, as shown in Table 1.

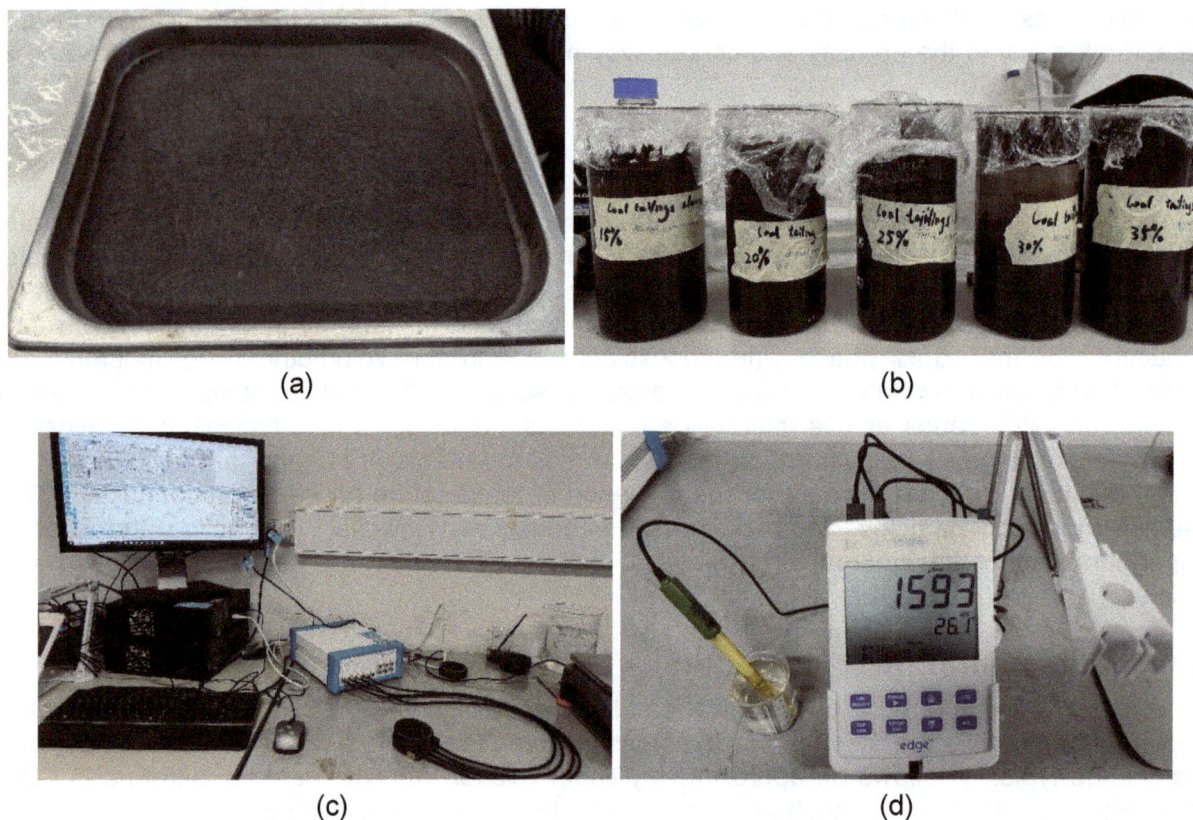

FIG 2 – Experimental procedures: sample preparation and conductivity testing: (a) fine coal tailings particles; (b) coal tailings slurry with different concentrations; (c) bulk electrical conductivity measurement; (d) pore water electrical conductivity measurement.

TABLE 1

XRD analysis results of coal tailings.

Minerals	Nominal wt%
Quartz	12.0
Anatase	0.5
Hematite	0.3
Siderite	1.5
K-Feldspar	3.2
Kaolinite	28.2
Illite/mica	2.9
Amorphous	51.5

In practice, the solid concentration of tailings slurry during pipeline transport is typically around 25 per cent by mass. To investigate the effect of solid concentration, tailings slurries were prepared at concentrations of 15 per cent, 20 per cent, 25 per cent, 30 per cent, and 35 per cent, as illustrated in Figure 2b. The experimental procedure is outlined here:

- Sodium chloride (NaCl) was dissolved in deionised water to prepare solutions with electrical conductivities of 0.001 S/m, 1 S/m, and 10 S/m. In addition, process water, with an electrical conductivity of approximately 0.158 S/m, was collected from the top of coal tailings slurry.

- Fine particle samples were combined with solutions of varying electrical conductivities to prepare tailings slurries with solid concentrations of 15 per cent, 20 per cent, 25 per cent, 30 per cent, and 35 per cent. For each concentration, four slurry samples were prepared using pore water with different conductivities: 0.001 S/m, 0.158 S/m, 1 S/m, and 10 S/m.

- Each sample was thoroughly mixed. A portion was taken and placed in the MFIA Impedance Analyser to measure the bulk electrical conductivity, as shown in Figure 2c. The actual solid concentration of each sample was also measured.

- The remaining sample was left to settle, and the supernatant was collected to remeasure the pore water conductivity, as shown in Figure 2d.

In most cases, the bulk conductivity is dominated by the electrical conductivity of the pore water, making it challenging to quantify the contribution from surface conduction. However, as shown in Equation 3, when the pore water conductivity approaches zero, the bulk conductivity of the slurry is primarily governed by surface conduction. Therefore, reducing the pore water conductivity as much as possible enables the investigation of surface conduction on particle surfaces. For this reason, NaCl solutions with conductivities ranging from 0.001 S/m to 10 S/m were prepared. This range allows for both an analysis of the influence of pore water conductivity on bulk conductivity and an estimation of the surface conduction. The results are presented in Figure 3.

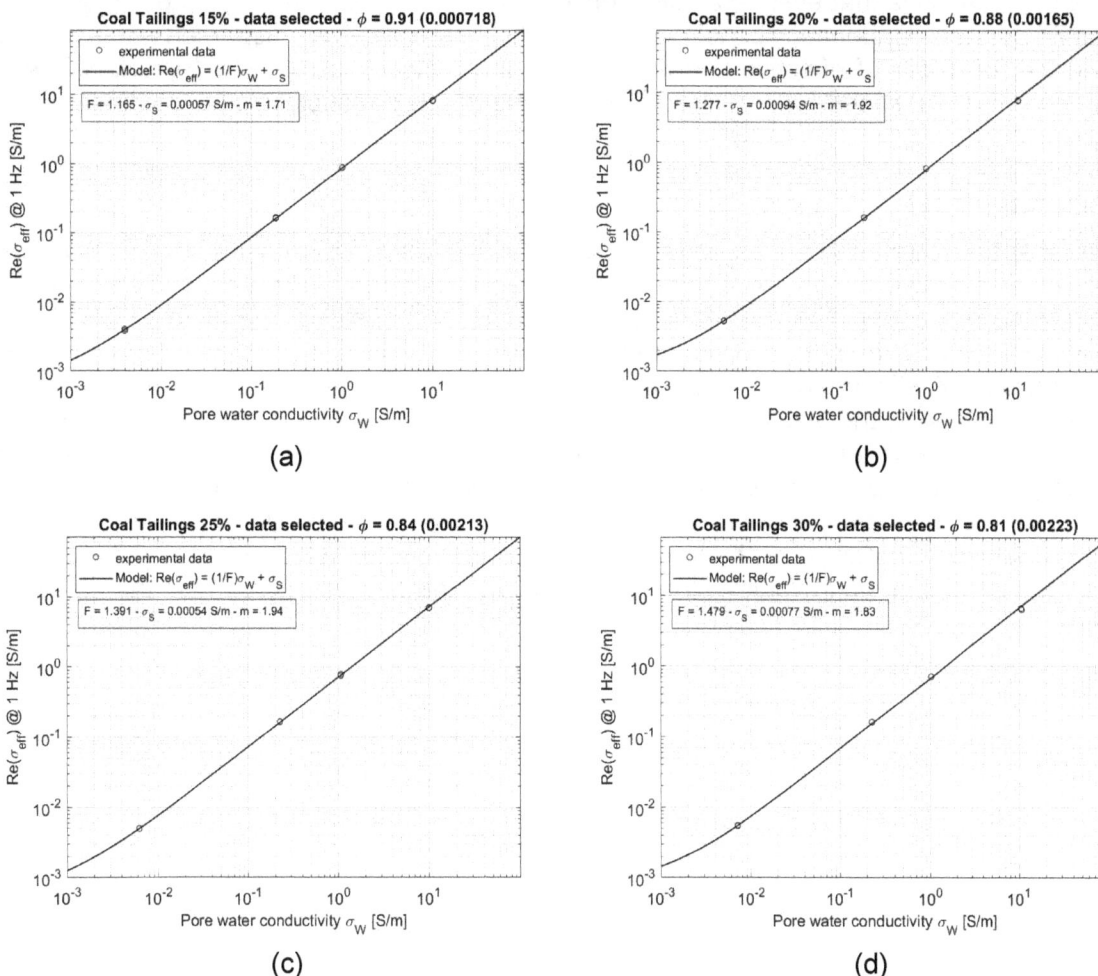

(a)

(b)

(c)

(d)

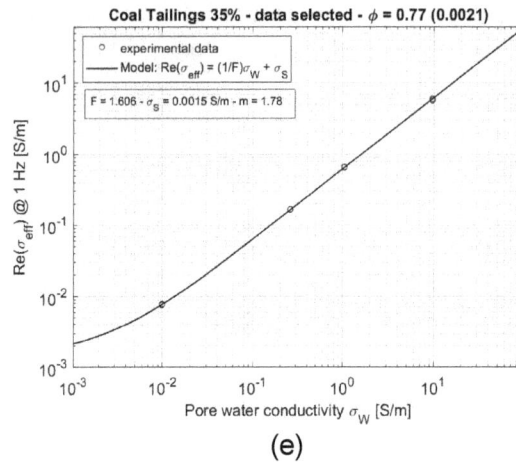

Coal Tailings 35% - data selected - $\phi = 0.77$ (0.0021)

experimental data
Model: $\mathrm{Re}(\sigma_{eff}) = (1/F)\sigma_W + \sigma_S$

$F = 1.606$ - $\sigma_S = 0.0015$ S/m - $m = 1.78$

(e)

FIG 3 – Relationship between bulk conductivity and pore water conductivity: (a) 15 per cent solids concentration; (b) 20 per cent solids concentration; (c) 25 per cent solids concentration; (d) 30 per cent solids concentration; (e) 35 per cent solids concentration.

As shown in the figure, the vertical axis represents the bulk conductivity, while the horizontal axis indicates the pore water conductivity. It can be observed that even when the pore water conductivity is very low (below 0.01 S/m), the tailings slurry does not exhibit a clear surface conduction effect. This suggests that within the solid concentration range of 15 per cent to 35 per cent, the bulk conductivity of the slurry remains predominantly controlled by the pore water conductivity, and the contribution from surface conduction associated with clay minerals can be considered negligible.

A possible explanation is that even at a solid concentration of 35 per cent, the porosity (φ) remains as high as 77 per cent. Under such conditions, clay particles are relatively dispersed within the slurry, preventing the formation of effective surface conduction pathways along particle surfaces. As the solid concentration increases further, the influence of surface conduction may gradually become more significant.

DAM BREAK TEST

A dam break apparatus was designed and constructed at the Long Pocket campus of The University of Queensland, as shown in Figure 4. The set-up includes an upstream and a downstream tank, both used to store tailings slurry. The upstream tank can store 1 m³ of slurry. A PVC pipe with 200 mm diameter is connected to the bottom of the upstream tank, with an ERT system installed at the middle section of the pipe. A sliding gate is positioned at the end of the pipe to control the flow of slurry. Before each experiment, the sliding gate is kept closed, and the slurry in the downstream tank is thoroughly stirred. The slurry is then pumped from the downstream tank to the upstream tank using a slurry pump and slurry pump hose. Once the sliding gate is opened, the slurry flows through the pipe into the downstream tank, and this entire process is continuously monitored by the ERT system. An ultrasonic sensor is mounted at the top of the upstream tank to monitor changes in slurry level. Based on the rate of change in volume, the flow velocity of the tailings slurry within the pipe can be estimated.

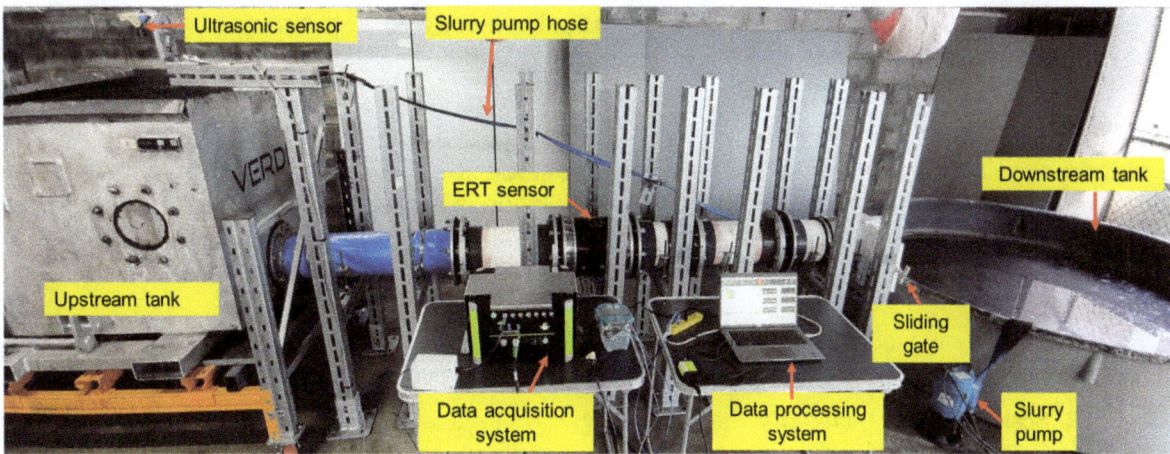

FIG 4 – Dam break apparatus.

The ERT system used in this study can generate 125 sets of raw data and reconstructing 125 frames of images per second, ensuring accurate capture of the conductivity distribution of the tailings slurry even under high flow velocity conditions. After each experiment, samples are taken to measure the solid concentration. The slurry is then diluted with process water to adjust the solid concentration, and the experiment is repeated. Figure 5 presents a selection of raw ERT measurements (2000 frames) under five different solid concentrations. The vertical axis represents the mean resistance, which is the average of multiple resistance measurements between electrode pairs. While it is directly related to the electrical resistance of the slurry, it does not represent the actual resistance value. The horizontal axis indicates the frame number. A clear separation can be observed between the resistance curves corresponding to different solid concentrations, indicating that the ERT system can detect variations in solid content.

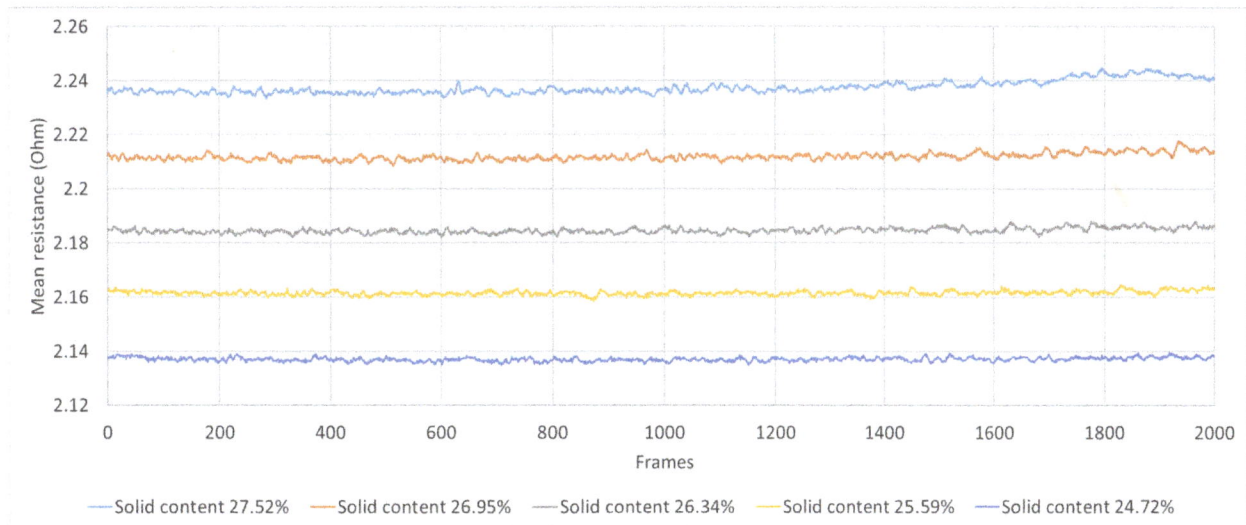

FIG 5 – Partial ERT measurement results of dam break test.

CONCLUSIONS

This paper presents the current progress in using an ERT system to monitor tailings slurry within pipelines, including the effects of pore water conductivity and clay surface conduction on the electrical conductivity of coal tailings slurry, as well as preliminary results from dam break experiments. The analysis indicates that the electrical conductivity of coal tailings slurry is primarily governed by the conductivity of the pore water. Under typical pipeline operating concentrations (15 per cent–35 per cent), the influence of clay minerals on slurry conductivity can be considered negligible. Furthermore, the dam break experiments demonstrate the effectiveness of the ERT system in distinguishing tailings slurries with different solid concentrations. Future work will focus on investigating the influence of temperature on ERT measurements. In addition, a series of pipe loop tests and field-scale trials will be conducted in the laboratory and at the mine site, respectively, to further evaluate the performance of the ERT system.

ACKNOWLEDGEMENTS

The authors would like to express their sincere gratitude to ACARP (Australian Coal Association Research Program) for funding this research. Appreciation is also extended to Stanwell Corporation for their support in providing the experimental site and samples.

REFERENCES

Bieberle, A, Härting, H U, Rabha, S, Schubert, M and Hampel, U, 2013. Gamma-Ray Computed Tomography for Imaging of Multiphase Flows, *Chemie Ingenieur Technik*, 85(7):1002–1011. https://doi.org/10.1002/cite.201200250

De Kretser, R, Scales, P J and Boger, D V, 1997. Improving clay-based tailings disposal: Case study on coal tailings, *AIChE Journal*, 43(7):1894–1903.

Dyakowski, T, Jeanmeure, L F C and Jaworski, A J, 2000. Applications of electrical tomography for gas–solids and liquid–solids flows — a review, *Powder Technology*, 112(3):174–192. https://doi.org/10.1016/S0032-5910(00)00292-8

Fu, Y, Horton, R, Ren, T and Heitman, J L, 2021. A general form of Archie's model for estimating bulk soil electrical conductivity, *Journal of Hydrology*, 597. https://doi.org/10.1016/j.jhydrol.2021.126160

Giguère, R, Fradette, L, Mignon, D and Tanguy, P A, 2008. Characterization of slurry flow regime transitions by ERT, *Chemical Engineering Research and Design*, 86(9):989–996. https://doi.org/10.1016/j.cherd.2008.03.014

Glover, P W J, 2010. A generalized Archie's law for n phases, *Geophysics*, 75(6):E247–E265. https://doi.org/10.1190/1.3509781

Kotzé, R, Adler, A, Sutherland, A and Deba, C N, 2019. Evaluation of Electrical Resistance Tomography imaging algorithms to monitor settling slurry pipe flow, *Flow Measurement and Instrumentation*, 68. https://doi.org/10.1016/j.flowmeasinst.2019.101572

Krupička, J and Matoušek, V, 2014. Gamma-ray-based measurement of concentration distribution in pipe flow of settling slurry: vertical profiles and tomographic maps, *Journal of Hydrology and Hydromechanics*, 62(2):126–132. https://doi.org/10.2478/johh-2014-0012

Field monitoring of soil covers for controlling moisture and oxygen ingress in wet climate waste rock dumps

C Zhang[1], L Tan[2], C Ptolemy[3] and D Williams[4]

1. Senior Research Fellow, Geotechnical Engineering Centre, University of Queensland, St Lucia Qld 4072. Email: chenming.zhang@uq.edu.au
2. Senior Research Assistant, Geotechnical Engineering Centre, University of Queensland, St Lucia Qld 4072. Email: lieven.tan@uq.edu.au
3. Senior Environmental Officer, Grange Resources, Burnie Tas 7320. Email: carl.ptolemy@grangeresources.com.au
4. Professor, Geotechnical Engineering Centre, School of Civil Engineering, University of Queensland, St Lucia Qld 4072. Email: d.williams@uq.edu.au

ABSTRACT

Acid mine drainage (AMD) is a significant environmental concern due to its ecological, ecotoxicological, and socio-economic impacts. Effective management and storage of potential acid-forming (PAF) materials during and after the life of a mine site present a major challenge. A key strategy for managing AMD is to control the net percolation of water and limit oxygen exchange, as these factors drive the formation of acid. This can be achieved by restricting oxygen ingress into PAF materials.

This study focuses on monitoring the 'as is' behaviour and performance of an existing incomplete soil cover system installed in a wet climate with an average annual rainfall of 1.9 m. The cover is currently incomplete and as such observes the clay like (C-Type) material barrier layer at varying depths, minimising net percolation and oxygen ingress into the underlying PAF waste. Above the C-Type layer, a non-acid-forming (NAF) material (A-type) is used as an armouring layer to mitigate evaporation, prevent erosion and root penetration through the C-Type layer.

A sensor array was deployed using sonic drilling and backfilled with the same material, extending to the PAF waste. These sensors were used to monitor moisture content, electrical conductivity (EC), temperature, and oxygen concentration profiles at various depths. Additionally, a weather station was installed to measure net inflow (via rainfall) and net outflow (via evapotranspiration).

Key findings from the study include:

- The clay like material barrier layer effectively intercepted water infiltration, limiting it from reaching the underlying PAF material.

- Much of the rainfall above the clay like layer was diverted horizontally through the NAF material and flowed horizontally, eventually appearing at the toe-drain.

- Oxygen (gaseous) content in the waterlogged waste remained relatively low, at less than 10 per cent.

This project provides valuable insights into the design of soil covers for AMD management in wet climates and presents the monitoring data collected to date.

INTRODUCTION

This study was engaged by Grange resources to the Geotechnical Engineering Centre at the University of Queensland (UQ) to perform long-term field monitoring on the performance of existing waste rock dump covers in controlling moisture and oxygen ingress in reducing acid mine drainage (AMD) risk.

AMD is an environmental issue caused by the exposure of sulfur-bearing minerals, like pyrite, to air and water during mining. This reaction produces sulfuric acid, which lowers the pH of surrounding water, making it highly acidic. The acid water often carries dissolved metals such as iron, copper, and lead, contaminating water bodies and harming the surrounding ecosystem. AMD occurs both during and after mining activities, continuing for years or decades post-closure. It can lead to the destruction of aquatic habitats, threaten biodiversity, and contaminate water supplies. Addressing AMD involves preventing water infiltration and oxygen ingress into potential acid forming wastes (PAF) and using neutralising agents or passive treatment systems to raise pH and remove contaminants (Jambor and Blowes, 2001).

AMD cover design typically involves multi-layered systems, including a base layer of impermeable material (like compacted clay), a drainage layer to direct water away, and a top layer for erosion control and vegetation. The cover also incorporates surface water management strategies and may include passive treatment systems to neutralise any acidic run-off. Long-term stability is crucial, and the design must withstand environmental processes like erosion and freeze-thaw cycles. Regular monitoring ensures the cover remains effective over time, with ongoing research into innovative materials and techniques to improve performance (Dold, 2003).

This field monitoring engagement aims to evaluate the performance of an incomplete constructed waste rock dump site located within a temperate climate region on its efficacy in AMD prevention. The region receives an average of 1942 mm of rainfall annually, with rain occurring consistently throughout the year. The region's high and consistent rainfall increases the risk of acidic run-off by promoting the oxidation of sulfur-bearing minerals in waste materials. Current waste rock dump characterisation of the site excluding chemical and synthetic barriers, involves the build-up of PAF classified as D-type waste, with a clay like material cover consisting of non-acid forming material (NAF) classified as C-Type, and the top armouring layer consists of high carbonate content NAF material classified as A-type to mitigate evaporation, prevent erosion and root penetration through the C-Type layer.

SET-UP AND INSTALLATION

Installation took place in May 2024 across five sites, with boreholes of 100 mm diameter drilled by sonic drilling. At each location, six Meter Group Teros-12 sensors were installed at various depths to provide comprehensive coverage of the stratigraphy, with the deepest sensor positioned just within the PAF wastes. Additionally, two gaseous oxygen sensors were installed at the same depth as the two deepest moisture sensors. The stratigraphic profile and sensor placement are depicted in Figure 1.

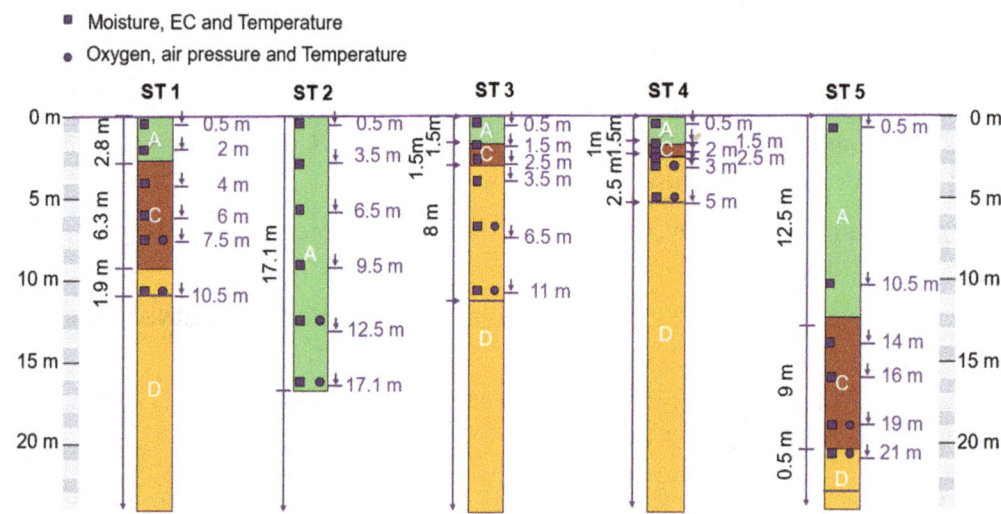

FIG 1 – Stratigraphy and sensor layout of each location.

Instrumentation and telemetry

Each borehole site was outfitted with a dedicated data logger acquisition system, employing 4G telemetry transmission for data transfer. In addition, ST3 and ST1 were equipped with UV and infrared sensors to measure solar radiation. At ST1, a weather station was also installed to monitor key environmental parameters, including temperature, humidity, pressure, wind speed, wind direction and rainfall (Figure 2).

FIG 2 – ST1 with weather station (left); sensor installation depicting PVC spacer support rods (right).

During the drilling process, materials were separated by layer and screened to remove large rocky debris. The screened material was then used to backfill the borehole. To ensure proper sensor placement, the depth of the borehole was measured, and sensors were secured to PVC sections at specified intervals (Figure 1). The installation process involved lowering the PVC support rods with attached sensors into the borehole as the drill casing was removed section by section and material was backfilled (Figure 2). To compensate for the loss of rocky material, additional backfill with similar characteristics was added to maintain the depth of the stratified layers.

For soil moisture readings, all reported moisture sensor measurements were fitted linearly between the same upper and lower boundary limit to obtain degree of saturation, as the sensor measures dielectric permittivity at 70 MHz, which effectively minimise the impact of mineralogy variations. Optical based oxygen sensors were used due to their low maintenance requirements for long-term operation with high accuracy at low concentrations, which required an adapter interface developed in-house to assist with long distance telemetry. Due to the varied conditions in which the oxygen sensors were located, post processing were required to filter low confidence oxygen telemetry marked by outlier temperature or barometric readings. This in conjunction with the adapter contributed to the gaps in oxygen telemetry readings.

FIELD STRATIFICATION PROFILE

Figure 3 illustrates the three material types present in the waste rock dump, based on core samples from ST1. From left to right, the sequence includes 2 m of NAF A-type armouring waste, followed by 6 m of clay like NAF C-Type material, and 2 m of PAF D-type waste.

FIG 3 – Image of sonic drill core samples for ST1: 3 m of NAF A-Type (left); 6 m of NAF C-Type (middle); 2 m of PAF D-Type Waste (right).

A summary profile of the layer thickness for A-type and C-Type material for all five locations are listed in Table 1.

TABLE 1

Stratified thickness of NAF A-type and C-Type layer summary.

	ST1	ST2	ST3	ST4	ST5
A-Type	2.8 m	17.1 m	1.5 m	1.5 m	12.5 m
C-Type	6.3 m	0 m	1.5 m	1 m	9 m

MONITORING OBSERVATION

ST1

Moisture at ST1

The degree of saturation was monitored throughout the recording period (Figure 4).

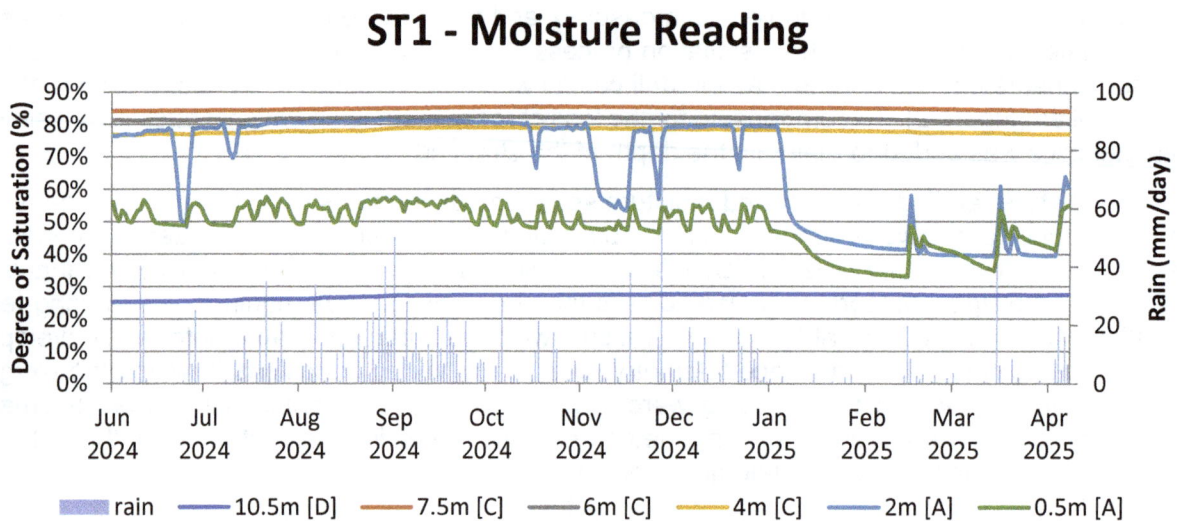

FIG 4 – Degree of saturation reading for ST1.

Water infiltration was detected to a minimum depth of 2 m within the NAF A-type armouring layer, particularly following a significant rainfall event in late November 2024, which recorded 90 mm of daily rainfall. Prior to this event, the ground was partially saturated to a depth of 2 m, but it quickly reached a peaked state during the rainfall. Between January and March 2025, during a period of minimal rainfall, a drying cycle was observed down to at least 2 m, likely caused by gravity drainage and evapotranspiration. Below 2 m, the ground moisture remained stable over the recorded period for sensors located within the NAF C-Type layer and the PAF D-type layer. The C-Type layer exhibited higher moisture readings compared to the sensor in the PAF layer.

Oxygen at ST1

Two oxygen sensors were installed for ST1 at depths of 7.5 m within the NAF C-Type layer and 10.5 m within the PAF. Both sensors recorded gaseous oxygen levels below 10 per cent, as illustrated in Figure 5.

ST1 - Gaseous Oxygen Level

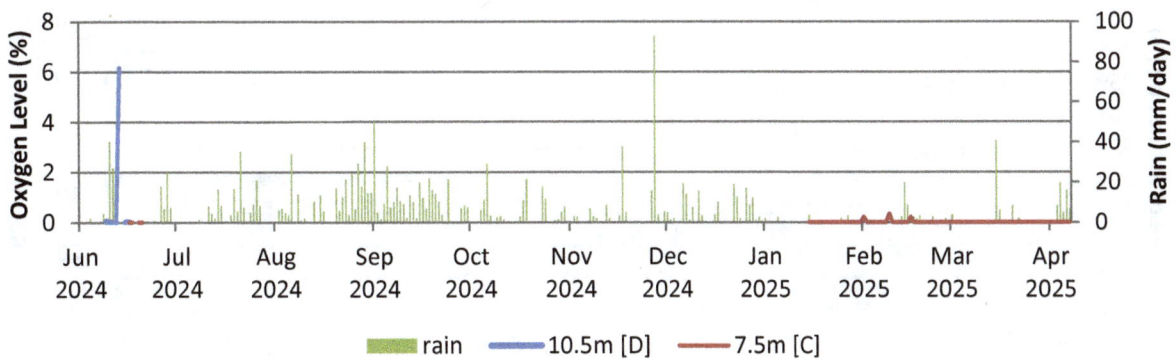

FIG 5 – Gaseous oxygen level for ST1.

Summary of data obtained at ST1

The observed water infiltration down to 2 m within the NAF A-type armouring layer, particularly following the significant rainfall event in November 2024, highlights the rapid infiltration capability of this layer. Prior to the event, the ground was partially saturated to 2 m, but the rainfall quickly led to a waterlogged state, demonstrating the armouring layer's high permeability. However, during the dry period between January and March 2025, a drying cycle was observed, indicating the armouring layer's limited moisture retention capacity.

Below 2 m, ground moisture remained stable, with the NAF C-Type layer showing higher moisture levels compared to the PAF D-type layer. The C-Type's layer's ability to effectively retain moisture also serves as a critical barrier to further water infiltration, preventing excessive moisture movement into the deeper layers. This underscores the C-Type layer's role in maintaining the stability of the moisture profile by limiting deeper infiltration and reinforcing its function as an effective water retention barrier.

The oxygen measurements in the NAF C-Type and PAF layers, with values consistently below 10 per cent, indicate low oxygen levels in these layers. These low oxygen readings suggest limited gas exchange due to the reduced permeability of the C-Type and PAF layers, further supporting the idea that the C-Type layer acts as an effective barrier to both water and gas movement.

Overall, the findings indicate that the armouring layer enables rapid infiltration and drainage, while the underlying C-Type layer effectively retains moisture and prevents deeper infiltration. The low oxygen levels further demonstrate the C-Type layer's role in restricting gas movement, contributing to moisture profile stability. Further monitoring is needed to evaluate the long-term performance of these layers in maintaining moisture and gas conditions.

ST2

A gap in telemetry was observed for ST2 between the period of September and October 2024, which was facilitated by the destruction of power cabling by the local wildlife.

Moisture at ST2

In contrast to other locations, the borehole drilled to a depth of 17.1 m for ST2 revealed only the NAF A-type armouring layer. During the observation period, the degree of saturation for ST2 showed minimal variation, with only slight increases recorded during the steady rainfall event from July to September 2024. A small, sharp increase in moisture was observed following the major rainfall event in November 2024, reaching down to 3.5 m, as shown in Figure 6.

ST2 - Moisture Reading

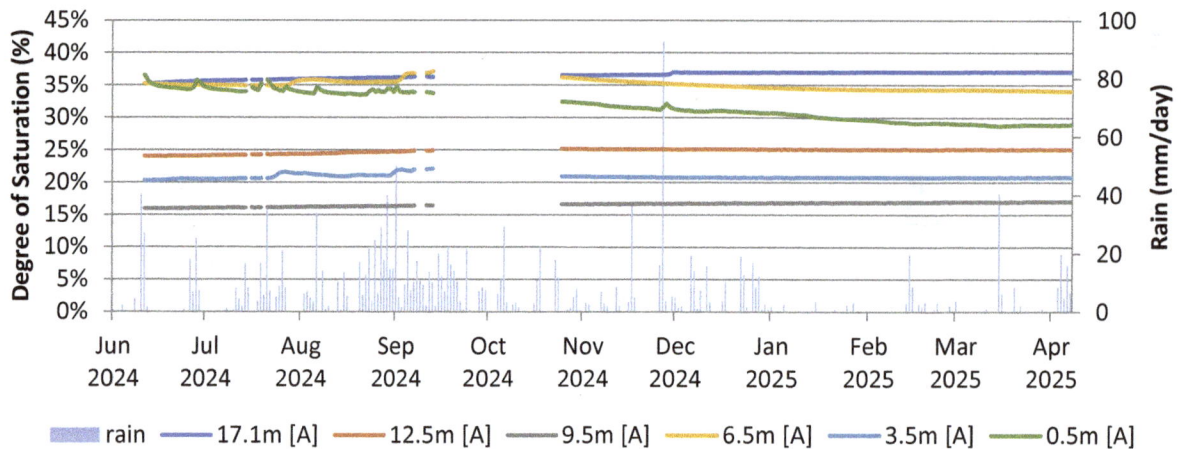

FIG 6 – Degree of saturation reading for ST2.

Oxygen at ST2

Two oxygen sensors were installed for ST2 at depths of 12.5 m and 17.1 m within the NAF A-type armouring layer. Initial oxygen readings were 17 per cent at 17.1 m and 4.5 per cent at 12.5 m. Over time, oxygen levels at both depths decreased, falling below 10 per cent, as shown in Figure 7.

ST2 - Gaseous Oxygen Level

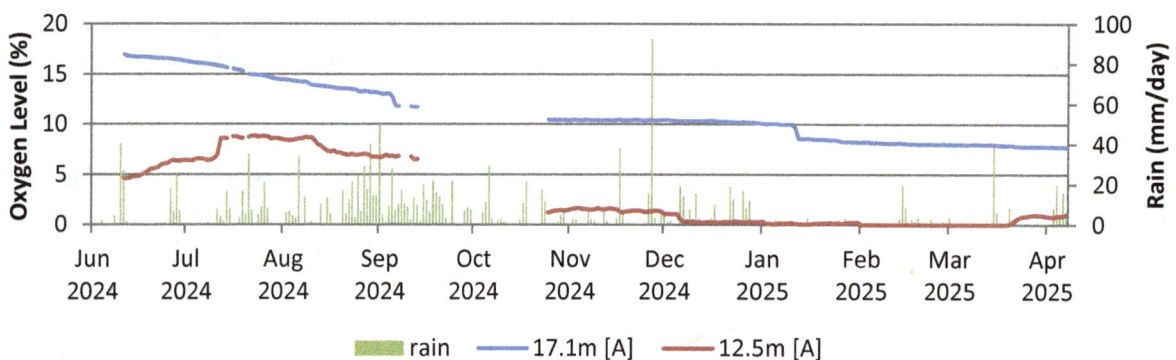

FIG 7 – Gaseous oxygen level for ST2.

Summary of data obtained at ST2

At ST2, the degree of saturation showed minimal variation over the recording period, with a slight steady increase during the rainfall period from July to September 2024, and a small, short rise following the major rainfall event in November 2024 extending to 3.5 m. This suggests the armouring layer facilitates rapid moisture infiltration with a limited capacity to retain moisture leading to quick drainage once rainfall stops.

Oxygen sensors at depths recorded levels which declined over time, falling below 10 per cent which indicates a likelihood of lack of aeration, which is ideal to maintain low oxygen levels to minimise a driver for AMD.

ST3

ST3 currently consists of a NAF A-type armouring layer approximately 1.5 m thick, with a NAF C-Type layer of 1.5 m in thickness located below. This C-Type layer is thicker than the 1 m NAF C-Type layer observed in ST4.

Moisture at ST3

Figure 8 presents the degree of saturation profile for ST3. Throughout the observation period, all sensors displayed distinct moisture dynamics in response to rainfall events. The surface sensor at 0.5 m exhibited the most significant fluctuations, with rapid moisture increases following rainfall. In contrast, sensors at 11 m and 6.5 m within the PAF showed minimal response, with moisture levels remaining relatively stable. The 11 m sensor displayed a gradual increase in moisture, indicating slow accumulation rather than rapid changes. The 3.5 m sensor exhibited moderate fluctuations with delayed responses to rainfall events.

ST3 - Moisture Reading

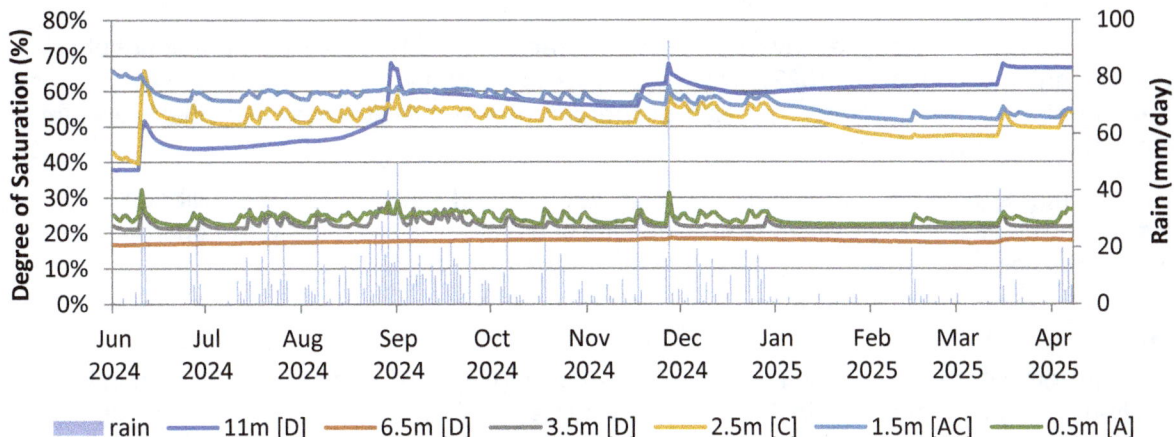

FIG 8 – Degree of saturation reading for ST3.

The 2.5 m sensor, located within the C-Type material, showed slower moisture cycles, reflecting higher water retention and slower recharge rates. Similarly, the 1.5 m sensor, positioned at the C-Type/A-type interface, demonstrated slower cycles, suggesting comparable moisture retention and recharge characteristics.

Oxygen at ST3

Two oxygen sensors were installed at depths of 6.5 m and 11 m within the PAF layer for ST3, with the measurements presented in Figure 9. The sensor at 6.5 m initially recorded an oxygen level of approximately 17 per cent, which gradually decreased over time, reaching around 12 per cent by November 2024. In contrast, the sensor at 11 m initially measured oxygen levels below 5 per cent, with a gradual increase observed, reaching a range of 10–15 per cent between January and April 2025.

ST3 - Gaseous Oxygen Level

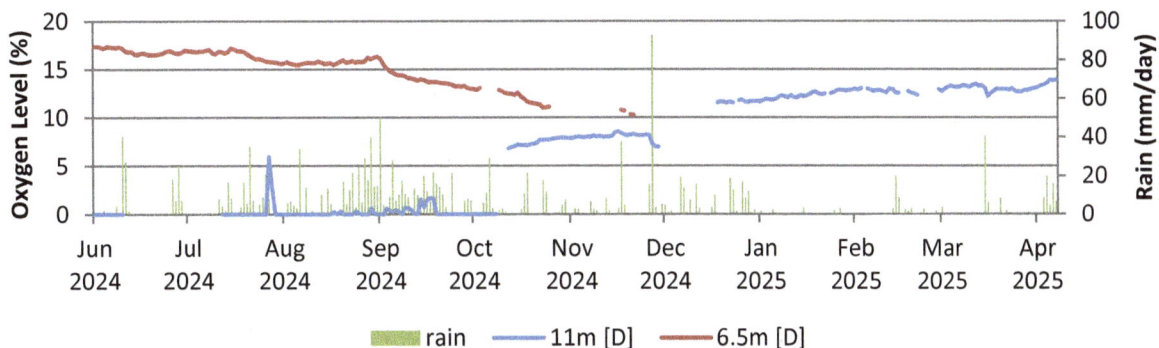

FIG 9 – Gaseous oxygen level for ST3.

Summary of data obtained at ST3

At the 0.5 m depth within the A-type gravel armouring layer, surface moisture exhibited rapid fluctuations following rainfall events, reflecting the high permeability and responsiveness characteristic of such materials. In contrast, sensors located at greater depths (11 m and 6.5 m) within the PAF Type D material displayed relatively stable moisture levels, with minimal sensitivity to short-term rainfall. The gradual increase in moisture observed at 11 m may be attributed to the finer grained backfill materials, which retain moisture more effectively and exhibit slower discharge rates. At the 3.5 m depth, the PAF sensor recorded slower and more gradual moisture changes, a trend also observed in intermediate-depth sensors at 2.5 m and 1.5 m, which are located within the C-Type layer and C-Type/A interface, respectively. Both sets of sensors exhibited slower moisture recharge and more gradual changes, indicating the moisture-retentive properties of these layers, which respond more gradually to rainfall events.

With respect to gaseous oxygen levels, a reduction was observed compared to the atmospheric level of 21 per cent (Lenton *et al*, 2022). However, due to inconsistent sensor behaviour, it remains unclear whether this reduction is a result of the soil profile characteristics or due to the inherent barrier effect of the bulk material.

These findings suggest that ST3 may not have yet reached a steady-state equilibrium, potentially due to disturbances during installation. Continued monitoring is necessary to confirm whether the system stabilises over time.

ST4

ST4 is currently characterised by a NAF A-type armouring layer of approximately 1.5 m in thickness, with a NAF C-Type layer of 1 m in thickness located below. This C-Type layer is slightly thinner than that of ST3, where the NAF C-Type layer measures approximately 1.5 m.

Throughout the observation period, ST4 experienced periodic power disruptions to the data acquisition system due to a faulty battery system, resulting in measurement and telemetry interruptions. These disruptions led to data gaps and timestamp misalignments, which in turn caused weak correlations between the recorded data and corresponding environmental events.

Moisture at ST4

Figure 10 illustrates that the degree of saturation for ST4 exhibited similar trends to ST3. The surface sensor demonstrated a rapid and dynamic response to water recharge and drying cycles, while deeper sensors within each stratified layer showed a slower response.

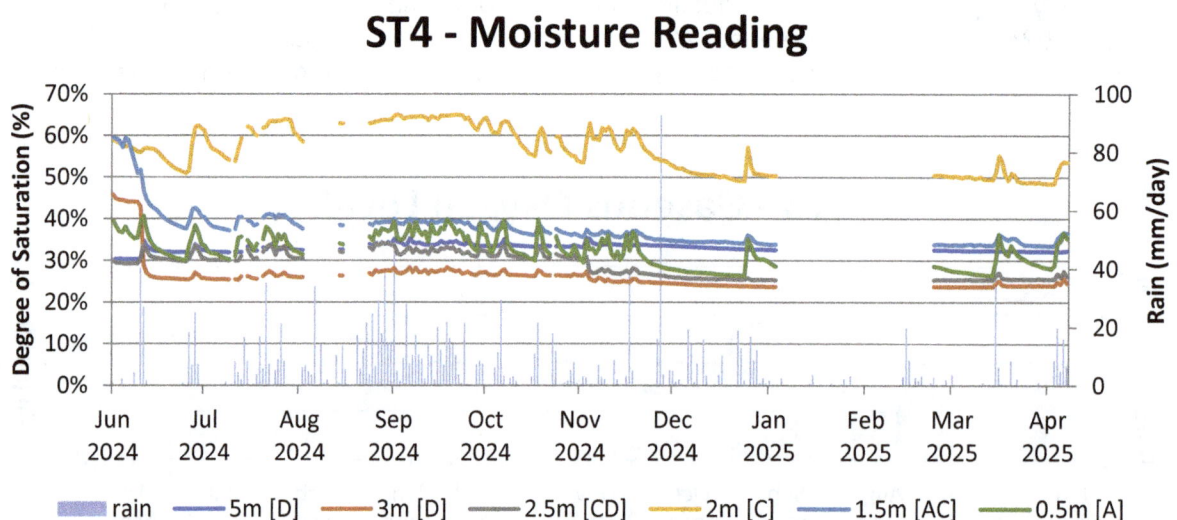

FIG 10 – Degree of saturation reading for ST4.

Oxygen at ST4

Two oxygen sensors were installed at ST4 at depths of 3 m and 5 m within the PAF layer, as shown in Figure 11. Initial oxygen levels at both depths were approximately 12 per cent, gradually decreasing until July 2024. At the 3 m depth, oxygen levels fluctuated between 10 per cent and 15 per cent from September to November 2024.

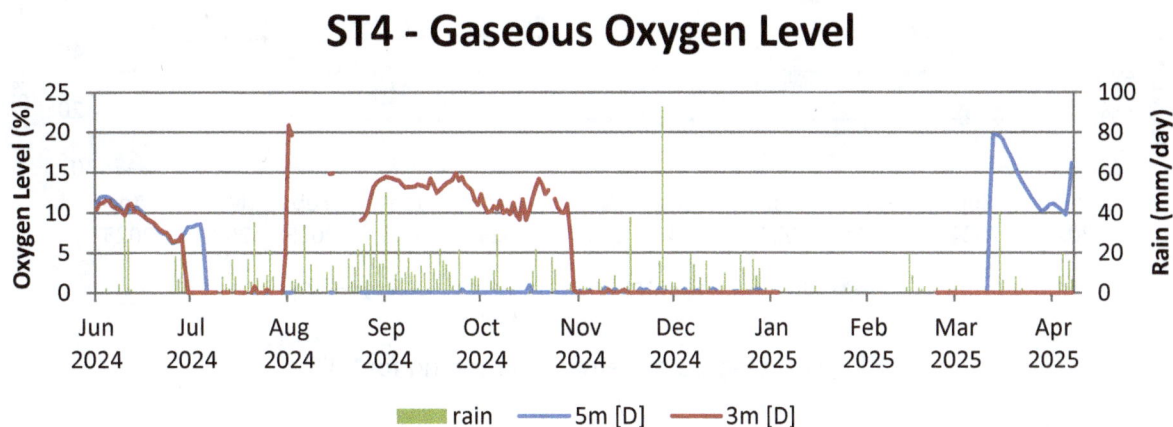

FIG 11 – Gaseous oxygen level for ST4.

Summary of data obtained at ST4

The response observed at ST4 closely mirrored that of ST3, where surface sensors reacted rapidly to environmental conditions. This is likely attributed to the gravelly composition of the armouring layer, which facilitates quicker infiltration. In contrast, the deeper sensors exhibited a more delayed and dampened response, consistent with the expected behaviour of compacted subsurface materials. The C-Type layer at ST4 appeared to effectively intercept and retain most of the infiltrating moisture, as evidenced by the minimal response recorded by sensors installed within the underlying PAF layer. These results suggest that the cover system is functioning as intended in reducing water percolation to the reactive material below.

However, limitations in the data set—namely power disruptions and timestamp inconsistencies—prevented accurate temporal correlation with rainfall events during the observation period. Additionally, sensor drift indicates that ST4 may not yet have reached hydrostatic equilibrium, likely due to post-installation soil disturbance. Continued monitoring is recommended to assess long-term stabilisation.

While gaseous oxygen data for ST4 were insufficient for definitive interpretation, the few available readings were below atmospheric levels, potentially indicating restricted oxygen diffusion within the profile. Further data collection is needed to confirm this trend and assess its implications for geochemical stability.

ST5

Moisture at ST5

Figure 12 illustrates the degree of saturation for ST5. Sensors within the NAF A-type armouring layer exhibited a rapid and dynamic response to rainfall events. In comparison, sensors located deeper in the NAF C-Type layer showed a more substantial recharge response, albeit with a similarly gradual discharge rate. Moisture sensor at 21 m, positioned within the PAF layer, demonstrated a more pronounced response to sustained rainfall events compared to isolated occurrences, with an estimated recharge of approximately 5 per cent. At 19 m below ground level (BGL), we observe a notable rise in moisture values towards the end of August 2024, with moisture telemetry indicating a slow increase at 16 m depth starting from June 2024 after the first rainfall event post installation.

ST5 - Moisture Reading

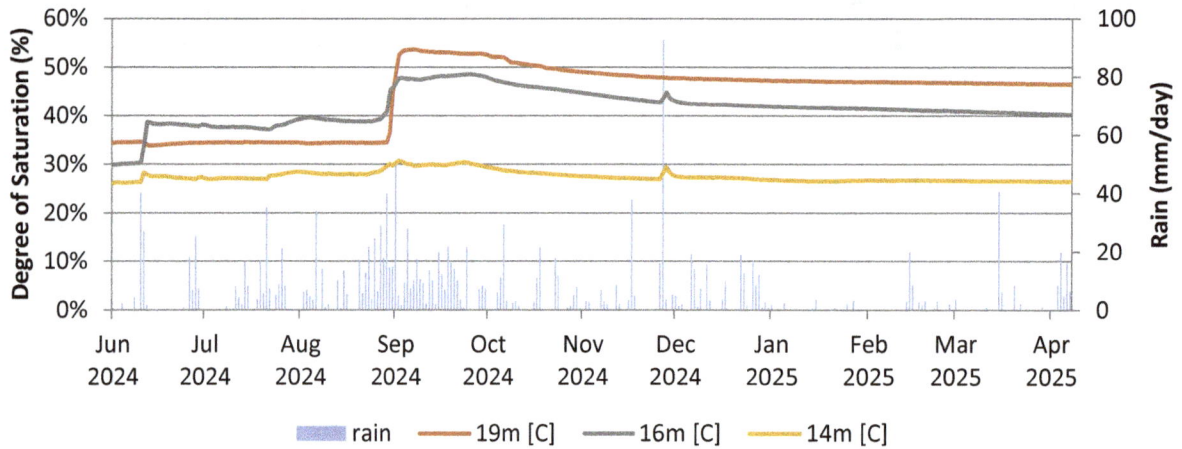

FIG 12 – Degree of saturation reading for ST5.

Oxygen at ST5

Two oxygen sensors installed at ST5 are positioned at depths of 19 m within the NAF C-Type layer and 21 m within the PAF. Both sensors recorded early gaseous oxygen levels below 10 per cent, as shown in Figure 13, with data available up to July 2024. No further measurements have been recorded since that time.

ST5 - Gaseous Oxygen Level

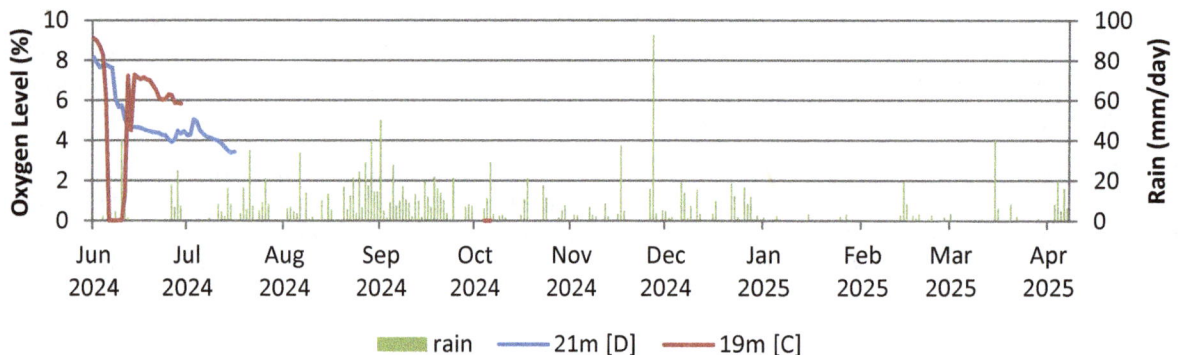

FIG 13 – Gaseous oxygen level for ST5.

Summary of data obtained at ST5

Due to the limited availability of gaseous oxygen data, definitive conclusions cannot be drawn; however, initial measurements indicate sub-atmospheric oxygen levels below 10 per cent, which may suggest a low-oxygen environment.

Moisture readings from surface sensors demonstrated expected performance, showing rapid responses to rainfall events. Sensors within the C-Type layer appeared effective in capturing and retaining moisture. This is correlated with a relatively low moisture measurement of the C-Type layer immediately post installation and only observing an initial sharp jump in moisture levels after the first initial rainfall event in June. Subsequent rainfall, water uptake and retention by the C-Type material is then observed at 16 m BGL, while the moisture level at 19 m BGL remained low. The delayed sharp rise at 19 m BGL in August could be attributed to the waterfront slowly percolating through the C-Type material finally reaching the area of measurement by the sensor at 19 m BGL. Although the sensor located at 14 m BGL was characterised to be within the C-Type layer, the response did not observe the same behaviour as sensors located at 16 m and 19 m depth, this could be attributed to a high comingling of the C-Type and A-type material at that depth. Although some moisture activity was detected by the sensor located within the PAF at a depth of 21 m, sensor drift observed at depths of 16 m and below suggests that conditions may not have reached equilibrium due to the disruption of the cover during instrumentation introducing a new equilibrium state. Thus, further monitoring is required to confirm stabilisation and assess long-term behaviour.

SUMMARY

The overall performance of the current cover designs appears effective in limiting water ingress into the underlying potentially acid-forming (PAF) layer. This is most clearly demonstrated at ST1, where the C-Type layer appears to intercept and shed infiltrating water. The waterlogged condition of the C-Type layer also corresponds with a low-oxygen environment within both the C-Type and PAF layers, indicating reduced gaseous oxygen diffusion.

While some moisture activity was recorded within the PAF layers at ST4, ST5, and ST6, the C-Type layer in these locations still appears to be functioning as intended by capturing and retaining moisture. Continued monitoring is necessary and is expected to be performed to the extent of instrument life to assess the long-term, stabilised performance of the cover system.

REFERENCES

Dold, B, 2003. Sulphidic mine waste management: The effects of climate and water flow on acid mine drainage, in *Mine Water and the Environment* (ed: D Zinke), pp 3–14 (Springer).

Jambor, L J and Blowes, D W, 2001. Environmental aspects of acid mine drainage, in *Environmental Geochemistry of Sulfide Oxidation*, pp 119–138 (Mineralogical Association of Canada).

Lenton, T M, Watson, A J, Daines, S J, Williams, H T and McClymont, E L, 2022. The role of oxygen in Earth's climate and biosphere, *Nature Reviews Earth & Environment*, 3(10):611–623.

Tailings and foundation characterisation

Key parameters affecting the compaction properties of filtered tailings

M F Barra[1], A C Matias[2], G Vizcarra[3] and A Heitor[4]

1. Senior Geotechnical Engineer, Vale S.A., Brazil. Email: fwbarra@gmail.com
2. Senior Geotechnical Engineer, Vale S.A., Brazil. Email: ana.carolina.p.matias@gmail.com
3. Specialist Geotechnical Engineer, Vale S.A., Brazil. Email: ginocalderon@hotmail.com
4. Lecturer in Geotechnical Engineering, School of Civil Engineering, University of Leeds, United Kingdom, Email: a.heitor@leeds.ac.uk

ABSTRACT

As an alternative to conventional tailings disposal, filtered tailings are becoming more common in the mining industry. This type of tailings is commonly disposed of in stacks, usually designed to be compacted. The compaction properties of mining tailings are critical for ensuring the stability, safety, and environmental sustainability of tailings storage facilities. This study investigates the key parameters influencing the compaction behaviour of iron ore tailings, focusing on particle size distribution, moisture content, compaction energy, and chemical composition. Particle size distribution plays a significant role, as finer particles tend to fill voids between coarser particles, enhancing density but potentially reducing permeability. Conversely, a well-graded particle distribution can improve compaction efficiency by optimising interparticle contact. Moisture content is another crucial factor, as it directly affects the soil's workability and compaction characteristics. Optimal moisture content, corresponding to the maximum dry density, is essential for achieving effective compaction, while deviations can lead to either insufficient compaction or excessive pore water pressure. Additionally, the mineralogical composition of tailings, including the presence of clay minerals, can affect compaction due to their water-absorbing and swelling properties. Tailings with a high percentage of fine particles often exhibit lower permeability and higher compressibility, complicating compaction efforts. Understanding these parameters is vital for designing efficient tailings management systems that minimise environmental risks, such as liquefaction or slope failure, while maximising storage capacity. This research highlights the interplay between these factors and provides deeper discussions on the behaviour of the materials, which can lead to future compaction strategies contributing to safer and more sustainable mining practices. The findings underscore the importance of tailored compaction approaches based on the specific characteristics of the tailings material.

INTRODUCTION

Iron ore beneficiation is a critical process in mining that enhances ore quality by removing impurities through stages including crushing, grinding, and magnetic separation (Meneses *et al*, 2024; Chapman, Gover and Ribbons, 2024). These processes generate tailings — waste materials comprising gangue minerals and residual impurities (Haynes *et al*, 2024). Effective tailings management is essential to mitigate environmental impacts and promote sustainable practices, with common disposal methods including dry stacking, paste disposal, and tailings dams.

A significant industry challenge lies in producing filtered tailings that meet geomechanical requirements cost-effectively, despite inherent variability in ore composition and operational conditions. Compacted filtered tailings stacks (CFTS) require dewatering to near-optimum moisture content and compaction to maximum density (Davies *et al*, 2010). However, fines and clay-sized particles complicate filtration efficiency, with technology selection heavily dependent on particle size distribution (PSD) and mineralogy.

The compaction properties of filtered tailings are pivotal for tailings storage facility (TSF) stability. Key parameters — moisture content, PSD, and consolidation characteristics dictate material behaviour. Filtered tailings, dewatered to reduce moisture, exhibit distinct compaction traits: moisture content governs density and shear strength (Meneses *et al*, 2024); PSD influences permeability and compressibility (Chapman, Gover and Ribbons, 2024); and consolidation affects long-term TSF stability (Rust and Rust, 2024).

Filtered tailings offer advantages over conventional slurry disposal, including improved geotechnical performance and reduced environmental risks (Davies, 2011; Fourie, 2012). Mechanically dewatered to a soil-like consistency, they enable dry stacking (Bussière, 2007), though their compaction depends critically on PSD, moisture, mineralogy, and energy input (Simms, 2017). This study examines these parameters to optimise compaction strategies, enhancing TSF safety and sustainability.

This study presents a comprehensive geotechnical characterisation of filtered tailings from iron ore processing plants, evaluating their compaction and mechanical properties for dry stack disposal applications. Through a series of laboratory tests, a comparison of the engineering behaviour of these tailings under conditions representative of their respective local climates and disposal environments. The research focuses on identifying critical material-specific parameters that influence stack stability, including moisture-density relationships, shear strength, and compressibility characteristics. By analysing tailings from distinct operational contexts, the study aims to develop tailored compaction strategies that account for regional variations in mineralogy, particle size distribution, and environmental conditions.

METHODOLOGY

Sample collection and preparation

To investigate the compaction behaviour of filtered iron ore tailings, seven samples were collected from distinct feed sources and processing streams within the Quadrilátero Ferrífero region (Minas Gerais, Brazil). These sources were selected to capture a representative range of mineralogical compositions and particle size distributions typical of industrial-scale operations. Furthermore, all the samples were collected after dewatering process, five of them being dewatered at filter press, one in vacuum disc filter and another by cycloning and sieving.

Sample preparation followed a standardised protocol to ensure consistency across tests:

- Air-drying was done to achieve stable moisture conditions suitable for laboratory testing. Sieving was conducted to remove particles larger than 2 mm, in accordance with geotechnical testing standards.

- Homogenisation to minimise variability and ensure uniformity across subsamples.

Particle size distribution analysis

The particle size distribution (PSD) of each sample was characterised using a dual-method approach:

- Sieve analysis (ASTM D6913; ASTM International, 2021c) was applied to the coarse fraction (>75 μm).

- Laser diffraction (ISO 13320:2020; International Organization for Standardization, 2020) was used for the fine fraction (<75 μm).

This combined methodology enabled comprehensive PSD profiling from clay-sized particles to coarse sands, facilitating accurate classification under the Unified Soil Classification System (USCS; ASTM D2487; ASTM International, 2021b).

Compaction testing

Compaction characteristics were evaluated using:

- Standard Proctor test (ASTM D698; ASTM International, 2021d), applying an energy input of 944 kJ/m³.

- Modified Proctor tests (ASTM D1557; ASTM International, 2021a), applying a higher compaction energy of 2700 kJ/m³. This test was conducted for one sample (AGL).

Samples were compacted at moisture contents within ±2 per cent of the estimated optimum moisture content (OMC). Each test condition was replicated three times to ensure statistical reliability and reproducibility of results.

Chemical analysis

The iron ore tailings samples were analysed using X-ray fluorescence (FRX) to determine their chemical composition. Powdered samples were pelletised and subjected to FRX to quantify major oxides such as Fe_2O_3, SiO_2, and Al_2O_3. The results, expressed in weight percent, provide insights into the elemental makeup of the samples, supporting mineralogical interpretation and geotechnical assessments.

Statistical analysis

A comprehensive statistical analysis was conducted to explore correlations and trends between material properties and compaction performance. The following methods were employed:

- Descriptive statistics to summarise key variables (eg mean, standard deviation, range). – Pearson correlation coefficients to assess linear relationships between particle size, moisture content, and compaction parameters (eg maximum dry density, OMC).

- Multiple linear regression analysis to quantify the influence of independent variables (eg fines content, mineralogy) on compaction outcomes.

- Validation of statistical models was performed using published data sets and field measurements from operational filtered tailings stacks.

This analytical framework enabled the identification of key predictors of compaction efficiency and provided a basis for extrapolating laboratory findings to field-scale applications.

RESULTS AND DISCUSSION

Physical characterisation

Grain size distribution

Figure 1 presents the particle size distributions of the analysed tailings samples, revealing significant variability across materials. The samples span from medium/fine sand (AGL, D_{50} = 0.2 mm) to predominantly silty material (MHF, D_{50} = 0.02 mm), reflecting the diverse processing methods employed at each source. Sample AGL's coarser distribution results from its gravity separation process, which preferentially retains larger particles, while MHF's finer characteristics stem from intensive grinding during its flotation-based concentration. Intermediate samples (eg MHA, SHA, MHT, BRC, SHF) show transitional distributions corresponding to their hybrid processing circuits. These fundamental differences in gradation will directly influence subsequent compaction behaviour, as demonstrated in the following sections.

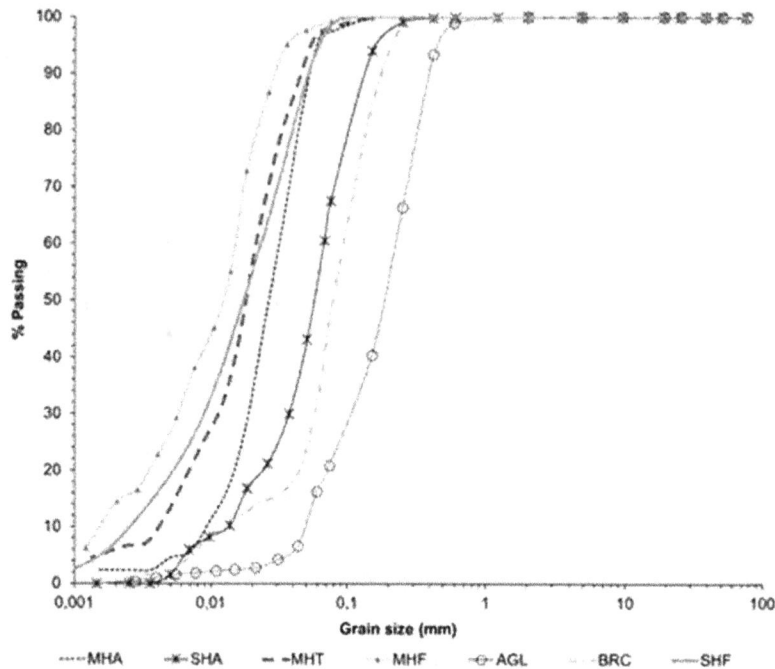

FIG 1 – Particle size distribution of studied tailings.

Table 1 summarises the physical, chemical and geotechnical properties of seven filtered tailings samples (MHA, SHA, MHT, MHF, AGL, BRC, SHF), revealing distinct compositional differences tied to their processing histories. Key trends emerge: samples MHA, MHT, MHF, and SHF exhibit high fines content (97–99 per cent) and classify as ML (silt) per the Unified Soil Classification System (USCS), while coarser samples AGL (80 per cent sand) and BRC (55 per cent sand) classify as SM (sandy silt). Notably, ultrafine content (<0.010 mm) varies significantly, from 2 per cent in AGL to 44 per cent in MHF, suggesting divergent dewatering challenges. The absence of measurable plasticity (LL/PL = "-") in all samples indicates non-cohesive behaviour, consistent with their dominant silt/sand compositions. These property variations—particularly in fines content (20–99 per cent) and gradation — will critically influence compaction performance, as explored in subsequent sections.

TABLE 1

Properties of testing materials.

Properties	MHA	MHT	MHF	SHA	SHF	AGL	BRC
Specific gravity, Gs	2.91	3.16	4.07	2.90	3.02	3.16	2.89
Fines content (%)	97	97	99	67	97	20	45
Ultrafine content (d < 0.010 mm) (%)	11	27	44	8	33	2	8
Liquid limit (%)	-	-	-	-	-	-	-
Plastic limit (%)	-	-	-	-	-	-	-
Plasticity index (%)	-	-	-	-	-	-	-
USCS*	ML	ML	ML	ML	ML	SM	SM
Grain size distribution (%)							
Sand (0.075 mm < d < 4.75 mm)	3	3	1	33	3	80	55
Silt (0.005 mm < d < 0.075 mm)	93	84	73	66	78	19	44
Clay (d < 0.005 mm)	4	13	26	1	19	1	1
Percentage of Iron (%)	10.5	24.6	39.9	11.3	16.9	-	-

* Unified Soil Classification System.

Additionally, the data from Table 1 reveals a clear trend linking the percentage of iron in the samples to both specific gravity and fines content. Samples with higher iron content, such as MHF and MHT, exhibit significantly higher specific gravities and also possess very high fines contents. This suggests that the presence of dense iron oxides like hematite and goethite contributes to both the increased mass per unit volume and the finer particle size distribution. In contrast, samples like AGL and BRC, which lack iron data and have lower specific gravities and fines contents, likely contain more quartz and less iron-rich phases. This correlation supports the interpretation that iron-rich tailings tend to be denser and finer, which has direct implications for their compaction behaviour and geotechnical performance.

Compaction

Figure 2 reveals that while MHF tailings achieves the highest dry density ($\gamma d \approx 2.3$ g/cm^3 at OMC), its 44 per cent ultrafines content (Figure 3, Table 1) results in elevated void ratios ($e \approx 0.8$), demanding Modified Proctor+ energy levels for effective compaction—increasing OPEX. This high void ratio also predisposes MHF to contractive behaviour and liquefaction risk, unlike coarser tailings (eg AGL, γd = 2 g/cm^3), which exhibit lower energy requirements. Thus, while MHF shows superior density, its geotechnical risks and operational costs necessitate careful evaluation.

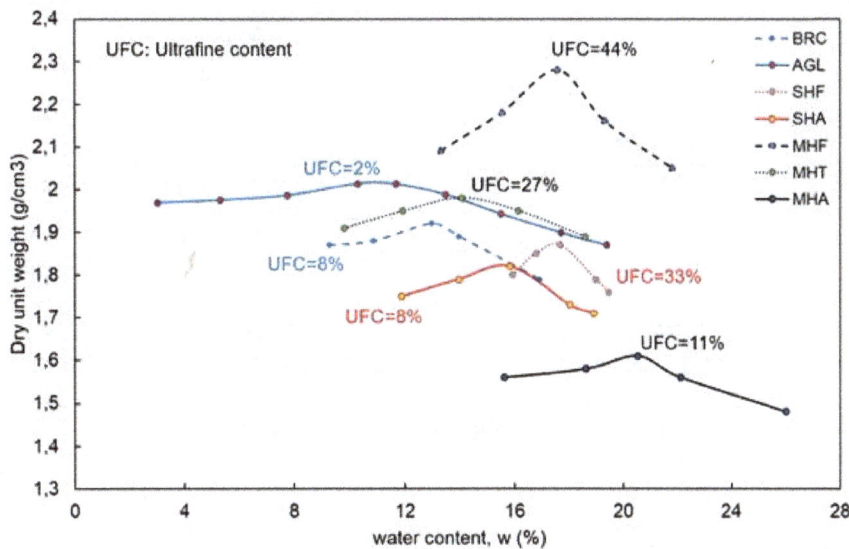

FIG 2 – Compaction curves of studied tailings.

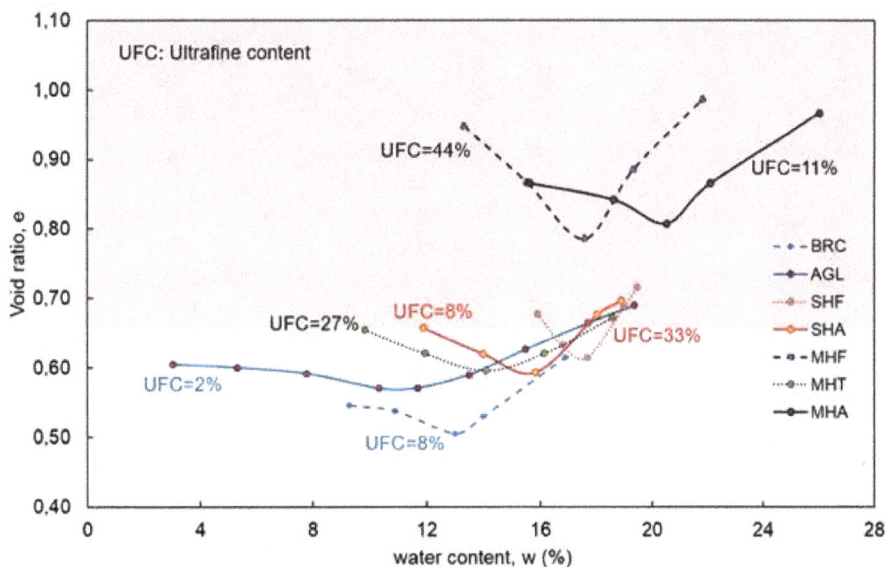

FIG 3 – Compaction curves of studied tailings in terms of void ratio.

Figure 4 demonstrates that exceeding optimum moisture content (OMC) leads to increased void ratios (Δe ≈ 0.10–0.20) at higher saturation levels (Sr> 85 per cent), indicating reduced compaction efficiency as pore water pressure inhibits particle rearrangement. Notably, MHF and SHA tailings reach their OMC at critically high saturation (Sr> 85 per cent), rendering them vulnerable to liquefaction. This behaviour contrasts with coarser tailings (eg AGL and BRC) achieving OMC at safer Sr (65–75 per cent), highlighting the need for: (1) strict moisture control for high-fines tailings; and (2) Modified Proctor energy to counteract saturation-induced density losses in operational conditions.

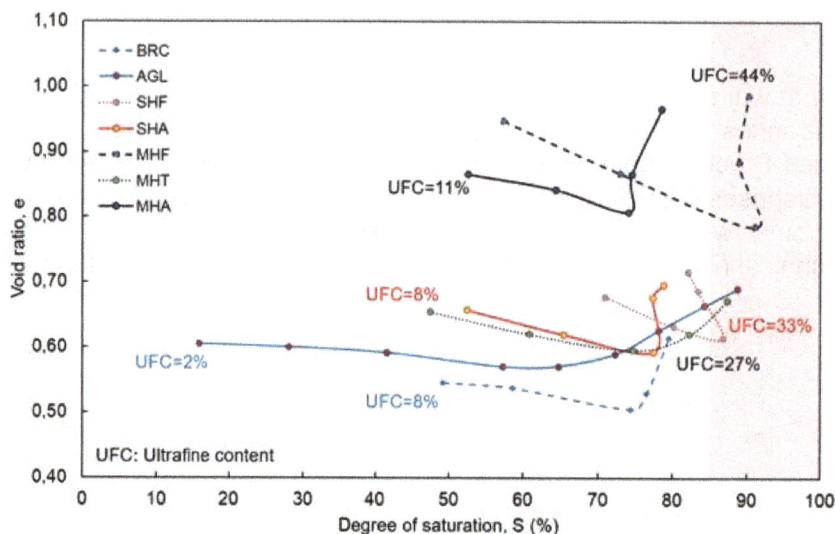

FIG 4 – Compaction curves of studied tailings in terms of void ratio and degree of saturation.

Figure 5 reveals a significant correlation (R^2> 0.95) between ultrafine content (%) and Fe concentration (%), indicating that iron-rich tailings fractions preferentially report to the ultrafine (<10 µm) fraction during processing. This empirical relationship enables operators to predict ultrafines content in real-time using routine Fe assays from the filtering plant. Such predictability is critical for: (1) anticipating compaction energy requirements for ultrafines; and (2) adjusting moisture targets to mitigate liquefaction risks in high-Fe tailings stacks.

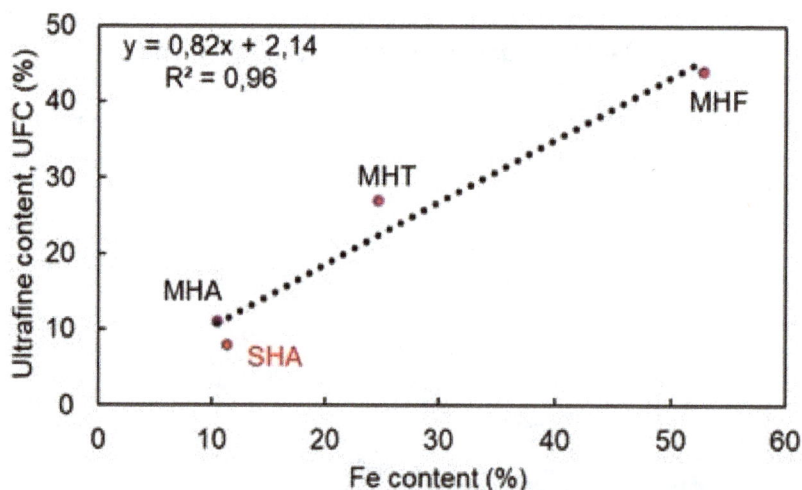

FIG 5 – Correlation between Fe content (%) and Ultrafine content (%).

Limitations

While this study provides valuable insights into the compaction behaviour of filtered iron ore tailings, several limitations should be acknowledged:

- The study focused exclusively on iron ore tailings, and the findings may not be directly applicable to other types of tailings (eg copper, gold).

- Similar approaches can be adopted for other types of tailings. However, it is important to note that different tailings behaviours may differ from those reported here due to their chemical composition and microstructure. Therefore, it is essential to conduct a comparative study to evaluate the behaviour of other tailings under compaction.

- The range of environmental and operational conditions tested was limited, and broader validation is needed to confirm the generalisability of the results.

- The laboratory tests were conducted under controlled conditions, which may not fully capture the complexities of field-scale operations.

Future research should aim to address these limitations by exploring a wider variety of tailings types and testing conditions, as well as conducting field trials to validate laboratory findings.

CONCLUSIONS

This study investigated the key parameters influencing the compaction behaviour of filtered iron ore tailings, with a focus on material-specific characteristics and their operational implications. The main findings and conclusions are as follows:

Material composition dictates compaction performance

Tailings with high ultrafine content (eg MHF, 44 per cent <10 µm) achieved the highest dry densities ($\gamma d \approx 2.3$ g/cm^3) but required Modified Proctor+ energy levels due to elevated void ratios ($e \approx 0.8$). These materials are prone to contractive behaviour and liquefaction, necessitating strict moisture control and potential stabilisation (eg per cent cement). Coarser tailings (eg AGL, 80 per cent sand) exhibited lower densities ($\gamma d \approx 2.0$ g/cm^3) but they are more operationally efficient.

Moisture-saturation balance is critical

High saturation levels (Sr> 85 per cent) at OMC, observed in fine tailings (MHF, SHA), reduced compaction efficiency and increased liquefaction risk. In contrast, coarser tailings (AGL, BRC) reached OMC at safer Sr (65–75 per cent), highlighting the need for tailored moisture management.

Predictive correlation for operational efficiency

The strong correlation ($R^2 = 0.96$) between Fe content and ultrafines enables real-time prediction of compaction energy requirements using routine plant assays, optimising operational planning and cost efficiency.

Recommendations for practice

For high-ultrafine tailings: Use Modified Proctor+ energy (≥ 2700 kJ/m^3) and thin lifts (0.2–0.3 m) and monitor Fe content to anticipate ultrafines and adjust moisture targets.

For all tailings: Prioritise saturation control (Sr <80 per cent) during compaction.

This study underscores that optimal tailings management requires balancing density, stability, and economics through material-specific compaction strategies, ultimately enhancing the safety and sustainability of filtered tailings storage.

REFERENCES

ASTM International, 2021a. ASTM D1557-21, Standard Test Methods for Laboratory Compaction Characteristics of Soil Using Modified Effort (56,000 kN·m/m^3 (2,700,000 ft·lbf/ft^3)), ASTM International, West Conshohocken.

ASTM International, 2021b. ASTM D2487-21, Standard Practice for Classification of Soils for Engineering Purposes (Unified Soil Classification System), ASTM International, West Conshohocken.

ASTM International, 2021c. ASTM D6913-21, Standard Test Methods for Particle-Size Distribution (Gradation) of Soils Using Sieve Analysis, ASTM International, West Conshohocken.

ASTM International, 2021d. ASTM D698-21, Standard Test Methods for Laboratory Compaction Characteristics of Soil Using Standard Effort (12 400 kN·m/m^3 (56,000 ft·lbf/ft^3)), ASTM International, West Conshohocken.

Bussière, B, 2007. Hydrogeotechnical properties of hard rock tailings from metal mines and emerging geoenvironmental disposal approaches, *Canadian Geotechnical Journal*, 44(9):1019–1052. https://doi.org/10.1139/T07-040

Chapman, P, Gover, S and Ribbons, J, 2024. Lessons learnt by applying a risk-based approach to existing and future tailings storage facilities, Proceedings of the 2024 Tailings and Mine Waste Conference, Westminster, CO, USA.

Davies, M P, 2011. Filtered dry stacked tailings—The fundamentals, in *Proceedings of Tailings and Mine Waste 2011*, pp 1–10.

Davies, M P, Lupo, J, Martin, T, McRoberts, E, Musse, M and Ritchie, D, 2010. Dewatered tailings practice – trends and observations, in *Proceedings Fourteenth International Conference on Tailings and Mine Waste*, pp 133–142 (AA Balkema: Netherlands).

Fourie, A B, 2012. Towards more sustainable tailings management, Keynote Lecture, International Seminar on Tailings Management, Santiago, Chile.

International Organization for Standardization, 2020. ISO 13320, Particle size analysis — Laser diffraction methods, 2nd edition, International Organization for Standardization, Geneva.

Meneses, B, Llano-Serna, M, Dressel, W, Coffey, J P and Gerritsen, T, 2024. A geotechnically derived screening method to assess the filterability of tailings, in *Paste 2024: Proceedings of the 26th International Conference on Paste, Thickened and Filtered Tailings* (eds: A B Fourie and D Reid), pp 115–128 (Australian Centre for Geomechanics: Perth). https://doi.org/10.36487/ACG_repo/2455_08

Simms, P, 2017. Compaction of tailings: A review, *Geotechnique*, 67(10):1–18. https://doi.org/10.1680/jgeot.16.P.200

Contractive behaviour in compacted earthfill and lessons for embankment stability

E Belcher[1]

1. Principal Engineer, Safe Dams and Tailings, Brisbane Qld 4000. Email: evan@safedams.net

ABSTRACT

This study addressed potential contractive behaviour in compacted earthfill. The transition from dilative to contractive behaviour has been given some attention lately. The question is typically applied to tailings and foundation materials. Less thought is given to the behaviour of compacted earthfill. However, yield stress can be reached within embankments as low as 10 m high. Failing to see this could increase the risk of embankment failure. This is an important topic where dams are incrementally raised. The topic was investigated at an existing Tailings Storage Facility. The main embankment comprised a starter dyke and three upstream raises. The hypothesis was that contractive behaviour was possible within the lower embankment. This was tested by estimating the yield stress of the compacted earthfill. Undisturbed and recompacted samples were tested. Parallel strength and consolidation tests were run. Results showed that local zones could become contractive. Yield stress was relatively consistent and typical values were in the range of 200 kPa to 300 kPa. This was an apparent preconsolidation stress due to compaction. The results were used to develop a hybrid strength model which considered both contractive and dilative behaviour. The model was then applied in an embankment stability assessment. Challenges included capturing evolution of the yield surface and treatment of pore pressures. The work is significant because contractive earthfill behaviour is typically overlooked. Key learnings relate to applying a hybrid strength model in embankment stability assessment. The method is generally suitable for risk-based design.

INTRODUCTION

Tailings storage facility (TSFs) embankments are often constructed from mechanically compacted earthfill. The aim of compaction is to improve the engineering properties of the material. This includes improving the strength, stiffness, and response to incremental loading. Compacted earthfill is generally constructed to achieve a dilative state. Dilative soils have a relatively stiff and strong loading response and may generate negative pore pressures during shearing.

However, initially dilative earthfill may become contractive over the life of a facility. Contractive behaviour is associated with relatively lower strength and stiffness – and contractive material may generate positive pore pressures when sheared. This is less favourable for embankment stability and must be adequately addressed in design.

A change in loading can cause material to transition from dilative to contractive. For example, this transition can be triggered by progressive saturation, tailings deposition, adding an embankment raise, or lowering the phreatic surface at closure. Authors have recognised similar mechanisms in major embankment failures (eg Mount Polley; Government of British Columbia, 2015). Foundation materials are known to be susceptible, though less emphasis is placed on compacted earthfill. As such, the potential transition to contractive behaviour is sometimes overlooked.

The author met this situation at an existing TSF. The TSF impoundment had been filled, operations had finished, and the facility was approaching closure. The main embankment was constructed from compacted earthfill and comprised a starter dyke with three upstream raises. The earthfill had been assigned an effective stress model in each successive raise design, which implied that undrained contraction was not a credible design condition. However, this assumption had not been directly verified, and any potential contractive behaviour would have implications for embankment stability.

To assess the risk, the design team developed a method to screen for potential contractive behaviour. The hypothesis was that contractive behaviour was possible within the lower embankment, which was tested by assessing the yield stress (p'y) of the material. The yield stress can be thought of as an apparent preconsolidation due to compaction and reflects the maximum vertical stress experienced to date. Contractive behaviour was flagged where saturated materials could be loaded near their yield stress.

The study found that contractive behaviour was possible within the lower embankment. Local zones of contractive material may develop at the upstream toe, and this was possible under various loading conditions. As such, material could transition from dilative to contractive behaviour over the life of the facility. This paper steps through the laboratory test methods, results, and screening procedure that were developed. The aim is to present a simple screening method that is usable for practising engineers.

Once potential contractive zones are flagged, the results can then be extended in various practical applications. For example, the impact of any contractive behaviour on embankment stability can be checked. This was also addressed in the current study. Strength test results were used to develop a hybrid strength model, which tried to capture a transition from dilative to contractive behaviour within increased loading. However, the method has several caveats and approximations. As such, this paper also details the development and application of the hybrid model. The aim is to highlight key lessons learned in applying a hybrid strength model in stability analyses. The current case study is included as a worked example.

BACKGROUND

Earthfill behaviour

Material behaviour can transition from dilative to contractive with a change in loading. This transition is associated with an approach to plastic yielding, which, in practical terms, is flagged where materials are loaded near their yield stress (eg Point A; Figure 1c). This simplification allows contractive behaviour to be screened using the over-consolidation ratio (OCR) – which is the ratio of *in situ* effective vertical stress ($\sigma'v$) to yield stress in one-dimensional compression.

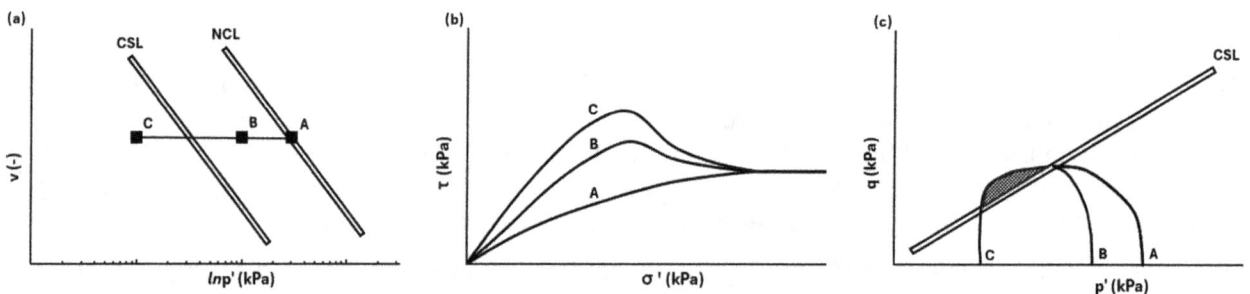

FIG 1 – Idealised behaviour of materials in undrained loading (after Budhu, 2011).

Mechanically compacting earthfill decreases the specific volume (v), which generally improves both the strength and stiffness. The relationship between specific volume, mean confining stress (p'), and earthfill behaviour can be explained with a critical state framework. Typical idealised behaviour during undrained loading is shown in Figure 1 (after Budhu, 2011).

States on the normal compression line (NCL) are normally consolidated (NC) (eg Point A; Figure 1a). Incremental loading will induce contractive behaviour and generate positive excess pore pressures. Peak strength and stiffness will be relatively low. States above the critical state line (CSL) are lightly over-consolidated (OC) (eg Point B; Figure 1a; OCR < 4). Peak strength is relatively higher but a transition towards contractive behaviour is possible near yielding.

States below the CSL are heavily OC (eg Point C; Figure 1a; OCR > 4). Loading induces a stiff and strongly dilative response up to yield (Atkinson, 2007). Peak strength and stiffness will be at their highest. As such, achieving a heavily OC state is generally the aim of mechanically compacting earthfill.

Yield stress can be estimated using standard consolidation tests, and the *in situ* vertical stress is generally approximated from depth, unit weight, and static pore pressure. Based on these values, an OCR around 1.5 can be used to mark the onset of transitional contractive behaviour. However, this is only a simplification used in engineering practice. Typical yield stress values start from around

150 kPa for mechanically compacted earthfill – which is comparable to stresses within many TSF embankments (eg > 10 m high). As such, potential contractive behaviour may be more common than generally considered, and this needs to be accounted for in any embankment stability assessment.

Stability assessment

Embankment stability is commonly assessed using limit equilibrium (LE) methods. Contractive behaviour is typically captured using an undrained strength model (eg Ladd, 1991). Dilative behaviour can be assigned an effective stress model (eg Mohr–Coulomb). However, the method cannot directly capture any transitional behaviour. Despite this, several workarounds are available. These include using a hybrid function (eg user-defined) or a strength-capped model (eg drained-undrained). This approach applies different strength models over different stress ranges (eg Brown and Gillani, 2016) and effectively caps the mobilised strength at the lesser of drained and undrained values. As such, transitional behaviour can be approximated. A hybrid function was used in the current assessment. This method is usable but has several limitations – several of which are highlighted in the current case study.

Case study

The current work is based on a real-world example. The case study is located on a mine site in western Queensland, Australia. The TSF embankment comprises a starter dyke and three upstream raises. The maximum embankment height is around 20 m. The starter dyke is around 8 m high and was constructed with compacted earthfill, which was generally intermediate plasticity silty clay. Fines content varied from 30 per cent to 80 per cent with minimal coarse sand and fine gravel. The earthfill was placed in 300 mm layers and compacted to a minimum density ratio (ie 98 per cent Standard Maximum Dry Density; SMDD).

Three upstream raises were made over the course of twenty years of operation, which significantly increased the stresses on the embankment. The TSF has now been filled and is awaiting closure. Cover and capping layers will be placed, and the phreatic surface will be permanently drawn down. This will change the embankment loading in future.

The compacted embankment has been historically assigned an effective stress model in stability analyses (ie shear-normal). The assumption implies dilative behaviour under ultimate loading conditions. However, this assumption had never been seriously challenged. As such, the assumption of dilative earthfill was checked in the current study. The hypothesis was that contractive behaviour was possible within the lower embankment. The first step was to assign a representative yield stress, which was addressed through detailed site investigations.

Site investigation

The site investigation comprised several auger-drilled boreholes. Samples of *in situ* embankment material (ie U75) were taken from a range of depths and locations. The intent was to capture any potential variation in yield stress due to *in situ* stress conditions. Material was also sampled from the base of the starter embankment. This zone has the highest vertical *in situ* stress and is most likely to host any potentially contractive material. Selected samples were submitted for testing, including consolidation, strength, and material characterisation tests. Consolidation test results informed the selection of a representative yield stress.

YIELD STRESS

Consolidation tests

Consolidation tests were conducted on *in situ* samples of embankment material to estimate a representative yield stress value. Several different tests were run as each has its own strengths and weaknesses. Samples were also checked for evidence of disturbance (eg Berre, Lunne and L'Heureux, 2021).

Conventional incremental loading (IL) tests were undertaken using Rowe cells (ASTM D2435-03; ASTM International, 2017) to achieve saturation under back-pressure. IL tests provide data at discrete loading increments. As such, a consolidation curve must be interpolated between the points.

This can make interpreting a yield stress value ambiguous. Controlled rate-of-strain strain (CRS) tests were run to address this (ASTM D4186-06; ASTM International, 2012). The tests provide a continuous data record, but the selected strain rate is known to impact the results (eg Maleksaeedi *et al*, 2018). The strain rate was varied to maintain excess pore pressures within a target range (ie 3 per cent to 15 per cent). Both the IL and CRS tests maintained a condition of zero lateral strain. Triaxial consolidation tests were also run (AS 1289.6.4.2; Standards Australia, 2016). Samples were isotopically consolidated and no strain rate was imposed. Isotropic test results will generally overestimate volumetric consolidation strain. In addition, mean stress (p') must be converted to an equivalent value in one-dimensional loading (σ'v). An at-rest earth pressure coefficient (K0) was assumed, and values were chosen based on published ranges (eg K0 = 0.4 to 0.7). However, the coefficient is not constant and will change as the test progresses (eg Llano-Serna *et al*, 2018).

Yield stress values were estimated using a range of methods, as selecting a representative value is subjective. Several graphic methods were trialled (after Paniagua *et al*, 2016; Umar and Sadrekarimi, 2017; Olek and Moskal, 2024). The author prefers to plot data in semi-logarithmic space and fit compression curves graphically (after Onitsuka *et al*, 1995). Yield stress results were also benchmarked using a work dissipation approach (Becker *et al*, 1987). This method can recover *in situ* stress conditions which provides additional confidence in sample integrity.

Consolidation results

Interpreted yield stress values generally ranged from 200 kPa to 300 kPa for the *in situ* samples. There was no consistent variation between the test methods. However, values varied by up to 10 per cent among the different methods of interpretation. As such, a representative value (ie 220 kPa) was conservatively assigned using the lower quartile of the data. There was also no consistent pattern with changes in fines content, plasticity, or location within the embankment section. This shows that assigning a single representative yield stress may be appropriate. Test results are summarised in Table 1.

TABLE 1

Summary of consolidation test results.

Sample	Test	Depth (m)	σ'v (kPa)	p'y (kPa)	OCR (-)
Sample I-1	Isotropic	6.8	130	200	1.5
Sample I-2	Isotropic	3.0	60	300	5.0
Sample I-3	Isotropic	6.0	120	300	2.5
Sample I-4	Isotropic	9.0	180	--	--
Sample I-5	Isotropic	5.3	110	200	1.8
Sample I-6	Isotropic	7.6	150	250	1.7
Sample I-7	Isotropic	10.3	200	250	1.3
Sample C-1	CRS	5.3	110	300	2.7
Sample C-2	CRS	3.3	70	230	3.3
Sample C-3	CRS	5.1	100	280	2.8
Sample R-1	Rowe	6.8	130	--	--
Sample R-2	Rowe	3.3	70	--	--
Sample R-3	Rowe	6.1	120	220	1.8
Sample R-4	Rowe	5.3	110	220	2.0
Sample R-5	Rowe	9.1	180	210	1.2

Rowe cell tests gave relatively low and consistent yield stress values. As such, these tests are recommended to target a relatively conservative result.

CONTRACTIVE BEHAVIOUR SCREENING

Screening methods

The representative yield stress (ie 220 kPa) was used to screen for any potentially contractive material. The screening was catered at the critical design section – which was translated onto a 1 m × 1 m grid in MS Excel. OCR values were calculated for every element. A threshold value was used to flag contractive behaviour (ie OCR ≤ 1.5). This value was benchmarked using triaxial test results – which included assessment of stress paths and pore pressure responses (incl Af coefficient) (after Head and Epps, 2014). Sensitivity was checked over a reasonable range and key assumptions were varied.

The screening considered two general bounding conditions. The first captured a rapid rise in the phreatic surface due to impoundment flooding. This increased the amount of saturated embankment material but decreased the effective stress. The second captured a lowering of the phreatic surface at closure, decreasing the amount of saturated material but increasing the effective stress.

Material above the phreatic surface was excluded. Contractive behaviour is unlikely during long-term unsaturated loading – as the material will have a relatively stiff and strong response. The positive effect of partially saturated loading was ignored in accordance with industry practice (eg after USACE, 2003).

Screening results

The screening successfully flagged zones of potential contractive and transitional material. Sensitivity trials confirmed that the zones were relatively persistent. A typical result is shown in Figure 2. The output shows potential contractive behaviour at the upstream toe (ie orange and red cells) – with shallower material flagged as generally dilative (ie green cells). Unsaturated material was excluded as noted above (ie grey cells).

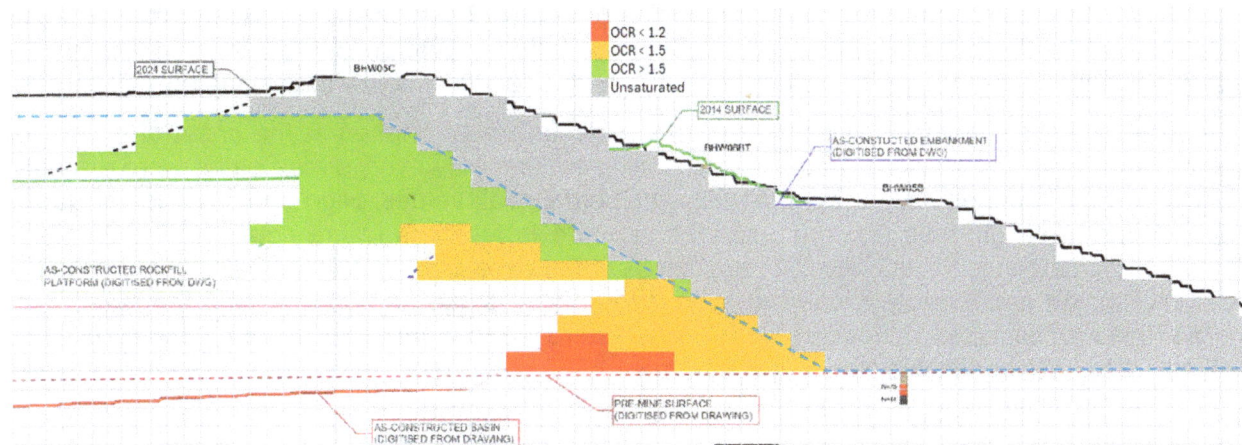

FIG 2 – Example of contractive behaviour screening (high phreatic surface).

Screening discussion

The screening results were generally robust, and flagged zones were not sensitive to changes in key assumptions. However, the screening only considers one-dimensional loading which is a simplification. Material behaviour is strongly dependent on biaxial loading. As such, the method can be extended into two dimensions by defining a yield curve and accounting for horizontal stresses (eg Muir Wood, 2007).

There is also a caveat around the evolution of the yield surface, which is not captured in subsequent stability assessments. For example, lowering the phreatic surface would increase the effective stress and cause more material to flag as contractive. However, the assigned yield stress remained constant in the hybrid strength model. This was a simplification. In reality, the material would consolidate in response to the new stress conditions, the yield curve would evolve, and the yield stress would increase. As such, the strength model would need to be updated in an iterative process. Despite this, the screening was able to flag potential contractive material, which was assigned a hybrid strength model in subsequent stability assessments.

HYBRID STRENGTH MODEL

A hybrid strength model was developed and applied. The model was assigned to any zones flagged as potentially contractive (ie OCR ≤ 1.5). The model capped the mobilised strength at the lesser of the drained and undrained values – as the aim was to capture a transition from dilative to contractive behaviour with increased stress.

Drained strength was assigned an effective stress envelope. This imposed a cap on undrained dilation at low confining stress. Undrained strength was assigned an undrained envelope (after Ladd, 1991). This was applied above the yield stress. Transitional strength was interpolated between them. The hybrid model was then applied as effective stress envelope in stability analyses (ie shear-normal). The implications are discussed later. Key parameters are summarised in Table 2. The parameters were informed by triaxial strength tests in accordance with industry standard methods (eg USACE, 2003).

TABLE 2

Summary of hybrid model strength parameters.

Model	Phi' (°) [1]	p'y (kPa) [2]	t/s [3]
Hybrid model	~27	220	0.25

(1) – Friction angle; 0–100 kPa; Non-linear; c' = 0 kPa;
(2) – Yield stress in one-dimensional compression;
(3) – Undrained strength ratio; NC value.

Strength tests

Triaxial strength tests were run on *in situ* samples (AS 1289.6.4.2; Standards Australia, 2016). All tests were run under isotropic consolidation (CIU) due to time constraints. Each sample was tested in a single stage, and stage pressures were chosen both above and below the estimated yield stress.

Low stress stages targeted the dilation cap. As such, failure was taken at the onset of dilation – which was marked by a decrease in excess pore pressure. However, stress paths were strongly dilative and the reported shear strengths are well below peak values. High stress stages targeted the undrained strength ratio (t/s) and used the peak deviator stress failure criterion. Results were above the typical range for NC silty clay, and stress paths were dilative even at high stage pressures (ie 400 kPa). An additional test was run at an even higher stage (ie 600 kPa) and the result was more in line with expected values (ie t/s = 0.22). These results suggest that the earthfill is still dilative well above the assigned yield stress (ie 220 kPa).

Test results and the hybrid strength model are shown in Figure 3. Historic data are included for reference. However, the results are from laboratory compacted samples – and define a practical lower bound to *in situ* conditions.

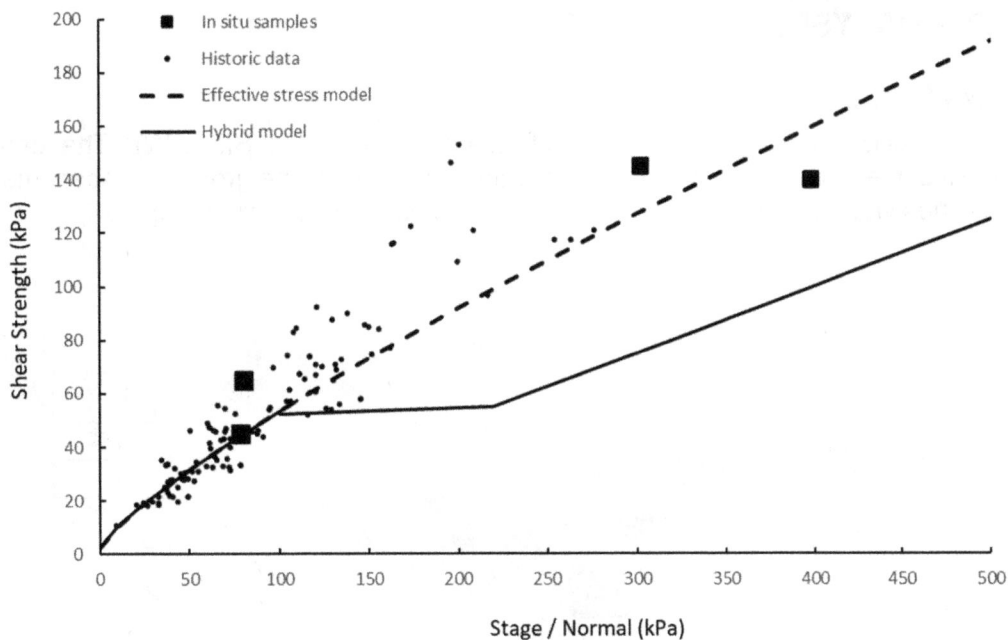

FIG 3 – Hybrid strength model including current and historic test data.

Strength model discussion

The hybrid strength model captured the transition from dilative to contractive behaviour with increased stress. This allowed the model to be applied in a standard LE stability assessment (eg shear-normal function). However, this application comes with a notable conservatism.

For example, the hybrid model includes a cap on dilation. This limits the mobilised strength at low confining stresses. The cap was imposed in line with industry practice (eg after Brown and Gillani, 2016). However, this is a conservative assumption and does not always capture real material behaviour. For example, the cap mostly eliminates the increase in undrained strength due to over-consolidation – which includes the mechanical compaction of earthfill.

The cap is typically justified via three general arguments. The first notes that undrained dilation generates negative excess pore pressure – and that this will increase the effective stress and mobilised peak strength at failure. The argument is that these negative pore pressures cannot be reliably sustained under field conditions, and as such, the positive effect of undrained dilation should be ignored. However, this is a somewhat inconsistent assumption. For example, undrained contraction generates positive excess pore pressures – which also need to be sustained in the field to generate a potential contractive response. The reasons to exclude negative but not positive pore pressures are not clear. This observation provides context for the imposed cap on dilation

The second argument notes that discrete slip planes may form during shearing – and that contractive behaviour may develop on a very local scale (after Atkinson, 2007). This is equivalent to a contraction of the yield curve during undrained dilation. While this is difficult to disprove in the field, no supporting evidence was found in the current test results. *In situ* samples showed typical OC behaviour, and stress paths were strongly dilative in low stress stages. Either way, it is not clear how the argument can be either validated or dismissed.

The third argument notes that over-consolidation can be destroyed by shrink-swell movements. For example, where shallow compacted earthfill could undergo net volumetric expansion over time. This could bring the material to an equivalent NC state. As such, the positive effect of any mechanical compaction should be ignored. However, this was checked and dismissed in the case study. Yield stress values were consistent with depth and shallow material was heavily OC (see Table 1). Shrink-swell tests showed the borrow material was only moderately reactive – and survey benchmarks were relatively static over the life of the facility. These observations show the argument is not valid for earthfill at the site. Regardless of the discussion above, the dilation cap was retained in subsequent stability analyses (ie Case 1). However, the impact of removing the cap was also assessed (ie Case 2).

STABILITY ANALYSES

Analysis methods

Embankment stability was assessed using standard methods (ie Slope/W). The general limit equilibrium (GLE) method was used. A critical section was taken at the greatest embankment height. Contractive zones were manually keyed in. An example output is shown in Figure 4.

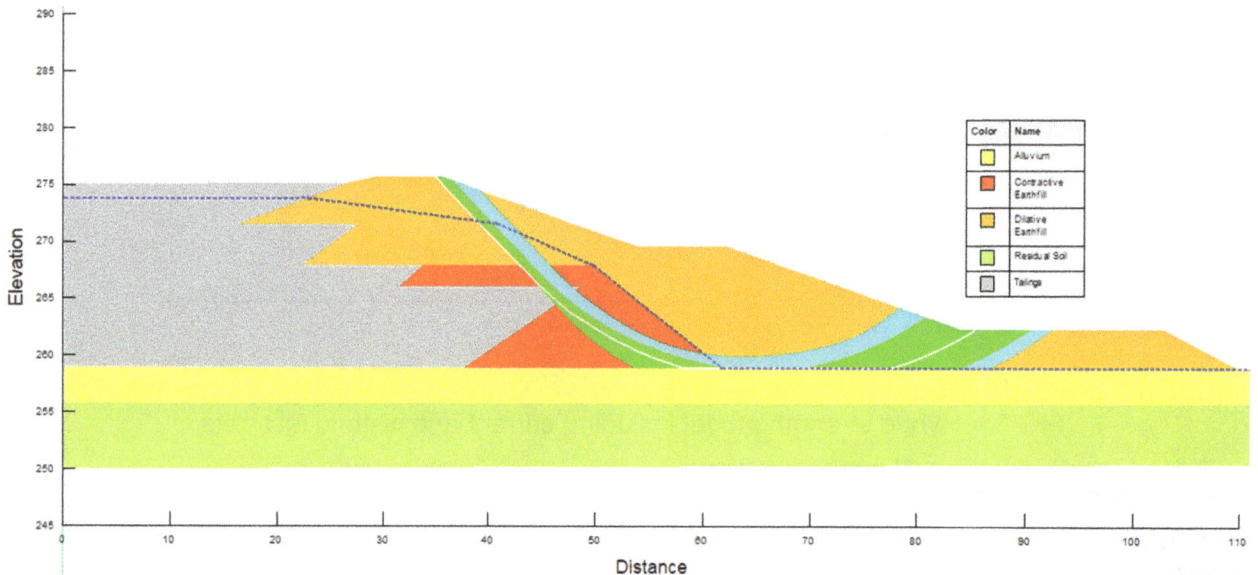

FIG 4 – Example of stability assessment output.

Two general cases were run. Case 1 assigned the hybrid model to material that screened as potentially contractive (ie Figure 4; red zone). Zones that were not contractive were given effective stress parameters (ie Figure 4; orange zone). Case 2 assigned an equivalent undrained model to the contractive material (ie Shansep). The model does not include a cap and allows peak strengths to develop through undrained dilation. This case was included to provide context for applying the cap in a compliance-based design. An effective stress envelope was given to the material above the phreatic surface in both cases (after USACE, 2003).

Stability results

Stability results are reported as Factor of Safety (FoS) values and are summarised in Table 3. Results showed that raising the phreatic surface was critical to stability (ie Case 1). This is because more material became saturated, which reduced the effective stress. However, relatively less material flagged as contractive. As such, the decrease in FoS may seem counterintuitive. The result was caused by applying an effective stress envelope (ie shear normal) to model undrained behaviour. For example, an undrained model would largely preserve strengths subject to a rise in the phreatic surface (ie Case 2). The increase in static pore pressure would be offset by a corresponding decrease in excess pore pressure at failure. Conversely, the effective stress envelope allowed strengths to reduce, and the effect was similar to double counting the pore pressure. This explanation was verified by comparing test results with equivalent mobilised strengths in the stability analyses. The comparison showed that the hybrid model consistently under-assigned strengths where the phreatic surface was raised. In addition, the impact was most pronounced at lower stresses where the imposed cap excluded any undrained dilation (ie Case 1). This approach mostly eliminates the increase in undrained strength due to over-consolidation, which highlights a key lesson learned.

TABLE 3

Summary of stability analysis results.

Case	Description	Phreatic surface	
		High	Low
Case 1	Hybrid model assigned to contractive zones including dilation cap	1.4	1.6
Case 2	Undrained model assigned to contractive zones excluding dilation cap	1.7	1.8

For example, in a compliance-based context, the design would be flagged as potentially non-compliant (eg Case 1 min FOS < 1.5 in long-term steady-state; after ANCOLD, 2019). However, this result is driven by the basic assumptions – being: (a) the application of an effective stress envelope to model undrained behaviour, and (b) the imposed cap on dilation. These are both conservative assumptions. As such, while it is necessary to check stability under both drained and undrained conditions, care should be taken when applying a combined hybrid model in compliance-based design.

This observation is consistent with the industry standard approach, where stability compliance is assessed under both drained and undrained conditions (eg ANCOLD, 2019). These represent general bounding cases. However, while different shear strength models may need to be applied under various loading conditions, assigned strengths are not necessarily capped at the lesser of all.

Conversely, some engineers assert that the standard approach has been superseded, and that strengths should be capped at worst-case values regardless of loading conditions (eg after Brown and Gillani, 2016). The current study doesn't necessarily disagree. Rather, the author recommends that existing standards are used to assess compliance (ie ANCOLD, 2019) and more remote assumptions and sensitivities are judged in a risk-based context (incl. hybrid and strength capped models). For example, a single threshold (eg FOS = 1.5) should not be seen as a hard distinction between safe and unsafe design. The result is only meaningful when the controlling assumptions are well understood.

CONCLUSIONS

This study addressed potential contractive behaviour in compacted earthfill. The paper outlined the laboratory test methods, results, and screening procedure that were developed. The screening was applied in a case study at an existing TSF. The hypothesis was that contractive behaviour was possible within the lower embankment. Results showed that local zones could transition from dilative to contractive under various loading cases. This result required embankment stability to be reassessed.

The transition from dilative to contractive behaviour was captured using a hybrid strength model, which capped the mobilised strength at the lesser of the drained and undrained values. However, this method is only a workaround and has several limitations, including both the inability to capture the evolution of the yield surface and the treatment of pore pressures. Despite this, the model can be applied as a useful check in a risk-based design.

ACKNOWLEDGEMENTS

The author would like to thank the reviewers for their contributions and discussion. The author acknowledges the mining company and the site as owners of the site investigation data.

DISCLAIMER

The author notes that site investigation data cited in this study is of a general nature only. Data have been selected, interpreted, and summarised for illustrative purposes. The data have been used in general accordance with fair use guidelines. All other information was available in the public domain.

ORIGINALITY

The author was responsible for technical reviews and managing the design program. The author declares that the published material is original research or otherwise cited. The author takes full responsibility for any personal opinions and commentary. No part of this work has been published in the past.

REFERENCES

ANCOLD, 2019. Guidelines on tailings dams planning, design, construction, operation and closure, Revision 1. Available from: <https://ancold.org.au/product/guidelines-on-tailings-dams-planning-design-construction-operation-and-closure-revision-1-july-2019/>

ASTM International, 2012. ASTM D4186-06 Standard test method for one-dimensional consolidation properties of saturated cohesive soils using controlled-strain loading. Available from: <https://store.astm.org/d4186-06.html>

ASTM International, 2017. ASTM D2435-03 Standard test method for one-dimensional consolidation properties of soils. Available from: <https://store.astm.org/d2435-03.html>

Atkinson, J, 2007. *The mechanics of soils and foundations*, second edition, pp 171–173 (Taylor and Francis Press).

Becker, D E, Crooks, J H A, Been, K and Jefferies, M G, 1987. Work as a criterion for determining in situ and yield stresses in clays, *Canadian Geotechnical Journal*, 24(4):549–564. https://doi.org/10.1139/t87-070

Berre, T, Lunne, T and L'Heureux, J-S, 2021. Quantification of sample disturbance for soft, lightly overconsolidated, sensitive clay samples, *Canadian Geotechnical Journal*, 59(2):300–303. https://doi.org/10.1139/cgj-2020-0551

Brown, B and Gillani, I, 2016. Common errors in the slope stability analyses of tailings dams, in *Proceedings of APSSIM 2016: First Asia Pacific Slope Stability in Mining Conference* (ed: P M Dight), pp 545–556 (Australian Centre for Geomechanics: Perth). https://doi.org/10.36487/ACG_rep/1604_36_Brown

Budhu, M, 2011. *Soil Mechanics and Foundations*, third edn, pp 332–336 (John Wiley and Sons).

Government of British Columbia, 2015. Report on Mount Polley tailings storage facility breach. Government of British Columbia, Independent Expert Engineering Investigation and Review Panel. Available from: <https://www.mountpolleyreviewpanel.ca/sites/default/files/report/ReportonMountPolleyTailingsStorageFacility Breach.pdf>

Head, K and Epps, R, 2014. *Manual of Soil Laboratory Testing Volume 3: Effective Stress Tests*, third edn, pp 10–12 (Whittles Publishing).

Ladd, C, 1991. Stability evaluation during staged construction, *Journal of Geotechnical Engineering*, 117(4):540–615.

Llano-Serna, M A, Farias, M M, Pedroso, D M, Williams, D J and Sheng, D, 2018. An assessment of statistically based relationships between critical state parameters, *Geotechnique*, 68(6):556–560. https://doi.org/10.1680/jgeot.16.T.012

Maleksaeedi, E, Nuth, M, Karray, M and Bonin, M, 2018. Application of a novel oedometer setup for performing constant-rate-or-strain (CRS) tests on soft soils, in *Proceedings of GeoEdmonton 2018, the 71st Canadian Geotechnical Conference and the 13th Joint CGS/IAH-CNC Groundwater Conference Transportation Geotechnique*, paper 217, 7 p (The Canadian Geotechnical Society and International Association of Hydrogeologists).

Muir Wood, D M, 2007. *Soil Behaviour and Critical State Soil Mechanics*, pp 65–76 (Cambridge University Press).

Olek, B S and Moskal, M, 2024. Comparison between methods for determining the effective vertical yield stress of intermediate fine-grained soils, *Scientific Reports*, 14(1):131. https://doi.org/10.1038/s41598-023-50026-2

Onitsuka, K, Hong, Z, Hara, Y and Yoshitake, S, 1995. Interpretation of oedometer test data for natural clays, *Soils and Foundations*, 35(3):61–70 (Japanese Geotechnical Society). https://doi.org/10.3208/sandf.35.61

Paniagua, P, L'Heureux, J-S, Yang, S L and Lunne, T, 2016. Study on the practices for preconsolidation stress evaluation from oedometer tests, in *Proceedings of the 17th Nordic Geotechnical Meeting (NGM)*, pp 547–555.

Standards Australia, 2016. AS 1289.6.4.2 Methods of testing soils for engineering purposes, Method 6.4.2: Soil strength and consolidation tests — Determination of compressive strength of a soil — Compressive strength of a saturated specimen tested in undrained triaxial compression with measurement of pore water pressure. Available from: <https://store.standards.org.au/product/as-1289-6-4-2-2016>

Umar, M and Sadrekarimi, A, 2017. Accuracy of determining pre-consolidation pressure from laboratory tests, *Canadian Geotechnical Journal*, 54(3):441–450. https://doi.org/10.1139/cgj-2016-0203

USACE, 2003. Engineering and Design: Slope Stability, Engineer Manual EM 1110-2-1902, US Army Corps of Engineers (USACE). Available from: <https://www.publications.usace.army.mil/Portals/76/Publications/Engineer Manuals/EM_1110-2-1902.pdf>

Evaluating shear strength and brittleness in mine tailings using variable rate shear vane apparatus

Y N Byrne[1], M Essien[2], I Entezari[3], R James[4] and N Pereira[5]

1. ConeTec, Burnaby BC V5A 4W2, Canada. Email: yasmin.byrne@conetec.com
2. ConeTec, Calgary AB T2C 2S4, Canada. Email: mary.essien@conetec.com
3. ConeTec, Burnaby BC V5A 4W2, Canada. Email: ientezari@conetec.com
4. BMA, Brisbane Qld 4000, Australia. Email: robert.james@bhp.com
5. Red Earth Engineering, Perth WA 6000, Australia.
   Email: nicolas.pereira@redearthengineering.com.au

ABSTRACT

Field vane shear testing (VST) is a widely used *in situ* test method for assessing undrained shear strength (S_u) in fine-grained plastic soils, including mine tailings. However, in lower-plasticity or coarser-grained materials, partial drainage can occur due to increased permeability, leading to elevated shear strength measurements. To mitigate this, VSTs are often performed at variable rotation rates, in excess of the ASTM D2573 (ASTM International, 2018) standard 0.1°/s, to ensure undrained shearing during the test. At the faster rates required to suppress drainage, it is common for failure to occur outside the specified time conditions of ISO 22476 (International Organization for Standardization (ISO), 2020) and AS1289.6.2.1 (Standards Australia, 2001).

This paper presents advancements in VST testing in tailings through two complementary studies: one driven by a large diverse data set and one material specific. The first study investigates post peak behaviour at fixed intervals of vane rotation to evaluate strain softening, using a large data set of standard-rate (0.1°/s) VSTs, screened to isolate undrained conditions. A novel stability criterion has also been developed to identify the point at which the rate of strength degradation stabilises; this value is defined as the post-peak shear strength. A pseudo-brittleness index based on post-peak strength was calculated to normalise strength loss and compare with brittleness derived from remoulded strength.

The second study examines variable-rate vane shear testing conducted in coalmine tailings in Australia since 2019, evaluating how rotation rate influences shear strength and the post-peak response in non-plastic materials. Together these studies aim to demonstrate the application of the vane shear test in tailings and refine the understanding of the post peak behaviour when compared to the remoulded strength.

INTRODUCTION

Field vane shear testing (VST) is a widely used *in situ* method for estimating undrained shear strength. Originally developed for use in saturated, plastic, cohesive fine-grained soils, the test is intended to capture both peak (S_u) and remoulded (S_{ur}) strengths under undrained conditions. As per ISO 22476 (ISO, 2020) the remoulded strength is defined as the lowest strength value after ten or more vane rotations (remoulding phase) as illustrated in Figure 1. In this paper, post-peak refers to any shear resistance following shear failure (failure is defined by the onset of softening) prior to the remoulding phase (Figure 1).

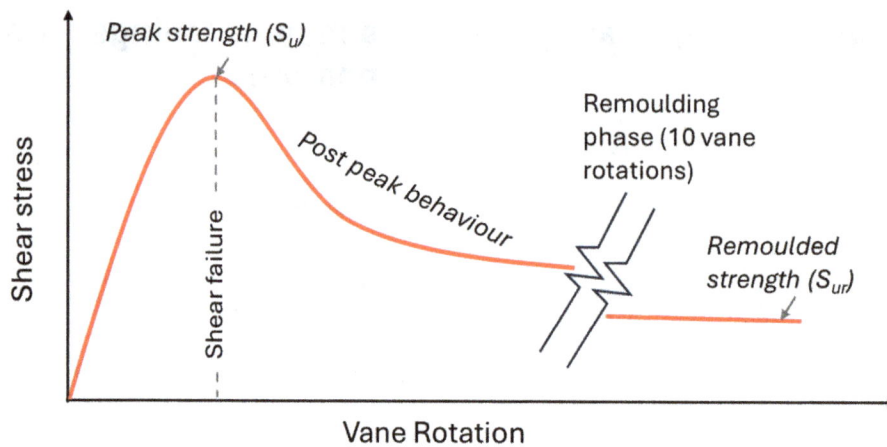

FIG 1 – Illustrative shear stress–rotation response from a field vane shear test.

In addition, VST data can be used to assess brittleness, defined as the relative loss of strength following peak. The Brittleness index (I_B) (Equation 1), as adopted in ICOLD Bulletin 194 (International Commission on Large Dams (ICOLD), 2023), provides a standardised measure for classifying tailings behaviour and guiding stability assessment:

$$I_B = \frac{S_u - S_{ur}}{S_u}$$

(1)

A higher brittleness index indicates a more pronounced strain-softening response, which is relevant for understanding post-failure deformation and mobility. In tailings materials, this behaviour has direct implications for assessing residual strength and the stability of tailings storage facilities (Locat and Demers, 1988; Jefferies and Been, 2006). In this paper, to explore strain softening, a *comparative* brittleness value is also calculated using a *post-peak stress* instead of the fully remoulded strength.

VST is increasingly conducted in mine tailings, including in higher permeability materials. Evidence for partial drainage during shearing has been seen in silts with plasticity indices up to 20 (Reid *et al*, 2023). To minimise the risk of overestimated shear strengths due to partially drained conditions, VST is increasingly conducted use faster rotation rates. These rates, typically selected based on the material's drainage characteristics, often exceed the standard rotation rate of 0.1°/s (ASTM D2573 (ASTM International, 2018) and AS1289.6.2.1-2001 (Standards Australia, 2001)). ASTM D2573 specifies 0.1°/s with a permissible variation is 0.05 to 0.12°/s.

This paper presents advancements in interpretating vane shear data in mine tailings through two complementary studies considering: one driven by a large, diverse data set and one material specific:

- The first study investigates post peak behaviour using a large data set of standard-rate VSTs, screened to isolate undrained conditions. Post-peak shear stress, measured at fixed intervals of vane rotation, is compared to the remoulded strength to evaluate strain softening. A novel stability criterion has also been developed to identify the point at which the rate of strength degradation stabilises; this value is defined as the post-peak shear strength.

- The second study examines 356 variable-rate *in situ* vane shear tests, conducted between 2019 and 2024, at four coal tailings storage facilities in Queensland, Australia. By comparing standard- and fast-rotation tests, this study investigates how changes in rotation rate influence the shear strength and post-peak behaviour in low-plasticity, fine-grained tailings. Adjacent CPTu soundings and pore pressure dissipation (PPD) tests were used to assess stratigraphic variability and evaluate drainage conditions.

BACKGROUND AND MOTIVATION

The drainage condition during a field vane test is governed by the soil's permeability, rotation rate, and the shear boundary geometry. In clays and other low permeability materials, vane testing at ISO standard rotation rates, typically produces undrained failure. However, in more permeable materials, such as low-plasticity silts and mine tailings, dissipation can occur during shearing, resulting in partially or fully drained conditions. This can result in elevated peak strength measurements that do not reflect the undrained response of the material (Blight, 1968; Reid, 2016; Hogan *et al*, 2025). Even modest drainage can significantly affect the measured strength, especially in contractive soils (Chandler, 1988). Heterogeneity, such as interbedded coarse materials or variable hydraulic conductivity, can further affect the drainage response at the scale of the test, while operational factors—such as a delay after vane insertion—may also allow drainage prior to shear mobilisation (Reid, 2016; Hogan *et al*, 2025).

To assess drainage conditions during VST, Blight's (1968) dimensionless time factor (T) (Equation 2) is typically used:

$$T = c_v t_f / D^2 \tag{2}$$

where t_f is the time to failure, D is the vane diameter, and c_v is the vertical coefficient of consolidation. Blight proposed a threshold of T < 0.02 for undrained conditions, while Chandler (1988) suggested an upper limit of T < 0.05. Values above these limits are associated with increasing degrees of partial or full drainage (T > 0.8) during shearing (Chandler, 1988).

In practice, the horizontal coefficient of consolidation (c_h), derived from CPTu dissipation tests is often used as a proxy for c_v. Ratios of c_h/c_v of between 1.0 and 1.5 are typically assumed (Hogan *et al*, 2022, 2025; Mundle, Esford and Julien, 2019; Reid, 2016), however these are based on empirical observations in homogeneous natural clays of marine, alluvial, deltaic, and lacustrine origin (Jamiolkowski *et al*, 1985). This assumption may not hold in stratified tailings deposits, where layering, hydraulic anisotropy, and depositional history can significantly affect consolidation behaviour.

Several recent VST studies have compared standard rotation rates (0.1°/s) with faster rates to evaluate the influence of rotation rate on peak shear strength in mine tailings (Reid, 2016; Mundle, Esford and Julien, 2019; Contreras and Harvey, 2021; Hogan *et al*, 2022; Harvey *et al*, 2023; Reid *et al*, 2023). These studies assess drainage conditions using T. At standard rotation rates, T often exceeds the undrained threshold, due to the permeability of the tailings tested. In contrast, fast rate tests — where the time to failure is reduced — typically achieve undrained conditions based on Harvey *et al* (2023) and Reid *et al* (2023) observe that peak shear strength obtained from fast rotation tests is lower than the standard rate tests, supporting the conclusion that undrained conditions have been achieved. An exception to this trend arises at very high rotation rates, where increased strength values have been reported. This behaviour is hypothesised to be due to viscous effects in rate-sensitive soils (Harvey *et al*, 2023; Chandler, 1988; Wilson *et al*, 2016). These studies demonstrate the importance of selecting rotation rates based on expected drainage behaviour in tailings materials (Hogan *et al*, 2025). As drainage conditions influence both peak, post-peak and remoulded responses, fast vane testing also offers an opportunity to better understand strain-softening and assess brittleness *in situ* — key themes explored in the following sections.

DATA-DRIVEN CASE STUDY – POST-PEAK ASSESSMENT

To explore post-peak behaviour and strain softening, a data set of 4336 *in situ* field vane shear tests was screened to target undrained conditions with minimal uncertainty. This data set required a CPT within a spatial distance of less than 10 m and a temporal gap of under 90 days between VST and paired CPT. Further screening criteria were selected to ensure consistency in soil conditions and to reduce the likelihood of partially drained or atypical responses, and included:

- a soil behaviour index Ic > 3 (silty clays and clays), (Robertson, 2009)
- VST depth below phreatic surface determined by CPTu
- testing was conducted at a standard rate of 0.1°/s
- low local variability in sleeve friction, fs (standard deviation <1 kPa) over a 15 cm window.

A total of 1006 tests were identified as suitable.

As guidance for tailings storage facilities evolves, there is increasing interest in whether a post-peak strength — rather than fully remoulded strength — could inform stability assessments under 'assumed triggered' conditions, where a factor of safety greater than 1.1 must still be demonstrated. In tailings, VSTs typically exhibit, an abrupt, rapid loss of strength immediately following failure, that with further shearing transitions into a gradual strength decay, through to the remoulded strength. This study considers usage of the post-peak response, prior to the remoulding phase, to evaluate whether a stable strength value can be identified prior to remoulding, and how this may relate to softening.

Initial analysis focused on assessing strength degradation as a function of angular rotation beyond the peak. Shear stress values were extracted at discrete rotation intervals — 60°, 90°, 120°, 150°, and 180°—and compared to both the peak and remoulded strengths. To normalise the strength loss across the data set, a pseudo-brittleness value ($I_{B\,(PP)}$) was calculated using this post-peak shear stress instead of the fully remoulded strength (Equation 3).

$$I_{B\,(PP)} = \frac{S_u - \tau(\theta_x)}{S_u}$$

(3)

Where $\tau(\theta_x)$ is the measured shear stress at rotation angle θ.

Figure 2 presents the pseudo-brittleness index at rotation angles of 60°, 120°, and 180° compared to brittleness index based on remoulded strength. These fixed-angle comparisons were intended to capture the progression of strength loss during post-peak shearing and to identify whether consistent trends emerged across the screened data set. While some correlation improved at larger rotation angles, the results remained scattered. This variability suggests that fixed-angle strength ratios cannot reliably capture brittleness across a range of materials, particularly in cases where strain softening is gradual, or the post-peak response extends over a wide deformation range.

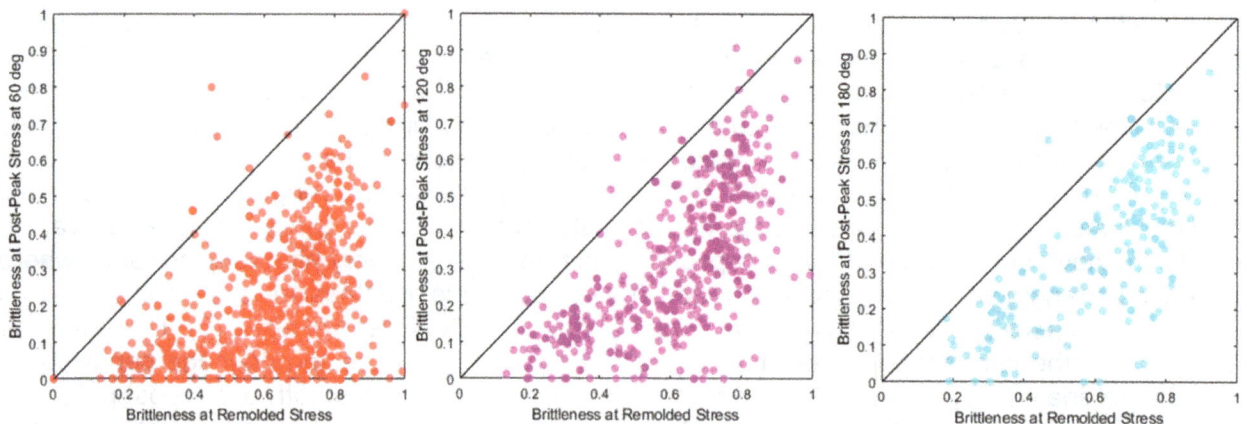

FIG 2 – Comparative brittleness index calculated at 60°, 120°, and 180° rotation angles compared to brittleness index based on remoulded strength.

Given the variability observed in fixed-angle comparisons, an alternative method was developed to identify a more representative post-peak strength. The shear stress rate of decay was considered stabilised when the change in normalised shear stress over a 20° rotation interval was less than 2 per cent of the peak value. The criterion is defined as:

$$\frac{\tau(\theta_2) - \tau(\theta_1)}{\tau_{peak}} < 0.02, \quad where \quad \theta_2 - \theta_1 = 20°$$

(4)

where: $\tau(\theta_x)$ is the measured shear stress at rotation angle θ and θ_2 and θ_1 are the start and end angles of the rotation interval. And 557 VST tests adequately met the post-peak stabilisation criteria.

This threshold was selected as a first-pass criterion to identify a point at which the rate of strength degradation had stabilised following rapid decay. Although preliminary and somewhat arbitrary, the method successfully isolated a post-peak strength where the rate of decay is stable, as illustrated in Figure 3 left, where an example data subset is shown indicating where the post-peak shear stress plateaus are identified. The corresponding shear stress value is referred to here as the stabilised post-peak strength. Figure 3 right presents a frequency distribution of the rotation angles at which stress stabilisation was detected, demonstrating the wide range of post-peak behaviour.

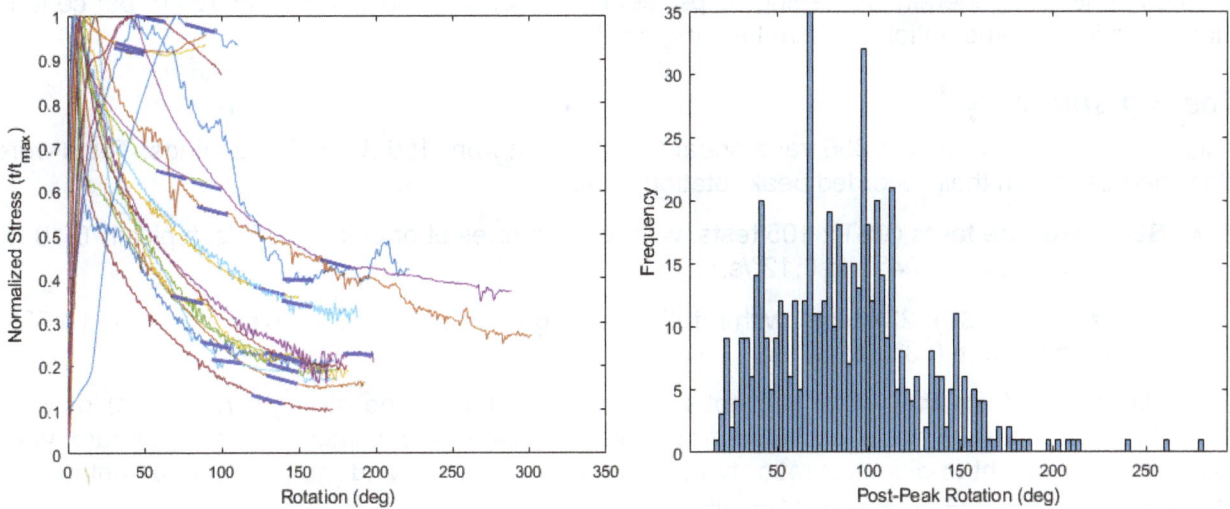

FIG 3 – (left) Example application of the stabilisation criterion over a sample data set with stabilised shear stress identified in blue. (right) Frequency distribution of rotation angles at which stabilised post-peak strength was identified.

Comparison of the stabilised post-peak strength against the remoulded strength (Figure 4) shows a general trend, where the post-peak strength (determined through stability criteria) is approximately twice the remoulded value. However, when corresponding brittleness values are compared, a significant amount of scatter remains.

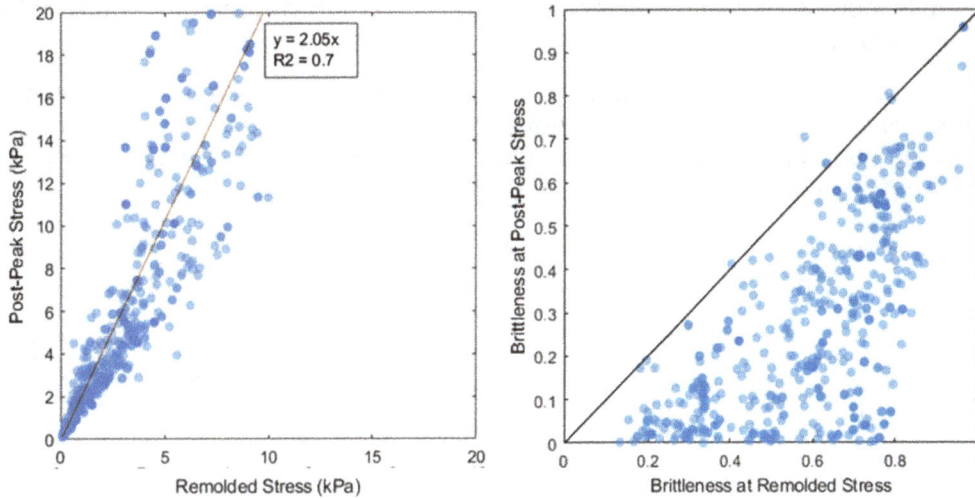

FIG 4 – left: Comparison between stabilised post-peak strength and remoulded strength. right: Comparison between brittleness determined by stabilised post-peak and remoulded strength.

MATERIAL-FOCUSED CASE STUDY – COAL TAILINGS

This study draws on a defined data set compiled from *in situ* testing campaigns conducted between 2019 and 2024 at four of BMA's coal tailings storage facilities in Queensland: Peak Downs, Blackwater, Saraji, and Goonyella Riverside. Only campaigns that incorporated fast vane shear testing (FVST) were included, allowing for evaluation of rate effects on strength and post-peak behaviour. These tests were generally accompanied by adjacent CPTu soundings and, in many cases, standard-rate vane tests for comparison. Stratigraphy at all sites was variable, often with interbedded fine and coarser layers observed. Vane testing targeted fine-grained zones within the deposits, where materials were identified as silty or clayey based on CPT soil behaviour type (SBT) – with an Ic (Robertson, 2009) typically ranging between 2.9 and 3.4.

Laboratory index testing indicated that the materials were generally of low plasticity, with varying fines contents but generally exceeding 50 per cent and plasticity indices around 10–15 per cent. In many locations the potential for micro layering exists.

Testing summary

This study includes a total of 356 vane shear tests, drawn from 159 distinct soundings. Tests were classified based on their recorded peak rotation rates:

- **Standard rate tests (VST):** 105 tests, with rotation rates at or below 0.12°/s, typically 0.10°/s, and ranging from 0.04°/s to 0.12°/s.

- **Fast tests (FVST):** 230 tests, with rotation rates greater than 0.12°/s, typically around 4.0°/s, and ranging from 0.3°/s to 31°/s.

An additional 21 tests that either did not achieve shear failure (no strength reduction) or were otherwise unsuccessful were excluded from further analyses. All tests utilised standard vane geometry (2:1 height-to-diameter ratio, typically 50 mm diameter) and included instrumentation to capture torque and rotation angle in real time.

The data set analysed includes many adjacent fast vane tests (FVST) and standard-rate vane tests (VST), conducted in twin soundings, enabling direct comparisons of rate effects within similar materials. However, this pairing approach was not consistently applied across the entire data set; at some locations, only fast or standard-rate tests were performed. 284 vane tests were conducted within 20 m of a CPTu sounding (typically with 5 m), and 123 included pore pressure dissipations (PPD) at or near vane depths to support interpretation.

Measured peak and remoulded shear strengths across the data set exhibited significant variability, reflecting the wide range of material types, drainage conditions, and strain rates tested. Even when shear strength is normalised with respect to depth, or when degrees of rotation to failure is considered, no trend is apparent.

All the tests generally adhered to ISO 22476-9 guidelines regarding vane geometry, test procedure, and insertion waiting times (ISO, 2020). However, the majority of tests (296 of 356) did not satisfy the ISO failure-duration criteria (60–180 seconds) due either to rapid rotation rates in FVSTs or partial drainage effects prolonging failure times in VSTs, as illustrated in Figure 5.

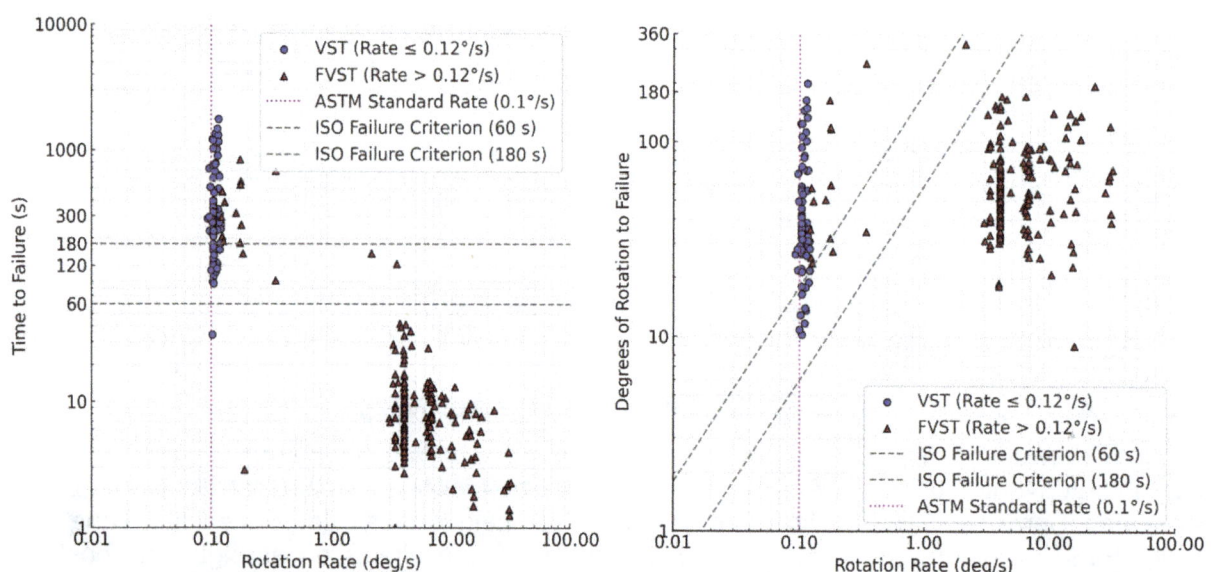

FIG 5 – Vane shear test time to failure (left) and degrees of rotation to failure (right) versus rotation rate for standard (VST ≤ 0.12°/s) and fast tests (FVST > 0.12°/s), showing ISO failure criteria (60 s to 180 s) and ASTM standard rate (0.1°/s).

Drainage assessment

Drainage conditions were evaluated using Blight's dimensionless time factor T. 55 per cent of the vane tests had adjacent CPTu, which included PPD suitable for the determination of T, within similar stratigraphic units and at comparable depths (within 1 m). The results of the drainage classification are presented in Figure 6. A standard rigidity index of 100 and an assumed $c_h/c_v = 1.0$ were applied to estimate consolidation coefficients, permitting comparison with other studies. Analysis of the PPD data indicated that standard-rate tests typically yielded $T > 0.05$ suggesting partially or fully drained conditions. In contrast, fast-rate tests commonly produced $T < 0.05$, consistent with undrained conditions as shown in Figure 6 (left). Review of the Ic and soil behaviour type demonstrate that this applies for both clay like and silt like tailings.

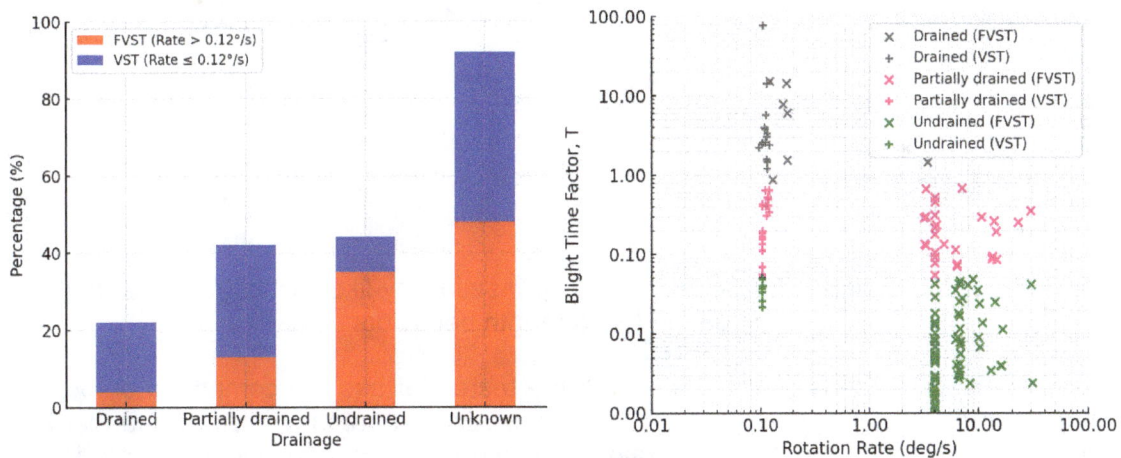

FIG 6 – Identification of drained (T > 0.8), partially drained (0.05 < T < 0.8) and undrained tests (T < 0.05) based on Blights' Factor T.

Paired VST and FVST tests

A total of 57 pairs of fast and standard-rate vane shear tests (FVST and VST) were analysed to assess the effect of rotation speed on shear strength determination. Contrary to expectations and findings in the literature, the data set exhibited a wide range of responses. Of the tests analysed, 65 per cent recorded a peak shear strength more than 10 per cent higher in the fast test (>0.12°/s) than in the standard-rate test. A similar trend was observed in the rotation angle to failure — with fast tests more frequently failing at higher rotation angles. In contrast, the remoulded strength showed no consistent trend with respect to rotation rate; results were nearly evenly split between tests where remoulded strength increased or decreased between the FVST and VST.

Material homogeneity was identified as a potential control on rate-dependent behaviour. In cases where CPT data indicated homogeneous, fine-grained, clay-like tailings — defined here as no change in SBT within ±0.5 m of the vane test depth — 75 per cent of fast tests produced a lower peak strength than the standard-rate test. A similar trend was observed in remoulded strength, with 85 per cent of fast tests showing a decrease. This is indicative of fully undrained behaviour during the FVST, where excess pore pressure is not allowed to dissipate. Where Blight's time factor (T) could be calculated, all fast tests were classified as undrained. In contrast, standard-rate tests in these zones exhibited a broader range of drainage conditions, from drained (7 per cent) to undrained (14 per cent), with the majority (57 per cent) falling within the partially drained category.

Figure 7 (left) shows a representative homogeneous example: a pair of vane tests conducted at 11.6 m depth in a material with Ic = 3.2. The FVST was performed at 10°/s, and the standard-rate test at 0.1°/s. Neither test satisfied the ISO 22476-9 (ISO, 2020) failure duration criteria; the FVST reached failure in seven seconds, while the VST required 1565 seconds. The impact of these durations is apparent when considered alongside the CPTu dissipation test (PPD) conducted at 11.4 m depth. While the dissipation response does not replicate the drainage mechanisms during vane shearing, it provides a valuable visual reference for relative drainage timescales. In this case, Blight's time factor —calculated using parameters derived from the adjacent PPD — indicates drained conditions for the VST (T = 3.059) and undrained conditions for the FVST (T = 0.015). These drainage classifications are consistent with both the dissipation behaviour observed in the PPD and the visual response of the vane tests: the VST shows a gradual buildup to peak, higher rotation at failure, and prolonged softening—indicative of drained conditions—while the FVST displays a sharper peak and rapid post-peak strength loss, characteristic of undrained shearing.

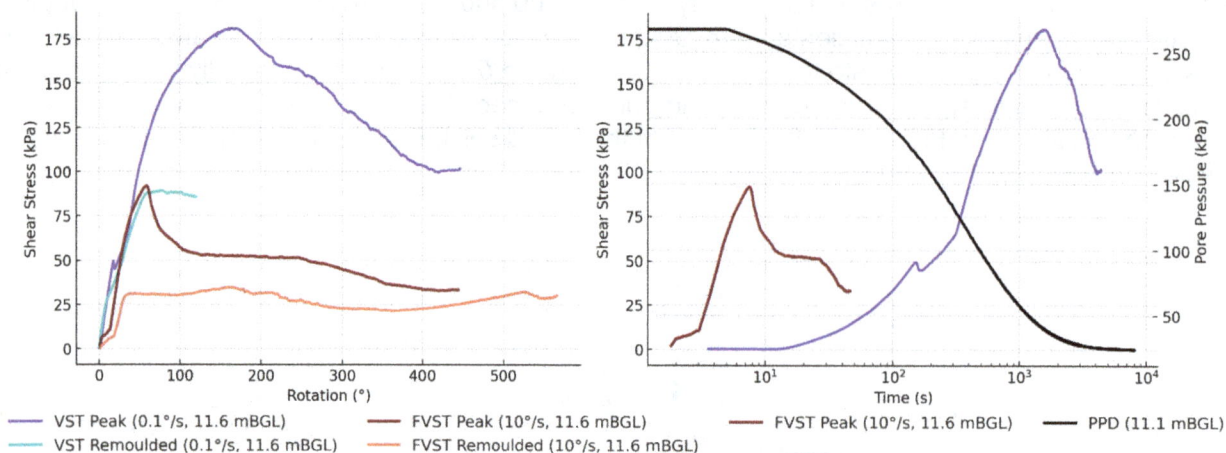

- VST Peak (0.1°/s, 11.6 mBGL)   FVST Peak (10°/s, 11.6 mBGL)   FVST Peak (10°/s, 11.6 mBGL)   PPD (11.1 mBGL)
- VST Remoulded (0.1°/s, 11.6 mBGL)   FVST Remoulded (10°/s, 11.6 mBGL)

FIG 7 – Comparison of paired FVST and VST response in clay-like tailings (Ic = 3.2) highlighting rate dependent strength behaviour.

Although the CPT suggests a clay-like material, the tested tailings are composed of low-plasticity fine-grained particles with higher permeability than typical natural clays. Under standard rotation rates, this higher permeability allows for drainage during shearing, resulting in elevated measured strengths. Increasing the rotation rate reduced the time to failure and more effectively suppressed drainage, enabling the FVST to capture a response more representative of undrained conditions. This highlights the importance of tailoring rotation rate to material permeability when interpreting VST results even in fine-grained tailings.

In contrast, tests conducted in heterogeneous tailings—identified based on multiple changes in soil behaviour type (SBT) within ±0.5 m of the vane test depth—showed a markedly different trend. In nearly all cases, the fast tests produced a higher peak shear strength than the standard-rate tests. These units typically contained thin interbedded layers of silt and clay or abrupt transitions in stratigraphy and were associated with greater scatter in both peak and remoulded strength response. The remoulded strength response in these zones was more variable; however, a majority of fast tests still showed higher remoulded strength compared to the slow tests.

Figure 8 illustrates this behaviour in a thinly layered unit, where the FVST, conducted at 7°/s, produced a higher peak shear strength (Su) than the standard-rate VST at 0.1°/s. Within a 1 m interval centred on the test depth, Ic values ranged from 2.5 to 3.3 (mean = 2.9), indicating variability in soil type. Although the VST reached failure at a lower rotation angle, the FVST failed much more rapidly and achieved an undrained condition (T = 0.02), whereas the VST response was classified as drained (T = 1.57).

Drainage classification in stratified or layered zones is inherently more uncertain. Hydraulic anisotropy and thin interbedded layers complicate the application of time factor approaches. The common assumption of a uniform c_h/c_v ratio, while convenient, may not be appropriate across highly variable tailings profiles, and could result in underestimation of drainage potential. Stratified layers are expected to result in a higher anisotropy. These issues merit further investigation, additionally, the impact of saturation has not been considered as part of this study.

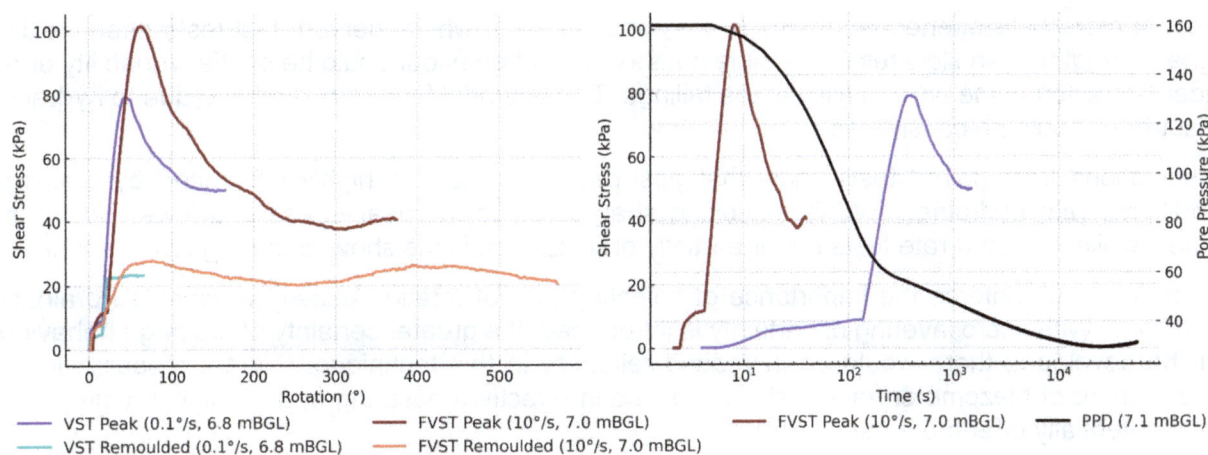

FIG 8 – Comparison of paired FVST and VST response in heterogeneous fine grained tailings highlighting rate dependent strength behaviour.

These layered zones are inherently more heterogeneous, and the increased potential for spatial variability is likely contributing to the inconsistent response observed. Unlike controlled academic studies that use multiple soundings, at multiple rotation rates at a single test depth, our study relied on paired tests in adjacent soundings—reflecting the constraints and realities of operational testing programs. While this approach is more commercially viable, it introduces limitations in resolving rate and drainage effects.

Though the current analysis of post-peak response in these case studies is limited to observational trends, the data provides valuable context for interpreting the broader data set of standard-rate VSTs. In particular, the paired tests from coal tailings illustrate how drainage conditions influence the shape of the post-peak response. Typically, in the FVSTs if undrained conditions were achieved, a sharper drop in shear strength following failure and lower post-peak values were observed compared to standard-rate VSTs performed under partially drained conditions. Both Figures 7 and 8 show this behaviour. In contrast, slower rotation tests often showed more gradual softening, likely because of drainage due to ongoing consolidation, as seen in Figure 7. While there is scatter in the results, these observations indicate the importance of understanding post-peak and strain softening behaviour rather than relying solely on remoulded strength values.

CONCLUSIONS

This paper presents a combined data-driven and material-focused evaluation of vane shear test behaviour in mine tailings, with emphasis on drainage conditions, rate effects, and post-peak response.

The data driven study evaluated post-peak strain softening using a screened data set of standard rate tests conducted under saturated, assumed undrained conditions. Comparisons of shear stress at fixed rotation angles revealed highly variable responses, indicating that a single rotation threshold cannot reliably capture or define post-peak behaviour. To address this, a rate-of-change criterion was introduced, defining the post-peak strength based on when the rate of strength degradation stabilises. Though the initial study is based on arbitrary criteria the approach shows promise as a practical means of characterising post-peak response rather than relying on remoulded strength alone.

From the coal tailings case study, there was a high degree of variability in response from the 356 vane shear tests analysed. The assessment incorporated adjacent CPTu and PPD data to better understand stratigraphic variability and drainage state. The PPDs demonstrate that even for tailings with a SBT of 'clay like', fast rotation rates are required to achieve undrained conditions. Of the 57 twinned fast and standard-rate tests, the majority of the fast tests found higher peak strengths, contrary to trends reported in prior studies. However, the heterogeneity of the material appears to be a key factor in this response.

Homogeneous, fine-grained tailings typically showed a reduction in peak and remoulded strength with increased rotation rate, consistent with suppressed drainage and an undrained response at fast rates. In more heterogeneous materials, the opposite trend was observed: fast tests often yielded higher strengths than slow tests. Possible causes of this behaviour could be spatial variability or an underestimation of the permeability of the tailings. The potential for partial drainage due to hydraulic anisotropy should be considered.

Observations from paired tests show that post-peak behaviour is highly influenced by drainage conditions, with undrained FVSTs exhibiting sharper drops in shear stress and lower post-peak values, while standard-rate tests under partially drained conditions showed more gradual softening.

These findings reinforce the importance of selecting rate of rotation based on expected drainage conditions. With micro layering uncertainty is introduced. If a greater certainty of drainage behaviour can be developed there would be increased reliability in this technique. Current research in the development of piezometer vanes, show promise in directly measuring c_v and Blight's time factor, could potentially offer more certainty.

REFERENCES

ASTM International, 2018. Standard Test Method for Field Vane Shear Test in Saturated Fine-Grained Soils, ASTM D2573/D2573M-18. https://doi.org/10.1520/d2573_d2573m-18

Blight, G E, 1968. A note on field vane testing of silty soils, *Canadian Geotechnical Journal*, 5(3):142–149. https://doi.org/10.1139/t68-014

Chandler, R J, 1988. The in situ measurement of the undrained shear strength of clays using the field vane, *Vane Shear Strength Testing in Soils: Field and Laboratory Studies*, pp 13–44 (ASTM International: West Conshohocken). https://doi.org/10.1520/STP10319S

Contreras, I A and Harvey, J W, 2021. The role of the vane shear test in mine tailings, in *Proceedings of the 25th International Conference on Tailings and Mine Waste*, pp 457–468.

Harvey, J W, Hogan, A A, Obeidat, D N, Contreras, I A and Kelly, S A, 2023. Establishing a site-specific standard of practice for field vane shear testing in mine tailings, in *Proceedings of Tailings and Mine Waste 2023*, pp 1493–1504 (The University of British Columbia: Vancouver).

Hogan, A A, Kelly, S A, Sharp, J T and DeJong, J T, 2025. A comprehensive review of field vane shear testing in mine tailings and recommendations for tailored standards, *Geotechnical Testing Journal*, 48(1). https://doi.org/10.1520/GTJ20240027

Hogan, A A, Kelly, S A, Stroeteboom, O and Robertson, P K, 2022. Modified field vane technology for improved reliability of undrained shear strength measurements in mine tailings, in *Proceedings of the 26th International Conference on Tailings and Mine Waste*, pp 262–273 (Colorado State University: Denver).

International Commission on Large Dams (ICOLD), 2023. Tailings Dam Safety – Monitoring, Instrumentation and Data Analysis, Bulletin 194, ICOLD.

International Organization for Standardization (ISO), 2020. Geotechnical investigation and testing – Field testing – Part 9: Field Vane Test (FVT and FVT-F), ISO 22476-9:2020 (International Organization for Standardization: Geneva).

Jamiolkowski, M, Ladd, C C, Germaine, J T and Lancellotta, R, 1985. New developments in field and laboratory testing of soils, in *Proceedings of the 11th International Conference on Soil Mechanics and Foundation Engineering*, pp 57–153.

Jefferies, M and Been, K, 2006. *Soil Liquefaction: A Critical State Approach* (Taylor and Francis).

Locat, J and Demers, D, 1988. Vane shear strength of sensitive clays, *Canadian Geotechnical Journal*, 25(4):799–806.

Mundle, C, Esford, F and Julien, M, 2019. Field shear vane testing in tailings, in *Proceedings of Tailings and Mine Waste 2019*, pp 1111–1123 (The University of British Columbia: Vancouver).

Reid, D, 2016. Effect of rotation rate on shear vane results in silty tailings, in *Proceedings of the Fifth Geotechnical and Geophysical Site Characterization*, pp 369–374 (Australian Geomechanics Society: St Ives).

Reid, D, Rodriguez, C, Fourie, A and Tiwari, B, 2023. Partial drainage effects during vane shear tests, with an emphasis on the measurement of remoulded strengths, Proceedings of Tailings and Mine Waste 2023.

Robertson, P K, 2009. Interpretation of cone penetration tests – A unified approach, *Canadian Geotechnical Journal*, 46(11):1337–1355. https://doi.org/10.1139/T09-065

Standards Australia, 2001. AS 1289.6.2.1-2001: Soil strength and consolidation tests – Determination of the shear strength of a soil – Field test using a vane.

Wilson, G W, Mundle, R, Contreras, M and Harvey, R, 2016. Viscous effects in rate-sensitive soils during vane shear testing, *Geotechnical Testing Journal*, 39(4):567–578.

Advancing vane shear testing – development and first field application of a piezometer vane (VSTu)

Y N Byrne[1], J Greig[2], A Small[3], J DeJong[4] and K Cator[5]

1. ConeTec, Burnaby BC, Canada. Email: yasmin.byrne@conetec.com
2. ConeTec, Burnaby BC, Canada. Email: jgreig@conetec.com
3. Klohn Crippen Berger, Fredericton NB, Canada. Email: asmall@klohn.com
4. UC Davis, Davis CA 95616, USA. Email: jdejong@ucdavis.edu
5. BHP, Legacy Assets, Saskatoon SK, Canada. Email: kendell.cator@bhp.com

ABSTRACT

This paper presents the development and initial field deployment of a prototype variable rate piezometer vane shear test device (VSTu or VR-VSTu) designed to measure porewater pressure (PWP) on the vane blade edge during insertion and shearing of soils or tailings in the field. The device was developed to address drainage uncertainties when conducting vane shear testing, in materials such as mine tailings, intermediate soils, and dredge materials. To ensure compatibility with existing VST results, the VSTu was designed to comply with standard vane dimensions, while integrating additional instrumentation. In addition to recording PWP and torque during shearing, the PWP response is recorded during insertion.

Observations of induced pore pressure, or lack thereof, during vane insertion and shearing can provide insights into whether materials are undrained, partially drained, or freely draining. By integrating PWP measurement during shearing, the VSTu has the potential to offer insights into the brittleness, sensitivity, consolidation, and large strain behaviour.

This paper presents results from the deployment of the VSTu at the East Kemptville tin tailings site in Nova Scotia, Canada. Field testing was conducted in both coarse and fine tailings, and results indicate the device can capture meaningful PWP responses during insertion and shearing, with implications for identifying drainage conditions and assessing undrained shear strength.

INTRODUCTION

Vane shear testing is widely used to measure the undrained shear strength of cohesive soils; however, current methods have limitations for partially draining or more permeable materials, such as mine tailings and dredge materials.

In coarse-grained materials, where achieving undrained conditions may not be possible even at high rotation rates (eg >4000°/min), the vane is likely measuring a drained or partially drained strength. While this does not represent undrained shear strength, the measured response may still be valuable. Research by Bezuidenhout and Torres-Cruz (2024) highlights the potential for drained vane test data to support state-based interpretations of tailings behaviour, including liquefaction assessment. Blindly interpreting peak torque as undrained shear strength without accounting for drainage conditions can result in inaccurate, overestimated strengths and unconservative geotechnical designs.

This highlights a critical need for an enhanced understanding of drainage potential and pore pressure response. Increasing the vane rotation speed is becoming common practice to minimise drainage potential; however, this approach also has limitations and unknowns related to selecting appropriate rotation rates and interpreting results. To address these uncertainties, this study introduces a prototype variable rate piezometer vane shear test device (VSTu), which integrates real-time porewater pressure (PWP) measurements during both insertion and shearing.

This paper presents the development, initial validation, and results of the first field application of the VSTu in tailings. The objectives include evaluating the tool's performance, demonstrating its compatibility with standard dimensions, and providing preliminary insights into drainage behaviour and soil response.

THE PIEZOMETER VANE

In situ vane shear testing, originally introduced in 1950 by Cadling and Odenstad and standardised internationally in the 1980s (ASTM, 1986; Chandler, 1988), is widely used to measure undrained shear strength in cohesive soils. However, standard vane shear protocols and procedures were originally developed and calibrated for use in low-permeability clays, where undrained conditions could be reliably assumed (Chandler, 1988). When applied to intermediate permeability materials such as silts, dredge materials, and mine tailings, the assumptions underpinning standard methods often breakdown. For example, Börgesson (1981) explicitly notes that the permeability and particle-size distribution of silts result in drainage conditions during vane shearing that cannot be clearly defined as either fully drained or fully undrained.

To counteract partial drainage effects, increasing the vane rotation speed has become common practice typically guided by Blight's (1968) method based on pore pressure dissipation criteria. However, increasing the vane rotation rate introduces further uncertainties regarding the identification of appropriate rotation speeds to capture an undrained response, while avoiding viscous effects (Hogan *et al*, 2022). Additionally, Morris and Williams (2011) raises questions in Blight's original interpretations of pore pressure dissipation, indicating that conventional assumptions about drainage following vane insertion may result in elevated pore pressures persisting from insertion.

Avoiding drainage may not always be necessary or possible—provided that drainage behaviour during testing is understood and measured. While this is a limitation of conventional vane testing tools, which do not record PWP during shearing, the VSTu prototype has the potential to inform on the drainage conditions throughout the test.

Earlier developments

The earliest published attempt to measure PWP during vane shear testing was by Wilson (1963), who used a hypodermic needle embedded in a vane blade during laboratory tests on saturated silt. This initiated early exploration into the link between shear-induced pore pressures and measured strength. More advanced instrumentation emerged in the 1980s and 1990s. Kimura and Saitoh (1983) conducted a comprehensive laboratory study on normally consolidated cohesive soils, directly, measuring PWP during both insertion and rotation. Their device used porous vane blade edges connected to an uphole directional valve and transducer, along with embedded pressure sensors in the soil, allowing continuous monitoring of pore pressure response throughout the test.

Charlie, Calderon and O'Neill (1994, 1995) later introduced the CSU (Colorado State University) Piezovane, a field-deployable vane shear device designed to assess liquefaction potential in saturated sands. The CSU Piezovane featured a pressure transducer embedded within the vane shaft, connected to four ports in the vane blade edges. While the blade dimensions conformed to ISO 22476-9 (International Organization for Standardization (ISO), 2020), to accommodate internal instrumentation the shaft diameter exceeded the ISO limit. The system collected high-frequency torque, rotation, and PWP data during rapid shearing (typically at 90°/s). The findings from Kimura and Saitoh (1983) revealed that high pore pressures were generated during vane insertion, with dissipation taking several hours—much longer than assumed in conventional interpretation. Interestingly, they found that pore pressure changes during rotation were comparatively small, indicating that insertion disturbance, not shearing, may dominate the drainage response in certain soils. The CSU Piezovane showed that contractive sands exhibited increased PWP during shear, while dilative sands showed decreases, allowing for *in situ* identification of liquefaction-susceptible layers—validated during post-earthquake investigations at Loma Prieta.

Both the Kimura and Saitoh device and the CSU Piezovane incorporated vane shaft or blade configurations with area ratio in excess of ISO 22476-9 (ISO, 2020), meaning that insertion disturbance and the associated excess PWP generation were likely greater than those produced by standard vanes. A thicker shaft displaces a greater volume of soil during insertion, leading to higher excess PWP and a more disturbed stress state around the vane prior to shearing (eg Kimura and Saitoh, 1983; Morris and Williams, 2011). This affects the comparability of measured strengths to conventional field vane tests. Additionally, the CSU Piezovane was developed specifically for sands and operated at significantly higher rotation speeds (up to 90°/s) than typically used in geotechnical practice.

Current prototype (VSTu)

The prototype VSTu is a variable-rate piezometer vane shear device developed to measure PWP during both insertion and shearing *in situ*. The patent pending system builds on ConeTec's existing vane shear testing platform, and includes an integrated pore pressure transducer, with pressure ports located directly at the vane edge. Figure 1 shows the blade with the pore pressure ports and the deployment rods passing through the motor that rotates the blade. Much like previous iterations the pore pressure transducer is set back from the vane blade, with a central port running through the centre of the vane allowing easy saturation and vane blade replacement. The design allows real-time measurement of induced PWP alongside torque and rotation angle.

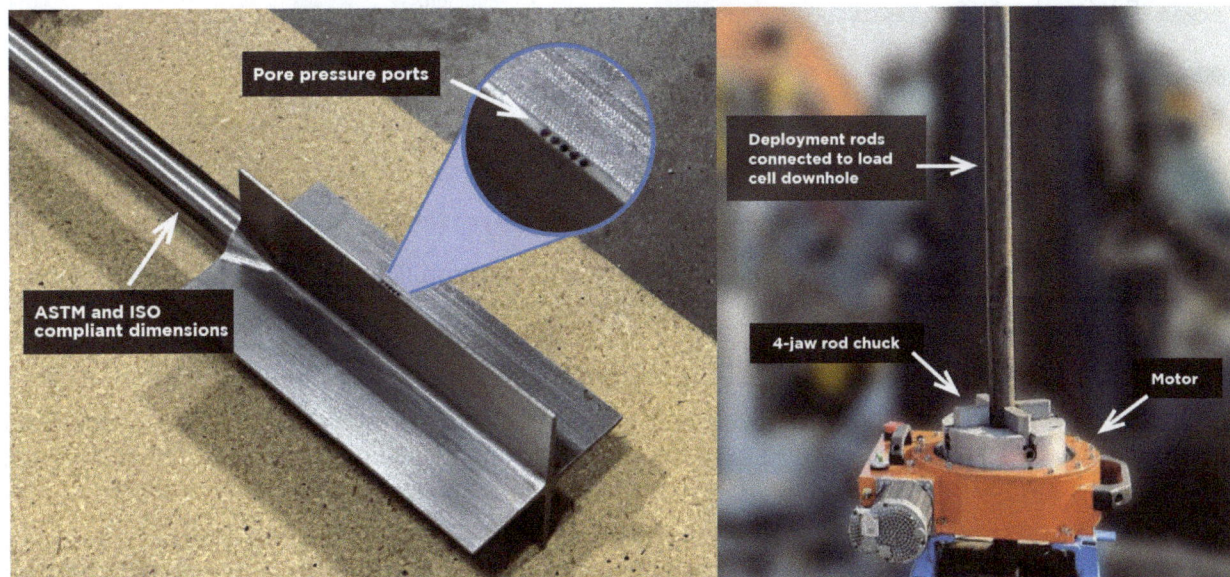

FIG 1 – Image of 40 × 80 mm piezometer vane blade, showing pore pressure ports.

The prototype was developed as a fast vane system, with rotation rates ranging from 5°/s to 30°/s, targeting potentially liquefiable materials where drainage is expected to occur under standard-rate vane shear testing. While the initial version does not accommodate the ASTM standard rate (0.1°/s) and speeds common for natural clays, a later prototype has been designed to support both fast and standard (slow) rotation speeds. The rotation speed range is applicable to intermediate permeability materials, such as mine tailings, silts, and dredge deposits. These materials were targeted to reduce uncertainty around the identification of a minimum vane rotation speed required to achieve undrained conditions, and to provide confirmation of whether a given test was, in fact, undrained. In addition, the system was intended to support investigation into how the *in situ* shear response of contractive and dilative materials differs.

Unlike earlier piezometer vane developments, the vane blade and shaft geometry of the prototype conform to ISO 22476-9 (ISO, 2020) dimensional requirements, including Area Ratio (R_a). The vane used in the initial field campaign had a R_a of 0.099 (9.9 per cent) This conformance ensures that data collected using the VSTu are directly comparable to conventional field vane and fast vane test results. As noted above, earlier instrumented vane designs incorporated oversized shafts; the resultant higher area ratios may have generated increased insertion disturbance compared to ISO-compliant vanes. In contrast, the VSTu maintains standard geometry while enabling pore pressure measurement.

Data validation demonstration

A system validation of the pore pressure measurement instrumentation was performed in a 20 m deep fluid-filled borehole, where the pore pressure transducer response was compared against expected hydrostatic pore pressure conditions. The transducer recorded stable and consistent PWP readings during lowering and static hold, confirming that the pore pressure measurement system was functioning correctly and capable of capturing changes in PWP as shown in Figure 2. This check was used to verify both the sensor response and the hydraulic continuity of the pressure port through the vane assembly. Each reading corresponded to the pore pressure that was expected at the depth of the reading, with the final reading corresponding to head of fluid in borehole.

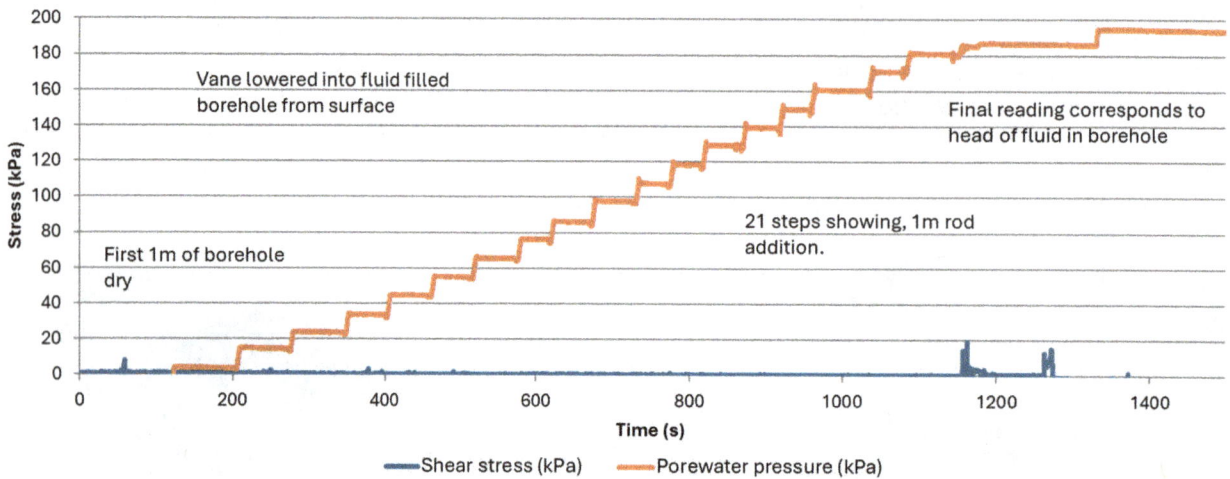

FIG 2 – PWP response during system validation in a 20 m deep fluid-filled borehole, confirming sensor performance.

THE SITE FOR THE FIRST APPLICATION

The first field application of the VSTu was at a closed tin mine in Nova Scotia, Canada, the East Kemptville site (EK). The site is located near the village of Kemptville, in south-western Nova Scotia, Canada and is owned and managed by Rio Algom Ltd., a wholly owned subsidiary of BHP.

From 1985 to 1992, the mine extracted tin from two on-site open pits and produced approximately 18.8 Mt of tailings. The site closed in 1992, and the reclamation activities were completed between 1992 and 1999. EK is currently in the Closure Active Care phase, during which time the site conducts routine operation, maintenance, and surveillance activities. During the milling and concentrating process, the EK tailings were separated into three streams: coarse tailings, fine tailings, and flotation tailings. The coarse tailings were rejected from the gravity separation process and hydraulically discharged using a single point discharge method. The fine tailings from the gravity separation process were combined with flotation tailings before being deposited into multiple tailings cells. The cells were designed to contain the fine tailings in the middle of the cell with perimeter coarse tailings containment dams. Internal rock fill dykes were designed to intersect with the perimeter coarse tailings dams, which created the cells for the purpose of impounding the fine tailings. The single discharge method was also used for the fine tailings mixture.

The finer tailings were deposited at the base of the facility and the coarse tailings placed over the finer tailings through hydraulic deposition and mechanical placement. Klohn Crippen Berger (KCB) has been involved at the site since 2018 as the engineer of record.

Geotechnical setting

The native soil strata at the EK site generally consists of organic deposits, overlying glacial till, which in turn overlie bedrock.

As noted above, the tailings impoundment contains three types of tailings: coarse tailings, fine tailings, and flotation tailings. The coarse tailings were free-draining sand, fine to medium sand-size material that originated from barren sand particles separated in the gravity separation process. The coarse tailings had an approximate fines content of 27 per cent. The fine tailings were finely ground material mainly composed of silt-size particles. The fine tailings and flotation tailings mixture had an estimated fines content as high as 96 per cent. The coarse and fine tailings were deposited in a tailings pile that has slope angles of 6H:1V and up to 22 m high. The liquefaction potential of the coarse and fine tailings was unknown. Previous investigation programs provided conflicting results.

In 2024, a comprehensive geotechnical investigation was conducted that included 31 boreholes, 12 seismic cone penetration tests (SCPTu), and seven fast vane shear tests (FVTs).

The primary objectives of the 2024 geotechnical investigation were to measure soil response from seismic tests to provide inputs for conducting Dynamic Site Response Analysis and to measure soil parameters and pore water pressure conditions to support planned updates of slope stability assessments.

Research questions

As part of the 2024 geotechnical investigation at the EK site, a piezometer vane testing plan was developed by KCB, ConeTec, and BHP, whereby the initial field deployment of ConeTec's VSTu prototype could be conducted. The piezometer vane testing plan was designed to allow a direct comparison of the results obtained using a standard vane to the results obtained with ConeTec's VSTu prototype.

The key objectives of the Piezometer Vane Testing Plan were to:

- Determine if insertion of the VR-eVSTu prototype can measure initial pore pressure generation.

- Determine if the VR-eVSTu prototype can provide information regarding whether the shearing is being conducted in an undrained manner.

- Determine the VR-eVSTu prototype rotation speed required to achieve undrained conditions and compare this to the predicted rotation speeds.

- Determine if rapid rotation has an influence on the shear strength results.

- Determine if the brittleness observed by the VR-eVSTu prototype is influence by the speed of the vane that was theoretically required to achieved undrained conditions.

Test locations and ground conditions

For the field application of the VSTu, a location was selected where there had already been two boreholes drilled with a piezometer installed and one SCPTu was advanced. Figure 3 shows the SCPTu profile with the coarse tailings clearly shown above the fine tailings (above a depth of 5.5 m).

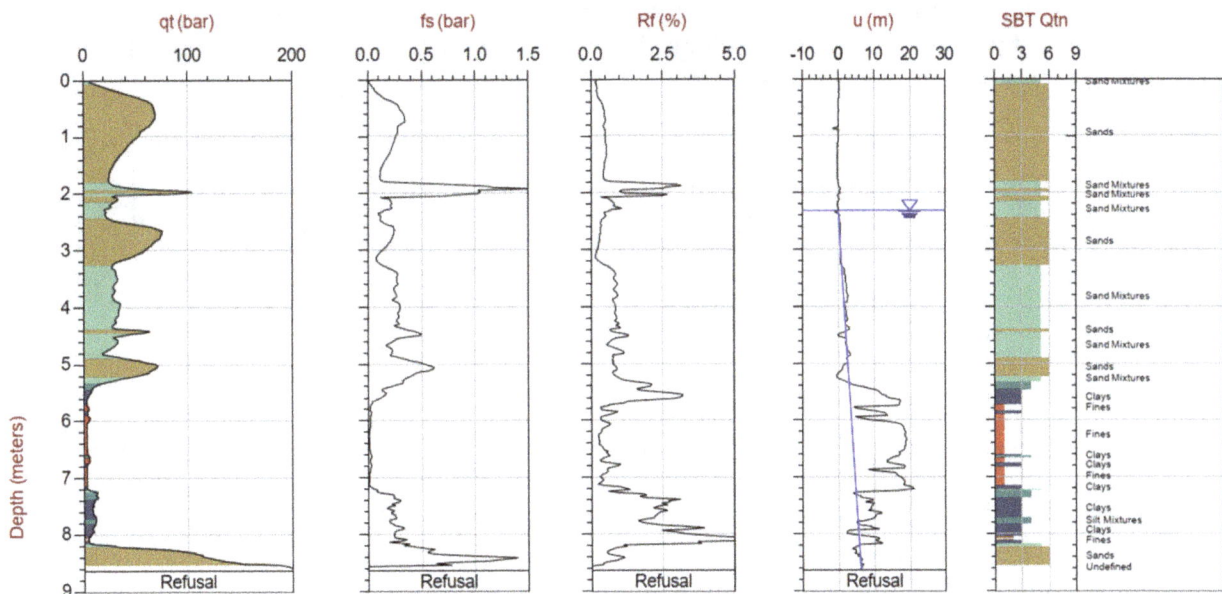

FIG 3 – SCPTu profile from East Kemptville Mine. All vane shear tests presented in this paper (VR-eVSTu and FVST) were conducted within 12 m of this sounding.

The water table at this location was at a depth of 2.27 m. The coarse tailings had a tip resistance above 20 bar and the dynamic pore water pressures below the water table were neutral. The fine tailings had a much lower tip resistance and sleeve friction and the dynamic pore pressures were positive. Below the fine tailings, the cone penetrated into the till and refused.

All vane tests presented in this paper (both VSTu and FVST) were conducted within 12 m of the SCPTu location.

THE METHODOLOGY

The *in situ* testing was conducted using the VSTu developed by ConeTec and applying a modified fast vane protocol tailored for intermediate permeability materials. These results of the nearby SCPTu and conventional fast vane tests were used to inform the selection of test depths and vane rotation speeds, which were specified by KCB. Ambient PWP was calculated from the phreatic surface identified in the SCPTu data.

- Sensor Saturation: Prior to every other test, the piezometer vane was saturated at the surface. Every second test was a direct push from the previous depth without retraction or saturation.

- Vane Insertion: The vane was pushed directly to the target depth using a cased rod system. PWP was continuously recorded during insertion. If excess pore pressure was observed at the end of insertion, a dissipation period was allowed before vane rotation. In all cases, excess pore pressure dissipated within the maximum permissible wait time specified by ISO 22476-9 (ISO, 2020) or ASTM D2573 (ASTM International, 2018).

- Peak Strength Test: The vane was rotated at a pre-specified constant rate between 6.7°/s and 12°/s, covering a total rotation of 360°, to capture the peak undrained shear strength.

- Remoulded Strength Phase: The vane was then rotated at a fixed rate of 30°/s for 4000° to determine the remoulded strength.

Torque, PWP, rotation angle, temperature, and inclination were recorded throughout the test sequence, including during insertion, peak strength testing, and the remoulded strength test. Rotation speeds were selected to allow comparison with existing fast vane tests and to explore potential rate effects across different material types.

THE RESULTS

In this study, piezometer vane test data were analysed against time rather than rotation, as no rotation occurs during the insertion and dissipation phases. For this initial field campaign, the data was divided into four phases: vane insertion, dissipation, peak test, and remoulded test. Figure 4 shows an example test with these phases annotated against time.

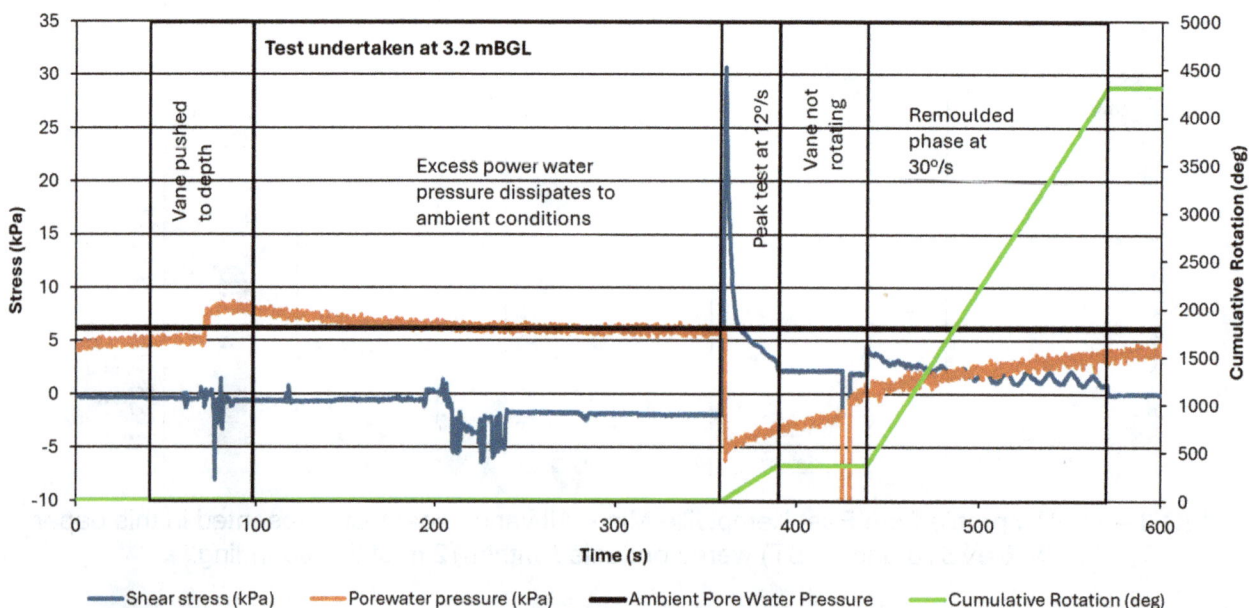

FIG 4 – Example piezometer test with insertion, dissipation, peak, and remoulded phases.

Comparison to conventional fast vane testing (FVST)

The peak shear strength results from the VSTu tests were compared to fast vane testing (FVST) conducted during the previous phase of ground investigation. For the peak tests, the VSTu was conducted at a rotation rates between 6.7–12°/s, and the FVST at 12°/s. These rates were selected based on assumed drainage behaviour of the tested materials. Figure 5 presents the results of peak strength (τ_{max}) and τ_{max}/σ_v', where τ_{max} is the peak shear strength and σ_v' is the vertical effective stress. Between 5.4 and 9.1 m depth, within the fine tailings—characterised as non-plastic and comprising over 75 per cent fines—there appears to be relatively good agreement between the two instruments. Based on the τ_{max}/σ_v' values, the shear response is interpreted to be undrained and the material is likely normally consolidated.

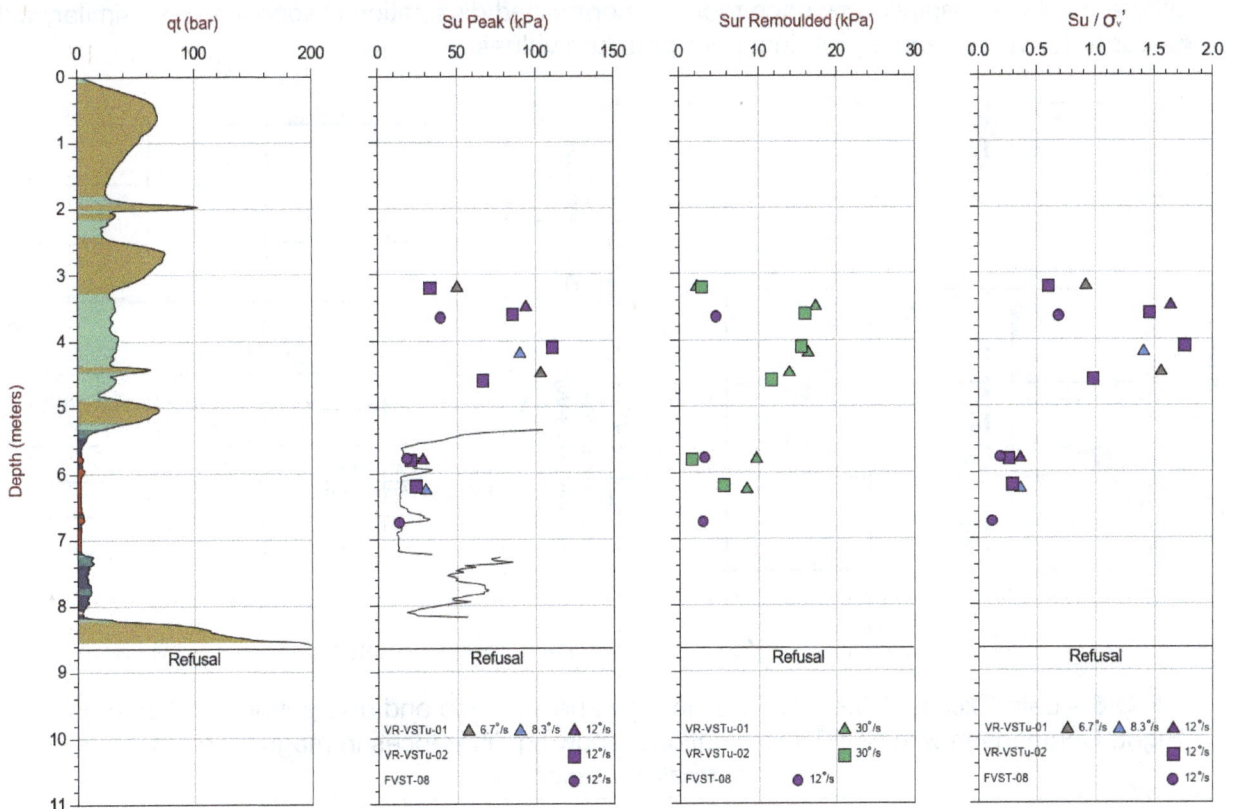

FIG 5 – Comparison of peak shear strength, remoulded shear strength and τ_{max}/σ_v' from VSTu and fast vane shear testing (FVST).

Results from the coarser tailings, between 3.2 and 4.5 m depth, show greater variability in τ_{max}. These materials exhibit a more heterogeneous fines content—above 27 per cent and a higher degree of stratigraphic variability, as shown in the SCPTu profile where the soil behaviour type fluctuates, indicating changes in material characteristics and permeability. The surrounding SCPTu soundings within 20 m also indicate greater spatial variability in response within the coarser tailings, further supporting the observed scatter in τ_{max} values. Based on τ_{max}/σ_v', it is likely that the tests in this interval captured a drained shear strength, rather than undrained response. In tailings with this composition, the rotation speeds used were not sufficient to prevent drainage during shearing.

There is the potential that allowing for PWP to dissipate – rather than commencing the vane rotation immediately – could have induced consolidation and resulted in elevated peak strengths. However, the dissipation wait time was always within ASTM and ISO requirements and comparable to wait times between insertion and peak from the FVST.

Piezometer vane results

There was variability in the observed piezometer vane response during both the vane insertion and the vane rotation.

In materials that are not free-draining, vane insertion is expected to generate excess PWP. Approximately half of the tests recorded such elevated pressures during insertion, however this was mixed between the fine and coarse tailings. The most significant response was noted in the fine tailings, where at 5.8 m below ground level (BGL), vane insertion generated approximately 20 kPa of excess PWP (Figure 6). This excess pressure was allowed to dissipate to ambient conditions before rotation commenced. The magnitude of pore pressure generated was lower than that reported by Kimura and Saitoh (1983), however the materials, experimental set-up and vane area ratios differed. It is expected that the thicker vanes shafts such as those in 1983 and 1994/95, would have generated greater excess pore pressures due to increased disturbance. In the adjacent SCPTu, a PPD (pore pressure dissipation) was performed at 5.85 m BGL; this is compared to the VSTu dissipation record (Figure 6). Although the magnitude of excess pore pressures differed because of the different volume displaced by each tool, the normalised dissipation responses were similar, with comparable T_{50} values and equilibrium pore pressure values.

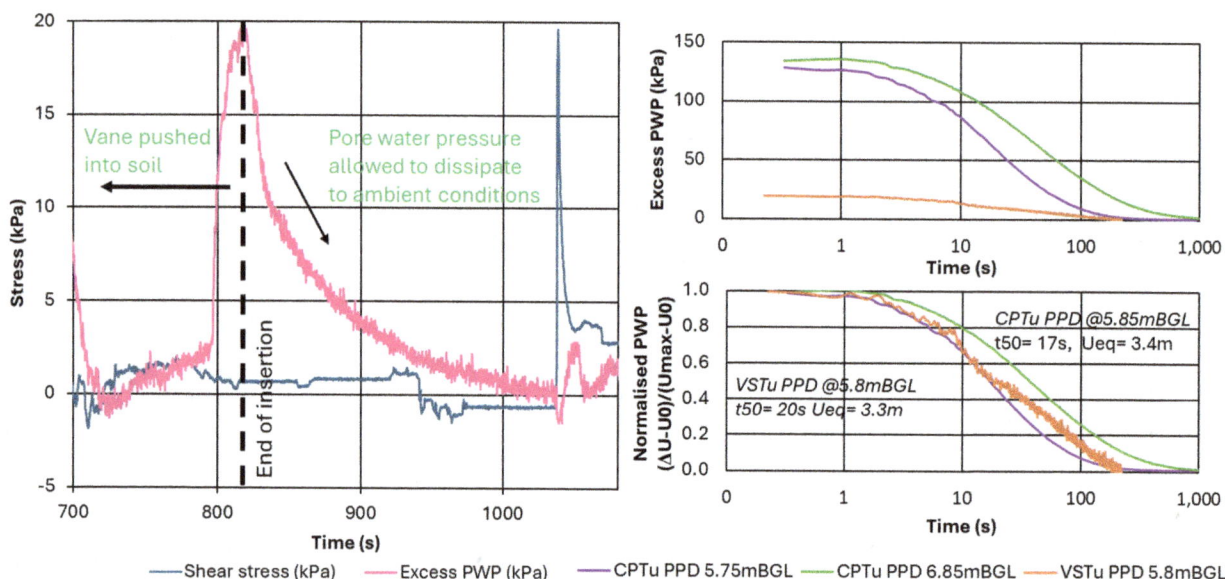

FIG 6 – Left: Excess PWP response during vane insertion and dissipation at 5.8 m BGL. Right: Comparison with SCPTu dissipations, showing differences in magnitude but a similar normalised response.

Although the PWP response during the peak test varied, a few trends can be observed from the limited data set. When Su/σ_v' exceeded 0.95, no excess porewater pressure was recorded during vane rotation—only the ambient pore pressure was observed (see Figure 7, left where no excess PWP was generated). In most cases, excess PWP was not generated prior to failure but instead occurred at or just after the peak, stabilising towards ambient conditions during post-peak shearing (see Figures 4 and 7, right). When excess pore pressures were generated during shearing, their magnitudes were generally lower than those observed during insertion. This trend is consistent with the observations of Kimura and Saitoh (1983).

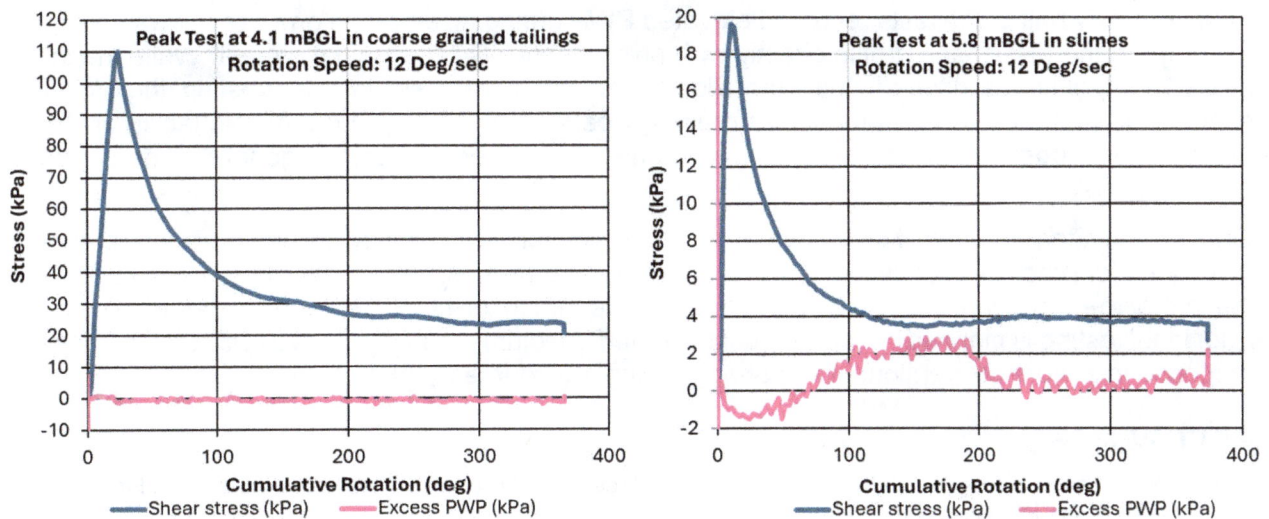

FIG 7 – Left: Example test in coarse tailings where no excess PWP was generated during shearing $(\tau_{max}/\sigma_v' =1.76)$. Right: Example test in fine tailings (slimes) showing typical excess PWP. $(\tau_{max}/\sigma_v' =0.26)$

Following the peak test, a remoulded test was conducted at a higher rotation speed of 30°/s. In the coarse tailings, the PWP during the remoulded phase generally remained stable or slowly returned toward ambient levels, largely unaffected by fluctuations in shear stress. However, in two tests conducted at 4.5 m BGL, a negative excess PWP developed during the remoulded test—despite no excess PWP having been generated during the preceding peak test. This suggests that the higher rotation rate may have induced an undrained response in those cases. In contrast, the fine tailings exhibited a more coupled behaviour, with PWP changes closely tracking variations in shear stress throughout the remoulded test.

Geotechnical implications

The results of the VSTu are being reviewed in detail and will be further compared to the other *in situ* tests that were conducted and the results of lab testing. An initial review of the results suggests the following:

- The VSTu prototype can provide an indication of whether the strength that is being measured is drained, undrained, or partially drained.

- For the undrained tests, the pattern of the shear strength versus rotation can provide an indication of potential brittleness of the tailings in terms of their undrained shear strength behaviour. Further research is required to equate the vane shear rotation with shear strain to better understand the brittleness.

- For the tests that did not rotate fast enough to obtain an undrained strength and were measuring a drained strength, the pattern of shear strength versus rotation might provide insights to the relative density of the tailings (state with respect to liquefaction potential). This is also an area of further research.

CONCLUSIONS

The piezometer vane prototype was successfully deployed in the field to measure PWP during insertion and shearing in tailings. An additional system validation test was conducted in a fluid-filled borehole, confirming the sensor response under known hydrostatic conditions. Since the prototype maintains ISO-compliant vane geometry, direct comparison with conventional fast vane test results is possible. In fine-grained slimes, the VSTu generally produced undrained responses consistent with fast vane tests conducted at similar rotation rates. In the coarser tailings, greater variability in Su and Su/σ_v' was observed, consistent with heterogeneity in SCPTu profiles and fines content. It is likely that partial drainage occurred in some tests, contributing to elevated shear strength values.

Approximately half of the tests recorded elevated PWP during insertion. In one case, it was possible to compare the dissipation profile directly with an adjacent SCPTu dissipation test. While the initial pressure magnitudes differed, the normalised responses were similar, suggesting the VSTu is capable of capturing meaningful *in situ* drainage behaviour. During shearing, variability in PWP response was observed, with both positive and negative excess pressures generated during peak and remoulded phases.

Given the limited data set, further research and validation are required to establish interpretation guidelines for practical application in permeable soils and tailings. Next steps include adjustments to the prototype hardware, including the possible integration of filter elements to refine response. Additional testing is planned in normally consolidated, medium-plasticity clays to establish a baseline response in a more conventional vane shear testing environment.

REFERENCES

ASTM International, 1986. ASTM D2573-86: Standard Test Method for Field Vane Shear Test in Saturated Fine-Grained Soils, ASTM International.

ASTM International, 2018. ASTM D2573/D2573M-18: Standard Test Method for Field Vane Shear Test in Saturated Fine-Grained Soils, ASTM International.

Bezuidenhout, A and Torres-Cruz, L A, 2024. Drained Laboratory Vane Shear Testing to Characterize Nonplastic Tailings, in Tailings and Mine Waste 2024 Conference.

Blight, G E, 1968. A study of the field vane shear test in clays, in *Proceedings of the Symposium on Shear Testing of Soils*, 361:78–107 (ASTM STP).

Börgesson, L, 1981. Mechanical properties of inorganic silt: Report 1 – Laboratory investigation, PhD thesis, Chalmers University of Technology, Gothenburg, Sweden.

Cadling, L and Odenstad, S, 1950. The vane borer: An apparatus for determining the shear strength of clay soils directly in the ground, in Royal Swedish Geotechnical Institute Proceedings No. 3.

Chandler, R J, 1988. The in-situ measurement of the undrained shear strength of clays using the field vane, in *Vane Shear Strength Testing in Soils: Field and Laboratory Studies*, 1014:13–44 (ASTM STP).

Charlie, W A, Calderon, P A and O'Neill, K, 1994. Liquefaction evaluation with the CSU Piezovane, *Geotechnical Testing Journal*, 17(3):329–339.

Charlie, W A, Calderon, P A and O'Neill, K, 1995. Estimating liquefaction potential of sand using the piezovane, in *Proceedings of the National Conference on Hydraulic Engineering*, pp 965–969 (ASCE).

Hogan, A A, Kelly, S A, Storteboom, O and Robertson, P K, 2022. Modified field vane technology for improved reliability of undrained shear strength measurements in mine tailings, in *Proceedings of Tailings and Mine Waste 2022 Conference,* pp 262–273.

International Organization for Standardization (ISO), 2020. ISO 22476-9:2020 – Geotechnical investigation and testing – Field testing – Part 9: Field Vane Test (FVT and FVT-F).

Kimura, T and Saitoh, K, 1983. Effect of disturbance due to insertion on vane shear strength of normally consolidated cohesive soils, *Soil and Foundations, Japanese Society of Soil Mechanics and Foundation Engineering*, 23(2), June.

Morris, P H and Williams, D J, 2011. A revision of Blight's model of field vane testing, *Canadian Geotechnical Journal*, 48(7):1089–1098.

Wilson, N E, 1963. Laboratory Vane Shear Tests and the Influence of Pore-Water Stresses, in *Laboratory Shear Testing of Soils,* ASTM Special Technical Publication No. 361:377–385.

Tailings dry soil mixing – a sustainable ground improvement technique for TSF operation

O Dudley[1] and L Moreno[2]

1. Senior Geotechnical Engineer, Red Earth Engineering, Brisbane Qld 4000.
   Email: oliver.dudley@redearthengineering.com.au
2. Geotechnical Engineer, Red Earth Engineering, Brisbane Qld 4000.
   Email: laura.moreno@redearthengineering.com.au

ABSTRACT

The stability of upstream raised Tailings Storage Facility (TSF) depends on the strength and consolidation of the tailings foundation. Best Available Technology (BAT) has recently been promoted throughout the tailings industry to mitigate the risk of tailings liquefaction failures. A common Australian practice is to achieve dilative tailings conditions via compaction of tailings with machinery to form a large, perimeter structural zone that forms the structural embankment to contain the tailings. The process of mechanical dewatering and compaction of the tailings is known as mud-farmed tailings. However, access for machinery to conduct mud farming or construction is often restricted due to soft, low-strength, non-conforming tailings. While traditional mud farming is an effective ground improvement method, its success relies on the precondition that previously treated tailings foundations meet compaction and strength requirements.

This paper introduces Tailings Dry Soil Mixing (TDSM), a ground improvement technique that blends soft, saturated tailings with dry tailings to accelerate dewatering and improve strength. The process promotes the reduction of moisture within the pore spaces and drying-induced consolidation of surface tailings. A case study is presented detailing methodology, machinery type, and before-and-after PANDA dynamic cone penetrometer test results, verifying improvement.

While this technique is suitable only under certain environmental and operational conditions, it presents a potential action to rectify unsuitable soft foundations early in tailings operations prior to subsequent raises, which cover the tailings at greater depths. The findings of this paper will provide insights into the practical application of Tailings Dry Soil Mixing as a tailings ground improvement technique, supporting safer and more resilient and sustainable TSF design and operation.

INTRODUCTION

The stability of upstream-raised Tailings Storage Facilities (TSFs) heavily depends on the *in situ* state of their tailings foundation and the perimeter structural zone. In recent years, global focus on TSF safety has intensified due to several high-profile failures, prompting the development and implementation of the Global Industry Standard on Tailings Management (GISTM; Global Tailings Review (GTR), 2020) to mitigate TSF-related risks throughout the facility life cycle, as well as the BAT principles derived from soil mechanics following the Mount Polley failure (Independent Expert Engineering Investigation and Review Panel (IEEIRP), 2015), which define three key components:

1. Eliminate surface water from impoundment.

2. Promote unsaturated conditions in the tailings with drainage provisions.

3. Achieve dilatant conditions throughout the tailings deposit by compaction.

One method aligned with these principles and commonly used in Australia is *mud farming*, which aims to produce dilatant, stable tailings conditions through drying and compaction processes.

At the study site, mud farming is the primary method used to promote dilatant behaviour and improve tailings foundation conditions for upstream raises. This is achieved using low-ground-pressure dozers to compact deposited tailings and enhance solar drying. While effective under favourable weather, the method is highly sensitive to environmental constraints, particularly rainfall and limited evaporation and tailings cycle times. Its success depends on tailings trafficability, which is often compromised in freshly deposited or saturated areas.

The site detailed in this paper is located in a tropical region characterised by high annual rainfall and humidity, particularly during the wet season. These climatic conditions can significantly limit the effectiveness of solar drying and impede surface drainage, resulting in prolonged consolidation times and an increased risk of failing quality control (QC) testing. In such cases, tailings may not meet the compaction and strength requirements for safe embankment construction, posing delays or safety concerns. Additionally, there is an added risk of the equipment bogging in tailings, limiting the effectiveness of mud farming machines. Operational pressures to maximise deposition rates may further reduce cycle times available for drying and consolidation, limiting improvement of the *in situ* foundation conditions.

This paper introduces Tailings Dry Soil Mixing (TDSM) as an alternative and adaptive ground improvement technique, designed to enhance drying consolidation and strength gain when environmental conditions limit traditional mud farming performance. TDSM involves the blending of soft, high moisture content (moisture content above the Plastic and Liquid Limit) with dry surface tailings material to reduce overall moisture content and accelerate consolidation. The process is proposed as a corrective action or supplementary treatment to improve substandard tailings mud farmed layers, especially during transition periods between wet and dry seasons. The overall aim is to meet future Life of Facility (LOF) stability criteria by forming a large perimeter structural zone, reducing the need for additional stability measures, such as downstream buttressing, and minimising the potential for large-strain consolidation and surface cracking.

This paper presents a case study showcasing the application of TDSM, including the methodology, equipment used, and field-testing results before and after implementation. The findings provide insight into the feasibility of TDSM as a sustainable, practical solution to improve TSF foundation conditions under challenging climatic and operational constraints.

TAILINGS DRYING INDUCED CONSOLIDATION

The tailings dewatering process involves reducing water content through physical processes such as:

- Sedimentation – particles settle after tailings are discharged, forming a soil layer. This is rapid in sub aqueous deposition but less relevant in sub ariel deposits.

- Consolidation – water is squeezed out of the soil skeleton due to self-weight or drainage, a process that can take significant time depending on the tailings depths and properties.

- Desiccation – evaporation at the surface creates suction, leading to water flow towards the surface and often resulting in shrinkage cracks that enhance evaporation.

- Evaporation – a major mechanism in arid climate influenced by salinity, environmental conditions and tailings properties.

While desiccation enhances drying through suction and surface cracking, its effectiveness is often limited in tropical climates. High humidity and frequent rainfall reduce evaporation rates and impede surface drainage. Studies (Fujiyasu, 1997; Li *et al*, 2012) indicate that additional limiting factors include:

- Formation of a surface crust, which hinders moisture migration.

- Decreasing hydraulic conductivity as the crust hardens.

- Suction exceeding the air entry value, slowing water loss during drying.

- Presence of an unsaturated zone beneath the crust.

- Climatic factors such as low temperature, high relative humidity, and reduced wind speed.

- Crack formation, which initially increases evaporation by exposing more surface area, may eventually stabilise the evaporation rate as the drying progresses.

Based on these factors, Australia's arid and semi-arid regions are among the most favourable environments for drying-induced consolidation, while wet tropical areas require tailored strategies.

Soil suction plays a key role in drying-induced consolidation. As surface water evaporates, matric suction increases, reducing water mobility and slowing evaporation. This leads to a progressive reduction in the evaporation rate as the soil dries and water becomes tightly bound within the pore structure (Lu and Likos, 2004). According to Iden and Durner (2008), this process can be divided into three stages:

Stage I (Atmosphere-Controlled): At high moisture content, soil evaporation proceeds at the potential rate dictated by atmospheric demand. Soil suction is low, and water is easily transported to the surface.

Stage II (Soil-Controlled): As water is depleted, soil suction increases, and water movement toward the surface slows. Evaporation becomes limited by the soil's ability to supply water.

Stage III (Residual Evaporation): Suction increases dramatically, water becomes tightly bound to particles, and evaporation rates become very low.

The Soil-Water Characteristic Curve (SWCC) is widely used to quantify the relationship between water content and matric suction. It describes how volumetric water content decreases as suction increases and is essential for predicting drying behaviour and water retention in tailings and engineered fills (Fredlund and Xing, 1994). Fine-grained materials, such as clayey tailings, exhibit higher suctions at a given moisture content compared to sandy soils, due to their smaller pore sizes and greater surface area (Khalili and Khabbaz, 1998).

In tropical climates, such as the site studied in this paper, high humidity and seasonal rainfall significantly reduce evaporation rates and limit the development of matric suction. Although high solar radiation can promote drying, wet season conditions—such as persistent cloud cover, saturated ground, and low vapor pressure deficit—hinder moisture loss, delaying the suction increases required to drive consolidation. These climatic constraints pose a challenge for surface densification methods like mud farming, which depend on drying-induced suction to achieve the required strength, and dry density targets necessary for safe tailings deposition and embankment construction.

Several studies have demonstrated that evaporation-induced suction can reach values exceeding 3000 kPa in fine-grained soils under intense drying conditions—sufficient to mobilise moisture and promote consolidation (Smits et al, 2012; Scanlon and Milly, 1994). However, the rate and magnitude of suction development are highly dependent on both soil texture and environmental conditions. Coarser soils tend to develop suction rapidly but lose moisture quickly, while finer soils exhibit a slower but more sustained increase in suction (Assouline et al, 2013), making them more responsive to prolonged drying periods.

These mechanisms are critical when evaluating the potential of TDSM. By blending dry tailings with soft, saturated material, TDSM enhances moisture gradients and encourages suction development, even under marginal drying conditions. This approach accelerates the transition from atmosphere-controlled to soil-controlled evaporation stages, improving consolidation and strength gain within shorter operational windows. As such, TDSM provides a viable and adaptable solution for improving tailings foundation conditions in TSFs affected by variable and climate-limited drying regimes.

TAILINGS GROUND IMPROVEMENT TECHNIQUES

Various ground improvement techniques have been developed and applied across the mining industry to enhance the mechanical behaviour of tailings, mitigate risks such as liquefaction and mud rushes, and improve overall structural performance of TSFs.

At the Kittilä Mine in Finland, Deep Soil Mixing (DSM) was employed to improve the mechanical properties of contractive tailings (Masengo et al, 2019). DSM involves the in situ mixing of binders (eg cement or lime) with soft soils to increase strength and reduce compressibility. At this site, the application of DSM significantly improved the strength and stiffness of the tailings, enabling safer and more stable TSF operations and demonstrating the technique's effectiveness in challenging geotechnical conditions.

The New Afton Mine addressed the risk of mud rushes by implementing a ground improvement strategy combining wick drains and consolidation loading (Adams and Friedman, 2017). Wick drains, or prefabricated vertical drains, accelerate consolidation by providing drainage pathways, while consolidation loading applies additional weight to compress the tailings. Together, these methods enhanced dewatering efficiency and tailings densification, increasing yield stress and overall material stability.

TAILINGS MUD FARMING AND PERFORMANCE CHALLENGES

Mud farming is a widely adopted method for achieving the target dry density of tailings to support planned storage capacity and future upstream embankment raises. The process involves the even distribution of tailings across the facility to optimise water recovery and maintain the design rate of rise. Common equipment used in mud farming includes CAT D6 low ground pressure (LGP) dozers (Dozer), high-and-wide (H&W) excavators, amphibious long-reach excavators (AMEX), and amphibious scroll-type machines (Scrolls).

Quality control for mud farming operations typically involves monitoring the following metrics:

- target tailings slurry solids content (by mass)
- target tailings pour thickness or rate of rise per deposition cycle and per annum
- dozer mud farmed length from embankment
- *in situ* dry density testing for mud farmed lots.

Despite its widespread use, mud farming presents several operational challenges. These include limited solar drying time, constraints on tailings deposition thickness, cycle times and water recovery schedules, and difficulties implementing corrective actions in continuously operated TSFs. Field observations indicate that mud-farmed lengths are often reduced during the wet season due to poor trafficability. In the dry season, non-operational cells—particularly those requiring upstream wall raises—are sometimes allowed to crust, after which dozers are used to track across the surface to extend the mud-farmed area as far as possible without becoming bogged.

Even with careful planning, equipment bogging has been observed during the early stages of cell deposition and mud farming operations. These delays increase equipment downtime and pose additional safety and operational risks. Furthermore, if dry density targets are not met, the *in situ* state of the tailings may shift from dilative to contractive under low effective stress conditions. This transition can reduce the factor of safety against slope instability and may result in large-strain consolidation and surface cracking and may require mitigation measures such as downstream buttressing.

TAILINGS CHARACTERISATION

To evaluate tailings segregation during deposition, surface samples were collected at distances of 10 m, 100 m, 150 m, 200 m, and 250 m from the spigot at two locations within the studied TSF. The Particle size distribution results show a clear trend: sand-sized particle content decreases with distance from the spigot, while fines content increases. The coarser particles tend to settle proximally, while finer silt and clay fractions are transported further across the beach. Figure 1 shows a decrease in sand content percentage and an increase in fines content percentage with distance away from the spigot location.

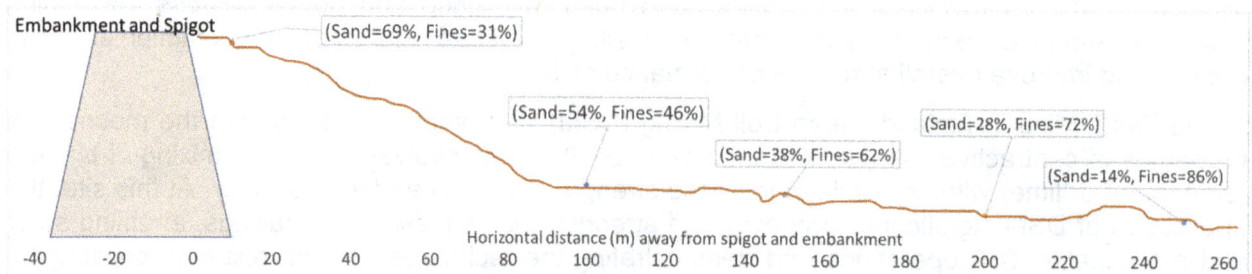

FIG 1 – Typical segregation of tailings particles at TSF studied.

This trend is consistent with field observations, where tailings particle size distribution varies based on spigot location and tailings beach slope, deposition methods, mud farming practices; and rainfall or pipe flushing events, which can remobilise fines and transport fines further downslope.

TAILINGS DRY SOIL MIXING (TDSM)

At a TSF site in northern Australia, mud farming equipment routinely became bogged within the mud farming lots—particularly between 100 m and 250 m from the perimeter embankment—during the wet season. The affected area, approximately 200 m by 200 m, is known to contain soft, unconsolidated, and sensitive tailings. TSF operators were tasked with recovering trafficability across as much of the mud farming floor as possible within this zone, despite the high likelihood of equipment bogging.

Figure 2 illustrates typical problematic tailings conditions beyond 150 m from the perimeter embankment. Figure 2a shows a surface crust formed by desiccated tailings, which can act as a capillary barrier and limit moisture movement to the surface, thereby reducing drying-induced consolidation. Beneath this crust, moisture content (MC) testing on sensitive tailings (Figure 2b) revealed MC = 41 per cent, confirming the poor trafficability conditions.

FIG 2 – Desiccated tailings crust overlying high-moisture sensitive tailings.

Following consultation with the Engineer of Record, the TSF operator implemented TDSM to improve trafficability and support future mud farming operations. The adopted TDSM methodology included:

- Blading dry surface tailings closes to embankment using a LGP dozer and pushing dry tailings toward the TDSM treatment area.

- Using a H&W excavator to perform TDSM along the front portion, closest to the perimeter wall.

- Continuing the mixing process with an AMEX excavator toward the rear of the TDSM treatment area.

The TDSM process involved breaking the existing surface crust and excavating to depths between 1.1 m and 1.9 m to retrieve soft, high-moisture tailings. These were then blended *in situ* with drier surface crust material and stockpile dry tailings material until a uniform moisture consistency was achieved. Upon completion, eight PANDA Dynamic Cone Penetrometer (DCP) tests (designated T1 to T8) were performed within the treated area to assess strength improvement. The area and test locations were also surveyed for record-keeping purposes.

The PANDA DCP is a lightweight device (20 kg), portable and easily transported by one person. The PANDA DCP can rapidly be deployed for testing, is quick to perform a test, can be undertaken by one person, and access areas inaccessible by low-pressure CPT rigs, making PANDA DCP a favourable tool for testing over soft tailings. The device consists of six main elements: hammer, instrumented anvil, rods, cones, Central Acquisition Unit (UCA), and Dialogue Terminal (TDD). The PANDA DCP measures dynamic cone tip resistance (q_d) and provides a strength profile with depth like a CPT plot. At present, the only recognised standards for PANDA testing are NF P 94–105 (Association Française de Normalisation (AFNOR), 1993) from France and NCh 3261–2012 (Instituto Nacional de Narmalizacion (INN), 2012) from Chile.

Summarised below is a timeline of TDSM works carried out at the back of mud farming lots:

- Month 0 – Ground proofing with dozers at back of mud farming lots to recover mud farming stable floor.

- Month 1 – TDSM with H&W excavator and AMEX long reach excavators at back of mud farming lots.

- Month 1 – Eight (8) PANDA DCP (designated T1-M1 to T8-M1) testing undertaken in the TDSM area following five days of TDSM.

- Month 1 to Month 3 – Cell non-operational and TDSM area allowed to dry and consolidate.

- Month 3 to Month 4 – Recommencement of tailings deposition in TSF cell and Scrolls begun cutting bleed drainage lines through TDSM area to facilitate drainage.

- Month 6 to Month 9 – tailings deposition ceased in cell and tailings allowed to dry and consolidate.

- Month 9 – Re-test the TDSM test locations carried out in Month 1. Eight PANDA DCP (designated T1-M9 to T8-M9) tests were conducted at the previous testing locations.

Local nearby Bureau of Meteorology (BOM) weather station data for the TDSM timeline mentioned above is detailed in Figure 3. The climatic conditions between Month 6 and Month 12 (dry season) appear favourable for drying-induced consolidation, characterised by consistently high evapotranspiration, very low to no rainfall, average maximum daily temperatures above 30°C, minimum relative humidity below 50 per cent, steady wind speeds, and high solar radiation.

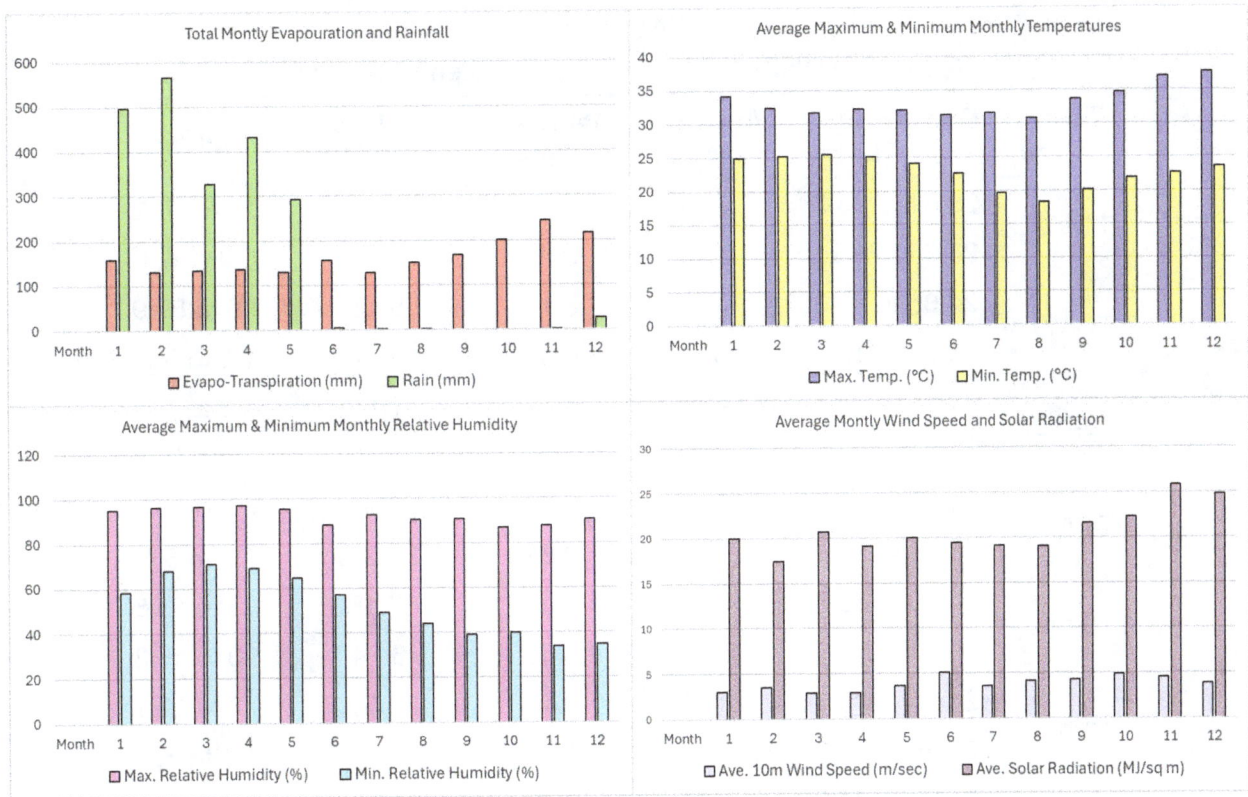

FIG 3 – Climatic conditions between Month 1 and Month 12 at TSF studied site (BOM Weather Station).

Table 1 presents the average PANDA dynamic cone tip resistance (q_d) recorded in Month 1, immediately following the TDSM implementation. Table 2 presents the q_d values measured in Month 9, after drying and consolidation.

TABLE 1

Results of PANDA DCP testing undertaken at end of TDSM Month 1.

Test location	Depth range (mRL)	Average q_d (MPa)	Standard deviation	Coefficient of variation	Range q_d (MPa)
T1-M1	24.56 to 21.85	0.62	0.26	41.4%	0.3 to 1.7
T2-M1	24.13 to 21.47	0.63	0.32	49.7%	0.16 to 1.8
T3-M1	24.08 to 21.37	0.64	0.16	25.6	0.2 to 1.2
T4-M1	23.82 to 21.11	0.40	0.18	44.2%	0.1 to 1.1
T5-M1	23.79 to 21.62	0.47	0.19	39.9%	0.2 to 1.2
T6-M1	23.72 to 20.68	0.49	0.19	39.5%	0.1 to 1.2
T7-M1	23.88 to 21.25	0.99	1.24	125.5%	0.1 to 5.24
T8-M1	23.87 to 21.66	0.66	1.03	156.1%	0.1 to 6.0

TABLE 2

Results of PANDA DCP testing undertaken in Month 9.

Test location	Depth range (mRL)	Average q_d (MPa)	Standard deviation	Coefficient of variation	Range
T1-M9	25.01 to 23.3	5.45	2.67	48.9%	0.9 to 18.2
	23.3 to 21.21	2.38	0.85	35.8%	0.2 to 6.1
T2-M9	24.85 to 23.3	14.14	8.14	57.6	0.6 to 59.9
	23.3 to 21.14	2.54	0.86	33.7	1.1 to 6.7
T3-M9	24.81 to 23.3	3.84	2.02	52.6%	0.9 to 7.9
	23.3 to 21.94	2.98	2.03	68.1%	1.0 to 15.8
T4-M9	24.77 to 23.3	24.77	3.25	92.9%	0.6 to 12.5
	23.3 to 21.21	2.42	0.83	34.3%	1.2 to 5.0
T5-M9	24.55 to 23.3	4.52	3.03	67.1%	1.3 to 13.6
	23.3 to 20.96	3.05	2.58	84.4	1.0 to 15.3
T6-M9	24.45 to 23.3	2.77	2.40	86.6%	0.3 to 12.1
	23.3 to 20.99	1.89	0.62	32.6%	1.1 to 3.7
T7-M9	24.32 to 23.3	1.86	0.53	28.3%	1.0 to 3.0
	23.3 to 20.61	2.05	0.84	41.1%	1.1 to 5.7
T8-M9	24.35 to 23.3	19.32	14.68	75.9%	1.4 to 98.7
	23.3 to 20.64	2.39	0.84	35.1%	0.6 to 4.7

Figure 4 illustrates the PANDA DCP test profiles of testing detailed in Table 1 (Month 1) and Table 2 (Month 9). The left graph shows results Month 1 after TDSM, while the right graph presents values Month 9 after TDSM. Dashed lines represent Month 1 data; solid lines correspond to Month 9. A clear improvement in strength is observed, with q_d increasing from <1 MPa to values exceeding 10 MPa at the surface and greater than 2 MPa average at depth in several locations. A CPT test (CPT-56), undertaken 20 m from T1 outside the TDSM zone prior to TDSM, was added to left hand graph to validate PANDA DCP results in TDSM zone. These results confirm the effectiveness of TDSM in enhancing tailings strength under partial drying conditions.

Based on the site-specific correlation developed for this location (Dudley and Llano-Serna, 2023), the estimated average *in situ* bulk density and undrained shear strength increased from γ = 16.4 kN/m^3 and Su = 7.4 kPa in Month 1 to γ = 19.1 kN/m^3 and Su = 57.0 kPa in Month 9, indicating substantial consolidation and strength gain over the monitoring period. The increase in cone tip resistance demonstrates measurable strength improvement due to the TDSM process and subsequent drying period. However, partial drainage and moisture variability at the time of testing may influence q_d values. A review of machine performance three months after the start-up of tailings deposition and mud farming operations, from Month 13 to Month 17, showed that the mud farming dozer is effectively traversing the entire 250 m extent (almost 100 per cent coverage) of the mud farming lots without bogging incidents and downtime. This is an improvement from 40 per cent to almost 100 per cent mud farming dozer coverage.

Future planned geotechnical investigations in the TDSM-treated area will aim to verify the performance of improved material and support its integration into the TSF ground model. *In situ* tests, such as cone penetration testing (CPT), vane shear tests, pore pressure dissipation tests, and mini-block sampling for laboratory characterisation, will be conducted to assess whether the treated tailings exhibit dilative or contractive behaviour under expected loading conditions. These results will inform the calibration of strength parameters and volumetric response in the numerical TSF models, and will be critical inputs for future stability analyses and raise design.

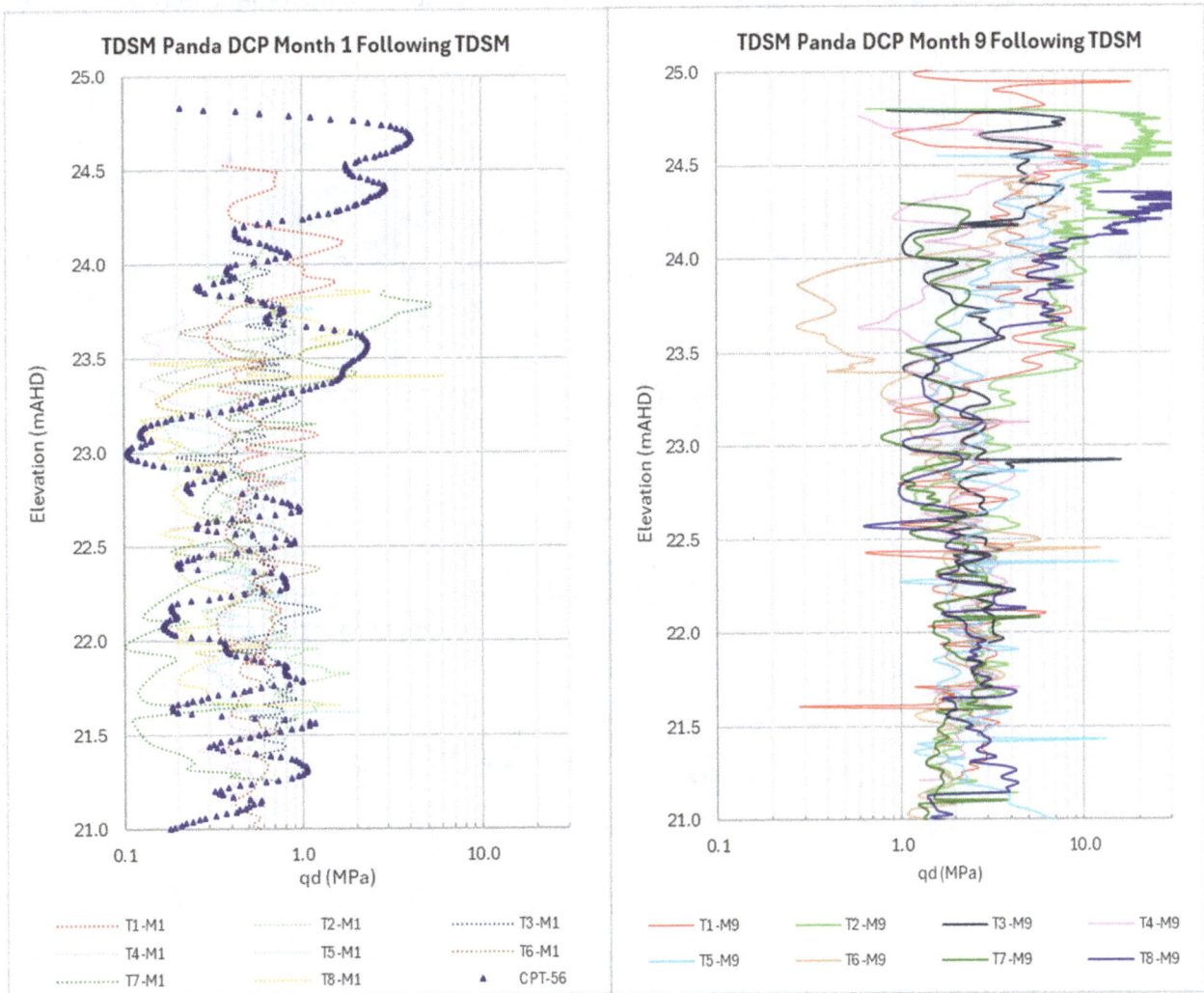

FIG 4 – Results of the PANDA DCP Undertaken in Month 1 and Month 9 following TDSM.

BENCHMARKING AGAINST UNTREATED TAILINGS

To benchmark the effectiveness of TDSM, a control area within the TSF—untreated by either mud farming or TDSM—was selected for comparison. This section of the facility is persistently saturated, with the tailings surface located near the phreatic level, limiting natural consolidation.

This control area was located outside the operational structural mud farming lots and therefore material will not affect the long-term stability of the embankment or TSF raises. As such, it had not undergone any ground improvement and provided a consistent baseline for evaluating strength changes attributable solely to TDSM.

In Month 2, three PANDA DCP tests (designated T10-M2 to T12-M2) were conducted in the control area, which had no history of ground improvement. The timeline of activities:

- Month 0: No previous mud farming beyond the extent of mud farming lots.

- Month 2: PANDA DCP testing T10-M2 to T12-M2 conducted at baseline; test locations surveyed for reference.

- Month 4 to Month 7: Scrolls commenced cutting bleed drainage lines.

- Month 12: PANDA DCP retesting at previous test location, T10-M12 to T12-M12.

Figure 5 illustrates the PANDA DCP test profiles of testing undertaken at the control test site. The data indicate negligible improvement in cone tip resistance between Month 2 and Month 12, confirming the lack of natural strength gain in the absence of ground improvement techniques.

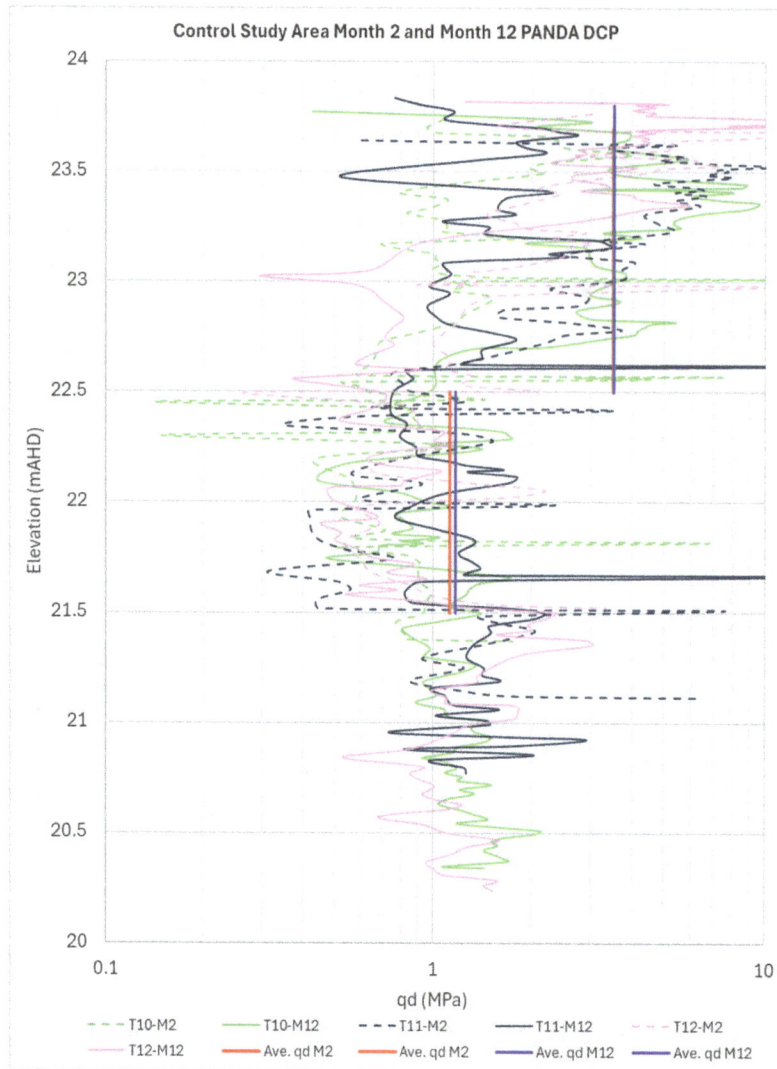

FIG 5 – PANDA DCP results in control test site.

CONCLUSION OF TAILINGS DRY SOIL MIX (TDSM) METHODOLOGY

The application of TDSM in a tropical TSF site, has demonstrated its potential as an effective ground improvement technique to recover soft tailings floor, improve mud farming machine trafficability and increase *in situ* density. The TDSM method enabled trafficability recovery in previously inaccessible areas and contributed to significant strength gains, as evidenced by PANDA DCP test results conducted over a 9-month period and the increase from 40 per cent dozer coverage to almost 100 per cent mud farming machine coverage of mud farming lots during start-up and cell operation. In contrast, the untreated control area showed negligible change in strength, supporting the conclusion that improvement was directly associated with TDSM.

TDSM enhances moisture redistribution and accelerates drying-induced consolidation. Beyond operational benefits, the resulting increase in density and shear strength reduces the likelihood of large-strain consolidation, surface cracking, and the need for downstream buttressing, all of which are critical to long-term TSF stability.

The following points are a proposed methodology for effective TDSM where the local climate conditions favour drying-induced consolidation:

- Tailings Dry Soil Mixing (TDSM) involves blending dry tailings and/or desiccated surface crust with soft, high-moisture, sensitive tailings using equipment such as H&W and AMEX excavators to the target depth. The mixed material should be allowed to dry over time, with repeated mixing as necessary to improve moisture uniformity, limit surface crusting, and enhance drying-induced consolidation and evaporation. Given the high initial moisture content of soft tailings, multiple mixing cycles may be required until a consistent and workable moisture condition is achieved.

- Record equipment used, mixing depth, and timing for each treatment zone.

- Conduct PANDA DCP or other suitable *in situ* strength testing (CPT, vane shear) upon completion of mixing. Record test coordinates and RL using differential GPS.

- Survey the full extent of the treated area for operational tracking.

- Use dozers to confirm stability of the treated surface prior to resuming mud farming operations.

Future planned geotechnical investigation of TDSM area will verify the performance of the TDSM material and characterise the material for incorporating into the TSF ground model. Verification testing will comprise *in situ* testing such as Cone Penetration testing, vane shear testing, pore pressure dissipation testing and mini block sampling for subsequent laboratory testing. These tests will be used to assess whether the treated tailings exhibit contractive or dilative behaviour under operational loads. This information will inform the calibration of material strength and stiffness parameters and will be critical for future stability assessments and TSF raise design.

Based on the results in this paper and experience with site and process of TDSM, the following conditions would ensure the effective TDSM process:

- **Non-operational cell:** TDSM to be undertaken in non-operational lots or TSF cell. This allows time for drying-induced consolidation of tailings.

- **Dry, Hot and Windy Environmental Conditions:** This is usually the dry season months in arid and tropical environments. These conditions ensure the surface of the dry soil mix dries and moisture evaporates which improves the dry density and *in situ* state of the material ie drying-induced consolidation.

- **Low Groundwater Conditions:** Tailings within saturated zones will not improve the drying of the tailings. Ensure that TDSM is conducted in areas with low groundwater levels within the non-operational cell. This can be verified by monitoring decant levels and reviewing RL readings from piezometers to confirm a low groundwater level (GWL) during the mixing operations. Test pits can be excavated within proposed TDSM area to confirm groundwater condition.

- **Adhere to or Enhance the Current Mixing Methodology:** Follow the established mixing methodology process detailed above. If possible, consider improvements to the methodology that could further enhance soil stability and ensure consistency across the treated area.

- **Implement Post-Mixing Quality Control (QC) Testing and Drying Periods:** Perform PANDA DCP testing after the mixing process and during mud farming operations to verify the effectiveness of the soil improvement.

- **Plan Future Geotechnical Investigation to target TDSM Area:** *In situ* and laboratory testing of TDSM material to verify the effectiveness of TDSM and characterise the strength to incorporate into TSF ground models.

REFERENCES

Adams, M and Friedman, D, 2017. Tailings impoundment stabilization to mitigate mudrush risk, in Proceedings of the 85th Annual Meeting of the International Commission on Large Dams (Knight Piésold Ltd).

Association Française de Normalisation (AFNOR), 1993. *NF P 94-105: Soil – Investigation and testing – Dynamic penetration test with variable energy – Determination of dynamic tip resistance qd*, Association Française de Normalisation, Paris, France.

Assouline, S, Narkis, K, Gherabli, R, Lefort, P and Silan, P, 2013. Soil evaporation from a drying soil surface: Diurnal dynamics and self-mulching effect, *Water Resources Research*, 49(7):3997–4008.

Dudley, O and Llano-Serna, M, 2023. Estimating in situ state of Tailings using panda dynamic cone penetrometer, in *Proceedings of the Mine Waste and Tailings Conference 2023*, pp 48–59 (The Australasian Institute of Mining and Metallurgy: Melbourne).

Fredlund, D G and Xing, A, 1994. Equations for the Soil-Water Characteristic Curve, *Canadian Geotechnical Journal*, 31(4):521–532.

Fujiyasu, Y, 1997. Evaporation behaviour of tailings, Doctoral dissertation, University of Western Australia, Department of Civil Engineering.

Global Tailings Review (GTR), 2020. Global Industry Standard on Tailings Management (GISTM) [online], Global Tailings Review. Available from: <https://globaltailingsreview.org/global-industry-standard/>

Iden, S C and Durner, W, 2008. Free-form estimation of the unsaturated hydraulic conductivity function using dual-step inverse modeling, *Vadose Zone Journal*, 7(4):1210–1220.

Independent Expert Engineering Investigation and Review Panel (IEEIRP), 2015. Report on Mount Polley Tailings Storage Facility Breach, Independent Expert Engineering Investigation and Review Panel, Province of British Columbia, January 30, 2015.

Instituto Nacional de Narmalizacion (INN), 2012. Tailings Deposits-Control of compaction with light dynamic penetrometer, Nch 3261–12, Santigo, Chile.

Keller Australia, nd. QAL Tailings Dam [online]. Available from: <https://www.keller.com.au/projects/qal-tailings-dam> [Accessed: 2 March 2025].

Khalili, N and Khabbaz, M H, 1998. A unique relationship for χ for the determination of the shear strength of unsaturated soils, *Géotechnique*, 48(5):681–687.

Li, A L, Been, K, Wislesky, I, Eldridge, T and Williams, D, 2012. Tailings initial consolidation and evaporative drying after deposition, in *Paste 2012: Proceedings of the 15th International Seminar on Paste and Thickened Tailings* (eds: R Jewell, A B Fourie and A Paterson), pp 25–42 (Australian Centre for Geomechanics: Perth).

Lu, N and Likos, W J, 2004. *Unsaturated Soil Mechanics* (Wiley).

Masengo, E, Julien, M R, Lavoie, P, Lépine, T, Nousiainen, J, Saukkoriipi, J, Piekkari, M and Karvo, J, 2019. Enhancement of contractive tailings using deep soil mixing techniques at Kittilä mine, in *Sustainable and Safe Dams Around the World* (eds: J Tournier, P Bennett and E Bibeau), pp 334–342 (Canadian Dam Association).

Scanlon, B R and Milly, P C D, 1994. Water and heat fluxes in desert soils, Field studies, *Water Resources Research*, 30(3):709–719.

Smits, K M, Sakaki, T, Limsuwat, A and Illangasekare, T H, 2012. Thermal conductivity of sands under varying moisture and porosity in drainage–wetting cycles, *Vadose Zone Journal*, 11(2).

High density 3D characterisation of tailings dams for engineering and safety performance

R Eddies[1], R Wood[2], S Sol[3] and D Valintine[4]

1. Solution Director, Fugro, Leidschendam, 2264 SG, Netherlands. Email: r.eddies@fugro.com
2. Consultant, Fugro, Houston, TX 77079, USA. Email: wrwood@fugro.com
3. Product Owner, Fugro, Leidschendam, 2264 SG, Netherlands. Email: s.sol@fugro.com
4. Manager Land Geophysics, Fugro, Houston, TX 77079, USA. Email: d.valintine@fugro.com

ABSTRACT

The historical impacts of tailings dam failures highlight the necessity for identification and evaluation of potential hazards throughout the entire life cycle. Linear structures such as tailings dams can experience localised failure. Subsurface models derived from intrusive measurements can leave significant uncertainty between 1D explorations. To explore the possibility of a paradigm shift in understanding internal dam characteristics, Codelco initiated a study at the El Teniente copper mine, Chile in partnership with Fugro to evaluate the effectiveness of 3D geophysical screening and dam model building for tailings dams. Seismic data were acquired at Dam A within Colihues and Cauquenes tailings storage facility and processed using adapted ambient noise tomography (ANT) techniques to build a dense 3D shear wave velocity model as a close proxy for a 3D stiffness model to a depth exceeding 100 m. The derived model was consistent with information from existing borehole logs and with 2D seismic data at shallow depths, demonstrating the significant depth coverage advantage of 3D ANT over conventional screening techniques. The velocity model showed that the internal dam structure was characterised by dense to very dense soils and soft rock. No adverse geological/geotechnical features were identified in the 3D velocity model. ANT provides a means for owner-operators to screen tailings dams and potential sites for new dams/dam extensions to build an early 3D geotechnical model that can inform timely risk management decisions. An ability to image the internal characteristics of a tailing dam and its foundation soils provides the means to better target localised zones of significance with subsequent intrusive geotechnical investigation. Integrating initial ANT screening and intrusive investigation assures owner-operators that the site has been effectively and efficiently characterised leading to the highest levels of confidence in the engineering performance and safety of the tailings facility. This approach founded on building early, dense ground models can also be readily extended to timely risk assessment of potentially onerous geotechnical conditions ahead of, for example, open pit operations.

INTRODUCTION

Critical risks associated with tailings dams and historical consequences of their failure, emphasise the need for targeted identification and assessment of potential hazards throughout a dam's entire life cycle. Linear structures such as tailings dams tend to suffer distress and/or fail locally and the traditional approach to assess such structures has tended to rely on local 1D measurements made in boreholes or with probes, supplemented by 2D MASW (multichannel analysis of surface waves) geophysical profiles that have limited depth penetration capabilities. Subsurface models resulting from these local measurements can contain significant uncertainty in the largely empty model space between 1D explorations and below the effective depth of characterisation of convention geophysical methods such as MASW. One of the most basic requirements in dam safety management is to understand the site-specific geomechanical and hydrogeological aspects of a dam – assumptions of unknown reliability are made today by geotechnical engineers in the absence of reliable, dense data characterising the engineered subsurface. To explore the possibility of a paradigm shift in understanding the internal characteristics of tailings dams, Codelco initiated a study in partnership with Fugro to evaluate the effectiveness of deep 3D geophysical screening for tailings dams.

STUDY SITE

Codelco's Colihues and Cauquenes tailings storage facility (TSF), is located at the El Teniente Mine located in the centre of Chile, approximately 90 km south of Santiago near the town of Rancagua. A number of structures make up the TSF and Dam A (Figure 1) (downstream construction) was selected for the screening study. Mining operations were ongoing during the data acquisition which were expected to provide sufficient ambient signal for the passive seismic investigation.

FIG 1 – Dam A (Colihues and Cauquenes tailings storage facility, El Teniente Mine).

From information provided by Codelco, the *in situ* bedrock beneath Dam A forms part of a volcanic-sedimentary sequence of Oligocene-Miocene age comprising basaltic to dacitic lavas, epiclastic and pyroclastic rocks. This is overlain by a morainic material upon which Dam A was constructed (Figure 2). Dam A is made up of various materials with fluvio-glacial material forming its upstream face and fluvial materials forming its downstream face. Dam A is approximately 1200 m wide and 80 m high. Note that the accuracy of the (as built) cross-section cannot be confirmed.

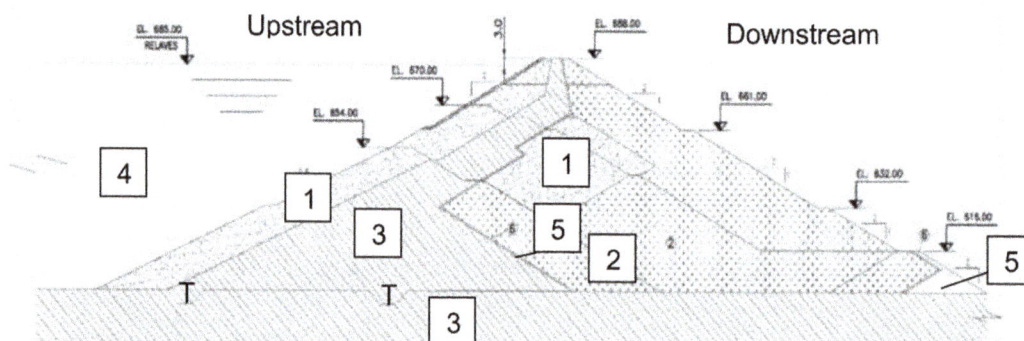

FIG 2 – Schematic as built cross-section, Dam A Notes: (1) Fluvio-glacial materials; (2) fluvial materials; (3) moraine; (4) tailings; (5) clay; (T) leak trench.

GEOPHYSICAL SCREENING

Adapted ambient noise tomography

Adapted ambient noise tomography (ANT) forms part of Fugro's solution to make the ground transparent for better data-driven decision-making. ANT (sometimes referred to as passive seismic interferometry) is a geophysical screening technology that from the capture and analysis of ambient seismic noise delivers a 3D volume of shear wave velocity that is directly related to small-strain stiffness, a key geotechnical parameter. Fugro have adapted the original methodology of Shapiro *et al* (2005) for engineering scale applications (as highlighted in Eddies, Wood and Staring, 2024) including the incorporation of active-source seismic data. Ambient seismic noise originates from both natural (ocean waves, wind) and anthropogenic (traffic, industrial processes) mechanisms. These mechanisms are frequency-dependent with a spectral boundary between natural and cultural noise around 1 Hz (Bonnefoy-Claudet, Cotton and Bard, 2006). With a focus on depths of 0–100 m for engineering purposes, the cultural noise range (1 Hz to 100 Hz) is most suitable. Cultural or anthropogenic seismic noise primarily originates from activities that occur at or near the surface, for example, traffic or industrial processes. Therefore, it is reasonable to assume that the noise wavefield mainly consists of surface waves. There are two types of surface waves: Rayleigh waves and Love waves. According to Yamanaka *et al* (1994), noise at frequencies above 1 Hz primarily consist of Rayleigh waves. These waves consist of different propagating modes in layered media (a typical subsurface), where the fundamental mode is typically the strongest. Fundamental mode surface waves are an approximate solution to the 2D wave equation with a frequency-dependent propagation velocity (Wapenaar *et al*, 2010). Therefore, reconstructing these Rayleigh waves from the recorded cultural noise is expected to be adequate to obtain a high-resolution shear wave velocity model of the subsurface (Picozzi *et al*, 2009).

The shear wave velocity (Vs) measured by shallow seismic techniques is directly related to the maximum (or initial) shear modulus (Gmax or G0) of soil or rock in geotechnical engineering. The relationship between shear wave velocity and the shear modulus is fundamental in the field of geotechnics, as it provides a non-destructive means to evaluate the stiffness of the ground materials, which is crucial for understanding soil behaviour under loading conditions. The shear modulus is a measure of the material's ability to resist shear deformation – it is especially important in the analysis and design of foundations, retaining structures, and slopes, as well as in the assessment of seismic site response. Shear wave velocity in soil is a useful indicator of soil stiffness and dynamic properties, and it varies depending on the soil type. Generally, denser and stiffer soils and rock have higher Vs values, while softer, looser soils have lower values (Table 1). Factors like confining pressure, void ratio, stress history, and aging also influence Vs.

TABLE 1

The NEHRP building code assigns one of six soil-profile types to a site, from hard rock (type A) to soft soils (types E or F), based on the mean shear wave velocity of top 30 m (FEMA, 2000).

Soil types	Rock/soil description	Average shear wave velocity (Vs) in top 30 m
A	Hard rock	>1500
B	Rock	760–1500
C	Dense soil/soft rock	360–760
D	Stiff soil	180–360
E	Soft soil	<180
F	Special soils requiring evaluation	

Typically, ambient noise sources provide low frequency content enabling the retrieval of lower frequency seismic signal to explore the subsurface beyond 100 m or more in 3D, significantly exceeding the limit of, for example, 2D MASW profiling techniques. Adapted ANT data acquisition involves the setting out of seismic sensors over a 2D surface grid that are sensitive to and can record low frequency ambient noise down to a frequency of 1 Hz over a period from days to weeks. The number of sensors can vary between a few hundred to several thousand as a function of the investigation objective. The ambient field is augmented using an active seismic source (such as a sledgehammer, small vibrating source or weight drop) that can sometimes provide higher frequency signals to improve near-surface information.

Data acquisition

Passive seismic data (from ambient noise sources) were acquired continuously on Dam A between April 29 and May 8, 2024, covering an area of ~145 m × 370 m. In addition, active seismic data (from sledgehammer blows) were also acquired within the period of passive seismic recording. Data acquisition involved 962 autonomous seismic sensors (nodes) set-up in a grid along the downstream face of the dam with a nominal spacing of about 7 m (Figure 3).

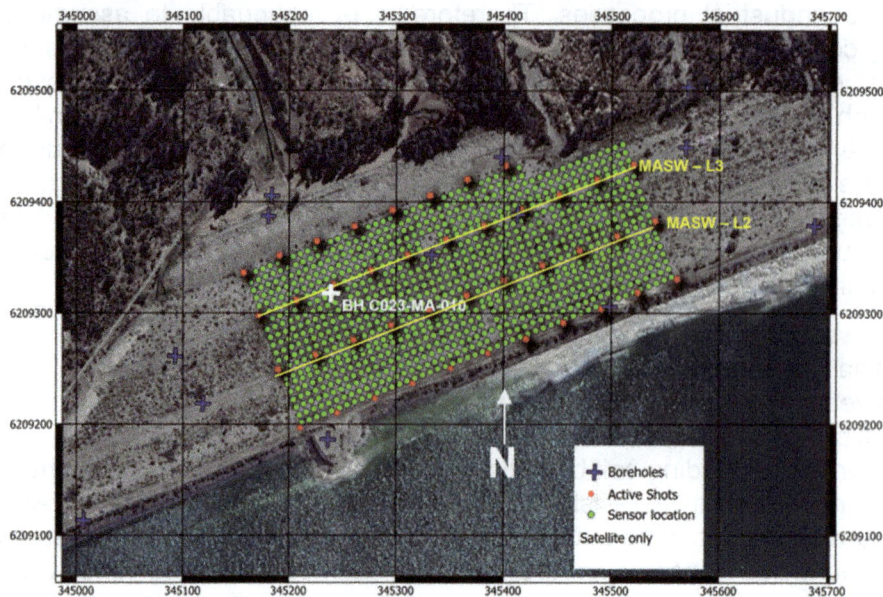

FIG 3 – Seismic sensor grid (green), active shot locations (red), boreholes and MASW lines, Dam A.

MASW profiles were acquired that derived shear wave velocities as a 2D profile (Figure 4). A reference borehole (BH CO23-MA-010) was located within the survey footprint to allow comparison between seismic velocities derived from 1D BH measurements and those from 2D MASW and 3D ANT. Due to the nature of the terrain, qualified roped-access teams deployed and topographically positioned the node array on the front face of the dam which comprised alluvial material including cobbles.

Data processing

The ambient noise passive seismic data were initially processed using a cross-correlation approach similar to the interferometric Multichannel Analysis of Surface Waves approach developed by Fugro (O'Connell and Turner, 2011) to estimate the Correlation Green's Functions (CGF) organised into virtual source gathers (so similar to a traditional active shot seismic record). The Green's function incorporates the properties of both the seismic source and the medium through which the seismic waves travel. It accounts for the geometry, material properties (such as elasticity) and any heterogeneities within the Earth's interior. When stacked over a relatively long amount of time surface waves that have been traversing the site emerge. An initial velocity model was developed from CGFs sampled over a coarse grid. The passive data contained seismic frequencies up to 16 Hz; the addition of active data increased the usable high frequency content to 40 Hz to ensure the initial model adequately represented the near surface (top few metres). Seismic interferometry was then applied to both phase velocity and time-domain (wavelet group velocity) dispersion analysis from the CGFs. The resulting travel time picks were projected onto a defined grid using travel time tomography. The result was a 3D (x, y, frequency) volume of group velocities. Using the initial velocity model, inter-receiver travel times were jointly inverted using a least-squares algorithm and a 3D (x, y, z) discretisation of the Vs model space.

RESULTS

The 3D shear wave velocity (Vs) model derived from adapted ANT comprising about 1 million data points is shown in Figure 4. Velocity values varied from about 300 m/s to 1300 m/s and represent an interval from surface to about 120 m depth. The 3D Vs model shows an increase of velocity with depth and the presence of lateral velocity/stiffness variations within the dam. The range of Vs was consistent with the presence of stiff/dense soils at low velocity through very dense soil/soft rock to rock at high velocity. Bedrock correlated well with the 900 m/s velocity contour (Figure 5). Localised velocity inversions were observed where velocity and therefore stiffness decreased with depth. However, the lowest velocities observed at Dam A below about 10 m depth were greater than

450 m/s equivalent to very dense/stiff soils and soft rock. Soft soils are characterised in commonly used classification schemes (Table 1) by shear wave velocities <180 m/s; such low velocities were not observed at the Dam A site. No adverse geological/geotechnical features were therefore identified in the 3D velocity model requiring further investigation.

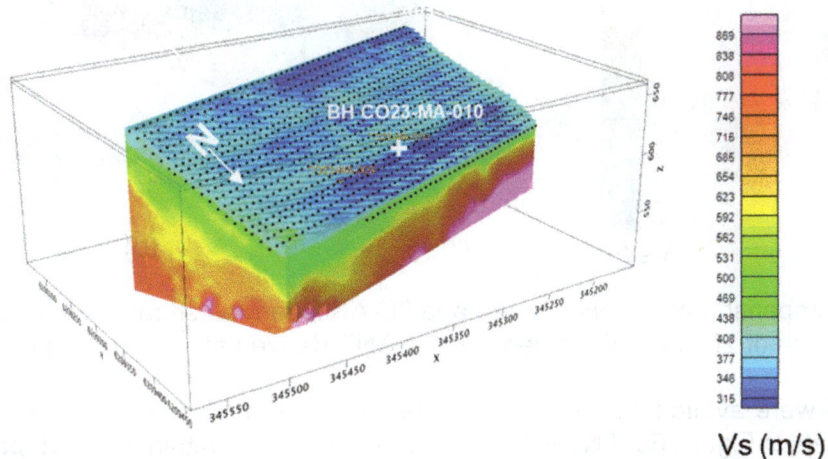

FIG 4 – 3D model of shear wave velocity derived from 3D ambient noise tomography (ANT), Dam A.

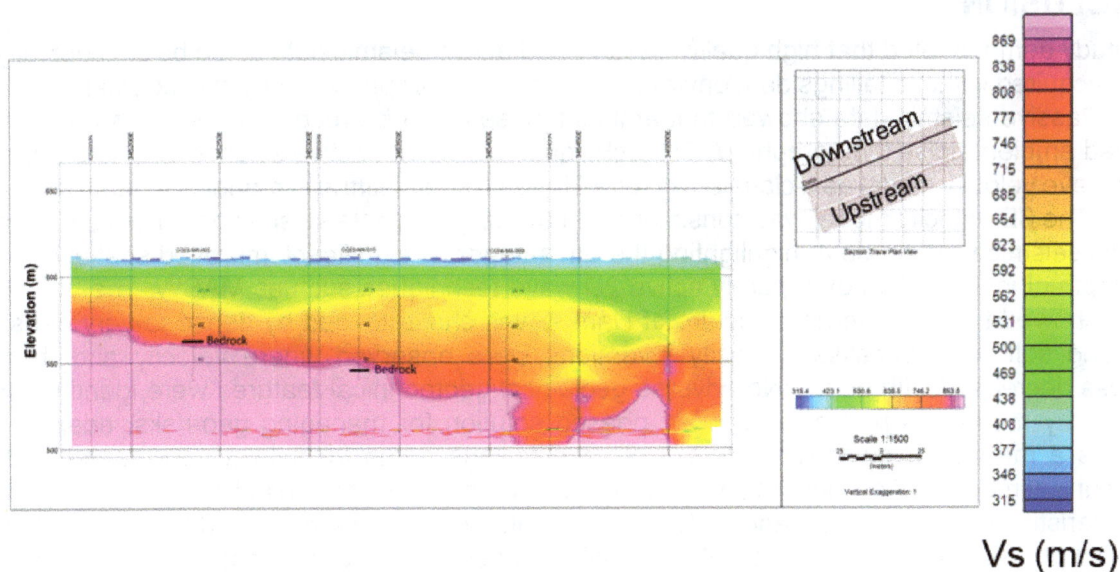

FIG 5 – Slice through 3D model of shear wave velocity derived from 3D ANT, Dam A showing correlation between 900 m/s velocity contour and bedrock encountered in boreholes.

Comparison 3D ANT Vs, 2D MASW Vs and 1D Borehole Vs

Borehole seismic (downhole Vs) data were available in BH CO23-MA-010 to a depth of about 35 m and are shown compared to nearby 3D ANT data in Figure 6a. The BH data show more rapid vertical velocity variations, but ANT velocities were found to be within 20 per cent of BH velocities. Note there is no implication that BH velocities were more accurate – downhole seismic in very coarse materials can be very heavily influenced by local conditions including the presence of high velocity boulders and also high seismic signal attenuation in such coarse materials giving rise to low frequency data and potential uncertainty in travel time picking and velocity estimation.

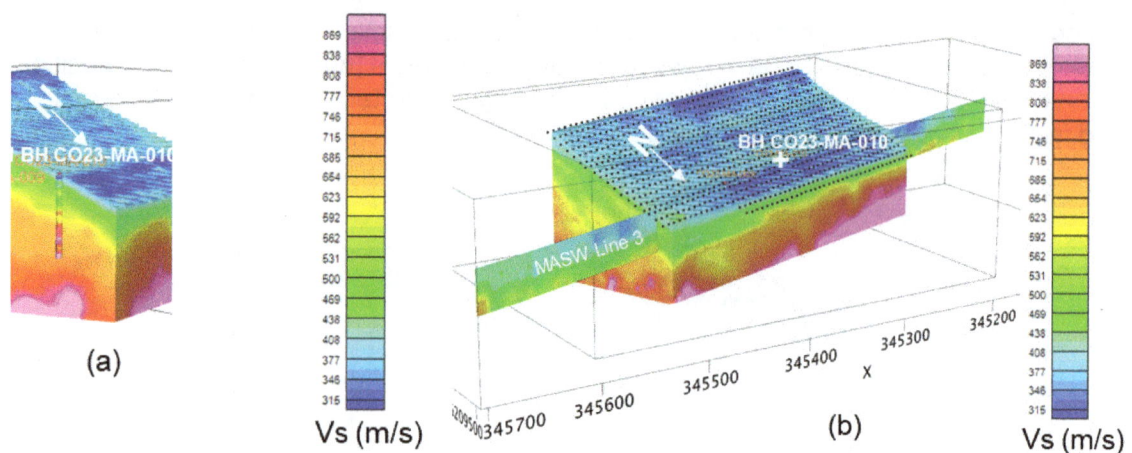

FIG 6 – (a) Comparison of 1D BH-derived and 3D ANT-derived shear wave velocity, Dam A;
(b) Comparison of 2D MASW-derived and 3D ANT-derived shear wave velocity, Dam A.

MASW data (Vs) were available as Line 3 to a depth of about 30 m and are shown compared to nearby 3D ANT data in Figure 6b. ANT velocities were found to be within about 15 per cent of MASW velocities but with significantly deeper investigation (factor of 3) and spatial coverage largely filling the 3D model space.

CONCLUSION

The study demonstrated that high quality passive and active seismic data could be successfully and safely acquired within a tailings dam environment with an appropriate safety management system in place. Passive seismic data showed that ambient noise could be recorded and transformed using adapted ambient noise tomography (ANT) techniques along with active seismic data to a dense 3D shear wave velocity model as a close proxy for a 3D small strain stiffness model to a depth exceeding 100 m. The 3D velocity model was consistent with existing borehole seismic data from the dam and with 2D seismic profiling data, highlighting the advantages of significant improved spatial coverage and depth of investigation over purely active 2D wave techniques such as MASW. The 3D velocity model showed that the internal structure of Dam A was characterised by dense to very dense/stiff soils and soft rock. Localised velocity inversions were observed where velocity and therefore stiffness decreased with depth. No adverse geological/geotechnical features were identified in the 3D velocity model. As part of a broader, phased solution for managing geo-risks, adapted ANT provides a means for owner-operators to screen tailings dams to build a very dense early 3D geotechnical model that can inform timely risk management decisions. An ability to image the internal characteristics of a tailing dam and its foundation soils provides the means to better visualise and target all geotechnical profiles of significance with subsequent intrusive geotechnical investigation. The fidelity and data point concentration of the subsurface geotechnical model permits much more reliable zonation of the site than can be achieved from interpolation between comparatively widely spaced information profiles (1D and 2D) obtained from conventional investigations. The shear wave velocity/shear stiffness derived from the ANT data can be imported directly into numerical analyses for performance-based design assessment. Integrating initial ANT screening and intrusive investigation significantly increases confidence for owner-operators that the site has been effectively and efficiently characterised leading to the highest levels of confidence in the engineering performance and safety of the tailings facility. This approach founded on building early, dense ground models can also be readily extended to timely risk assessment of potentially onerous geotechnical conditions ahead of, for example, open pit operations.

ACKNOWLEDGEMENTS

The authors express their sincere thanks to Codelco management for their practical support to undertake the investigation at the El Teniente mine and for the permission to publish the preliminary results of the study.

REFERENCES

Bonnefoy-Claudet, S, Cotton, F and Bard, P Y, 2006. The nature of noise wavefield and its applications for site effects studies: A literature review, *Earth-Science Reviews*, 79(3–4):205–227.

Eddies, R, Wood, R and Staring, M, 2024. Early Screening for Improved Management of Geo-risks, Proceedings of the 7th International Conference on Geotechnical and Geophysical Site Characterization, Barcelona, 18–21 June 2024.

Federal Emergency Management Agency (FEMA), 2000. FEMA 368:2000 – NEHRP Recommended Provisions for Seismic Regulations for New Buildings and Other Structures, 2000 edn, developed by the Federal Emergency Management Agency (FEMA).

O'Connell, D R O and Turner, J, 2011. Interferometric Multichannel Analysis of Surface Waves (IMASW), *Bulletin of the Seismological Society of America*, 101(5):2122–2141.

Picozzi, M, Parolai, S, Bindi, D and Strollo, A, 2009. Characterization of shallow geology by high-frequency seismic noise tomography, *Geophysical Journal International*, 176(1):164–174.

Shapiro, N M, Campillo, M, Stehly, L and Ritzwoller, M H, 2005. High resolution surface wave tomography from ambient seismic noise, *Science*, 307:1615–1618.

Wapenaar, K, Draganov, D, Snieder, R, Campman, X and Verdel, A, 2010. Tutorial on seismic interferometry: Part 1— Basic principles and applications, *Geophysics*, 75(5):75A195–75A209.

Yamanaka, H, Takemura, M, Ishida, H and Niwa, M, 1994. Characteristics of long period microtremors and their applicability in exploration of deep sedimentary layers, *Bulletin of the Seismological Society of America*, 84(6):1831–1841.

Considerations for stone column quality control and design intent verification for dam stabilisation

J Eldridge[1], R Dellamea[2] and J Rola[3]

1. Consultant, SRK Consulting (Australasia) Pty Ltd, West Perth WA 6005.
   Email: jeldridge@srk.com.au
2. Senior Consultant, SRK Consulting (Australasia) Pty Ltd, West Perth WA 6005.
   Email: rdellamea@srk.com.au
3. Principal Consultant, SRK Consulting (Australasia) Pty Ltd, West Perth WA 6005.
   Email: jrola@srk.com.au

ABSTRACT

Recent catastrophic tailings dam failures have driven the identification and need for dam stabilisation solutions worldwide in an effort to improve factors of safety under a post-seismic loading condition, where contractive materials are present within the structural foundation zone. Although not common place, specific circumstances (ie limited footprint, lack of buttress material etc) may drive the use of ground improvement (GI) as part of a wider dam stabilisation solution.

GI techniques for dam stabilisation typically aim to create a structural block to cut-off problematic foundation layers (ie soft/liquefiable materials). This, in combination with a berm/buttress can provide a satisfactory stabilisation solution. Stone columns (SCs) are a GI technique that has been used successfully in a number of dam stabilisation projects.

To confirm that SC construction achieves the design intent, robust construction quality control (QC) is necessary. Some common considerations for SC QC include assessment of stone column construction depth, termination amperage, stone column diameter, pre-drill depth (if required) and the actual installed soil replacement ratio. As well, it is common to conduct cone penetration tests (CPTs) following SC construction to confirm the achieved undrained shear strength of the densified soil between the columns.

This paper discusses the key considerations for SC QC for dam stabilisation and presents how these QC considerations can be implemented using data from a recent tailings dam stabilisation project, including identification of some challenges encountered during construction. The implementation of a detailed SC QC program during construction allowed early identification of remediation works/areas. The paper discusses how potential non-conformances were assessed and addressed without causing significant project delays and/or potentially unrealistic remediation works with a design intent verification example.

INTRODUCTION

Recent catastrophic tailings dam failures have driven the identification and need for dam stabilisation solutions worldwide in an effort to improve factors of safety under a post-seismic loading condition, where contractive materials are present within the structural foundation zone. A common and universal dam stabilisation approach, that has proven effective, is the placement of buttress fill material to increase the resisting forces. Although not common place, specific circumstances (ie limited footprint, lack of buttress material etc) may drive the use of GI as part of a wider dam stabilisation solution.

GI techniques for dam stabilisation typically aim to create a structural block to cut-off problematic foundation layers (ie soft/liquefiable materials). This, in combination with a berm/buttress can often provide a satisfactory stabilisation solution. Some common GI techniques used for dam stabilisation include deep soil mixing, cement bentonite slurry walls, dynamic compaction, jet-grouting and SCs. This paper will focus on SCs.

SCs are a GI technique that has been used successfully in a variety of dam stabilisation projects. SCs are an *in situ* strengthening technique which, for dam stabilisation, aims to create a composite material with increased shear capacity providing an overall improvement in the resisting forces across a problematic unit.

Unlike the thin lift placement of buttress fill, where each layer can be visually assessed and tested to full depth, SCs are constructed below the ground surface where visual confirmation is not possible and testing less direct. Therefore, a robust construction QC program is necessary to confirm that SC construction achieves the design requirements. This paper discusses some key considerations for SC QC and presents a method (with a worked example using real data) that can be used to verify if the design intent is achieved when QC data indicate a key design requirement has not been met.

BACKGROUND – STONE COLUMNS

SCs are an *in situ* strengthening technique, which for dam stabilisation, aims to create a composite material with increased shear capacity providing an overall improvement in the resisting forces across a problematic unit. Like many GI techniques, SCs are typically used to reduce settlement and increase load-bearing capacity, and are particularly effective at improving slope stability and reducing liquefaction risk by densification and increasing shear strength within a soil.

SCs are vertical inclusions that may be constructed using stone or sand. They are typically installed in a grid pattern through loose and soft soils/problematic units. The vertical inclusions have higher stiffness, shear strength and generally improved drainage characteristics when compared to the existing *in situ* soil. Therefore, the area improved by the SCs may see improved bearing capacity, reduced total and differential settlements, and accelerated consolidation (due to the additional drainage capacity of the granular material).

Once constructed, the SCs create a strong 'shear key' (also referred to as a GI block) of non-liquefiable material within the existing foundation stratum, and placement of a berm over the GI block acts as a buttress to the slope of the dam while also increasing the strength of the block as of result of added vertical effective stress.

Construction

A typical SC construction sequence is presented in Figure 1. SCs are constructed with a vibrating probe which penetrates the loose/soft soils via cavity expansion to the target improvement depth, typically densifying the surrounding soils. An auger can also be used to create a pilot hole approaching the target improvement depth when hard soils are present above the target unit. For the dry bottom feed SC construction method, upon reaching the target depth, material is fed through the top of the probe to the base of the hole, filling the void left as the probe is raised in short intervals (ie 0.5 m etc). The probe is then lowered and vibrated into the placed stone to densify and build the SC in intervals. Generally, stone is placed into each interval and vibrated until a specified compaction amperage is achieved. Depending on the soil encountered, this may result in differing amounts of stone required at each depth interval (See Figure 1). Very loose or soft soils generally require a higher stone volume, resulting in an increased column diameter over the specific interval. Dense soils may achieve the compaction amperage with less stone, and therefore have a reduced column diameter.

As SCs are constructed below the ground surface and visual confirmation is not possible, operators and offsiders manually record the stone loaded into the probe (ie the number of hopper buckets) at each interval to track the volume of stone used and allow estimation of the constructed SC diameter across the depth interval. SCs are generally most effective when the soft soil is confined to a relatively thin layer, but can also be implemented for thick units.

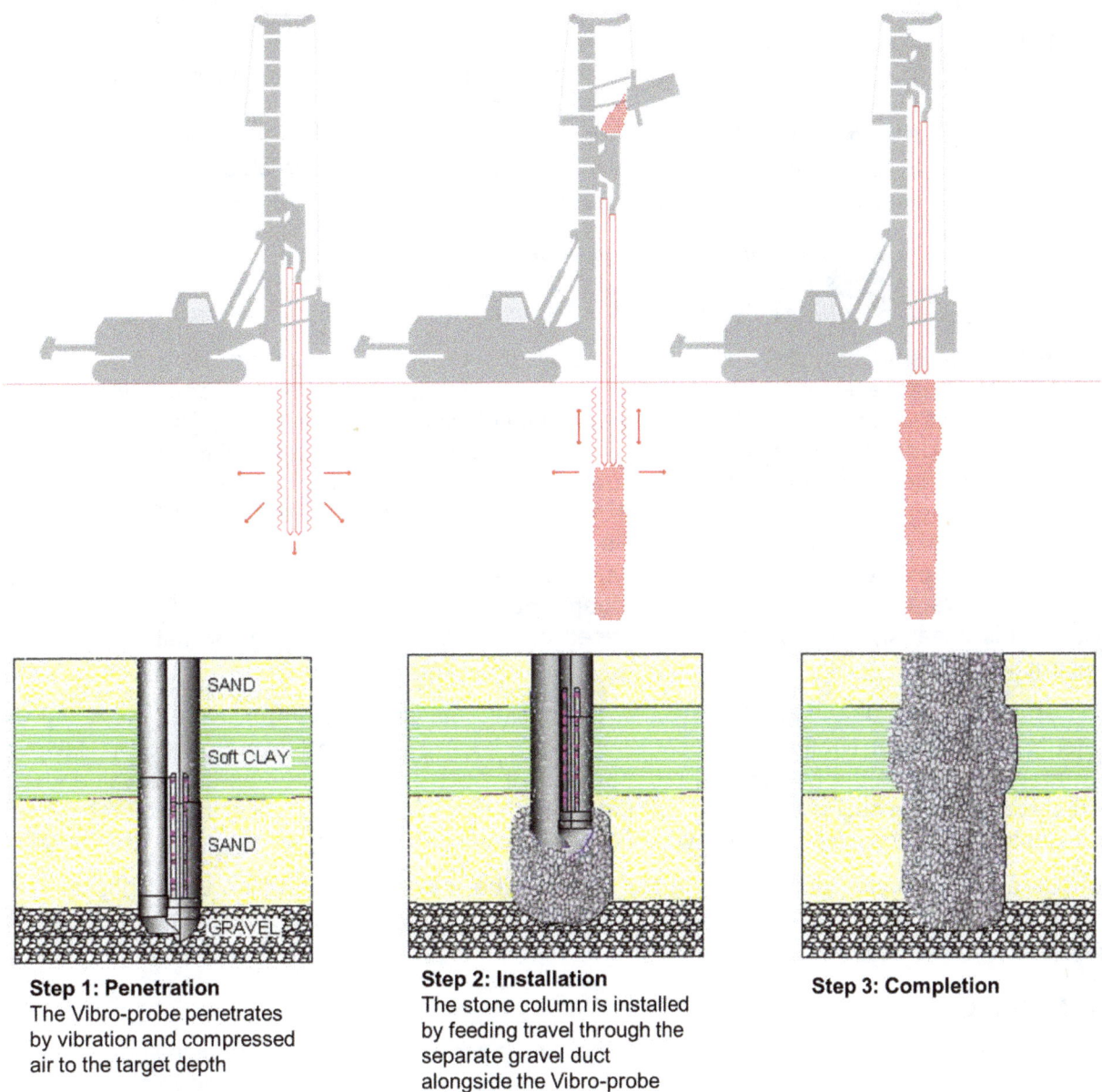

Step 1: Penetration
The Vibro-probe penetrates
by vibration and compressed
air to the target depth

Step 2: Installation
The stone column is installed
by feeding travel through the
separate gravel duct
alongside the Vibro-probe

Step 3: Completion

FIG 1 – Typical stone column installation sequence (Menard Oceania, 2025).

Due to the vibratory effort of the probe, SC construction may also lead to densification and strength improvement for soils surrounding the vertical inclusion. For incorporation in design, on-site trials are necessary to confirm the improvement in the surrounding soils for each stratum encountered, noting that fine grained soils in between adjacent SCs may not densify as part of construction.

Pore pressure generation

Excess pore water pressure (PWP) is generated during SC construction in loose/soft saturated soils. For dam stabilisation projects, the propagation of excess PWP from SC construction in brittle/liquefiable soils is a risk that must be well understood as it could trigger contractive undrained shearing. Well instrumented field trials can be conducted to inform excess PWP generation from SC construction as an input to the overall design/construction staging. The following are key considerations:

- SC trials should be undertaken in similar ground conditions as the proposed GI block, but sufficiently far away as to not influence the critical infrastructure.

- Instrumentation (ie vibrating wire piezometers) should be installed within each critical unit at various spacings from the SC trial to track PWP generation/dissipation spatial and time dependency.

Design

When designing SCs for dam stabilisation, a homogenised GI block is commonly adopted for two-dimensional slope stability analyses using limit equilibrium methods. This allows the establishment of equivalent material properties for the composite ground (ie soil and SCs). The homogenised GI block strength, $s_{composite}$, can be calculated using Equation 1 (Composite block shear strength (US Department of Transportation Federal Highway Administration (US DoT FHWA), 2017).

$$s_{composite} = RR \times s_{stone\ columns} + (1 - RR) \times s_{soil} \qquad (1)$$
$$s: shear\ strength$$
$$RR: replacement\ ratio\ (\%)$$

The design s_{soil} is typically developed using a statistical representation of the interpreted data at a selected percentile value commensurate with the risk of the project. As s_{soil} is dependent on each unit encountered, implementation of the homogenised block requires estimating different composite strength parameters for each unit, and therefore, a good understanding of the strength of the foundation soils before and after SC construction is paramount. This can be estimated from cone penetration testing. A trial can be conducted including CPTs both before and after SC installation to consider potential strength improvement in the soil surrounding the SCs. The timing of post-SC CPTs should consider excess PWP generation/dissipation as this will impact the strength estimate. In the absence of trials, it is commonplace to disregard improvement or densification of the soil between SCs. In this case, s_{soil} would be defined based on the soil shear strength estimates prior to SC construction.

With an overall aim of achieving $s_{composite}$, the design specifies the target depth, RR and stone material providing high confidence in achieving the $s_{stone\ columns}$ component of Equation 1. However, s_{soil} may vary due to the innate nature of soils and therefore the adopted percentile value should suitably consider the observed variability of the improved soils and overall risk of the project to ensure the adopted strength is achieved following SC construction.

As SCs are constructed below the ground surface where visual confirmation is not possible and testing less direct, a robust QC program is necessary to confirm that SC construction achieves the design requirements (ie target depth, RR etc).

QUALITY CONTROL

Similar to density testing of compacted earth fill being a proxy for achieving a design shear strength, a variety of QC measures may be implemented during SC construction to increase confidence in the constructed GI block achieving the design $s_{composite}$.

The following sections discuss suggested QC for dry bottom feed SCs for dam stabilisation. While not exhaustive, this list presents the critical controls required for design verification.

Set-out of column

Column locations are typically set-out by a surveyor and actual installation locations are surveyed following completion. This ensures the correct number of columns are installed in the correct locations based on the design RRs.

Grading, strength and crushing resistance of stone

Grading, strength and crushing resistance of the stone used for SC construction is monitored with ongoing particle size distribution, point load tests and aggregate crushing value (ACV) testing respectively throughout construction. This ensures the strength properties adopted for the stone material in the composite shear block strength calculation (Equation 1) remain consistent and compliant with design.

Depth of pre-drilling

Where required, pre-drilling depth is tracked by operators for each column. It is important to ensure column construction depth equals or exceeds the pre-drill depth to not leave disturbed foundation soils below constructed columns.

Depth of column

Column depth is tracked with the software integrated into the SC rigs for comparison against pre-drill depth and design target depth. Design target depth may not always be specified by depth below ground level as uncertainty may exist regarding the extent of the loose/soft layer being improved. Instead target depth may be specified by termination at a competent material underlying the unit being improved. In this instance it is also important to track power consumption at the termination depth.

Power consumption (amperage) during penetration and compaction

Based on the GI contractor's experience and SC trials on-site, it is common to define an expected range of termination amperage and compaction amperage. Termination amperage is used to discern when a competent or strong material is encountered and is typically related to target installation depth. Similarly, compaction amperage is used to discern when sufficient stone has been placed and vibrated within each depth interval as the SC is installed. Achieving compaction amperage assists in ensuring target stone column diameters are achieved.

Number of hopper buckets used per depth interval

The number of hopper buckets deposited at each depth interval of the SC is counted and tracked by an offsider. The weight and/or volume of stone per bucket is measurable and known, therefore allowing estimation of the volume of stone installed within each depth interval. This can be converted to a column diameter estimate at each depth interval. In addition to the offsider's manual count of hopper buckets, buckets are also counted electronically by the operator who presses a button within the SC rig each time a bucket is deposited into the probe. This provides both electronic and physical records of the amount of stone installed within each SC.

Stone consumption over depth

With the volume estimates per depth interval, comparison can be made of the expected stone volume to be installed at each depth interval against the actual installed volume. The expected volume is calculated based on the design target SC diameter and the depth interval. As previously discussed, and shown in Figure 1, actual installed SC diameter can vary depending on the soil unit. However, design diameters are usually provided as a minimum diameter. Volume greater than the expected volume (and hence greater diameter) is considered volume overconsumption and volume less than design considered underconsumption. Volume overconsumption and underconsumption should also be tracked over the full depth of the column.

Diameter of column

Volume overconsumption/underconsumption is usually accompanied by an estimate of the average diameter of the column. Generally, this is provided as an average diameter over the full depth of the column as this is readily estimated from the QC data. A limit may be set for SC diameter below the design diameter which requires remediation (eg average diameter <80 per cent of design diameter).

Improvement of soil between columns

Following SC construction, and after sufficient time has passed for excess PWP to dissipate, CPTs can be conducted to confirm the strength of soils between SCs. Generally, it is expected that soil shear strength will improve though this is not always true. The estimated shear strength from CPTs can be compared to the adopted design values to confirm the adequacy of the SCs alongside the previously discussed controls.

Summary

Adequately installed SCs by the above quality controls would:

- have stone consistent with the adopted design properties
- have greater installation depth than pre-drill depth

- be installed to the design target depth (or greater)

- have achieved the termination amperage at the base depth and achieved compaction amperage at each depth interval

- have installed the expected volume of stone as a minimum

- have achieved the design diameter

- where considered in the design, have improved the surrounding soils in line with trial estimates.

However, in instances where some of the above quality controls are not met or have conflicting outcomes, it may be useful to conduct design intent verifications to avoid unnecessary or unrealistic remediation works.

DESIGN INTENT VERIFICATION

The GI block is comprised of SCs at specified RR/diameters adopted based on the design composite shear strength value (see Equation 1). Therefore, the overall design intent is for the GI block to achieve this composite shear strength value with the specified RR and SC diameter acting as parameters that can be measured to provide confidence that the design intent is achieved based on QC data.

Where the specified RR/SC diameters are not achieved based on the QC data, additional remedial SCs could be constructed to increase the overall RR to the specified value. However, remedial SCs may not be required provided the design GI block composite shear strength value is achieved. Verification of this can be undertaken using post-SC CPTs and estimation of SC diameters per depth interval. Ideally, post-SC CPTs should be undertaken after excess pore pressures generated by SC installation have dissipated. However, depending on the soil units encountered this may be difficult and pressure to demobilise equipment may drive testing to be conducted earlier. In these instances, consideration into the impact of these excess pore pressures on strength estimation is prudent.

Figure 2 illustrates the inputs for estimation of the composite shear strength when accompanied by post-SC CPTs.

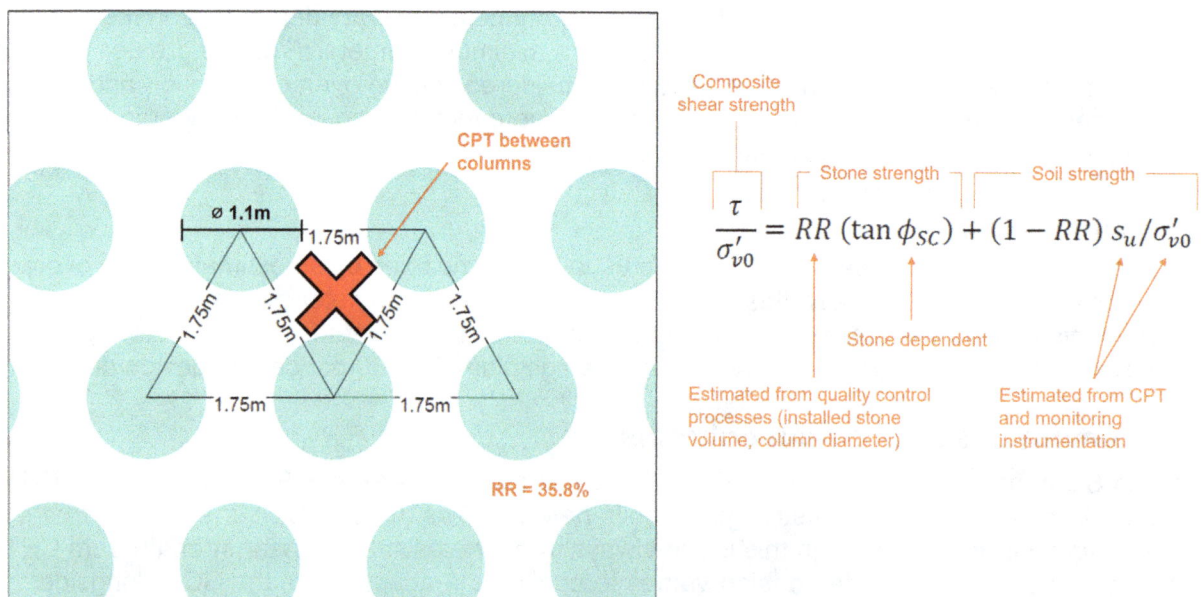

FIG 2 – Equivalent strength estimation.

The RR is calculated from estimates of installed SC diameter (with depth) for the three SCs surrounding the CPT and compared against the portion of soil within the same three SCs as presented in Equation 2 (Replacement ratio calculation (US DoT FHWA, 2017)).

$$RR = \frac{A_{stone\ columns}}{A_{total}}$$

(2)

Estimation of the SC diameter per depth interval requires crosschecking the electronic and manual QC records for inconsistencies, and interpretation of the stone deposition intervals based on the software-tracked movements of the probe and hopper buckets used during construction. This is time consuming and cannot necessarily be automated due to the nature of the vibration probe, compaction process and the resultant data output. As a result, estimation of these diameters is typically reserved for SCs adjacent to a CPT from which strength verifications are taken. Diameter estimates per depth interval allow estimation of the RR (see Equation 2) with depth of the CPT, which then enables calculation of the composite shear strength with depth for comparison against design values.

As discussed in the Background – Stone columns section and shown in Figure 1, it is common for the SC construction diameter to increase and decrease depending on the strength of the soil encountered. As the composite GI block shear strength is a function of stone strength and soil strength components (see Figure 2), it is possible that with sufficient soil strength the adopted design strengths can be achieved despite a reduced stone strength component due to reduced SC diameter (ie reduced RR). In this sense, SCs constructed within a dense/strong material that may result in lower than design diameters (including correspondingly low RRs), may still be proven adequate by achieving the design composite strength provided a higher soil strength in the area is confirmed. If the design composite strength is achieved, remedial SCs are not required.

DESIGN INTENT VERIFICATION EXAMPLE

For a recent dam stabilisation project in Australia, the authors defined technical specifications for SCs, in collaboration with the ground improvement contractor, in accordance with the controls suggested in this paper. The specifications were used to assess SC compliance on a daily basis for the construction of approximately 6500 SCs at a target diameter of 1.1 m in 9 months.

During construction, there was an area where the constructed SCs met the termination amperage, pre-drilling and installation depth criteria, confirming compliance with the design installation depth. Furthermore, compaction amperage throughout construction was within the expected range. However, as shown in Figure 3, volume consumption and average SC diameter in the area suggested numerous non-conforming SCs were constructed.

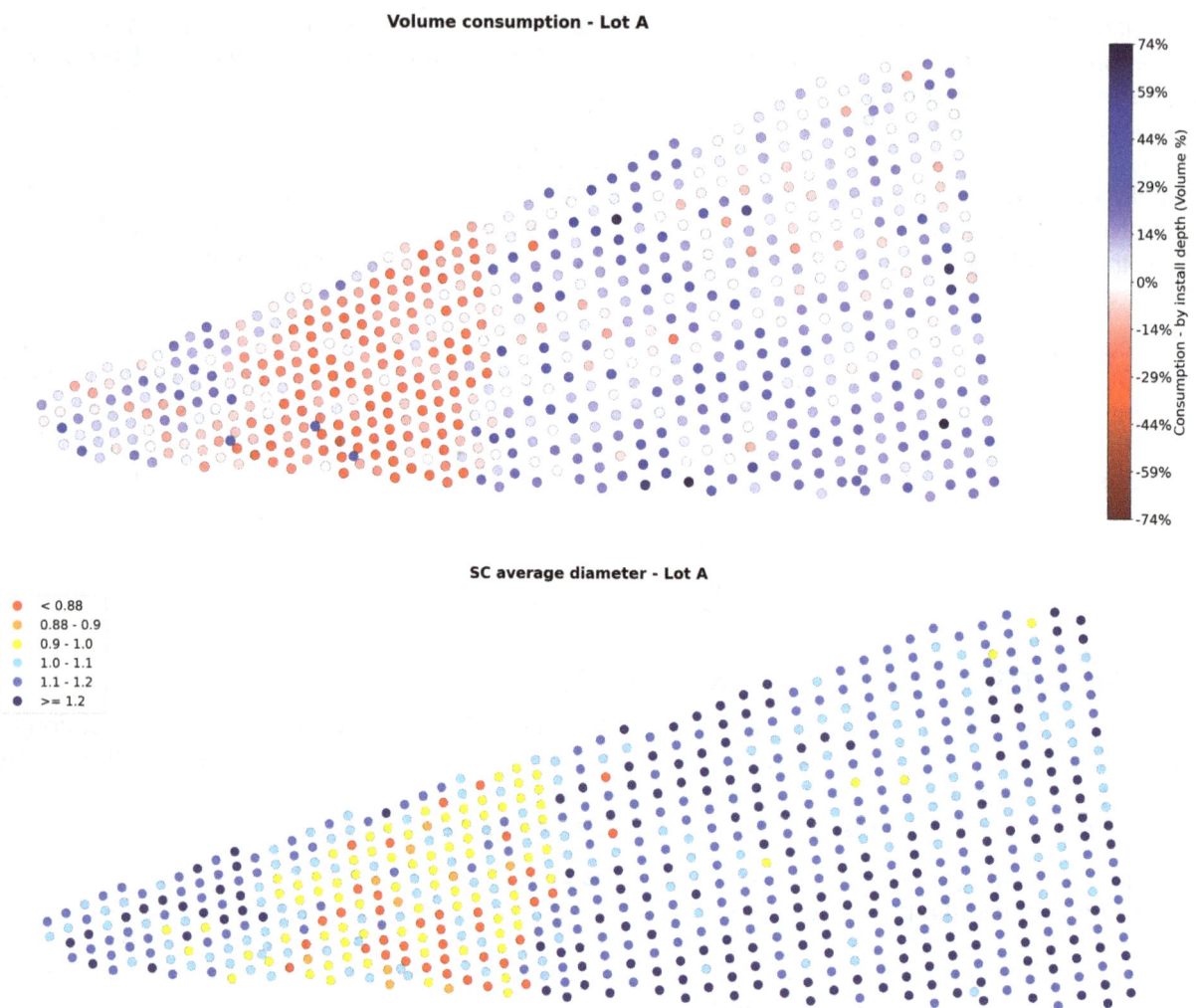

FIG 3 – Stone column volume consumption and diameter heatmaps.

For volume consumption, blue represents volume overconsumption – SCs where a greater volume of stone was installed than estimated from design depth and diameter. Red represents columns with volume underconsumption (volume less than design).

It can be observed that there is a large cluster of volume underconsumption, suggesting that the design SC diameter and RR has not been achieved in this area. This is reinforced by the heatmap of average SC diameter where it is observed that a similar area of SCs are below the design target diameter of 1.1 m when considering average SC diameter over the total depth of the column.

When remediating SCs for failing the average diameter or volume criteria, the appropriate number of additional remedial SCs to reach the design RR for that area is estimated. In order to reach the design RR for the area shown in Figure 3, it was calculated that approximately 70 remedial SCs would have been required (equating to approximately 30 per cent more SCs than design). While being a costly exercise purely due to rework, it may also be challenging to achieve the specified remedial SC diameter (0.9 m on this project). The soil surrounding the non-conforming SCs may have already densified as a result of the constructed SCs, increasing the mechanical effort required to construct the remedial SCs. This increases the cost and time effort of constructing the remedial SCs significantly, and there is a risk that the remedial SCs are also non-conforming (ie not achieving a diameter of 0.9 m).

The authors considered that the low diameter SCs constructed may be a result of locally higher soil strength. As discussed in the Design intent verification section, if the soil strength is sufficiently higher than the design estimates, lower than design SC diameters (and hence RRs) may still provide sufficient shear strength, achieving the design intent. Despite extensive site investigation campaigns, there were gaps in the reliance information for this area. To confirm if remediation works were required in the area, further design intent verification was undertaken.

The design intent verification was completed as follows.

Conduct additional CPTs within the area of non-conforming SCs. Note that CPTs were already planned to be undertaken as part of QC as suggested in this paper. This meant that comparing results against CPTs conducted within conforming areas was also possible. The CPTs conducted across these areas are presented on the volume consumption heatmap in Figure 4.

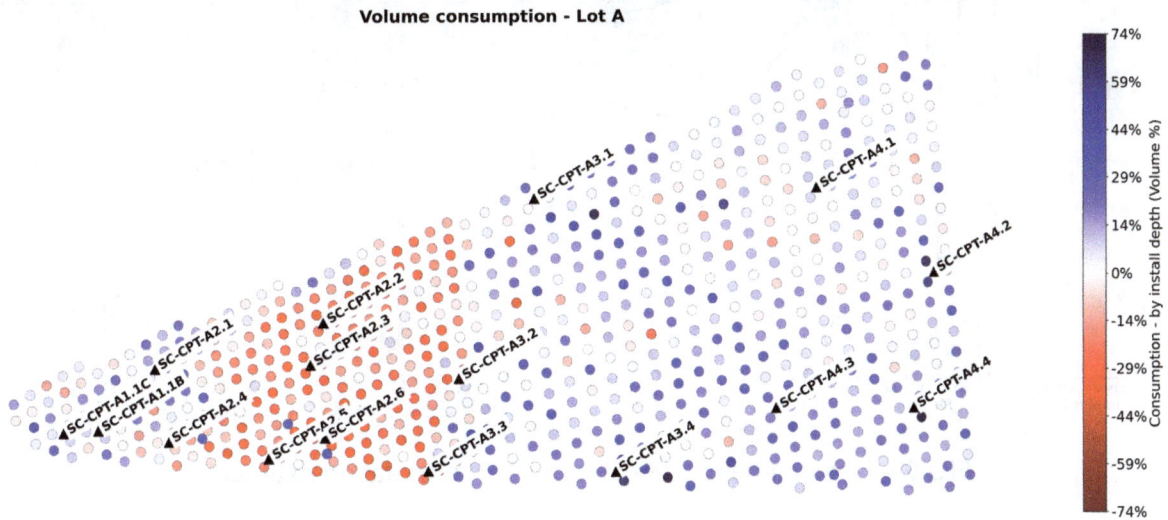

FIG 4 – Cone penetration test plan.

Estimate SC diameter with depth for the three SCs adjacent to each CPT. As discussed in Design intent verification, the process of estimating SC diameter for each interval of stone deposition/compaction can be cumbersome and therefore this interpretation was only completed for SCs immediately adjacent to the post-construction CPTs.

Calculate RR with depth based on the SC diameter with depth estimations. The estimated SC diameter for each depth interval for each of the three SCs surrounding each CPT can be used to calculate $A_{stone\ columns}$ for input into Equation 2. A_{total} is calculated for the triangular area connecting the three SCs based on the SC spacing.

Estimate composite GI block shear strength using Equation 1. Soil shear strength (Figure 2) was estimated from CPT based on Robertson's (2022) liquefied strength approach. Figure 5 presents an example of the CPT interpretation for one of the CPTs within the non-conforming area of SCs. Diameter data from three adjacent SCs is used to inform RR surrounding the CPT.

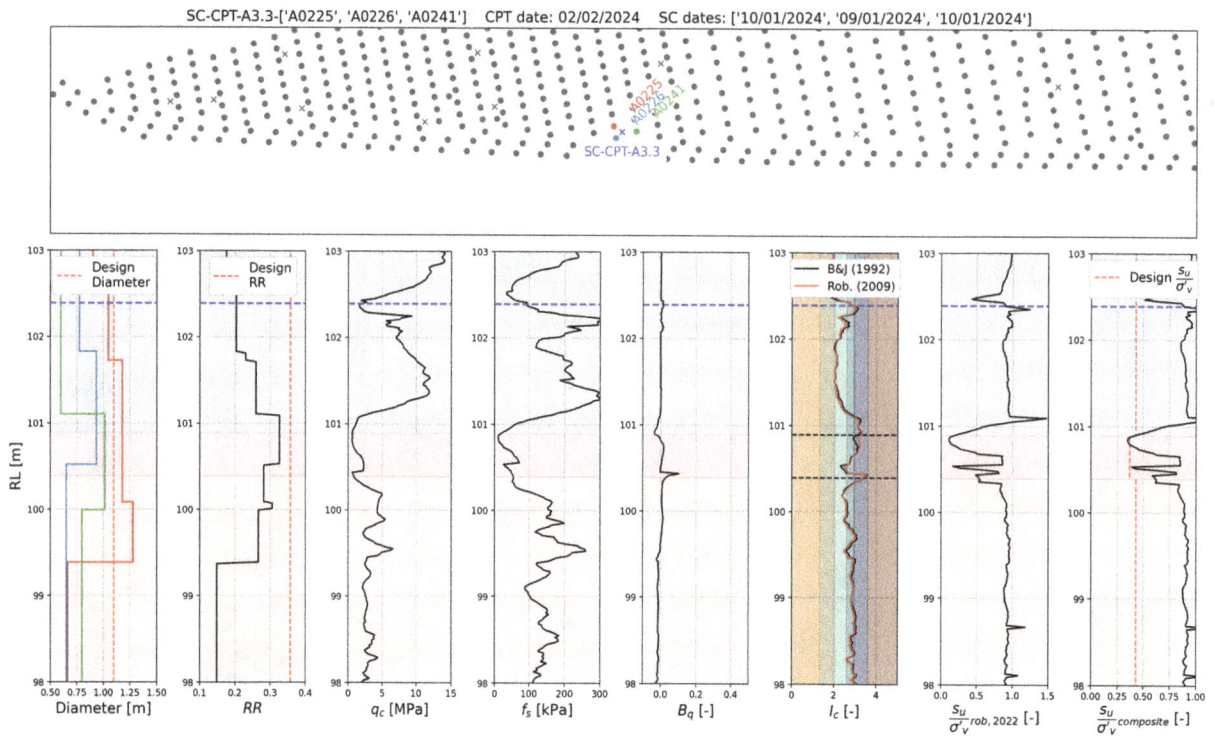

FIG 5 – Design intent verification CPT interpretation example.

Compare the estimated composite GI block shear strength by depth with the adopted design shear strength for each material unit. Design diameter, RR and composite GI block shear strength are plotted with dashed lines on their respective plots in Figure 5 (Plots 1, 2, and 8).

Comparison was made on an individual basis (ie for each individual CPT) and for the construction lot as a whole. For the CPT presented in Figure 5, it is observed that the estimated liquefied shear strength of the soil between RL 103 m and RL 101 m is generally greater than 0.5. As a result, the composite GI block shear strength was estimated to be greater than the design value of 0.43 despite the below design RR installed over this depth interval. Similarly, for the critical soil unit between RL 100.9 m and RL 100.4 m the estimated composite GI block shear strength was estimated to be equal to or greater than the design value of 0.37 despite the below design RR installed.

It was estimated that for this lot, only 7 per cent of data points within the critical material unit were below the design composite GI block shear strength. Furthermore, the non-compliant data points were localised, comprising of thin sub-layers that did not form a continuous layer across the improvement area. As a result, it was concluded that the design intent had been achieved for the SCs installed within this lot and remedial SCs were not required.

CONCLUSIONS

Specific circumstances such as limited footprint or lack of buttress material may drive the use of GI as part of a wider dam stabilisation solution. SCs are an *in situ* strengthening technique that has been used successfully in a variety of dam stabilisation projects. SCs aim to create a composite material with increased shear capacity providing an overall improvement in the resisting forces across a problematic unit.

SCs are constructed below the ground surface where visual confirmation is not possible and testing less direct. Therefore, a robust QC program is necessary to confirm that SC construction achieves the design requirements. Key considerations for SC QC are presented alongside a method that can be used to verify if the design intent is achieved when QC data indicate a key design requirement has not been met. Design intent verification was used to successfully confirm satisfactory performance of SCs on a recent project in Australia, in line with the design intent, avoiding costly remediation works.

ACKNOWLEDGEMENTS

The authors would like to acknowledge Menard Oceania for their support and collaboration on the project. The authors also acknowledge the support of the dam owners to undertake the project and share the learnings documented in this paper.

REFERENCES

Menard Oceania, 2025. Stone columns [online]. Available from: <https://menardoceania.com.au/soil-expert-portfolio/stone-columns/> [Accessed: 18 March 2025].

Robertson, P K, 2022. Evaluation of flow liquefaction and liquefied strength using the cone penetration test: an update, *Canadian Geotechnical Journal*, 59(4):620–624. https://doi.org/10.1139/cgj-2020–0657

US Department of Transportation Federal Highway Administration (US DoT FHWA), 2017. Ground Modification Methods Reference Manual – Volume 1, FHWA-NHI-16–027 [online]. Available from: <https://highways.dot.gov/sites/fhwa.dot.gov/files/FHWA-NHI-16–027.pdf> [Accessed: 18 March 2025].

A comparison of uniform intact and reconstituted specimens of silty-sand tailings

V Kunasegaram[1], D Reid[2], A Fourie[3] and C Vulpe[4]

1. Research Fellow, The University of Western Australia, Crawley WA 6009.
   Email: vijayakanthan.kunasegaram@uwa.edu.au
2. Principal Engineer/Adjunct Research Fellow, Red Earth Engineering/The University of Western Australia, Crawley WA 6009. Email: david.reid@uwa.edu.au
3. Professor, The University of Western Australia, Crawley WA 6009.
   Email: andy.fourie@uwa.edu.au
4. Senior Lecturer, The University of Western Australia, Crawley WA 6009.
   Email: cristina.vulpe@uwa.edu.au

ABSTRACT

The critical state line (CSL) represents an essential tool in the static liquefaction susceptibility assessment of mine tailings. Although it has been shown that the CSL can be consistently and reliably obtained for sandy silt tailings through triaxial testing of moist tamped reconstituted specimens, there is ongoing debate whether the reconstituted specimens reproduce the actual CSL of the *in situ* material. Some previous research has highlighted uncertainty in replicating *in situ* behaviour due to fabric effects, particularly particle orientation and assembly. While addressing the observations of fabric, this paper presents a comparison of triaxial tests carried out on specimens of silty sand tailings trimmed from a high-quality intact block against reconstituted specimens of material taken from the same block using the moist tamping (MT) preparation method. The intact block used in these tests was homogeneous, with only a slight variation of particle size distribution with depth. All the specimens trimmed from the intact block tended towards the inferred MT CSL. Contrary to some available literature, there was no significant variation of the critical state shearing behaviour of intact and reconstituted specimens at the same void ratio and effective stresses, in both drained and undrained loading. This study verifies that MT can provide accurate definition of critical state conditions for *in situ* material that is uniform.

INTRODUCTION

Tailings Storage Facility (TSF) failures pose significant environmental, economic, and safety risks, often leading to catastrophic consequences such as dam collapses, extensive downstream contamination, and loss of life. Understanding the underlying causes of these failures is crucial for improving the design, construction, and long-term stability of TSFs. Among the various failure mechanisms, static liquefaction has emerged as a leading cause, often resulting in sudden and extensive collapses of tailings dams. Static liquefaction is a complex phenomenon where the geotechnical and material characteristics of a tailings embankment is coupled with hydraulic conditions. Unlike dynamic liquefaction, which is induced by seismic activity, static liquefaction can occur due to gradual loading, rainfall infiltration, or construction-induced internal stress changes, making it particularly difficult to predict timeously.

Quantifying susceptibility to static liquefaction is often dealt with by reference to the Critical State Line concept (CSL) using reconstituted samples, because of the difficulties in obtaining high-quality undisturbed samples, particularly with low-plasticity silts or silty sand particles. Relying on such critical state parameters of homogeneous materials, and indiscriminate use of those parameters in more sophisticated numerical simulations to replicate the *in situ* behaviour of large-scale TSFs with variable depositional histories, could lead to significant discrepancies, owing to macro or micro layering caused by subaqueous deposition, as observed by Baziar and Dobry (1995).

Subaqueous deposition and associated hydrodynamic processes influence the formation of distinct sedimentary layers with varying geotechnical properties. Hydraulically placed fills are deposited in a slurry form and variations in material properties and water content can lead to the formation of localised zones of higher or lower density and strength. This variability can significantly influence the mechanical behaviour and stability of the fill, affecting settlement patterns, permeability, and overall structural integrity. Homogeneity in a hydraulically placed fill is extremely rare as revealed in previous fabric studies of multiple localised layers within a small block sample (Baziar and Dobry, 1995; Reid *et al*, 2018; Reid and Fanni, 2022; Chang, Heymann and Clayton, 2011).

Significant layering can create hydraulic discontinuities, influencing drainage, pore water pressure dissipation, and the potential for liquefaction or shear failure. However, decades of investigations on silty tailings reconstitution methods and the use of resulting CSLs (Høeg, Dyvik and Sandbækken, 2000; Chang, 2009; Chang, Heymann and Clayton, 2011) have placed less emphasis on fabric studies or mechanical assessment of layering other than the visual observations. These studies also concluded that none of the reconstitution methods provide a reasonable reproduction of *in situ* fabric and undrained shearing behaviour of predominately silty tailings. Discrepancies related to *in situ* layering observed in these studies are discussed in Reid and Fanni (2022). Recent studies (Reid and Fanni, 2022; Reid, Fourie and Fanni, 2022) with careful consideration of *in situ* layering as small as 20 mm intervals vertically, dictates that the intact block samples of predominately silty tailings generally tended towards the moist tamped (MT) CSL and layering has a more significant influence on the end states of intact block samples than those of the homogeneous MT specimens.

Concurrent with ongoing arguments regarding the best reconstitution method and the associated importance of *in situ* layering, inherent variability, and the impact of particle rearrangement, the moist tamping (MT) method is frequently used in engineering practice for TSFs as demonstrated by its widespread use in recent significant TSF failure investigations (Jefferies *et al*, 2019; Robertson *et al*, 2019). Furthermore, MT has emerged as a widely used technique for reconstituting disturbed soil/tailings due to its ability to control density and moisture content with precision.

This paper explores the application of the MT technique to reconstitute a uniform soil deposit, reviewing its effectiveness in simulating MT reconstituted counterparts of intact specimens and assessing the mechanical behaviour of intact and MT specimens.

BLOCK SAMPLING AND MATERIAL PREPARATION

The TSF, which provided a high-quality block sample for this study, is a storage facility for nickel ore tailings and is mainly composed of silt-sized non-plastic tailings. As part of the site investigation and sampling, the undisturbed blocks were extracted from several locations on the TSF beach at shallow depth raging from the surface to up to 2 m depth. Six block samples were collected from areas of the TSF that were dry enough to access and retrieve blocks (which were partially saturated) and this study mainly investigates the findings from one of the recovered blocks. Extracted block samples were carefully wrapped on-site, with several alternating layers of cellophane and aluminium foil. After that, they were packed in airtight containers and cushioned with polyurethane foam to transport to the laboratory.

Upon arriving at the University of Western Australia Geotechnical Testing Laboratory, the block sample was unwrapped and inspected for visual indications of layering. It is noted that the block appeared to be homogeneous on visual inspection. Then, small sub-samples with sufficient volume to enable wet sieving were collected vertically, from as little as 20 mm layers from the block surface as shown in Figure 1a. The variation of percent passing through 75 μm sieve is illustrated along the depth in Figure 1a. The result of wet sieving indicates that the block was acceptably homogenous with an average of 14 per cent fines throughout the depth, except a significant layering observed closer to the bottom of the block. Thanks to the uniformity of the block sample, it was possible to obtain a few gradationally homogeneous intact specimens for triaxial compression tests in this study.

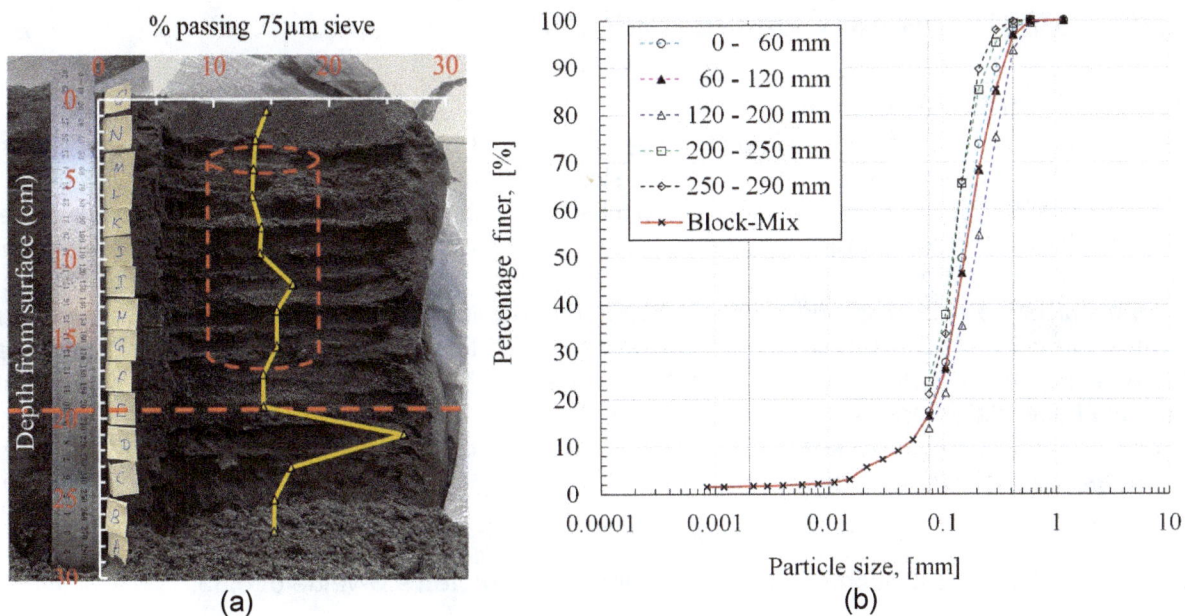

FIG 1 – Sub-sampling of material from block to assess homogeneity: (a) Variation of % passing 75 µm sieve along the depth; and (b) Particle size distribution of composite bulk and a range of block samples.

Composite bulk sample for reconstitution

Leftover materials from the block specimen preparation and trimming process were collected in a sealable storage bag after being thoroughly mixed to create a composite bulk sample for reconstitution. Index properties of the composite bulk sample are given in Table 1, and the gradation of the composite bulk sample from Sedigraph and wet sieving is illustrated in Figure 1b. A portion of the intact block sample was also used to confirm the gradational homogeneity of the block at an interval of 60 mm vertically and the results are plotted in Figure 1b. In this investigation, the Sedigraph results below 75 µm were omitted due to the quantity of fines required from the high-quality block sample for the hydrometer analysis. Grading curves obtained for 0–60 mm and 60–120 mm, match quite closely with the bulk mix and the variations are considered insignificant. However, the gradations of remaining three sub layers have small variations with a maximum of ±7 per cent for particles passing 75 µm from the gradation of the bulk mix. Based on sieve analysis and index properties, the material is non-plastic and predominantly silt-sized, with a specific gravity (G_s) of 3.22.

TABLE 1

Index properties of composite bulk sample.

Test type	Property	Unit	Value
Pycnometer	Gs	-	3.22
Wet sieving	P<75 µm	%	17
Sedigraph	P<38 µm	%	9
Sedigraph	D60	µm	195
Sedigraph	D10	µm	40
	Liquid limit	%	22
Atterberg limits	Plastic limit	%	Non-plastic
	Plasticity index	%	Non-plastic

TESTING METHODS AND SAMPLE RECONSTITUTION

Intact block specimens

Upon completion of layering investigations, the mechanical properties of the tailings were studied using a series of triaxial compression (TX-C) tests. The block was first trimmed with a scalpel and then an internally lightly greased stainless-steel tube with a sharp cutting edge was carefully inserted into the block to prepare the intact specimens. In total, four intact specimens were obtained for triaxial testing, each having a diameter of 72 mm and approximately 144 mm height. To prevent possible sample disturbance at the top and bottom during trimming and to maintain homogeneity, the centre portion of the upper 190 mm uniform section was selected to prepare the cylindrical specimens as indicated in Figure 1a. Once trimmed to the desired height, the specimen was extruded directly onto the triaxial base platen and secured with a membrane and top cap under a small value of suction applied through the valves to keep the *in situ* granular structure. After that, general triaxial testing procedures were followed as follows:

1. Flushing the specimen with deionised deaired water to remove entrapped air.

2. Back pressure saturation to achieve a minimum Skempton's B-value of 0.95.

3. Isotropic consolidation to the desired stress state.

4. Drained or undrained shearing of the specimen until the inferred critical state and finally.

5. End-of-test freezing (Sladen and Handford, 1987) of the specimen to estimate the void ratio as accurately as possible.

Shearing rates of 0.03 mm/min and 0.1 mm/min were adopted for drained and undrained tests, respectively. The same procedure was followed for all the tests conducted in this study and a summary of intact block tests is described in Table 2.

TABLE 2
Summary of intact block triaxial compression tests.

Test ID	Test type	Initial states		Consolidated states				End of shearing states		
		e_o	Ψ_o	p'_c (kPa)	q_c (kPa)	e_c	Ψ	p'_c (kPa)	q_c (kPa)	e_{cs}
BL-2a	CID	0.83	0.11	901	3	0.73	0.01	1886	2957	0.63
BL-2b	CIU	0.87	0.05	206	4	0.80	-0.02	375	572	0.80
BL-2c	CID	0.86	0.00	51	3	0.84	-0.02	109	178	0.85
BL-2d	CIU	0.88	0.16	902	3	0.75	0.03	709	1076	0.75

Moist tamping

One popular approach for reconstituting tailings specimens in laboratory settings is the Moist Tamping (MT) method. Recent extensive TSF failure investigations indicate that the MT is used in parallel with continuous debates on the ideal reconstitution mechanism (Jefferies *et al*, 2019; Robertson *et al*, 2019). Moist tamping involves compacting granular soils at a controlled moisture content, which helps in minimising segregation and achieving uniformity in the specimen structure. Considering the homegeneity of the block samples, the MT method was adopted as a reconstitution technique in this study. In total, seven MT specimens were prepared to determine the CSL of the composite bulk sample. Ladd's (1978) under-compaction technique was used to produce the specimens in eight layers at a gravimetric water content of approximately 5 per cent. Layer masses were estimated with caution to prepare loose contractive specimens to eliminate shear banding at high strains. Oversized highly lubricated end platens were used to allow uniform specimen deformation throughout the specimen height. Three MT specimens (MT-2a, MT-2b and MT-2d) were targeted to closely mimic the *in situ* density of trimmed block specimens to directly compare the responses of block and MT specimens. A summary of MT tests conducted in this study is described in Table 3.

TABLE 3

Summary of MT samples.

Test ID	Test type	Initial states		Consolidated states				End of shearing states		
		e_o	Ψ_o	p'_c (kPa)	q_c (kPa)	e_c	Ψ	p'_c (kPa)	q_c (kPa)	e_{cs}
MT-2a	CID	0.88	0.02	51	5	0.86	0.00	100	145	0.85
MT-2b	CIU	0.89	0.07	202	6	0.85	0.03	94	137	0.85
MT-2c	CID	0.89	0.14	702	6	0.80	0.05	1425	2173	0.67
MT-2d	CIU	0.88	0.16	903	6	0.79	0.07	434	654	0.79
MT-2e	CID	0.88	0.06	202	5	0.83	0.01	398	591	0.78
MT-2f	CIU	0.87	0.03	103	5	0.85	0.01	69	72	0.85
MT-2g	CID	0.88	0.11	502	5	0.80	0.03	979	141	0.70

RESULTS AND DISCUSSION

MT samples

Observed undrained and drained shearing responses of MT specimens are illustrated in Figures 2 and 3, respectively. MT specimens in this study were prepared slightly looser than the block specimens to exhibit contractive behaviour and sheared to inferred critical state conditions distinguished by insignificant variations in mean effective stress, deviator stress, shear-induced pore pressure, and/or volumetric strain. All MT specimens showed signs of contraction except the specimen 'MT-2a', which was initially contractive, however slightly dilative beyond an axial strain of 8 per cent as shown in Figure 3b. The outcomes of MT tests suggest an average critical state friction ratio (M_{tc}) of 1.47 when considering all MT specimens. Stress paths of each MT specimen and corresponding inferred CSL are plotted in a state diagram along with the initial state of intact specimens in Figure 4. A random mean effective stress of 10 kPa was used to indicate the relative position of intact samples with respect to the CSL. Observed curvature along the inferred critical states of MT specimens led to the selection of a power law fit, as proposed by Li and Wang (1998). Referring to the relative location of MT CSL, the initial state parameter (Ψ_0) of loose MT specimens (MT-2a to MT-2g) varies in a range from 0.02 to 0.16 as given in Table 3.

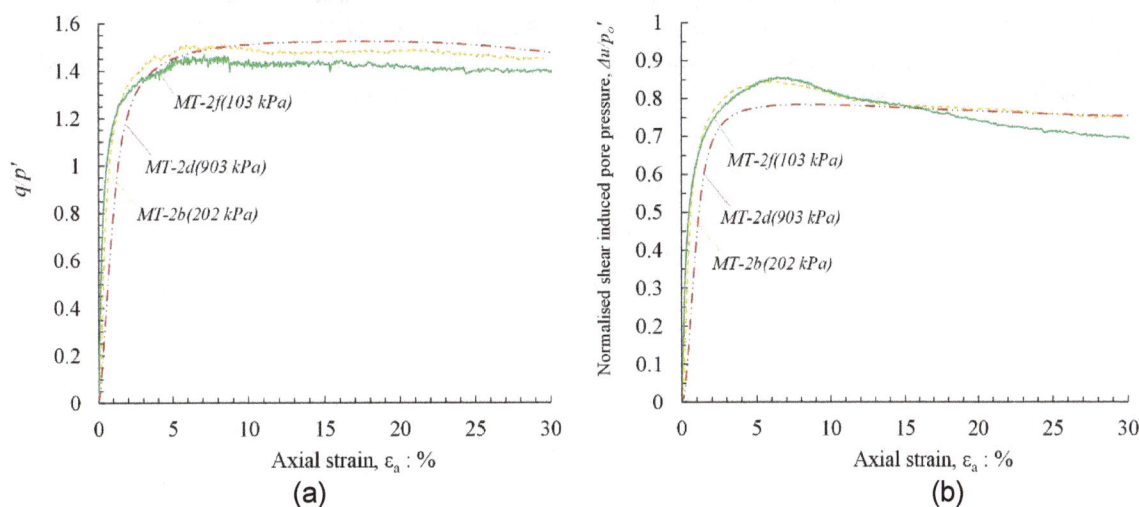

FIG 2 – Results of MT CIU tests: (a) normalised stress-strain data; (b) normalised shear-induced pore pressure.

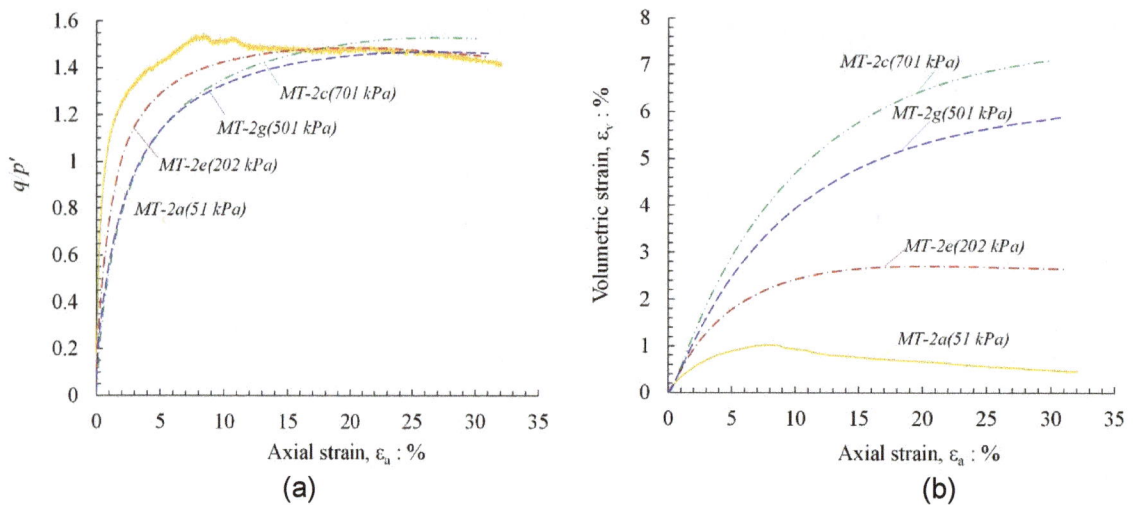

FIG 3 – Results of MT CID tests: (a) normalised stress–strain data; (b) volumetric strain.

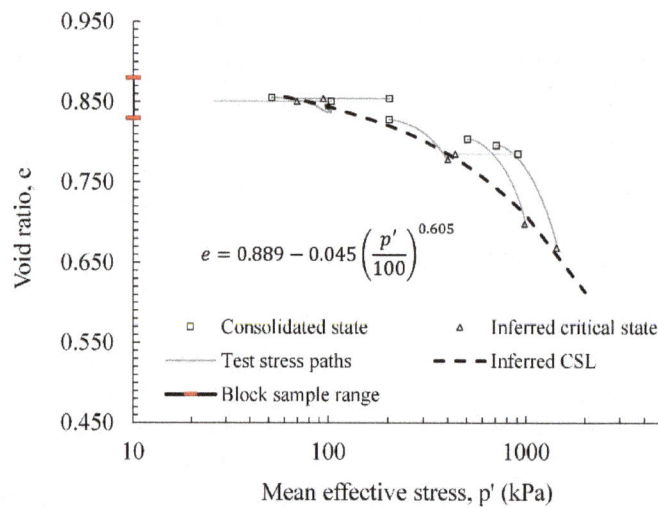

$$e = 0.889 - 0.045 \left(\frac{p'}{100} \right)^{0.605}$$

□ Consolidated state   △ Inferred critical state
— Test stress paths   – – Inferred CSL
— Block sample range

FIG 4 – MT test state diagram and inferred CSL.

Intact block samples

Drained and undrained shearing responses of intact block specimens are illustrated in Figures 5 and 6, respectively. Additionally, MT reconstituted counterparts that were consolidated to identical initial stress states are also drawn in Figures 5 and 6 to enable direct comparison between intact and MT specimens. The outcomes of intact block tests indicate an average critical state friction ratio (M_{tc}) of 1.56 based on the end of shearing state when considering all intact specimens. The observed average M_{tc} from block specimens is slightly higher than that of the MT results.

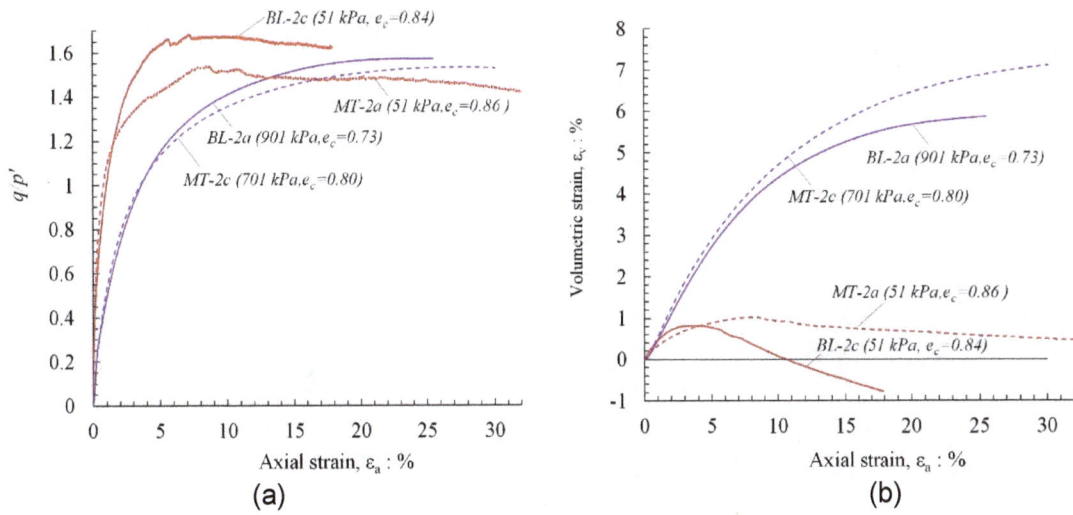

FIG 5 – Results of intact block CID tests compared with MT results: (a) normalised stress–strain data; (b) volumetric strain.

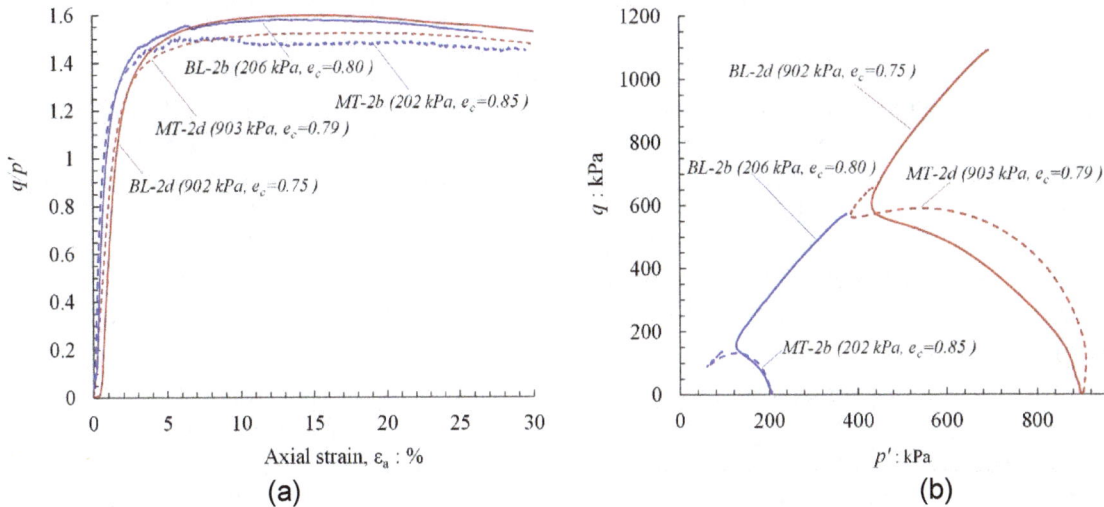

FIG 6 – Results of intact block CIU tests compared with MT results: (a) normalised stress–strain data; (b) q-p state diagram.

Intact block sample BL-2c and reconstituted counterpart MT-2a exhibit a very similar stress-strain response up to a q/p' of 1.2 as shown in Figure 5a. However, the intact specimen exhibits an early dilative behaviour at small strains and relatively higher rate of dilatancy compared to the reconstituted specimen, as seen in Figure 5b. The observed deviation in shearing response beyond a q/p' of 1.2, and the lower peak strength of MT-2a could be attributed to relatively higher state parameter (Ψ) of MT-2a ($\Psi = 0.00$) compared to BL-2c ($\Psi = -0.02$). Comparable to the drained test results, the undrained response of intact block specimens also exhibited a consistent shearing response with a similar stress strain relationship as shown in Figure 6a. Significant differences between intact and MT exist in the q-p' state plot, illustrated in Figure 6b. Referring to Figure 6, at higher strains, intact specimens BL-2b ($\Psi = -0.02$) and BL-2d ($\Psi = 0.03$) changed from their initial contractive phase to dilatative. MT samples MT-2b ($\Psi = 0.03$) and MT-2d ($\Psi = 0.07$), which were at looser states than the intact also exhibited a similar phase transition. However, the strain level at which the transition occurs was higher than that of intact specimens as expected for a specimen with higher Ψ. Overall, the MT and intact specimens achieve a distinct critical state friction angle at large strain, but at various stress states.

Stress paths of intact block specimens are illustrated along with the inferred MT CSL in Figure 7. The exclusion of MT stress paths in Figure 7 improves the comparability between intact specimens and MT CSL. It is observed that all the block specimens trend towards the MT CSL and indicates a close fit, meaning that the MT reconstitution method can be effectively used to replicate the CSL of the uniform intact block.

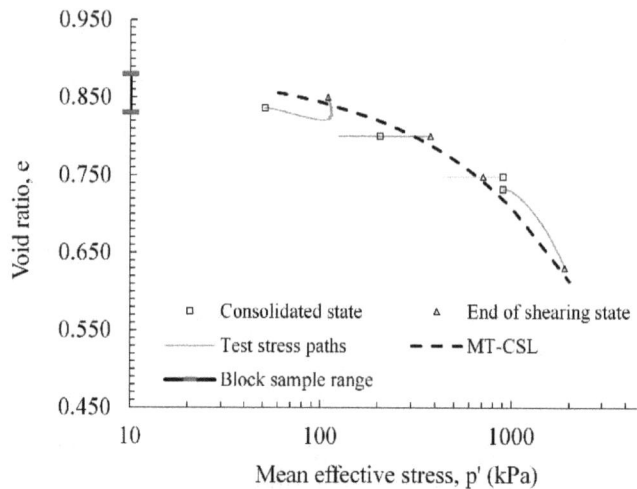

FIG 7 – Intact block test state diagram with MT CSL for comparison.

CONCLUSIONS

A detailed examination of an undisturbed block sample was conducted to study the presence of layering within a block sample obtained from a TSF. The suitability of the MT reconstitution method to replicate undisturbed material CSL was evaluated using a series of uniform homogeneous block specimens and MT reconstituted specimens. The study reveals that the intact block and MT specimens tend towards a very similar CSL. Small discrepancies seen in the shearing responses of MT and block specimens may be caused by variations in the state parameters prior to shearing. Overall, MT reconstituted specimens effectively capture the shearing behaviour and critical states of homogeneous intact specimens.

The findings of this study thus negate the criticisms often levelled against the use of MT for preparing specimens of silty or sandy tailings for testing to determine the critical state locus. This conclusion applies to relatively uniform zones of a TSF. Subsequent work is planned to investigate how representative MT tests on blended specimens are for determining the critical state locus of a layered tailings deposit.

ACKNOWLEDGEMENTS

The authors would like to gratefully acknowledge funding awarded from the Amira P1217 Evaluation of Tailings Storage Facilities monitoring technologies Project, funded through Amira Global by Gold Fields, Agincourt Resources, Boliden, Anglo American, Lundin Mining, Rio Tinto, Independence Group, CMOC, BHP, LKAB, ArcelorMittal, 3vGeomatics, GroundProbe, Institute of Mine Seismology, CGG Services UK, Sercel, Canary Systems and Loupe Geophysics.

REFERENCES

Baziar, M H and Dobry, R, 1995. Residual strength and large-deformation potential of loose silty sands, *J Geotech Engng*, 121(12):896–906.

Chang, N, 2009. The effect of fabric on the behaviour of gold tailings, PhD thesis, University of Pretoria, Pretoria, South Africa.

Chang, N, Heymann, G and Clayton, C, 2011. The effect of fabric on the behaviour of gold tailings, *Géotechnique*, 61(3):187–197. https://doi.org/10.1680/geot.9.P.066

Høeg, K, Dyvik, R and Sandbækken, G, 2000. Strength of undisturbed versus reconstituted silt and silty sand specimens, *J Geotech Geoenviron Engng*, 126(7):606–617.

Jefferies, M, Morgenstern, N R, Van Zyl, D V and Wates, J, 2019. Report on NTSF Embankment Failure, Cadia Valley Operations, for Ashurst Australia.

Ladd, R, 1978. Preparing test specimens using undercompaction, *Geotech Test J*, 1(1):16–23.

Li, X S and Wang, Y, 1998. Linear representation of steady-state line for sand, *J Geotech Geoenviron Engng*, 124(12):1215–1217.

Reid, D and Fanni, R, 2022. A comparison of intact and reconstituted samples of a silt tailings, *Géotechnique*, 72(2):176–188.

Reid, D, Fanni, R, Koh, K and Orea, I, 2018. Characterisation of a subaqueously deposited silt iron ore tailings, *Geotechnique Letters,* 8(4):278–283.

Reid, D, Fourie, A and Fanni, R, 2022. Layering–the missing factor in fabric studies?, in Proceedings of the 20th International conference on Soil Mechanics and Geotechnical Engineering.

Robertson, P K, de Melo, L, Williams, D J and Wilson, G W, 2019. Report of the Expert Panel on the technical causes of the Failure of Feijão Dam, I.

Sladen, J A and Handford, G, 1987. A potential systematic error in laboratory testing of very loose sands, *Can Geotech J*, 24(3):462–466.

Triaxial laboratory shear vane test – a tool to help improve the understanding of the brittleness of tailings materials

M Llano-Serna[1], H Joer[2], R Rekowski[3] and W van Rhyn[4]

1. Principal Engineer, Red Earth Engineering a Geosyntec Company, Brisbane Qld 4000. Email: marcelo.llano@redearthengineering.com.au
2. Director, GTI Perth, Perth WA 6077. Email: hjoer@gtiperth.com.au
3. Senior Principal Engineer, Stantec, Brisbane Qld 4000. Email: robert.rekowski@stantec.com
4. Tailings Engineer, WSP, Perth WA 6000. Email: wil.vanrhyn@wsp.com

ABSTRACT

The International Commission on Large Dams (ICOLD), in Bulletin No. 194 (2022), has provided updated guidance on assessing brittle and ductile behaviour. ICOLD recognised different opinions within the geotechnical engineering profession on applying the classical Bishop's brittleness index and highlighted that conventional approaches may not directly address the strain aspects related to brittleness. The Global Industry Standard on Tailings Management (GISTM; Global Tailings Review (GTR), 2020) requires that brittle failure modes be identified and addressed with conservative criteria. In ICOLD words, *there is currently very little guidance in the technical literature* [to guide brittle behaviour identification]. ICOLD also provided stress-strain curves of ductile and brittle materials. This paper explores the development and early work of a laboratory testing technique that aims to provide insights into tailings' brittle behaviour. An approach that integrates elements of stress-strain, such as the triaxial test, is combined with the mini-vane shear test. This combination of triaxial and vane shear tests is named the Trivane. It involves a vane shear mechanism inside a triaxial chamber, allowing for shear strength measurements using a vane shear under controlled confining pressures. The test enables a more refined prediction of undrained shear strength for future dam construction stages. Current field vane and laboratory vane shear testing only allow testing at surface conditions in the case of conventional mini-vane laboratory testing and as-built conditions in the case of field vane testing. The study describes the development of the tool and presents early results in well-behaved Kaolin clay and the initial test trials with tailings materials. The paper summarises the learnings and future work required to help bridge the gap highlighted by ICOLD.

INTRODUCTION

The International Commission on Large Dams (ICOLD), in Bulletin No. 194 (ICOLD, 2022) define brittleness as the tendency of a material to suddenly lose strength without first undergoing significant plastic deformation, typically occurring at a relatively low degree of strain. Brittleness has historically been defined using the brittleness index (IB) concept as defined by Bishop (1967). IB is a normalised parameter that varies from zero to one and is expressed as:

$$IB = \frac{\tau_p - \tau_r}{\tau_p}$$

where τ_p is the peak (or yield) undrained shear strength value and τ_r is the residual undrained shear strength value. A value of zero means no strength loss (ie a ductile material with no brittleness) and a value of one means the residual strength undrained strength loss is zero. Experimental techniques to assess IB in the context of shearing mechanisms are various, and the relationship between the quantities that can be measured is unclear. For example, Mitchell and Soga (2005) present the conceptual relationship between peak, critical and residual strengths on an idealised triaxial stress-strain curve, see Figure 1. The conceptual triaxial stress-strain shown in Figure 1 is consistent with the mechanism shown in the stress-strain curves of ductile and brittle materials introduced by ICOLD (2022). However, although there is a relative consensus in the tailings industry regarding the concept illustrated in Figure 1, the reality of triaxial testing techniques to assess peak and residual quantities is not straightforward. That is because reaching a sufficient strain inside a triaxial testing device to reach true residual undrained shear strength is difficult when residual strength lies beyond 30 per cent vertical strain.

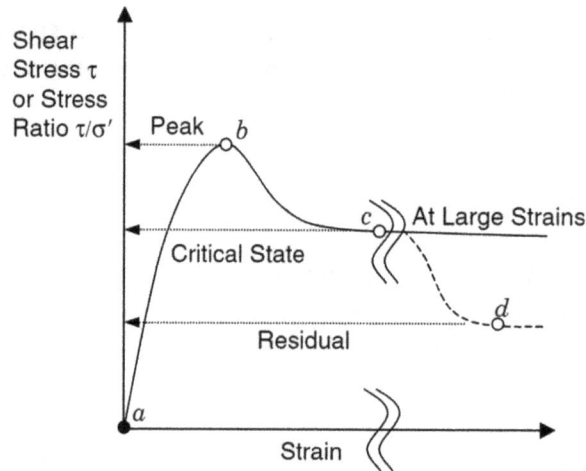

FIG 1 – A conceptual stress-strain curve for peak, critical and residual strengths (Mitchell and Soga, 2005).

Because of this key limitation in conventional triaxial testing, engineers use complementary testing techniques such as vane shear testing (VST) to assess residual strengths. VST can be undertaken using the laboratory miniature vane shear test (LVT) (ASTM D4648M, 2024) or the field vane shear test (FVT) (ASTM D2573/D2573M, 2018). Challenges arise because, despite advantages, VST techniques possess some significant drawbacks for design engineers in the tailings dam industry. For example, on the one hand, the conventional FVT is an *in situ* testing technique, which means that the method can only be applied after the dam has been operational and tailings have been deposited. On the other hand, LVT methods do not account for confinement stress. It means that test results from LVT represent tailings deposited near the surface but do not account for the gain of undrained shear strength due to confinement.

MOTIVATION

Richards (1988) is a comprehensive reference that compiles technical knowledge, best practices and research on VST. Although the book is almost four decades old, it is still used to date as a key reference. The introductory overview by Richards summarised research and development topics still unresolved in 2025. A summary of the topics highlighted by Richards and that are relevant to ICOLD (2022) in the context of brittleness characterisation includes:

- Continue investigations of the failure mode in the VST.

- Obtain the stress-strain curve, and relate it to the VST for the interpretation of results.

- Extend comparisons among the different *in situ* tests and the standard laboratory tests to the full range of applicable soil types in engineering practice.

TRIVANE

A standard triaxial apparatus was modified to incorporate an automatic laboratory shear vane apparatus (Trivane), see Figure 2. The laboratory shear vane apparatus was mounted onto a newly fabricated crossbar to support the additional weight of the shear vane apparatus. During Trivane testing the torque measurement is undertaken outside and above the triaxial cell. In the Trivane the vane is located in the centre of the sample.

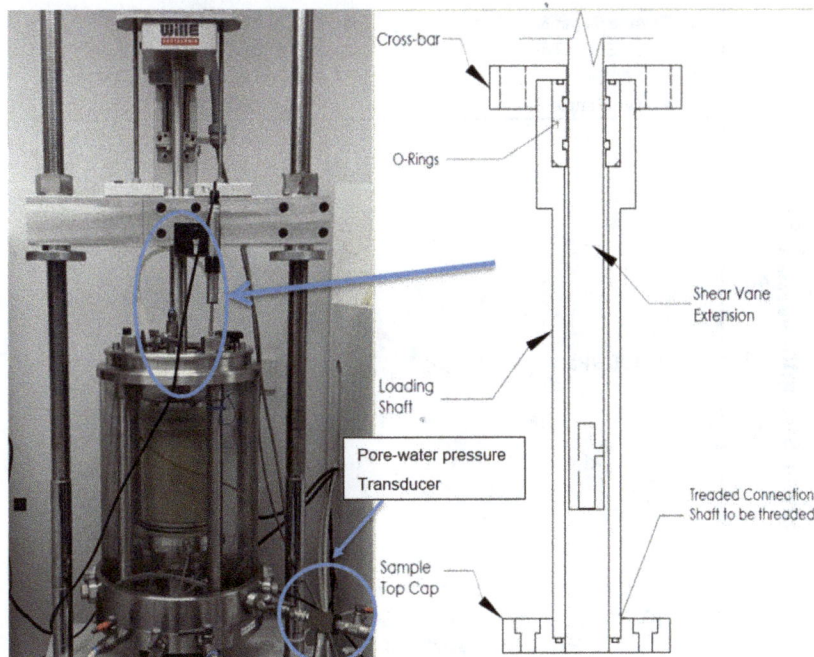

FIG 2 – Trivane apparatus with loading shaft cross-section.

The Trivane allows to:

- Measure pore-water pressures in the same manner as during conventional triaxial testing.

- Apply confinement stress and consolidate samples under isotropic and anisotropic conditions for subsequent VST testing.

- Measure torque versus degree of rotation during VST testing for a range of rates of rotation as recommended by Harvey and Hogan (2023).

Rod friction has been found to be an integral component of the process that requires correction for the Trivane tests. A custom made calibrated rod without blades, with a length and diameter similar to the shear vane is used. Rod friction testing has been completed at various confining stresses ranging between 25 kPa and 100 kPa. The rod friction has been found to be non-constant during shearing and generally increases with confinement stress and rotation angle. Results have been summarised in van Rhyn (2023).

The Trivane sample diameter is nominally 100 mm. Vane height and diameter of 25.4 mm and 12.7 mm are conventionally adopted. The adopted diameter dimensions follow Dzuy and Boger (1985) recommendations. The vane penetration depth for tests is done at a depth of 70 mm for the top of the vane to penetrate a minimum depth equal to the height of the vane, see ASTM D4648-16 (2024).

EARLY RESULTS

Kaolin samples were used to commission the Trivane. The Kaolin used has a liquid limit of 62 per cent and a plasticity index of 21 per cent. Specific gravity is 2.61 and the clay content (<2 µm) is 20 per cent. Commissioning Trivane tests were generally conducted at a rotation rate of 6°/min. Simple shear and direct shear testing was performed on the Kaolin at various confinement stresses of between ~20 kPa and 100 kPa. The results in Figure 3 include a comparison of peak shear strength measured with the various apparatuses during commissioning. The Trivane results are seen to compare very well with both simple shear and direct shear test results. Power law strength envelopes were developed for the Trivane, simple shear and direct shear test results. R^2 vales showed excellent fitting curve results with minor differences associated with shearing mechanisms. Details of the commissioning campaign are documented by van Rhyn (2023).

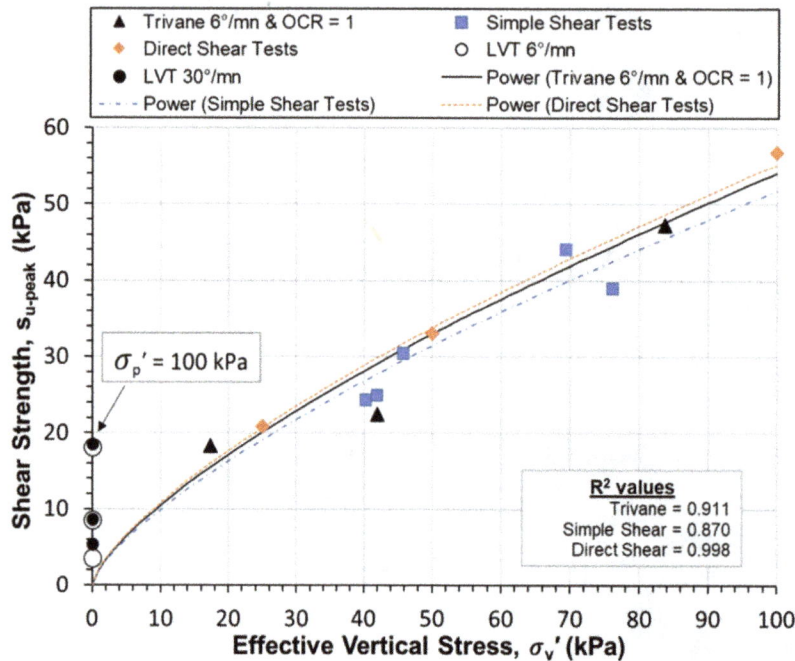

FIG 3 – Trivane test results compared with test results using direct shear, simple shear test and LVT. All tests were completed on Kaolin samples.

The relationship between the Over Consolidation Ratio (OCR) and the shear strength ratio measured in Kaolin samples is presented in Figure 4. A strong correspondence was found between test results and the relationship proposed by Chandler (1988). The trendline presented from Chandler was obtained from FVT performed on clays at various sites.

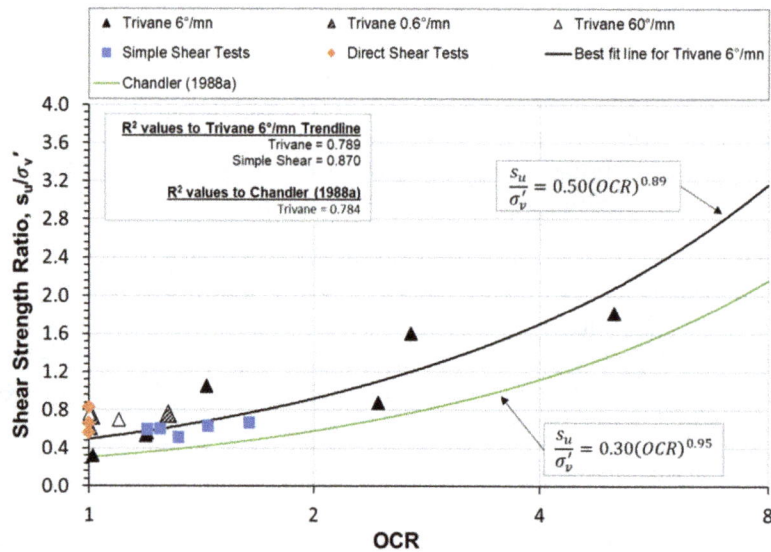

FIG 4 – Effective stress ratio versus OCR on Kaolin samples.

The Trivane has been used successfully in two tailings projects, see Figures 5 and 6. The first project included Bauxite tailings (Figure 5). For the Bauxite tailings various FVT were undertaken at various depths and locations using two rotation rates (60°/min and 120°/min). Samples from the Bauxite tailings were recovered from an adjacent area where the FVT was completed and sent to the laboratory for subsequent VST testing. Samples were remoulded at the laboratory. LVT and Trivane tests were completed. The Trivane sample was isotopically confined at 100 kPa. The FVT were expected to be completed at confinement stresses between 100 kPa and 150 kPa nominally. The preliminary investigation into the correspondence between the Trivane tests and the FVT showed agreement, and minor differences are attributed to differences in the rotation rate and the fact that not the same sample was tested using both techniques. The Trivane sample was isotropically consolidated while the field test was undertaken at a location where the stress state would be more representative of a k_0 stress condition – this key difference further explains differences in peak shear strengths estimated. A comparison between LVT, FVT and Trivane results revealed that LVT testing is inadequate for strength tailings characterisation at depths different than surface level.

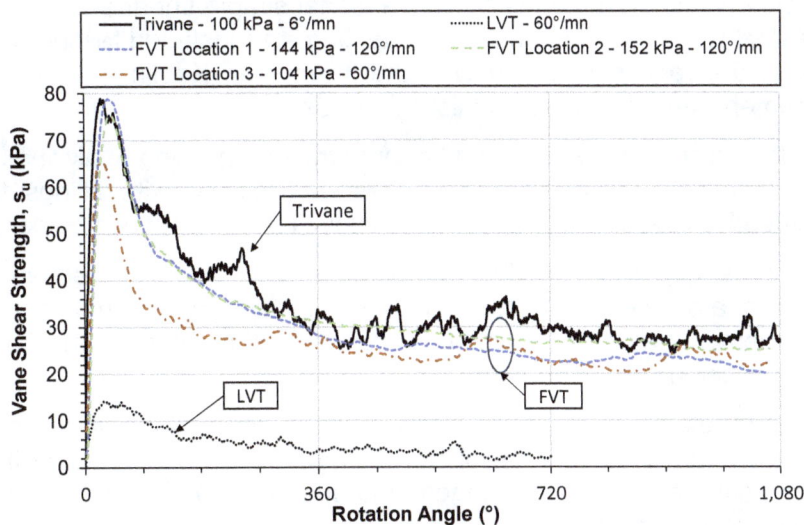

FIG 5 – Comparison between field and Trivane results in bauxite tailings.

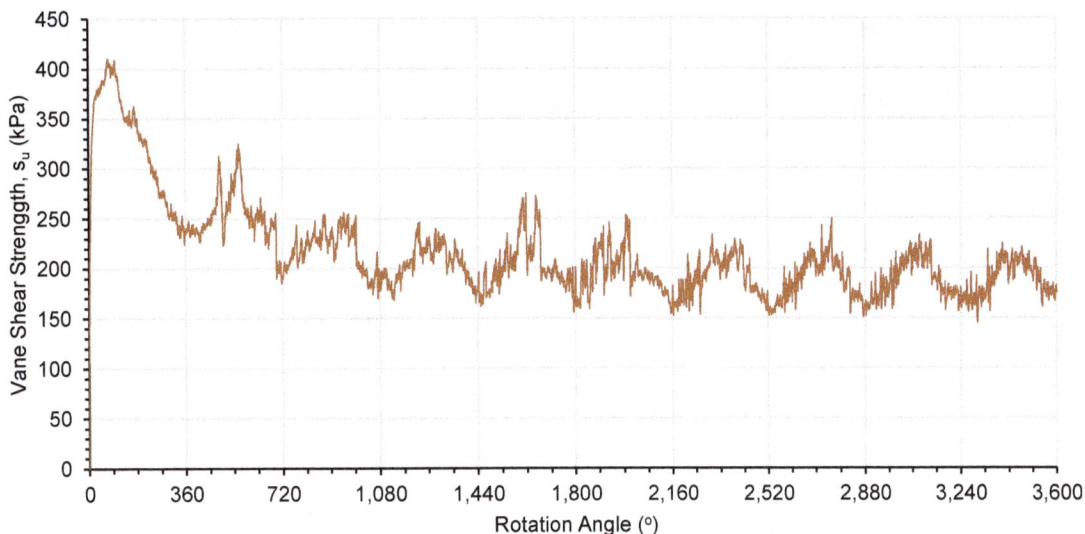

FIG 6 – Trivane results in Rare Earth Element tailings.

Figure 6 shows the Trivane results of a Rare Earth Element tailings at a confinement stress of 175 kPa. The test was undertaken to inform a design in the pre-feasibility study stage of a greenfield site, where tailings were not yet deposited for FVT. Trivane testing allowed the identification of peak and residual strengths for design. Previous conventional triaxial testing in the Rare Earth Element tailings did not allow the determination of residual strengths. The Rare Earth Element testing results indicated semi-typical anomalies in Trivane data, particularly the oscillations around the residual value. The anomalies and methods to address them are well documented in FVT by Buttling and Burges (2023).

CONCLUSIONS

The Global Industry Standard on Tailings Management (GISTM; GTR, 2020) requires that brittle failure modes be identified and addressed with conservative criteria. In ICOLD words, *there is currently very little guidance in the technical literature* [to guide brittle behaviour identification]. ICOLD also provided stress-strain curves of ductile and brittle materials. This paper explores the development and early work of a laboratory testing technique that aims to provide insights into tailings' brittle behaviour.

A device that combines triaxial testing with vane shear testing is presented here. Early test results include a commissioning campaign using well-behaved Kaolin. The commissioning stage demonstrated adequacy of the Trivane to determine peak strengths in Kaolin when compared with tests in the same material when undertaking conventional simple shear and direct shear testing at various confinement stresses. The commissioning testing also included testing at various OCRs. A comparison between the results obtained with the Trivane in Kaolin prepared at various OCR showed good agreement with historically published trends.

The potential to capture a similar degree of brittleness when comparing peak and residual strengths in field and laboratory settings was demonstrated. Test results in Bauxite tailings demonstrated how the Trivane can be used to breach the gap identified by ICOLD in relation to brittleness identification.

The Trivane has been successfully used to characterise a Rare Earth Element tailings to inform strength parameters for a pre-feasibility study stage, where tailings were not yet deposited for FVT. The last application demonstrates the potential to complement methods to predict undrained shear strength for future dam construction stages.

More testing is needed to standardise results. For example, the development of standard procedures are needed to inform: i) the determination of rod friction, ii) how conventional triaxial testing and Trivane compare for brittle tailings, iii) the range of material applicability and iv) determination of the optimum rate of rotation prior to shearing.

REFERENCES

ASTM International, 2018. ASTM D2573/D2573M-16: Standard test method for field vane shear test in saturated fine-grained soils, West Conshohocken, ASTM International.

ASTM International, 2024. ASTM D4648M-10: Standard test method for laboratory miniature vane shear test for saturated fine-grained soil, West Conshohocken, ASTM International.

ASTM International, 2024. ASTM D4648M-16: Standard test method for laboratory miniature vane shear test for saturated fine-grained clayey soil, West Conshohocken, ASTM International.

Bishop, A W, 1967. Progressive failure-with special reference to the mechanism causing it, in *Proc Geotech Conf*, 2:142–150).

Buttling, S and Burgess, J, 2023. Anomalies in field vane shear testing data, in Proceedings of the 14th Australia and New Zealand Conference on Geomechanics, Cairns 2023 (ANZ2023).

Chandler, R J, 1988. The in-situ measurement of the undrained shear strength of clays using the field vane, in *Vane shear strength testing in soils: field and laboratory studies* (ed: A F Richards), 1014:13–44.

Dzuy, N Q and Boger, D V, 1985. Direct yield stress measurement with the vane method, *Journal of Rheology*, 29(3):335–347.

Global Tailings Review (GTR), 2020. Global Industry Standard on Tailings Management (GISTM) [online], Global Tailings Review. Available from: <https://globaltailingsreview.org/global-industry-standard/>

Harvey, J W and Hogan, A, 2023. Establishing a site-specific standard of practice for field vane shear testing in mine tailings, in *Proceedings of Tailings and Mine Waste 2023* (ed: R Kuitunen), (University of British Columbia). Available from: <https://open.library.ubc.ca/media/stream/pdf/59368/1.0438115/3> [Accessed: 5 May 2025].

International Committee on Large Dams (ICOLD), 2022 Bulletin No. 194 Tailings Dam Safety.

Mitchell, J K and Soga, K, 2005. *Fundamentals of soil behavior*, vol 3:558 (New York: John Wiley and Sons).

Richards, A F (ed), 1988. *Vane shear strength testing in soils: Field and laboratory studies,* vol 1014 (ASTM International).

van Rhyn, W, 2023. Trivane test – improving correspondence between lab and field vane tests, Curtin University.

Paradigm shifts in geotechnical *in situ* testing equipment and methods – to make them 'work' in tailings materials: especially in very soft stuff

A J McConnell[1] and M K D Chapman[2]

1. Founder, Insitu Geotech Services Pty Ltd (IGS), Brisbane Qld 4014. Email: allan@insitu.com.au
2. Managing Director, Insitu Geotech Services Pty Ltd (IGS), Brisbane Qld 4014. Email: mark@insitu.com.au

ABSTRACT

Most *in situ* testing tools and methods evolved originally for use in testing natural soils. But tailings are not natural soils. The authors will highlight, explain and discuss significant shifts in equipment and test methodology that have taken place (or are taking place) for three of the most popular *in situ* tests used to characterise tailings materials: (1) The most-used test, the CPT, evolved in the 1950s to 1990s, originally in Holland and is the subject of detailed, in places very rigorous, international standards; (2) The *in situ* Vane Shear Test, deemed by some geotechnical practitioners as the basic, almost picture-perfect way to determine undrained shear strength of cohesive soils, appeared in geotechnical practice in 1948 and has changed little since; it is also governed by international standards; and (3) the Flat Plate Dilatometer – the DMT – that first appeared in 1977 and evolved into a meaningful tool by 1980; in its basic form it is described in a 2001 International Society for Soil Mechanics and Foundation Engineers (ISSMGE) Committee TC16 Report; and it is covered by international standards. These three tests, the equipment and the test methodologies, have evolved significantly in the past decade, and are still evolving, as they are being adapted to give better data for use in tailings characterisation. In some cases this involves significant design, calibration, or test-method paradigm shifts. Certainly, simply, or even rigorously, 'following the standards' does in many instances lead to useless or near-useless data. This can cause problems – old practices are hard to break – what is the proper practice? – much of this is not written in the standards. Much of this is a work in progress.

INTRODUCTION AND EXPLANATION

This paper is primarily about tailings materials that have consistency and shear strength in the range very soft (VS) to soft (S) as defined by AS 1726–2017 Table 11 shown and highlighted in Figure 1.

CONSISTENCY TERMS FOR COHESIVE SOILS

Consistency	Field guide to consistency	Indicative undrained shear strength kPa
Very Soft (VS)	Exudes between the fingers when squeezed in hand	≤12
Soft (S)	Can be moulded by light finger pressure	>12 and ≤25
Firm (F)	Can be moulded by strong finger pressure	>25 and ≤50
Stiff (St)	Cannot be moulded by fingers	>50 and ≤100
Very Stiff (VSt)	Can be indented by thumb nail	>100 and ≤200
Hard (H)	Can be indented with difficulty by thumb nail	>200
Friable (Fr)	Can be easily crumbled or broken into small pieces by hand	—

FIG 1 – Table 11 from AS 1726–2017 – authors' highlighting.

In fact some discussion will be about the lower half of the Very Soft range, which for expediency here in this paper will sometimes be termed 'ooze' – ie material with shear strength of (say) 6 kPa or less. While there are deposits of natural soils with strength in this ooze range, they are not as common as with materials stored in Tailings Storage Facilities (TSFs), where their strength and other properties are typically critical to design and planning.

These very low strengths evolve from the relative very short history of the deposits; if you are lucky, materials may be normally consolidated; often they are under-consolidated.

CONE PENETRATION TESTS (CPTs)

CPT is the most used geotechnical test for characterisation of TSF materials, because:

- The test has a long history of success in natural soft and very soft (and much stiffer) soils.

- The test is relatively rapid, allowing larger numbers of tests at lower cost than other options.

- The test is semi-continuous, typically with readings every 10 mm or 20 mm of depth, making it excellent for stratum recognition and property profiling.

- There are many correlations in the literature to help with interpreting the data (but it relies heavily on such published correlations); note that most of these are for natural soils.

- CPT has become habit because of convenience and familiarity.

So what are the problems with CPTs in testing tailings ooze?

There are two significant issues, both related to the low strength of these materials:

1. Calibration and sensitivity issues, which impact all parameters measured, chiefly q_c and f_s.

2. A design issue reducing reliability/repeatability of very low sleeve friction (f_s) measurements.

Calibration issues

Most CPT cones are not sensitive enough by manufacture nor calibrated in the very low ranges required to reliably profile these very soft materials. This problem exists in the ooze being focused on here but in fact extends above that range into the realms of very soft and soft stuff. A nice illustration of this, with notes by the authors, comes from research by Lunne *et al* (2018), Figure 2.

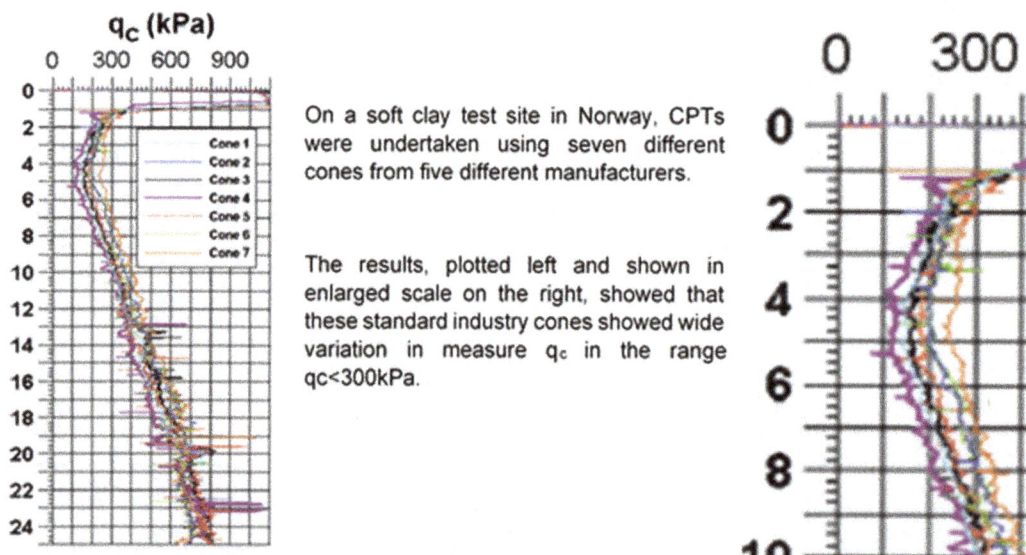

On a soft clay test site in Norway, CPTs were undertaken using seven different cones from five different manufacturers.

The results, plotted left and shown in enlarged scale on the right, showed that these standard industry cones showed wide variation in measure q_c in the range $q_c < 300$ kPa.

FIG 2 – Norwegian test site – showing normal CPTs' inability to properly measure q_c at very low values.

Something of great consequence is that CPT cone manufacturers do not themselves typically (properly or at all) calibrate this very low part of the test range. Hence, if they care about it, it is up to CPT Operators to do this themselves. This is despite the two most recent ISO CPT standards (dated 2012 and 2022) clearly both specifying such calibration, see Figures 3 and 4.

ISO 22476-1:2012(E)

A.2.2 Calibration of cone resistance and sleeve friction

Incrementally axially loading and unloading the cone and the friction sleeve calibrate the cone resistance and sleeve friction. The calibrations of cone resistance and sleeve friction can be carried out separately, but the other sensors should be checked individually to ensure that the applied load does not influence them. The calibration is carried out for various measuring ranges, with special emphasis on those ranges relevant for the forthcoming tests.

FIG 3 – Excerpt from ISO 22476–1:2012 (authors' underlining).

ISO 22476-1:2022(E)

B.2.2 Measuring intervals for calibration

The measuring intervals for calibration of the sensors of the cone penetrometer should be selected to cover the measuring intervals of interest. The measuring intervals should include zero load. Considerations for the measuring intervals of interest include expected ground conditions and operational setting.

FIG 4 – Excerpt from ISO 22476–1:2022 (authors' underlining).

The compression cone design issue that spoils tiny f$_s$ values

Figure 5 shows a snap-shot explanation of the design issue discussed under this heading. This is discussed in much more detail in McConnell and Wassenaar (2022).

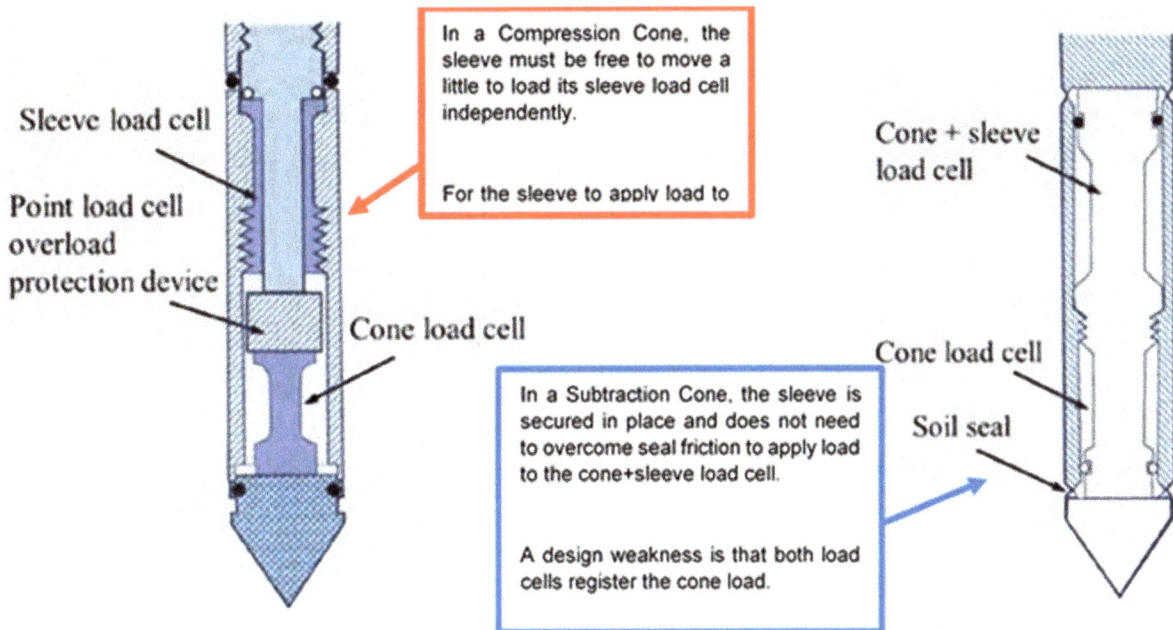

Sleeve load cell

Point load cell overload protection device

In a Compression Cone, the sleeve must be free to move a little to load its sleeve load cell independently.

For the sleeve to apply load to

Cone load cell

In a Subtraction Cone, the sleeve is secured in place and does not need to overcome seal friction to apply load to the cone+sleeve load cell.

A design weakness is that both load cells register the cone load.

Cone + sleeve load cell

Cone load cell

Soil seal

FIG 5 – CPT seal friction explanation. From Robertson and Cabal (2024); modified and annotated by the authors.

Referring to the captions in Figure 5. This matter was discussed in the paper by McConnell and Wassenaar (2022). That paper describes one successful solution to this problem; now available to all.

Referring back to the research and paper by Lunne *et al* (2018), mentioned previously, another figure from that paper is shown as Figure 6; with accompanying notes by the authors.

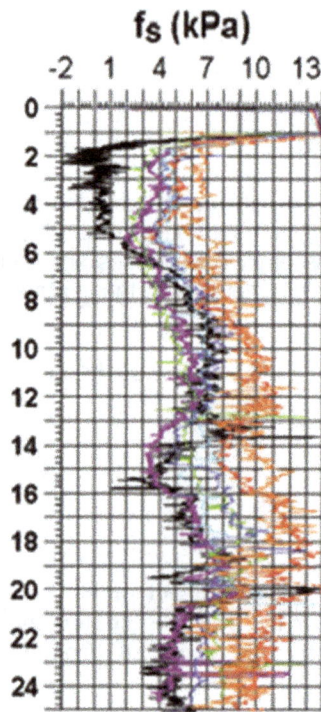

On a soft clay site in Norway, CPTs were undertaken using seven different cones from five different manufacturers. The results show significant variation in attempts to measure f_s in the lowest ranges. Obviously one or all were incorrect.

A very wide variation was reported, and one cone measured negative values for the top few metres of penetration.

FIG 6 – Norwegian test site – showing inability of a normal CPT to reliably measure f_s at very low values.

The issue of friction from sleeve seals was discussed in a paper by Santos, Barwise and Alexander (2014). That paper explained experiments they had made trialling calibrations of sleeves of 'everyday' compression cones, both with and without the dirt seals in place. Figure 7 is from that paper and clearly show the inhibiting effect of seal friction on measurement of f_s values.

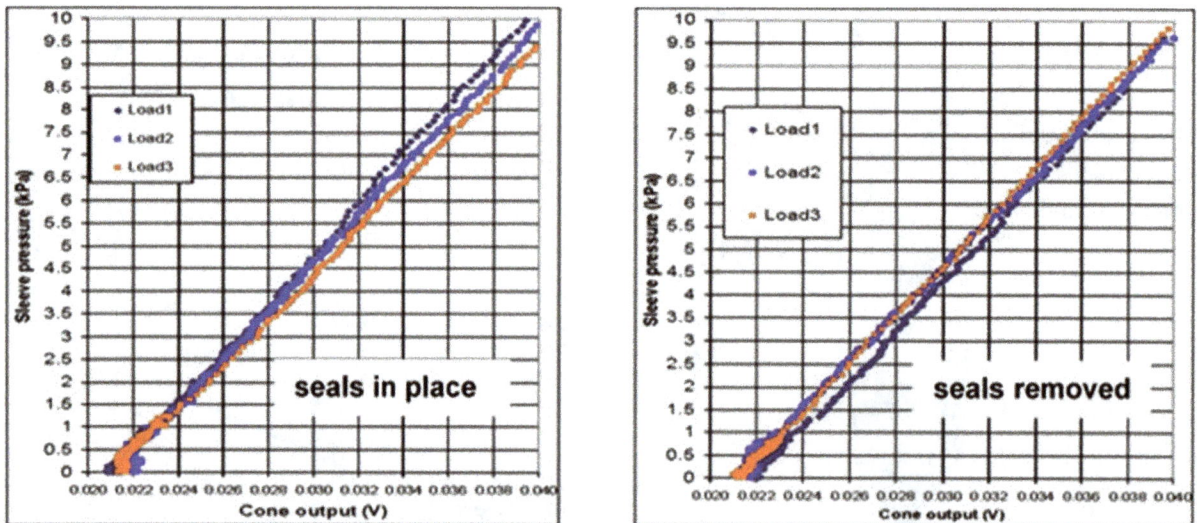

FIG 7 – Figure from Santos, Barwise and Alexander (2014) showing the influence of seal friction.

Shown below is a typical example of the inability of a normal compression cone to measure f_s values in very soft tailings. Cone resistance q_c was clearly measurable in this example, but significant zones of $f_s=0$ were indicated. Note that this meant, on this occasion, that Soil Behaviour Type – SBT (Robertson, 2016) – for these zones could not be interpreted, as shown in Figure 8.

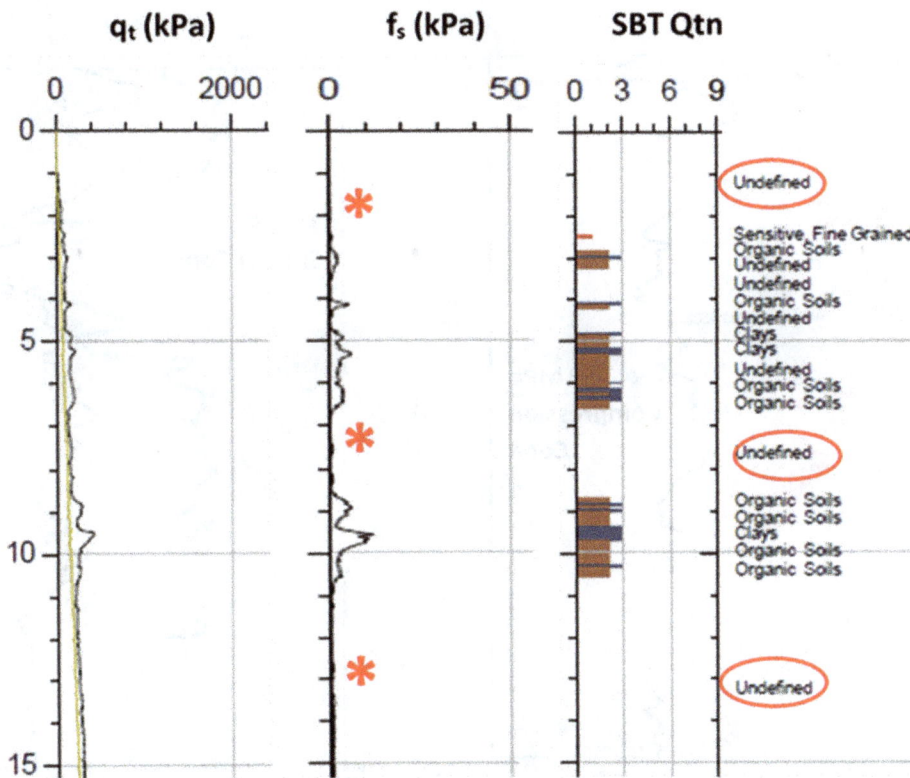

FIG 8 – Confidential – a typical compression cone test in very soft ooze-like tailings.

In the CPT profession it is quite common to simply accept such problematic data – *'because it can't be bettered – it's normal'*. This is an unhappy paradigm that has now been shifted, at least for some.

Solutions – *to make CPTs useful in ooze (especially in TSFs)*

A paradigm-shifting CPT has been described in McConnell and Wassenaar (2022). Referring to annotated Figure 5, this new cone's design, developed by the authors and a manufacturer in The Netherlands (Geomil), relies on the fact that subtraction cones do not develop friction in dirt seals behind the friction sleeve, as the sleeve is not free to move. A subtraction cone with more compressible materials in the cone body, and very sensitive load cells, has been developed, and is now available commercially in both 3 MPa and 10 MPa q_c capacities. Examples of the use of this cone type compared to normal compression cones are shown in the following, Figures 9 and 10. This cone design can now be used in tailings-ooze, with consistent and repeatable f_s results.

Note that calibration of this cone type is absolutely critical, as in a subtraction cone (refer Figure 5) both load cells must be near-perfectly matched; even small differences will lead to a major shift in f_s and consequently in determination of SBT. It's not just a design issue, it is a 'life of cone' matter.

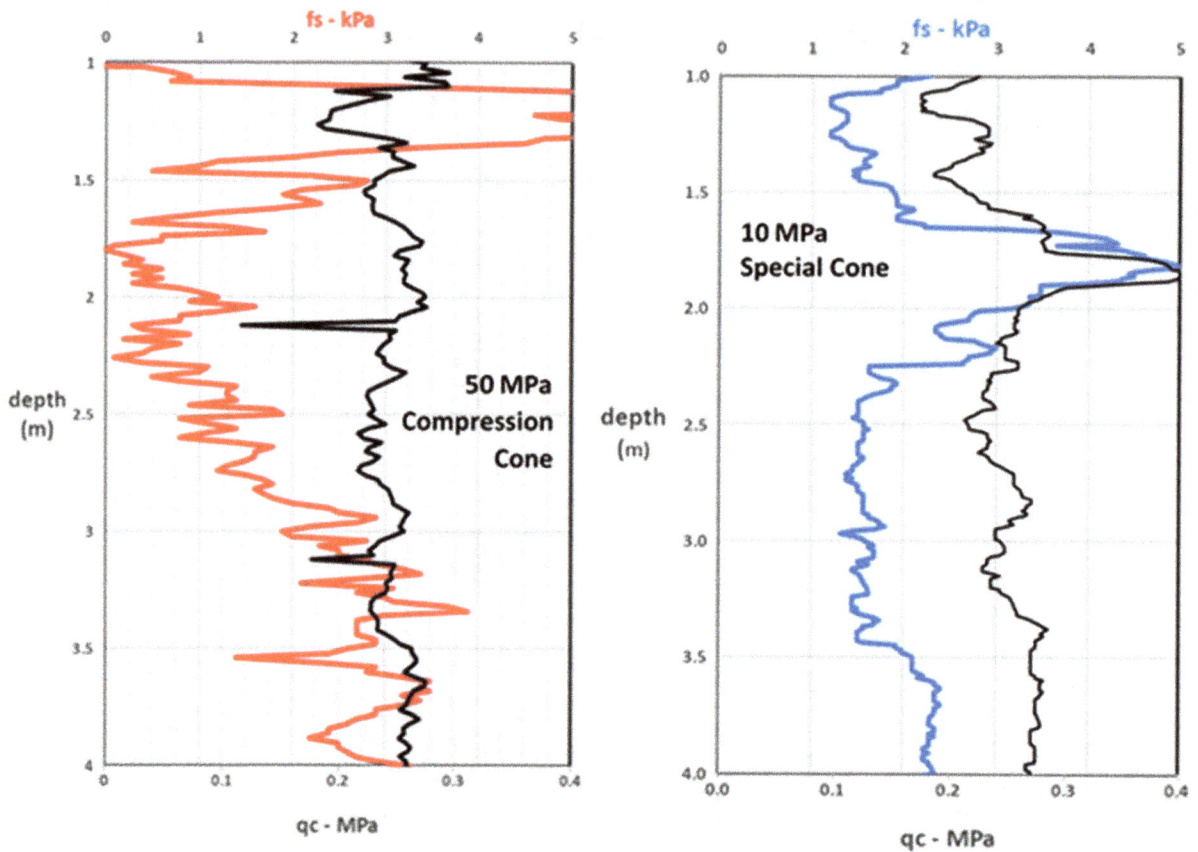

FIG 9 – Confidential – side-by side tests in a red mud dam.

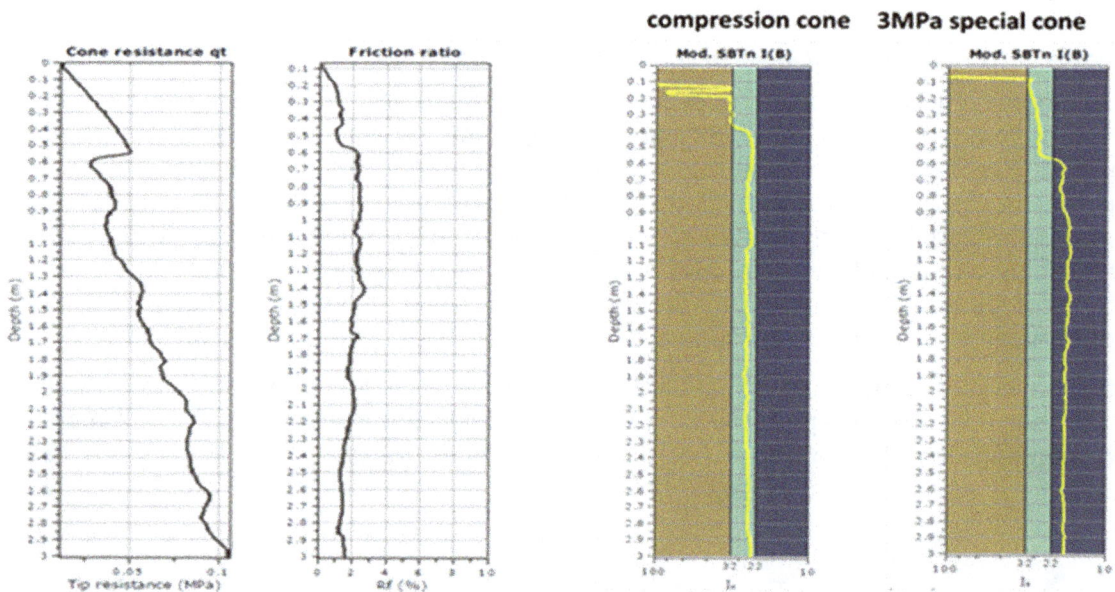

FIG 10 – Confidential – side-by side tests in a TSF – showing SBT transition.

It is readily observable in Figure 9 that f_s measurements in the left hand figure (in red) bear no relationship to the low q_c values being measured (black). Whereas in the right hand figure, in blue f_s versus black q_c, a clear-cut relationship can be seen, such as should be expected. These tailings are very soft, with q_c values measured of around 250 kPa. This better f_s data leads to quite different interpretations of Soil Behaviour Type (SBT). In Figure 10 a critical SBT shift is seen.

VANE SHEAR TESTS (VST)

The general principles of direct-push *in situ* vane shear testing is shown in the two parts of Figure 11. The vane is pushed to depth as per the right-hand part of the figure; then it is rotated by rods inside casing, with torque measured, as per the left hand part of the figure.

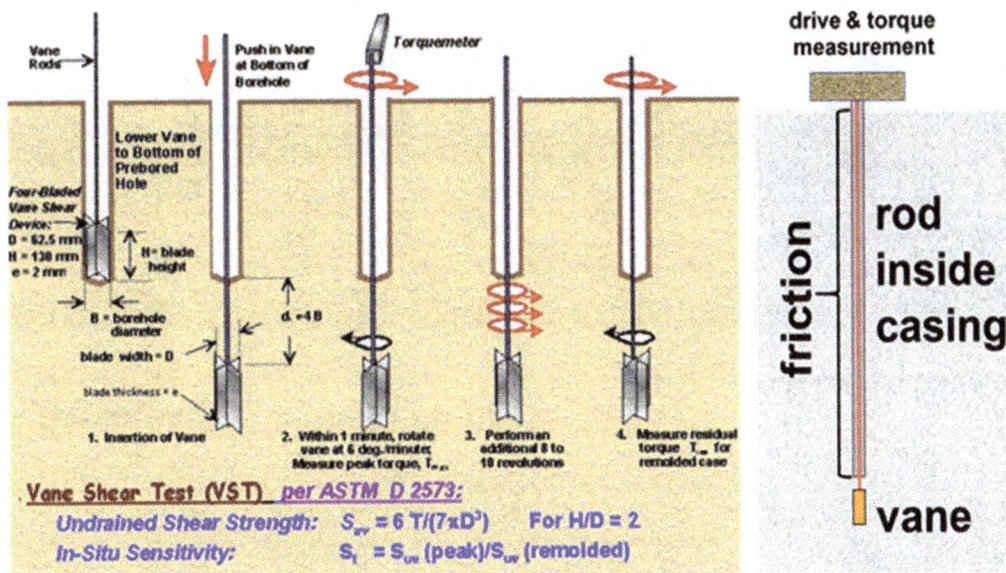

FIG 11 – Principles of direct-push vane (LHS after Mayne, 2013) – RHS prepared for this paper.

Historically: (a) the test has been limited to testing of clay materials; (b) rotation and torque measurements have been made by devices above ground; (c) rotation speeds have been directed by standards to be very slow (usually 6° per min); (d) differing vane sizes are chosen to suit the material being tested, ranging typically from D × H (mm) of 50 × 100 to 100 × 200.

So what are the differences with VSTs in testing tailings ooze?

There are two significant issues, both related to the nature of these materials; discussed in the following: (a) the effect of even small rod friction on results; (b) testing in partially draining materials.

Errors associated with even very small rod friction

Rotation torque required to overcome friction between the rods inside casing, while possibly being small, can be significant compared to the torque required to actually turn the vane itself to shear very soft ooze. Standards specify, and common-sense dictates, that corrections must be made to allow for this friction. But these corrections are not 100 per cent certain or uniform, and uncertainty in them can swamp the test data, making that data overall uncertain.

This problem can be overcome by the use of an alternative vane shear test drive system that eliminates any need for friction corrections. In recent years equipment has been developed commercially by (at least) one company in The Netherlands (A P van den Berg) that makes this possible, as shown in Figure 12.

Using such equipment, vane shear tests can be undertaken that will consistently and repeatably measure *in situ* shear strengths in the very soft range, including materials with shear strength <5 kPa. Consistent and repeatable measurement in such conditions was pretty-much unheard of until this equipment was developed. The concept and method are now industry best practice.

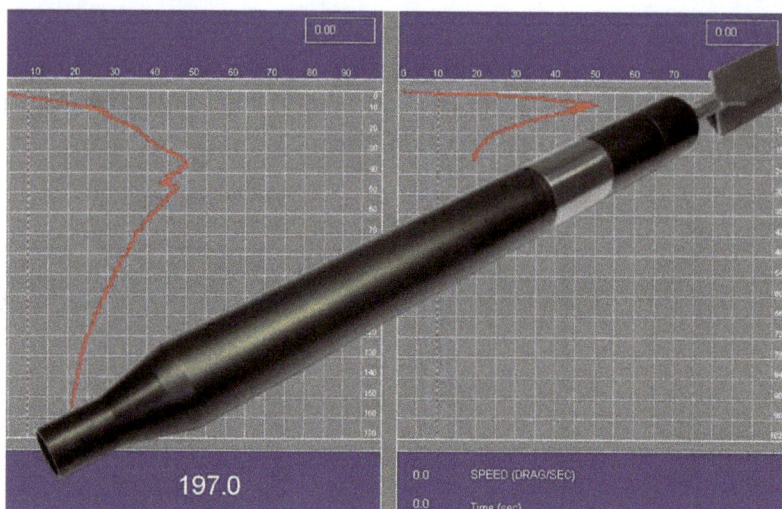

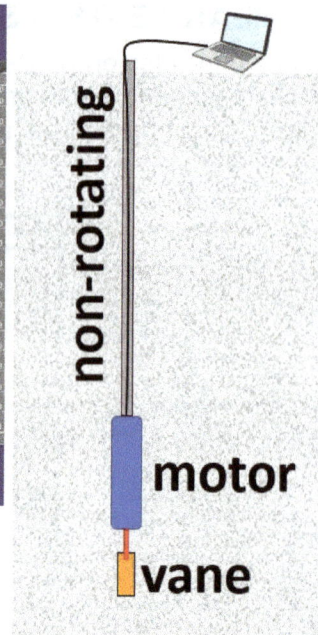

FIG 12 – LHS the equipment (from A P van den Berg website <https://www.apvandenberg.com/>) – RHS prepared for this paper.

The two test plots in Figure 13 are from different Australian TSFs tested using the system described above.

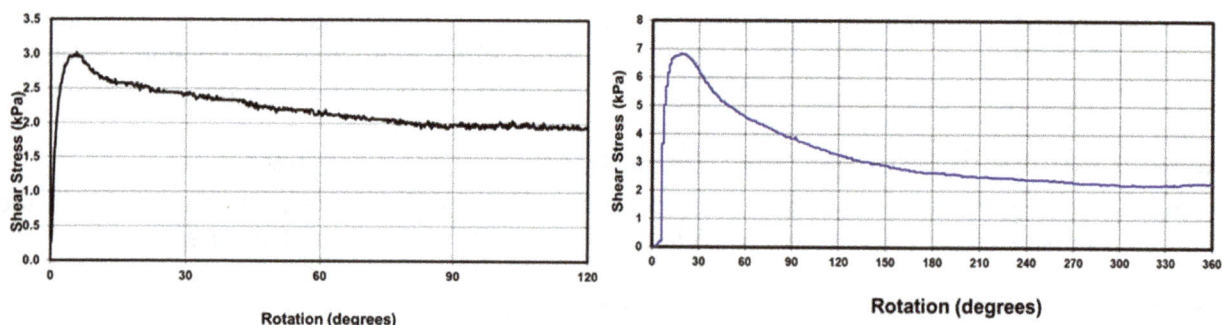

FIG 13 – Two tests on TSFs using the equipment discussed above.

Testing in partially draining materials – 'the times they are a changing'

Vane shear tests have been traditionally limited to testing clay-type soils. Test rotation speeds have been dictated by standards, and, to some extent, by custom. Almost all tests done 'properly' have been rotated at 6° per min or similar. Because the tests were made in clay soils the results have always been deemed to be of undrained shear strength, S_u.

Tailings engineers have to deal with materials that are very often faster draining than natural clay soils and, as a consequence, vane shear tests made in these materials at the traditional slow rotation speed have yielded drained or partially drained shear strength. Sometimes quite a mish-mash of data has arisen that seems to fit no proper geotechnical pattern.

Tailings engineers want data, and typically they are seeking undrained strength of their faster-draining ooze-like materials. Many have recently challenged the paradigm of the very slow test rotation speeds and have experimented with, and taken into reasonably common practice, significantly higher rotation speeds; and results have been good. Basically the idea is simple – test at a higher rotation speed so that drainage does not occur during the test.

Because such fast testing is not according to standards, this is a decision made by the tailings engineer and is typically decided on-site, by experimenting with different speeds. Fortunately the latest modern testing equipment can accommodate this. The equipment described previously can test up to 360° per min, some 60 times faster than dictated by standards. The test in Figure 14 was undertaken in an Australian TSF by rotating at 360° per min. This means one full 360° rotation takes one min, instead of the previous one hr required.

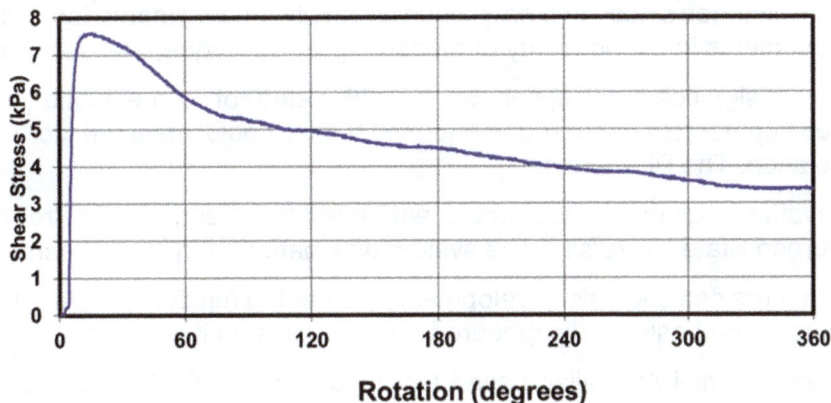

FIG 14 – Vane Shear Test on a TSF with rotation speed 360° per min.

This is all pretty new stuff, it is not covered by any existing standards, and ideas and theories can vary from tailings engineer to tailings engineer. In other words it is 'a work in progress'.

Notwithstanding the 'work in progress' nature of this, A P van den Berg (at least) has developed equipment that permits testing at even much higher rotation speeds than discussed above.

The very new equipment (Figure 15) can test at rotation speeds ranging from 60°–3600° per min. At its highest speed this is some 600 times faster than the standards.

FIG 15 – The 'latest' VST equipment – designed to rotate at from 60° to 3600° per min.

Something old is new again – the Piezovane

The vane shear test principles and the device itself were created by Skempton around 1948 and published in *Geotechnique*. They have in their customary form changed very little in 76 years.

The most significant changes since 1948 have been relatively recent automation of the blade rotation system, and improvements in methodology of measuring blade-torque, as described previously.

In 1987 however, a significant exception to this 76 years of non-evolution was reported in *Geotechnique*; development of vane shear equipment that could measure pore pressure during insertion and/or rotation. The Piezovane.

The new and innovative equipment and process emanated from research by Prof Wayne A Charlie and others at Colorado State University. The system was patented by the researchers in 1992.

But, as far as the authors can see, this development was not too much appreciated; the 'Piezovane', evolved, got patented, and basically disappeared from practice until recently.

About 2–3 years ago a small Australian-based testing contractor, CPTS, developed a piezo-vane device, and very recently another contractor, Conetec, has developed one (<https://www.conetec.com/services/in-situ-testing/piezovane>).

The former, referenced in a 2021 paper by Wentzinger and Keulemans, and the latter, announced very recently, have pore pressure access to a transducer through holes through the edges of the blades, very similar to the 1992-patented device shown in Figure 16.

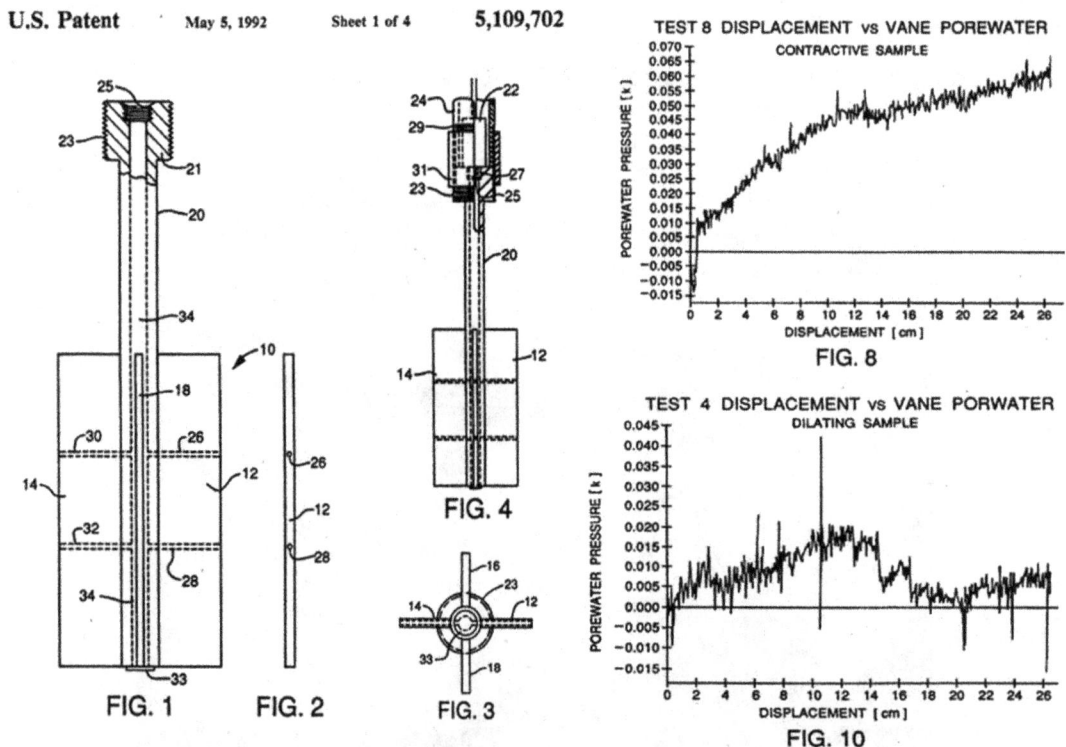

FIG 16 – Selected figures from the 1992 patent documentation (from Charlie and Butler, 1992).

It seems that this old/new technique in one form or another has potential to evolve into a functional tool, though at the time of writing it has certainly not yet become part of routine geotechnical practice.

But 'watch this space'.

FLAT PLATE DILATOMETER TESTING (DMT) AND THE MEDUSA DMT REVOLUTION

The original mechanical DMT – still in use today

The Flat Plate Dilatometer (DMT) was introduced into site characterisation by (the late) Dr Silvano Marchetti in 1977 (Marchetti, 1980). The equipment and test technique are both deceptively simple-looking, but via correlations and algorithms a range of important engineering parameters can be determined.

For readers who are not familiar with this test, it is described in detail in a 2001 ISSMGE TC16 Report; and it is covered by international standards. The general principles are shown in Figure 17, taken from the TC16 Report. A ready way to view the TC16 Report is via <https://www.marchetti-dmt.it/>. This an excellent website that provides much background and research data, and also provides detailed information on the correlations and algorithms involved in the standard test's interpretation.

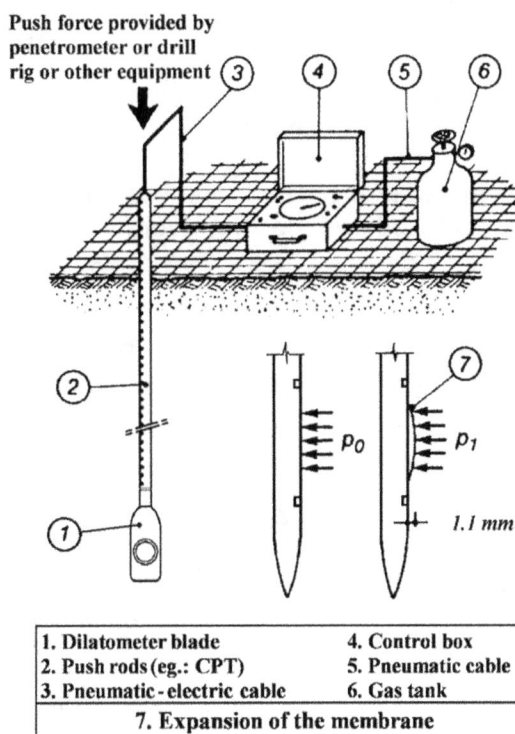

1. Dilatometer blade	4. Control box
2. Push rods (eg.: CPT)	5. Pneumatic cable
3. Pneumatic - electric cable	6. Gas tank
7. Expansion of the membrane	

FIG 17 – Principles of the original Flat Plate Dilatometer.

The authors' company has been using DMT in its original form, with several evolving improvements, since about 2004. Over that time it has become the second-most-common *in situ* test by the company, after CPT.

However, the original DMT is: (a) essentially manual-operator-dependant; (b) subject to potential errors when testing at greater depths, due to gas flow and pressure-transmission delays; (c) not really suited to testing tailings ooze on a day-to-day basis, due to sensitivity and repeatability issues.

The latest development – the Medusa DMT – now well-suited to tailings

The latest evolution of the equipment, ie the Medusa DMT, first released by Marchetti in 2018, is seen by some as one of the most important developments in *in situ* geotechnical testing. Certainly now DMT testing is suited to testing in tailings ooze, it has been used for such on several recent occasions.

The Medusa DMT was developed to basically overcome the weaknesses listed as just discussed, and as such it is another paradigm shift in tailings ooze testing. The Medusa's principles and a typical 'basic' result plot from a tailings project are shown in Figure 18.

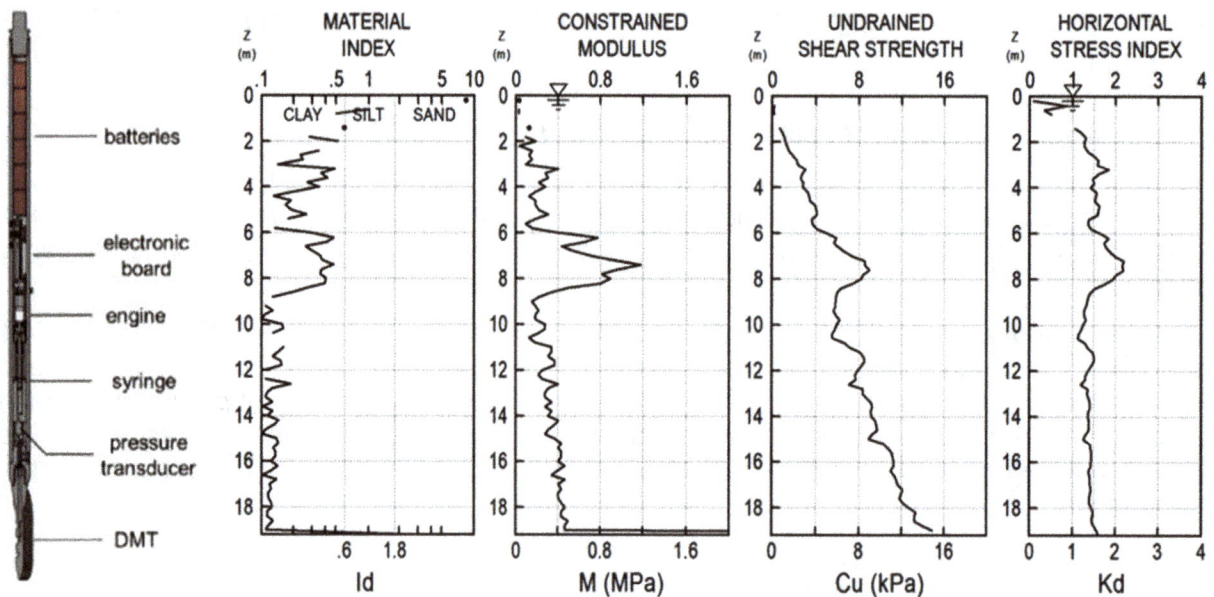

FIG 18 – LHS The Medusa DMT principles – RHS is a typical 'basic' output from a test in tailings ooze.

The Medusa DMT is a fully automated probe. The probe has an onboard electronic control system, rechargeable batteries and a motorised syringe that hydraulically controls the expansion of the membrane. The blade has the same dimensions as the standard DMT. Connected to a computer, real time results can be viewed by the operator.

This new and improved technology not only provides the parameters derived from the standard DMT test but also offers some considerable advantages over the normal mechanical device:

- Almost completely removes any operator influence/bias over the test data. It is very much more responsive/repeatable in soft materials.

- Due to the electronic control of the hydraulic inflation of the DMT membrane the Medusa enables short or long duration horizontal stress dissipation tests automatically and autonomously. Dissipation tests were theoretically possible but labour-intensive using the original mechanical DMT equipment. These are now essentially automated and thus able to be made 'normal practice' using the Medusa DMT.

- Control of the test and assessment of the dissipation test data allows evaluation of whether test results are undrained, drained or partially drained. Understanding this data can help better determine penetration speeds and or rotation speeds for other testing such as CPT and vane shear tests.

The Medusa DMT is a very powerful tool, it provides a greater understanding of *in situ* test data and therefore *in situ* conditions than has previously been possible.

While the technology is now stable and robust, the uses of the Medusa DMT, beyond the well-established uses of the original DMT, are a work in progress. Much is being made of the device's ability to very quickly repeat what are known as 'A Readings', several times per second if needed, and over long periods, leading to short or long-term data on change of total horizontal stress, which equates to measurements of pore pressure dissipation. Kelly, Chapman and Chamberlain (2024) has compared dissipation tests results by Medusa DMT with conventional dissipation test results by CPT piezocone, and have achieved very comparable outcomes. On one occasion a problematic soil's permeability properties were able to be determined by Medusa DMT where they could not be determined by CPT piezocone.

Long-term tests are possible – a recent real example is shown in Figure 19 (from a TSF) that ran autonomously for 68 hrs – a test undertaken to measure pore pressure dissipation rate, and also projected to establish coefficient-of-earth-pressure-at-rest K_0 (the latter at research level with very encouraging results). Determination of K_0 is a valuable work in progress; the authors know no other site investigation test to establish this important parameter.

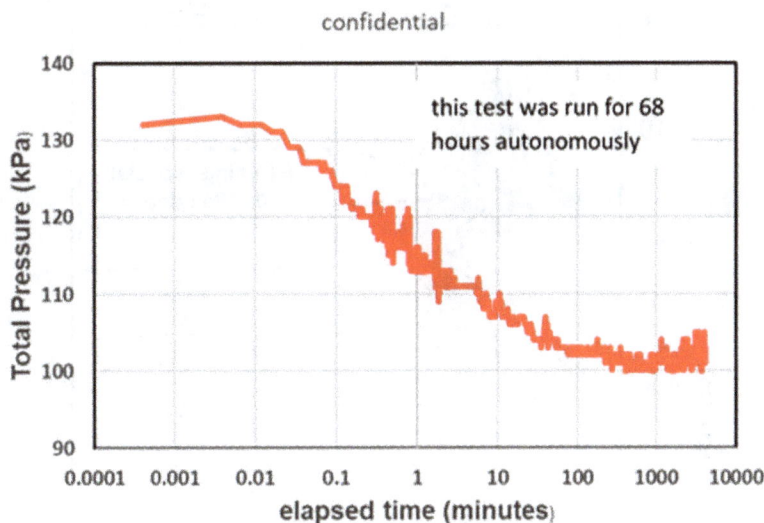

FIG 19 – Medusa DMT – Long-term 'Repeated A' dissipation test.

SUMMARY AND CONCLUSION – WHAT'S THE POINT OF ALL OF THIS?

Summary

Most *in situ* testing tools and methods evolved originally for use in testing natural soils. But tailings are not natural soils.

The authors have highlighted, explained and discussed significant shifts in equipment and test methodology that have taken place (or are taking place) for three of the most popular *in situ* tests used to characterise tailings materials:

1. The most-used test, the CPT.

2. The *in situ* vane shear test, deemed by some geotechnical practitioners as the basic, almost picture-perfect way to determine undrained shear strength.

3. The Flat Plate Dilatometer – the DMT – describing its latest iteration, the Medusa DMT.

While some of this is a work in progress, things are definitely changing, and any geotechnical practitioner who works in (or wants to work in) the tailings industry needs to come on board with these changes.

CONCLUSION

Figure 20 shows the sort of high quality data that can be achieved by taking account of the matters discussed here. This is not a lone example.

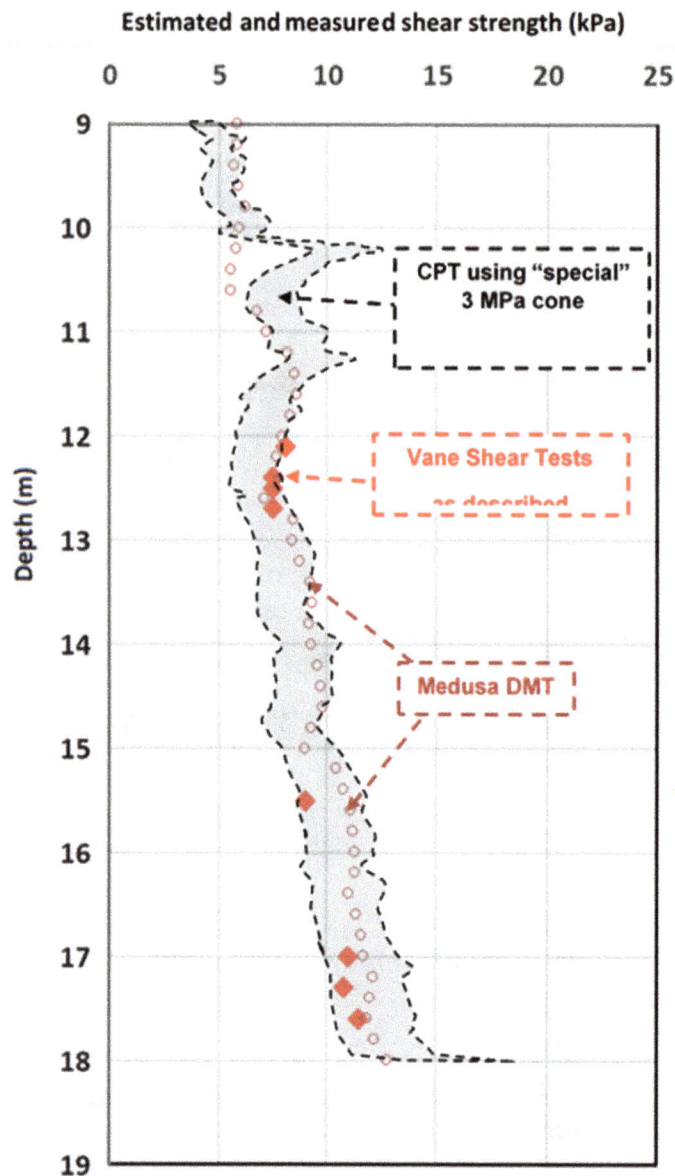

FIG 20 – Illustration of test data from a red mud dam – 'special' 3 MPa CPT, VST and Medusa DMT.

REFERENCES

Australian Standards (AS), 2017. AS 1726:2017 Geotechnical site investigations.

Charlie, W A and Butler, L W, 1992. Method for determining liquefaction potential of cohesionless soils, United States Patent Number 5,109,702

International Organization for Standardization (ISO), 2022. ISO 22476–1:2022 (and its predecessor dated 2012) Geotechnical investigation and testing — Field testing — Part 1: Electrical cone and piezocone penetration test.

International Society for Soil Mechanics and Foundation Engineers (ISSMGE), 2001. TC-16 report, The Flat Dilatometer Test in Soil Investigations.

Kelly, R, Chapman, M and Chamberlain, S, 2024. Interpretation of thermal and pore pressure dissipation tests, AGS Symposium, Brisbane.

Lunne, T, Strandwick, S, Kasin, K and L'Heurex, J S, 2018. Effect of cone penetrometer type on CPTU results at a soft clay test site in Norway CPT'18 Conference, Delft University of Technology, The Netherlands.

Marchetti, D, 2018. Dilatometer and Seismic Dilatometer Testing Offshore: Available Experience and New Developments, *Geotech Testing J,* 41(5):967–977.

Marchetti, S, 1980. In situ Tests by Flat Dilatometer, *J Geotech Eng Div*, 106(GT3):299–321.

Mayne, P W, 2013. Power Point Presentation at Geotechnical Seminar, University of Newcastle, NSW, Australia.

McConnell, A and Wassenaar, E, 2022. An innovative new 3MPa CPT – to detect and measure very small f_s values, in Fifth International Symposium on Cone Penetration Testing (CPT'22), Bologna, Italy, 8–10 June 2022.

Robertson, P K and Cabal, K L, 2024. Guide to Cone Penetration Testing for Geotechnical Engineering, 7th edn.

Robertson, P K, 2016. Cone penetration test (CPT)-based soil behaviour type (SBT) classification system – an update, *Canadian Geotechnical Journal*, 53(12).

Santos, R S, Barwise, A and Alexander, M, 2014. Improved CPT sleeve friction sensitivity in soft soils, Las Vegas, USA, in Third International Symposium on Cone Penetration Testing.

Skempton, A W, 1948. Vane tests in the alluvial plane of the River Forth near Grangemouth, *Geotechnique,* 1(2).

Wentzinger, B and Keulemans, Y, 2021. The effect of vane shear rotation speed on the estimation of tailings undrained strength, in Mine Waste and Tailings Conference, Brisbane.

The power of geometallurgical characterisation – how to turn legacy tailings into assets

L Nicholls[1], A Parbhakar-Fox[2], R Valenta[3] and E Wightman[4]

1. PhD Candidate, Sustainable Minerals Institute, University of Queensland, St Lucia Qld 4072. Email: l.nicholls@uq.edu.au
2. Associate Professor, Sustainable Minerals Institute, University of Queensland, St Lucia Qld 4072. Email: a.parbhakarfox@uq.edu.au
3. Institute Director, Sustainable Minerals Institute, University of Queensland, St Lucia Qld 4072. Email: r.valenta@uq.edu.au
4. Associate, Sustainable Minerals Institute, University of Queensland, St Lucia Qld 4072. Email: e.wightman@uq.edu.au

ABSTRACT

Copper mining is responsible for generating approximately 46 per cent of the world's tailings, amounting to nearly 3.7 Bt annually. Many critical minerals, including cobalt (Co), tellurium (Te), selenium (Se), molybdenum (Mo), rhenium (Re), bismuth (Bi), indium (In), and arsenic (As), as well as strategic elements such as sulfur (S), lead (Pb), silver (Ag), and zinc (Zn), are commonly associated with copper (Cu) mineralisation and subsequently remain in Cu tailings. Recovering these elements as secondary resources for the green energy transition has the potential to generate side-stream revenue, enhance portfolio diversification, and financially support site rehabilitation efforts. However, a fundamental challenge for the industry is determining the most effective methodologies for identifying, characterising, and prioritising mine waste repositories, based on their critical mineral content and reprocessing potential.

To address this challenge, a comprehensive multi-scale characterisation approach is necessary, utilising a systematic top-down methodology to optimise resource allocation and expenditure. This study presents a structured roadmap for waste characterisation, demonstrated through a case study on Cu tailings within the North-west Queensland Minerals Province. The approach integrates data collection across multiple scales—from repository-scale assessments down to mineral and elemental analysis—and evaluates findings in relation to reprocessing technologies. The workflow incorporates geophysics, geology, mineralogy, geochemistry, and geometallurgy, presented as a tailings characterisation toolbox. Specific techniques highlighted in this case study include a Loupe EM survey, geochemical analyses, facies classification, MLA mineralogical assessments, and geometallurgical test work. Key misconceptions about tailings composition and recoverable elements are addressed, highlighting overlooked value in mine waste. This research provides a structured methodology for assessing reprocessing potential and full-value mining opportunities, offering practical guidance for industry stakeholders aiming to maximise resource efficiency while supporting sustainable mining practices.

INTRODUCTION

Globally, an estimated 282.5 Bt of tailings are stored across ~3400 active tailings storage facilities (TSFs) and over 8500 legacy sites (ICMM, 2019). Historical mining practices, driven by market demand and processing inefficiencies, have resulted in the prioritisation of mining high-grade, marketable metals, leaving lower-grade material and metals that were previously considered less valuable in waste dumps and tailings dams (Lottermoser, 2010). These waste streams contain an estimated US$3.4 trillion worth of critical, strategic, and precious minerals (Minerals Research of Western Australia, 2023).

Mine waste valorisation is the process of waste reclamation for value recovery (Lottermoser, 2011; Hudson-Edwards and Dold, 2015). It presents a Circular Economy approach to transforming waste liabilities into assets, with the combined benefit of reducing the waste volume and environmental footprint (Bellenfant et al, 2013; Tayebi-Khorami et al, 2019; Kinnunen et al, 2022). Yet despite its economic and environmental benefits, mine waste reprocessing remains outside mainstream mining business models (Kinnunen and Kaksonen, 2019; Zinck *et al*, 2019). Key barriers perceived by the industry include technical, economic, environmental, policy, and social challenges (Figure 1).

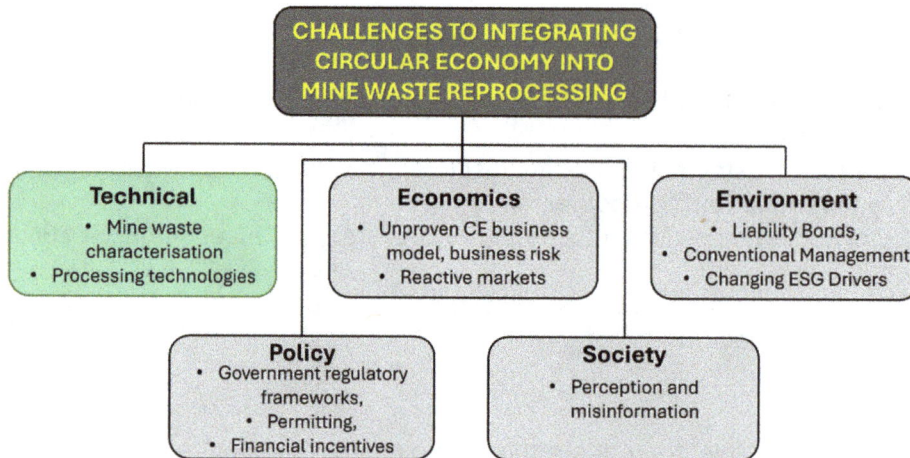

FIG 1 – Summary of challenges to integrating CE into mine waste reprocessing in the mining industry. Providing data for the green box is the focus of this paper. Summarised from Kinnunen and Kaksonen (2019), Tayebi-Khorami *et al* (2019) and Zinck *et al* (2019).

Enhancing technical capabilities through advanced characterisation approaches is necessary for understanding material composition, assessing environmental risks, and identifying value recovery opportunities. This is particularly important for legacy tailings, where limited data on material provenance, deposition history, and weathering effects hinder reprocessing efforts. There is a need for more case studies to demonstrate the value addition provided by detailed characterisation, which will inform and build confidence in the mining industry that waste valorisation can be a viable component of mainstream mining (Bellenfant *et al*, 2013; Kinnunen and Kaksonen, 2019; Parbhakar-Fox and Baumgartner, 2023). This paper and case study contributes to bridging the gap between research and industry application, through demonstrating a multiscale, top-down approach to characterisation that is both cost-effective and provides the required data to inform reprocessing decisions.

CHARACTERISATION APPROACH

Characterisation of mine waste has conventionally targeted specific management outcomes, such as AMD management, land rehabilitation or reprocessing for value extraction, however, rarely encompasses the breadth of all potential uses and management options as a whole material of value in a circular mining economy. A comprehensive characterisation approach is required which includes environmental risk and valorisation options using an integrated geometallurgical framework.

The Characterisation Toolbox for Tailings (Figure 2) has been developed as a collation of techniques which can be used at various stages of tailings characterisation depending on intent (ie recovery, recycle, remediation). This deductive framework prioritises low cost, high quantity data in the macro-scale before narrowing the focus to targeted micro-scale analyses (Parbhakar-Fox, 2019). The advantage of this staged approach is that key data is collected to inform go/no-go checkpoints on projects, making the characterisation economically justifiable.

Key to this toolbox is that techniques of each disciple (geology, metallurgy, geoenvironmental) occur in parallel and findings are integrated. This allows for collection of comprehensive geometallurgical data sets to assess value and environmental risk simultaneously.

CHARACTERISATION TOOLBOX FOR TAILINGS

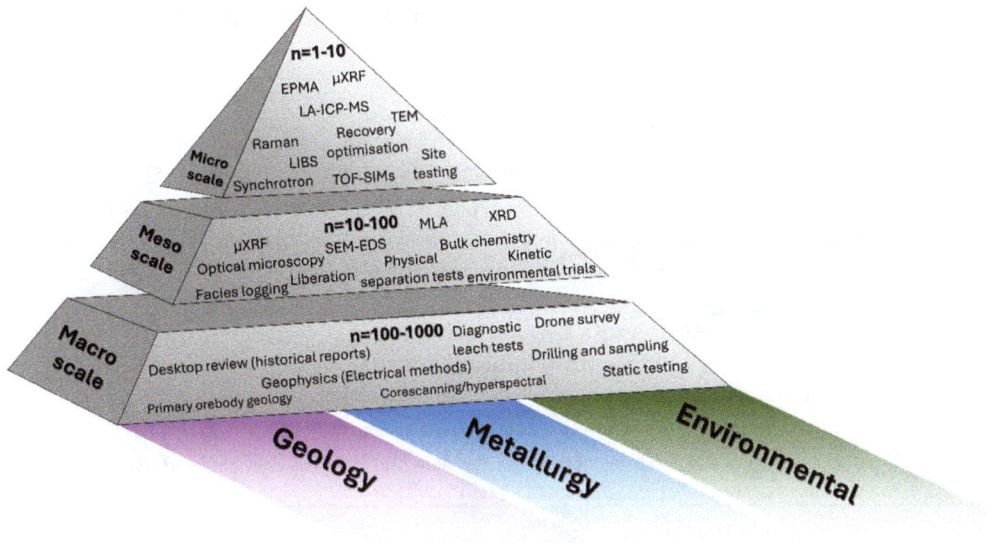

FIG 2 – Characterisation Toolbox for Tailings follows a deductive framework integrating geology, metallurgy and environmental test work to inform tailings reprocessing opportunities.

An applied roadmap of critical questions was developed to narrow the focus of the characterisation to the specific site requirements and goals, streamlining the process (Figure 3). This enables industry to undertake the staged characterisation required to make informed decisions on reprocessing potential.

Roadmap of critical questions
to guide tailings management within a CE context.

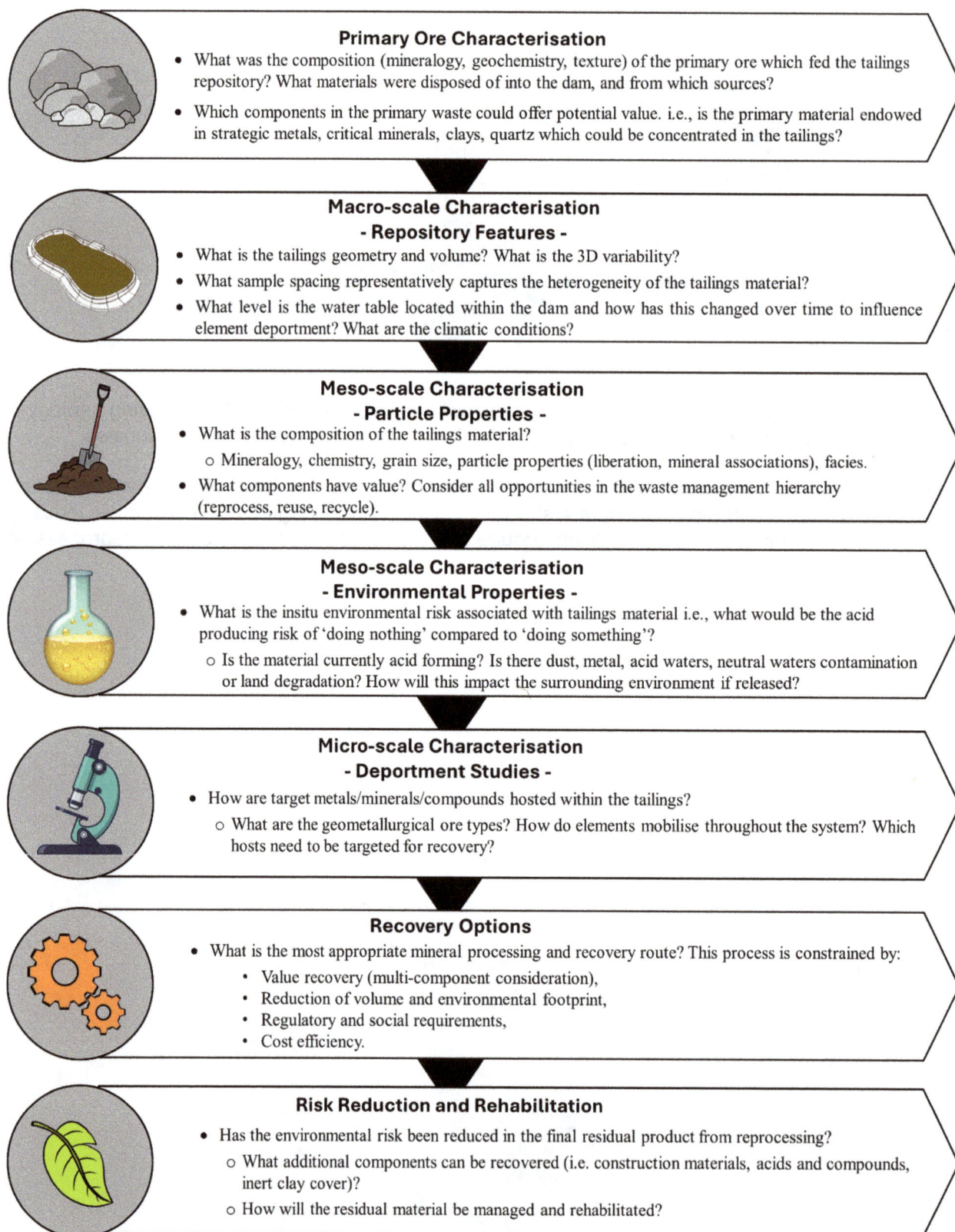

Primary Ore Characterisation

- What was the composition (mineralogy, geochemistry, texture) of the primary ore which fed the tailings repository? What materials were disposed of into the dam, and from which sources?
- Which components in the primary waste could offer potential value. i.e., is the primary material endowed in strategic metals, critical minerals, clays, quartz which could be concentrated in the tailings?

Macro-scale Characterisation
- Repository Features -

- What is the tailings geometry and volume? What is the 3D variability?
- What sample spacing representatively captures the heterogeneity of the tailings material?
- What level is the water table located within the dam and how has this changed over time to influence element deportment? What are the climatic conditions?

Meso-scale Characterisation
- Particle Properties -

- What is the composition of the tailings material?
  - Mineralogy, chemistry, grain size, particle properties (liberation, mineral associations), facies.
- What components have value? Consider all opportunities in the waste management hierarchy (reprocess, reuse, recycle).

Meso-scale Characterisation
- Environmental Properties -

- What is the insitu environmental risk associated with tailings material i.e., what would be the acid producing risk of 'doing nothing' compared to 'doing something'?
  - Is the material currently acid forming? Is there dust, metal, acid waters, neutral waters contamination or land degradation? How will this impact the surrounding environment if released?

Micro-scale Characterisation
- Deportment Studies -

- How are target metals/minerals/compounds hosted within the tailings?
  - What are the geometallurgical ore types? How do elements mobilise throughout the system? Which hosts need to be targeted for recovery?

Recovery Options

- What is the most appropriate mineral processing and recovery route? This process is constrained by:
  - Value recovery (multi-component consideration),
  - Reduction of volume and environmental footprint,
  - Regulatory and social requirements,
  - Cost efficiency.

Risk Reduction and Rehabilitation

- Has the environmental risk been reduced in the final residual product from reprocessing?
  - What additional components can be recovered (i.e. construction materials, acids and compounds, inert clay cover)?
  - How will the residual material be managed and rehabilitated?

FIG 3 – Roadmap to link key research outcomes with industry application to guide tailings characterisation.

CASE STUDY – NORTH-WEST QUEENSLAND MINERALS PROVINCE (NWQMP)

To demonstrate the benefits of undertaking a multi-scale and comprehensive characterisation approach, a copper mine with legacy flotation tailings was selected as the case study site within the NWQMP. This site is one of 29 operating and historical copper mines in the North-west Queensland Minerals province (Figure 4), many of which have associated Co mineralisation (Spatial and Graphic Services, 2019). Two main deposit types host Co in the NWQMP: sedimentary-hosted copper deposits, including the world-class Mt Isa Copper Mine, as well as the Capricorn Copper, Mount Oxide and Lady Annie deposits, and deposits related to the IOCG class of deposits, including Ernest Henry, Rocklands, Eloise and Osborne deposits (Geological Survey of Queensland, 2011). Cobalt often occurs as a refractory trace element in pyrite, which is suppressed during Cu flotation and rejected to tailings (Jefferson Montoya *et al*, 2024). As a result, there is significant potential for secondary cobalt resources within copper tailings in the region, with estimates exceeding 200 kt in the NWQMP (Golev, 2022). The clustering of these deposits in the region gives credence to the concept of a centralised reprocessing plant in a hub-and-spoke model, which diversify the region's interests into secondary resources mining and a Co production centre for Queensland (Degeling, 2020; The Next Economy, 2025).

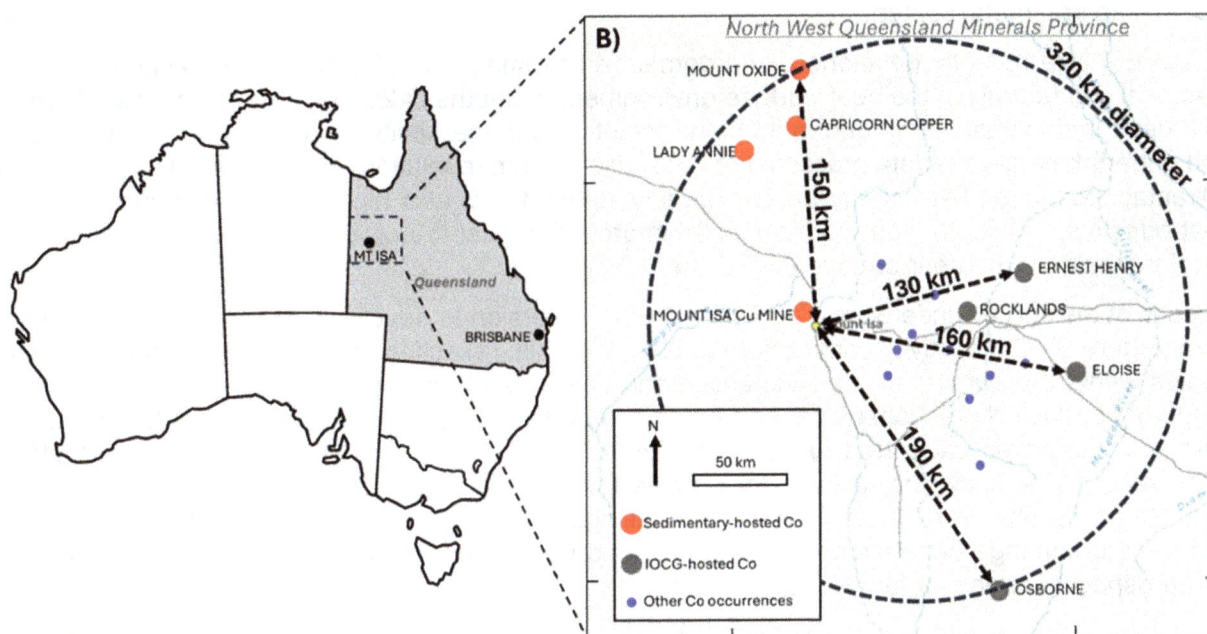

FIG 4 – Map and location of the NWQMP, showing the major operating and historical copper mines and Co occurrences in the region within 200 km of Mt Isa.

CRITICAL MINERAL ENDOWMENT – COBALT FOCUS

Cobalt is a critical metal essential for electric vehicle (EV) batteries, superalloys, hard metals, ceramics, pigments, catalysts, and permanent magnets (Hitzman *et al*, 2017; Savinova *et al*, 2023). Its high corrosion resistance, thermal stability, ferromagnetism, and energy storage capacity make it invaluable in modern technology (Mudd *et al*, 2013; Visual Capitalist, 2020). Despite emerging alternative battery chemistries, demand for Co is projected to double by 2030 due to its superior energy density and longevity in transportable batteries (Ryu *et al*, 2021; Cobalt Institute, 2024). However, supply risks are significant, with ~70 per cent of global Co production concentrated in the Democratic Republic of the Congo (DRC) and 74 per cent of refining occurring in China, raising geopolitical and ethical concerns (Cobalt Institute, 2024). As a result, Co is classified as a critical mineral in most OECD countries (Hitzman *et al*, 2017; European Commission, 2020; US Geological Survey (USGS), 2020; Australian Government, 2023; International Energy Agency, 2024).

Ninety-eight per cent of Co is mined as a by-product of Cu or Ni deposits, primarily from sediment-hosted Cu-Co, magmatic Ni-Cu, and Ni-Co laterite deposits (Hitzman *et al*, 2017; Savinova *et al*, 2023). Grades range from 0.02 to 0.92 wt per cent Co, averaging 0.2 wt per cent Co (Slack, Kimball and Shedd, 2017; Dehaine *et al*, 2021). Cobalt occurs in various mineral phases, including sulfides, arsenides, oxides, carbonates, silicates, selenides, and sulfates, often substituting for Fe, Cu, Mn, or Ni in minerals such as pyrite, chalcopyrite, and Mn-oxides (Young, 1957; Dehaine *et al*, 2021). While Co can be hosted in solid solution with pyrite up to 3 wt per cent (Kuyvenhoven and Townley, 2019), weathering can mobilise it into secondary phases under acidic conditions (Ziwa, Crane and Hudson-Edwards, 2021). The feasibility of Co recovery depends on its mineralogical deportment, necessitating detailed characterisation of geochemical distribution, mineral hosts, and fluid flow within tailings (Dehaine *et al*, 2021). This is especially necessary in legacy tailings environments, where material heterogeneity and extensive weathering significantly impact composition and metal distribution. A multiscale characterisation approach is therefore essential for determining reprocessing potential and optimising recovery strategies.

MACRO-SCALE – REPOSITORY MAPPING

Electrical geophysical methods are typically employed to map tailings dam walls, due to the high conductivity contrast between the surrounding country rock and the highly conductive tailings material (Campbell and Fitterman, 2000; Martín-Crespo *et al*, 2018; Nikonow, Rammlmair and Furche, 2019; Georgieva, Stoyanov and Dimovski, 2021). Conductivity responses in tailings are driven by many factors, including salinity and acidity, clay content and grain size, degree of saturation and sulfide and metal content.

Loupe EM is a recently developed time-domain electromagnetic (TEM) system designed to map electrical conductivity in the near-surface environment to depths of 25 to 40 m (Street *et al*, 2018). It was used in this case study due to its high resolution and the ability to acquire multi-channel data and undertake real-time data processing to optimise system resolution and DOI, which is particularly advantageous in the heterogeneous tailings environment. It is also highly portable (two-man team), cost-effective, quick to capture and with improved signal-to-noise processing for separating interference from the near-surface environment.

Results of the survey underwent a 1D layered earth inversion to develop a 3D volume of the tailings dam (Figure 5). Conductivity contrasts were clearly visible, reflecting the presence of standing water bodies within the dam, grain size variability and salinity and acidity related to metal content. The range of conductivity responses varied from 0.1 to 5000 mS/m, with an average of 142 mS/m. The tailings volume was calculated to be 7 725 729 m³. Based on a dry density of 1.54 to 2.05 t/m³, depending on porosity, the tailings dry particle tonnage was estimated to be between 11.9 and 15.9 Mt. Being able to map the profile of the tailings dam and provide an updated estimate of the volume and tonnage of a historic material provided important information on the size (economics) and prospective areas for further characterisation.

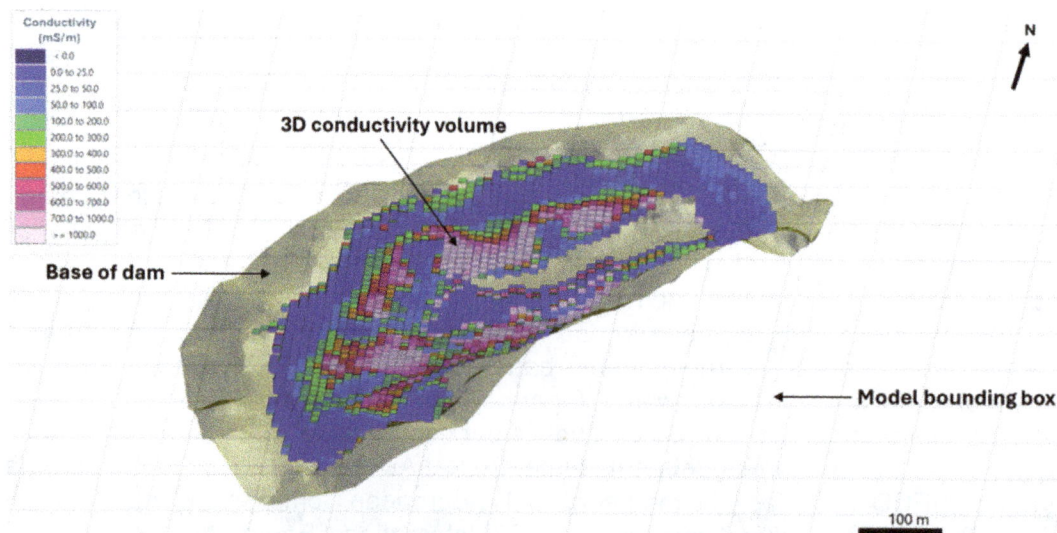

FIG 5 – 3D volume of 1D layered earth inversion model for legacy tailings case study in NWQMP. Model by Reid (2024).

MESO-SCALE – FACIES AND MINERAL MAPPING

Key to determining the mining and reprocessing methods is understanding the spatial distribution of mineralogy and Co hosts in the system. Facies are distinct units within a tailings dam that exhibit similar texture, mineralogy, and geochemistry, determined by both the depositional environment and weathering processes. Facies are mapped on a 10 to 100 cm scale. Twelve facies were logged across the study site (Figure 6), highlighting the high heterogeneity of the material. Salient features included efflorescent salts, bright green processing precipitates, red hematitic residues, yellow and orange iron (oxy)-hydroxides and black sulfides.

Facies A   **Facies B**   **Facies C**   **Facies D1**

Facies D2   **Facies D3**   **Facies D4**   **Facies E**

Facies F   **Facies G**   **Facies H1**   **Facies H2**

FIG 6 – Hand sample images of 12 facies identified in the legacy tailings, highlighting the highly heterogeneous mature of the material.

The tailings were enriched in critical metals (Australian Government, 2024), including As, Bi, Co, Mn, Se and Sb, up to ~10–1000 times the average crust abundance. Some strategic (Cu, S), precious (Ag) and other elements (Fe, C, Pb, Tl) were also relatively enriched. Cobalt returned an average of 322 ppm and a maximum of 1930 ppm, while Cu returned an average of 4700 ppm and a maximum grade of 2.6 per cent.

The tailings mineralogy (n=31) was dominated by quartz and pyrite, accounting for 37 wt per cent and 12 wt per cent of the total modal mineralogy (Figure 7). Other major phases included clays (13 wt per cent), feldspars (6 wt per cent), sulfates (6 wt per cent) and oxides (3 wt per cent). Critical minerals were predicted to mainly reside in pyrite, as well as minor chalcopyrite, oxides and sulfates and trace Co-sulfides such as cobaltite, alloclasite and carrollite. Tailings particle size distribution (PSD) from MLA analysis (n=10) returned an average P_{80} of 97 µm, with individual samples ranging from 65 to 169 µm. Comparatively, the pyrite PSD was much finer with an average P_{80} of 46 µm, and a range of 22 to 80 µm, while the Co-sulfides had a P_{80} of 17 µm. Multi-addition NAG pH (n=43) ranged between pH 2.53 and 8.48 and the EC ranged from 250.3 µS to 4540 µS. Tailings that were highly oxidised had minimal acid producing potential, while tailings that were deeper and in more anaerobic conditions retained the primary sulfide minerals.

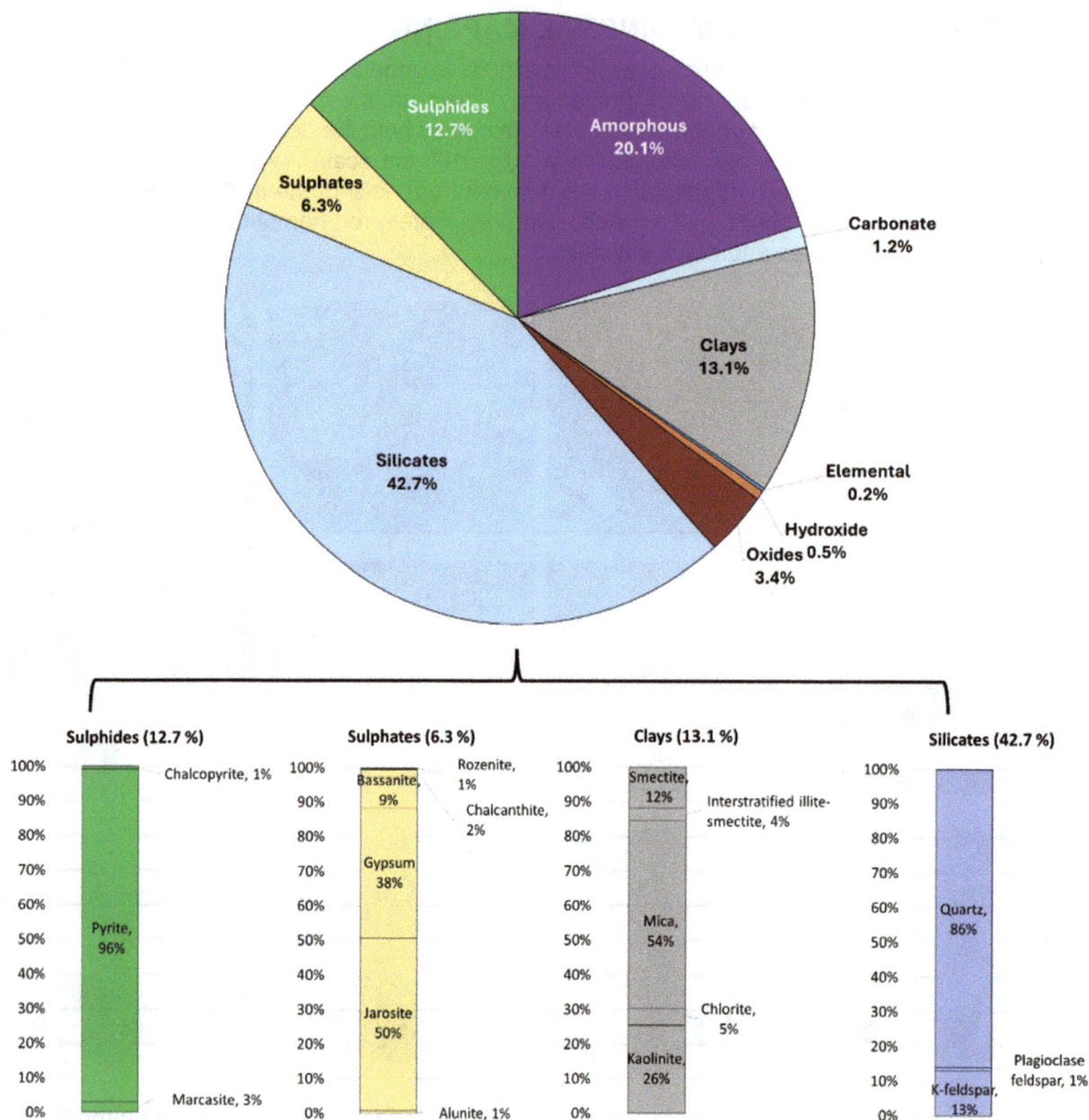

FIG 7 – Hand sample images of 12 facies identified in the legacy tailings, highlighting the highly heterogeneous mature of the material.

MICRO-SCALE – ELEMENT DEPORTMENT

Although mineralogical analyses are important for understanding the bulk composition of the material and predicting how these would respond to reprocessing technologies, tracking element deportment within these minerals is critical to targeting highly endowed minerals for the highest recovery returns. Originally at this case study site Co was assumed to occur in pyrite. However, studies using LA ICP-MS and XANES at the Australian Synchrotron revealed that up to 16 per cent of Co occurred in Co-sulfides, 40 per cent of Co occurred in leachable phases (Figure 8) and up to 44 per cent remained in pyrite or other phases, including up to 29 per cent of Co in the slimes fraction (<5 µm).

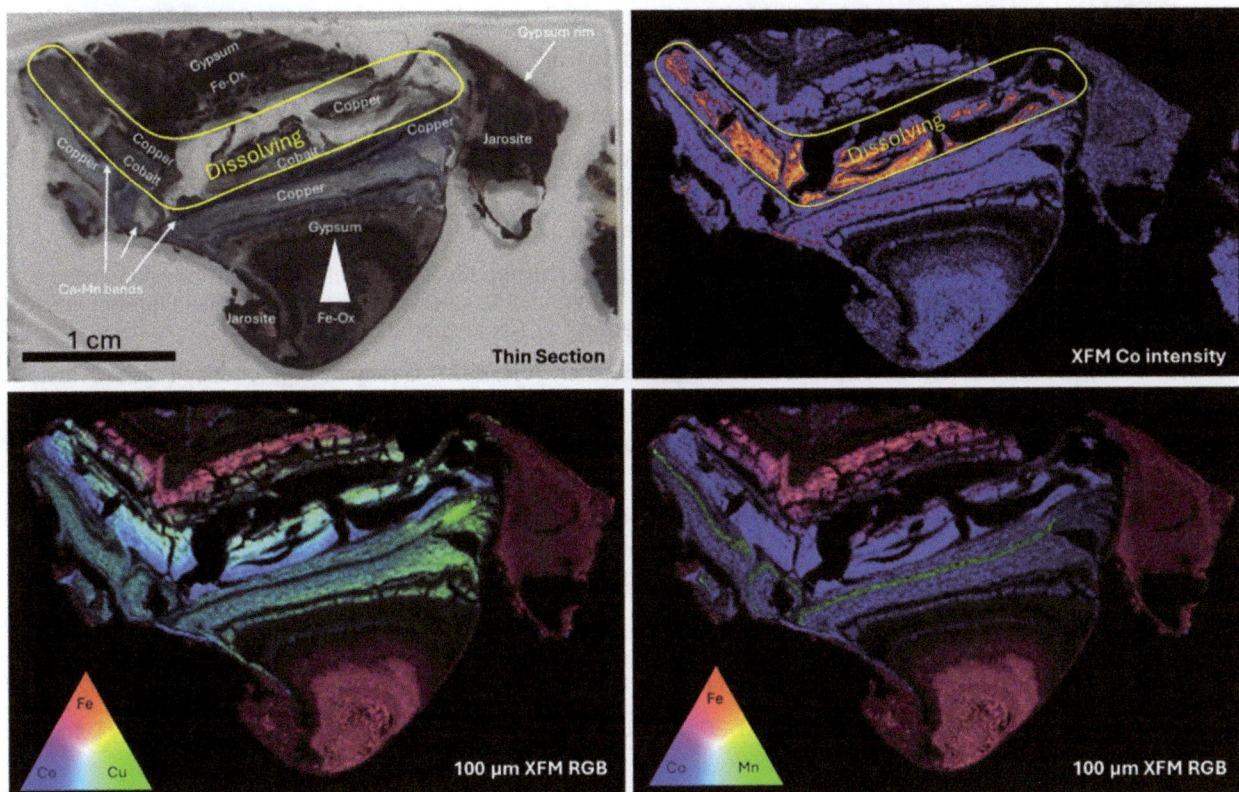

FIG 8 – XFM scans of Co, Cu, Mn, Fe in tailings sample, showing processing precipitates and water soluble features endowed in Co and Cu.

Cobalt deportment was influenced by primary ore, original processing methods and weathering to result in a highly dynamic environment, where Co had been mobilised and reprecipitated overtime in different phases. This information opened up opportunities for reprocessing to consider low cost *in situ* leaching with acid or microbiology, and utilisation of the existing flotation infrastructure to recover additional value not just from Co, but also Cu, As, S and Ag. This would effectively desulfurise the tailings to reduce acid producing potential and environmental legacy.

CONCLUSION

The findings from this study highlight the importance of a comprehensive, multi-scale characterisation approach to uncover and assess the potential of these secondary resources. By integrating geophysical, geological, and geometallurgical techniques, this approach can optimise resource identification, increase the efficiency of reprocessing strategies, and support the development of sustainable mining practices. Through such methodologies, the mining industry can move towards a Circular Economy, transforming waste liabilities into profitable and environmentally responsible assets, while also addressing the challenges posed by global supply chain risks and resource scarcity. Ultimately, these efforts can contribute to a more sustainable and diversified resource portfolio, benefiting both the mining sector and society.

ACKNOWLEDGEMENTS

The authors would like to thank the Geological Survey of Queensland in the Department of Natural Resources and Mines, Manufacturing, and Regional and Rural Development for funding the research costs of this project. Thank you to Loupe EM geophysics, Sasha Aivazpourporgou and James Reid for the data processing and inversions. Thank you to CSIRO for XRD analysis and the ANSTO for XFM beamtime.

REFERENCES

Australian Government, 2023. Critical minerals strategy 2023–2030, Department of Industry Science and Resources. Available from: <https://www.ga.gov.au/scientific-topics/minerals/critical-minerals>

Australian Government, 2024. Australia's Critical Minerals List and Strategic Materials List 2024, Critical Minerals Office.

Bellenfant, G, Guezennec, A G, Bodenan, F, D'Hugues, P and Cassard, D, 2013. Reprocessing of mining waste: combining environmental management and metal recovery?, in *Proceedings of the Eighth International Seminar on Mine Closure*, pp 571–582 (Australian Centre for Geomechanics: Perth).

Campbell, D and Fitterman, D, 2000. Geoelectrical Methods for Investigating Mine Dumps, in Proceedings of the International Conference on Acid Rock Drainage (ICARD2000), Denver, Colorado, 21–24 May 2000.

Cobalt Institute, 2024. Cobalt Market Report 2023, Cobalt Institute.

Degeling, H, 2020. Queensland's new economy minerals initiative [online], AusIMM Bulletin (The Australasian Institute of Mining and Metallurgy: Melbourne). Available from: <https://www.ausimm.com/bulletin/bulletin-articles/queenslands-new-economy-minerals-initiative/> [Accessed: 17 September 2021].

Dehaine, Q, Tijsseling, L T, Glass, H J, Törmänen, T and Butcher, A R, 2021. Geometallurgy of cobalt ores: A review: *Minerals Engineering*, vol 160.

European Commission, 2020. Critical Raw Materials Resilience: Charting a Path towards greater Security and Sustainability, Brussels, European Commission.

Geological Survey of Queensland, 2011. North-West Queensland Mineral and Energy Report, Department of Employment, Economic Development and Innovation: Brisbane.

Georgieva, G, Stoyanov, N and Dimovski, S, 2021. Geophysical Approach for Mapping Zones of low Permeability Endangering the Stability of a Tailings dam Wall, pp 1–5.

Golev, A, 2022. Cobalt in focus: what is Queensland's potential? [online]. Available from: <https://geoscience.data.qld.gov.au/blog/cobalt-focus-what-queenslands-potential> [Accessed: 13 Sep 2024].

Hitzman, M, Bookstrom, A, Slack, J and Zientek, M, 2017. Cobalt—Styles of Deposits and the Search for Primary Deposits, Virginia, US Geological Survey (USGS).

Hudson-Edwards, K A and Dold, B, 2015. Mine waste characterization, management and remediation, *Minerals*, 5(1):82–85

International Council on Mining and Metals (ICMM), 2019. Global Tailings Review [online]. Available from: <https://globaltailingsreview.org/> [Accessed: 21 Oct 2021].

International Energy Agency (IEA), 2024. Global Critical Minerals Outlook 2024 [online]. Available from: <https://iea.blob.core.windows.net/assets/ee01701d-1d5c-4ba8-9df6-abeeac9de99a/GlobalCriticalMineralsOutlook2024.pdf> [Accessed: Aug 2024].

Jefferson Montoya, M, Jones, T R, Curtis-Morar, C, Parbhakar-Fox, A, Forbes, G and Forbes, E, 2024. Impact of Pyrite Textures Prevalence on Flotation Performance in Mount Isa Copper Orebodies, *Mineral Processing and Extractive Metallurgy Review*, pp 1–15.

Kinnunen, P H M and Kaksonen, A H, 2019. Towards circular economy in mining: Opportunities and bottlenecks for tailings valorization, *Journal of Cleaner Production*, 228:153–160.

Kinnunen, P, Karhu, M, Yli-Rantala, E, Kivikytö-Reponen, P and Mäkinen, J, 2022. A review of circular economy strategies for mine tailings, *Cleaner Engineering and Technology*, 8:100499.

Kuyvenhoven, R and Townley, B, 2019. Challenges and opportunities for cobalt recovery at copper plants, in Flotation'19 Conference, Cape Town, April 2019.

Lottermoser, B G, 2011. Recycling, reuse and rehabilitation of mine wastes, *Elements*, 7:405–410.

Lottermoser, B, 2010. Mine Wastes Characterization, Treatment and Environmental Impacts, Heidelberg (Springer: Berlin).

Martín-Crespo, T, Gómez-Ortiz, D, Martín-Velázquez, S, Martínez-Pagán, P, De Ignacio, C, Lillo, J and Faz, Á, 2018. Geoenvironmental characterization of unstable abandoned mine tailings combining geophysical and geochemical methods (Cartagena-La Union district, Spain), *Engineering Geology*, 232:135–146.

Minerals Research of Western Australia (MRIWA), 2023. Alternative Use of Tailings and Waste [online]. Available from: <https://www.mriwa.wa.gov.au/minerals-research-advancing-western-australia/focus-areas/alternative-use-of-tailings-and-waste/> [Accessed: 10 Nov 2024].

Mudd, G M, Weng, Z, Jowitt, S M, Turnbull, I D and Graedel, T, E, 2013. Quantifying the recoverable resources of by-product metals: The case of cobalt, *Ore Geology Reviews*, 55:87–98.

Next Economy, The, 2025. Mount Isa future ready economy roadmap, in Mount Isa City Council, Queensland.

Nikonow, W, Rammlmair, D and Furche, M, 2019. A multidisciplinary approach considering geochemical reorganization and internal structure of tailings impoundments for metal exploration, *Applied Geochemistry*, 104:51–59.

Parbhakar-Fox, A and Baumgartner, R, 2023. Action versus reaction – how geometallurgy can improve mine waste management across the life-of-mine, *Elements*, 19:371–376.

Parbhakar-Fox, A, 2019. Reinventing the Wheel: the Environmental Geometallurgy Matrix and its Supporting Tools, in Procemin Geomet 2019 Conference, Santiago, Chile, 20–22 November 2019.

Reid, J, 2024. 1D layered earth inversion model for Capricorn Copper Loupe EM data, in Mira Geoscience, Australia.

Ryu, H-H, Sun, H H, Myung, S-T, Yoon, C S and Sun, Y-K, 2021. Reducing cobalt from lithium-ion batteries for the electric vehicle era, *Energy and Environmental Science*, 14:844–852.

Savinova, E, Evans, C, Lèbre, É, Stringer, M, Azadi, M and Valenta, R K, 2023. Will global cobalt supply meet demand? The geological, mineral processing, production and geographic risk profile of cobalt, *Resources, Conservation and Recycling*, 190:106855.

Slack, J F, Kimball, B E and Shedd, K B, 2017. Cobalt, ch F, *Critical mineral resources of the United States—Economic and environmental geology and prospects for future supply* (eds: K J Schulz, J H DeYoung Jr, R R Seal II and D Bradley), (US Geological Survey).

Spatial and Graphic Services, 2019. Queensland's major mineral, coal and petroleum operations and resources, in Geological Survey of Queensland, Department of Natural Resources.

Street, G, Duncan, A, Fullagar, P and Tresidder, R, 2018. LOUPE – A Portable EM Profiling System, *ASEG Extended Abstracts*, pp 1–3.

Tayebi-Khorami, M, Edraki, M, Corder, G and Golev, A, 2019. Re-thinking mining waste through an integrative approach led by circular economy aspirations, *Minerals*, 9:286.

US Geological Survey (USGS), 2020. Mineral commodity summaries 2020 (US Geological Survey).

Visual Capitalist, 2020. Ethical Supply: The Search for Cobalt Beyond the Congo [online]. Available from: <https://www.visualcapitalist.com/ethical-supply-the-search-for-cobalt-beyond-the-congo/> [Accessed: 12 April 2021].

Young, R, 1957. The geochemistry of cobalt, *Geochimica et Cosmochimica Acta*, 13:28–41.

Zinck, J, Tisch, B, Cheng, T and Cameron, R, 2019. Mining Value from Waste Initiative: Towards a Low Carbon and Circular Economy, *REWAS*, pp 325–332 (Springer International Publishing).

Ziwa, G, Crane, R and Hudson-Edwards, K A, 2021. Geochemistry, Mineralogy and Microbiology of Cobalt in Mining-Affected Environments, *Minerals*, 11:22.

Lessons learned – construction of earthfill embankment over lacustrine foundation material

A Parsi[1] and W Herweynen[2]

1. Geotechnical Engineer, GHD, Burnie Tas 7320. Email: amin.parsi@ghd.com
2. Dams and Geotechnical Engineer, GHD, Hobart Tas 7000. Email: wesley.herweynen@ghd.com

ABSTRACT

Constructing earthfill embankments over soft Lacustrine deposits presents significant geotechnical difficulties due to the foundation's low strength, high plasticity, and vulnerability to deformation under load. This paper examines the case of a tailings dam embankment at a mine site in Tasmania, built over deep Lacustrine sediments overlain by Fluvioglacial deposits. To address seepage risks posed by the permeable overburden, a clay-filled cut-off trench was installed. However, construction activities encountered considerable challenges including groundwater ingress, deformation of trench sidewalls, and difficulty achieving compaction within the low strength Lacustrine foundation.

Despite repeated trials, conventional stabilisation techniques proved ineffective. The paper details the cut-off trench construction process and assesses the geotechnical performance of the foundation materials during and after construction. Observations from the project underscore the limitations of traditional earthwork methods in such settings and highlight the importance of tailoring construction techniques to site-specific geological conditions. The findings offer practical lessons and design considerations for future embankment projects constructed on similar low strength soil profiles.

INTRODUCTION

The project involves the staged construction of a tailings storage facility located in western Tasmania. The facility is being progressively raised to accommodate ongoing tailings production. The storage is developed using a zoned earthfill embankment constructed by the downstream method, incorporating a low-permeability core, drainage filters, and rock fill shoulders. A typical section of the embankment geometry is presented in Figure 1.

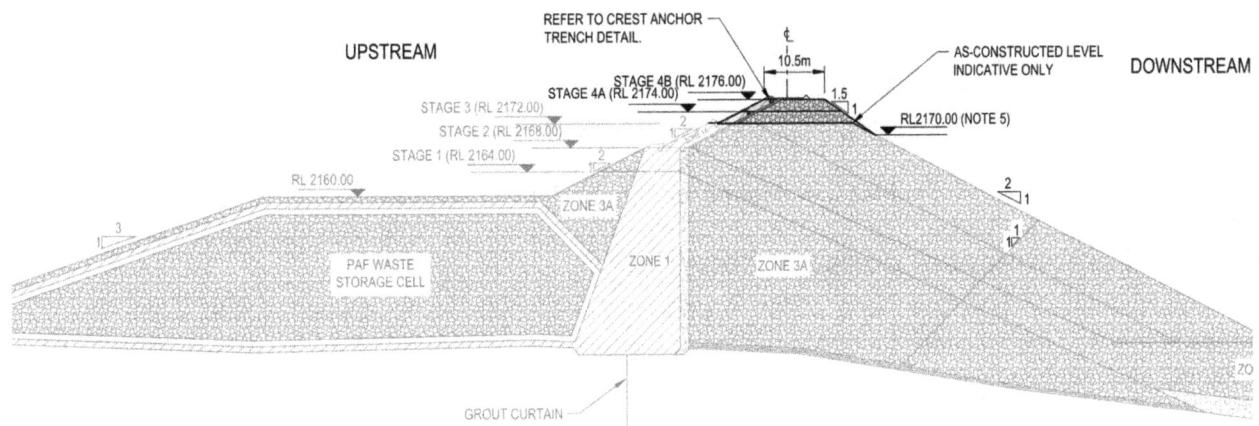

FIG 1 – Typical section of the general embankment geometry for existing raises.

A key challenge in the recent raise design was the presence of soft Lacustrine deposits beneath sections of the proposed embankment footprint. In previous stages, these deposits were removed to ensure foundation stability; however, due to cost and constructability considerations, and also the deep depth of the Lacustrine layer, the design adopted a strategy to retain the Lacustrine material *in situ* in selected areas. This change necessitated additional geotechnical investigation and analysis to assess the implications for constructability, long-term stability and deformation. The first proposed assessment was the construction of a trial embankment over this area.

This paper presents a case study of the construction trial over Lacustrine material, providing valuable lessons for similar future projects.

SITE CONDITION AND GEOLOGY

The site is underlain by a complex sequence of geological materials, primarily composed of peat topsoil, Fluvioglacial deposits, Lacustrine clays and silts, glacial till, and Neoproterozoic bedrock. The Fluvioglacial deposits typically consist of sand, gravel, and cobbles with variable fines content generally found from the surface up to 2 m thick. Lacustrine units are under the Fluvioglacial. This material are soft clay and silt, generally found between three and 13 m thick in the areas of interest.

The Lacustrine material was classified as low to intermediate plasticity clay (CL-CI) with high *in situ* moisture content and relatively low undrained shear strength, making it susceptible to cyclic softening and potential strength loss under seismic loading. Portions of the foundation where these materials could not be economically removed were subject to numerical deformation assessments to evaluate their influence on embankment performance.

Groundwater is present at shallow depths across the site, contributing to construction challenges related to excavation and compaction. Several geotechnical investigations between 1992 and 2024 provided a robust understanding of the foundation conditions, including CPTu and triaxial testing, plate load testing, and laboratory characterisation of foundation and construction materials. This geotechnical framework informed the design and risk assessment processes.

CONSTRUCTION CHALLENGES

To assess the constructability of the embankment over the site geology, the construction of a 50 m long trial embankment was proposed. The trial embankment comprises a 5 m high rock fill (Zone 3A) embankment and a clay cut-off trench (Zone 1) on the upstream. A typical section of the trial embankment is presented in Figure 2.

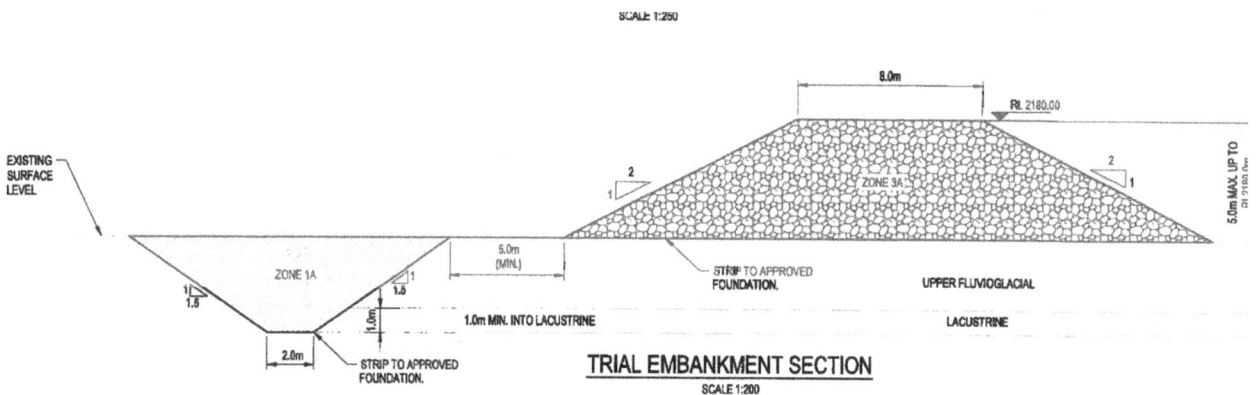

FIG 2 – Trial embankment typical section.

Rock fill embankment construction trial

An approximately 50–60 m long virgin foundation was cleared to construct the Zone 3A trial embankment. The cleared foundation was noted to be weathered argillite with signs of some Fluvioglacial materials present (Figure 3).

FIG 3 – Trial embankment cleared foundation

Zone 3A material was placed directly over the excavated foundation with no filter material or additional treatments undertaken on the cleared foundation. Zone 3A placement methodology on the Trial embankment was similar to the previous raises of the embankment including 500 mm placement thickness for each layer, compacted with a vibrating smooth drum roller. Prior to placement of Zone 3A material, a few vibrating wire piezometers were installed in the both Fluvioglacial and Lacustrine foundation material.

The intent of the Zone 3A placement trial was to demonstrate constructability of this material on the glacial foundations (Fluvioglacial underlayed by Lacustrine). The trial Zone 3A embankment was constructed 5 m high, with 16 m wide crest, replicating the ultimate heigh of the dam in this area. The trial appeared a success and there was no evidence of liquefaction of foundation materials during the construction. Survey monuments were installed on the trial embankment crest to monitor movements.

Cut-off trench construction trial

The embankment section of interest in this study was a starter dam, requiring a low-permeability vertical barrier to control seepage through the foundation. As such, a cut-off trench was deemed necessary to control the foundation seepage. Given the presence of deep and soft Lacustrine material in the foundation, concerns were raised about the constructability of such a trench and the performance of clay backfill in these conditions. To address these concerns, a trial cut-off trench construction was proposed.

The trial aimed to assess whether clay material could be placed and compacted effectively over low strength Lacustrine soils, and to evaluate the behaviour of the trench under construction conditions. The trench was excavated through overlying Fluvioglacial material and keyed into the Lacustrine layer, targeting a minimum depth of 1 m into the Lacustrine foundation (refer to Figures 2 and 4).

FIG 4 – Cut-off trench trial construction.

The excavation was approximately 3 m wide and up to 2.5 m deep. A few test pits in this area indicated that the Lacustrine layer is five to 7 m thick in this area.

During excavation, significant groundwater ingress was observed once the Lacustrine was encountered, causing rapid sidewall deformation and challenges in maintaining trench stability.

Several trials were conducted on the cut-off trench to compact Zone 1A material.

The first attempt was to place the Zone 1 with 150 mm layer thickness as per the technical specifications for the previous raises of the embankment. This placement technique caused significant deformation and bow waving up to 150 mm on the placed layers as more roller passes were undertaken.

Figure 5 presents the significant moisture presents in the trench as soon as the first layer placement started.

FIG 5 – water ingress into the trench as soon as the material placement began.

A test pit was undertaken on the first Zone 1A layer which showed the foundation floor of the trench have moved up by 150 mm when checked using the excavator GPS.

Following the floor heaving, the Zone 1A layer was removed due to contamination.

The second attempt was to place a 500 mm thick layer of Zone 1 to form a bridging layer. This 500 mm thick layer was compacted using two passes of a static pad foot roller which showed slight movement post compaction. An additional 400 mm thick Zone 1A layer was placed which had six passes form a static and two passes on vibrating pad foot roller. This layer showed signs of deformation in isolated regions, but no significant bow weaving was noticed. The region which showed no signs of deformation had additional 300 mm layer of Zone 1A material placed which was compacted using vibrating pad foot roller. As this additional 300 mm layer was being compacted, minor movement was visible.

Nuclear Density Gauge Tests (NDT) were also undertaken on the compacted 400 mm and 300 mm layers. The results show the minimum required density was not met, suggesting adequate compaction was not achieved.

Hence, the cut-off trench construction trial was considered unsuccessful. So, alternative approaches such as slurry wall and sheet piles are to be considered and assessed.

PERFORMANCE MONITORING AND OBSERVATIONS

As previously mentioned, vibrating wire piezometers were installed in both the Fluvioglacial and Lacustrine foundations beneath the trial embankment prior to construction. Also, a few deformation monitoring prisms were installed on the crest of the trial embankment after construction completion.

Pore pressure and movement data has been gathered over the 12 months since construction.

Pore pressures as measured by the piezometers have stabilised since construction and there is no evidence of excess pore pressure remaining. Survey results from the four settlement monuments installed on the crest of the trial embankment all show <10 mm of settlement across a six-month period. Although the data set is limited, the magnitude of settlement appears insignificant.

LESSONS LEARNED

The trial embankment construction over low strength Lacustrine deposits offered critical insights into both the performance of traditional embankment construction techniques and the limitations encountered under challenging geotechnical conditions.

One of the primary takeaways was the importance of assessing foundation conditions early, especially where soft, compressible soils such as Lacustrine clays are present.

The trial highlighted that while the rock fill shoulders (Zone 3A) could be placed successfully on glacial soils with no signs of foundation liquefaction or instability, the cut-off trench presented significant difficulties. This is because the top thin layer of Fluvioglacial was not removed for the construction of Zone 3A, so this layer acted as a bridging layer between Lacustrine and the embankment material. This is wile the attempt to construct a clay cut-off trench directly over the Lacustrine foundation encountered instability, trench wall deformation, and inadequate compaction – despite multiple variations in placement and compaction methodology.

Monitoring data indicated minimal long-term settlement and stable pore pressure in the foundations beneath the trial embankment, suggesting that the fill itself did not overstress the Lacustrine soils. However, the unsuccessful trench trial highlighted the need for alternative seepage barriers, such as slurry walls or sheet piles, better suited to these conditions.

CONCLUSION

The key conclusions and recommendations for similar projects include:

- Removing any low strength foundation material if economically suits the project is the writer's recommendation.

- Undertaking early constructability trials when working with low strength foundations is highly recommended.

- Direct construction over low strength clay foundation material in the case study was failed due to excessive deformations and lack of proper compaction.

- Even the presence of a thin high strength material (in this case Fluvioglacial) could act as a bridging layer and enhance construction and embankment stability and deformation.

- Instrumentations monitoring data of the trial embankment significantly helped the design and provided a more clear vision for the risk assessment.

- A deformation analysis calibrated using the monitoring instruments is highly recommended to assess the long-term behaviour of the embankment and foundation.

- Consider non-conventional seepage control systems where trench stability and compaction are compromised.

ACKNOWLEDGEMENTS

The author would like to acknowledge Wesley Herweynen for his valuable contributions throughout the project. Appreciation is also extended to the GHD design and site engineering team for their dedicated efforts during the planning, execution, and monitoring of the trial embankment. Their collaboration was essential to the insights gained from this work.

Numerical investigation on the effect of corners to static liquefaction triggering

N Pereira[1], D Reid[2], S Prabhu[3] and F Urbina[4]

1. Senior Tailings Engineer, Red Earth Engineering – A Geosyntec Company, Perth WA 6000. Email: nicolas.pereira@redearthengineering.com.au
2. Principal Engineer, Red Earth Engineering – A Geosyntec Company, Perth WA 6000. Email: david.reid@redearthengineering.com.au
3. Tailings Engineer, Red Earth Engineering – A Geosyntec Company, Perth WA 6000. Email: sprabhu@redearthengineering.com.au
4. Senior Tailings Engineer, Red Earth Engineering – A Geosyntec Company, Perth WA 6000. Email: felipe.urbina@redearthengineering.com.au

ABSTRACT

The use of numerical methods to assess the risk of static liquefaction in tailings storage facility (TSF) slopes has increased significantly over the past decade. Much of this impetus has been driven by a series of major TSF failures attributed to static liquefaction, and finite difference (FD) models being used as part of the post-failure investigations. The FD models carried out have included both two and three-dimensional analyses, depending on capabilities of the available code/constitutive model, the geometry of the problem, and the spatial distribution of contributory soil layers. For valley TSFs and other constrained structures there is general agreement on the importances of considering the three-dimensional nature of the problem. For paddock-style TSFs, common in regions like the Goldfields of Western Australia, two-dimensional (plane strain) analysis is often deemed sufficient due to the typically long, uniform slopes. However, anecdotal evidence of some failures (eg TVA Kingston) suggest that static liquefaction may initiate in corner areas where slope angles are lower, a somewhat counterintuitive observation.

This paper investigates the effects of corner geometry on uniform paddock-style TSFs using FLAC3D FD analyses. Tailings are modelled as loose, contractive material incrementally placed, with the NorSand constitutive model and parameters from Cadia tailings for comparison with previous relevant works. To investigate the effect of corner geometries on the potential for the static liquefaction triggering, the current study conducts stress-deformation analysis of an idealised TSF with an upstream-raised slope under three scenarios: (i) a conventional two dimensional (2D) plane strain analysis; (ii) a three-dimensional (3D) analysis using the same geometry extended in the third dimension with an 'inner' corner configuration (Scenario 2); and (iii) a 3D analysis incorporating an 'outer' corner geometry (Scenario 3).

The results of the analysis suggest that, although instability is triggered later in inward corners compared to outward corners and plane strain conditions, the delay is not substantial. While there is a general notion that inward corners are significantly more stable, this seems to be the critical condition in the context of static liquefaction, although the final stable height of the TSF is still comparable across all three scenarios. This was primarily the result of looser *in situ* states and lower Lode angles in the inward corner model, resulting in lower instability stress ratios. The implications and limitations of the study are discussed.

INTRODUCTION

Statically-triggered flow liquefaction represents one of the largest risks for the perimeter embankment stability of tailings storage facilities (TSFs) as evidenced by a series of recent failures featuring such behaviour (Morgenstern *et al*, 2016; Jefferies *et al*, 2019; Robertson *et al*, 2019; Arroyo and Gens, 2021). These events, the work of investigation panels afterwards, and a general increase in availability of numerical tools and computational power have all led to a greater use of stress-deformation based methods to investigation the stability of TSFs, including with respect to potential statically-triggered flow liquefaction (Reid *et al*, 2022a; Sottile *et al*, 2022; Babaki and Tannant, 2024). Similarly, efforts in the laboratory element characterisation – a key input to the development of such numerical models – has seen similar increases over the same period (Reid *et al*, 2021, 2023; Fanni, Reid and Fourie, 2024).

Given the complexity of stress-deformation modelling, wherever possible it is useful to simplify three dimensional problems to two-dimensional plane strain analyses. While some situations, such as TSFs in narrow valleys, demand three dimensional analyses, a significant number of TSFs feature relatively long uniform embankments and therefore two-dimensional analyses are generally seen as appropriate. While even relatively uniform embankments often have corners, these are typically not seen as an area of specific focus or that necessitates the use of three-dimensional models, as corners generally have shallower slopes than the plane-strain portion of the perimeter embankment.

An important nuance related to corners in TSFs that are otherwise relatively uniform in relation to the stress conditions *in situ* near to the corners – specifically, how does such geometry affect *in situ* stress conditions and do these geometry effects increase, or decrease, the propensity for static triggering in corner area. For context, in stress-deformation numerical modelling state of practice, the stress ratio ($\eta=q/p'$) is used to characterise the *in situ* stresses and the *in situ* state parameters (ψ) allows to check whether η has reached the estimated instability stress ratio (η_{IL}), which represents the onset of softening for a given element. Important to this comparison is the process for estimating the *in situ* value of η_{IL}, as there is evidence it can be affected by principal stress angle (Reid, 2020; Fanni, Reid and Fourie, 2022), which, to the authors knowledge, is rarely included in such assessments. Less controversially, it now seems state of practice to 'scale' η_{IL} on the basis of the value of Lode angle *in situ* – as it reduces critical state friction ratio (*M*), so it appears to also reduce η_{IL} (Jefferies *et al*, 2019; Fanni *et al*, 2024). To the authors knowledge, there has been little investigation into the effects of three dimensional analyses on the evolution of Lode angle *in situ*, something which may be a major factor in differences seen at corner locations to the uniform plane strain portions of embankments.

While the impact of three dimensional geometry on the results of limit equilibrium (LE) stability has been assessed (Firincioglu and Ercanoglu, 2021; Kumar, Choudhary and Burman, 2022), the authors are aware of little work examining the potential for corners – in otherwise relatively uniform TSF embankment/foundation geometry – to affect the outcomes of a static liquefaction assessment or triggering potential. One anecdotal example would be the failure of the TVA Kingston ash pond (AECOM, 2009) appears to have initiated at a corner. However, given the likely process of subaqueous deposition of coal ash in that portion of the pond, it seems plausible that triggering might be the result of particularly loose material having been deposited in the corner rather than any specific geometric effect, noting that at TVA Kingston deposition was occurring at the opposite end of the deposit in early development of the facility.

To investigate the effect of corner geometries on the potential for the static liquefaction triggering, the current study conducts stress-deformation analysis of an idealised TSF with an upstream-raised slope under three scenarios: (i) a conventional two dimensional (2D) plane strain analysis (ii) a three-dimensional (3D) analysis using the same geometry extended in the third dimension with an 'inner' corner configuration (Scenario 2), and (iii) a 3D analysis incorporating an 'outer' corner geometry (Scenario 3). Each model is 'constructed' incrementally in stages to reflect the tailing deposition and to ensure consistency in loading conditions for the different scenarios. During the construction process, the *in situ* tailings are continuously interrogated to identify the elements that reach a trigger condition of $\eta>\eta_{IL}$. Where this occurs, softening is initiated, with zones reaching liquefied strengths after 10 per cent strain. The model is subsequently solved, and if stable, the deposition is continued.

NUMERICAL MODEL DEVELOPMENT

General development and geometry

For the current investigation, the idealised 2D TSF cross-section used by Reid *et al* (2022a) was used as reference as: (i) the model, being already developed and established, enabled a useful 2D 'baseline' for the current work, (ii) the model simulates loose tailings in an upstream-raised TSF, a relevant scenario for statically-triggered flow liquefaction and (iii) utilised the Cadia TSF NorSand parameters developed by Jefferies *et al* (2019). The 2D model geometry used in the current study, adapted from Reid *et al* (2022) is presented in Figure 1. The model was developed in FLAC3D v9.3.

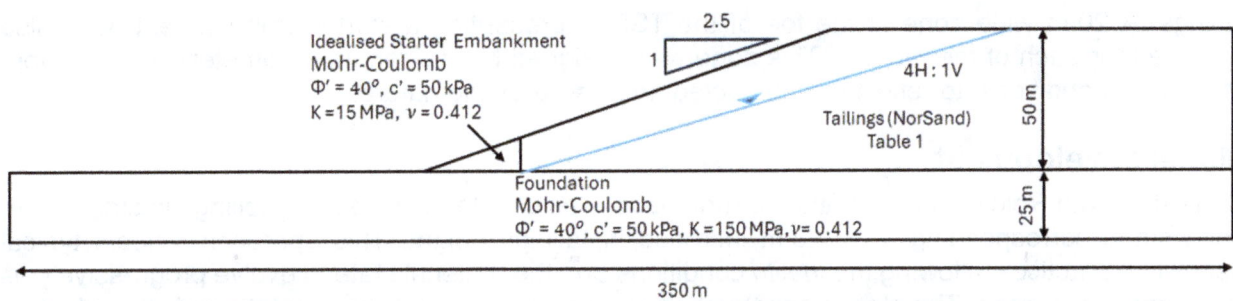

FIG 1 – Two-dimensional model at the end of deposition.

The geometry in Figure 1 represents the plane strain 'baseline' case (Scenario 1) and is used as a reference for comparison with the other scenarios in the current work. This geometry was then incorporated into an 'inner' and 'outer' corner arrangement as presented in Figure 2.

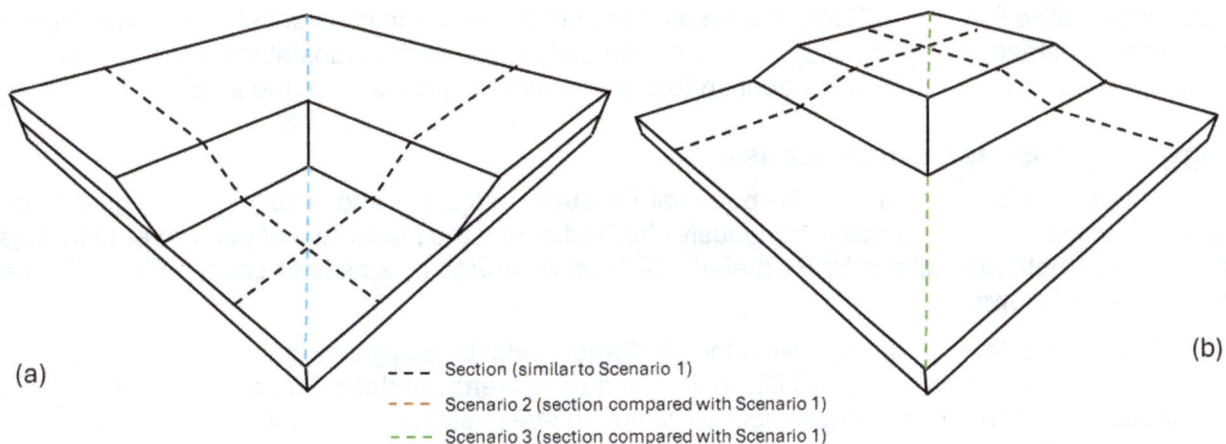

FIG 2 – Three-dimensional model – (a) inner corner (Scenario 2) (b) outer corner (Scenario 3).

Constitutive model and calibration

Based on Reid *et al* (2022), the NorSand constitutive model was adopted to represent the tailings in both the 2D and 3D models. In addition, the parameter set developed and used by the Cadia investigation panel (Jefferies *et al*, 2019) was also considered. The summary of the parameters is provided in Table 1.

TABLE 1

Cadia tailings NorSand parameter summary.

Parameter	Parameter value
CSL definition	$e_{cs} = 0.906 - 0.355\,(p'/100)^{0.119}$
Critical-state friction ratio under triaxial compression conditions, Mtc	1.5
Volumetric coupling parameter, N	0.3
State dilatancy constant, χ	8
Plastic hardening, *H*	50–450Ψ
Elastic shear modulus, G	$17\ \text{MPa} \cdot (p'/100)^{0.76}$
Poisson's ratio, v	0.2

Similar to Reid *et al* (2022), the foundation was assumed to be stiff rock in all the scenarios to exclude foundation-related deformations from the analysis and focus the assessment on the behaviour of the tailings. A 20 m wide zone at the toe of the TSF, representing a starter embankment, was also assumed for each of the models. This zone was assigned fixed strength parameters, independent of phreatic conditions, to reflect its compacted and free-draining nature.

Model development

Consistent with Reid *et al* (2022), the current model was developed by placing tailings in stages with each stage corresponding to a 2.5 m increase in tailings height. This approach reflects typical deposition practices, allowing the *in situ* conditions and the material state to evolve progressively as the tailings level rises. The staged construction ensures a more realistic simulation of stress history and density development within the tailings mass. A seed state parameter (ψ) of +0.06 is used for the tailings which is also consistent with the state used by Reid *et al* (2022) in their model.

The phreatic surface in the model was also developed in an incremental manner, alongside the tailings with the pore pressure in the zones increasing gradually with the deposition process. The phreatic surface developed in a 4H:1V slope from the upstream toe of the starter embankment. Importantly, unlike Reid *et al* (2022), this paper does not focus on the constant shear drained trigger mechanism. Instead, it emphasises the potential for tailings to reach an unstable state as a result of ongoing deposition, with particular attention to the influence of geometry on the stability.

Triggering assessment process

For all the scenarios considered, the potential for static triggering and resulting flow liquefaction failure of the idealised slope was continuously checked during the placement of each layer of tailings based on the methods outlined by Reid *et al* (2022) as illustrated by a pseudo-code in Figure 3. The steps were as follows:

1. Place a new layer of tailings using the NorSand model by assigning the properties presented in Table 1 and a seed ψ of +0.06, then adjust the phreatic surface accordingly. The model is subsequently brought to equilibrium allowing the stress states to develop.

2. Interrogate zones below the phreatic surface to check if $\eta > \eta_{IL}$. This check includes scaling of η_{IL} from triaxial compression condition to the relevant *in situ* stress condition on the basis of the Lode angle of each element, adopting the relationship between M and Lode angle proposed by Jefferies and Shuttle (2011). The triaxial compression η_{IL} versus ψ relationship is presented in Figure 4 based on NorSand simulations.

3. If the criteria of $\eta > \eta_{IL}$ for any zone is met, the zones are switched to Mohr–Coulomb, maintaining its current shear strength and allowed to soften. The softening is carried out such that the liquefied strength of the tailings is reached within an additional 10 per cent strain. The liquefied strength as a function of ψ as defined by Jefferies and Been (2015), which is used in the model, is outlined in Figure 5.

4. As elements soften, and brittle stress redistribution occurs in the zones. The model is run to see if additional zones exceed the criteria of $\eta > \eta_{IL}$.

5. Steps 2–4 are repeated until either (i) the model achieves equilibrium, and no new elements are triggered, in which case Step 1 is repeated for the next layer or (ii) the onset of instability due to flow liquefaction occurs, represented by the model not achieving equilibrium.

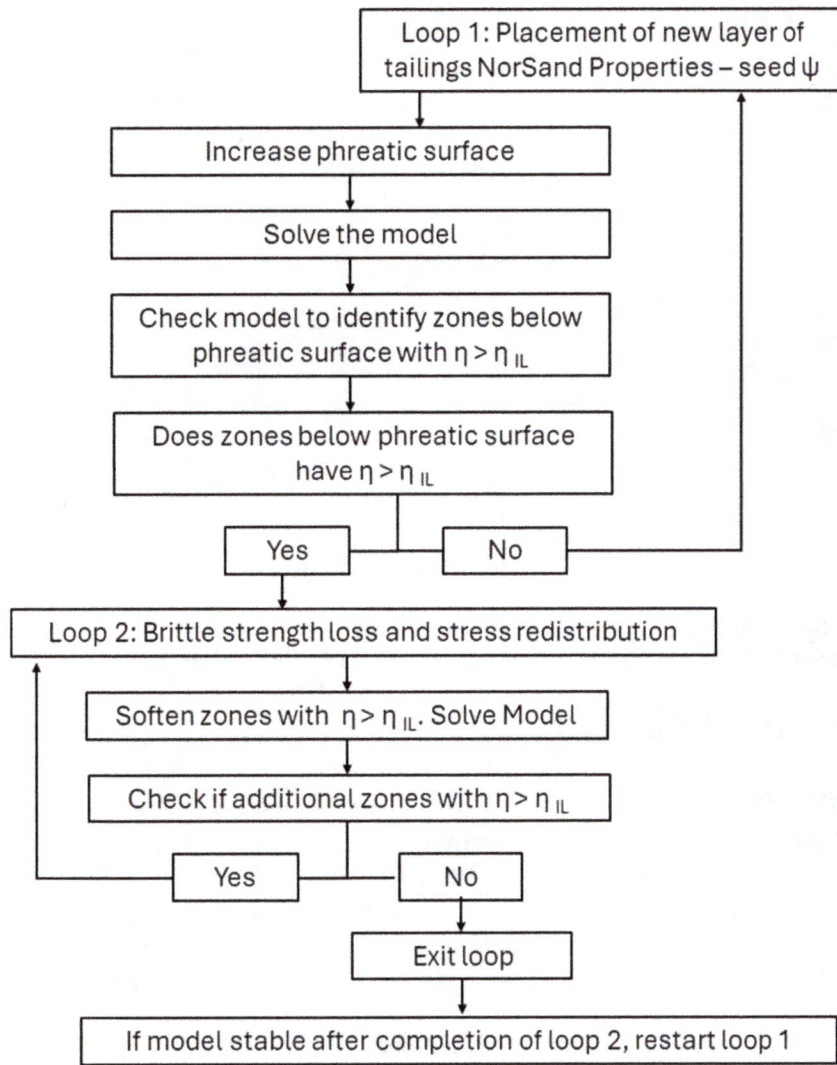

FIG 3 – Pseudo-code of the construction process and triggering assessment.

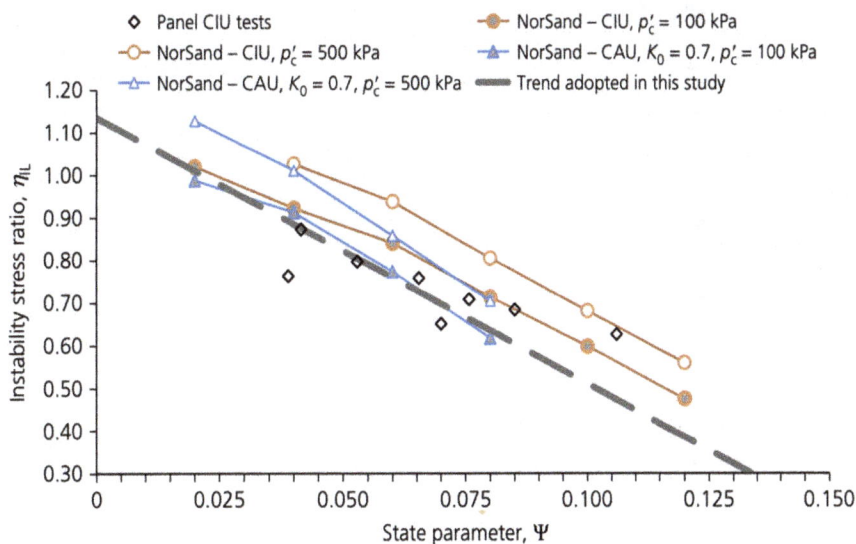

FIG 4 – Trends between instability stress ratio and state parameter, numerical simulations and experimental data for TX-C conditions. CAU, anisotropically consolidated undrained; CIU, isotropically consolidated undrained (Reid *et al*, 2022).

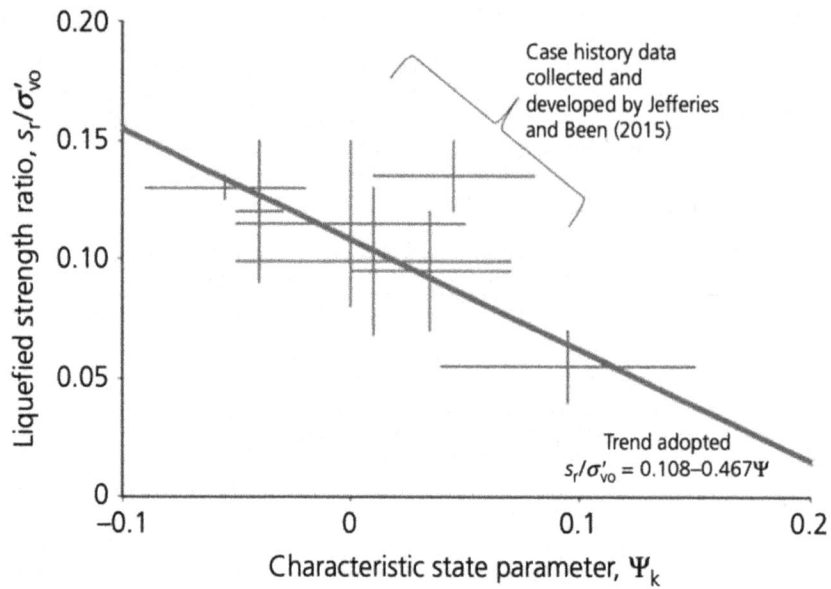

FIG 5 – Flow liquefaction case histories developed by Jefferies and Been (2015) and the trend assumed for NorSand models in the current work (Reid *et al*, 2022).

NUMERICAL MODEL RESULTS

Maximum height achieved

The outcomes of the three scenarios are summarised in Table 2, which indicates that none of the models reached the final tailings height of 50 m. Scenario 2 becomes unstable by the end of stage 14, reaching a maximum height of 35 m. Instability is triggered in Scenario 1 and Scenario 3 by the end of stage 15, reaching a maximum height of 37.5 m. The reasons behind this difference in achieved height and triggering conditions are examined in more detail subsequently.

TABLE 2

Summary of model outcomes for the three scenarios.

Model	Max. height achieved (m)	Area of triggering	Average α, (near slope) (degrees)	Average Intermediate stress ratio b (near slope)
Plane strain (Scenario 1)	37.5	Bottom layer of tailings, close to the toe	42	0.06
Inner corner (Scenario 2)	35.0	Bottom layer tailings, close to the toe	35	0.28
Outer corner (Scenario 3)	37.5	Saturated zone close to the slope face	23	0.14

In situ conditions at triggering

The results of the models are presented as contours in two and three dimensions, where relevant, in Figure 6. For a direct comparison, the results are plotted at the end of stage 13, prior to the onset of instability for all the scenarios, when the height of the tailings is 32.5 m.

To further illustrate the difference between the *in situ* stress conditions in the three scenarios Figure 7, presents the values of b and ψ plotted against distance from the toe for the lower portion of the tailings.

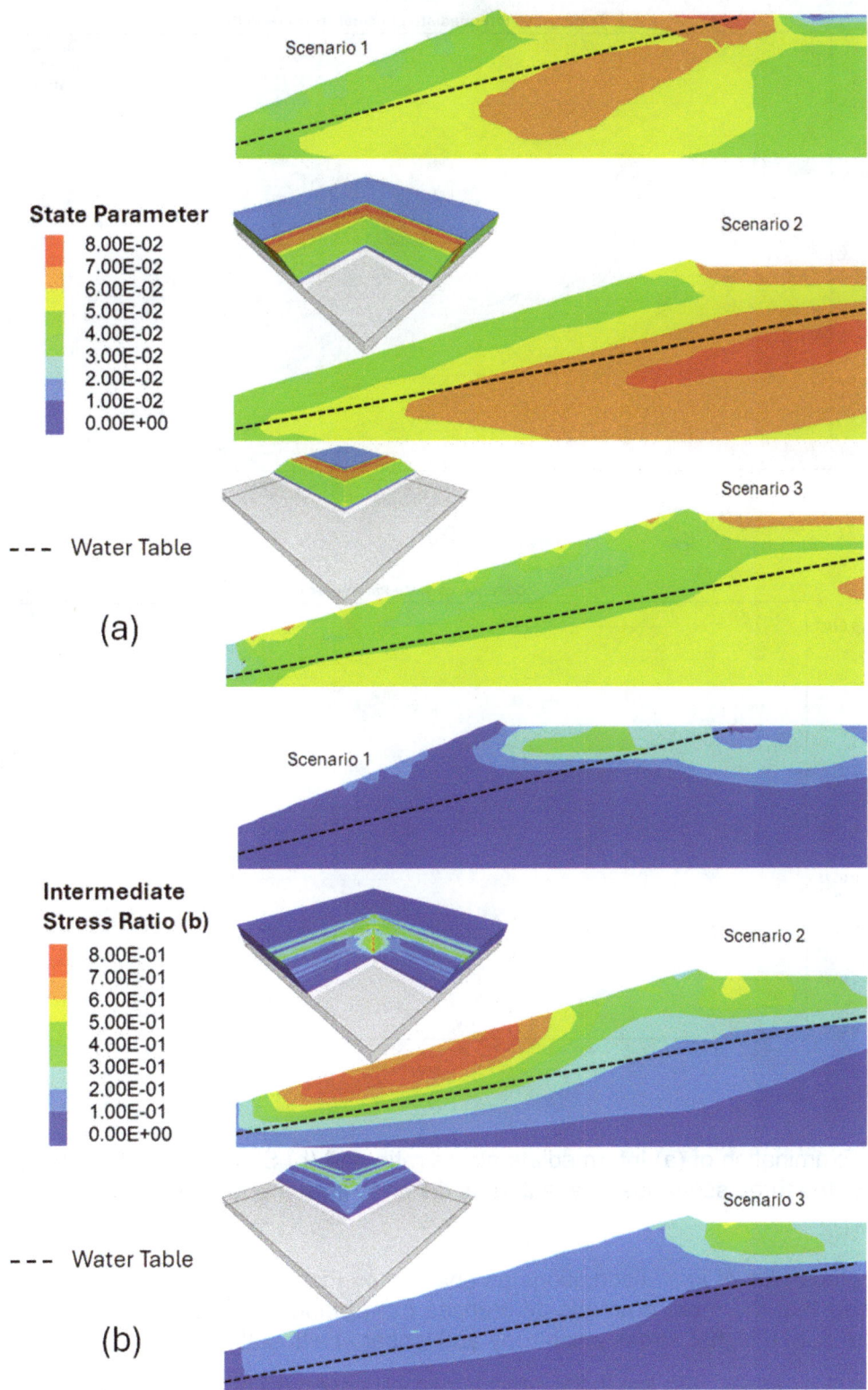

FIG 6 – Contour plots for the three scenarios (a) state parameter (b) intermediate stress ratio (stage 13).

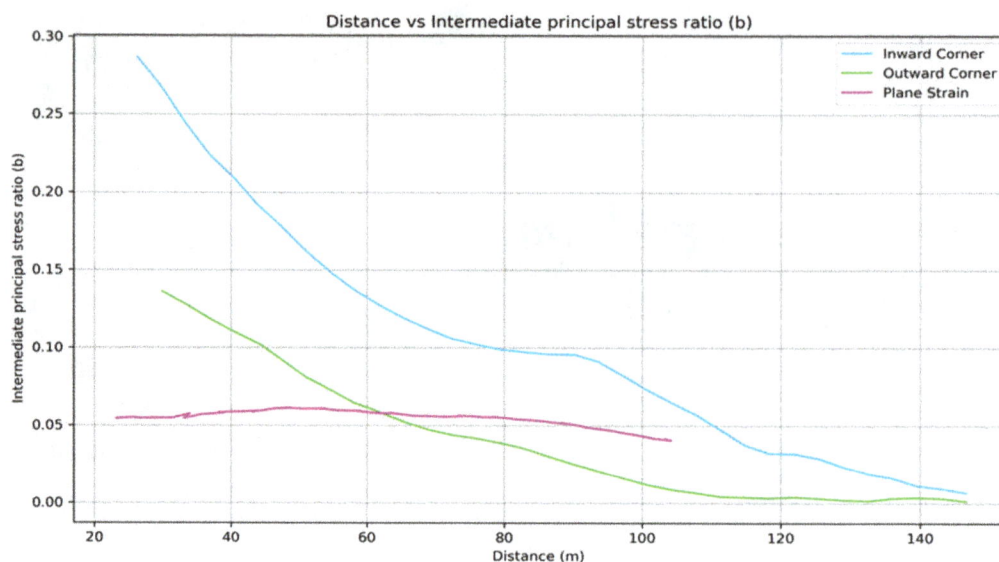

(a)

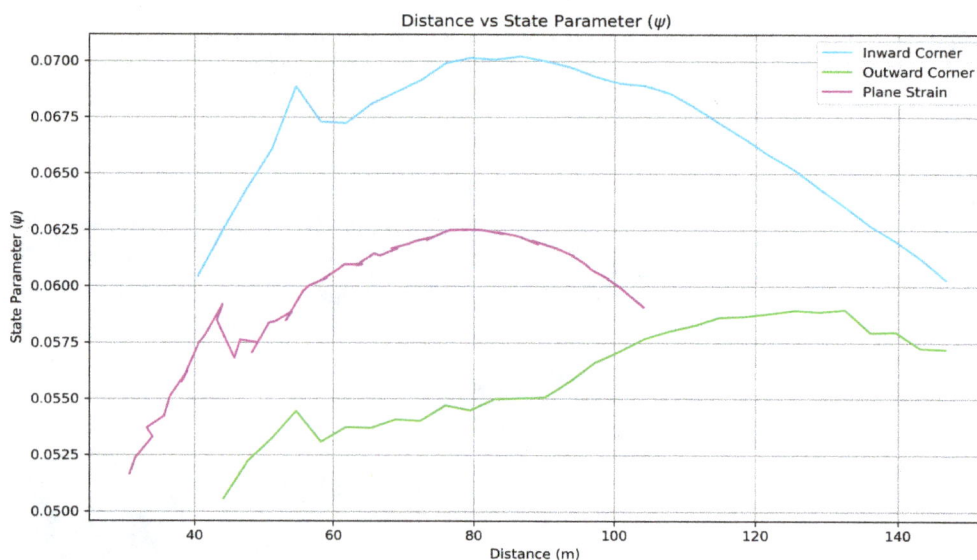

(b)

FIG 7 – Examination of (a) Intermediate stress ratio and (b) State parameter after Stage 13 placement of the three scenarios, presented as distance from the toe for the bottom portion of the model.

The direct comparison of *in situ* conditions as enabled by Figure 6 provides some useful insights – it can be seen that the most dramatic differences from plane strain conditions are at the inward corner, where both a higher stress anisotropy and looser *in situ* state occur.

Ψ values slightly denser than the initial seed ψ can be observed near the slope, pointing to some degree of shear densification during the construction process. This effect is more pronounced in Scenarios 3 and 1 respectively. Scenario 2 presents states looser than initial. It is again emphasised that all models adopted the same seed ψ – the differences seen within the models is therefore a result of the stress conditions and the prediction on consolidation behaviour made by the NorSand constitutive model under these conditions.

Greater variation in intermediate stress ratio (b) can be observed in Scenario 2 across the entire section. However, the low values of b observed in Scenario 1 do not align with those presented by Reid, Fanni and Fourie (2022), pointing to conditions under the slope being closer to triaxial than expected. This suggest that the implementation of NorSand in FLAC3D v9.3 might differ from the original user defined model (UDM) developed by Shuttle and Jefferies (2018), preventing stress anisotropy to be captured correctly.

Principal stress angle

The triggering process carried out in the current models did not include the potential effects of principal stress angle (α). This was based on principal stress angle typically not being included in published examples of numerical models investigating static liquefaction. Alternatively, there is a significant body of experimental evidence suggesting that for sands and tailings principal stress angle is likely to affect η_{IL} (Reid, 2020; Fanni, Reid and Fourie, 2022; Reid, Fanni and Fourie, 2022) – although in some cases, when using moist tamping sample preparation, the effects appear to be negligible (Fanni, Reid and Fourie, 2024). Given the potential for principal stress angle playing a role, the values of alpha along the centreline of each model are presented in Figure 8, further investigation of this potential effect on the outcomes is planned for subsequent investigations of the same idealised models.

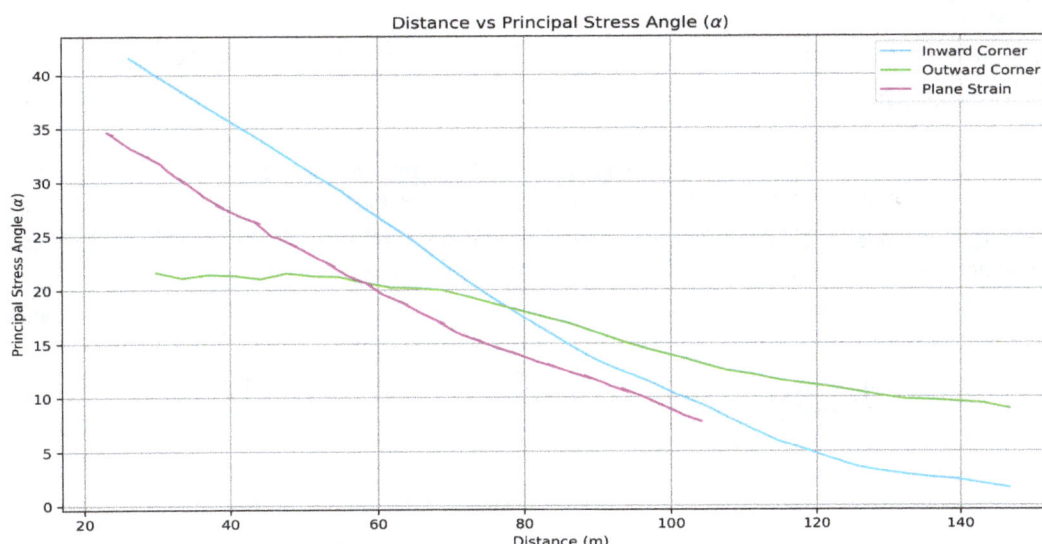

FIG 8 – Principal stress angle (α) variation from the toe of each scenario for the bottom portion of the model.

CONCLUSIONS

In many TSFs, the embankment geometry is sufficiently uniform that two-dimensional plane strain analyses are considered sufficient. This greatly simplifies the process of stress-deformation, which is of benefit given the existing range of challenges and uncertainties in such modelling cases. However, it is common for relatively uniform TSF embankments to feature corners of various geometries. There appears to be little investigation of the effects of such corners, at least in the context of static liquefaction triggering in a stress-deformation framework.

The current paper undertook an investigation of the effects of corner geometry on the potential for statically-triggering flow liquefaction by extending previous analyses of an idealised slope in plane strain conditions to include two three dimensional models, comprising an inner and outer corner geometry in an otherwise identical slope. The assessment was made by comparing *in situ* η to η_{IL} as the material was incrementally deposited in the model, where η_{IL} included the effects of Lode angle 'scaling' based on current state of practice techniques. The conclusions that can be drawn this modelling exercise are outlined as follows:

The inward corner (Scenario 2) showed the most dramatic increase in b and Lode angle, along with a looser state parameter, when compared to plane strain conditions. This combination leads to a lower η_{IL}, allowing the onset of the triggering condition at a lower maximum height, when compared to the other scenarios.

Both the outwards corner (Scenario 3) and plane strain (Scenario 1) trigger at the same maximum height, however the mechanisms are vastly different. Triggering initiates near the bottom for

Scenario 2, similarly to that of the inwards corner. In the case of Scenario 3, failure is initiated in the saturated area closest to the slope face.

While the effect of principal stress angle was not included in the analyses outlined, given the increased principal stress angle seen in the inwards corner model, this may warrant further investigation.

Finally, while some differences were seen between the three scenarios, the results did not strongly suggest an increased propensity for triggering at outer corners such as might have been suggested by the observations of the TVA Kingston failure. However, given the low values of b observed in Scenario 1 when compared to similar works (Reid *et al*, 2022), further investigation and inspection of the implementation of NorSand is required to confirm its reliability. It must be recognised that the results outlined herein are largely dependent on the behaviour of the NorSand constitutive model as implemented in FLAC3D v9.3. Comparison of stress-path element tests and the model performance would serve as a useful verification to these outcomes.

REFERENCES

AECOM, 2009. Root cause analysis of TVA Kingston dredge pond failure on December 22, 2008.

Arroyo, M and Gens, A, 2021. Computational analyses of Dam I at the Corrego de Feijao mine in Brumadinho – Final Report.

Babaki, A P and Tannant, D D, 2024. Numerical evaluation of static liquefaction–induced flowslide causing the Edenville dam failure, Michigan, *Journal of Geotechnical and Geoenvironmental Engineering*, 150.

Fanni, R, Reid, D and Fourie, A B, 2022. Effect of principal stress direction on the instability of sand under the constant shear drained stress path, *Géotechnique*, 74(9):875–891. https://doi.org/10.1680/jgeot.22.00062

Fanni, R, Reid, D and Fourie, A, 2025. Drained and undrained behaviour of a sandy silt gold tailings under general multiaxial conditions, *Géotechnique*, in press. https://doi.org/10.1680/jgeot.23.00186

Firincioglu, B S and Ercanoglu, M, 2021. Insights and perspectives into the limit equilibrium method from 2D and 3D analyses, *Engineering Geology*, 281:105968.

Jefferies, M and Been, K, 2015. *Soil Liquefaction: A Critical State Approach*, 2nd edition (CRC Press: Boca Raton).

Jefferies, M and Shuttle, D, 2011. On the operating critical friction ratio in general stress states, *Géotechnique*, 61:709–713.

Jefferies, M, Morgenstern, N R, Van Zyl, D V and Wates, J, 2019. Report on NTSF Embankment Failure, Cadia Valley Operations, for Ashurst Australia.

Kumar, S, Choudhary, S S and Burman, A, 2022. Recent advances in 3D slope stability analysis: a detailed review, *Modeling Earth Systems and Environment*.

Morgenstern, N R, Vick, S G, Viotti, C B and Watts, B D, 2016. Fundão Tailings Dam Review Panel: Report on the immediate causes of the failure of the Fundão Dam.

Reid, D, 2020. On the effect of anisotropy on drained state liquefaction triggering, *Géotechnique Letters*, 10:393–397.

Reid, D, Dickinson, S, Mital, U, Fanni, R and Fourie, A, 2022. On some uncertainties related to static liquefaction triggering assessments, in *Proceedings of the Institution of Civil Engineers – Geotechnical Engineering*, 175:181–199.

Reid, D, Fanni, R and Fourie, A, 2022. Assessing the undrained strength cross-anisotropy of three tailings types, *Géotechnique Letters*, 12:1–7.

Reid, D, Fourie, A B, Ayala, J L, Dickinson, S, Ochoa-Cornejo, F, Fanni, R, Garfias, J, Viana da Fonseca, A, Ghafghazi, M, Ovalle, C, Riemer, M, Rismanchian, A, Olivera, R and Suazo, G, 2021. Results of a critical state line test round robin program, *Géotechnique*, 71:616–630.

Reid, D, Fourie, A, Dickinson, S, Shanmugarajah, T, Fanni, R, Smith, K, Garfias, J, Yuan, B, Ghafghazi, M and Duyvesty, A, 2023. Results of a dilatancy round robin, *Geotechnical Research*, 12(1):2–28. https://doi.org/10.1680/jgere.24.00008

Robertson, P K, de Melo, L, Williams, D J and Wilson, G W, 2019. Report of the Expert Panel on the technical causes of the Failure of Feijão Dam, I.

Shuttle, D and Jefferies, M, 2018. NorSand Implementation as UDM.

Sottile, M, Sfriso, A, Lizcano, A, Ledesma, O and Cueto, I A, 2022. Flow Liquefaction Triggering Analyses of a Tailings Storage Facility by means of a Simplified Numerical Procedure, 20th International Conference on Soil Mechanics and Geotechnical Engineering (ICSMGE).

Rock fill breakdown under high stress values

S Quintero[1], C Zhang[2], S Wijeweera[3], Z Ci[4] and D Williams[5]

1. Senior Research Technologist, The University of Queensland, St Lucia Qld 4072.
   Email: s.quintero@uq.edu.au
2. Senior Research Fellow, The University of Queensland, St Lucia Qld 4072.
   Email: chenming.zhang@uq.edu.au
3. Geotechnical Engineer, BHP, Brisbane Qld 4000. Email: saduni.wijeweera@bhp.com
4. Geotechnical Engineer, Evolution Mining, Brisbane Qld 4000.
   Email: ci.ooi@evolutionmining.com
5. Emeritus Professor, The University of Queensland, St Lucia Qld 4072.
   Email: d.williams@uq.edu.au

ABSTRACT

The use of rock fill materials in the construction of embankments and storage dams is a standard practice at mine sites worldwide. These materials offer a cost-effective and geotechnically reliable solution when proper construction methodologies are followed. However, due to their unbound nature, rock fill structures are particularly susceptible to water infiltration, whether from rainfall or seepage within storage facilities. This infiltration can significantly impact the material's behaviour, potentially leading to increased settlement, elevated pore water pressure, and even structural failure, which pose critical risks to mining operations.

This study investigates the particle breakdown of a Type 2.3 road base rock fill under high vertical stress and varying moisture content. A large-scale, high-stress consolidation apparatus was utilised to evaluate the performance of the material under vertical stresses of up to 10 MPa. Tests were conducted on samples with three distinct moisture levels, representing conditions from dry to fully saturated. Particle size distribution analyses were performed before and after each test to quantify the extent of material breakdown.

The findings reveal that, contrary to expectations, this particular rock fill material did not demonstrate increased particle breakdown when fully saturated under high-stress conditions. Instead, the rate of particle breakdown remained consistent regardless of moisture content, even under substantial applied stresses.

These results offer valuable insights into the strength and durability of compacted rock fill materials. Such information is critical for optimising the design and ensuring the long-term stability of geotechnical structures in mining environments.

INTRODUCTION

The use of rock fill in civil engineering projects is widely recognised for its high permeability, strength, and durability. It is particularly effective in stabilising steep embankments and providing natural drainage in road construction. While rock fill offers superior stability and drainage, it requires careful handling due to its large size and weight (Dharmawardene, 2004).

In mining construction, rock fill is extensively relied upon for embankment and dam construction (Sitharam and Hegde, 2016) and as backfill in underground mining stopes (Ratnayaka, Brandt and Johnson, 2009). This material can be locally sourced, is cost-effective, and possesses mechanical properties that far exceed the load-bearing requirements for geotechnical designs (Australian National Committee on Large Dams Incorporated (ANCOLD Inc), 2019).

The mining industry is currently under intense scrutiny, driving companies to pursue the goal of 'zero harm to people and the environment,' as outlined in the Global Industry Standard on Tailings Management released in 2020 (Global Tailings Review (GTR), 2020). This standard emphasises rigorous safety protocols, environmental protection measures, and continuous monitoring to prevent accidents and mitigate ecological impact.

While rock fill is a reliable material, it requires continuous testing, monitoring, and maintenance after construction (Oke and Hashemi, 2021). Key technical challenges include long-term settlement and loss of structural integrity, which can result from moisture content fluctuations, transitions from full to partial saturation, and particle breakage under high confinement stress. Long-term issues in rock fill dams may include settlement, cracking, and leakage due to creep deformation, weathering, and cyclic wetting/drying, which can ultimately lead to structural failure (Cheng *et al*, 2023; Zhou *et al*, 2020; Bauer, 2021).

In tailings dams, the constant presence of water keeps the phreatic line close to the surface of the ponds. This water level is crucial to the stability of the dam, as seepage-induced damage can lead to dam failure (Dong, Sun and Li, 2017). The added weight of the saturated tailings below the phreatic line further increases the load on the dam, resulting in slow consolidation, reduced shear and effective stress, and heightened risks of dam instability and failure (Lyu *et al*, 2019).

Conventional testing methods often fail to accurately represent the material used on-site, as they typically involve a smaller, scrapped fraction that disregards the voids created by larger coarse particles excluded in laboratory tests. Common maximum particle sizes in rock fill can exceed 19 mm, with some particles reaching up to 60 mm (Zerui, 2017). In contrast, conventional consolidation and direct shear tests typically use particles of up to 4.75 mm, depending on the mould size. Rock fill generally contains little to no fine particles (below 75 microns).

This study aims to investigate the impact of moisture content variations under high stress conditions, which can cause particle breakage and significant settlement in rock fill material. Large-scale consolidation and shear frame tests were employed to simulate particle sizes comparable to those found on-site. The test regime will assess key rock fill properties, such as particle breakage through particle size distributions, settlement variations via consolidation tests, density changes, mechanical loss during shear stress tests, and the effects of moisture content across a range from dry to fully saturated conditions.

MATERIALS AND METHODS

Material used

The testing regime for this project was conducted on a single sample type: a Type 2.3 road base material sourced from a quarry in south-east Queensland. This material is highly reliable, widely available, and frequently utilised in civil projects, particularly for drainage systems, embankments, and dams.

This quarry material, sourced from an open pit mine, is comparable to hard rock fill material often found in waste rock dumps at mine sites and used in geotechnical structures. This hard rock material from waste rock dumps is similar to the Type 2.3 road base material used in this study. Like quarry material, waste rock dumps are typically the most visually prominent landforms left after open pit mining and are susceptible to erosion (Department of Mines, Industry Regulation and Safety, 2021). The stability of these dumps is primarily controlled by the rock fill, assuming the natural slope has been cleared of vegetation and soft materials. The strength of the rock fill in the waste dumps depends on the degree of weathering and fracturing of the rock, which is indicated by the particle size distribution of the rock fill (Williams and Walker, 1984).

Rock fill is frequently used in the construction of the outer shells of water-retaining dams due to its inherent strength, which allows for the safe and economical construction of steep slopes. Typically, rock fill is sourced by blasting and possibly crushing hard rock from a quarry. The rock fill may also need to be screened to remove oversized boulders that are difficult to place, and washed to remove excessive fines, ensuring that the rock fill is free-draining. Rock fill primarily consists of angular to sub-angular particles of durable, fresh rock, and its behaviour is influenced by various factors, including its mineralogical composition, particle grading, size, and shape. This process is similar to the method used to obtain road base material in pavement structures.

Another advantage of using locally sourced quarry material in this study is the ability to conduct a neutral analysis of a representative material that has been proven reliable for the stability of geotechnical structures, regardless of its location. The term Type 2.3 refers to a specific strength and particle size range that the material must meet. This could also incentivize the characterisation of waste rock, with the aim of standardising its use as rock fill in geotechnical structures.

Material mineralogy

To obtain the mineralogy of the material used in this study, a high-end spectrometer Bruker PUMA Series 2 from the Fire laboratory in the University of Queensland. An X-ray fluorescent basic elements study was carried out on the Type 2.3 road base. Table 1 gives the elements and concentration found.

TABLE 1

XRF test mineralogy.

Basic element	Concentration (%)
Silicon – Si	56.8
Aluminium – Al	15.8
Iron – Fe	11.4
Potassium – K	8.5
Calcium – Ca	4.1
Magnesium – Mg	1.6
Titanium – Ti	1
Sulfur – S	0.3
Manganese – Mn	0.2
Strontium – Sr	0.1
Chromium – Cr	0.1
Rubidium – Rb	0.1

The major elements found are also those that can be generally found in rock fill mineralogy. Silicon and calcium being the must predominant elements that can be related to rock fill (Oyedotun, 2018).

Test methodology

To assess the initial properties of the material, particle size distribution and standard compaction tests were performed. These tests provided data on the initial size distribution and determined the optimum moisture content (OMC) and maximum dry density (MDD) of the sample.

The study was divided into two main scenarios:

- Scenario 1: Fully Saturated Sample Under Confinement

  This scenario examined the behaviour of the sample under fully saturated conditions, simulating its location below the phreatic line and subjected to large confinement loads. For this phase, the sample was maintained at a constant target moisture content of 13.4 per cent.

  Three subsamples were prepared at varying dry densities to represent different levels of dam stability, with the MDD indicating the most stable condition.

  Large shear tests were conducted under normal loads of 50 kPa, 100 kPa, and 200 kPa.

  Key parameters analysed in this scenario were the friction angle and changes in dry density.

- Scenario 2: Particle Breakage at High Confining Pressures

  The second part of the study focused on the extent of particle breakage in the rock fill under high confining pressures, with varying moisture contents. Unlike most previous studies, which primarily examine fully saturated or completely dry samples, this research included partially saturated samples, reflecting conditions similar to those expected during construction (near the OMC).

Three moisture content levels were investigated:

1. Dry condition (no added moisture).
2. Partial saturation (target moisture content of 6.4 per cent, equivalent to OMC).
3. Full saturation (target moisture content of 13.4 per cent).

Each sample prepared at these target moisture contents underwent single-stage vertical consolidation tests under two stress levels:

1. 5 MPa, approximating a depth of 250 m.
2. 10 MPa, approximating a depth of 500 m.

These tests aimed to evaluate the settlement behaviour and particle breakage under realistic *in situ* loading conditions, offering insights into how moisture content and confining pressure influence the stability and integrity of rock fill materials.

Particle size distribution tests

The particle size distribution, shown in Figure 1, was performed on the received sample to determine its initial particle size range.

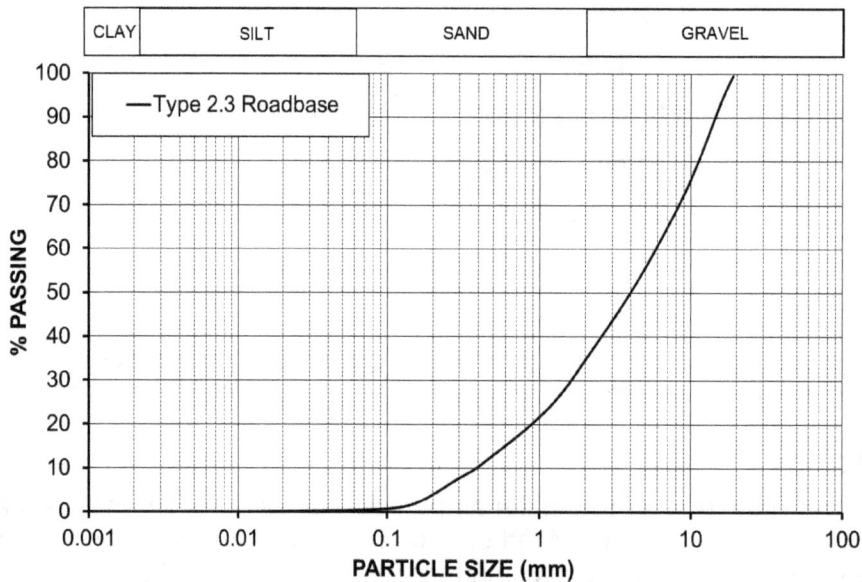

FIG 1 – Original dry PSD on Type 2.3 road base.

The PSD analysis indicates that the sample is predominantly gravel, with over 60 per cent of particles exceeding 2.36 mm in size. Additionally, the sample contains minimal fine material, with less than 1 per cent passing the 75 µm sieve. Based on these characteristics, the material can be classified as sandy gravel. Figure 2 illustrates the physical appearance of the sample.

FIG 2 – Type 2.3 material.

Standard compaction test

Figure 3 presents the curve for the standard compaction test conducted using the 19 mm particle size equivalent mould. The MDD is 2175 kg/m^3, and the OMC is 6.5 per cent.

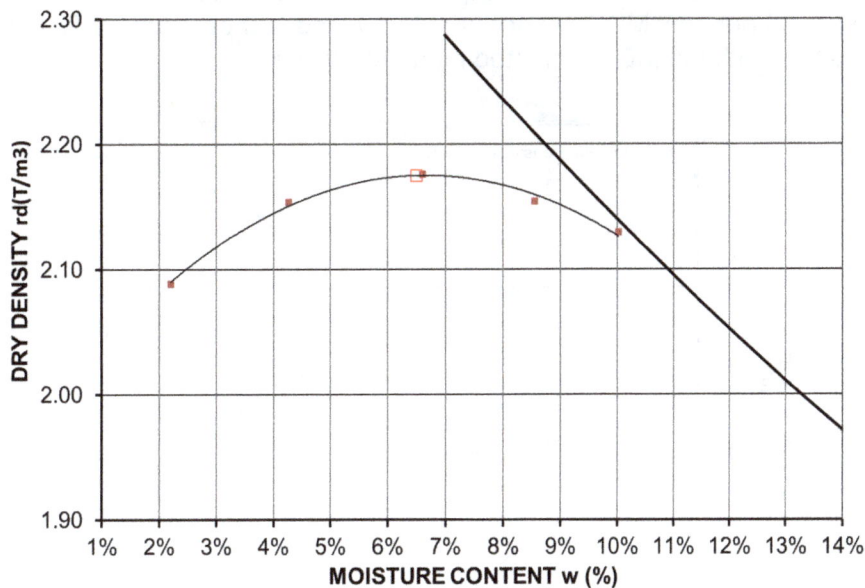

FIG 3 – Standard compaction test results.

Large shear stress test

Large shear test was conducted on samples measuring 150 mm × 150 mm, with a sample height of 15 mm to maintain a 1:1 aspect ratio (Figure 4). This configuration allows for the inclusion of coarser particles in the sample without causing tilting of the sample height during testing. This specialised equipment is located in the Geomechanics Laboratories at the University of Queensland.

FIG 4 – Large Shear frame.

Sample preparation and density values

The sample was prepared with a target moisture content (MC) of 13.4 per cent. Three different dry density values were used to prepare each sample. Based on the MDD of 2175 kg/m³, the selected target densities were 80 per cent MDD, 90 per cent MDD, and 100 per cent MDD. Figure 5 illustrates the achieved densities, with the MDD indicated by a dotted line.

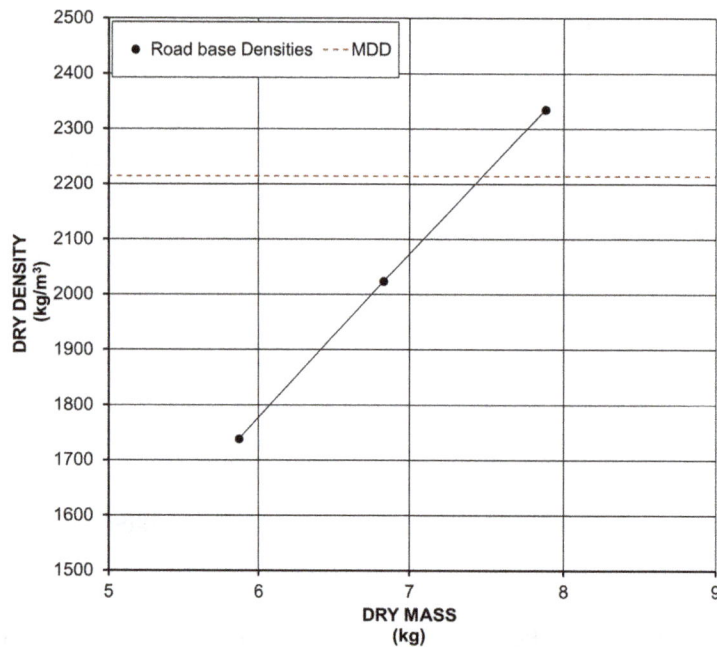

FIG 5 – Achieved dry density values per sample.

Large scale high stress oedometer test

The loading frame used for these consolidation tests was a custom-built high-stress oedometer located in the Geomechanics Laboratories at the University of Queensland. This apparatus can apply hydraulic pressures of up to 20 MPa on samples with diameters of 150 mm and heights of up to 150 mm. These large-scale dimensions allow for the inclusion of coarser particles in the samples, enabling a more accurate representation of the maximum particle size used in rock fill on-site. Figure 6 shows the loading frame and sample size.

FIG 6 – High stress oedometer.

The primary parameters for analysis after each test include the void ratio and density changes, the index of compressibility, sample settlement, and an indirect calculation of hydraulic conductivity.

Following the consolidation tests, each sample was collected for further particle size distribution analysis. These tests were then compared to the initial dry particle size distribution of the sample when first received, providing quantifiable insights into particle breakage across different moisture contents.

Sample preparation and moisture content values

The material was prepared with a consistent target dry density of 2175 kg/m^3, corresponding to the sample's MDD. Three moisture content levels were evaluated: a dry sample at 0 per cent MC, a partially saturated sample with a target OMC of 6.5 per cent, and a fully saturated sample at 13.4 per cent MC. The resulting total mass variations for each moisture content is given in Figure 7.

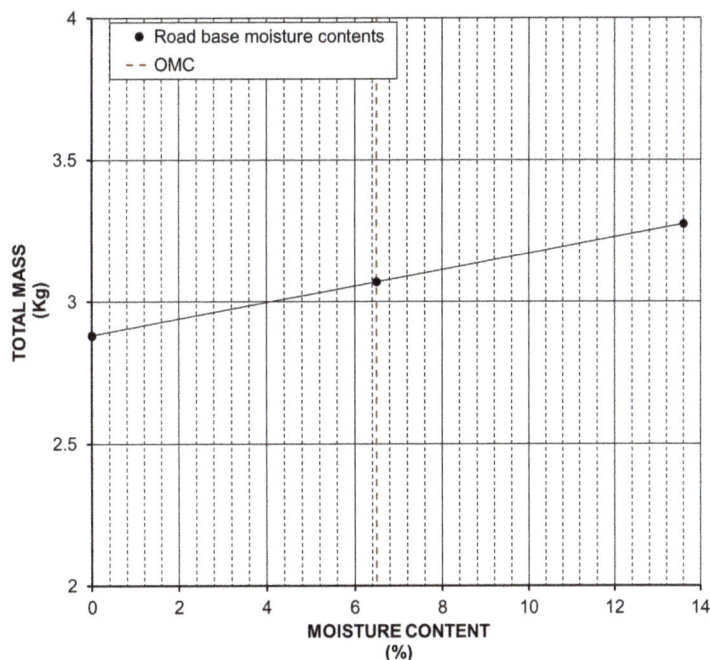

FIG 7 – Achieved moisture contents per sample.

TEST RESULTS

Influence of density on mechanical parameters of the material

The materials prepared at loose density (80 per cent MDD), medium density (90 per cent MDD) and high density (100 per cent MDD) on the large shear frame were subject to vertical loads of 50 kPa, 100 kPa and 200 kPa, respectively. The tests were done on single stage at a shearing rate of 1 mm/min.

The first parameter to analyse is the settlement of all samples and all loads. Figure 8 shows an expected behaviour for loose material with very little density of 1 mm. The denser the sample becomes, the greater the dilatancy. This is a common behaviour for dense granular material. The highest dilatancy was evident at the 50 kPa load for the highly dense sample.

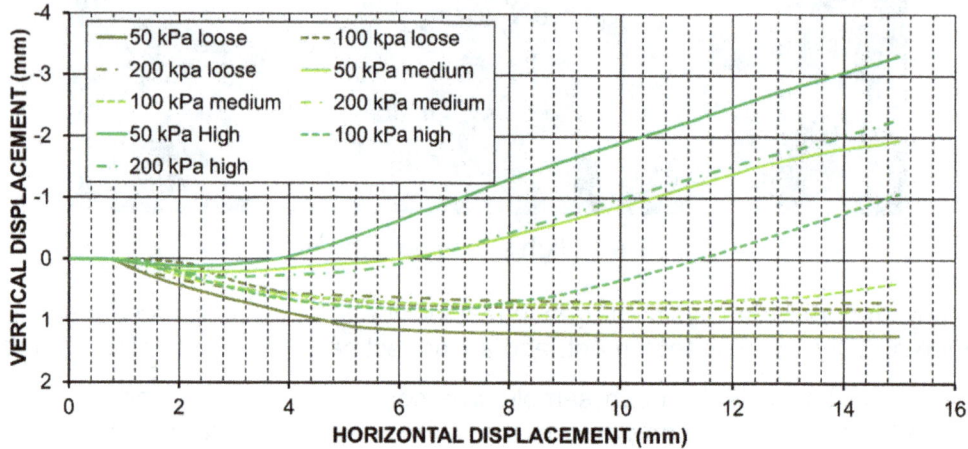

FIG 8 – Vertical displacement of shearing tests.

The dilatancy is an indication that as the material is denser, meaning less voids between particles, the coarser particles overlap with each other as the sample is subject to a shearing strength. This results in the particles having to move on top of each other causing in an apparent expansion of the sample indicating a loss in the density. This also means there is likely more particle breakage when the sample is dilating. At a dry density of 100 per cent MDD, the dilatancy occurs on all loads, meaning there is some particle breakage from low vertical loads. For the settlement of 80 per cent MDD, the particle still has voids so there is only consolidation of the material. It is minimum with not larger than 1.2 mm in a sample of 150 mm of initial height. The 90 per cent sample shows varying results. On one side, the 50 kPa shows the greatest dilatancy where the sample initially consolidates and then it dilates before the peak shear occurs (around 4 mm). The vertical loads of 100 kPa and 200 kPa are larger enough to continue to consolidate the sample as it shears, implying the sample is not dense enough to dilate.

Figure 9 shows the mechanical parameters of the shear strength test. It is worth noting that there is an increasing apparent cohesion that plateaus at 100 kPa. This cohesion is given by the water filling up the voids between particles.

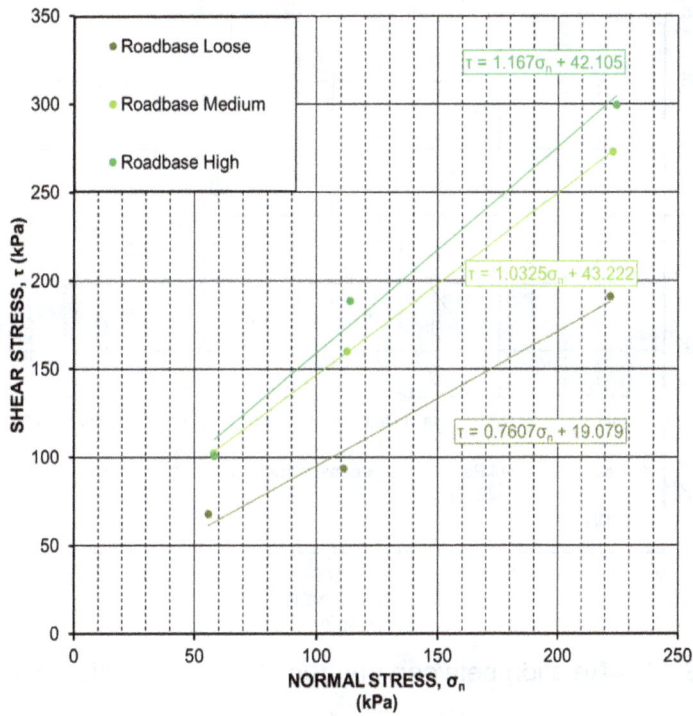

Legend in figure:
- Roadbase Loose
- Roadbase Medium
- Roadbase High

$\tau = 1.167\sigma_n + 42.105$

$\tau = 1.0325\sigma_n + 43.222$

$\tau = 0.7607\sigma_n + 19.079$

FIG 9 – Friction angle changes of the direct shear tests.

Table 2 shows the test results for each large shear test. the friction angle increases as expected to almost a rock like material. This is despite the high moisture content. The proximity between the friction angle values between the medium density and high-density sample indicate that the mechanical strength is not lineal but rather logarithmic or polynomial, reaching plateau with further vertical stress increments.

TABLE 2

Shear stress parameters.

Sample type	Corrected normal stress (kPa)	Corrected max. Shear stress (kPa)	Cohesion (kPa)	Friction angle (°)	Initial measured dry density (kg/m^3)	Final dry density (kg/m^3)
Low density	55.77	68.01	19.079	37.26	1738.3	1822.7
	111.11	93.85			1738.3	1844.7
	222.09	191.28			1738.3	1688.3
Medium density	57.75	102.23	43.222	45.92	2023.7	2067.4
	112.29	160.09			2023.7	2099.6
	222.96	273.13			2023.7	1906.8
High density	58.27	100.82	42.105	49.41	2335.0	2401.6
	113.69	188.70			2335.0	2425.2
	224.57	299.54			2335.0	2189.2

Figure 10 shows a comparison between the initial dry density and the dry density reached at peak shear stress. This is done assuming a constant friction angle per vertical load, per sample. It can be observed that the dry density changes are smaller at peak, however a decrease of the dry density is evident due to the dilatancy occurred during the test. This also shows that the initial consolidation, before shearing is not large enough to prevent the sample from lowering its dry density below its maximum designed dry density (MDD).

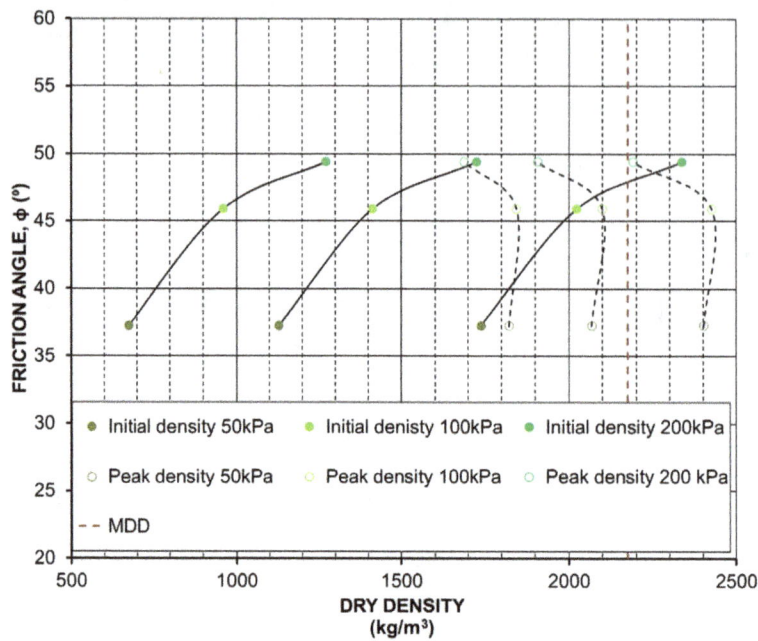

FIG 10 – Relation between dry density versus friction angle.

Influence of moisture content on the sample settlement and particle size

The consolidation tests proposed on this study have two major components to analyse. One is the particle breakage after consolidation for each moisture content, and the second is the effect of the moisture content in the settlement for each consolidation cycle.

Particle breakage after consolidation

Figure 11 shows a summary of all PSD tests after each consolidation cycle for all moisture contents plus the original sample PSD.

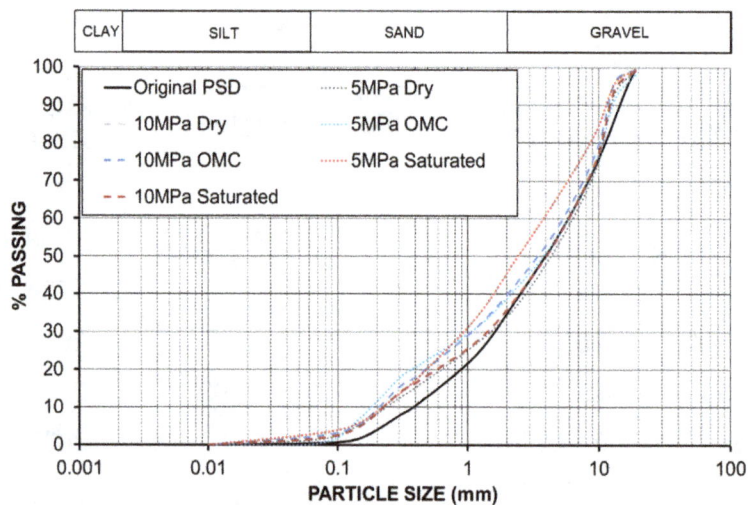

FIG 11 – PSD tests for all moisture contents.

Overall, the particle breakage regardless of the consolidation still warrants that the material is kept within its range of gravel size. The fines do increase on all samples but remains below 5 per cent of the total sample. This means the sample still complies with the PSD ranges required for construction. The highest particle breakdown occurs at 2.36 mm and below.

Table 3 shows the material passing below specific particle percentages compared to the original PSD. The percentage change in effective particle stress increased as the applied stress rose from 5 MPa to 10 MPa. Notably, the finer particles experienced the highest levels of breakdown, with the percentage change being most pronounced in this size range, while coarser particles showed little to no degradation.

TABLE 3

PSD of sample under various loads and moisture conditions.

Particle percentage relations	5 MPa			10 MPa		
	Dry	OMC	Saturated	Dry	OMC	Saturated
D10	50	50	40	50	50	42.5
D30	6.25	18.75	28.13	37.5	31.25	28.75
D50	0	0	0	35.9	12.8	0
D60	0	0	0	36.6	10	0
D90	0	0	0	12.3	6.2	0.7

Consolidation test results

A total of six single-stage high-stress consolidation tests were conducted, with three tests performed at 5 MPa and three at 10 MPa. Each test was run for approximately 24 hrs, although full consolidation was generally achieved within the first hr of load application.

As illustrated in Figure 12, the sample prepared under partial saturation (with a moisture content near 6.5 per cent, corresponding to the OMC) exhibited the greatest settlement. This behaviour was consistent across both 5 MPa and 10 MPa loading conditions. The samples exhibiting the least settlement under 5 MPa loading were the fully saturated and driest samples, indicating that the level of saturation significantly influences settlement behaviour.

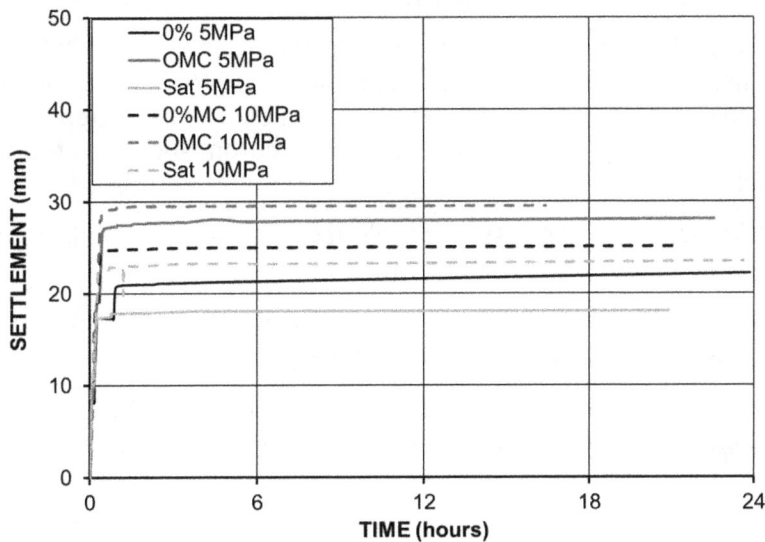

FIG 12 – Settlement for high stress oedometer tests.

Similarly, Figure 13 demonstrates that the sample prepared at the OMC achieved the highest dry density. This suggests that at OMC, particles have sufficient mobility to rearrange and minimise void spaces effectively.

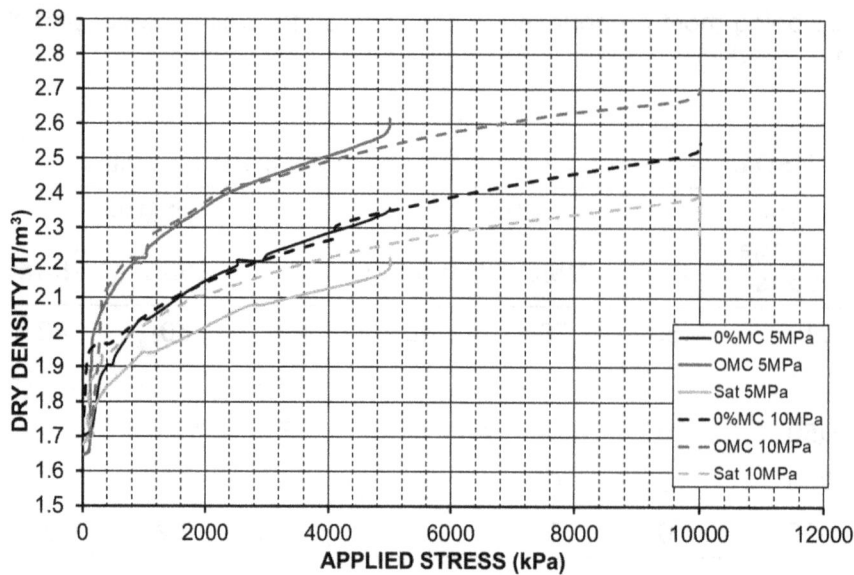

FIG 13 – Dry density results for consolidation tests.

CONCLUSIONS

This study investigated the effect of various moisture contents—targeting 0 per cent, 6.5 per cent, and 13.5 per cent, representing dry, partially saturated, and fully saturated states—on a rock fill sample under varying confining pressures. The aim was to determine the optimal conditions for compacting rock fill. A high-stress consolidation cell capable of applying loads up to 10 MPa was utilised to simulate real *in situ* loading conditions without requiring excessive scalping of the samples to fit the mould.

The key findings and recommendations are as follows:

- Effect of Density on Friction Angle: Samples with lower densities exhibit a lower friction angle, as expected. There is a significant increase in friction angle for samples at 90 per cent of the MDD, after which it remains constant at higher densities. For saturated samples, this indicates that the mechanical integrity of the rock fill is only compromised when the density drops below 90 per cent MDD.

- Dilatancy and Dry Density: The loss of dry density due to particle dilatancy during shearing becomes more pronounced at the highest dry density.

- Particle Breakdown: Particle breakdown predominantly occurs in the coarse sand and fine gravel size ranges, especially around 2.36 mm. Larger particles remain largely intact.

- Partially Saturated Samples: Partial saturation leads to the most significant particle breakage and settlement under confining pressures. This suggests that higher consolidation occurs in partially saturated samples, as more voids allow particles to rearrange under confinement.

- Long-Term Monitoring Recommendations: Since partial saturation aligns closely with the OMC used during construction, proper long-term monitoring of settlement and moisture changes is recommended for rock fill structures. If MDD drops below 90 per cent under saturation, the mechanical stability of the material can significantly deteriorate. It is critical to target rock fill placement around the phreatic line for stability.

- Settlement Under Compaction Pressures: Samples compacted under 5 MPa and 10 MPa show the greatest settlement at OMC compared to dry and fully saturated samples.

- Void Ratio Behaviour: At 5 MPa loading, samples at OMC achieved the lowest void ratio. Under 10 MPa loading, fully saturated samples exhibited the lowest void ratio.

- Particle Size Distribution: Overall, the tested rock fill samples were considered well-graded, except for fully saturated samples under 10 MPa, which deviated from this grading.

ACKNOWLEDGEMENT

The author would like to thank and acknowledge the prompt contribution from Dr. Sergio Zarate Galindo from the Fire group of the University of Queensland to carry out the XRF testing in such short notice.

REFERENCES

Australian National Committee on Large Dams Incorporated (ANCOLD Inc), 2019. Guidelines On Tailings Dams - Planning, Design, Construction, Operation And Closure, revision 1, ANCOLD Inc. Available from: <https://ancold.org.au/product/guidelines-on-tailings-dams-planning-design-construction-operation-and-closure-revision-1-july-2019/>

Bauer, E, 2021. Long-Term Behavior of Coarse-Grained Rock fill Material and Their Constitutive Modeling, in *Dam Engineering – Recent Advances in Design and Analysis* (eds: Z Fu and E Bauer), (Rijeka: IntechOpen).

Cheng, J, Ma, G, Zhang, G, Chang, X and Zhou, W, 2023. A theoretical model for evaluating the deterioration of mechanical properties of rock fill materials, *Computers and Geotechnics,* 163:105757. https://doi.org/10.1016/j.compgeo.2023.105757

Department of Mines, Industry Regulation and Safety, 2021. Waste Rock Dumps Guidelines, Government of Western Australia, Australia.

Dharmawardene, W, 2004. The use of rock fill – some considerations, presented at the 57th Canadian Geotechnical Conference.

Dong, L, Sun, D and Li, X, 2017. Theoretical and Case Studies of Interval Nonprobabilistic Reliability for Tailing Dam Stability, *Geofluids,* 2017(1):8745894. https://doi.org/10.1155/2017/8745894

Global Tailings Review (GTR), 2020. Global Industry Standard on Tailings Management (GISTM) [online], Global Tailings Review. Available from: <https://globaltailingsreview.org/global-industry-standard/>

Lyu, Z, Chai, J, Xu, Z, Qin, Y and Cao, J, 2019. A Comprehensive Review on Reasons for Tailings Dam Failures Based on Case History, *Advances in Civil Engineering,* 2019:4159306. https://doi.org/10.1155/2019/4159306

Oke, J and Hashemi, A, 2021. In situ backfill monitoring database, in *Paste 2021: Proceedings of the 24th International Conference on Paste, Thickened and Filtered Tailings* (eds: A B Fourie and D Reid), pp 353–368 (Australian Centre for Geomechanics, Perth). https://doi.org/10.36487/ACG_repo/2115_29

Oyedotun, T D T, 2018. X-ray fluorescence (XRF) in the investigation of the composition of earth materials: a review and an overview, *Geology, Ecology and Landscapes,* 2(2):148–154. https://doi.org/10.1080/24749508.2018.1452459

Ratnayaka, D D, Brandt, M J and Johnson, M K, 2009. *Twort's Water Supply* (Elsevier).

Sitharam, T G and Hegde, A, 2016. Stability analysis of rock fill tailing dam: an Indian case study, *International Journal of Geotechnical Engineering,* 11(4):332–342. https://doi.org/10.1080/19386362.2016.1221574

Williams, D and Walker, L, 1984. *Laboratory and Field Strength of Mine Waste Rock,* presented at the Conference on Geomechanics Perth.

Zerui, B M, 2017. An Innovative Approach to Determine Particle Size Distribution for Rock fill, Dams and Water 1 for the Future.

Zhou, X, Chi, S, Jia, Y and Shao, X, 2020. A new wetting deformation simulation method based on changes in mechanical properties, *Computers and Geotechnics,* 117:103261. https://doi.org/10.1016/j.compgeo.2019.103261

Distinguishing between the different types of post-peak strength loss

D Reid[1]

1. Principal Tailings Engineer, Red Earth Engineering, Perth WA 6000.
   Email: david.reid@redearthengineering.com.au

ABSTRACT

The propensity for loose soils and tailings to undergo a post-peak loss of strength (ie strain softening) is a fundamental aspect of tailings storage facility (TSF) failures involving slope instability. The loss of strength that can occur in geomaterials – often dramatic, and with little warning, can lead to flow slides and thus significant damage to the surrounding area. Indeed, for this reason much current analysis of tailings and foundation soil mechanical behaviour is focused on determining brittleness, ie the magnitude of post-peak strength loss.

Factors contributing to the post-peak loss of strength of geomaterials include shear-induced excess pore water pressure (SIEPWP) in contractive soils during undrained shearing, or frictional softening where platy clay particles align to form a 'slickenslide' with lower effective frictional strength. While both mechanisms have seen considerable discussion, experimental evidence, and implementation in the analysis of slope failures, there remains much ambiguity as to which mechanism is potentially relevant in a given situation. Further, there appears to be no consensus as to if these two mechanisms interact in some cases, and if so when and how, with this having considerable importance in some slope models where the potential for a soil to 'stay' weakened could influence the potential future stability of a structure.

The current paper carries out a review of historical laboratory testing and other experimental evidence to examine the interaction of frictional softening and shear induced SIEPWP in post-peak strength loss of geomaterials. The relevance of this distinction on some recent TSF failures and their post-failure investigation process is outlined. Experimental techniques to assess which mechanism is relevant and in what proportion are discussed. Finally, the potential for post-peak reconsolidation to lead to an increase in strength for bonded materials is examined through additional experiments.

INTRODUCTION

The strain softening nature of loose tailings and some foundation soils has been a major factor in many of the recent tailings storage facility (TSF) failures of the past decade. Indeed, strain softening behaviour, in the form of qualitative stress-strain plots and ranges of Brittleness Index I_B (Bishop, 1967), now forms integral guidance for the stability analysis of TSFs (ICOLD, 2023). Investigation of the brittleness of tailings now forms regular engineering practice for such facilities. An idealised stress-strain curve featuring stain softening is presented as Figure 1 for illustration purposes.

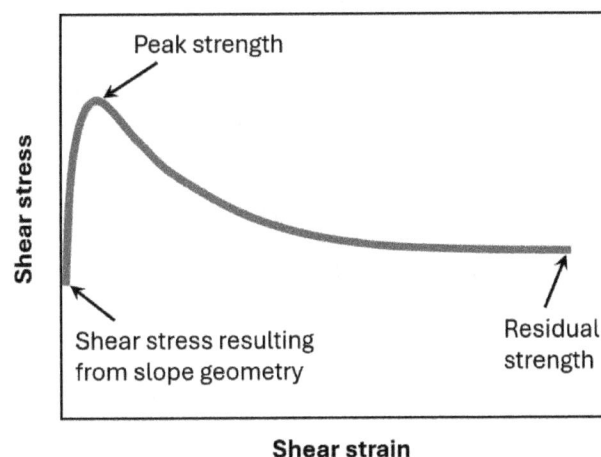

FIG 1 – Idealised stress-strain curve showing brittle strength loss.

While strain softening of tailings and soils noted in Figure 1 is ubiquitous, an important distinction must be drawn between two forms of strain softening, demarcated as follows for the purpose of the current paper:

- *Type I:* Contractive undrained shearing of loose soils, where the reduction in shear strength with strain is a direct result of shear-induced excess pore water pressure (SIEPWP) as a contractive soil is prevented from reducing in volume owing to an inability to discharge pore fluid in a time frame relevant to the shearing process. There is no suggestion of a reduction in friction angle in such cases – indeed, conceptualising the contractive undrained shearing process, say the peak strength observed, with a reduced friction angle can lead to much conceptual difficulty (eg as discussed by Jefferies and Been, 2015). Importantly, were Type I strain softening to occur in a discrete portion of a TSF and failure of the slope did not occur, there is every reason to believe that strength gain would occur over time as SIEPWP dissipated – indeed, this is exactly what is shown by the available experimental evidence that is discussed subsequently.

- *Type II:* Drained frictional softening of soils made up of a significant proportion of clay particles, wherein there is an unambiguous reduction in the friction angle of the soil owing to the creation of a 'slickenslide' where clay particles have aligned. This form of soil behaviour has undergone significant historical laboratory investigation (eg Lupini, Skinner and Vaughan, 1981) and has been the leading cause of a number of slope failures (Alonso and Gens, 2006a, 2006b, 2006c). Importantly, there is little evidence that in most cases the reduction in friction angle will subsequently 'heal' over time (Mesri and Huvaj-Sarihan, 2012; Stark and Hussain, 2010). That is, once drained strain softening has occurred, there is unlikely to be a significant improvement in the future.

The two types of strain softening outlined are distinct and well documented in the geotechnical literature. However, the author sees frequent examples where strain softening behaviour is being discussed in geotechnical practice and it is unclear which type is being considered, leading to uncertainty and inconsistency. This is particularly important given the increased advocation for performance-based design – if there is an intention to capture the behaviour of a TSF over a period of decades, it becomes increasingly important what assumptions will be made regarding the long-term behaviour of isolated regions of strain-softened material. Further, while the two distinct cases have been well studied in isolation, there is arguably overlap in these two cases – specifically, the large-strain behaviour of clays during undrained shearing, and whether frictional softening may become involved. The lack of full understanding of some of these issues by the geotechnical profession has also been discussed (Thakur *et al*, 2014).

The purpose of the current paper is as follows:

- to explicitly outline the two types of strain softening

- highlight areas of potential overlap of the two mechanisms and available data to enable investigating this

- examining examples in recent case histories that provide an indication as to how these processes are accounted for in current engineering practice

- outline some tests carried out to expand our understanding of these issues with respect to the potential for reconsolidation-based strength gain of cemented soils.

DISTINCT TYPE I AND TYPE I STRAIN SOFTENING

Purpose

To provide greater context and detail for the reader on the mechanics of the two forms of strain softening discussed herein, examples of test data and first provided.

Type I – shear-induced excess pore water pressure

Strain softening as a result of shear-induced excess pore water pressure is first demonstrated by means of an undrained triaxial compression test on a gold tailings carried out by the author for demonstration purposes. Figure 2a shows the conventional stress-strain response of a brittle soil, while also showing the SIEPWP developing during shearing, owing to the contractive nature of the soil and the restriction of drainage in an undrained test. Figure 2b shows mobilised friction angle (ie frictional strength developed at that point of the test) against axial strain for the same test. Presenting the test data in this manner clearly demonstrates the increasing mobilised friction as the soil tends towards critical – clearly, no reduction in friction angle is seen, or is responsible for the strain softening response. The strain softening response in such cases is solely a result of the significant contractile response of loose soils. This form of behaviour is observed in a wide range of loose soils, from sands to clays.

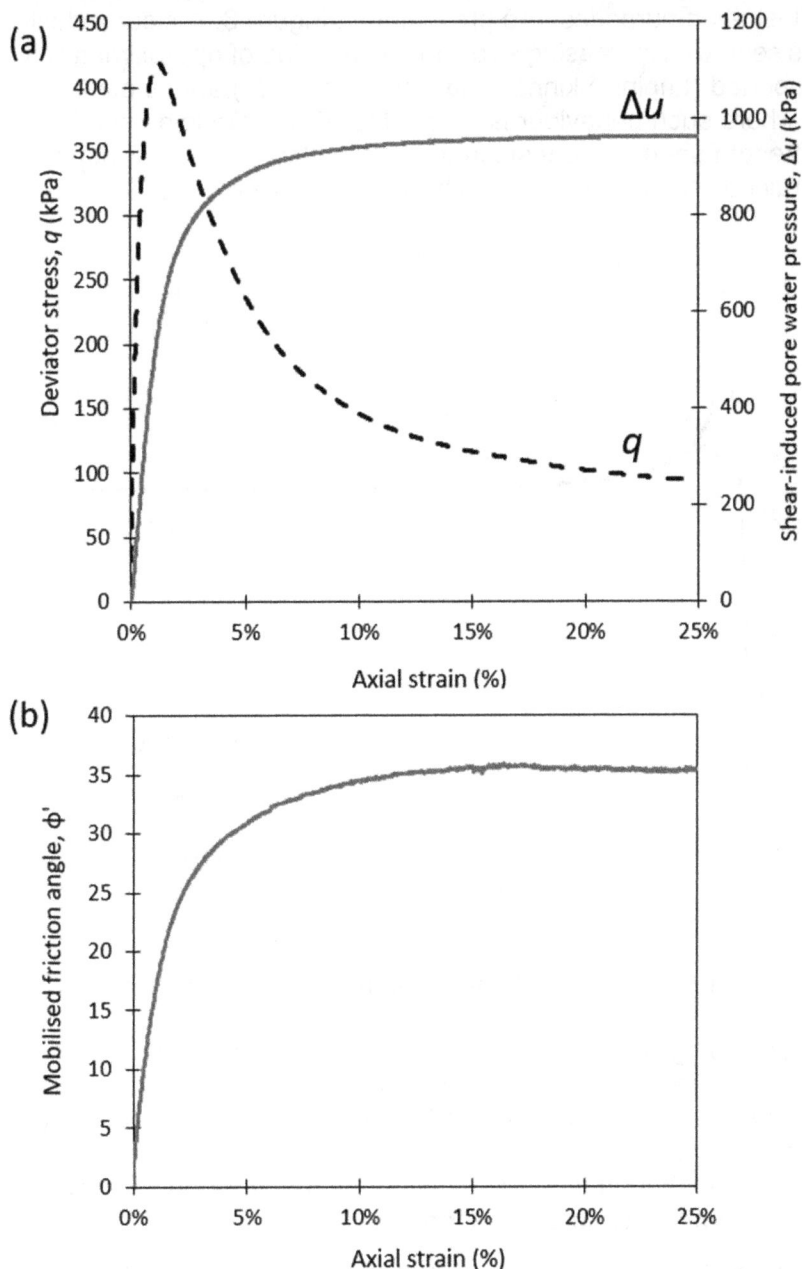

FIG 2 – Undrained triaxial response of loose gold tailings: (a) deviator stress and SIEPWP; and (b) mobilised friction angle.

A less common consideration for soils that behave such as that in Figure 2 is what would occur *in situ* if an element or zone of material underwent such a response, but where the overall slope did not fail – that is, despite the loss of strength and resulting brittle load shedding in a small part of the slope the overall system remained stable. In such a case, the significant SIEPWP would then have time to dissipate. Vulpe and Fourie (2025) present results examining this behaviour, showing how post-shearing reconsolidation (ie allowing dissipation of SIEPWP) then leads to an increase in the long-term strength of the element. As will be seen subsequently, this increase in strength of small zones that undergo strain softening, and then are allowed time for SIEPWP dissipation, is not commonly included in modelling of TSFs.

Type II – drained frictional strain softening

Drained strain softening – ie where excess pore pressure development is clearly excluded from the process – is perhaps best outlined by means of drained ring shear tests. In such a test, with very low shear rates and open drainage boundaries, clays can be sheared to high strains to examine the evolution of their friction angle without SIEPWP. The results of a typical test of this type, carried out by the author for illustration purposes, are presented in Figure 3. A clear reduction in the mobilised friction from ~18° is seen with increasing strain to a final value of approximately 13°. Such behaviour has been widely reported (Lupini, Skinner and Vaughan, 1981) and is commonly incorporated into numerical models where such behaviour is suspected (Potts, Dounias and Vaughn, 1990). Finally, while the loss of strength seen in the tests in the kaolin clay in Figure 3 appears relatively modest, the potential for drained strength loss is much greater in overconsolidated soils and/or soils with structure

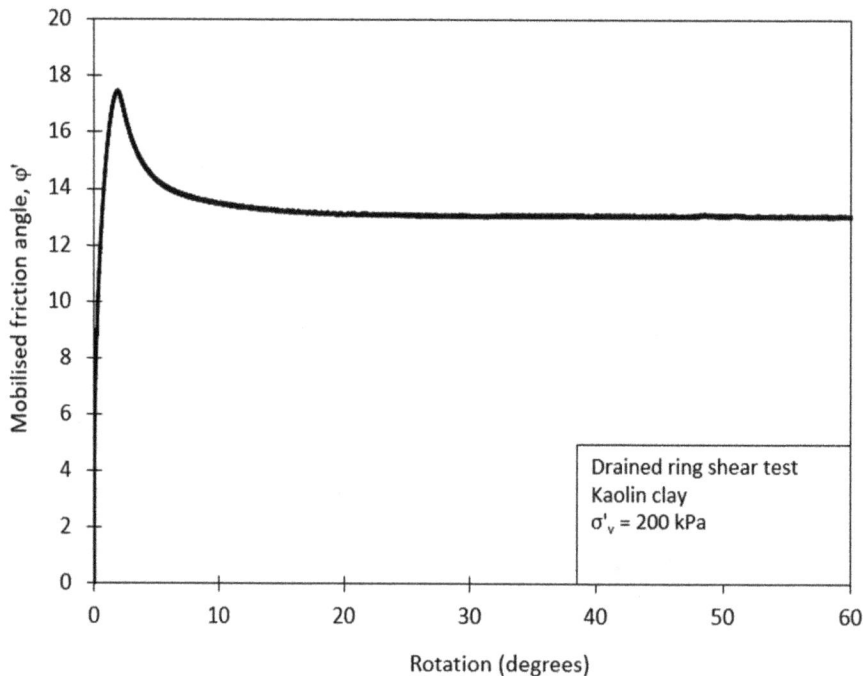

FIG 3 – Drained ring shear results on kaolin clay.

An important note regarding drained strain softening relates to the range of soils that can exhibit such a response. Given the mechanism involved in clay particle alignment, detailed reviews and testing of a range of soil suggest that about 40 per cent clay-sized particles (<2 μm) are required such that a reduction in friction angle in the manner outlined in Figure 3 will be observed (eg Lupini, Skinner and Vaughan, 1981). This is a useful observation to take cognisance of, as it guides which *in situ* soils may be susceptible to such a response.

In contrast to the potential for long-term strength recovery (and, indeed, even increase) to soils having undergone undrained shearing, available evidence suggests that the long-term strength recovery of soils having undergone drained frictional strain softening is unlikely in most cases (Mesri and Huvaj-Sarihan, 2012). For example, Stark and Hussain (2010) carried out a detailed study at a range of stresses, showing that only at relatively low stresses (<100 kPa) was any strength recovery with time seen, and indeed this strength recovery was seen to be lost quickly with any subsequent shear displacement. Recovery of strength is therefore rarely included in models developed to assess slope behaviour with drained frictional strain softening.

INVESTIGATION OF AREAS OF POTENTIAL OVERLAP

General

The previous examples were selected as they were distinct, unambiguous examples of the two forms of strain softening. These were useful to 'set the scene' for these two mechanisms. However, further insight can be gained through examination of other, more bespoke forms of testing. Specifically, while there is clearly no doubt that loose/'wet' clays can exhibit SIEPWP, the potential evolution of the friction angle of such materials at high strain during undrained shearing is less clear (eg Thakur et al, 2014). Indeed, as will be seen subsequently, whether such a process occurs can be quite relevant to how some numerical modelling processes would be undertaken involving such soils.

CVRS ring shear tests results

Stark and Contreras (1996) present development of a constant volume ring shear (CVRS)s that combined the use of the constant volume technique as per the commonly-used DSS (Dyvik et al, 1987) while also enabling shearing to much larger strains consistent with ring shear testing. Such a test is ideal for exploring the questions relevant to the current paper. Tests carried out using this or similar devices with sufficiently reliable results to investigate frictional response at large strain during undrained shearing that could be located by the author are summarised in Table 1. Of the tests identified, the Drammen clay and BCC materials are sufficiently clayey that drained frictional softening would be expected – at least in a conventional ring shear test under drained conditions.

TABLE 1
CVRS literature review summary.

Reference	Material	% <2 μm	Plasticity index
Stark and Contreras (1996)	Drammen clay	60–70	60–70
Michaud (2017)	Saint-Jude clayey silt	24	24
Stark (2021)	Bootlegger Cove clay (BCC)	4–20	4–20

The results presented in Figure 4a show the distinct strain softening in each test in terms of undrained strength ratio – with I_B in excess of 0.5 for all the tested samples. Significant SIEPWP was seen in all tests. The higher undrained strengths for BCC are a result of this being an intact overconsolidated sample, rather than a normally consolidated specimen as per the other tests. Turning to the mobilised friction angle with shear displacement in Figure 4b, decrease is seen in all tests. However, this decrease is relatively modest and does not appear consistent with expectations for the drained residual friction angle of Drammen and BCC, in particular. Further, the modest decrease in mobilised friction angle confirms that it is the contraction and SIEPWP that is the dominant source of post-peak strength loss. This suggests that were the strain softening seen in Figure 4a to occur in isolated portions of a foundation (and this did not result in slope failure) that in the long-term, strength recovery would be likely as SIEPWP dissipated.

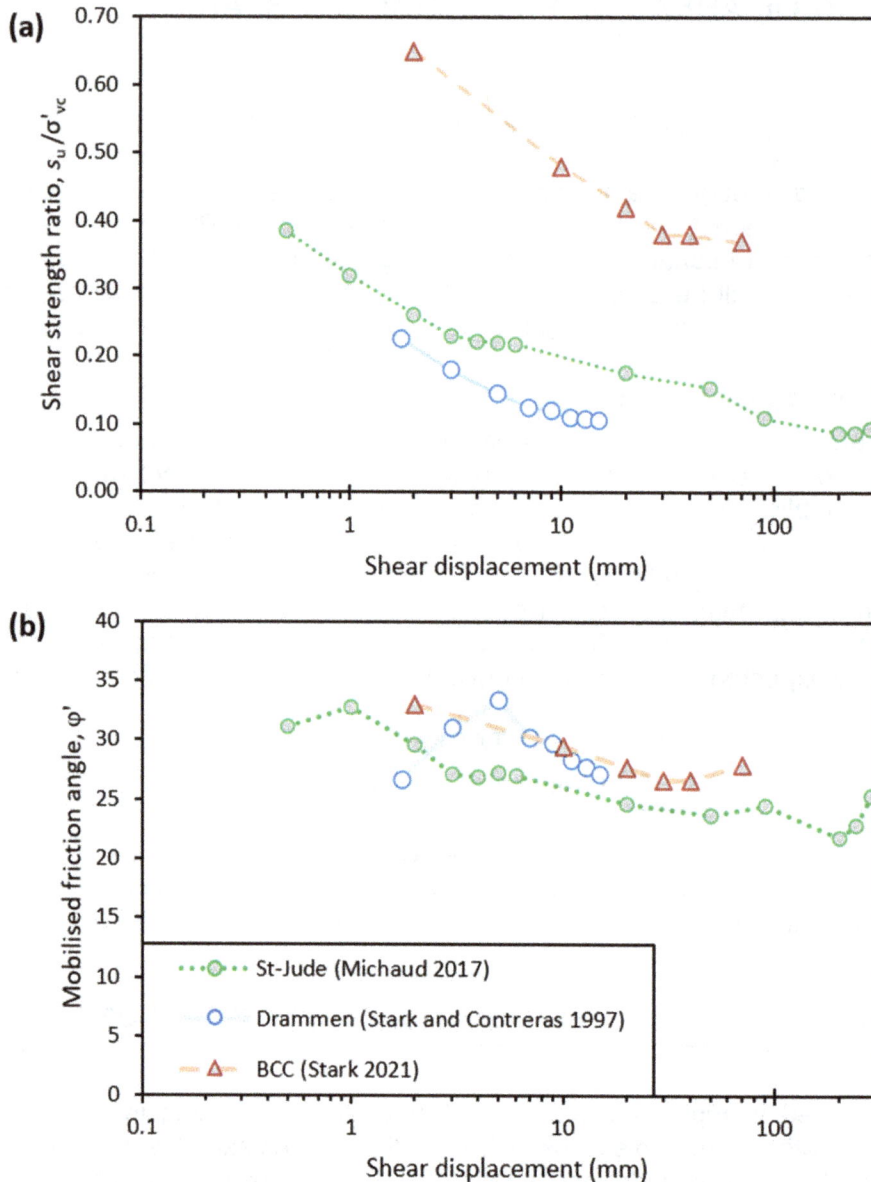

FIG 4 – Reinterpretation of CVRS testing of clays published by Michaud (2017), Stark and Contreras (1996) and Stark (2021) as: (a) shear strength ratio versus shear displacement; and (b) mobilised friction angle versus shear displacement.

Multistage DSS test results

The CVRS system is a relatively unique device which, while seeing some increased application (eg Mmbando *et al*, 2022; Reid, Tiwari and Fourie, 2023), suffers from potential difficulties related to side friction. Further, to the author's knowledge, none of the available CVRS studies on clayey soils involve reconsolidation after undrained shearing to assess whether the strength loss is permanent or could recover – a crucial understanding, if numerical stress deformation analyses of decades-long TSF development lives are to be reasonable captured with clayey foundations that undergo some amount of strain softening.

An alternative means to examine the potential evolution of friction angle at high strains during undrained shearing of clays is the 'multistage' DSS test (MSDSS). This involves shearing a sample within the DSS, under strain control, both forward and backward repeatedly. This partially overcomes the strain limitations of the DSS system, similar to what is attempted in a 'reversible' direct shear test. Zabolotnii, Morgenstern and Wilson (2022). made use of such tests to estimate the potential strain softening characteristics of the Upper GLU at Mount Polley (discussed subsequently).

A series of MSDSS tests carried out on kaolin clay by Reid, Fanni and Fourie (2023) are presented in Figure 5 as undrained shear strength ratio versus shear strain. The multi-stage shearing loops are evident to approximate limits of ± 25 per cent strain (the practical one directional strain limits of the DSS) The results show an increasing reduction in shear stress with strain. Further, the behaviour of the specimens after reconsolidating – ie allowing SIEPWP to dissipate – are shown, suggesting significant recovery in strength. In other words, the soil does not 'stay' weakened, consistent with the triaxial tests on tailings by Vulpe and Fourie (2024). This is consistent with the CVRS rests on clayey soils presented earlier, suggesting that SIEPWP was the dominant source of strain softening.

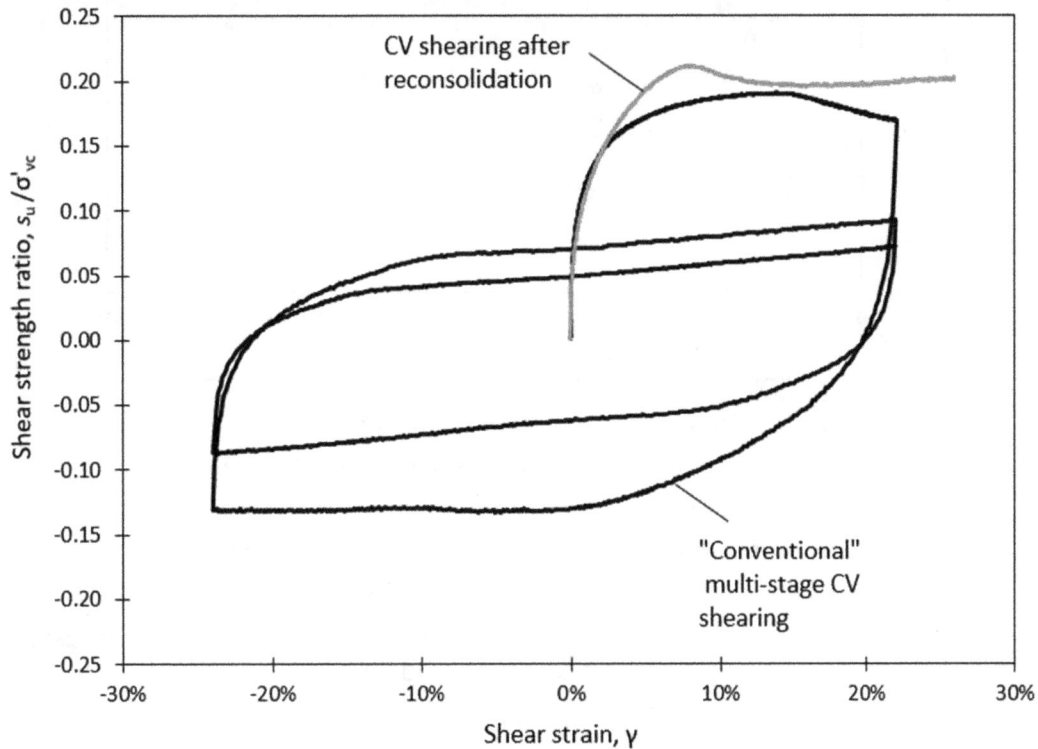

FIG 5 – Example of a MSDSS test on normally consolidated reconstituted kaolin clay, after Reid, Fanni and Fourie (2023).

The results presented in Figures 5 and 6 relate solely to the undrained shear strengths observed in the kaolin during MSDSS shearing. However, mobilised friction angle can be calculated for the shearing in a MSDSS in a similar manner to that of ring shear tests. This is presented in Figure 7 for the kaolin MSDSS tests, drawing a distinction between the initial shearing stage and that carried out after the significant initial shear strain and subsequent reconsolidation. This comparison suggests that the multistage shearing process – carried out under undrained conditions – is resulting in a reduction in friction angle. Included in Figure 7 are the peak and residual friction angles for normally consolidated kaolin as previously outlined in Figure 3, showing reasonable agreement to the values obtained from the drained ring shear test. Therefore, while the predominant mechanism for strength loss in the undrained MSDSS shearing of kaolin was SIEPWP – and therefore strength recovery after reconsolidation – a frictional softening process also appears to have occurred.

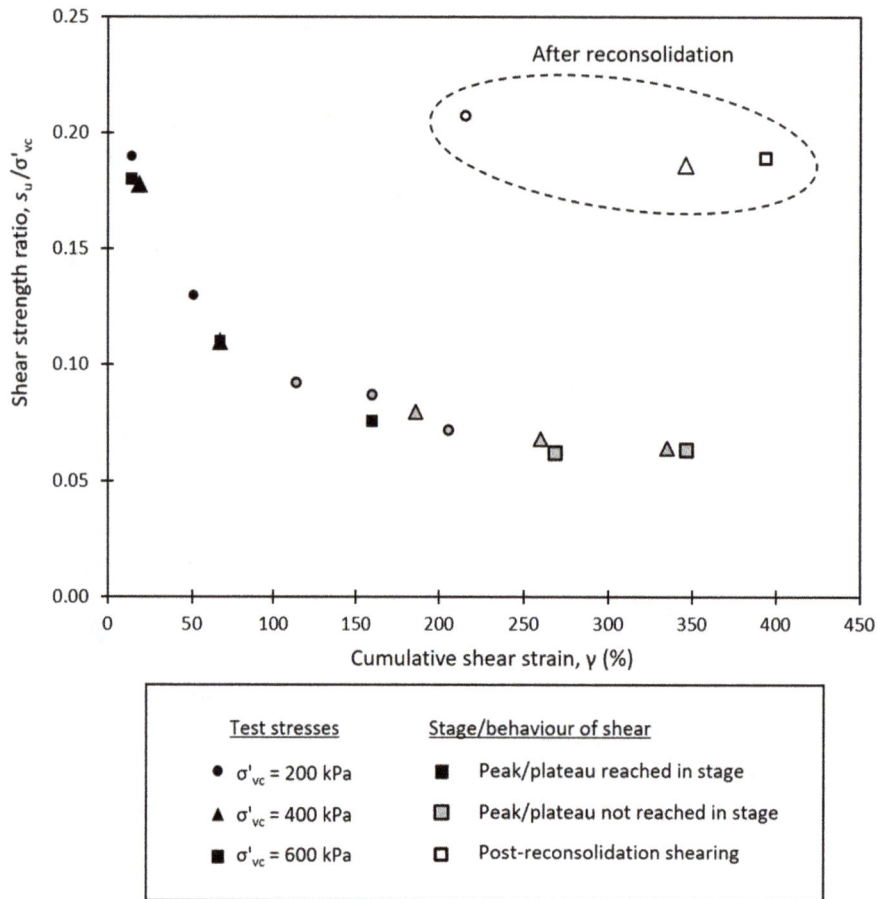

FIG 6 – Summary of MSDSS tests by Reid, Fanni and Fourie (2023), including measure shear strengths after reconsolidation.

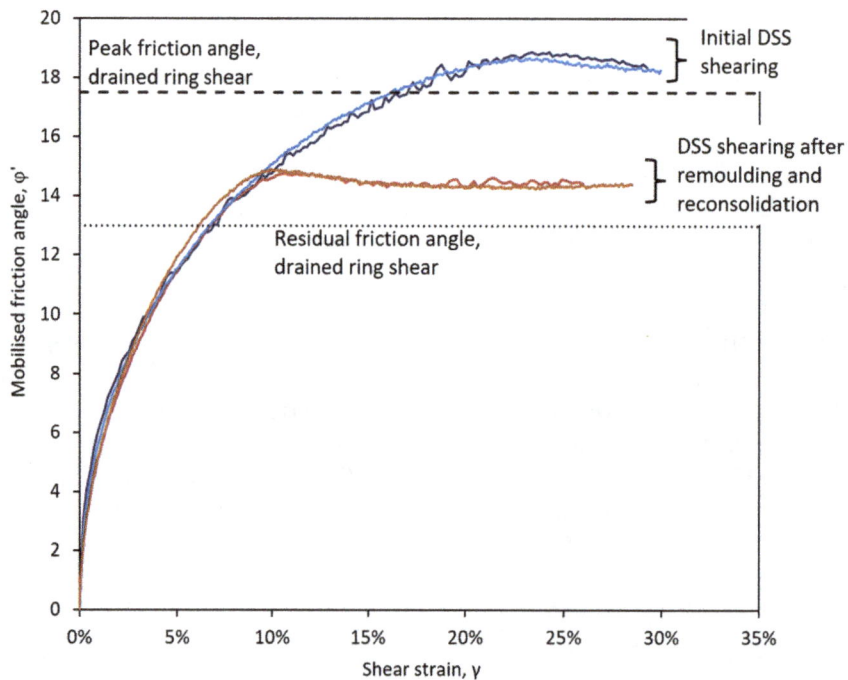

FIG 7 – Mobilised friction angle during MSDSS tests on kaolin, initial shearing and shearing after significant shear strain and subsequent reconsolidation.

INTERPRETATION OF RECENT FAILURE BACK ANALYSES

General

Many of the recent TSF failures included soils and/or tailings that were inferred to have undergone strain softening, and the investigations included models of the structures over a sufficiently long period of time that if strength recovery was viewed as a plausible mechanism, it could be implemented. It is instructive to examine the interpretation of the soil behaviour in each of these cases in the context of the current paper, as is carried out subsequently.

Aznalcóllar

The Aznalcóllar dam failed in 1998 resulting a catastrophic flow of pyritic acid tailings in a nearby river. Investigation details of the failure are presented by Alonso and Gens (2006a, 2006b, 2006c). The primary geotechnical cause of the slope failure being construction of downstream embankments onto a low permeability, overconsolidated brittle (in drained shearing) clay foundation. It is emphasised that the clay foundation was overconsolidated, and thus contractive undrained shear is not relevant to this failure. It therefore serves as a useful example of drained frictional softening. Importantly, in modelling exercises examining the behaviour of the foundation, the material unambiguously 'stays' weakened over time (Zabala and Alonso, 2011), consistent with the lack of meaningful strength recovery seen after the reduction of drained strengths.

Fundão slimes

An important aspect of the lateral extrusion hypothesis (Morgenstern *et al*, 2016) for the triggering of the Fundão failure was the strain softening of slimes inferred to have underlay the loose sand tailings. The strain softening of the slimes, and resulting deformations, was integral to the modelled lateral-extrusion driven increase in stress ratio in the overlying loose sand. The assumptions around the slimes strain softening are of relevance to this current discussion.

While element testing was carried out on some samples of slimes, it appears the primary input to the strength characteristics of this material were taken from a back analysis of a historic trial embankment construction failure. The inferred back-analysed strengths being a peak undrained strength ratio of 0.22 and residual value of 0.07 – it being crucial to note that these strengths are presented as undrained strengths, and there is no discussion of frictional softening of the slimes of which the author is aware. Indeed, the typical index properties of the slimes are inconsistent with drained frictional softening. Another important distinction for the interested reader who may refer to the source material is that although Morgenstern *et al* (2016) used a friction angle input to characterise the slimes, this was simply for convenience, and the values of friction angle were calculated based on the relevant undrained strength ratios.

Morgenstern *et al* (2016) modelled two primary scenarios involving strain softening of the slimes: (i) that the slimes underwent strain softening shortly prior to the failure; and (ii) that the slimes in their entirety underwent strain softening almost immediately after their deposition, and remained strain softened for a periods of years up until the failure occurred.

It is suggested that hypothesis (ii) appears implausible based on previous discussions and the index properties of the slimes – undrained strain softening would certainly 'heal' to some degree over the period of years relevant to the TSF life and modelling exercise, and there is no reason to suspect drained frictional softening could occur in the slimes. While the distinction between scenarios (i) and (ii) may have little importance in the overall lateral extrusion hypothesis, given similar stress states were produced in the overlying sands in either case, the inclusion of scenario (ii) in the modelling work is of interest in one of the theses of this paper – that there appears to be significant uncertainty and variability in how the longer-term behaviour of strain softened fine-grained soils is modelled.

Mount Polley Upper GLU

As noted previously, the undrained strain softening characteristics of the Upper GLU relevant to the Mount Polley failure were modelled by (Zabolotnii *et al*, 2022) taking input from MSDSS tests on the Upper GLU foundation layer. These showed significant undrained strain softening. It is again emphasised that the DSS is a constant volume test and thus significant vertical effective stress reduction occurs during shearing equivalent to the SIEPWP that would develop in an undrained test.

While the results of the MSDSS tests on the Upper GLU are uncontroversial, what is of interest here is how the soil, after strain softening in the numerical model, was inferred to behaved in the longer term when modelled. Strain softening of portions of the Upper GLU were first seen in the modelling of Zabolotnii, Morgenstern and Wilson (2022) in Stage 7, approximately two years prior to the failure. However, despite this strain softening being primarily driven by SIEPWP, the areas that had undergone strain softening were assumed to stay softened for the remainder of the modelled life of the facility.

Regarding the decision to allow strain softened zones to remain in such a condition, Reid, Fanni and Fourie (2023) carried out MSDSS tests on kaolin clay with additional 'healing' periods of reconsolidation and found that strength recovery did occur (discussed previously, Figures 5–7). Zabolotnii, Morgenstern and Wilson (2023) acknowledged this likely behaviour of the Upper GLU and further outlined model details that indicate this assumption had negligible effect on the overall outcomes of their modelling of the failure. Therefore, what is important here is not the specifics of Mount Polley and the behaviour of the Upper GLU, but more broadly how such behaviour should be accounted for in other modelling exercises where the decisions around the longevity of softening could be more important. It is emphasised that the benefits of performance based design and stress-deformation modelling of the entire life of TSFs is widely advocated.

Feijão tailings

A recent work involving the modelling of the Feijão TSF failure (Zhu, Zhang and Puzrin, 2024) assumed, based on inferred bonding of the tailings (Robertson *et al*, 2019) that once a portion of the tailings underwent strain softening that the material would remain in a weakened condition indefinitely. Also, unlike some previous cases where there was potential ambiguity as to whether this assumption was explicitly made or the logic underpinning it, the authors of this work clearly hold to this view (Zhu, Zhang and Puzrin, 2025) despite objections as to its feasibility (Reid *et al*, 2025). Notwithstanding alternative views regarding the potential for bonding of the Feijão tailings (Arroyo and Gens, 2021), the question of whether a bonded soils would regain strength following strain softening and subsequent reconsolidation cannot, to the author's knowledge, be answered based on any available experimental data. For this reason, a laboratory test program was designed to further investigate this topic as outlined subsequently.

Cementation and post-shearing strength gain

Rationale, methods, and materials

Considering the views of Zhu *et al* (2024, 2025) with respect to the hypothesised behaviour of bonded soils, a set of experiments were designed and carried out to allow an initial investigation into the behaviour of cemented soils after shearing, and whether, after post-peak strain softening and subsequent reconsolidation any strength recovery occurred.

The initial experimentation to this end water carried out two soil/cementing agent combinations for testing in MSDSS-type tests:

1. A mixture of 20 per cent kaolin clay and 80 per cent silica fine sand (20K80SFS) used frequently as a research soil at UWA. This material was selected as it can be prepared to a loose, brittle soil with slurry deposition. Portland cement at either 0.25 or 1.0 per cent concentration to dry solids mass was used. Tests were carried out adopting the same technique of Reid, Fanni and Fourie (2023) on kaolin as previously outlined where, after multistage shearing, reconsolidation was carried out followed by a second episode of undrained shearing. Samples were prepared as a loose, non-segregating slurry at 90 per cent GWC, then a mass of Portland cement was added to produce the desired concentration, and the slurry with cement was then thoroughly mixed. The mixture was then poured either into a 100 mm diameter stainless steel column or directly into a DSS mould, depending on the target vertical effective stress for the sample/test.

2. SFS sand, prepared using moist tamping (MT) with 10 per cent gypsum concentration to dry solid mass. Gypsum was adopted as there is evidence it can create a more brittle cementation than Portland cement (Ismail *et al*, 2002), therefore making it potentially a more suitable cementing agent in the context of the current discussion. Samples were prepared using MT with the dry gypsum cement having been thoroughly mixed through the moist material. After tamping into the DSS apparatus the specimen was consolidated to 50 kPa vertical effective stress and then flushed with water.

The 20K80SFS material was tested with identical procedures to that of Reid, Fanni and Fourie (2023) on kaolin as previously outlined. The SFS with gypsum was sheared in a single direction with four episodes of reconsolidation, followed by multistage shearing, and a final additional reconsolidation episode. The addition reconsolidation episodes were possible in the SFS with gypsum owing to its much faster post-peak decrease in strength compared to 20K80SFS with Portland cement. The concentration of the two cementing agents was at the lower range typically seen in such studies, as the purpose of the testing was to investigate the 'light' bonding suggested by some as being relevant to Feijão.

Results – 20K80SFS with Portland cement

The 20K80SFS are first summarised as consolidated void ratio against consolidated vertical effective stress in Figure 8, illustrating the looser void ratios achieved with cementation – consistent with typical behaviour in cement-treated clayey slurries, and with the idealised behaviour of bonded soils that may similarly exist at looser states than those without bonding.

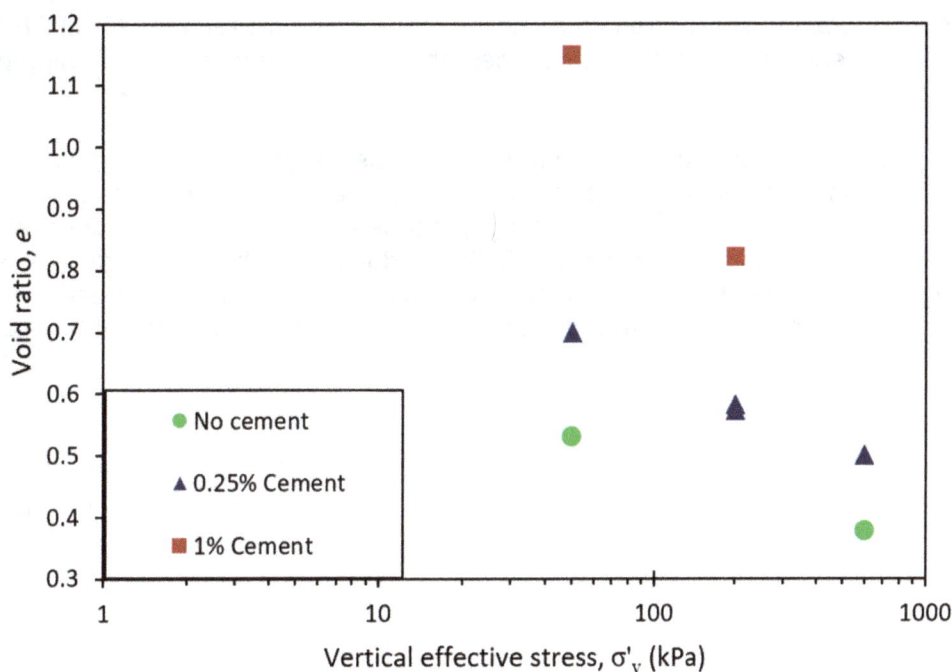

FIG 8 – Consolidation characteristics of 20K80SFS: without cement, 0.25 per cent cement and 1 per cent cement as void ratio versus vertical effective stress.

The peak undrained shear strength for all the 20K80SFS tests carried out are summarised in Figure 9 against consolidated void ratio. The looser states achieved with Portland cement addition are clear, as is the significantly higher shear strength at a particular density resulting from cement addition. Again, this response is consistent with the typically postulated behaviour of bonded soils, suggesting that the results of the testing may be of some relevance to the previous discussions related to the potential for strength gain in bonded soils after strain softening.

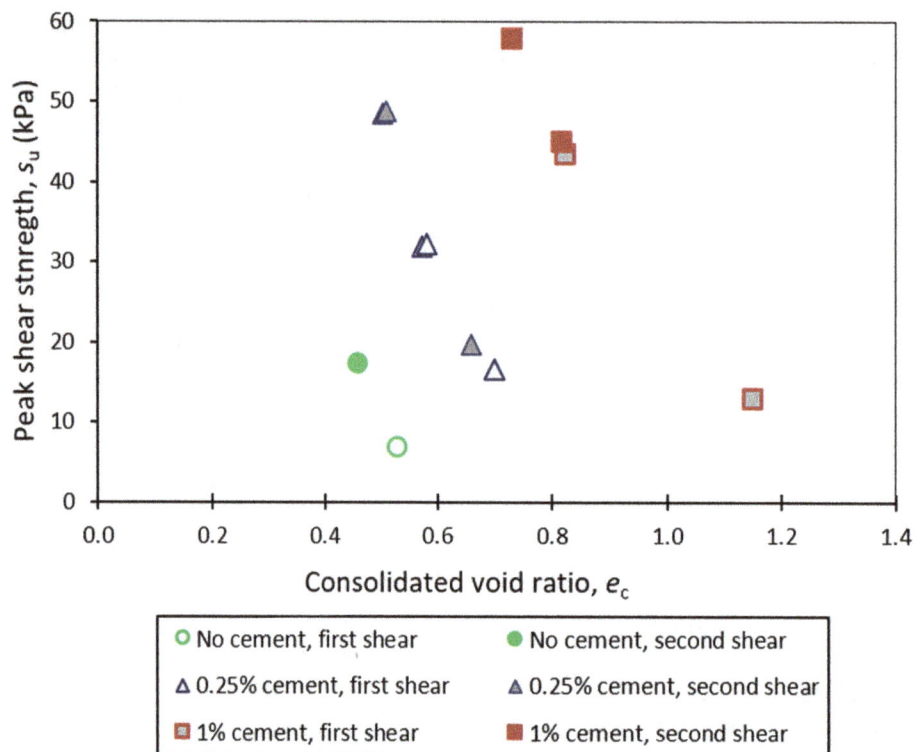

FIG 9 – Peak undrained shear strength against consolidated void ratio for 20K80SFS with no cement, 0.25 per cent cement and 1 per cent cement in initial shearing and following post-shear reconsolidation.

The 20K80SFS test results are summarised in Figure 10 as the shear strength ratio obtained at the end of each increment of shearing against cumulative shear strain. The results are separated between that of the initial multi-stage shearing process and the final undrained shearing that occurred after reconsolidation was permitted. Both a reduction in shear strength with strain in the initial shearing, and significant reconsolidation strength gain, are evidence. These results are not-supportive of the suggesting that cemented soils will not regain shear strength over time through the dissipation of SIEPWP.

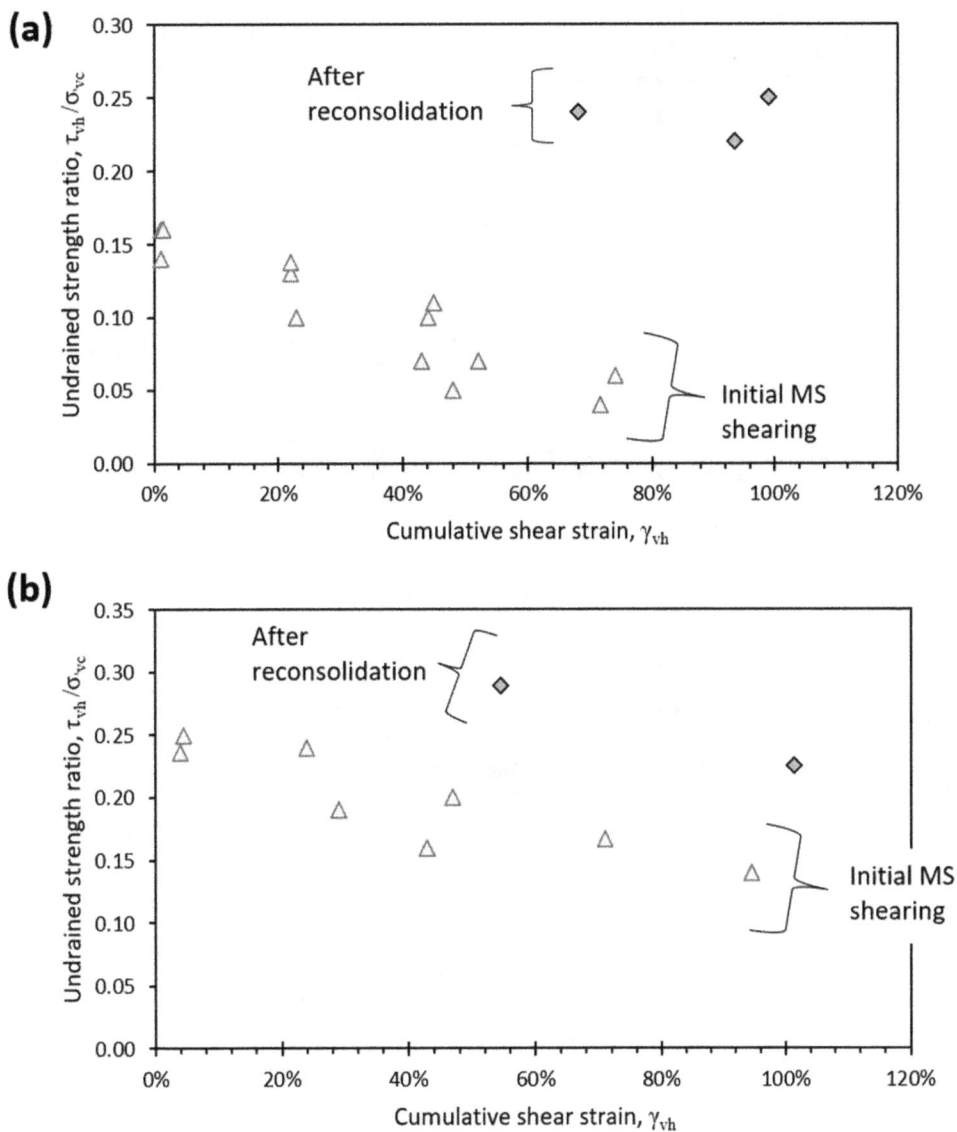

FIG 10 – Undrained shear strength with cumulative shear strain, and undrained shear strength following reconsolidation: (a) 0.25 per cent cement and (b) 1.0 per cent cement.

Results – SFS with gypsum

The results of the MSDSS test carried out on SFS with 10 per cent gypsum is summarised in Figure 11, showing the MSDSS stress-strain response for all stages and a summary of the peak shear strength of each stage against consolidated void ratio for each stage. The results clearly show that after each post-peak strain softening episode, reconsolidation results in a subsequent increase in shear strength. Most importantly, after the longest shearing increment, the subsequent shear strength (Point 5) is significantly increased, to far higher than all the previous stages. This increase is clearly a result of the consolidation densification that occurs as SIEPWP is allowed to dissipate. Further, this happens despite the tendency of Point 5, presumably owing to significant damage to the gypsum cementation bonds, to shift closer to the untreated trend in void ratio – shear strength space (ie Figure 11b). This suggests that the density increase from consolidation is sufficient to outweigh the likely destruction of gypsum-based bonding from large shear strains.

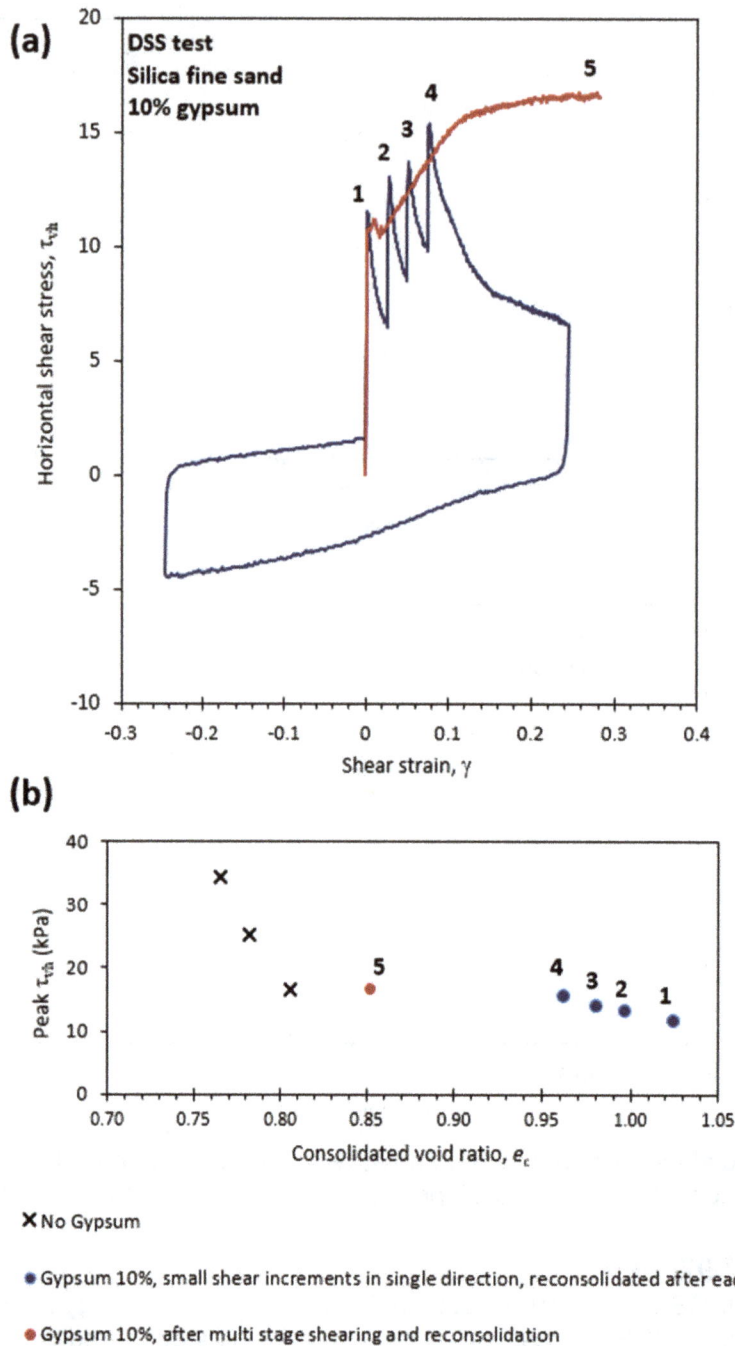

FIG 11 – Results of MSDSS test on SFS with 10 per cent gypsum: (a) MSDSS stress-strain response, with each peak shear strength value indicated; and (b) summary of peak shear strength against consolidated void ratio for various MSDSS stages, and including reference data on untreated SFS for comparison.

CONCLUSIONS

The strain softening behaviour of many loose tailings and some brittle foundation soils have been major contributors to a series of recent TSF failures. While much progress has been made in the understanding and classification of the potential for such strain softening, some uncertainty remains with respect to which form of strain softening is under consideration in some design scenarios – that driven primarily by SIEPWP in loose soils, or drained frictional softening in clayey soils. This distinction can be important given the need to simulate the long-term behaviour of soils within and below TSFs over a period of decades in many numerical stress-deformation analyses. Further, there remains significant uncertainty as to if, and to what degree, soil that has strain softened could potential regain strength if reconsolidation is able to occur.

The current study examined these issues by means of a literature review of the two distinct mechanisms that can drive strain softening, followed by a more detail examination of CVRS and MSDSS tests that can elucidate the mechanical response of soils where both SIEPWP and drained frictional softening may be occurring. MSDSS tests were also used to highlight the increase in strength seen with reconsolidation is permitted, even in clayey soils for which drained frictional softening can occur. Finally, owing to uncertainty around the potential for post-reconsolidations strength gain in bonded soils, tests were carried out on two soils with difference cementing agents to examine their behaviour in MSDSS tests. These showed that for both combinations tested, while sufficient shearing may be able to damage bonds, reconsolidation still results in significant increase in subsequent strength owing to an increase in density that can outweigh the potential losses from bond breakage.

ACKNOWLEDGEMENTS

The author thanks Riccardo Fanni for his useful advice on different techniques to produce suitable cementation.

REFERENCES

Alonso, E E and Gens, A, 2006a. Aznalcóllar dam failure, Part 1: Field observations and material properties, *Géotechnique*, 56:165–183.

Alonso, E E and Gens, A, 2006b. Aznalcóllar dam failure, Part 2: Stability conditions and failure mechanism, *Géotechnique*, 56:185–201.

Alonso, E E and Gens, A, 2006c. Aznalcóllar dam failure, Part 3: Dynamics of the motion, *Géotechnique*, 56:203–210.

Arroyo, M and Gens, A, 2021. Computational analyses of Dam I at the Corrego de Feijao mine in Brumadinho – Final Report.

Bishop, A W, 1967. Progressive failure with special reference to the mechanism causing it, in *Proceedings of the Geotechnical Conference*, pp 142–150, Oslo, Norway.

Dyvick, R, Berre, T, Lacasse, S and Raadim, 1987. Comparison of truly undrained and constant volume direct simple shear tests, *Geotechnique*, 37(1):3–10.

International Committee on Large Dams (ICOLD), 2023. Bulletin 194 – Tailings Dam Safety.

Ismail, M A, Joer, H A, Sim, W H and Randolph, M F, 2002. Effect of cement type on shear behavior of cemented calcareous soil, *Journal of Geotechnical and Geoenvironmental Engineering*, 128:520–529.

Jefferies, M G and Been, K, 2015. *Soil Liquefaction – A Critical State Approach*, second edition (CRC Press).

Lupini, J F, Skinner, A E and Vaughan, P R, 1981. The drained residual strength of cohesive soils, *Géotechnique*, 31:181–213.

Mesri, G and Huvaj-Sarihan, N, 2012. Residual Shear Strength Measured by Laboratory Tests and Mobilized in Landslides, *Journal of Geotechnical and Geoenvironmental Engineering*, 138:585–593.

Michaud, H, 2017. Comportement de l'argile de Saint-Jude sous cisaillement annulaire à volume constant, MSc thesis, Département de génie civil et de génie des eaux, Université Laval, Québec, 185 p.

Mmbando, E, Fourie, A, Reid, D, O'Loughlin, C, Gao, J and Wang, Y, 2022. Residual strength based on CPT sleeve friction and a constant volume ring shear device, in *Proceedings of the 26th International Conference on Tailings and Mine Waste*, pp 297–306 (University of British Columbia).

Morgenstern, N R, Vick, S G, Viotti, C B and Watts, B D, 2016. Fundão Tailings Dam Review Panel: Report on the immediate causes of the failure of the Fundão Dam.

Potts, D M, Dounias, G T and Vaughn, P R, 1990. Finite element analysis of progressive failure of Carsington embankment, *Geotechnique*, 40(1):79–101.

Reid, D, Arroyo, M, Jefferies, M, Fanni, R, Fourie, A, Gens, A and Mánica, M, 2025. Creep deformation does not explain the Brumadinho disaster, *Communications, Earth and Environment*, 6.

Reid, D, Fanni, R and Fourie, A B, 2023. Discussion of 'Mechanism of failure of the Mount Polley Tailings Storage Facility', *Canadian Geotechnical Journal*, 60(7):1095–1098.

Reid, D, Tiwari, B and Fourie, A B, 2023. Large strain shearing behaviour of untreated and polymer treated clayey silt slurry specimens, in *Paste 2023: Proceedings of the 25th International Conference on Paste, Thickened and Filtered Tailings* (eds: G W Wilson, N A Beier, D C Sego, A B Fourie and D Reid), pp 676–685 (University of Alberta: Edmonton, and Australian Centre for Geomechanics: Perth). https://doi.org/10.36487/ACG_repo/2355_52

Robertson, P K, de Melo, L, Williams, D J and Wilson, G W, 2019. Report of the Expert Panel on the technical causes of the Failure of Feijão Dam, I.

Stark, T D, 2021. Constant volume ring shear specimen trimming and testing, *Geotechnical Testing Journal*, 44(5).

Stark, T D and Contreras, I A, 1996. Constant volume ring shear apparatus, *Geotechnical Testing Journal*, 19(1):3–11.

Stark, T D and Hussain, M, 2010. Drained residual strength for landslides, GeoFlorida 2010.

Thakur, V, Jostad, H P, Amundsen, H A and Degago, S A, 2014. How well do we understand the undrained strained softening response in soft sensitive clays?, *Landslides in Sensitive Clays: Natural and Technological Hazards Research*, 36:ch 23 (Springer).

Vulpe, C and Fourie, A B, 2025. Effect of shear-induced consolidation on the behaviour of a loose silt tailings, *Canadian Geotechnical Journal*, 62:1–10. https://doi.org/10.1139/cgj-2024-0083

Zabala, F and Alonso, E E, 2011. Progressive failure of Aznalcóllar dam using the material point method, *Géotechnique*, 61:795–808.

Zabolotnii, E, Morgenstern N R and Wilson, G W, 2022. Mechanism of Failure of the Mount Polley Tailings, *Canadian Geotechnical Journal*, 59(8):1503–1518. https://doi.org/10.1139/cgj-2021-0036

Zabolotnii, E, Morgenstern N R and Wilson, G W, 2023. Reply to the discussion by Reid *et al* on 'Mechanism of failure of the Mount Polley Tailings Storage Facility', *Canadian Geotechnical Journal*, 60(7):1099–1101. https://doi.org/10.1139/cgj-2023-0014

Zhu, F, Zhang, W and Puzrin, A M, 2024. The slip surface mechanism of delayed failure of the Brumadinho tailings dam in 2019, *Communications Earth and Environment*, 5:33. https://doi.org/10.1038/s43247-023-01086-9

Zhu, F, Zhang, W and Puzrin, A M, 2025. Reply to: Creep deformation does not explain the Brumadinho disaster, *Communications Earth and Environment*, 6:181. https://doi.org/10.1038/s43247-025-02068-9

NorSand calibration using similarity-based optimisation – a pilot study

H Shen[1], L Kirsten[2] and H Brandao[3]

1. Senior Geotechnical Engineer, Klohn Crippen Berger, Brisbane Qld 4000.
   Email: hshen@klohn.com
2. Senior Geotechnical Engineer, Associate, Klohn Crippen Berger, Brisbane, Qld, 4000.
   Email: lkirsten@klohn.com
3. Senior Geotechnical Engineer, Klohn Crippen Berger, Brisbane, Qld, 4000.
   Email: hbrandao@klohn.com

EXTENDED ABSTRACT

The NorSand (NS) critical state soil model (Jefferies, 1993) has become a widely adopted framework in geotechnical engineering for characterising tailings, sand embankments (eg sand dams), and foundation materials. It is also instrumental in evaluating static liquefaction susceptibility and simulating deformation. Over the years, NS has undergone refinements, notably by Shuttle and Jefferies (2016), further solidifying its role as an industry standard.

NS calibration relies on three core theoretical pillars:

1. The Critical State Locus (parameterised by Γ, λ, or a power-law fit).

2. The Taylor-Bishop interlocking strength model (parameters M and N).

3. The State-Dilatancy Relationship (parameter X).

In addition, the NS-specific plastic hardening modulus H (governing plastic stiffness) is typically derived through iterative forward modelling.

Despite these theoretical foundations, uncertainty in soil properties—whether arising from variability in triaxial testing, sample preparation, or parameter estimation—often leads to repeated trial-and-error calibrations. These uncertainties are clearly accounted for by the two round robin testing programmes on the critical state locus (Reid *et al*, 2021) and dilatancy parameters (Reid *et al*, 2025). However, these minor 'tuning' iterations during model calibration can become rather tedious.

To address this challenge, we introduce a semi-automated optimisation framework that explicitly accounts for parameter uncertainty and evaluates the similarity between measured and simulated triaxial responses. This approach does not replace engineering judgment but rather aims to accelerate and streamline the calibration process, reducing its repetitive nature.

The proposed framework consists of three key steps:

1. NS Parameter Derivation:

   NS input parameters are extracted based on the three fundamental theorems. As a case study, we focus on Fraser River Sand, a well-characterised material in literature. Baseline parameters were adopted from Jefferies and Been's (2016) prior calibrations. See Table 1 and Figures 1 to 3 for reference.

2. Defining Parameter Uncertainty Ranges:

   Parameter uncertainty is quantified based on the distribution of selected critical state points and maximum dilatancy data. For this preliminary implementation, we estimate uncertainty using standard error, which allows the search space to reflect plausible variability in M, N, X, and ψ_0. See Table 1 and Figures 1 to 3 for reference.

3. Similarity Assessment via Optimisation:

   A similarity metric—Frechet Distance from Alt and Godau (1995)—is used to compare measured triaxial curves with simulations generated by NorTXL program (Jefferies and Been, 2016) across the uncertainty space. For tractability, this study limits the search space to three dimensions (N, X, ψ_0). However, the framework is extendable to higher-dimensional optimisation and will be further developed in future work. See Figures 4 and 5 for reference.

This pilot study demonstrates that parameter uncertainty can be explicitly visualised and optimised against triaxial measurements using similarity metrics, helping reduce calibration subjectivity and iteration time. Further work on quantitative soil model calibration is ongoing.

TABLE 1

NS parameters and the uncertainties for Fraser River Sands.

Method \ Parameters	a	b	c	M_{tc}	N	χ_{tc}
Best fit	1.028	0.107	0.29	1.47	0.53	4.87
Calibration from Jefferies and Been (2016)	1.01	0.087	0.38	1.47	0.5	5
Standard error of the calibration by Jefferies and Been (2016)	0.044	0.049	0.188	0.02	0.05	0.25
Optimisation range	N/A this time, but ψ_0 perturbed by ±0.01			1.47	0.45, 0.5, 0.55	4.75, 5, 5.25

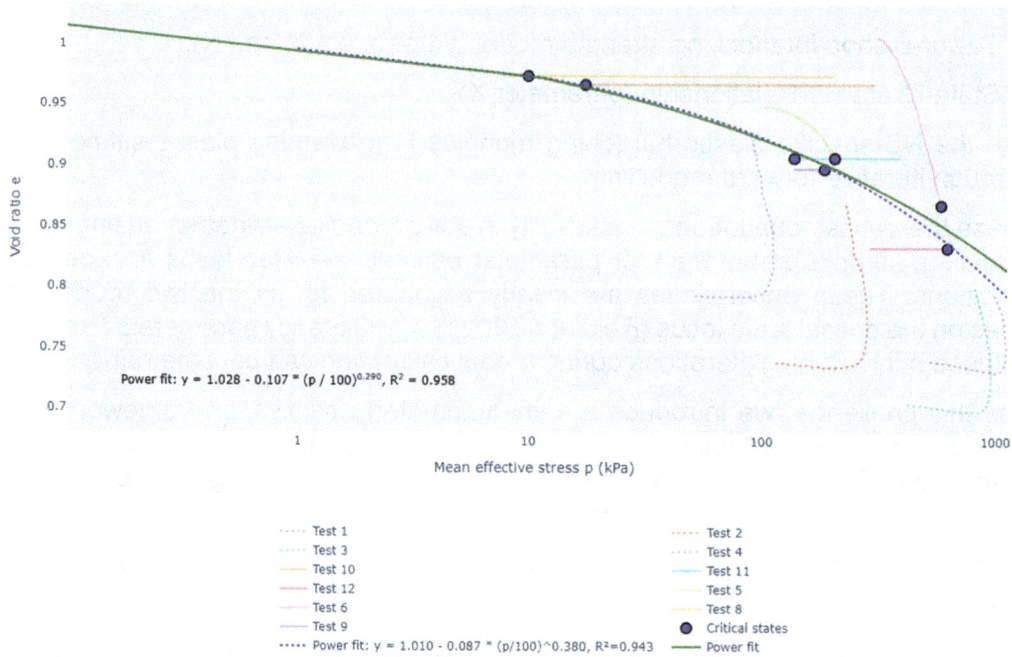

Power fit: y = 1.028 - 0.107 * (p / 100)$^{0.290}$, R^2 = 0.958

Test 1
Test 3
Test 10
Test 12
Test 6
Test 9
Power fit: y = 1.010 - 0.087 * (p/100)^0.380, R²=0.943

Test 2
Test 4
Test 11
Test 5
Test 8
Critical states
Power fit

FIG 1 – Critical state locus of Fraser River Sands, green solid line: the best fit from the least squares method, blue dash line: calibration by Jefferies and Been (2016).

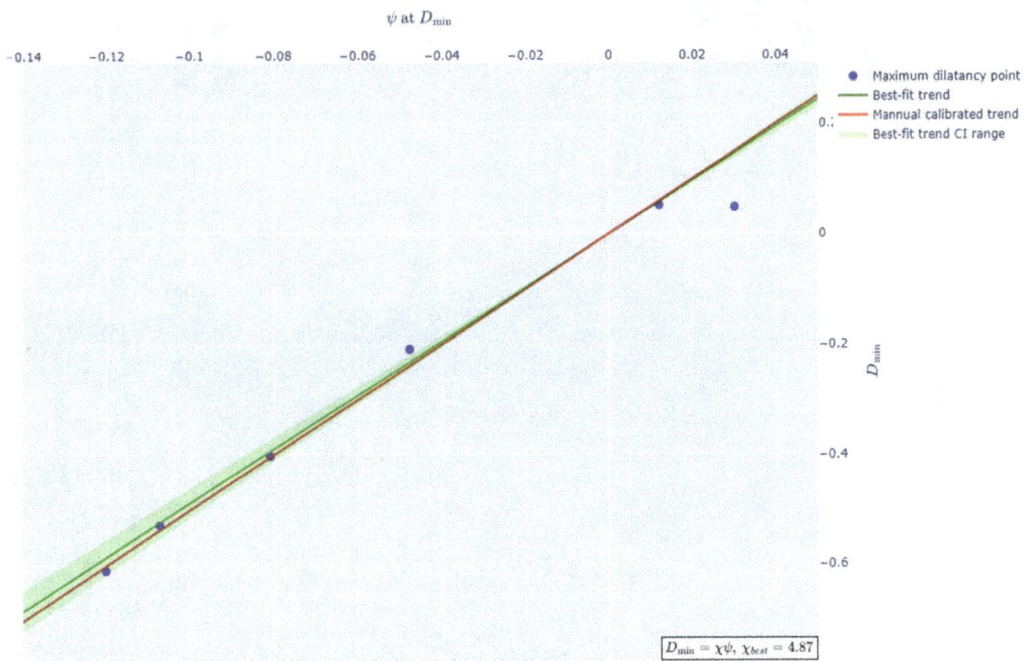

FIG 2 – State-dilatancy relation of Fraser River Sands: red trend calibrated by Jefferies and Been (2016).

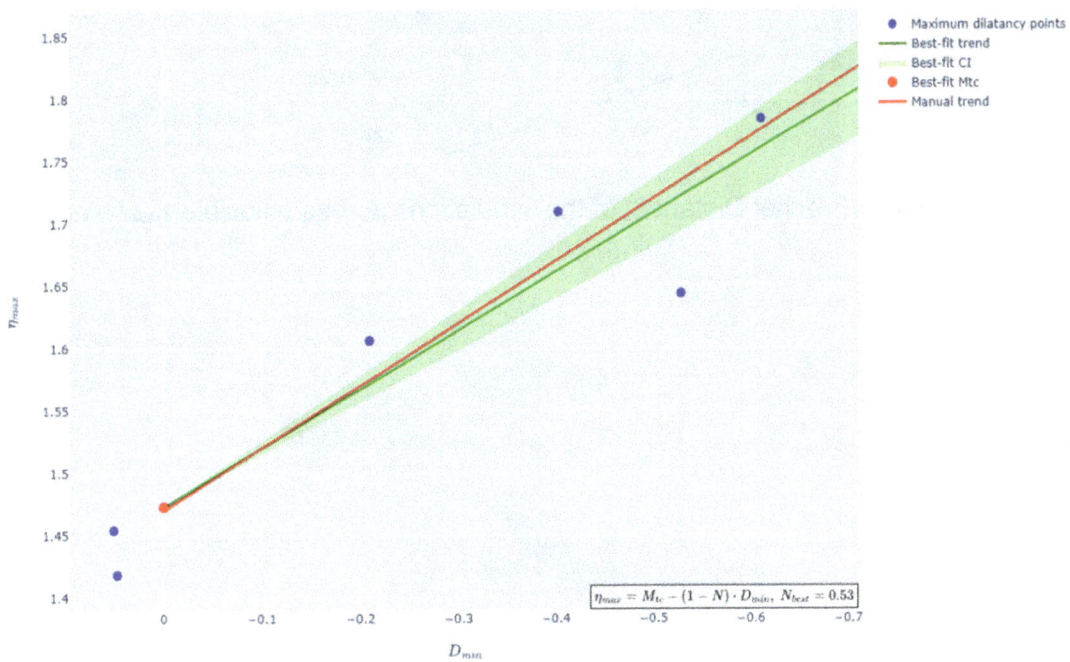

FIG 3 – Stress-dilatancy relation of Fraser River Sands: red trend calibrated by Jefferies and Been (2016).

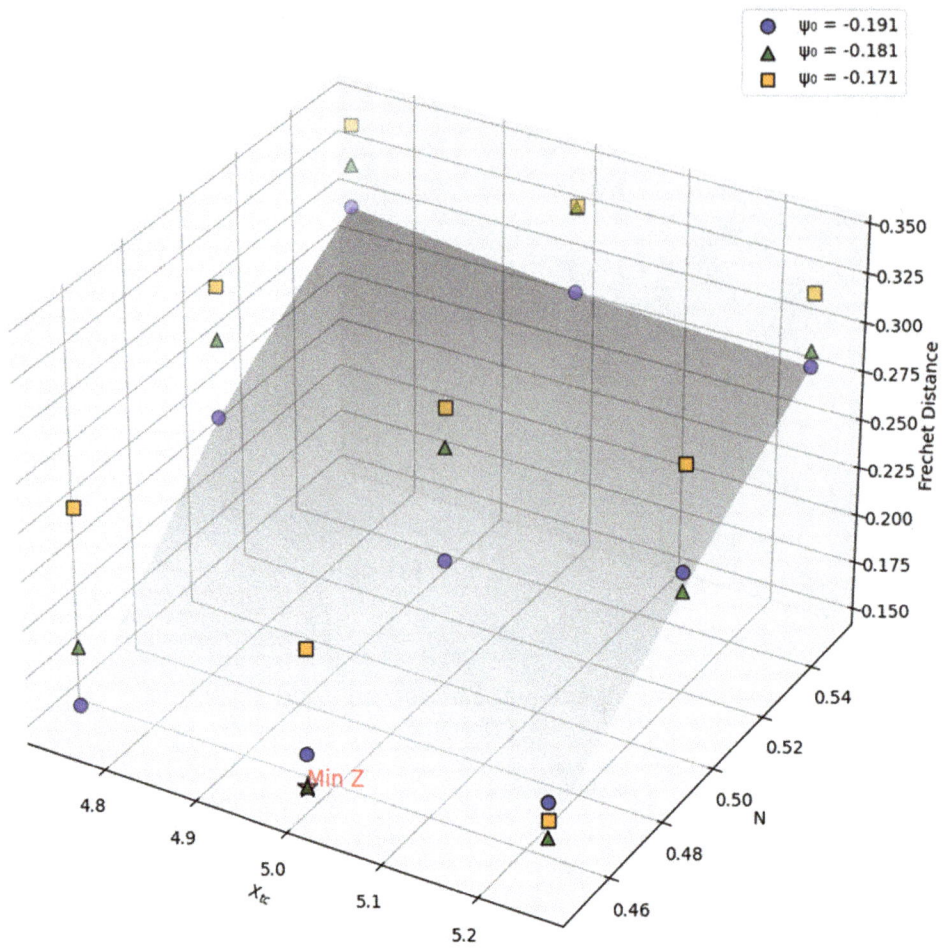

FIG 4 – Similarity (Frechet Distance) of the simulations across plausible X_{tc}, N, and ψ_0.

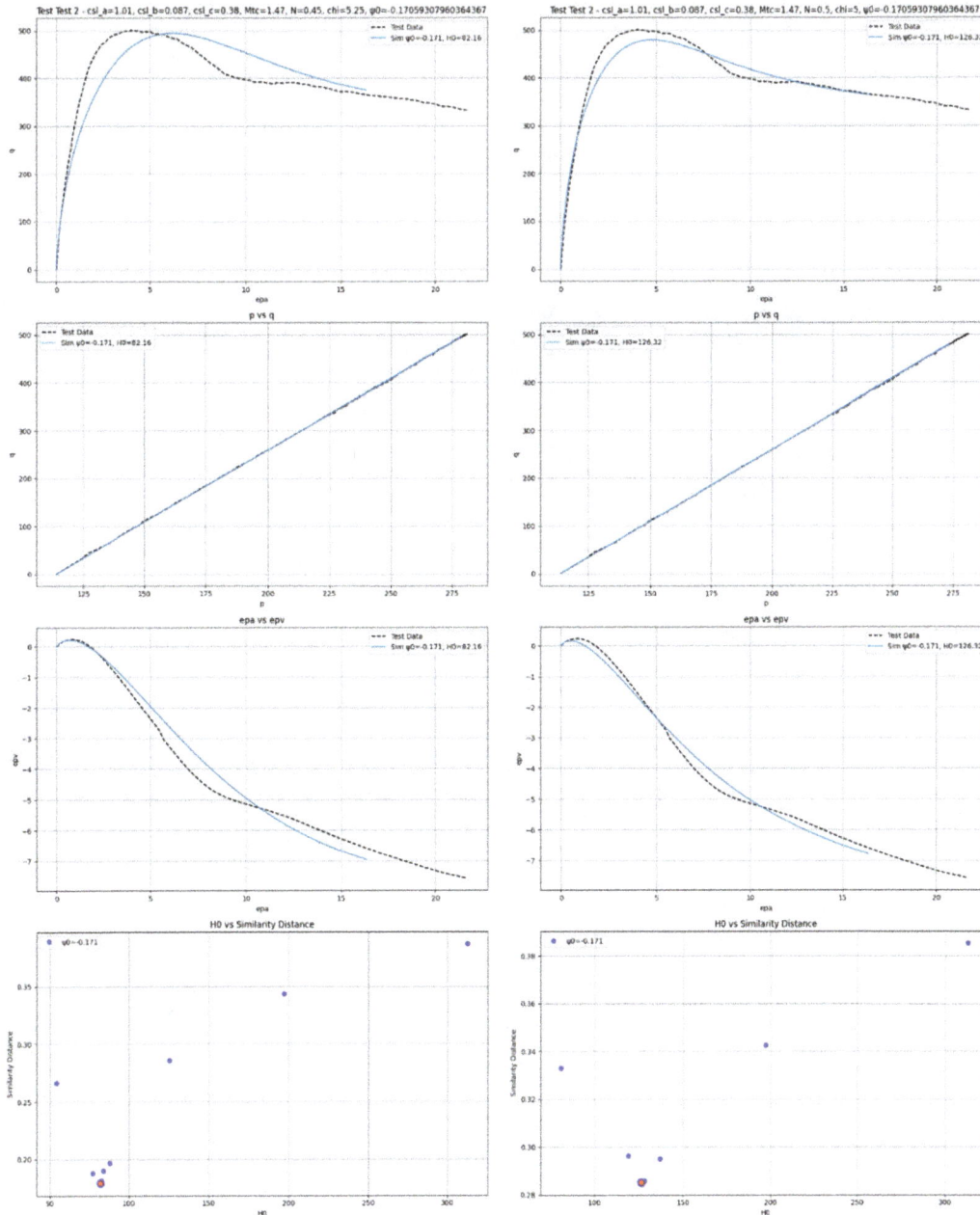

FIG 5 – Typical results from two sets of X_{tc}, N, and ψ_0 (Note only pre-peak stress and strains are compared in the similarity measurement).

REFERENCES

Alt, H and Godau, M, 1995. Computing the Fréchet distance between two polygonal curves, *International Journal of Computational Geometry and Applications*, 5(1–2):75–91.

Jefferies, M and Been, K, 2016. Soil Liquefaction: A Critical State Approach, 2nd edition (CRC Press). https://doi.org/10.1201/b19114

Jefferies, M, 1993. Nor-Sand: a simple critical state model for sand, *Géotechnique* 43:91–103. https://doi.org/10.1680/geot.1993.43.1.91

Reid, D, Fourie, A, Ayala, J L, Dickinson, S, Ochoa-Cornejo, F, Fanni, R, Garfias, J, Viana da Fonseca, A, Ghafghazi, M, Ovalle, C, Riemer, M, Rismanchian, A, Olivera, R and Suazo, G, 2021. Results of a critical state line testing round robin programme, *Géotechnique*, 71(7):616–630. https://doi.org/10.1680/jgeot.19.P.373

Reid, D, Fourie, A, Dickinson, S, Shanmugarajah, T, Fanni, R, Smith, K, Garfias, J, Yuan, B, Ghafghazi, M and Duyvestyn, A, 2025. Results of a dilatancy round robin, *Geotechnical Research*, 12(1):2–28. https://doi.org/10.1680/jgere.24.00008

Shuttle, D and Jefferies, M, 2016. Determining silt state from CPTu, *Geotech Res*, 3:90–118. https://doi.org/10.1680/jgere.16.00008

Connecting tailings from the laboratory to *in situ* through numerical modelling

K Song¹, S H Lines² and M Llano-Serna³

1. Tailings Engineer, Red Earth Engineering a Geosyntec Company, Brisbane Qld 4000.
   Email: kobi.song@redearthengineering.com.au
2. Senior Tailings Engineer, Red Earth Engineering a Geosyntec Company, Brisbane Qld 4000.
   Email: scott.lines@redearthengineering.com.au
3. Principal Tailings Engineer, Red Earth Engineering a Geosyntec Company, Brisbane Qld 4000.
   Email: marcelo.llano@redearthengineering.com.au

ABSTRACT

One of the challenges of geotechnical practitioners is to take the laboratory testing results and convert them into practicable outcomes that can be applied. A reason for this, is the difficulty to know that the laboratory sample is representative of the soil in the field. Just the removal process during a site investigation results in extensive disturbance, even when the most careful procedures are followed. The stress conditions change, densification can occur depending on the method, followed by lifting the sample and transporting it on roads that are often not silky smooth. Then once in the laboratory, the problems start; there are limitations of the testing apparatuses and sample preparation, should the sample be moist tamped or will slurry deposition better represent the field conditions?

This paper seeks to address some of these aspects by linking triaxial testing to the cone penetration testing (CPTu) conducted *in situ*. This is done through large-strain modelling using the particle finite element method (G-PFEM) with an extended version of the Clay and Sand Model (CASM). The workflow involved calibration of the CASM on the triaxial results and then simulating the CPTu using those calibrated soil parameters. The calibration was possible in part due to obtaining high-quality mini-block sampling (MBS). A subsequent comparison between MBS results and those of reconstituted samples further emphasises the importance of minimising disturbance through such actions as sampling technique. This is followed by an examination of the *in situ* state parameter (Ψ) and the role it plays in the stress distribution at the time of failure. Furthermore, recommendations and learnings are provided for future projects and applications. The results show that calibration of numerical models using high-quality samples has significant benefits when compared with calibration of results using reconstituted samples.

INTRODUCTION

CPTu is a widely adopted *in situ* technique for characterising soft, fine-grained soils such as mine tailings. Its ability to provide continuous profiles (easily repeatable) of cone resistance (qt), sleeve friction (fs), and pore pressure (u2) makes it particularly useful in assessing geotechnical parameters and soil behaviour in stratified, weak, and sensitive deposits.

It is common for a significant geotechnical project to have a component of both laboratory testing and field testing to improve understanding of the potential behaviour of the soil. A challenge for practitioners is how to link the data obtained in a laboratory to the potential behaviour of the soil in the field. Currently, most interpretation of CPTu data is done with empirical methods to form this link.

This study applies G-PFEM, introduced by Monforte *et al* (2017a, 2017b, 2018), to interpret CPTu penetration in soft tailings. G-PFEM uses numerical modelling to simulate cone penetration. The constitutive model implemented in G-PFEM is CASM (Clay and Sand Model). CASM has gained attention for the usage of modelling tailings from well-documented tailings dam failures such as Brumadinho, Arroyo and Gens (2022). CASM is a critical state model and is capable of capturing both contractive and dilative responses in a unified manner. G-PFEM has been used previously for the interpretation of CPTu's by Bernardo *et al* (2024). In this work, the tailings parameters were calibrated using triaxial (TX) testing undertaken in MBS. The sampling was undertaken at a location adjacent to the CPTu. Parameters calibrated on the laboratory results were used to simulate CPTu with G-PFEM, including qt, fs, and u2. The G-PFEM results were then compared against the CPTu data next to the MBS location. Empirical normalised soil behaviour type (SBTn) charts were used to evaluate the classification consistency between the numerical and *in situ* data. The discussion focuses on discrepancies in cone tip resistance, pore pressure response, and the role of constitutive and interface assumptions in influencing the results. A comparison between MBS and laboratory reconstituted samples using state-of-the-art sample preparation methods at various laboratories revealed that there are still challenges in replicating *in situ* conditions in laboratory settings.

BACKGROUND

A geotechnical investigation was undertaken at a bauxite tailings facility in Northern Australia. The investigation included both CPTu and an extensive laboratory campaign, which included classification testing in addition to more advanced geotechnical testing such as TX, oedometer and simple shear. Laboratory testing was undertaken on samples recovered from adjacent locations to CPTu borings using MBS as well as on reconstituted samples.

The tailings storage facility (TSF) experiences a tropical monsoonal climate and with approximately 2 m of rainfall each year, mostly between December and March. It is a relatively young TSF, constructed in 2018 and having only undergone several wall raises. It is approximately 1000 Ha and is separated into two cells, with one cell being raised each year. The design life is 25 years, with a series of 12 to 13 upstream raises anticipated, totalling a raise of 20 m. A total of nine to ten million dry tons per annum of tailings are deposited into the TSF.

SITE INVESTIGATION

The tailings layer was investigated with the CPTu using a compression cone with a capacity of 25 MPa. Figure 1 displays the *in situ* ground profile of interest (4–6 mbgl), the phreatic surface is not shown but is 3.6 mbgl, inferred using *in situ* dissipation testing. It is noted that the atypical 'soft' bauxite tailings experienced a cone tip resistance (q_c) of typically <1 MPa and the friction ratio (R_f) <0.5, indicative of sensitive–contractive behaviour.

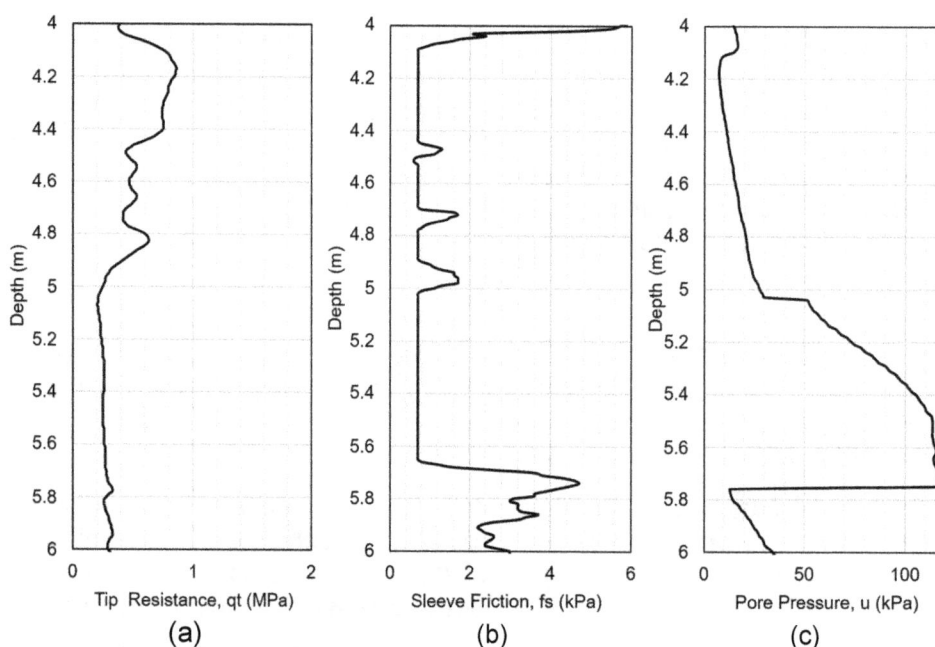

FIG 1 – CPTu profile of the investigated location.

MBSs were recovered below the phreatic surface from 5.0–5.6 mbgl, oedometer tests and consolidated isotropic undrained (CIU) TX were conducted. Bag samples were also recovered to undertake laboratory remoulding testing. CIU TX laboratory testing included end of testing freezing and moist tamping preparations and as Jefferies and Been (2016) recommendations. Laboratory testing was undertaken at various laboratories to check reproducibility, Reid *et al* (2020) has reported significant variation on results as a function of laboratory practices when undertaking critical state soil testing for tailings.

LABORATORY TESTING

Six CIU TX tests were conducted with trimmed MBSs. The initial void ratios of each sample were measured prior to the tests and varied marginally from 1.26 to 1.29. Sample quality indexes for samples were assessed following recommendations by Lunne, Berre and Strandvik (1997) and Berre, Lunne and L'Heureux (2022) with the ratings of very good to excellent. The TX stress trajectories are presented in Figure 2.

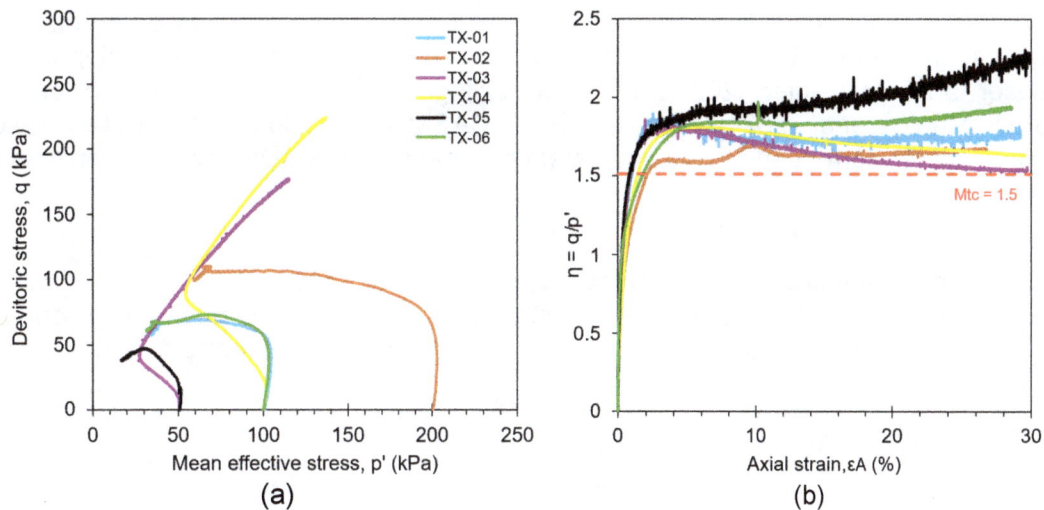

FIG 2 – Laboratory TX results on MBS tailings samples, (a) Cambridge p' – q Plot; (b) η – εA Plot.

TX-02, TX-05 and TX-06 showed dilation and strain hardening. TX-01, TX-03 and TX-04 showed a low to moderate brittleness. A stress trajectory of 100 kPa mean effective confining stress in TX-01 was considered to be the most representative of field conditions (80 kPa) and was selected for further modelling.

Figure 3 displays the difference between the high-quality MBS and the reconstituted samples (using moist tamping) stress-strain behaviour. The MBSs continue to show contractive behaviour after 5 per cent deformation. In comparison, all three tests of the reconstituted samples show a dilative response after the initial contractive stage.

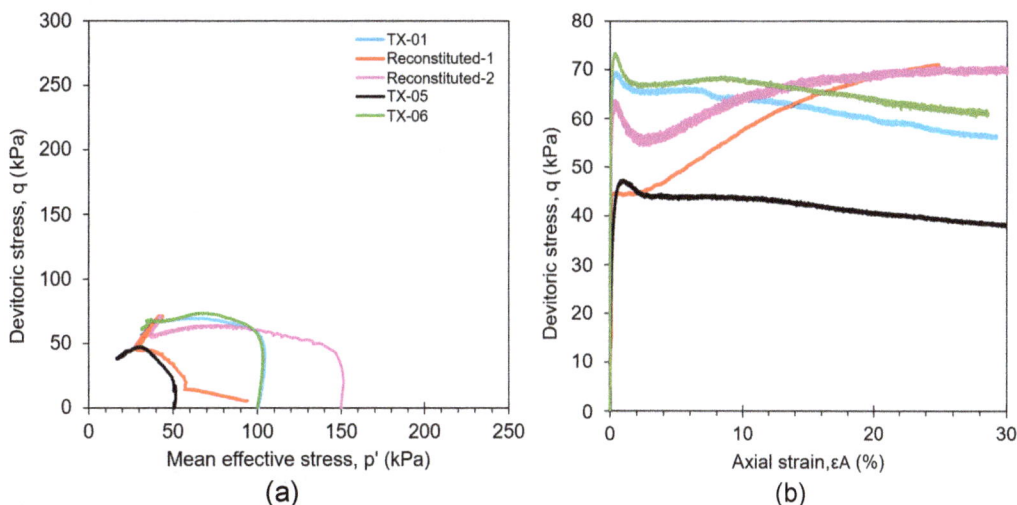

FIG 3 – Laboratory TX results showing MBS and reconstituted (a) Cambridge p'- q Plot; (b) q – εA Plot.

These materials' behaviour difference implies that the reconstituted lab samples aren't capturing some of the brittle behaviour evident in the MBS. These are challenges are hypothesized to relate with the loss of *in situ* fabric such as bonding and cementation. Such aspects of reconstituting samples has previously been demonstrated by Reid, Fanni and Fourie (2023). This result can make interpretation of the strength more challenging, especially when using reconstituted sample as it introduces the possibility of either over- or underestimating the materials residual strength. Another consideration is the limitations of the testing apparatus which may influence the stress-strain behaviour.

NUMERICAL MODELLING

CASM is a elastoplastic critical state model proposed by Yu (1998) and has been reported to be particularly well-suited for modelling the non-linear soil behaviour of cone penetration in saturated soils (Hauser and Schweiger, 2021). Because tailings are frequently found to be normally consolidated, Mánica *et al* (2021) recommends a CASM formulation that imposes ensures all constant stress paths correspond to a unique state parameter value, given by Equation 1 as follows:

$$\xi = (\lambda - \kappa)lnr\left[1 - \left(\frac{\eta}{M}\right)^n\right] \tag{1}$$

where η is known as the stress ratio; λ is the slope of the CSL, M is the stress ratio, and κ is the recompression line slope. Parameters r (spacing ratio) and n (stress-state coefficient) are the additional CASM parameters. r and n are defined by Equations 2 and 3 respectively:

$$r = \frac{p_0'}{p_x'} \tag{2}$$

$$\left(\frac{\eta}{M}\right)^n = 1 - \frac{\xi}{\xi_R} \tag{3}$$

where p_0' is the initial mean effective stress, p_x' is the reference mean effective stress used for normalisation, and ξ_R is a positive reference state parameter that denotes the vertical distance between the CSL and a reference consolidation line, $\xi_R = (\lambda - \kappa)lnr$.

The variation of the yield surface depends on a function of the stress invariants p' and q, λ, M, the pre-consolidation pressure as well as CASM parameters r, and n, shown in Equation 4:

$$f = \left(\frac{q}{Mp'}\right)^n + \frac{\xi}{\xi_R} - 1 \tag{4}$$

Parameter calibration

TX-06 was simulated at the element level using CASM. The stress path and the stress-strain plot are shown in Figure 4.

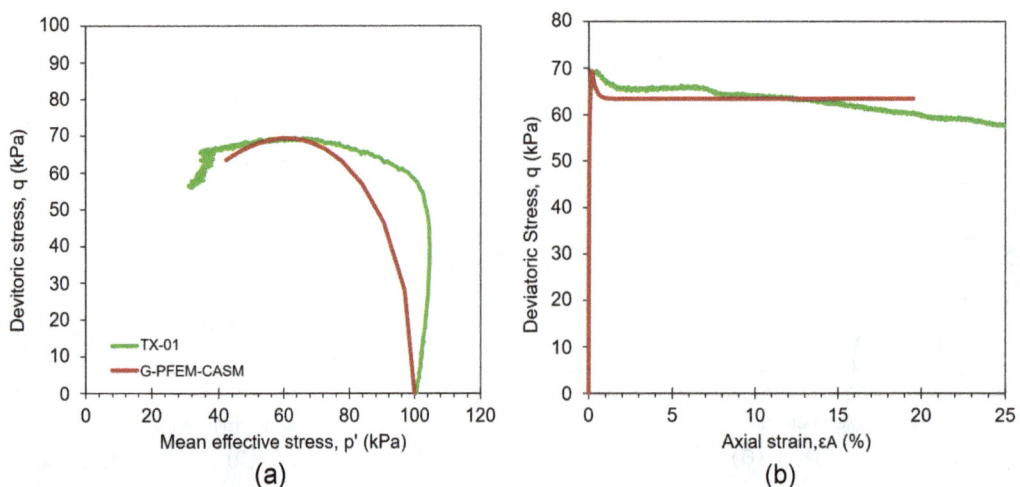

FIG 4 – G-PFEM-CASM modelling result versus laboratory TX-01 result, (a) Cambridge p-q' Plot; (b) Stress – Strain Plot.

The swelling and isotropic compression slope, κ^* and λ^*, are defined in the εvol-p' space, which are modified based on the initial void ratio and are defined in Equations 5 and 6 as follows:

$$\lambda^* = \frac{\lambda}{1+e_0} \tag{5}$$

$$\kappa^* = \frac{\kappa}{1+e_0} \tag{6}$$

The spacing ratio and the stress ratio were calibrated iteratively. The calibration indicates good agreement between the measured and modelled peak undrained shear strength ratio. Both the model and the test reached peak undrained shear strength at less than 1 per cent axial strain. Divergence between the model and the test include the residual undrained shear strength (in contrary to the model, TX-01 does not show a constant residual undrained shear strength). Moreover, the rate of strength loss is much more rapid in the model than in TX-01. The calibrated parameters utilised for the CPTu penetration simulation are shown in Table 1.

TABLE 1

CASM parameters used in Pocket G-PFEM simulations.

λ^*	κ^*	n	r	M	ϕ_{tc}	OCR
0.0301	0.0018	2.5	2.0	1.5	37.7	1

RESULTS OF A SIMULATION OF CPTU PENETRATION

CPTu profile simulation

In G-PFEM the soil formation around the cone during penetration is simulated using a fully coupled hydromechanical model with a single axisymmetric mesh. To minimise the boundary effects, the computational domain is defined as an axisymmetric soil column with a depth 15 times the cone diameter. The cone is modelled as a rigid body with a standard apex angle of 60° and a base diameter of 35.7 mm, consistent with the *in situ* CPTu test cone specifications.

The vertical boundaries are constrained in the horizontal direction, while the bottom boundary is fixed in both directions. The top boundary is free to deform, and a non-penetrable contact interface is defined between the cone and the tailings domain, as shown in Figure 5a. The mesh size is generated with triangular elements. The CPTu penetration simulation set-up is presented in Figure 5b.

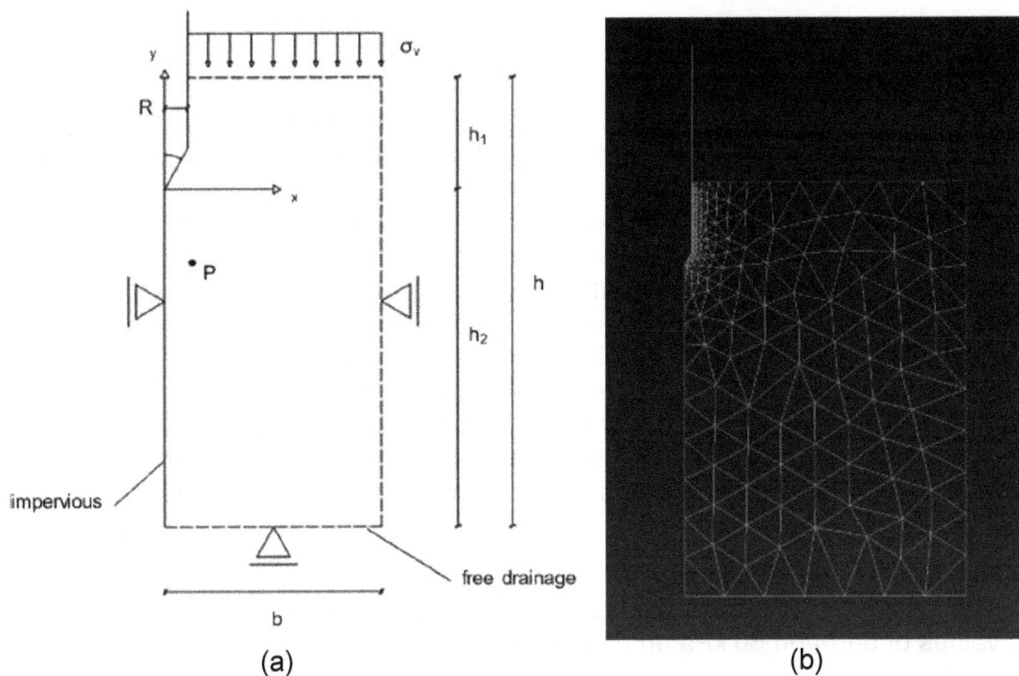

(a)   (b)

FIG 5 – G-PFEM model (a) conceptual set-up (Hauser and Schweiger, 2021); (b) G-PFEM Mesh.

The simulated cone resistance (qt) was compared with field CPTu measurements to assess the model performance, see Figure 6. The general shape and trend of the q_t profile with depth were successfully reproduced; however, the magnitude of the simulated q_t values was consistently higher than the field data across the entire depth range, with the value being overestimated by 20 per cent to 30 per cent. This overestimation may be attributed to the use of a rough cone model, which assumes a no-slip interface between the cone and the surrounding soil. This results in full shear stress transfer at the contact surface and can lead to artificially elevated resistance values, as it neglects potential interface softening, remoulding, or partial slip that occurs under real field conditions. Further investigations are necessary to confirm this hypothesis. Llano-Serna and Contreras (2019) have demonstrated that the role of cone roughness in analytical and numerical cone models is less prevalent than historically assumed, see for example Houlsby (1982).

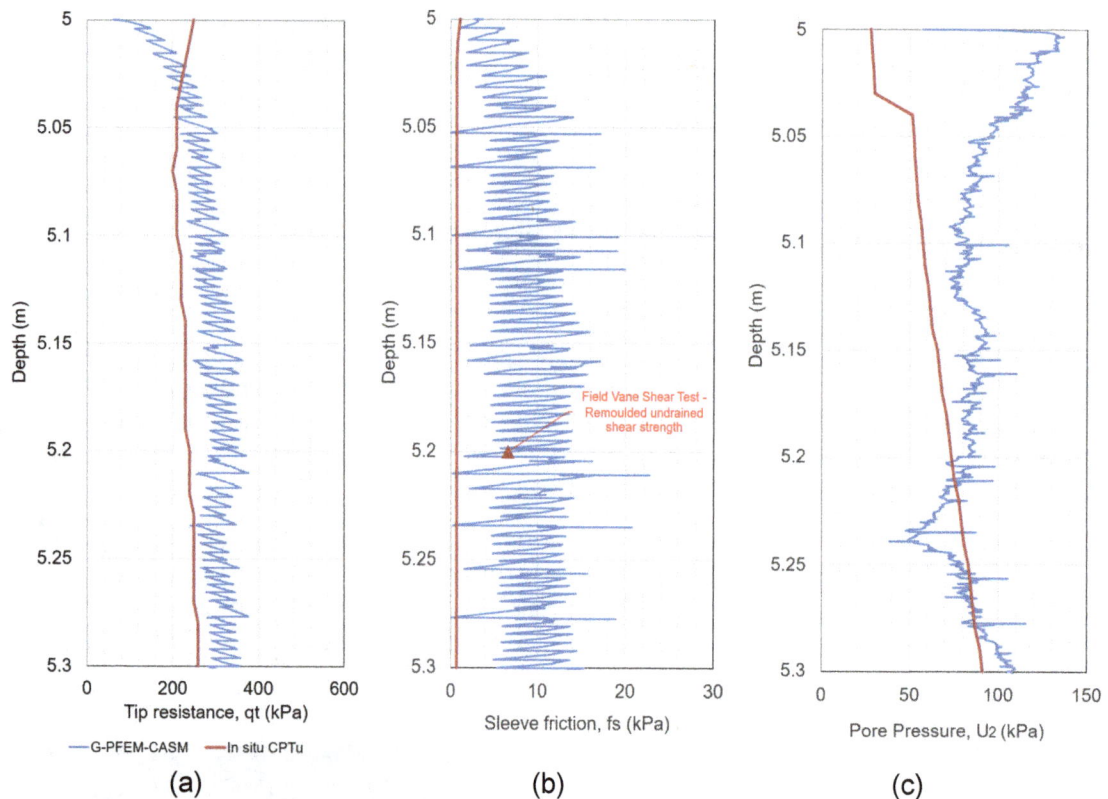

FIG 6 – G-PFEM modelled results versus CPTu results, (a) cone tip resistance versus depth, (b) sleeve friction versus depth, and (c) dynamic pore pressure at the u2 position.

Figure 6b shows that the G-PFEM modelled f_s oscillates between 5 kPa and 10 kPa, this is significantly larger than the *in situ* f_s results. According to McConnell and Wassenaar (2022), using a high-capacity compression cone (25 MPa) compared to a more sensitive cone (3 MPa), the f_s value may be underestimated. The field vane shear test (VST) residual undrained shear strength obtained between 5.2 to 5.7 mbgl ranged from 6 to 7.5 kPa, which is greater than the CPTu cone measured value but approximately equal to the modelled G-PFEM value. The assessment undertaken herein does not allow us to conclude the nature of the oscillation in the modelled f_s trace. It is hypothesised that a finer mesh could help mitigate the instability of results. However, finer meshes take considerably longer to run the simulations and render this kind of assessment less feasible for a consulting practice.

The simulated pore pressure response (u_2) showed good agreement with the field data, see Figure 6c. The magnitude of excess pore pressure during penetration was relatively consistent with measured values of between 50 kPa and 100 kPa.

Soil behaviour type prediction

The Soil Behaviour Type (SBT) classification results from the numerical simulation were compared against *in situ* CPTu data using three established methods: the Schneider *et al* (2008) chart, the Robertson (1990) normalised B_q plot, and the Q_t-F_r chart proposed by Robertson (1990). Results of the comparison are presented in Figure 6.

The following key observations are made:

- The Schneider *et al* (2008) chart is presented in Figure 7a. Both CPTu and G-PFEM simulated results show consistency as tailings were predicted to behave between silts and low I_r clays and transitional soils. However, as it is noted that q_t was been overestimated using G-PFEM.

- Robertson (1990) normalised B_q plot is presented in Figure 7b. Field data clusters around zones 2 and 3 (organic soils-peats and clays clay to silty clay respectively). The modelled results plot broadly within zone 3. Differences are attributed to the numerical model overestimation of the cone resistance that results in an overestimated normalised cone resistance Q_t.

- Robertson (1990) Q_t-F_r chart is presented in Figure 6c. The results show that field data is classified as sensitive, fine grained (zone 1) and zone 3. The model data is classified between zone 3 and zone 4 (silt mixtures clayey silt to silty clay). The use of a CPTu tool with inadequate sensitivity to capture fs value is highlighted in this application.

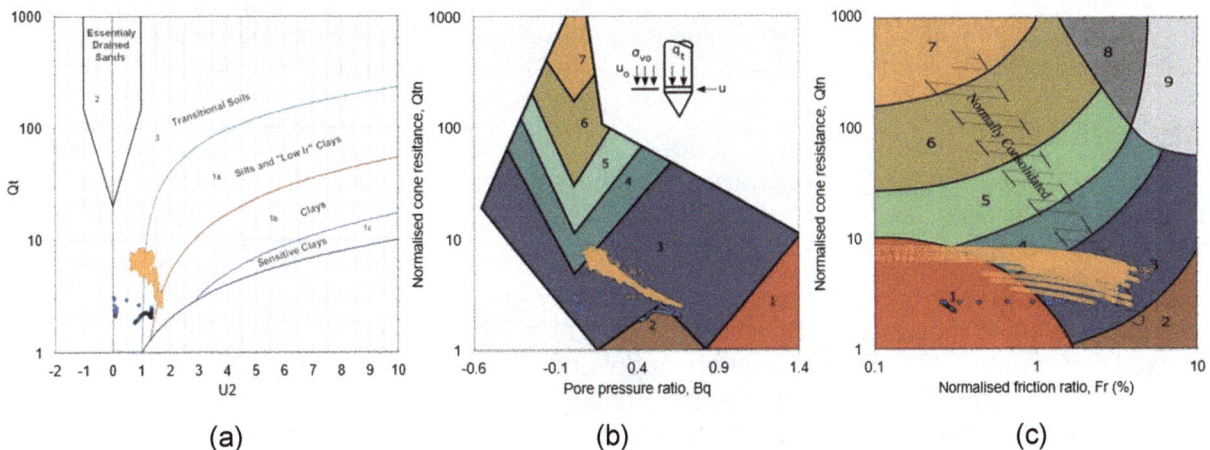

(a) (b) (c)

FIG 7 – Comparison of G-PFEM modelled results and *in situ* CPTu based SBT results:
(a) Schneider *et al* (2008) correlation plot, Orange: G-PFEM-CASM Result, Blue: *In situ* CPTu based Result; (b) Robertson (1990) normalised B_q plot; (c) Robertson (1990) Q_t-F_r chart. Soil Behaviour Type (Robertson, 1990): 1. Sensitive fine-grained; 2. Organic; 3. Clay; 4. Silt-mixtures; 5. Sand-mixtures; 6. Sand; 7. Gravelly sand to sand; 8. Very stiff sand to clayey sand; 9. Very stiff fine-grained.

DISCUSSION

This study assessed the capability of Pocket G-PFEM, combined with the CASM constitutive model, to simulate CPTu penetration in soft, fine-grained tailings and aid data quality control and interpretation. Although still in the development stage, the numerical approach has shown potential in capturing key features of cone penetration behaviour under large deformation and undrained conditions for soft tailings.

The laboratory testing section results highlighted the importance of high-quality sampling using MBS techniques for subsequent testing and numerical modelling calibration. Laboratory testing was undertaken at well-known, reputable facilities using state of art remoulding techniques. The results indicated that remoulding of tailings did not result in the same kind of brittleness recorded from MBS. These differences are attributed to fabric anisotropy (as documented by Reid, Fanni and Fourie, 2023).

Furthermore, numerical modelling results were enhanced due to high-quality sampling for model calibration. The calibration process is dependent on experience and iterative fitting of the calibration parameters, which can be time consuming. There are some inherent disadvantages when undertaking the calibration, for example, testing on remoulded tailings is conventional practice in the industry. It is difficult to understand if testing on remoulded tailings can replicate the same degree of brittleness that tailings in the field can develop.

An interesting future development for G-PFEM–CASM could be to undertake reverse engineering for modelling. For example, use the physical quantities from CPTu as input parameters for G-PFEM and allow the model to produce CASM parameters that fit the CPT trace. This exercise could be replicated for other constitutive relationships.

The simulation produced pore pressure (u_2) responses that generally aligned well with the field CPTu data. However, a notable overprediction of cone resistance (q_t) was observed throughout the full depth range. This discrepancy has the potential of an impact SBT classification. In the example shown here, the simulated data clustered in zones typically associated with less sensitive materials. It is hypothesised that differences could be attributed to: i) temperature equalisation not being undertaken during the field investigation; and ii) models adopted to capture the roughness interface between the cone and the tailings. The CASM implementation does not include temperature effects and a rough cone was assumed for modelling.

The most notable contribution of the modelling presented was the capacity to infer CPT sleeve friction. Highly sensitive cones such as those recommended by McConnell and Wassenaar (2022) might not be available in all scenarios. In the case study presented, the data were processed after the site investigation was completed, and G-PFEM–CASM was useful to improve interpretation.

CONCLUSION

High-quality MBSs were used to undertake a site characterisation. Laboratory testing focused on TX testing. TX results were used to calibrate the CASM constitutive relationship. G-PFEM was used to model CPT penetration of sensitive tailings at depths of between 5 m and 5.3 m.

Field CPT data was used to benchmark G-PFEM model results using conventional CPT classification charts such as those proposed by Schneider et al (2008) and Robertson (1990). The benchmarking was used to determine differences between model and site measurements in cone tip resistance of up to 30 per cent attributed to the numerical strategy to model cone roughness and temperature equalisation practices during cone penetration. The sleeve friction measured in situ was found to be inaccurate and less than expected due to a number of reasons (ie cone sensitivity). Residual strength was interpreted using G-PFEM–CASM and confirmed by the field VST results.

The results of this study demonstrate that calibrating the G-PFEM–CASM simulation framework using TX laboratory data can offer a relatively effective means of replicating CPTu behaviour in soft, fine-grained tailings. Despite inherent simplifications and model limitations, the simulated penetration profile shows agreement with in situ pore pressure trends and captures the general evolution of cone resistance.

The approach represents a promising direction for integrating laboratory test data into advanced numerical simulations of in situ testing, ultimately enhancing the interpretive power of CPTu results in complex ground conditions.

ACKNOWLEDGEMENTS

The authors are grateful for the assistance provided by the research team at UPC and CIMNE, especially by Marcos Arroyo, Lluís Monforte and Laurin Hauser, both in providing access to Pocket G-PFEM and by giving us their valuable support during its utilisation.

REFERENCES

Arroyo, M and Gens, A, 2022. Computational Analyses of Dam I Failure at the Corrego de Feijao Mine in Brumadinho, Final Report for VALE SA, August 2021, International Center for Numerical Methods in Engineering (CIMNE).

Bernardo, K, Tasso, N, Sottile, M G and Sfriso, A, 2024. A method to estimate the state parameter from CPTu soundings using Pocket G-PFEM, Barcelona, Spain, 7th International Conference on Geotechnical and Geophysical Site Characterization.

Berre, T, Lunne, T and L'Heureux, J-S, 2022. Quantification of sample disturbance for soft, lightly overconsolidated, sensitive clay samples, *Canadian Geotechnical Journal*, 59(2):300–303.

Hauser, L and Schweiger, F H, 2021. Numerical study on undrained cone penetration in structured soil using G-PFEM, *Computers and Geotechnics*, 133.

Houlsby, G T, 1982. Theoretical analysis of the fall cone test, *Géotechnique*, 32(2):111–118.

Jefferies, M and Been, K, 2016. *Soil Liquefaction A Critical State Approach*, second edition (CRC Press).

Llano-Serna, M and Contreras, L-F, 2019. The effect of surface roughness and shear rate during fall-cone calibration, *Geotechnique*, 70(4):332–342.

Lunne, T, Berre, T and Starndvik, S, 1997. Sample Disturbance effects in soft low plastic Norwegian Clay, Symposium on Recent Developments in Soil and Pavement Mechanics.

Mánica, M A, Arroyo, M, Gens, A and Monforte, L, 2021. Application of a critical state model to the Merriespruittailings dam failure.

McConnell, A and Wassenaar, E, 2022. An innovative new 3MPa CPT—to detect and measure very small fs values, in *Cone Penetration Testing*, pp 197–202 (CRC Press).

Monforte, L, Arroyo, M, Carbonell, M J and Gens, A, 2017a. Numerical simulation of undrained insertion problems in geotechnical engineering with the Particle Finite Element Method (PFEM), *Computers and Geotechnics*, 144–156.

Monforte, L, Carbonell, J M, Arroyo, M and Gens, A, 2017b. Performance of mixed formulations for the particle finite element method in soil mechanics problems, *Computational Particle Mechanics*, 269–284.

Monforte, L, Carbonell, J M, Arroyo, M and Gens, A, 2018. Coupled effective stress analysis of insertion problems in geotechnics with the Particle Finite Element Method, *Computers and Geotechnics*, 114–129.

Reid, D, Fanni, R and Fourie, A, 2023. Linking laboratory quasi-steady state strengths to field scale performance of tailings, 8th International Symposium on Deformation Characteristics of Geomaterials (ISDCG2023).

Reid, D, Fourie, A, Juan, A and Dickinson, S, 2020. Results of a critical state line testing round robin programme, *Geotechnique*, 71(7):616–630.

Robertson, P K, 1990. Soil classification using cone penetration test, *Canadian Geotechnical Journal*, 151–158.

Schneider, J A, Randolph, F M, Mayne, W P and Ramsey, R N, 2008. Analysis of Factors Influencing Soil Classification Using Normalized Piezocone Tip Resistance and Pore Pressure Parameters, *Journal of Geotechnical and Geoenvironmental Engineering*, 1569–1586.

Yu, H S, 1998. CASM: A unified state parameter model for clay and sand, *International Journal for Numerical and Analytical Methods in Geomechanics*, 621–653.

Numerical validation of equivalent strength methods – a case study on a stone column-stabilised tailing dam

M Sottile[1], J Rola[2], A Hubaut[3], A Sfriso[4] and L Roldan[5]

1. Principal Consultant, SRK Consulting, Argentina. Email: msottile@srk.com.ar
2. Principal Consultant, SRK Consulting, Perth, Australia. Email: jrola@srk.com.au
3. Design Manager, Menard Oceania, Sydney, Australia. Email: ahubaut@menard.com.au
4. Corporate Consultant, SRK Consulting, Argentina. Email: asfriso@srk.com.ar
5. Senior Consultant, SRK Consulting, Argentina. Email: lroldan@srk.com.ar

ABSTRACT

The complexity of the 3D arrangement of stone columns often requires simplifications into 2D models using equivalent strength parameters for design purposes. This study employs a simple expression to compute the equivalent block strength and investigates its validity by comparing factors of safety from 2D analyses with those from 3D numerical strip models that account for a realistic interaction between stone columns and surrounding soil.

Information of a real dam tailings stabilisation project is used. The modelling is done for three representative embankment sections under varied replacement ratios and soil conditions. Results show that the Shear Strength Reduction (SSR) factors calculated from 3D finite element simulations are in acceptable agreement with the Factors of Safety (FoS) from 2D limit equilibrium methods using equivalent strength, with maximum differences of 5 per cent. Moreover, an overall agreement was found in the failure mechanisms, which minor differences that can be attributed to the role of stress distribution and soil-column interactions in 3D models. These results confirm the reliability of the equivalent strength approach for practical 2D designs.

INTRODUCTION

There are many methods available to improve the stability of earth dams, including berms and buttresses; excavation and replacement; *in situ* densification (eg vibro-compaction, vibro-replacement, dynamic-compaction, compaction-grouting); *in situ* strengthening (eg stone columns, jet grouting, dynamic replacement, concrete piling, deep soil mixing); drainage (eg strip drains, stone columns, trenches). The selection of the method relies on several factors, including soil type and properties, groundwater conditions, project requirements, depth and extent of weak layers, time constraints, environmental and constructability constraints, availability of equipment and materials, cost-effectiveness, and experience (Mitchell and Gallagher, 1998).

Stone columns consist of vertical inclusions of clean crushed rock that are installed in a grid using the vibro-replacement method. They increase soil density, improve shear strength, and provide drainage for pore pressure dissipation; therefore, they are particularly effective in improving slope stability and reducing the liquefaction risk. This technique has been used successfully to reinforce the foundation of several dams, eg Salmon Lake Dam (Luehring *et al*, 2001); Mormon Island Auxiliary Dam (Nickell, Allen and Ledbetter, 1994) and Steel Creek Dam (Keller, Castro and Rogers, 1987).

Designing stone column arrangements often involves simplifying 3D models to 2D using equivalent strength parameters. This study validates these parameters for stone column-stabilised tailing dams by comparing 2D and 3D models. Numerical analyses were conducted on three cross-sections of an existing tailings embankment, using the Shear Strength Reduction (SSR) method in Plaxis 3D and 2D limit equilibrium (LE) methods, assessing the impact of different stone column replacement ratios (RR).

LITERATURE REVIEW

Various methods have been developed to model stone columns (SCs), including simplified 2D plane-strain approaches and detailed 3D finite element (FE) simulations. The following review offers a comprehensive overview of the methodologies currently employed in SC modelling. It highlights the use of equivalent strength techniques for stability as well as consolidation-focused approaches, particularly in the context of dam reinforcement.

Castro (2017) reviewed various modelling techniques for both ordinary and geosynthetic-encased stone columns, emphasizing numerical approaches such as unit cell analysis, cylindrical ring models, and three-dimensional full or partial simulations. The study highlighted the importance of considering critical column length and installation effects, particularly in non-encased columns. A key insight was that the column critical length is primarily dependent on footing dimensions rather than column length alone, and different simplified geometrical models exhibit varying levels of suitability for different analytical purposes.

Chakraborty and Sawant (2022) conducted numerical simulations to evaluate the performance of embankments on liquefiable soils, specifically focusing on the effectiveness of stone columns in mitigating settlements induced by liquefaction. They employed the equivalent plane strip concept to model the cylindrical stone columns as continuous strips, utilising equivalence relationships for permeability and stiffness parameters. Their parametric study indicated that although stone columns substantially reduce excess pore pressure and heaving at the embankment toe, their effectiveness decreases with increasing cyclic load amplitudes.

Ng and Tan (2015) introduced a simplified homogenisation method for stone column designs called the Equivalent Column Method (ECM), which estimates equivalent stiffness and permeability. ECM allows for straightforward numerical implementation and maintains accuracy in settlement predictions, unlike complex finite element constitutive models. Design charts for various area replacement ratios and permeability values were developed to provide practical insights for embankment and large tank foundation improvements.

Tan, Tjahyono and Oo (2008) developed a plane-strain conversion approach for modelling stone column-reinforced ground, allowing axisymmetric unit cells to be transformed into equivalent plane-strain models for 2D numerical analysis. Two conversion methods were proposed: one based on permeability matching and the other on column width adjustment. Their study demonstrated that both methods yield reasonable long-term consolidation settlements under linear-elastic material modelling, whereas for elastoplastic materials, the width-matching method produced more reliable results.

Zhang, Han and Ye (2014) conducted a numerical analysis on the stability of deep-seated slopes in embankments supported by stone columns over soft clay. They compared two modelling approaches: the column-wall method and the equivalent area method. Their findings indicated that the equivalent area method consistently produced continuous critical slip surfaces, whereas the column-wall method exhibited discontinuous failure zones. The authors recommended applying a correction factor to the equivalent area model under short-term conditions to ensure its accuracy in slope stability assessments.

CASE STUDY

Dam description

The dam analysed in this study is a 26 m high downstream-raised structure, originally constructed in the 1980s as a water retention facility. It spans approximately 800 m across a shallow valley, which was previously covered with 6 to 8 m of hydraulically deposited tailings. The dam consists of three primary sections: the crest, the berm, and the toe, each playing a distinct role in the structural integrity of the embankment (Figure 1).

The site has been extensively investigated through *in situ* testing, including 206 cone penetration tests (CPTu), 37 boreholes (BH), and 35 test pits (TP), along with additional geophysical and laboratory investigations. Rola *et al* (2024) and Sottile, Crocker and Roldan (2024) provide a thorough discussion on the ground model elaboration for the unreinforced condition of the facility.

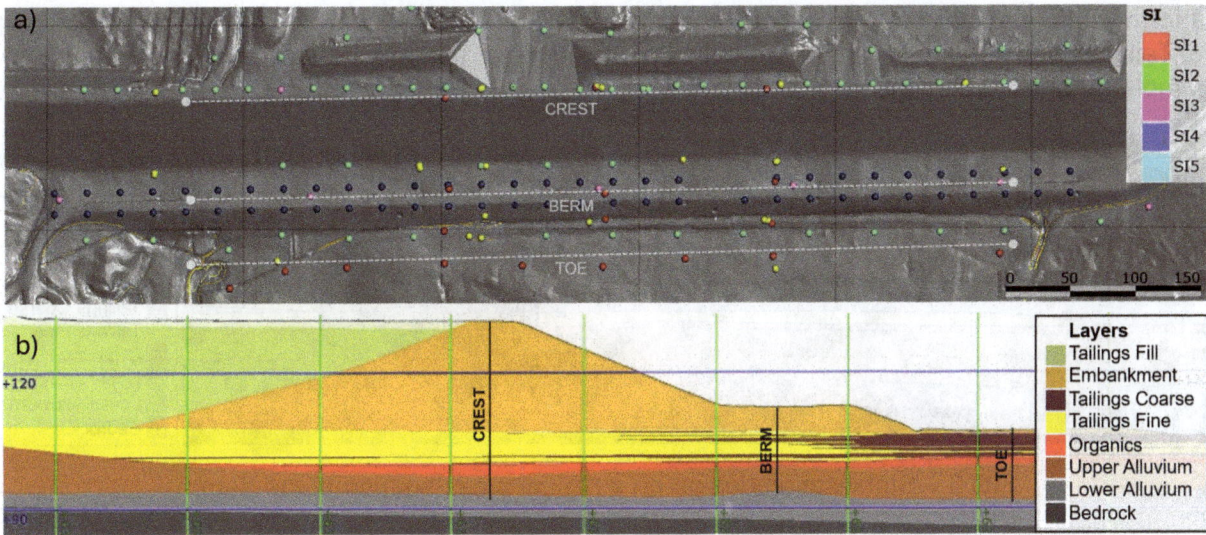

FIG 1 – (a) Plain view of the dam; (b) Typical dam cross-section at the centre of the valley.

Geotechnical units

The subsurface stratigraphy has distinct layers, each affecting dam stability and foundation behaviour. Figure 2 shows a CPT-BH pair at the crest of the dam. The main geotechnical units are:

1. Embankment Fill, a compacted clayey material forming the main structure of the dam, with moderate strength and low pore pressure response during the cone penetration.

2. Tailings, lying beneath the embankment and extending laterally across the dam area, is composed of coarser and finer portions with differences in sleeve friction measurements.

3. Organic Layer, a very soft clayey layer with high organic content with very low tip resistance and high pore pressure response during cone penetration.

4. Upper Alluvium, a natural clay layer with low strength, with occasional higher values due to the presence of sand or gravel inclusions.

5. Lower Alluvium, a stiff gravelly clay, identified by a sharp increase in cone tip resistance, up to refusal, and low pore pressure ratios.

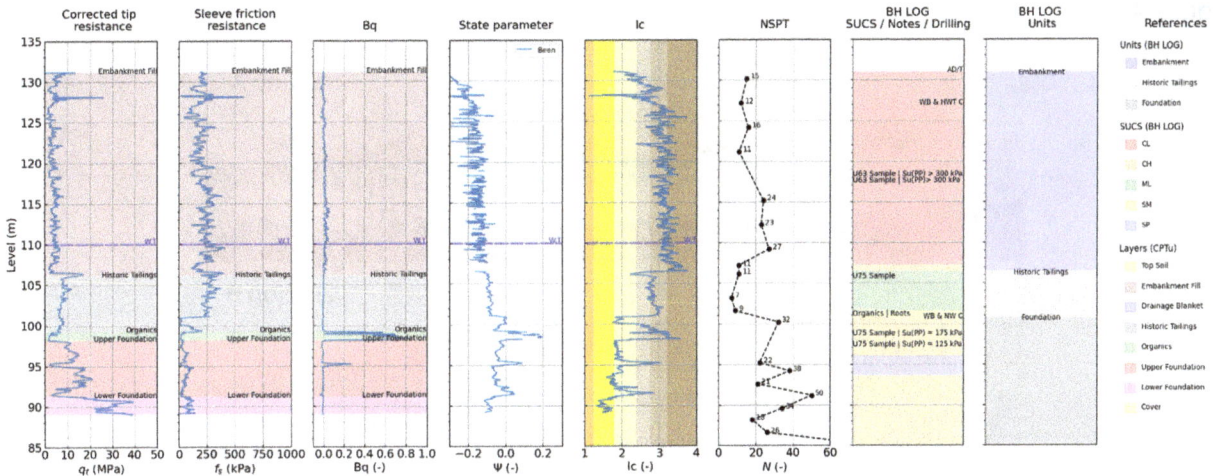

FIG 2 – Example of layer identification using CPTu and BH logs located at the dam crest.

Stabilisation design

The proposed dam stabilisation includes constructing a working platform which also serves as a safety buttress, improving the ground near the toe area with stone columns, and adding a buttress. The necessary GI width and replacement ratio differ by section to ensure the minimum Factor of Safety (FoS) is achieved. A stepped buttress up to RL +117/121 m is battered to the levelling platform with a slope of 1V:2.5H. Short SCs with an RR of 22.7 per cent and a width of 4 m through the Tailings and Organics layers were incorporated near the toe as a mitigation measure. An example for one section is shown in Figure 3.

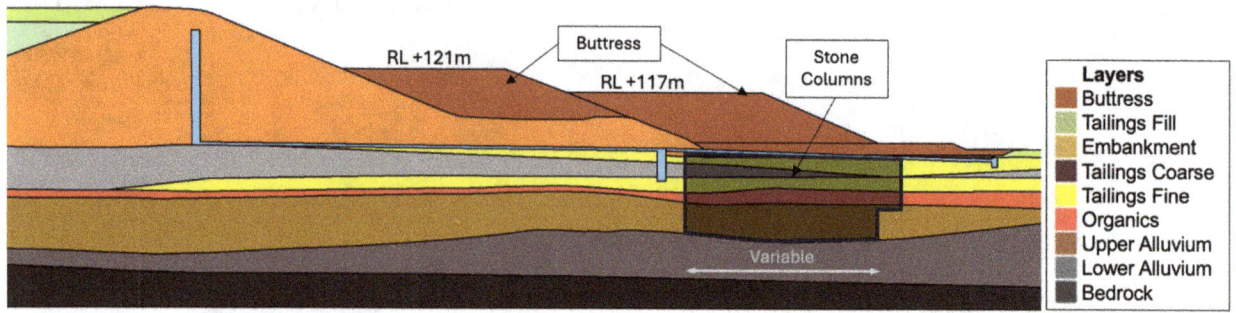

FIG 3 – Stabilisation design on a typical cross-section of the dam.

Equivalent strength method

The expression used in this study correspond to the equivalent area method, which has been used by Zhang, Han and Ye (2014). The equivalent strength is computed based on the proportion of the improved area occupied by the stone columns (Replacement Ratio, RR). The friction angle of the stone columns ($\phi'_{sc} = 40°$) and the undrained shear strength of the surrounding soil (s_u/σ'_{v0}) are combined using a weighted average as follows:

$$\frac{\tau_f}{\sigma'_{v0}} = RR \ \tan(\phi'_{sc}) + (1 - RR)\frac{s_u}{\sigma'_{v0}}$$

This formula is applied independently for each geotechnical layer (eg Tailings Coarse, Tailings Fine, Organics, Upper Alluvium) considering peak and residual USR.

NUMERICAL MODELLING APPROACH

Geometry

Three cross-sections of the dam were modelled: Section A with RR = 30.4 per cent; Section B with RR = 35.8 per cent; and Section C with RR = 22.7 per cent. These sections were chosen to represent spatial variability in geotechnical units and replacement ratios. Their cross-sections are shown in Figure 4.

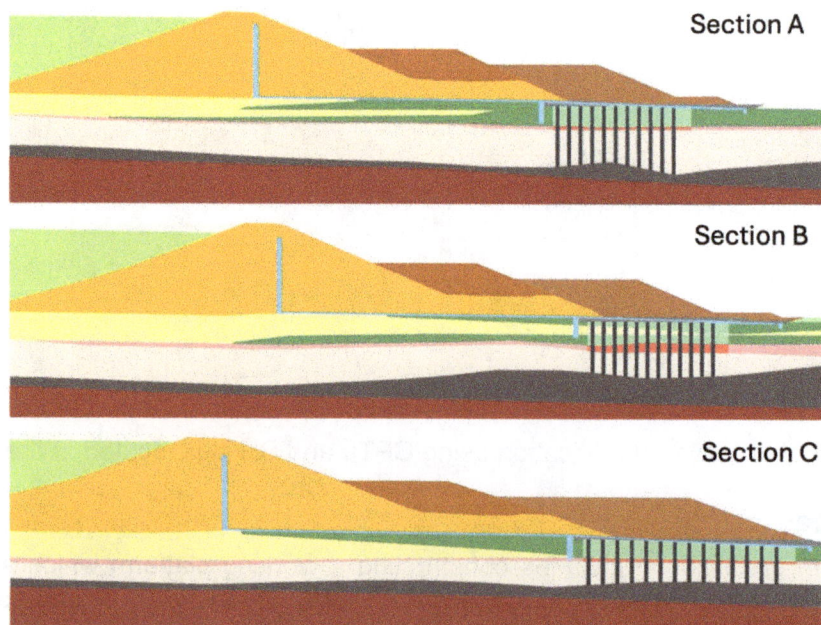

FIG 4 – Geometry of the three selected cross-sections for this study: Section A, B and C.

Constitutive models and material parameters

Two constitutive models were employed in the 3D finite element (FE) numerical analyses:

1. Hardening Soil with Small-Strain Stiffness (HSS): for all materials during construction stages, for dilative materials (Lower Alluvium, Embankment, Buttress) in the safety analyses and for stone columns in the 3D models; Table 1 summarises the most relevant parameters of all units, which were determined using laboratory/*in situ* geotechnical testing data and trial pads executed near the toe of the dam.

2. SHANSEP Mohr–Coulomb (MC): Applied only in the safety analyses within the 3D FE model to represent peak and residual USR of contractive materials (Tailings Coarse, Tailings Fine, Organic and Upper Alluvium) in the zones outside the GI area and in-between SCs to account for the local effects of stone column installation (Roldan *et al*, 2024). Table 2 presents the calibrated parameters.

TABLE 1
HSS model parameters adopted in FE analysis.

Material	γ_{sat} (kN/m)	E_{50}^{ref} (MPa)	E_{oed}^{ref} (MPa)	E_{ur}^{ref} (MPa)	m (-)	G_0^{ref} (MPa)	$\gamma_{0.7}$ (-)	c' (kPa)	ϕ' (°)
Tailings Coarse	15.5	5.0	3.7	70	0.5	70	1E-04	0	35
Tailings Fine	15.5	5.0	3.7	70	0.5	70	1E-04	0	33
Organics	17.5	2.0	2.0	32	0.8	70	3E-04	0	28
Upper Alluvium	18.5	40.0	40.0	120	0.5	150	2E-04	0	28
Lower Alluvium	19.5	80.0	80.0	250	0.5	265	2E-04	0	34
Embankment	20.5	60.0	60.0	175	0.5	180	2E-04	5	30
Buttress Berm	19.5	42.0	42.0	105	0.5	125	2E-04	5	30
Buttress Toe	20.5	60.0	60.0	175	0.5	180	2E-04	2	40
Tailings Fill	15.5	5.0	3.7	70	0.5	70	1E-04	0	33
Stone Columns	20.0	140.0	140.0	360	0.2	360	1E-04	0	38

TABLE 2
SHANSEP parameters – Peak and residual USR.

Material	Outside SCs [A, B]		In-between SCs [A]		Equivalent Peak USR [B]			Equivalent Residual USR [B]		
	Peak USR [-]	Res USR [-]	Peak USR [-]	Res USR [-]	RR = 22.7%	RR = 30.4%	RR = 35.8%	RR = 22.7%	RR = 30.4%	RR = 35.8%
Tailings Coarse	0.25	0.10	0.53	0.53	0.60	0.62	0.64	0.60	0.62	0.64
Tailings Fine	0.27	0.14	0.27	0.20	0.40	0.44	0.47	0.35	0.39	0.43
Organic	0.23	0.09	0.23	0.11	0.37	0.42	0.45	0.28	0.33	0.37
Upper Alluvium [T]	0.62[1]	0.43[1]	0.25	0.20	0.38	0.43	0.46	0.35	0.39	0.43
Upper Alluvium [C]	0.25	0.20	N/A	N/A	N/A	N/A	N/A	N/A	N/A	N/A
Lower Alluvium	-	0.43	N/A	N/A	N/A	N/A	N/A	N/A	N/A	N/A

Notes: [A] = Used in 3D FE analyses; [B] = used in 2D LE analyses. Upper Alluvium location at [C] = Crest area and [T] = Toe area, where the effect of OCR is incorporated directly on strength ratios [1] (ie SHANSEP-m exponent is set to 0). N/A applies for units that are not affected by the SCs installation.

In the 2D limit equilibrium (LE) analyses, the models employed were:

1. Mohr–Coulomb for the dilative soil units, with effective strength parameters as per the HSS model (ie c' and ϕ').

2. SHANSEP (ie s_u/σ'_{v0} peak and residual) for the contractive soil units outside the zone of influence of the GI installation (Outside SCs – Table 2).

3. SHANSEP with equivalent undrained strength parameters for the contractive units influenced by SC installation (ie Tailings, Organics, and Upper Alluvium) and calculated for different replacement ratios (22.7 per cent, 30.4 per cent, and 35.8 per cent) using the equivalent strength equation presented above (Equivalent Peak/Residual USR – Table 2).

Mesh

Plaxis 3D was used to create strip models of each section. The model width equalled the SC spacing based on the corresponding RR. The geometry and mesh for section B are illustrated in Figure 5; an analogous approach was followed for the other two sections.

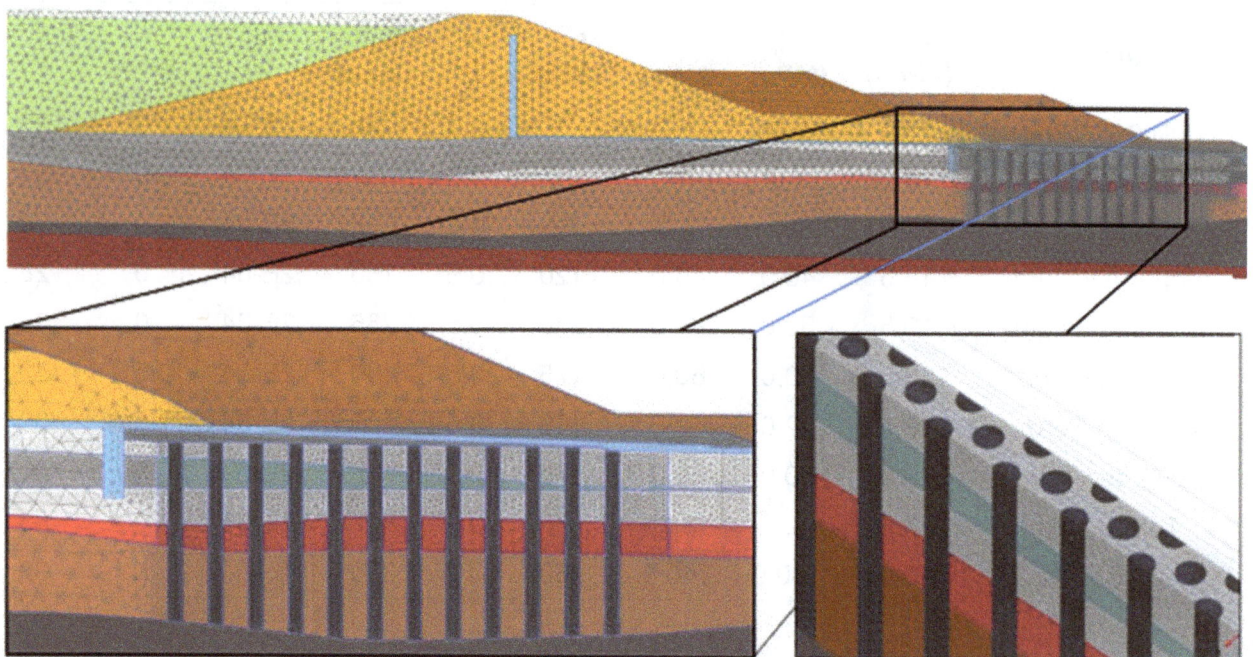

FIG 5 – Geometry and mesh – Example for Section B.

Modelling sequence

The following construction sequence was modelled for each 3D-strip section (Figure 6):

Phase 0 – Initial Phase. A K0-procedure was performed; a $K_0 = 1.0$ was adopted for soils between the SCs to account for installation effects. This strategy allows the stresses at the ground improvement area to be initialised in a simplified but realistic way, while avoiding a complex and detailed SC installation sequence.

Phase 1 – Nil Phase. A drained plastic phase was performed to recalculate the initial stresses to achieve equilibrium.

Phase 2 – Buttress construction. A drained plastic phase was performed to activate the final geometry of the buttress.

Phase 3 – Peak strength analyses. SHANSEP peak undrained materials were assigned to corresponding materials (3a), and safety analyses were conducted subsequently (3b).

Phase 4 – Residual strength analyses. SHANSEP residual undrained materials were assigned to corresponding materials (4a), and safety analyses were conducted subsequently (4b).

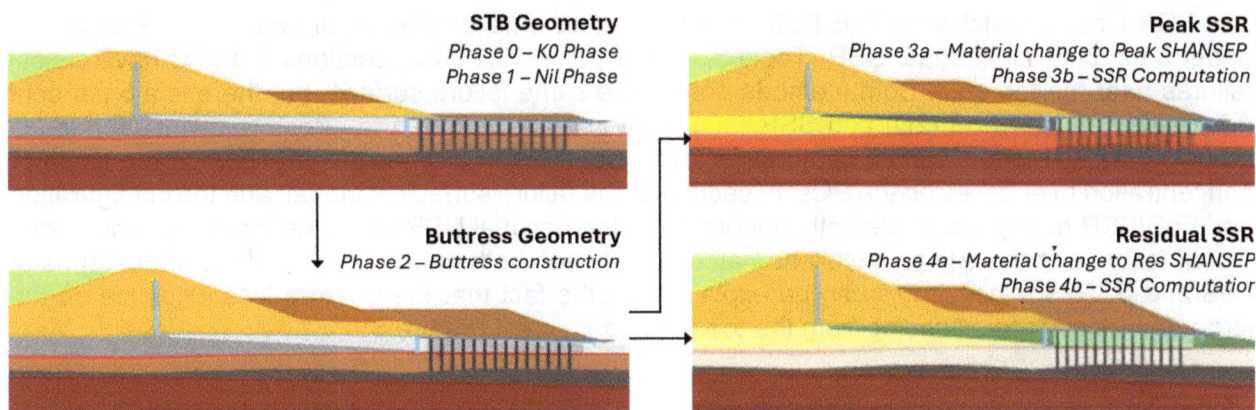

FIG 6 – Modelling sequence. Example for Section C.

RESULTS

Table 3 shows the SSR values from 3D models and the FoS from 2D LE analyses using equivalent strengths for GI. Figures 6 and 7 show a comparison of the failure mechanisms for peak and residual USR cases, respectively. For the case of the 3D FE model, deviatoric strains contours at failure are shown; these correspond to a non-converged stages, and so, they should only be interpreted to understand the location of the failure surface.

TABLE 3

Peak and Residual SSR and FoS values for each section.

Section	RR (%)	Peak USR		Residual USR	
		Plaxis 3D	LE 2D Equivalent	Plaxis 3D	LE 2D Equivalent
A	30.4	1.64 L	1.72 L	1.07 G	1.10 G
B	35.8	1.78 G	1.80 G	1.10 G	1.11 G
B	22.7	1.49 L	1.55 L	1.13 G	1.10 G

Notes: L = Local failure; G = Global failure.

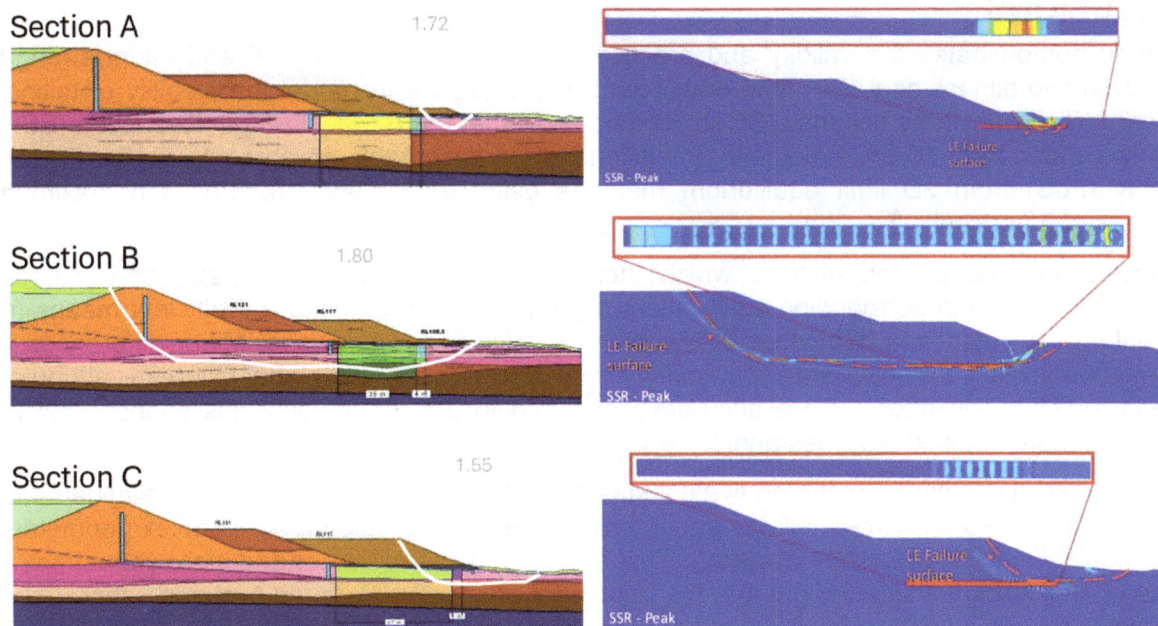

FIG 7 – Comparison of failure surfaces between 2D LE analyses using equivalent strength (left) and 3D FE modelling using the full geometry (right). Peak USR case. Section A, B and C.

The SSR closely matched the LE FoS, with differences smaller than 5 per cent for peak USR and under 3 per cent for residual USR (Table 3). In the peak USR case, sections A and C have critical failures near the toe area. Both methods predict the same failure surface, but there is a 5 per cent difference between FoS (2D) and SSR (3D). This can be attributed to stress re-distribution in 3D models that LE models cannot capture; for example, rotation of principal stresses on the slope or concentration of stresses on the SCs. In section B, the failure surface is global, and the configuration and FoS/SSR nearly match perfectly (Figure 7). In the residual USR case, the three sections form a global failure and there is an excellent matching between both methods. The difference in terms of FoS and SSR is minimal; this can be explained by the fact that the GI area less influence on the larger failure surface extending from the toe to the dam crest (Figure 8).

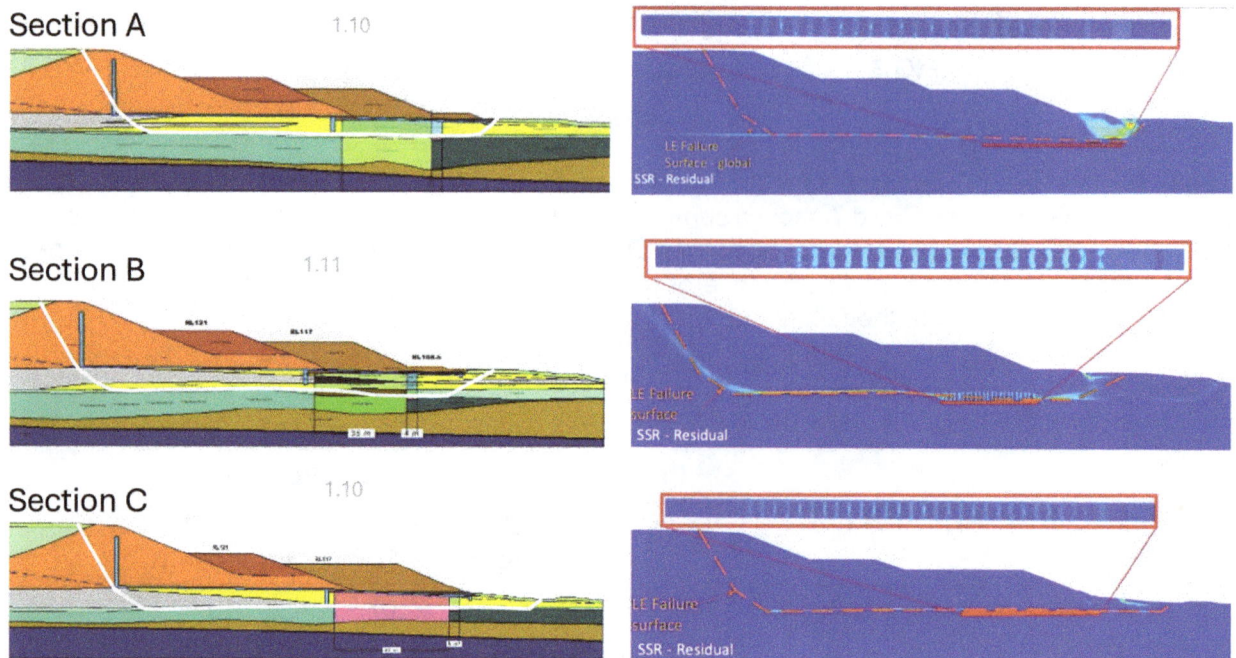

FIG 8 – Comparison of failure surfaces between 2D LE analyses using equivalent strength (left) and 3D FE modelling using the full geometry (right). Residual USR case. Section A, B and C.

CONCLUSIONS

This study demonstrates the validity and practical effectiveness of using an equivalent strength approach in two-dimensional (2D) limit equilibrium (LE) analyses to model the stability of tailings dams reinforced with stone columns (SCs). Results show that the Shear Strength Reduction (SSR) factors calculated from 3D finite element simulations are in acceptable agreement with the Factors of Safety (FoS) from 2D limit equilibrium methods using equivalent strength, with maximum differences of 5 per cent.

The equivalent strength formulation, which incorporates the replacement ratio and material parameters of SCs and surrounding soils, offers a robust and computationally efficient means for representing the composite behaviour of reinforced ground in 2D analyses. Minor differences in failure mechanisms were observed —particularly related to local versus global failures— attributed to stress redistribution effects more accurately captured in 3D simulations; this aspect shall be studied with more detailed analyses and/or be validated with instrumentation.

Overall, the results confirm that the equivalent strength method is a reliable tool for preliminary and detailed design of SC-reinforced tailings dams, enabling engineers to simplify complex geometries without significantly compromising accuracy. This approach facilitates safer and more cost-effective designs while retaining critical insight into the behaviour of improved ground systems under varied conditions.

ACKNOWLEDGEMENTS

The authors thank SRK Consulting and Menard Oceania for their support and contributions to this study. The collaboration of Nicolas Tasso on the modelling works is also highly appreciated.

REFERENCES

Castro, J, 2017. Modelling Stone Columns, *Materials*, 10(7):782. https://doi.org/10.3390/ma10070782

Chakraborty, A and Sawant, V A, 2022. Numerical Simulation of Earthen Embankment Resting on Liquefiable Soil and Remediation Using Stone Columns, *International Journal of Geomechanics*, 22(11):04022205. https://doi.org/10.1061/(ASCE)GM.1943-5622.0002559

Keller, T O, Castro, G O and Rogers, J H, 1987. Steel Creek Dam foundation densification, in Proceedings of the Symposium on Soil Improvement – A Ten-Year Update, vol 12, ASCE Geotechnical Special Publication.

Luehring, R, Snorteland, N, Stevens, M and Mejia, L, 2001. Liquefaction mitigation of a silty dam foundation using vibro-stone columns and drainage wicks: a case history at Salmon Lake Dam, Proceedings of 21st USSD Annual Meeting, USA.

Mitchell, J K and Gallagher, P M, 1998. Engineering guidelines on ground improvement for civil works structures and facilities, US Army Corp of Engineering Division Directorate of Civil Works.

Ng, K S and Tan, S A, 2015. Simplified homogenization method in stone column designs, *Soils and Foundations*, 55(1):154–165. https://doi.org/10.1016/j.sandf.2014.12.012

Nickell, J S, Allen, M G and Ledbetter, R H, 1994. Seismic remediation for liquefiable gravels: Mormon Island auxiliary dam and others in Northern California, Proceedings of the International Workshop on Remedial Treatment of Liquefiable Soils, Japan.

Rola, J, Sottile, M, Rivas, N, Roldan, L and Sfriso, A, 2024. Development of a 3D ground model to design the stabilisation of a dam founded on weak liquefiable ground, Proceedings of the 7th International Conference on Geotechnical and Geophysical Site Characterization (ISC), Spain.

Roldan, L, Sottile, M, Sfriso, A and Rola, J, 2024. Ground improvement between stone columns: a performance study on various soil types using in-situ testing, Proceedings of the XVIII ECSMGE 2024, Portugal.

Sottile, M, Crocker, J and Roldan, L, 2024. Interpretation of CPTu data using machine learning techniques to develop the ground model of a dam, Proceedings of the 7th International Conference on Geotechnical and Geophysical Site Characterization (ISC), Spain.

Tan, S A, Tjahyono, S and Oo, K K, 2008. Simplified Plane-Strain Modeling of Stone-Column Reinforced Ground, *Journal of Geotechnical and Geoenvironmental Engineering*, 134(2):185–194. https://doi.org/10.1061/(ASCE)1090-0241(2008)134:2(185)

Zhang, Z, Han, J and Ye, G, 2014. Numerical investigation on factors for deep-seated slope stability of stone column-supported embankments over soft clay, *Engineering Geology*, 168:104–113. https://doi.org/10.1016/j.enggeo.2013.11.004

Effect of mud farming and drying on shear strength of bauxite tailings

S Srinivasan[1], D Reid[2] and M Kyaw[3]

1. Geotechnical Engineer, Alcoa, Pinjarra WA 6208. Email: shriram.srinivasan@alcoa.com
2. Principal Tailings Engineer, Red Earth Engineering, Perth WA 6000.
   Email: david.reid@redearthengineering.com.au
3. Senior Geotechnical Engineer, Alcoa, Pinjarra WA 6208. Email: min.kyaw@alcoa.com

INTRODUCTION

Mud farming is commonly used to increase the density of red muds and subsequent mechanical behaviour of red muds. Successful drying appears likely to produce an over consolidated residue that may be dilative at low effective stresses and/or have far higher undrained shear strengths with increasing effective stresses compared to material that has not undergone drying and farming. This allows miners to effectively plan and safely increase tailing stack heights lowering future costs on buttressing and other remediation measures. However, while surficial examination of red mud farming benefits has been examined frequently – for example, surficial vane shear tests for operation monitoring – there is less direct experimental evidence as to the increase in undrained shear strength resulting from different magnitudes of drying and mud farming. The current study conducted at Alcoa's Wagerup Residue Storage Area in Western Australia presents an experimental comparison of the mechanical behaviour of residue muds having undergone different magnitudes of farming and drying: (i) samples dried over a winter period; and (ii) a block sample that was allowed to dry over the summer in a fallow cell.

TESTING METHOD AND MATERIALS

The mechanical response of each material under a range of vertical effective stresses was first characterised by means of direct simple shear (DSS) testing on samples trimmed from block samples and 86 mm piston samples. DSS specimens were obtained by means of advancing a sharp cutting ring of 70 mm into the tube or block and carefully trimming material with a scalpel as the ring advanced, as shown in Figure 1. For samples obtained from the winter block or piston samples, the material was clearly at or very close to saturation during sampling, and therefore specific efforts to saturate the samples prior to testing were not necessary. Alternatively, the block sample that had dried over summer was in an unsaturated state after trimming. Therefore, the trimmed specimens, within rings, were placed within a caustic pore fluid 'bath' to inundate them for a period of one week prior to testing. Within the bath a small bedding load of 5 kPa was maintained on the samples with weights.

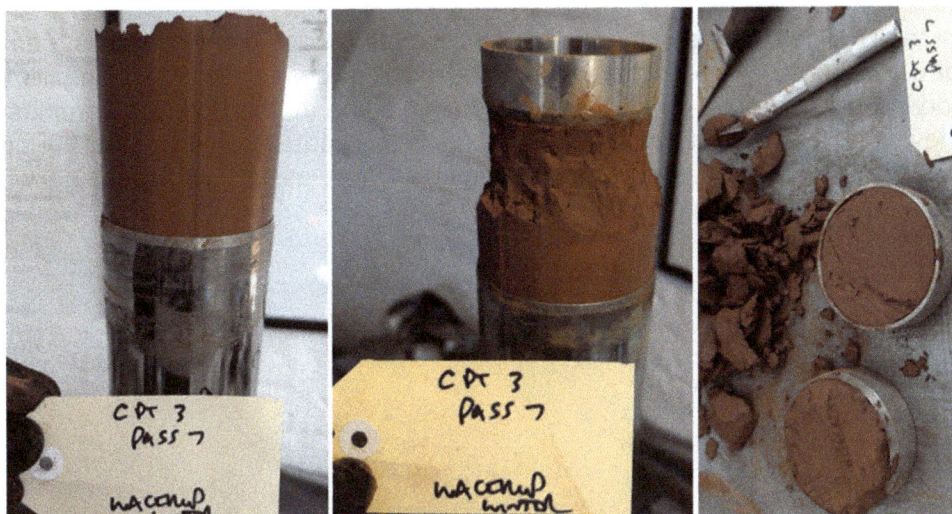

FIG 1 – Extrusion and trimming samples from CPT piston samples.

To enable measurement of the shear strength of the material at lower solids contents, the trimmings from block and tube samples were re-slurried and used in two ways:

1. Poured into a tube and then consolidated to 50 kPa vertical effective stress with weights. After completion of one-dimensional consolidation, the sample was extruded and trimmed for DSS testing.

2. Poured directly into the DSS mould and then consolidated to a target vertical effective stress. This approach was needed when consolidating to stresses below 50 kPa, as use of samples from consolidation tubes would have resulted in over consolidated specimens.

The preparation of a slurry sample with one-dimensional consolidation is outlined in Figure 2.

FIG 2 – Sample preparation from a slurry using one-dimensional consolidation in a tube.

Regardless of the sample preparation method, once within the DSS apparatus samples were consolidated to the target vertical effective stress in stages, then sheared under constant volume conditions at approximately 5 per cent strain per hr. At the end of the test each sample was dried in an oven to obtain the final mass of water and mass of solids and salt.

RESULTS

The results of the DSS tests are summarised as uncorrected solids content against peak undrained shear strength in Figure 3. This format is selected as it is commonly used on-site to track the progress of drying and its effect on strength. Further, plotting all the different samples together in this format shows that they exhibit similar relationships in this context. This suggests relatively similar materials in the context of stress at a given density, which assists in assessing what may be contributing to differences shown subsequently.

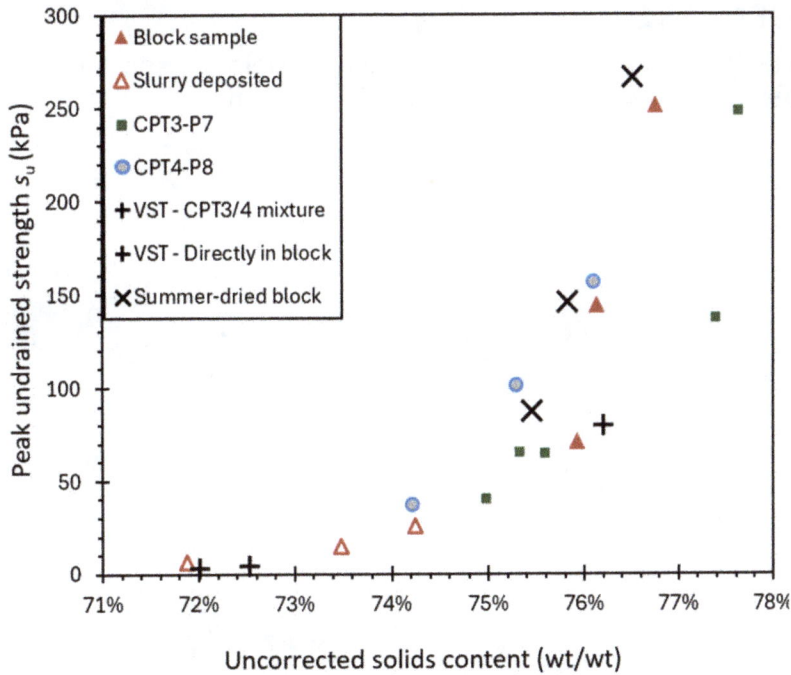

FIG 3 – Uncorrected solids content versus peak undrained shear strength.

The results of the test program are presented as consolidated vertical effective stress versus peak undrained strength ratio in Figure 4. This format allows an examination of the evolution of shear strength with increasing overburden pressure (ie subsequently deposited tailings), and can indicate which samples received significant overconsolidation effects from drying and amphirolling. The results clearly show that the summer-dried block achieved far greater overconsolidation effects as a result of the additional drying that was achieved, which the CPT piston samples show overconsolidation but of a significantly lower magnitude. Once reaching a vertical effective stress of 1000 kPa, all samples have converged to the typical ~0.25 ratio for the material when prepared in the laboratory in a normally consolidated state from a slurry.

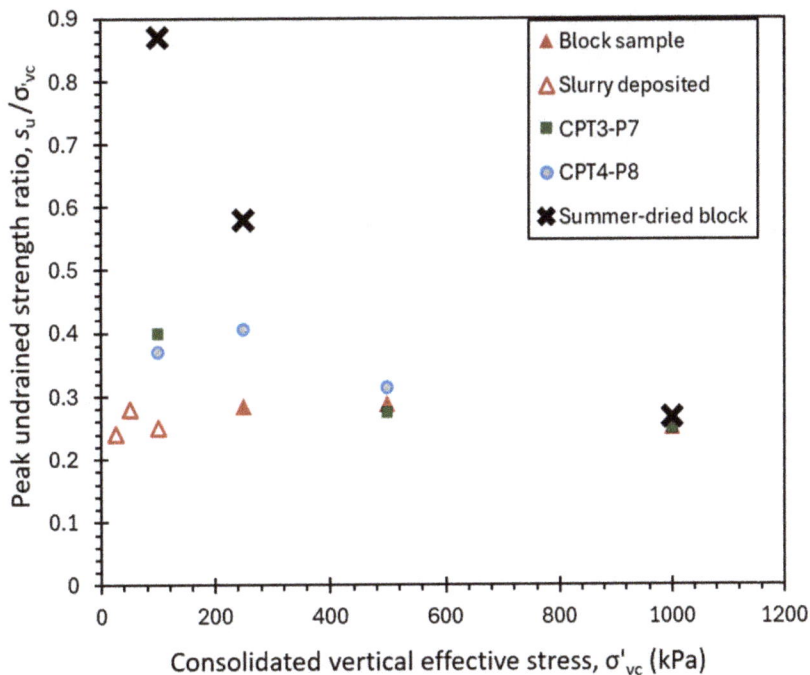

FIG 4 – Undrained strength ratio against consolidated vertical effective stress.

ACKNOWLEDGEMENTS

The authors would like to acknowledge Alcoa for their approval to present this paper and for providing access to the Wagerup Residue Storage Area for this research.

Strength, liquefaction and cone penetration test results in unsaturated silty tailings

Y Wang[1,2], A Russell[3] and R Rekowski[4]

1. Research Associate, University of New South Wales, Sydney NSW 2052.
   Email: yanzhi.wang@unsw.edu.au
2. Geotechnical Engineer, Stantec, Sydney NSW 2065. Email: alex.wang2@stantec.com
3. Professor, University of New South Wales, Sydney NSW 2052. Email: a.russell@unsw.edu.au
4. Senior Principal Tailings Engineer, Stantec, Brisbane Qld 4000.
   Email: robert.rekowski@stantec.com

ABSTRACT

The behaviour of two silty tailings are investigated, focusing on their potential for static liquefaction and how they can be characterised using the cone penetration test (CPT) under unsaturated conditions. The study emphasises loose states and high degrees of saturation. The tailings are first characterised using triaxial tests. The consideration of the closed system condition is novel and critical, as rapid instability and deformation can occur, preventing the escape of air and water from the tailings. A bounding surface plasticity model is adapted for the closed system condition and calibrated using triaxial test results, achieving good correlation. Simulations are conducted to identify factors influencing instability and to model rising water tables under constant total stress in the field. CPTs are performed on the tailings in a calibration chamber, correlating cone resistance to tailings states for both saturated and unsaturated conditions. A cavity expansion analysis, incorporating the bounding surface model, is performed and compared with CPT results. A linear relationship between cavity wall pressure and cone resistance is identified, enabling the creation of interpretation charts for unsaturated tailings.

INTRODUCTION

Tailings Storage Facilities (TSF)s fail catastrophically far too often. In many cases the tailings impounded inside a TSF reduce in strength with or without the addition of significant external disturbances or loads. Initially liquefaction was considered to happen only in fully saturated states, but later findings show that they can also happen, although are less likely, when unsaturated. Tailings can experience a reduction in effective stress and strain soften during closed-system loading, attaining a very low residual strength. The conditions required for it to occur are not well understood although there is evidence of it in poorly graded materials containing predominantly sand sized particles when the degree of saturation (S_r) is as low as 90 per cent.

EXPERIMENTAL CHARACTERISATION OF THE TAILINGS

A series of laboratory tests were performed on a gold tailings and a copper tailings. The gold tailings originated from Australia, and the copper tailings from the USA. Both tailings are sandy silts. Triaxial tests, including constant suction and closed system shearing, are performed. A variety of void ratios were also used.

Closed system triaxial tests were conducted on both tailings. The samples were prepared under the same procedure as the constant suction tests. Water was flushed from the bottom of the sample after it was installed onto the apparatus to increase the degree of saturation. Closed system shearing was imposed by increasing axial stress while keeping cell pressure constant. The sample volume change was tracked by monitoring the pore air pressure continuously through a sensor connected to the top of the sample. Boyle's law was then used to obtain the pore air volume change, which was also the volume change of the entire sample.

STABILITY OF THE TAILINGS FROM CONSTITUTIVE MODELLING

A bounding surface plasticity model was used to simulate the tailings' behaviours. The model was calibrated for the two tailings using constant suction and closed system triaxials.

To assess the susceptibility of static liquefaction it is necessary to determine whether the materials in question will be contractive or dilative on shearing. The approach adopted by the industry is to determine the state parameter (ψ) which is the difference between the *in situ* voids ratio and the voids ratio at the critical state for the same mean effective stress conditions (Been and Jeffries, 1985). Currently the industry generally relies on correlations with the results of piezocone penetration tests (PCPTs) or PCPTs with seismic measurements due to the general belief in the difficulty of obtaining representative samples from the field (Sottile, Kerguelen and Sfriso, 2019). Some of the CPT correlations are empirical, some are based on cavity expansion theory, and their reliability is likely to vary for different tailings materials. Also, the different methods can produce quite different estimates of state parameter for the same material.

Efforts have been undertaken to improve the quality of sampling of soft, low plasticity cohesive soils by using the mini block sampler, a modified Sherbrooke sampler developed by the Norwegian Geotechnical Institute (Lines *et al*, 2023). The sampler carves (essentially by overcoring) a cylindrical block specimen of 160 mm diameter using three annular cutters, that carve a 50 mm wide slot around the specimen. When a sample of around 300 mm in length has been carved, three horizontal blades are activated to cut the base of the specimen. The horizontal cutters provide support to the specimen when it is lifted to the ground surface. This technique is considered effective in appropriate materials down to borehole depths of up to around 15 m. Unfortunately, the technique would not be expected to be successful in materials with little cohesion such as silty fine sands.

Advanced laboratory tests conducted diligently are considered by the industry to be able to define the relationship between critical state void ratio and mean effective stress. An ability to recover a high-quality sample for subsequent laboratory determination of void ratio would provide an independent and direct measure of state parameter and would be useful for site specific calibration of PCPT correlations.

METHODOLOGY

The sampling device – design objectives

A significant challenge facing high fidelity determination of the *in situ* voids ratio of a very loose/loose cohesionless sample is densification of the sample during sample recovery and perhaps to a lesser extent during transportation and handling in the laboratory. The novel and advanced element of this sampler is its ability to positively retain all constituents (air, water and soil) of the sample recovered from the subsurface within a sealed chamber of calibrated volume (ie the volume of the sample at the time it is taken is known). As the laboratory determination of voids ratio only requires the measurement of the mass of the individual constituents within a known volume, any disturbance to the soil structure during transportation and handling does not affect the calculation of voids ratio.

The cutting shoe has been carefully designed to minimise disturbance during insertion, and the calibrated sampling chamber has been designed to be as short as possible with a polished internal surface to minimise friction. In materials containing hard, gravel sized particles, there is risk of damage to the cutting shoe, however this is hardened to minimise this risk. The sample shoe is unlikely to be damaged when the sampler is used in fine tailings. The upper sample collection chamber has a slightly increased internal diameter to further relieve internal friction. A primary design objective of the sampler, therefore, is to minimise disturbance during sampling, particularly densification within the sample tube due to internal friction between the sample and the device.

Many tailings facility operators prefer not to allow personnel onto the impounded tailings for Health and Safety reasons, requiring, where possible, any site characterisation activities to be conducted from remotely operated vehicles. Therefore, the second primary design objective of the sampler was for it to be deployable from the base of a drilled borehole or by pushing from the surface with a conventional or remotely operated CPT system. Maximising sample quality requires that the sampler can be advanced, closed-ended, and opened remotely at the chosen depth to recover the sample.

Finally, sufficient material must be retained in the sealed sampling chamber to allow determination of the Specific Gravity of the sample.

The sampling device – primary components

To be effective the sampling device must positively retain all constituents within a known volume of sample. There is a ball valve at the end of the sample tube which is opened during sampling and then closed after completion of sampling. The sample to be used for the void ratio determination is sealed within the throat of the ball valve. The throat of the ball valve has a known, calibrated volume of the order of 55 cm^3. The ball valve assembly is detachable from the main body of the sampler so that the sample remains sealed until it is eventually opened for testing in the laboratory. Any number of sealed samples can be taken on a site using additional ball valve assemblies that are attached prior to each sampler deployment. Alternatively, an accurate weighing scale can be mobilised to the site to determine the wet weight of the sample immediately after recovery. The sampler includes a chamber above the ball valve for collecting additional material for laboratory testing on reconstituted samples. The sampling device is illustrated schematically in Figure 1.

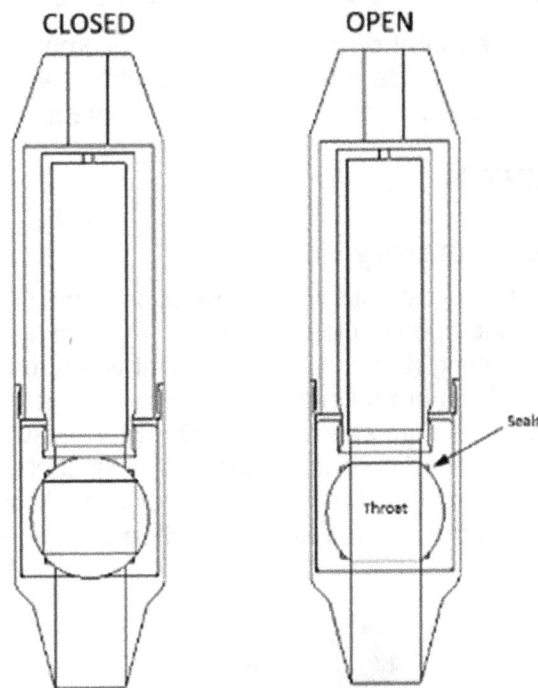

FIG 1 – Schematic illustration – void ratio sampler.

The sampler incorporates a hydraulically operated piston that rotates the valve via a connecting rod. Water is used as the pressurising fluid and pressure is applied from the surface through a hydraulic hose inside the rods used to deploy the sampler and by using the deployment rods themselves as a hydraulic conduit. The sampler ball valve is normally closed, such that during deployment material cannot enter the sampling chambers.

The sampling device – operation procedure

The ability to deploy the sampler with the ball valve closed means that the sampler can either be deployed into the base of a drilled hole using a drill rig, like other high quality sampling devices, or pushed from the surface to a predetermined sampling depth using a CPT thrust machine or similar. The nose of the sample shoe is fitted with a plastic plug to prevent material entering the shoe during pushing from the surface. For testing from the base of the borehole, the operation principle allows the sampler to be penetrated closed beyond the reasonable zone of disturbance at the base of the borehole caused by drilling operations.

Pushing the tube closed ended may cause some compaction immediately below the tube and therefore the ball valve is opened a short distance above the desired elevation for void ratio determination. Once the valve has been opened with hydraulic pressure, the sampler is advanced at a slow and steady rate with material passing through the throat of the ball valve into the chamber above.

At the end of the sampling stroke, the ball valve is closed with hydraulic pressure completely sealing the material within the valve. Thus, all constituents of the sample (air, water and soil) are retained as intended. After recovery of the sampler to the surface the closed ball valve is detached, material in the upper chamber collected and preserved for transportation to the laboratory and a replacement ball valve assembly fitted ready for the next sampling operation. It is expected that layers requiring void ratio sampling would be first identified in a PCPT using available correlations for state parameter. The sampler can be deployed by any typical geotechnical drilling or CPT rig.

On return to the geotechnical laboratory, the ball valve assembly is first cleaned externally, dried and then weighed before opening. After opening the ball valve, the sample in the throat is collected and weighed, taking care not to lose any of the soil or water constituents. Alternatively, an accurate weighing scale can be mobilised to the site to determine the wet weight of the sample immediately after recovery. Bulk density, moisture content and specific gravity can be determined using standard laboratory tests, allowing calculation of the voids ratio and degree of saturation.

After testing, the sample from the throat of the ball valve is recombined with the disturbed sample taken from the upper chamber to provide material for laboratory testing on reconstituted samples such as shear tests to determine the critical state line for the material.

RESULTS AND DISCUSSION

Prototype testing in the laboratory

A test rig was developed to demonstrate satisfactory operation of the valve and determine indicative forces required to rotate it from the open to closed position. The test rig allowed a cylindrical sample to be reconstituted by Moist Tamping (Ladd, 1978) such that when rotated the valve sampled from the centre of the cylinder of soil. Iron ore tailings from a mine in Brazil were used for the laboratory tests. This material was a fine to medium sand (D_{90} 0.533 mm, D_{10} 0.067 mm) with Specific Gravity of 2.98. Tests were performed at four different nominal moisture contents as wet as practicable for the moist tamping method (maximum moisture content 26 per cent). Visual inspection of the sampler after valve actuation showed that the sample had been sheared smoothly in the expected curved profile (Figure 2).

FIG 2 – Sample in the throat of the valve.

The four samples were reconstituted by careful moist tamping. The valve was actuated using a triaxial test loading frame allowing forces and displacements to be measured and recorded.

Wet unit weight and void ratio values were determined for the whole sample prior to valve actuation and for the material sampled within the throat of the valve. For the four tests, the average difference in void ratio between the taken sample and the initial sample was 0.007 with a maximum difference of 0.015, with the void ratio in the valve always smaller than for the whole. Void ratios of specimens were between 0.63 to 0.68. For wet unit weight, the ratio between wet unit weight of the whole sample to that of the valve averaged 0.996, with the greatest difference being a ratio of 0.991.

Field trials

Field trials have been performed at Fugro's test site in Nootdorp, Netherlands using a 20 t thrust CPT track truck, Figure 3. The CPT profile is shown in Figure 4. Samples were recovered at 2.6 m and 3 m depth in a silty fine sand (cone resistance q_c ~400 kPa) and at 5 m depth in a sandy silt (cone resistance q_c ~ 300 kPa), shown in Figure 5. Specific Gravities were in the range 2.50 to 2.66 and *in situ* moisture contents were in the range 38 per cent to 71 per cent. Atterberg limits were not determined. Although these sample depths were relatively shallow, to investigate loose soils, the principle of the sampler allows tests to be performed at any depth that the CPT rig is capable of pushing to and when deployed with a drilling rig, the deployment depth is only limited by the length of hydraulic hose mobilised to site.

FIG 3 – CPT track truck.

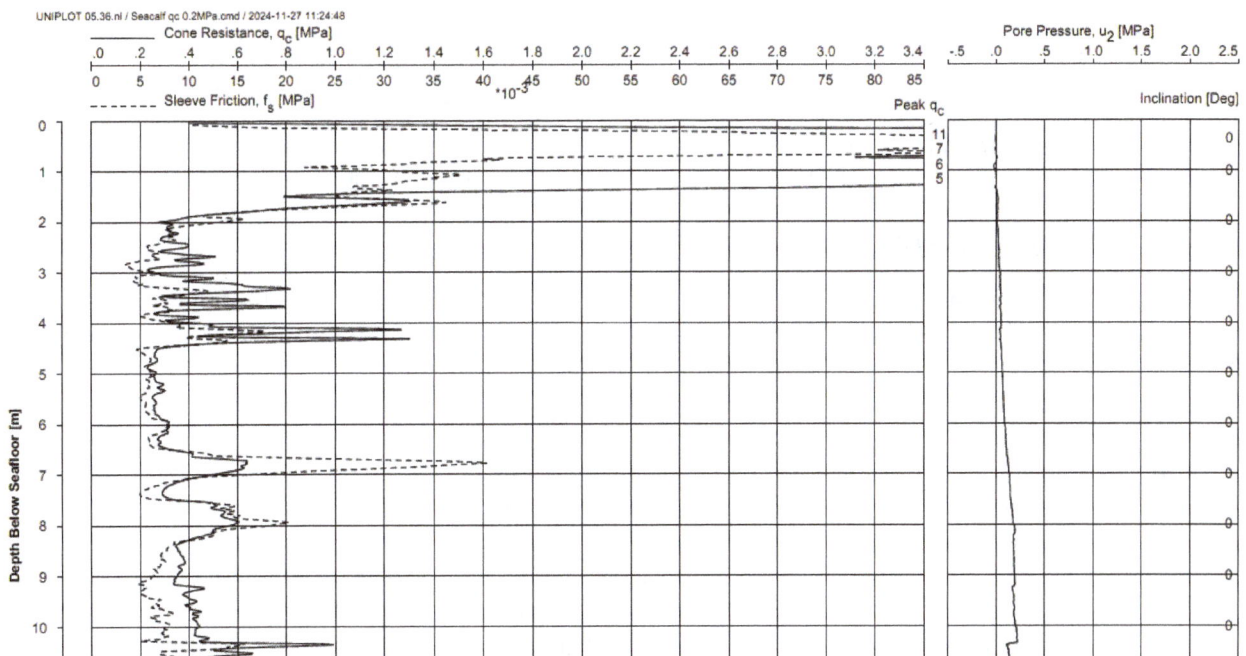

FIG 4 – Test site cone penetration test profile.

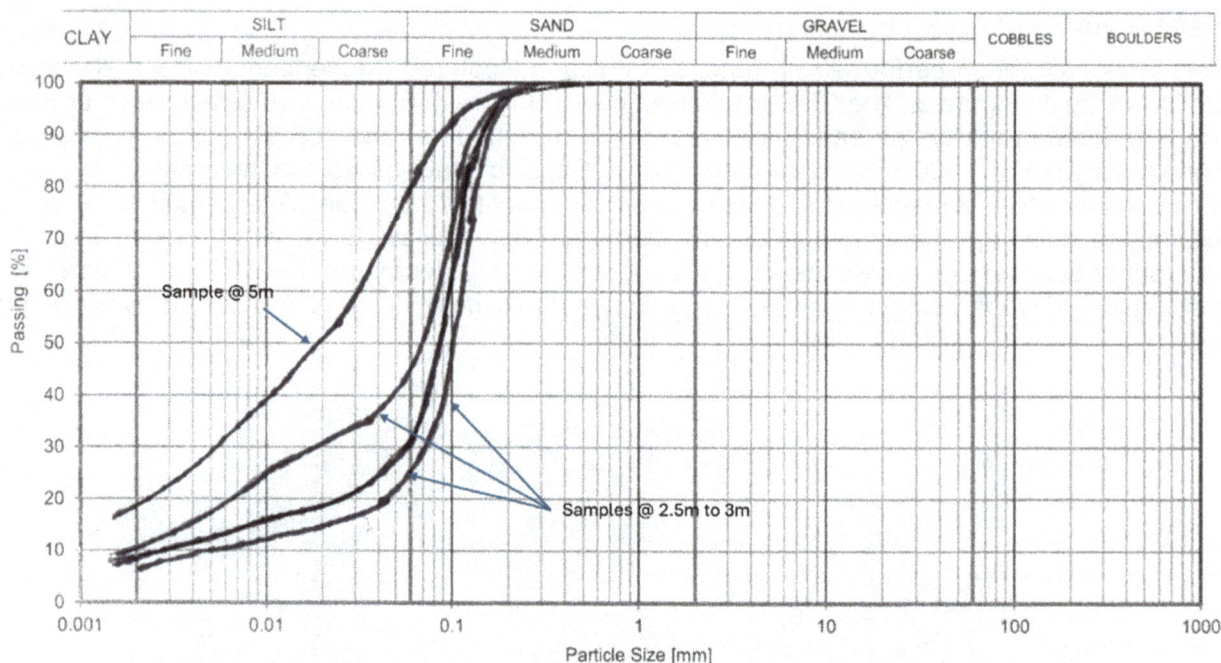

FIG 5 – Sample particle size distribution curves.

Test results on the samples obtained from the throat of the ball are summarised in Table 1.

TABLE 1

Laboratory test results.

Sampling date	Depth (m)	Bulk density (Mg/m³)	Dry density (Mg/m³)	Particle density (Mg/m³)	Water content (%)	Void ratio	Degree of saturation (%)
28 Nov 24	3.0	1.78	1.28	2.64	39.4	1.065	97.5
31 Jan 25	5.0	1.49	0.88	2.49	70.6	1.843	95.4
17 Feb 25	2.5	1.73	1.24	2.66	40.1	1.149	92.8
17 Feb 25	2.5	1.59	1.03	2.62	54.4	1.543	92.4

The soils in the Netherlands in the location of the test site are known to be organic and the void ratios and specific gravities quoted above fall within the range of established experience.

FUTURE DEVELOPMENTS

Further field trials on a mine tailings facility are planned in mid-2025. A large displacement finite element numerical analysis of sampler penetration is being undertaken to better understand the sampling process and refine the design of the cutting shoe.

CONCLUSION

A sampling device that can be deployed at the base of a borehole or pushed from the surface using CPT equipment has been designed and developed and demonstrated to be effective in recovering a high-quality sample for the purposes of *in situ* void ratio and moisture content determination in the laboratory. Deployment using CPT equipment will facilitate use of the sampling device from remotely operated vehicles that can be safely deployed on the impounded tailings deposits. Sufficient material can be recovered to allow testing on reconstituted samples, for example to investigate the critical state line. The device overcomes the current depth limitation of the mini block sampler, will be successful in sampling low cohesion/cohesionless materials and is capable of being deployed using remotely operated equipment, a key HSE consideration for working on tailings impoundments.

Such void ratio determinations will allow increased confidence in the assessment of static liquefaction potential in tailings materials and other saturated loose cohesionless soils via an independent assessment of State Parameter *in situ*. Reliable determinations of *in situ* void ratio will allow improved site-specific correlations to be made for 'State Parameter' to be developed for PCPT data, building the industry knowledge base for such relationships.

ACKNOWLEDGEMENTS

The authors kindly acknowledge Fugro's permission to publish this paper.

REFERENCES

Been, K and Jefferies, M G, 1985. A state parameter for sand, *Geotechnique*, 35:99–112.

International Council on Mining and Metals (ICMM), 2020. Global Industry Standard on Tailings Management (GISTM), ICMM: London.

Ladd, R S, 1978. Preparing Test Specimens by Undercompaction, *Geotechnical Testing Journal*, 1(1):16–33. http://dx.doi.org/10.1520/GTJ10364J

Lines, S H, Llano-Serna, M, Rekowski, R and Ludlow, W, 2023. Identification and analysis of a complex layer within the footprint of a tailings storage facility foundation utilising mini-block sampling, in *Proceedings of the Mine Waste and Tailings Conference*, pp 60–79 (The Australasian Institute of Mining and Metallurgy: Melbourne).

Morgenstern, N R, Vick, S G and Van Zyl, D, 2015. Report on Mount Polley Tailings Storage Facility Breach, report by the Independent Expert Engineering Investigation and Review Panel, Government of British Columbia. Available from: <www.mountpolleyreviewpanel.ca/final-report>

Robertson, P K, de Melo, L, Williams, D and Wilson, G W, 2019. Report of the Expert Panel on the Technical Causes of the Failure of Feijao Dam I, commissioned by Vale S.A. Available from: <www.b1technicalinvestigation.com>

Sottile, M, Kerguelen, A and Sfriso, A, 2019. A Comparison of Procedures for Determining the State Parameter of Silt-like Tailings, in *Proceedings of the 17th African Regional Conference on Soil Mechanics and Geotechnical Engineering*, pp 1–5 (ISSMGE).

Calibration of N_{kt} value for iron ore tailings

J Zhang[1] and W Dressel[2]

1. Senior Tailings/Geotechnical Engineer, Red Earth Engineering, Perth WA 6000.
   Email: sam.zhang@redearthengineering.com.au
2. Principal, Red Earth Engineering, Perth WA 6000.
   Email: waldo.dressel@redearthengineering.com.au

ABSTRACT

It is common practice to undertake Cone Penetration Testing (CPT) supplemented by electric Vane Shear Testing (eVST) to assess the *in situ* undrained shear strength of tailings using the equation $s_u = (q_t - \sigma_{v0})/N_{kt}$. However, a common issue of this practice is the calibrated N_{kt} value could be exceptionally low, leading to unconservative estimation of undrained shear strength.

Some research has suggested that undertaking eVST on some tailings, eg iron ore tailings, under the standard/suggested Rate of Rotation of 6° per minute (AS 1289.6.2.1-2001; Standards Australia, 2013) may be too slow to achieve a fully undrained condition. A (partially) drained condition could result in a higher measured shear strength of the tailings, and thus an exceptionally low calibrated N_{kt} value.

This paper presents a case study of calibrating N_{kt} value for iron ore tailings using the results of CPT and eVST campaigns designed to overcome the above-mentioned shortcomings.

Key considerations in the design of the CPT and eVST campaigns include:

- The eVST was undertaken at varied rotation rates higher than 6° per minute to investigate the effect of drainage conditions on the shear strength of iron ore tailings.

- Shelby tube sampling was undertaken at the locations of the eVST with subsequent laboratory testing carried out, aiming to confirm the degree of saturation and particle segregation for the tested locations.

Key findings of this case study include:

- A correlation can be established between calculated N_{kt} values and Blight's Time Factor T.

- This correlation appears to be insensitive to factors such as dilatancy and particle segregation.

- A T of 0.05 represents the transition between drained and undrained conditions, where the tested shear strength of the iron ore tailings is the lowest. Therefore, the calculated N_{kt} value corresponding to T of 0.05 may be adopted as design N_{kt} value.

INTRODUCTION

To estimate the peak undrained shear strength (s_u) of fine-grained soils from Cone Penetration Testing (CPT) data, the following empirical correlation has been widely used (Lunne, Robertson and Powell, 2002):

$$s_u = \frac{(q_t - \sigma_{v0})}{N_{kt}} \tag{1}$$

Where:

q_t is cone resistance corrected for pore pressure effects

σ_{v0} is total *in situ* vertical stress

N_{kt} is an empirical cone factor

A successful application of Equation 1 is heavily dependent on a reasonable N_{kt} value. Estimation of N_{kt} value can be achieved through proposed theoretical or empirical correlations with parameters such as rigidity index (I_r) (Teh, 1987), plasticity index (PI) (Aas *et al*, 1986), and pore pressure ratio (B_q) (Karlsrud, Lunne and Brattlien, 1996).

To accurately account for site-specific factors such as stress history and particle segregation, N_{kt} value may also be directly calibrated based on the results of *in situ* electric Vane Shear Tests (eVST).

However, recent research suggests that fully undrained conditions may not always be achievable in eVST carried out on certain tailings under the suggested rate of rotation of 6° per minute by AS 1289.6.2.1-2001 (Standards Australia, 2013). (Partially) drained conditions could lead to higher measured shear strengths for the tailings (Reid, 2016; Reid *et al*, 2023), potentially resulting in exceptionally low calculated N_{kt} values.

This paper examines a case study on calibrating N_{kt} value for iron ore tailings accounting for drainage conditions, using the results of CPT and eVST campaigns designed to address the aforementioned limitations.

TESTING PROGRAM

This study investigates two iron ore tailings storage facilities (TSFs), both of which serve the same iron ore mine in Western Australia (WA). TSF #1 has been decommissioned, while TSF #2 is currently active. For confidentiality reasons, further details about the mine site and its location cannot be disclosed.

Field testing program

Two separate CPT campaigns supplemented by eVST were carried out for the TSFs by different CPT contractors, with key details summarised in Table 1.

TABLE 1

Summary of CPT and eVST.

TSF	TSF #1	TSF #2
CPT cone area (cm²)	15	10
CPT penetration rate (mm/s)	20 or 40	5, 10, 20 or 30
eVST vane size	Double tapered (45°, 45°) 80 (H) × 40 (D) mm	Rectangular 100 (H) × 50 (D) mm
eVST rate of rotation (°/min)	6	6, 12 or 24
eVST rod friction correction	Uphole load cell + slip coupler or downhole load cell	Uphole load cell + slip coupler

All CPT were carried out with pore water pressure measurement (CPTu). Some downhole seismic testing (SCPTu) was carried out for TSF #1.

To investigate the effect of drainage conditions on the strength of the iron ore tailings, the CPT and eVST were carried out at varied rates.

Each eVST was paired with and carried out directly adjacent to a corresponding CPT.

Laboratory testing program

Shelby tube samples were collected from the same locations and depths as the eVST. Laboratory testing on these samples included classification tests, oedometer tests and triaxial tests. Additionally, bulk samples were also collected from the tailings surface for classification tests.

The degree of disturbance to the Shelby tube samples was assessed using the oedometer test results, following the method proposed by Berre, Lunne and L'Heureux (2022). The results indicate that the majority of the samples are of very good to excellent, or good to fair quality.

PROPERTIES AND *IN SITU* CONDITIONS OF TAILINGS

While drainage conditions are the primary focus of this study, the properties and *in situ* conditions of the tailings, such as classification, dilatancy, saturation, and microstructure, may also influence the results of eVST. These factors will be briefly discussed herein.

Classification

As shown in Figure 1, the tailings samples are generally classified as clayey silt to sandy silt with low to high plasticity, and significant variations in the particle size distribution (PSD) and Atterberg Limits (AL) are evident.

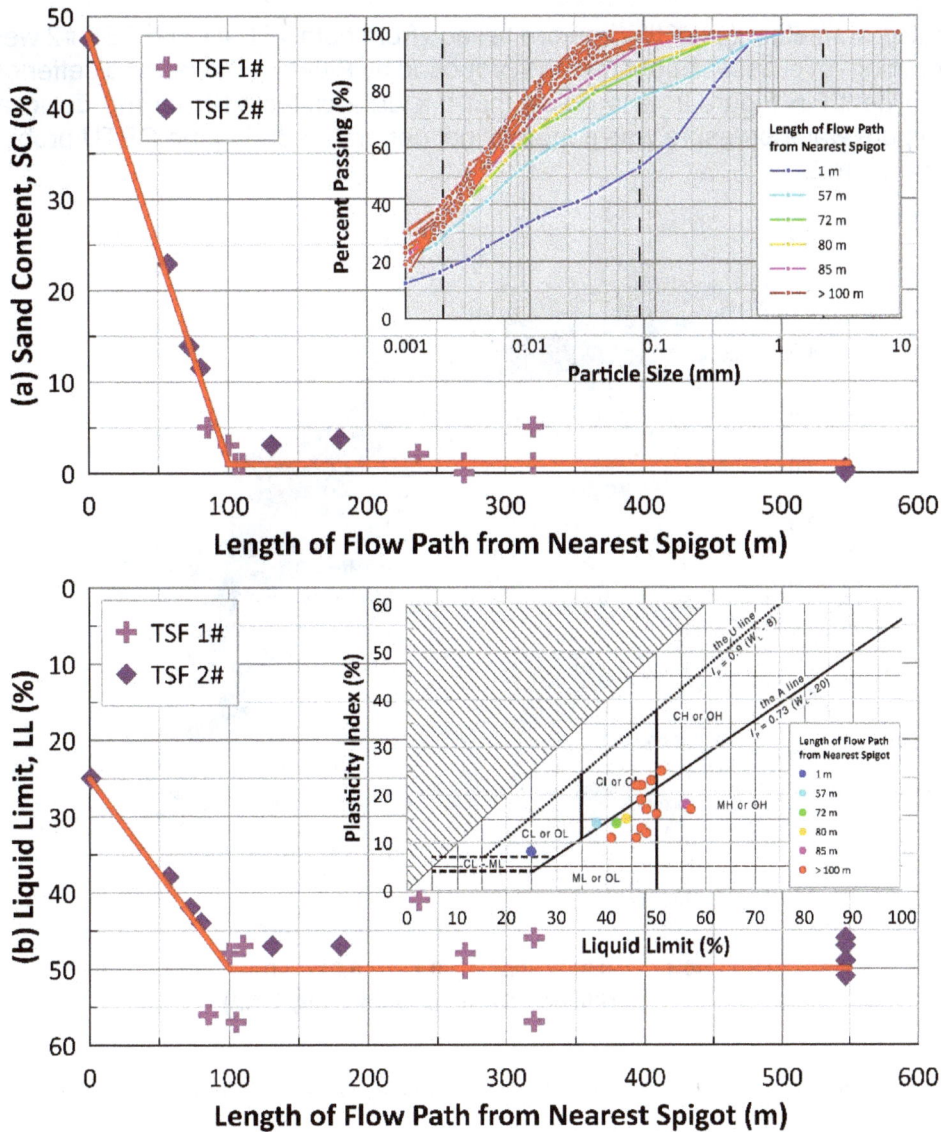

FIG 1 – (a) Particle size distribution, and (b) Atterberg Limits of iron ore tailings.

The sand content (SC) and liquid limit (LL) of the tailings are plotted against the length of the flow path from the nearest spigot. Significant particle segregation is evident within the first 100 m, as the SC decreases sharply from approximately 47 per cent to 0 per cent, while the LL rises from around 25 per cent to 50 per cent. Beyond 100 m, particle segregation ceases, and the PSD and AL of the tailings become relatively stable.

There is no obvious deviation in classification between the tailings from TSF #1 and TSF #2.

Dilatancy

Given the site's arid and tropical climate, significant desiccation occurs on the tailings surface when tailings deposition is paused. Consequently, a certain depth of surficial tailings is expected to be over-consolidated due to desiccation (Dong, Lu and Fox, 2020), despite the absence of tailings harvesting at TSF #1 and TSF #2.

Over-consolidated tailings may exhibit dilative behaviour, where their undrained strength exceeds drained strength. In contrast, normally consolidated tailings tend to exhibit contractive behaviour, with undrained strength lower than drained strength (Ayala, Fourie and Reid, 2023). Therefore, to investigate the influence of drainage conditions on the strength of the iron ore tailings, it is important to identify the transition between the dilative and contractive tailings.

The *in situ* state parameters (ψ) of the iron ore tailings from both TSF #1 and TSF #2 were estimated using the screening-level assessment method proposed by Plewes, Davies and Jefferies (1992) and plotted against depth in Figure 2. It is noted that the data collected from the desiccated tailings, where excess pore water pressure was generally not generated during the CPTU probing, has been excluded from Figure 2.

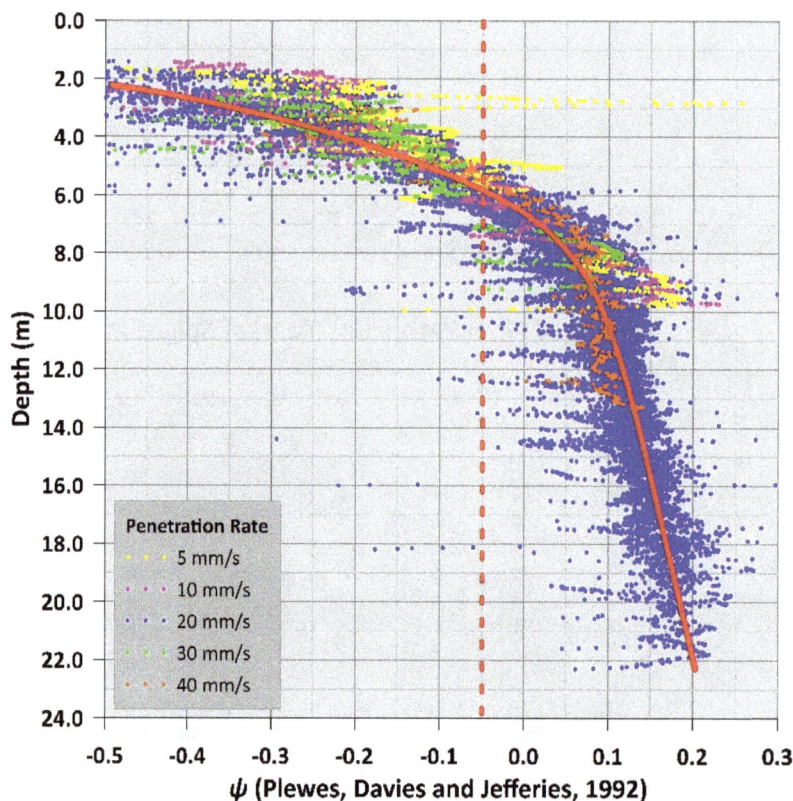

FIG 2 – Estimated *in situ* state parameters of tailings (Plewes, Davies and Jefferies, 1992).

Regardless of penetration rates, the ψ of the iron ore tailings increases with depth, reaching -0.05 at approximately 6 m, which marks the transition between dilative and contractive behaviour (Shuttle and Cunning, 2008). This suggests that the iron ore tailings shallower than 6 m exhibit dilative behaviour, whereas those deeper than 6 m are contractive.

Saturation

All eVST were carried out below the desiccated tailings. The degrees of saturation (s_r) at the eVST locations were estimated using bulk density (ρ_b), specific gravity (G_s), and moisture content (MC) tested from the Shelby tube samples. The results indicate that the s_r values at the eVST locations were generally above 90 per cent, implying that the eVST were primarily governed by saturated soil mechanics, with the effect of matric suction being generally negligible.

Microstructure

The modified normalised small-strain rigidity index (K_G^*) of the tailings from TSF #1, calculated based on the SCPTu results using the method proposed by Robertson (2016), generally ranges between 100 and 330, indicating that the tailings are generally young and uncemented. Although no SCPTu was carried out for TSF #2, given that its tailings are younger than those in TSF #1, it is reasonable to consider them similarly young and uncemented.

DRAINAGE CONDITIONS

Cone penetration testing

To evaluate the drainage conditions of the CPT, the normalised penetration velocity (V), as proposed by Finnie and Randolph (1994), are calculated using Equation 2:

$$V = \frac{v \times d}{c_v} \tag{2}$$

Where:

v is CPT penetration rate

d is cone diameter

c_v is vertical coefficient of consolidation, estimated from oedometer test results

Regardless of penetration rate, the V values of all CPT exceed 30, suggesting that fully undrained conditions were generally achieved across all CPT (Randolph and Hope, 2004).

This finding is further supported by the minimal variations in cone resistance observed in CPT carried out at different penetration rates within the same location.

Electric vane shear testing

The *in situ* shear strength of the tailings (s_{fv}) were calculated based on the eVST results, following ASTM D2573/D2573M-18 (ASTM International, 2018), using Equation 3:

$$s_{fv} = \frac{12 \cdot T_{max}}{\pi D^2 \cdot \left(\frac{D}{\cos(i_T)} + \frac{D}{\cos(i_B)} + 6H \right)} \tag{3}$$

Where:

T_{max} is the maximum value of measured torque corrected for apparatus and rod friction

D and H are the width and height of vane blade, respectively

i_T and i_B are the angles of the taper at the top and bottom of vane

The drainage conditions of the eVST are evaluated using the time factor (T) proposed by Blight (1968). This factor is calculated using Equation 4:

$$T = \frac{c_v t_f}{D^2} \tag{4}$$

Where:

t_f is time to failure

c_v is estimated from the result of the oedometer test conducted on the corresponding Shelby tube sample, collected near the eVST location at similar coordinates and depth

The values of s_{fv} are plotted against T in Figure 3. The data points are separately grouped by facility, rate of rotation, depth, and length of flow path, to examine the influence of these factors on s_{fv}.

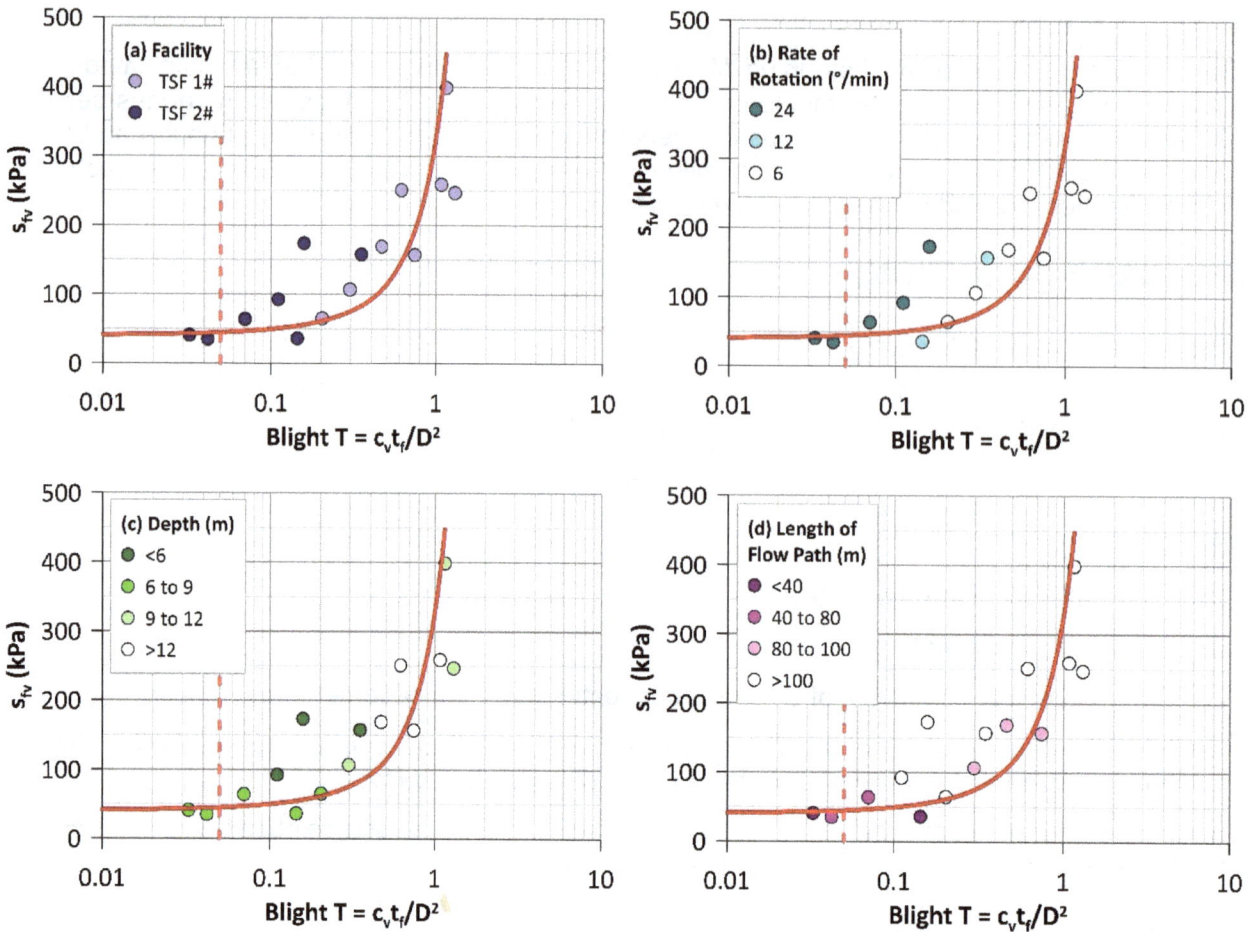

FIG 3 – s_{fv} from eVST versus Blight T, grouped by: (a) facility; (b) rate of rotation; (c) depth; and (d) length of flow path.

The following observations can be made:

- The value of s_{fv} generally decreases with T, stabilising at approximately 40 kPa as T drops below 0.05. This trend aligns with the findings of Chandler (1988), who suggested that fully undrained condition can be considered achieved where $T < 0.05$.

- The data points from TSF #1 and TSF #2 predominantly follow the same trend line.

- eVST carried out at a rotation rate of 6° per minute generally resulted in s_{fv} values exceeding 40 kPa and T values above 0.05, indicating partially drained conditions. In contrast, eVST performed at a rotation rate of 24° per minute may achieve $T < 0.05$, thereby reaching fully undrained conditions.

- No obvious viscosity effects can be observed under the maximum rotation rate of 24° per minute used in this study, as the s_{fv} values from depths greater than 6 m do not exhibit an obvious increasing trend as T decreases.

- The data points from eVST conducted at depths greater 6 m exhibit less deviation from the trend line, whereas those from eVST conducted at shallower depths tend to shift slightly above it. This suggests that the dilative tailings at depths less than 6 m may have higher s_{fv} compared to deeper contractive tailings under identical drainage conditions, likely contributing to the scatters observed in the trend line.

- Despite the significant particle segregation observed within the first 100 m of the flow path, no clear influence of the change in tailings classification on the trend line can be observed.

Calibration of N_{kt}

Derived from Equation 1, N_{kt} can be calculated using Equation 5:

$$N_{kt} = \frac{(q_t - \sigma_v)}{s_{fv}} \tag{5}$$

To obtain a N_{kt} value representative of fully undrained conditions using Equation 5, both the q_t from CPT and s_{fv} from eVST shall correspond to fully undrained conditions. While fully undrained conditions have been achieved across all CPT, this is not the case for eVST.

To address this issue, the calculated N_{kt} using Equation 5 are plotted against T in Figure 4. The data points are separately grouped by facility, rate of rotation, depth, and length of flow path, to examine the influence of these factors on N_{kt} values.

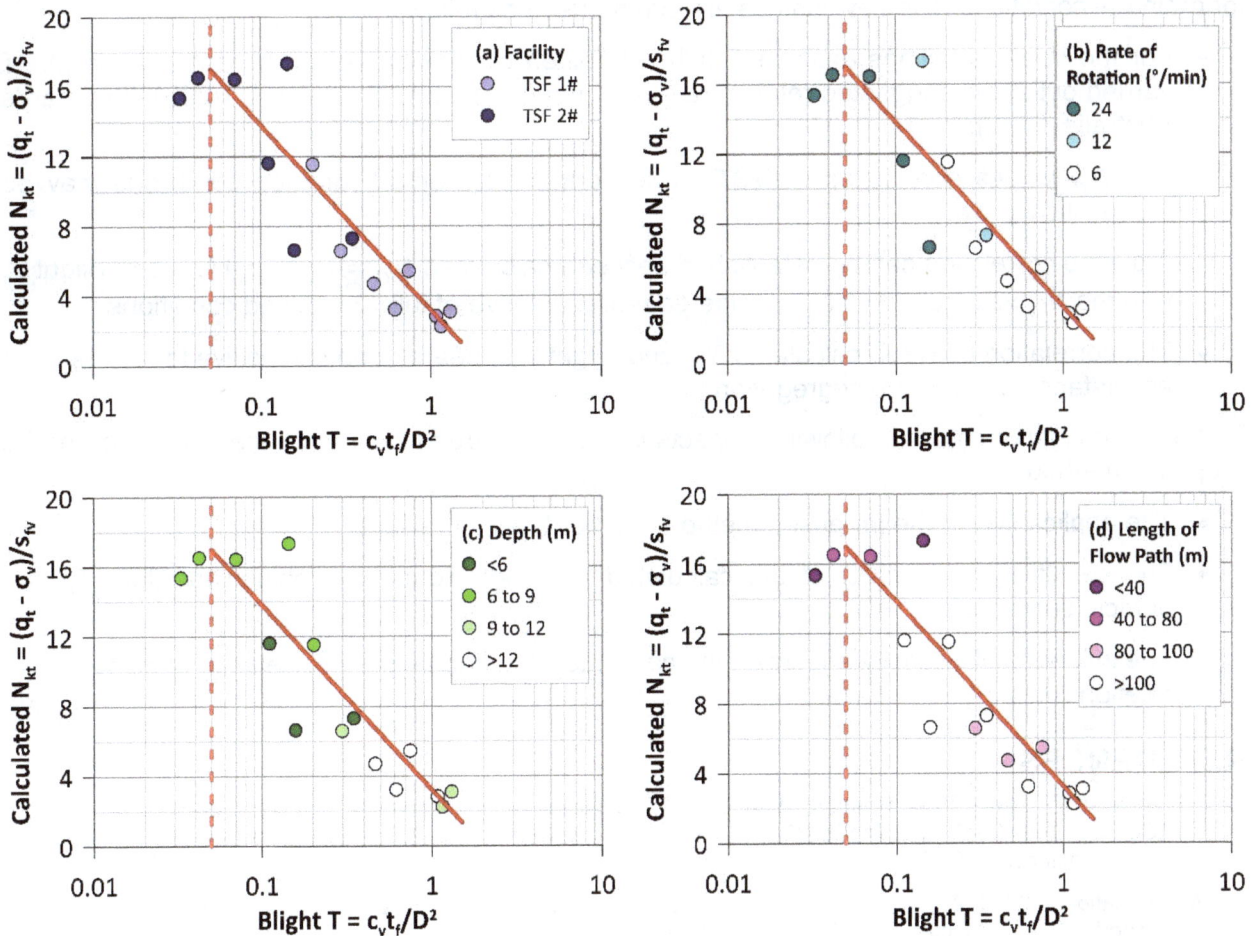

FIG 4 – Calculated N_{kt} versus Blight T, grouped by: (a) facility; (b) rate of rotation; (c) depth; and (d) length of flow path.

The following observations can be made:

- The calculated N_{kt} value generally increases as T decreases, following a linear trend in a semi-logarithmic space. The Pearson Correlation Coefficient between the calculated N_{kt} value and $\log T$ is calculated as -0.904, suggesting a very strong negative correlation.

- The calculated N_{kt} value approaches a value of 17 as T reaches 0.05. This suggests that an N_{kt} value of 17 can be considered representative of fully undrained conditions.

- The data points from TSF #1 and TSF #2 largely align with the same trend line.

- The N_{kt} values calculated from eVST carried out at a rotation rate of 6° per minute generally range from 2 to 12, with corresponding T values between 0.2 and 1.3. This suggests that unrealistically low N_{kt} values may result from eVST carried out under partially drained conditions.

- The data points from the dilative tailings at depths less than 6 m integrate well with those from deeper contractive tailings, exhibiting less scatters compared to Figure 3c. This is likely due to the q_t from the dilative tailings being influenced in a similar manner as the s_{fv}, resulting in these two analogous effects cancelling out when combined in Equation 5.

- The correlation between calculated N_{kt} and T appears unaffected by the significant particle segregation observed within the first 100 m of the flow path.

CONCLUSIONS

This paper presents a case study on calibrating the N_{kt} value for iron ore tailings while accounting for drainage conditions. The key findings are summarised as follows:

- For the iron ore tailings examined, fully undrained conditions cannot be achieved in eVST carried out at the suggested rotation rate of 6° per minute by AS 1289.6.2.1-2001 (Standards Australia, 2013).

- The N_{kt} values derived from eVST carried out under partially drained conditions may be unrealistically low.

- A simple correlation can be established between calculated $N_{kt} = (q_t - \sigma_v)/s_{fv}$ and Blight T, with the N_{kt} value at $T = 0.05$ considered representative of fully undrained conditions.

- The correlation between calculated N_{kt} and Blight T appears to be insensitive to factors such as dilatancy and particle segregation.

Further investigation into the following aspects is recommended to enhance the application of the proposed method:

- The applicability of the proposed method to other types of tailings.

- The sensitivity/insensitivity of the calculated N_{kt} and Blight T correlation to dilatancy and particle segregation.

- The influence of viscosity effects on the calculated N_{kt} and Blight T correlation at higher eVST rotation rates.

REFERENCES

Aas, G, Lacasse, S, Lunne, T and Hoeg, K, 1986. Use of in situ tests for foundation design on clay, in *Proceedings of the ASCE Specialty Conference In Situ '86, Use of In Situ tests in Geotechnical Engineering*, pp 1–30 (American Society of Civil Engineers (ASCE)).

ASTM International, 2018. ASTM D2573/D2573M-18, Standard test method for field vane shear test in saturated fine-grained soils (ASTM International: West Conshohocken).

Ayala, J, Fourie, A and Reid, D, 2023. A Unified Approach for the Analysis of CPT Partial Drainage Effects within a Critical State Soil Mechanics Framework in Mine Tailings, *Journal of Geotechnical and Geoenvironmental Engineering*, 149(6):04023036. https://doi.org/10.1061/JGGEFKG.TENG-10915

Berre, T, Lunne, T and L'Heureux, J-S, 2022. Quantification of sample disturbance for soft, lightly overconsolidated, sensitive clay samples, *Canadian Geotechnical Journal*, 59(2):300–303. https://doi.org/10.1139/cgj-2020-0551

Blight, G E, 1968. A note on field vane testing of silty soils, *Canadian Geotechnical Journal*, 5(3):142–149. https://doi.org/10.1139/t68-014

Chandler, R J, 1988. The in-situ measurement of the undrained shear strength of clays using the field vane, in *Vane Shear Strength Testing in Soils: Field and Laboratory Studies* (ed: A F Richards), pp 19428–2959. https://doi.org/10.1520/STP10319S

Dong, Y, Lu, N and Fox, P J, 2020. Drying-Induced Consolidation in Soil, *Journal of Geotechnical and Geoenvironmental Engineering*, 146(9):04020092. https://doi.org/10.1061/(ASCE)GT.1943-5606.0002327

Finnie, I and Randolph, M, 1994. Punch-through and liquefaction induced failure of shallow foundations on calcareous sediments, in *Seventh International Conference on the Behaviour of Offshore Structures*, pp 217–230 (Pergamon: Boston).

Karlsrud, K, Lunne, T and Brattlien, K, 1996. Improved CPTU interpretations based on block samples, in *Publikasjon-Norges Geotekniske Institutt*, pp 195–201.

Lunne, T, Robertson, P K and Powell, J J M, 2002. *Cone penetration testing in geotechnical practice* (Taylor and Francis Group). https://doi.org/10.1201/9781482295047

Plewes, H, Davies, M and Jefferies, M, 1992. CPT based screening procedure for evaluating liquefaction susceptibility, in *Proceedings of the 45th Canadian Geotechnical Conference*, pp 1–9.

Randolph, M and Hope, S, 2004. Effect of cone velocity on cone resistance and excess pore pressures, in *Proceedings of the IS Osaka – Engineering Practice and Performance of Soft Deposits*, pp 147–152 (Yodogawa Kogisha Co, Ltd: Osaka).

Reid, D, 2016. Effect of rotation rate on shear vane results in a silty tailings, in *Proceedings of Geotechnical and Geophysical Site Characterization*, pp 369–374.

Reid, D, Rodriguez, C, Fourie, A and Tiwari, B, 2023. Partial drainage effects during vane shear tests, with an emphasis on the measurement of remoulded strengths, in Proceedings of Tailings and Mine Waste Conference 2023 (The University of British Columbia). https://doi.org/10.14288/1.0438118

Robertson, P K, 2016. Cone penetration test (CPT)-based soil behaviour type (SBT) classification system — an update, *Canadian Geotechnical Journal*, 53(12):1910–1927. https://doi.org/10.1139/cgj-2016-0044

Shuttle, D A and Cunning, J, 2008. Reply to the discussion by Robertson on 'Liquefaction potential of silts from CPTu', *Canadian Geotechnical Journal*, 45(1):142–145. https://doi.org/10.1139/T07-119

Standards Australia, 2013. AS 1289.6.2.1-2001 (Rec:2013) – Soil strength and consolidation tests – determination of the shear strength of a soil – field test using a vane.

Teh, C L, 1987. An analytical study of the cone penetration test, PhD thesis, University of Oxford.

The Life of Mine | Mine Waste and Tailings Conference 2025 | Brisbane, Australia | 29–30 July 2025

Tailings dam breach and runout analysis

Comparative analysis of dam breach modelling using the Eulerian approach and geometrical debris flows

P Abbasimaedeh[1], C Hogg[2] and P Atmajaya[3]

1. Principal Geotechnical Engineer, CMW Geoscience, Perth WA 6155.
   Email: pouyana@cmwgeo.com
2. Senior Principal Tailings Engineer, CMW Geoscience, Perth WA 6155.
   Email: chrish@cmwgeo.com
3. Senior Geotechnical Engineer, CMW Geoscience, Perth WA 6155.
   Email: prasudia@cmwgeo.com

ABSTRACT

Tailings dam breaches present significant environmental and safety challenges due to the complex behaviour of tailings materials and the high consequences of failure. The current study presents a comparative analysis of two widely used dam breach modelling techniques: the Eulerian hydrodynamic approach using a numerical method and the geometrical debris flow approach. A case study from a proposed gold mining Tailings Storage Facility in Western Australia was used to evaluate breach formation, flow propagation, and deposition patterns under a sunny-day failure scenario. Breach geometries were estimated using empirical methods, with a released tailings volume of approximately 3.9 Mm^3 and a peak flow rate of up to 442 m^3/s. The numerical model, configured with the Shallow Water Equations and Exner-Layer Model, captured flow dynamics, inundation extents, and velocity profiles, predicting a runout distance of 1.5 km. Meanwhile, the geometrical debris flow model provided detailed insights into tailings runout and final deposition, with a maximum extent of 1.3 km and a spatial distribution of released material. While the Eulerian approach offered superior outputs for emergency response planning through dynamic discharge behaviour and terrain interaction, the geometrical model was more suitable for long-term deposition assessment. The findings demonstrate that both models have distinct advantages, and their integration enhances the robustness of tailings breach assessments.

INTRODUCTION

Tailings dam breaches pose severe environmental, social, and economic risks due to the sudden and often catastrophic release of vast volumes of mine waste. Historically, numerous tailings dam failures have led to catastrophic losses, reinforcing the need for robust risk assessment and breach modelling approaches. Statistical data indicate that tailings dams fail at a rate over 100 times higher than water retaining dams, with an estimated 1.2 per cent annual failure rate, compared to 0.01 per cent for conventional dams (Azam and Li, 2010).

In 1962, the Huogudu failure in China released 3.3 Mm^3 of tailings, resulting in 171 fatalities, while the 1965 El Cobre failure in Chile caused 200 deaths following the release of 0.35 Mm^3 of tailings. In the United States, the 1972 Buffalo Creek disaster in West Virginia caused 125 fatalities when a coal tailings dam collapsed, unleashing 0.5 Mm^3 of waste. The 1985 Strava disaster in Italy, one of the deadliest tailings dam failures in Europe, saw 0.2 Mm^3 of tailings devastate downstream communities, claiming 268 lives. More recently, the 2008 Toashi tailings dam failure in China resulted in 277 fatalities, following the release of 0.19 Mm^3 of tailings, marking one of the worst tailings-related disasters of the 21st century. The 2019 Brumadinho failure, which engulfed entire settlements with 12 Mm^3 of tailings, remains among the deadliest in history. Major disasters such as the 2015 Fundão dam failure in Brazil, which released 43 million cubic metres (Mm^3) of mine waste, and the 2019 Brumadinho collapse, which claimed 267 lives, underscore the devastating consequences of tailings dam failures (Morgenstern, Vick and Van Zyl, 2016).

Unlike water-retaining dam failures, tailings dam breaches involve the mobilisation of fine-grained mining residues, resulting in longer runout distances with hyper-concentrated, mud, or debris flows, while the breach shape and dimensions can vary significantly due to the complex erosion and sediment transport processes unique to tailings materials. While traditional dam breach models typically assume Newtonian fluid behaviour, tailings dam failures involve additional complexities such as progressive failure mechanisms, embankment erosion, highly variable breach geometries and especially non-Newtonian material behaviour (Jeyapalan, Duncan and Seed, 1983).

The Global Industry Standard on Tailings Management (GISTM), introduced in 2020 (Global Tailings Review, 2020) following the Brumadinho disaster, now mandates tailings dam break assessment for all TSFs to enhance risk assessment and emergency planning (Canadian Dam Association (CDA), 2021). However, modelling tailings dam failures remains a challenge due to the complex interplay between fluid dynamics, sediment transport, breach formation processes, and geotechnical engineering. Traditional water dam breach models, based on Eulerian hydrodynamic principles, struggle to capture the non-Newtonian behaviour of tailings, particularly in cases involving liquefaction-induced failure and rapid flow transformation, so new computational approaches incorporating debris flow modelling and two-phase flow dynamics are being explored to improve predictive accuracy (Sreekumar et al, 2024).

There are several numerical modelling techniques used in tailings dam breach modelling, however, hydrodynamic (Eulerian) models and geometrical debris flow models are the models the current study focused on (Brunner, 2014). On the other hand, geometrical debris flow modelling, represented by models such as Muk3D (MineBridge Software Inc., 2024), provides an alternative approach by focusing on mass movement dynamics rather than fluid mechanics. Unlike HEC-RAS 2D, which primarily tracks water movement, debris flow simulates breach widening, material erosion, and debris transport using a three-dimensional framework (Hungr et al, 2005; Medina, Hürlimann and Bateman, 2008). Recent research by Rodrigues (2024) and the author's experience in recent projects have expanded the use of geometrical debris flow modelling for large-scale tailings dam breach predictions, highlighting its advantages in capturing complex debris-laden flows and post-failure material redistribution.

This study conducts a comparative analysis of the Eulerian hydrodynamic approach and the geometrical debris flow approach to assess their effectiveness in modelling breach formation, flow propagation, inundation mapping, and sediment transport in tailings dam failures. Using a real-world tailings dam case study in Western Australia, this research explored the strengths, limitations, and practical applications of both models, offering insights into their suitability for TSF breach risk assessment and emergency response planning. By integrating hydrodynamic simulations with debris flow modelling, this research enhances predictive methodologies for tailings dam breach analysis and contributes to the development of more effective and TSF management strategies, thus improving safety and minimising environmental risks.

MATERIALS AND METHODS

Case study

The case study focuses on a proposed TSF in the Pilbara Region of Western Australia (Figure 1), which will serve as a critical infrastructure of an active gold mining operation. The facility consists of a two-cell paddock storage system strategically positioned adjacent to an existing TSF to enhance tailings management plus operational and construction efficiencies. A new open pit excavation is also planned to the west of the northern cell of the proposed TSF, raising concerns about the potential release of tailings material into the pit during both the excavation and operational phases and Closure. This situation highlighted the need for a comprehensive breach assessment to evaluate the risks associated with tailings mobility and containment stability under different failure scenarios.

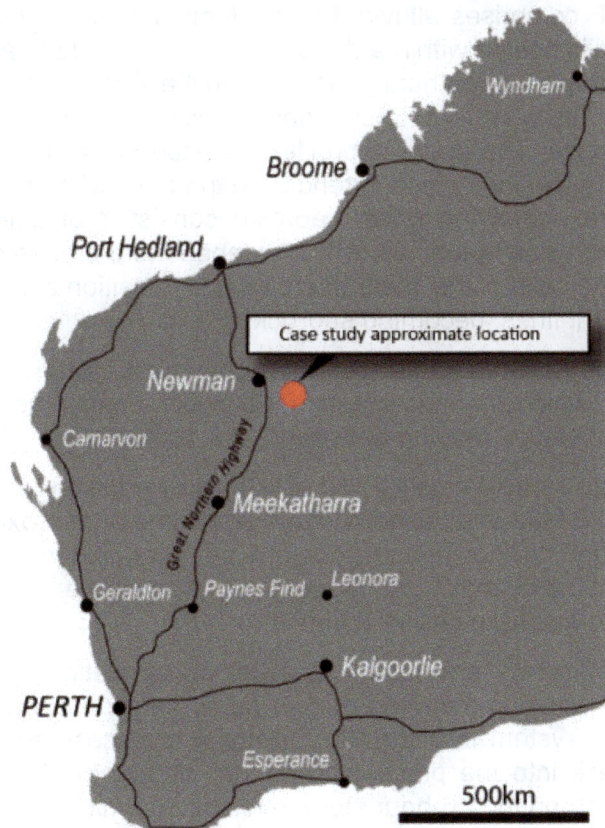

FIG 1 – Case study location in Western Australia.

The perimeter containment embankment structure of the TSF will have an initial crest elevation of RL 594.0 mAHD (Stage 1), which is planned to be raised once to a final crest elevation of RL 598.0 mAHD (Stage 2), reaching a maximum height of approximately 17 m. The embankment is engineered with crest widths of between 57.5 m (Stage 1) and 36 m (Stage 2), 1(V):2(H) slope on the upstream face, and 1(V):3(H) slope on the downstream face, to provide stability. The embankment will be zoned into three distinct layers: Zone A, an upstream low-permeability clay barrier; Zone B, a traffic-compacted mine waste zone; and Zone C, a downstream waste dump zone. The Stage 1 embankment will incorporate a cut-off trench 1.5 m to 2.5 m deep with 4 m wide base founded on a Ferricrete layer in order to reduce seepage losses. Raising of the embankment will utilise downstream construction technique. The aerial photo and site plan for the proposed embankment and mining site are presented in Figure 2.

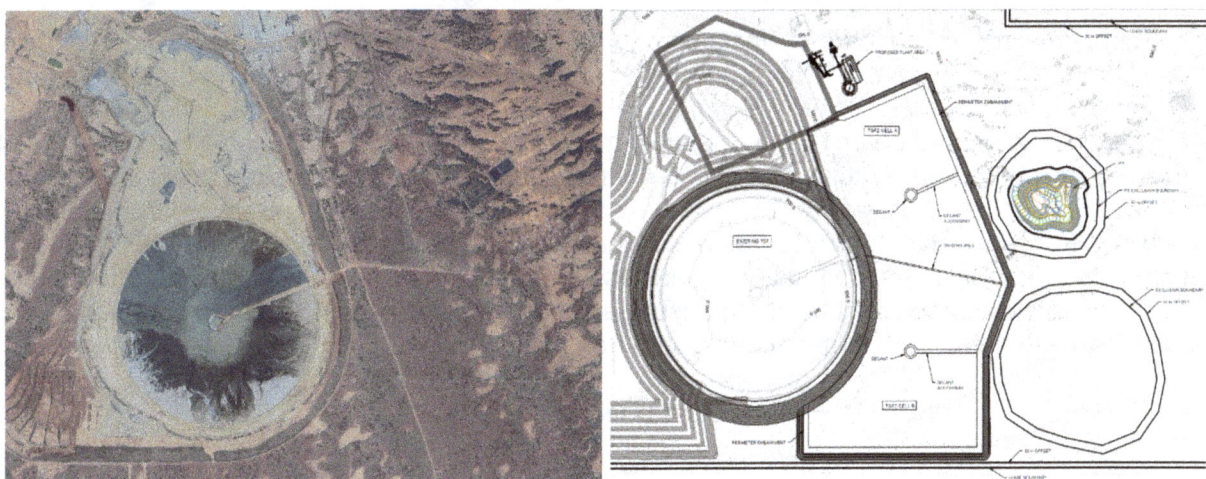

FIG 2 – The aerial photo (left) and site plan (right) of the case study TSF.

The foundation of the TSF comprises alluvial fan on dominantly flat topography. A shallow near-surface supergene horizon is hosted within a duricrust, which extends to approximately 10 m below ground levels (mbgl). The depth of weathering, as noted in the adjacent open pit to the TSF, extends to approximately 60 mbgl. The weathered rock (saprolite) was described as comprising of an upper saprolite and a lower saprolite. The upper saprolite consisted of sandy, mainly kaolinitic clay (60–70 per cent fines content by weight), which extended to the base of complete oxidation at between 25 mbgl and 35 mbgl. The underlying lower saprolite consisted of smectitic and kaolinitic clay (50 per cent fines), and sand-sized particles, which extended to the base of weathering at between 50 mbgl and 60 mbgl. At the base of the lower saprolite is a transition zone comprising highly friable saprolite with 20–30 per cent fines, becoming saprock.

Geotechnical investigations have characterised the subsurface material as dense to very dense, exhibiting low permeability, which is a critical factor for ensuring containment efficiency, and borehole permeability tests indicated values ranging between 2.2×10^{-6} m/s and 2.6×10^{-7} m/s.

The TSF is designed to store approximately 27 Mt of tailings over the target of 5.5 years life assuming an ore processing rate of 4.5 Mt/a and based on storage volume of approximately 19.4 Mm^3, tailings *in situ* density of 1.4 t/m^3 (dry), tailings beach slope of 1 per cent, and minimum operational and total embankment freeboards of 0.3 m and 0.5 m, respectively. The northern cell will store approximately 7.6 Mm^3 of tailings, and the southern cell 11.8 Mm^3.

To enhance operational efficiency and minimise water accumulation, the TSF will incorporate a decant system featuring a pontoon-mounted decant pump positioned within a 60 m diameter rock ring decant structure. This system facilitates the removal of excess surface water, which will be subsequently recycled back into the processing circuit. Hydrological analysis indicated that the probable maximum precipitation for a 6-hour storm event is 855 mm.

The facility has been designed with sufficient storage capacity to accommodate Probable Maximum Precipitation (PMP) events. While rainy-day failure modes were not identified as critical, there were notable concerns regarding internal instability mechanisms that could contribute to structural distress over time. The primary focus of the breach modelling, therefore, was on sunny-day failure scenarios where gradual internal weakening and potential static liquefaction pose the greatest risks to embankment integrity.

Breach geometry

The geometry of a tailings dam breach is a critical factor influencing the magnitude and impact of a failure event. It affects the volume of released tailings, peak outflow rates, and downstream inundation potential. The breach process typically unfolds in two phases. The first phase involves the discharge of the supernatant pond, carrying eroded tailings and dam fill materials. The second phase follows as flowable tailings are released due to liquefaction or the progressive slumping of unsupported materials, leading to a widespread flow of tailings.

The characteristics of the breach, including its shape and dimensions, were estimated using empirical models and historical failure data. Various parametric models, including those developed by Froehlich (2008), Xu and Zhang (2009), and Pierce, Thornton and Abt (2010), were considered to estimate breach parameters. In this study, the MacDonald and Langridge-Monopolis (1984) method, which assumed Newtonian behaviour for released material, was applied to estimate breach characteristics under both sunny-day and worst-case failure scenarios. Some parameters like breach development time and total released volume were calculated based on TSF geometry, construction methodology and industry regular assumptions. The breach was assumed to develop through the entire embankment height to ensure a conservative assessment of the potential tailings release. The estimated breach geometry is summarised in Table 1.

TABLE 1

Assumed breach geometry summary for case study.

Parameter	Worst case scenario	Sunny day scenario	Unit
Breach width (avg.)	19	16	m
Breach height	17	17	m
Breach base width	11	8	m
Breach top width	28	25	m
Breach development time	~4	~4	hr
Released tailings volume	4.77	3.89	Mm^3
Total tailings storage volume	11.8	11.8	Mm^3
Bulk density	2.55	2.55	t/m^3

The estimated breach development time was approximately four hours, reflecting the progressive nature of downstream tailings dam failures. Embankment erosion and tailings mobilisation contribute to the widening and deepening of the breach over time. The tailings storage volume modelled in a breach was 11.8 Mm^3, with a bulk density of 2.55 t/m^3. In the worst-case failure scenario, the released tailings volume was calculated to be 4.77 Mm^3, while in the sunny-day failure scenario, it was slightly lower at 3.89 Mm^3. These estimates were derived based on breach geometry, basin characteristics, and geotechnical assessments of tailings liquefaction potential.

The estimated runout flow rates were calculated by considering breach volume, development time, and hydrograph shapes. The average tailings flow rate (QF) was determined to be 221 m^3/s in the worst case and 180 m^3/s in the sunny-day scenario, while the peak tailings flow rate (QP) reached 442 m^3/s and 361 m^3/s, respectively. Using the Rico, Benito and Díez-Herrero (2008) empirical equation for breach peak discharge, maximum flow rates of 2060 m^3/s (worst case) and 1891 m^3/s (sunny day) was estimated, suggesting that breach progression can generate high-magnitude flow waves. The predicted runout distance was estimated at 2 km in the worst-case scenario and 1.5 km in the sunny-day failure, reflecting the potential extent of downstream tailings inundation (Seddon, 2010). A summary of peak flow and run-out distance is presented in Table 2.

TABLE 2

The summary of peak flow and runout distance for failure scenarios.

Parameter	Worst case scenario	Sunny day scenario	Unit
Average flow rate	221	180	m^3/s
Peak flow rate	442	361	m^3/s
Max flow rate*	2060	1891	m^3/s
Estimated runout distance	2	1.5	km

* Rico, Benito and Díez-Herrero (2008).

Numerical Eulerian modelling approach

HEC-RAS 2D, developed by the US Army Corps of Engineers (USACE, 2021), is a widely used semi-physically based hydrodynamic model that applies the Saint-Venant equations to simulate unsteady flow conditions in breached dams and allows users to define breach parameters manually or through a simplified physical approach that dynamically computes breach widening and erosion rates. The recent addition of a non-Newtonian solver to HEC-RAS 2D has attempted to address this limitation (Scholtz and Chetty, 2021), but challenges remain in accurately modelling rheological effects and sediment entrainment (Gildeh, Tourchi and Yu, 2020).

The HEC-RAS 2D model was employed to simulate the breach evolution and downstream flood propagation of the TSF under a sunny-day failure scenario. This scenario assumes internal instability leading to a full embankment breach without external triggers such as seismic activity or extreme precipitation, and that released material is fully liquified. Given the proximity of the breach to the proposed open pit, the simulation aimed to assess the potential extent of tailings mobility and deposition. The breach was modelled as extending down to the foundation level, representing a worst-case scenario in terms of material release and downstream impact.

To capture the complex flow behaviour of the released tailings-laden water, the Shallow Water Equations with the Exner-Layer Model (SWE-ELM) solver was utilised. This solver provides enhanced capabilities for modelling erosion and sediment transport within breach zones, improving the accuracy of flood propagation predictions. Unlike traditional Eulerian models that assume Newtonian fluid behaviour, the SWE-ELM approach incorporates dynamic interactions between water flow and sediment movement. However, no correction factor for non-Newtonian flow was applied in this study, as the model focused on bulk flow characteristics rather than detailed rheological effects.

To ensure numerical stability and convergence, the Courant number was maintained below 1, preventing excessive numerical diffusion and ensuring a stable solution. Additionally, the maximum iteration count was set to 20, allowing for refined calculations while balancing computational efficiency. The PARDISO matrix solver was employed to optimise computational performance, reducing processing time while maintaining numerical accuracy.

The numerical model was developed using a high-resolution LiDAR-derived digital elevation model with 1 m contours, ensuring a precise representation of the TSF topography and downstream floodplain. A computational mesh with an average cell size of 5 m in the potential flow downstream area was employed, with localised refinements in critical areas such as the breach location and anticipated high-velocity flow zones. The full breach development was simulated over a four hour duration, reflecting a progressive failure mechanism consistent with downstream embankment erosion and liquefaction-driven tailings release. The estimated volume of released tailings under this scenario was 3.89 Mm³, as determined from breach geometry calculations.

Hydraulic roughness was defined using Manning's coefficient (n) set to 0.3, which aligns with hydrological studies on sediment-laden flood events. A schematic of the HEC-RAS 2D numerical model is presented in Figure 3.

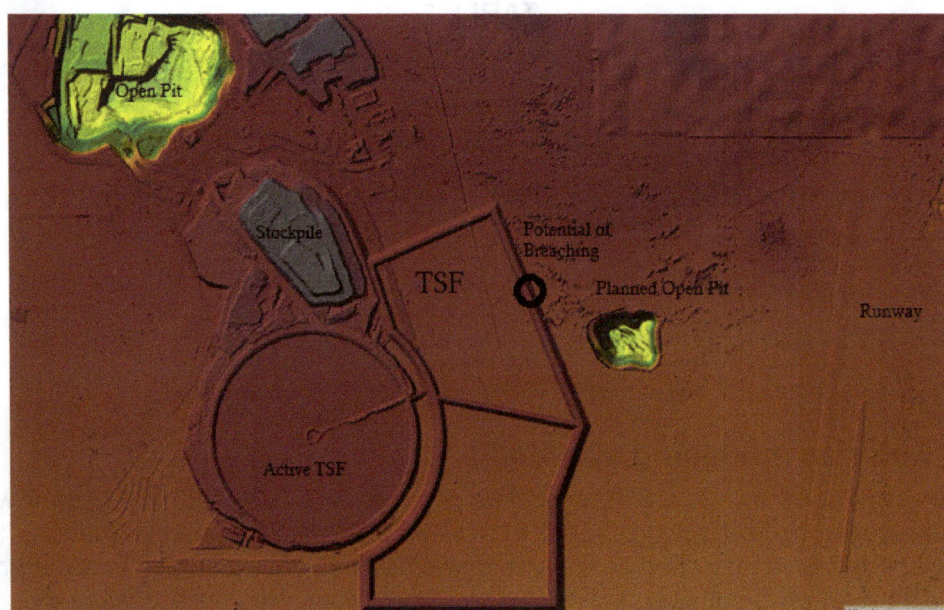

FIG 3 – Schematic of HEC-RAS 2D model covering planned adjacent open pit and TSFs.

Geometrical debris flow modelling approach

The geometrical debris flow modelling approach was implemented using Muk3D, a three-dimensional numerical tool specifically designed for simulating the runout behaviour of failures from earth structures such as tailings dams. Unlike Eulerian hydrodynamic models that primarily capture fluid motion, Muk3D focuses on mass movement dynamics, allowing for a detailed assessment of runout volume, deposition patterns, and tailings mobility.

For this study, Muk3D was used to estimate the runout volume and deposition extent following a sunny-day failure scenario, ensuring that the simulated tailings mobility aligned with the expected breach hydrograph. The modelling was performed using Tailings+ that allows advanced tailings flow simulations based on material-specific rheological and depositional properties. The computational domain was divided into three primary grids with high accuracy of 5 m cell meshing, representing the downstream floodplain, the embankment structure, and the tailings pond, ensuring a refined resolution in critical areas affecting breach development and runout behaviour.

Geometrical models typically lack detailed hydrodynamic interactions, making them less suitable for cases where fluid-structure coupling and sediment-water interactions are significant. The breach was assumed to initiate at the top of the foundation level, consistent with the assumptions in the Eulerian numerical modelling approach, with breach width and elevation defined according to the parameters provided in the breach geometry calculations. The breach location was aligned with the numerical model, ensuring comparability between both methodologies.

The embankment failure was modelled with a slope of 1(V):2(H), reflecting the anticipated failure mechanism based on geotechnical assessments. The runout slope was iteratively adjusted within the model to achieve a total released volume of 3.9 Mm³, ensuring that the simulated deposition matched the expected post-failure tailings distribution. The final modelled slope ranged from 0 per cent to 1 per cent, representing the natural gradients of the downstream floodplain and the anticipated flow resistance as tailings propagated outward from the breach zone. A schematic Muk3D model used in the current study is presented in Figure 4.

FIG 4 – Schematic model for geometrical debris flow in Muk3D.

RESULTS and DISCUSSION

The results of Eulerian hydrodynamic modelling and geometrical debris flow modelling are presented in the following sections separately. The HEC-RAS 2D model, based on the SWE-ELM solver, focuses on the hydrodynamics of the breach-induced flood wave, providing estimates of peak discharge, flow velocities, and inundation extents. Meanwhile, the Muk3D model, a geometrical debris flow simulator, is tailored for estimating runout volumes and predicting the final distribution of tailings deposits.

Results from Eulerian hydrodynamic simulations

The HEC-RAS 2D simulation results illustrated rapid flood wave propagation immediately following the breach, characterised by an abrupt rise to peak flow due to the sudden release of fluidised tailings. The model predicts a total discharge of 3.6 Mm³ within a four hour period, which aligns closely with the calculated estimates presented in Tables 1 and 2. The outflow initially exhibits high-energy, momentum-driven behaviour, resulting in a rapid surge, before progressively dissipating as the material disperses downstream. The initial outflow rate is at its peak during this phase, as the stored material is released under maximum head pressure, leading to the most significant drop in volume within a short time frame. The steepness of the curve in this early stage highlights the criticality of the initial two hours in emergency response, as this is when the highest flow rates and the most severe downstream impacts are likely to occur. Figure 5 illustrates the discharge rate over time for the proposed breaching scenario, highlighting the transition from the initial surge to a gradual attenuation phase, which is influenced by terrain interaction, flow resistance, and sediment deposition.

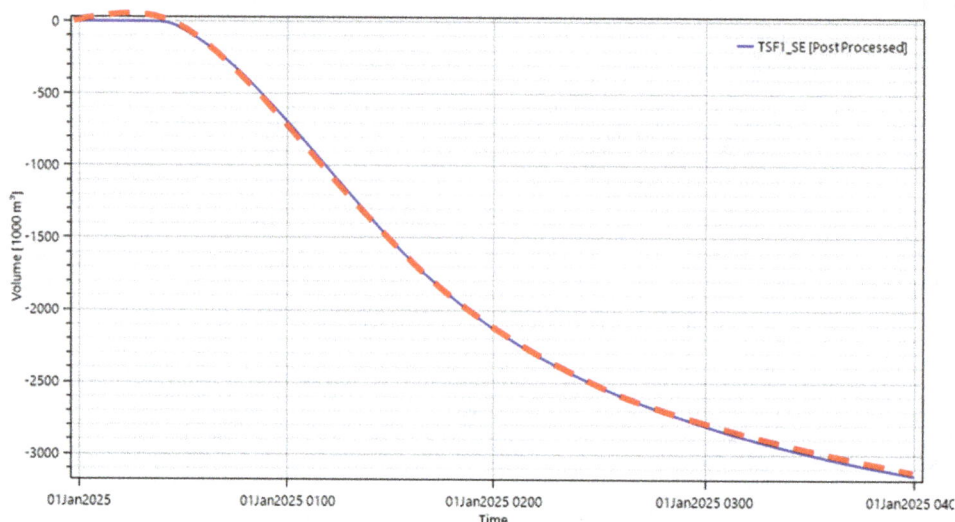

FIG 5 – The volume discharge versus time.

The results of the numerical flow calculations, incorporating terrain interactions and predefined flow parameters, are presented in Figure 6. The simulation indicates that the released tailings followed the natural topographic gradients, allowing a portion of the material to flow into the open pit, where it accumulated at a maximum depth of approximately 12 m at the lowest elevation within the pit. The remaining material was fully transported into the downstream flow area, spreading across the terrain. After four hours, the minimum depth of deposition in the farthest downstream region (south of the TSF) was estimated at ≤0.1m. Despite the relatively thin deposition in these areas, the simulation results confirm the presence of released material, highlighting the model's sensitivity to detecting low-thickness sediment deposition over extended distances.

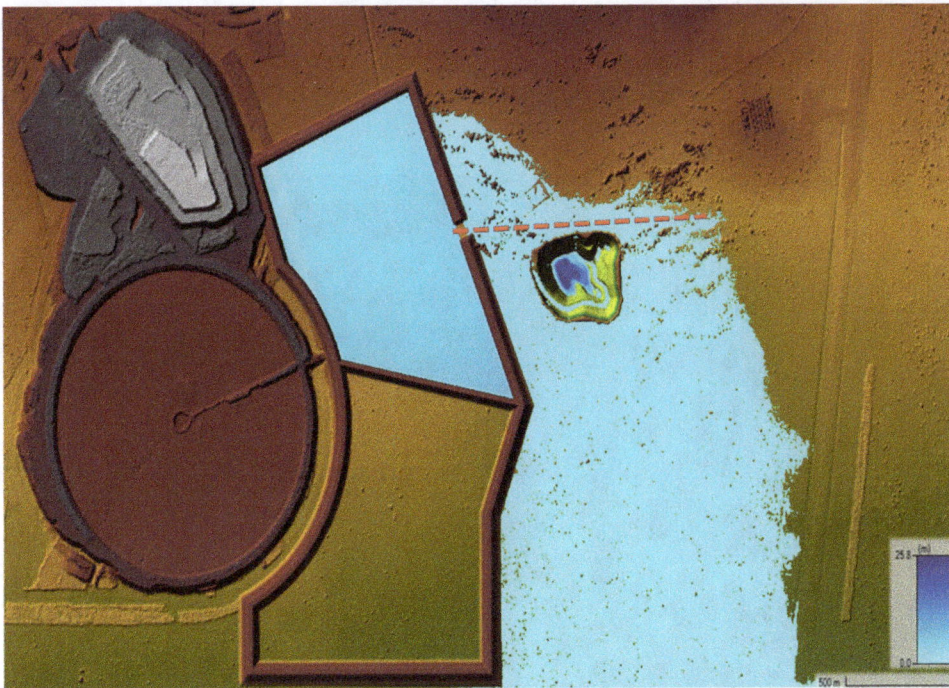

FIG 6 – Released material maximum depth and flow section.

Figure 7 presents the relative level variations of the released material over the flow path shown as a red dashed line in Figure 6, capturing the dynamic evolution of tailings surface elevation. The maximum and minimum RL levels of the released material are approximately RL 593.0 m and RL 585.5 m, respectively. The graph exhibits a sharp initial drop in material surface elevation, indicative of a high-energy, sudden outflow triggered by the breach formation and the rapid release of tailings.

As the tailings continue to propagate downstream, the surface elevation gradually declines, reflecting a progressive attenuation of flow energy due to sediment deposition, topographic variations, and hydraulic resistance. This trend signifies a transition from an initial high-velocity surge to a more stable, gravity-driven flow regime, emphasizing the crucial role of terrain morphology in controlling tailings mobility and final deposition patterns.

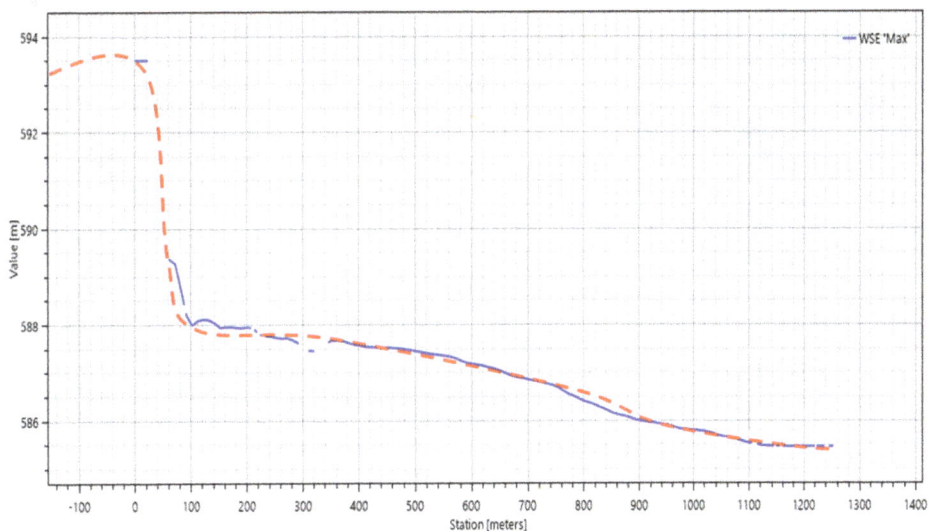

FIG 7 – Flow relative level alongside the flow section.

Flow velocities were highest in proximity to the breach, exceeding 12 m/s, before gradually decreasing as the inundation wave propagated over a 1.5 km runout distance. Near the breach, maximum flow depths reached an average of 3 m, highlighting the significant volume of tailings mobilised during the failure event. The progressive reduction in velocity suggests energy dissipation through topographic interaction, flow resistance, and sediment deposition, shaping the extent and distribution of the released material. Figure 8 presents the distribution of released material velocity in the maximum loading case.

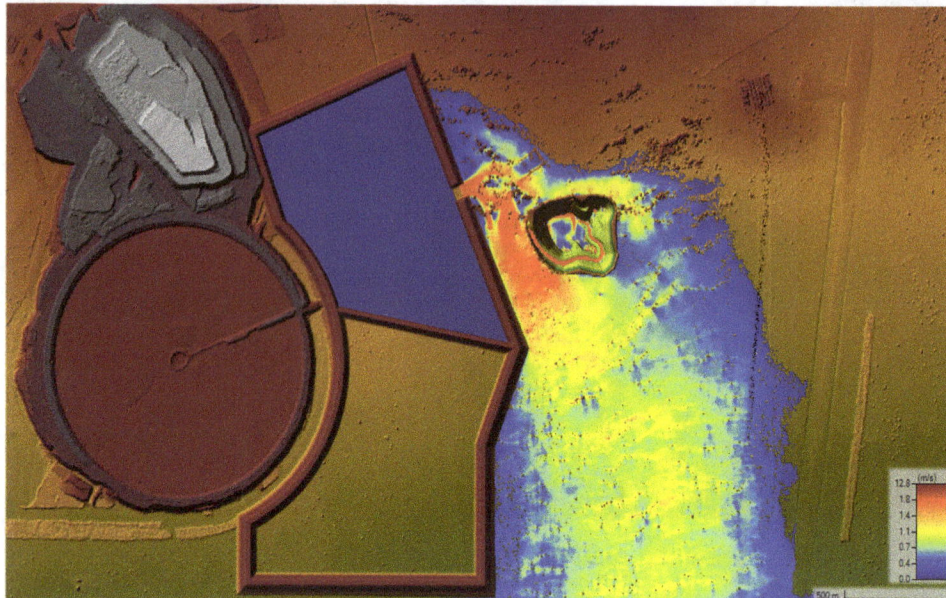

FIG 8 – The results of released material velocity contours.

Figure 9 presents the arrival time for a 0.3 m flow depth resulting from the TSF breach, simulated numerically. The legend suggests that the earliest flow reaches approximately 0.29 hrs (around 17 mins), while the latest arrival occurs at 3.96 hrs, demonstrating a time-dependent flow expansion influenced by topography and material properties.

FIG 9 – Arrival time for a 0.3 m flow depth.

Results of debris flow simulations

The geometrical debris flow model provided a more gradual release and runout profile, reflecting the progressive redistribution of tailings across the downstream landscape. The model estimated a total runout volume of 3.9 Mm³, closely matching the anticipated release from breach calculations. Unlike the numerical model, which focuses on flood propagation, geometrical flow provided detailed predictions of where the released tailings would ultimately settle, identifying high-deposition areas near the breach and progressively thinning layers further downstream. The embankment failure was simulated with a slope of 1(V):2(H), with breach width and elevation parameters matching those used in the numerical model. The breach location was consistent between both modelling approaches, allowing for direct comparisons of flood wave characteristics and material transport behaviour.

Figure 10 presents the simulation results, with the red line marking the extent of material released from the breach area. The model predicts a controlled, geometrically constrained flow pattern, where tailings primarily move downslope following terrain gradients and natural topographic features. Unlike fluid dynamics-based models that account for momentum-driven flow, turbulence, and viscosity effects, this simulation employs a geometric and empirical deposition approach, resulting in a more structured and predictable flow path. The debris flow model estimates a maximum runout distance of 1.3 km aligned with the breach failure direction. Only a small portion of the material entered the open pit, while most of the released tailings remain confined behind the bund wall surrounding the pit, highlighting the influence of terrain constraints on flow mobility and deposition patterns.

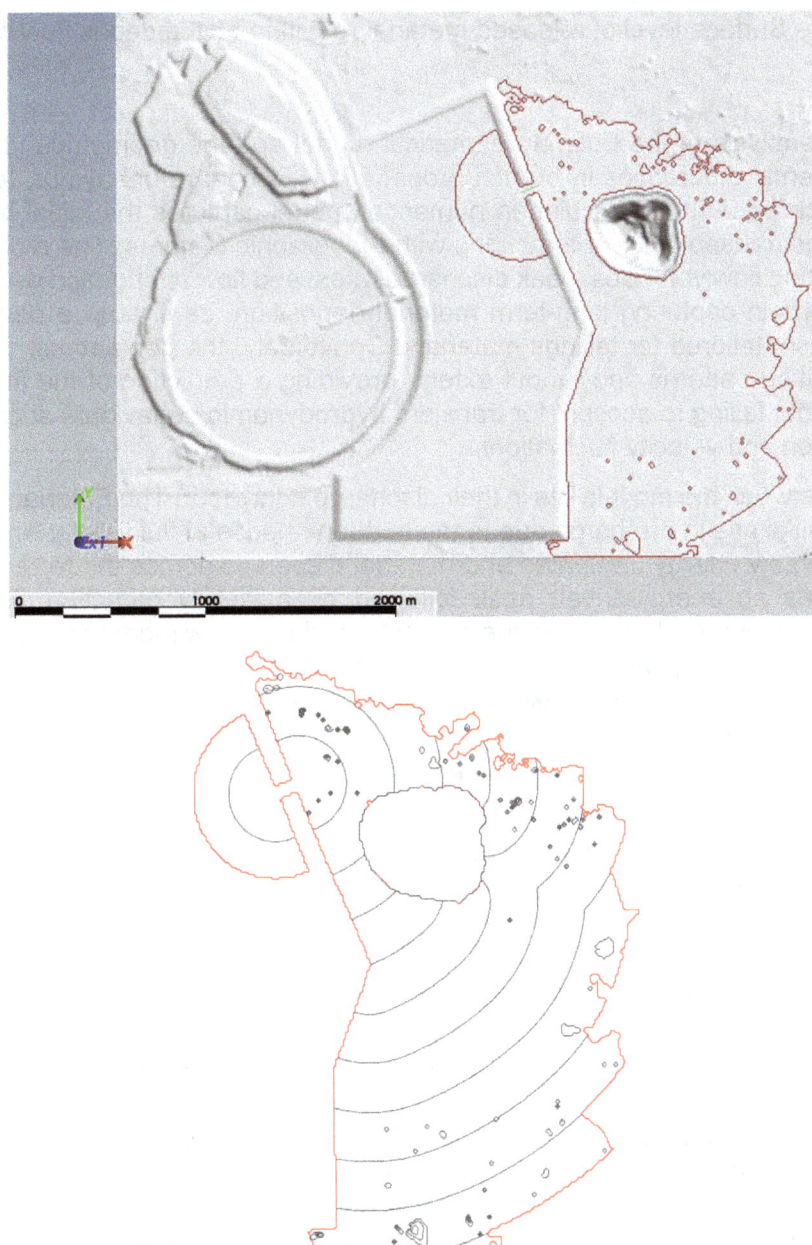

FIG 10 – Results of released tailing in the terrain area (up); results of released material and 1 m contour line (bottom).

Figure 11 presents the released material surface and terrain profile along the same section used in the numerical model. The simulation results indicate that no sudden drop in material elevation occurred at the initial breach stage, suggesting a gradual flow onset rather than an abrupt collapse. The released material surface exhibits a consistent slope, transitioning smoothly from RL 592 m to RL 585.5 m, reflecting a steady deposition pattern. The minimum elevation result aligns well with the numerical model's predictions, confirming the model's accuracy in capturing final deposition levels. However, the predicted deposition extends approximately 50 m farther than the numerical model outcome, potentially due to differences in terrain resolution, flow assumptions, or the influence of secondary flow dynamics not fully captured in the empirical deposition approach.

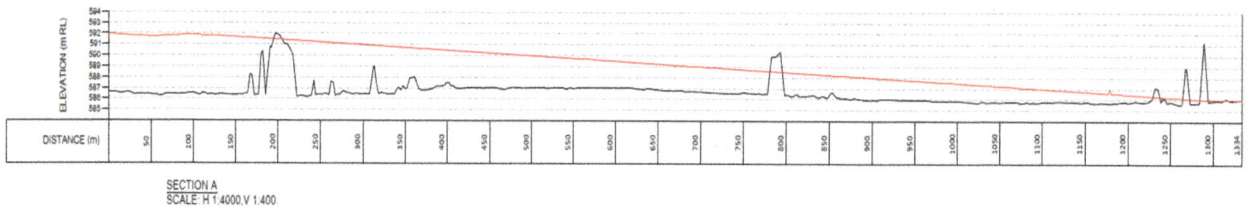

FIG 11 – Surface level of released material resulting by the debris flow model.

Discussion

The comparative analysis of the Eulerian numerical model and the geometrical debris flow model highlights fundamental differences in breach propagation, discharge behaviour, terrain sensitivity, and final deposition patterns. The Eulerian numerical model captures the initial high-energy flood surge, rapid tailings transport, and interactions with topographic features. The model provides high accuracy in predicting flow velocities, peak discharge rates, and flow attenuation over time. However, it exhibits limitations in capturing long-term material deposition, as it lacks a dedicated sediment transport mechanism tailored for tailings materials. Conversely, the geometrical debris flow model focuses on deposition patterns and runout extent, providing a prediction of the final distribution of released material but failing to account for transient hydrodynamic behaviours such as momentum-driven flow evolution and velocity fluctuations.

A key distinction between the models lies in their discharge behaviour. The Eulerian numerical model predicts a sharp initial rise in discharge due to the sudden release of fluidised tailings, followed by a gradual decline as flow energy dissipates and material disperses across the terrain. The discharge curve demonstrates an energy-driven peak followed by a steady reduction, which aligns with observed dam breach events. In contrast, the geometrical debris flow model does not provide a time-dependent discharge curve, as it assumes a steady-state mass flow governed by geometric spreading principles. This difference makes the Eulerian model more applicable for real-time hazard assessments and emergency response planning, while the geometrical debris flow model is effective for long-term impact assessments and runout prediction.

Terrain sensitivity further differentiates the two models. The Eulerian numerical model dynamically adjusts flow velocities and depths based on terrain interaction, allowing for the simulation of material accumulation in depressions, backflow into pits, and localised ponding. This capability is critical for accurately predicting the extent of inundation in complex landscapes. The geometrical debris flow model, however, follows a more rigid spreading pattern based on predefined angles and geometric constraints, leading to an underestimation of material accumulation in enclosed depressions or restricted flow areas.

The strengths of both models suggest that an integrated approach offers a more comprehensive framework for tailings dam breach analysis. The Eulerian numerical model excels in predicting breach hydrographs, velocity distribution, and flood wave propagation, making it an essential tool for emergency response and risk mitigation. The geometrical debris flow model provides valuable insights into the final deposition of tailings, allowing engineers to assess long-term environmental impacts and develop remediation strategies. By combining both methodologies, a more complete assessment of breach risks, material movement, and sediment deposition can be achieved, ensuring a balanced approach to real-time hazard management and post-failure recovery planning. The comparative inundation map resulting from the numerical and geometrical models is presented in Figure 12.

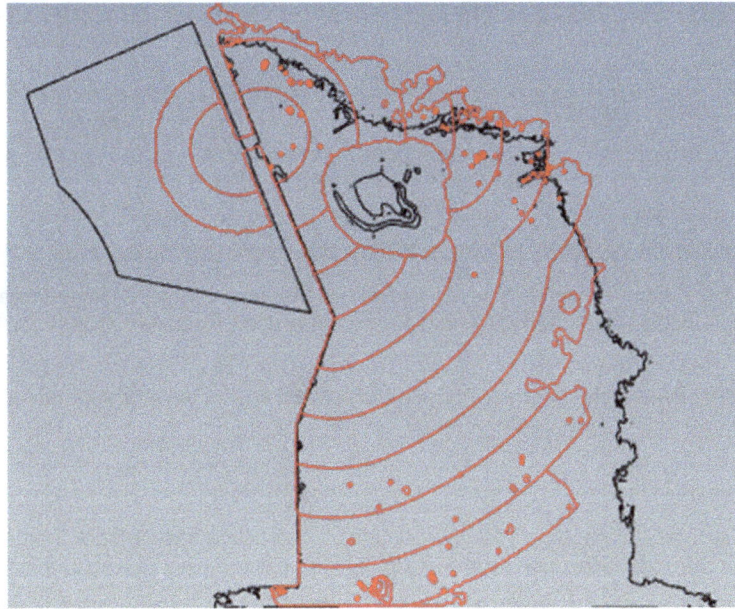

FIG 12 – The inundation map resulting from geometrical debris flow (red line) and numerical model (black line).

CONCLUSION

This study compared two modelling approaches, Eulerian hydrodynamic modelling and geometrical debris flow simulation, to evaluate their effectiveness in predicting breach behaviour, tailings runout, and deposition patterns following a tailings dam failure. Using a real-world case study from Western Australia, both models were applied to a sunny-day failure scenario, allowing for direct comparison of their capabilities and limitations in a consistent framework.

Key findings from the analysis include:

- Eulerian hydrodynamic modelling provided detailed insights into breach development, peak discharge, flow velocity, arrival time, and inundation extent.

- Geometrical debris flow modelling estimated tailings runout extent and final deposition, offering a clear spatial understanding of post-failure material distribution.

- Muk3D lacked transient flow dynamics but produced accurate topography-driven results aligned with observed terrain influences.

- Both models were analysed based on similar total released volumes (~3.9 Mm^3), with close agreement on final deposition elevation and runout distances (1.3–1.5 km).

- The Eulerian model will be better suited for emergency response planning, while the geometrical model will assist in long-term environmental impact assessment.

Overall, this comparative analysis highlights the value of integrating both modelling approaches to enhance the tailings dam breach assessments. Combining the strengths of hydrodynamic flow prediction with robust deposition mapping enables a more holistic understanding of TSF failure consequences, supporting better-informed decision-making for risk mitigation and management strategies.

ACKNOWLEDGEMENT

The author would like to extend sincere thanks to Darren Edward for his valuable contribution in modelling on this paper. Appreciation is also extended to CMW Geoscience for their support throughout the preparation of this work.

REFERENCES

Azam, S and Li, Q, 2010. Tailings dam failures: A review of the last hundred years, *Geotechnical News*, 28(4):50–53.

Brunner, G, 2014. Using HEC-RAS for Dam Break Studies, August issue, US Army Corps of Engineers, Hydrologic Engineering Center.

Canadian Dam Association (CDA), 2021. Technical Bulletin: Application of Dam Safety Guidelines to Mining Dams, Canadian Dam Association, Toronto, Canada.

Froehlich, D C, 2008. Embankment dam breach parameters and their uncertainties, *Journal of Hydraulic Engineering*, 134(12):1708–1720. https://doi.org/10.1061/(ASCE)0733-9429(2008)134:12(1708)

Gildeh, M M, Tourchi, A and Yu, X, 2020. Numerical modelling of tailings dam breach using a coupled Eulerian-Lagrangian method, *Computers and Geotechnics*, 119:103383. https://doi.org/10.1016/j.compgeo.2019.103383

Global Tailings Review, 2020. Global Industry Standard on Tailings Management (GISTM), International Council on Mining and Metals (ICMM), United Nations Environment Programme (UNEP) and Principles for Responsible Investment (PRI), August 2020.

Hungr, O, Evans, S G, Bovis, M and Hutchinson, J N, 2005. A review of the classification of landslides of the flow type, *Environmental and Engineering Geoscience*, 11(3):221–238.

Jeyapalan, J K, Duncan, J M and Seed, H B, 1983. Analysis of flow failures of mine tailings dams, *Journal of Geotechnical Engineering*, 109(2):150–171.

MacDonald, T C and Langridge-Monopolis, J, 1984. Breaching Characteristics of Dam Failures, *Journal of Hydraulic Engineering*, 110(5):567–586.

Medina, V, Hürlimann, M and Bateman, A, 2008. Application of FLAT Model for debris-flow hazard assessment, *Natural Hazards and Earth System Sciences*, 8:923–938. https://doi.org/10.5194/nhess-8-923-2008

MineBridge Software Inc., 2024. Muk3D: 3D Earth Structures Modelling Solution, MineBridge Software Inc. Available from: <https://www.muk3d.com>

Morgenstern, N R, Vick, S G and Van Zyl, D, 2016. Fundão tailings dam review panel: Report on the immediate causes of the failure of the Fundão Dam, BHP Billiton and Vale, Brazil.

Pierce, M W, Thornton, C I and Abt, S R, 2010. Predicting Peak Outflow from Breached Embankment Dams, *Journal of Hydrologic Engineering*, 15(5):338–349.

Rico, M, Benito, G and Díez-Herrero, A, 2008. Floods from tailings dam failures, *Journal of Hazardous Materials*, 154:79–87.

Rodrigues, B O, 2024. Assessment of the runout behavior of hypothetical failures of tailings stacks through empirical and hydrodynamic methods, Universidade Federal de Minas Gerais, School of Engineering, Postgraduate Program in Sanitation, Environment and Water Resources, Belo Horizonte, Brazil.

Scholtz, S and Chetty, K, 2021. HEC-RAS, 2D non-Newtonian flow modelling for mine tailings dam breach analysis, Proceedings of the International Mine Water Association Conference 2021.

Seddon, K D, 2010. Approaches to estimation of runout distances of liquefied tailings, in *Proceedings of the First International Seminar on the Reduction of Risk in the Management of Tailings and Mine Waste (Mine Waste 2010)* (eds: A B Fourie and R Jewell), pp 63–70 (Australian Centre for Geomechanics: Perth).

Sreekumar, U, Kheirkhah Gildeh, H, Mohammadian, A, Rennie, C and Nistor, I, 2024. Tailings dam breach outflow modelling: A review, *Mine Water and the Environment*, 43:563–587. https://doi.org/10.1007/s10230-024-01015-y

US Army Corps of Engineers (USACE), 2021. HEC-RAS River Analysis System User's Manual Version 6, Institute for Water Resources, Hydrologic Engineering Center, Davis, CA, USA.

Xu, Y and Zhang, L M, 2009. Breaching parameters for earth and rockfill dams, *Journal of Geotechnical and Geoenvironmental Engineering*, 135(12):1957–1970. https://doi.org/10.1061/(ASCE)GT.1943-5606.0000162

Proof-of-concept probabilistic inundation model based on probabilistic distribution of tailings dam breach release volumes

G Bullard[1], M McKellar[2], V Rojanschi[3] and B Russell[4]

1. Civil Engineer, BGC Engineering Inc., Halifax NS B3J 3N6, Canada.
   Email: gbullard@bgcengineering.ca
2. Junior Intern, BGC Engineering Inc., Toronto ON M5H 3T9, Canada.
   Email: mmckellar@bgcengineering.ca
3. Principal Water Resources Engineer, BGC Engineering Inc., Calgary AB T2P 3L8, Canada.
   Email: vrojanschi@bgcengineering.ca
4. Principal Geotechnical Engineer, BGC Engineering Inc., Vancouver BC V6Z 0C8, Canada.
   Email: brussell@bgcengineering.ca

ABSTRACT

Water and tailings dams are among the largest engineered structures on earth. Consequences of failure can be catastrophic to communities, people, infrastructure, and the environment. Dam breach and inundation studies of credible failure modes are routinely undertaken to support emergency response planning and preparedness, to estimate populations at risk, support dam consequence classification, and support the development of risk management solutions. All of which is challenging due to the multiple scenarios used in the analysis (varying failure modes, pond levels or breach geometries), none of which are typically associated with a probability of failure. Furthermore, is it typical for the tailings dam breach and runout analysis to be conducted on conservative assumptions, which are appropriate to support emergency response or dam consequence classification; however, these assumptions may not define appropriately the downstream risk given their low likelihood of occurrence.

A previously published paper has developed a risk screening-level tool, which applies probability distributions of tailings dam breach volumes based on the failure mode. The current work, and focus of this paper, is to advance that work and develop a screening-level probabilistic runout model that provides a likelihood to downstream elements at risk. This model has been calibrated to two historical tailings dam breach failures: Feijao and Merrispruit. The ultimate goal of this work is to identify the failure modes associated with the highest downstream risk, which will inform the assumptions to be used in a detailed tailings dam breach assessment to support risk-informed decision-making in design.

INTRODUCTION

Water and tailings dams are among the largest engineered structures on earth. Their primary uses include hydroelectric power generation, agricultural and domestic water storage, and mine waste management. Impoundment volumes behind such dams represent a hazard to downstream communities, including people, infrastructure, and the environment. Tailings dam breaches have caused nearly 3000 fatalities in the past century (Santamarina, Torres-Cruz and Bachus, 2019). Recent tailings dam failures include the 2019 failure in Corrego do Feijao in Brumadinho, Brazil causing over 250 deaths (Rotta *et al*, 2020) and the Luming Mine tailings breach in March, 2020 threatening the source of drinking water for Tieli City (population 68 000) (Wise Uranium Project, 2021). The examples above highlight the potential consequences of dam breach failures and the importance of understanding and communicating this consequence to all stakeholders.

Dam breach and inundation studies of hypothetical failure scenarios are undertaken to determine the appropriate dam consequence classification, to develop risk management solutions for dam failure modes, and to support emergency planning (Figure 1). There are a number of guidance documents to support dam breach and inundation studies, and dam consequence classification, including the Canadian Dam Association guidelines (CDA, 2013, 2019), Global Tailings Review (GTR, 2020), and the US Mine Safety and Health Administration (MSHA) classification system. This guidance is to base dam breach assessments on so called 'credible failure scenarios', which are often based on conservative assumptions leading to results representing 'worst-case' scenarios that manifest the most extreme consequence.

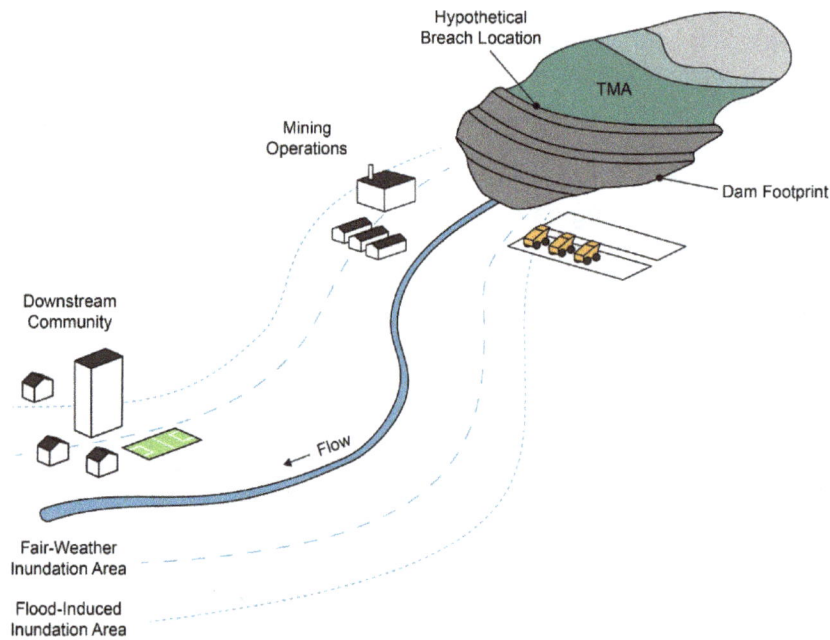

FIG 1 – Schematic of inundation maps for flood-induced and fair-weather dam breach scenarios.

Best practice guidance for the tailings dam industry includes requirements to perform and regularly update risk assessments, including identification and characterisation of potential failure modes (GTR, 2020; and International Commission of Large Dams (ICOLD, 2022). Previous work by Porter, Russel and Watts (2023) and Bale *et al* (2024) have developed a framework for a screening-level tool, called Probability of Failure (PoF), to support the selection of tailings dam breach volumes to assist with screening-level assessments and dam breach assessments.

This paper expands on the work by Porter, Russel and Watts (2023) and Bale *et al* (2024) and takes the probability distribution of volume for a given failure mode to estimate the runout extents for the range of probabilities using a screening-level runout model. This will provide a spatial estimate of the consequence based on a certain failure mode. The current study explores the application of the Progressive Debris-Flow routing and inundation model to tailings dam breaches by calibrating to two historical events: Feijao and Merriespruit.

This work is another piece to a larger vision of creating a tool, which is meant to complement expert-based quantitative or semi-quantitative risk assessment methods and accelerate the adoption of risk-informed decision-making in all phases of tailings dam management. The last component of this vision would be to combine the results from the present study with other exposure models, where life loss and economic loss could be estimated and compared against established risk tolerance criteria.

METHODS

Model description

The screening-level runout model selected for this study was developed to model post-wildfire debris flows (also a non-Newtonian flow) in California called the Progressive Debris-Flow routing and inundation model (ProDF) (Gorr *et al*, 2022). The model employs an iterative variation of the multiple flow direction (MFD) routing algorithm to predict the spread of non-Newtonian flows across a landscape. This algorithm distributes the flow volume to all downslope neighbouring cells based on the slope of each cell. Flow depth at each point along the flow path is calculated using empirical relationships that relate volume, topographic slope, and flow properties.

The model uses an iterative process to refine the flow routing surface. In the first iteration, flow is routed based on the original topography. After each iteration, a small fraction of the computed flow depth is added to the routing surface, representing the flow surface rather than the topography. This iterative process is repeated for a user-defined number of iterations, allowing the model to simulate lateral spreading of the flow. Flow routing stops when the shear stress at the base of the flow falls below a user-defined yield strength, ensuring that the model only routes flow-through areas where the flow can physically move.

The final output of ProDF is the extent of flow inundation and peak-flow depth at each point along the flow path. This model does not estimate sediment deposition thickness or flow velocity.

Input parameters

The ProDF model requires several key input parameters to accurately simulate flow inundation:

- topography
- three inputs defining flow behaviour
- breach initiation point
- bounds on the model extent (ie estimate on how far the runout will travel)
- source volume.

Topographic data from AW3D30 DSM data map published by the Japan Aerospace Exploration Agency (JAEA, 1997) was used to cover each of the model extents. This data set has a 30 m spatial resolution and 5 m height accuracy. A higher resolution topography would provide less uncertainty in the results; however, given that this is a screening-level tool designed to be used for a global inventory, a publicly available source was selected to be consistent with the future use cases of this tool.

The three inputs to define the flow behaviour are: flow density, yield strength, and a flow resistance coefficient. A constant flow density of 2000 kg/m^3 was used for the event calibration. The yield strength and flow resistance coefficient can vary widely and must be calibrated for the model using observed data from past tailings dam breach events to ensure the model accurately reproduces the extent and depth of tailings dam breach inundation.

The coordinates of the estimated initiation point are entered into the model using the latitude and longitude. In this paper, only historical case studies were investigated and so the bounds on the model extent were known. If this were not the case, the user can provide a best guess of the extents. If the extent is too small, the runout will end at the edge of the extent provided and a larger extent will increase model run-times. For reference, running the probabilistic model with one set of flow resistance parameters takes approximately 5–10 mins.

The final input is the source volume. The original model uses a deterministic code to evaluate the runout (ie one source volume). This study has adjusted the code to incorporate numerous source volumes to represent the range in release volumes which could occur based on a specific failure mode (eg overtopping). This study now allows the user to input 100 source volumes, which has been selected to represent each percentile of release volume based on the release volume tool developed by Bale *et al* (2024).

RESULTS

Model calibration

The yield strength and flow resistance coefficients were calibrated using two historical case studies: Feijao and Merriespruit. These case histories were selected as these failures have been well documented and they represent a range of release volumes and failure modes. These failure cases were also used in the previous work conducted by Bale *et al* (2024) and so preliminary results describing the probabilistic release volume distributions have been completed. A key finding in a previous study by Ghahramani *et al* (2022) was that multiple sets of rheological parameters could produce comparable output results and that the best-fit parameters were non-transferable between the four models. Therefore, the calibrated yield stress was intentionally not consistent with previous studies and should not be compared directly.

To perform the calibration, the yield stress range was inputted between 10–1000 Pa and flow resistance coefficient between 10–60 s^{-1}m m$^{-0.5}$. These values were based on the range considered from historical tailings events and those used in the calibration of the debris flow models by Gorr *et al* (2022). Based on the results of the simulation, the range of values were narrowed to determine the plausible values that provided the closest match to the known failure inundation area. The closeness of fit was evaluated by estimating the percentage of overlap between the 'wet' historical inundation area and the 'wet' simulation area, the percentage of false negatives ('wet' historical inundation area and a 'dry' simulation area) and false positives ('dry' historical inundation area and a 'wet' simulation area).

The calibration reveals a narrow range of values for the flow resistance that could effectively represent the liquefied tailings. For the model comparison, the flow resistance was held at a constant 10 s^{-1}m m$^{-0.5}$, an average value of the initial calibration results of the historical events. The calibrated value of yield strength for the Feijao failure simulation was 46 Pa. This value on the lower end of the range for this wet and very mobile liquefaction failure through the channelised creek valley that was said to behave as a high-energy debris flow (de Lima *et al*, 2020). For Feijao, the closeness of fit parameters from the inundation extent based on the flow resistance parameters above are: 75 per cent overlap, 24 per cent false negative and 20 per cent false positive.

The calibrated yield strength input for the Merriespruit failure was 730 Pa, a higher value considering that a significant proportion of the outflow was free water from the facility that travelled over the near flat and unconfined terrain (Wagener, 1997). The calibrated values for the Merriespruit failure fall within the range of yield stresses, 55–1000 Pa, presented in Ghahramani *et al* (2022). For Merriespruit, the closeness of fit parameters are: 97 per cent overlap, 2 per cent false negative, 87 per cent false positive.

Model runout

The calibration range of yield strength and flow resistance values could effectively represent the inundation extents of liquefied tailings, which typically have a sediment concentration of 20–50 per cent by volume. Now, the calibrated inundation model can be combined with the output form the probabilistic release volume tool, described by Bale *et al* (2024).

The original ProDF model was adapted to be able to output an inundation map, which shows the inundation extent associated with each release volume in the probabilistic range (eg the 5th percentile release volume out of the probabilistic range). The calibrated model of both the Feijao failure and the Merriespruit failure are shown in Figure 2. The inundation extent for the higher percentile volume outflows are shown in dark shades (purples), and the inundation extent for the lower percentile volume outflows are shown in lighter colours (yellow and orange). The distribution of inundation extents can be compared with the estimated historical runout, which is shown in red.

In the Feijao simulation, near the breach location, there is a small discrepancy between the modelled and estimated historical inundation extent due to the coarse topography not being representative of the dam structures prior to failure (Figure 2a). This allows the simulation to backfill into the impoundment, which is not considered an accurate representation. Between the upstream source and downstream extent of the historical event, there is good correlation between all the simulations and the estimated historical extent. Relative to the used probabilistic release volume range, the known failure outflow of 9.7 Mm^3 (79 per cent of the total stored volume, outlined in red (Adria *et al*, 2023)) is related to the range's 99 per cent percentile. Since 100 simulations were conducted, only one inundation model (shown in deep purple) had a larger inundation extent than the historical event.

In the Merriespruit simulation, there is a significantly smaller runout distance due to the flatter topography than the Feijao case and due to the smaller release volume (Figure 2b). Approximately 95 per cent of the inundation models have an extent that exceeds the historical event. The maximum simulated inundation extent is as much as 1 km beyond the historical event, in some places. This aligns with the work done by Bale *et al* (2024), who reports that, based on the generated probability distributions curves, there was a 96 per cent probability for the inundation extent to exceed the actually observed inundation area; this was due to the actually released volume being at the low end of the probable release range. The inundation extent for the 50th percentile (median condition) would have had a significantly greater extent than the historical event.

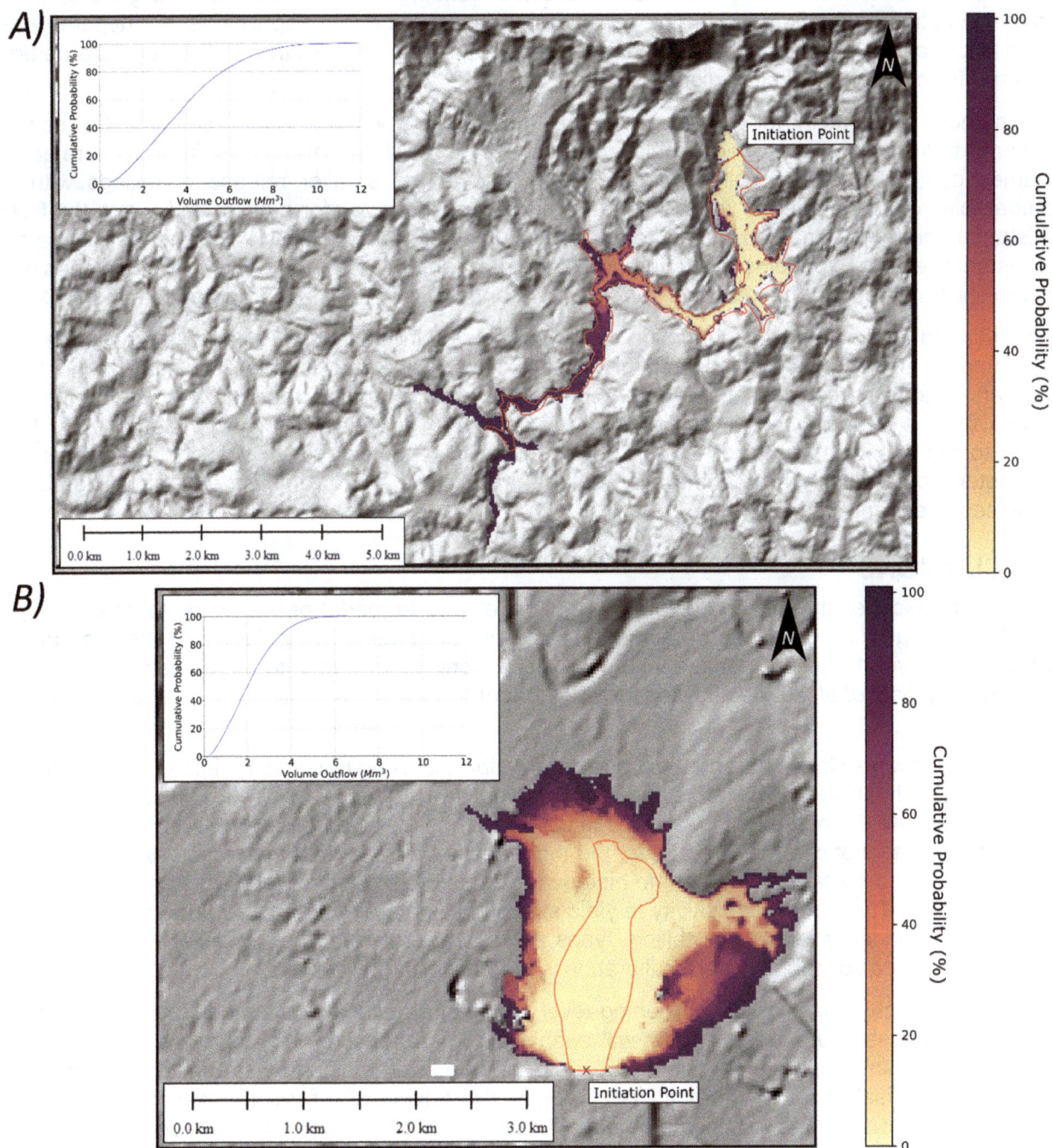

FIG 2 – Simulation of the: (a) 2019 Feijao failure; and (b) 1994 Merriespruit dam failure for a probable distribution of outflow volume based on an estimated release result from the PoF tool (Bale *et al*, 2024). Darker colours represent the extent of larger volume outflow and lighter colours represent the extent of smaller outflows. This distribution is based on calibrated inputs of flow parameters to the known inundation extents for both events (shown in red).

DISCUSSION

Adjusting the yield stress and flow resistance parameters allowed ProDF models to reproduce the bulk behaviours of the historical failures at Feijao and Merriespruit. The authors recognise the challenge in applying a new model that has never been used to model the runout of a tailings dam breach event, especially to hypothetical future failure scenarios, where calibration is not possible. Continued work aims to take this proof of concept and continue calibrating to multiple historical tailings dam failures with varying failure modes, release volumes, and tailings properties. The goal is to develop an understanding of the relationship between physical properties that can be measured *in situ* or in a laboratory (eg. water content, grain size distribution, minerology, yield stress etc) and the model's resistance parameters. From that understanding, a user-guide can be generated to provide direction on the flow resistance parameters that could be most applicable to a hypothetical future failure at another facility, based on the further understanding of the flow resistance parameters gathered from the study of case histories.

The authors also acknowledge that varying the release volume could change the solids concentration of the slurry, which could impact the flow resisting parameters (ie smaller release volumes could have lower solids concentrations due to the smaller release of tailings with a comparable volume of release water volume). This means that even for a calibrated event, the flow resistance parameters may only apply to release volumes of approximately the same magnitude as the historical event. Expanding the calibration of the resistance parameter to more case histories will help to shed light on the role of solids concentration and the flow resistance parameters in ProDF.

Using this reduced complexity inundation model as a screening-level tool will enable practitioners a visual understanding of the likely extents of a potential failure and quick comparison in the inundation between various percentiles of release volumes and failure modes (eg 50th percentile release volume from overtopping and 95th percentile release from foundation failure). The ability of the model to apply simple inputs of topography, outflow volume, location of breach and three flow parameters to rapidly generate inundation extents makes ProDF a promising tool for tailings dam risk assessment applications.

CONCLUSIONS

This paper describes the ability of ProDF, a tool developed to model debris flows, to estimate the runout of tailings dam breaches. The model has been adapted to be paired with the probability distribution curves for failure release volumes, which were generated with the screening-level tool developed by Bale *et al* (2024), allowing for 100 inundation extents to be generated in approximately 5 mins.

The model's flow resistance parameters have been calibrated using Feijao and Merriespruit historical failures. For both failures, a flow resistance parameter of 10 $s^{-1}m\,m^{-0.5}$ was used. The calibrated yield strengths were 46 Pa for the Feijao failure and 730 Pa for the Merriespruit failure. With the calibrated values of flow resistance and yield stress, the model estimated the total inundation extent of these two failures with a percent overlap to the historical extents of 75 per cent for Feijao and 97 per cent for Merriespruit. The model's sensitivity to yield stress highlights the need for further calibration with additional historical failures and tailing types. This will enhance the model's applicability to hypothetical future scenarios, where calibration is not possible.

Ultimately, the integration of this screening-level tool with other exposure models, such as those estimating life loss and economic loss, will enable a comprehensive risk assessment framework. This approach can accelerate adoption of risk-informed decision-making by streamlining the means of risk characterisation, helping practitioners identify risk drivers at a facility or within a portfolio of facilities, to prioritise efforts to mitigate potential impacts related to tailings dam breaches.

ACKNOWLEDGEMENTS

The authors would like to acknowledge BGC Engineering Inc., who provided funding through Research and Development to complete this work, Sophia Zubrycky for introducing the ProDF model and Matt Lato for his endless support and mentorship.

REFERENCES

Adria, D A M, Ghahramani, N, Rana, N M, Violeta, M, McDougall, S, Evans, S G and Take, W A, 2023. Insights from the Compilation and Critical Assessment of breach and Runout Characteristics from Historical Tailings Dam Failures: Implications for Numerical Modelling, *Mine Water and the Environment*, 42:650–669. https://doi.org/10.1007/s10230-023-00964-0

Bale, S, Coia, V, Russell, B and Clohan, D, 2024. Probability Distributions of Tailings Dam Breach Volumes by Failure Mode as Part of a Risk Screening-Level Tool, in *Proceedings Tailings and Mine Waste 2024*, pp 897–909.

Canadian Dam Association (CDA), 2013. Dam Safety Guidelines 2007 (2013 edition).

Canadian Dam Association (CDA), 2019. Technical Bulletin: Application of Dam Safety Guidelines to Mining Dams.

de Lima, R E, de Lima Picanço, J, da Silva, A F and Acordes, F A, 2020. An anthropogenic flow type gravitational mass movement: the Córrego do Feijão tailings dam disaster, Brumadinho, Brazil, *Landslides*, 17(12):2895–2906.

Ghahramani, N, Chen, H J, Clohan, D, Liu, S, Llano-Serna, M, Rana, N M, McDougall, S, Evans, S G and Take, W A, 2022. A benchmarking study of four numerical runout models for the simulation of tailings flows, *Sci Total Environ*, 827:154245. https://doi.org/10.1016/j.scitotenv.2022.154245

Global Tailings Review (GTR), 2020. Global Industry Standard on Tailings Management (GISTM) [online], Global Tailings Review. Available from: <https://globaltailingsreview.org/global-industry-standard/>

Gorr, A N, McGuire, L A, Youberg, A M and Rengers, F K, 2022. A progressive flow-routing model for rapid assessment of debris-flow inundation, *Landslides*, 19:2055-2073. https://doi.org/10.1007/s10346-022-01890-y

ICOLD, 2022. Tailings Dam Safety, Bulletin 194.

Japan Aerospace Exploration Agency (JAEA), 1997. AWD30 DSM data map [Database]. Available from: <http://www.eorc.jaxa.jp/ALOS/en/aw3d30/data/index.htm>

Porter, M, Russell, B and Watts, B, 2023. Using Statistics and Judgement to Generate Screening-level Estimates of Tailings Dam Risk, in *Proceedings Tailings and Mine Waste 2023*, pp 1217–1229.

Rotta, L H S, Alcantara, E, Park, E, Negri, R G, Lin, Y N, Bernardo, N, Mendes, T S G and Filho, C R S, 2020. The 2019 Brumadinho tailings dam collapse: possible cause and impacts of the worst human and environmental disaster in Brazil, *Int J Appl Earth Obs Geoinf*, 90:102119. https://doi.org/10.1016/j.jag.2020.102119

Santamarina, J C, Torres-Cruz, L A and Bachus, R C, 2019. Why coal ash and tailings dam disasters occur, *Sci*, 364(6440):526–528.

Wagener, F, 1997. The Merriespruit slimes dam failure: overview and lessons learned, *J South Afr Inst Civ Eng*, 39(3):11–15.

Wise Uranium Project, 2021. Chronology of major tailings dam failures viewed March 2021. Available from: <https://www.wise-uranium.org/mdaf.html>

Impact of topography accuracy and resolution on the results of dam breach analysis inundation simulations

V Nell[1], A Ng[2] and A Ellithorpe[3]

1. VP Products and Solutions, PhotoSat, Vancouver V4E 1T2, Canada.
   Email: veronique.nell@photosat.ca
2. Head of Production, PhotoSat, Vancouver V4E 1T2, Canada. Email: allan.ng@photosat.ca
3. Product Manager, PhotoSat, Vancouver V4E 1T2, Canada.
   Email: angela.ellithorpe@photosat.ca

ABSTRACT

Dam breach analyses (DBA) are crucial for ensuring the safety and environmental protection of tailings storage facilities (TSF) by predicting potential inundation areas and informing emergency response plans.

Topography of downstream areas is a primary input to DBA inundation simulations. This paper examines how the resolution and accuracy of topographic data impact the results of these simulations, demonstrating that lower resolution data does not necessarily lead to more conservative outcomes, contrary to previous assumptions. Additionally, it highlights the ground features that become visible only with high-resolution input data, underscoring the advantages of using higher resolution data in dam breach analyses.

INTRODUCTION

A downstream topographic surface is a crucial input to a Dam Breach Analysis (DBA) inundation simulation. These areas are large, often stretching hundreds of kilometres downstream. Previous studies have suggested that lower resolution topography would result in 'larger run-out areas, greater flow volumes and more rapid flow streams' (Halliday and Arenas, 2019).

The Canadian Dam Association (CDA) bulletin (CDA, 2021) suggests that the 'best possible' topographic data should be used. Available data ranges from low-resolution, low-accuracy Shuttle Radar Topography Mission (SRTM) data to high-resolution, high-accuracy data from LiDAR or high-resolution optical satellites. The SRTM data is publicly available and processed topography from this data comes at a much lower price than topography produced from LiDAR or high-resolution satellite imagery.

When choosing topographic data for an inundation study, the following factors should be considered:

- Topography resolution – distance between each point in the elevation grid.
- Vertical and horizontal accuracy – how close to reality is each point in the grid.
- Area – the topographic surface needs to encompass the full area that would be affected by the flow or tailings discharge.
- Date – archived data is often available but the topographic surface needs be representative of the current state of the downstream area.

PhotoSat examined the impact of resolution and accuracy of the topographic data to determine how variation would affect the results of the DBA inundation simulation. We compared:

- maximum depth of inundation
- total length of inundation path
- extent of inundation area
- flood time arrival
- depth-velocity product.

SITE SELECTION

For this study, PhotoSat chose a site using the following criteria:

- Data availability – both SRTM and archive high-resolution optical satellite imagery were available with collection dates that were representative of current ground conditions.

- Terrain variety – the river channel varied in width and included smaller channels branching off from the main river channel.

- Land-use variety – the hypothetical inundation area includes urban, agricultural, and rural areas.

Figure 1 shows an orthorectified photo of the area. It is important to note that the area used for this comparison does not lie in a potential dam breach inundation area and the results are only intended to illustrate the impact that topography resolution and accuracy can have on the results of a simulation.

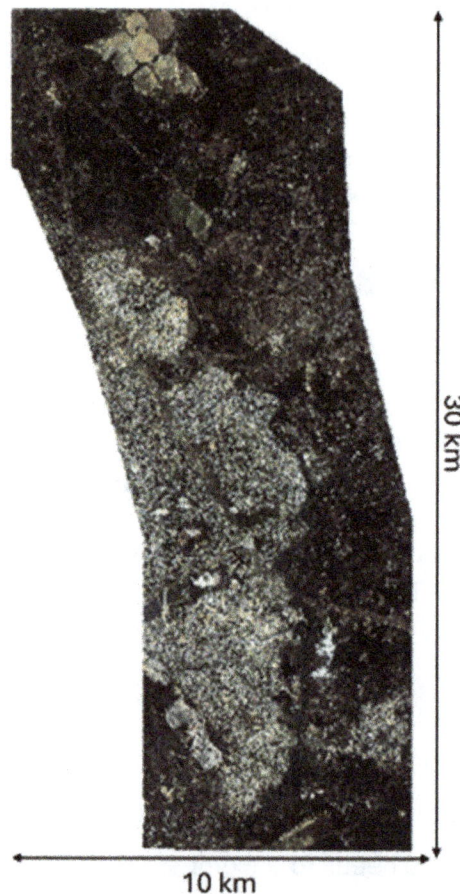

FIG 1 – Study area (WorldView imagery, courtesy of MAXAR).

COMPARISON OF TOPOGRAPHY

Resolution

The resolution of a topographic surface is the distance between each point in the survey grid. In other words, it is the horizontal grid spacing of the elevation survey grid. It is important to consider that higher resolution does not necessarily correlate with higher accuracy.

Topographic survey resolution and image resolution are often misinterpreted as being one and the same but that is not the case. The image resolution refers to the size of the orthophoto's pixels. For example, a 1 m survey can be derived from a 30 cm of 50 cm satellite photo (stereo pair).

SRTM topography typically has a resolution of 10 m to 30 m (30 m in the geographic region chosen for this study), whereas high-resolution surveys can have resolutions of 1 m or better.

Figure 2 illustrates how different survey grid resolutions approximate the true ground surface of a hypothetical surface. Since there is a lower density of survey points, the topographic surface is less detailed and is therefore a coarser approximation of the true ground surface.

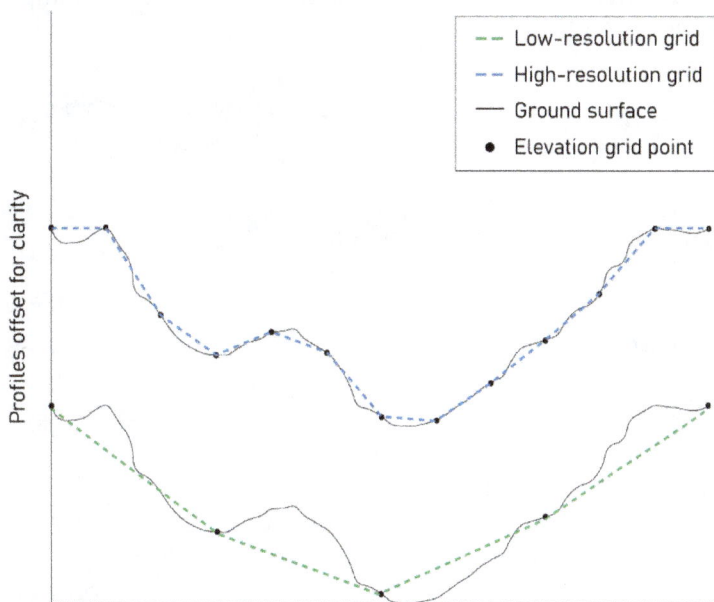

FIG 2 – High-resolution versus low-resolution topographic surface.

In the context of a DBA simulation, the resolution of the topographic surface will directly impact the cell size of the flow model as the cell size must be equal to or greater than the resolution of the survey grid. As shown in Figure 3, larger cell sizes leave gaps that cannot be filled in the simulation whereas smaller cell sizes allow the flow to follow the surface of the inundated area more closely and therefor lead to a more plausible representation of what might occur in a real-world event.

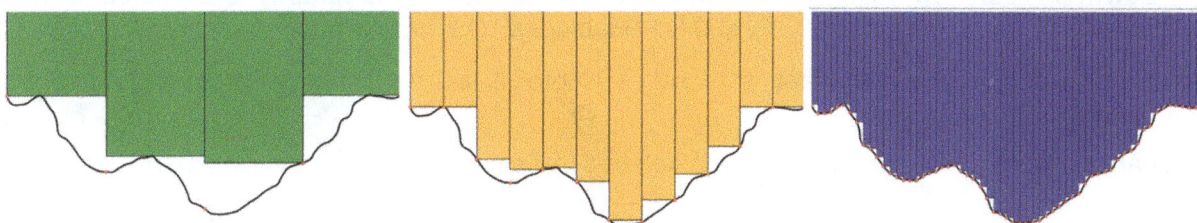

FIG 3 – Cell size for a low, medium and high-resolution elevation grid.

Note that high-resolution topographic surfaces and smaller cell sizes lead to larger file sizes which will increase the time needed to run the simulation.

Accuracy

The accuracy of a topographic surface is defined both horizontally and vertically. It should be reported using standard methodology such as the accuracy standards published by the United States Geological Survey (USGS).

PhotoSat produces high-accuracy surveys from satellite imagery with horizontal and vertical accuracies better than 0.2 m Root Mean Square Error (RMSE) and 0.3 m Linear Error with 90 per cent confidence (LE90). These accuracies are reported using ASPRS and USGS standards (Positional Accuracy Standards for Digital Geospatial Data) and consistent across the survey area. They are supported by published accuracy reports (Mitchell, 2016; ASPRS, 2014).

Lower accuracy topographic surfaces such as SRTM or other off-the-shelf Digital Terrain Models (DTM) often have inconsistent accuracies across the area and can contain errors including:

- Offsets: when the whole elevation grid appears to be shifted in a specific direction.

- Tilts: when false slopes appear in the surface.

- Vertical errors: when small, localised errors create false features in the terrain.

Figure 4 illustrates, using a hypothetical profile, how a vertical error can impact the channel depth and shape in the topographic surface.

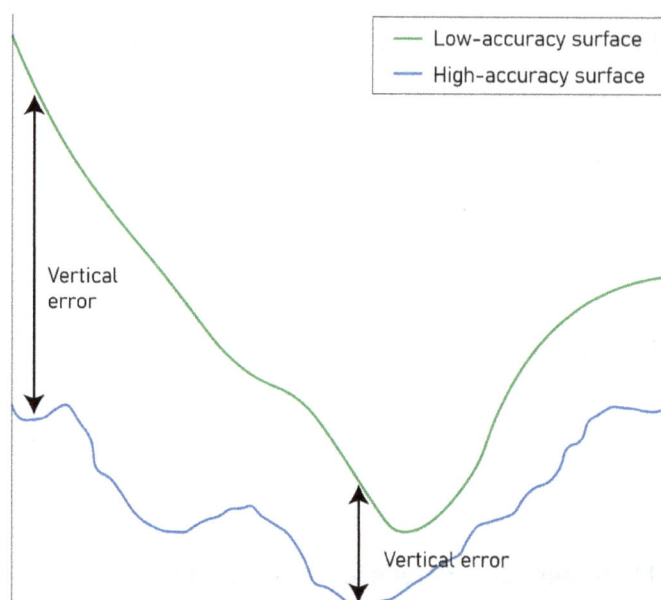

FIG 4 – Effect of vertical error on channel depth.

Data sets used in this study

For this comparison, we chose:

- Low-resolution, low-accuracy data set: SRTM data. SRTM topography is available for most areas in the world between 60°N and 60°S latitudes. The data is publicly available and free.

- High-resolution, high-accuracy data set: a PhotoSat survey produced from high-resolution optical satellite imagery.

The characteristics of the two data sets are detailed in Table 1.

TABLE 1

Topography characteristics.

Characteristics	Low resolution, low accuracy	High resolution, high accuracy
Source	Shuttle Radar Topography Mission (SRTM)	High-resolution satellite imagery (WorldView, courtesy of MAXAR) processed through PhotoSat's proprietary geophysical processing algorithms
Elevation grid spacing (topography resolution)	30 m	1 m
Vertical accuracy	2 m–3 m	< 0.2 m
(RMSE, USGS reporting standard)	2000 (various dates)	August 10 and 27, 2019

Figures 5 and 6 show a comparison of the SRTM topography and the PhotoSat survey over the same area.

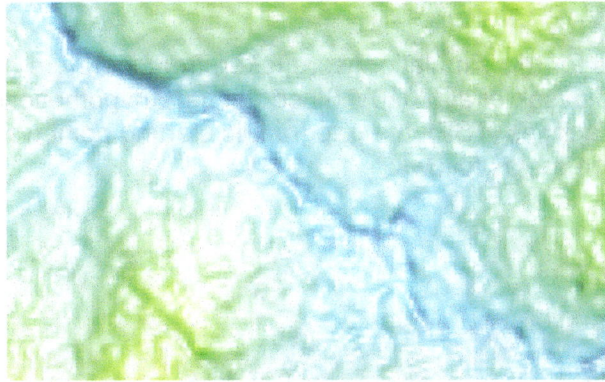

FIG 5 – Example of SRTM topographic surface.

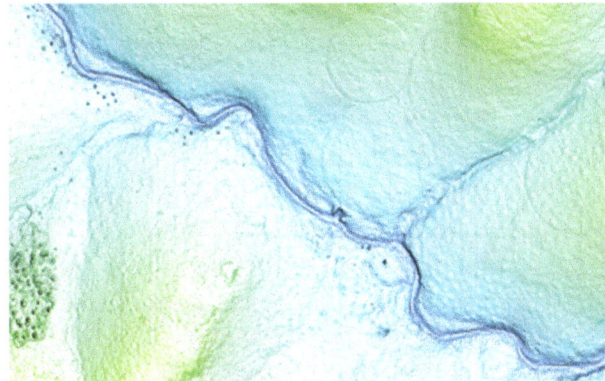

FIG 6 – Example of PhotoSat high-accuracy survey.

Figure 7 illustrates how the river channel width and depth are affected by the resolution and accuracy of the topographic surface. In both cases, the surface represents the top of the water surface as it was at time of capture, not the below-water ground surface. However, the high-resolution, high-accuracy PhotoSat survey shows a more detailed representation of the shape of the channel. The narrower, deeper channel in the high-resolution topographic surface will constrain the flow to this narrower channel. The elevation offsets of up to 6 m will also have an impact on the results of the simulation.

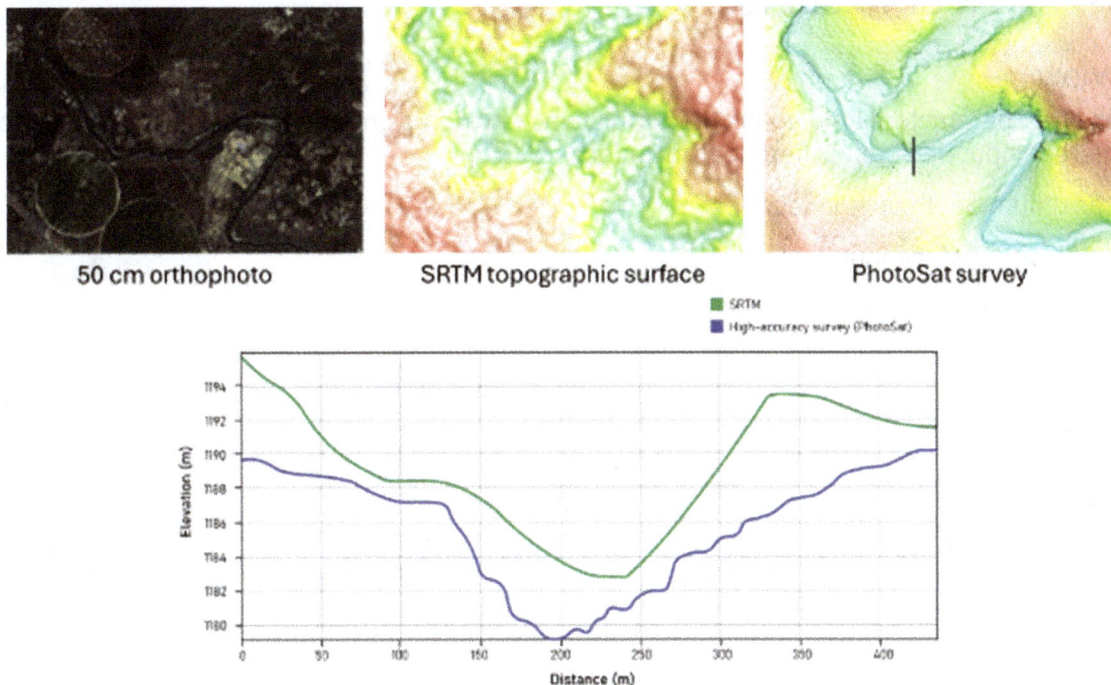

50 cm orthophoto   SRTM topographic surface   PhotoSat survey

FIG 7 – Comparison of topographic surfaces and channel profiles (at black line).

SIMULATION INPUTS

The input parameters used for this simulation were taken from literature and based on a real-world dam breach scenario (Petkovšek *et al*, 2020). Since this is a hypothetical inundation area with no upstream tailings storage facilities (TSF), the geographic starting point was arbitrarily selected. The parameters of the simulation are as follows:

- Duration: 7 hrs.

- Maximum flow: 2500 m³/s.

- Manning's roughness: 0.06.

- Riverbed conditions: dry.

FIG 8 – Hydrograph used for the simulation.

The inundation simulations were run using the Hydronia RiverFlow2D software. Since the resolution of SRTM is 30 m, an initial simulation cell size of 30 m was chosen for the comparison. The high-accuracy survey has a resolution of 1 m which allowed simulations with smaller cell size to be done and compared.

For each topographic surface, three simulations were run using three different values for concentration by volume (Cv); 0.25, 0.38 and 0.5.

COMPARISON OF RESULTS

Overview

With all other parameters held equal, the simulations showed significantly differences results for the two different input data set. Differences included:

- maximum depth of inundation

- total length of the inundation path

- extent of the inundated area

- flood arrival time.

Maximum depth of inundation

Figures 9 and 10 show the difference in predicted inundation depth at different points in the downstream area. The depth difference between the SRTM and high-resolution simulations was as much as 6 m with a depth of 9 m with the high-resolution data set and 15 m with the low-resolution data set. A possible explanation for these large differences is localised vertical errors in the SRTM data resulting in artificially created holes in the topographic surface.

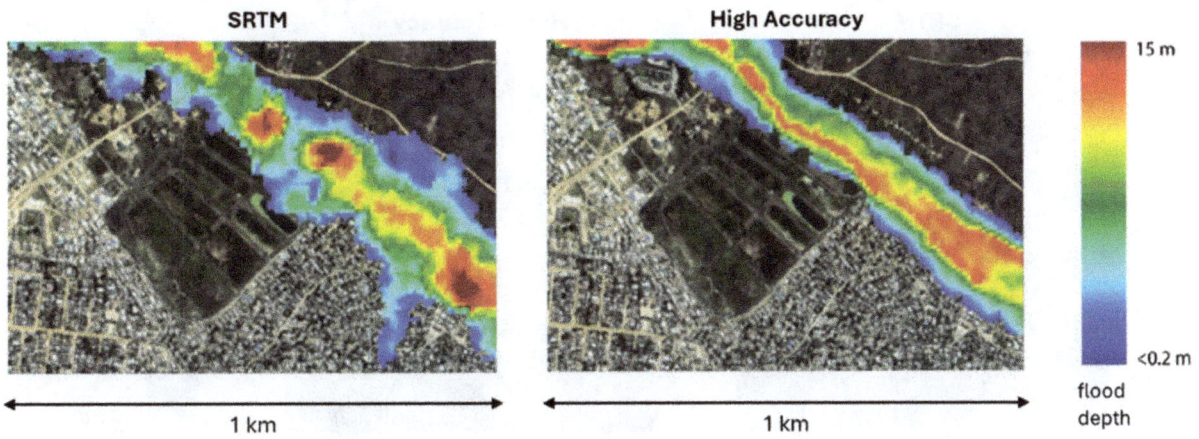

FIG 9 – Comparison of predicted maximum flow depth, urban area (Cv = 0.38).

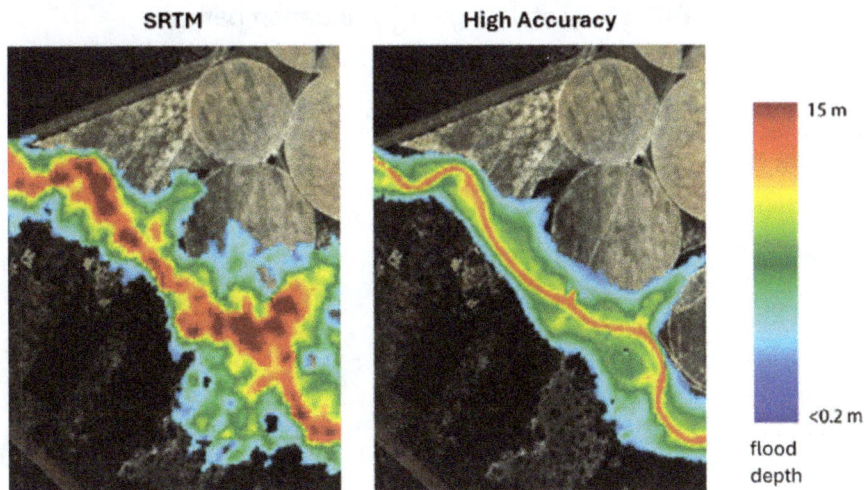

FIG 10 – Comparison of predicted maximum flow depth, flat area (Cv = 0.25).

Total length of inundation path

The results of the simulations showed significant differences in the total flood length. Figure 11 shows the end of the tailings run-out area in the simulation using the least flowable material (Cv = 0.5). The simulations showed that the inundation path extended an additional 1.5 km down the channel when using the high-resolution, high-accuracy topographic surface, which contradicts the thought that a lower resolution data set would lead to a longer flood length. This may be caused by artificial 'holes' in the low-resolution SRTM topographic surface that get 'filled' with the simulated outflow leading to less outflow continuing downstream.

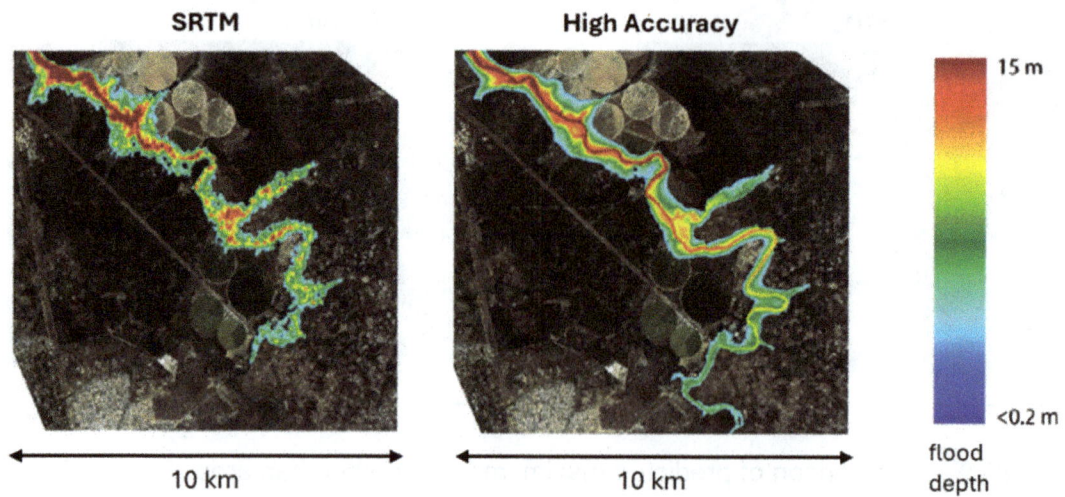

FIG 11 – Total length of inundation path.

Extent of inundation area

The simulations showed significant differences in the location and size of the area affected by the inundation. Figures 12 and 13 illustrates the difference in zones affected by the inundation in an urban area.

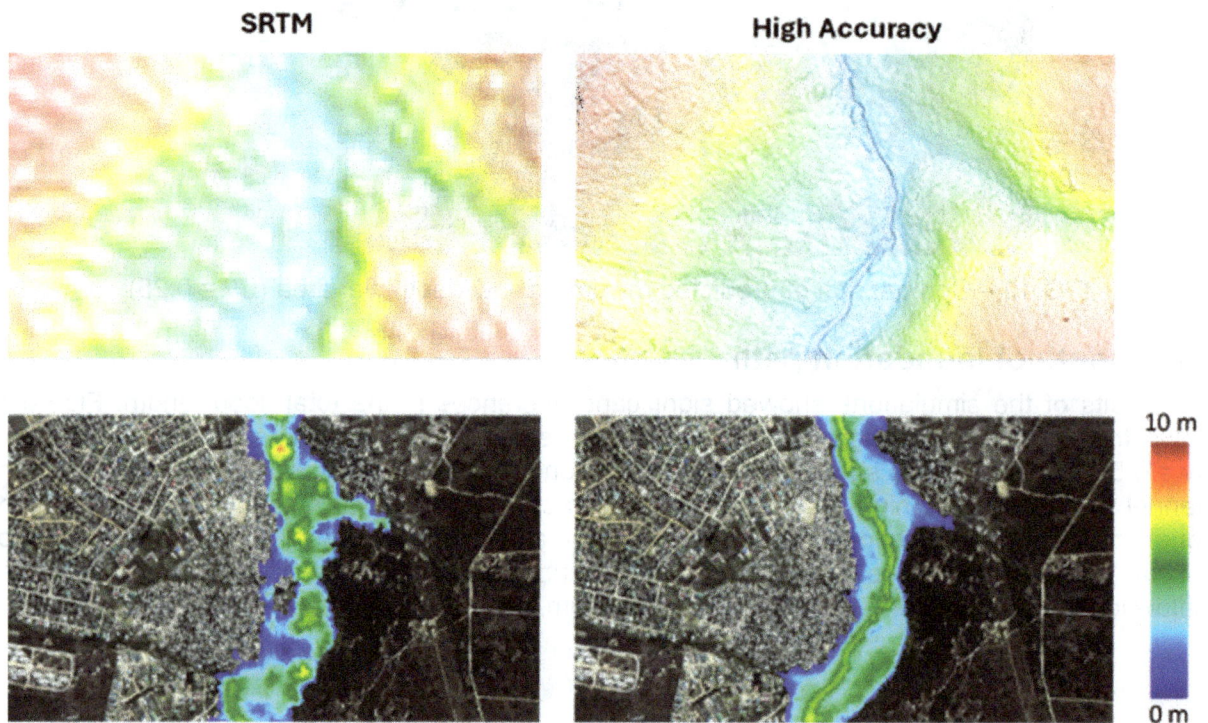

FIG 12 – SRTM versus high-accuracy area and depth.

Figure 13 shows the areas only affected when using the SRTM data set in magenta, and the areas only affected when using the high-accuracy data set in blue.

FIG 13 – Differences in affected areas.

The structure counting shown in Figure 14 illustrates how the difference in results could lead to the misguided emergency response planning and the misallocation of emergency response efforts.

FIG 14 – Structures in affected areas.

Flood time arrival

Using a Cv of 0.25, the difference in flood arrival time between the low-resolution data set and the high-resolution data set was 7 hrs at the end of predicted inundation channel, with the low-resolution SRTM simulation arriving after the high-resolution simulation. The SRTM simulation consistently arrived after the high-resolution data simulation over the length of the channel, including in urban areas. This contradicts the previously accepted idea the lower resolution topographic data would lead to a faster flood arrival time. This may have a significant impact on emergency planning.

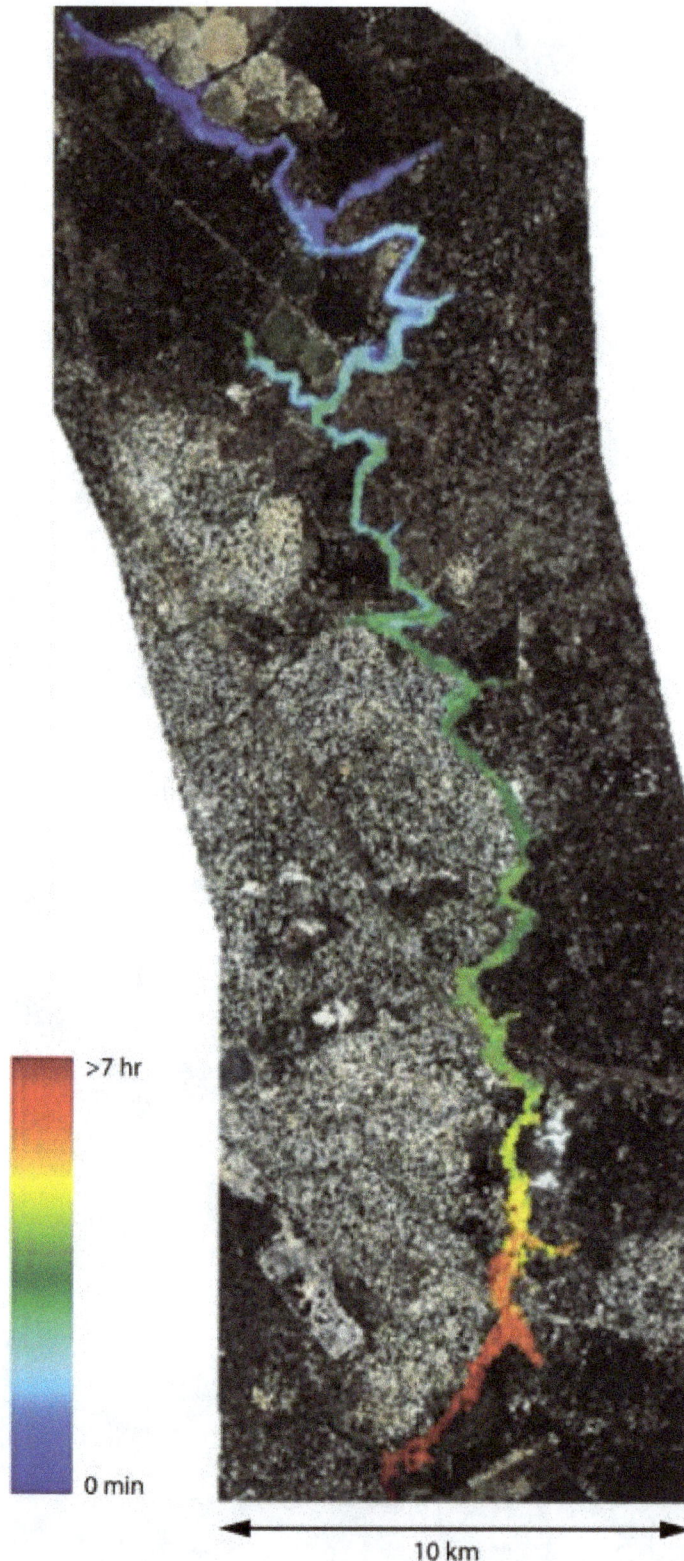

FIG 15 – Difference in arrival time (high-resolution input leading to a faster arrival time).

CONCLUSION

This study showed that both resolution and accuracy of the input topography had significant implications for the results of a DBA inundation simulation. The resolution and accuracy of the topography led to significant differences in maximum depth of the predicted flow, total length of the inundation channel, extent of the inundation area and flood arrival time. These differences have important implications for emergency planning, emergency response, and environmental remediation planning.

The findings of this study align with the CDA's (Canadian Dam Association, 2021) recommendation to use the highest quality topography available. Using low quality topography runs the risk of introducing significant errors into the inundation simulation.

ACKNOWLEDGEMENTS

PhotoSat would like to thank Dr Violeta Martin of Knight Piésold for her support, and particularly her advice in selecting suitable parameters for the simulations.

REFERENCES

ASPRS, 2014. Positional Accuracy Standards for Digital Geospatial Data, edition 1, version 1.0, November 2014.

Canadian Dam Association (CDA), 2021. Technical Bulletin: Tailings Dam Breach Analysis.

Halliday, A and Arenas, A, 2019. Impacts of Topography Quality on Dam Breach Assessment, Golder Associates.

Mitchell, G, 2016. Accuracy assessment – WorldView 3, Garlock Fault, 1GCP, 13 cm RMS E, 22 cm LE90.

Petkovšek, G, Hassan, M, Lumbroso, D and Roca, M, 2020. A Two-Fluid Simulation of Tailings Dam Breaching, *Mine Water and the Environment,* 40:1–15. https://doi.org/10.1007/s10230-020-00717-3

Identifying the best fit distributions for tailings dam breach parameters

L Newby[1] and K Warman[2]

1. Principal Tailings Engineer, WSP, Perth WA 6000. Email: leilani.newby@wsp.com
2. Water Resources Engineer, WSP, Brisbane Qld 4000. Email: kyle.warman@wsp.com

ABSTRACT

The adequacy of tailings dam failure impact assessments is significantly affected by the accuracy of the predicted breach runout hydrograph used when modelling the failure. This often results in overly optimistic or pessimistic inundation maps. The state of practice for tailings dam breach assessment (TDBA) is thus evolving into a risk-based approach, which considers the potential magnitude of the breach hydrograph, based on a sensitivity of the key input parameters, to confirm the credibility of the dam breach model.

The physical process of tailings dam breaches is multifaceted, depending both on the dam failure scenario and the failure modes. This study focuses on the failure modes that may lead to a loss of containment due to a physical breach of the tailings dam, namely collapse and overtopping. The Canadian Dam Association has defined four cases of TDBA, based on the flow of the supernatant pond and the flow of tailings due to liquefaction (Canadian Dam Association, 2021). Both the failure mode and TDBA case should be considered in a risk-based approach.

The main objective of this study is to identify the best-fit probability distribution of tailings dam breach parameters using a combination of tailings dam failure and earthen embankment dam failure databases. Four parameters were selected for evaluation: (i) side slope of trapezoidal breach, (ii) height of breach, (iii) bottom width of breach, and (iv) formation time. To identify the best-fit distributions, a correlation and comparative analysis is undertaken for various combinations of TDBA Cases and failure modes.

The output from this study can be used to perform a probabilistic dam breach analysis, using software which can perform a Monte Carlo simulation to randomly sample various breach parameters about the best-fit distribution. This will contribute to the development of probabilistic dam breach inundation maps instead of deterministic mapping.

INTRODUCTION

A TSF failure is defined as a physical breach resulting in the uncontrolled release of stored tailings and water. Currently, there are few published guidelines available pertaining specifically to TSF breach analyses, with most of the empirical equations and dam breach formation models in use today being based on historical failure data of earthen embankment dams (Fontaine and Martin, 2015). The methodology followed for a TSF breach analysis is thus generally based on the aforementioned equations, which calculates the breach size and breach formation time presented in the form of a breach outflow hydrograph.

The outflow hydrograph resulting from a breach is used to determine the extent of inundation, peak flow arrival times, velocities and depths at various downstream locations (Froehlich, 1987). Generation of a dam breach outflow hydrograph requires an estimation of the time of formation (tf) and the peak flow (Qpeak) of the breach. These are determined from the breach parameters, as illustrated in Figure 1.

The accuracy of the predicted breach runout hydrograph used when modelling the failure significantly affects tailings dam failure impact assessments. Supporting literature indicates that a wide range of results can be expected based on the empirical equations used (Davies and Martin, 2000b). This, combined with the large degree of uncertainty associated with applying these deterministic methods to TSFs, can often result in overly optimistic or pessimistic inundation maps, as illustrated in this paper.

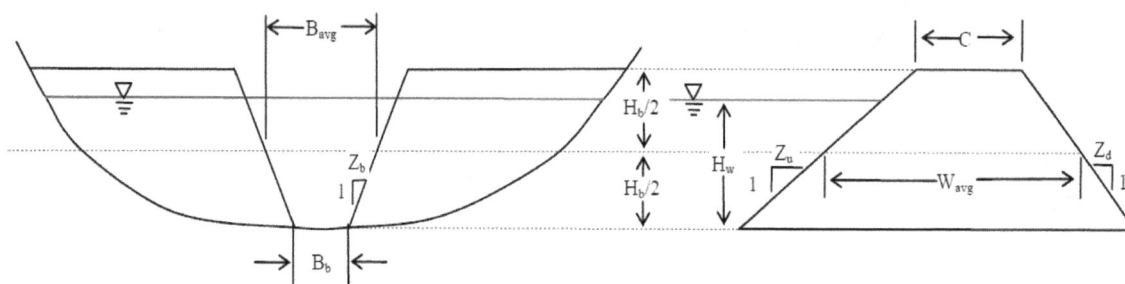

FIG 1 – Indicative breach parameters (State of Colorado Department of Natural Resources Division of Water Resources, 2010).

A STATISTICAL COMPARISON OF EMPIRICAL EQUATIONS

The main objective of this study is to identify the best-fit probability distribution of tailings dam breach parameters using a combination of tailings dam failure and earthen embankment dam failure databases. For simplicity purposes, best fit probability distributions are limited to the following statistical distributions: uniform; normal; log-normal and triangular (Goodell, 2019).

Four parameters were selected for evaluation. To account for variations in the size of the dams considered, it was necessary to consider parameter ratios in some instances. These ratios are considered to be representative of the standalone parameter. A summary of the parameters selected, and the corresponding parameter is presented in Table 1.

TABLE 1

Dam breach parameters for evaluation.

Database parameter	Analysed parameter	Unit
Side slope of trapezoidal breach (Zb)	Side slope of trapezoidal breach (Zb)	m/m
Height of breach (H_B)	Breach to dam height ratio (H_B/H_D)	m/m
Formation time (t_f)	Formation time (t_f)	hours
Bottom breach width B_B	Bottom breach width to dam height ratio (B_B/H_D)	m/m

As an initial step, the probability distribution of earth fill dam breach parameters was evaluated using a dam failure database of 3861 observations (Bernard-Garcia and Mahdi, 2020). To identify the best-fit distributions, a correlation analysis was conducted, followed by a comparative analysis per failure mode. The following three failure modes were considered based on the data available: overtopping, internal erosion and 'other/undefined'.

A similar approach was followed to evaluate the probabilistic distribution fitting of tailings dam breach parameters considering a tailings dam breach database of 26 events (Bowker and Chambers, 2017). Based on the available data, the breach parameters were grouped into the following failure modes:

- Overtopping: the breach starts at the crest of the TSF, usually due to either poor operation of the TSF or an extreme flood event.

- Collapse: the breach originates at some point below the crest of the TSF due to either piping, erosion, or some other form of wall instability.

- Liquefaction: the breach is caused by inadequate undrained shear strength of the embankment and contained tailings material.

Breach to dam height ratio (H_B/H_D)

The analyses of the breach to dam height ratio considered a database of 328 parameters, of which 313 were applicable to earthen embankment water retaining dams and 15 applicable to TSFs (of these, three correspond to downstream construction and the remaining to upstream construction methods). The data distribution according to each failure mode is presented in Tables 2 to 4.

TABLE 2

Breach to dam height ratio: water dams.

Parameter	Overtopping	Internal erosion	Other	All
Minium (a)	0.04	0.05	0.02	0.02
Maximum (c)	5.33	2.14	1.38	5.33
Mean (μ)	0.92	0.94	0.78	0.91
Standard Deviation (σ)	0.42	0.30	0.37	0.39
Log Mean (μ)	0.82	0.85	0.68	0.81
Log Standard Deviation (σ)	1.71	1.70	2.21	1.82
Mode (b)	1.00	1.00	1.00	1.00
Count	202.00	85.00	26.00	313.00

TABLE 3

Breach to dam height ratio: TSFs.

Parameter	Overtopping	Liquefaction	All
Minium (a)	0.68	0.29	0.29
Maximum (c)	0.87	1.00	1.00
Mean (μ)	0.75	0.72	0.74
Standard Deviation (σ)	0.08	0.29	0.24
Log Mean (μ)	0.74	0.65	0.69
Log Standard Deviation (σ)	1.11	1.61	1.50
Mode (b)	0.75	1.00	1.00
Count	4.00	10.00	15.00

TABLE 4

Breach to dam height ratio: Combination of TSFs and Water dams.

Parameter	Overtopping	Other	All
Minium (a)	0.04	0.02	0.02
Maximum (c)	5.33	2.14	5.33
Mean (μ)	0.91	0.89	0.90
Standard Deviation (σ)	0.42	0.32	0.38
Log Mean (μ)	0.82	0.77	0.80
Log Standard Deviation (σ)	1.70	1.99	1.81
Mode (b)	1.00	1.00	1.00
Count	206.00	121.00	328.00

A frequency distribution plot, considering a frequency interval of 0.02, is presented graphically in Figure 2. As indicated graphically, the ratio of breach to dam height corresponds to a value of 1.0 for approximately 50 per cent of instances for water dams whereas TSFs have slightly more variance with only 20 per cent of instances resulting in a breach to dam height ratio of 1.0. It is noted that in some instances of water dam breaches, the ratio of breach height to dam height exceeds 1.0. This could either be due to erosion of the downstream embankment or errors in measuring the historical data.

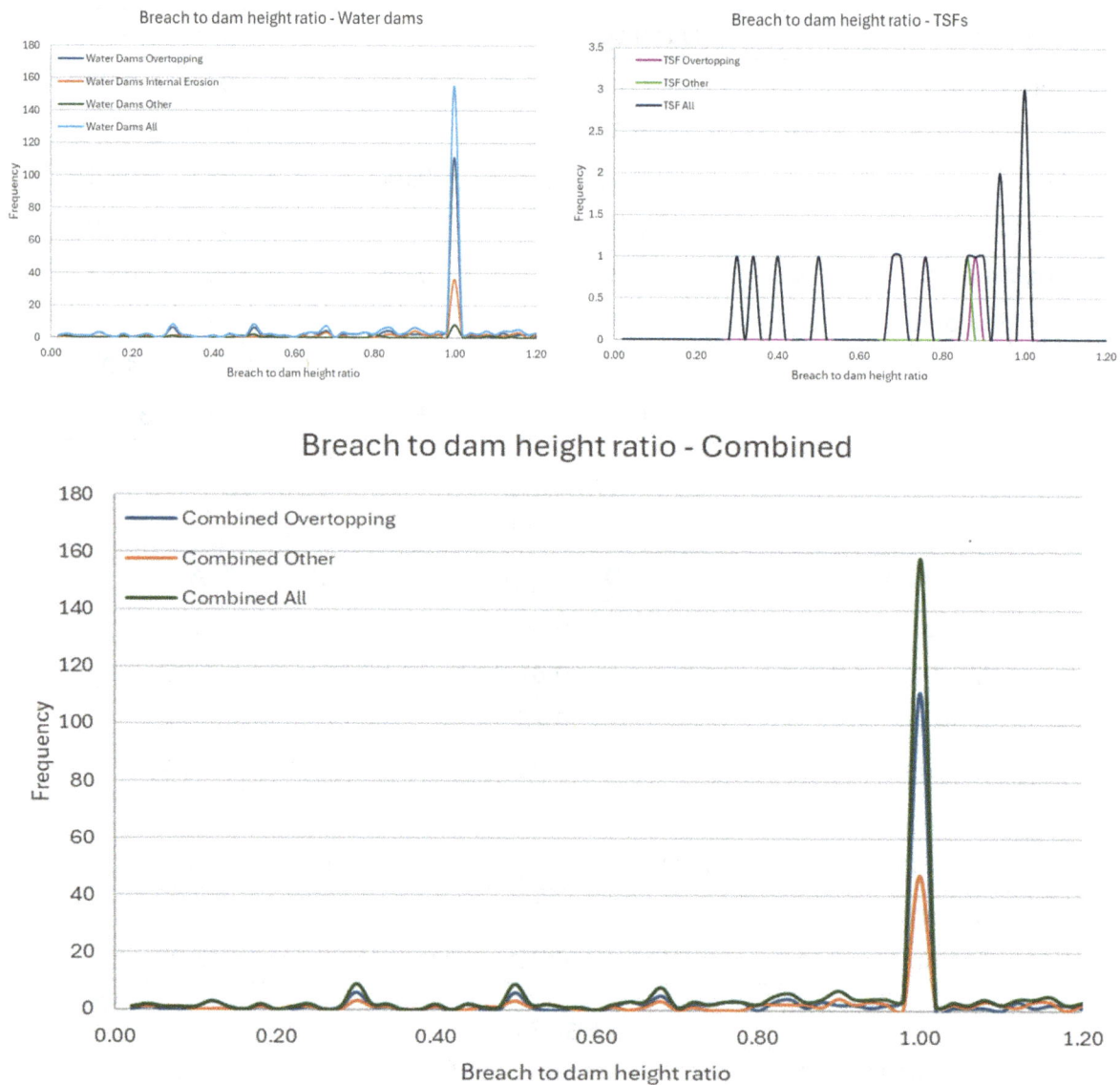

FIG 2 – Frequency distribution of breach to dam height ratios.

Based on the graphs, a uniform distribution can be discounted. The data was therefore evaluated for the best fit of normal, log normal and triangular distributions. A presentation of this evaluation according to the coefficient of best fit is presented in Table 5.

The analysis indicates that breach to dam height ratio of a water dam follows a log-normal distribution in all instances, regardless of failure mode, whereas all instances of a TSF failure mode correlate more strongly to a triangular distribution. However, since triangular distributions tend to be suitable for a scenario with limited data available, as is the case for the TSF failure database, an argument could be made that once more data becomes available the breach to dam height ratio for TSF failures shall similarly trend towards a log-normal distribution.

TABLE 5
Coefficient of best fit for ratio of breach to dam height.

Parameter	Normal distribution	Log-normal distribution	Triangular distribution
Water Dams – Overtopping	0.37	0.59	0.04
Water Dams – Internal Erosion	0.41	0.57	0.04
Water Dams – Other	0.22	0.51	0.03
Water Dams – All	0.37	0.56	0.04
TSF – Overtopping	0.04	0.12	0.56
TSF – Liquefaction	0.08	0.75	0.98
TSF – All	0.08	0.57	0.97
Combined – Overtopping	0.38	0.60	0.04
Combined – Other	0.34	0.50	0.04
Combined – All	0.37	0.58	0.04

The exception to this rationale would be the case of an overtopping TSF failure. In the instances evaluated, the tailings within the TSF were classified as non-liquefiable (Case 1B). In these instances, the breach height is less likely to extend the full height of the dam and is more driven by the volume of the pool and the height of the pool and possibly saturated tailings retained by the TSF. Based on the limited information currently available, both a uniform distribution (with a minimum of 0.68 H_B/H_D and a maximum of 0.87 H_B/H_D) as well as a triangular distribution (with a mode of 0.75 H_B/H_D) would be suitable in this instance.

Bottom breach width to dam height ratio

The analyses of the bottom breach width dam height ratio were completed as per the above, considering a database of 127 parameters, of which 112 were applicable to earthen embankment water retaining dams and 15 applicable to TSFs (of these, three correspond to downstream construction and the remaining to upstream construction methods). The result of the analysis is presented in Table 6, indicating that a log-normal distribution is the best fit in all instances.

TABLE 6

Coefficient of best fit for side slope of trapezoidal breach.

Parameter	Normal distribution	Log-normal distribution	Triangular distribution
Water Dams – Overtopping	0.28	0.87	0.03
Water Dams – Internal Erosion	0.42	0.86	0.00
Water Dams – Other	0.67	0.60	0.71
Water Dams – All	0.36	0.83	0.03
TSF – Overtopping	0.34	0.84	0.03
TSF – Liquefaction	0.07	0.83	0.30
TSF – All	0.16	0.89	0.25
Combined – Overtopping	0.29	0.89	0.02
Combined – Other	0.35	0.85	0.06
Combined – All	0.34	0.85	0.03

Side slope of trapezoidal breach and time of breach formation

The analyses of the side slope of the trapezoidal breach, assumed to be equal on both sides, considered a database of 181 parameters, of which 170 were applicable to earthen embankment water retaining dams and 11 applicable to TSFs. The time of formation considered a database of 83 water dam parameters and five TSF parameters (liquefaction only, of which two were upstream constructed and three downstream constructed). The results are presented in Tables 7 and 8 respectively.

TABLE 7

Coefficient of best fit for side slope of trapezoidal breach.

Parameter	Normal distribution	Log-normal distribution	Triangular distribution	Uniform distribution
Water Dams – Overtopping	0.01	0.19	0.26	0.07
Water Dams – Internal Erosion	0.43	0.44	0.00	0.33
Water Dams – All	0.15	0.34	0.09	0.25
TSF – Overtopping	0.01	0.03	0.33	0.71
TSF – Liquefaction	0.01	0.02	0.19	0.43
TSF – All	0.04	0.01	0.11	0.17
Combined – Overtopping	0.01	0.18	0.29	0.07
Combined – Other	0.40	0.43	0.00	0.37
Combined – All	0.14	0.33	0.08	0.25

TABLE 8

Coefficient of best fit for time of breach formation.

Parameter	Normal distribution	Log-normal distribution	Triangular distribution	Uniform distribution
Water Dams – Overtopping	0.43	0.34	0.40	0.32
Water Dams – Internal Erosion	0.46	0.31	0.31	0.11
Water Dams – All	0.47	0.40	0.40	0.31
TSF – Liquefaction	0.02	0.16	0.14	0.23
TSF – All	0.02	0.16	0.07	0.47
Combined – All	0.45	0.42	0.39	0.35

In both instances, the data from TSF failures correspond best with a uniform distribution. It is expected that this is as a result of limited data, rather than a true reflection of the best fit probability distribution to the data. Data analysed from the side slope of trapezoidal water dam breaches generally follow a log-normal distribution, and the time of formation of a water dam breach follows a normal distribution. It is recommended that these distributions be adapted in the absence of more data for TSF failures.

CASE STUDY

TSF geometry

The above probability distributions were applied to a hypothetical TSF to examine the variations in the extent of the release volume, using the probabilistic distributions described above compared to traditional deterministic methods.

The TSF under consideration is a valley impoundment dam with an upstream constructed embankment wall, and liquefiable tailings consistent with a 1A failure case. A summary of the geometrical properties of the TSF is presented in Table 9.

TABLE 9

TSF geometric parameters.

Parameter	Value
Volume of impounded tailings	35 000 000 m³
Dam Height	50 m
Crest Length	1500 m
Crest Width	25 m
Crest Slope	1: 3
Total volume of supernatant water	1 500 000 m³

As noted by Davies and Martin (2000a) the key failure mode of upstream constructed embankment TSFs is a static/transient load induced liquefaction failure. Additionally, based on various other current guidelines, it should be assumed that the tailings are liquefiable unless evidence proves otherwise. Thus overtopping, collapse and liquefaction were considered as credible failure modes for the TSF, with a liquefaction failure expecting to result in the most severe impact downstream and thus selected for further analysis.

Release volume

Sensitivity analyses conducted by others on Tailings Storage Facilities (TSF) breach parameters indicate that the largest variation to the extent of inundation is a result of variation in the release volume – ie uncertainty associated with the size of the breach hydrograph. A record of historical TSF failures around the world compiled by Bowker and Chambers (2017) indicates that the volume released represented as a percentage of the total containment volume ranges from 2 per cent to 100 per cent. A sensitivity analysis on the amount of volume released is outside the scope of this paper.

The expected release volume was determined by 3-dimensional modelling of the expected release during a breach based on a conical failure shape. This method takes both the volume of free water and the geometry of the TSF into consideration and will result in different release volumes for breaches located on various flanks of the TSF, compared to deterministic equations which does not necessarily consider TSF geometry.

The conical failure shape was calculated assuming a slope of 3.5 per cent from the toe of the TSF for the final post liquefied residual slope of the tailings. This was correlated with the slope of the natural ground below the TSF, which is assumed to remain intact in case of a breach. As the hypothetical TSF is a valley impoundment, the slope of the natural ground was steeper than the recommended post-failure liquefaction angle, which indicates that should a liquefaction failure occur a significant volume of the liquefied tailings will flow downstream. The expected release volume modelled equates to approximately 28 million m³ (around 75 per cent of the total storage volume). A corresponding stage versus storage curve was used as input to the analysis.

Deterministic breach model

Traditional deterministic methods were used to conduct a preliminary dam breach analysis. The following empirical equations were considered in the analyses:

- Froehlich (1995, 2008, 2016)

- MacDonald and Langridge-Monopolis (US Army Corps of Engineers (USACE), 2014)

- Von Thun and Gillette (1990)

- USBR (Chow, 1959).

The range of hydrographs obtained using the above equations is presented in Figure 3. A breach height of 50 m was assumed, resulting in an average breach width range of 40 m–180 m and a side slope range of 1:0.3–1:3.

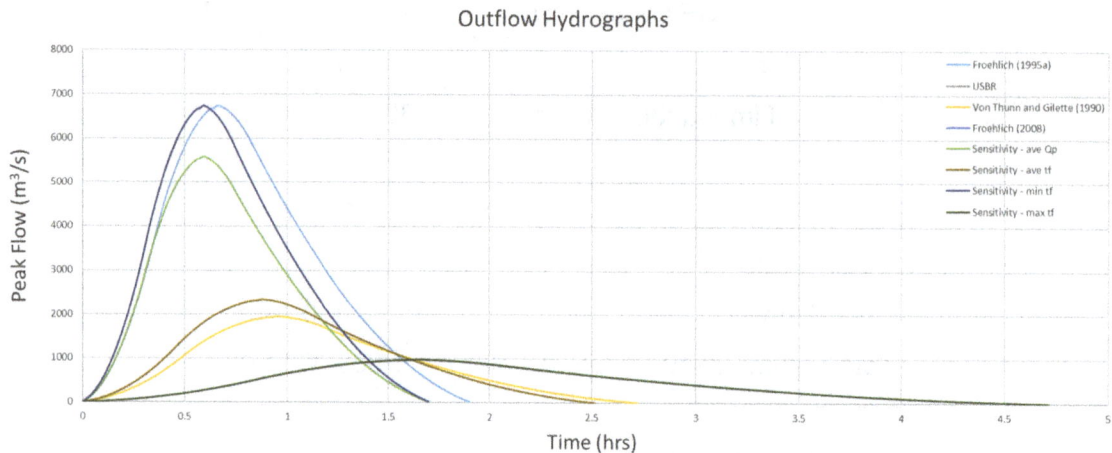

FIG 3 – Breach hydrographs from empirical equations.

An average breach width of 150 m and a side slope ratio of 1:1.5 was selected for the breach analysis. A single mixed breach outflow with an averaged solids concentration was assumed and modelled as a non-Newtonian fluid, considering typical rheological parameters of gold tailings, to determine the downstream inundation extents.

Downstream flood routing of the breached tailings material was completed using the hydraulic modelling software HEC-RAS (Hydrologic Engineering Centre River Analysis System). HEC-RAS considers the viscosity and yield stress relationships, as determined by the rheology of the tailings, and couples water and sediment flow by using full dynamic wave momentum equations to represent non-Newtonian flows. Additionally, the model analyses flow in two dimensions since discharge of tailings material under a breach event is expected to flow in both lateral and longitudinal directions.

Probabilistic breach model

A probabilistic dam breach analyses was performed using the software McBreach. McBreach uses a Monte Carlo simulation to randomly sample various breach parameters about a predefined statistical distribution (Goodell, 2019). This is performed in conjunction with HEC-RAS.

Best fit statistical distributions were chosen based on the range of breach parameters calculated in the preceding section. The following statistical distributions were assigned:

- Breach height: The breach height was assigned a log-normal distribution with a mode of 50 m. Note that an absolute maximum of 52 m was specified, as constrained by the embankment height.

- Bottom width of breach: The assigned distribution is a log-normal distribution with a mode of 150 m. The absolute minimum and maximum values were adjusted to fit the geometrical constraints of the TSF embankment.

- Side slopes of breach: The side slopes of the breach were assigned a log-normal distribution with a mode of 1:1.4.

- Time of formation: The time of formation was assigned a normal distribution.

Results

Statistical convergence of the computed peak discharges was achieved at 8000 realisations. The following exceedance probabilities were determined: 1 per cent; 5 per cent; 50 per cent; and 95 per cent. An exceedance probability inundation map overlaid by a deterministically determined inundation map of the 3D modelled failure is presented in Figure 4.

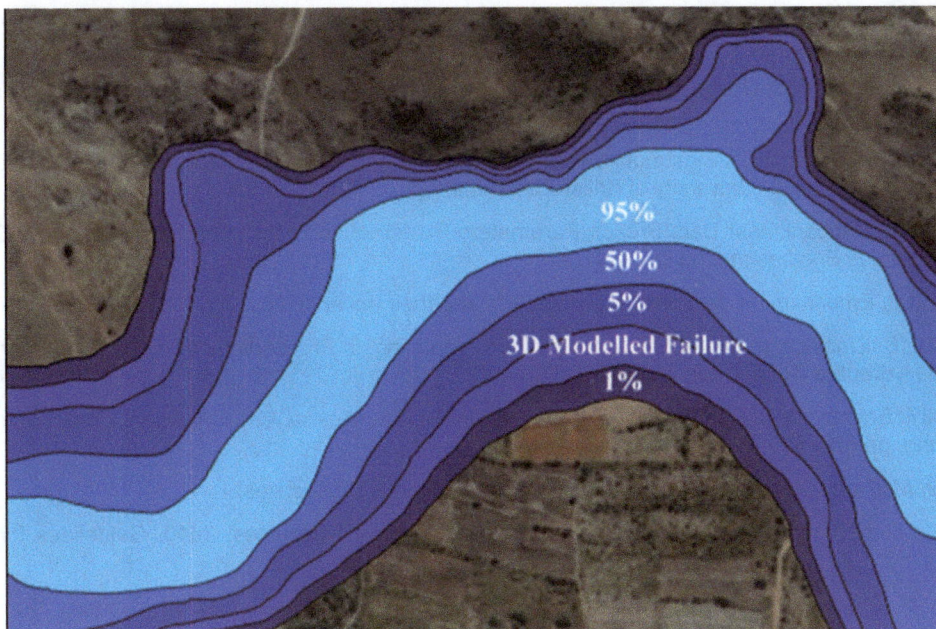

FIG 4 – Comparison inundation map.

Results from the probabilistic analyses indicate that the 3D modelled release volume falls between the 5 per cent and 1 per cent probability of exceedance. It is important to note that the maximum velocity, flow depth and flood arrival times for various locations downstream will change based on the breach hydrograph selected. A noteworthy finding from this case study is that the quickest time to formation and thus the most conservative flood arrival time corresponds to the 95 per cent exceedance probability.

Therefore, instead of selecting the most conservative release volume and corresponding breach outflow hydrograph, as is current practice, a probabilistic analysis enables us to prioritise which parameters are the most important for specific purposes (ie emergency response planning, selection of remediation measures downstream etc) and select a probability of exceedance accordingly.

CONCLUSION – AN ARGUMENT FOR THE PROBABILISTIC DAM BREACH MODEL

The breach hydrograph is typically determined using deterministically computed breach parameters. Due to uncertainty regarding these parameters, it is common practice to assume the most conservative scenario which often results in large inundation extents and poor input for decision-making.

To account for these uncertainties a probabilistic approach to TSF breach analysis is required. The state of practice for tailings dam breach assessment should thus evolve into a risk-based approach, which considers the potential magnitude of the breach hydrograph, based on a sensitivity of the key input parameters, to confirm the credibility of the dam breach model.

The output from this study can be used to perform a probabilistic dam breach analysis, using software which can perform a Monte Carlo simulation to randomly sample various breach parameters about the best-fit distribution. This will contribute to the development of probabilistic dam breach inundation maps instead of deterministic mapping.

REFERENCES

Bernard-Garcia, M and Mahdi, T, 2020. A worldwide historical dam failure's database, *Borealis*, v1. https://doi.org/10.5683/SP2/E7Z09B

Bowker, L N and Chambers, D, 2017. TSF Failures 1915–2017 [online], *Centre for Science in Public Participation*. Available from: <http://www.csp2.org/tsf-failures-1915-2017> [Accessed: 13 March 2025].

Canadian Dam Association, 2021. Technical Bulletin: Tailings Dam Breach Analysis

Chow, V, 1959. *Open Channel Hydraulics* (McGraw Hill Classic Publisher).

Davies, M and Martin, T, 2000a. Mine Tailings Dams: When Things Go Wrong, in *Tailings Dams 2000, Association of State Dam Safety Officials,* pp 261–273 (US Committee on Large Dams).

Davies, M and Martin, T, 2000b. Upstream constructed tailings dams – A review of the basics, *Tailings and Mine Waste,* pp 3–15.

Fontaine, D and Martin, V, 2015. Tailings mobilization estimates for dam breach studies, in *Proceedings of Tailings and Mine Waste*, pp 343–365 (University of British Columbia).

Froehlich, D C, 1995. Embankment Dam Breach Parameters Revisited, in *Proceedings of 1995 ASCE Conference on Water Resources Engineering*, pp 887–891.

Froehlich, D C, 2008. Embankment dam breach parameters and their uncertainties, *Journal of Hydraulic Engineering.*

Froehlich, D C, 2016. Empirical model of embankment dam breaching, in Proceedings from the international conference on Fluvial Hydraulics (River Flow).

Froehlich, D, 1987. Embankment Dam Breach Parameters, in *Proceedings of ASCE National Conference on Hydraulic Engineering*, pp 570–575.

Goodell, C, 2019. *McBreach User's Manual,* version 5.0.7 (Kleinschmidt Associates).

State of Colorado Department of Natural Resources Division of Water Resources, 2010. Guidelines for Dam Breach Analysis.

US Army Corps of Engineers (USACE), 2014. Using HEC-RAS for Dam Break Studies: Training Document No. 39 (Hydrologic Engineering Centre).

Von Thun, J L and Gillette, D R, 1990. Guidance on Breach Parameters, in Internal Memorandum, US Dept of the Interior, Bureau of Reclamation.

Common mistakes when trying to arrange emergency preparedness for tailings storage facility failure

O Wennstrom[1]

1. Technical Director – Crisis and Emergency Management, GHD, Perth WA 6000.
Email: olle.wennstrom@ghd.com

ABSTRACT

The Global Industry Standard on Tailings Management (GISTM) requires tailings storage facility (TSF) owners to *'provide immediate response to save lives, supply humanitarian aid and minimise environmental harm'* (GISTM Requirement 13.4, International Council on Mining and Metals (ICMM), 2020). This requires owners to create emergency preparedness, which comprises a suitable emergency management and emergency response organisation, an understandable emergency preparedness and response plan and the capability to, based on the plan, execute an emergency response.

Emergency response is defined by the United Nations International Strategy for Disaster Reduction (UNISDR, 2010) as the tactical response to mitigate the impact of an emergency, whilst emergency management is defined as *'the organisation and management of resources for dealing with all aspects of emergencies'* (Australian Institute for Disaster Resilience, 2023).

There are various standards, guidelines, and documents advising on how to create emergency preparedness for a TSF. However, perhaps because of the number of documents, there have been numerous attempts at creating the desired level of emergency preparedness that have fundamentally failed. The majority of these failed, or non-compliant attempts, can be traced back to one or more common categories of errors.

This paper presents an overview of these common errors, or pitfalls, describes why they most likely occur and provides guidance on how to avoid them when arranging emergency preparedness, composing an emergency preparedness and response plan, and developing the associated training.

The advice given in this paper is based on many reviews by the author of existing and proposed plans, as well as observing many training and exercise activities. The observations in this paper are informed by the author's experience of working with emergency response and emergency management.

This paper delivers an easy-to-apply model for avoiding common pitfalls and advice on where to find the descriptions of *'what good looks like.'*

INTRODUCTION

The Global Industry Standard on Tailings Management (GISTM), was launched in August 2020. A great initiative, in which emergency response and long-term recovery was given a chapter.

To support the integration of GISTM into the International Council on Mining and Metals' (ICMM) existing assurance and validation processes, the Conformance Protocols for the Global Industry Standard on Tailings Management (ICMM, 2021a) were released in May 2021. In the same month, the ICMM Tailings Management: Good practice guide (ICMM, 2021b) was released, focusing on primarily technical issues and recommendations, good practice for design, construction, operation, and closure.

During the last few years, many companies have taken quantum leaps to improve their emergency preparedness and response capability. Some have produced admirable results, many are just meeting the target of compliance, and yet again quite a few are without doubt unsatisfactory.

With the intent of learning from mistakes, avoiding repeats, and improving the quality of emergency preparedness and response capability. This paper presents an overview of these common errors or pitfalls, describes why they most likely occur, provides guidance on how to avoid them when arranging emergency preparedness, composing an emergency preparedness and response plan, and developing the associated training.

The advice given in this paper is based on many reviews by the author of existing and proposed plans, as well as observing many training and exercise activities. The observations are informed by the author's extensive experience of working with emergency response and emergency management.

REQUIREMENTS

GISTM does in its *Topic V: Emergency Response and Long-Term Recovery* present four principles on emergency response and five principles on recovery. Let us start with analysing what these requirements actually ask for.

Requirement 13.1

As part of the Tailings Management System, use best practices and emergency response expertise to prepare and implement a site-specific tailings facility Emergency Preparedness and Response Plan (EPRP) based on credible flow failure scenarios and the assessment of potential consequences.

Test and update the EPRP at all phases of the tailings facility life cycle at a frequency established in the plan, or more frequently if triggered by a material change either to the tailings facility or to the social, environmental, and local economic context. Meaningfully engage with employees and contractors to inform the EPRP and co-develop community-focused emergency preparedness measures with project-affected people.

Requirement 13.2

Engage with public sector agencies, first responders, local authorities and institutions and take reasonable steps to assess the capability of emergency response services to address the hazards identified in the tailings facility EPRP, identify gaps in capability and use this information to support the development of a collaborative plan to improve preparedness.

Requirement 13.3

Considering community-focused measures and public sector capacity, the Operator shall take all reasonable steps to maintain a shared state of readiness for tailings facility credible flow failure scenarios by securing resources and carrying out annual training and exercises. The Operator shall conduct emergency response simulations at a frequency established in the EPRP but at least every three years for tailings facilities with potential loss of life.

Requirement 13.4

In the case of a catastrophic tailings facility failure, provide immediate response to save lives, supply humanitarian aid and minimise environmental harm.

Summary of Principle 13

In summary the whole Principle 13 requires the Operator to: *'In the case of a catastrophic tailings facility failure, provide immediate response to save lives, supply humanitarian aid and minimise environmental harm'*, as clearly stated in 13.4. 13.1 to 13.3 are guidance on how to reach to that capability. Hence, the emergency preparedness arrangements that are not able to deliver the above stated, are not compliant.

OBSERVATIONS

When reviewing EPRP it is not often that perfect samples come across one's desk. Potentially because they are good and do not require review. The common ones are the mediocre to substandard ones, which can in general be divided under following groups:

- 'We should not be in this situation!'
- 'Call a friend!'
- 'There are principles!'

These three categories will be further developed in the following sections.

'We should not be in this situation!'

This category is often initially characterised by an underlying error in who the EPRP is written for and what it aims to achieve. Plans in this category tend to be written by and for tailings storage facility (TSF) team members. The 'Tailings Team' are normally responsible for the operation and maintenance of the TSF, as described in Figure 1, in which maintaining the integrity of the dam is a crucial part. Hence the facility's Operations Maintenance and Surveillance (OMS) Manual describes the required efforts to achieve this. Consequently, the OMS is written for the Tailings Team and describes how to avoid a TSF failure.

FIG 1 – Responsibility for operation and maintenance of the TSF rests with the 'Tailing Team'.

If the mission of maintaining the integrity fails, an emergency caused by the TSF failure might occur. An emergency is defined as 'an event, actual or imminent, which endangers or threatens to endanger life, property or the environment, and which requires a significant coordinated response' (Emergency Management Australia, 1998).

While the emergency is caused by the TSF and its failure, it will almost certainly not occur within the TSF itself. The emergency is elsewhere, as described in Figure 2. An emergency operation, in response to a failure of a TSF or dam is often a multi-disciplinary, multi-organisational operation. Impacting multiple jurisdictions and will be managed by the sites Incident Management Team in coordination with the Public Emergency Services, and the Local Government.

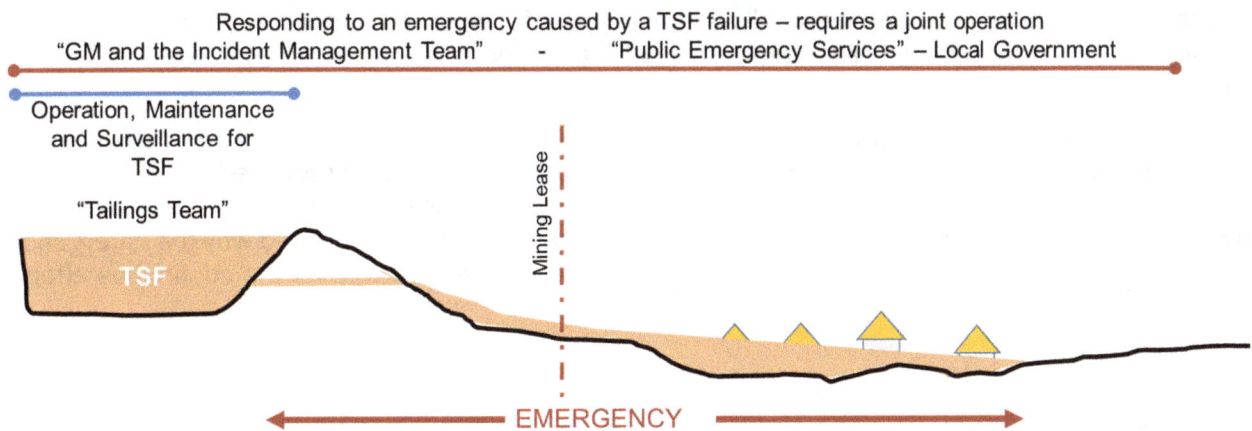

FIG 2 – Responding to an emergency caused by a TSF failure – a joint operation.

If the incorrect assumption is made, at the outset of the EPRP development, that the plan is to guide the 'TSF team' the plan is often directed at how to prevent the failure of the dam. In most cases this is already well described in the OMS. An EPRP in this category becomes a repeat of that document, often with little or no additional guidance.

The error can in some cases be exacerbated if the EPRP is aimed at the wrong target group, the TSF team instead of the Incident Management Team (IMT). An emergency caused by a TSF failure is many times one of the worse kind of emergencies a mining operation can meet. Whilst it may cause extensive threat to life, property, and the environment, it may most certainly cause an *'abnormal and unstable situation that threatens the organisation's strategic objectives, reputation or viability'* (BS 11200:2014; British Standards, 2014), which is the definition of a crisis.

It is highly unlikely that the TSF team will be best suited to lead the Operator's response to the emergency caused by the failure, coordinate with the Public Emergency Services, and manage the crisis that may arise out of the emergency.

'Call a friend!'

The second category that deserves a chapter of its own in this paper is the EPRPs dominated by elaborate call-trees and flow diagrams.

Communication is a tool for command, control, and coordination, but it will on its own not mitigate any emergency. An emergency operation can simplistically be partitioned into three main blocks in terms of communication:

1. Activation phase.
2. Operational phase.
3. Termination and handover phase.

During the activation phase it is essential to not only contact the correct recipients but also to convey the information to enable each party of the operation to activate and conduct initial planning for the response correctly. The information volume that will enable this is a conveyance of the three W's: *'What has/is going to happen, When, and Where?'* Hence a call tree with only phone numbers is not very helpful. It needs to comprise a clear instruction on how the 'Three W's' shall be conveyed.

Once in the operational phase communication must align with the Command and Control organisation. This should normally be outlined in the Operator's general emergency management arrangements, as it is for the Public Emergency Services in their respective standing plans and arrangements.

An emergency response operation, if it is to be successful, will require a clear strategy, developed into tactics and techniques. The communication must be able to convey these elements between the designated role holders in the organisation.

During the operational phase, information is often sent to the community, both directly and via news media. These contacts must also follow the chain of command, and any contradictions or ambiguities, must be avoided. The responsibility for this type of communication is often defined in legislation.

At the end of an emergency operation, the organisations must not lose the control they have gained. GISTM points out the responsibility for the Operator to have a capability for medium to long-term recovery. In most jurisdictions, the local government has a responsibility as well for such recovery, hence coordination is appropriate. Delivering a clear understanding of when and why the emergency operation is now going into a recovery operation, and who is responsible for the continuous efforts to return to normal conditions.

'There are principles!'

The third category often appears to be written by the Operators legal and/or public relations department. These plans provide lengthy explanations of the principal responsibilities of organisations that are likely to be involved in a joint emergency response to the impact of a TSF failure. They are characterised by a high level of detachment from the actual operation, which the plan is meant to guide.

This type of material is useful as a legal background document for the person writing the EPRP, to ensure it is based on legal and contractual platforms correctly. However, this background information has very limited application during an ongoing emergency response operation.

AVOIDING THE THREE COMMON ERRORS

General

The EPRP should be developed by a professional in that field, just as a specialist designs a TSF. Below are some hopefully useful tips on how to avoid the three common errors described previously in this paper.

'We should not be in this situation!'

Define who the EPRP is to be written for and what purpose it is to serve. This requires a knowledge of both the Operator's general and site-wide emergency management arrangements, as well as the legal framework pertaining to emergency management for the state where the operation is located. Remember that the EPRP is always the Operator's document, developed in conjunction with external stakeholders, but not intended to serve their needs as a plan for their emergency response.

Each organisation participating in the joint operation requires its own plan, as a plan is written for a specific target group, with its own organisation, competency, professional language, and needs. Sharing plans with other organisations has limited value, but keeping the coordination partners informed and up to date on your plan is beneficial and in line with the GISTM concept of shared state of readiness.

Clearly separate the OMS from the EPRP, as illustrated in Figure 3, avoid duplications, but align the two documents, especially pertaining to Operational Conditions, the criteria for activation of the EPRP, and mobilisation of the IMT.

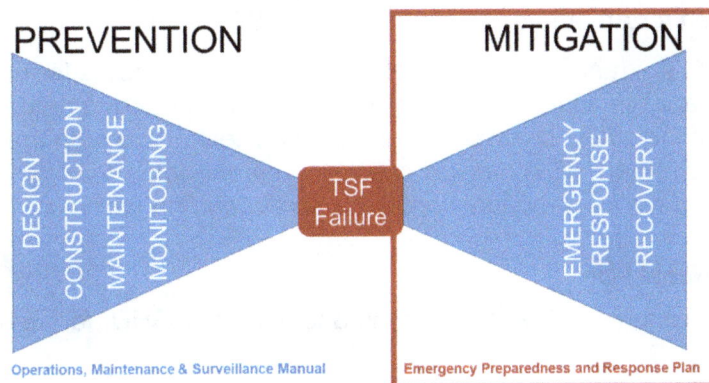

FIG 3 – The EPRP should describe the response to the effects of an emergency.

'Call a friend!'

View communication as a tool in command and control. Ensure, especially during the activation phase of an emergency, that all communication is clear and unambiguous. A functional way of doing this is using pre-scripted alert messages and conveying information on *'The three W's.'*

This is an example of such a rescripted message:

> *'This is a message from XXX company at this location.*
>
> *A failure of the tailings storage facility is imminent and expected within the next XX hours.*
>
> *We recommend an immediate activation of relevant and preplanned emergency services and police to our preplanned rendezvous point.*
>
> *Please advise on a suitable email address. I am sending you a map of the potential inundation area and our estimated timeline.'* [Deliver an inundation map and a timeline].

To facilitate communication in the Operational Phase a template for situation reporting and the exchange of Liaison Officers are powerful tools.

'There are principles!'

Ensure that the EPRP is developed based on the principles, not as a description of them. Remember that the IMT normally have other roles and in case of a major emergency, are to spring into action, in a somewhat different and many times challenging role. They need a good manual pointing out what to do, not the framework underpinning it.

CONCLUSIONS

GISTM Requirements 13.1 and 13.2 serve as guides for developing the EPRP. The EPRP must be aimed at the Operator's organisation that is to manage any major emergency internally, and coordinate with external organisations, such as Public Emergency Services and Local Government.

A response to an emergency caused by a TSF failure is most likely going to be a multi-disciplinary, multi-organisational operation, impacting multiple jurisdictions. Hence, it requires the Operator's entire Chain of Command to be activated.

To meet the above, the EPRP should be written for the IMT, and guide the IMT Leader and the team on *'What to do in the unlikely event of a TSF failure.'* The guidance on maintenance and operation of the TSF, both in normal operation and in critical stages, should already be in the OMS.

In developing a functional EPRP, based on Requirements 13.1, the Operator shall *'use best practices and emergency response expertise.'* Avoid the three most common errors, namely: *'We should not be in this situation!,' 'Call a friend!,' 'There are principles!'*

GISTM 13.3 delivers guidance on the competency that is to be gained and how to achieve this. The IMT and its IMT Leader must gain a relevant level of competency to be able to plan and execute a joint response operation for an emergency caused by a TSF failure.

If the EPRP offers good guidance to the Operator's organisation, that is to respond to and manage the emergency, and that organisation is correctly set-up, with proper competency acquired through training and exercise of the whole chain of command, the likelihood is high that the organisation can deliver of GISTM Requirement 13.4: *'In the case of a catastrophic tailings facility failure, provide immediate response to save lives, supply humanitarian aid and minimise environmental harm'.*

ACKNOWLEDGEMENTS

Andrew White, Senior Technical Director – Dams and Tailings, GHD, for valuable information and guidance.

John Phillips, Senior Technical Director – Tailings, GHD, for valuable inputs to the development of the concept of emergency preparedness for TSF failure.

REFERENCES

Australian Institute for Disaster Resilience (AIDR), 2020. Flood Emergency Planning for Disaster Resilience, first edition, Australian Disaster Resilience Handbook Collection. Available from: <https://knowledge.aidr.org.au/media/8266/aidr_handbookcollection_flood-emergency-planning_2020.pdf>

British Standards, 2014. BS 11200:2014; Crisis Management – Guidance and good practice. https://doi.org/10.3403/30274343U

International Council on Mining and Metals (ICMM), 2021a. Conformance Protocols: Global Industry Standard on Tailings Management; May 2021.

International Council on Mining and Metals (ICMM), 2021b. Tailings Management: Good practice guide; May 2021.

International Council on Mining and Metals (ICMM), 2020. The Global Industry Standard on Tailings Management (GISTM); March 2020.

Emergency Management Australia, 1998. Australian Emergency Manual Series, Part 1, The Fundamentals, Manual 3, Australian Emergency Management Glossary.

United Nations International Strategy for Disaster Reduction (UNISDR), 2 February 2010. United Nations International Strategy for Disaster Reduction (UNISDR) Secretariat Evaluation - Final report. Available from: <https://www.unisdr.org/files/12659_UNISDRevaluation2009finalreport.pdf>

Author index

www.ingramcontent.com/pod-product-compliance
Lightning Source LLC
Chambersburg PA
CBHW061102210326
41597CB00021B/3957